Lecture Notes in Computer Science 14009

The series Lecture Notes in Computer Science (LNCS), including its subseries Lecture Notes in Artificial Intelligence (LNAI) and Lecture Notes in Bioinformatics (LNBI), has established itself as a medium for the publication of new developments in computer science and information technology research, teaching, and education.

LNCS enjoys close cooperation with the computer science R & D community, the series counts many renowned academics among its volume editors and paper authors, and collaborates with prestigious societies. Its mission is to serve this international community by providing an invaluable service, mainly focused on the publication of conference and workshop proceedings and postproceedings. LNCS commenced publication in 1973.

Luca Calatroni · Marco Donatelli ·
Serena Morigi · Marco Prato ·
Matteo Santacesaria
Editors

Scale Space and Variational Methods in Computer Vision

9th International Conference, SSVM 2023
Santa Margherita di Pula, Italy, May 21–25, 2023
Proceedings

Editors
Luca Calatroni
CNRS, Université Côte d'Azur
Sophia-Antipolis, France

Marco Donatelli
University of Insubria
Como, Italy

Serena Morigi
University of Bologna
Bologna, Italy

Marco Prato
University of Modena and Reggio Emilia
Modena, Italy

Matteo Santacesaria
University of Genova
Genova, Italy

ISSN 0302-9743 ISSN 1611-3349 (electronic)
Lecture Notes in Computer Science
ISBN 978-3-031-31974-7 ISBN 978-3-031-31975-4 (eBook)
https://doi.org/10.1007/978-3-031-31975-4

This Springer imprint is published by the registered company Springer Nature Switzerland AG
The registered company address is: Gewerbestrasse 11, 6330 Cham, Switzerland

Preface

The International Conference on Scale Space and Variational Methods in Computer Vision (SSVM) series was born in 2007 in Ischia, from the fusion of the biannual Conference on Scale Space and the workshop on Variational, Geometric and Level Set Methods (VLSM). Since then, it has become a major event in the mathematical imaging community. Following the tradition of biannual events taking place at beautiful, remote places, the 9th edition of this conference (SSVM 2023, https://eventi.unibo.it/ssvm2023) took place from May 21–25, 2023 in the small town of Santa Margherita di Pula, on the Sardinian coast in Italy.

The conference provides a platform for state-of-the-art scientific exchange on topics related to computer vision and image analysis, including diverse themes such as 3D-vision, convex and non-convex variational modeling, image analysis, inverse problems in imaging, optimization methods in imaging, machine learning in imaging, PDEs in image processing, registration, restoration and reconstruction, scale-space methods, and segmentation. The 57 contributions (24 selected for oral and 33 for poster presentation) in this proceedings volume demonstrate a strong and lively community. They underwent a rigorous double-blind peer review process similar to that of a high-ranking journal in the field.

Following the tradition of previous SSVM conferences, four outstanding researchers were invited to give a keynote presentation: Coloma Ballester (Universitat Pompeu Fabra, Spain), Michael Bronstein (University of Oxford, UK), Francesca Odone (Università di Genova, Italy) and Jean-Christophe Pesquet (CentraleSupélec, France).

We would like to thank all those who contributed to the success of this conference. We are grateful to the authors for their contributions. We are indebted to the reviewers for their commitment, hard work, and enthusiasm that made this event possible at the highest scientific standards. We thank the Italian INdAM-GNCS, the Universities of Bologna and Cagliari and the I3S Laboratory in Sophia-Antipolis for financial support. Lastly, we thank the steering committee and the organizers of all the previous SSVM conferences.

April 2023

Luca Calatroni
Marco Donatelli
Serena Morigi
Marco Prato
Giuseppe Rodriguez
Matteo Santacesaria

Organization

Conference Chairs

Luca Calatroni	CNRS, Université Côte d'Azur, France
Marco Donatelli	Università dell'Insubria, Italy
Serena Morigi	Università di Bologna, Italy
Marco Prato	Università di Modena e Reggio Emilia, Italy
Giuseppe Rodriguez	Università di Cagliari, Italy
Matteo Santacesaria	Università di Genova, Italy

Program Committee

Andrés Almansa	CNRS, Université Paris Descartes, France
Jean-François Aujol	Institut de Mathématiques de Bordeaux, France
Coloma Ballester	Universitat Pompeu Fabra, Spain
Martin Benning	Queen Mary University of London, UK
Marcelo Bertalmío	Spanish National Research Council, Spain
Davide Bianchi	Harbin Institute of Technology, China
Laure Blanc-Féraud	CNRS, Université Côte d'Azur, France
Kristian Bredies	Universität Graz, Austria
Michael Breuß	BTU Cottbus, Germany
Tatiana A. Bubba	University of Bath, UK
Alessandro Buccini	Università di Cagliari, Italy
Martin Burger	FAU Erlangen-Nürnberg, Germany
Daniela Calvetti	Case Western Reserve University, USA
Valentina Candiani	Università di Genova, Italy
Antonin Chambolle	Université Paris-Dauphine, France
Raymond Honfu Chan	Chinese University of Hong Kong, China
Ke Chen	University of Liverpool, UK
Julie Delon	Université Paris Cité, France
Agnès Desolneux	CNRS, ENS Paris-Saclay, France
Yiqiu Dong	TU Denmark, Denmark
Remco Duits	TU Eindhoven, The Netherlands
Fabio Durastante	Università di Pisa, Italy
Vincent Duval	Inria, France
Matthias Ehrhardt	University of Bath, UK
Peter Elbau	Universität Wien, Austria

Paul Escande	CNRS, Aix-Marseille Université, France
Jalal Fadili	CNRS, ENSICAEN, France
Caterina Fenu	Università di Cagliari, Italy
Guy Gilboa	Technion, Israel
Markus Grasmair	NTNU, Norway
Martin Holler	Universität Graz, Austria
Sung Ha Kang	Georgia Tech, USA
Yuri Korolev	University of Bath, UK
Arjan Kuijper	Fraunhofer IGD, Germany
Alessandro Lanza	Università di Bologna, Italy
François Lauze	Københavns Universitet, Denmark
Carole Le Guyader	INSA Rouen, France
Damiana Lazzaro	Università di Bologna, Italy
Jan Lellman	Universität zu Lübeck, Germany
Tony Lindeberg	KTH, Sweden
Elena Loli Piccolomini	Università di Bologna, Italy
Dirk Lorenz	TU Braunschweig, Germany
Simon Masnou	Université Claude Bernard Lyon 1, France
Jan Modersitzki	Universität zu Lübeck, Germany
Michael Möller	Universität Siegen, Germany
Subhadip Mukherjee	University of Bath, UK
Nicholas Nelsen	California Institute of Technology, USA
Michael Ng	Hong Kong Baptist University, China
Thomas Oberlin	Université de Toulouse, France
Peter Ochs	Universität Tübingen, Germany
Nicolas Papadakis	CNRS, Université de Bordeaux, France
Kostas Papafitsoros	Queen Mary University of London, UK
Marcelo Pereyra	Heriot-Watt University, UK
Jean-Christophe Pesquet	CentraleSupélec, France
Thomas Pock	TU Graz, Austria
Federica Porta	Università di Modena e Reggio Emilia, Italy
Yvain Quéau	CNRS, ENSICAEN, France
Julien Rabin	CNRS, ENSICAEN, France
Luca Ratti	Università di Bologna, Italy
Giuseppe Rodriguez	Università di Cagliari, Italy
Martin Rumpf	Universität Bonn, Germany
Otmar Scherzer	Universität Wien, Austria
Christoph Schnörr	Universität Heidelberg, Germany
Samuli Siltanen	Helsingin Yliopisto, Finland
Emmanuel Soubies	CNRS, Université Toulouse, France
Gabriele Steidl	TU Berlin, Germany
Julián Tachella	CNRS, ENS de Lyon, France

Xue-Cheng Tai	Hong Kong Baptist University, China
Silvia Villa	Università di Genova, Italy
Joachim Weickert	Universität des Saarlandes, Germany
Pierre Weiss	CNRS, Université Toulouse, France
Martin Welk	UMIT Tyrol, Austria
Xiaoqun Zhang	Shanghai Jiao Tong University, China

External Referees

Laetitia Chapel	Université de Bretagne Sud, France
Rita Fioresi	Università di Bologna, Italy
Patrizio Frosini	Università di Bologna, Italy
Jan Gerken	TU Berlin, Germany
Andreas Hauptmann	Oulun Yliopisto, Finland
Johannes Hertrich	TU Berlin, Germany
Martin Huska	Università di Bologna, Italy
Samuel Hurault	Institut de Mathématiques de Bordeaux, France
Sebastian Neumayer	EPFL Lausanne, Switzerland
Alasdair Newson	Télécom Paris, France
Romain Petit	Università di Genova, Italy
Pascal Peter	Universität des Saarlandes, Germany
Yann Traonmilin	Institut de Mathématiques de Bordeaux, France

Invited Speakers

Coloma Ballester	Universitat Pompeu Fabra, Spain
Michael Bronstein	University of Oxford, UK, and Twitter
Francesca Odone	Università di Genova, Italy
Jean-Christophe Pesquet	Université Paris-Saclay, France

Sponsoring Institutions

Dipartimento di Matematica, Università di Bologna, Italy
Dipartimento di Matematica e Informatica, Università di Cagliari, Italy
Laboratoire I3S, Université Côte d'Azur, France
Istituto Nazionale di Alta Matematica - Gruppo Nazionale per il Calcolo Scientifico, Italy

Contents

Machine and Deep Learning in Imaging

Inverse Problems in Imaging

Explicit Diffusion of Gaussian Mixture Model Based Image Priors

Martin Zach[1(✉)], Thomas Pock[1], Erich Kobler[2], and Antonin Chambolle[3]

[1] Graz University of Technology, 8010 Graz, Austria
{martin.zach,pock}@icg.tugraz.at
[2] Universitätsklinikum Bonn, 53127 Bonn, Germany
kobler@uni-bonn.de
[3] Université Paris-Dauphine-PSL, 75775 Paris Cedex 16, France
antonin.chambolle@ceremade.dauphine.fr

Abstract. In this work we tackle the problem of estimating the density f_X of a random variable X by successive smoothing, such that the smoothed random variable Y fulfills $(\partial_t - \Delta_1) f_Y(\,\cdot\,, t) = 0$, $f_Y(\,\cdot\,, 0) = f_X$. With a focus on image processing, we propose a product/fields-of-experts model with Gaussian mixture experts that admits an analytic expression for $f_Y(\,\cdot\,, t)$ under an orthogonality constraint on the filters. This construction naturally allows the model to be trained simultaneously over the entire diffusion horizon using empirical Bayes. We show preliminary results on image denoising where our model leads to competitive results while being tractable, interpretable, and having only a small number of learnable parameters. As a byproduct, our model can be used for reliable noise estimation, allowing blind denoising of images corrupted by heteroscedastic noise.

Keywords: Diffusion Models · Empirical Bayes · Gaussian Mixture · Blind Denoising

1 Introduction

Consider the practical problem of estimating the probability density $f_X : \mathcal{X} \to \mathbb{R}$ of a random variable X in $\mathcal{X}$, given a set of data samples $\{x_i\}_{i=1}^N$ drawn from f_X.[1] This is a challenging problem in high dimension (e.g. for images of size $M \times N$, i.e. $\mathcal{X} = \mathbb{R}^{M \times N}$), due to extremely sparsely populated regions. A fruitful approach is to estimate the density at different times when undergoing a diffusion process. Intuitively, the diffusion equilibrates high- and low-density regions over time, thus easing the estimation problem.

[1] For notational convenience, throughout this article we do not make a distinction between the *distribution* and *density* of a random variable.

Supplementary Information The online version contains supplementary material available at https://doi.org/10.1007/978-3-031-31975-4_1.

L. Calatroni et al. (Eds.): SSVM 2023, LNCS 14009, pp. 3–15, 2023.
https://doi.org/10.1007/978-3-031-31975-4_1

Let Y_t (carelessly) denote the random variable whose distribution is defined by diffusing f_X for some time t. We denote the density of Y_t by $f_Y(\,\cdot\,,t)$, which fulfills the diffusion equation $(\partial_t - \Delta_1)f_Y(\,\cdot\,,t) = 0$, $f_Y(\,\cdot\,,0) = f_X$. The empirical Bayes theory [13] provides a machinery for reversing the diffusion process: Given an instantiation of the random variable Y_t, the Bayesian least-squares estimate of X can be expressed solely using $f_Y(\,\cdot\,,t)$. Importantly, this holds for all positive t, as long as f_Y is properly constructed.

In practice we wish to have a parametrized, trainable model of f_Y, say f_θ where θ is a parameter vector, such that $f_Y(x,t) \approx f_\theta(x,t)$ for all $x \in \mathcal{X}$ and all $t \in [0,\infty)$. Recent choices [17,18] for the family of functions $f_\theta(\,\cdot\,,t)$ were of practical nature: Instead of an analytic expression for f_θ at any time t, authors proposed a time-conditioned network in the hope that it can learn to behave as if it had undergone a diffusion process. Further, instead of worrying about the normalization $\int_\mathcal{X} f_Y(\,\cdot\,,t) = 1$ for all $t \in [0,\infty)$, usually they directly estimate the *score* $-\nabla_1 \log f_Y(\,\cdot\,,t) : \mathcal{X} \to \mathcal{X}$ with some network $s_\theta(\,\cdot\,,t) : \mathcal{X} \to \mathcal{X}$. This has the advantage that normalization constants vanish, but usually, the constraint $\partial_j(s_\theta(\,\cdot\,,t))_i = \partial_i(s_\theta(\,\cdot\,,t))_j$ is not enforced in the architecture of s_θ. Thus, $s_\theta(\,\cdot\,,t)$ is in general not the gradient of a scalar function (the negative-log-density it claims to model).

In this paper, we pursue a more principled approach. Specifically, we leverage Gaussian mixure models (GMMs) to represent the popular product/fields-ofexperts (FoE) model [6,15] and show that under an orthogonality constraint of the associated filters, the diffusion of the model can be expressed analytically.

2 Background

In this section, we first emphasize the importance of the diffusion process in density estimation (and sampling) in high dimensions. Then, we detail the relationship between diffusing the density function, empirical Bayes, and denoising score matching [19].

2.1 Diffusion Eases Density Estimation and Sampling

Let f_X be a density on $\mathcal{X} \subset \mathbb{R}^d$. A major difficulty in estimating f_X with parametric models is that f_X is extremely sparsely populated in high dimensional spaces[2], i.e., $d \gg 1$. This phenomenon has many names, e.g. the curse of dimensionality or the manifold hypothesis [1]. Thus, the learning problem is difficult, since meaningful gradients are rare. Conversely, let us for the moment assume we have a model $\tilde{f}_X$ that approximates f_X well. In general, it is still very challenging to generate a set of points $\{x_i\}_{i=1}^I$ such that we can confidently say that the associated empirical density $\frac{1}{I}\sum_{i=1}^I \delta(\,\cdot\, - x_i)$ approximates $\tilde{f}_X$. This is because, in general, there does not exist a procedure to directly draw from

[2] Without any reference to samples $x_i \sim f_X$, an equivalent statement may be that f_X is (close to) zero almost everywhere (in the layman—not measure-theoretic—sense).

$\tilde{f}_X$, and (modern) Markov chain Monte Carlo (MCMC) relies on the estimated gradients of $\tilde{f}_X$ and, in practice, only works well for unimodal distributions [18].

The isotropic diffusion process or heat equation

$$(\partial_t - \Delta_1) f(\cdot, t) = 0 \text{ with initial condition } f(\cdot, 0) = f_X \tag{1}$$

equilibrates the density in f_X, thus mitigating the challenges outlined above. Here, ∂_t denotes $\frac{\partial}{\partial t}$ and $\Delta_1 = \operatorname{tr}(\nabla_1^2)$ is the Laplace operator, where the 1 indicates application to the first argument. We detail the evolution of f_X under this process and relations to empirical Bayes in Sect. 2.2.

Learning $f(\cdot, t)$ for $t \geq 0$ is more stable since the diffusion "fills the space" with meaningful gradients [17]. Of course, this assumes that for different times t_1 and t_2, the models of $f(\cdot, t_1)$ and $f(\cdot, t_2)$ are somehow related to each other. As an example of this relation, the recently popularized noise-conditional score-network [18] shares convolution filters over time, but their input is transformed through a time-conditional instance normalization. In this work, we make this relation explicit by considering a family of functions $f(\cdot, 0)$ for which $f(\cdot, t)$ can be expressed analytically.

For *sampling*, $f(\cdot, t)$ for $t > 0$ can help by gradually moving samples towards high-density regions of f_X, regardless of initialization. To utilize this, a very simple idea with relations to simulated annealing [8,18] is to have a pre-defined time schedule $t_T > t_{T-1} > \ldots > t_1 > 0$ and sample $f(\cdot, t_i)$, $i = T, \ldots, 0$ (e.g. with Langevin dynamics [14]) successively.

2.2 Diffusion, Empirical Bayes, and Denoising Score Matching

In this section, similar to the introduction, we again adopt the interpretation that the evolution in (1) defines the density of a smoothed random variable Y_t. That is, Y_t is a random variable with probability density $f_Y(\cdot, t)$, which fulfills $(\partial_t - \Delta_1) f_Y(\cdot, t) = 0$ and $f_Y(\cdot, 0) = f_X$. It is well known that the Green's function of (1) is a Gaussian (see e.g. [2]) with zero mean and variance $\sigma^2(t) = 2t\mathrm{Id}$. In other words, for $t > 0$ we can write $f_Y(\cdot, t) = G_{0,2t\mathrm{Id}} * f_X$, where

$$G_{\mu,\Sigma}(x) = |2\pi\Sigma|^{-1/2} \exp\left(-\frac{1}{2}(x-\mu)^\top \Sigma^{-1} (x-\mu)\right). \tag{2}$$

Thus, the diffusion process constructs a (linear) *scale space in the space of probability densities.* In terms of the random variables, $Y_t = X + \sqrt{2t}N$ where $N \sim \mathcal{N}(0, \mathrm{Id})$. Next, we show how to estimate the corresponding instantiation of X which has "most likely" spawned an instantiation of Y_t using empirical Bayes.

In the school of empirical Bayes [13], we try to estimate a clean random variable given a corrupted instantiation, using only knowledge about the corrupted density. In particular, for our setup, Miyasawa [10] has shown that the Bayesian minimum mean-squared-error (MMSE) estimator $\hat{x}_{\mathrm{EB}}$ for an instantiation y_t of Y_t is

$$\hat{x}_{\mathrm{EB}}(y_t) = y_t + \sigma^2(t) \nabla_1 \log f_Y(y_t, t), \tag{3}$$

which is also known as Tweedie's formula [3]. Raphan and Simoncelli [12] extended the empirical Bayes framework to arbitrary corruptions and coined the term non-parametric empirical Bayes least-squares (NEBLS).

Recently, (3) has been used for parameter estimation [18,19]: Let $\{x_i\}_{i=1}^I$ be a dataset of I samples drawn from f_X and let Y_t be governed by diffusion. Then, both the left- and right-hand side of (3) are *known*—in expectation. This naturally leads to the loss function

$$\min_\theta \int_{(0,\infty)} \mathbb{E}_{(x,y_t)\sim f_{X\times Y_t}} \|x - y_t - \sigma(t)^2 \nabla_1 \log f_\theta(y_t, t)\|^2 \,\mathrm{d}t \tag{4}$$

for estimating θ such that $f_\theta \approx f_Y$. Here, $f_{X\times Y_t}$ denotes the joint distribution of real and degraded points. This learning problem is known as denoising score matching [7,18,19].

3 Methods

In this section, we introduce a patch and convolutional model to approximate the prior distribution of natural images. For both models, we present conditions such that they obey the diffusion process.

3.1 Patch Model

Patch-based prior models such as expected patch log-likelihood (EPLL) [20] typically use GMMs to approximately learn a prior for natural image patches. Throughout this section, we approximate the density of image patches $p \in \mathbb{R}^a$ of size $a = b \times b$ as a product of GMM experts, i.e.

$$\tilde{f}_\theta(p, t) = Z(\{k_j\}_{j=1}^J)^{-1} \prod_{j=1}^J \psi_j(\langle k_j, p\rangle, w_j, t), \tag{5}$$

in analogy to the product-of-experts model [6]. $Z(\{k_j\}_{j=1}^J)$ is required such that $\tilde{f}_\theta$ is properly normalized. Every expert $\psi_j : (\mathbb{R} \times \Delta^L \times [0,\infty)) \to \mathbb{R}^+$ for $j = 1, \ldots, J$ models the density of associated filters k_j for all diffusion times t by an one-dimensional GMM with L components of the form

$$\psi_j(x, w_j, t) = \sum_{l=1}^L w_{jl} G_{\mu_l, \sigma_j^2(t)}(x). \tag{6}$$

The weights of each expert $w_j = (w_{j1}, \ldots, w_{jL})^\top$ must satisfy the unit simplex constraint, i.e., $w_j \in \Delta^L$, $\Delta^L = \{x \in \mathbb{R}^L : x_l \geq 0, \sum_{i=1}^L x_l = 1\}$. Although not necessary, we assume for simplicity that all ψ_j have the same number of components and the discretization of their means μ_l over the real line is shared and fixed a priori (for details see Sect. 4.1). Further, the variances of all components of each expert are shared and are modeled as

$$\sigma_j^2(t) = \sigma_0^2 + c_j 2t,$$

where σ_0 is chosen to support the uniform discretization of the means μ_l and $c_j \in \mathbb{R}_{++}$ are constants, to reflect the convolution effect of the diffusion process. Thus, we can summarize the learnable parameters $\theta = \{(k_j, w_j)\}_{j=1}^J$.

Next, we detail how the diffusion process changes the variance $\sigma_j^2(t)$ of each expert. In particular, we exploit two well-known properties of GMMs: First, the product of GMMs is again a GMM, see e.g. [16]. This allows us to work on highly expressive models that enable efficient *evaluations* due to factorization. Second, we use the fact that there exists an analytical solution to the diffusion equation if f_X is a GMM: The associated Green's function is a Gaussian with isotropic covariance $2t\mathrm{Id}$. Hence, diffusion amounts to the convolution of two Gaussians for every component due to the linearity of convolution. Using previous notation, if X is a random variable with normal distribution $\mathcal{N}(\mu_X, \Sigma_X)$, then Y_t follows the distribution $\mathcal{N}(\mu_X, \Sigma_X + 2t\mathrm{Id})$. In particular, the mean remains unchanged, thus it is sufficient to only adapt the variances in (5) with the diffusion time. We call our model Gaussian mixture diffusion model (GMDM) due to the combination of GMMs and the diffusion framework.

First, the following theorem establishes the exact form of (6) as a GMM on $\mathbb{R}^a$. We denote with $\hat{l}(j)$ a fixed but arbitrary selection from the index set $\{1, \dots, L\}$ for each $j \in \{1, \dots, J\}$.

Theorem 1. *$\tilde{f}_\theta(\cdot, 0)$ is a homoscedastic GMM on $\mathbb{R}^a$ with L^J components, precision matrix*

$$(\Sigma_a)^{-1} = \frac{1}{\sigma_0^2} \sum_{j=1}^J (k_j \otimes k_j) \text{ and means } \mu_{a,\hat{l}} = \Sigma_a \sum_{j=1}^J k_j \mu_{\hat{l}(j)}. \tag{7}$$

Proof. By definition,

$$\prod_{j=1}^J \psi(\langle k_j, p\rangle, w_j, 0) = \prod_{j=1}^J \sum_{l=1}^L \frac{w_{jl}}{\sqrt{2\pi\sigma_0^2}} \exp\left(-\frac{1}{2\sigma_0^2}(\langle k_j, p\rangle - \mu_l)^2\right). \tag{8}$$

The general component of the above reads as

$$(2\pi\sigma_0^2)^{-\frac{J}{2}} \left(\prod_{j=1}^J w_{j\hat{l}(j)}\right) \exp\left(-\frac{1}{2\sigma_0^2} \sum_{j=1}^J (\langle k_j, p\rangle - \mu_{\hat{l}(j)})^2\right). \tag{9}$$

To find $(\Sigma_a)^{-1}$, we complete the square: Motivated by $\nabla_p \|p - \mu_a\|_{\Sigma_a^{-1}}^2 / 2 = \Sigma_a^{-1}(p - \mu_a)$ we find $\nabla_p\big(\frac{1}{2\sigma_0^2} \sum_{j=1}^J (\langle k_j, p\rangle - \mu_{\hat{l}(j)})^2\big) = \frac{1}{\sigma_0^2} \sum_{j=1}^J (k_j \otimes k_j)p - k_j \mu_{\hat{l}(j)}$ and the theorem immediately follows.

The next theorem establishes a tractable analytical expression for the diffusion process under the assumption of pair-wise orthogonal filters, that is

$$\forall (i,j) \in \{1, \dots, J\}^2 \quad \langle k_j, k_i\rangle = \begin{cases} 0 & \text{if } i \neq j, \\ \|k_j\|^2 & \text{else.} \end{cases} \tag{10}$$

This assumption does limit expressiveness since it does not allow over-complete models. However, the idea of orthogonal transformations and the resulting decomposing conditional models have recently been successfully used in the context of wavelet score-networks [5].

Theorem 2 (Patch diffusion). *Under assumption* (10), $\tilde{f}_\theta(\cdot, t)$ *satisfies the diffusion equation* $(\partial_t - \Delta_1)\tilde{f}_\theta(\cdot, t) = 0$ *if* $\sigma_j^2(t) = \sigma_0^2 + \|k_j\|^2 2t$.

Proof. Assuming (10), the Eigendecomposition of the precision matrix can be trivially constructed. In particular, $(\Sigma_a)^{-1} = \sum_{j=1}^J \frac{\|k_j\|^2}{\sigma_0^2}(\frac{k_j}{\|k_j\|} \otimes \frac{k_j}{\|k_j\|})$, hence $\Sigma_a = \sum_{j=1}^J \frac{\sigma_0^2}{\|k_j\|^2}(\frac{k_j}{\|k_j\|} \otimes \frac{k_j}{\|k_j\|})$. As discussed in Sect. 2.2, Σ_a evolves as $\Sigma_a \mapsto \Sigma_a + 2t\mathrm{Id}_a$ under diffusion. Equivalently, for all $j = 1, \dots, J$ Eigenvalues, $\frac{\sigma_0^2}{\|k_j\|^2} \mapsto \frac{\sigma_0^2 + 2t\|k_j\|^2}{\|k_j\|^2}$. Recall that σ_0^2 is just $\sigma_j^2(0)$. Thus, $\tilde{f}(\cdot, t)$ satisfies the diffusion equation if $\sigma_j^2(t) = \sigma_0^2 + \|k_j\|^2 2t$.

Corollary 1. *With assumption* (10) *the potential functions* $\psi_j(\cdot, w_j, t)$ *in* (5) *model the marginal distribution of the random variable* $Z_{j,t} = \langle k_j, Y_t \rangle$. *In addition,* $\tilde{f}_\theta$ *is normalized when* $Z(\{k_j\}_{j=1}^J)^{-1} = \prod_{j=1}^J \|k_j\|^2$.

Proof. Consider one component of the resulting homoscedastic GMM: $\hat{Y}_t \sim \mathcal{N}(\mu_{a,\hat{l}}, \Sigma_a + 2t\mathrm{Id}_a)$. The distribution of $\hat{Z}_{j,t} = \langle k_j, \hat{Y}_t \rangle$ is (see e.g. [4] for a proof) $\hat{Z}_{j,t} \sim \mathcal{N}(k_j^\top \mu_{a,\hat{l}}, k_j^\top(\Sigma_a + 2t\mathrm{Id}_a)k_j) = \mathcal{N}(\mu_{\hat{l}(j)}, \sigma_0^2 + 2t\|k_j\|^2)$. The claim follows from the linear combination of the different components.

We note that (7) only specifies a covariance matrix if $J = a$, otherwise the matrix is singular. In the case $J < a$, we restrict the analysis to the subspace $\mathrm{span}(\{k_1, \dots, k_J\})$. In particular, we also assume that the diffusion process does not transport density out of this subspace.

3.2 Convolutional Model

To avoid the extraction and combination of patches in patch-based image priors and still account for the local nature of low-level image features, we describe a convolutional GMDM next. The following analysis assumes vectorized images $x \in \mathbb{R}^n$ with n pixels; the generalization to higher dimensions is straightforward. In analogy to the patch-based model of the previous section, we extend the FoE model [15] to our considered diffusion setting by accounting for the diffusion time t and obtain[3]

$$f_\theta(x, t) \propto \prod_{i=1}^n \prod_{j=1}^J \psi_j((K_j x)_i, w_j, t). \tag{11}$$

Here, each expert ψ_j models the density of convolution *features* extracted by convolution kernels $\{k_j\}_{j=1}^J$ of size $a = b \times b$. $\{K_j\}_{j=1}^J \subset \mathbb{R}^{n \times n}$ are the corresponding

[3] For simplicity, we discard the normalization constant Z, which is independent of t.

matrix representations and all convolutions are cyclic, i.e., $K_j x \equiv k_j *_n x$, where $*_n$ denotes a 2-dimensional convolution with cyclic boundary conditions. Further, $w_j \in \triangle^L$ are the weight the components of each expert ψ_j as in Eq. (6). As in the patch model, it is sufficient to adapt the variances $\sigma_j^2(t)$ by the diffusion time as the following analysis shows.

By definition for $t = 0$, we have

$$f_\theta(x, 0) = \prod_{i=1}^{n} \prod_{j=1}^{J} \sum_{l=1}^{L} \frac{w_{jl}}{\sqrt{2\pi\sigma_0^2}} \exp\left(-\frac{((K_j x)_i - \mu_l)^2}{2\sigma_0^2}\right). \tag{12}$$

First, we expand the product over the pixels

$$f_\theta(x, 0) = \prod_{j=1}^{J} \sum_{\hat{l}(i)=1}^{L^n} (2\pi\sigma_0^2)^{-\frac{n}{2}} \overline{w}_{j\hat{l}(i)} \exp\left(-\frac{\|(K_j x) - \mu_{\hat{l}(i)}\|^2}{2\sigma_0^2}\right) \tag{13}$$

using the index map $\hat{l}(i)$ and $\overline{w}_{j\hat{l}(i)} = \prod_{i=1}^{I} w_{j\hat{l}(i)}$. Further, expanding over the features results in

$$f_\theta(x, 0) = \sum_{\hat{\imath}(i,j)=1}^{(L^n)^J} (2\pi\sigma_0^2)^{-\frac{nJ}{2}} \overline{\overline{w}}_{\hat{\imath}(i,j)} \exp\left(-\frac{1}{2\sigma_0^2} \sum_{j=1}^{J} \|(K_j x) - \mu_{\hat{\imath}(i,j)}\|^2\right), \tag{14}$$

where $\overline{\overline{w}}_{\hat{\imath}(i,j)} = \prod_{j=1}^{J} \prod_{i=1}^{I} w_{\hat{\imath}(i,j)}$ Observe that Eq. (14) again describes a homoscedastic GMM with precision $\Sigma^{-1} = \frac{1}{\sigma_0^2} \sum_{j=1}^{J} K_j^\top K_j$ and means $\tilde{\mu}_{\hat{\imath}(i,j)} = \Sigma \frac{1}{\sigma_0^2} \sum_{j=1}^{J} K_j^\top \mu_{\hat{\imath}(i,j)}$. Due to the assumed boundary conditions, the Fourier transform $\mathcal{F}$ diagonalizes the convolution matrices: $K_j = \mathcal{F}^* \operatorname{diag}(\mathcal{F} k_j) \mathcal{F}$. Thus, the precision matrix can be expressed as

$$\Sigma^{-1} = \mathcal{F}^* \operatorname{diag}\left(\sum_{j=1}^{J} \frac{|\mathcal{F} k_j|^2}{\sigma^2}\right) \mathcal{F} \tag{15}$$

where we used $\mathcal{F}\mathcal{F}^* = \mathrm{Id}$, $\bar{z}z = |z|^2$ and $|\cdot|$ denotes the complex modulus acting element-wise on its argument. We assume that the spectra of k_j have disjoint support, i.e.

$$\Gamma_i \cap \Gamma_j = \emptyset \;\text{ if }\; i \neq j, \tag{16}$$

where $\Gamma_j = \operatorname{supp} \mathcal{F} k_j$. Note that, in analogy to the pair-wise orthogonality of the filters in the patch model (10), from this immediately follows that $\langle \mathcal{F} k_j, \mathcal{F} k_i \rangle = 0$ when $i \neq j$. In addition, we assume that the magnitude is constant over the support, i.e.

$$|\mathcal{F} k_j| = \xi_j \mathbb{1}_{\Gamma_j}, \tag{17}$$

where $\mathbb{1}_A$ is the characteristic function of the set A.

Theorem 3 (Convolutional Diffusion). *Under assumptions* (16) *and* (17), *$f_\theta(\,\cdot\,, t)$ satisfies the diffusion equation $(\partial_t - \Delta_1) f_\theta(\,\cdot\,, t) = 0$ if $\bar{\sigma}_j^2(t) = \sigma_0^2 + \xi_j^2 2t$.*

Proof. In analogy to Theorem 2, with Eq. (15) $\mathcal{F}^* \operatorname{diag}\left(\sum_{j=1}^J \frac{\sigma^2}{|\mathcal{F}k_j|^2}\right)\mathcal{F} \mapsto \mathcal{F}^* \operatorname{diag}\left(\frac{\sigma^2+2t\sum_{j=1}^J |\mathcal{F}k_j|^2}{\sum_{j=1}^J |\mathcal{F}k_j|^2}\right)\mathcal{F}$ under diffusion. The inner sum decomposes as

$$\frac{\sigma_0^2 + 2t\sum_{j=1}^J |\mathcal{F}k_j|^2}{\sum_{j=1}^J |\mathcal{F}k_j|^2} = \sum_{j=1}^J \frac{\sigma_0^2 + 2t|\mathcal{F}k_j|^2}{|\mathcal{F}k_j|^2} \tag{18}$$

using (16), and with (17) the numerator reduces to $\sigma_0^2 + 2t\xi_j^2$.

4 Numerical Results

4.1 Numerical Optimization

For all experiments, ψ_j is a $L = 125$ component GMM, with equidistant means μ_l in the interval $[-\gamma, \gamma]$, where we chose $\gamma = 1$. To support the uniform discretization of the means, the shared standard deviation of the experts is $\sigma_0 = \frac{2\gamma}{L-1}$. Assuming zero-mean filters of size $b \times b$, we use $J = b^2 - 1$ filters. The filters are initialized by independently drawing their entries from a zero-mean Gaussian distribution with standard deviation b^{-1}. We avoid simplex projections by replacing w_j with learnable parameters ζ_j, from which w_j are computed using a soft-argmax $w_{jl} = \frac{\exp \zeta_{jl}}{\sum_{l=1}^L \exp \zeta_{jl}}$, and initialize $\zeta_{jl} = \frac{0.1\sqrt{\alpha}}{1+\alpha\mu_l^2}$, where $\alpha = 1000$.

For the numerical experiments, f_X reflects the distribution of rotated and flipped $b \times b$ patches from the 400 gray-scale images in the BSDS 500 [9] training and test set, with each pixel in the interval $[0, 1]$. We optimize the parameters $\theta = \{(k_j, \zeta_j)\}_{j=1}^J$ in (4) using the iPALM algorithm [11] with respect to a randomly chosen batch of size 3200 for 100 000 steps. We approximate the infinite-time diffusion process by uniformly drawing $\sqrt{2t}$ from the interval $[0, 0.4]$. We detail how we ensure the orthogonality of the filters during the iterations of iPALM in the next section.

Enforcing Orthogonality. Let $K = [k_1, k_2, \dots, k_J] \in \mathbb{R}^{a \times J}$ denote the matrix obtained by horizontally stacking the filters. We are interested in finding

$$\operatorname{proj}_{\mathcal{O}}(K) = \underset{M \in \mathcal{O}}{\arg\min} \|M - K\|_F^2 \tag{19}$$

where $\mathcal{O} = \{X \in \mathbb{R}^{a \times J} : X^\top X = D^2\}$, $D = \operatorname{diag}(\lambda_1, \lambda_2, \dots, \lambda_J)$ is diagonal, and $\|\cdot\|_F$ is the Frobenius norm. Since $\operatorname{proj}_{\mathcal{O}}(K)^\top \operatorname{proj}_{\mathcal{O}}(K) = D^2$ we can represent it as $\operatorname{proj}_{\mathcal{O}}(K) = OD$ with O semi-unitary ($O^\top O = \mathrm{Id}$). Other than positivity, we do not place any restrictions on $\lambda_1, \dots, \lambda_J$, as these are related to the precision in our model. Thus, we rewrite the objective

$$\operatorname{proj}_{\mathcal{O}}(K) = \underset{\substack{O^\top O = \mathrm{Id}_J \\ D = \operatorname{diag}(\lambda_1, \dots, \lambda_J)}}{\arg\min} \left\{ \|OD - K\|_F^2 = \|K\|_F^2 - 2\langle K, OD\rangle_F + \|D\|_F^2 \right\} \tag{20}$$

Algorithm 1: Algorithm for orthogonalizing a set of filters K.

Input : $K = [k_1, \ldots, k_J] \in \mathbb{R}^{a \times J}$, $B \in \mathbb{N}$, $D^{(1)} = \mathrm{Id}_J$
Output: $O^{(B)} D^{(B)} = \mathrm{proj}_{\mathcal{O}}(K)$
1 **for** $b \in 1, \ldots, B-1$ **do**
2 $\quad U^{(b)} P^{(b)} = D^{(b)} K^\top$ // Polar decomposition
3 $\quad O^{(b+1)} = U^{(b)}$
4 $\quad D^{(b+1)}_{i,i} = \left(((O^{(b+1)})^\top K)_{i,i}\right)_+$

where $\langle \cdot, \cdot \rangle_F$ is the Frobenius inner product.

We propose the following alternating minimization scheme for finding O and D. The solution for the reduced sub-problem in O can be computed by setting $O = U$, using the polar decomposition of $DK^\top = UP$, where $U \in \mathbb{R}^{J \times a}$ is semi-unitary ($U^\top U = \mathrm{Id}_a$) and $P = P^\top \in \mathbb{S}^a_+$. The sub-problem in D is solved by setting $D_{i,i} = \left((O^\top K)_{i,i}\right)_+$. The algorithm is summarized in Algorithm 1, where we have empirically observed fast convergence; $B = 3$ steps already yielded satisfactory results. A theoretical analysis of the algorithm is presented in the supplemental material.

To visually evaluate whether our learned model matches the empirical marginal densities for any diffusion time t, we plot them in Fig. 1. At the top, the learned 7×7 orthogonal filters k_j are depicted, the associated learned potential functions $-\log \psi_j$ are shown below. Indeed, they match the empirical marginal responses

$$h_j(z, t) = -\log \mathbb{E}_{p \sim f_X} \delta(z - \langle k_j, p \rangle) \tag{21}$$

visualized at the bottom almost perfectly even at extremely low-density tails. This supports the theoretical argument that the diffusion process eases the problem of density estimation outlined in the introductory sections.

4.2 Sampling

A direct consequence of Corollary 1 is that our model admits a simple sampling procedure: The statistical independence of the components allows drawing random patches by $Y_t = \sum_{j=1}^J \frac{k_j}{\|k_j\|^2} Z_{j,t}$, where $Z_{j,t}$ is sampled from the one-dimensional GMM ψ_j. The samples in Fig. 2 indicate a good match over a wide range of t. However, for small t the generated patches appear slightly noisy, which is due to an over-smooth approximation of the sharply peaked marginals around 0. This indicates that the (easily adapted) discretization of μ_l is not optimal.

4.3 Image Denoising

To exploit our patch-based prior for whole-image denoising, we define the EPLL of a noisy image $y \in \mathbb{R}^n$ with variance $\sigma^2(t) = 2t$ as

$$e(y, t) = \sum_{j=1}^{\tilde{n}} p_j^{-1} \log \tilde{f}_\theta(P_j y, t). \tag{22}$$

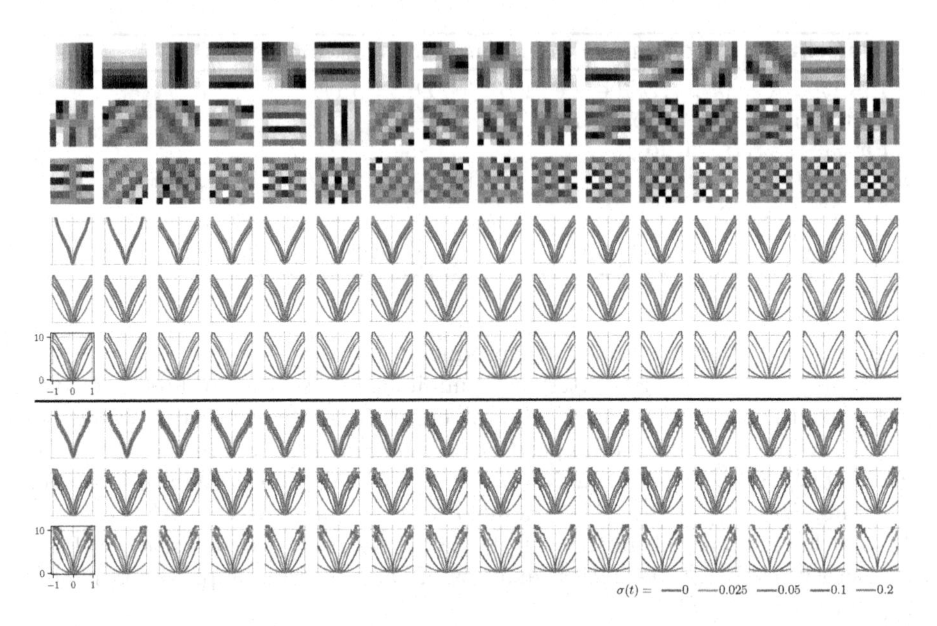

Fig. 1. Learned filters k_j (top, the intervals show the values of black and white respectively, amplified by a factor of 10) and potential functions $-\log \psi_j$ (middle). On the bottom, the empirical marginal filter response histograms are drawn.

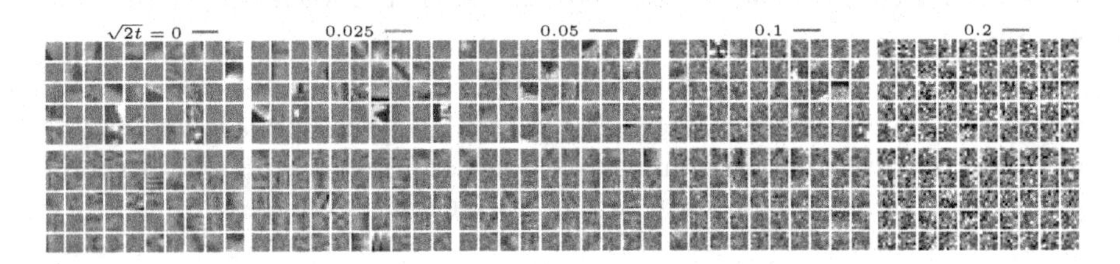

Fig. 2. Samples from the random variable Y_t (top) and generated patches (bottom).

Here, $\tilde{n}$ denotes the total number of overlapping patches (e.g. for $n = 4 \times 4$ and $a = 3 \times 3$, $\tilde{n} = 4$), $P_i : \mathbb{R}^n \to \mathbb{R}^a$ denotes the patch-extraction matrix for the i-th patch and $p_i = \left(\sum_{j=1}^{\tilde{n}} P_j^\top P_j\right)_{i,i}$ counts the number of overlapping patches to compensate for boundary effects. We consider two inference methods: Empirical Bayes-patch averaging (EB-PA)

$$\hat{x}_{\text{EB-PA}}(y,t) = y + \sigma^2(t)\nabla_1 e(y,t) = y + 2t \sum_{j=1}^{\tilde{n}} p_j^{-1} P_j^\top \nabla_1 \log \tilde{f}_\theta(P_j y, t) \tag{23}$$

computes patch-wise MMSE estimates and combines them by averaging. To bridge the local-global gap and allow patch cross-talk, we propose to find $\arg\min_x \frac{\lambda}{2}\|x - y\|^2 - e(x, \bar{t})$ with a proximal gradient continuation (PGC) scheme utilizing an empirical Bayes step size rule and a predefined schedule for $\bar{t}^{(k)}$ that

Table 1. Quantitative denoising results. Reference numbers are taken from [20].

255σ	FoE [15]	GMM-EPLL [20]	GMDM ($b = 7 \mid b = 15$)			
			EB-PA		PGC	
15	30.18	31.21	30.00	30.34	30.37	30.69
25	27.77	28.72	27.47	27.81	28.13	28.39
50	23.29	25.72	24.61	24.96	25.32	25.50
100	16.68	23.19	22.14	22.74	23.07	23.12
#Params	648	819 200	8352	78 400	8352	78 400

approaches 0 starting from $\bar{t}^{(0)} = t$. The algorithm reduces to

$$\hat{x}_{\text{PGC}}^{(k+1)} = \frac{\hat{x}_{\text{EB-PA}}(\hat{x}_{\text{PGC}}^{(k)}, \bar{t}^{(k)}) + 2\lambda \bar{t}^{(k)} y}{1 + 2\lambda \bar{t}^{(k)}} \tag{24}$$

where $\hat{x}_{\text{PGC}}^{(0)} = y$ and we found λ via grid-search on the training data.

We show a quantitative comparison on the 68 test images from [9] against FoE [15] and GMM-EPLL [20] in Table 1. The reference methods were chosen since they share some methodology in terms of density estimation and have roughly the same computational complexity. Especially for high noise levels, our method shows competitive performance given the relatively small number of parameters.

We want to emphasize that these results were obtained when applying the patch-based prior "convolutionally" (in the sense of (23), i.e. we consider all overlapping patches). In analogy to the gain observed when comparing the product-of-experts to the FoE model [15, Tab. 3 vs. Tab. 5], we expect significant performance improvements when utilizing our proposed convolutional prior (11).

4.4 Noise Estimation and Blind Image Denoising

The construction of our model allows us to interpret $\tilde{f}_\theta(\,\cdot\,, t)$ as a time-conditional likelihood density. Thus, it can naturally be used for noise estimation: Fig. 3 shows the expected negative-log density[4] $\mathbb{E}_{p \sim f_X, \eta \sim \mathcal{N}(0,\text{Id})} l_\theta(p + \sigma\eta, t)$ over a range of σ and t. The noise estimate $\sigma \mapsto \arg\min_t \mathbb{E}_{p \sim f_X, \eta \sim \mathcal{N}(0,\text{Id})} l_\theta(p + \sigma\eta, t)$ perfectly matches the identity map $\sigma \mapsto \sqrt{2t}$.

We demonstrate that we can utilize *one* model for *noise estimation* and *heteroscedastic blind denoising* in Fig. 4: For all overlapping patches $P_j y$ in the corrupted image, we estimate the noise level through $\hat{t}_j = \arg\max_t \tilde{f}_\theta(P_j y, t)$ and reconstruct the image with one empirical Bayes step

$$\hat{x}_{\text{blind}}(y) = y + 2 \sum_{j=1}^{\tilde{n}} \hat{t}_j p_j^{-1} P_j^\top \nabla_1 \log \tilde{f}_\theta(P_j y, \hat{t}_j). \tag{25}$$

[4] For visualization purposes, we normalized the negative-log density to have a minimum of zero over t: $l_\theta(x, t) = -\log \tilde{f}_\theta(x, t) - (\max_t \log \tilde{f}_\theta(x, t))$.

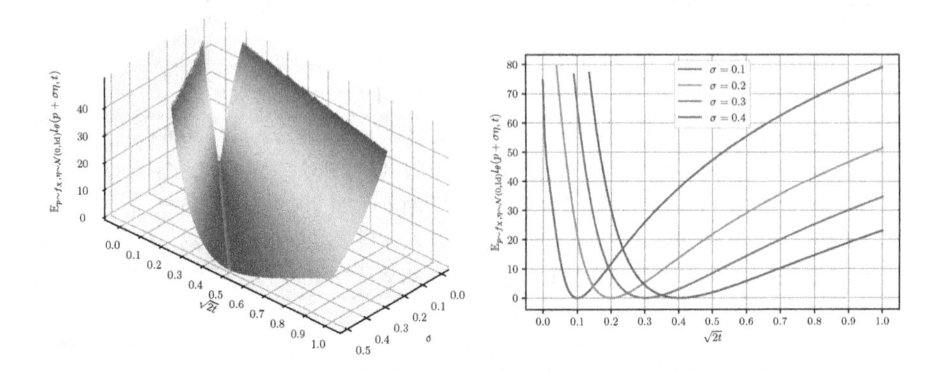

Fig. 3. Expected normalized negative-log density along with the noise estimate —— $\sigma \mapsto \arg\min_t \mathbb{E}_{p\sim f_X,\eta\sim\mathcal{N}(0,\mathrm{Id})} l_\theta(p+\sigma\eta,t)$;, —— $\sigma \mapsto \sqrt{2t}$; (left) and the slices at $\sigma \in \{0.1, 0.2, 0.3, 0.4\}$ (right).

Fig. 4. Blind denoising, from left to right: Image corrupted by heteroscedastic Gaussian noise in a checkerboard-pattern (standard deviation 0.1 and 0.2), noise estimation map, EB-PA denoising result, and absolute difference to the reference image.

In the difference image (Fig. 4, right), the checkerboard-like noise map is hardly visible, indicating that the noise estimation is robust also when confronted with little data.

5 Conclusion

In this paper, we introduced GMDMs as products of GMMs on filter responses that allow for an explicit solution of the diffusion equation of the associated density. Our explicit formulation enables learning of FoE-like image priors simultaneously for all diffusion times using denoising score matching. Our numerical results demonstrated that GMDMs capture the statistics of natural image patches well for any noise level and hence are suitable for heteroscedastic (blind) image denoising. In future work, we plan to extend the numerical evaluation to the convolutional model and apply our framework to challenging inverse problems in medical imaging.

References

1. Bengio, Y., Courville, A., Vincent, P.: Representation learning: a review and new perspectives. IEEE Trans. Pattern Anal. Mach. Intell. **35**(8), 1798–1828 (2013)

2. Cole, K., Beck, J., Haji-Sheikh, A., Litkouhi, B.: Heat Conduction Using Greens Functions. CRC Press, Boca Raton (2010)
3. Efron, B.: Tweedie's formula and selection bias. J. Am. Stat. Assoc. **106**(496), 1602–1614 (2011)
4. Gut, A.: An Intermediate Course in Probability. Springer, New York (2009). https://doi.org/10.1007/978-1-4419-0162-0
5. Guth, F., Coste, S., Bortoli, V.D., Mallat, S.: Wavelet score-based generative modeling. In: Advances in Neural Information Processing Systems (2022)
6. Hinton, G.E.: Training products of experts by minimizing contrastive divergence. Neural Comput. **14**(8), 1771–1800 (2002)
7. Hyvarinen, A.: Estimation of non-normalized statistical models by score matching. J. Mach. Learn. Res. 14 (2005)
8. Kirkpatrick, S., Gelatt, C.D., Vecchi, M.P.: Optimization by simulated annealing. Science **220**(4598), 671–680 (1983)
9. Martin, D., Fowlkes, C., Tal, D., Malik, J.: A database of human segmented natural images and its application to evaluating segmentation algorithms and measuring ecological statistics. In: Proceedings of the International Conference on Computer Vision, vol. 2, pp. 416–423 (2001)
10. Miyasawa, K.: An empirical bayes estimator of the mean of a normal population. In: Bulletin of the International Statistical Institute, pp. 161–188 (1961)
11. Pock, T., Sabach, S.: Inertial proximal alternating linearized minimization (iPALM) for nonconvex and nonsmooth problems. SIAM J. Imag. Sci. **9**(4), 1756–1787 (2016)
12. Raphan, M., Simoncelli, E.P.: Least squares estimation without priors or supervision. Neural Comput. **23**(2), 374–420 (2011)
13. Robbins, H.: An empirical bayes approach to statistics. In: Proceedings of the Berkeley Symposium on Mathematical Statistics and Probability, pp. 157–163 (1956)
14. Roberts, G.O., Tweedie, R.L.: Exponential convergence of langevin distributions and their discrete approximations. Bernoulli **2**(4), 341–363 (1996)
15. Roth, S., Black, M.J.: Fields of experts. Int. J. Comput. Vision **82**(2), 205–229 (2009)
16. Schrempf, O., Feiermann, O., Hanebeck, U.: Optimal mixture approximation of the product of mixtures. In: Proceedings of the International Conference on Information Fusion, vol. 1, pp. 85–92 (2005)
17. Song, Y., Ermon, S.: Generative modeling by estimating gradients of the data distribution. In: Advances in Neural Information Processing Systems, vol. 32 (2019)
18. Song, Y., Sohl-Dickstein, J., Kingma, D.P., Kumar, A., Ermon, S., Poole, B.: Score-based generative modeling through stochastic differential equations. In: Proceedings of the International Conference on Learning Representations (2021)
19. Vincent, P.: A connection between score matching and denoising autoencoders. Neural Comput. **23**(7), 1661–1674 (2011)
20. Zoran, D., Weiss, Y.: From learning models of natural image patches to whole image restoration. In: Proceedings of the International Conference on Computer Vision, pp. 479–486 (2011)

Efficient Neural Generation of 4K Masks for Homogeneous Diffusion Inpainting

Karl Schrader(✉), Pascal Peter, Niklas Kämper, and Joachim Weickert

Mathematical Image Analysis Group, Faculty of Mathematics and Computer Science, Campus E1.7, Saarland University, 66041 Saarbrücken, Germany
{schrader,peter,kaemper,weickert}@mia.uni-saarland.de

Abstract. With well-selected data, homogeneous diffusion inpainting can reconstruct images from sparse data with high quality. While 4K colour images of size 3840×2160 can already be inpainted in real time, optimising the known data for applications like image compression remains challenging: Widely used stochastic strategies can take days for a single 4K image. Recently, a first neural approach for this so-called mask optimisation problem offered high speed and good quality for small images. It trains a mask generation network with the help of a neural inpainting surrogate. However, these mask networks can only output masks for the resolution and mask density they were trained for. We solve these problems and enable mask optimisation for high-resolution images through a neuroexplicit coarse-to-fine strategy. Additionally, we improve the training and interpretability of mask networks by including a numerical inpainting solver directly into the network. This allows to generate masks for 4K images in around 0.6 s while exceeding the quality of stochastic methods on practically relevant densities. Compared to popular existing approaches, this is an acceleration of up to four orders of magnitude.

Keywords: Image Inpainting · Diffusion · Partial Differential Equations · Data Optimisation · Deep Learning

1 Introduction

Inpainting-based image compression [12] is surprisingly simple: During encoding, only a carefully optimised subset of all pixel locations and their values is stored. In the decoding phase, the missing information is reconstructed in a lossy way by some inpainting method. Optimising the inpainting data, the so-called mask, is essential for obtaining a good reconstruction: Even with a simple method such as homogeneous diffusion inpainting [6], results can be surprisingly good.

This work has received funding from the European Research Council (ERC) under the European Union's Horizon 2020 research and innovation programme (grant agreement no. 741215, ERC Advanced Grant INCOVID).

Supplementary Information The online version contains supplementary material available at https://doi.org/10.1007/978-3-031-31975-4_2.

L. Calatroni et al. (Eds.): SSVM 2023, LNCS 14009, pp. 16–28, 2023.
https://doi.org/10.1007/978-3-031-31975-4_2

Simple model-driven approaches for this optimisation problem require a compromise between reconstruction quality and speed [4,17]. Fast and qualitatively convincing results rely on sophisticated algorithms and implementations [7]. A first attempt to combine both advantages within a deep learning setting has been proposed recently by Alt et al. [2]. It is simple, fast, and approaches the quality of widely used model-driven optimisation strategies. It trains a mask generator network with the help of a neural inpainting surrogate. Extensions [21] allow to optimise not only the positions, but also the pixel values of the known data.

While offering a very good combination of quality and speed for greyscale images of sizes up to 256×256 and colour images up to 128×128, it is inflexible: The mask network can only generate masks for the density and resolution it was trained for. Furthermore, a naïve extension to high-resolution images requires a prohibitive amounts of compute power during training: At a resolution of 3840×2160, 4K colour images have around 25 million values, two to three orders of magnitude more than previously considered.

Our Contribution. With the first coarse-to-fine approach for neural mask generation, we address the weak points of the neural approach while preserving its high speed and increasing the quality of the generated masks. We partition the image into patches, and generate masks for each. Mask pixel budgets are assigned to each patch using an optimality result of Belhachmi et al. [4]. This constitutes the first transfer of their findings to the discrete setting which does not involve dithering, a step that usually leads to substantial quality losses. Our approach matches the quality of widely used stochastic mask optimisation strategies on 4K images of size 3840×2160, while being up to four orders of magnitude faster. With a new training process, we solve the challenging problem of integrating model-based inpainting directly into the network. This removes the previous need for surrogate solvers and yields an overall more transparent and reliable architecture. Additionally, it greatly reduces the number of trainable weights and speeds up the training.

Related Work. There are many approaches for mask optimisation. The taxonomy below is an extension of the one of Peter et al. [21].

1. *Analytic Approaches.* For the continuous setting, Belhachmi et al. [4] were able to show that optimal masks can be derived from the Laplacian magnitude (modulus) of the image. To discretise this result, the Laplacian magnitude has been dithered so far. We apply this result in a novel way to estimate optimal patch densities in Sect. 4. Since analytic approaches do not require any inpaintings, they are very fast. However, dithering comes with significant compromises and typically results in low quality [17].
2. *Nonsmooth Optimisation Strategies.* Finding a good inpainting mask that minimises the mean squared error of the reconstruction can be phrased as a nonsmooth bilevel optimisation problem. It can be solved with primal-dual methods [5,14,18]. While being able to achieve high quality, they do not allow the user to specify a target density as for to our approach. In addition, they produce nonbinary masks and require binarisation as a postprocessing step

which leads to a loss of quality. This can be alleviated by applying tonal optimisation [17], which optimises not only the position, but also the value of the mask pixels. While this can introduce small errors to the known data, it is often beneficial for the overall quality of the reconstruction.

3. *Sparsification Methods.* This class of methods starts with a full mask and successively remove mask pixels with a low impact on reconstruction quality. One widely used implementation is *probabilistic sparsification* by Mainberger et al. [17]. It allows a target density to be specified and produces binary masks directly. However, it requires an inpainting to be computed in every step which results in long runtimes, especially on large images.
4. *Densification Methods.* They start with an empty mask and successively add those pixels to it which increase reconstruction quality the most [9,12]. Combined with ideas like finite elements [7], they constitute a fairly new but promising class of methods.
5. *Relocation Methods.* Greedy optimisation strategies can get stuck in local minima. To escape from them, *nonlocal pixel exchange* [17] tests if moving some randomly selected mask pixels into the unknown regions leads to a better reconstruction. Successful moves are kept, while unsuccessful ones are reverted. Given sufficient time, this method can produce very good masks. It is usually applied as a postprocessing tool.
6. *Neural Methods.* Works such as [8,20] learn the inpainting operator along with a mask generation network. As this results in an opaque model, we focus on well-understood homogeneous diffusion inpainting. For it, [2,21] have shown that a mask generation network can be successfully trained with the help of an inpainting surrogate, but focus on small images of sizes up to 256×256. Opposed to our model, new densities and resolutions require a new model to be trained. Their framework also allows for the training of a tonal optimisation network with the same strategy. While this is also the case for our approach, we focus on the spatial case due to space limitations. Inference with neural mask generation is very fast since it requires no inpaintings.

These approaches can be summarised as follows: Category 1 is very fast as it requires no inpaintings, but offers low quality. Higher quality is be achieved by Categories 2–4 at the expense of speed, as they all need to calculate many inpaintings. Given sufficient iterations, Category 5 can improve any mask further. Previous neural methods from Category 6 offer good quality and speed, but are not as flexible as model-driven approaches. We will exploit optimality results from Category 1 without sacrificing quality through dithering to alleviate the shortcomings of Category 6.

Organisation of the Paper. Section 2 reviews inpainting with homogeneous diffusion and model-driven spatial optimisation. Section 3 presents our improvements to the baseline neural mask generation architecture, followed by our new coarse-to-fine mask generation strategy in Sect. 4. Section 5 provides empirical evidence for our patch density estimation and compares our neural approach against model-driven ones. We end with conclusions and an outlook in Sect. 6.

2 Homogeneous Diffusion Inpainting

Let us consider a continuous greyscale image $f : \Omega \to \mathbb{R}$ defined on a rectangular image domain Ω which is only known on a mask $K \subset \Omega$. The goal of inpainting is to restore the inpainting region $\Omega \setminus K$ with a homogeneous diffusion process that propagates information; see e.g. [13]. For homogeneous diffusion inpainting [6], the reconstruction solves the partial differential equation

$$(1 - c)\Delta u - c(u - f) = 0 \tag{1}$$

where $c : \Omega \to \{0, 1\}$ is a binary confidence function which indicates if a point $\boldsymbol{x}$ is part of the mask. Known values are specified by $c(\boldsymbol{x}) = 1$ and are preserved. Missing data are indicated by $c(\boldsymbol{x}) = 0$ and are inpainted with the steady state $\Delta u = 0$ of the homogeneous diffusion equation, where $\Delta = \partial_{xx} + \partial_{yy}$ is the Laplacian. We assume reflecting boundary conditions at the image boundary $\partial\Omega$. Nonbinary masks are also well-defined within the mathematical framework, but are hard to compress and only used as an intermediate step in this work. For colour images, each channel is considered independently, since homogeneous diffusion is a linear operator.

While more sophisticated inpainting models exist [12,23], homogeneous diffusion is attractive for inpainting-based compression: Its simplicity allows for an analytic continuous theory for inpainting data selection [4] and enables fast implementations [15]. Furthermore, its complete lack of parameters make it easy to use, and it can yield high-quality reconstruction for well optimised inpainting data. These properties make it a suitable candidate for mask optimisation.

In the discrete setting, we consider digital colour images $\boldsymbol{f} \in \mathbb{R}^{3n_x n_y}$ with resolution $n_x \times n_y$. The inpainting Eq. (1) is discretised with finite differences, leading to a linear system of equations. The reconstruction $\boldsymbol{u}$ can then be computed by a suitable numerical solver such as the conjugate gradient method; see e.g. [24]. Mask optimisation strategies are designed to generate discrete masks $\boldsymbol{c} \in \{0, 1\}^{n_x n_y}$ such that the corresponding reconstruction $\boldsymbol{u}$ approximates the original image well. They are constrained by the mask density $\|\boldsymbol{c}\|_1/(n_x n_y)$ where $\|.\|_1$ is the 1-norm, which should match the desired target density d.

2.1 Model-Based Mask Optimisation

A mask optimisation strategy that does not require any inpaintings has been proposed by Belhachmi et al. [4]. They show that for optimal masks, the local density should increase with the Laplacian magnitude. In Sect. 4 we use this result to predict a suitable density for image regions. However, the application of these results to discrete images relies on dithering. The commonly used Floyd-Steinberg dithering [11] is fast and simple, but suffers from a directional bias and generates masks of relatively low quality.

A method which produces better masks but remains simple is *probabilistic sparsification* [17]. It starts with a full mask and gradually removes pixels until the target density d is reached. In every iteration, a fraction p of mask pixels is

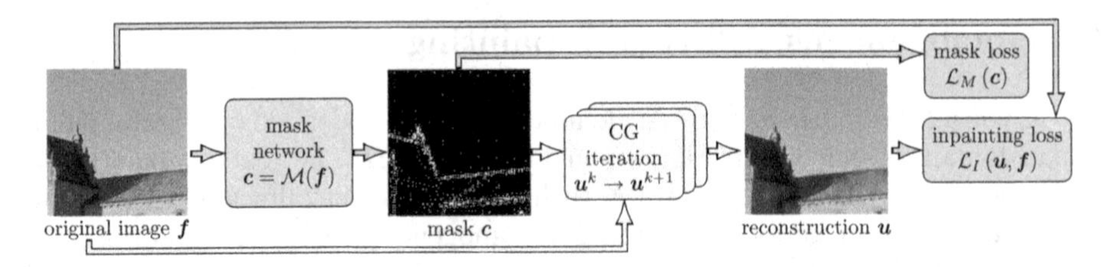

Fig. 1. Our Mask Optimisation Architecture. The mask network and its losses are coloured in blue, our newly introduced sequence of CG iterations is given in yellow.

removed. Then, an inpainting is computed, and the fraction q which resulted in the largest loss of quality is added back.

We also consider *nonlocal pixel exchange* [17] as a post-processing method. In each step, it moves a fraction p of mask pixels into the unknown image area. It then inpaints with the new mask to check for improvements: Exchanges which increase inpainting quality are kept, while unsuccessful ones are reverted. This is repeated for k cycles with $\|\boldsymbol{c}\|_1$ iterations each. Nonlocal pixel exchange can help greedy approaches like probabilistic sparsification escape from poor local minima at the cost of significant runtime.

Both probablisitic sparsification and nonlocal pixel exchange require an inpainting for each iteration, with the other operations being quick in comparison. As such, their runtime is primarily determined by the number and the speed of the inpainting operations. A naïve CPU-based inpainter using a conjugate gradient solver takes around 3.5 seconds per 4K colour image on a contemporary PC. At this speed, 5 cycles of nonlocal pixel exchange for a mask with 5% known data would take around 12 weeks. To enable meaningful comparisons, we have developed implementations of probablistic sparsification and nonlocal pixel exchange that use the very fast GPU-based inpainting of Kämper et al. [15]. With it, an iteration of either method requires only ≈ 6 ms, a speedup of two orders of magnitude. To the best of our knowledge, our implementations are the fastest available.

3 Neural Mask Generation

We follow the basic network architecture from Peter et al. [21] with a mask generator network and an inpainting approximator. The latter is used to train the mask network by evaluating the reconstruction quality and is discarded during inference. An overview of our architecture can be found in Fig. 1. While we retain the mask network as is, we completely replace the original inpainting approximator.

Mask Network. The mask network receives the original image $\boldsymbol{f}$ and generates a mask $\boldsymbol{c} = \mathcal{M}(\boldsymbol{f})$. Its output is restricted to $[0, 1]$ by applying a sigmoid activation function. Additionally, the network outputs are rescaled if they exceed the desired density d. The goal of the network is to produce masks such that the corresponding inpainted image $\boldsymbol{u}$ is close to the original $\boldsymbol{f}$. To this end, we minimise their mean squared error (MSE) $\mathcal{L}_I(\boldsymbol{u}, \boldsymbol{f}) = \frac{1}{n_x n_y} \|\boldsymbol{u} - \boldsymbol{f}\|_2^2$ where $\|.\|_2$ is the Euclidean norm. In addition, we require a mechanism to encourage the generation of binary masks. To this end, we penalise the inverse variance of the mask using $\mathcal{L}_M(\boldsymbol{c}) = \alpha(\sigma_{\boldsymbol{c}}^2 + \varepsilon)^{-1}$ where ε is a small numerical constant to avoid division by 0, and α balances variance loss $\mathcal{L}_M$ and inpainting loss $\mathcal{L}_I$.

Inpainting Approximator. To train the mask network with an inpainting loss $\mathcal{L}_I$, we need to approximate the inpainting process inside the network. Previous approaches [2,21] facilitate this by including a surrogate inpainting network which receives the original $\boldsymbol{f}$ and mask $\boldsymbol{c}$, and is trained to find a reconstruction $\boldsymbol{u}$ which solves the discrete version of the inpainting Eq. (1):

$$(\boldsymbol{I} - \boldsymbol{C})\boldsymbol{A}\boldsymbol{u} - \boldsymbol{C}(\boldsymbol{u} - \boldsymbol{f}) = \boldsymbol{0} \tag{2}$$

Here, the matrix $\boldsymbol{C} = \operatorname{diag}(\boldsymbol{c})$ contains the mask entries on the diagonal, and $\boldsymbol{A}$ applies a finite difference discretisation of the Laplacian Δ with reflecting boundary conditions. This approach significantly increases the total number of weights and adds complexity to the architecture. Furthermore, it decreases interpretability as the surrogate is, apart from its loss function, a black box.

We propose to replace it by a sufficient number of iterations of a successful numerical solver. As homogeneous diffusion leads to a linear system of equations with a symmetric system matrix, we can use the conjugate gradient (CG) method [24]. It offers convergence guarantees while remaining simple and efficient. As each iteration is differentiable, we can backpropagate through the solver to train the mask network. By introducing this well-understood numerical solver, we have reduced the total number of weights by half compared to [21] while increasing interpretability. Section 5.2 confirms that this improves inpainting approximation quality greatly and the quality of generated masks slightly.

Mask Generation in Practice. As the generated masks are not guaranteed to be binary, we need to binarise them. To this end, Peter et al. [21] perform weighted coinflips at each mask pixel and then choose the best-performing mask out of 30 attempts. We found that rounding leads to a comparable quality for our masks. It is less time intensive and involves no randomness.

4 Coarse-to-Fine Approach for Mask Generation

Simply partitioning a large image into patches and generating masks with equal density for each leads to suboptimal results: Textured regions receive too few, while homogeneous areas receive too many mask pixels. As such, we require a fast and simple mechanism that estimates suitable densities for each patch while taking the whole image into account.

To this end, we estimate a good patch density using the average Laplacian magnitude per patch, similar to the approach by Belhachmi et al. [4]. They have shown that for optimal continuous masks, the local mask density should

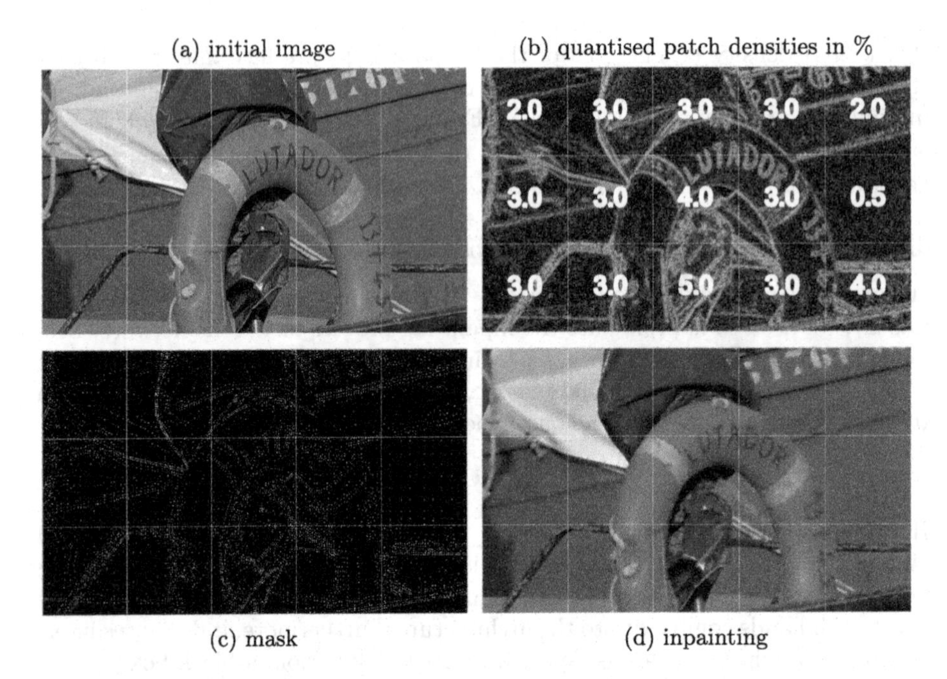

Fig. 2. Stages of the Mask Generation Process. (a) Initial image with subdivision into patches. (b) Quantised patch densities, and Laplacian magnitude with logarithmic dynamic compression for better visibility in the background. (c) Binary mask with 3% known data. (d) Inpainted result.

increase with the Laplacian magnitude. The discrete approximation of this result through dithering requires significant compromises. We, however, aggregate the Laplacian magnitude over an image patch to estimate the optimal density and thus avoid dithering completely. To the best of our knowledge, we are the first to apply the optimality result this way. This motivates the following algorithm, which is also visualised in Fig. 2:

1. Compute the Laplacian magnitude of the luma channel for every pixel.
2. Rescale it such that its global mean matches the target density.
3. Compute the target patch densities as the mean of the rescaled Laplacian magnitude per patch. Section 5.2 confirms that these target densities correlate well with high-quality masks.
4. Quantise the patch densities to values for which pre-trained mask networks are available and generate masks for every patch.
5. Assemble the mask patches into a mask for the whole image.

As an additional benefit, this methodology allows to generate masks with arbitrary densities. This is achieved by selecting densities per patch out of the available ones such that their mean approximates the desired average well.

Fig. 3. Our 12 test images of size 3840×2160. Photos by J. Weickert.

5 Experiments

After mentioning the technical details in Sect. 5.1, we show the improvements made by introducing a CG solver into the network in Sect. 5.2. Additionally, we provide empirical evidence for our patch density estimation in Sect. 5.3, and compare the mask generation performance for 4K images in Sect. 5.4.

5.1 Experimental Setup

For our mask generator, we use the same small multiscale context aggregation network [22] with ≈ 2.9 million parameters and settings as in [21] to facilitate direct comparisons. It is based on a U-Net architecture with four scales and uses blocks of parallel dilated convolutions with different dilation rates. All mask networks are trained with the losses described in Sect. 3. The mask variance loss weight is set to $\alpha = 0.01$. The surrogate network in Sect. 5.2 shares the mask network architecture and uses the squared residual of the discrete inpainting Eq. (2) as its loss. All networks are trained for 100 epochs with a batch size of 8 and the Adam optimiser [16] with a learning rate of $5 \cdot 10^{-5}$. Performance is evaluated on an AMD Ryzen 7 5800X CPU and an Nvidia RTX 3090 GPU. For comparisons against [21], we trained on a subset of 100,000 images from ImageNet [10] sampled by Dai et al. [8]. Tests are performed on the full BSDS500

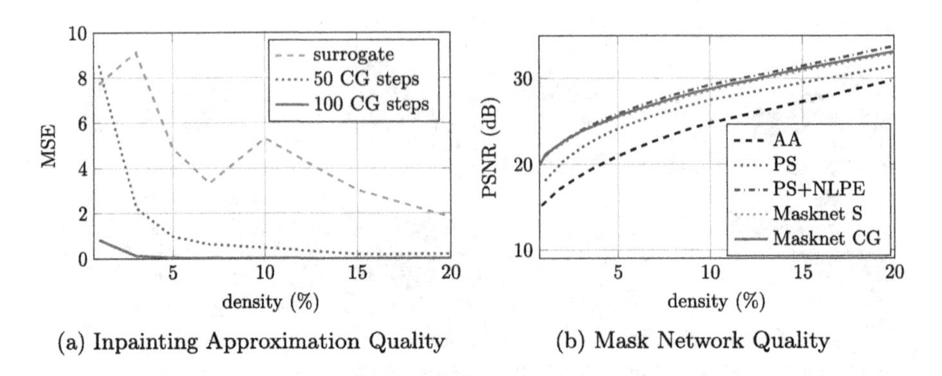

(a) Inpainting Approximation Quality (b) Mask Network Quality

Fig. 4. Comparison of Different Inpainting Approximators. (a) Distance between the inpainting approximations and a converged inpainting. The MSEs decrease in a nonmonotone fashion due to the varying quality of different mask and inpainting networks. (b) Comparison of mask networks trained with either a surrogate inpainter (Masknet S) or 100 CG iterations (Masknet CG). Even though 100 CG iterations are a much better inpainting approximator, the quality of the mask network trained with them is only slightly improved. Both methods are better than the analytic approach (AA) and probablistic sparsification (PS), and are very close to PS with added nonlocal pixel exchange (PS+NLPE).

dataset [3]. We used 128×128 centre crops for both. The networks used for 4K inpainting were trained on 100,000 random patches of size 120×120 from the high-resolution image dataset Div2K [1]. Tests are performed on 12 representative 4K images photographed by one of the authors; see Fig. 3. Our selection of pre-trained networks is optimised for target densities $\leq 10\%$, as those are practically relevant for compression. The set contains networks for different densities in the range from 0.5% to 80%. For 1%–15%, we use increments of 1%. For 15%–25% we increment by 2%, and by 5% for densities between 25% and 50%. Finally, in the range of 50%–80% we have steps of 10%. Reference inpaintings are computed using a CG solver which is stopped after a relative decrease of the Euclidean norm of the residual of 10^{-6}.

For comparisons against model-driven approaches, we use the analytic approach (AA) by Belhachmi et al. [4], and probabilistic sparsification (PS) alone or combined with nonlocal pixel exchange (PS+NLPE). PS uses candidate fractions $p = 0.3$ and $q = 0.005$. NLPE runs for 5 cycles. The other NLPE parameters were optimised individually for the different target resolutions.

5.2 Quality of Inpainting Approximations

We measure the quality of different inpainting approximators by comparing against inpaintings with our reference solver. To avoid a bias against the surrogate inpainting networks, we test on binarised masks generated by their corresponding mask networks. In Fig. 4(a), we see that 100 CG iterations are a better

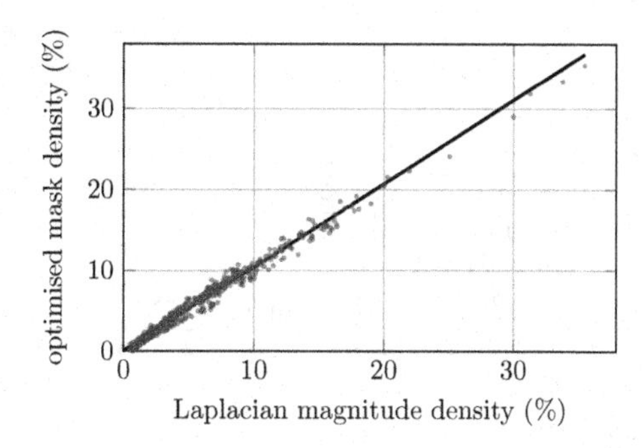

Fig. 5. Patch Density Correlation between Laplacian Magnitude and High Quality Mask. Comparison of the mean rescaled Laplacian magnitude for each patch of the image *lofsdalen* against patch densities of a high quality mask, both with a total density of 5%. A correlation coefficient of 0.994 confirms the strong connection.

approximator than the surrogate network across all densities. Additionally, they allow for faster training: On image batches, one forward and backward pass takes only 1.9 ms per image, while the surrogate requires 5.6 ms. Nevertheless, the quality of the trained mask network is only slightly improved by using CG as an inpainting approximator as can be seen in Fig. 4(b). This makes sense as the reconstruction error is about 3000 times larger than the approximation error for 100 CG steps and all densities, making small deviations insignificant.

5.3 Justification of Mask Pixel Distribution

In Fig. 5 we show that the rescaled Laplacian magnitude is a good predictor of optimal mask densities at a patch level. There we compare against the patch densities of a well optimised mask generated with the approach from Chizhov and Weickert [7]. The relationship between the patch densities is almost linear, and the mean absolute error between densities is only 0.5%. While dithering the rescaled Laplacian magnitude leads to low-quality masks, its aggregation over a patch avoids this and produces a good coarse scale density estimate.

5.4 High-Resolution Mask Generation

The tests on 4K images in Fig. 6(a) show that our masknet trained with 100 CG iteration is superior to AA and PS for all densities. In addition, we are even able to outperform PS+NLPE on densities smaller than 8%. This range is practically relevant for inpainting-based compression.

In Fig. 6(b) we compare the speed of the different methods. The runtime of PS+NLPE scales with the required inpaintings, taking between 30 min and 8.5 h depending on the density. In contrast, our neural approach takes only about 0.6 seconds per image. It is even quicker for the lowest densities where

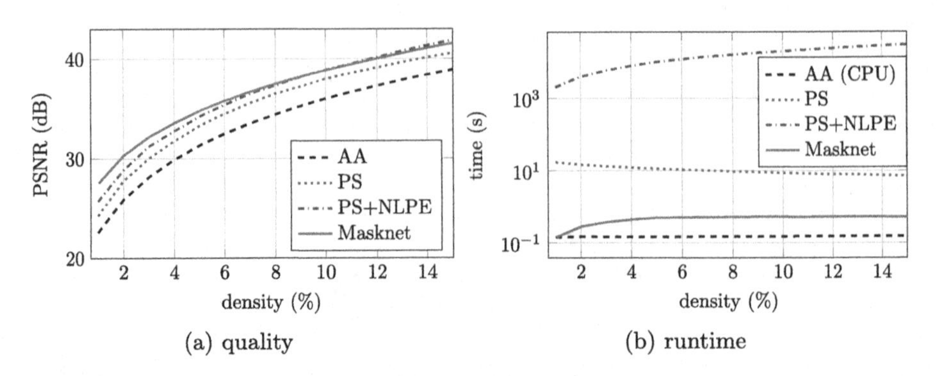

Fig. 6. Spatial Optimisation. Our masknet outperforms the faster analytic approach, but also the slower PS across all densities. It even beats the significantly slower PS+NLPE for densities $\leq 8\%$. Time measurements exclude input/output operations.

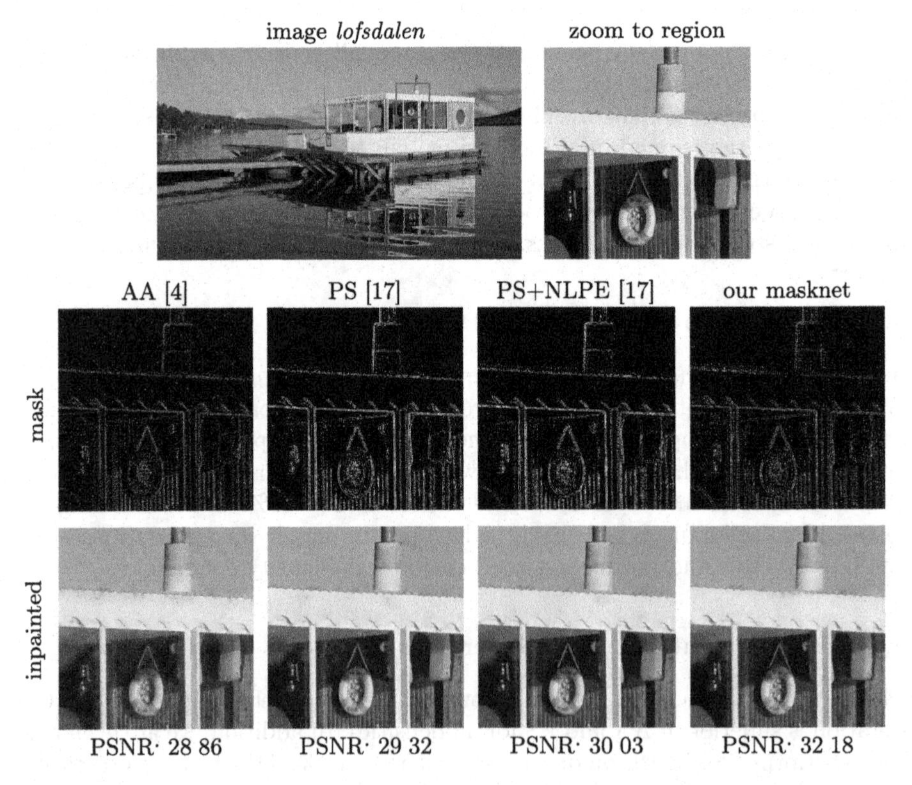

Fig. 7. Visual Comparison for 4% Mask Density on *lofsdalen*. PSNRs are for the whole image. Notice the discoloured sky for PS and PS+NLPE. The full images are available in the supplementary material.

fewer of the pre-trained mask networks are active and the batches per net are larger. Our masknet is also more than 10 times faster than the qualitatively worse PS. The analytic approach requires no inpaintings, and even a CPU-based implementation is significantly faster than our neural method. Still, its quality is inferior by a large margin. In Fig. 7, the most noticeable differences are slightly blurry edges for AA, and a discoloured sky for PS and PS+NLPE.

6 Conclusions

Our coarse-to-fine approach for mask generation is orders of magnitude faster than qualitatively similar approaches on 4K images. This shows that the estimation of patch densities using the optimality result by Belhachmi et al. [4] works well, and our mask networks are able to outperform dithering. Performance of the neural approach relative to PS+NLPE even grew with the increase in resolution. Our experiments suggest that the coarse-to-fine approach produces local problems that are easier and faster to solve in high quality. This is in line with transform-based codecs like JPEG [19], which also use a block structure to enable parallelisation and reduce complexity.

In the future, we will investigate these coarse-to-fine approaches for mask generation further using various local solvers. This avenue of research becomes even more desirable given the ever-increasing parallelisation capabilities of modern GPUs, rapidly increasing image resolutions, and the fact that reasonably simple solvers such as CG cannot offer optimal linear complexity.

References

1. Agustsson, E., Timofte, R.: NTIRE 2017 challenge on single image super-resolution: dataset and study. In: Proceedings of the 2017 IEEE Computer Society Conference on Computer Vision and Pattern Recognition Workshops, Honolulu, vol. 1, pp. 1122–1131 (2017)
2. Alt, T., Peter, P., Weickert, J.: Learning sparse masks for diffusion-based image inpainting. In: Pinho, A.J., Georgieva, P., Teixeira, L.F., Sánchez, J.A. (eds.) IbPRIA 2022. LNCS, vol. 13256, pp. 528–539. Springer, Cham (2022). https://doi.org/10.1007/978-3-031-04881-4_42
3. Arbelaez, P., Maire, M., Fowlkes, C., Malik, J.: Contour detection and hierarchical image segmentation. IEEE Trans. Pattern Anal. Mach. Intell. **33**(5), 898–916 (2011)
4. Belhachmi, Z., Bucur, D., Burgeth, B., Weickert, J.: How to choose interpolation data in images. SIAM J. Appl. Math. **70**(1), 333–352 (2009)
5. Bonettini, S., Loris, I., Porta, F., Prato, M., Rebegoldi, S.: On the convergence of a linesearch based proximal-gradient method for nonconvex optimization. Inverse Prob. **33**(5), 055005 (2017)
6. Carlsson, S.: Sketch based coding of grey level images. Signal Process. **15**, 57–83 (1988)
7. Chizhov, V., Weickert, J.: Efficient data optimisation for harmonic inpainting with finite elements. In: Tsapatsoulis, N., Panayides, A., Theocharides, T., Lanitis, A., Pattichis, C., Vento, M. (eds.) CAIP 2021. LNCS, vol. 13053, pp. 432–441. Springer, Cham (2021). https://doi.org/10.1007/978-3-030-89131-2_40

8. Dai, Q., Chopp, H., Pouyet, E., Cossairt, O., Walton, M., Katsaggelos, A.K.: Adaptive image sampling using deep learning and its application on X-ray fluorescence image reconstruction. IEEE Trans. Multimedia **22**(10), 2564–2578 (2019)
9. Daropoulos, V., Augustin, M., Weickert, J.: Sparse inpainting with smoothed particle hydrodynamics. SIAM J. Appl. Math. **14**(4), 1669–1704 (2021)
10. Deng, J., Dong, W., Socher, R., Li, L.J., Li, K., Fei-Fei, L.: ImageNet: a large-scale hierarchical image database. In: Proceedings of the 2009 IEEE Computer Society Conference on Computer Vision and Pattern Recognition, Miami, pp. 248–255 (2009)
11. Floyd, R.W., Steinberg, L.: An adaptive algorithm for spatial grey scale. In: Proceedings of the Society of Information Display, vol. 17, pp. 75–77 (1976)
12. Galić, I., Weickert, J., Welk, M., Bruhn, A., Belyaev, A., Seidel, H.P.: Image compression with anisotropic diffusion. J. Math. Imaging Vis. **31**(2–3), 255–269 (2008)
13. Guillemot, C., Le Meur, O.: Image inpainting: overview and recent advances. IEEE Signal Process. Mag. **31**(1), 127–144 (2014)
14. Hoeltgen, L., Setzer, S., Weickert, J.: An optimal control approach to find sparse data for Laplace interpolation. In: Heyden, A., Kahl, F., Olsson, C., Oskarsson, M., Tai, X.C. (eds.) Energy Minimisation Methods in Computer Vision and Pattern Recognition. Lecture Notes in Computer Science, vol. 8081, pp. 151–164. Springer, Heidelberg (2013). https://doi.org/10.1007/978-3-642-40395-8_12
15. Kämper, N., Weickert, J.: Domain decomposition algorithms for real-time homogeneous diffusion inpainting in 4K. In: Proceedings of the 2022 IEEE International Conference on Acoustics, Speech and Signal Processing, Singapore, pp. 1680–1684 (2022)
16. Kingma, D.P., Ba, J.: Adam: a method for stochastic optimization. In: Proceedings of the 3rd International Conference on Learning Representations, San Diego (2015)
17. Mainberger, M., et al.: Optimising spatial and tonal data for homogeneous diffusion inpainting. In: Bruckstein, A.M., ter Haar Romeny, B., Bronstein, A.M., Bronstein, M.M. (eds.) SSVM 2011. LNCS, vol. 6667, pp. 26–37. Springer, Heidelberg (2012). https://doi.org/10.1007/978-3-642-24785-9_3
18. Ochs, P., Chen, Y., Brox, T., Pock, T.: iPiano: inertial proximal algorithm for nonconvex optimization. SIAM J. Imag. Sci. **7**(2), 1388–1419 (2014)
19. Pennebaker, W.B., Mitchell, J.L.: JPEG: Still Image Data Compression Standard. Springer, New York (1992)
20. Peter, P.: A Wasserstein GAN for joint learning of inpainting and its spatial optimisation. arXiv:2202.05623 [eess.IV] (2022)
21. Peter, P., Schrader, K., Alt, T., Weickert, J.: Deep spatial and tonal data optimisation for homogeneous diffusion inpainting. arXiv:2208.14371 [eess.IV] (2022)
22. Vašata, D., Halama, T., Friedjungová, M.: Image inpainting using Wasserstein generative adversarial imputation network. In: Farkaš, I., Masulli, P., Otte, S., Wermter, S. (eds.) ICANN 2021. LNCS, vol. 12892, pp. 575–586. Springer, Cham (2021). https://doi.org/10.1007/978-3-030-86340-1_46
23. Weickert, J., Welk, M.: Tensor field interpolation with PDEs. In: Weickert, J., Hagen, H. (eds.) Visualization and Processing of Tensor Fields, pp. 315–325. Springer, Heidelberg (2006). https://doi.org/10.1007/3-540-31272-2_19
24. Wendland, H.: Numerical Linear Algebra: An Introduction. Cambridge University Press, Cambridge (2017)

Theoretical Foundations for Pseudo-Inversion of Nonlinear Operators

Eyal Gofer(✉) and Guy Gilboa

Technion - Israel Institute of Technology, 32000 Haifa, Israel
{eyal.gofer,guy.gilboa}@ee.technion.ac.il
https://www.vision-and-sensing.com/

Abstract. The Moore-Penrose inverse is widely used in physics, statistics, and various fields of engineering. It captures well the notion of inversion of linear operators in the case of overcomplete data. In data science, *nonlinear operators* are extensively used. In this paper we characterize the fundamental properties of a pseudo-inverse (PI) for nonlinear operators.

The concept is defined broadly. First for general sets, and then a refinement for normed spaces. The PI for normed spaces yields the Moore-Penrose inverse when the operator is a matrix. We present conditions for existence and uniqueness of a PI and establish theoretical results investigating its properties, such as continuity, its value for operator compositions and projection operators, and others. Analytic expressions are given for the PI of some well-known, non-invertible, nonlinear operators, such as hard- or soft-thresholding and ReLU. Finally, we analyze a neural layer and discuss relations to wavelet thresholding.

Keywords: pseudo-inverse · generalized inverse · inverse problems · nonlinear operators

1 Introduction

Operator inversion is a fundamental mathematical problem concerning various fields in science and engineering. The vast majority of research devoted to this topic is concerned with generalized inversion of *linear operators*. However, *nonlinear operators* are extensively used today in numerous domains, and specifically in machine learning and image processing. Given that many nonlinear operators are not invertible, it is highly instrumental to formulate the generalized inverse of nonlinear operators and to analyze its properties.

Generalized inversion of linear operators has been studied extensively since the 1950's, following the paper of Penrose [10]. As noted in [1,2], that work rediscovered, simplified, and made more accessible the definitions first made by Moore [9], in what is referred to today as *the Moore-Penrose inverse*, often also called the matrix pseudo-inverse. A detailed overview of various generalized inverse definitions and properties is available in the book of Ben-Israel and Greville [2].

Supplementary Information The online version contains supplementary material available at https://doi.org/10.1007/978-3-031-31975-4_3.

L. Calatroni et al. (Eds.): SSVM 2023, LNCS 14009, pp. 29–41, 2023.
https://doi.org/10.1007/978-3-031-31975-4_3

The research related to generalized inversion of nonlinear operators has been very scarce. In [13] the notion of *nonlinear PI* is given, for the first time, to the best of our knowledge. It is in the context of least square estimation in image restoration. This topic, however, is not further developed. Characteristics of the PI and issues such as existence and uniqueness are not discussed. The most comprehensive study on generalized inversion of nonlinear operators is in the work of Dermanis [4]. While focused on geodesy, he defines inversion broadly. We significantly extend the initial results of Dermanis and attempt to establish a general theory, relevant to data science. In control applications (see [6,8]), pseudo-inversion is used for the design of controllers for nonlinear dynamics. The inversion is meant to approximately cancel the dynamic of the plant. In these studies, inversion is discussed in a rather narrow applied context.

The goal of this paper is to define a broad and general notion of PI of nonlinear operators in various settings and to analyze its mathematical properties. The main contributions of the paper are: 1. We explain and illustrate the general concept of pseudo-inversion of nonlinear operators for sets, as well as its fundamental properties. It relies on the first two axioms of Moore and Penrose and generally is not unique. 2. For nonlinear operators in normed spaces a stronger definition can be formulated, based on best approximate solution, which directly coincides with Moore-Penrose in the linear setting. We show that although the first Moore-Penrose axiom is implied, the second one is still required explicitly, unlike the linear case. 3. Certain theoretical results are established, such as conditions for existence and uniqueness, continuity of the inverse of a continuous operator over a compact set, and various settings in which the inverse of an operator composed by several simpler operators can be inferred. 4. Finally, analytic expressions of the PI for some canonical functions are given, as well as explicit computations of a neural-layer inversion and relations to wavelet thresholding. Due to space limitations, **some details and many proofs are omitted**, and may be found in the supplementary material.

Some Definitions and Notations. For an operator $T : V \to W$ between sets V and W, $T(V')$ denotes the image of $V' \subseteq V$ by T, and $T^{-1}(W')$ denotes the preimage of $W' \subseteq W$ by T. The restriction of T to a subset $V' \subseteq V$ is denoted by $T|_{V'}$. The composition of operators $T_1 : V_2 \to V_3$ and $T_2 : V_1 \to V_2$ is denoted by $T_1 \circ T_2$ or $T_1 T_2$. The set of all operators from V to W is denoted by W^V, and we write $|V|$ for the cardinality of the set V. An operator $T : V \to W$ is *idempotent* or a *generalized projection* iff $T \circ T = T$. We write $\arg\min_{v \in V}\{f(v)\}$ or $\arg\min\{f(v) : v \in V\}$ for the set of elements in V that minimize a function $f : V \to \mathbb{R}$. The closed ball with center a and radius r is denoted by $\bar{B}(a, r)$, and the n-dimensional sphere in $\mathbb{R}^{n+1}$ is denoted by S^n. The notation $\mathbb{R}^+$ specifies non-negative reals. The indicator function of an event A is denoted by $\mathbb{I}\{A\}$.

2 The Moore-Penrose Properties and Partial Notions of Inversion

The Moore-Penrose (MP) inverse is defined in the linear domain as follows.

Definition 1 (Moore-Penrose pseudo-inverse). *Let V and W be finite-dimensional inner-product spaces over $\mathbb{C}$. Let $T : V \to W$ be a linear operator with the adjoint*

$T^* : W \to V$. The PI of T is a linear operator $T^\dagger : W \to V$, which admits the following identities:

$$\text{MP1: } TT^\dagger T = T \qquad \text{MP3: } (TT^\dagger)^* = TT^\dagger$$
$$\text{MP2: } T^\dagger TT^\dagger = T^\dagger \qquad \text{MP4: } (T^\dagger T)^* = T^\dagger T$$

Penrose [10] showed that this PI exists uniquely for every linear operator. It may be calculated, for example, using singular value decomposition (SVD, see, e.g., [2]). The same work by Penrose showed that the PI is involutive, namely, $T^{\dagger\dagger} = T$. He also showed [11] that it yields a *best approximate solution* (BAS) to the equation $Tv = w$. That is, for every $w \in W$, $\|Tv - w\|_2$ is minimized over $v \in V$ by $v^* = T^\dagger(w)$, and among all such minimizers, it uniquely has the smallest L_2 norm.

Let us now examine how the Moore-Penrose scheme can be adapted to nonlinear operators. A direct application, unfortunately, does not work. Recall that the adjoint of a linear operator satisfies $\langle Tv, w\rangle = \langle v, T^* w\rangle$ for every $v \in V$, $w \in W$. As can be easily shown, the properties of the inner product restrict T to be linear. This means that the adjoint operation as defined here does not extend to nonlinear operators, and that MP3–4, which involve the adjoint, need to be replaced. Thus, MP1–2 are extended in a way that is suitable for nonlinear operators.

It should be noted that any possible subset of the four MP properties (and indeed other possible properties) may serve as the basis for the definition of a different type of inverse. These various inverses have been studied intensively for linear operators (see [2]). In keeping with the notation of [2], the MP PI is a $\{1, 2, 3, 4\}$-inverse. For any subset $\{i, j, \ldots, k\} \subseteq \{1, 2, 3, 4\}$ one may also talk of $T\{i, j, \ldots, k\}$, the set of all $\{i, j, \ldots, k\}$-inverses of the operator T. The same notations will be used here for a nonlinear operator T as well. We first examine $\{1, 2\}$-inverses of nonlinear operators, namely, inverses that satisfy MP1–2.

3 The $\{1,2\}$-Inverses of Nonlinear Operators

It is important to note that V and W are no longer required to be inner-product or even vector spaces, and they may in fact be any two general nonempty sets. The following lemma pinpoints the nature of a $\{1, 2\}$-inverse.

Lemma 1. *Let $T : V \to W$ and $T^\ddagger : W \to V$, where V and W are nonempty sets. The statement $T^\ddagger \in T\{1, 2\}$ is equivalent to the following: $\forall w \in T(V)$, $T^\ddagger(w) \in T^{-1}(\{w\})$, and $\forall w \notin T(V)$, $T^\ddagger(w) = T^\ddagger(w')$ for some $w' \in T(V)$.*

Proof. Assume that $T^\ddagger \in T\{1, 2\}$. If $w \in T(V)$, then $w = T(v)$ for some $v \in V$. By MP1, $w = T(v) = TT^\ddagger T(v) = TT^\ddagger(w)$, so $T^\ddagger(w) \in T^{-1}(\{w\})$. If $w \notin T(V)$, let $w' = T(T^\ddagger(w))$, so $w' \in T(V)$, and by MP2, $T^\ddagger(w') = T^\ddagger(w)$.

In the other direction, for every v we have $T(v) \in T(V)$, therefore $T^\ddagger(T(v)) \in T^{-1}(\{T(v)\})$, implying that $TT^\ddagger T(v) = T(v)$, satisfying MP1. As for MP2, if $w \in T(V)$, we have $T^\ddagger(w) \in T^{-1}(\{w\})$, so $TT^\ddagger(w) = w$, and thus $T^\ddagger TT^\ddagger(w) = T^\ddagger(w)$. If $w \notin T(V)$, then $T^\ddagger(w) = T^\ddagger(w')$ for some $w' \in T(V)$, and we already know that $T^\ddagger TT^\ddagger(w') = T^\ddagger(w')$. As a result, $T^\ddagger TT^\ddagger(w) = T^\ddagger(w)$, completing the proof. □

Lemma 1 provides a recipe for constructing a $\{1,2\}$-inverse $T^\ddagger$ of an operator T. First, for each element w in the image of T, define $T^\ddagger(w)$ as some arbitrary element in its preimage $T^{-1}(\{w\})$. Second, for any element w not in the image of T, select some arbitrary element in $T(V)$, and use its (already defined) inverse as the value for $T^\ddagger(w)$.

Thus, MP1–2 leave us with two degrees of freedom in defining the inverse. One is in selecting a subset V_0 of V, which contains exactly one source of each element in $T(V)$. It is easy to see that this set is exactly $T^\ddagger T(V)$. The other is an arbitrary mapping P_0 from $W \setminus T(V)$ to $T(V)$, which may be extended to a mapping from W to $T(V)$ by defining it as the identity mapping on $T(V)$. This mapping clearly equals $TT^\ddagger$. The following theorem summarizes the resulting picture.

Theorem 1. *Let $T: V \to W$ be an operator, let $V_0 \subseteq V$ contain exactly one source for each element in $T(V)$, and let $P_0 : W \to T(V)$ satisfy that its restriction to $T(V)$ is the identity mapping. Then the following hold. 1. The restriction $T|_{V_0}$ is a bijection from V_0 onto $T(V)$. 2. The function $T^\ddagger = (T|_{V_0})^{-1}P_0$ is a $\{1,2\}$-inverse of T that uniquely satisfies the combined requirements MP1–2, $TT^\ddagger = P_0$, and $T^\ddagger T(V) = V_0$. 3. Applying the construction of part 2 to $T^\ddagger$ with the set $W_0 = T(V)$ and mapping $Q_0 = T^\ddagger T$, yields $T^{\ddagger\ddagger} = (T^\ddagger|_{W_0})^{-1}Q_0 = T$. 4. The constructions of parts 2 and 3 generalize the MP PI for linear operators. Specifically, if $V = \mathbb{C}^n$, $W = \mathbb{C}^m$, and T is linear, then picking $V_0 = T^\dagger T(V)$ and $P_0 = TT^\dagger$ yields $T^\ddagger = (T|_{V_0})^{-1}P_0 = T^\dagger$ and picking $W_0 = T^\dagger(V)$ and $Q_0 = T^\dagger T$ yields $T^{\ddagger\ddagger} = (T^\ddagger|_{W_0})^{-1}Q_0 = T^{\dagger\dagger} = T$. 5. The functions $TT^\ddagger$ and $T^\ddagger T$ are idempotent. 6. If T is a bijection of V onto W, then $T^\ddagger = T^{-1}$ is the only $\{1,2\}$-inverse of T.*

Proof. 1. Immediate, since V_0 contains exactly one source for each element in $T(V)$. 2. If $w \in T(V)$ then P_0 maps it to itself, and $(T|_{V_0})^{-1}$ then maps it to one of its sources. If $w \notin T(V)$ then $(T|_{V_0})^{-1}P_0$ maps it to the inverse of an element in $T(V)$. By Lemma 1, $T^\ddagger$ satisfies MP1–2. We have that $TT^\ddagger = T(T|_{V_0})^{-1}P_0 = P_0$ and $T^\ddagger T(V) = (T|_{V_0})^{-1}P_0T(V) = (T|_{V_0})^{-1}T(V) = V_0$. As for uniqueness, Lemma 1 describes the degrees of freedom in defining a $\{1,2\}$-inverse. For $w \in W \setminus T(V)$, $TT^\ddagger(w)$ is clearly the element in $T(V)$ whose inverse is associated with w. Otherwise, $w \in T(V)$ and $T^\ddagger T(V) = V_0$. Since V_0 contains exactly one source for w, there is no choice in defining its inverse. 3. First we need to verify that $W_0 \subseteq W$ contains exactly one source under $T^\ddagger$ for each element in $T^\ddagger(W)$. We have by parts 1 and 2 that $T^\ddagger(W) = V_0$ and $W_0 = T(V)$ has exactly one source under $T^\ddagger$ for each element in V_0. We also have that $Q_0 = T^\ddagger T$ maps V to $T^\ddagger(W)$ and its restriction to $T^\ddagger(W)$ is the identity mapping. By part 2 we thus have that defining $T^{\ddagger\ddagger} = (T^\ddagger|_{W_0})^{-1}Q_0$ complies with MP1–2 and satisfies $T^\ddagger T^{\ddagger\ddagger} = Q_0$ and $T^{\ddagger\ddagger}T^\ddagger(W) = W_0$. To show that $T^{\ddagger\ddagger} = T$ we can use the uniqueness of the inverse shown in part 2. It is clear that $T \in T^\ddagger\{1,2\}$ by Lemma 1. In addition, $T^\ddagger T = Q_0$ by definition, and $TT^\ddagger(W) = T(V) = W_0$, and we are done. 4. Let $V = \mathbb{C}^n$, $W = \mathbb{C}^m$, let $T(v) = Av$ for a complex m by n matrix A, and let $A^\dagger$ be its MP inverse. We set $V_0 = A^\dagger A(V)$ and $P_0 = AA^\dagger$. Thus, $P_0 : W \to A(V)$, and for every $w = Av$, $P_0(w) = AA^\dagger Av = Av = w$, as required. In addition, $A(V_0) = AA^\dagger A(V) = A(V)$, so every element in $A(V)$ has at least one source in V_0. To show that there is exactly one source, it is enough to show that V_0 and $A(V)$, which are both vector spaces, have the same dimension. Let $A = U_1 \Sigma U_2^*$

be the singular value decomposition of A, where U_1 and U_2 are unitary and Σ is a generalized diagonal matrix. Then the MP inverse of A is given by $A^\dagger = U_2\Sigma^\dagger U_1^*$, where $\Sigma^\dagger_{ji} = \Sigma^{-1}_{ij}\mathbb{I}\{\Sigma_{ij} \neq 0\}$. Thus $AA^\dagger = U_1D_1U_1^*$ and $A^\dagger A = U_2D_2U_2^*$, where $D_1 = \Sigma\Sigma^\dagger$ and $D_2 = \Sigma^\dagger\Sigma$ are diagonal matrices with only ones and zeros, and the same trace. As a result, both $P_0(W)$ (which equals $A(V)$) and V_0 have an equal dimension that is the common trace of D_1 and D_2, as we wanted to show. The fact that this construction in fact yields $A^\dagger$ follows since $A^\dagger = A^\dagger AA^\dagger = AP_0 = (A|_{V_0})^{-1}P_0$, where the last equality holds since A is a bijection from V_0 to $A(V) = P_0(W)$. As for $A^{\dagger\dagger}$ (part 3), both our inverse and the MP inverse yield A (see [10], Lemma 1), so they coincide again. 5. Directly from MP1–2, we have that $TT^\ddagger TT^\ddagger = TT^\ddagger$ and $T^\ddagger TT^\ddagger T = T^\ddagger T$. 6. By Lemma 1, there are no degrees of freedom in defining $T^\ddagger$, since $W \setminus T(V)$ is empty and there is a single source for every element. On the other hand, it is clear that T^{-1} satisfies MP1–2. □

It is instructive to consider the symmetry, or lack thereof, between T and $T^\ddagger$. The roles of T and $T^\ddagger$ are completely symmetric in MP1–2, and it is therefore possible to define $T^{\ddagger\ddagger} = T$ if only MP1–2 have to be satisfied. The construction given in part 2 of Theorem 1 does not restrict the degrees of freedom allowed by MP1–2, yet embodies them (through V_0 and P_0) in a way that is not necessarily symmetric in the roles of T and $T^\ddagger$. The specific construction for $T^{\ddagger\ddagger}$ in part 3 aims to preserve the symmetry between T and $T^\ddagger$ and achieves the desired involution property of the inverse, namely, $T^{\ddagger\ddagger} = T$.

Ultimately, Theorem 1 describes T as a composition of an endomorphism of V, $T^\ddagger T$, and a bijection, $T|_{V_0}$, where $V_0 = T^\ddagger T(V)$. Symmetrically, $T^\ddagger$ is a composition of an endomorphism of W, $TT^\ddagger$, and a bijection $T^\ddagger|_{W_0}$, which is the inverse of the bijection that is part of T. These compositions, $T = T(T^\ddagger T)$ and $T^\ddagger = T^\ddagger(TT^\ddagger)$, are inherent in MP1–2. The endomorphisms above are also idempotent, or generalized projections. It should be noted that such operators do not require a metric, like conventional metric projections. The full scheme is depicted in Fig. 1.

If one seeks to assign a particular $\{1, 2\}$-inverse for every operator from V to W and indeed also from W to V, then that is always possible by following the recipe of part 2 of Theorem 1 or Lemma 1. In addition, if this assignment is a bijection $F : W^V \to V^W$, then the involution property may be satisfied for all operators by defining the inverse of $F(T)$ for $T : V \to W$ as T. Note that this definition is proper since F is a bijection, and that $T^\ddagger \in T\{1, 2\}$ implies $T \in T^\ddagger\{1, 2\}$ since the roles of the operator and the inverse in MP1–2 are symmetrical.

The existence of a bijection between W^V and V^W (equivalently, $|W^V| = |V^W|$) is a necessary condition for all operators to have an involutive $\{1, 2\}$-inverse. For example, if $|V^W| < |W^V|$, then there are two different operators $T_1, T_2 \in W^V$ that are assigned the same inverse in $T^\ddagger \in V^W$, and clearly the inverse of $T^\ddagger$ cannot be equal to both.

4 A Pseudo-Inverse for Nonlinear Operators in Normed Spaces

We now examine the concept of PI, which applies to normed spaces and coincides with MP1–4 for matrices. For this reason we will use the notation $T^\dagger$ for this type of inverse as well. We note that this definition does not use the adjoint operation and can be applied to nonlinear operators.

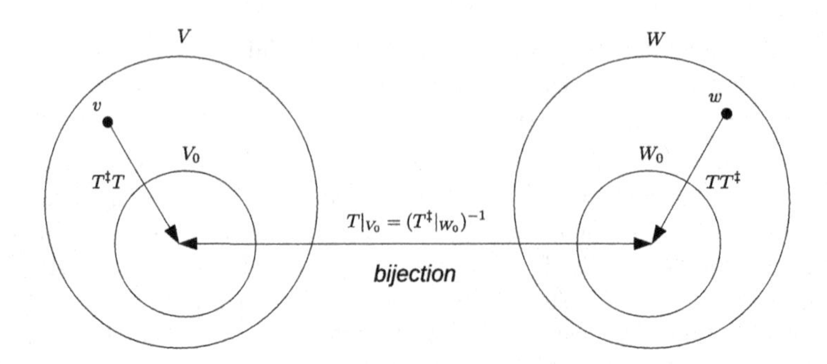

Fig. 1. A symmetric depiction of the $\{1, 2\}$-inverse scheme, as detailed in Theorem 1. The inverse 'projects' a point w to $W_0 = T(V)$ and then maps it to V_0 using $T|_{V_0}^{-1}$ (equivalently, $T^{\ddagger}|_{W_0}$), which is a proper bijection from W_0 onto V_0. In the other direction, T equivalently first 'projects' a point v onto V_0, where T is then a proper bijection onto W_0. The generalized projections are done by $T^{\ddagger}T$ in V and by $TT^{\ddagger}$ in W.

Definition 2 (Pseudo-inverse for normed spaces). *Let V and W be subsets of normed spaces over F ($\mathbb{R}$ or $\mathbb{C}$), and let $T : V \to W$ be an operator. A PI of T is an operator $T^{\dagger} : W \to V$ that admits the following identities:*

$$\text{BAS: } \forall w \in W, T^{\dagger}(w) \in \arg\min\{\|v\| : v \in \arg\min_{v' \in V}\{\|T(v') - w\|\}\}$$

$$\text{MP2: } T^{\dagger}TT^{\dagger} = T^{\dagger}$$

The calculation of $T^{\dagger}(w)$ for a given $w \in W$ can be translated into two consecutive minimization problems: first, find $m_w = \min_{v \in V}\{\|T(v) - w\|\}$, and then minimize $\|v\|$ for $v \in V$ s.t. $\|T(v) - w\| = m_w$. Note that the definition implicitly requires that the minima exist. Any solutions are then also required to satisfy MP2.

The BAS property is modeled after the best approximate solution property of the MP inverse. It replaces MP3 and MP4, so the definition no longer relies on the adjoint operation. In addition, BAS directly implies MP1, making it redundant. This means that a PI according to Definition 2 is in particular a $\{1, 2\}$-inverse. Property MP2 is not implied by BAS, and neither is the involution property.

More concretely, let V and W be subsets of normed spaces, let $T : V \to W$, and let $S : W \to V$ satisfy BAS w.r.t. T. Then it may be shown that 1. It holds that $S \in T\{1\}$. 2. For every $w \in T(V)$, $STS(w) = S(w)$, namely, MP2 is satisfied on $T(V)$. 3. In general, S does not necessarily satisfy MP2. 4. It might be impossible to define involutive PIs for all operators in W^V, even when all these operators have PIs.

The nonlinear PI has the desirable elementary property that for any bijection T, T^{-1} is its unique valid PI. By itself, however, Definition 2 implies neither existence nor uniqueness. In what follows we will seek interesting scenarios where this is indeed the case. One such important scenario is the original case of matrices, discussed by Penrose. As shown by him [11], for a linear operator T, the BAS property is uniquely satisfied by the MP inverse. Since the MP inverse also satisfies MP2, we can state the following:

if V and W are finite-dimensional inner-product spaces over $\mathbb{C}$, and $T : V \to W$ is a linear operator, then Definition 2 w.r.t. the induced norms is uniquely satisfied by the MP inverse.

Focusing on the image of the operator (equivalently, if the operator is onto) allows for easier characterization of existence and uniqueness. Specifically, let V and W be subsets of normed spaces, and let $T : V \to W$ and $E \subseteq T(V)$. It may be shown that a PI $T^\dagger : E \to V$ exists iff $A_w = \arg\min\{\|v\| : v \in T^{-1}(\{w\})\}$ is well-defined (minimum exists) for every $w \in E$. A PI $T^\dagger : E \to V$ exists uniquely iff A_w is a singleton for every $w \in E$.

With the following restriction, the PI may be uniquely defined beyond the image. Assume that for every $w \in T(V)$, $\arg\min\{\|v\| : T(v) = w\}$ is a singleton, and let $W' = \{w \in W : \arg\min\{\|w_0 - w\| : w_0 \in T(V)\}$ is a singleton$\}$. Then it may be shown that a PI $T^\dagger : W' \to V$ exists and it is unique.

We now consider the important case of continuous operators. The next theorem shows that a PI always exists over the whole range if the domain of the operator is compact. Conditions for uniqueness and continuity of the inverse are also given.

Theorem 2 (Continuous operators over a compact set). *Let V and W be subsets of normed spaces where V is compact, and let $T : V \to W$ be continuous. Then 1. A PI $T^\dagger : W \to V$ exists. 2. If we assume further that W is a Hilbert space, T is injective, and $T(V)$ is convex, and define $T^\dagger : W \to V$ as $T^\dagger = T^{-1} \circ P_{T(V)}$, where $P_{T(V)}$ is the projection onto $T(V)$ and $T^{-1} : T(V) \to V$ is the real inverse of T, then $T^\dagger$ is continuous and is the unique PI of T.*

Proof. 1. For every $w \in W$, the function $g_w : V \to \mathbb{R}$ defined as $g_w(v) = \|T(v) - w\|$ is continuous. By the extreme value theorem, it attains a global minimum m_w on V. The set $G_w = \{v \in V : \|T(v) - w\| = m_w\} = g_w^{-1}(\{m_w\})$ is closed, as the preimage of a closed set by a continuous function. As a subset of V, G_w is also compact, and the function $h : G_w \to \mathbb{R}$, defined as $h(v) = \|v\|$ is continuous and attains a minimum μ_w on G_w, again by the extreme value theorem. The set of elements in G_w with minimal norm will be denoted by H_w. By definition, the BAS property can be satisfied by defining $T^\dagger(w) \in H_w$. However, this must be done while also satisfying MP2. By Lemma 1, MP1–2 are satisfied if we choose $T^\dagger(w) \in T^{-1}(\{w\})$ for $w \in T(V)$, and $T^\dagger(w) = T^\dagger(w')$ for some $w' \in T(V)$ for $w \notin T(V)$. For $w \in T(V)$ we pick $T^\dagger(w)$ arbitrarily in H_w, and indeed $T^\dagger(w) \in H_w \subseteq G_w = T^{-1}(\{w\})$. For $w \notin T(V)$, picking $T^\dagger(w) = T^\dagger T(v)$ for some arbitrary $v \in H_w$ would thus guarantee that we satisfy MP1–2, and it remains only to show that $T^\dagger T(v) \in H_w$ if $v \in H_w$. Now, it holds that $H_w = \{u \in V : \|T(u) - w\| = m_w$ and $\|u\| = \mu_w\}$. By MP1 we have that $\|T(T^\dagger T(v)) - w\| = \|T(v) - w\| = m_w$, so $T^\dagger T(v) \in G_w$ and it is left to show that $\|T^\dagger T(v)\| = \mu_w$. It holds that $T^\dagger T(v)$ is a source of $T(v)$ under T with minimal norm. Since v is also a source of $T(v)$, $\|T^\dagger T(v)\| \leq \|v\| = \mu_w$. However, the minimal norm of elements in G_w is μ_w, so $\|T^\dagger T(v)\| = \mu_w$, and we are done. 2. Since T is continuous and V is compact, then $T(V)$ is compact, and therefore, closed. The projection $P_{T(V)}$ is thus well defined, by the Hilbert projection theorem, and it is continuous.[1] The operator T is a bijection from V to $T(V)$, so the real inverse $T^{-1} : T(V) \to V$ is well defined.

[1] This latter result, given in [3] for spaces over $\mathbb{R}$, is easily extended to spaces over $\mathbb{C}$ as well.

Since V is compact and T is continuous, it is well known that T^{-1} is also continuous. This holds since the preimage of any closed (and hence compact) set in V by T^{-1} is its image by T; this image is compact, since T is continuous, and hence closed. Define $T^\dagger : W \rightarrow V$ as $T^\dagger = T^{-1} \circ P_{T(V)}$. This operator is continuous as the composition of two continuous operators. There is a unique PI of T defined on W, and to satisfy the BAS property, this PI must coincide with $T^\dagger$, and thus this unique PI must be $T^\dagger$. □

One may show examples of continuous operators with a unique PI that nevertheless have a discontinuity even on the image. For $V = [-r, r]$ (with, say, $r \geq 4$) and $W = \mathbb{R}$, the PI of $T(v) = (v-2)^3 - (v-2)$ has a discontinuity since some points in the image have more than one source (Fig. 2). The existence and uniqueness of a PI of T on the image is guaranteed. The only violation of the assumptions of part 2 of Theorem 2 is that T is not injective.

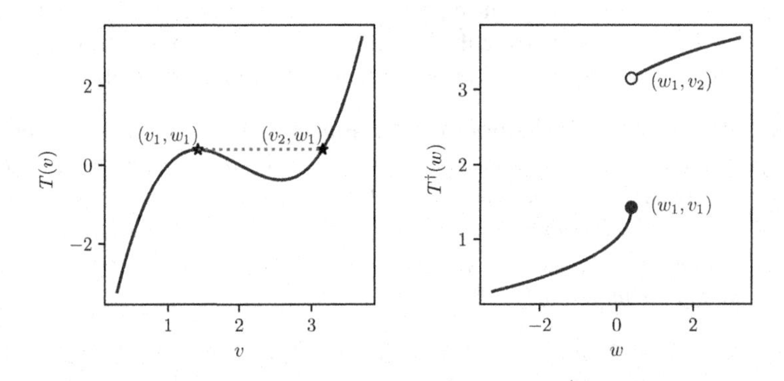

Fig. 2. An example of a discontinuity in a unique PI due to local extrema, with $T(v) = (v-2)^3 - (v-2)$. The PI value at $w_1 = T(v_1)$ is v_1, which of the two sources of w_1 has the smaller norm. For any $\epsilon > 0$, $T(v_1) + \epsilon$ has a single source that is greater than v_2.

The impact of Theorem 2 may be extended to non-compact domains V that are a countable union of compact sets $V_1 \subseteq V_2 \subseteq \ldots$, for example, $\mathbb{R}^n = \cup_{m=1}^{\infty} \bar{B}(0, m)$. For every point $w \in W$ one may relate the sequence $(T|_{V_n})^\dagger(w)$ to $T^\dagger(w)$. Specifically, let $T : V \rightarrow W$, where $V = \cup_{n=1}^{\infty} V_n$ and $V_1 \subseteq V_2 \subseteq \ldots$, denote $T_n = T|_{V_n}$, and assume a PI $T_n^\dagger : W \rightarrow V_n$ exists for every n. If for every $w \in W$ there is $N = N(w) \in \mathbb{N}$ s.t. $n \geq N$ implies $T_n^\dagger(w) = T_N^\dagger(w)$, then it may be shown that $T^\dagger : W \rightarrow V$, defined as $T^\dagger(w) = \lim_{n\rightarrow\infty} T_n^\dagger(w)$, is a PI for T. Furthermore, if the PI for each T_n is unique, then $\lim_{n\rightarrow\infty} T_n^\dagger$ is the unique PI for T.

Projections. For considering metric projections, the following concept will be key.

Definition 3. *A nonempty subset C of a metric space V, with a metric d, is a Chebyshev set, if for every $v \in V$, there is exactly one element $v' \in C$ s.t. $d(v, v') = \inf_{u \in C} d(v, u)$. The projection operator onto C, $P_C : V \rightarrow V$, may thus be uniquely defined at v by setting $P_C(v) = v'$.*

By the Hilbert projection theorem (see, equivalently, [12], Theorem 4.10), in a Hilbert space, every nonempty, closed, and convex set is a Chebyshev set. We mention a stronger characterization, namely, that a normed space is rotund and reflexive iff each of its nonempty closed convex subsets is a Chebyshev set (see [7, p. 436] for more details).

Let V be a subset of a Hilbert space, and let $1 \leq n \in \mathbb{N}$, $0 \in C_n \subseteq C_{n-1} \cdots \subseteq C_1 \subseteq V$, where $C_1, \ldots, C_n$ are closed and convex. Let $P_{C_i} : V \to V$ be the projection onto C_i for $i = 1, \ldots, n$, and define $P = P_{C_n} \circ \cdots \circ P_{C_1}$. Then it may be shown that P_{C_n} is the unique PI of P. For $n = 1$, we have that P_{C_1} is its own unique PI.

5 High-Level Properties of the Nonlinear Inverse

The purpose of this section is to describe the inverse for complex cases using simpler components. We will consider both $\{1, 2\}$-inverses and PIs.

Product Operator. Let $T_i : V_i \to W_i$, where V_i and W_i are sets for $1 \leq i \leq n$, and let $T_i^\ddagger \in T_i\{1, 2\}$. Define the product operator $\Pi_i T_i : \Pi_i V_i \to \Pi_i W_i$ as $(\Pi_i T_i)((v_1, \ldots, v_n)) = (T_1(v_1), \ldots, T_n(v_n))$. It may be shown that 1. The operator $(\Pi_i T_i)^\ddagger : \Pi_i W_i \to \Pi_i V_i$, defined $(\Pi_i T_i)^\ddagger((w_1, \ldots, w_n)) = (T_1^\ddagger(w_1), \ldots, T_n^\ddagger(w_n))$, is a $\{1, 2\}$-inverse of $\Pi_i T_i$. In addition, the involution property is preserved, namely, $\Pi_i T_i \in (\Pi_i T_i^\ddagger)\{1, 2\}$. 2. For every i, assume further that V_i and W_i are subsets of normed spaces and $T_i^\ddagger$ is also a PI of T_i. Consider $\Pi_i V_i$ and $\Pi_i W_i$ as subsets in the normed direct product spaces, each equipped with a norm that is some function of the norms of components, where that function is strictly increasing in each component (e.g., the sum of component norms). Then $\Pi_i T_i^\ddagger$ is also a PI of $\Pi_i T_i$.

Remark 1. In the above statement, the norm of each product space may be the L_p norm of the vector of component norms, for any $1 \leq p < \infty$, which is strictly increasing in each component. In particular, if each V_i and W_i is $\mathbb{R}$ with the absolute value norm, and $\sigma : \mathbb{R} \to \mathbb{R}$ has PI $\sigma^\dagger : \mathbb{R} \to \mathbb{R}$, then $T : \mathbb{R}^n \to \mathbb{R}^n$, the entrywise application of σ, has PI $(\sigma^\dagger, \ldots, \sigma^\dagger)$ for L_p, $1 \leq p < \infty$.

Composition with Bijections. We will need the following definition.

Definition 4 (Norm-monotone operators). *Let V, W be subsets of normed spaces. An operator $T : V \to W$ is norm-monotone if for every $v_1, v_2 \in V$, $\|v_1\| \leq \|v_2\|$ implies $\|T(v_1)\| \leq \|T(v_2)\|$.*

Let V, V_1, W, and W_1 be sets, let $T : V \to W$, and let $S_1 : W \to W_1$ and $S_2 : V_1 \to V$ be bijections. It may be shown that 1. If $T^\ddagger \in T\{1, 2\}$, then $S_2^{-1} T^\ddagger S_1^{-1} \in (S_1 T S_2)\{1, 2\}$. 2. Assuming further that V, V_1, W, and W_1 are subsets of normed spaces, aS_1 is an isometry for some $a \neq 0$, and S_2^{-1} is norm-monotone, it holds that if $T^\ddagger$ is a PI of T, then $S_2^{-1} T^\ddagger S_1^{-1}$ is a PI of $S_1 T S_2$. Consequently, if V and W are normed spaces, and $T : V \to W$ has a PI $T^\dagger : W \to V$, then for any scalars $a, b \neq 0$ and vector $w_0 \in W$, $b^{-1} T^\dagger (a^{-1}(w - w_0))$ is a PI of $aT(bv) + w_0$.

Compositions. Let U, V, W be sets, and let $T : U \to V$ and $S : V \to W$. It may be shown that 1. If $S_1^\ddagger \in S|_{T(U)}\{1, 2\}$ and $T^\ddagger \in T\{1, 2\}$, then $T^\ddagger S_1^\ddagger \in (ST)\{1, 2\}$. Consequently, if $S^\ddagger \in S\{1, 2\}$ and T is onto V, then $T^\ddagger S^\ddagger$ is a $\{1, 2\}$-inverse of ST, but if T is not onto, that does not necessarily hold. 2. Assuming that U, V, and W are subsets of normed spaces, and letting $T^\dagger$ and $S_1^\dagger$ be PIs of T and $S|_{T(U)}$, respectively, there is a setting where ST has a unique PI, and it is not $T^\dagger S_1^\dagger$.

A specific composition of particular interest is that of a projection applied after an operator, even (perhaps especially) if the operator does not have a PI. One such case is that of a neural layer, which will be considered in Sect. 6.

Theorem 3 (Left-composition with projection). *Let $T : V \to W$, where V and W are subsets of normed spaces, let $\emptyset \neq C \subseteq T(V)$ be a Chebyshev set, and let $P_C : W \to W$ be the projection operator onto C. 1. For $w \in W$, write $B_w = \arg\min\{\|v\| : v \in V,\ P_C T(v) = P_C(w)\}$, and let $W' \subseteq W$ be the set of elements $w \in W$ for which B_w is well-defined. Then $w \in W' \setminus C$ implies $P_C(w) \in W' \cap C$, and the following recursive definition yields a PI $(P_C T)^\dagger : W' \to V$:*

$$(P_C T)^\dagger(w) = \begin{cases} \text{some } v \in B_w, & w \in W' \cap C \\ (P_C T)^\dagger(P_C(w)), & w \in W' \setminus C\,. \end{cases}$$

2. If, in addition, T is continuous, W is a Hilbert space, and every closed ball is compact in V, then the PI of part 1 is defined on all of W ($W' = W$).

Proof. 1. For any $w \in W$, BAS necessitates that

$$(P_C T)^\dagger(w) \in \arg\min\{\|v\| : v \in V \text{ minimizes } \|P_C T(v) - w\|\}\,. \tag{1}$$

Since $C \subseteq T(V)$, there is some v_0 s.t. $T(v_0) = P_C(w)$, and thus $P_C T(v_0) = P_C(w)$. Since C is a Chebyshev set, $v \in V$ minimizes $\|P_C T(v) - w\|$ iff $P_C T(v) = P_C(w)$, and the BAS requirement becomes

$$(P_C T)^\dagger(w) \in \arg\min\{\|v\| : v \in V,\ P_C T(v) = P_C(w)\} = B_w\,. \tag{2}$$

If $w \in W' \cap C$, then our definition satisfies BAS and also MP2, since

$$(P_C T)^\dagger (P_C T)(P_C T)^\dagger(w) = (P_C T)^\dagger(P_C(w)) = (P_C T)^\dagger(w). \tag{3}$$

If $w \in W' \setminus C$, then by definition of B_w, $B_{P_C(w)} = B_w$. Thus, $B_{P_C(w)}$ is well-defined, so $P_C(w) \in W' \cap C$ and $(P_C T)^\dagger(P_C(w))$ is already defined. It satisfies BAS since it is in $B_{P_C(w)}$, and also satisfies MP2 since

$$(P_C T)^\dagger (P_C T)(P_C T)^\dagger(w) = (P_C T)^\dagger (P_C T)(P_C T)^\dagger(P_C(w)) \tag{4}$$

$$= (P_C T)^\dagger(P_C(w)) = {P_C}^\dagger(w)\,. \tag{5}$$

2. A Chebyshev set in a Hilbert space is closed and convex, so P_C is continuous [3]. Therefore, $P_C T$ is continuous as well. Thus, for any $w \in C$, $A = (P_C T)^{-1}(\{w\})$ is closed. There is some $v_0 \in V$ s.t. $P_C T(v_0) = w$, so $A_1 = A \cap \{v \in V : \|v\| \leq \|v_0\|\} \neq \emptyset$. Since A_1 is compact, the norm function attains a minimum m_w on that set by the extreme value theorem, and $B_w = \{v \in V : \|v\| = m_w\}$ is well-defined. For any $w \in W$, $B_w = B_{P_C(w)}$ is thus also well-defined, so $W' = W$. □

6 Test Cases

In this section we consider nonlinear PIs for specific cases. Recall that a valid PI $T^\dagger$ of an operator $T : V \to W$ should satisfy BAS and MP2.

Some One-Dimensional Operators. In Table 1 we give the PIs for some simple operators over $\mathbb{R}$, with the L_2 norm (the only norm up to scaling). The domain of the PI shows where it may be defined. We write $\mathrm{sgn}_\epsilon(v) = \min\{1, \max\{-1, v/\epsilon\}\}$ for $\epsilon > 0$, so $\mathrm{sgn}(v) = \lim_{\epsilon \to 0+} \mathrm{sgn}_\epsilon(v)$, pointwise. To be clear, $\mathrm{sgn}(v) = \mathbb{I}\{v > 0\} - \mathbb{I}\{v < 0\}$.

Table 1. PIs of some one-dimensional operators, $T : \mathbb{R} \to \mathbb{R}$.

$T(v)$	$T^\dagger(w)$	Domain of $T^\dagger$	Unique
v^2	$\pm\sqrt{\max\{w, 0\}}$	$\mathbb{R}$	–
$(v-a)^2, a \neq 0$	$a - \mathrm{sgn}(a)\sqrt{\max\{w, 0\}}$	$\mathbb{R}$	+
$\max\{v, 0\}$ (ReLU)	$\max\{w, 0\}$	$\mathbb{R}$	+
$v \cdot \mathbb{I}\{\lvert v\rvert \geq a\}, a \geq 0$	$\mathrm{sgn}(w) \cdot \mathbb{I}\{\lvert w\rvert > \frac{a}{2}\} \max(a, \lvert w\rvert\}$	$\mathbb{R}$	+
$\mathrm{sgn}(v) \max\{\lvert v\rvert - a, 0\}, a \geq 0$	$\mathrm{sgn}(w)(\lvert w\rvert + a)$	$\mathbb{R}$	+
$\tanh(v)$	$\mathrm{arctanh}(w)$	$(-1, 1)$	+
$\mathrm{sgn}(v)$	0	$[-\frac{1}{2}, \frac{1}{2}]$	+
$\mathrm{sgn}_\epsilon(v), \epsilon > 0$	$\epsilon\, \mathrm{sgn}_1(w)$	$\mathbb{R}$	+
$\exp(v)$	$\log(w)$	$(0, \infty)$	+
$\sin(v)$	$\arcsin(\min\{1, \max\{-1, w\}\})$	$\mathbb{R}$	+

A Neural Layer. We consider the PI of a single layer of a trained multilayer perceptron. Let $V = \mathbb{R}^n$ and $W = \mathbb{R}^m$, let A be an m by n matrix, and let $\sigma : \mathbb{R} \to \mathbb{R}$ be a transfer function (we consider either tanh or ReLU). With some abuse of notation, we will also write $\sigma : W \to W$ for the elementwise application of σ to a vector in W. Given $v \in V$, a neural layer operator $T : V \to W$ is defined as $T(v) = \sigma(Av)$, which is continuous for both choices of σ and smooth for tanh. We consider the L_2 norm for both spaces and make the simplifying assumption that $A(V) = W$ (thus $m \leq n$).

Note that a $\{1, 2\}$-inverse is readily obtained for this setting. In particular, for $\sigma =$ ReLU, we have that $A^\dagger \sigma \in (\sigma A)\{1, 2\}$, since the PIs of A and σ are also $\{1, 2\}$-inverses, and entrywise ReLU is its own PI. We next consider PIs.

Claim 1. *Let $T : V \to W$, where $V = \mathbb{R}^n$, $W = \mathbb{R}^m$, $m \leq n$, $T = \sigma A$, and A is a full-rank matrix. 1. If $\sigma = \tanh$, then T has a unique PI, $T^\dagger : (-1, 1)^m \to V$, defined as $T^\dagger(w) = A^\dagger(\mathrm{arctanh}(w))$. 2. Let $\sigma = \tanh$, let $1 < k \in \mathbb{N}$, and define $C_k = [-1 + 1/k, 1 - 1/k]^m$ and $T_k = P_{C_k} T$. Then $\lim_{k \to \infty} T_k = T$, and we may define a valid PI $T_k^\dagger : W \to V$. 3. For $\sigma =$ ReLU, we can recursively define a valid PI $T^\dagger : W \to V$, where for $w \in W \setminus [0, \infty)^m$ we have $T^\dagger(w) = T^\dagger(\sigma(w))$, and for $w \in [0, \infty)^m$, $T^\dagger(w)$ is the solution to the quadratic program $\min \|v\|^2$ s.t. $(Av)_i = w_i, \forall i\ w_i > 0$, and $(Av)_i \leq 0, \forall i\ w_i = 0$.*

Wavelet Thresholding. Let $V = W = \mathbb{R}^n$, let A be an n by n matrix representing the elements of an orthonormal wavelet basis, and let $\sigma : \mathbb{R} \to \mathbb{R}$ be some thresholding operator. Wavelet thresholding [5] takes as input a vector $v \in \mathbb{R}^n$, which represents a possibly noisy image. The image is transformed into its wavelet representation $Av \in W$, and then the thresholding operator σ is applied entrywise to Av. The result will be written as $\sigma(Av)$ or $(\sigma A)(v)$. Finally, an inverse transform is applied, and the result, $A^{-1}\sigma(Av)$, is the denoised image. For the thresholding operator, we will be particularly interested in hard thresholding, $\xi_a(x) = x \cdot \mathbb{I}\{|x| \geq a\}$, and soft thresholding, $\eta_a(x) = \text{sgn}(x)\max\{|x| - a, 0\}$, where $a \in \mathbb{R}^+$ in both cases.

Claim 2. *Let $T : V \to W$, where $T = \sigma A$. 1. If σ has a PI $\sigma^\dagger : \mathbb{R} \to \mathbb{R}$, then $T^\dagger = A^{-1}\sigma^\dagger$ is a PI of T, where all the PIs are w.r.t. the L_2 norm. 2. If $\sigma = \xi_a$ for any $a \geq 0$, then $T^\dagger T = A^{-1}\sigma A$. Namely, wavelet hard thresholding is equivalent to applying T and then its PI. In contrast, for $\sigma = \eta_a$ (soft thresholding), $T^\dagger T = A^{-1}\sigma A$ does not hold for $a > 0$.*

We note that denoising in general is ideally idempotent, and so is $T^\dagger T$ for any T, since $T^\dagger T T^\dagger T = T^\dagger T$ by MP2.

7 Conclusion

This work attempts to establish a broad and coherent theory for the PI of nonlinear operators. We give a procedure to construct an inverse satisfying the first two MP axioms in a very general setting, along with some essential properties. A stricter notion is defined for normed spaces through a minimization problem, as well as MP2. Such a minimization can be thought of as the limit of Tikhonov regularization and applies to regularized loss minimization. Explicit nonlinear PIs are given for cases relevant to learning and image processing. Our analysis of pseudo-inversion of compound operators may pave the way for efficient numerical algorithms to solve the nonlinear PI problem.

Acknowledgements. We acknowledge support by grant agreement No. 777826 (NoMADS), by the Israel Science Foundation (Grant No. 534/19), by the Ministry of Science and Technology (Grant No. 5074/22) and by the Ollendorff Minerva Center.

References

1. Baksalary, O.M., Trenkler, G.: The Moore-Penrose inverse: a hundred years on a frontline of physics research. Eur. Phys. J. H **46**(1), 1–10 (2021)
2. Ben-Israel, A., Greville, T.N.: Generalized Inverses: Theory and Applications, vol. 15. Springer, New York (2003). https://doi.org/10.1007/b97366
3. Cheney, W., Goldstein, A.A.: Proximity maps for convex sets. Proc. Am. Math. Soc. **10**(3), 448–450 (1959)
4. Dermanis, A.: Generalized inverses of nonlinear mappings and the nonlinear geodetic datum problem. J. Geodesy **72**(2), 71–100 (1998)
5. Donoho, D.L., Johnstone, I.M.: Adapting to unknown smoothness via wavelet shrinkage. J. Am. Stat. Assoc. **90**(432), 1200–1224 (1995)

6. Liu, Y., Zhu, J.J.: Singular perturbation analysis for trajectory linearization control. In: 2007 American Control Conference, pp. 3047–3052. IEEE (2007)
7. Megginson, R.E.: An Introduction to Banach Space Theory, vol. 183. Springer, New York (1998). https://doi.org/10.1007/978-1-4612-0603-3
8. Mickle, M.C., Huang, R., Zhu, J.J.: Unstable, nonminimum phase, nonlinear tracking by trajectory linearization control. In: Proceedings of the 2004 IEEE International Conference on Control Applications, 2004, vol. 1, pp. 812–818. IEEE (2004)
9. Moore, E.H.: On the reciprocal of the general algebraic matrix. Bull. Am. Math. Soc. **26**, 394–395 (1920)
10. Penrose, R.: A generalized inverse for matrices. Math. Proc. Camb. Philos. Soc. **51**(3), 406–413 (1955)
11. Penrose, R.: On best approximate solutions of linear matrix equations. Math. Proc. Camb. Philos. Soc. **52**(1), 17–19 (1956)
12. Rudin, W.: Real and Complex Analysis. McGraw-Hill Inc., New York (1987)
13. Zervakis, M.E., Venetsanopoulos, A.N.: Iterative least squares estimators in nonlinear image restoration. IEEE Trans. Signal Process. **40**(4), 927–945 (1992)

A Frame Decomposition of the Funk-Radon Transform

Michael Quellmalz[1(✉)], Lukas Weissinger[2], Simon Hubmer[2], and Paul D. Erchinger[1]

[1] Technische Universität Berlin, Institute of Mathematics, MA 4-3, Straße des 17. Juni 136, 10623 Berlin, Germany
quellmalz@math.tu-berlin.de, erchinger@campus.tu-berlin.de
[2] Johann Radon Institute Linz, Altenbergerstraße 69, 4040 Linz, Austria
{lukas.weissinger,simon.hubmer}@ricam.oeaw.ac.at
https://www.tu.berlin/imageanalysis, https://www.ricam.oeaw.ac.at/

Abstract. The Funk-Radon transform assigns to a function defined on the unit sphere its integrals along all great circles of the sphere. In this paper, we consider a frame decomposition of the Funk-Radon transform, which is a flexible alternative to the singular value decomposition. In particular, we construct a novel frame decomposition based on trigonometric polynomials and show its application for the inversion of the Funk-Radon transform. Our theoretical findings are verified by numerical experiments, which also incorporate a regularization scheme.

Keywords: Funk-Radon Transform · Frame Decompositions · Inverse and Ill-Posed Problems · Numerical Analysis · Tomography

1 Introduction

The *Funk-Radon transform* assigns to a function $f\colon \mathbb{S}^2 \to \mathbb{C}$ defined on the two-dimensional *unit sphere* $\mathbb{S}^2{:=}\{\boldsymbol{\xi} \in \mathbb{R}^3 : \|\boldsymbol{\xi}\| = 1\}$ its integrals along all great circles of the sphere, i.e.,

$$Rf(\boldsymbol{\xi}){:=}\frac{1}{2\pi}\int_{\boldsymbol{\xi}^\top \boldsymbol{\eta}=0} f(\boldsymbol{\eta})\,\mathrm{d}\boldsymbol{\eta}\,, \qquad \forall \boldsymbol{\xi} \in \mathbb{S}^2\,, \tag{1}$$

where $\mathrm{d}\boldsymbol{\eta}$ denotes the arclength on the great circle perpendicular to $\boldsymbol{\xi}$. Tracing back to works of Funk [17] and Minkowski [36] in the early twentieth century, it is also known as Funk transform, Minkowski-Funk transform or spherical Radon transform. It has found applications in diffusion MRI [43,49], radar imaging [52], Compton camera imaging [48], photoacoustic tomography [24], and geometric tomography [18, Chap. 4]. Besides analytic inversion formulas, e.g., [4,17,21,28], the numerical reconstruction of functions given its Funk-Radon transform can be done using mollifier methods [34,44], the eigenvalue decomposition [23], or discretization on the cubed sphere [5]. Generalizations have been developed for various non-central sections of the sphere [2,39,45,46] or for derivatives [28,42].

L. Calatroni et al. (Eds.): SSVM 2023, LNCS 14009, pp. 42–54, 2023.
https://doi.org/10.1007/978-3-031-31975-4_4

In this paper, we are interested in *frame decompositions* (FDs) of the Funk-Radon transform. Originally developed in the framework of wavelet-vaguelette decompositions [1,10,11,14,15,31,33] and then extended to biorthogonal curvelet and shearlet decompositions [6,8], FDs are generalizations of the singular value decomposition (SVD) [12,16,25–27,50]. In particular, they allow SVD-like decompositions of bounded linear operators also in those cases when the SVD itself is either unknown, its computation is infeasible, or its structure is unfavourable. More precisely, given a bounded linear operator $A\colon X \to Y$ between real or complex Hilbert spaces X and Y, an FD of A is a decomposition of the form

$$Ax = \sum_{k=1}^{\infty} \sigma_k \langle x, e_k \rangle_X \tilde{f}_k , \qquad \forall\, x \in X . \tag{2}$$

Here, the sets $\{e_k\}_{k\in\mathbb{N}}$ and $\{f_k\}_{k\in\mathbb{N}}$ form frames over X and Y, respectively, and $\{\tilde{f}_k\}_{k\in\mathbb{N}}$ denotes the dual frame of the frame $\{f_k\}_{k\in\mathbb{N}}$; see Sect. 2 below. The main requirement on e_k and f_k is that they satisfy the quasi-singular relation

$$\overline{\sigma_k}\, e_k = A^* f_k , \qquad \forall\, k \in \mathbb{N} , \tag{3}$$

where $\overline{\sigma_k}$ denotes the complex conjugate of the coefficient $\sigma_k \in \mathbb{C}$ and A^* the adjoint of A. Using the FD (2) it is possible to compute (approximate) solutions of the linear equation $Ax = y$, and to develop filter-based regularization schemes as for the SVD [12,26,27]. However, the question remains whether frames satisfying (3) can be found. While this is possible by geometric considerations for some examples [11,14,15,25], an explicit construction "recipe" exists in case that A satisfies the stability condition

$$c_1 \|x\|_X \leq \|Ax\|_Z \leq c_2 \|x\|_X , \qquad \forall\, x \in X , \tag{4}$$

for some constants $c_1, c_2 > 0$ and a Hilbert space $Z \subseteq Y$. In this case, one can start with an arbitrary frame $\{f_k\}_{k\in\mathbb{N}}$ over Y with the additional property

$$a_1 \|y\|_Z^2 \leq \sum_{k=1}^{\infty} \alpha_k^2 |\langle y, f_k \rangle_Y|^2 \leq a_2 \|y\|_Z^2 , \qquad \forall\, y \in Y , \tag{5}$$

for coefficients $0 \neq \alpha_k \in \mathbb{R}$ and some constants $a_1, a_2 > 0$. Then, one defines

$$e_k := \alpha_k A^* f_k ,$$

which results in a frame $\{e_k\}_{k\in\mathbb{N}}$ over X which satisfies (3) with $\overline{\sigma_k} = 1/\alpha_k$ [26]. In case that Z and Y are Sobolev spaces, frames $\{f_k\}_{k\in\mathbb{N}}$ satisfying (5) can often be found (e.g., orthonormal wavelets [9]), which has resulted in FDs of the classic Radon transform [26,27]. On the other hand, while the Funk-Radon transform satisfies a stability property of the form (4), see Theorem 2 below, frames which satisfy (5) are more difficult to find. The standard candidate would be spherical harmonics, which, however, already are the eigenfunctions of the Funk-Radon transform, and thus offer no further insight.

Hence, in this paper we consider a different approach for constructing FDs, which was originally outlined in [26]. This approach is still based on the stability property (4), but instead of (5) it only requires that

$$\|y\|_Y \leq \|y\|_Z \,, \qquad \forall y \in Z \,, \tag{6}$$

that $Z \subseteq Y$ is a dense subspace of Y, and that the frame functions f_k are elements of Z, i.e., $\|f_k\|_Z < \infty$. In this case, one can build an alternative FD of A similar to (2), which can then be used to compute the (unique) solution of the linear operator equation $Ax = y$ for any y in the range $\mathcal{R}(A)$ of A, and to develop stable reconstruction approaches in case of noisy data y^δ.

The aim of this paper is to show that the above approach is applicable to the Funk-Radon transform. In particular, we construct an FD using trigonometric functions, which have the advantage of their fast computation outperforming spherical harmonics, cf. [35,53]. For this, we first review some background on frames and FDs in Sect. 2. Then, in Sect. 3 we show that all required properties are satisfied for the Funk-Radon transform with a suitable choice of the functions f_k. Furthermore, we provide explicit expressions for the frame functions e_k, leading to an FD and a corresponding reconstruction formula. Finally, in Sect. 4 we consider the efficient implementation of our derived FD and evaluate its reconstruction quality on a number of numerical examples.

2 Background on Frames and Frame Decompositions

In this section, we review some background on frames and FDs, collected from the seminal works [7,9] and the recent article [26], respectively.

Definition 1. *A sequence $\{e_k\}_{k\in\mathbb{N}}$ in a Hilbert space X is called a* frame *over X, if and only if there exist* frame bounds $0 < B_1, B_2 \in \mathbb{R}$ *such that there holds*

$$B_1 \|x\|_X^2 \leq \sum_{k=1}^{\infty} |\langle x, e_k \rangle_X|^2 \leq B_2 \|x\|_X^2 \,, \qquad \forall x \in X \,. \tag{7}$$

For a given frame $\{e_k\}_{k\in\mathbb{N}}$, one can consider the *frame (analysis) operator* F and its adjoint *(synthesis)* operator F^*, which are given by

$$F : X \to \ell_2(\mathbb{N}) \,, \qquad x \mapsto (\langle x, e_k \rangle_X)_{k\in\mathbb{N}} \,,$$

$$F^* : \ell_2(\mathbb{N}) \to X \,, \qquad (a_k)_{k\in\mathbb{N}} \mapsto \sum_{k=1}^{\infty} a_k e_k \,.$$

Due to (7) there holds $\sqrt{B_1} \leq \|F\| = \|F^*\| \leq \sqrt{B_2}$. Furthermore, one can define

$$Sx := F^* F x = \sum_{k=1}^{\infty} \langle x, e_k \rangle_X \, e_k \,, \qquad \text{and} \qquad \tilde{e}_k := S^{-1} e_k \,.$$

Since S is continuously invertible with $B_1 I \leq S \leq B_2 I$, the functions $\tilde{e}_k$ are well-defined, and the set $\{\tilde{e}_k\}_{k\in\mathbb{N}}$ forms a frame over X with frame bounds B_2^{-1}, B_1^{-1} called the *dual frame* of $\{e_k\}_{k\in\mathbb{N}}$. Furthermore, it can be shown that

$$x = \sum_{k=1}^{\infty} \langle x, \tilde{e}_k \rangle_X e_k = \sum_{k=1}^{\infty} \langle x, e_k \rangle_X \tilde{e}_k , \qquad \forall x \in X .$$

In general, this decomposition is not unique, which is a key difference between frames and bases, but it can be understood as the "most economical" one [9].

Next, we consider FDs. For this, we use the following

Assumption 1. *The operator $A\colon X \to Y$ between the Hilbert spaces X, Y is bounded, linear, and satisfies condition* (4) *for some constants $c_1, c_2 > 0$, where the Hilbert space $Z \subseteq Y$ is a dense subspace of Y satisfying* (6)*. Furthermore, the set $\{f_k\}_{k\in\mathbb{N}}$ forms a frame over Y with frame bounds $C_1, C_2 > 0$, and $\{\tilde{f}_k\}_{k\in\mathbb{N}}$ denotes the dual frame of $\{f_k\}_{k\in\mathbb{N}}$. Moreover, the functions f_k are elements of Z, i.e., $\|f_k\|_Z < \infty$, and $E\colon Z \to Y$, $z \mapsto z$, denotes the* embedding operator.

Now, the key idea for constructing an FD of A is the suitable choice of a frame $\{e_k\}_{k\in\mathbb{N}}$ over X based on the frame $\{f_k\}_{k\in\mathbb{N}}$, as outlined in

Proposition 1. **([26, Lem. 4.5]).** *Let $A\colon X \to Y$ and let Assumption 1 hold. Then the set $\{e_k\}_{k\in\mathbb{N}}$, where the functions e_k are defined as*

$$e_k := A^* L f_k , \qquad \text{where} \qquad L := (EE^*)^{-1/2} , \tag{8}$$

form a frame over X with frame bounds $B_1 = c_1^2 C_1$ and $B_2 = c_2^2 C_2$, where C_1 and C_2 are the frame bounds of $\{f_k\}_{k\in\mathbb{N}}$, and c_1 and c_2 are as in Assumption 1.

The choice (8) for the functions e_k leads us to the following

Proposition 2. **([26, Lem. 4.6]).** *Let $A\colon X \to Y$, let Assumption 1 hold, and let the functions e_k be defined as in* (8)*. Then for all $x \in X$ there holds*

$$LAx = \sum_{k=1}^{\infty} \langle x, e_k \rangle_X \tilde{f}_k , \qquad \text{and} \qquad Ax = L^{-1} \left(\sum_{k=1}^{\infty} \langle x, e_k \rangle_X \tilde{f}_k \right) , \tag{9}$$

where the second equality uses the fact that $L^{-1} = (EE^*)^{1/2}$ is continuous.

Next, we consider the solution of the linear operator equation $Ax = y$ using the FD of A from above. Summarizing results from Lemma 4.7, Theorem 4.8, and Remark 4.2 in [26], which are essentially consequences of (9), we obtain

Theorem 1. *Let $A\colon X \to Y$ be a bounded linear operator, let Assumption 1 hold, and let the functions e_k be defined as in* (8)*. Then for any $y \in \mathcal{R}(A) \subseteq Z$,*

$$A^\dagger y := \sum_{k=1}^{\infty} \langle Ly, f_k \rangle_Y \tilde{e}_k ,$$

is the well-defined, unique solution of $Ax = y$, and $\|A^\dagger y\|_X \leq \sqrt{C_2/B_1}\, \|Ly\|_Y$.

3 Frame Decompositions of the Funk-Radon Transform

The eigenvalue decomposition of the Funk-Radon transform (1), which is also an FD, is due to [36]. Denoting by P_ℓ the ℓ-th Legendre polynomial, we have

$$RY_\ell^m = P_\ell(0)\, Y_\ell^m = \begin{cases} \frac{(-1)^{\ell/2}(\ell-1)!!}{\ell!!}\, Y_\ell^m, & \ell \text{ even}, \\ 0, & \ell \text{ odd}, \end{cases} \tag{10}$$

where the eigenfunctions are the *spherical harmonics* Y_ℓ^m [37] of degree $\ell \in \mathbb{N}_0$ and order $m = -\ell, \dots, \ell$, which form an orthonormal basis of $L^2(\mathbb{S}^2)$. From its definition in (1), we see that Rf is even for any $f\colon \mathbb{S}^2 \to \mathbb{C}$, i.e., $Rf(\boldsymbol{\xi}) = Rf(-\boldsymbol{\xi})$ for all $\boldsymbol{\xi} \in \mathbb{S}^2$. Conversely, Rf vanishes for odd functions $f(\boldsymbol{\xi}) = -f(-\boldsymbol{\xi})$, so we can expect to recover only even functions f from their Funk-Radon transform.

The spherical *Sobolev space* $H^s(\mathbb{S}^2)$, $s \in \mathbb{R}$, can be defined as the completion of $C^\infty(\mathbb{S}^2)$ with respect to the norm [3]

$$\|f\|_{H^s(\mathbb{S}^2)} := \sum_{\ell=0}^{\infty} \sum_{m=-\ell}^{\ell} (\ell + \tfrac{1}{2})^{2s} |\langle f, Y_\ell^m \rangle_{L^2(\mathbb{S}^2)}|^2, \tag{11}$$

where $\langle f, g\rangle_{L^2(\mathbb{S}^2)} := \int_{\mathbb{S}^2} f(\boldsymbol{\xi}) g(\boldsymbol{\xi})\, \mathrm{d}\boldsymbol{\xi}$, and we denote by $H^s_{\text{even}}(\mathbb{S}^2)$ its restriction to even functions, which is the span of spherical harmonics Y_ℓ^m with even degree $\ell \in 2\mathbb{N}_0$. The Sobolev spaces are nested, i.e., $H^s(\mathbb{S}^2) \subsetneq H^r(\mathbb{S}^2)$ whenever $s > r$, and we have $H^0(\mathbb{S}^2) = L^2(\mathbb{S}^2)$.

Theorem 2. *Let $s \geq 0$. The Funk-Radon transform R defined in (1) extends to a continuous and bijective operator from $X = H^s_{\text{even}}(\mathbb{S}^2)$ to $Z = H^{s+1/2}_{\text{even}}(\mathbb{S}^2)$ that satisfies (4) with the bounds $c_1 = \sqrt{1/2}$ and $c_2 = \sqrt{2/\pi}$. Furthermore, it also extends to a continuous and self-adjoint operator from $H^s_{\text{even}}(\mathbb{S}^2)$ to $H^s_{\text{even}}(\mathbb{S}^2)$.*

Proof. The bijectivity of $R\colon X \to Y$ is due to [47, § 4]. From Theorem 3.13 in [40], we know that c_1 and c_2 in (4) are characterized by

$$c_1 \left(\ell + \tfrac{1}{2}\right)^{-\frac{1}{2}} \leq \frac{(\ell-1)!!}{\ell!!} = |P_\ell(0)| \leq c_2 \left(\ell + \tfrac{1}{2}\right)^{-\frac{1}{2}}, \qquad \forall \ell \in 2\mathbb{N}_0. \tag{12}$$

Analogously to the proof of Lemma 3.2 in [22], we can see that the sequence $2\mathbb{N}_0 \ni \ell \mapsto (\ell+1/2)^{1/2}\, (\ell-1)!!/\ell!!$ is increasing and converges to $2/\pi$ for $\ell \to \infty$. Therefore, it is bounded from below by its value $1/2$ for $\ell = 0$ and from above by its limit $2/\pi$. Furthermore, $R\colon X \to X$ is self-adjoint since its eigenvalues $P_\ell(0)$, cf. (10), are real. □

In the following, we describe a trigonometric basis on the sphere that allows us to obtain a novel FD of the Funk-Radon transform R. We start with the *spherical coordinate transform*

$$\phi(\lambda, \theta) := (\cos\lambda \sin\theta, \sin\lambda \sin\theta, \cos\theta), \qquad \forall \lambda \in [0, 2\pi),\ \theta \in [0, \pi], \tag{13}$$

which is one-to-one except for $\theta \in \{0, \pi\}$ corresponding to the north and south pole. We assume λ to be 2π-periodic, and define the trigonometric basis functions

$$b_{n,k} \colon \mathbb{S}^2 \to \mathbb{C}\,, \; b_{n,k}(\phi(\lambda,\theta)) := \frac{e^{in\lambda}\sin(k\theta)}{\pi\sqrt{\sin\theta}}\,, \qquad \forall n \in \mathbb{Z},\; k \in \mathbb{N}, \tag{14}$$

which are well-defined and continuous on $\mathbb{S}^2$ since $b_{n,k}$ vanishes for $\theta \to 0$ and $\theta \to \pi$. Trigonometric bases on $\mathbb{S}^2$ bear the advantage of their simple and fast computation [53], and form the foundation of the double Fourier sphere method [35,51]. Note that our functions (14) slightly differ from the ones in [35] as we take $\sin(k\theta)$ in order to avoid singularities at the poles.

Lemma 1. *The sequence $\{b_{n,k}\}_{n\in\mathbb{Z},k\in\mathbb{N}}$ forms an orthonormal basis of $L^2(\mathbb{S}^2)$.*

Proof. Let $n, n' \in \mathbb{Z}$ and $k, k' \in \mathbb{N}$. Since the integral on $\mathbb{S}^2$ in spherical coordinates (13) reads as $\sin(\theta)\,d\theta\,d\lambda$, we have

$$\langle b_{n,k}, b_{n',k'}\rangle_{L^2(\mathbb{S}^2)} = \frac{1}{\pi^2}\int_0^\pi \int_0^{2\pi} e^{i(n-n')\lambda}\sin(k\theta)\sin(k'\theta)\,d\lambda\,d\theta = \delta_{n,n'}\delta_{k,k'}\,,$$

which shows the orthonormality. The completeness follows from the completeness of $\{e^{in\cdot}\}_{n\in\mathbb{Z}}$ in $L^2([0,2\pi])$ and of $\{\sin(k\cdot)\}_{k\in\mathbb{N}}$ in $L^2([0,\pi])$. □

For $\boldsymbol{\xi} = \phi(\lambda,\theta) \in \mathbb{S}^2$, its antipodal point is $-\boldsymbol{\xi} = \phi(\pi+\lambda, \pi-\theta)$. Hence, we obtain the symmetry relation $b_{n,k}(-\boldsymbol{\xi}) = (-1)^{n+k+1} b_{n,k}(\boldsymbol{\xi})$, which implies that the sequence $\{b_{n,k}\}_{n\in\mathbb{Z},k\in\mathbb{N},n+k \text{ odd}}$ is an orthonormal basis of $L^2_{\text{even}}(\mathbb{S}^2)$.

Lemma 2. *The basis functions $b_{n,k}$, $n \in \mathbb{Z}$, $k \in \mathbb{N}$ with $n+k$ odd are well-defined elements of $H^1_{\text{even}}(\mathbb{S}^2)$. In particular, we also have $b_{n,k} \in H^{1/2}_{\text{even}}(\mathbb{S}^2)$.*

Proof. By [41, § 5.2], the $H^1(\mathbb{S}^2)$ Sobolev norm of $f \in C^1(\mathbb{S}^2)$ can be written as

$$\|f\|^2_{H^1(\mathbb{S}^2)} = \sum_{i=1}^{3} \|[\nabla_{\mathbb{S}^2} f]_i\|^2_{L^2(\mathbb{S}^2)} + \tfrac{1}{4}\|f\|^2_{L^2(\mathbb{S}^2)}\,, \tag{15}$$

where $[\nabla_{\mathbb{S}^2} f]_i$ denotes the i-th coordinate of the surface gradient given in spherical coordinates by

$$\nabla_{\mathbb{S}^2} f(\phi(\lambda,\theta)) = \phi(\lambda, \tfrac{\pi}{2}+\theta)\,\partial_\theta f(\phi(\lambda,\theta)) + \phi(\tfrac{\pi}{2}+\lambda, \tfrac{\pi}{2})\,\tfrac{1}{\sin\theta}\,\partial_\lambda f(\phi(\lambda,\theta))\,.$$

Let $n \in \mathbb{Z}$ and $k \in \mathbb{N}$. For $f = b_{n,k}$, we have

$$\partial_\theta b_{n,k}(\phi(\lambda,\theta)) = \frac{e^{in\lambda}}{\pi}\left(\frac{k\cos(k\theta)}{\sqrt{\sin\theta}} - \frac{\cos(\theta)\,\sin(k\theta)}{2\,\sin^{3/2}\theta}\right)$$

and

$$\partial_\lambda b_{n,k}(\phi(\lambda,\theta)) = in\,\frac{e^{in\lambda}}{\pi}\,\frac{\sin(k\theta)}{\sqrt{\sin\theta}}\,.$$

Employing the facts that sine and cosine functions, in particular the components of ϕ, are bounded by one and that $|a+b|^2 \leq 2(|a|^2+|b|^2)$ for $a, b \in \mathbb{R}$, we obtain

$$\begin{aligned}|[\nabla_{\mathbb{S}^2} b_{n,k}(\phi(\lambda,\theta))]_i|^2 &\leq 2\left|\frac{k\cos(k\theta)}{\pi\sqrt{\sin\theta}} - \frac{\cos(\theta)\sin(k\theta)}{2\pi\sin^{3/2}\theta}\right|^2 + 2\frac{n^2\sin^2(k\theta)}{\pi^2\sin^3\theta}\\ &\leq \frac{4k^2}{\pi^2\sin\theta} + \frac{\sin^2(k\theta)}{\pi^2\sin^3\theta} + \frac{2n^2\sin^2(k\theta)}{\pi^2\sin^3\theta}.\end{aligned}$$

Hence, we have

$$\begin{aligned}\|[\nabla_{\mathbb{S}^2} b_{n,k}]_i\|^2_{L^2(\mathbb{S}^2)} &= \int_0^\pi\int_0^{2\pi} |[\nabla_{\mathbb{S}^2} b_{n,k}]_i|^2 \sin(\theta)\,\mathrm{d}\lambda\,\mathrm{d}\theta\\ &\leq \int_0^\pi\int_0^{2\pi}\left(\frac{4k^2}{\pi^2} + (1+2n^2)\frac{\sin^2(k\theta)}{\pi^2\sin^2\theta}\right)\mathrm{d}\lambda\,\mathrm{d}\theta\\ &= 8k^2 + (1+2n^2)2k,\end{aligned}$$

where the last equality follows from the the integration formula [38, Ex. 1.15] of the $(k-1)$th Fejér kernel. Finally, we conclude from (15) and Lemma 1 that

$$\|b_{n,k}\|^2_{H^1(\mathbb{S}^2)} = \sum_{i=1}^3 \|[\nabla_{\mathbb{S}^2} b_{n,k}]_i\|^2_{L^2(\mathbb{S}^2)} + \tfrac{1}{4}\|b_{n,k}\|^2_{L^2(\mathbb{S}^2)} \leq 6k(4k+1+2n^2) + \tfrac{1}{4}$$

is finite. The claim follows as $H^1(\mathbb{S}^2)$ is continuously embedded in $H^{1/2}(\mathbb{S}^2)$. □

Theorem 3. *Let* $E\colon H^{1/2}_{\mathrm{even}}(\mathbb{S}^2) \to L^2_{\mathrm{even}}(\mathbb{S}^2)$ *denote the embedding operator, and set* $L{:=}(EE^*)^{-1/2}$. *Then*

$$e_{n,k}{:=}RLb_{n,k}, \qquad (n,k) \in J{:=}\{(n,k) \in \mathbb{Z}\times\mathbb{N} : n+k \text{ odd}\},$$

is a frame in $L^2_{\mathrm{even}}(\mathbb{S}^2)$, *and for any* $g \in H^{1/2}_{\mathrm{even}}(\mathbb{S}^2)$, *the unique solution* $f \in L^2_{\mathrm{even}}(\mathbb{S}^2)$ *of the inversion problem of the Funk-Radon transform* $Rf = g$ *satisfies*

$$f = R^\ddagger g{:=} \sum_{(n,k)\in J} \langle Lg, b_{n,k}\rangle_{L^2(\mathbb{S}^2)}\tilde{e}_{n,k}, \tag{16}$$

where $\tilde{e}_{n,k}$ *is the dual frame of* $e_{n,k}$. *It holds that* $\|R^\ddagger g\|_{L^2(\mathbb{S}^2)} \leq 2\|Lg\|_{L^2(\mathbb{S}^2)}$.

Proof. From Theorem 2, Lemmas 1 and 2, we see that Assumption 1 is satisfied with $X = Y = L^2_{\mathrm{even}}(\mathbb{S}^2)$ and $Z = H^{1/2}_{\mathrm{even}}(\mathbb{S}^2)$. The claim follows by Theorem 1. □

Remark 1. The definition (11) of the Sobolev spaces yields that

$$E^*f = EE^*f = \sum_{\ell=0}^\infty\sum_{m=-\ell}^{\ell} (\ell+\tfrac{1}{2})^{-1}\langle f, Y_\ell^m\rangle_{L^2(\mathbb{S}^2)}\, Y_\ell^m, \qquad \forall f \in L^2(\mathbb{S}^2).$$

Hence, the self-adjoint operator L from Theorem 3 is a multiplication operator with respect to the spherical harmonics, and we have for all $g \in H^{1/2}(\mathbb{S}^2)$ that

$$Lg = (EE^*)^{-1/2} g = \sum_{\ell=0}^{\infty} \sum_{m=-\ell}^{\ell} (\ell + \tfrac{1}{2})^{1/2} \langle g, Y_\ell^m \rangle_{L^2(\mathbb{S}^2)} \, Y_\ell^m \,. \tag{17}$$

Denoting by $\Delta_{\mathbb{S}^2}$ the Laplace-Beltrami operator on $\mathbb{S}^2$, we obtain by its eigenvalue decomposition [3, p. 121] that $Lg = (-\Delta_{\mathbb{S}^2} + 1/4)^{1/4} g$ for $g \in H^{1/2}(\mathbb{S}^2)$. Furthermore, combining (10) and (17), it follows that

$$RLg = \sum_{\ell=0}^{\infty} \sum_{m=-\ell}^{\ell} P_\ell(0) \, (\ell + \tfrac{1}{2})^{1/2} \langle g, Y_\ell^m \rangle_{L^2(\mathbb{S}^2)} \, Y_\ell^m \,, \tag{18}$$

and since $|P_\ell(0)|$ decays as $(\ell + \frac{1}{2})^{-1/2}$ by (12), we obtain $RLg \in H^{1/2}(\mathbb{S}^2)$.

4 Numerical Results

Next, we discuss the use of regularization for our FD of the Funk-Radon transform, outline the main steps of its implementation, and present numerical results. First, note that noisy data $g^\delta := Rf + \delta$ does not necessarily belong to the space $H^{1/2}(\mathbb{S}^2)$, and thus $R^\ddagger g^\delta$ is not well-defined. This necessitates regularization, for which we consider stable approximations of $R^\ddagger g^\delta$ defined by

$$f_\alpha^\delta := R^\ddagger U_\alpha g^\delta \,, \qquad \text{and} \qquad LU_\alpha g^\delta := \sum_{\ell=0}^{\infty} \sum_{m=-\ell}^{\ell} \sigma_\ell h_\alpha(\sigma_\ell^2) \langle g^\delta, Y_\ell^m \rangle_{L^2(\mathbb{S}^2)} \, Y_\ell^m \,,$$

where $\sigma_\ell := (\ell + \frac{1}{2})^{-1/2}$ and $R^\ddagger$ is given in (16). This amounts to a filtering of the coefficients σ_k of L via a suitable *filter function* h_α approximating $s \mapsto 1/s$ as in the SVD case [13]. In our numerical experiments, we investigate the positive influence of the Tikhonov filter functions $h_\alpha(s) = 1/(\alpha + s)$ on the reconstruction quality. Note that the choice $h_\alpha(s) = 1/\sqrt{s}$ implies $U_\alpha = L^{-1}$, which by Remark 1 basically amounts to a smoothing operator; cf., e.g., [30].

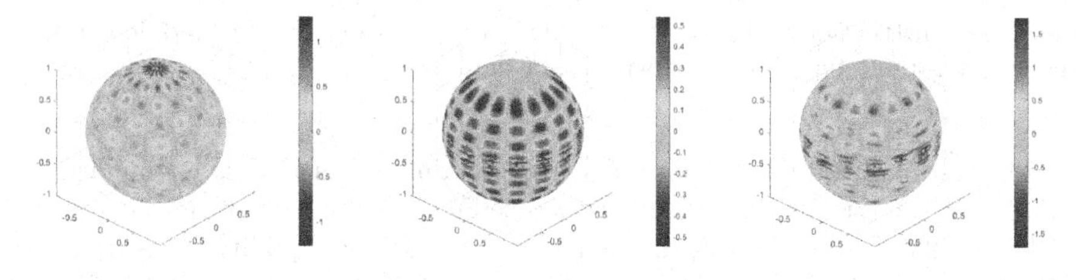

Fig. 1. Ingredients of the FD: b_{nk} (left), e_{nk} (middle), $\tilde{e}_{nk}$ (right) for $n = k = 8$.

For discretization of the problem, we approximate functions on the sphere with a Chebyshev-type quadrature, i.e., quadrature points such that all weights are equal; see [20]. In our computations, we use quadrature points (spherical design) being exact up to degree 200 taken from [19]. All computations involving spherical harmonics are done using the NFSFT (Non-equispaced fast spherical Fourier transform) software [29,32]. Note that the dual frame functions of an FD can be pre-computed and stored, such that the computational effort of computing $R^{\ddagger}g$ according to (16) for any new measurement g amounts to one application of the operator L (or LU_α in the noisy case), computation of the inner products $\langle Lg, b_{n,k}\rangle_{L^2_{\text{even}}(\mathbb{S}^2)}$ and a summation over these. As starting point for the computation, we only use a finite set of frame functions $\{b_{n,k} : (n,k) \in J, |n| \le N, k \le N\}$ for some $N \in \mathbb{N}$. The frame functions $e_{n,k} = RLb_{n,k}$ are evaluated at the quadrature nodes via the spherical harmonic decomposition (18) up to degree $\ell \le 100$. The dual frame functions $\tilde{f}_{n,k}$ are computed using the matrix representation of the linear operator S with respect to the basis $b_{n,k}$, see also [27, Chap. 5]. Note that the inversion of this matrix may itself be an ill-conditioned problem requiring regularization. Examples of the involved frame functions are depicted in Fig. 1. All computations are performed using Matlab R2022b.

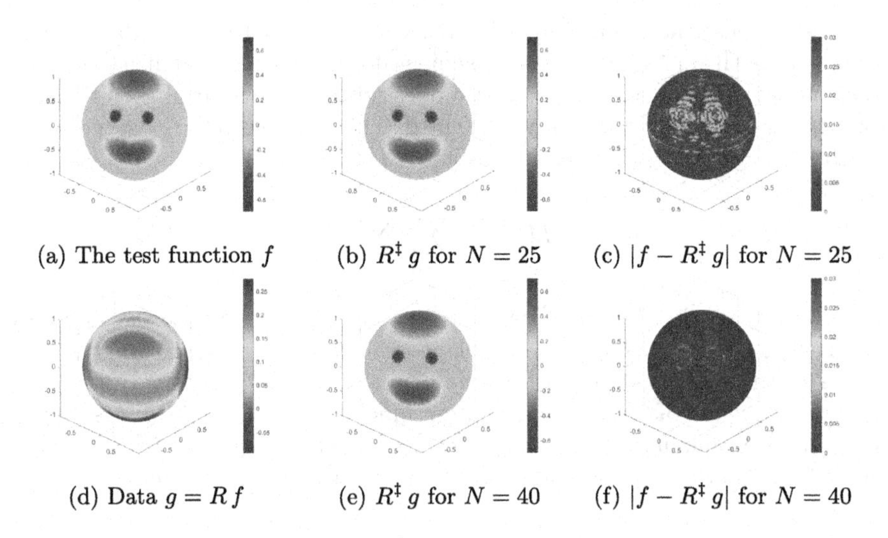

(a) The test function f (b) $R^{\ddagger}g$ for $N = 25$ (c) $|f - R^{\ddagger}g|$ for $N = 25$

(d) Data $g = Rf$ (e) $R^{\ddagger}g$ for $N = 40$ (f) $|f - R^{\ddagger}g|$ for $N = 40$

Fig. 2. Reconstruction evaluation for the Chebyshev-type quadrature for different numbers of dual-frame functions used with exact data.

The test function used in our numerical experiments is a linear combination of radially symmetric, quadratic splines, whose Funk-Radon transform is computed explicitly [23, Lem. 4.1], to prevent inverse crimes. Our quality measure is the relative reconstruction error, i.e., $\|f - R^{\ddagger}g\|_{L^2(\mathbb{S}^2)}/\|f\|_{L^2(\mathbb{S}^2)}$. In Fig. 2, we see that increasing the number of used frames highly improves reconstruction quality, reducing the relative reconstruction error from 0.023 for $N = 25$ to 0.006 for $N = 40$. For noisy data shown in Fig. 3, the regularization parameter α is chosen such that the relative reconstruction error is minimized. In the case of $N = 25$ and

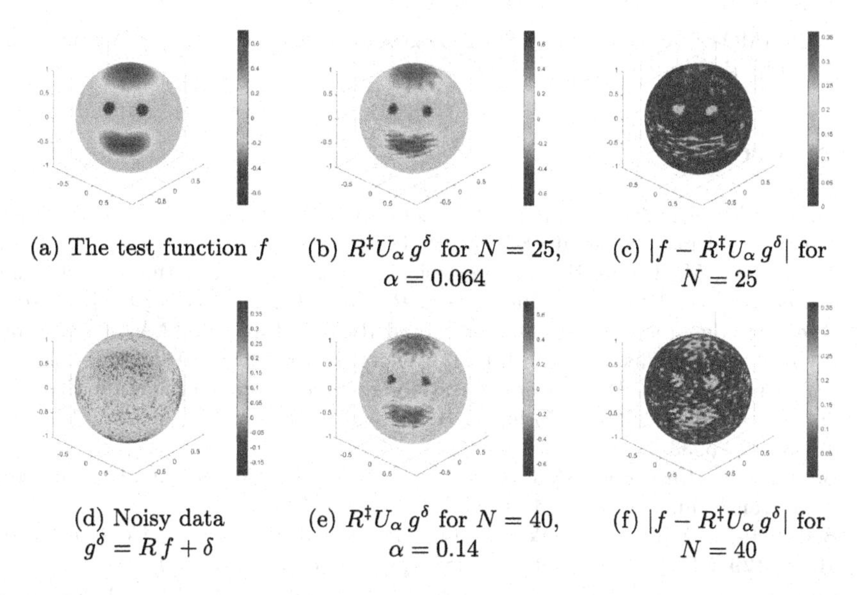

(a) The test function f (b) $R^{\ddagger}U_{\alpha}\,g^{\delta}$ for $N = 25$, $\alpha = 0.064$ (c) $|f - R^{\ddagger}U_{\alpha}\,g^{\delta}|$ for $N = 25$

(d) Noisy data $g^{\delta} = Rf + \delta$ (e) $R^{\ddagger}U_{\alpha}\,g^{\delta}$ for $N = 40$, $\alpha = 0.14$ (f) $|f - R^{\ddagger}U_{\alpha}\,g^{\delta}|$ for $N = 40$

Fig. 3. Reconstruction evaluation for the Chebyshev-type quadrature for different numbers of dual-frame functions used with Gaussian noise δ, noise level 20%.

noise level 20%, the error for the non-regularized solution (i.e., $\alpha = 0$) is 0.269, while the error for the regularized solution with parameter $\alpha = 0.076$ reduces to 0.222. However, we see that the increment of the number of frame functions actually results in a loss of reconstruction quality to an error value of 0.332 (optimally regularized). This can be explained by the regularization effect of the truncation itself: more frame functions result in a less stable reconstruction, but yield a higher accuracy in case of exact data. Note that all specific error values in the noisy case are insignificantly varying for the specific realization of the randomly generated Gaussian noise δ.

5 Conclusion

In this paper, we derived a novel frame decomposition of the Funk-Radon transform utilizing trigonometric basis functions $b_{n,k}$ on the unit sphere and suitable embedding operators in Sobolev spaces. This decomposition does not involve the spherical harmonics and leads to an explicit inversion formula for the Funk-Radon transform. In our numerical examples, we obtained promising reconstruction results even in the case of very large noise by including regularization. While the regularization itself currently uses a spherical harmonics expansion of the operator L, in our future work we aim to apply other forms of regularization avoiding the computationally expensive spherical harmonics entirely.

Acknowledgements. This work was funded by the Austrian Science Fund (FWF): project F6805-N36 (SH) and the German Research Foundation (DFG): project

495365311 (MQ) within the SFB F68: "Tomography Across the Scales". LW is partially supported by the State of Upper Austria.

References

1. Abramovich, F., Silverman, B.W.: Wavelet decomposition approaches to statistical inverse problems. Biometrika **85**(1), 115–129 (1998)
2. Agranovsky, M., Rubin, B.: Non-geodesic spherical funk transforms with one and two centers. In: Bauer, W., Duduchava, R., Grudsky, S., Kaashoek, M.A. (eds.) Operator Algebras, Toeplitz Operators and Related Topics. OTAA, vol. 279, pp. 29–52. Springer, Cham (2020). https://doi.org/10.1007/978-3-030-44651-2_7
3. Atkinson, K., Han, W.: Spherical Harmonics and Approximations on the Unit Sphere: An Introduction. Springer, Heidelberg (2012). https://doi.org/10.1007/978-3-642-25983-8
4. Bailey, T.N., Eastwood, M.G., Gover, A., Mason, L.: Complex analysis and the Funk transform. J. Korean Math. Soc. **40**(4), 577–593 (2003)
5. Bellet, J.-B.: A discrete Funk transform on the cubed sphere. J. Comput. Appl. Math. **429**, 115205 (2023). https://doi.org/10.1016/j.cam.2023.115205
6. Candes, E.J., Donoho, D.L.: Recovering edges in ill-posed inverse problems: optimality of curvelet frames. Ann. Stat. **30**(3), 784–842 (2002). https://doi.org/10.1214/aos/1028674842
7. Christensen, O.: An Introduction to Frames and Riesz Bases. Applied and Numerical Harmonic Analysis, Birkhäuser, Cham (2016)
8. Colonna, F., Easley, G., Guo, K., Labate, D.: Radon transform inversion using the shearlet representation. Appl. Comput. Harmon. Anal. **29**(2), 232–250 (2010). https://doi.org/10.1016/j.acha.2009.10.005
9. Daubechies, I.: Ten Lectures on Wavelets. Society for Industrial and Applied Mathematics, Philadelphia (1992). https://doi.org/10.1137/1.9781611970104
10. Dicken, V., Maass, P.: Wavelet-Galerkin methods for ill-posed problems. J. Inverse Ill-Posed Probl. **4**(3), 203–221 (1996). https://doi.org/10.1515/jiip.1996.4.3.203
11. Donoho, D.L.: Nonlinear solution of linear inverse problems by Wavelet-Vaguelette decomposition. Appl. Comput. Harmon. Anal. **2**(2), 101–126 (1995). https://doi.org/10.1006/acha.1995.1008
12. Ebner, A., Frikel, J., Lorenz, D., Schwab, J., Haltmeier, M.: Regularization of inverse problems by filtered diagonal frame decomposition. Appl. Comput. Harmon. Anal. **62**, 66–83 (2023). https://doi.org/10.1016/j.acha.2022.08.005
13. Engl, H.W., Hanke, M., Neubauer, A.: Regularization of Inverse Problems. Kluwer Academic Publishers, Dordrecht (1996)
14. Frikel, J.: Sparse regularization in limited angle tomography. Appl. Comput. Harmon. Anal. **34**(1), 117–141 (2013). https://doi.org/10.1016/j.acha.2012.03.005
15. Frikel, J., Haltmeier, M.: Efficient regularization with wavelet sparsity constraints in photoacoustic tomography. Inverse Probl. **34**(2), 024006 (2018). https://doi.org/10.1088/1361-6420/aaa0ac
16. Frikel, J., Haltmeier, M.: Sparse regularization of inverse problems by operator-adapted frame thresholding. In: Dörfler, W., et al. (eds.) Mathematics of Wave Phenomena, pp. 163–178. Springer, Cham (2020). https://doi.org/10.1007/978-3-030-47174-3_10
17. Funk, P.: Über Flächen mit lauter geschlossenen geodätischen Linien. Math. Ann. **74**(2), 278–300 (1913). https://doi.org/10.1007/BF01456044

18. Gardner, R.J.: Geometric Tomography, 2nd edn. Cambridge University Press, Cambridge (2006). https://doi.org/10.1017/CBO9781107341029
19. Gräf, M.: Quadrature rules on manifolds. https://www.tu-chemnitz.de/~potts/workgroup/graef/quadrature
20. Gräf, M., Potts, D.: On the computation of spherical designs by a new optimization approach based on fast spherical Fourier transforms. Numer. Math. **119**, 699–724 (2011). https://doi.org/10.1007/s00211-011-0399-7
21. Helgason, S.: Integral Geometry and Radon Transforms. Springer, New York (2011). https://doi.org/10.1007/978-1-4419-6055-9
22. Hielscher, R., Potts, D., Quellmalz, M.: An SVD in spherical surface wave tomography. In: Hofmann, B., Leitão, A., Zubelli, J.P. (eds.) New Trends in Parameter Identification for Mathematical Models. TM, pp. 121–144. Springer, Cham (2018). https://doi.org/10.1007/978-3-319-70824-9_7
23. Hielscher, R., Quellmalz, M.: Optimal mollifiers for spherical deconvolution. Inverse Probl. **31**(8), 085001 (2015). https://doi.org/10.1088/0266-5611/31/8/085001
24. Hristova, Y., Moon, S., Steinhauer, D.: A Radon-type transform arising in photoacoustic tomography with circular detectors: spherical geometry. Inverse Probl. Sci. Eng. **24**(6), 974–989 (2016). https://doi.org/10.1080/17415977.2015.1088537
25. Hubmer, S., Ramlau, R.: A frame decomposition of the atmospheric tomography operator. Inverse Probl. **36**(9), 094001 (2020). https://doi.org/10.1088/1361-6420/aba4fe
26. Hubmer, S., Ramlau, R.: Frame decompositions of bounded linear operators in Hilbert spaces with applications in tomography. Inverse Probl. **37**(5), 055001 (2021). https://doi.org/10.1088/1361-6420/abe5b8
27. Hubmer, S., Ramlau, R., Weissinger, L.: On regularization via frame decompositions with applications in tomography. Inverse Probl. **38**(5), 055003 (2022). https://doi.org/10.1088/1361-6420/ac5b86
28. Kazantsev, S.G.: Funk-Minkowski transform and spherical convolution of Hilbert type in reconstructing functions on the sphere. Sib. Èlektron. Mat. Izv. **15**, 1630–1650 (2018). https://doi.org/10.33048/semi.2018.15.135
29. Keiner, J., Kunis, S., Potts, D.: Using NFFT3 - a software library for various nonequispaced fast Fourier transforms. ACM Trans. Math. Softw. **36**, Article 19, 1–30 (2009). https://doi.org/10.1145/1555386.1555388
30. Klann, E., Ramlau, R.: Regularization by fractional filter methods and data smoothing. Inverse Probl. **24**(2), 025018 (2008)
31. Kudryavtsev, A.A., Shestakov, O.V.: Estimation of the loss function when using Wavelet-Vaguelette decomposition for solving Ill-posed problems. J. Math. Sci. **237**(6), 804–809 (2019). https://doi.org/10.1007/s10958-019-04206-z
32. Kunis, S., Potts, D.: Fast spherical Fourier algorithms. J. Comput. Appl. Math. **161**, 75–98 (2003). https://doi.org/10.1016/S0377-0427(03)00546-6
33. Lee, N.: Wavelet-Vaguelette decompositions and homogeneous equations. ProQuest LLC, Ann Arbor, thesis (Ph.D.)-Purdue University (1997)
34. Louis, A.K., Riplinger, M., Spiess, M., Spodarev, E.: Inversion algorithms for the spherical Radon and cosine transform. Inverse Probl. **27**(3), 035015 (2011). https://doi.org/10.1088/0266-5611/27/3/035015
35. Mildenberger, S., Quellmalz, M.: Approximation properties of the double Fourier sphere method. J. Fourier Anal. Appl. **28**(2), 1–30 (2022). https://doi.org/10.1007/s00041-022-09928-4
36. Minkowski, H.: Sur les corps de largeur constante. Matematiceskij Sbornik **25**(3), 505–508 (1905). https://mi.mathnet.ru/sm6643

37. Müller, C.: Spherical Harmonics. Springer, Aachen (1966)
38. Plonka, G., Potts, D., Steidl, G., Tasche, M.: Numerical Fourier Analysis. Birkhäuser, Cham (2018). https://doi.org/10.1007/978-3-030-04306-3
39. Quellmalz, M.: A generalization of the Funk-Radon transform. Inverse Probl. **33**(3), 035016 (2017). https://doi.org/10.1088/1361-6420/33/3/035016
40. Quellmalz, M.: Reconstructing functions on the sphere from circular means. Dissertation, Universitätsverlag Chemnitz (2019). https://nbn-resolving.org/urn:nbn:de:bsz:ch1-qucosa2-384068
41. Quellmalz, M.: The Funk-Radon transform for hyperplane sections through a common point. Anal. Math. Phys. **10**(3), 1–29 (2020). https://doi.org/10.1007/s13324-020-00383-2
42. Quellmalz, M., Hielscher, R., Louis, A.K.: The cone-beam transform and spherical convolution operators. Inverse Probl. **34**(10), 105006 (2018). https://doi.org/10.1088/1361-6420/aad679
43. Rauff, A., Timmins, L.H., Whitaker, R.T., Weiss, J.A.: A nonparametric approach for estimating three-dimensional fiber orientation distribution functions (ODFs) in fibrous materials. IEEE Trans. Med. Imaging **41**(2), 446–455 (2022). https://doi.org/10.1109/TMI.2021.3115716
44. Riplinger, M., Spiess, M.: Numerical inversion of the spherical Radon transform and the cosine transform using the approximate inverse with a special class of locally supported mollifiers. J. Inverse Ill-Posed Probl. **22**(4), 497–536 (2013). https://doi.org/10.1515/jip-2012-0095
45. Rubin, B.: On the spherical slice transform. Anal. Appl. **20**(3), 483–497 (2022). https://doi.org/10.1142/S021953052150024X
46. Salman, Y.: Recovering functions defined on the unit sphere by integration on a special family of sub-spheres. Anal. Math. Phys. **7**(2), 165–185 (2016). https://doi.org/10.1007/s13324-016-0135-7
47. Strichartz, R.S.: L^p estimates for Radon transforms in Euclidean and non-Euclidean spaces. Duke Math. J. **48**(4), 699–727 (1981)
48. Terzioglu, F.: Recovering a function from its integrals over conical surfaces through relations with the Radon transform. Inverse Probl. **39**(2), 024005 (2023). https://doi.org/10.1088/1361-6420/acad24
49. Tuch, D.S.: Q-ball imaging. Magn. Reson. Med. **52**(6), 1358–1372 (2004). https://doi.org/10.1002/mrm.20279
50. Weissinger, L.: Realization of the frame decomposition of the atmospheric tomography operator. Master's thesis, JKU Linz (2021). https://lisss.jku.at/permalink/f/n2r1to/ULI_alma5185824070003340
51. Wilber, H., Townsend, A., Wright, G.B.: Computing with functions in spherical and polar geometries II. The disk. SIAM J. Sci. Comput. **39**(3), C238–C262 (2017). https://doi.org/10.1137/16M1070207
52. Yarman, C.E., Yazici, B.: Inversion of the circular averages transform using the Funk transform. Inverse Probl. **27**(6), 065001 (2011). https://doi.org/10.1088/0266-5611/27/6/065001
53. Yee, S.Y.K.: Studies on Fourier series on spheres. Mon. Weather Rev. **108**(5), 676–678 (1980). https://doi.org/10.1175/1520-0493(1980)108⟨0676:SOFSOS⟩2.0.CO;2

Prony-Based Super-Resolution Phase Retrieval of Sparse, Multidimensional Signals

Robert Beinert(✉) and Saghar Rezaei

Technische Universität Berlin, Institute of Mathematics,
Straße des 17. Juni 136, 10623 Berlin, Germany
robert.beinert@tu-berlin.de, s.rezaei@campus.tu-berlin.de
https://tu.berlin/imageanalysis/

Abstract. Phase retrieval consists in the recovery of an unknown signal from phaseless measurements of its usually complex-valued Fourier transform. Without further assumptions, this problem is notorious to be severe ill posed such that the recovery of the true signal is nearly impossible. In certain applications like crystallography, speckle imaging in astronomy, or blind channel estimation in communications, the unknown signal has a specific, sparse structure. In this paper, we exploit these sparse structure to recover the unknown signal uniquely up to inevitable ambiguities as global phase shifts, transitions, and conjugated reflections. Although using a constructive proof essentially based on Prony's method, our focus lies on the derivation of a recovery guarantee for multidimensional signals using an adaptive sampling scheme. Instead of sampling the entire multidimensional Fourier intensity, we only employ Fourier samples along certain adaptively chosen lines. For two-dimensional signals, an analogous result can be established for samples in generic directions. The number of samples here scales quadratically to the sparsity level of the unknown signal.

Keywords: Phase retrieval · uniqueness guarantees · sparse signals · Prony's method · adaptive sampling

1 Introduction

Phase retrieval is one of the major challenges in many imaging tasks in physics and engineering. For instance, phase retrieval is an essential component of the imaging techniques: ptychography [17], crystallography [18], speckle imaging [16], diffraction tomography [9]. Although there are many different problem formulations summarized as "phase retrieval", the main task consists in the recovery of an image or signal form the magnitudes of usually complex-valued

Supported by the Federal Ministry of Education and Research (BMBF, Germany) [grant number 13N15754].

L. Calatroni et al. (Eds.): SSVM 2023, LNCS 14009, pp. 55–67, 2023.
https://doi.org/10.1007/978-3-031-31975-4_5

measurements. If we think at the application in optics, the measurements often correspond to the magnitudes of the Fourier or Fresnel transform. Although the Fourier and Fresnel transform are invertible, the loss of the phase turns the imaging task into an severe ill-posed inverse problem. The central challenge is here the strong ambiguousness, which has been studied for the continuous as well as for the discrete setting, see for instance [5,10,12]. To overcome the ambiguousness, different a priori informations like support constraints or non-negativity as well as additional measurements have been studied [3,4,8,14,15].

In this paper, we are interested in the recovery of multidimensional, sparse signals, which are modeled as complex measure

$$\mu := \sum_{n=1}^{N} c_n \, \nu(\cdot - T_n),$$

where ν is a known structure. Choosing ν as Dirac measure, we obtain point measures or spike functions. But other choices like Gaussians are reasonable too and lead to more regular functions. We here consider the Fourier phase retrieval problem, meaning that we want to recover c_n and T_n from samples of the Fourier intensity— the magnitude of the Fourier transform— of μ. Our focus lies on the derivation of uniqueness guarantees, i.e. on assumptions allowing the determination of c_n and T_n up to unavoidable ambiguities. Since the transitions T_n may lie on a continuous domain, our problem can be interpreted as super-resolution phase retrieval [1]. This kind of problem arises in applications like crystallography [18], speckle imaging in astronomy [16] as well as in blind channel estimation in communication [2].

Methodology and Relation to the Literature To show that phase retrieval of sparse, multidimensional signals is possible in principle, we rely on the Prony-based techniques in [6], where super-resolution phase retrieval is studied for one-dimensional signals. For the generalization to multidimensional signals, we employ an adaptive sampling strategy, where the Fourier intensity of the unknown signal is measured along adaptively chosen or along generic lines. This sampling method traces back to [20], where the recovery of sparse signals from their (complex-valued) Fourier transform is studied. Our main idea is to use the specific sampling setup to reduce the multidimensional phase retrieval problem into a series of one-dimensional problems, to solve these one-dimensional instances as in [6], and to combine the extracted informations to solve the multidimensional instance. The considered super-resolution phase retrieval problem has also been studied in [1], where a greedy-like algorithm is proposed to solve the problem numerically. In difference to [1], where the entire Fourier domain is sampled, we show that the unknown D-dimensional signal is completely determined by $\mathcal{O}(N^2)$ measurements on $2D - 1$ lines, where N denotes the sparsity level.

Contribution The contribution of this paper is the derivation of a recovery guarantee for super-resolution phase retrieval of multidimensional signals. The main theorem shows that, under mild assumptions like collision-freeness, the unknown signal is completely determined by equispaced samples on a few lines. All in all, we here require $\mathcal{O}(DN^2)$ samples of the Fourier intensity to recover a D-dimensional signal composed of N components. This shows that, at least theoretically, it is enough to use a space sampling setup to recover a sparse signal. Although the proofs are constructive, the focus lies on the theoretical uniqueness guarantee since the applied Prony method is known to be unstable for noisy measurements.

Outline Before considering the super-resolution phase retrieval problem, we briefly introduce the needed concepts like the Fourier transform of measures and Prony's method in Sect. 2. In Sect. 3, we define the considered problem in more details and discuss the unavoidable ambiguities. Section 4 is devoted to the one-dimensional sparse phase retrieval problem, which we generalize to the multidimensional setting in Sect. 5. The numerical simulation in Sect. 6 show that the constructive proofs can be implemented to recover the unknown signal at least in the noise-free setting. A final discussion is given in Sect. 7.

2 Preleminaries

2.1 Fourier Transform of Measures

Subsequently, the unknown signals are characterized as complex measures. For this, let $\mathcal{B}(\mathbb{R}^D)$ denote the Borel σ-algebra of the Euclidean space $\mathbb{R}^D$, and $\mathcal{M}(\mathbb{R}^D)$ the space of all regular, finite, complex measures. Every considered signal is then interpreted as complex-valued mapping $\mu\colon \mathcal{B}(\mathbb{R}^D) \to \mathbb{C}$ with $\mu \in \mathcal{M}(\mathbb{R}^D)$. Recall that $\mathcal{M}(\mathbb{R}^D)$ is the dual space of $C_0(\mathbb{R}^D)$, which consists of all continuous functions $\phi\colon \mathbb{R}^D \to \mathbb{C}$ where $\phi(x)$ vanishes for $\|x\| \to \infty$. Every measure $\mu \in \mathcal{M}(\mathbb{R}^D)$ hence defines a continuous, linear mapping via

$$\phi \mapsto \langle \mu, \phi \rangle := \int_{\mathbb{R}^D} \phi(x) \mathrm{d}\mu(x).$$

The convolution of two measures $\mu, \nu \in \mathcal{M}(\mathbb{R}^D)$ is indirectly defined as

$$\langle \mu * \nu, \phi \rangle := \iint_{\mathbb{R}^D \times \mathbb{R}^D} \phi(x + y) \mathrm{d}\nu(y) \mathrm{d}\mu(x), \quad \phi \in C_0(\mathbb{R}^D).$$

The Fourier transform on $\mathcal{M}(\mathbb{R}^D)$ is given by $\mathcal{F}\colon \mathcal{M}(\mathbb{R}^D) \to C_{\mathrm{b}}(\mathbb{R}^D)$ with

$$\mathcal{F}[\mu](\omega) := \hat{\mu}(\omega) := \int_{\mathbb{R}^D} \mathrm{e}^{-\mathrm{i}\langle \omega, x \rangle} \mathrm{d}\mu(x), \quad \omega \in \mathbb{R}^D,$$

where $C_{\mathrm{b}}(\mathbb{R}^D)$ consists of all bounded, continuous functions on $\mathbb{R}^D$. Notice that $\mathcal{F}$ is continuous and injective. Further, the *Fourier convolution theorem* states

$$\mathcal{F}[\mu * \nu] = \hat{\mu}\hat{\nu}.$$

2.2 Prony's Method

The Fourier intensities of a sparse signal on $\mathbb{R}$ are essentially given by a non-negative exponential sum $E\colon \mathbb{R} \to \mathbb{R}$ of the form

$$E(\omega) := \sum_{\ell=-L}^{L} \gamma_\ell \, \mathrm{e}^{-\mathrm{i}\omega\tau_\ell} = \gamma_0 + \sum_{\ell=1}^{L} \left(\gamma_\ell \, \mathrm{e}^{-\mathrm{i}\omega\tau_\ell} + \bar{\gamma}_\ell \, \mathrm{e}^{\mathrm{i}\omega\tau_\ell}\right) \tag{1}$$

with $\gamma_\ell = \bar{\gamma}_{-\ell} \in \mathbb{C} \setminus \{0\}$ and $\tau_\ell = -\tau_{-\ell} \in \mathbb{R}$ for $\ell = 0, \dots, L$. Note that $\gamma_0 \in \mathbb{R} \setminus \{0\}$ and $\tau_0 = 0$. For pairwise distinct τ_ℓ, the parameters γ_ℓ and τ_ℓ may be recovered by Prony's method [13,21,22]. In a nutshell, for $h > 0$ with $h\tau_\ell < \pi$, $\ell = 1, \dots, L$, and for $M \geq 4L + 1$, we define the Prony polynomial

$$\Lambda(z) := \prod_{\ell=-L}^{L} \left(z - \mathrm{e}^{-\mathrm{i}h\tau_\ell}\right) = \sum_{k=0}^{2L+1} \lambda_k \, z^k, \quad z \in \mathbb{C}, \tag{2}$$

and observe

$$\sum_{k=0}^{2L+1} \lambda_k \, E(h(k+m)) = \sum_{\ell=-L}^{L} \gamma_\ell \, \mathrm{e}^{-\mathrm{i}hm\tau_\ell} \, \Lambda(\mathrm{e}^{-\mathrm{i}h\tau_\ell}) = 0$$

for $m = 0, \dots, M - 2L - 1$. Due to $\lambda_{2L+1} = 1$, we may compute the remaining λ_ℓ by solving an equation system, whose solution is, in fact, unique. Knowing λ_ℓ, we determine the roots of Λ and the frequencies τ_ℓ. The coefficients γ_ℓ are then given by the over-determined Vandermonde-type system

$$\sum_{\ell=-L}^{L} \gamma_\ell \, \mathrm{e}^{-\mathrm{i}hm\tau_\ell} = E(hm), \quad m = 0, \dots, M. \tag{3}$$

For our numerical experiments, we use the so-called Approximative Prony Method (APM) by Potts and Tasche [21], which is based on the above consideration but is numerically more stable.

Algorithm 1. (APM, [21]).
Input: $L \in \mathbb{N}$, $M \geq 4L + 1$, $h > 0$ with $h\tau_\ell < \pi$, $(E(hm))_{m=0}^{M}$.

1. Compute the right singular vector $(\lambda_\ell)_{\ell=0}^{2L+1}$ to the smallest singular value of

$$(E(h(k+m)))_{m,k=0}^{M-2L-1,2L+1}.$$

2. Compute the roots $(z_\ell)_{\ell=-L}^{L}$ of $\Lambda(z)$ in (2) in order $z_\ell = \bar{z}_{-\ell}$.
3. Compute the least-square solution $(\gamma_\ell)_{\ell=-L}^{L}$ of (3).
4. Set $\tau_\ell := h^{-1} \arg z_\ell$.

Output: γ_ℓ, τ_ℓ, $\ell = -L, \dots, L$.

Instead of the exact number of terms $2L+1$, the method can be applied with an upper estimate for L. In this case, z_ℓ not lying on the unit circle and terms with small γ_ℓ can be neglected.

3 The Phase Retrieval Problem

Originally, phase retrieval means the recovery of an unknown signal only from the intensities of its Fourier transform. In the following, we consider phase retrieval for sparse signals, which are superpositions of finitely many transitions of one known structure. Moreover, signal and structure are modeled as complex measures. Thus, the true signal $\mu \in \mathcal{M}(\mathbb{R}^D)$ is a superposition of finitely many transitions $\nu_{T_n} := \nu(\cdot - T_n)$ with $T_n \in \mathbb{R}^D$ and $\nu \in \mathcal{M}(\mathbb{R}^D)$. Using the convolution and the Dirac measure δ, the considered *phase retrieval problem* has the following form: recover the coefficients $c_n \in \mathbb{C} \setminus \{0\}$ and the translations $T_n \in \mathbb{R}^D$ of the *structured signal*

$$\mu := \sum_{n=1}^{N} c_n \, \nu_{T_n} = \nu * \Big(\sum_{n=1}^{N} c_n \, \delta_{T_n} \Big) \tag{4}$$

with $\nu \in \mathcal{M}(\mathbb{R}^D)$ from samples of its (squared) Fourier intensity

$$|\hat{\mu}(\omega)|^2 = |\hat{\nu}(\omega)|^2 \sum_{n=1}^{N} \sum_{k=1}^{N} c_n \bar{c}_k \, \mathrm{e}^{-\mathrm{i}\langle \omega, T_n - T_k \rangle} = |\hat{\nu}(\omega)|^2 \sum_{\ell=-L}^{L} \gamma_\ell \, \mathrm{e}^{-\mathrm{i}\langle \omega, \tau_\ell \rangle}. \tag{5}$$

The (indexed) families of translates and coefficients are henceforth denoted by $\mathfrak{T} := [T_1, \ldots, T_N]$ and $\mathfrak{C} := [c_1, \ldots, c_N]$, where the index of c_n and T_n always corresponds to each other. Conceivable structures are the Dirac measure δ, in which case μ may be interpreted as Dirac signal, or a Gaussian, in which case μ may be interpreted as ordinary function via its density function. The considered phase retrieval problem is never uniquely solvable.

Lemma 1. (Trivial Ambiguities). *For every $\mu \in \mathcal{M}(\mathbb{R}^D)$, global phase shifts, transitions, and conjugated reflections have the same Fourier intensity, i.e.*

$$|\hat{\mu}| = |\mathcal{F}[\mathrm{e}^{\mathrm{i}\alpha} \mu]| = |\mathcal{F}[\mu(\cdot - x_0)]| = |\mathcal{F}[\overline{\mu(-\cdot)}]|, \quad \alpha \in \mathbb{R},\ x_0 \in \mathbb{R}^D.$$

Since the statement can be established using standard computation rules of the Fourier transform, the proof is omitted. Besides these so-called trivial ambiguities, there may occur further non-trivial ambiguities.

Lemma 2. (Non-Trivial Ambiguities). *Let $\mu = \mu_1 * \mu_2$ be the convolution of $\mu_1, \mu_2 \in \mathcal{M}(\mathbb{R}^D)$. The signal $\mu_1 * \overline{\mu_2(-\cdot)}$ then has the same Fourier intensity.*

The statement immediately follows form the Fourier convolution theorem. As an immediate consequence, we may conjugate and reflect the structure part ν and the location part $\sum_{n=1}^{N} c_n \, \delta_{T_n}$ of (4) independently of each other. Therefore, we may only hope to recover μ up to the following ambiguities.

Definition 1. (Inevitable Ambiguities). The signal in (4) can only be recovered up to a global phase shift of all coefficients in $\mathfrak{C}$; a global shift of all transitions in $\mathfrak{T}$; and the conjugation of $\mathfrak{C}$ together with the reflection of $\mathfrak{T}$.

In general, the number of non-trivial ambiguities can be immense. For example, under the assumption that the transitions $\mathfrak{T}$ are equispaced, there may exists up to 2^{N-2} non-trivial ambiguities [5], which can be characterized as in Lemma 2. To get rid of the non-trivial ambiguities, we have to assume that the transitions $\mathfrak{T}$ are somehow unregular.

Definition 2. (Collision-Freeness). A family $\mathfrak{T}:=[T_1, \dots, T_N] \subset \mathbb{R}^D$ is called *collision-free* if the differences $T_n - T_k$ are pairwise distinct for all $n \neq k$.

4 Sparse Phase Retrieval on the Line

Phase retrieval of structured signals on the line has been well studied [6,7,19, 23]. Under certain assumptions, the recovery from equispaced measurements is principally possible.

Theorem 1. (Phase Retrieval, *[6,* **Thm 3.1]).** *Let μ be of the form* (4) *with $\hat{\nu}(\omega) \neq 0$, $\omega \in \mathbb{R}$, collision-free $T_1 < \cdots < T_N$, and $|c_1| \neq |c_N|$. Further, choose $h > 0$ such that $h(T_n - T_k) < \pi$ for n, k. Then μ can be uniquely reconstructed from $|\hat{\mu}(hm)|$, $m = 0, \dots, 2\,N(N-1)+1$, up to inevitable ambiguities.*

The constructive proof leads to a two-step algorithm: First, the parameters γ_ℓ and τ_ℓ of (5) are determined using Prony's method. Second, the hidden relation between the indices n, k and ℓ is revealed.

Algorithm 2. (Phase Retrieval on the Line, [6]**).**
Input: $N \in \mathbb{N}$, $M \geq 2\,N(N-1)+1$, $h > 0$ with $h(T_n - T_k) < \pi$, $(|\hat{\mu}(hm)|^2)_{m=0}^{M}$.

1. Apply APM with $E(hm) := |\hat{\mu}(hm)|^2 / |\hat{\nu}(hm)|^2$ to determine $(\tau_\ell)_{\ell=-L}^{L}$ and $(\gamma_\ell)_{\ell=-L}^{L}$ in (5). Assume that $(\tau_\ell)_{\ell=-L}^{L}$ is ordered increasingly.
2. Set $\mathfrak{D} := [\tau_k : k = 0, \dots, N(N-1)/2]$ and $L := N(N-1)/2$.
3. Set $T_1 := 0$, $T_N := \tau_L$, $T_{N-1} := \tau_{L-1}$. Find ℓ^* with $|T_N - T_{N-1}| = \tau_{\ell^*}$. Set $c_1 := |\gamma_L \bar{\gamma}_{L-1} / \gamma_{\ell^*}|^{1/2}$, $c_N := \gamma_L / \bar{c}_1$, $c_{N-1} := \gamma_{L-1} / \bar{c}_1$. Remove τ_0, τ_{ℓ^*}, τ_{L-1}, τ_L from $\mathfrak{D}$.
4. Initiate the lists $\mathfrak{T} := [T_1, T_N, T_{N-1}]$ and $\mathfrak{C} := [c_1, c_N, c_{N-1}]$.
5. For the maximal remaining τ_{L^*} in $\mathfrak{D}$, find ℓ^* with $\tau_{L^*} + \tau_{\ell^*} = T_N$. Compute $d_{\mathrm{r}} := \gamma_{L^*} / \bar{c}_1$ and $d_{\mathrm{l}} := \gamma_{\ell^*} / \bar{c}_1$.
   a) If $|c_N \bar{d}_{\mathrm{r}} - \gamma_{\ell^*}| < |c_N \bar{d}_{\mathrm{l}} - \gamma_{L^*}|$, add τ_{L^*} to $\mathfrak{T}$ and d_{r} to $\mathfrak{C}$.
   b) Otherwise add τ_{ℓ^*} to $\mathfrak{T}$ and d_{l} to $\mathfrak{C}$.
   Remove all $|S - S'|$ with $S, S' \in \mathfrak{T}$ from $\mathfrak{D}$, and repeat until $\mathfrak{D}$ is empty.

Output: $\mathfrak{C}$, $\mathfrak{T}$.

5 Sparse Phase Retrieval on the Real Space

To extend the phase retrieval procedure from the line to the real space, we combine Theorem 1 and Algorithm 2 with the sampling strategy in [20]. Instead of sampling the whole Fourier domain, we only require the Fourier intensity $|\hat{\mu}|$

sampled along adaptively chosen lines in $\mathbb{R}^D$. Initially, we consider the setup where we have given samples along the Cartesian axes spanned by the unit vectors e_d, $d = 1, \dots, D$, and where we choose additional sampling lines accordingly. Since the condition $|c_1| \neq |c_N|$ in Theorem 1 strongly depends on the geometry of the transitions T_n, we henceforth assume that the absolute values $|\mathfrak{C}|$ of the coefficients are distinct.

Theorem 2. (Phase Retrieval on the Real Space). *Let μ be of the form (4) with $\hat{\nu}(\omega) \neq 0$, $\omega \in \mathbb{R}^D$, collision-free $\mathfrak{T}^{e_d}$ for every $d = 1, \dots, D$, and distinct $|\mathfrak{C}|$. Further, choose $h > 0$ such that $h \|T_n - T_k\| < \pi$ for all n, k. Then there exist $\theta_d \in \mathbb{R}^d$, $d = 1, \dots, D-1$, with $\|\theta_d\| = 1$ such that μ can be uniquely reconstructed from*

$$\{|\hat{\mu}(hm\, e_d)|, : m = 0, \dots, 2N(N-1)+1, d = 1, \dots, D\}$$

and the adaptive samples

$$\{|\hat{\mu}(hm\, \theta_d)| : m = 0, \dots, 2N(N-1)+1, d = 1, \dots, D-1\}$$

up to inevitable ambiguities.

Proof. Initially, we consider equispaced Fourier samples along an arbitrary line. For $\zeta \in \mathbb{R}^D$ with $\|\zeta\| = 1$, such samples have the form

$$|\hat{\mu}(hm\,\zeta)|^2 = |\hat{\nu}(hm\,\zeta)|^2 \left| \sum_{n=1}^{N} c_n \, \mathrm{e}^{-\mathrm{i}hm\langle \zeta, T_n \rangle} \right|^2.$$

Essentially, these samples are the squared Fourier intensities of a point measure on the line with coefficient $c_n^\zeta := c_n$ and transitions $T_n^\zeta := \langle \zeta, T_n \rangle$. Without loss of generality, we henceforth assume that the families $\mathfrak{C}^\zeta := [c_n^\zeta]$ and $\mathfrak{T}^\zeta := [T_n^\zeta]$ are ordered such that $0 = T_1^\zeta < \cdots < T_N^\zeta$. If $\mathfrak{T}^\zeta$ is collision-free, and $\mathfrak{C}^\zeta$ fulfils $|c_1^\zeta| \neq |c_N^\zeta|$, the assumption of Theorem 1 are satisfied, and $\mathfrak{C}^\zeta$ and $\mathfrak{T}^\zeta$ can be

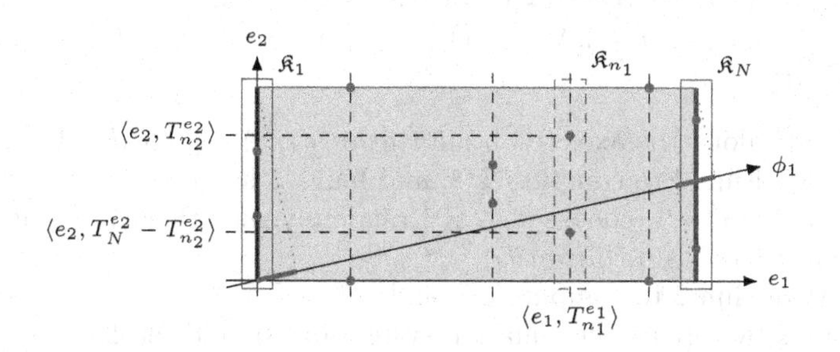

Fig. 1. Construction of the candidate sets $\mathfrak{K}_{n_1}$ in the two-dimensional setup. Condition (7) is fulfilled if the projection of the left and right edge onto the line spanned by ϕ_1 —indicated by the red regions— are disjoint.

determined by Algorithm 2 up to trivial ambiguities. Where the global phase shifts and additional transitions are unproblematic, the conjugated reflection ambiguity has to be resolved. Owing to the distinct absolute values $|\mathfrak{C}|$, we can identify the transitions $T_n^{e_d}$ with each other. More precisely, for every $n_1 \in \{1, \dots, N\}$, we find unique indices $n_2, \dots, n_D$ such that $|c_{n_1}^{e_1}| = \dots = |c_{n_D}^{e_D}|$. Up to a global shift, the true transitions T_n of (4) are contained in

$$\tilde{\mathfrak{T}} := \mathfrak{K}_1 \cup \dots \cup \mathfrak{K}_N. \tag{6}$$

with the candidate sets

$$\mathfrak{K}_{n_1} := \{(T_{n_1}^{e_1}, R_{n_2}^{e_2}, \dots, R_{n_D}^{e_D}) : R_{n_d}^{e_d} \in \{T_{n_d}^{e_d}, (T_N^{e_d} - T_{n_d}^{e_d})\}\}$$

where $|c_{n_1}^{e_1}| = \dots = |c_{n_D}^{e_D}|$. The construction of $\tilde{\mathfrak{T}}$ is schematically illustrated in Fig. 1. Notice that the candidate sets $\mathfrak{K}_{n_1}$ are the vertices of $(D-1)$-dimensional cubes lying on shifted versions of the hyperplane orthogonal to e_1. Therefore, we find ϕ_d with $\|\phi_d\| = 1$ such that $\{e_1, \phi_1, \dots, \phi_{D-1}\}$ form a basis and that

$$\langle \phi_d, S_1 \rangle \le \langle \phi_d, S_N \rangle \quad \text{for all} \quad S_1 \in \mathfrak{K}_1, S_N \in \mathfrak{K}_N, \tag{7}$$

see also Fig. 1. Moreover, ϕ_d may be chosen so that $\mathfrak{T}^{\phi_d}$ is collision-free, and Algorithm 2 can be applied to recover $\mathfrak{C}^{\phi_d}$ and $\mathfrak{T}^{\phi_d}$. Due to (7), the conjugated reflection ambiguity for $\mathfrak{T}^{\phi_d}$ can be resolved, and the true transitions are given by

$$\left\{ \left((e_1 | \phi_1 | \dots | \phi_D)^* \right)^{-1} (T_{n_1}^{e_1}, T_{n_2}^{\phi_1}, \dots, T_{n_D}^{\phi_{D-1}})^* : n_1 = 1, \dots, N \right\}, \tag{8}$$

where $*$ denotes the conjugation and transposition, and where the indices are uniquely given by $|c_{n_1}^{e_1}| = |c_{n_2}^{\phi_1}| = \dots = |c_{n_D}^{\phi_{D-1}}|$. The reflection of $\mathfrak{T}^{e_1}$ would yield the conjugated reflection of μ, which concludes the proof. □

Similarly to the recovery guarantee on the line (Theorem 1), the proof is constructive and summerizes to the following reconstruction method, which we will also use for the numerical simulations in Sect. 6.

Algorithm 3. (Phase Retrieval on the Real Space).
Input: $D \in \mathbb{N}$, $N \in \mathbb{N}$, $M \ge 2N(N-1)+1$, $h > 0$ with $h\|T_n - T_k\| < \pi$, adaptive samples of $|\hat{\mu}|^2$.

1. Sample $|\hat{\mu}|^2$ along the axes to obtain $(|\hat{\mu}(hm\, e_d)|^2)_{m=0}^M$ for $d = 1, \dots, D$.
2. Apply Algorithm 2 to compute $\mathfrak{T}^{e_d}$, and build $\tilde{\mathfrak{T}}$ in (6).
3. For $d = 1, \dots, D-1$, choose $\theta_d \in \mathbb{R}^D$ randomly such that (7) is satisfied.
4. Sample $|\hat{\mu}|^2$ to obtain $(|\hat{\mu}(hm\, \theta_d)|^2)_{m=0}^M$.
5. Apply Algorithm 2 to compute $\mathfrak{C}^{\theta_d}$ and $\mathfrak{T}^{\theta_d}$.
6. Compute $\mathfrak{T}$ by solving the equation system in (8), and set $\mathfrak{C} := \mathfrak{C}^{e_1}$.

Output: $\mathfrak{C}$, $\mathfrak{T}$.

The statement of Theorem 2 remains valid if the unit vectors e_d are replaced by an arbitrary basis ψ_d. Similarly to (8), the candidate set $\tilde{\mathfrak{T}}$ can then be determined by

$$\tilde{\mathfrak{T}} := \big((\psi_1 | \dots | \psi_D)^*\big)^{-1} (\mathfrak{K}_1 \cup \cdots \mathfrak{K}_N). \tag{9}$$

One of the key assumptions in Theorem 2 is that the coordinates of $\mathfrak{T}$ with respect to the considered directions are collision-free. For a given set of transitions $\mathfrak{T}$, this is holds true for almost all directions $\phi \in \mathbb{R}^D$, i.e. up to a Lebesgue null set.

Lemma 3. *Let $\mathfrak{T} := [T_1, \dots, T_N] \subset \mathbb{R}^D$ be collision-free, then the family $\mathfrak{T}^\theta := [\langle \theta, T_n \rangle : T_n \in \mathfrak{T}]$ is collision-free for almost all $\theta \in \mathbb{R}^D$.*

Proof. Let $T_{i_1}, T_{i_2}, T_{i_3}, T_{i_4}$ be arbitrary points in $\mathfrak{T}$, where only T_{i_2} and T_{i_3} may coincide. The dimension of the subspace $\{\theta \in \mathbb{R}^D : \langle \theta, T_{i_1} - T_{i_2} \rangle = \langle \theta, T_{i_3} - T_{i_4} \rangle\}$ may be $D - 1$ at the most since $\mathfrak{T}$ is collision-free. Thus the family $\mathfrak{T}^\theta$ can only contain collisions for θ lying in the union of finitely many lower-dimensional subspaces, which gives the assertion. □

Against the background of Lemma 3, and since D generic vectors form a basis, the sampling along adaptively chosen lines in the two-dimensional setup can be replaced by the sampling along arbitrary generic lines.

Theorem 3. (Phase Retrieval in 2D). *Let μ be of the form* (4) *with $\hat{\nu}(\omega) \neq 0$, $\omega \in \mathbb{R}^2$, collision-free $\mathfrak{T}$, and distinct $|\mathfrak{C}|$. Further, let ψ_1, ψ_2, ψ_3 be generic vectors in $\mathbb{R}^2$. Then μ can be uniquely reconstructed from*

$$\big\{ |\hat{\mu}(hm\, \psi_d)| : m = 0, \dots, 2N(N-1)+1, d = 1, \dots, 3 \big\}$$

up to inevitable ambiguities.

Proof. To establish the statement, we adapt the proof of Theorem 2, where ψ_1, ψ_2 play the role of e_1, e_2, and ψ_3 the role of the adaptive direction ϕ_1. Due to Lemma 3, the families $\mathfrak{T}^{\phi_d}$ are collision-free for generic lines; so the application of Algorithm 2 is unproblematic. Considering the construction of $\mathfrak{K}_{n_1}$, we notice that the candidates are contained in a parallelogram, where $\mathfrak{K}_1$ and $\mathfrak{K}_N$ lie on opposite edges, see Fig. 1 for an illustration. If the coordinates of $\mathfrak{K}_1$ and $\mathfrak{K}_N$ satisfy (7) with respect to the third direction ψ_3, we can apply the procedure in the proof of Theorem 2 to recover μ. Otherwise, we interchange the role of ψ_1 and ψ_2. In this case, we obtain a second set of candidates in the same parallelogram. The new sets $\mathfrak{K}'_1$ and $\mathfrak{K}'_N$, however, lie on the remaining to two edges such that ψ_3 now fulfils (7), and μ can be recovered. □

6 Simulations

To substantiate the theoretical observations, we apply the constructive proof (Algorithm 3) to a minor, synthetic example. Our goal is recover the sparse

Table 1. Transitions and coefficients of the true sparse signal μ. Algorithm 3 recovers $\mathfrak{T}$ up to a maximal absolute error of $6.982 \cdot 10^{-8}$, and $\mathfrak{C}$ up to a maximal absolute error of $2.898 \cdot 10^{-5}$.

n	1	2	3	4	5
T_n^*	(27.374, 27.258)	(13.065, 32.008)	(8.847, 37.665)	(0.000, 13.874)	(23.876, 0.000)
c_n	7.293 + 5.115i	30.665 + 2.258i	2.740 + 22.286i	1.576 + 49.834i	17.400 + 46.587i

structure in Fig. 2a, which consists of five sources. Each source corresponds to a Gaussian with standard derivation 1/2 and to a complex coefficient. For repeatability, the exact locations and coefficients are given in Table 1. The Fourier intensity of the true signal is shown in Fig. 2b. Instead of sampling the whole Fourier domain, we only use equispaced samples on three predefined lines, which are depicted as red lines in Fig. 2b. The first two lines correspond to the Cartesian axes, and the third to the angle 0.143π. On each line, we take 100 equispaced, noise-free samples, which is slightly more than the 83 samples to employ the one-dimensional, sparse phase retrieval in Algorithm 2. The sampling distance is $h := \max\{\|T_k - T_n\|\}/2\pi \approx 0.0387$.

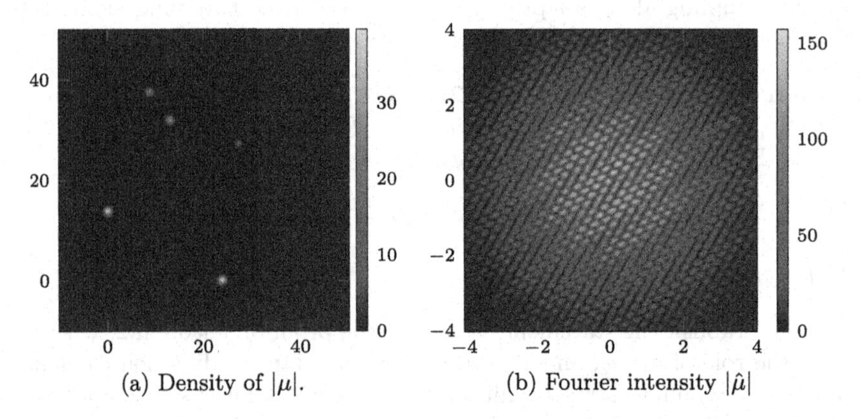

(a) Density of $|\mu|$. (b) Fourier intensity $|\hat{\mu}|$

Fig. 2. Absolute value of the complex-valued density of the true μ and the corresponding Fourier intensity $|\hat{\mu}|$. The red lines indicate the predefined sampling direction for Algorithm 3.

Applying Algorithm 3 here yields an accurate approximation of the true transitions $\mathfrak{T}$. To compare the recovered values $\mathfrak{T}'$ with the true ones, we conjugate and reflect the recovered signal such that both signals have the same orientation. Furthermore, we shift both signals so that the leftmost and lowermost vectors lie on the Cartesian axes. The recovered values are shown in Fig. 3 and nearly coincide with the true values. The maximal absolute error after eliminating the shift and conjugated reflection ambiguity is given by

$$\max_{n=1,\dots,5}\{|T_n - T_n'|\} = 6.982 \cdot 10^{-8}.$$

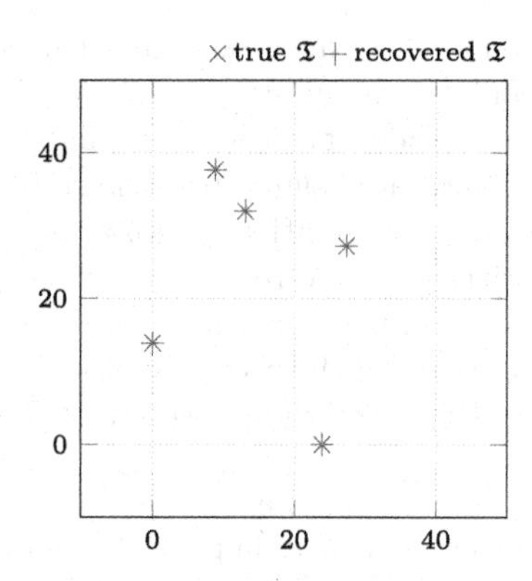

Fig. 3. True and recovered transitions $\mathfrak{T}$ and $\mathfrak{T}'$ after removing the trivial shift and conjugated reflection ambiguity.

The maximal absolute error for the coefficients is here

$$\max_{n=1,\dots,5} \{|c_n - \mathrm{e}^{\mathrm{i}\alpha}\, c_n'|\} = 2.898 \cdot 10^{-5},$$

where we choose the global phase shift α such that the maximal absolute error is minimized.

This first numerical example shows that a sparse, two-dimensional signal can be completely recovered using only samples on three predefined lines instead of sampling the whole Fourier domain. In all fairness, Prony's method is known to be very sensitive to noise and to become unstable if the frequencies— here $\langle \zeta, T_n - T_k \rangle$— almost coincide. Since Prony's method has to recover $2N(N-1)+1$ frequencies for an N-sparse signal, N cannot be increased very far. The development of a stable algorithm for the proposed sampling scheme, which can deal with additional noise, thus remains open for further research.

7 Conclusion

The focus of this paper is the sparse phase retrieval problem as it occurs in speckle imaging and crystallography. Combining the one-dimensional phase retrieval approach in [6,19] with the adaptive sampling strategy from [20], we derive a Prony-based recovery method for D-dimensional, sparse signals. This constructive method is the key ingredient to establish recovery guarantees for collision-free signals. Instead of sampling the Fourier intensity on the entire D-dimensional domain, we only require samples from $2D-1$ adaptively chosen or generic lines. This leads to a sampling complexity of $\mathcal{O}(DN^2)$, where D is the dimension of the signal and N the sparsity level, i.e. the number of shifted components. The N^2 is accounted for by Prony's method, which we use to determine the parameters of the given Fourier intensity —an exponential sum with quadratic structure. Since the unknown signal consists of merely $(D+2)N$ real parameters, the question arises whether the signal can also be uniquely recovered using less measurements. This question is left for further research. During the numerical simulations, we show that the Prony-based method can, in principle, be applied to recover the wanted signal from noise-free measurements. Since Prony's method is, however, very sensitive to noise, one of our next steps is to derive a more stable algorithm for the recovery of specific structured, multidimensional signals. A first step in this direction could be to transfer the numerical methods from [11], dealing with the recovery of sparse signals from complex Fourier samples along radial lines, to the phase retrieval setting.

Acknowledgements. The authors would like to thank the anonymous reviewers for their valuable comments and suggestions to improve the manuscript

References

1. Baechler, G., Kreković, M., Ranieri, J., Chebira, A., Lu, Y.M., Vetterli, M.: Super resolution phase retrieval for sparse signals. IEEE Trans. Signal Process. **67**(18), 4839–4854 (2019). https://doi.org/10.1109/TSP.2019.2931169
2. Barbotin, Y., Vetterli, M.: Fast and robust parametric estimation of jointly sparse channels. IEEE J. Emerg. Sel. Topics Power Electron. **2**(3), 402–412 (2012). https://doi.org/10.1109/JETCAS.2012.2214872
3. Beinert, R.: Non-negativity constraints in the one-dimensional discrete-time phase retrieval problem. Inf. Inference **6**(2), 213–224 (2017). https://doi.org/10.1093/imaiai/iaw018
4. Beinert, R.: One-dimensional phase retrieval with additional interference intensity measurements. Results Math. **72**(1–2), 1–24 (2017). https://doi.org/10.1007/s00025-016-0633-9
5. Beinert, R., Plonka, G.: Ambiguities in one-dimensional discrete phase retrieval from Fourier magnitudes. J. Fourier Anal. Appl. **21**(6), 1169–1198 (2015). https://doi.org/10.1007/s00041-015-9405-2
6. Beinert, R., Plonka, G.: Sparse phase retrieval of one-dimensional signals by Prony's method. Front. Appl. Math. Stat. **3**, 5 (2017). https://doi.org/10.3389/fams.2017.00005
7. Beinert, R., Plonka, G.: Sparse phase retrieval of structured signals by Prony's method. PAMM **17**(1), 829–830 (2017). https://doi.org/10.1002/pamm.201710382
8. Beinert, R., Plonka, G.: Enforcing uniqueness in one-dimensional phase retrieval by additional signal information in time domain. Appl. Comput. Harmon. Anal. **45**(3), 505–525 (2018). https://doi.org/10.1016/j.acha.2016.12.002
9. Beinert, R., Quellmalz, M.: Total variation-based reconstruction and phase retrieval for diffraction tomography. SIAM J. Imaging Sci. **15**(3), 1373–1399 (2022). https://doi.org/10.1137/22M1474382
10. Bendory, T., Beinert, R., Eldar, Y.C.: Fourier phase retrieval: uniqueness and algorithms. In: Boche, H., Caire, G., Calderbank, R., März, M., Kutyniok, G., Mathar, R. (eds.) Compressed Sensing and its Applications. ANHA, pp. 55–91. Springer, Cham (2017). https://doi.org/10.1007/978-3-319-69802-1_2
11. Dossal, C., Duval, V., Poon, C.: Sampling the Fourier transform along radial lines. SIAM J. Numer. Anal. **55**(6), 2540–2564 (2017). https://doi.org/10.1137/16M1108807

12. Grohs, P., Koppensteiner, S., Rathmair, M.: Phase retrieval: uniqueness and stability. SIAM Rev. **62**(2), 301–350 (2020). https://doi.org/10.1137/19M1256865
13. Hildebrand, F.B.: Introduction to Numerical Analysis, 2nd edn. Dover Publications, New York (1987)
14. Klibanov, M.V., Kamburg, V.G.: Uniqueness of a one-dimensional phase retrieval problem. Inverse Probl. **30**(7), 075004, 10 (2014). https://doi.org/10.1088/0266-5611/30/7/075004
15. Klibanov, M.V., Sacks, P.E., Tikhonravov, A.V.: The phase retrieval problem. Inverse Prob. **11**(1), 1–28 (1995). https://doi.org/10.1088/0266-5611/11/1/001
16. Knox, K.T.: Image retrieval from astronomical speckle patterns. J. Opt. Soc. Am. **66**(11), 1236–1239 (1976). https://doi.org/10.1364/JOSA.66.001236
17. Konijnenberg, A., Coene, W., Pereira, S., Urbach, H.: Combining ptychographical algorithms with the Hybrid Input-Output (HIO) algorithm. Ultramicroscopy **171**, 43–54 (2016). https://doi.org/10.1016/j.ultramic.2016.08.020
18. Millane, R.P.: Phase retrieval in crystallography and optics. J. Opt. Soc. Am. A **7**(3), 394–411 (1990). https://doi.org/10.1364/JOSAA.7.000394
19. Plonka, Gerlind, Potts, Daniel, Steidl, Gabriele, Tasche, Manfred: Numerical Fourier Analysis. ANHA, Springer, Cham (2018). https://doi.org/10.1007/978-3-030-04306-3
20. Plonka, G., Wischerhoff, M.: How many Fourier samples are needed for real function reconstruction? J. Appl. Math. Comput. **42**(1–2), 117–137 (2013). https://doi.org/10.1007/s12190-012-0624-2
21. Potts, D., Tasche, M.: Parameter estimation for exponential sums by approximate Prony method. Signal Process. **90**(5), 1631–1642 (2010). https://doi.org/10.1016/j.sigpro.2009.11.012
22. Prony, R.: Essai expérimental et analytique sur les lois de la dilatabilité des fluides élastiques et sur celles de la force expansive de la vapeur de l'eau et de la vapeur de l'alkool, á différentes températures. Journal de l'École polytechnique. **2**, 24–76 (1795)
23. Ranieri, J., Chebira, A., Lu, Y.M., Vetterli, M.: Phase retrieval for sparse signals: uniqueness conditions (2013). https://doi.org/10.48550/ARXIV.1308.3058

Limited Electrodes Models in Electrical Impedance Tomography Reconstruction

Francesco Colibazzi[1], Damiana Lazzaro[1](✉), Serena Morigi[1], and Andrea Samorè[1,2]

[1] Department of Mathematics, University of Bologna, Bologna, Italy
{francesco.colibazzi2,damiana.lazzaro,serena.morigi}@unibo.it
[2] School of Biomedical Engineering, University of Sydney, Sydney, Australia
andrea.samore@sydney.edu.au

Abstract. We state the Limited Electrode problem in Electrical Impedance Tomography, and propose solutions inspired by the application of compressed sensing techniques and deep learning strategies on the raw boundary impedance data. These strategies allow to recover the target reconstruction quality while using a relatively low number of nonlinear measurements, assuming sparsity-gradient conductivity. This would help reducing modelling costs and computational power, thus enhancing applicability of EIT.

1 Introduction

In Electrical Impedance Tomography (EIT), the electrical conductivity σ inside a domain Ω is estimated from current and voltage measurements (I_F, V_M) at its boundary $\partial\Omega$. This imaging technique has been mainly used for medical imaging, industrial monitoring, geological studies, and, more recently, biotechnology [7] and structural health monitoring [11]. The voltage and current values are measured through a finite number n of electrodes attached to the boundary of the object. The number of electrodes used is a trade-off between accuracy, measurement time and processing time. We expect, in general, that having more electrodes and thus more measurements, results in a higher reconstruction quality. However, increasing their number is not straightforward due to the added complexities and costs involved both in the instrumentation and in the time required to set up the system. Also, increasing the number of electrodes can lead to reduce their size, and thus increase contact impedance. Moreover, the more components are present in a system, the easier it is to have a few of them malfunctioning, especially electrodes, thus reducing reliability.

It is thus clear that obtaining high quality EIT reconstructions with a reduced number of electrodes (and thus of measurements) would be of great help to reduce the costs and increase the reliability of EIT in practical applications. It would be even more useful to know the sufficient number of measurements to recover an optimal σ, under sparsity conditions on the unknown conductivity.

L. Calatroni et al. (Eds.): SSVM 2023, LNCS 14009, pp. 68–80, 2023.
https://doi.org/10.1007/978-3-031-31975-4_6

In this work we formulate this problem naming it EIT-Limited Electrodes (EIT-LE) problem, and denoting with LE Model (LEM) the numerical methods proposed for its solution. There seems to be a natural connection with the basic ideas of Compressed Sensing (CS) theory which deals with the recovery of a sparse signal from a small number of linear measurements. In a well-posed context, the CS theory, under appropriate hypotheses, identifies a relationship between the minimum number of measurements necessary to obtain a "good reconstruction" of the signal, its size and its sparsity in some domain. In this paper, we instead address the compressed sensing recovery problem in a setting where the observations are nonlinear and we are concerned with sparse recovery principles for ill-posed inverse problems. Suppose we know the minimum number of measurements required to obtain an optimal σ^* reconstruction, but we are under the more realistic assumption of having fewer measurements (electrodes) available. It would be desirable to apply a method (LEM) which nevertheless allows to obtain a reconstruction of quality comparable to σ^*.

In this work we investigate two different LEMs to the EIT-LE problem. The first approach involves the applicability of CS techniques by exploiting the gradient sparsity of the conductivity distribution, which leads to a nonlinear variational problem. The second proposal is a learned residual approach on raw boundary impedance data. Both strategies will recover a reconstruction quality obtainable with a high number of electrodes (and thus measurements) while using for acquisitions a relatively low number of electrodes.

This paper is organized as follows. In Sect. 2 the EIT reconstruction inverse problem is introduced and the EIT-LE is formulated in Sect. 3. A variational CS-based LEM is proposed in Sect. 4 and a network-based LEM is described in Sect. 5. The performance of the two different proposals is discussed in Sect. 6 and conclusions are drawn in Sect. 7.

2 Preliminaries on EIT Reconstructions

Following the so-called *complete electrode model* [10], one can fully describe the behavior of the voltages on the electrodes E_i at the object/domain boundary $\partial\Omega_{E_i}$, given the conductivity of the medium σ and a set of injected currents $I_F = [I_1, \ldots, I_n]$. This process constitutes the EIT forward model and is described by a nonlinear forward operator denoted by $F(\sigma; I_F)$. Since the continuous forward operator is Fréchet differentiable, F' is a matrix, called the Jacobian of F and in the following denoted by J. The numerical computation of the EIT forward model can be obtained by employing Finite Element (FE) techniques using a discrete approximation Ω_h of the domain Ω, represented by a triangulated mesh composed by n_T triangles.

Given a set of voltage and current measurements, the goal of the EIT inverse problem is to compute the approximate conductivity σ^* such that the voltage potentials predicted through the forward model are as close to the actual measurements V_M as possible, which reads as

$$\sigma^* \in \arg\min_{\sigma \in \mathbb{R}^{n_T}} \|r(\sigma)\|_2^2, \quad r(\sigma) := F(\sigma; I_F) - V_M. \tag{1}$$

The underlying optimization problem is hard to solve, not only as the boundary currents depend non-linearly on the conductivity which makes the optimization problem non-convex, but also for the well-known sensitivity of the solutions to small voltage perturbations. To reduce the ill-posedness while keeping high accuracy reconstructions, the nonlinear least squares EIT problem (1) is usually regularized by adding a penalty term and solving the regularized nonlinear minimization problem by the Gauss-Newton iterative method, reported in the Appendix to the intended purpose of use in the proposed numerical solution of the EIT-LE problem. Many other regularization strategies have been proposed for this highly ill-posed inverse problem [4,6,9]. We investigate how suit these regularization methods to deal with the challenge EIT-LE problem with a low number of data/measurements.

The quality of the conductivity reconstruction σ^* is strictly related to the quantity and the quality of the acquired measurements V_M. In this simplified context we are aware of neglecting many other factors which could affect the reconstruction accuracy, i.e., the mismodeling of the domain, misplacement of electrodes. The data acquisition depends on the stimulation pattern. In our work the collected measurements V_M are obtained using the adjacent acquisition protocol: adjacent electrodes for the injections and adjacent electrodes for the measurements. When a pair of electrodes is chosen as driving pair, it is usually excluded from the available measuring pairs. Each measurement of $V_M \in \mathbb{R}^{n_M}$, with $n_M = n \times (n-3)$, is characterized by the pair (I_k, V_ℓ), $k = 1, \ldots, n, \ell = 1, \ldots, n-3$, where I_k indicates the current injection at the couple of electrodes (E_k, E_{k+1}), and $V_\ell \in V_M$ represents the potential measure between the electrodes $(E_\ell, E_{\ell+1})$. As the measured data is unavoidably noisy, we assume the following noisy forward observation model

$$F(\sigma; I_F) = V_M + \eta, \tag{2}$$

where $\eta \in \mathbb{R}^{n_M}$ is a vector $\eta \sim \mathcal{N}(0, \bar{n}\bar{V}_M)$ of Gaussian distributed measurements characterized by zero-mean, and noise level $\bar{n}$, with $\bar{V}_M$ the voltage average value.

3 EIT-Limited Electrodes Problem Setup

Let $\sigma_L \in \mathbb{R}^{n_T}$ be the conductivity recovered by an EIT setup with n_L electrodes and stimulation pattern (I_F^L, V_M^L). A more accurate reconstruction $\sigma_H \in \mathbb{R}^{n_T}$ is expected to be obtained with n_H electrodes, and stimulation pattern (I_F^H, V_M^H), with $n_H > n_L$, and, consequently, a larger number of measurements $|V_M^H| >> |V_M^L|$.

Limited Electrodes Model (LEM)
Under the assumption of gradient-sparsity of $\sigma \in \mathbb{R}^{n_T}$, given a limited number of measurements n_M^L collected from a low-resource EIT configuration (I_F^L, V_M^L) with n_L electrodes, the objective of a Limited Electrode Model is to recover a

reconstruction $\sigma^* \in \mathbb{R}^{n_T}$ *from* $V_M^L \in \mathbb{R}^{n_M^L}$ *measures even if* $n_M^L << n_T$*, with an error bounded by:*

$$\|\sigma^* - \sigma\| \leq C(\|\eta\|, \|\nabla\sigma\|_0, n_T, n_M^L), \tag{3}$$

where $C > 0$ *is a small constant, and the* ℓ_0 *pseudonorm* $\|x\|_0$ *measures the gradient sparsity counting the nonzero value in* x*.*

The LEM will be considered successful if it reconstructs σ^* of better quality with respect to σ_L obtained by solving the inverse optimization problem (1), from measurements $V_M = V_M^L$ and injections $I_F = I_F^L$. Furthermore, the LEM will be considered as optimal, in case it reconstructs a distribution σ^* close to σ_H.

We finally remark that, due to the ill-posedness of the EIT inverse problem, the straightforward interpolation of n_M^L on the boundary $\partial\Omega$ to obtain interpolated n_M^H measurements for solving (1), produces unacceptable reconstructions.

4 A Compressed Sensing Approach to EIT-LE Problem

Compressed sensing allows signals to be sampled far below the rate traditionally prescribed [5]. Most of the theory developed for compressed sensing signal recovery assumes that samples are taken using linear measurements in a well-posed context. In this paper, we instead address the compressed sensing recovery problem in a setting where the observations are nonlinear and we are concerned with sparse recovery principles for nonlinear ill-posed inverse problems.

In the linear case, CS theory relies on the sufficient restricted isometry property (RIP) of the linear measurement operator, that can be interpreted in terms of the Lipschitz property of the operator itself and its inverse. The classical CS theory has been extended in [2] to arbitrary linear operators with bounded inverse considering in particular the linearized EIT problem. In [2] the authors show that the electrical conductivity may be stably recovered from a number of linearized EIT measurements proportional to the sparsity of the signal with respect to a wavelet basis, up to a log factor. Preliminary theoretical results in [1,3,8] paved the way for addressing EIT using CS strategies even in the more complex nonlinear setup.

In this section we address a preliminary proposal in this direction. In addition we show experimentally that the number of measurements n_M needed for an optimal recovery are proportional to the gradient sparsity of the unknown conductivity distribution.

Let $\sigma \in \mathbb{R}^{n_T}$ be defined in $\Omega_h \subset \mathbb{R}^2$. The data of the reconstruction problem are represented in spatial domain by the set of potential measurements V_M^L obtained by a limited EIT configuration with n_L electrodes. Let n_M^L be the number of measurements using an n_L-electrodes configuration, and n_M^H be the ones obtained by the n_H-electrodes configuration, with $n_M^L << n_M^H$. Considering $\mathcal{S} \in \mathbb{R}^{n_M^L \times n_M^H}$ to be an acquisition-dependent projection matrix which allows to select a restricted number of potential measurements, then the compressed measurement model, named in this context *LE acquisition model* $F_{\mathcal{S}} : \mathbb{R}^{n^H_M} \rightarrow \mathbb{R}^{n^L_M}$, reads as

$$F_{\mathcal{S}}(\sigma; I^H_F) = V^L_M, \tag{4}$$

with $F_{\mathcal{S}}(\sigma; I^H_F) = \mathcal{S}(F(\sigma; I^H_F))$. In a more realistic scenario in which the measurements $V^L_M \in \mathbb{R}^{n^L_M}$ are corrupted by noise according to (2), the CS reconstruction problem can be stated as the following constrained minimization:

$$\sigma^* \in \arg\min_{\sigma \in \mathbb{R}^{n_T}} R(\sigma) \quad \text{subject to} \quad \|\mathcal{S}(F(\sigma; I^H_F)) - V^L_M\|_2^2 \leq \eta^2, \tag{5}$$

where $R(\sigma)$ is a sparsifying function. In the present context we have assumed that the conductivity distribution, reconstructed from each set of measurements, varies only in spatially localized small subregions of Ω, which consequently leads to imposing the sparsity of the gradient of the distribution. Therefore our reconstruction method chooses the convex, sparsity inducing function as $R(\sigma) := \|\nabla\sigma\|_1$, namely, it minimizes the total variation of the conductivity distribution. Sparse reconstruction can be performed by converting problem (5) into an unconstrained minimization problem and then solving it by applying the Regularized Gauss-Newton (RGN) method illustrated in Appendix with residual $r(\sigma) = \mathcal{S}(F(\sigma; I^H_F)) - V^L_M$. The matrix $\mathcal{S}$ has a block rectangular structure with $n_L \times n_H$ blocks, each block is either an Identity matrix of dimension $n_L - 3$, in case the injection is part of the acquired measurements, either a zero block $\mathbf{0}_{n_L-3}$, otherwise. The position of the nonzero blocks in a row depends on the selected acquisition protocol.

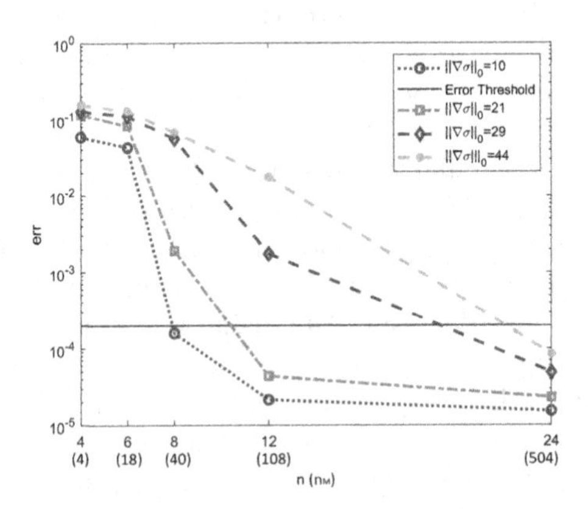

Fig. 1. Plot of the error reconstruction for an increasing number of electrodes n (and measurements n_M) for four samples σ with different gradient-sparsity.

The relation between the sparsity factor $\|\nabla\sigma\|_0$ and the number of nonlinear measurements n_M, a well-known result for linear CS, will be the subject of future investigation. By the way of illustration, we plot in Fig. 1, for four conductivity samples, the relative reconstruction errors $err := \|\sigma - \sigma^*\|/\|\sigma\|$, as the number of electrodes n (and thus the number of measurements n_M) varies. Each sample is representative of a different gradient-sparsity level, charaterized by $||\nabla\sigma||_0 = \{10, 21, 29, 44\}$. Using $n = \{4, 6, 8, 12, 24\}$ electrodes, then the corresponding measurements are $n_M = \{4, 18, 40, 108, 504\}$, respectively. The horizontal blue line indicates the error threshold below which the reconstruction can be considered exact (corresponding to a PSNR=70). As expected, as the gradient-sparsity decreases, which corresponds to increasing values of $||\nabla\sigma||_0$, the number of measurements necessary to obtain a perfect reconstruction increases. It is also important to note that, after a certain threshold, the reconstruction improvement in the face of an increase in number of measurements is no longer significant.

5 A Learned Approach to EIT-LE Problem

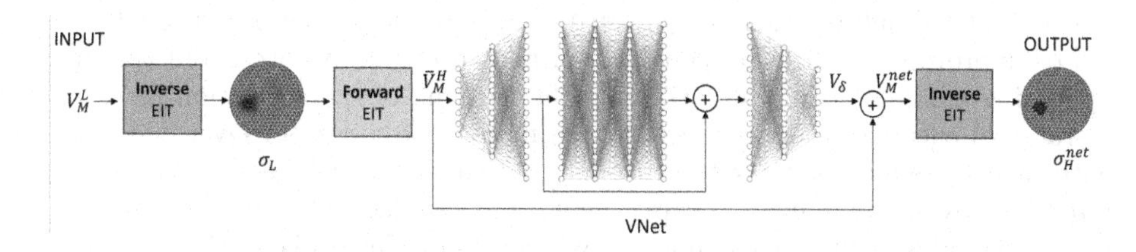

Fig. 2. Learned LEM

A schematic of the EIT-LE framework is shown in Fig. 2. Given the voltage measurements V_M^L collected using n_L electrodes according to the stimulation pattern (I_F^L, V_M^L), the reconstructed σ_L is computed by solving (1) using RGN-TV or RGN-TIK algorithm (see Appendix for details). Then by applying the forward EIT model $F(\sigma_L; I_F^H)$, with I_F^H virtual current injections, we get n_M^H virtual measurements $\bar{V}_M^H$. These measurements differ from V_M^H obtained by an EIT configuration (I_F^H, V_M^H) with n_H actual electrodes settings. We propose a residual network named VNet which learns a correction factor V_δ to recover a set of measurements V_M^{net}

$$V_M^{net} = V_\delta + \bar{V}_M^H. \tag{6}$$

This residual strategy estimates the quantization noise caused by the acquisition using a limited number of measurements. The improved conductivity reconstruction σ_H^{net} is finally computed by solving the inverse problem (1).

5.1 VNet Architecture

VNet is a deep fully-connected neural network (FCNN) built to provide an ad hoc efficient solution to the EIT-LE problem. In particular, VNet is an autoencoder network with an encoder stack to learn high-level features, a latent space which is composed of the higher dimensional hidden layers, and a decoder module to reconstruct high-resolution voltage measurements.

Following a supervised learning-based approach, we denominate f_Θ, with parameters Θ, a predefined function that, given the input measurements $\bar{V}_M^H$, computes the updated measurement vector V_M^{net}, according to the proposed architecture VNet. We aim at learning the optimal vector Θ by minimizing a loss function $\mathcal{L}$ on a training data set containing N measures-reconstructions pairs $(V_{M_i}^H, \sigma_H^i)$, $i = 1, \ldots, N$, where $V_{M_i}^H$ are the potential differences obtained by a configuration with n_H electrodes. In particular, the learning-based method is modeled as the following optimization problem

$$\min_{\Theta} \frac{1}{N} \sum_{i=1}^{N} \mathcal{L}(\sigma_i^*, \sigma_H^i) \quad \textbf{subject to} \quad \sigma_i^* = f_\Theta(V_{M_i}^H),$$

$$\mathcal{L}(\sigma_i^*, \sigma_H^i) = \|V_{M_i}^{net}(\Theta) - V_{M_i}^H\|_2^2 + \|\sigma_i^* - \sigma_H^i\|_2^2.$$

The considered autoencoder is composed of fully connected layers, dropout layers to improve generalization capabilities, and residuals to both avoid the disappearing gradients effect and allowing the learning of the correction term. In a standard encoder-decoder structure, the number of nodes per layer decreases with each subsequent layer of the encoder, and increases back in the decoder. As a result of extensive numerical experimentations, we decided instead to increase the number of neurons in each hidden layer in the encoder, and compress it in the decoder. The final hyperparameters (neurons and layers) configuration is tuned by optimizing the model capacity via pruning technique: the neurons which have no impact on the performance of the network are trimmed during the training. The encoder module consists of three layers, initialized with n_M^H values, which learn an array of upsampling weights by increasing them up to a certain number m. Then the latent space module has four layers each of the same size and finally the decoder has three layers with the last of dimension down to n_M^H. Each layer i is a standard linear operator, with weight coefficients $\theta_{i,j}$, which transforms the input features x_j into output features y_i, and reads as

$$y_i = \phi(\sum_{j=1}^{m} \theta_{i,j} x_j + b_i), \quad \phi(z) = PReLU(z) := \begin{cases} z, & \text{if}\, z \geq 0 \\ \beta z, & \text{otherwise} \end{cases}$$

where ϕ is the Parametric Rectified Liner Unit (PReLU) activation function for each layer, with learnable parameter β, except for the last layer where ϕ is the Hyperbolic Tangent (Tanh) function. Since PReLU has a learnable coefficient for the negative part of features, it can avoid the "dead features" caused by zero gradients in ReLU. The Tanh activation function is used to limit the values in $V_\delta \in [-1, 1]$, and improve the stability of training. Since the correction factor for each voltage is expected to be very small, it is important to have an activation function which is continuous and differentiable near the origin.

A dropout operator with probability $p = 0.2$ is inserted in the latent block to randomly zero some of the elements of the input layer using samples from a

Table 1. Average results on the entire testing set: (a)-(b) Data-PC; (c)-(d) Data-S.

	PSNR	SSIM
σ_6	20.30	0.694
σ_{12}^{net}	21.05	0.676
σ_{24}	26.10	0.836

(a)

	PSNR	SSIM
σ_8	22.07	0.741
σ_{16}^{net}	23.16	0.765
σ_{16}	25.25	0.848

(b)

	PSNR	SSIM
σ_6	24.35	0.668
σ_{12}^{net}	25.16	0.697
σ_{12}	27.81	0.763

(c)

	PSNR	SSIM
σ_{12}	27.81	0.763
σ_{24}^{net}	28.90	0.777
σ_{24}	32.27	0.890

(d)

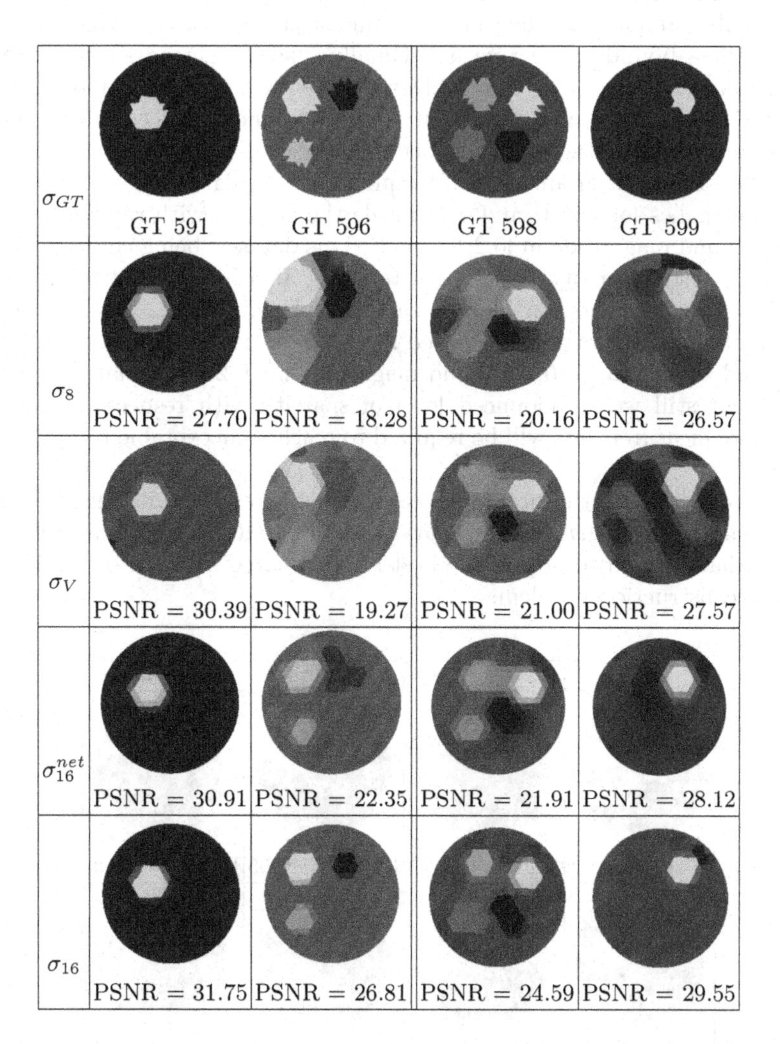

Fig. 3. Example 1: (first row) σ_{GT}, samples 591, 596, 598, 599 of the Data-PC dataset; (second row) reconstructions using RGN-TV from V_M^8 measurements; (third row) results of CS-based LEM; (fourth row) results of learned LEM; (fifth row) reconstruction using RGN-TV from V_M^{16} measurements.

Bernoulli distribution. The idea is that instead of letting layers learn the underlying mapping, we let the network fit the residual mapping.

6 Numerical Experiments

The following numerical experiments show the performance of the proposed EIT-LE frameworks for computing gradient-sparse reconstructions of synthetically built 2D EIT samples. All the examples simulate a circular tank slice of unitary radius with a boundary ring with n equally-spaced electrodes. We built two different datasets, the first, named Data-PC, with a few separated anomalies characterized by constant conductivity, and a second dataset, named Data-S, with much more smooth anomalies, occasionally overlapped.

In Data-PC each ground-truth sample σ_{GT} has a random number from 1 to 4 of anomalies localized randomly inside the domain with random radius in $[0.15, 0.25]$ and magnitude in $[0.2, 2]\Omega m^{-1}$. This dataset, being characterized by a strong gradient sparsity, lends itself to being almost perfectly reconstructed by a few measurements. In Data-S each sample σ_{GT} has a random number between 10 and 20 anomalies localized randomly inside the domain, eventually overlapped, radius in $[0.10, 0.15]$ and magnitude in $[1, 2]$. This dataset presents a lower but still well-pronounced level of sparsity with respect to Data-PC, hence more measurements will be required to reach an acceptable reconstruction accuracy.

For both datasets the homogeneous conductivity of the background liquid is set to be $\sigma_0 = 1.0\Omega m^{-1}$. The parameter α in (RGN-Tik) and (RGN-TV) has been hand-tuned to obtain the best approximated solution of the ill-posed inverse reconstruction problem.

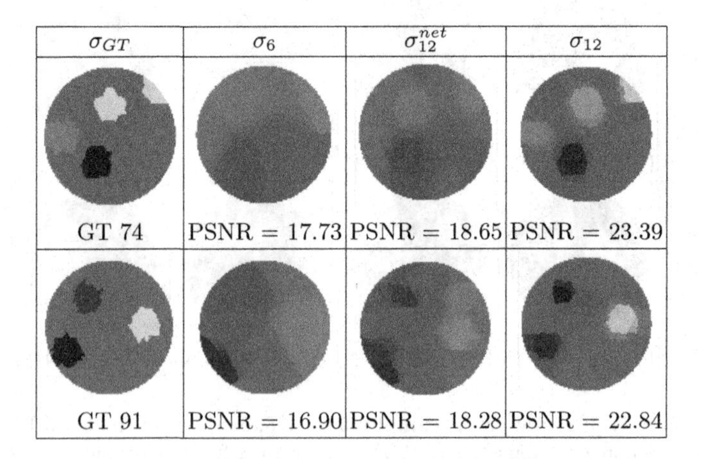

Fig. 4. Example 2: Reconstructions of samples 74 and 91 from Data-PC dataset, obtained using n_M^6 measurements (second column), the Learned LEM framework (third column), and n_M^{12} measurements (fourth column)

Example 1: CS-based Variational LEM vs Learned LEM Network
Given a set of $n_M^L = 40$ voltage measurements (V_M^L, I_F^L), with $n_L = 8$ electrodes, we compare the reconstructions obtained by solving the CS-based model (5), with those computed by the learned LEM, using RGN-TV for the inverse EIT, on the Data-PC training set of 500 samples, a validation set with 100 samples, and a test set with 10 samples. Figure 3 illustrates the reconstructions starting from noise-free measurements (in the first two columns) and two reconstructions obtained from noisy measurements according to the degradation model (2) with $\eta = 5 \times 10^{-3}\, \bar{V}_M\, rand(n_M)$, (in the third and fourth columns), corresponding to SNR=46dB of the measurements. The higher the gradient sparsity in the conductivity distribution, the more significant the gain in reconstruction accuracy, measured with the Peak SNR metric. Results from the learned LEM show a better separability of the anomalies and less artifacts with respect to the ones from the CS-based LEM. Despite the ill-posedness of the involved EIT inverse problems, both approaches allow to reconstruct the conductivity distributions from very few measurements under gradient-sparsity conditions, even in presence of noise.

Example 2: Learned LEM applied to Data-PC
In this example we consider the application of the learned LEM to the Data-PC dataset, using RGN-TV algorithm to solve (1). We tested this approach in two different settings: using $n_L = 6$ and $n_L = 8$ electrodes to obtain results qualitatively comparable with those obtained by $n_H = 12$ and $n_H = 16$ electrodes, respectively. In both cases the initial measurements were perturbed to obtain an average SNR of 46 dB. Figure 4 illustrates, for two different samples of the dataset, the reconstructions σ_{12}^{net}, results of the learned LEM framework, compared with σ_6 (second column), obtained by a voltage measurement (I_F^6, V_M^6), and σ_{12} (fourth column), obtained by (I_F^{12}, V_M^{12}). The corresponding average metrics over the testing dataset are summarized in Table 1 (a). A similar comparison has been performed for reconstructions σ_{16}^{net} compared to those obtained by (I_F^8, V_M^8) and by (I_F^{16}, V_M^{16}), and the corresponding average results are summarized in Table 1 (b).

Example 3: Learned LEM applied to Data-S
We applied the learned LEM to the Data-S dataset, using RGN-Tik algorithm to solve (1). The complexity of the dataset Data-S demands for more measurements n_M to obtain acceptable results. We tested two settings: using $n_L = 6$ electrodes (corresponding to $n_M^6 = 18$ measurements) and $n_L = 12$ electrodes ($n_M^{12} = 108$ measurements) to obtain results qualitatively comparable with those obtained by $n_H = 12$ and $n_H = 24$ electrodes, respectively. In both cases the initial measurements were perturbed to obtain an average SNR of 46dB.

In Fig. 5 the reconstructions for two different samples of the dataset Data-S are shown: the reconstructions σ_{12}^{net}, results of the learned LEM framework, compared with σ_6, obtained by a voltage measurement (I_F^6, V_M^6) and σ_{12}, obtained by (I_F^{12}, V_M^{12}), in the second and fourth columns, respectively. The corresponding average metrics PSNR and SSIM over the testing dataset are summarized

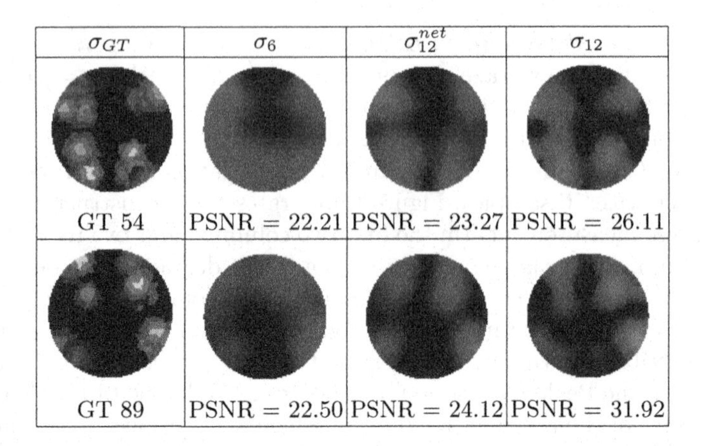

Fig. 5. Example 3: Reconstructions of samples 54 and 89 in the Data-S dataset obtained using n_M^6 measurements (second column), the Learned LEM (third column), and n_M^{12} measurements (fourth column).

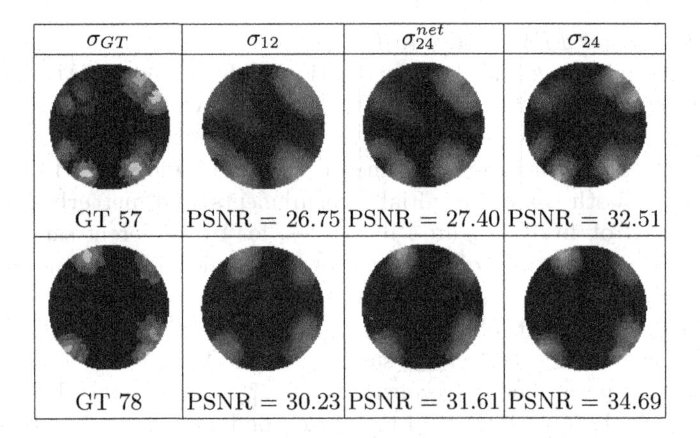

Fig. 6. Example 3: Reconstructions of samples 57 and 78 in Data-S dataset obtained using n_M^{12} measurements (second column), the Learned LEM framework (third column), and n_M^{24} measurements (fourth column).

in Table 1 (c). In a similar settings, Fig. 6 compares the reconstructions for σ_{12}, σ_{24}^{net}, and σ_{24}, and Table 1 (d) summarizes the associated average PSNR and SSIM values.

7 Conclusion and Future Work

Obtaining accurate EIT reconstructions is generally dependent on the amount of data available and therefore on the number of electrodes present in the acquisition system. The increase in the number of these stimulation/acquisition elements is however associated with non-negligible technical difficulties and high costs. This work introduces the EIT-Limited Electrodes Model for the recovery of high-quality reconstructions from a limited number of measurements. Under the assumption of gradient-sparsity, we propose a variational LEM and a neural network-based LEM. Numerical results show a significant improvement in the quality of the reconstructed conductivity distribution both for the simulated datasets Data-PC, and Data-S, which present a different level of gradient-sparsity. This paves the way to LEM approaches which approximate reconstructions of high quality even if starting from a low profile setup with a few electrodes; thus reducing costs and widening the applicability of this imaging technique. The obtained results can be easily applied to any stimulation pattern. Future developments will include a theoretical investigation on the CS nonlinear measurements.

7.1 Appendix: regularized Gauss-Newton method for the inverse EIT problem

Given a nonlinear residual $r(\sigma)$, the ill-posedness of problem (1) can be alleviated by adding to the cost function a regularization term $R(\sigma)$, which stabilizes the solution

$$\sigma^* \in \arg\min_{\sigma \in \mathbb{R}^{n_T}} \|r(\sigma)\|_2^2 + \frac{\alpha}{2} R(\sigma), \tag{7}$$

with $\alpha > 0$ the regularization parameter. An approximated solution of the unconstrained regularized nonlinear minimization problem (7) can be obtained by applying the iterative Regularized Gauss-Newton (RGN) method, which, starting from an initial guess $\sigma^{(0)}$, performs a line search along the direction $p^{(k)}$ to obtain the new conductivity iterate $\sigma^{(k+1)} = \sigma^{(k)} + p^{(k)}$, following the search direction $p^{(k)}$ from the current iterate.

For the choice $R(\sigma) := \|\nabla\sigma\|_2^2$, the functional in (7) is the classical generalized nonlinear Tikhonov model, a standard choice in EIT [4], and the (RGN) method can be applied with $p^{(k)}$ determined by solving the linear normal equations

$$(J(\sigma^{(k)})^T J(\sigma^{(k)}) + \alpha L^T L)p = J(\sigma^{(k)})^T r(\sigma) - \alpha L^T L \sigma^{(k)}, \tag{RGN-Tik}$$

where $L^T L \in \mathbb{R}^{n_T \times n_T}$ denotes the Laplacian. In case the total variation regularizer $R(\sigma) = \|\nabla\sigma\|_1$ is considered, we can resort to the discretization presented in [4], with $R(\sigma) \approx D^T E^{-1} D$, with D denoting the discrete gradient, and $E = 1/diag(\sqrt{(D\sigma)^2 + \epsilon^2})$, $\bar{D} = \sqrt{E}D$. The direction $p^{(k)}$ is then obtained by solving the linear normal equations

$$(J(\sigma^{(k)})^T J(\sigma^{(k)}) + \alpha \bar{D}^T \bar{D})p = J(\sigma^{(k)})^T r(\sigma) - \alpha \bar{D}^T \bar{D} \sigma^{(k)}. \tag{RGN-TV}$$

The linear systems in RGN-Tik and RGN-TV are solved by a few iterations of the conjugate gradient method.

References

1. Alberti, G.S., Santacesaria, M.: Calderón's inverse problem with a finite number of measurements. Forum Math. Sigma. **7**, e35 (2019). https://doi.org/10.1017/fms.2019.31
2. Alberti, G.S., Santacesaria, M.: Infinite dimensional compressed sensing from anisotropic measurements and applications to inverse problems in PDE. Appl. Comput. Harmon. Anal. **50**, 105–146 (2021). https://doi.org/10.1016/j.acha.2019.08.002
3. Blumensath, T.: Compressed sensing with nonlinear observations and related nonlinear optimization problems. IEEE Trans. Inf. Theory. **59**(6), 3466–3474, 2245716 (2013). https://doi.org/10.1109/TIT.2013
4. Borsic, A., et al.: In Vivo impedance imaging with total variation regularization. IEEE Trans. Med. Imaging. **29**(1), 44–54 (2010). https://doi.org/10.1109/TMI.2009.2022540
5. Candès, E.J., Romberg, J.K., Tao, T.: Stable signal recovery from incomplete and inaccurate measurements. Commun. Pure Appl. Math. **59**(8), 1207–1223 (2006). https://doi.org/10.1002/cpa.20124
6. Colibazzi, F., Lazzaro, D., Morigi, S., Samoré, A.: Learning nonlinear electrical impedance tomography. J. Sci. Comput. **90**(1), 1–23 (2021). https://doi.org/10.1007/s10915-021-01716-4
7. Cortesi, M., et al.: Development of an electrical impedance tomography set-up for the quantification of mineralization in biopolymer scaffolds. Physiol. Measur. **42**(6), 064001 (2021). https://doi.org/10.1088/1361-6579/ac023b
8. Harrach, B.: Uniqueness and Lipschitz stability in electrical impedance tomography with finitely many electrodes. Inverse Prob. **35**(2), 024005 (2019). https://doi.org/10.1088/1361-6420/aaf6fc
9. Jauhiainen, J., et al.: Relaxed Gauss-Newton methods with applications to electrical impedance tomography. SIAM J. Imaging Sci. **13**(3), 1415–1445 (2020). https://doi.org/10.1137/20M1321711
10. Somersalo, E., Cheney, M., Isaacson, D.: Existence and uniqueness for electrode models for electric current computed tomography. SIAM J. Appl. Math. **52**(4), 1023–1040 (1992). https://doi.org/10.1137/0152060
11. Tallman, T.N., Smyl, D.J.: Structural health and condition monitoring via electrical impedance tomography in self-sensing materials: a review. Smart Mater. Struct. **29**(12), 123001 (2020). https://doi.org/10.1088/1361-665X/abb352

On Trainable Multiplicative Noise Removal Models

Mahipal Jetta(✉), Utkarsh Singh, and Padmaja Yinukula

Mahindra University, Hyderabad, India
mahipal.jetta@mahindrauniversity.edu.in

Abstract. In most of the real world imaging applications like synthetic aperture radar (SAR), microscope and laser images, multiplicative noise is more prevalent than the additive Gaussian noise. In our present work, we have focused on multiplicative noise removal models. Standard diffusion based multiplicative Gamma noise removal models produce highly smoothed images. In addition, their restoration performance mainly depends on the proper choice of the built-in parameters and the diffusion coefficient. In this paper, we study various learning frameworks, including our own, regarding the applicability of the models in case of denoising of highly corrupted images. We show through numerical experiments that the considered trainable models perform better than the state-of-the-art PDE models in terms of peak-signal-to-noise ratio (PSNR).

Keywords: Nonlinear Diffusion · Deep Learning · Multiplicative Noise

1 Introduction

Multiplicative noise appears in synthetic aperture radar (SAR), microscope and laser images, which hinders the user to perform high level image processing tasks such as image understanding and segmentation. Therefore, pre-processing of such images is a necessary requirement.

The relationship among the corrupted image f, the original image g and the noise η is in general modeled as [2]

$$f = g\eta. \tag{1}$$

The problem here is to estimate g, given f. Here the noise η is assumed to follow Gamma distribution with mean 1:

$$p(\eta) = \begin{cases} \frac{L^L}{\Gamma(L)}\eta^{L-1}exp(-L\eta) & \text{for } \eta > 0, \\ 0 \quad \text{for } \eta = 0, \end{cases}$$

where L denotes the number of looks and $\Gamma(.)$ is the Gamma function.

The major difficulty in addressing this problem is the corruption due to multiplicative type noise as it destructs the underlying image depending on its

L. Calatroni et al. (Eds.): SSVM 2023, LNCS 14009, pp. 81–93, 2023.
https://doi.org/10.1007/978-3-031-31975-4_7

intensity values, that is, the brighter pixels will be corrupted heavily than the darker pixels.

Several approaches have been proposed in the literature to address this problem, see, for example, [10] and the references therein. They are broadly classified into the following: (i) Partial Differential Equations (PDE) based approaches, (ii) Variational methods, (iii) Nonlocal methods, and (iv) Transformation based approaches.

In this article, we focus on the PDE based approaches. The speckle reduction anisotropic diffusion (SRAD) was developed in [11], and it is based on a spatial domain filter proposed by Lee *et al.* [5]. This model incorporated the noise statistics into the diffusion coefficient instead of the image gradient to better identify the edge locations, and hence achieved reasonable edge preservation while removing the noise. Inspired by this work, many second order nonlinear diffusion filters have been proposed in the literature [7,8,10,12]. Most of these models use only gradient or its equivalent operator information to remove the multiplicative noise.

Recently, Shan *et al.* [9] proposed a diffusion model that uses both the gradient and the gray level of the image in the diffusion coefficient. This model takes the following form

$$\begin{cases} u_\tau = div(c(u_\sigma, |\nabla u_\sigma|)\nabla u) \ \ \forall(\mathbf{x},\tau) \in \Omega \times (0,T) \\ u(\mathbf{x},0) = f(\mathbf{x}), \quad \forall \mathbf{x} \in \Omega \times \{\tau = 0\} \\ \frac{\partial u}{\partial \mathbf{n}} = 0 \quad \forall(\mathbf{x},\tau) \in \partial\Omega \times (0,T) \end{cases} \tag{2}$$

where $\Omega \subset \mathbf{R}^2$, $T > 0$ denotes fixed time, f is the noisy image with the intensity range in $[0,1]$, and the diffusion coefficient $c(u_\sigma, |\nabla u_\sigma|) = \left(\frac{u_\sigma}{M}\right)^\alpha \frac{1}{1+|\nabla u_\sigma|^\beta}$. Here α, β and σ are positive constants, $u_\sigma = G_\sigma * u$, which is the convolution between the Gaussian function with standard deviation σ and u, and $M = \max_{\mathbf{x}\in\Omega} |u_\sigma(\mathbf{x},\tau)|$. This choice of the diffusion coefficient $c(.)$ helps in protecting the edges while removing the noise, besides offering theoretical advantage regarding the well-posedness.

Majee *et al.* [6] developed a well-posed telegraph diffusion model for better restoration of the noisy image. This model has the following form

$$\begin{cases} u_{\tau\tau} + \gamma u_\tau - div(c(u_\sigma, |\nabla u_\sigma|)\nabla u) = 0 \ \ \forall(\mathbf{x},\tau) \in \Omega \times (0,T) \\ u(\mathbf{x},0) = f(\mathbf{x}), \quad u_\tau(\mathbf{x},0) = 0 \ \ \forall \mathbf{x} \in \Omega \\ \frac{\partial u}{\partial \mathbf{n}} = 0 \quad \forall(\mathbf{x},\tau) \in \partial\Omega \times (0,T) \end{cases} \tag{3}$$

where $c(u_\sigma, |\nabla u_\sigma|) = \frac{2|u_\sigma|^\nu}{(M_\sigma)^\nu + |u_\sigma|^\nu} \cdot \frac{1}{1+\left(\frac{|\nabla u_\sigma|}{K}\right)^2}$. Here γ, $K > 0$ are constants, and $\nu \geq 1$. This model uses the benefit of both wave equation and diffusion equation in preserving the highly oscillatory patterns in the image.

It is important to note here that the performance of the above mentioned models depend heavily on the choice of the underlying parameters. Therefore, in this paper, we adopt a trainable nonlinear diffusion filter, which has originally been developed to remove additive noise. Here we train various models to obtain optimal values for the underlying parameters to remove the multiplicative Gamma noise. The details of these trainable models are given in the following section.

This paper is organized as follows: In Sect. 2, we provide the details of the trainable nonlinear diffusion filters for multiplicative noise removal. The experimental set up and numerical comparisons with existing works are provided in Sect. 3. Finally, conclusions constitute Sect. 4.

2 Trainable Models for Multiplicative Noise Removal

2.1 TNRD Model

The trainable nonlinear reaction diffusion (TNRD) has been used to remove additive Gaussian noise from an image [3]. The parameters of this model cannot be directly applied to remove the multiplicative noise as they will interpret noise as edges and hence the noise will be enhanced. Therefore, we adopt this model and train the parameters appropriately to remove the multiplicative noise. The details are given as follows.

Greedy Training. Consider the TNRD model

$$\begin{cases} u_t = u_{t-1} - \left(\sum_{i=1}^{N_k} \bar{k}_i^t * \phi_i^t(k_i^t * u_{t-1}) + \lambda^t (u_{t-1} - f_n) \right) \\ u|_{t=0} = f_n \end{cases} \tag{4}$$

where f_n is the initial image corrupted with multiplicative Gamma noise, u_t is the image obtained at t^{th} iteration, $\bar{k}_i$ indicates the point reflection of the kernel k_i which, in case of 2D, is the rotation of k_i by 180^0, '$*$' indicates convolution, ϕ_i is an influence function which helps in distinguishing between noisy and prominent regions, and N_k denotes the number of filters. A typical role of these kernels is to better estimate the spatial derivatives of the image.

The goal of TNRD is to obtain optimal kernels, influence functions and parameter λ to achieve the best possible filtered image. Here we set up a learning framework to obtain these parameters. In this connection, we take a data set containing S number of sample images and corrupt them with multiplicative Gamma noise. We then consider the loss function

$$\ell(\phi_i^t, k_i^t, \lambda^t) = \frac{1}{2} \sum_{s=1}^{S} ||u_t^{(s)} - f^s||_2^2, \tag{5}$$

with $u_t^s = u_{t-1}^s - \left(\sum_{i=1}^{N_k} \bar{k}_i^t * \phi_i^t(k_i^t * u_{t-1}^s) + \lambda^t (u_{t-1}^s - f_n^s) \right)$, and $f^{(s)}$ is the ground truth (clean) image. The parameters $(\phi_i^t, k_i^t, \lambda^t)$, $i = 1, 2, 3, .., N_k$, are obtained by minimizing $\ell(.)$ using an optimization algorithm, for example, L-BFGS or the Gradient-descent method, for each stage t. A typical optimization algorithm involves the computation of gradient of the loss function. We refer to Chen *et al.* [3] and its supplementary material to see the calculation of $\nabla \ell$.

It is important to note that each kernel k_i is expressed as a linear combination of Discrete-Cosine transform (DCT) basis, and each influence function ϕ_i is expressed as a weighted sum of Gaussian radial basis functions. The gradient of the loss function $\ell(.)$ will then have components as derivatives of ℓ with respect to λ, scalars involved in each kernel approximation and the weights involved in approximation of influence function. We then employ an optimization algorithm, precisely L-BFGS algorithm, to find these parameters which minimize $\ell(.)$ over the entire samples in the given data set.

These optimal parameters obtained at each stage will then be employed at the corresponding stage to update the given input image f_n.

Joint Training. After greedy training where loss is evaluated and minimised at every stage, the trained coefficients, $\Theta_t = (\phi_i^t, k_i^t, \lambda^t)$, are taken as initial values in the joint training framework.

The objective of joint training is to optimise the parameters of all the T stages simultaneously. It is formulated as

$$\mathcal{L}(\Theta_{1,\ldots,T}) = \sum_{s=1}^{S} l\left(u_T^{(s)}, u_{gt}^{(s)}\right) = \sum_{s=1}^{S} \frac{1}{2} ||Tu_T^{(s)} - u_{gt}^{(s)}||_2^2$$

It should be noted that unlike in greedy training, the loss is only affected by the output of the final stage T. The derivations of the gradient of the loss function with respect to Θ_t have been explained thoroughly in the supplementary material [3]. To ensure continuity, the term needed to find the gradient have been mentioned below.

$$\frac{\partial l}{\partial \Theta_t} = \frac{\partial u_t}{\partial \Theta_t} . \frac{\partial l}{\partial u_t}$$

Here, $\frac{\partial u_t}{\partial \Theta_t}$ have already been mentioned in the greedy training section, and $\frac{\partial l}{\partial u_t}$ is found through standard back-propagation:

$$\begin{aligned} \frac{\partial l}{\partial u_t} &= \frac{\partial u_{t+1}}{\partial u_t} . \frac{\partial l}{\partial u_{t+1}} \\ &= \frac{\partial u_{t+1}}{\partial u_t} \cdots \frac{\partial u_T}{\partial u_{T-1}} . e \end{aligned}$$

where

$$e = \frac{\partial l}{\partial u_T} = T^{\mathcal{T}} (Tu_T - u_{gt})$$

$$\frac{\partial u_{t+1}}{\partial u_t} = P_T^{\mathcal{T}} . \left(\left(1 - \lambda^{t+1}\right) I - \sum_{i=1}^{N_k} K_i^{t+1^{\mathcal{T}}} . \Lambda_i . \left(\overline{K}_i^{t+1}\right)^{\mathcal{T}} \right)$$

$$\Lambda_i = \operatorname{diag}\left(\phi_i^{t+1'}(z_1), \ldots, \phi_i^{t+1'}(z_p)\right) \text{ with } z = k_i^{t+1} * u_{tp}.$$

Here, the (padding) matrix P implies that the image array is being padded to compensate for the loss in border pixels upon convolution, and K_i is a sparse matrix made with elements from kernel k_i which is populated with values corresponding to u_{t+1}. The u_{tp} denotes the padded version of u_t.

2.2 L-TNRD Model

The multiplicative noise removal can also be addressed by converting it into additive noise removal problem by applying the log transformation.

In this regard, we consider the logarithm of both noisy and corresponding clean images in the training data set and apply the (additive noise removal) TNRD model with this transformed data set. We then obtain the restored image in the original domain by taking the exponential of the image obtained in log domain. Here we call this approach as L-TNRD model.

2.3 Feng *et al* Model

Here the authors [4] have incorporated the speckle statistics into the energy functional, and developed a learning framework which was derived based on a proximal gradient descent method. The diffusion process takes the following form for a sample image u^s.

$$u_{t+1}^s = \frac{\tilde{u}_{t+1}^s + \sqrt{\tilde{u}_{t+1}^{s^2} + 8(1+2\lambda^{t+1})\lambda^{t+1} f^2}}{2(1+2\lambda^{t+1})} \tag{6}$$

where

$$\tilde{u}_{t+1}^s = u_t^s - \left(\sum_{i=1}^{N_k} \bar{k}_i^t * \phi_i^t (k_i^t * u_{t-1}^s) \right). \tag{7}$$

Then the parameters of this model are learned by minimizing (5) with (6) and (7).

2.4 Proposed Model

We consider the Aubert *et al.* [1] model to derive the diffusion-reaction process. The main reason for considering this model is that it is derived based on the modeling of the multiplicative noise.

The main challenge here is to ensure stability of the diffusion-reaction process as there is a chance of u becoming closer to zero which eventually leads to training failure. To avoid such a situation, we add a small positive quantity ϵ to u^2 in the denominator. The diffusion-reaction process is given by

$$u_t^s = u_{t-1}^s - \left(\sum_{i=1}^{N_k} \bar{k}_i^t * \phi_i^t (k_i^t * u_{t-1}^s) + \lambda^t \frac{u_{t-1}^s - f_n^s}{((u_{t-1}^s)^2 + \epsilon)} \right) \tag{8}$$

The architecture of this process is shown in Fig. 1.

The gradient of loss are same as in case of TNRD model except the following changes:

$$\frac{\partial u_t}{\partial \lambda^t} = \left(\frac{u_{(t-1)p} - f_{np}}{u^2_{(t-1)p} + \epsilon} \right)^\tau$$
$$\frac{\partial u_{t+1}}{\partial u_t} = \left(1 - \lambda^{t+1} \left(\frac{2f-u}{u^3+\epsilon}\right)\right) I - \sum_{i=1}^{N_k} K_i^{t+1^\tau} .\Lambda_i. \left(\overline{K}_i^{t+1}\right)^\tau$$

The other components of the gradient remain same as in case of the TNRD model.

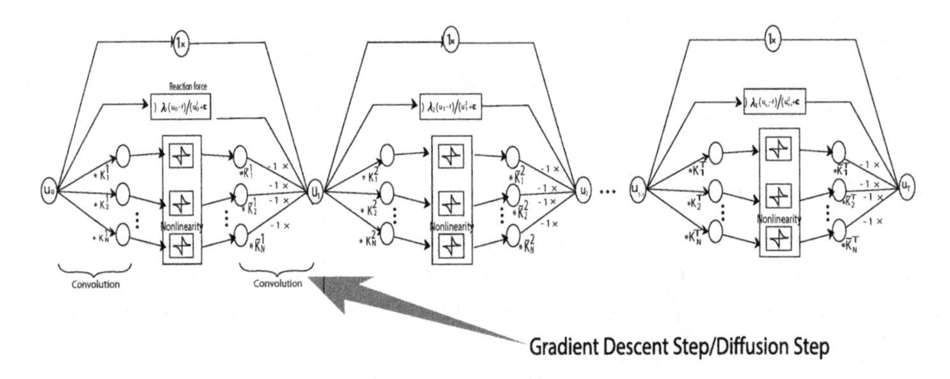

Fig. 1. The architecture of the proposed model.

3 Numerical Experiments

In this section, we compare the performance of the trained multiplicative noise removal models with the two state-of-the-art PDE models (2) and (3).

We have implemented these models, assuming the intensity range in [0, 1], with suitable parameters so that we get their optimal performance. Apart from these parameters, we set $\gamma = 1$, $\nu = 1$ and $\sigma = 1$ in both the PDE based models wherever applicable. In case of trainable models, we have used the same data set as considered in [3]. The trainable models have been trained with 5×5 kernels and the influence function has been approximated with 63 Gaussian radial basis functions over $[-310, 310]$. The number of weights in the kernel approximation has taken to be 24 as the filter size is 25. The training has been performed up to 8 stages. The same parameters have been used in the all considered trainable models. It is important to note that the performance with greedy training alone is reported here in all the numerical simulations.

To compare the performance of all the considered methods, we have taken three natural images and corrupted them with Gamma noise with $(L = 1)$. We can see from the Figs. 2, 3 and 4 that the trainable nonlinear diffusion models produced better quality filtered images when compared with the PDE based models. The same has been confirmed in terms of image quality metric (PSNR), see Table 1. They can also better preserve the important details to a reasonable extent, see the edges of the restored images in Figs. 2, 3 and 4.

The log-transformation based method, in general, does not preserve the mean of the initial image [1]. Therefore, the performance of the L-TNRD model is inferior to all the other methods considered in this paper, see Table 3. Hence, we focus mainly on the other three learned models. The better performance of these learned models is attributed to their learning ability of non-standard kernels and influence functions. The characteristic influence (ϕ) and penalty functions (ρ) by the proposed model are shown in Fig. 5. This mixture of filter response dependent convex and concave type penalty functions helps in smoothing and enhancing the features. Similar profiles have been observed in the multiplicative noise removal TNRD and Feng *et al.* models. It is striking that all the three learned models are lacking the rotation-invariance and producing some artifacts in the restored image.

We have then considered the evaluation of the three learned models in terms of restored image quality. In this regard, we have considered a sample of 68 images and corrupted them with multiplicative Gamma noise with $L = 1$, and performed the denoising process with all the three trainable models. The average of PSNR of the obtained filtered images is more or less the same, see Table 2. Also, in the same table, the maximum PSNR (Max) row and the Variance row show that we cannot confirm one trainable model is superior to the other two considered models. The same has been confirmed when worked with individual images, refer Table 3.

Table 1. Comparison of various models

	Image	Shan *et al.*			Majee *et al.*			TNRD	Feng *et al.*	Proposed
		α	β	PSNR	α	β	PSNR	PSNR	PSNR	PSNR
$L = 1$	Cameraman	1.1	0.6	20.25	0.5	2.5	20.91	21.08	20.91	20.94
	Pirate	0.6	0.3	22.37	0.5	1.5	22.46	23.21	23.25	23.23
	Lena	1.5	0.5	21.19	0.5	2	21.41	22.21	22.08	22.10

The mean performance of the trained models, taken over a sample of 68 images is provided in (2).

Table 2. Mean-Performance statistics in terms of PSNR

Metric	TNRD	Feng et al.	Proposed
Mean	21.90	21.96	21.91
Max	27.35	27.39	27.67
Variance	5.59	5.83	5.97

Fig. 2. Denoising performance of the models. Noisy image is obtained by corrupting the original *Lena* image with multiplicative Gamma noise with $L = 1$.

(a) Noisy PSNR: 12.22

(b) Shan *et al.* PSNR:20.25

(c) Majee *et al.* PSNR:20.91

(d) TNRD PSNR:21.08

(e) Feng *et al.* PSNR:20.91

(f) Proposed PSNR:20.94

Fig. 3. Denoising performance of the models. Noisy image is obtained by corrupting the original *Cameraman* image with multiplicative Gamma noise with $L = 1$.

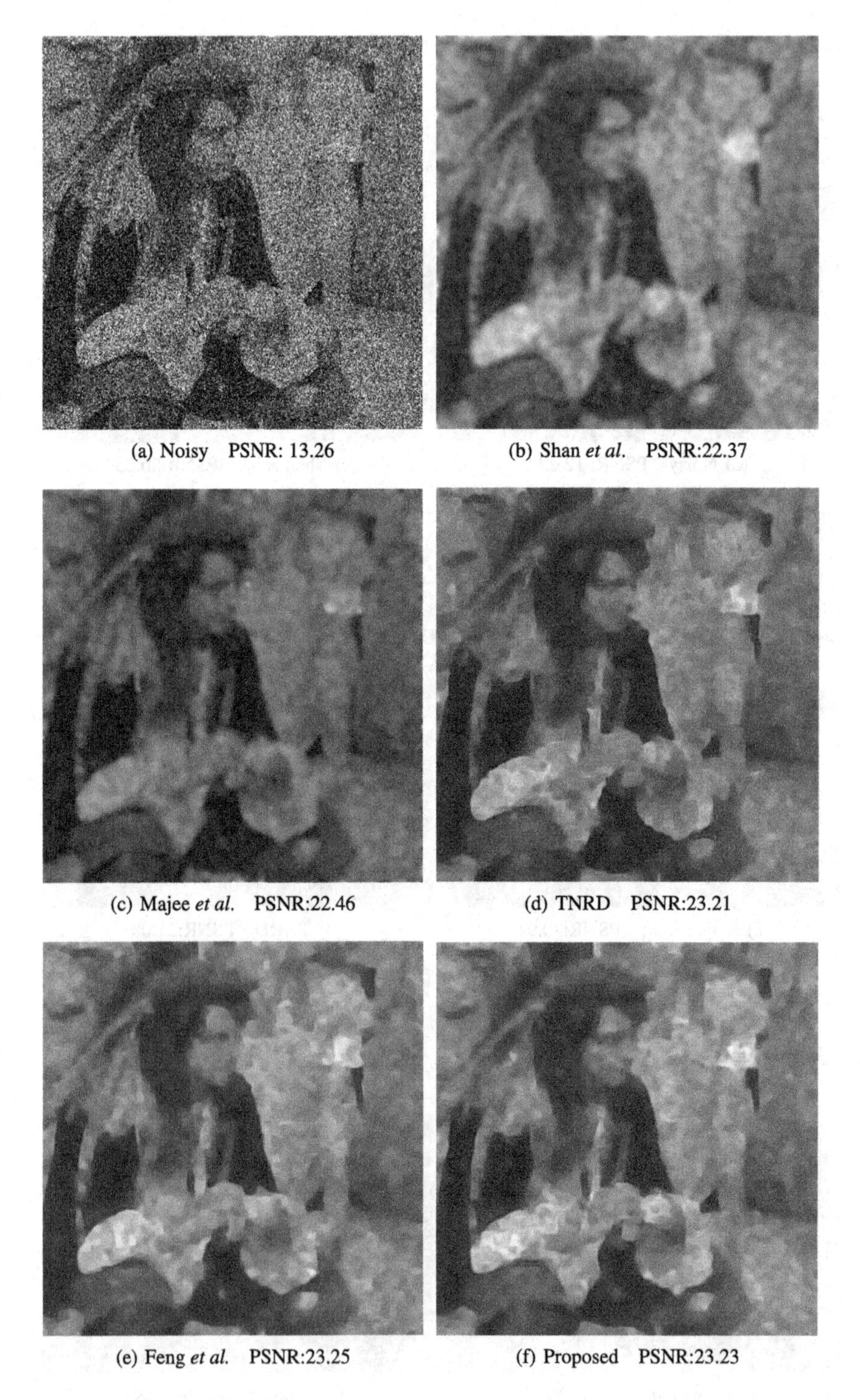

Fig. 4. Denoising performance of the models. Noisy image is obtained by corrupting the original *Pirate* image with multiplicative Gamma noise with $L = 1$.

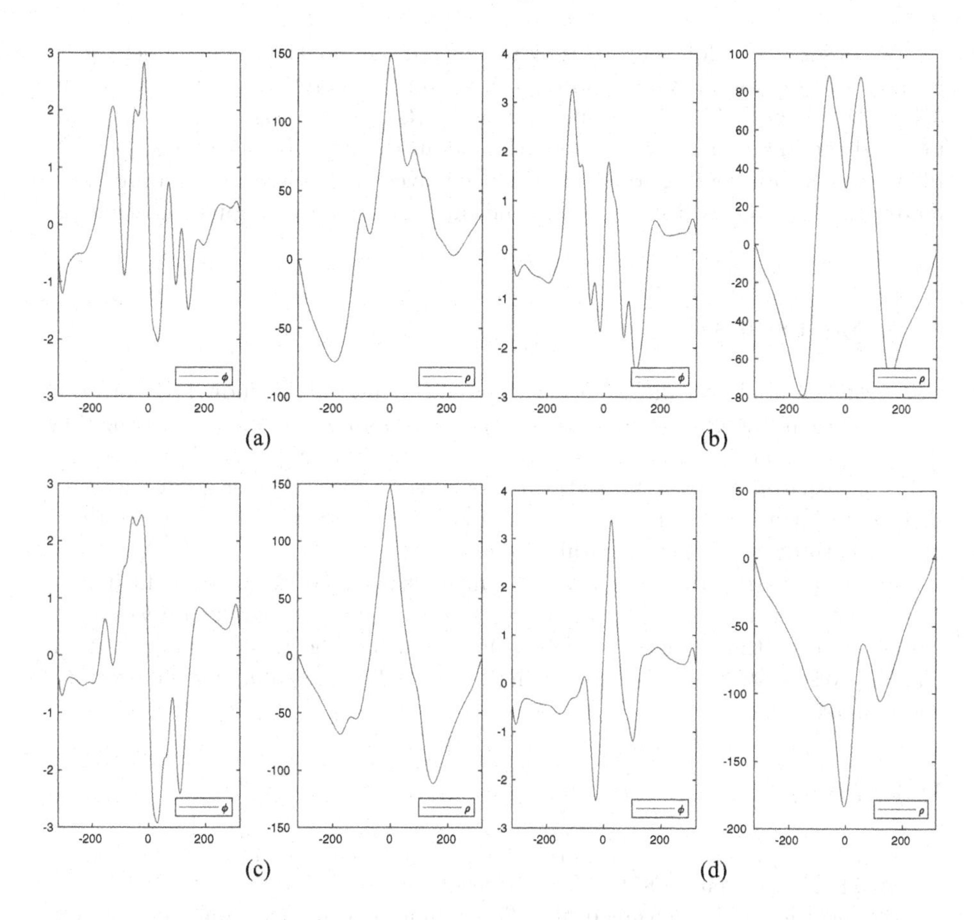

Fig. 5. Type of influence and the corresponding potential functions, learned by the proposed model.

Table 3. Trained models performance in terms of PSNR

Image	TNRD	L-TNRD	Feng et al.	Proposed
penguin	23.25	10.04	**23.62**	23.33
wolf	23.91	11.34	**24.51**	23.96
soldiers	23.32	13.22	**23.38**	23.33
mountaineer	23.99	11.36	24.01	**24.02**
snake	23.86	12.64	23.90	**23.97**
vase	26.83	13.78	26.92	**26.93**
elephants	**25.22**	14.34	25.13	25.16
surfer	24.02	14.96	**24.07**	23.92
airplane	27.36	13.47	**27.86**	27.77
camel	26.11	14.55	25.14	**26.20**

We observe the following through numerical simulations which were performed on a system with configuration: Intel© core^{TM} i5-7500 CPU @ 3.40 GHz × 4 with 7.7 GiB memory. The TNRD model takes 2.64 s for testing, and produces paint brush type artefacts in the filtered image. The Feng *et al.* model consumes 8.90 s for testing as it involves the update through proximal mapping. The proposed model takes the same amount of computation time as TNRD.

4 Conclusions

The trainable nonlinear diffusion-reaction models have been studied to learn the underlying parameters to remove multiplicative Gamma noise. A new trainable diffusion-reaction process has been developed by unfolding the Aubert *et al.* diffusion-reaction process. These trained models produced better quality filtered images, in terms of PSNR, when compared with the two state-of-the-art PDE based multiplicative noise removal models.

We also observe that the trainable models produce some artifacts in the filtered image while restoring highly corrupted images. In our future work, we wish to improve the trainable models by incorporating the gray-level information and the noise statistics into the learning framework while enforcing the stability constraints.

References

1. Aubert, G., Aujol, J.F.: A variational approach to removing multiplicative noise. SIAM J. Appl. Math. **68**(4), 925–946 (2008)
2. Bioucas-Dias, J.M., Figueiredo, M.A.T.: Multiplicative noise removal using variable splitting and constrained optimization. IEEE Trans. Image Process. **19**(7), 1720–1730 (2010). https://doi.org/10.1109/TIP.2010.2045029
3. Chen, Y., Pock, T.: Trainable nonlinear reaction diffusion: a flexible framework for fast and effective image restoration. IEEE Trans. Pattern Anal. Mach. Intell. **39**(6), 1256–1272 (2017). https://doi.org/10.1109/TPAMI.2016.2596743
4. Feng, W., Chen, Y.: Speckle reduction with trained nonlinear diffusion filtering. J. Math. Imaging Vision **58**(1), 162–178 (2017)
5. Lee, J.S.: Digital image enhancement and noise filtering by use of local statistics. IEEE Trans. Pattern Anal. Mach. Intell. **2**, 165–168 (1980)
6. Majee, S., Ray, R.K., Majee, A.K.: A gray level indicator-based regularized telegraph diffusion model: application to image despeckling. SIAM J. Imaging Sci. **13**(2), 844–870 (2020)
7. Mishra, D., Chaudhury, S., Sarkar, M., Soin, A.S., Sharma, V.: Edge probability and pixel relativity-based speckle reducing anisotropic diffusion. IEEE Trans. Image Process. **27**(2), 649–664 (2018). https://doi.org/10.1109/TIP.2017.2762590
8. Ramos-Llordén, G., Vegas-Sánchez-Ferrero, G., Martin-Fernandez, M., Alberola-López, C., Aja-Fernández, S.: Anisotropic diffusion filter with memory based on speckle statistics for ultrasound images. IEEE Trans. Image Process. **24**(1), 345–358 (2015)

9. Shan, X., Sun, J., Guo, Z.: Multiplicative noise removal based on the smooth diffusion equation. J. Math. Imaging Vision **61**(6), 763–779 (2019)
10. Yao, W., Guo, Z., Sun, J., Wu, B., Gao, H.: Multiplicative noise removal for texture images based on adaptive anisotropic fractional diffusion equations. SIAM J. Imaging Sci. **12**(2), 839–873 (2019)
11. Yu, Y., Acton, S.T.: Speckle reducing anisotropic diffusion. IEEE Trans. Image Process. **11**(11), 1260–1270 (2002)
12. Zhou, Z., Guo, Z., Dong, G., Sun, J., Zhang, D., Wu, B.: A doubly degenerate diffusion model based on the gray level indicator for multiplicative noise removal. IEEE Trans. Image Process. **24**(1), 249–260 (2015). https://doi.org/10.1109/TIP.2014.eps

Surface Reconstruction from Noisy Point Cloud Using Directional G-norm

Guangyu Cui(✉), Ho Law, and Sung Ha Kang

Georgia Institute of Technology, Atlanta, GA, USA
{gcui8,hlaw}@gatech.edu, kang@math.gatech.edu

Abstract. We propose a method to reconstruct surface from noisy point cloud data, by obtaining a clean zero level set of their signed distance function (SDF). Due to the noise in the given point cloud, the distance function is oscillatory and lacks smoothness, especially near the data point. We denoise the SDF using a modified G-norm along the tangential direction of the input point cloud data. While there is abundant work for obtaining a smooth surface from a noisy data, our contribution is to provide an insight into how noisy data corrupts reconstruction of surfaces, through extracting the noisy component from the distance function directly. We apply Augmented Lagrangian Method to optimize the objective energy function and solve the subproblems. We present various numerical results to validate the proposed method.

Keywords: Point Cloud · Surface Reconstruction · G-norm · Decomposition

1 Introduction

Reconstructing surfaces from point cloud data has been extensively studied for decades. The problem formulation can be described as follows: the input is point cloud data $\mathcal{P}$, typically consisting of the locations of points sampled from the interested surface. The objective is to generate a surface that minimizes the total distance to the point cloud while capturing meaningful geometric and topological property of the underlying ground truth surface.

In real application, raw point cloud data often cannot represent the original surface accurately. Some typical issues are: non-uniform sampling which cause lack of dense samples on high curvature areas of the surface; noise which may cause over-fitting; outliers which may cause false correspondences; misalignment and missing data which can be considered as a higher level artifact and it needs the model to correspondingly have high level prior information to properly recover the surface [3].

The surface reconstruction techniques can be separated into two major categories: explicit and implicit methods. Explicit methods, such as Delaunay triangulations [12], alpha-shape [4] and Voronoi diagrams [2], aim to interpolate most

Research is supported in part by Simons Foundation 584960.

L. Calatroni et al. (Eds.): SSVM 2023, LNCS 14009, pp. 94–106, 2023.
https://doi.org/10.1007/978-3-031-31975-4_8

of the points with some post-processing procedure in order to resolve over-fitting issue. The objective of implicit method is to directly construct the surface, typically as zero-level set of distance function to the point cloud. As pointed out by Carr [5], determining the outward normal direction of point cloud is crucial in implicit methods. This can be categorized into global and local approaches. Local approaches such as using moving least square to evaluate normal direction is comprehensively reviewed by Cheng [7]. As for global approaches, one of the state of the art method proposed by Kazhdan [11] approximated normal directions a priori in the input data, and solved Poisson equations to construct a signed distance function to the point cloud.

Mean-curvature motion of level set function [16] well suits the challenges of surface reconstruction problem, since it inherently contains the curvature smoothing term that counteracts robustly against overfitting, and the formation of level set function automatically handles the consistency of outward surface normal direction globally. One of fundamental work was proposed by Zhao [19], where the authors computed the unsigned distance function d to the point cloud, and evolved the zero-level surface via geodesic active contour [6]. Numerous improved models have been studied for different types of regularizations. In [13], a ridge and corner preserving model was studied. In [14], the authors defined the surface via a collection of anisotropic Gaussians centered at each entry of the input point cloud, and used TVG-L1 model for minimization. In [9], authors studied Euler's elastica energy as regularization term, and proposed schemes based on operator splitting method and augmented Lagrangian method to minimize their proposed functional.

There is also work to reconstruct surface via zero level set using Machine Learning techniques, e.g., [8,18] and many others. In this work, we focus on training free approach.

Our contribution is that we proposed a novel norm on the noisy distance function space, that would naturally absorb the noise along a specific direction and give us a cleaner zero level-set. We denoise the SDF using a modified G-norm along the tangential direction of the input point cloud data. Our method is easy to implement as our algorithm consists of a few steps without excessive parameters or pre-or-post processing steps.

This paper is organized as follows: In Sect. 2, we provide a review on some basic knowledge and models. The proposed method is introduced in Sect. 3, followed by the proposed numerical scheme and details in Sect. 4. The experiments are shown in Sect. 5.

2 Notation and Motivation

We first review the computation of surface reconstruction from level-set approaches [16], and discuss the noise effect on the zero level surface.

2.1 Review on Computation of SDF

Let $\mathcal{P} \subset \Omega$ be the set of input point cloud data, subsampled from an unknown C^2 simple closed surface Γ_0. To compute the signed distance function, Zhao et al. [19,20] proposed to solve the following Eikonal Equation:

$$\begin{aligned} |\nabla d(x)| &= 1 \quad \text{for } x \in \Omega, \\ d(x) &= 0 \quad \text{for } x \in \mathcal{P}. \end{aligned} \tag{1}$$

The authors used active-contour formulation, where the estimated surface is defined as the zero contour of level set function $\phi(\mathbf{x})$, i.e. $\Gamma = \{\mathbf{x} \mid \phi(\mathbf{x}) = 0\}$, and minimized the following energy, for $1 \leq p \leq \infty$:

$$E_p(\Gamma) = \left(\int_\Gamma d_0^p(\mathbf{x}) d\sigma\right)^{\frac{1}{p}} = \left(\int_\Omega d_0^p(\mathbf{x}) |\nabla \phi(\mathbf{x})| \delta(\phi(\mathbf{x})) d\mathbf{x}\right)^{\frac{1}{p}} \tag{2}$$

Using Euler-Lagrange Equation, the authors solved a PDE and obtain the level set function $\phi(\mathbf{x})$. Under such setting, $\phi(\mathbf{x})$ can be considered an estimation of SDF, and we use this formulation in this paper.

2.2 Effect of Noise on SDF

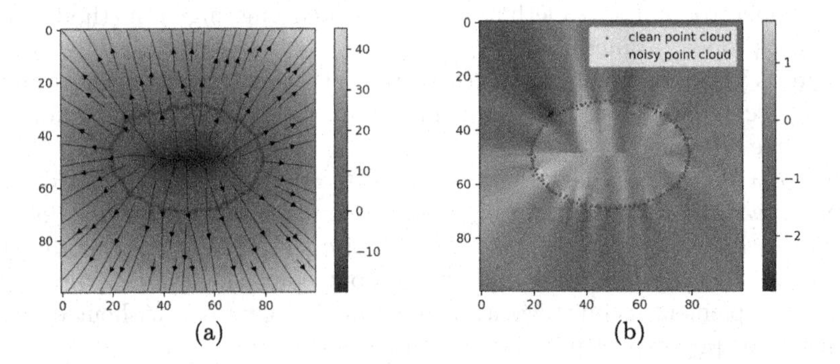

Fig. 1. (a) SDF to the noisy point cloud (red). (b) Two sets of point cloud are marked as blue for clean point cloud and red for noisy point cloud. The difference between the clean and the noisy SDFare presented as surface. (Color figure online)

Due to the noise in the point cloud, the computed SDF typically lacks smoothness, as shown in Fig. 1(a), where the blue contour is the surface found by locating the zero level set of the SDF computed using the method in [19]. In Fig. 1(b), comparing the clean and the noisy point clouds, we observe that noise is an oscillation away from the zero level set, of which the gradient seems to be perpendicular to the clean surface $\mathcal{P}$. This motivates us to remove the oscillation through decomposition [15], and use G-norm in tangential direction.

3 The Proposed Method: Directional G-norm Based Surface Reconstruction

We decompose the noisy distance function d into two functions ϕ and n, such that $d : \Omega \to \mathbb{R}$ and $d = \phi + n$ on Ω. This $\phi \in C(\Omega; \mathbb{R})$ ideally represents a function with smoother gradient, and the function n, the noise perpendicular to the tangential direction of the surface Γ_0. We propose the following directional G-norm based surface reconstruction energy:

$$E(\phi, n; \vec{w}_\sigma, f) = \|\overrightarrow{w}_\sigma \cdot \nabla\phi\|_2^2 + \lambda\|n\|_{*,f}^2 \quad \text{subject to } d = \phi + n. \tag{3}$$

The first term $\vec{w}_\sigma$ denotes the tangential direction along which we denoise the SDF. We define the direction $\vec{w}_\sigma$ by

$$\vec{w}_\sigma = G_\sigma * \left(\begin{bmatrix} 0 & 1 \\ -1 & 0 \end{bmatrix} \cdot \nabla d \right), \tag{4}$$

where G_σ is Gaussian kernel with variance σ and $*$ denotes element-wise convolution. This Gaussian kernel helps to stabilize the computation of tangential vector field and weaken the small scale oscillation.

We call the second norm to be the directional G-norm, which is used to capture oscillations. For a scale adjusting function $f : \Omega \to \mathbb{R}$ to be explained, the directional G-norm is defined as

$$\|n\|_{*,f} = \inf_{g \in C^1(\Omega;\mathbb{R})} \{\|fg\|_2 \mid n = \vec{w}_\sigma \cdot \nabla g\}, \tag{5}$$

on the space:

$$\mathcal{N}(\vec{w}_\sigma) = \{n : \Omega \to \mathbb{R} \mid \exists g \in C^1(\Omega; \mathbb{R}) \text{ such that } n = \vec{w}_\sigma \cdot \nabla g\}. \tag{6}$$

We consider the norm in the negative Sobolev space as in [10,17]. This model extends from Meyer's G-space [15], which includes all n such that $n = \nabla \cdot \nabla g$. The divergence of gradient of g helps to capture the noise in all direction, while the proposed modified G-norm in (5) includes only noise in direction that is perpendicular to $\vec{w}_\sigma$. Since the noise originates from the point cloud $\mathcal{P}$ around the zero level set of d, the scale of the oscillation changes depending on the distance from the point cloud in Ω. We chose a $\epsilon > 0$, and use $f(x) = (|d(x) - \min_{z\in\Omega} d(z)| + \epsilon)^{-1}$ to properly capture the scale differences to improve the denoising effect of $\|\cdot\|_{*,f}$.

In (3), the first norm $\|\vec{w}_\sigma \cdot \nabla\phi\|_2$ is small when ϕ is smooth along direction $\vec{w}_\sigma$, and the second norm $\|n\|_{*,f}$ is small when ϕ has high-frequency oscillations perpendicular to $\vec{w}_\sigma$. With $\lambda \ll 1$, the smoothness induced by $\|\vec{w}_\sigma \cdot \nabla\phi\|_2$ is emphasised, forcing the ϕ to be smooth as (3) is minimized. If λ is large, then n becomes less tolerant to lower frequency components, and they go to the smooth part ϕ. The parameter λ balances these effects.

3.1 Minimisation Method

To minimize (3), we apply Augmented Lagrangian Method (ALM). The energy is rewritten as follows:

$$E(\phi, g, Z; \vec{w}_\sigma, f) = \|\vec{w}_\sigma \cdot \nabla\phi\|_2^2 + \lambda\|fg\|_2^2 - \langle Z, \phi + \vec{w}_\sigma \cdot \nabla g - d\rangle + \frac{\beta}{2}\|\phi + \vec{w}_\sigma \cdot \nabla g - d\|_2^2 \tag{7}$$

This leads us to the iteration scheme

$$\phi^{k+1} = \underset{\phi}{\operatorname{argmin}} E(\phi, g^k, Z^k; \vec{w}_\sigma, f) \tag{8}$$

$$g^{k+1} = \underset{g}{\operatorname{argmin}} E(\phi^{k+1}, g, Z^k; \vec{w}_\sigma, f) \tag{9}$$

$$Z^{k+1} = Z^k - \beta(\phi^{k+1} + \vec{w}_\sigma \cdot \nabla g^{k+1} - d). \tag{10}$$

The subproblem (8), with expansion and completing the square, becomes

$$\phi^{k+1} = \underset{\phi}{\operatorname{argmin}} \int_\Omega (\vec{w}_\sigma \cdot \nabla\phi)^2 + \frac{\beta}{2}\left(\frac{1}{\beta}Z^k - (\phi + \vec{w}_\sigma \cdot \nabla g^k - d)\right)^2 \mathrm{d}x, \tag{11}$$

using Euler-Lagrange equation leads to the following PDE:

$$2\nabla \cdot ((\vec{w}_\sigma \cdot \nabla\phi)\vec{w}_\sigma) + \beta\phi = \beta\left(\frac{1}{\beta}Z^k - (\vec{w}_\sigma \cdot \nabla g^k - d)\right). \tag{12}$$

The subproblem (9) can be explicitly written as:

$$g^{k+1} = \underset{g}{\operatorname{argmin}} \int_\Omega \lambda(fg)^2 + \frac{\beta}{2}\left(\frac{1}{\beta}Z^k - (\phi + \vec{w}_\sigma \cdot \nabla g^k - d)\right)^2 \mathrm{d}x. \tag{13}$$

Applying Euler-Lagrange equation again gives us:

$$\beta\nabla \cdot ((\vec{w}_\sigma \cdot \nabla g)\vec{w}_\sigma) + 2\lambda fg = -\beta\nabla \cdot \left(\left(\frac{1}{\beta}Z^k - (\phi^{k+1} - d)\right)\vec{w}_\sigma\right). \tag{14}$$

In next section, we discuss the details to solve (12) and (14) in discrete setting.

4 Numerical Scheme

Assume the domain Ω is discretized into Cartesian grid of size $h \times w$, denoted by $\mathbb{Z}_{[1,h]\times[1,w]} = \{(x, y) \in \mathbb{Z}^2 \mid 1 \leq x \leq w, 1 \leq y \leq h\}$. We vectorize the noisy distance function d, the smooth function ϕ and the noise function n by vertically stacking columns in their matrices from left to right into one vector, denoted as $\hat{d}$, $\hat{\phi}$ and $\hat{n}$ respectively. This process is denoted as vec$(\cdot)$.

Then, we discretize $\frac{\mathrm{d}}{\mathrm{d}x}$ and $\frac{\mathrm{d}}{\mathrm{d}y}$ as $\mathcal{D}_x = I_{w\times w} \otimes D_h \in \mathbb{R}^{hw\times hw}$ and $\mathcal{D}_y = D_w \otimes I_{h\times h} \in \mathbb{R}^{hw\times hw}$ respectively, where $\otimes$ denotes the Kronecker product, and assuming Neumann boundary condition, $D_k = -1$ if $i = j$ and $i \neq 1$ or k; 1 if

$j = i+1$ and $i \neq 1$ or k; 0 otherwise. Thus, gradient operator and divergence operator are discretized as $\mathcal{D} = \begin{bmatrix} \mathcal{D}_x \\ \mathcal{D}_y \end{bmatrix}$ and $\mathcal{D}^T$ respectively. For vector field $\vec{w}_\sigma = (w_{\sigma,x}, w_{\sigma,y})$, we discretize it into $\vec{w}_\sigma := W = \left[\text{diag}(w_{\sigma,x})\ \text{diag}(w_{\sigma,y})\right] \in \mathbb{R}^{hw \times 2hw}$.

Here diag$(\cdot)$ means a diagonal matrix with the input vector being the diagonal entries in that order. We discretize the auxiliary function by

$$f := F = \text{diag}\left(\text{vec}(f(\mathbb{Z}_{[1,h]\times[1,w]}))\right) \tag{15}$$

where $f(\mathbb{Z}_{[1,h]\times[1,w]}) \in \mathbb{R}^{h\times w}$(i.e. applying f to lattice graph of shape $h \times w$).

4.1 Linear Systems

Using the discretized operator $\mathcal{D}$ and vector field W, the discretized version of the first subproblem (12) becomes

$$(2\mathcal{D}^T W^T W \mathcal{D} + \beta I)\hat{\phi} = \beta\left(\frac{1}{\beta}Z^k - W\mathcal{D}\hat{g}^k + \hat{d}\right). \tag{16}$$

It is clear that the matrice $2\mathcal{D}^T W^T W\mathcal{D} + \beta I$ is sparse and symmetric positive definite. We use methods for positive definite coefficient matrices, for example Conjugate Gradient Descent.

For the second subproblem (14), with discretized F shown in (15), it becomes

$$(\beta\mathcal{D}^T W^T W\mathcal{D} + 2\lambda F)\hat{g} = -\beta\mathcal{D}^T W^T\left(\frac{1}{\beta}Z^k - (\hat{\phi}^{k+1} - \hat{d})\right). \tag{17}$$

Since $\mathcal{D}^T W^T W\mathcal{D}$ is semi-positive definite, and λ is chosen to be small for stronger denoising effect, as we discuss in previous section, $\beta\mathcal{D}^T W^T W\mathcal{D} + 2\lambda F$ may be ill-conditioned. Thus we use Tikhonov regularization: for some $\kappa > 0$, we instead solve the regularized linear system

$$(\beta\mathcal{D}^T W^T W\mathcal{D} + 2\lambda F + \kappa I)\hat{g} = -\beta\mathcal{D}^T W^T\left(\frac{1}{\beta}Z^k - (\hat{\phi}^{k+1} - \hat{d})\right) \tag{18}$$

After solving for $\hat{\phi}^{k+1}$ and $\hat{g}^{k+1}$, we update Z^{k+1} in (10) by

$$Z^{k+1} = Z^k - \beta(\hat{\phi}^{k+1} + W^T\hat{g}^{k+1} - \hat{d}). \tag{19}$$

Our minimisation scheme solves (16), (18) and (19) iteratively, until the convergence condition is met: reaching a maximum iteration number or $\|\hat{\phi}^{k+1} - \hat{\phi}^k\| < \epsilon$ for some small positive ϵ. We set the initial condition to be $\hat{\phi}^0 = \hat{d}$, $\hat{g}^0 = 0$ everywhere, and Z^0 a random matrix from normal distribution with mean 0 and variance 1. We set the initial condition to be $\hat{\phi}^0 = \hat{d}$, $\hat{g}^0 = 0$ everywhere, and Z^0 a random matrix from normal distribution with mean 0 and variance 1.

5 Numerical Results

In this section, we first present (i) our surface reconstruction result and the energy change to validate the algorithm. Secondly, we (ii) compare denoising result with and without using directional information, to illustrate the necessity of using directional G-norm. Next, we (iii) vary the noise level to test our model's denoising capability. We also present (iv) the effect of choosing the parameter of the tangential vector field. We (v) include a preliminary result of applying this method to a 3D point cloud data set.

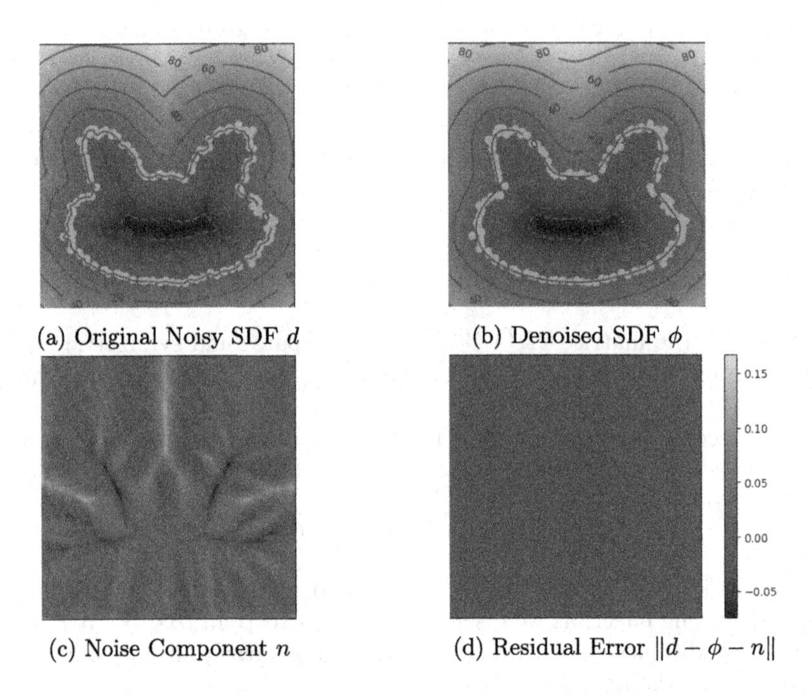

(a) Original Noisy SDF d (b) Denoised SDF ϕ (c) Noise Component n (d) Residual Error $\|d - \phi - n\|$

Fig. 2. Decomposition Result of a Rabbit's Head. Notice that (b) shows the zero level contour captures the shape well and smoothly. (d) shows the residual error is small.

1D Surface Reconstruction: Figure 2 shows our reconstruction result from noisy point cloud data sampled from a rabbit's head. Figure 2 (a) and (b) show the noisy SDF d and the denoised result ϕ, superposed with contour plots, and Fig. 2 (c) and (d) show the noise component n and residual error $d-\phi-n$. In (d), the average residual error $\mathbb{E}[\|d - \phi - n\|]$ is 7.04×10^{-5}. The zero level set of the denoised SDF is smooth and able to capture the underlying shape. In (c), it shows how noise perturbs the SDF; oscillatory components is added perpendicular to the surface, with particularly large magnitude at the corners. The computation of SDF enforces $|\nabla d| = 1$ almost everywhere causing discontinuity in first order derivative of d along lines passing through corners. The first term in (3) enforces the first order smoothness along $\vec{w}_\sigma$, and, with the constraint $d = \phi + n$, n captures larger oscillations around the corners.

Figure 3 shows the change of each term $||\vec{w}_\sigma \cdot \nabla\phi||$, $||n||_{*,f}$ and the total energy of the ALM equation (7) for the rabbit example in Fig. 2. As expected, the norm $||\vec{w}_\sigma \cdot \nabla\phi||$ and the total energy decrease until they converge. The norm $||n||_{*,f}$ increases during the iteration, as it capture more oscillation by the directional G-norm, i.e. higher variance noise, thus the norm value increases.

With vs Without Directional Denoising: We compare the result using G-norm, with and without directional denoising. Figure 4 shows an experiment of using the proposed directional G-norm versus denoising in all direction [10,17], with the same set of parameters λ, β, κ and the convergence conditions. This replaces the first term in equation (7) with $||\nabla\phi||$, and $n = \vec{w}_\sigma \cdot \nabla g$ with $\nabla \cdot \nabla g$ in Eq. 6. In Fig. 4c (c), it shows considering noise in all directions smooths the surface and behaves poorly at maintaining sharp corners. In addition, Euler Lagrange Equation leads to a fourth order PDE when solving g-subproblem, of which the coefficient matrix is more ill-conditioned. This results in a larger residual error and longer computation time. The mean error and computation

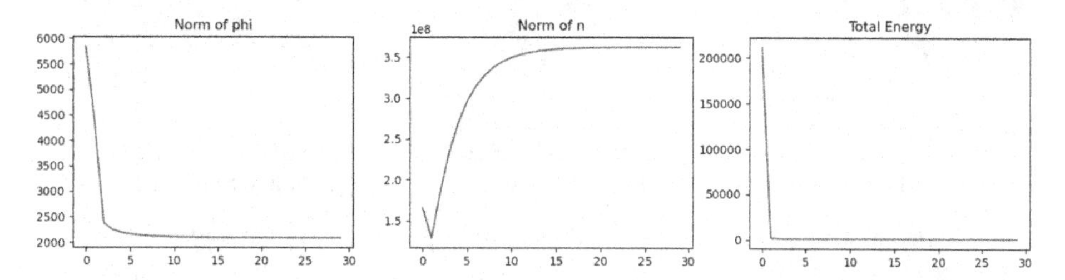

Fig. 3. Change of norms and energy for rabbit's head example. From left to right: the first figure shows the norm $||\vec{w}_\sigma \cdot \nabla\phi||$; the second shows the norm $||n||_{*,f}$; and the last shows the total energy calculated by Equation (7).

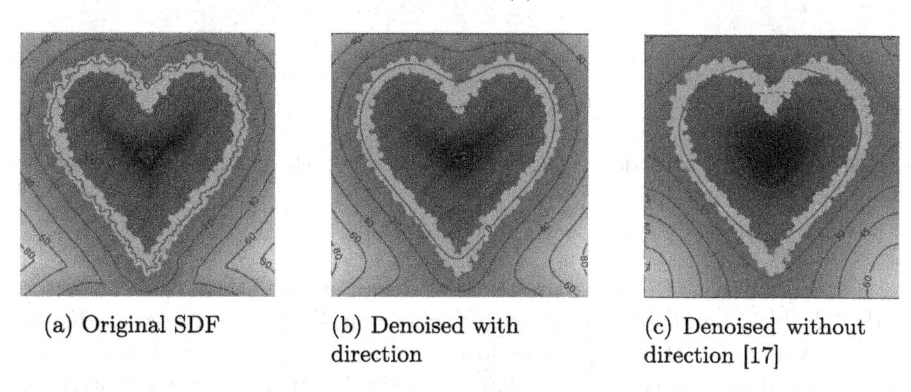

(a) Original SDF

(b) Denoised with direction

(c) Denoised without direction [17]

Fig. 4. Comparison between using and not using $\vec{w}_\sigma$ to denoise. (c) shows that without using directional denoising, the corners are more rounded.

time of using the proposed directional G-norm are 1.89×10^{-3} and 72 seconds; without a specific direction, they surge to 3.18×10^{-2} and 375 seconds.

Stable against high level of noise: We test our algorithm's performance on different noise level, and results are presented in Fig. 5. We first sample points $\mathcal{P}$ from a clean heart shape, then added noise: for each $p \in \mathcal{P}$, $p \leftarrow p + v$, where v is a random unit vector with magnitude sampled from $\mathcal{N}(0, \sigma^2)$. In Fig. 5, from left to right, the noise levels are increasing, $\sigma = 1, 3, 5$ respectively. The proposed method reconstructs the surface well.

Relation of $\vec{w}$ and λ: We explore the influence of the choice of the vector field $\vec{w}$ to the behavior of the weight λ in (3). In Fig. 6, denote the initial surface to be Γ_0, the corresponding SDF to be ϕ_0, and the denoised surface via minimizing (3) to be $\Gamma_\sigma[\lambda]$. Here σ is the convolution variance, and the parameter λ on the G-norm.

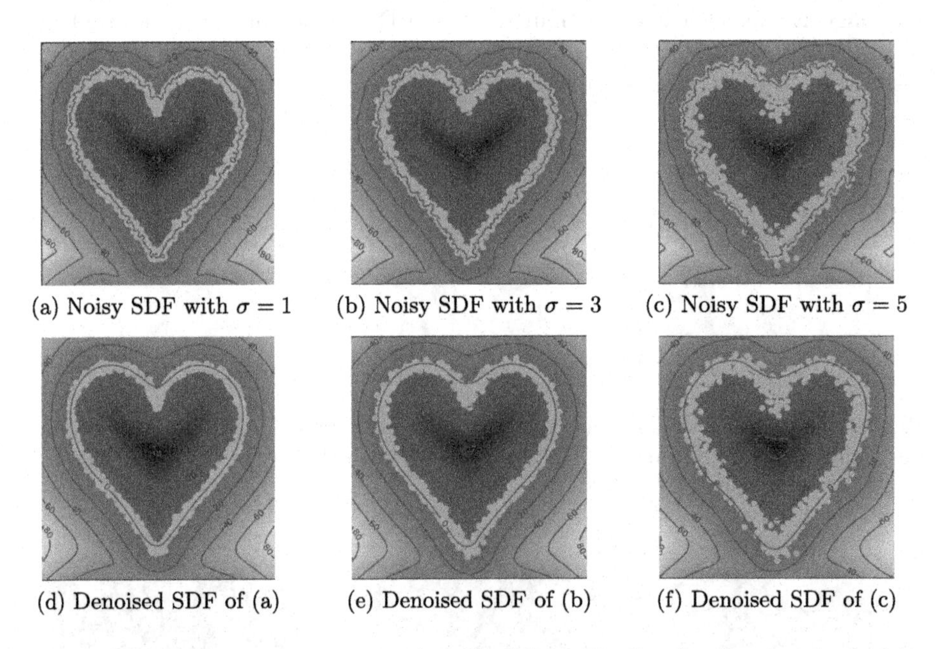

(a) Noisy SDF with $\sigma = 1$ (b) Noisy SDF with $\sigma = 3$ (c) Noisy SDF with $\sigma = 5$

(d) Denoised SDF of (a) (e) Denoised SDF of (b) (f) Denoised SDF of (c)

Fig. 5. Surface reconstruction example with different levels of noise. Our method is robust against different levels of noise.

This experiment reveals two properties of our model. First, the tangent vector field of $\Gamma_\sigma[0]$ is parallel to $\vec{w}_\sigma|_{\Gamma_\sigma[0]}$ everywhere on $\Gamma_\sigma[0]$. And as we gradually increase λ in $\Gamma_\sigma[\lambda]$, the surface should converge to Γ_0, the original surface. In Fig. 6 (a) and (b), we choose $\sigma_1 = 0.6$ and $\sigma_2 = 0.2$ to create two vector fields $\vec{w}_{\sigma_1}$ and $\vec{w}_{\sigma_2}$ respectively, and apply our model with varying value of λ. In both Fig. 6 (a) and (b), the denoised surface $\Gamma_{\sigma_i}[\lambda]$ converges to $\Gamma_{\sigma_i}[0]$ as $\lambda \to 0$. In our model, $\vec{w}_\sigma$ serves as a guideline for transforming Γ_0 to the target surface

$\Gamma_\sigma[0]$ of which the tangent vector field is parallel to $\vec{w}_\sigma$. Second, from Γ_0 to $\Gamma_\sigma[0]$, λ controls how much details are kept in $\phi_\sigma[\lambda]$ through the decomposition. Figure 6 (c) (d) shows zoom-in detail of how small scale oscillation is gradually removed from Γ_0 respecting $\Gamma_\sigma[0]$, since $\|\cdot\|_{*,f}$ is defined with respect to $\vec{w}_\sigma$.

Surface Reconstruction: We generalize the proposed method for 3D surface reconstruction. On a two-dimensional manifold $\mathcal{M}$, we construct a frame field $\mathcal{W} = (\vec{w}_N, \vec{w}_1, \vec{w}_2)$ where $\vec{w}_N$ approximates outward surface normal, and $\vec{w}_1, \vec{w}_2$ approximate the first and the second principal direction on the tangential plane of $\mathcal{M}$ [1]. We model noise as oscillation on the tangent plane $T_\mathcal{M}$, and extend ϕ's smoothness constraint onto the surface $\|\nabla_{T_\mathcal{M}}\phi\|_2$. We reformulate the directional g-norm (5) in $\mathbb{R}^3$ as:

$$\|n\|_* = \inf_{g_1,g_2\in(\Omega;\mathbb{R})} \left\{ \| \sqrt{\|g_1\|_2^2 + \|g_2\|_2^2} \mid n = \nabla_{\vec{w}_1} g_1 + \nabla_{\vec{w}_2} g_2 \right\}. \tag{20}$$

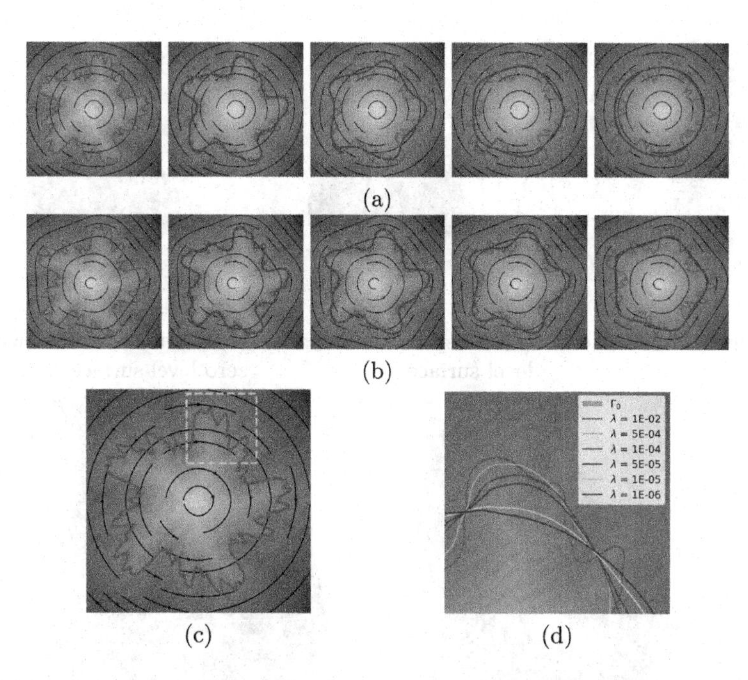

Fig. 6. (a) Original surface Γ_0 (red) and denoised surfaces $\Gamma_{\sigma_1}[\lambda]$ (blue), with λ gradually decreasing, the vector field $\vec{w}_{\sigma_1}$ (black) with $\sigma_1 = 0.6$, and the background represents the corresponding SDF $\phi_{\sigma_1}[\lambda]$. (b) The same set of experiment as (a) except with $\sigma_2 = 0.2$. Notice the blue curve keeps the general shape more. (c) Enlarged version of (a), with yellow dotted square indicating zoom-in area. (d) Zoom-in of (a), with multiple lines indicating surfaces with different λ. (Color figure online)

The energy (3) becomes:

$$E(\phi, n; \mathcal{W}) = \frac{1}{2}\|\nabla_{T_\mathcal{M}}\phi\|_2^2 + \frac{\lambda}{2}\|n\|_*^2 \quad \text{subject to } d = \phi + n. \tag{21}$$

Here ϕ represents the smoothed version with respect to the oscillation on its tangent planes. To implement ALM on the 3D model, we introduce the discretized version of $\nabla_{T_{\mathcal{M}}}$. Discretize the 3D domain Ω into Cartesian grid of size $h \times w \times l$. Denote $W_N, W_1, W_2 \in \mathbb{R}^{hwl \times 3hwl}$ as the discretized version of $\vec{w}_N, \vec{w}_1, \vec{w}_2$ as in section 4, define $P = I - {W_N}^T W_N \in \mathbb{R}^{3hwl \times 3hwl}$, the projection operator onto the surfaces orthogonal to $\vec{w}_N$. Let $\mathcal{D} = \begin{bmatrix}\mathcal{D}_x^T & \mathcal{D}_y^T & \mathcal{D}_z^T\end{bmatrix}^T \in \mathbb{R}^{3hwl \times hwl}$ be the 3-D differential operator, the discretized version of the energy (21) becomes:

$$\begin{aligned}\mathcal{E}(\hat{\phi}, \hat{g_1}, \hat{g_2}, Z) = &\frac{1}{2}\|P\mathcal{D}\hat{\phi}\|_2^2 + \frac{\lambda}{2}(\|\hat{g}_1\|^2 + \|\hat{g}_2\|^2) \\ &- \langle Z, \hat{\phi} + W_1\mathcal{D}\hat{g}_1 + W_2\mathcal{D}\hat{g}_2 - \hat{d}\rangle + \frac{\beta}{2}\|\hat{\phi} + W_1\mathcal{D}\hat{g}_1 + W_2\mathcal{D}\hat{g}_2 - \hat{d}\|_2^2.\end{aligned} \tag{22}$$

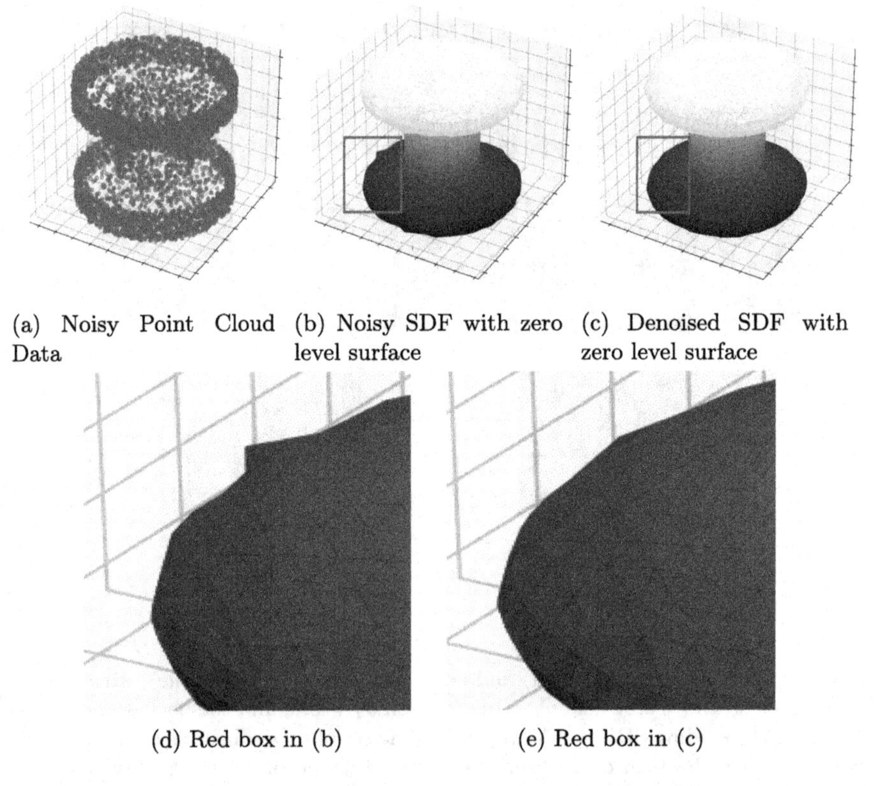

(a) Noisy Point Cloud Data (b) Noisy SDF with zero level surface (c) Denoised SDF with zero level surface

(d) Red box in (b) (e) Red box in (c)

Fig. 7. Denoising Experiment with synthetic dumb bell surface. (a) point cloud sampled from a dumbbell surface with noise, (b) surface corresponding to the noisy SDF, (c) denoised SDF, (d)(e) are zoom-in of (b)(c).

We use similar numerical scheme as in Sect. 4. Figure 7 shows reconstruction result on a synthetic dumbbell point cloud. (a) and (b) show the noisy given point cloud and its SDF. Zoom-ed images in (d) and (e) show that oscillation in the given surface is denoised in (e). Figure 8 shows reconstruction results of Standford bunny and armadillo. With the directional G-norm formulation, the reconstruction in (c) shows smoother SDF. The parameters λ, β control the scale of the oscillation, which enable our method to provide a smooth surface reconstruction, while preserving the geometric details of the object.

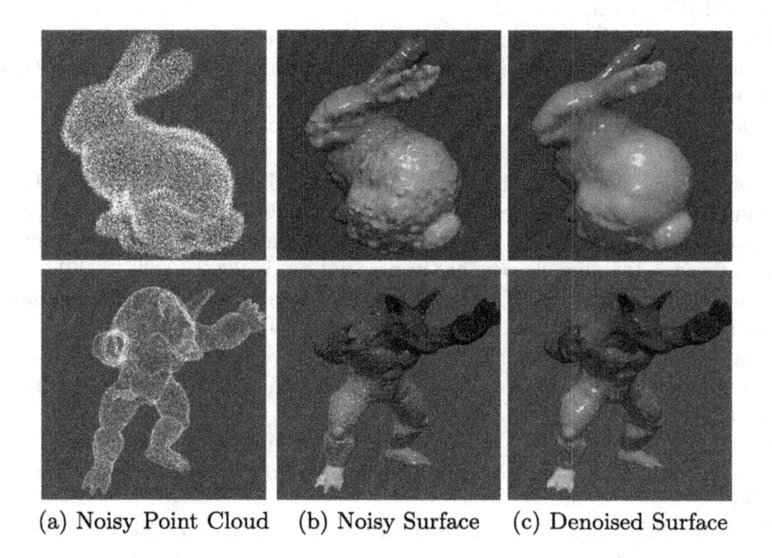

(a) Noisy Point Cloud (b) Noisy Surface (c) Denoised Surface

Fig. 8. First row: Surface reconstruction result of Stanford bunny on a $128 \times 128 \times 128$ grid, we used $\sigma = 12$ for Gaussian kernel, with $\lambda = 1, \beta = 1$ for regularity terms; second row: armadillo on a $256 \times 256 \times 256$ grid, with $\sigma = 8$, $\lambda = 1, \beta = 10$.

6 Conclusion

We present a method using directional G-norm to denoise the SDF for surface reconstruction of noisy point cloud. Our model respect tangential direction provided by the vector field $\vec{w}_\sigma$, and only take out oscillations from the perpendicular direction. It provides the potential for preserving corners, and for better understanding of surface texture; An efficient numerical scheme (ALM) is developed for our model. In our experiment setting, our scheme usually converges in a few iterations; Our framework is flexible in the sense that other energies that are commonly used in image processing can be easily migrated into our formulation.

References

1. Albin, E., Knikker, R., Xin, S., Paschereit, C.O., D'Angelo, Y.: Computational assessment of curvatures and principal directions of implicit surfaces from 3d scalar data. In: Floater, M., Lyche, T., Mazure, M.-L., Mørken, K., Schumaker, L.L. (eds.) MMCS 2016. LNCS, vol. 10521, pp. 1–22. Springer, Cham (2017). https://doi.org/10.1007/978-3-319-67885-6_1

2. Amenta, N., Bern, M., Kamvysselis, M.: A new voronoi-based surface reconstruction algorithm. In: Proceedings of the 25th Annual Conference on Computer Graphics and Interactive Techniques, pp. 415–421 (1998)
3. Berger, M., et al.: A survey of surface reconstruction from point clouds. In: Computer Graphics Forum, vol. 36, pp. 301–329. Wiley Online Library (2017)
4. Bernardini, F., Mittleman, J., Rushmeier, H., Silva, C., Taubin, G.: The ball-pivoting algorithm for surface reconstruction. IEEE Trans. Visual Comput. Graph. **5**(4), 349–359 (1999)
5. Carr, J.C., et al.: Reconstruction and representation of 3d objects with radial basis functions. In: Proceedings of the 28th Annual Conference on Computer Graphics and Interactive Techniques, pp. 67–76 (2001)
6. Caselles, V., Kimmel, R., Sapiro, G.: Geodesic active contours. Int. J. Comput. Vision **22**(1), 61–79 (1997)
7. Cheng, Z.Q., Wang, Y., Li, B., Xu, K., Dang, G., Jin, S.: A survey of methods for moving least squares surfaces. In: VG/PBG@ SIGGRAPH, pp. 9–23 (2008)
8. Gropp, A., Yariv, L., Haim, N., Atzmon, M., Lipman, Y.: Implicit geometric regularization for learning shapes. arXiv preprint arXiv:2002.10099 (2020)
9. He, Y., Kang, S.H., Liu, H.: Curvature regularized surface reconstruction from point clouds. SIAM J. Imag. Sci. **13**(4), 1834–1859 (2020)
10. Huska, M., Kang, S.H., Lanza, A., Morigi, S.: A variational approach to additive image decomposition into structure, harmonic, and oscillatory components. SIAM J. Imag. Sci. **14**(4), 1749–1789 (2021)
11. Kazhdan, M., Bolitho, M., Hoppe, H.: Poisson surface reconstruction. In: Proceedings of the Fourth Eurographics Symposium on Geometry Processing, vol. 7 (2006)
12. Kolluri, R., Shewchuk, J.R., O'Brien, J.F.: Spectral surface reconstruction from noisy point clouds. In: Proceedings of the 2004 Eurographics/ACM SIGGRAPH Symposium on Geometry Processing, pp. 11–21 (2004)
13. Lai, R., Tai, X.C., Chan, T.F.: A ridge and corner preserving model for surface restoration. SIAM J. Sci. Comput. **35**(2), A675–A695 (2013)
14. Liang, J., Park, F., Zhao, H.: Robust and efficient implicit surface reconstruction for point clouds based on convexified image segmentation. J. Sci. Comput. **54**(2), 577–602 (2013)
15. Meyer, Y.: Oscillating Patterns in Image Processing and Nonlinear Evolution Equations: The Fifteenth Dean Jacqueline B. Lewis Memorial Lectures. American Mathematical Society, USA (2001)
16. Osher, S., Sethian, J.A.: Fronts propagating with curvature-dependent speed: algorithms based on Hamilton-Jacobi formulations. J. Comput. Phys. **79**(1), 12–49 (1988)
17. Osher, S., Solé, A., Vese, L.: Image decomposition and restoration using total variation minimization and the h. Multiscale Model. Simul. **1**(3), 349–370 (2003)
18. Park, J.J., Florence, P., Straub, J., Newcombe, R., Lovegrove, S.: DeepSDF: learning continuous signed distance functions for shape representation. In: Proceedings of the IEEE/CVF Conference on Computer Vision and Pattern Recognition, pp. 165–174 (2019)
19. Zhao, H.K., Osher, S., Fedkiw, R.: Fast surface reconstruction using the level set method. In: Proceedings IEEE Workshop on Variational and Level Set Methods in Computer Vision, pp. 194–201. IEEE (2001)
20. Zhao, H.: A fast sweeping method for Eikonal equations. Math. Comput. **74**, 603–627 (2004)

Regularized Material Decomposition for K-edge Separation in Hyperspectral Computed Tomography

Francesca Bevilacqua[1(✉)], Yiqiu Dong[2], and Jakob Sauer Jørgensen[2]

[1] Department of Mathematics, University of Bologna, Bologna, Italy
francesca.bevilacqu8@unibo.it
[2] Department of Applied Mathematics and Computer Science, Technical University of Denmark, Kgs. Lyngby, Denmark
{yido,jakj}@dtu.dk

Abstract. Hyperspectral computed tomography is a developing technique that exploits the property of materials to attenuate X-rays in different quantities depending on the specific energy. It allows to not only reconstruct the object, but also to estimate the concentration of the materials which compose it. The objective of the present study is to obtain an accurate material decomposition from noisy few-projection data. A preliminary comparative study of reconstruction methods based on material decomposition is performed, employing a phantom composed of materials with similar attenuation profiles with characteristic K-edges separated by only 2, 4 and 6 keV. It is found that a one-stage method encompassing both material decomposition and tomographic reconstruction in a single variational model performs better than a more conventional two-stage approach. It is further found that better modelling of noise through use of a weighted least-squares data fidelity improves reconstruction and material separation, as does the use total variation and L1-norm regularization.

Keywords: Hyperspectral CT · Material Decomposition · Variational Methods · K-edge Imaging

1 Introduction

Spectral computed tomography (CT) is an evolving technique which exploits the dependence of the attenuation coefficients of the materials with respect to the used energy. Compared to conventional CT, spectral CT employs a photon-counting detector that records the energy of individual photons. In particular, hyperspectral CT is a special case of spectral CT, in which we have many (e.g. more than five) energy channels, and fine energy resolution, ca 1 keV. This means we can model it as a linear process within each energy bin and work with a fine grid of discrete energy dependent data. In this way it is easier to distinguish

L. Calatroni et al. (Eds.): SSVM 2023, LNCS 14009, pp. 107–119, 2023.
https://doi.org/10.1007/978-3-031-31975-4_9

materials that have similar attenuation coefficients in an energy range, but different in others. Using the energy dependent data, the material decomposition process allows to identify, quantify and differentiate the presence of the several materials inside the reconstructed object. Extensive work on material decomposition is present in literature, however mostly focused on medical imaging, has few energy bins and few materials, and typically material separation is not based on K-edges or at most one material has a K-edge.

Different strategies to reconstruct material-specific images from the photon counts can be divided into two main categories: two-stage and one-stage methods. The two-stage methods separate the process of material decomposition and image reconstruction. One way is called image-based methods, which first reconstruct the images according to each energy bin and then perform material decomposition from reconstructed images, e.g. the work in [1,2]. The other way is called projection-based methods, which first decompose the multi-energy projections to each material, then perform image reconstruction independently. [14,15]. The one-stage methods basically formulate these two steps as one problem and solve it directly. They take the photon counts as the input and directly reconstruct the set of material volumes, [4,17,18].

When we apply variational methods to reconstruct material-specific images, the choice of data-fidelity term and regularization terms is important. Several regularization techniques have been applied to hyperspectral CT. For example, regularization by denoising [3], total variation (TV) [4,6,17], Non Local Total Variation (NLTV) [18], and the point-wise separation regularizer for dual energy CT [5], etc.

In this paper, we focus on material decomposition for distinguishing materials that have high atomic number with similar attenuation coefficients and K-edges in the considered energy range. Further, we consider the case with a small number of projections in order to keep the scan time reasonable. Photon counting detectors, especially with high energy resolution, are slow and have limits on flux they can handle. Therefore, a small number of projections ensure that the scan time is acceptable.

Here we perform some preliminary comparative study with simulated data to establish which methods are more promising. It would set the stage for further study. Specifically, we seek to address three questions.

(1) Is it worthwhile employing the more involved one-stage method or is the simpler two-stage method sufficient?
(2) How to model the noise, i.e. which data-fidelity term to employ?
(3) Which regularizer to use for sparse materials: TV or L1?

The paper is structured as follows. In Sect. 2 the mathematical model of hyperspectral CT is presented. Based on one-stage strategy, Sect. 3 summarize several choices on data-fidelity terms and regularization terms that are used in numerical tests. In Sect. 4 the setup of the experiments is explained, while 5 shows the numerical results. Finally, Sect. 6 concludes the paper.

2 Hyperspectral Computed Tomography

Assume that the attenuation coefficient $\hat{\mu}_{ej}$ of the object at the pixel j and the energy index e can be written as

$$\hat{\mu}_{ej} = \sum_{m=1}^{N_m} \mu_{em} x_{mj}, \quad e \in \{1, \dots, N_e\}, \quad j \in \{1, \dots, t\} \tag{1}$$

where N_m is the number of the materials, N_e is the number of energy channels, t the pixels in the object, μ_{em} denotes the mass-attenuation coefficient of the material m at the energy index e, and x_{mj} denotes the concentration of the material m at the pixel j. The attenuation profiles $\{\mu_{em}\}$ are known, and the final aim is to find the concentration maps for all the materials, i.e., $\{x_{mj}\}$.

Based on the Beer's Law [11], the X-ray intensity before and after illumination through the object follows

$$I_{ie} = S_{ie} \exp\left(-\sum_{j=1}^{t} a_{ij} \hat{\mu}_{ej}\right) = S_{ie} \exp\left(-\sum_{j=1}^{t} a_{ij} \left(\sum_{m=1}^{N_m} \mu_{em} x_{mj}\right)\right), \tag{2}$$

where I_{ie} denotes the expected photon count at the detector pixel i for the energy level e, S_{ie} is the incoming source intensity, i.e. the number of photons for the energy index e that are expected to be detected in pixel i of the detector if there was no attenuating object, and a_{ij} represents the projection coefficient for the detector pixel i and the object pixel j. The corresponding matrix formulation is

$$I = S\, e^{-A\hat{M}^T} = S\, e^{-A(M\,X)^T} = S\, e^{-A\,X^T M^T} \tag{3}$$

where $I = (I_{ie}), S = (S_{ie}) \in \mathbb{R}_+^{d \times N_e}$, $M = (\mu_{em}) \in \mathbb{R}_+^{s \times N_m}$, $\hat{M} = (\hat{\mu}_{em}) \in \mathbb{R}_+^{N_e \times t}$, $X = (x_{mj}) \in \mathbb{R}_+^{N_m \times t}$, $A = (a_{ij}) \in \mathbb{R}_+^{d \times t}$, d is the number of the detector pixels and the product between S and $e^{-A\hat{M}^T}$ is considered point-wise.

Due to the photon-counting system in the detector, the data is affected by the Poisson noise, then the forward model reads

$$Y_{ie} \sim \text{Poisson}\,(I_{ie}) \quad \text{with } I = S\, e^{-A\,X^T M^T}, \tag{4}$$

where $Y = (Y_{ie}) \in \mathbb{R}_+^{d \times N_e}$.

3 One-Stage Methods

For one-stage methods, the reconstruction can be obtained by solving an optimization problem in the form

$$\min_{X \in [0,1]^{N_m \times t}} F_Y(X) + \lambda R(X), \tag{5}$$

where $F_Y(X)$ is the data-fidelity term that is based on the forward model and the noise distribution, $R(X)$ is the regularization term that promotes certain properties on the reconstruction, and $\lambda > 0$ is the regularization parameter that balances the effect of $F_Y(X)$ and $R(X)$. In contrast to [5], we allow the mix of multiple materials at the same pixel, and the element in X represents the percentage of a certain material at a certain pixel. Therefore, we introduce the constraint that all elements in X are bounded in $[0, 1]$. In principle one could also enforce that within each pixel these should sum to 1 but we do not exploit that here.

The data-fidelity term, $F_Y(X)$, takes into account the noise statistics that affects the data. In the case of Poisson noise, based on the Maximum Likelihood approach, the data-fidelity term becomes the generalized Kullback-Leibler (KL) divergence between $S\,e^{-AX^TM^T}$ and Y [12,13]. For computational simplicity, we can use the weighted least square (WLS) to approximate the KL divergence [10]. By first introducing $V = (v_{ie}) \in \mathbb{R}^{d\times s}$ with $v_{ie} = -\log(Y_{ie}/S_e)$, and the weight matrix $W = (w_{ie}) \in \mathbb{R}^{d\times s}$ with $w_{ie} = Y_{ie}/S_e$, WLS can be expressed as

$$F_V(X) = \|A(\tilde{M}X)^T - V\|_W^2, \tag{6}$$

where $\|\cdot\|_W^2 = \sum_{i,e} w_{ie} \cdot_{ie}^2$. In our numerical experiments, we compare the reconstruction results by using WLS with the ones by using the most common choice, least square (LS) data-fidelity term, i.e.,

$$F_V(X) = \|A(MX)^T - V\|^2, \tag{7}$$

where $\|\cdot\|$ denotes the Frobenius norm.

The choice of the regularization term, $R(X)$, depends on the prior information on X. We will introduce different regularization to different materials depending on their properties and consider the sum of them, each with a specific regularization parameter. In particular, we will consider L1 regularization for materials that are present with spikes and total variation (TV) regularization for the piece-wise constant materials. The TV and L1 regularization term on the material m, i.e., $X_m = (x_{m1}, \cdots, x_{mt})^T \in \mathbb{R}^t$, are defined respectively as

$$\begin{aligned} \mathrm{TV}_m(X) &= \|\nabla X_m\|_{2,1}, \\ \mathrm{L1}_m(X) &= \|X_m\|_1, \end{aligned} \tag{8}$$

where $\|\cdot\|_{2,1}$ on the gradient of X_m gives the 2D isotropic TV of X_m.

4 Experimental Setup

In this section we explain the setup for the material decomposition experiments for hyperspectral CT that will be carried out. With the focus on materials with high atomic number and similar attenuation coefficients, we consider an object X with five different materials: Yb, Lu, Ta, Os and H_2O. In particular, we choose the first four materials due to their attenuation profiles, which are approximately all in the same range and the most important is that they are characterized by a jump (K-edge) 2, 4, and 6 keV apart from each other, see Fig. 1 (*Left*).

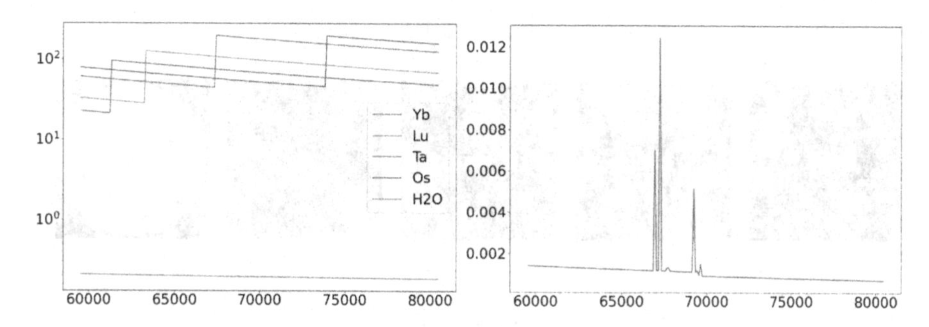

Fig. 1. *Left.* Attenuation profiles of the five materials obtained by multiplying the mass attenuation coefficients with the density of each material. *Right.* Normalized source spectrum. The spectrum is shown only in the considered energies, but it is normalized over all the energy range (from 1 to 121 KeV). The x-axes represent the energy level in eV.

As introduced in the previous section we consider the possibility that more than one material can be present in a pixel of the object. For this reason, we constructed phantoms in which some materials are present both alone and mixed with others. Furthermore, we assume that all materials are surrounded by water. Figure 2 shows two phantoms, i.e. the concentration maps of the materials Yb, Lu, Ta, Os and H_2O respectively. Note that in phantom 2 the third material is present in very small quantities. The phantoms in Fig. 2 as well as all reconstructions in the following sections are shown in the range $[0, 1]$. To simplify, the materials Yb, Lu, Ta, Os and H_2O will be addressed as material 1,2,3,4 and 5, respectively.

We focus here on an X-ray energy range surrounding the K-edges i.e. from 60 keV to 80 keV. In reality, during the CT acquisition, all the energies in the considered spectrum contribute to the process; but the spectral CT detector is only able to distinguish photons with a limited energy resolution, for example 1 keV. In order to mimic this process, and especially to avoid the inverse crime, we simulate data on a finer grid of energies, followed by binning (summing) data into 1 keV energy bins. Specifically, we considered a fine energy grid between 59550 and 80450 eV with a width of 0.1 keV. Figure 1 (*Right*) shows the normalized source spectrum for the fine grid of considered energies. After applying the Beer's Law and the noise (as in (2)) the photons are binned into 21 energy bins centered around 60, 61, ..., 79 and 80 KeV.

The data was generated considering a phantom with $t = 256^2$ pixels (pixel size 4.6875^2 μm^2) and a fan-beam setup with 45 projections in the range $[0, 2\pi]$, 256 detector pixels (pixel size 7.8125 μm), a distance of 0.5 cm between the source and the detector and 0.3 cm between the source and the center of rotation. The experiments are carried out with few projections (only 45) and a short exposure time, this leads to noisy data with reduced information content. From the data

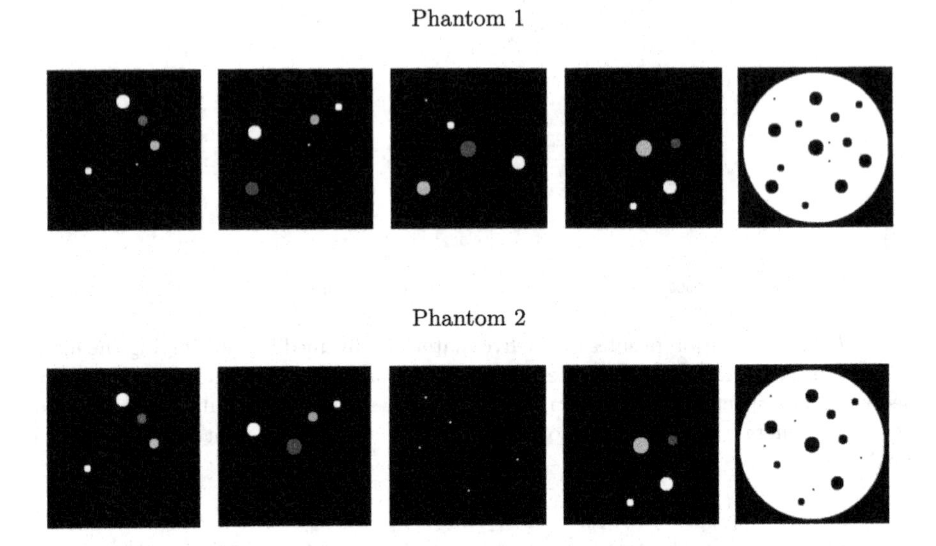

Fig. 2. Considered phantoms: concentration maps of the materials Yb, Lu, Ta, Os and H_2O (from left to right).

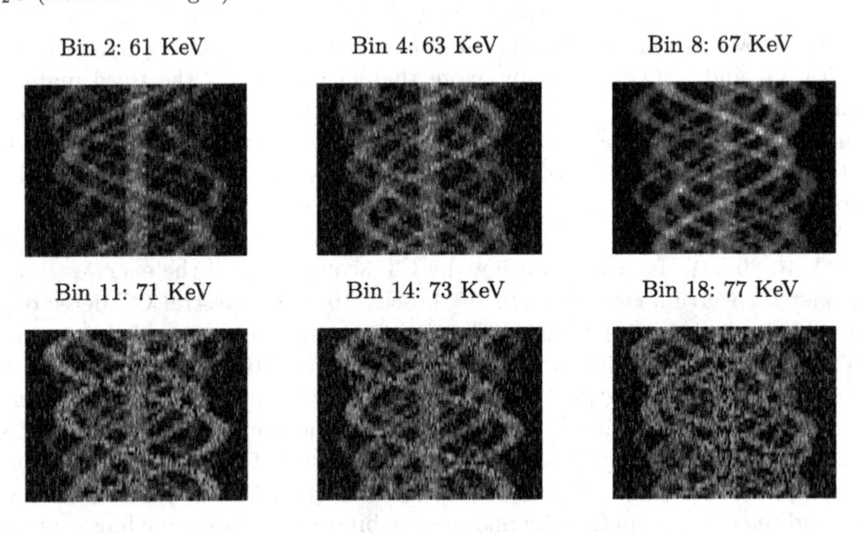

Fig. 3. Sinogram data for some energy bins. Each row of the sinogram corresponds to a projection.

shown in Fig. 3, one can easily note how the different materials and K-edges, together with the noise levels based on the spectrum, lead to different looking sinograms over the energy bins.

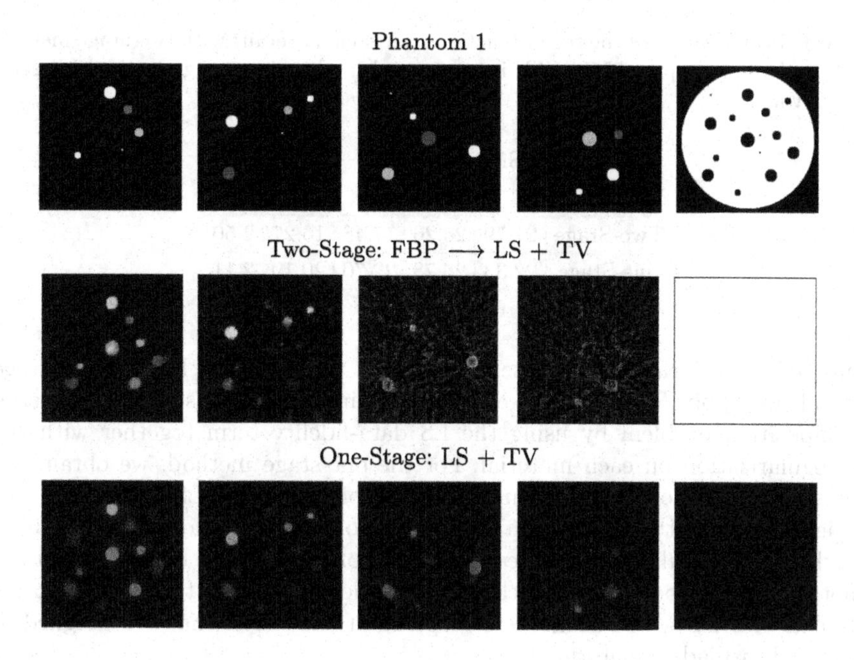

Fig. 4. Reconstruction with two-stage method (using FBP in the first step and LS + TV in the second) and the one-stage method.

Reconstruction quality is addressed qualitatively by visual inspection and quantitatively by PSNR (in dB) on each material [16].

All the experiments are performed using the Core Imaging Library (CIL) [7,8], an open-source Python framework for tomographic imaging that allows us to generate the data and reconstruct it using both standard techniques such as filtered backprojection (FBP) and advanced ones that incorporate some regularizations. To solve the minimization problem in (5) we use the primal dual hybrid gradient (PDHG) algorithm which is available in CIL. In addiction, the SpekPy toolkit [9] is used to calculate and manipulate the x-ray tube spectra, while the XrayDB library provides the attenuation profiles of the materials in Fig. 1.

5 Numerical Results

In this section, we explain and discuss the numerical results obtained in three experiments. Each experiment tries to address one of the questions listed in Sect. 1.

5.1 One-stage Vs Two-Stage Method

In the introduction section we briefly explained the differences between two-stage and one-stage methods. In this experiment we use phantom 1 given in Fig. 1 to

Table 1. PSNR values of the reconstructions (on each material) with two-stage method (FBP in the first step and LS + TV in the second) and the one-stage method (directly LS + TV).

	PSNR (in dB)				
Material	1	2	3	4	5
Two-Stage	21.79	26.26	17.48	16.27	2.50
One-Stage	22.33	26.78	22.70	20.10	2.11

compare the performance of these two types of the methods. For the two-stage method, we apply FBP on each energy data first and then solve the material decomposition problem by using the LS data-fidelity term together with the TV regularization on each material. For the one-stage method, we obtain the reconstruction by solving the minimization problem (5) with LS (7) and TV (8). In Fig. 4 we show the reconstruction results from both methods, while Table 1 lists the PSNR results. One can see that the second step of the two-stage method is not able to compensate the lack of information present in the energy images obtained with FBP, as we can see by looking at the circles that are assigned to material 1 instead of material 4.

5.2 LS Vs WLS

In this experiment we compare the results obtained with the one-stage methods by using LS defined in (7) and WLS given in (6) as data-fidelity terms. LS is the most commonly used data-fidelity term, and it potentially assumes that the noise model is additive Gaussian. WLS is an approximation of the data-fidelity term coming from Poisson noise model. In Fig. 5 we show the phantom and the reconstructions from both LS and WLS together with the TV regularization on each material, while Table 2 contains the PSNR values of the decomposed materials in all the cases. The test is done for two levels of noise and the different data are obtained by halving the incoming source spectrum S. The regularization parameters are chosen manually to give the highest PSNR. It is obvious that WLS gives better reconstructions than LS. In particular, for the lower noise case, the reconstruction from WLS is almost the same as the ground truth and its PSNR values are more than 4 dB higher than from LS (except for material 5). For the higher noise case, LS cannot recognise material 3 and 4 and considers that most of the circles are made by material 1. But WLS, whose PSNR values are significantly higher, detects correctly most of them except that it has a bit difficulty to distinguish material 1 and 4. This might happen because the two materials have similar attenuation coefficients for energies between 62 and 74 KeV. Furthermore, as we can see in Fig. 1, this energy interval corresponds to higher values of the energy spectrum which makes the data obtained from this energy range more powerful than the others. Regarding material 5, with its small attenuation profile compare to the other materials, neither method is able to recognize it and it will therefore need to be further investigated in the future.

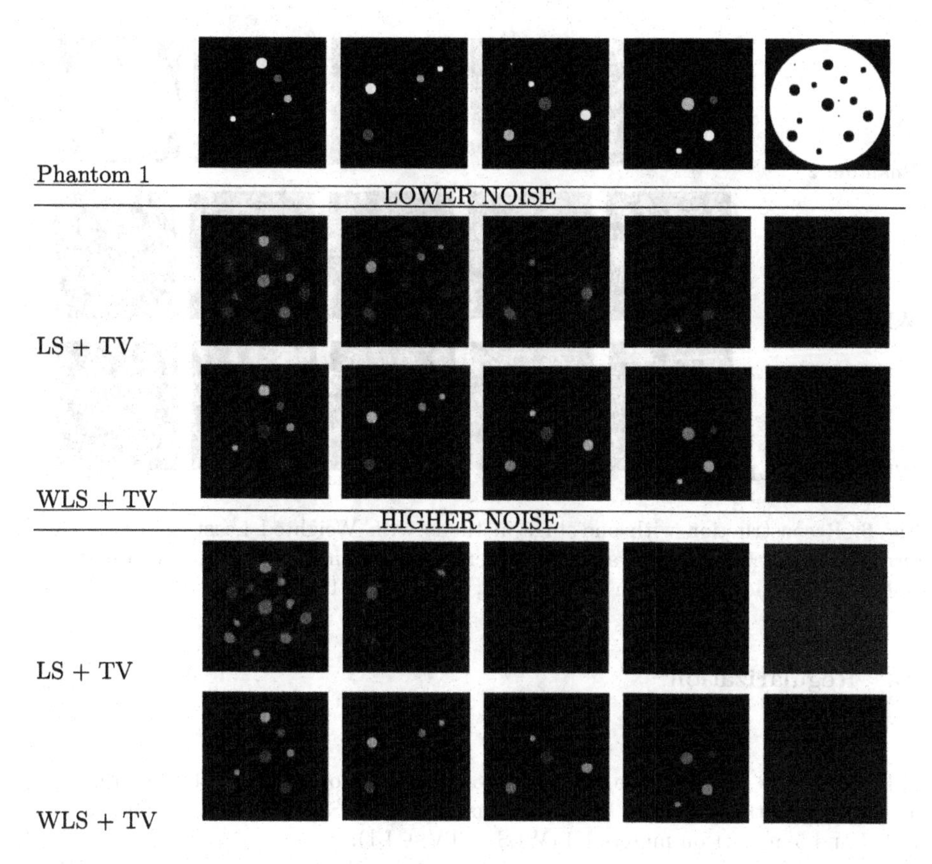

Fig. 5. Reconstruction from one-stage methods with LS and WLS data-fidelity term and TV regularization for two different noise levels.

Table 2. PSNR values of the reconstructions obtained with the TV regularization on each material and two data-fidelity terms: WLS and LS. The table reports the results for two noise levels.

		PSNR (in dB)				
Material		1	2	3	4	5
LOWER NOISE	LS + TV	22.33	26.78	22.70	20.10	2.11
	WLS + TV	28.99	30.03	29.40	27.22	2.32
HIGHER NOISE	LS + TV	21.2039	22.7809	19.6872	19.2476	3.5111
	WLS + TV	25.9380	27.0200	24.9557	23.7285	2.6190

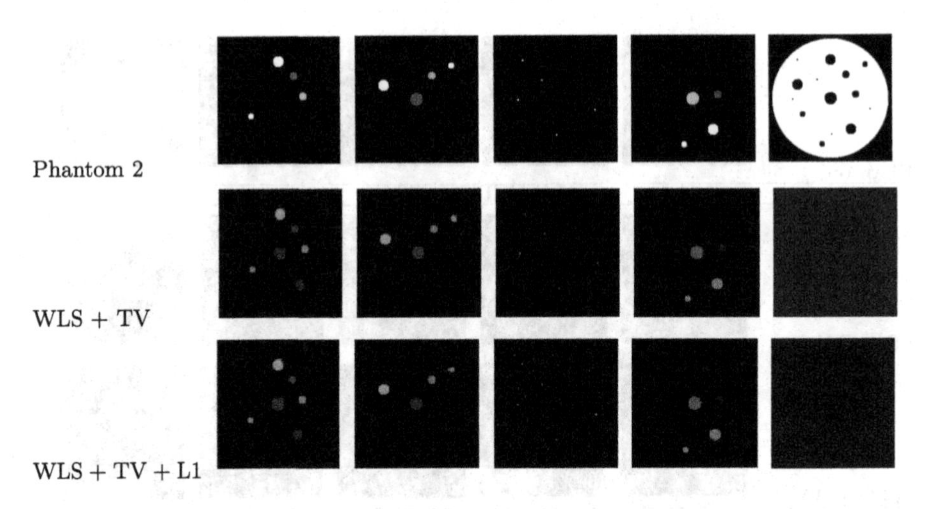

Fig. 6. Reconstruction with one-stage methods with Weighted Least Square fidelity term and different regularizers on the materials: in the first case TV on each material and in the second case TV on material 1,2, and 5 and L1 on material 3.

5.3 Regularization

Table 3. PSNR values of the reconstructions for Phantom 2 with WLS and different regularizers on the materials: TV on each material (WLS + TV) and TV on material 1, 2, 4 and 5 and L1 on material 3 (WLS + TV + L1).

	PSNR (in dB)				
Material	1	2	3	4	5
WLS + TV	26.32	27.20	34.85	23.90	2.56
WLS + TV + L1	26.40	26.48	29.68	23.87	2.23

In previous experiments, TV is used for all materials, since we suppose that the materials are distributed in a piece-wise constant way across the object. In reality, the materials may be located in different manners. For example, in Phantom 2 the third material is present as spikes. In this experiment, we consider a regularizer on each material depending on its properties: TV on material 1,2,4 and 5 and L1 on material 3, and we compare its results with the one using TV on all materials. Both the TV and L1 parameters are set manually to give the highest PSNR; the one chosen for L1 is three orders of magnitude smaller than for TV. Figure 6 shows the visual results, while Table 3 list the PSNR value for each material. We can see that the reconstruction are similar for material 1,2,4 and 5, while, for material 3, the images present some differences. This is also reflected by PSNR that are similar with the exception of the third material. To better

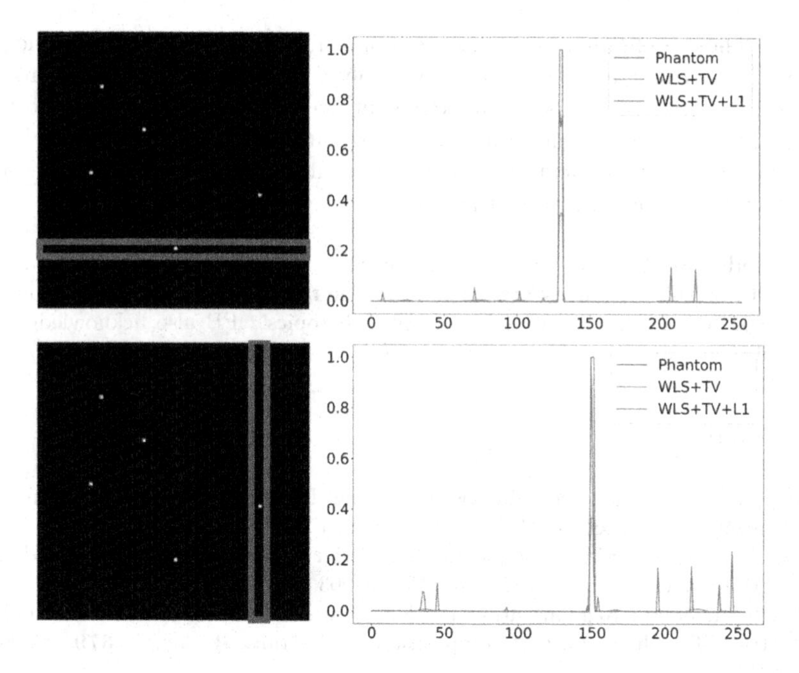

Fig. 7. Line plots of some crucial rows and columns of the reconstructions using WLS+TV+L1 and WLS+TV.

highlight these differences, in Fig. 7 one can observe the line plots of one crucial row and one column of material 3 for the WLS+TV and WLS+TV+L1 reconstructions together with the groundtruth. The results obtained with WLS+TV does not contain noise, but the values on the small dots are equal or less than 0.5 while they should be around 1. On the other hand, WLS+TV+L1 reconstructs the dots with the right values so they are more visible compared to the TV case, but some small spikes appear in the images, making it noisier. These results perfectly reflect the behavior of the two regularizers: TV leads to smoother images with shorter jumps and higher values of PSNR, while the L1 feature of sparsifying the object lowers the PSNR results but identifies correctly the values of the small dots.

6 Conclusion

In the context of hyperspectral CT, we conducted a preliminary study for the material decomposition of materials with high atomic number with K-edges close to each other in the considered energy range. From the comparison between the two-stage and one-stage methods, we choose the latter and performed more test to analyse the different data-fidelity and regularization terms. The weighted least squares data fidelity increases considerably the quality of the reconstruction, while the choice of the regularizer depends on the property of the single material. These preliminary findings are promising, as they lead towards the possibility of performing accelerated energy-resolved imaging with better separation of materials. They set the stage for our continued research, as our future work includes further development of dedicated regularizers to separate materials reliably especially for small features, including an investigation of why materials 1 and 4 blend and how to resolve the issue.

Acknowledgements. This work was supported by the Villum Investigator grant no. 25893 and the Villum Synergy grant no. VIL50096 from The Villum Foundation, and by the ex60 project "Funds for selected research topics". FB also acknowledges the "National Group for Scientific Computation (GNCS-INDAM)".

References

1. Feng, J., et al.: Image-domain based material decomposition by multi-constraint optimization for spectral CT. IEEE Access **8**, 155450–155458 (2020)
2. Tao, S., et al.: Material decomposition with prior knowledge aware iterative denoising (MD-PKAID). Phys. Med. Biol. **63**, 195003 (2018)
3. Perelli, A. et al.: Regularization by denoising sub-sampled Newton method for spectral CT multi-material decomposition. Phil. Trans. R. Soc. A. **379**, 20200191 (2021)
4. Mory, C., et al.: Comparison of five one-step reconstruction algorithms for spectral CT. Phys. Med. Biol. **63**, 235001 (2018)
5. Gondzio, J., et al.: Material-separating regularizer for multi-energy X-ray tomography. Inverse Prob. **38**, 025013 (2022)
6. Hu, Y. et al.: Spectral computed tomography with linearization and preconditioning. SIAM Journal on Scientific Computing. SIAM J. Sci. Comput. **41**(5), S370–S380 (2019)
7. Jørgensen, J.S., et al.: Core imaging library part I: a versatile python framework for tomographic imaging. Phil. Trans. R. Soc. A. **379**, 20200192 (2021)
8. Papoutsellis, E., et al.: Core imaging library - Part II: multichannel reconstruction for dynamic and spectral. Phil. Trans. R. Soc. A. **379**, 20200193 (2021)
9. Poludniowski, G., et al.: Technical note: spekPy v2.0-a software toolkit for modeling x-ray tube spectra. Med. Phys. **48**(7), 3630–3637 (2021)
10. Hansen, P. C., Jørgensen, J. S., Lionheart, B.: Computed tomography: algorithms, insight, and just enough theory. In: SIAM (2021)
11. Buzug, T. M.: Computed Tomography From Photon Statistics to Modern Cone-Beam CT. Springer, Cham (2008). https://doi.org/10.1007/978-3-540-39408-2
12. Resmerita, E., et al.: Joint additive Kullback-Leibler residual minimization and regularization for linear inverse problems. Math. Methods Appl. Sci. **30**(13), 1527–44 (2007)
13. Kullback, S., et al.: On information and sufficiency. Ann. Math. Stat. **22**(1), 79–86 (1951)
14. Alvarez, R.E., et al.: Energy-selective reconstructions in X-ray computerized tomography. Phys. Med. Biol. **21**, 733–44 (1976)
15. Schlomka, J.P., et al.: Experimental feasibility of multi-energy photon-counting K-edge imaging in pre-clinical computed tomography. Phys. Med. Biol. **53**, 4031 (2008)

16. Horé, A., Ziou, D.: Image quality metrics: PSNR vs. SSIM. In: 2010 20th International Conference on Pattern Recognition, pp. 2366–2369, Istanbul, Turkey (2010). https://doi.org/10.1109/ICPR.2010.579
17. Schmidt, T.G., et al.: A spectral CT method to directly estimate basis material maps from experimental photon-counting data. IEEE Trans. Med. Imaging **36**(9), 1808–1819 (2017)
18. Liu, J., et al.: TICMR: total image constrained material reconstruction via nonlocal total variation regularization for spectral CT. IEEE Trans. Med. Imaging **35**(12), 2578–2586 (2016)

Quaternary Image Decomposition with Cross-Correlation-Based Multi-parameter Selection

Laura Girometti, Martin Huska, Alessandro Lanza(✉), and Serena Morigi

Department of Mathematics, University of Bologna, Bologna, Italy
{laura.girometti2,martin.huska,alessandro.lanza2,serena.morigi}@unibo.it

Abstract. We propose a two-stage variational model for the additive decomposition of images into piecewise constant, smooth, textured and white noise components. The challenging separation of noise from texture is successfully achieved by including a normalized whiteness constraint in the model, and the selection of the regularization parameters is performed based on a novel multi-parameter cross-correlation principle. The two resulting minimization problems are efficiently solved by means of the alternating directions method of multipliers. Numerical results show the potentiality of the proposed model for the decomposition of textured images corrupted by several kinds of additive white noises.

Keywords: Variational image decomposition · Whiteness · Auto- and cross-correlation · Automatic parameter selection

1 Introduction

An important problem in image analysis is to separate different features in images. However, when the image is noisy, the decomposition process becomes challenging, especially in the separation of the textural component.

In the last two decades many papers were published on image decomposition, addressing modelling and algorithmic aspects and presenting the use of image decomposition in cartooning, texture separation, denoising, soft shadow/spot light removal and structure retrieval - see, e.g., the recent works [7–9] and the references therein. Given the desired properties of the image components all the valuable contributions to this problem rely on a variational-based formulation which minimizes the sum of different energy norms: total variation (TV) semi-norm, [14], L^1-norm, G-norm [1,12], approximation of the G-norm by the $div(\mathrm{L}^p)$-norm [15] and by the H^{-1}-norm [13], homogeneous Besov spaces [5], to model the oscillatory component of an image. A balanced combination of TV semi-norm and L^2-norm for separating the piecewise constant and smooth

Supplementary Information The online version contains supplementary material available at https://doi.org/10.1007/978-3-031-31975-4_10.

L. Calatroni et al. (Eds.): SSVM 2023, LNCS 14009, pp. 120–133, 2023.
https://doi.org/10.1007/978-3-031-31975-4_10

components is proposed in [6]. The intrinsic difficulty with these minimization problems comes from the numerical intractability of the considered norms [3,16], from the tuning of the numerous model parameters, and, overall, from the complexity of extracting noise from a textured image, given the strong similarity between these two components. This paper aims to give robust and efficient answers to these problems by exploiting statistical image characterizations and advanced numerical optimization algorithms.

We present a two-stage variational approach for the decomposition of a given (vectorized) image $f \in \mathbb{R}^N$ into the sum of four characteristic components:

$$f = c + s + t + n, \quad \text{with:} \quad o := t + n, \; u := c + s + t, \tag{1}$$

where c, s, t and n represent the cartoon or geometric (piecewise constant), the smooth, the texture and the noise components, respectively, whereas the introduced composite components o and u indicate the oscillatory and noise-free parts of f, respectively. The cartoon, smooth, and oscillatory components c, s, o are separated in the proposed first stage, with the three parts well-captured by using a non-convex TV-like term for c, a quadratic Tikhonov term for s and Meyer's G-norm for o. The estimated oscillatory component is then separated into texture and white noise parts t, n in the second stage by exploiting the noise whiteness property. This allows to deal effectively with a large class of important noises such as, e.g., those characterized by Gaussian, Laplacian and uniform distributions, which can be found in many applications. In the first stage, the two free regularization parameters are selected based on a novel multi-parameter cross-correlation principle which extends the single-parameter criterion recently proposed in [7]. The first and second stage optimization problems are efficiently solved by means of the alternating directions method of multipliers (ADMM).

In Sect. 2, we begin by recalling some preliminary definitions. The proposed quaternary two-stage variational decomposition model is introduced and motivated in Sect. 3. The cartoon-smooth-oscillating components separation approach (Stage I) is analysed in Sect. 4.1 - modelling insights - and in Sect. 4.2 - resolvability conditions. The multi-parameter selection is discussed in Sect. 5, while in Sect. 6 the computational ADMM framework is detailed. In Sect. 7 we demonstrate the ability of the proposed decomposition model to separate the desired image features and conclusions are drawn in Sect. 8.

2 Preliminaries and Notations

We recall some notions and definitions which will be useful in the rest of the paper. Let us consider two non-zero images in matrix form $x, y \in \mathbb{R}^{h \times w}$,

$$x = \{x_{i,j}\}_{(i,j)\in\Omega}, \; y = \{y_{i,j}\}_{(i,j)\in\Omega}, \; \Omega := \{0, \ldots, h-1\}\times\{0, \ldots, w-1\}. \tag{2}$$

Upon the assumption of suitable boundary conditions for x, y, the *sample normalized cross-correlation* of the two images x and y and the *sample normalized auto-correlation* of image x are the two matrix-valued functions ρ : $\mathbb{R}^{h\times w} \times \mathbb{R}^{h\times w} \to \mathbb{R}^{h\times w}$ and $\varphi : \mathbb{R}^{h\times w} \to \mathbb{R}^{h\times w}$ defined by

$$\rho(x,y) = \{\rho_{l,m}(x,y)\}_{(l,m)\in\Omega}, \quad \varphi(x) = \{\varphi_{l,m}(x)\}_{(l,m)\in\Omega}, \tag{3}$$

with scalar components $\rho_{l,m}(x,y)$ and $\varphi_{l,m}(x)$ given by

$$\rho_{l,m}(x,y) = \frac{1}{\|x\|_2\|y\|_2} \sum_{(i,j)\in\Omega} x_{i,j}\, y_{i+l,j+m}, \quad (l,m)\in\Omega, \tag{4}$$

$$\varphi_{l,m}(x) = \rho_{l,m}(x,x) = \frac{1}{\|x\|_2^2} \sum_{(i,j)\in\Omega} x_{i,j}\, x_{i+l,j+m}, \quad (l,m)\in\Omega, \tag{5}$$

respectively, where $\|\cdot\|_2$ in (4)–(5) denotes the Frobenius matrix norm and index pairs (l,m) are commonly called *lags*. It is well known that the sample normalized cross- and auto-correlations satisfy $\rho_{l,m}(x,y),\ \varphi_{l,m}(x,y) \in [-1,1]\ \forall (l,m) \in \overline{\Theta}$.

We introduce the following non-negative scale-independent scalar measure of correlation $\mathcal{C} : \mathbb{R}^{h\times w} \times \mathbb{R}^{h\times w} \to \mathbb{R}_+$ between the images x and y:

$$\mathcal{C}(x,y) := \frac{1}{N}\,\|\rho(x,y)\|_2^2\,. \tag{6}$$

Finally, we recall the Meyer's characterization of highly oscillating images - the component o in (1) - in the G-space, dual of BV-space [12], endowed with the G-norm defined in the discrete setting by

$$\|o\|_{\mathrm{G}} = \inf\left\{ \|g\|_\infty \;\middle|\; o = \mathrm{div}(g),\ g = (g^{(1)}, g^{(2)}) \in \mathbb{R}^{h\times w} \times \mathbb{R}^{h\times w} \right\}, \tag{7}$$

where $\|g\|_\infty := \max_{i,j} |g_{i,j}|$, with $|g_{i,j}| = \sqrt{(g_{i,j}^{(1)})^2 + (g_{i,j}^{(2)})^2}$. The G-space is very good to model oscillating patterns such as texture and noise, characterized by zero-mean functions of small G-norm [2].

3 Proposed Two-Stage Variational Decomposition Model

An observed image f, composed as in (1), is separated into cartoon, smooth, texture and noise components by solving the following two-stage model:

- **Stage I**: Given the observation f, compute estimates $\widehat{c}$, $\widehat{s}$, $\widehat{o}$ of the cartoon, smooth and oscillatory components c, s, o in (1) by solving

$$\{\widehat{c}, \widehat{s}, \widehat{o}\} \in \mathop{\arg\min}_{\substack{c,\,s,\,o\,\in\,\mathbb{R}^N \\ c+s+o=f}} \mathcal{J}_1(c,s,o;\gamma_1,\gamma_2,a_1), \tag{8}$$

$$\mathcal{J}_1(c,s,o;\gamma_1,\gamma_2,a_1) = \gamma_1 \sum_{i=1}^{N} \phi\left(\|(\mathrm{D}c)_i\|_2\,; a_1\right) + \frac{\gamma_2}{2}\,\|\mathrm{H}s\|_2^2 + \|o\|_G, \tag{9}$$

with penalty function ϕ defined in (13) and G-norm defined in (7).

- **Stage II:** Given the estimate $\widehat{o}$ from Stage I, compute estimates $\widehat{t}$, $\widehat{n}$ of the texture and white noise components t, n in (1) by solving

$$\{\widehat{t}, \widehat{n}\} \in \arg\min_{\substack{t,\, n \,\in\, \mathbb{R}^N \\ t\, +\, n\, =\, \widehat{o}}} \left\{ \mathcal{J}_2(t, n; a_2, \alpha) = \sum_{i=1}^{N} \phi\left(\|(\mathrm{D}t)_i\|_2 \,; a_2\right) + \imath_{\mathcal{W}_\alpha}(n) \right\}, \quad (10)$$

where γ_1, γ_2 in (8) are positive parameters, $\mathrm{D} := (\mathrm{D}_h; \mathrm{D}_v) \in \mathbb{R}^{2N \times N}$, with $\mathrm{D}_h, \mathrm{D}_v \in \mathbb{R}^{N \times N}$ finite difference operators discretizing the first-order horizontal and vertical partial derivatives of an image, respectively, and where the discrete gradient of image x at pixel i is denoted by $(\mathrm{D}x)_i := ((\mathrm{D}_h\, x)_i\,;\, (\mathrm{D}_v\, x)_i) \in \mathbb{R}^2$. The matrix H discretizes the second-order horizontal, vertical and mixed partial derivatives, with $(\mathrm{H}x)_i := ((\mathrm{H}_{hh}\, x)_i\,;\, (\mathrm{H}_{hv}\, x)_i\,; (\mathrm{H}_{vh}\, x)_i\,;\, (\mathrm{H}_{vv}\, x)_i) \in \mathbb{R}^4$ denoting the vectorized Hessian at pixel i. The function $\imath_{\mathcal{W}_\alpha} : \mathbb{R}^N \to \overline{\mathbb{R}} := \mathbb{R} \cup \{+\infty\}$ in (10) is the indicator function of the set $\mathcal{W}_\alpha \subset \mathbb{R}^N$, namely $\imath_{\mathcal{W}_\alpha} = 0$ for $x \in \mathcal{W}_\alpha$, $\imath_{\mathcal{W}_\alpha} = +\infty$ for $x \notin \mathcal{W}_\alpha$. Therefore, the target white noise component n or, equivalently, the residue image $\widehat{o} - t$ of Stage II model (10), must belong to the parametric set $\mathcal{W}_\alpha$, referred to as the *normalized whiteness set* with $\alpha \in \mathbb{R}_{++}$ called the *whiteness parameter*, and defined as follows:

$$\begin{aligned} \mathcal{W}_\alpha &:= \left\{ n \in \mathbb{R}^{h \times w} : \; -w_\alpha \leq \varphi_{l,m}(n) \leq w_\alpha \;\; \forall\, (l,m) \in \overline{\Theta}_0 := \overline{\Theta} \setminus \{(0,0)\} \right\} \\ &= \left\{ n \in \mathbb{R}^{h \times w} : \; -w_\alpha\, n^T n \leq \left(n \star n\right)(l,m) \leq w_\alpha\, n^T n \;\; \forall\, (l,m) \in \overline{\Theta}_0 \right\}. \end{aligned} \quad (11)$$

Motivated by the asymptotic distribution of the sample normalised auto- correlation $\varphi(n)$ (see [7]), a natural choice for the non-negative scalar w_α is

$$w_\alpha = \frac{\alpha}{\sqrt{N}}\,, \quad (12)$$

where the whiteness parameter α allows to directly set the probability that the sample normalized auto-correlation of a white noise realization at any given non-zero lag falls inside the whiteness set.

The parametric function $\phi(\,\cdot\,; a) : \mathbb{R}_+ \to \mathbb{R}_+$ in (8) is a re-parameterized and re-scaled version of the minimax concave (MC) penalty, namely a simple piecewise quadratic function defined by:

$$\phi(t; a) \;=\; \begin{cases} -\dfrac{a}{2}\, t^2 + \sqrt{2a}\, t & \text{for } \; t \in \left[\, 0, \sqrt{2/a} \,\right), \\ 1 & \text{for } \; t \in \left[\, \sqrt{2/a}, +\infty \,\right), \end{cases} \quad (13)$$

with $a \in \mathbb{R}_+$ called the *concavity parameter* of penalty ϕ. In fact, since $a = -\min_{t \neq 0} \phi''(t; a)$, it represents a measure of the degree of non-convexity of ϕ.

4 On the Decomposition Stage I

4.1 Insights on the Effect of Different Norms

Given the desired properties of components c, s, t and n, an ideal decomposition model is characterized by the minimization of energies expressed in terms of norms: $\sum_i \phi(\|(\mathrm{D}u_1)_i\|;a) =$ is minimal for $x = c$, $\|\mathrm{H}x\|_2$ is minimal for $x = s$, $\|x\|_G$ is minimal for $x = t$, $\|\varphi(x)\|_\infty := \max_{(l,m)\neq(0,0)} |\varphi_{(l,m)}(x)|$ is minimal for $x = n$.

In Table 1, we provide an insight on this conjecture, for synthetic images with different features illustrated in Fig. 1: piecewise constant scalar fields with sharp edges (u_1); smooth-gradient scalar fields with varying frequency of oscillations according to parameter z, (u_2); realization of Gaussian noise with increasing standard deviations σ, (u_3).

The noise component n is perfectly captured by $\|\varphi(u_3)\|_\infty$ making it minimal and constant for any values of standard deviation σ, while H^{-1}-norm and G-norm values increase as the noise level increases. If the term $\|\varphi(u_3)\|_\infty$ is not present in the model, then the noise component n is absorbed by the G-norm.

The cartoon component c is well detected by minimal values of $\sum_i \phi(\|(\mathrm{D}u_1)_i\|;a) =$ in the first block of Table 1 for all z values but the last column where the image no longer looks like a piecewise constant image, but it looks more like a texture.

Similarly the smooth component s is separated by a minimal $\|\mathrm{H}s\|_2$ in the second block of Table 1, for all z but the last column where the image has a pronounced textural component.

For what concern the texture component t, both the H^{-1}-norm and G-norm capture oscillations and they decrease for increasing frequency z, which is good, they both increase for increasing oscillation amplitude, which is not so good. However, G-norm is independent from image dimension, while H^{-1}-norm increases proportionally with dimension. Moreover, H^{-1}-norm recognizes as texture only the fine oscillatory patterns (see last column), while G-norm remains limited and small for different scale repeated patterns.

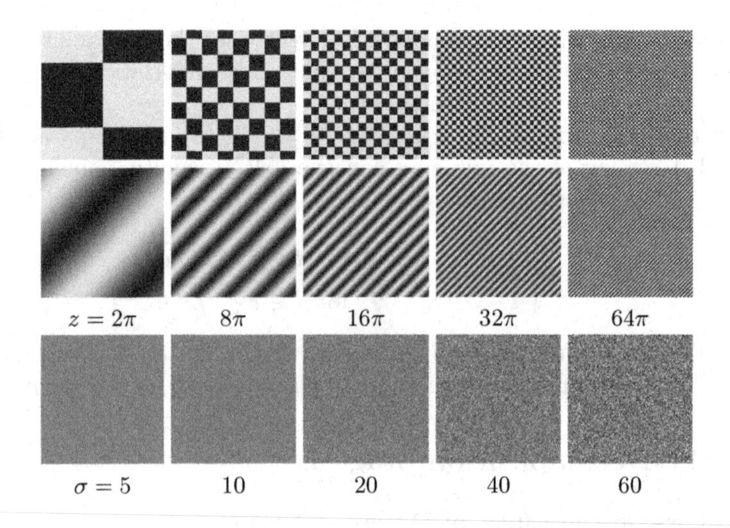

Fig. 1. Sample images used to evaluate model norms.

Table 1. Model norms evaluated for images u_1, u_2 and u_3 in Fig. 1

Chessboard	$z =$	2π	8π	16π	32π	64π
	$\sum_i \phi(\|(\mathrm{D}u_1)_i\|; a) =$	793	3124	6120	11728	21408
	$\|\mathrm{H}u_1\|_2 =$	7526	16307	24100	36516	57708
	$\|u_1\|_{\mathrm{H}^{-1}} =$	472098	100009	49882	27215	17810
	$\|u_1\|_{\mathrm{G}} =$	4520	1102	583	365	226
	$\|\varphi(u_1)\|_\infty =$	0.98	0.98	0.96	0.92	0.92
Diagonal Stripes	$z =$	2π	8π	16π	32π	64π
	$\sum_i \phi(\|(\mathrm{D}u_2)_i\|; a) =$	39999	39999	39999	39999	39999
	$\|\mathrm{H}u_2\|_2 =$	28.05	448	1786	7031	26367
	$\|u_2\|_{\mathrm{H}^{-1}} =$	364147	82381	40500	20218	10392
	$\|u_2\|_{\mathrm{G}} =$	2753	715	373	199	106
	$\|\varphi(u_2)\|_\infty =$	1.00	1.00	1.00	1.00	1.00
Noise Image	$\sigma =$	5	10	20	40	60
	$\sum_i \phi(\|(\mathrm{D}u_3)_i\|; a) =$	39999	39999	39999	39999	39999
	$\|\mathrm{H}u_3\|_2 =$	4454	8884	17906	35574	53572
	$\|u_3\|_{\mathrm{H}^{-1}} =$	945	1989	4029	7649	12289
	$\|u_3\|_{\mathrm{G}} =$	16.67	33.42	72.16	134	197
	$\|\varphi(u_3)\|_\infty =$	2.06e-2	2.20e-2	1.99e-2	2.42e-2	2.02e-2

4.2 Analysis of the Model

The goal of this section is to shortly analyze the Stage I optimization problem (8)–(9). To facilitate the analysis, first we note that, after replacing in (8)–(9) the explicit definition of the G-norm of the oscillating component $o = \mathrm{D}^T g$, and exploiting the additive image formation model (1) (from which $s = f - c - o$), problem (8)–(9) is equivalent to the following one:

$$\{\widehat{c}, \widehat{g}\} \in \arg\min_{c \in \mathbb{R}^N,\, g \in \mathbb{R}^{2N}} \widetilde{\mathcal{J}}_1(c, g; \gamma_1, \gamma_2, a_1), \tag{14}$$

$$\widetilde{\mathcal{J}}_1(c, g; \gamma_1, \gamma_2, a_1) = \gamma_1 \sum_{i=1}^{N} \phi\left(\|(\mathrm{D}c)_i\|_2 ; a_1\right) + \frac{\gamma_2}{2} \left\|\mathrm{H}(f - c - \mathrm{D}^T g)\right\|_2^2 + \|g\|_\infty \tag{15}$$

$$\widehat{o} = \mathrm{D}^T \widehat{g}, \quad \widehat{s} = f - \widehat{c} - \widehat{o}. \tag{16}$$

In the following Propositions 1, 2 (see proofs in the Supplementary Material) we analyze the optimization problem (14)–(15), with focus on convexity and coerciveness of the cost function $\widehat{\mathcal{J}}_1$, and afterward, on existence of solutions to the problem. To simplify notations, we introduce the total optimization variable $x := (c; g) \in \mathbb{R}^{3N}$.

Proposition 1. *For any $f \in \mathbb{R}^N$ and any $\gamma_1, \gamma_2, a_1 \in \mathbb{R}_{++}$, the function $\widetilde{\mathcal{J}}_1$ in* (15) *is proper, continuous, bounded from below by zero, convex and coercive in g,* *non-convex and non-coercive in* c*, hence non-convex and non-coercive in* x*. However, the function* $\widetilde{\mathcal{J}}_1$ *admits global minimizers.*

Proposition 2. *For any* $f \in \mathbb{R}^N$ *and any* $\gamma_1, \gamma_2, a_1 \in \mathbb{R}_{++}$*, the function* $\widetilde{\mathcal{J}}_1$ *in (15) is constant along straight lines in its domain* $\mathbb{R}^{3N}$ *of direction defined by the vector* $d := \left(1_N; 0_{2N}\right)$*. Hence, any constrained model of the form*

$$\{\widehat{c}, \widehat{g}\} \in \underset{(c;g)\,\in\,\mathcal{A}_{q_1,q_2}}{\arg\min}\;\; \widetilde{\mathcal{J}}_1(c, g; \gamma_1, \gamma_2, a_1)\,, \tag{17}$$

with $\mathcal{A}_{q_1,q_2} \subset \mathbb{R}^{3N}$ *one among the infinity of* $(3N-1)$*-d affine feasible sets*

$$\mathcal{A}_{q_1,q_2} = \left\{x \in \mathbb{R}^{3N} :\; q_1^T x \,=\, q_2 \text{ with } q_1^T d \neq 0\right\}, \quad q_1 \in \mathbb{R}^{3N}, \; q_2 \in \mathbb{R}, \tag{18}$$

admits solutions and the solutions are equivalent to those of the unconstrained model (14)–(15) in terms of (minimum) cost function value.

5 Multi-parameter Selection via Cross-Correlation

Selecting the regularization parameter of a two-terms (two-components) variational decomposition model based on the cross-correlation between the two estimated components has been proposed in the recent work [7], and referred to as the Cross-Correlation Principle (CCP). However, in the proposed Stage I decomposition model (8) we have three terms/components and two free regularization parameters. Hence, we propose an extension of the CCP in [7] to this more complicated case and refer the extended principle as Multi-Parameter CCP (MPCCP). The MPCCP for Stage I model is formulated as follows:

$$\text{Select } \; (\gamma_1, \gamma_2) = (\widehat{\gamma}_1, \widehat{\gamma}_2) \; \text{ such that } \; \{\widehat{\gamma}_1, \widehat{\gamma}_2\} \in \underset{\gamma_1, \gamma_2\,\in\,\mathbb{R}_{++}}{\arg\min}\; C(\gamma_1, \gamma_2)\,, \tag{19}$$

where the multi-parameter cross-correlation scalar measure $C : \mathbb{R}^2_{++} \to \mathbb{R}_+$ reads

$$\begin{aligned} C(\gamma_1, \gamma_2) &= \alpha_{c,o}\, \mathcal{C}\left(\widehat{c}(\gamma_1, \gamma_2), \widehat{o}(\gamma_1, \gamma_2)\right) + \alpha_{s,o}\, \mathcal{C}\left(\widehat{s}(\gamma_1, \gamma_2), \widehat{o}(\gamma_1, \gamma_2)\right) \\ &\quad + \alpha_{c,s}\, \mathcal{C}\left(\widehat{c}(\gamma_1, \gamma_2), \widehat{s}(\gamma_1, \gamma_2)\right)\,, \end{aligned} \tag{20}$$

with $\widehat{c}(\gamma_1, \gamma_2)$, $\widehat{s}(\gamma_1, \gamma_2)$ and $\widehat{o}(\gamma_1, \gamma_2)$ the (γ_1, γ_2)-dependent solution components of Stage I model (8), with the cross-correlation scalar measure $\mathcal{C}(\cdot, \cdot)$ defined in (6) and with the cross-correlation weights $\alpha_{c,o}, \alpha_{s,o}, \alpha_{c,s} \in [0, 1]$ allowing to tune the contributions of the three cross-correlations between c and o, s and o, c and s, respectively, to the introduced ternary cross-correlation among c, s, o. In particular, in the first example of the numerical section we will provide evidence that a good choice for the weights is $\alpha_{c,o} = \alpha_{s,o} = 0.5$, $\alpha_{c,s} = 0$. The idea encoded by the MPCCP in (19)–(20) is to select the pair (γ_1, γ_2) in Stage I model (8) which leads to "as separate as possible" solution components $\widehat{c}(\gamma_1, \gamma_2)$, $\widehat{s}(\gamma_1, \gamma_2)$, $\widehat{o}(\gamma_1, \gamma_2)$, with separability measured by cross-correlation.

The cost function $C(\gamma_1, \gamma_2)$ we aim to minimize in the MPCCP in (19)–(20) can be evaluated for any given pair (γ_1, γ_2) (by first solving numerically the associated Stage I model and then directly applying the function C by computing the three cross-correlations) but does not have an explicit form and, hence, the derivatives can not be calculated. A theoretical analysis of the properties of function C is therefore a very hard task and is out of the scope of this paper. However, the numerical tests that we will present seem to indicate that C is continuous (if not differentiable) and unimodal with a unique global minimizer. Hence, by applying some convergent zero-order minimization algorithm - i.e., relying on evaluations of the cost function C but not of its derivatives - for unimodal functions of two variables, MPCCP could be made fully automatic.

6 An ADMM-Based Numerical Solution for Stage I

In this section we consider the solution of the minimization problem (14)–(15) by using ADMM. We introduce two auxiliary variables $h \in \mathbb{R}^{2N}$, $z \in \mathbb{R}^{2N}$ and solve the following equivalent constrained problem:

$$\begin{aligned} \{\widehat{c}, \widehat{g}, \widehat{h}, \widehat{z}\} \in \arg\min_{\substack{c\in\mathbb{R}^N,\\ g,h,z\in\mathbb{R}^{2N},}} & \left\{ \gamma_1 \sum_{i=1}^{N} \phi\left(\|h_i\|_2\,; a_1\right) + \frac{\gamma_2}{2} \left\|\mathrm{H}(f - c - \mathrm{D}^T g)\right\|_2^2 + \|z\|_\infty \right\} \\ \text{s. t.} \qquad & h = \mathrm{D}c, \;\; z = g. \end{aligned} \tag{21}$$

To solve (21), we define the augmented Lagrangian function

$$\begin{aligned} \mathcal{L}(c, g, h, z, \lambda_1, \lambda_2; \gamma_1, \gamma_2) = {} & \gamma_1 \sum_{i=1}^{N} \phi\left(\|h_i\|_2; a_1\right) + \frac{\gamma_2}{2} \|\mathrm{H}(f - c - \mathrm{D}^T g)\|_2^2 + \|z\|_\infty \\ & - \langle \lambda_1, h - \mathrm{D}c \rangle + \frac{\beta_1}{2} \|h - \mathrm{D}c\|_2^2 - \langle \lambda_2, z - g \rangle + \frac{\beta_2}{2} \|z - g\|_2^2, \end{aligned} \tag{22}$$

where $\lambda_1 \in \mathbb{R}^{2N}$ is the vector of Lagrange multipliers associated with the linear constraints $h = \mathrm{D}c$, $\lambda_2 \in \mathbb{R}^{2N}$ is the vector of Lagrange multipliers associated with the linear constraints $z = g$, and $\beta_1, \beta_2 \in \mathbb{R}_{++}$ are the ADMM penalty parameters. Given the previously computed (or initialized for $k = 0$) vectors $c^{(k)}$, $g^{(k)}$, $h^{(k)}$, $z^{(k)}$, $\lambda_1^{(k)}$, $\lambda_2^{(k)}$ and the regularization parameters, the k-th iteration of the proposed ADMM-based iterative scheme reads as follows:

$$\begin{aligned} \left(c^{(k+1)}, g^{(k+1)}\right) \in \arg\min_{c\in\mathbb{R}^N, g\in\mathbb{R}^{2N}} & \left\{ \frac{\gamma_2}{2} \|\mathrm{H}(f - c - \mathrm{D}^T g)\|_2^2 - \left\langle \lambda_1^{(k)}, h^{(k)} - \mathrm{D}c \right\rangle \right. \\ & \left. + \frac{\beta_1}{2} \|h^{(k)} - \mathrm{D}c\|_2^2 - \left\langle \lambda_2^{(k)}, z^{(k)} - g \right\rangle + \frac{\beta_2}{2} \|z^{(k)} - g\|_2^2 \right\}, \end{aligned} \tag{23}$$

$$h^{(k+1)} \in \arg\min_{h\in\mathbb{R}^{2N}} \sum_{i=1}^{N} \phi\left(\|h_i\|_2; a_1\right) + \frac{\beta_1}{2\gamma_1} \left\| h - \left(\mathrm{D}c^{(k+1)} + \frac{1}{\beta_1} \lambda_1^{(k)} \right) \right\|_2^2, \tag{24}$$

$$z^{(k+1)} \in \arg\min_{z\in\mathbb{R}^{2N}} \frac{1}{\beta_2}\|z\|_\infty + \frac{1}{2}\left\|z - \left(g^{(k+1)} + \frac{1}{\beta_2}\lambda_2^{(k)}\right)\right\|_2^2, \tag{25}$$

$$\begin{pmatrix}\lambda_1^{(k+1)}\\ \lambda_2^{(k+1)}\end{pmatrix} = \begin{pmatrix}\lambda_1^{(k)} - \beta_1\left(h^{(k+1)} - \mathrm{D}c^{(k+1)}\right)\\ \lambda_2^{(k)} - \beta_2\left(z^{(k+1)} - g^{(k+1)}\right)\end{pmatrix}. \tag{26}$$

To solve (23), we apply the first-order optimality conditions which lead to solve the following linear system:

$$\mathrm{A}\begin{pmatrix}c\\ g\end{pmatrix} = \begin{pmatrix}\gamma_2\mathrm{H}^T\mathrm{H}f + \beta_1\mathrm{D}^T\left(h^{(k)} - \frac{1}{\beta_1}\lambda_1^{(k)}\right)\\ \gamma_2\mathrm{D}\mathrm{H}^T\mathrm{H}f + \beta_2 z^{(k)} - \lambda_2^{(k)}\end{pmatrix}, \tag{27}$$

where

$$\mathrm{A} = \begin{bmatrix}\gamma_2\mathrm{H}^T\mathrm{H} + \beta_1\mathrm{D}^T\mathrm{D} & \gamma_2\mathrm{H}^T\mathrm{H}\mathrm{D}^T\\ \gamma_2\mathrm{D}\mathrm{H}^T\mathrm{H} & \gamma_2\mathrm{D}\mathrm{H}^T\mathrm{H}\mathrm{D}^T + \beta_2\mathrm{I}\end{bmatrix}, \tag{28}$$

which can be solved via a sparse Cholesky solver. Problem (24) is equivalent to N two-dimensional problems which can be efficiently solved in closed form [8]. In particular, rewriting (24) for each $h_i \in \mathbb{R}^2$, $i = 1,\ldots,N$

$$h_i^{(k+1)} \in \arg\min_{h\in\mathbb{R}^2} \phi\left(\|h\|_2; a_1\right) + \frac{\beta_1}{2\gamma_1}\left\|h - \left(\left(\mathrm{D}c^{(k+1)}\right)_i + \frac{1}{\beta_1}\lambda_{1,i}^{(k)}\right)\right\|_2^2, \tag{29}$$

and satisfying the convexity condition $\beta_1 > a\gamma_1$ (see, e.g., [11]), the closed form in [8] can be applied directly. Problem (25) is not separable but it is equivalent to the proximity operator of the mixed $\ell_{\infty,2}$ norm, which also admits a closed form efficient solution [4]. Finally, the ADMM solution of Stage II in (10) is similar to that in [10] and is described in the Supplementary Material.

7 Numerical Examples

In this section we present some preliminary results on the performance of the proposed two-stage decomposition model when applied to images synthetically corrupted by additive white noise of different types among uniform (AWUN), Gaussian (AWGN) and Laplacian (AWLN). We consider three test images of size 200×200 and 256×256 pixels - *geometric*, *coast* and *skyscrapers* - which contain flat regions, neat edges and textures.

Example 1. In this example, we evaluate the performance of Stage I of our decomposition framework on the *geometric* image to showcase the decomposition performance together with the cross-correlation parameter selection. The parameter a_1 (and similarly a_2 for Stage II) can be estimated by directly imposing the abscissa of the transient point $\bar{a}_1 = \sqrt{2/a_1}$ in (13) as $a_1 = 2/\bar{a}_1^2$. The rationale is to penalize equally every salient discontinuity in c, therefore the value $\bar{a}$ should aim to estimate the minimal nonzero salient gradient of c. According to this strategy, for this example we used $\bar{a}_1 = 0.4$. Moreover, the ADMM algorithm was initialized by $h^{(0)} = Df$, $z^{(0)} = \lambda_1^{(0)} = \lambda_2^{(0)} = 0$. The first row of Fig. 2 reports the input image

f, and the resulting components $\widehat{c}$, $\widehat{s}$, $\widehat{o}$, respectively. Each component is visualized in range [0.05,0.95] to demonstrate better both the characteristics as well as reconstruction for each component. In the second row of Fig. 2 we report the cross-correlation surfaces, namely $C(\gamma_1, \gamma_2)$ in (20) with weights $\alpha_{c,o} = \alpha_{s,o} = 0.5$, $\alpha_{c,s} = 0$ - used to select $\widehat{\gamma}_1, \widehat{\gamma}_2, \mathcal{C}(\widehat{c}(\cdot), \widehat{s}(\cdot)), \mathcal{C}(\widehat{s}(\cdot), \widehat{o}(\cdot))$ and $\mathcal{C}(\widehat{c}(\cdot), \widehat{o}(\cdot))$ for varying γ_1 and γ_2. In the latter three surfaces the minimum cross-correlation is represented by a solid dot, while the minimum $\min(C(\gamma_1, \gamma_2)) = C(\widehat{\gamma}_1, \widehat{\gamma}_2) = 5.31 \times 10^{-8}$ by a diamond marker. In the third row of Fig. 2, from second to last column, we report the associated Signal-to-Noise Ratio (SNR) surfaces for each component for varying γ_1, γ_2 and the mean SNR surface

$$SNR_{mean} = (SNR(\widehat{c}(\gamma_1, \gamma_2)) + SNR(\widehat{s}(\gamma_1, \gamma_2)) + SNR(\widehat{o}(\gamma_1, \gamma_2)))/3.$$

Black dot indicates the maximum SNR attained, while the asterisk marks the maximum $\max(SNR_{mean})$. It can be noticed that the parameter pair $(\widehat{\gamma}_1, \widehat{\gamma}_2)$ selected by the proposed MPCCP is close to the pair (γ_1, γ_2) yielding a SNR equal to the maximum $\max(SNR_{mean})$. This indicates that our MPCCP works well for this example. In particular, the SNR values attained by each reconstructed component $\widehat{c}$, $\widehat{s}$, $\widehat{o}$ are $SNR(\widehat{c}(\widehat{\gamma}_1, \widehat{\gamma}_2)) = 47.16$, $SNR(\widehat{s}(\widehat{\gamma}_1, \widehat{\gamma}_2)) = 52.44$ and $SNR(\widehat{o}(\widehat{\gamma}_1, \widehat{\gamma}_2)) = 36.26$.

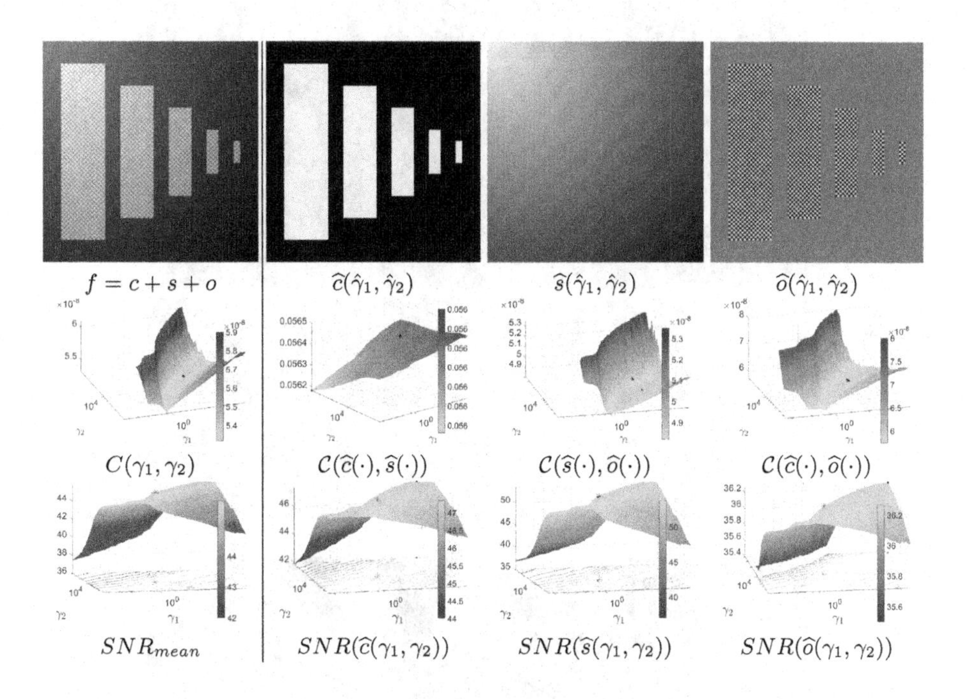

Fig. 2. Decomposition results of *geometric* image based on cross-correlation parameter selection for Stage I.

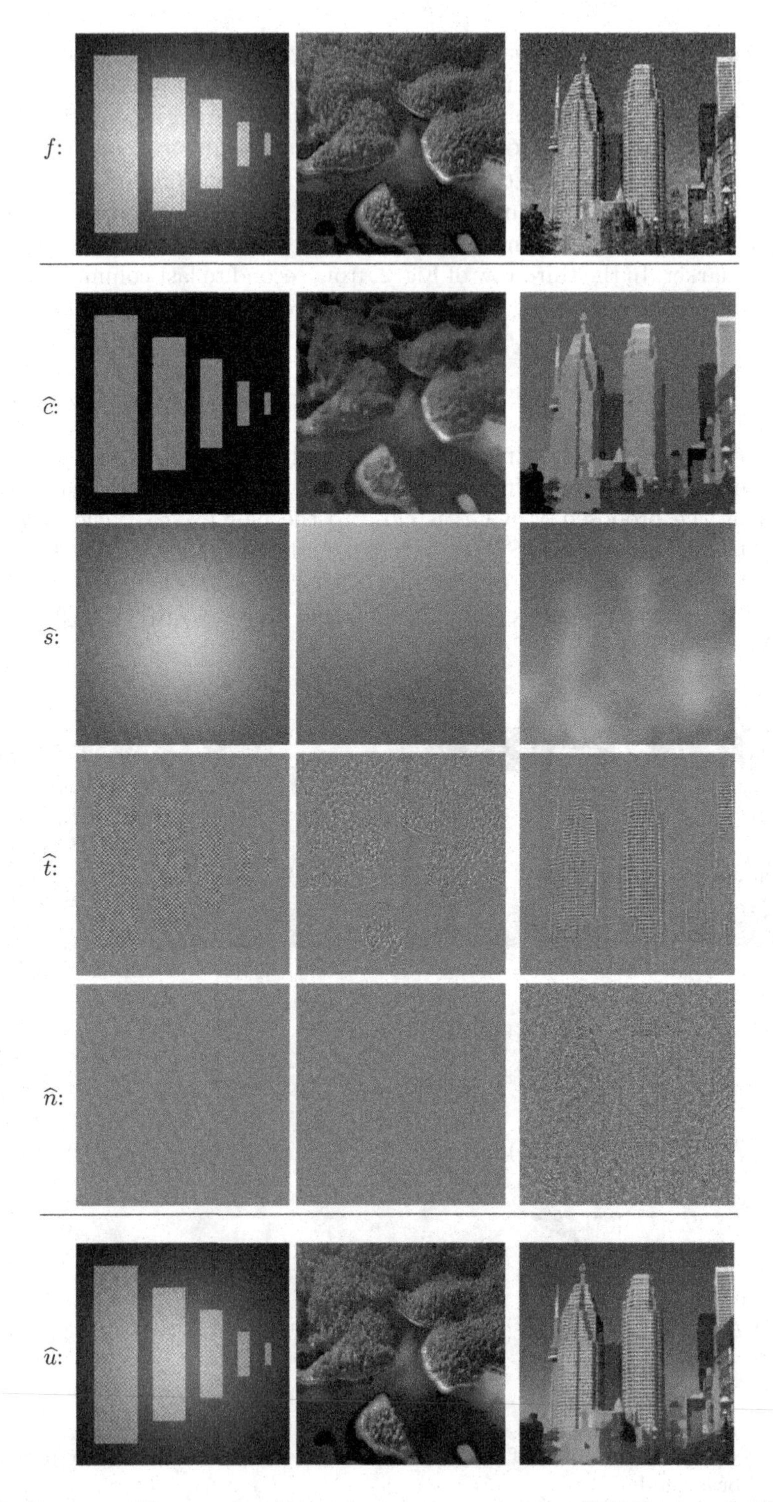

Fig. 3. Decomposition results of test images corrupted by AWLN, $\sigma = 6$, AWUN, $\sigma = 10$, and AWGN, $\sigma = 25$, from left to right respectively.

Example 2. In this example we present the results of applying Stage I + Stage II to *geometric* - where we changed the smooth component to a spotlight effect - *coast* and *skyscraper* images, corrupted with a realization of AWLN with standard deviation $\sigma = 6$, AWUN with $\sigma = 10$, AWGN with $\sigma = 25$, respectively. Stage I used $\bar{a}_1 = 0.4$ with $h^{(0)} = Df$, $z^{(0)} = \lambda_1^{(0)} = \lambda_2^{(0)} = 0$ initialization for *geometric* image, while $\bar{a}_1 = 0.1$ with zero initialization of the above variables for the two photographic images. For Stage II, we used $\bar{a}_2 = 20$, for *geometric*, $\bar{a}_2 = 0.3$ for *coast* and $\bar{a}_2 = 0.7$ for *skyscraper*. In the first row of Fig. 3 we report the observed images f used as input for Stage I. The oscillatory component $\widehat{o}$ was then used as input for Stage II. From second to fifth row of Fig. 3 we report the resulting components $\widehat{c}$, $\widehat{s}$, $\widehat{t}$ and $\widehat{n}$, where $\widehat{s}$, $\widehat{t}$ and $\widehat{n}$ have been set its mean to 0.5 for visualization purposes. In the last row of Fig. 3 we report images $\widehat{u} = \widehat{c} + \widehat{s} + \widehat{t}$ which represents the denoised version of the degraded image f.

Remark 1. Due to the nature of G-space, it may seem more natural to model the texture component t by the G-norm, i.e. to use the following model for Stage II:

$$\{\widehat{t}_2, \widehat{n}_2\} \in \arg\min_{t,n \in \mathbb{R}^N} \{\|t\|_G + \imath_{\mathcal{W}_\alpha}(n)\} \text{ s.t.: } t + n = \widehat{o}. \tag{30}$$

In Fig. 4, the decomposition results $\widehat{t}_2$, $\widehat{n}_2$, obtained by solving problem (30) for $\widehat{o}$ of *geometric* test image, are reported. In the second column of Fig. 4, the horizontal cross-sections are shown: the ground truth (dashed black), and the solutions of Stage II models (10) and (30) (solid red and blue, respectively). From these results, it is clear that the G-norm is not suitable to separate texture from noise. In particular, for localized textures, like the one in Fig. 4, some noise can be included in the texture component without changing its G-norm value.

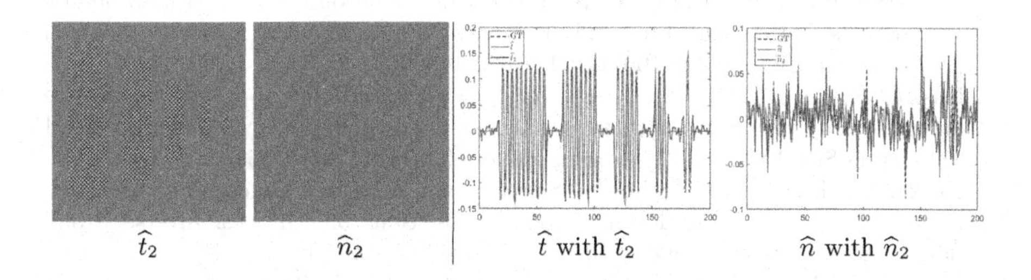

Fig. 4. Stage II model (30): decomposition results (left) and central horizontal cross-sections (right) of texture and noise with $\widehat{t}$ and $\widehat{n}$ from the first column of Fig. 3.

8 Conclusions

In this paper we presented a two-stage variational model for the additive decomposition of images corrupted by additive white noise into cartoon, smooth, texture and noise components. We also proposed a novel multi-parameter selection criterion based on cross-correlation which, together with the usage of a whiteness constraint for the noise component, makes the model context- and noise-unaware. Some numerical examples are presented which indicate the good quality decomposition results achievable by the proposed approach.

Acknowledgments. This work was supported in part by the National Group for Scientific Computation (GNCS-INDAM) and in part by MUR RFO projects.

References

1. Aujol, J.F., Aubert, G., Blanc-Feraud, L., Chambolle, A.: Image decomposition into a bounded variation component and an oscillating component. J. Math. Imaging Vis. **22**(1), 71–88 (2005). https://doi.org/10.1007/s10851-005-4783-8
2. Aujol, J.F., Chambolle, A.: Dual norms and image decomposition models. J. Math. Imaging Vis. **63**(1), 85–104 (2005). https://doi.org/10.1007/s11263-005-4948-3
3. Aujol, J.F., Kang, S.H.: Color image decomposition and restoration. J. Vis. Commun. Image Represent. **17**(4), 916–928 (2006). https://doi.org/10.1016/j.jvcir.2005.02.001
4. Eriksson, A., Isaksson, M.: Pseudoconvex proximal splitting for L-infinity problems in multiview geometry. In: Proceedings of Computer Vision and Pattern Recognition (CVPR 2014), pp. 4066–4073 (2014). https://doi.org/10.1109/CVPR.2014.518
5. Garnett, J.B., Le, T.M., Meyer, Y., Vese, L.: Image decompositions using bounded variation and generalized homogeneous besov spaces. Appl. Comput. Harmon. Anal. **23**(1), 25–56 (2007). https://doi.org/10.1016/j.acha.2007.01.005
6. Gholami, A., Hosseini, S.M.: A balanced combination of Tikhonov and total variation regularizations for reconstruction of piecewise-smooth signals. Signal Process. **93**(7), 1945–1960 (2013). https://doi.org/10.1016/j.sigpro.2012.12.008
7. Girometti, L., Lanza, A., Morigi, S.: Ternary image decomposition with automatic parameter selection via auto- and cross-correlation. Adv. Comput. Math. **49**(1) (2023). https://doi.org/10.1007/s10444-022-10000-4
8. Huska, M., Kang, S.H., Lanza, A., Morigi, S.: A variational approach to additive image decomposition into structure, harmonic and oscillatory components. SIAM J. Imaging Sci. **14**(4) (2021). https://doi.org/10.1137/20M1355987
9. Huska, M., Lanza, A., Morigi, S., Selesnick, I.: A convex-nonconvex variational method for the additive decomposition of functions on surfaces. Inverse Probl. **35**(12) (2019). https://doi.org/10.1088/1361-6420/ab2d44
10. Lanza, A., Morigi, S., Sciacchitano, F., Sgallari, F.: Whiteness constraints in a unified variational framework for image restoration. J. Math. Imaging Vis. **60**(9), 1503–1526 (2018). https://doi.org/10.1007/s10851-018-0845-6
11. Lanza, A., Morigi, S., Sgallari, F.: Convex image denoising via non-convex regularization with parameter selection. J. Math. Imaging Vis. **56**(2), 195–220 (2016). https://doi.org/10.1007/s10851-016-0655-7
12. Meyer, Y.: Oscillating Patterns in Image Processing and Nonlinear Evolution Equations: The Fifteenth Dean Jacqueline B. Lewis Memorial Lectures. American Mathematical Society, USA (2001)
13. Osher, S., Solé, A., Vese, L.: Image decomposition and restoration using total variation minimization and the H^{-1}-norm. SIAM J. Multiscale Model. Simul. **1**(3), 349–370 (2003). https://doi.org/10.1137/S1540345902416247

14. Rudin, L., Osher, S., Fatemi, E.: Nonlinear total variation based noise removal algorithms. Physica D **60**(1–4), 259–268 (1992). https://doi.org/10.1016/0167-2789(92)90242-F
15. Vese, L., Osher, S.: Modeling textures with total variation minimization and oscillating patterns in image processing. J. Sci. Comput. **19**(1–3), 553–572 (2003). https://doi.org/10.1023/A:1025384832106
16. Wen, Y.W., Sun, H.W., Ng, M.: A primal-dual method for the Meyer model of cartoon and texture decomposition. Numer. Linear Algebra Appl. **26**(2) (2019). https://doi.org/10.1002/nla.2224

Machine and Deep Learning in Imaging

EmNeF: Neural Fields for Embedded Variational Problems in Imaging

Danielle Bednarski(✉) and Jan Lellmann

Institute of Mathematics and Image Computing, University of Lübeck, Lübeck, Germany
{bednarski,lellmann}@mic.uni-luebeck.de

Abstract. We propose a model-driven neural fields approach for solving variational problems. The approach can be applied to a variety of problems with convex, 1-homogeneous regularizer and arbitrary, possibly non-convex, data term. Our strategy is to embed the non-convex energy into a higher-dimensional space, reaching a convex primal-dual formulation. Instead of using classical gradient-descent based optimization algorithms, we propose training multiple fields representing the primal and dual variables in order to solve the problem.

Keywords: Neural fields · Functional lifting · Embedding · GANs

1 Motivation

We consider variational image processing problems with energies of the form

$$\inf_{u \in U} F(u), \quad F(u) := \underbrace{\int_\Omega \rho(x, u(x))\,\mathrm{d}x}_{\text{data term}} + \underbrace{\int_\Omega \eta(\nabla u(x))\,\mathrm{d}x}_{\text{regularizer}}, \tag{1}$$

where $u : \Omega \to \Gamma$, $\Omega \subset \mathbb{R}^n$ open and bounded, and $\Gamma \subset \mathbb{R}$ compact. Let U be a suitable solution space and assume that $\rho : \Omega \times \Gamma \to \overline{\mathbb{R}}$ is proper, non-negative and possibly non-convex with respect to u, and $\eta : \mathbb{R}^n \to \mathbb{R}$ is non-negative, one-homogeneous and convex. Typically, ρ will encode the data term, while η defines the regularizer.

Problems of this form are found in many imaging applications, such as image denoising, segmentation and stereo vision [16]. Classical gradient-based optimization algorithms, such as (non-convex) PDHG [2,9,18], can be used to find a minimizer. However, if the energy is non-convex, these algorithms might converge to a local instead of a global minimum. Different convexification strategies, such as discrete Markov random fields [6] and continuous current- [12] and measure-based [19] lifting strategies, have been proposed in order to handle non-convexities. The idea of these strategies is to reformulate the given energy in a higher-dimensional space, such that this high-dimensional energy is convex and that its global minimizer maps to a global minimizer of the original problem.

L. Calatroni et al. (Eds.): SSVM 2023, LNCS 14009, pp. 137–148, 2023.
https://doi.org/10.1007/978-3-031-31975-4_11

A downside of the these embedding – or "lifting" – approaches is the increased size of the embedded problem. Typically, one has to discretize the problem over at least one additional dimension, typically the range of the solution, and weigh the increase in runtime against discretization artefacts due to a coarse grid. Although special discretization strategies can decrease the number of required grid points in the additional dimension [10,11], they considerably complicate implementation and impact flexibility.

Neural networks have gained great popularity, since they have proven to be quite successful at solving imaging tasks. Early imaging networks were trained in a data-driven manner and many of the prevailing networks are still trained in this fashion. In recent years, model-driven networks such as physics-informed neural networks (PINNs) [4] and neural fields [20] have emerged. PINNs overcome data shortage by embedding knowledge of the physical model – described via partial differential equations – into the learning process. Neural fields typically are coordinate-based networks, which parameterize physical properties over space and/or time, most prominently neural radiance fields (NeRFs) for view synthesis [8].

Outline and Contribution. In this work, we propose EmNeF, a neural field approach for solving problems of the form (1). In Sect. 2, we recall the calibration-based embedding method [12], which transforms a non-convex primal problem (1) into a higher-dimensional, convex, primal-dual problem of the form

$$\inf_{\nu \in \mathcal{U}} \sup_{\varphi \in \mathcal{K}} \mathcal{F}(\nu, \varphi), \quad \mathcal{F}(\nu, \varphi) = \int_{\Omega \times \Gamma} \langle \varphi, (D\nu) \rangle . \tag{2}$$

In Sect. 3, we propose a neural field based approach to determine implicit solutions of the primal-dual problem (2). Similar to the strategy used for generative adversarial networks (GANs) [5], we train individual neural fields for the primal and dual variables. During training, the fields are updated in an alternating manner; while one field aims to minimize the energy (2) with respect to ν, the other two fields aim to maximize the energy with respect to φ. As the approach finds an implicit representation of ν, the solution can be extracted on an arbitrarily fine grid during inference.

2 Embedding

Denote by $\mathbf{1}_u : \Omega \times \Gamma \to \{0,1\}$ the indicator of the subgraph of a function u,

$$\mathbf{1}_u(x,t) = \begin{cases} 1, & u(x) \geq t, \\ 0, & u(x) < t. \end{cases} \tag{3}$$

The authors of [1,3,12] derive an embedding method based on the complete graph of u, which is described as the measure-theoretic boundary of $\mathbf{1}_u$. They show that energies of the form (1) with $u \in \mathrm{SBV}(\Omega; \mathbb{R})$ can equivalently be written as

$$F(u) = \sup_{\varphi \in \mathcal{K}} \int_{\Omega \times \Gamma} \langle \varphi, D\mathbf{1}_u \rangle =: \mathcal{F}(\mathbf{1}_u), \tag{4}$$

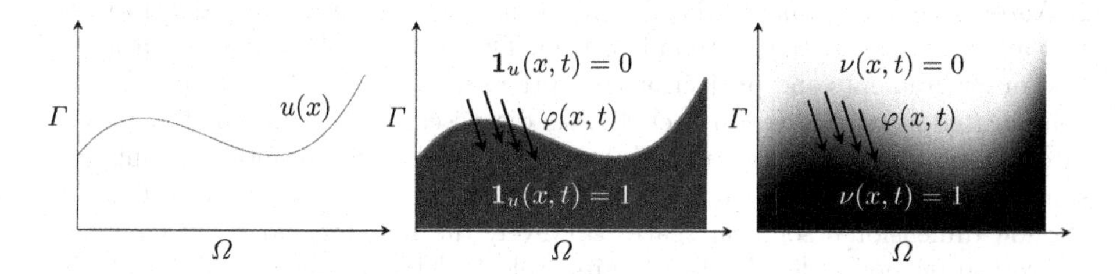

Fig. 1. Calibration-based embedding. Instead of minimizing a primal problem over a function u **(left)**, the problem is reformulated in terms of the characteristic function of the subgraph $\mathbf{1}_u$ **(middle)**. The resulting formulation is a primal-dual problem, which can be understood as the maximum flux of vector fields φ through the complete graph. This primal-dual problem becomes convex, if one extends the space of feasible solutions to the convex hull of the subgraph indicator functions, i.e., ν **(right)**.

where $D\mathbf{1}_u$ is the distributional derivative of $\mathbf{1}_u$. Denoting by η^* the Fenchel conjugate with respect to the second variable, the constraint set is given as

$$\mathcal{K} := \{(\varphi_x, \varphi_t) \in C_c^1(\Omega \times \Gamma; \mathbb{R}^n \times \mathbb{R}) : \varphi_t + \rho(x,t) \geq \eta^*(x, \varphi_x(x,t))\}. \quad (5)$$

Note that the Fenchel conjugate of a convex, one-homogeneous function is the indicator function of a convex set [13, Cor. 13.2.1], i.e., it will only assume values in $\{0, +\infty\}$. Thus, under our assumption that the regularizing integrand η is one-homogeneous, convex, and independent of x, the constraint decouples into

$$\varphi_t + \rho(x,t) \geq 0 \quad \text{and} \quad \varphi_x(x,t) \in \text{dom}(\eta^*), \quad (6)$$

where $\text{dom}(\eta^*) := \{p \in \mathbb{R}^n | \eta^*(p) < +\infty\}$.

A convex problem is obtained by relaxing the set of feasible primal solutions from the set of indicator functions to the set of (pointwise) non-increasing functions mapping to the interval $[0,1]$. The final formulation of the embedded primal-dual problem, formulated on continuous domain and range, is

$$\inf_{\nu \in \mathcal{C}} \sup_{\substack{\varphi_x \in \mathcal{K}_x \\ \varphi_t \in \mathcal{K}_t}} \int_{\Omega \times \Gamma} \left\langle \begin{pmatrix} \varphi_x \\ \varphi_t \end{pmatrix}, D\nu \right\rangle, \quad (7)$$

$$\begin{aligned}
\mathcal{C} := \{\nu \in \text{BV}_{\text{loc}}(\Omega \times \Gamma; [0,1]) : \quad & \nu(x,t) = 1 \quad \forall t \leq \Gamma_{\min}, & (8)\\
& \nu(x,t) = 0 \quad \forall t > \Gamma_{\max}, & (9)\\
& \nu(x, \cdot) \text{ non-increasing.}\}, & (10)\\
\mathcal{K}_x := \{\varphi_x \in C_0(\Omega \times \Gamma; \mathbb{R}^d) : \quad & \varphi_x(x,t) \in \text{dom}(\eta^*) \quad \forall (x,t) \in \Omega \times \Gamma\}, & (11)\\
\mathcal{K}_t := \{\varphi_t \in C_0(\Omega \times \Gamma; \mathbb{R}) : \quad & -\varphi_t(x,t) \leq \rho(x,t) \quad \forall (x,t) \in \Omega \times \Gamma\}. & (12)
\end{aligned}$$

A central result [12, Thm. 3.1] states that if ν^* is a global minimizer of (7), then for almost every $s \in [0,1)$ the characteristic function $\mathbf{1}_{\{\nu^* > s\}}$ is in fact the characteristic function of a subgraph of a minimizer of (1). Therefore, solving the convex problem (7) allows to find a global minimizer of the non-convex problem (1). For a visualization of the embedding method, see Fig. 1.

A Note on Optimization. In [12], the problem is discretized in a straightforward way and solved using the PDHG algorithm. The latter involves the calculation of first-order gradients and projections onto the sets $\mathcal{C}, \mathcal{K}_t$ and $\mathcal{K}_x$ – which are not complicated as we will see in Sect. 4.1. As discussed in [12, Thm. 3.1], any non-integral solution $\nu^* = \arg\min_{\mathcal{C}} \mathcal{F}(\nu)$ can be thresholded to obtain an integral solution and then projected onto a global minimizer $u^* = \arg\min_{U} F(u)$ in the low-dimensional solution space. However, the number and exact choice of discretization points for Γ plays a vital role and has a great influence on the quality of the solution; see also Fig. 4b.

In [10,11], the authors propose a more involved "sublabel-accurate" discretization, which preserves information on local minimal of the data term in between the chosen labels. This discretization helps to greatly improve results for fewer labels. However, the implementation is more intricate (projections onto the Fenchel conjugate of the data term) and, to date, there is no approach for mapping non-integral global minimizers of the embedded problem to global minimizers of the original problem. Therefore, the theoretical results presented in [12, Thm. 3.1] do not apply directly.

With our neural fields based approach, we hope to combine classical results and modern techniques in order to overcome the dimensionality curse of the straightforward discretization used in [12]. Due to the nonlinear architecture of the network, we lose convexity of the problem. Therefore, convergence is much less clear and an important issue for future work.

3 EmNeF

In the neural field approach, spatial information is encoded by determining the weights $\theta \in \mathbb{R}^p$ of a model $\boldsymbol{F}_\theta : \mathbb{R}^n \to \mathbb{R}^d$ that directly translates $\mathbb{R}^n$-valued coordinates into $\mathbb{R}^d$-valued quantities, such as physical properties encoding appearance. This stands in contrast to classical convolutional neural networks, where the output of the network typically consists in multidimensional grids of samples of said quantities. Similarly, existing methods for solving (7) have focused on discretizing the primal and dual variables on various grids or using finite elements [7,11].

Here, we aim to solve problem (7) with a neural field approach: we aim to learn an implicit, parametric representation of the optimization variables. Therefore, we introduce three different neural fields – one field for the primal variable $\boldsymbol{F}^{\nu}_{\theta^\nu} : \mathbb{R}^{n+1} \to \mathbb{R}$ and two fields for the dual variables $\boldsymbol{F}^{\varphi_x}_{\theta^x} : \mathbb{R}^{n+1} \to \mathbb{R}^n$ and $\boldsymbol{F}^{\varphi_t}_{\theta^t} : \mathbb{R}^{n+1} \to \mathbb{R}$.

Training. Instead of training (solving) on a fixed grid, we sample random coordinates. In the terminology of neural networks, these random points are batches and not only help to speed up single iterations, but also allow to learn the ν, φ_x and φ_t implicitly, i.e., in a continuous manner. Later, in the inference (evaluation) step, this is especially crucial with respect to $\nu(\cdot, t)$ so as to avoid discretization artefacts related to the embedding method.

Before passing the coordinates to the neural fields, we pass them through a Fourier feature mapping as proposed in [17]. It has been shown that random Fourier features (RFF) improve the capability of networks with low-dimensional input coordinates to learn high-frequency functions [17,21].

The RFF encoding $\gamma : \mathbb{R}^n \to \mathbb{R}^m$ transforms the input x according to

$$\gamma(x) = [a_1 \cos(b_1^\top x), a_1 \sin(b_1^\top x), ..., a_{\frac{m}{2}} \cos(b_{\frac{m}{2}}^\top x), a_{\frac{m}{2}} \sin(b_{\frac{m}{2}}^\top x)], \tag{13}$$

before passing it to the main model $\boldsymbol{F}_\theta$. Here $m \gg n$ and $b_i \in \mathbb{R}^n$ are random coefficients drawn from the multivariate normal distribution $\mathcal{N}(0, 2\pi\sigma^2 I)$. In our experiments, we set all $a_i = 1$ and use randomly chosen but fixed b_i for all three neural fields.

In each training step, the primal and dual fields are updated alternatingly in order to iteratively approximate a solution of (7). While we perform a gradient descent step on the primal field $\boldsymbol{F}_{\theta^\nu}^\nu : \mathbb{R}^{n+1} \to \mathbb{R}$, the inf − sup structure requires to perform gradient ascent steps on the dual fields $\boldsymbol{F}_{\theta^x}^{\varphi_x} : \mathbb{R}^{n+1} \to \mathbb{R}^n$ and $\boldsymbol{F}_{\theta^t}^{\varphi_t} : \mathbb{R}^{n+1} \to \mathbb{R}$. For a batch $\boldsymbol{X} = \{\boldsymbol{x}_i \in \Omega^h \times \Gamma^h | i = 1, ..., k\}$, this amounts to the following primal and dual loss functions:

$$\text{primal:} \quad L_\nu(\theta^\nu; \boldsymbol{X}) = \frac{1}{k} \sum_{i=1}^{k} \boldsymbol{F}_{\theta^x}^{\varphi_x}(\boldsymbol{x}_i)^\top \nabla_x^h \boldsymbol{F}_{\theta^\nu}^\nu(\boldsymbol{x}_i) + \boldsymbol{F}_{\theta^t}^{\varphi_t}(\boldsymbol{x}_i) \nabla_t^h \boldsymbol{F}_{\theta^\nu}^\nu(\boldsymbol{x}_i), \tag{14}$$

$$\text{dual:} \quad L_x(\theta^x; \boldsymbol{X}) = \frac{1}{k} \sum_{i=1}^{k} -\boldsymbol{F}_{\theta^x}^{\varphi_x}(\boldsymbol{x}_i)^\top \nabla_x^h \boldsymbol{F}_{\theta^\nu}^\nu(\boldsymbol{x}_i), \tag{15}$$

$$L_t(\theta^t; \boldsymbol{X}) = \frac{1}{k} \sum_{i=1}^{k} -\boldsymbol{F}_{\theta^t}^{\varphi_t}(\boldsymbol{x}_i) \nabla_t^h \boldsymbol{F}_{\theta^\nu}^\nu(\boldsymbol{x}_i). \tag{16}$$

In order to approximate $\nabla_x^h \boldsymbol{F}_{\theta^\nu}^\nu$ and $\nabla_t^h \boldsymbol{F}_{\theta^\nu}^\nu$, we also evaluate $\boldsymbol{F}_{\theta^\nu}^\nu$ on neighboring points of $\boldsymbol{Q}$ and use finite differences. The constraints are hardcoded into the network as described in the next chapter.

Inference. For inference, which in the classical approach amounts to evaluating/interpolating the solution ν after the optimization problem has been solved, we evaluate the primal neural field $\boldsymbol{F}_{\theta^\nu}^\nu : \mathbb{R}^{n+1} \to \mathbb{R}$ on a uniform grid $\Omega^h \times \Gamma^h$ in $\Omega \times \Gamma$, with a particularly fine resolution with respect to the range $\Gamma = [\gamma_0, \gamma_1]$. Following the central result in [12] [Thm. 3.1], we choose a threshold $s \in [0, 1)$ and estimate the solution u^* of the original problem (1) from the (approximate) solution ν^* of the relaxed problem (7) as follows (Fig. 2):

$$u(x) = \int_{\gamma_0}^{\gamma_1} \mathbf{1}_{\nu^*(x,\cdot)>s}(t)\, dt. \tag{17}$$

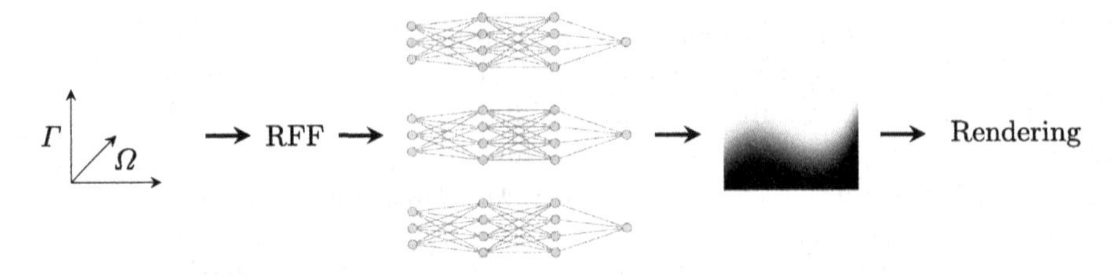

Fig. 2. EmNeF Coordinates are passed through a random Fourier feature mapping (RFF). In the training/optimization stage, multiple primal (dual) neural fields are iteratively updated to minimize (maximize) the energy (7). The inference/final interpolation can be performed on an arbitrarily fine grid for an optimal resolution of the lifted solution ν^*. Thresholding of ν^* results in an approximate minimizer u^* of (1).

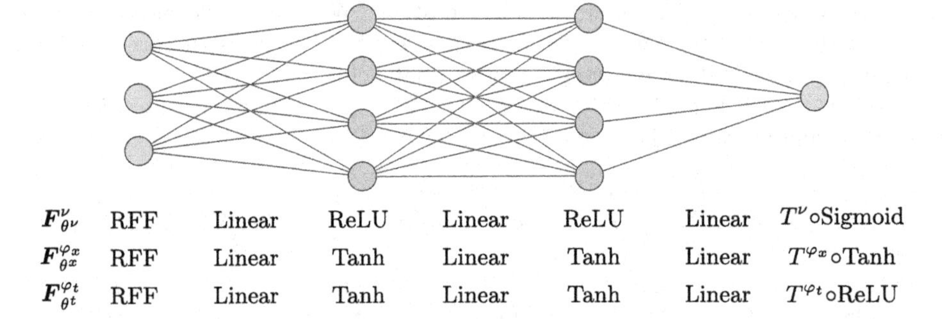

Fig. 3. Network architecture. All networks consist of six layers. The transformations T^* are further described by (22), (24) and (18).

4 Network Architecture

All networks have two hidden layers configured as shown in Fig. 3. The RFF encoding expands each coordinate vector from $\mathbb{R}^3$ to $\mathbb{R}^{256}$. This dimension is kept in the following layers, only the last layer reduces the output to $\mathbb{R}^1$ $(\boldsymbol{F}^{\nu}_{\theta^\nu}, \boldsymbol{F}^{\varphi_t}_{\theta^t})$ and $\mathbb{R}^2$ $(\boldsymbol{F}^{\varphi_x}_{\theta^x})$.

4.1 Implementation of the Constraints

We opted for implementing the constraints $\mathcal{C}, \mathcal{K}_x$, and $\mathcal{K}_t$ as hard constraints by adjusting the output of the neural fields. Adding suitable Lagrange multipliers is also feasible, but would result in a more complicated primal-dual structure and require additional variables for the multipliers.

Primal Constraint $\mathcal{C}$. The embedded solution space is defined by (8)–(10). We employ a sigmoid activation function as the last layer of $\boldsymbol{F}^{\theta^\nu}_{\nu}$ and define an additional transformation in order to ensure that the boundary constraints are satisfied. The monotonicity condition (10) is enforced implicitly, as the multipliers in φ_t are unbounded from above. The boundary condition (8)–(9) is enforced by sampling grid points from a space with slightly enlarged range $\Omega \times \Gamma^{\pm}$, with $\Gamma^{\pm} \supset \Gamma$, and applying the following pointwise transformation to the output of $\boldsymbol{F}^{\nu}_{\theta^{\nu}}$:

$$T^{\nu} \circ \boldsymbol{F}^{\nu}_{\theta^{\nu}}(x,t) := \begin{cases} 1, & \text{if } t \leq \Gamma_{\min}, \\ 0, & \text{if } t \geq \Gamma_{\max}, \\ \boldsymbol{F}^{\nu}_{\theta^{\nu}}(x,t), & \text{otherwise.} \end{cases} \tag{18}$$

Dual Constraint $\mathcal{K}_x$. In our experiments, we use the isotropic total variation regularizer defined as

$$TV(u) := \sup_{\substack{\psi \in C^1_c(\Omega,\mathbb{R}^n) \\ \|\psi(x)\|_2 \leq 1}} \int_{\Omega} \lambda \langle \psi, Du \rangle . \tag{19}$$

In the embedded problem, the regularizer is implemented via constraint (11). In order to calculate the (pointwise) Fenchel conjugate η^* of the regularizer, we use the following established connection between the support σ_{Ψ} and indicator δ_{Ψ} function of some convex set Ψ [14, Prop. 11.3, Example 11.4a)]:

$$\eta^*(\cdot) = (\sigma_{\Psi}(\cdot))^* = \delta_{\Psi}(\cdot). \tag{20}$$

This, together with the conjugacy rules listed on [14, p. 475], leads to the following equivalence:

$$\varphi_x(x,t) \in \operatorname{dom}\{\eta^*\} \quad \Leftrightarrow \quad \|\varphi_x(x,t)\|_2 \leq \lambda. \tag{21}$$

In order to enforce the constraint on the output of $\boldsymbol{F}^{\varphi_x}_{\theta^x}$, we choose a Tanh activation function as the last layer. This guarantees that the values of $\boldsymbol{F}^{\varphi_x}_{\theta^x}$ are restricted to $[-1,1]$. For anisotropic total variation this would suffice in case of $\lambda = 1$; for a more isotropic variant, we additionally perform the transformation

$$T^{\varphi_x} \circ \boldsymbol{F}^{\varphi_x}_{\theta^x}(x,t) := \begin{cases} \lambda \frac{\boldsymbol{F}^{\varphi_x}_{\theta^x}(x,t)}{\|\boldsymbol{F}^{\varphi_x}_{\theta^x}(x,t)\|_2}, & \text{if } \frac{1}{\lambda}\|\boldsymbol{F}^{\varphi_x}_{\theta^x}(x,t)\|_2 > 1, \\ \boldsymbol{F}^{\varphi_x}_{\theta^x}(x,t), & \text{otherwise,} \end{cases} \tag{22}$$

which enforces (21) by projecting onto the scaled unit ball with radius λ.

Dual Constraint $\mathcal{K}_t$. In the embedded problem, the data term is implemented via constraint (12). In order to enforce the constraint on the output of $\boldsymbol{F}^{\varphi_t}_{\theta^t}$, we choose a ReLU activation function as the last layer, which guarantees that $\boldsymbol{F}^{\varphi_t}_{\theta^t}$ assumes values in $[0, +\infty)$. Note that

$$\boldsymbol{F}^{\varphi_t}_{\theta^t} \geq 0 \quad \Leftrightarrow \quad \boldsymbol{F}^{\varphi_t}_{\theta^t} - \rho \geq -\rho \quad \Leftrightarrow \quad -(\boldsymbol{F}^{\varphi_t}_{\theta^t} - \rho) \leq \rho. \tag{23}$$

Comparing this to the required constraint (12), one sees that the constraint can be enforced by simply transforming the output of the last ReLU layer in the neural field $\boldsymbol{F}^{t}_{\theta^t}$ according to

$$T^{\varphi_t} \circ \boldsymbol{F}^{\varphi_t}_{\theta^t}(x,t) := \boldsymbol{F}^{\varphi_t}_{\theta^t}(x,t) - \rho(x,t). \tag{24}$$

5 Numerical Results

In the following, we investigate a proof-of-concept implementation of our method. We used the same network architecture for all experiments and trained on 2500 randomly chosen grid points per batch. For each experiment, we manually selected the variance parameter σ associated to the RFF encoding, the regularization parameter λ, the number of epochs, as well as the step size used for the SGD optimizer. The experiments were run on Ubuntu 18.04.6 LTS, Python 3.8, GeForce RTX 2070, CUDA 11.1, Intel(R) Core(TM) i7-8700 CPU @ 3.20 GHz and 64 GB of RAM. Training time was approximately 3–8 min per run.

5.1 Denoising

Consider the Rudin-Osher-Fatemi denoising problem

$$F(u) = \int_{\Omega} (u(x) - f(x))^2 \,\mathrm{d}x + \lambda TV(u). \tag{25}$$

While convex in itself and thus not the primary target of any convex relaxation method, this allows us to use a classical non-smooth convex solver (here CVXPY) in order to compute a ground truth. The results in Fig. 4a show that our neural field approach approximates the ground truth. A certain amount of additional blur is introduced, which we attribute to the RFF step, since retraining the model using starting weights from the previous run and iteratively increasing the variance σ does remove some of the additional noise.

One major challenge when working with embedding methods is the discretization of the label space Γ. The quality of the results is heavily influenced by the number and specific choice of labels, see Fig. 4b. Our new method represents the solution implicitly and, therefore, does not suffer from this issue.

In a related experiment, we test how the new approach handles missing data by masking parts of the data term

$$F(u) = \int_{\Omega} w(x)(u(x) - f(x))^2 \,\mathrm{d}x + \lambda TV(u). \tag{26}$$

We solved the problem with isotropic TV and respectively with and without the masking term, using the exact same parameters. Results can be seen in Fig. 4c.

5.2 Stereo Matching

A typical application for embedding/lifting approaches is the non-convex stereo matching problem. We assume that the two input images I_1 and I_2 are rectified, i.e., the epipolar lines in the images align, so that the unknown – but desired – displacement of points between the two images is restricted to the x_2 axis and can be modeled as a scalar function u. The variational problem is defined as

$$F(u) = \int_{\Omega} \max\left\{|I_1(x_1, x_2 + u(x)) - I_2(x_1, x_2)|, 0.15\right\} \mathrm{d}x + \lambda TV(u). \tag{27}$$

(a) **Ground truth.** Comparison of result from our neural fields-based approach **(center)** with results of a classic CVXPY-generated solution **(left)** on a convex problem. The difference `u_nf - u_cvx` is shown on the **right**. Our approach leads to a slightly blurrier solution, which we attribute to the RFF features.

(b) **Discretization.** One advantage of our new approach is that solutions are represented implicitly. In the embedding method, this is of special interest when considering the label axis Γ. By using 2, 7, 100 **(left to right)** labels during inference, we can easily demonstrate the discretization artefacts induced by a coarse grid with respect to Γ (the colormap is changed, such that the discrimination between values is easier). The coarser the discretization, the more cartoon-like are the solutions. Using classical optimization approaches, one has to weigh an increase in runtime against these discretization artefacts. In our approach, choosing a fine grid during inference does not increase the run time substantially.

(c) **Inpainting.** We solved an inpainting problem by masking parts of the data term. The input data is shown **(left)** and the masked area is marked by a rectangle. Next to it, is the solution of the EmNeF approach without **(center)** and with **(right)** the masking term.

Fig. 4. EmNeF applied to the convex denoising problem (25). In Fig. 4a and 4b, we trained for 50000 epochs, primal learning rate 0.3, dual learning rate 0.83, $\sigma = 4.3$ and $\lambda = 10^{-4}$. In Fig. 4c, we trained for 25000 epochs, primal learning rate 0.3, dual learning rate 0.83, $\sigma = 1$ and $\lambda = 5 \cdot 10^{-4}$.

(a) **Comparison to classical sublabel-accurate relaxation. Left:** Input image pair; red lines mark the sections shown in the profile plots in Fig. 5b.**Center:** results obtained by solving a sublabel-accurate discretization of (2) with the PDHG algorithm [10]. **Right:** results from the proposed approach based on neural fields. Although the EmNeF results are a little blurry, they show the potential of the proposed method: Without much fine-tuning, one can achieve comparable results to established algorithms. In addition, the EmNeF approach can easily be adapted to fit different data terms, whereas classical optimization strategies require to implement projections onto epigraphs, i.e., the conjugate of the data term, which is not a trivial problem and might require further approximation steps.

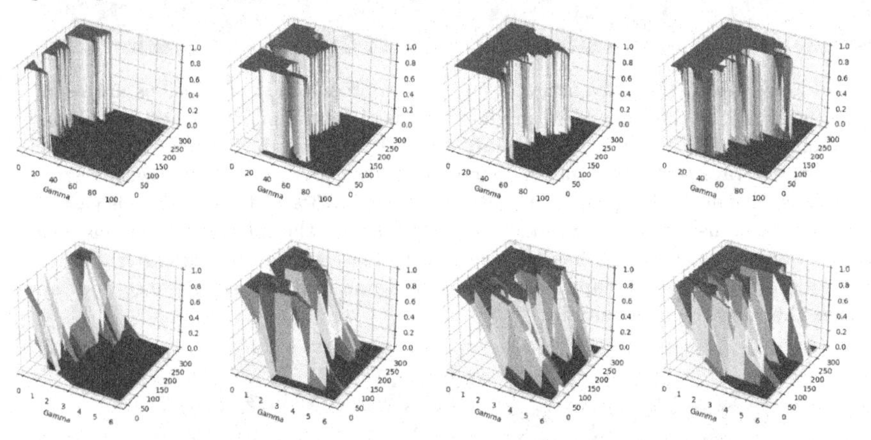

(b) **Integral solutions.** The figures depict sections over the x_2 and Γ axis of the solutions in Fig. 5a with fixed $x_1 \in \{10, 50, 90, 110\}$. Our new approach **(top)** learns implicit representations of the solution, therefore we can choose an arbitrary fine discretization with respect to the Γ axis during inference; here we chose 100 labels. The edges of the jumps are clearly defined, which shows that the learning approach has a bias towards integral solutions. The classical approach **(bottom)** provides explicit solutions but requires a full discretization with respect to Γ. The obtained solutions are less integral.

Fig. 5. EmNeF applied to the nonconvex stereo matching problem (27) on the "Umbrella" data set [15]. We optimize over 100000 epochs, with a primal learning rate of 0.2 and dual learning rate of 0.8. We set $\sigma = 4.3$, $\lambda = 9e-05$ and assume that the maximal shift is 30 pixels.

In [10,11], a sublabel-accurate discretization of the embedding method (2) is introduced. We used the two associated libraries `prost` and `sublabel_relax` provided on github to compute a ground truth. The results can be found in Figs. 5a and 5b.

The most striking difference between our solution and the ground truth is the smoothness of our results. We assume that it is linked to the choice of variance σ in the RFF encoding step. We observe that, in the beginning, a high variance hinders the neural fields at learning data relevant information. A possible solution and part of future work could be to adaptively choose σ.

6 Outlook

In this work, we considered variational problems with 1-homogeneous, convex regularizer and nonconvex data term. We used an established embedding method to express the problem in a higher-dimensional space, such that the resulting energy is convex and the global minimizer relates to the global minimizer of the original problem. Instead of using classical non-smooth first-order optimization for solving the primal-dual problem, we proposed to learn a representation of the solution by training multiple neural fields in a GAN-inspired way.

We believe that the numerical results presented in this work are very promising: although theory is still mostly open, the approach empirically allows to obtain solutions comparable to those of classical solvers, with a surprisingly simple implementation that can easily be adapted to different applications by changing the pointwise implementation of the (original, low-dimensional) data term and adapting the parameters σ and λ. In comparison, classical primal-dual optimization strategies for solving the embedded problem require to implement projections onto epigraphs, e.g., of the conjugate of the data term, which is non-trivial and can require further approximation steps.

Future work could be concerned with a more analytical study of the approach such as convergence behavior, adaptive strategies for choosing σ, refining the network architecture, and investigating how well the continuous solution space is approximated.

Acknowledgements. The authors acknowledge support through DFG grant LE 4064/1-1 "Functional Lifting 2.0: Efficient Convexifications for Imaging and Vision" and NVIDIA Corporation.

References

1. Alberti, G., Bouchitté, G., Dal Maso, G.: The calibration method for the Mumford-shah functional and free-discontinuity problems. Calc. Var. Partial. Differ. Equ. **16**(3), 299–333 (2003). https://doi.org/10.1007/s005260100152
2. Chambolle, A., Pock, T.: A first-order primal-dual algorithm for convex problems with applications to imaging. J. Math. Imaging Vis. **40**(1), 120–145 (2011). https://doi.org/10.1007/s10851-010-0251-1

3. Chambolle, A.: Convex representation for lower semicontinuous envelopes of functionals in L1. J. Convex Anal. **8**(1), 149–170 (2001)
4. Cuomo, S., Di Cola, V.S., Giampaolo, F., Rozza, G., Raissi, M., Piccialli, F.: Scientific machine learning through physics-informed neural networks: where we are and what's next. J. Sci. Comput. **92**(3), 88 (2022). https://doi.org/10.1007/s10915-022-01939-z
5. Goodfellow, I., et al.: Generative adversarial networks. Commun. ACM **63**(11), 139–144 (2020). https://doi.org/10.1145/3422622
6. Ishikawa, H.: Exact optimization for Markova random fields with convex priors. IEEE Trans. Pattern Anal. Mach. Intell. **25**(10), 1333–1336 (2003). https://doi.org/10.1109/TPAMI.2003.1233908
7. Laude, E., Möllenhoff, T., Moeller, M., Lellmann, J., Cremers, D.: Sublabel-accurate convex relaxation of Vectorial multilabel energies. In: Leibe, B., Matas, J., Sebe, N., Welling, M. (eds.) ECCV 2016. LNCS, vol. 9905, pp. 614–627. Springer, Cham (2016). https://doi.org/10.1007/978-3-319-46448-0_37
8. Mildenhall, B., Srinivasan, P.P., Tancik, M., Barron, J.T., Ramamoorthi, R., Ng, R.: NeRF: representing scenes as neural radiance fields for view synthesis. Commun. ACM **65**(1), 99–106 (2021). https://doi.org/10.1145/3503250
9. Moöllenhoff, T., Strekalovskiy, E., Moeller, M., Cremers, D.: The primal-dual hybrid gradient method for semiconvex splittings. SIAM J. Imag. Sci. **8**(2), 827–857 (2015). https://doi.org/10.1137/140976601
10. Möllenhoff, T., Laude, E., Möller, M., Lellmann, J., Cremers, D.: Sublabel-accurate relaxation of nonconvex energies. CoRR abs/1512.01383 (2015). https://doi.org/10.1109/CVPR.2016.428
11. Möllenhoff, T., Cremers, D.: Sublabel-accurate discretization of nonconvex free-discontinuity problems. In: Proceedings of the IEEE International Conference on Computer Vision. pp. 1183–1191 (2017). https://doi.org/10.1109/ICCV.2017.134
12. Pock, T., Cremers, D., Bischof, H., Chambolle, A.: Global solutions of variational models with convex regularization. SIAM J. Imag. Sci. **3**(4), 1122–1145 (2010). https://doi.org/10.1137/090757617
13. Rockafellar, R.T.: Convex Analysis, vol. 28. Princeton University Press (1970)
14. Rockafellar, R.T., Wets, R.J.: Variational Analysis, vol. 317. Springe, Heidelberg (2009). https://doi.org/10.1007/978-3-642-02431-3
15. Scharstein, D., et al.: High-resolution stereo datasets with subpixel-accurate ground truth. In: Jiang, X., Hornegger, J., Koch, R. (eds.) GCPR 2014. LNCS, vol. 8753, pp. 31–42. Springer, Cham (2014). https://doi.org/10.1007/978-3-319-11752-2_3
16. Scherzer, O., Grasmair, M., Grossauer, H., Haltmeier, M., Lenzen, F.: Variational Methods in Imaging, Applied Mathematical Sciences, vol. 167. Springer, New York (2009). https://doi.org/10.1007/978-0-387-69277-7
17. Tancik, M., et al.: Fourier features let networks learn high frequency functions in low dimensional domains. Adv. Neural. Inf. Process. Syst. **33**, 7537–7547 (2020)
18. Valkonen, T.: A primal-dual hybrid gradient method for nonlinear operators with applications to MRI. Inverse Prob. **30**(5), 055012 (2014)
19. Vogt, T.: Measure Valued Variational Models With Application In Image Processing. Ph.D. thesis, Universität zu Lübeck (2019)
20. Xie, Y., et al.: Neural fields in visual computing and beyond. In: Computer Graphics Forum, vol. 41, pp. 641–676. Wiley Online Library (2022). https://doi.org/10.1111/cgf.14505
21. Zheng, J., Ramasinghe, S., Lucey, S.: Rethinking positional encoding. arXiv preprint arXiv:2107.02561 (2021)

GenHarris-ResNet: A Rotation Invariant Neural Network Based on Elementary Symmetric Polynomials

Valentin Penaud--Polge(✉), Santiago Velasco-Forero, and Jesus Angulo

Mines Paris, PSL University, Center for Mathematical Morphology (CMM), Fontainebleau, France
{valentin.penaud_polge,santiago.velasco,jesus.angulo}@minesparis.psl.eu

Abstract. In this paper, we propose a rotation invariant neural network based on Gaussian derivatives. The proposed network covers the main steps of the Harris corner detector in a generalized manner. More precisely, the Harris corner response function is a combination of the elementary symmetric polynomials of the integrated dyadic (outer) product of the gradient with itself. In the same way, we define matrices to be the self dyadic product of vectors composed with higher order partial derivatives and combine the elementary symmetric polynomials. A specific global pooling layer is used to mimic the local pooling used by Harris in his method. The proposed network is evaluated through three experiments. It first shows a quasi perfect invariance to rotations on Fashion-MNIST, it obtains competitive results compared to other rotation invariant networks on MNIST-Rot, and it obtains better performances classifying galaxies (EFIGI Dataset) than networks using up to a thousand times more trainable parameters.

Keywords: Elementary Symmetric Polynomials · Structure Tensor · Harris Corner Detector · Gaussian Derivatives · Neural Network · Galaxy Morphologies

1 Introduction

Deep neural networks have been successfully applied across a large number of areas, including but not limited to computer vision [8], remote sensing [32], biology [23]. In many of these areas the data contains geometric properties or symmetries that can be exploited when designing neural networks. Prior to the Deep Learning (DL) era, *rotation invariance* in computer vision were mainly inspired by the Harris Corner Detector (HCD) [12], through local features extraction [21]. The main idea behind the HCD is to extract features based on the local geometry of the image. To do so the trace and the determinant of a matrix called *structure tensor*, defined to be the locally averaged outer product of the gradient with

This work was granted access to the HPC resources of IDRIS under the allocation 2022-[AD011013367] made by GENCI.

L. Calatroni et al. (Eds.): SSVM 2023, LNCS 14009, pp. 149–161, 2023.
https://doi.org/10.1007/978-3-031-31975-4_12

itself, are used to define a feature called the corner response function. Then this feature function is used to detect corners and edges depending where the function takes its local extrema. In DL, the use of intrinsic properties of invariance and equivariance from the data is an active area of research [3,6,17,28]. Two of the main motivations to study equi/in-variance in neural networks are quite intuitive: i) Objects in natural images are not always oriented, scaled or localized in the same way unless the environment of acquisition is highly constrained. ii) It has already shown outstanding results with the example of *translation equivariance* whose most known representatives are Convolutional Neural Networks (CNN). In order to fix the principal notions of the paper, we recall the definition of *equivariance* and *invariance*. Let $f(\cdot)$ be a function and $g(\cdot)$ a transformation. f is equivariant, respectively invariant, to g if and only if $g(f(\cdot)) = f(g(\cdot))$, respectively $g(f(\cdot)) = f(\cdot)$. If f is equi/in-variant to all the transformations of a group $\mathcal{G}$, then f is said to be equi/in-variant to $\mathcal{G}$. In this paper, we focus on the invariance to the group of rotations. Our contribution in this paper is threefold: First, we generalize the corner response function to higher orders of derivation using structure tensor with higher derivatives, using the entire set of elementary symmetric polynomials and not only the determinant and the trace. Second, we propose and validate the use of top-k max/min pooling as feature extractor. Third, we show that the proposed approach can be used in the context of DL with the *GenHarris-ResNet*.

2 Related Work

The structure tensor is well known in the image processing community [4]. Its famous application remains the HCD [12]. Several generalizations of this structure tensor have been proposed. An approach involves applying the complete HCD method to higher dimensional data, for instance in [15], a space-time interest point detector is proposed for moving corners, object collisions and more generally changes in movements. A generalization of the structure tensor to higher-order tensors is proposed in [29], where the outer product of the gradient vector is applied several times instead of only once. With this high-order tensor, a contrast function is defined whose extrema is used to detect edges and junctions. Another generalized structure tensor was proposed in [2] which corresponds to the classical structure tensor but in a harmonic system of coordinates. In [22], a theory allowing local characterisation of multiple orientations was presented. It supposes that an image is a linear combination of several images carrying, each of them, an orientation. It can be noted that the order of derivation used in the generalized structure tensor is directly linked to the number of orientations to be estimated. This generalization of the structure tensor is the closest to ours and the only difference comes from factors weighting the components of the tensor which allow us to obtain interesting mathematical properties. Rotation invariance has been addressed through various paradigms in DL. The most known approaches are spatial transformer networks [14] and group equivariant convolutional networks [6]. Spatial transformers are used to predict a transformation applied to the input tensor in order to use only a properly transformed section of it. The second approach is intrinsically, by its mathematical construction, invariant to a group of rotations. The principle of group equivariant convolutional networks is to extend convolution to other types of transformations and not only translation. Thus, the filters of a group equivariant convolutional network not only runs through all the possible translation in the input tensor but also through a (restricted) discrete group of rotations. Nevertheless, this approach has the advantage of data efficiency as the learned filters are common to all the rotations of the chosen group. In order to reduce the computation resources, a modified but similar approach has been proposed in [19]. Instead of keeping all the rotated copies of the tensor from a layer to another, a pooling operation along the rotation axis is applied to the output tensor of each layer. In order to keep the orientation information, the pooling does not only return the values of the tensor but also the angle value of the orientation map it has been pooled from. In [11], a CNN using constrained filters is proposed to obtain rotation invariance as the filters used in this network are steerable. But a discretization of the angles is still mandatory even though the angle sampling can be finer than [6,11] without interpolation. Finally, a recently published paper [28] proposes to use linear combinations of differential invariants as the filters. The used differential invariants are derived using the moving frame method and consist in normalized polynomials of first and second order Gaussian derivatives which makes this network the closest to ours among the mentioned ones.

3 Mathematical Preliminaries

In this section we present the mathematical foundation of the Gaussian Elementary Symmetric Polynomials (ESPs) Layer, key element of the architecture of a GenHarris-ResNet. We will first recall basic knowledge about Gaussian derivative kernels. Then we will present a simple generalization of the structure tensor to higher order of derivation, which we call *Generalized Structure Tensor for Higher Derivatives* (GSTHD). Finally we will recall how to compute the ESPs through the power sums and the Girard-Newton's identities without the direct use of the eigenvalues.

3.1 Gaussian Partial Derivatives

The one-dimensional centered Gaussian kernel and its pth derivative are respectively given by

$$G(x,\sigma) = \frac{1}{(2\pi\sigma^2)^{\frac{1}{2}}} e^{\frac{-x^2}{2\sigma^2}} \; ; \; G_p(x,\sigma) = \frac{\partial^p G(x,\sigma)}{\partial x^p}, \tag{1}$$

where σ represents the standard deviation (or scale). Gaussian derivatives of any order can be expressed as a Hermite polynomial multiplied by a Gaussian kernel, and as Hermite polynomials are easily expressible, we will use them to formulate the isotropic Gaussian derivative kernels without any use of derivation operations. The one-dimensional Hermite polynomial of order p is given by the following equation:

$$H_p(x) = \sum_{i=0}^{\lfloor p/2 \rfloor} \frac{(-1)^i p!}{i!(p-2i)!} (2x)^{p-2i} . \tag{2}$$

The relation between the one-dimensional Gaussian derivative of order p and the Hermite polynomial of order p is given by

$$G_p(x,\sigma) = \left(-\frac{1}{\sqrt{2}\sigma}\right)^p \sqrt{2\pi}\sigma H_p\left(\frac{x}{\sqrt{2}\sigma}\right) G\left(\frac{x}{\sqrt{2}},\sigma\right)^2 . \tag{3}$$

This relation between Hermite polynomials and Gaussian derivatives has been used several times in papers applying Gaussian derivative kernels to CNNs [13, 25]. From this, we can define the two-dimensional isotropic separated Gaussian derivative kernel of order $p+q$:

$$G_{p,q}(x_1,x_2,\sigma) = G_p(x_1,\sigma) G_q(x_2,\sigma) . \tag{4}$$

The convolution of a Gaussian derivative kernel $G_{p,q}(\cdot,\cdot,\sigma)$ with an image I with be denoted $L_{p,q}^{(\sigma)}(\cdot) := G_{p,q}(\cdot,\sigma) * I(\cdot)$.

3.2 Generalized Structure Tensor for Higher Derivatives

The structure tensor is defined to be the integrated dyadic product of the Gaussian gradient with itself. Before defining the proposed GSTHD, we first recall the definition of the dyadic (outer) product and an associated, simple but useful, lemma which is the key to the rotational equivariance property.

Dyadic Product and Orthogonal Matrices. Given two vectors u, v n-dimensional in $\mathbb{R}^n$ with $u = (u_1, ..., u_n)$ and $v = (v_1, ..., v_n)$, the dyadic product of u with v, written as $u \otimes v$, is defined by

$$\forall i,j \in [\![1,n]\!],\ (u \otimes v)_{i,j} = u_i v_j \tag{5}$$

The following lemma will be taken into consideration to define the vectors used to construct the GSTHD as it informs us how the dyadic product of a vector with itself behaves when an orthogonal transformation is applied to the vector. We will denote by $\mathcal{O}^n(\mathbb{R})$, the set of real orthogonal $n \times n$ matrices, i.e., $\forall P \in \mathcal{O}^n(\mathbb{R})$, $P^T = P^{-1}$.

Lemma 1. *$\forall u, v \in \mathbb{R}^n$, if $\exists P \in \mathcal{O}^n(\mathbb{R})$ s.t. $u = Pv$, then*

$$u \otimes u = (Pv) \otimes (Pv) = P(v \otimes v) P^{-1}.$$

Therefore if we define the vector used to compute the GSTHD in a way that a rotation applied to the input image (or tensor) implies an orthogonal transformation of the vector, then the ESPs can be used as rotational equivariant features as they are similarity invariants.

Generalized Structure Tensor for Higher Derivatives. The two main requirements to properly define the GSTHD are: i) to be a generalization of the well known tensor structure when considering first order derivatives and ii) to take into account Lemma 1. For an order $k \in \mathbb{N}$, we propose to use the following vector defined at every coordinate of the image:

$$\mathcal{L}_k^{(\sigma)} = \left(\sqrt{\binom{k}{p}} L_{k-p,p}^{(\sigma)}\right)_{p \in [\![0,k]\!]} \tag{6}$$

where $\binom{k}{p} = \begin{cases} \frac{k!}{p!(k-p)!} & \text{if } 0 \leq p \leq k \\ 0 & \text{otherwise} \end{cases}$.

We can immediately observe that the case $k = 1$ gives the gradient as required by the first requirement. The GSTHD of order k, scale σ and integrated at scale σ_{int}, denoted as $\mathcal{M}_k^{(\sigma)}$ is, pointwise (or pixelwise) defined as follow:

$$\mathcal{M}_k^{(\sigma)} = G(\cdot, \sigma_{int}) * \left(\mathcal{L}_k^{(\sigma)} \otimes \mathcal{L}_k^{(\sigma)}\right) \tag{7}$$

Before taking into account the second requirement stated above, let us define the notations when a rotation is applied to the input image $I : \mathbb{R}^2 \to \mathbb{R}$. We will denote as (x_1, x_2) the system of coordinates attached to I. If a rotation of angle θ is applied to I, the coordinates also transform themselves into a rotated ones (x_1', x_2') through the relation

$$\begin{pmatrix} x_1' \\ x_2' \end{pmatrix} = \begin{pmatrix} \cos(\theta) & \sin(\theta) \\ -\sin(\theta) & \cos(\theta) \end{pmatrix} \begin{pmatrix} x_1 \\ x_2 \end{pmatrix}. \tag{8}$$

In the rest of this subsection, every mathematical object expressed in this rotated system of coordinates will hold an apostrophe. As an example $\mathcal{L'}_k^{(\sigma)}$ corresponds to $\mathcal{L}_k^{(\sigma)}$ expressed with the rotated coordinates. The operator describing a rotation of angle θ applied to an image will be denoted $R_\theta(.)$, i.e., $I' = R_\theta(I)$.

The following result allows us to use Lemma 1, i.e., answers to the second requirement.

Proposition 1. *Let I be the input image, $\forall \theta \in [-\pi, \pi], \forall k \in \mathbb{N}$:*

$$I' = R_\theta(I) \implies \exists \mathcal{P}_k(\theta) \in \mathcal{O}^{k+1}(\mathbb{R}) \mid \mathcal{L'}_k^{(\sigma)} = \mathcal{P}_k(\theta)\mathcal{L}_k^{(\sigma)} \text{ and } \mathcal{M'}_k^{(\sigma)} = \mathcal{P}_k(\theta)\mathcal{M}_k^{(\sigma)}\mathcal{P}_k(\theta)^T.$$

More precisely $\forall i, j \in [\![0, k]\!]$,

$$(\mathcal{P}_k(\theta))_{i,j} = \frac{\sqrt{\binom{k}{i}}}{\sqrt{\binom{k}{j}}} \sum_{r=max(0,i+j-k)}^{min(i,j)} \binom{i}{r}\binom{k-i}{j-r}(-1)^{i-r}\sin(\theta)^{i+j-2r}\cos(\theta)^{k+2r-i-j}. \tag{9}$$

3.3 Elementary Symmetric Polynomials

This section relies on basic mathematical knowledge [18]. We showed in Proposition 1 that applying a rotation to the input image/tensor only results in an orthogonal change of basis of $\mathcal{M}_k^{(\sigma)}$ at each point of the image. This means that using elementary symmetric polynomials, which are similarity invariants of these matrices, leads to rotational equivariant features. In this subsection we recall the definition of the ESPs and how to compute them without directly knowing the spectrum of the studied matrices.

Let $M \in M_n(\mathbb{R})$ be a diagonalizable real matrix and let $\Lambda = (\lambda_1, ..., \lambda_n)$ be its eigenvalues. For all k in $[\![1, n]\!]$, the Elementary Symmetric Polynomials are defined as

$$e_k(\lambda_1, ..., \lambda_n) = \sum_{1 \leq i_1 < ... < i_k \leq n} \prod_{j=1}^{k} \lambda_{i_j}. \tag{10}$$

The two mostly used ESPs are the determinant (e_n) and the trace (e_1) of a matrix as it is the case, for example, in the Harris corner detector [12] where the corner response is defined to be $R = e_2\left(\mathcal{M}_1^{(\sigma)}\right) - \beta\left(e_1\left(\mathcal{M}_1^{(\sigma)}\right)\right)^2$. As we propose a generalization to the Harris corner detector for deep learning by using matrices of higher dimensions than two, we will use the entire set of ESPs of each matrix. Nevertheless, as it is not suitable to define a layer computing directly the eigenvalues of some matrices, we propose to use Girard-Newton's identities to avoid this problem. To this end, we recall the definition of the power sums of Λ of order k:

$$p_k(\lambda_1, ..., \lambda_n) = \sum_{i=1}^{n} \lambda_i^k. \tag{11}$$

These quantities have two major advantages: First, they are easy to compute in the case of a diagonalizable matrix, $p_k(\Lambda) = \text{trace}\left(M^k\right)$, and second, they can be used to compute the ESPs using the Girard-Newton's identities:

$$e_k(\Lambda) = \frac{1}{k} \sum_{i=1}^{k} (-1)^{i-1} e_{k-i}(\Lambda) p_i(\Lambda). \tag{12}$$

Finally, it can be highlighted that using a Gaussian kernel to integrate the matrices $\mathcal{M'}_k^{(\sigma)}$ does not alter the result of Proposition 1 as the matrices of each pixel of a rotated image are all transformed using the matrix $\mathcal{P}_k(\theta)$.

4 GenHarris-ResNet

As stated in Sect. 1, the GenHarris-ResNet aims to generalize, in a deep learning approach, the HCD. As highlighted above, the HCD can be divided into two main stages: the computation of a feature function (a corner response function) and the selection of interest points using the feature function as a decision criterion. We present in this section how we propose to adapt these two processes to deep

Algorithm 1. ESP Layer Algorithm

Given an input tensor I and the hyperparameters derivation order N, a list of scale values Σ and an integration scale σ_{int}

1. Create empty list E
2. For σ in Σ
   (a) For $i = 0$ to N
      i. Compute the $\mathcal{L}_i^\sigma$ channelwise.
      ii. Compute the outer product of $\mathcal{L}_i^\sigma$ with itself.
      iii. $\mathcal{M}_i^\sigma \leftarrow$ Convolution of each element of the resulting matrix across all the channels with $G(\cdot, \sigma_{int})$
      iv. Compute the homogenized absolute ESPs $\{|e_1|, |e_2|^{\frac{1}{2}}, ..., |e_{i+1}|^{\frac{1}{i+1}}\}$ of $\mathcal{M}_i^\sigma$ using the power sums and the Girard-Newton's identities.
      v. Add $e_1, (e_2)^{\frac{1}{2}}, ..., (e_{i+1})^{\frac{1}{i+1}}$ to E
3. Return E stacked into a tensor.

learning: one with the definition of a feature function block, composed of an ESP layer, two linear combinations and a normalization. And the other with a global extrema pooling layer.

In the same way Harris and Stephens used the trace and the determinant of the structure tensor to obtain the corner response function, we propose to use the elementary symmetric polynomials of the GSTHD and to combine them. More precisely, the ESP layer has three hyperparameters: a maximal order of derivation N, a list of scale values Σ and an integration σ_{int}. Given an input tensor I, the ESP Layer performs Algorithm 1. It can be highlighted that no parameters are learned in the ESP layer, it only computes ESPs in a differentiable manner. Once the ESPs are computed, two convolutional layers using filters of size 1×1 and a layer normalization are used to combine them. The combination of these layers constitutes a feature function block and several blocks in cascade with residual connections constitute the feature function part of the GenHarris-ResNet. After computing the corner response function, Harris and Stephens determine the interest points (corners and edges) through a MaxPooling and a directional MinPooling. In other words, Harris and Stephens proposed to use the local extrema of the corner response function. In our GenHarris-ResNet, we propose to use a global pooling layer which returns the values of the top $2K$ global maxima and global minima, i.e., the K highest values and the K lowest values. Note that global pooling changes the rotation equivariance property of the feature function part of the network into rotation invariance [3]. Finally, a dense layer ends the GenHarris-ResNet in the case of a classification task. The overall architecture of the GenHarris-ResNet is illustrated in Fig. 1.

5 Experiments

This section contains three parts: i) A rotation invariance study using Fashion-MNIST. ii) A comparison with previous works on MNIST-Rot. iii) The

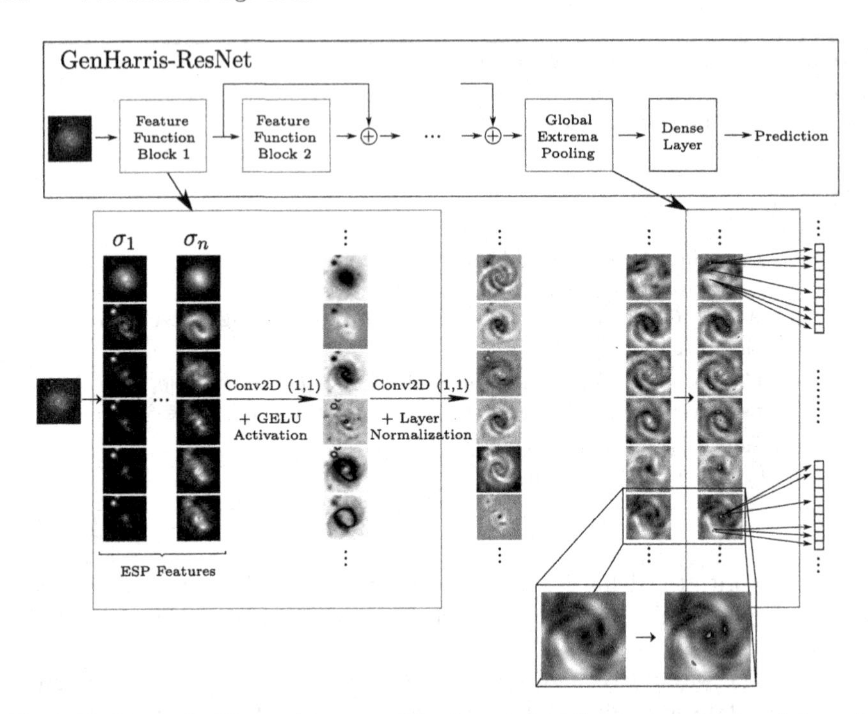

Fig. 1. GenHarris-ResNet architecture: The first part of the proposed network is equivariant to rotation and is composed of several feature function blocks arranged in a ResNet configuration. Then, the feature maps are given to the global extrema pooling layer which brings the invariance property. A dense layer ends the network.

application of the GenHarris-ResNet to a real-images problem: the classification of galaxy morphologies using the EFIGI Dataset [1]. An overall ablation study is realized in the first and last experiments.

Rotation Invariance. Fashion-MNIST is used to study the effects of the number of feature function blocks used in the GenHarris-ResNet and of the value taken by K in the global extrema pooling layer. We also test the invariance to rotations of the proposed model. We trained several GenHarris-ResNets on Fashion-MNIST without any data augmentation. The considered models are composed of either two or three feature function blocks (denoted as $L = 2$ or $L = 3$ in Table 1). The feature function blocks had a maximal order of derivation N values two and a list of scales values $[1.0, 2.0, 3.0]$. Several values of K for the global extrema pooling were tested. The rotation invariance of the trained GenHarris-ResNets has been tested on rotated versions of the entire test dataset of Fashion-MNIST using bilinear interpolation. The rotations applied to the test dataset have been discretized with an angle step value of $10°$, giving 36 rotated versions of the samples. Three quantities have been used to evaluate the networks: i) The accuracy at $\theta = 0$, denoted Acc_0, to evaluate the classification performance of the networks. ii) The mean accuracy deviation $\mathbb{E}\left(|\mathrm{Acc}_0 - \mathrm{Acc}\left(\theta\right)|\right)$.

Table 1. Test accuracy, rotation invariance and number of parameters of several configurations of the GenHarris-ResNet trained on Fashion-MNIST

Network	Acc_0	$\mathbb{E}(\lvert\text{Acc}_0 - \text{Acc}(\theta)\rvert)$	$\mathbb{E}(\text{Cons Error}(\theta))$	nb param
Vanilla CNN	91.88%	68.31%	76.21%	12.4k
$L = 2$, GlobalMaxPooling	87.76%	0.99%	5.63%	3.8k
$L = 2$, $K = 1$	87.92%	1.10%	5.55%	4.1k
$L = 2$, $K = 2$	88.40%	1.27%	5.15%	4.7k
$L = 2$, $K = 10$	89.03%	0.99%	4.94%	9.9k
$L = 2$, $K = 20$	88.62%	0.83%	4.58%	16.3k
$L = 3$, $K = 2$	89.70%	0.82%	4.62%	6.5k
$L = 3$, $K = 10$	90.26%	1.00%	3.94%	11.6k
$L = 3$, $K = 20$	89.54%	1.04%	4.23%	18k

iii) The *mean consistency error*, denoted $\mathbb{E}(\text{Cons Error}(\theta))$. We recall that the consistency is the percentage of predictions that did not change after applying the transformation, even the wrong predictions are taken into account. What is called here the consistency error is in the contrary, the percentage of predictions that changed. Table 1 shows the performances of the GenHarris-ResNets compared to the performances of a classical three layers CNN as a reference for comparison. The results show that despite the fact that the CNN slightly outperforms the GenHarris-ResNets when no rotation is applied to the test dataset, the proposed networks show a nearly perfect invariance to rotations even though no transformed data has been used during training. It should be highlighted that at 90°, 180° and 270°, the consistency values of all the GenHarris-ResNets are above 99.9%, implying that the non perfect invariance may come from the interpolation method used to rotate the images, the low resolution of the images and the discretization of the Gaussian derivative kernels. It is observable that for a fixed value of K, using three feature function blocks increases the performance of the GenHarris-ResNet, when applied to Fashion-MNIST. The better performances obtained when using the global extrema pooling compared to global max pooling show that using only one value to describe one image is not the best way to summarize feature maps to discriminate properly the classes. The increase of K from one to ten in the global extrema pooling layer leads to a small increase of the performance of the network until it reaches $K = 20$ and decreases. This suggests that an optimal number of values exists to summarize the feature maps when using extreme values. Using less values than the optimal one may not be enough to summarize correctly the feature maps and using more extreme values may cause too much redundancy while unnecessarily increasing the number of trainable parameters.

Comparison on MNIST-Rot. In order to compare the performance of the GenHarris-ResNet to other methods dealing with rotation invariance in deep

Table 2. Comparison of the test error obtained on MNIST-Rot when training on MNIST with the five best results from the paper [26]. Networks: RP_RF_1_32 [10], Spherical CNN [7], ORN-8 (ORAlign) [31], RED-NN [26], Covariant CNN [27].

Network	RP_RF_1	Spherical CNN	ORN-8	RED-NN	Covariant CNN	Ours
% Test error	12.20	6.00	16.24	2.05	17.21	3.83
# parameters	1M	68k	969k	42k	7k	11.6k

learning, we tested it on MNIST-Rot [16] as it represents the standard dataset for this task. Nevertheless, we believe that evaluating the invariance of a model to a transformation should be done without including transformed samples in the training dataset to truly appreciate the intrinsic invariance property. We chose the configuration of the GenHarris-ResNet that obtained the best accuracy in the previous experiment and we trained it from scratch on the training set of MNIST without any data augmentation and we used 2k images of the MNIST-rot training dataset for validation. We compare the performance of our model with networks of the literature trained in similar conditions. The test error obtained on MNIST-Rot is presented in Table 2 together with published results. Even though the GenHarris-ResNet does not have the lowest error value, it challenges the state of art methods while having a low number of trainable parameters.

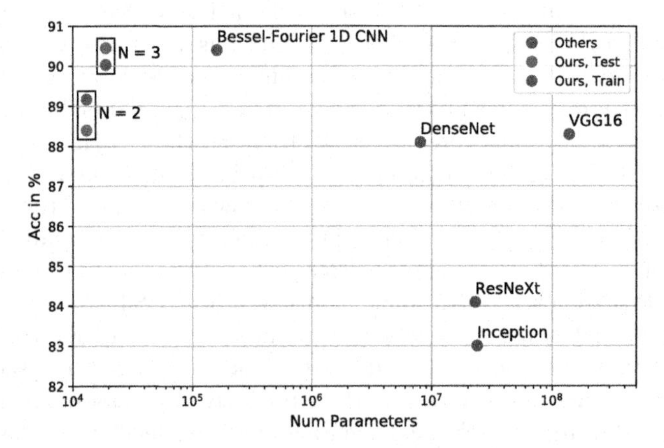

Fig. 2. Test accuracy values of related work networks on EF-4: blue points. Test (respectively training) accuracy values of the proposed networks: red (respectively green) points. It can be pointed out that the model 'Bessel-Fourier 1D CNN' [5] (Color figure online) has similar performances to the proposed model but does not hold invariance to translations due to the use of Bessel-fourier moments (its use is mainly restricted to images displaying centered objects).

Table 3. Confusion matrix of the test predictions obtained for the model N = 3 with the following label indexing: 0: Elliptical, 1: Irregular, 2: Lenticular, 3: Spiral.

		Predicted labels			
		0	1	2	3
True labels	0	27	0	1	1
	1	0	21	0	4
	2	11	0	36	7
	3	2	7	9	314

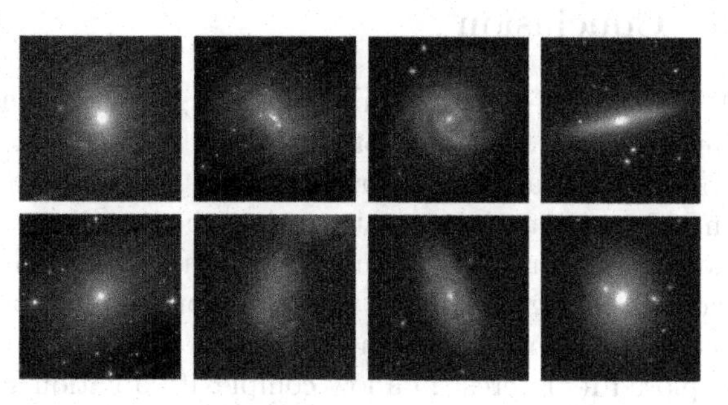

Fig. 3. Samples from the four classes of EF-4. From left to right: Elliptical, irregular, spiral, lenticular.

Galaxy Morphologies. This last experiment aims to apply the GenHarris-ResNet to a classification task on real images. As the feature function block uses the local structure of images to compute its feature maps, we applied the GenHarris-ResNet to the classification of galaxy morphologies to demonstrate its ability to capture the geometry of objects in images. It also serves to complete an overall ablation study of the layer, through the study of the effect of N, the maximal order of derivation in the ESP Layer. We used the EF-4 configuration, named after its number of classes, of the EFIGI dataset [20]. See Fig. 3 for examples of images of each class. The EF-4 dataset was randomly split in a training/validation set using 90% of the images and a testing set using the remaining samples. Two GenHarris-ResNet composed of three feature function blocks have been used for this task: one using only feature function blocks with $N = 2$ and another using $N = 3$ in the last two blocks. We found better performances using $K = 30$ for the second order GenHarris-ResNet and $K = 50$ for the third order GenHarris-ResNet. The test accuracy values obtained are shown in Fig. 2 together with the performances obtained by other networks [5,20]. As the random split of the EF-4 dataset is surely not the same between our networks and the others, we also give the accuracy values obtained on the training set to show that the results obtained by our networks are not due to a lucky composition of the test set. The results presented in Fig. 2 show that the GenHarris-ResNets have competitive performances compared with the other networks while using 10 to 1000 times less trainable parameters. The results also show that, in the case of this dataset, using a maximal order of derivation equals to three allows to obtain better performances than using $N = 2$. We highlight that this result might not hold in general because the use of $N = 3$ requires a value of σ_{int}, in (7), high enough so that the matrices $\mathcal{M}_k^\sigma$ are well defined. Finally, the confusion matrix of the third order GenHarris-ResNet obtained on the test dataset is given in Table3 for future comparisons.

6 Conclusion

We proposed a low-parameter neural network architecture which allows us to introduce the paradigm of the Harris Corner Detector to DL. The motivation was to define a new network using feature maps that are equivariant to rotations in a rigorous mathematical way, i.e., the invariance property is intrinsic and does not depend on data augmentation: the generalization to orientations not considered during training is total. The proposed approach extended the structure tensor to higher Gaussian derivatives and used a larger set of ESPs. We have explored its interest in a low complexity ablation study which revealed a nearly perfect invariance to rotations. The GenHarris-ResNet also showed competitive performances with related approaches for rotation invariance when evaluated on MNIST-Rot and outperformed networks using millions of trainable parameters on the task of galaxy classification based on their morphology.

Code, Data and Proofs Availability. The code, the EF-4 dataset splits and detailed proofs (based on [9,24,30]) are available at https://github.com/Penaud-Polge/GenHarrisResNet.

References

1. Baillard, A., Bertin, E., De Lapparent, V., et al.: The EFIGI catalogue of 4458 nearby galaxies with detailed morphology. Astron. Astrophy. **532**, A74 (2011)
2. Bigun, J., Bigun, T., Nilsson, K.: Recognition by symmetry derivatives and the generalized structure tensor. IEEE PAMI **26**(12), 1590–1605 (2004)
3. Bronstein, M.M., Bruna, J., Cohen, T., Veličković, P.: Geometric deep learning: Grids, groups, graphs, geodesics, and gauges. arXiv preprint arXiv:2104.13478 (2021)
4. Brox, T., Weickert, J., Burgeth, B., Mrázek, P.: Nonlinear structure tensors. Image Vis. Comput. **24**(1), 41–55 (2006)
5. Camacho-Bello, C.J., Gutiérrez-Lazcano, L., et al.: Rotation-invariant image classification using a novel 1D CNN and multichannel accurate Bessel-Fourier moments. Boletín de Ciencias Básicas e Ingenierías **10**(3), 1–4 (2022)
6. Cohen, T., Welling, M.: Group equivariant convolutional networks. In: International Conference on Machine Learning, pp. 2990–2999. PMLR (2016)
7. Cohen, T.S., Geiger, M., Köhler, J., Welling, M.: Spherical CNNs. In: 6th International Conference on Learning Representations, ICLR 2018, Vancouver, BC, Canada, 30–May 3 April 2018, Conference Track Proceedings (2018)
8. Feng, D., Haase-Schütz, C., et al.: Deep multi-modal object detection and semantic segmentation for autonomous driving: Datasets, methods, and challenges. Trans. Intell. Transp. Syst. **22**(3), 1341–1360 (2020)
9. Florack, L.M., et al.: Scale and the differential structure of images. Image Vis. Comput. **10**(6), 376–388 (1992)
10. Follmann, P., Bottger, T.: A rotationally-invariant convolution module by feature map back-rotation. In: 2018 IEEE WACV, pp. 784–792. IEEE (2018)
11. Graham, S., Epstein, D., Rajpoot, N.: Dense steerable filter CNNs for exploiting rotational symmetry in histology images. IEEE Trans. Med. Imaging **39**(12), 4124–4136 (2020)

12. Harris, C., Stephens, M., et al.: A combined corner and edge detector. In: Proceedings of the Alvey Vision Conference, pp. 23.1-23.6 (1988)
13. Jacobsen, J.H., Van Gemert, J., Lou, Z., Smeulders, A.W.: Structured receptive fields in CNNs. In: IEEE CVPR, pp. 2610–2619 (2016)
14. Jaderberg, M., Simonyan, K., Zisserman, A., et al.: Spatial transformer networks. Adv. Neural. Inf. Process. Syst. **28**, 2017–2025 (2015)
15. Laptev, I., Lindeberg, T.: Space-time interest points. In: IEEE CVPR, vol. 1, pp. 432–439 (2003)
16. Larochelle, H., Erhan, D., et al.: An empirical evaluation of deep architectures on problems with many factors of variation. In: 24th ICML, pp. 473–480 (2007)
17. Lindeberg, T.: Scale-covariant and scale-invariant Gaussian derivative networks. J. Math. Imaging Vis. **64**(3), 223–242 (2022)
18. Macdonald, I.G.: Symmetric Functions and Hall Polynomials. Oxford University Press (1998)
19. Marcos, D., Volpi, M., Komodakis, N., Tuia, D.: Rotation equivariant vector field networks. In: IEEE CVPR, pp. 5048–5057 (2017)
20. Martinazzo, A., Espadoto, M., Hirata, N.S.: Deep learning for astronomical object classification: a case study. In: VISIGRAPP, pp. 87–95 (2020)
21. Mikolajczyk, K., Tuytelaars, T., et al.: A comparison of affine region detectors. Int. J. Comput. Vision **65**(1), 43–72 (2005)
22. Mühlich, M., Aach, T.: A theory of multiple orientation estimation. In: Leonardis, A., Bischof, H., Pinz, A. (eds.) ECCV 2006. LNCS, vol. 3952, pp. 69–82. Springer, Heidelberg (2006). https://doi.org/10.1007/11744047_6
23. Naylor, P., Laé, M., Reyal, F., Walter, T.: Nuclei segmentation in histopathology images using deep neural networks. In: 14th ISBI, pp. 933–936. IEEE (2017)
24. Park, W., Leibon, G., Rockmore, D.N., Chirikjian, G.S.: Accurate image rotation using hermite expansions. IEEE TIP **18**(9), 1988–2003 (2009)
25. Penaud-Polge, V., Velasco-Forero, S., Angulo, J.: Fully trainable Gaussian derivative convolutional layer. In: 29th IEEE ICIP (2022)
26. Salas, R.R., Dokladal, P., Dokladalova, E.: A minimal model for classification of rotated objects with prediction of the angle of rotation. J. Vis. Commun. Image Represent. **75**, 103054 (2021)
27. Salas, R.R., Dokladalova, E., Dokladal, P.: Rotation invariant CNN using scattering transform for image classification. In: 2019 IEEE ICIP, pp. 654–658. IEEE (2019)
28. Sangalli, M., Blusseau, S., Velasco-Forero, S., Angulo, J.: Moving frame net: SE(3)-equivariant network for volumes. In: NeurIPS Workshop on Symmetry and Geometry in Neural Representations, pp. 81–97. PMLR (2023)
29. Schultz, T., Weickert, J., Seidel, H.P.: A higher-order structure tensor. In: Laidlaw, D., Weickert, J. (eds) Visualization and Processing of Tensor Fields, pp. 263–279. Springer, Berlin (2009). https://doi.org/10.1007/978-3-540-88378-4_13
30. Silván-Cárdenas, J.L., Escalante-Ramírez, B.: The multiscale hermite transform for local orientation analysis. IEEE TIP **15**(5), 1236–1253 (2006)
31. Zhou, Y., Ye, Q., Qiu, Q., Jiao, J.: Oriented response networks. In: IEEE CVPR, pp. 519–528 (2017)
32. Zhu, X.X., et al.: Deep learning in remote sensing: a comprehensive review and list of resources. IEEE Geosci. Remote Sens. Mag. **5**(4), 8–36 (2017)

Compressive Learning of Deep Regularization for Denoising

Hui Shi(✉), Yann Traonmilin, and Jean-François Aujol

Univ. Bordeaux, Bordeaux INP, CNRS, IMB, UMR 5251, 33400 Talence, France
hui.shi@u-bordeaux.fr

Abstract. Solving ill-posed inverse problems can be done accurately if a regularizer well adapted to the nature of the data is available. Such regularizer can be systematically linked with the distribution of the data itself through the maximum a posteriori Bayesian framework. Recently, regularizers designed with the help of deep neural networks (DNN) received impressive success. Such regularizers are typically learned from large datasets. To reduce the computational burden of this task, we propose to adapt the compressive learning framework to the learning of regularizers parametrized by DNN. Our work shows the feasibility of batchless learning of regularizers from a compressed dataset. In order to achieve this, we propose an approximation of the compression operator that can be calculated explicitly for the task of learning a regularizer by DNN. We show that the proposed regularizer is capable of modeling complex regularity prior and can be used for denoising.

Keywords: Regularization · Compressive learning · Denoising

1 Introduction

We consider the denoising problem, i.e. finding an accurate estimate $\hat{x}$ of the original signal $x \in \mathbb{R}^d$ from the observed noisy signal $y \in \mathbb{R}^d$:

$$y = x + \epsilon, \tag{1}$$

where the noise ϵ (assumed to be additive white Gaussian noise of standard deviation σ) is independent of x. Recovering x from its degraded version y is an ill-posed problem and we needs to use additional (prior) information about the unknown signal x to obtain meaningful solutions. Common strategies [2] for solving inverse problems often define an estimator which minimizes

$$\hat{x} \in \arg\min_x F(x) + \lambda R(x), \tag{2}$$

where F is the data fidelity term making the solution consistent with the observation y and R is the regularization term weighted by $\lambda > 0$ that incorporates

This work was partly funded by ANR project EFFIREG - ANR-20-CE40-0001.

L. Calatroni et al. (Eds.): SSVM 2023, LNCS 14009, pp. 162–174, 2023.
https://doi.org/10.1007/978-3-031-31975-4_13

the prior information. The choice of R depends on the statistics of the signal of interest which is not always available in real-life applications.

The maximum a posteriori (MAP) Bayesian framework provides a useful tool to interpret such methods. The MAP estimator is given by:

$$\hat{x}_{\mathrm{MAP}} \propto \arg\min_{x} \|y - x\|_2^2 - \lambda log(\mu(x)) \quad (3)$$

where μ denotes a prior probability law (of density $\mu(\cdot)$) of the unknown data x. In this context, the regularizer is related to the prior distribution of the data, i.e., $R(x) = -log(\mu(x))$.

It is not an easy task to accurately estimate a prior model, especially in high-dimensional spaces. Classical Bayesian approaches, e.g. in image processing, rely on explicit priors such as total variation or Gaussian mixture models (GMM) [22] trained on a database of image patches. Recently, researchers propose to use DNN to design the regularizer. Methods such as the total deep variation [11], adversarial regularizers [12,15], as well as the Plug & Play approach and its extensions [9,21] deliver remarkably accurate results. However, such models are typically learned from large datasets. Estimating their parameters from such a large-scale dataset is a serious computational challenge.

Compressive Learning. One possibility to reduce the computational resources of learning consists in using the compressive learning (CL) framework [4,5,7]. The main idea of CL, coined as sketching, is to compress the whole data collection into a fixed-size representation, a so-called *sketch* of data, such that enough information relevant to the considered learning task is captured. Then the learned parameters are estimated by minimizing a non-linear least-square problem built with the sketch. The size of sketch m is chosen proportional to the intrinsic complexity of the learning task. Meanwhile, the cost of inferring the parameters of interest from the sketch does not depend on the number of data in the initial collection but on the number of parameters we want to estimate. Hence, it is possible to exploit arbitrarily large datasets in the sketching framework without demanding more computational resources.

During the sketching phase, a huge collection of n d-dimensional data vectors $X = \{x_i\}_{i=1}^n$ is summarized into a single m-dimensional ($m \ll n$) vector $\hat{z}$ with:

$$\hat{z} = \frac{1}{n}\sum_{i=1}^{n} \Phi(x_i) = \mathcal{S}(\hat{\mu}_n), \quad (4)$$

where $\hat{\mu}_n := \frac{1}{n}\sum_{i=1}^{n} \delta_{x_i}$ the empirical probability distribution of the data, δ_{x_i} is the Dirac measure at x_i and the function $\Phi : \mathbb{R}^d \rightarrow \mathbb{R}^m$ is called the feature map (typically random Fourier moments). The operator $\mathcal{S}$ is a linear operator on measures μ defined by $\mathcal{S}\mu := \mathbb{E}_{X\sim\mu}\Phi(X)$. An estimate of a distribution μ (or of distributional parameters θ of interest) is computed by solving:

$$\mu_\theta^* = \arg\min_{\mu_\theta} \|\hat{z} - \mathcal{S}\mu_\theta\|_2^2. \quad (5)$$

In practice, this "sketch matching" problem can be solved by greedy compressive learning Orthogonal Matching Pursuit (OMP) algorithm and its extension Compressive Learning-OMP with replacement [10]. When the distribution μ is a GMM in high-dimension with flat tail covariances, the problem can also be solved by the Low-Rank OMP algorithm (LR-OMP). It was shown that the prior model learned with LR-OMP can be used to perform image denoising [20].

These greedy algorithms are suitable for any sketching operator $\mathcal{S}$ and any distribution density μ, as long as the sketch $\mathcal{S}\mu$ and its gradient $\nabla_\theta \mathcal{S}\mu$ with respect to the distributional parameters θ of interest have a closed-form expression: the core of these OMP-based algorithms is computing the expression of $\mathcal{S}\mu$ and $\nabla_\theta \mathcal{S}\mu$. However, real-life data needs to be modeled with more complex distributions. In this case, the sketching feature map may not have a closed form. This limits the possible use of the sketching framework in practice.

In this paper, our goal is to recover a good approximation of the probability distribution of any unknown data from its sketch (i.e. beyond GMM). As neural networks (NN) have great expressive power [8,14], we propose to tackle the problems by adapting the sketching to NN. More precisely, we propose to define the regularizer R_θ parametrized by a DNN f_θ (precisely a ReLU network) as

$$R_\theta(\cdot) = \|f_\theta(\cdot)\|_2^2. \tag{6}$$

Such a regularization corresponds to the parametric distribution density $\mu_\theta \propto e^{-\|f_\theta(\cdot)\|_2^2}$. Thus it can be viewed as a generalized Gaussian distribution, where the bilinear form induced by the covariance matrix is replaced by a network. Due to the fact that NN have good generalization properties, the proposed regularization should be capable of encoding complex probability distributions. Unfortunately, a direct practical application of existing tools is not possible as closed-form expressions of $\mathcal{S}\mu$ are not available for sketching operator $\mathcal{S}$ based on random Fourier features.

Contributions and Outline. In this work, we show the feasibility of learning regularizers parametrized by a DNN from a compressed database. Once the network is trained, the regularizer can be used for denoising. To do so, we propose to approximate the sketching operator $\mathcal{S}$ by a discrete version $\mathcal{S}_d$ whose feature map can be calculated with closed-form expressions, and such that the approximation still permits to apply the sketch matching estimation method. The approximation is performed on a grid of the domain where the data is located.

To find an estimate of the distribution μ_θ, we adapt the sketch matching problem with our approximate sketching operator in the following way:

$$\theta^* \in \arg\min_{\theta \in \Theta} \|\mathcal{S}_d \mu_\theta(p) - \hat{z}\|_2^2, \tag{7}$$

where Θ is a set were the DNN inducing the regularizer $R_\theta(\cdot) = \|f_\theta(\cdot)\|_2^2$ can be parametrized (i.e. weights and bias). This problem can be solved practically with gradient descent based methods.

As we do not need the original dataset during the training process, the training procedure does not need to build batches of data. As a consequence, *each gradient descent iteration in the training incorporates information from the whole original database.* Once the empirical sketch has been computed (in a single pass, possibly in parallel), the dataset can be removed from memory. This reduces the memory complexity of the learning task. Moreover, the Jacobian $\nabla \mathcal{S}\mu_\theta$ can be computed efficiently with back-propagation.

Our approach overcomes the limits of greedy learning algorithms of the original sketching framework: regardless of the complexity of the data distribution, the proposed sketching operator allows us to always have a closed-form expression of $\mathcal{S}_d\mu_\theta$. Thus, the sketching is no longer limited to the distribution densities for which the Fourier transform is explicit.

As a result, the learned regularizer can be used to solve inverse problems. The effectiveness of the proposed scheme is tested on synthetic examples and real dataset. Due to the limitation of our approximation of the sketching operator (the dependence on training points), the feasibility is illustrated on 2-D and 3-D data with possibly complex distributions. Our work thus opens the broader open question of designing closed-form sketching operators in high dimension.

The rest of this article is organized as follows. We start by introducing the sketching framework, ReLU networks and some related works in Sect. 2. In Sect. 3, we describe the proposed framework: the adaptation of the compressive learning framework to the learning of regularizers parameterized by ReLU networks. Section 4 illustrates the performance of the proposed methods on both synthetic data and real-life data. Finally, conclusions are drawn in Sect. 5.

2 Background, Related Works

We suppose that data samples x_i are modeled as independent and identically distributed random vectors having an unknown probability distribution with density $\mu \in \mathcal{D}$ ($\mathcal{D}$ is the set of probability measures over $\mathbb{R}^d$). We define the linear sketching operator $\mathcal{S}$ that maps μ to the m-dimensional sketch vector z:

$$\begin{aligned} \mathcal{S} &: \mathcal{D} \to \mathbb{C}^m \\ z &= \mathcal{S}\mu := \int_{\mathbb{R}^d} \mu(x)\Phi(X)dx. \end{aligned} \tag{8}$$

When the transformation (sketching feature map) $\Phi(\cdot)$ is built with random frequencies of the Fourier transform, the l-th component of the sketch is

$$z_l = \int_{\mathbb{R}^d} e^{-j<\omega_l,x>}\mu(x)dx, \quad \text{for} \quad l = 1, \dots, m, \tag{9}$$

where $\{\omega_l\}_{l=1}^m \in \mathbb{R}^d$ are frequencies drawn at random. Taking a statistical perspective, the components z_l can be seen as samples of the characteristic function of $\mu_{\mathcal{X}}$. Accordingly, given a dataset $X = \{x_i\}_{i=1}^n$, the empirical sketch $\hat{z}$ can be computed from the samples of the database as

$$\hat{z}_l = \frac{1}{n}\sum_{i=1}^{n} e^{-j<\omega_l,x_i>}, \quad \text{for} \quad l = 1, \dots, m. \tag{10}$$

The compression ratio r is m/nd. It was shown [5,6,10] that when the probability distribution μ has a low dimension structure, e.g. a GMM, one can recover it (with high probability) from enough randomly chosen samples of its Fourier transform. The required size of the sketch is typically of the order of the number of parameters we need to estimate.

ReLU Network. A ReLU network, denoted by f_θ, is defined as a fully connected, feed-forward network (multi-layer perceptrons) with rectified linear unit (ReLU) activations. This activation has grown in popularity in feed-forward networks due to the success of first-order gradient based heuristic algorithms and the improvement in convergence to the approximated function for training [13].

Related Works. The sketching framework has been successfully applied to parametric models including GMMs [5,10,20], K-means clustering [5,10] and classification [17]. These methods are limited to the models for which the sketch function has a closed form. In our work, we apply the sketching to neural networks to encode more complex probability distributions. The sketching has been used for neural networks once [18]. In their work, the authors combine the sketching with generative networks to generate data samples, while in our work, we aim at learning a deep regularizer for solving the inverse problem. In addition, the authors of [18] proposed to approximate the sketching map by Monte-Carlo sampling. In our approach, we propose to do the approximation with a discrete sketching operator. The sketching framework mentioned above focus on data-independent approximation, i.e. the sketches are obtained by averaging random features. In [1], the authors propose to perform the sketching based on a Nyström approximation, the latter is data-dependent and shows empirically better performances for K-means clustering and Gaussian modeling.

3 Proposed Method

In this Section, we explain how we adapt the sketching framework to estimate regularizations by DNN. We start by explaining why there are no explicit closed-form expressions of the sketching function available in the context of prior parametrized by DNN. Intuitively, since ReLU networks define piecewise affine functions, we can indeed express a ReLU network f_θ as:

$$f_\theta(x) = \sum_{\gamma=1}^{N_R} \mathbf{1}_{R_\gamma}(x)(W_\gamma x + b_\gamma), \tag{11}$$

where $\mathbf{1}_{R_\gamma}$ is the indicator function of each of the N_R affine regions R_γ, with parameters (W_γ, b_γ).

Given a dataset X, we aim at learning, from only the sketch z, an approximation μ_θ for the probability distribution μ generating X. We consider a regularizer of the form $R_\theta(\cdot) = \|f_\theta(\cdot)\|_2^2$ which corresponds to parametric densities of the form $\mu_\theta(\cdot) \propto e^{-R_\theta(\cdot)}$. Ideally, with the definition in (9), the sketch would have to be calculated as

$$\begin{aligned} z_l &= \int_{\mathbb{R}^d} e^{-j<\omega,x>} e^{-\|f_\theta(x)\|_2^2} dx \\ &= \int_{x_d} \cdots \int_{x_1} e^{-j\sum_{p=1}^d \omega_p x_p} e^{-\sum_{p=1}^d (\sum_i^{N_R} \mathbf{1}_{R_i}(x)((W_i x)_p + b_{i_p}))^2} dx_1 \cdots dx_d. \end{aligned} \tag{12}$$

However, to the best of our knowledge, there is no analytic expression of such Fourier transform (Fourier transform on polygons). To tackle this issue, we consider approximating the continuous Fourier transform on a set of discrete points.

To be specific, we define an approximation $\mathcal{S}_d : \mathbb{R}^d \to \mathbb{C}^m$ of the sketching operator $\mathcal{S}$ such that $\mathcal{S}_d\mu_\theta(\omega) \approx \mathcal{S}\mu(\omega)$ for a given frequency ω. The approximated sketch $\tilde{z}$ then has components:

$$\tilde{z}_l = |\Delta\Omega| \sum_{p_i \in \Omega} e^{-j<\omega_l,p_i>} e^{-\|f_\theta(p_i)\|_2^2}, \tag{13}$$

where p_i is a point in the d-dimensional cell Ω with volume $|\Delta\Omega|$. Of course, the major pitfall of this approximation is the limitation for applications in high dimension as the number of points is exponential with respect to the dimension d. The required boundedness (or approximate boundedness such as in the Gaussian case) of the data is a valid assumption in many practical applications in signal and image processing.

As a consequence, given N points $\{p_i\}_{i=1}^N$ on the grid where the dataset X lives and the empirical sketch defined as (4), we consider a ReLU network sketch matching problem as finding the network parameters θ^* in the set Θ of possible parametrizations, such that

$$\theta^* = \arg\min_{\theta \in \Theta} \|\mathcal{S}_d\mu_\theta - \hat{z}\|_2^2. \tag{14}$$

With the discretization, if μ_θ is differentiable at point p_i, the gradient of $\mathcal{S}_d\mu_\theta$ with respect to the parameters θ can be computed easily by using the automatic differentiation. Note that the discretization is used only in the estimation of the regularizer from the sketch. It thus only impacts the calculation time and memory requirement of the estimation of the regularizer and not the size of the compressed dataset itself.

Denoising with Prior. With the learned regularization term, we solve the variational problem (3) which yields the minimization of:

$$G(x) = \|x - y\|_2^2 + \lambda\|f_\theta(x)\|_2^2. \tag{15}$$

The optimization problem can be solved by gradient descent based methods. Similarly, we can compute the gradient by using automatic differentiation. Also, note that this denoising method can easily be extended to other linear inverse problems, such as interpolation and deconvolution by including the corresponding forward measurement operator.

4 Experiments

Synthetic Data. We first test our framework with 2-D and 3-D synthetic data. To illustrate the advantage of the compressive learning framework in terms of (computational) learning times, the used training datasets are made of $n = 10^6$ samples which are generated from: a spiral with parameters: the radius of circular curve $R = 0.3$ to 1, spiral length $L = 2\pi$; and a zero-mean GMM of 2 Gaussians. The source code is available at [19] which contains parts of code taken from the Python Compressive Learning toolbox [16]. To train the network, we use the Adam optimizer with a learning rate of value 10^{-6}. The number of points on the grid is set to $N = 20^d$ where d denotes the data dimension. For comparison, we propose to learn the regularizer on the non compressed dataset using the same network with the following learning objective function:

$$\theta' = \arg\min_{\theta} \sum_{i=1}^{N} \left\| \|f_\theta(p_i)\|_2^2 - dist_i \right\|_2^2, \tag{16}$$

where $dist_i = \min_j \|x_j - p_i\|_2^2$ is the distance between the data x_j and its nearest grid point. This objective function imposes a regularizer that approximates the function "distance to the model". Note that for non compressive learning, as we do not go through a (implicit) model of the density, we explicitly give the distance value, which is not necessary when using our proposed sketched method.

For the 2-D experiments, the network f_θ is designed as a ReLU network with 3 fully connected hidden layers with 64, 128, and 256 neurons in each layer respectively. Figure 1 shows the experimental results for 2-D synthetic data. The first column shows the training spiral and GMM samples. The results shown in the *2nd* column are models learned from 4000-fold compressed dataset while producing comparable results to those learned from a non-compressed dataset (*3rd* column), indicating that our approach achieves its goal: efficiently learn prior probability densites from compressed datasets (which will be evaluated when used as a regularization on real audio data). In fact, we match the distribution density of data directly in the compressive method which is not trivial for the non-compressed method.

Table 1 shows the needed learning times (in hours) with respect to the different sketch size $m = 50, 100, 1000, 5000$, i.e. with compression ratio $r = 40000, 20000, 2000, 400$ respectively. We see that training the same dataset, the non compressed learning takes much longer (20 times) than the compressive learning. Meanwhile, the proposed compressive learning approach is capable of recovering good approximations of the probability distribution of sample data.

Denoising Results. The learned regularizers are evaluated on the denoising of white Gaussian noise. The noisy dataset is made of 500 samples generated with noise level σ^2. We choose the optimal hyper-parameter values (the learning rate η and the regularizer parameter λ) for each model. Figure 2 visually illustrates the 2-D denoising results using regularizers learned from the compressed dataset

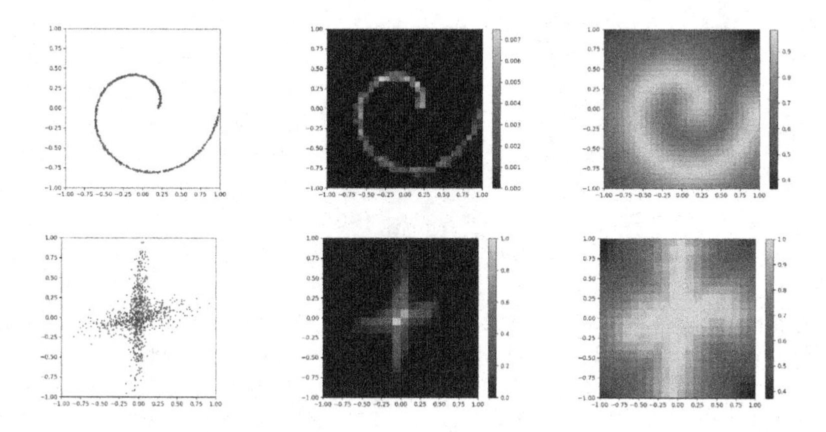

Fig. 1. The distribution densities of sample data (left) learned from compressed dataset with sketching (compression ratio $r = 4000$) (middle) and original dataset (right).

Table 1. Table of leaning times with respect to the compression ratio: sketch used for training is r times smaller than the original dataset (results in bold).

	r	sketching				non compressed
		40000	20000	2000	400	
Time	Spiral	0.19h	0.23h	0.28h	0.72h	**28.7h**
	GMM	0.23h	0.18h	0.24h	0.73h	**28.5h**

4000 times smaller (left) and the original dataset (right). Figure 3 shows the 3-D denoising results with different noise levels.

We assess the effect compression ratio, *i.e.* the sketch size, in Table 2 and use the signal-to-noise ratio (SNR) to evaluate the effectiveness of our method. We generate 10 different noisy datasets of 500 samples for each data type with

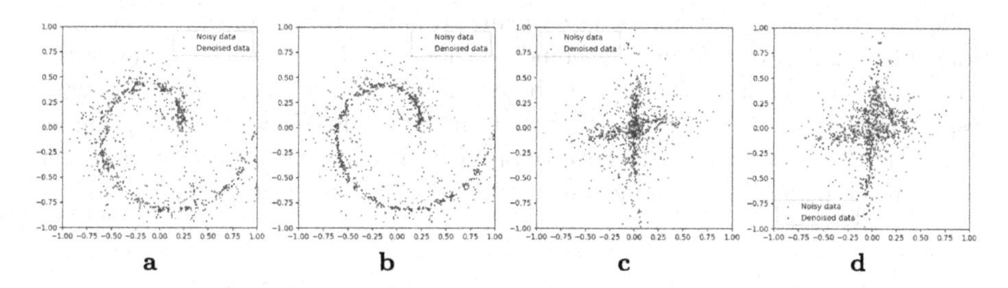

Fig. 2. Denoising results ($\sigma^2 = 0.15$) with regularizers learned with the compressed dataset with compression ratio $r = 4000$ (**a**, **c**) and the original dataset (**b**, **d**).

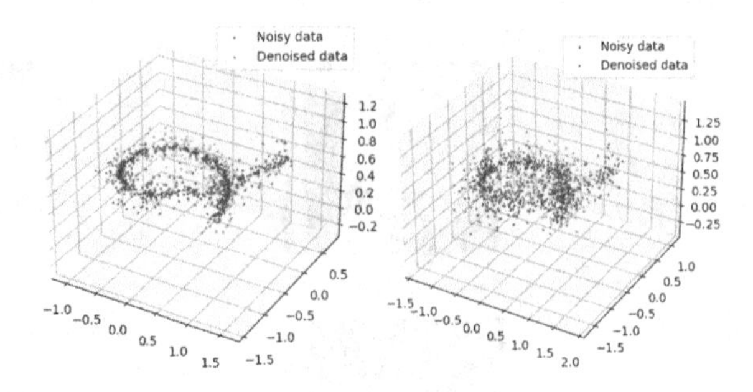

Fig. 3. 3-D Denoising results ($\sigma^2 = 0.1$ (left) and $\sigma^2 = 0.2$ (right)) with densities learned from the proposed method with a compression ratio $r = 3000$.

noise level $\sigma^2 = 0.15$. Table 2 shows the average gain in SNR with respect to the priors learned with different compression ratios. It is shown that the proposed approach is capable to learn distributions from databases with high compression ratios, and that regularizers trained from the compressed datasets have similar denoising performance compared to regularizers trained on their original non compressed datasets. When we reduce the sketch size, the denoising performance drops slightly.

Table 2. Table of reconstruction loss with respect to the compression ratio: sketch used for training is r times smaller than the original dataset.

	r	sketching				non compressed
		40000	20000	2000	400	
SNR Gain	Spiral	1.37	1.89	1.92	2.31	2.61
	GMM	0.80	0.86	1.57	1.82	1.77

The number of frequencies used to compute sketches affects memory storage, while the number of grid points used during training process affects the learning time. The number of grid points should be chosen well if we want to control the learning time well, since the number of grid points N grows exponentially with respect to the dimension d. Experiments from Fig. 4 also show that overparameterized networks (with more layers or more neurons per layer) can achieve better results while using less learning time and fewer necessary grid points.

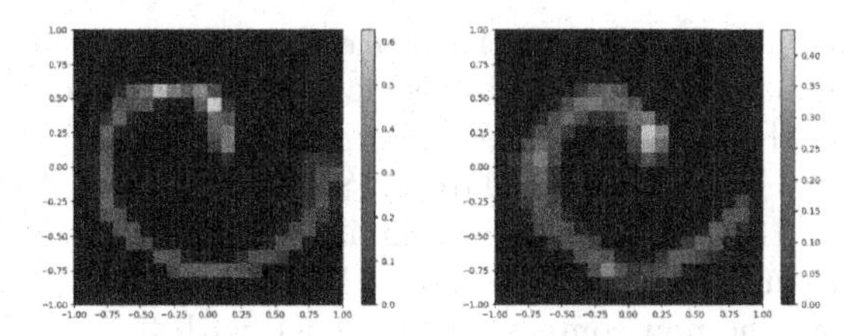

Fig. 4. Regularizers learned via the proposed method with different learning parameters. Using the same number of grid points, we have better result when the network has more neurons. (Left) 3 hidden layers with 64, 128, 256 neurons in each layer. (Right) 3 hidden layers with 64, 128, 192 neurons in each layer.

Robustness to Noise During Training. We train the regularizer on a compressed dataset of data samples generated with noise level $\sigma_{train}^2 = 0.15$ using the same network and training procedure as described above. Figure 5 shows the learned distribution density and the denoising result. The result shows that the regularizer trained with a compressed noisy dataset has good denoising performance. This illustrates that our approach is robust to low noise level. This is easy to understand due to the fact that adding Gaussian noise corresponds to a convolution of the density with a Gaussian kernel, which does not change the shape of the distribution if small enough. It is even possible to add a deconvolution term to the distribution parameter estimation if the noise level in the training dataset is known.

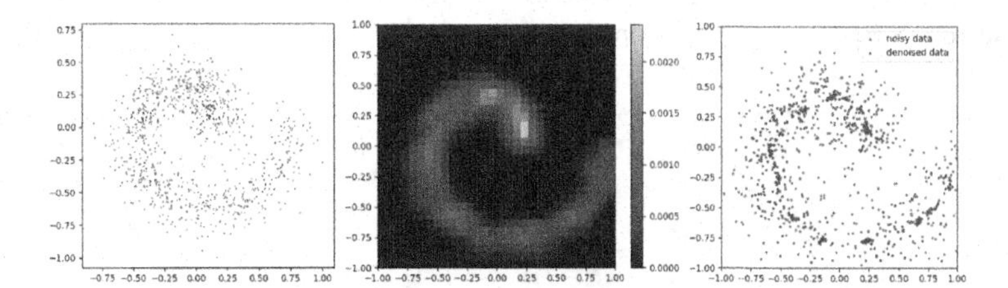

Fig. 5. Denoising result (right, $\sigma^2 = 0.2$) with density (middle) learned from the compressived noisy dataset (left, $\sigma_{train}^2 = 0.15$).

Application to Audio Denoising. We perform experiments on recorded musical notes (monophonic 16kHz audio snippets) from the NSynth dataset [3]. The training data is an extracted 0.125s audio recorded from an acoustic guitar. After filtering the normalized audio data s by two 4th-order Butterworth low-pass filters h_1 and h_2 with a cutoff frequency of 1.5kHz and 3.75kHz, three frequency responses are constructed with $s_1 = h_1 * s$, $s_2 = h2 * (s - s_1)$, and $s_3 = s - s1 - s2$. Then the frequency responses are concatenated, hence the training set is of dimension 2000×3; i.e. 2000 samples in dimension 3. The regularizer is learned from a sketch of size $m = 200$, *i.e.* the dataset is compressed by a factor of 30. Once the regularizer is learned, it is used to denoise the audio corrupted by Gaussian white noise of noise levels $\sigma^2 = 0.1$ and $\sigma^2 = 0.2$.

Figure 6 and 7 show the audio denoising results with different noise levels. In the two cases, we gain more than 1 dB on SNR in the case of small noise and more than 2.5 dB in case of large noise. Similar denoising results (gain of 1 dB in the small noise and 1.94 dB in the case of large noise) are obtained from the priors learned from the non compressed approach with 3 times slower training time. The results show that, in addition to the original low-pass filtering effect, denoising can be achieved in low dimensions even when temporal consistency between individual samples is not guaranteed. These first feasibility results for solving inverse problem using regularizers learned from sketch are promising, especially if it is possible to extend this frramework to high dimension.

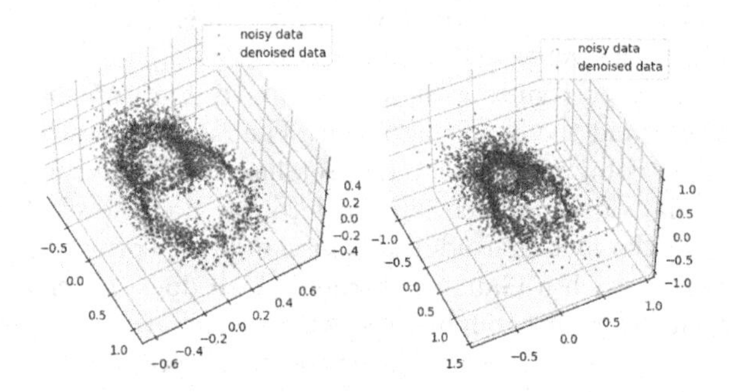

Fig. 6. Audio denoising performances for different noise levels: (left) $\sigma^2 = 0.1$, SNR is 10.03 for noisy data and 11.39 for denoised data; (right) $\sigma^2 = 0.2$, SNR is 4.01 for noisy data and 6.64 for denoised data.

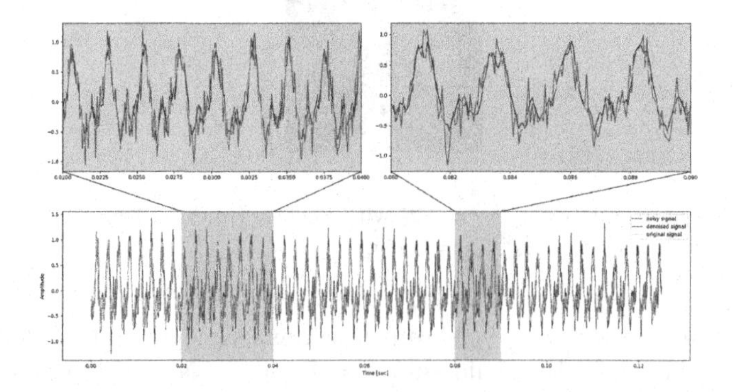

Fig. 7. Audio denoising performance for noise level $\sigma^2 = 0.2$.

5 Conclusions

In this work, we illustrate the feasibility of adapting the compressive learning framework to the learning of a regularizer parameterized by a DNN. We achieve this by approximating the original sketching operator with a discrete one. With the proposed approximated sketching operator, the "sketch matching" problem can be solved with a gradient based algorithm. In addition, we define a new parametrization of the regularizer to solve the inverse problem. The regularizer is defined as the squared ℓ^2-norm of a ReLU network and learned with a compressive dataset instead of the original dataset. It gathers the advantages of sketching which reduces the learning cost and of NN which have great expressive power. Experiment results on 2-D/3-D synthetic data and audio data show that our method accomplishes the objective of compressive learning, illustrating the potential of sketched NN. However, our method relies on a discretization of the domain on which the data resides, which limits its use in high-dimensional domains (*e.g.*, for image denoising). Future works will be needed to overcome this limitation. We want to design a fast sketching operator that avoids such discretization. This leads to a major open question: can we find a sketching operator such that any distribution parametrized by a DNN can be estimated from the sketch? Does such a sketching operator exist? A positive answer to these questions could drastically change the way deep priors are trained.

References

1. Chatalic, A., Carratino, L., De Vito, E., Rosasco, L.: Mean Nyström embeddings for adaptive compressive learning. In: International Conference on Artificial Intelligence and Statistics, pp. 9869–9889. PMLR (2022)
2. Demoment, G.: Image reconstruction and restoration: overview of common estimation structures and problems. IEEE Trans. Acoust. Speech Signal Process. **37**(12), 2024–2036 (1989)
3. Engel, J., et al.: Neural audio synthesis of musical notes with wavenet autoencoders. In: International Conference on Machine Learning, pp. 1068–1077. PMLR (2017)
4. Gribonval, R., Blanchard, G., Keriven, N., Traonmilin, Y.: Compressive statistical learning with random feature moments. Math. Stat. Learn. **3**(2), 113–164 (2021)
5. Gribonval, R., Blanchard, G., Keriven, N., Traonmilin, Y.: Statistical learning guarantees for compressive clustering and compressive mixture modeling. Math. Stat. Learn. **3**(2), 165–257 (2021)
6. Gribonval, R., Chatalic, A., Keriven, N., Schellekens, V., Jacques, L., Schniter, P.: Sketching datasets for large-scale learning (long version). arXiv preprint arXiv:2008.01839 (2020)
7. Gribonval, R., Chatalic, A., Keriven, N., Schellekens, V., Jacques, L., Schniter, P.: Sketching data sets for large-scale learning: keeping only what you need. IEEE Signal Process. Mag. **38**(5), 12–36 (2021)
8. Hornik, K., Stinchcombe, M., White, H.: Multilayer feedforward networks are universal approximators. Neural Netw. **2**(5), 359–366 (1989)
9. Hurault, S., Leclaire, A., Papadakis, N.: Gradient step denoiser for convergent plug-and-play. arXiv preprint arXiv:2110.03220 (2021)
10. Keriven, N., Bourrier, A., Gribonval, R., Pérez, P.: Sketching for large-scale learning of mixture models. Inf. Inference **7**(3), 447–508 (2018)
11. Kobler, E., Effland, A., Kunisch, K., Pock, T.: Total deep variation: a stable regularization method for inverse problems. IEEE Trans. Pattern Anal. Mach. Intell. **44**(12), 9163–9180 (2021)

12. Lunz, S., Öktem, O., Schönlieb, C.B.: Adversarial regularizers in inverse problems. In: Advances in Neural Information Processing Systems, vol. 31 (2018)
13. Nair, V., Hinton, G.E.: Rectified linear units improve restricted Boltzmann machines. In: International Conference on Machine Learning (2010)
14. Pan, X., Srikumar, V.: Expressiveness of rectifier networks. In: International Conference on Machine Learning, pp. 2427–2435. PMLR (2016)
15. Prost, J., Houdard, A., Almansa, A., Papadakis, N.: Learning local regularization for variational image restoration. In: Elmoataz, A., Fadili, J., Quéau, Y., Rabin, J., Simon, L. (eds.) SSVM 2021. LNCS, vol. 12679, pp. 358–370. Springer, Cham (2021). https://doi.org/10.1007/978-3-030-75549-2_29
16. Schellekens, V.: Pycle: a python compressive learning toolbox (2020). https://doi.org/10.5281/zenodo.3855114
17. Schellekens, V., Jacques, L.: Compressive classification (machine learning without learning). arXiv preprint arXiv:1812.01410 (2018)
18. Schellekens, V., Jacques, L.: Compressive learning of generative networks. arXiv preprint arXiv:2002.05095 (2020)
19. Shi, H.: https://github.com/shihui1224/sketching-deep-regu-for-denoising
20. Shi, H., Traonmilin, Y., Aujol, J.F.: Compressive learning for patch-based image denoising. SIAM J. Imag. Sci. **15**(3), 1184–1212 (2022)
21. Venkatakrishnan, S.V., Bouman, C.A., Wohlberg, B.: Plug-and-play priors for model based reconstruction. In: IEEE Global Conference on Signal and Information Processing, pp. 945–948. IEEE (2013)
22. Zoran, D., Weiss, Y.: From learning models of natural image patches to whole image restoration. In: International Conference on Computer Vision, pp. 479–486. IEEE (2011)

Graph Laplacian and Neural Networks for Inverse Problems in Imaging: GraphLaNet

Davide Bianchi[1], Marco Donatelli[2], Davide Evangelista[3], Wenbin Li[1(✉)], and Elena Loli Piccolomini[3]

[1] Harbin Institute of Technology, Shenzhen, Shenzhen, China
{bianchi,liwenbin}@hit.edu.cn
[2] University of Insubria, Como, Italy
marco.donatelli@uninsubria.it
[3] University of Bologna, Bologna, Italy
{davide.evangelista5,elena.loli}@unibo.it

Abstract. In imaging problems, the graph Laplacian is proven to be a very effective regularization operator when a good approximation of the image to restore is available. In this paper, we study a Tikhonov method that embeds the graph Laplacian operator in a ℓ_1–norm penalty term. The novelty is that the graph Laplacian is built upon a first approximation of the solution obtained as the output of a trained neural network. Numerical examples in 2D computerized tomography demonstrate the efficacy of the proposed method.

1 Introduction

We consider ill-posed inverse problems in imaging. Images can be well represented as ordered vectors in $\mathbb{R}^n$ where each entry is the intensity of a pixel. Given an ill-posed linear operator $K : \mathbb{R}^n \to \mathbb{R}^m$, see [9,16], a fixed $\boldsymbol{x}_{gt} \in \mathbb{R}^n$ and $\boldsymbol{y} := K\boldsymbol{x}_{gt}$, we want to recover a good approximation of the ground-truth image $\boldsymbol{x}_{gt}$ from a noisy observation $\boldsymbol{y}^\delta$ of $\boldsymbol{y}$, that is

$$\boldsymbol{y}^\delta := K\boldsymbol{x}_{gt} + \boldsymbol{\eta}, \tag{1}$$

where $\boldsymbol{\eta}$ is a random and unavoidable perturbation. By passing to a least-square formulation, we need to compute

$$\operatorname*{argmin}_{\boldsymbol{x}\in\mathbb{R}^n} \|K\boldsymbol{x} - \boldsymbol{y}^\delta\|_2^2. \tag{2}$$

The ill-posedness of K and the presence of noise measurements compel us to regularize the model problem, and we reformulate (2) in the following variational form

$$\operatorname*{argmin}_{\boldsymbol{x}\in\mathbb{R}^n} \frac{1}{2}\|K\boldsymbol{x} - \boldsymbol{y}^\delta\|_2^2 + \lambda\|L\boldsymbol{x}\|_1, \tag{3}$$

L. Calatroni et al. (Eds.): SSVM 2023, LNCS 14009, pp. 175–186, 2023.
https://doi.org/10.1007/978-3-031-31975-4_14

where $L : \mathbb{R}^n \to \mathbb{R}^s$ is a linear operator such that $\ker(L) \cap \ker(K) = \{\mathbf{0}\}$. The above Eq. (3) belongs to the framework of generalized Tikhonov methods. Here, $\frac{1}{2}\|K\boldsymbol{x} - \boldsymbol{y}^\delta\|_2^2$ measures the fidelity of the reconstruction, $\|L\boldsymbol{x}\|_1$ is a penalty term which regularizes the problem, and λ is the regularization parameter that provides a trade-off between the fidelity and the regularization.

In general, the image to be recovered has near-constant intensities in most areas of its region, and discontinuities appear at the interface between different areas. To take advantage of this feature, the linear operator L of differential type is developed to provide regularization in Eq. (3) [10]. For example, in [8], the authors introduce graph-based non-local operators for the image denoising problem; in [5], this approach is further investigated in the context of image segmentation; while in [2,4], it is applied to image deblurring and computerized tomography problems.

The graph-based operator L has the capability of capturing image features by exploiting the affinity between image pixels. In particular, it is proposed to take $L = \Delta$, i.e., the graph Laplacian associated to a suitable graph, built upon a first approximation, $\Psi(\boldsymbol{y}^\delta)$, of the ground-truth image, where $\Psi : \mathbb{R}^m \to \mathbb{R}^n$ denotes a chosen reconstruction method for the model problem. The weights of the graph Laplacian Δ are determined on the first approximation $\Psi(\boldsymbol{y}^\delta)$. Empirically, it is known that the better the approximation $\Psi(\boldsymbol{y}^\delta)$ is, in terms of closeness to the ground-truth image, the higher quality one can achieve in the recovered solution of Eq. (3). As a result, the construction of $\Psi(\boldsymbol{y}^\delta)$ plays an important role and it deserves further study.

In this work, we consider a two-step method inspired by [4] for solving the minimization problem (3), where $L = \Delta$. The first step of the method computes an initial approximation $\Psi(\boldsymbol{y}^\delta)$ of $\boldsymbol{x}_{gt}$, which is used to construct the graph Laplacian Δ. The main novelty of this paper concerns the computation of a reliable $\Psi(\boldsymbol{y}^\delta)$. This step is performed by resorting to a simple neural network architecture. The neural network is not able to reconstruct the details of $\boldsymbol{x}_{gt}$, but it provides a good graph to construct the operator Δ which is close to what would be obtained with $\boldsymbol{x}_{gt}$. The combination of the variational model with a neural network leads to a *data and knowledge driven* approach which provides a robust reconstruction. The overall algorithm produces final solutions of outstanding quality in 2D computerized tomography.

The paper is organized as follows. In Sect. 2, we introduce the graph Laplacian and explain how to build it from a given image. In Sect. 3, we provide a mathematical definition of the neural network and discuss its training process. Section 4 presents our overall algorithm named graphLaNet. Section 5 and Section 6 provide the setup and results of our numerical experiment in 2D computerized tomography. Lastly, Sect. 7 draws the conclusion.

2 Building the Graph Laplacian

For an introduction to modern graph theory, we refer to [11]. We consider as a (finite) graph any triple (P, w, ν) where

- P is a finite set of elements called *nodes*;
- $w : P \times P \to [0, +\infty)$ is a symmetric function with zero diagonal, called *edge-weight* function;
- $\nu : P \to (0, +\infty)$ is a node measure, that here we assume to be constant for each node:

$$\nu(p) = \bar{\nu}, \qquad \forall p \in P.$$

We denote $\mathcal{C}(P)$ the set of real-valued functions on P, that is $\mathcal{C}(P) := \{\boldsymbol{x} : P \to \mathbb{R}\}$, and with $\deg \in \mathcal{C}(P)$ the *degree* function such that

$$\deg(p) := \sum_{q \in P} w(p, q).$$

Given a graph (P, w, ν), the graph Laplacian Δ is the linear operator on $\mathcal{C}(P)$ whose action is defined via

$$\begin{aligned} \Delta \boldsymbol{x}(p) &:= \frac{1}{\bar{\nu}} \sum_{q \in P} w(p, q)(\boldsymbol{x}(p) - \boldsymbol{x}(q)) \\ &= \frac{1}{\bar{\nu}} \left(\deg(p)\boldsymbol{x}(p) - \sum_{q \in P} w(p, q)\boldsymbol{x}(q) \right). \end{aligned} \tag{4}$$

Indicating with W the *adjacency* matrix associated to w, that is

$$(W)_{p,q} := w(p, q),$$

and with D the diagonal matrix associated to the degree function, then the operator Δ in (4) can be rewritten as

$$\Delta = \frac{1}{\bar{\nu}}(D - W).$$

In particular, notice that Δ is uniquely defined by $w(\cdot, \cdot)$ and $\bar{\nu}$.

If we think of P as the set of pixels of an image $\boldsymbol{x}$, then an image can be seen as a function $\boldsymbol{x} \in \mathcal{C}(P)$. In a natural way we make a one-to-one association between pixels and points of a discretized rectangle $P \subset \mathbb{R}^2$, endowed with the metric $\|\cdot\|_\infty$. If we have $n_1 \times n_2$ pixels, then $\mathcal{C}(P) \cong \mathbb{R}^{n_1 \cdot n_2}$.

Suppose now that we have given a fixed $\tilde{\boldsymbol{x}} \in \mathcal{C}(P)$, which is the realization of a reconstruction method $\Psi : \mathbb{R}^m \to \mathbb{R}^n$, that is

$$\tilde{\boldsymbol{x}} := \Psi(\boldsymbol{y}^\delta).$$

Then we can build a graph Laplacian Δ^Ψ from $\tilde{\boldsymbol{x}}$ in the following way: Define

$$\begin{aligned} w(p, q) &:= \begin{cases} \mathrm{e}^{-\frac{|\tilde{x}(p) - \tilde{x}(q)|^2}{\sigma^2}}, & \text{if } p \neq q \text{ and } \|p - q\|_\infty \leq R, \\ 0, & \text{otherwise}, \end{cases} \\ \bar{\nu} &:= \|W\|_F, \end{aligned}$$

where $\|\cdot\|_F$ is the Frobenius norm, and σ, R are positive constants. Let us observe that we will make explicit in the notation the dependence of the graph Laplacian Δ^{Ψ} to the reconstruction method Ψ. The choice of the Gaussian kernel as edge-weight function is pretty common, see [5,14], and it has a theoretical explanation based on the relationship between the graph Laplacian and the discretization of the Laplacian-Beltrami operator on manifolds, see e.g. [17].

3 Our Neural Network

To introduce the neural architecture used in our framework, we need some preliminary notations. Fixed $L \in \mathbb{N}$ we define:

$$\Theta := \{\mathcal{W}, b, \mathcal{S}\}$$

where $\mathcal{W} = \{W^l \in \mathbb{R}^{\nu_{l+1/2} \times \nu_l}\}_{l=1}^L$ and $\mathcal{S} = \{S_{l,k} \in \mathbb{R}^{\nu_{l+1/2} \times \nu_k} \mid l > k\}_{l,k=1}^L$ are collections of matrices, and $b := \{\boldsymbol{b}^l \in \mathbb{R}^{\nu_{l+1/2}}\}_{l=1}^L$ is a collection of vectors. W^l and $S_{l,k}$ are called weight matrices and skip connections, respectively, while $\boldsymbol{b}^l$ are called bias vectors and Θ is called the *set of parameters*. Finally, we indicate with $\rho := \{\rho^l : \mathbb{R}^{\nu_{l+1/2}} \to \mathbb{R}^{\nu_{l+1}}\}_{l=1}^L$ a set of possible nonlinear functions and we call *neural network architecture* the pair $\mathcal{A} := (\Theta, \rho)$.

Definition 1. *We define the parametric family of* neural network *reconstructors* $\Phi_\Theta : \mathbb{R}^m \to \mathbb{R}^n$*, with architecture* $\mathcal{A}$ *and pre-processing layer* $\Gamma : \mathbb{R}^m \to \mathbb{R}^{\nu_1}$*, as*

$$\mathcal{F}_{\mathcal{A}} := \{\Phi_\Theta : \mathbb{R}^m \to \mathbb{R}^n \mid \Theta \in \mathbb{R}^s\}$$

where, for any $\boldsymbol{y} \in \mathbb{R}^m$*, the action of each* Φ_Θ *is given by*

$$\begin{cases} \Phi_\Theta(\boldsymbol{y}) := \boldsymbol{x}, \\ \boldsymbol{x} := \boldsymbol{x}^{L+1} \in \mathbb{R}^n, \\ \boldsymbol{x}^{l+1} := \rho^l(W^l \boldsymbol{x}^l + \boldsymbol{b}^l + \sum_{k=1}^{l-1} S_{l,k} \boldsymbol{x}^k) \in \mathbb{R}^{\nu_{l+1}} \quad \forall l = 1, \ldots, L, \\ \boldsymbol{x}^1 := \Gamma(\boldsymbol{y}) \in \mathbb{R}^{\nu_1}. \end{cases} \tag{5}$$

The pre-processing layer Γ has been used, for example, in an application of single photon emission computerized tomography (SPECT) [1], and it helps to improve the stability of the resulting network [7].

The architecture of the neural network used in this work is a modified version of the U-net [15], whose detailed structure is depicted in Fig. 1. U-net is a convolutional neural network, originally designed for biomedical image segmentation, which consists of a contracting path and a symmetric expanding path so that it has a U-shaped architecture. The network is trained for 50 epochs with the Adam optimization algorithm [12] and fixed step size of 10^{-3}. It is trained to minimize the mean squared error (MSE) between the output and the processed input with no additional modifications such as regularizers or the use of Dropout layers. The batch size is set to the maximum value our GPU could handle, that is 16. The training is executed on an NVIDIA RTX A4000 GPU card, requiring approximately 30 min to be completed. More detail on the network and its training can be found in [7].

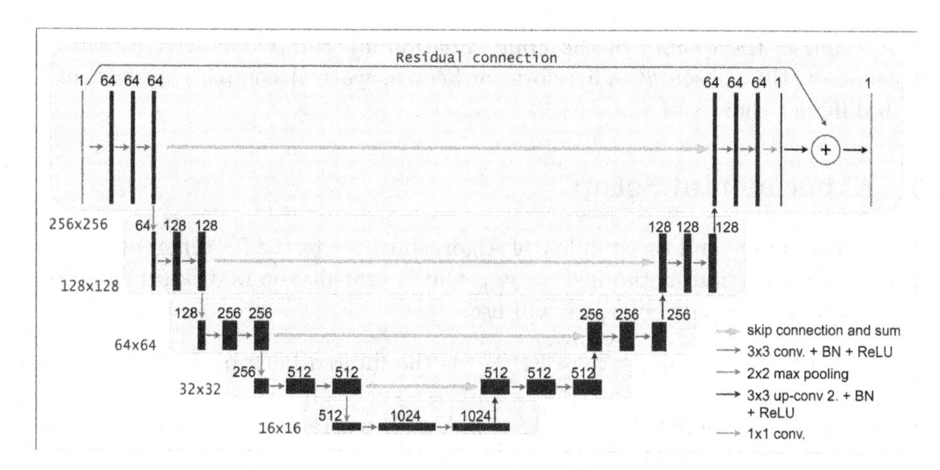

Fig. 1. A diagram of the architecture used in the experiments.

4 GraphLaNet

In [2] it is observed that the quality of the solution of (3) strongly depends on the details of $\boldsymbol{x}_{gt}$ which can be approximated by the initial reconstruction $\Psi(\boldsymbol{y}^\delta)$, from which the graph Laplacian Δ^Ψ is built following the procedure described in Sect. 2. We propose to generate $\Psi(\boldsymbol{y}^\delta)$ with a neural network Φ_Θ.

Given a reconstruction method $\Psi : \mathbb{R}^m \to \mathbb{R}^n$ for the model problem (2), we define Algorithm 1 below.

Input : $\boldsymbol{y}^\delta$, λ, Ψ;
Output: $\boldsymbol{x}_\lambda^\Psi$;

1 Compute $\Psi(\boldsymbol{y}^\delta)$;
2 Build Δ^Ψ from $\Psi(\boldsymbol{y}^\delta)$ as described in Section 2;
3 Solve $\boldsymbol{x}_\lambda^\Psi := \underset{\boldsymbol{x}\in\mathbb{R}^n}{\operatorname{argmin}} \frac{1}{2}\|K\boldsymbol{x} - \boldsymbol{y}^\delta\|_2^2 + \lambda\|\Delta^\Psi \boldsymbol{x}\|_1$.

Algorithm 1: Two-step method to solve the problem (3), where $L = \Delta^\Psi$ is constructed from a first approximation $\Psi(\boldsymbol{y}^\delta)$ of $\boldsymbol{x}_{gt}$.

Algorithm 1 is a two-step method. Step 1 computes a first approximation of $\boldsymbol{x}_{gt}$, which is used in Step 2 to construct the graph Laplacian Δ^Ψ. Finally, Step 3 solves the problem (3) where L is the graph Laplacian Δ^Ψ computed in the previous step. To solve the minimization problem in Step 3 we use the algorithm in [4], while other algorithms could be used as well.

When the chosen reconstruction method is a neural network, that is $\Psi = \Phi_\Theta$, we call Algorithm 1 graphLaNet. It is the combination of the graph Laplacian and a neural network in a variational problem of the form (3).

Of course, the quality of the approximation $\boldsymbol{x}_\lambda^\Psi$ computed by Algorithm 1 depends on the chosen Ψ. Therefore, in Sect. 6, we will compare Algorithm 1 with different choices of Ψ.

5 Experimental Setup

Both the accuracy and the stability of Algorithm 1 are tested for different choices of the reconstruction methods $\Psi = \Psi_{(\cdot)}$, which provide the first approximation of the solution. In particular, we will use:

1. $\Psi_{FBP}(\boldsymbol{y}^\delta) := \text{FBP}(\boldsymbol{y}^\delta)$, where FBP$(\cdot)$ is the filtered back projection, see for example [16].
2. $\Psi_{Tik}(\boldsymbol{y}^\delta) := \frac{1}{2}\underset{\boldsymbol{x}\in\mathbb{R}^N}{\text{argmin}}\|K\boldsymbol{x} - \boldsymbol{y}^\delta\|_2^2 + \mu\|\nabla\boldsymbol{x}\|_2^2$, where ∇ is a discretization of the gradient, see for example [2].
3. $\Psi_{Net}(\boldsymbol{y}^\delta) := \Phi_\theta(\boldsymbol{y}^\delta)$, where Φ_Θ is the trained neural network described in Sect. 3. In particular, we set the pre-processing layer to be the filtered back projection operator, i.e., $\Gamma = \text{FBP}$.
4. $\Psi_{gt}(\boldsymbol{y}^\delta) \equiv \boldsymbol{x}_{gt}$, where $\boldsymbol{x}_{gt}$ is the ground-truth.

Let us recall that our graphLaNet proposal implements the reconstruction method Ψ_{Net} at point 3. We also underline that the method 4 sets the ground-truth $\boldsymbol{x}_{gt}$ as a "first approximation" of the solution. This is clearly unfeasible in practice, but we will consider it as a baseline for our experiments to show the behavior of the best possible theoretical solution to the model problem (3).
The quantitative results of our experiments will be measured as reconstruction relative error (RRE), defined as

$$\text{RRE} = \frac{\|\boldsymbol{x}_{gt} - \boldsymbol{x}\|^2}{\|\boldsymbol{x}_{gt}\|^2},$$

peak signal-to-noise ratio (PSNR), defined as

$$\text{PSNR} = 20\log_{10}\left(\frac{255}{\|\boldsymbol{x}_{gt} - \boldsymbol{x}\|}\right),$$

and structural similarity (SSIM) [18].

5.1 Few-View CT Reconstruction

We tested our algorithm on a few-view CT reconstruction setup, where the matrix K represents an under-determined Radon transform matrix computed from a parallel beam geometry, in an angular range $[0°, 180°]$ and angular step of 1.5°, i.e. 120 projections. Each datum $\boldsymbol{x}$ has dimension 256×256 pixels and we compute the sinogram $\boldsymbol{y} = K\boldsymbol{x}_{gt}$ with dimension 363×120 pixels. The noisy version of $\boldsymbol{y}$ is computed as:

$$\boldsymbol{y}^\delta = \boldsymbol{y} + \boldsymbol{\eta}, \qquad \boldsymbol{\eta} \sim \sigma^2 \cdot \mathcal{N}(0, I).$$

5.2 The Data Set

To validate the procedure described in Algorithm 1, we chose the COULE data set [6], which is composed of several random ellipses of different shape and color. For the training of the neural network Φ_Θ in Sect. 3, the dataset is divided into a training set and a test set, both of 400 images. Assuming an a-priori information on the noise distribution, we set the standard deviation σ of all the input images $\boldsymbol{y}_i^\delta$ to 0.7.

5.3 Choice of Regularization Parameters

The λ parameter in Algorithm 1 controls the amount of regularization. There are several automatic choice rules to compute it, such as generalized cross validation or discrepancy principle [9]. However, it is well known that these rules are not completely reliable. Since we have access to a collection of ground-truth data, and both the noise statistics and the overall structure of each image in the dataset are similar, we computed, for each method, a parameter $\lambda > 0$ minimizing the relative error between the computed solution and the ground-truth $\boldsymbol{x}_{gt}$ over the training dataset. We have finally set λ as the mean of the values computed for each image of the data set. Similarly, the couple of parameters (λ, μ), required to compute the $\boldsymbol{x}_\lambda^{\Psi_{Tik}}$ solution, is obtained by minimizing the relative error surface.

6 Numerical Results

We report here the results obtained from the experiment described in Sect. 5. In the first experiment we fixed for the test set the standard deviation of the noise to $\sigma = 0.7$, the same value used for the training of the neural network.

In Table 1 we show the values of the considered RRE, PSNR and SSIM metrics obtained on the test set by running the experiments over all the images in the test set (400 images) and by reporting the average and standard deviation of the metrics. As it is clear, Algorithm 1 always improves the overall performances of each initial reconstruction method Ψ. The best results are obtained by graphLaNet, by far, that is Algorithm 1 combined with Ψ_{Net}. In the bottom of the Table row we show, just as comparison, the best limit-values obtained by Algorithm 1 when fed with the ground-truth image $\boldsymbol{x}_{gt} \equiv \Psi_{gt}(\boldsymbol{y}^\delta)$ as initial reconstruction. Observe how the SSIM value of $\boldsymbol{x}_\lambda^{\Psi_{Net}}$ is very close to the SSIM limit-value of $\boldsymbol{x}_\lambda^{\Psi_{gt}}$. The graphLaNet algorithm always achieves the best outcomes, considerably improving the quality of the reconstructions. From Fig. 2, showing the results on one particular image of the test set, it is evident that the starting image obtained with FBP is corrupted by streaking artifacts which are inherited by the final $\boldsymbol{x}_\lambda^{\Psi_{FBP}}$. Comparing the results from Tikhonov and the neural network, we notice that $\Psi_{Tik}(\boldsymbol{y}^\delta)$ is a bit more blurred than $\Psi_{Net}(\boldsymbol{y}^\delta)$. In the final output, some noise is present especially on the dark background of

$\boldsymbol{x}_\lambda^{\Psi_{Tik}}$, whereas $\boldsymbol{x}_\lambda^{\Psi_{Net}}$ is more homogeneous. The metrics relative to these two approaches in Table 1 confirm that the approach using neural network is more effective.

Table 1. Reported values (average and standard deviation) of the metrics for the considered experiments, which have been run all over the 400 images belonging to the test set. Inside the brackets we indicate the change rate of the metrics, from Ψ to $\boldsymbol{x}_\lambda^{\Psi}$.

$\Psi/\boldsymbol{x}_\lambda^{\Psi}$	RRE	PSNR	SSIM
$\Psi_{FBP}(\boldsymbol{y}^\delta)$	0.206 ± 0.025	24.678 ± 0.566	0.4464 ± 0.017
$\boldsymbol{x}_\lambda^{\Psi_{FBP}}$	0.136 ± 0.017	28.269 ± 0.620	0.715 ± 0.016
	(-54%)	$(+14\%)$	$(+60\%)$
$\Psi_{Tik}(\boldsymbol{y}^\delta)$	0.184 ± 0.021	25.618 ± 0.918	0.732 ± 0.015
$\boldsymbol{x}_\lambda^{\Psi_{Tik}}$	0.059 ± 0.009	35.356 ± 0.833	0.961 ± 0.003
	(-32%)	$(+38\%)$	$(+31\%)$
$\Psi_{Net}(\boldsymbol{y}^\delta)$	0.080 ± 0.010	32.889 ± 1.64	0.709 ± 0.027
$\boldsymbol{x}_\lambda^{\Psi_{Net}}$ (graphLaNet)	$\mathbf{0.0464 \pm 0.006}$	$\mathbf{37.632 \pm 0.695}$	$\mathbf{0.975 \pm 0.002}$
	(-58%)	$(+14\%)$	$(+37\%)$
$\boldsymbol{x}_\lambda^{\Psi_{gt}}$	0.011 ± 0.002	49.935 ± 0.445	0.976 ± 0.002

In the following experiments, we test the methods on data with noise standard deviation σ in the range $[0, 1]$. In Fig. 3 we report the results (as the mean on the test set) in terms of PSNR (on the left) and SSIM (on the right) obtained with different values of σ. We underline that the performance of graphLaNet remains consistently better all along the interval $[0, 1]$, compared to Algorithm 1 combined with other reconstruction methods Ψ. In particular, it is remarkable how the SSIM values of graphLaNet keep being close to the SSIM limit-values of $\boldsymbol{x}_\lambda^{\Psi_{gt}}$.

Finally, in Fig. 4 we show a direct visualization of the stability of $\boldsymbol{x}_\lambda^{\Psi_{Tik}}$ and the graphLaNet, $\boldsymbol{x}_\lambda^{\Psi_{Net}}$, from one test image by computing the error image $|\boldsymbol{x}_\lambda^{\Psi_{(\cdot)}} - \boldsymbol{x}_{gt}|$ for the two reconstructions, and varying the amount of noise in $\boldsymbol{y}^\delta$. Let us remark that the neural network is not re-trained for taking into account the different intensities of σ.

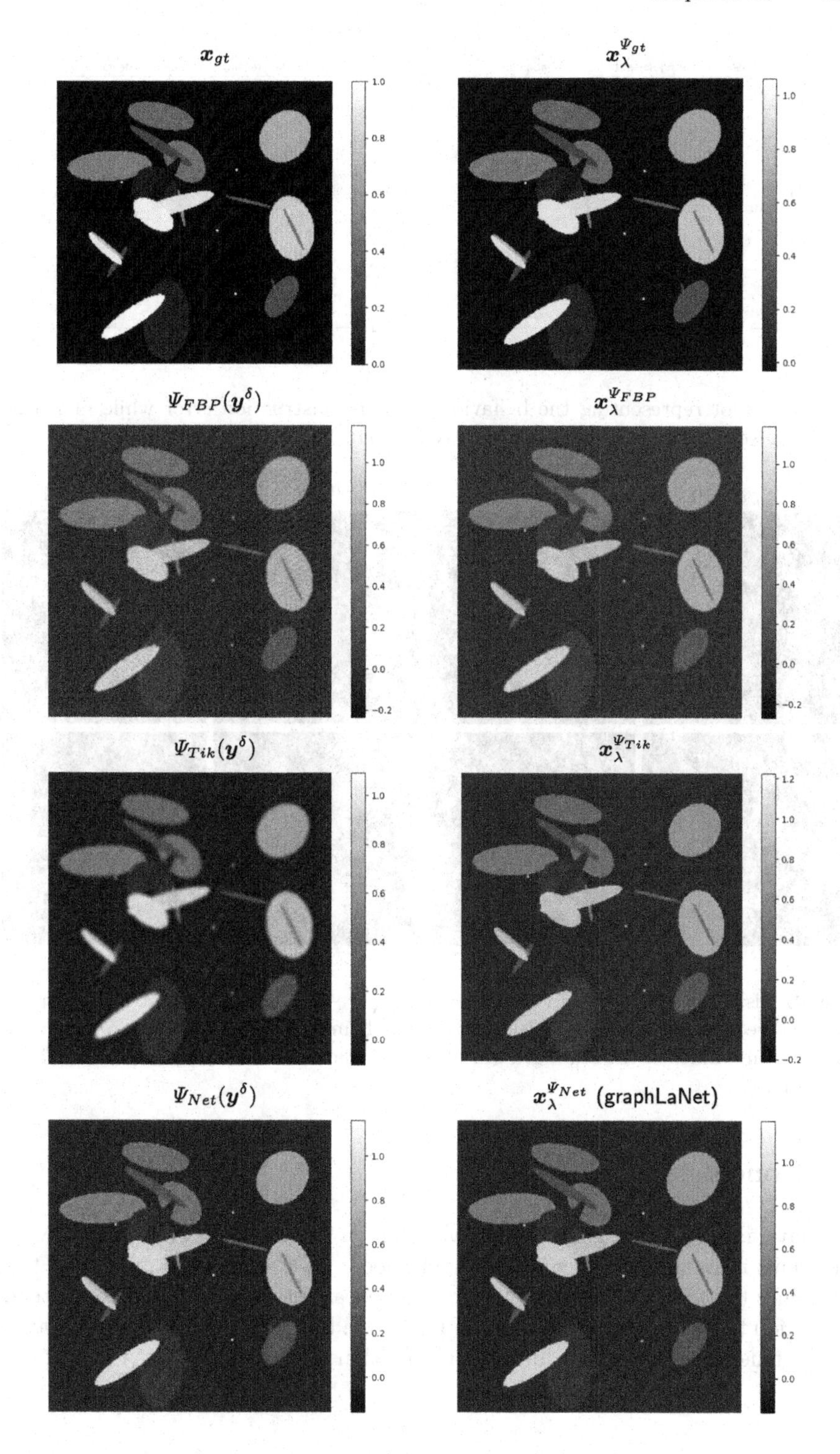

Fig. 2. Visual inspection of the reconstructions from one test image y^δ.

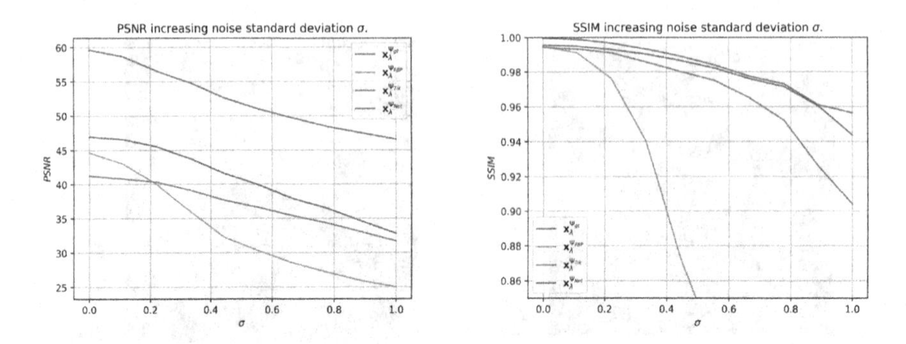

Fig. 3. A plot representing the behavior of the reconstruction error while increasing the noise standard deviation σ. *Left:* PSNR. *Right:* SSIM.

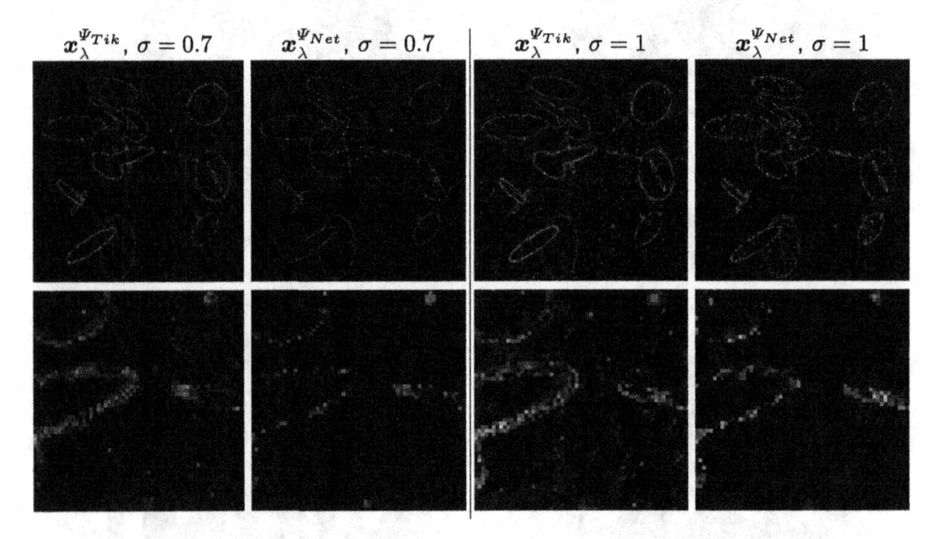

Fig. 4. Visualization of the difference image $|x_\lambda^{\Psi_{(\cdot)}} - x_{gt}|$ from one test image for two different reconstructions. In the top row the whole image, in the bottom row a zoom-in for different values of the standard deviation σ of the noise in the test set ($\sigma = 0.7$ and $\sigma = 0.1$).

7 Conclusion

We have developed the graphLaNet algorithm for the solutions of ill-posed inverse problems in imaging science. The graphLaNet is an ℓ^2-ℓ^1 Tikhonov method that combines the tools of graph Laplacian and neural networks. The Tikhonov regularization term is weighted by a graph Laplacian built upon a first approximation of the true image, and such first approximation is constructed by a neural network.

The outcomes from some preliminary numerical experiments are extremely promising. As shown in Sect. 6, the graphLaNet generates solutions with improving qualities for visualization and better performances in the metrics of RRE, PSNR and SSIM. In addition, the recovered solutions by graphLaNet have remarkable stability with respect to random noise contamination, in particular in terms of the SSIM metric.

That said, there is still work to be done, in particular concerning the rigorous proofs of convergence and stability results, which will be addressed in an upcoming paper. Nevertheless, those excellent preliminary experiments open the doors to several further research directions. An interesting direction regards the possibility to learn the edge-weight function of the graph Laplacian as well, where the penalty term should be built by embedding the graph Laplacian into a neural network architecture [3,13].

Acknowledgments. We want to thank the authors of [6] to share the dataset and the authors of [4] to share the code. Davide Bianchi is supported by NSFC (grant no. 12250410253). Wenbin Li is supported by Natural Science Foundation of Shenzhen (grant no. JCYJ20190806144005645) and NSFC (grant no. 41804096). Marco Donatelli is partially supported by GNCS (project 2022 "Tecniche numeriche per lo studio dei problemi inversi e l'analisi delle reti complesse").

References

1. Ao, W., Li, W., Qian, J.: A data and knowledge driven approach for SPECT using convolutional neural networks and iterative algorithms. J. Inverse Ill-Posed Probl. **29**(4), 543–555 (2021). https://doi.org/10.1515/jiip-2020-0056
2. Bianchi, D., Buccini, A., Donatelli, M., Randazzo, E.: Graph Laplacian for image deblurring. ETNA **55**, 169–186 (2022). https://doi.org/10.1553/etna_vol55s169
3. Bianchi, D., Lai, G., Li, W.: Uniformly convex neural networks and non-stationary iterated network Tikhonov (iNETT) method. Inverse Prob. (2023). https://doi.org/10.1088/1361-6420/acc2b6
4. Buccini, A., Donatelli, M.: Graph Laplacian in $\ell^2 - \ell^q$ regularization for image reconstruction. In: 2021 21st International Conference on Computational Science and Its Applications (ICCSA), pp. 29–38. IEEE (2021). https://doi.org/10.1109/ICCSA54496.2021.00015
5. Calatroni, L., van Gennip, Y., Schönlieb, C.-B., Rowland, H.M., Flenner, A.: Graph clustering, variational image segmentation methods and Hough transform scale detection for object measurement in images. J. Math. Imaging Vision **57**(2), 269–291 (2016). https://doi.org/10.1007/s10851-016-0678-0
6. Evangelista, D., Morotti, E., Loli Piccolomini, E.: COULE dataset (2021). https://www.kaggle.com/datasets/loiboresearchgroup/coule-dataset
7. Evangelista, D., Morotti, E., Piccolomini, E.L.: RISING: a new framework for model-based few-view CT image reconstruction with deep learning. Comput. Med. Imaging Graph. **103**, 102156 (2023). https://doi.org/10.1016/j.compmedimag.2022.102156
8. Gilboa, G., Osher, S.: Nonlocal operators with applications to image processing. Multiscale Model. Simul. **7**(3), 1005–1028 (2009). https://doi.org/10.1137/070698592

9. Hansen, P.C.: Rank-deficient and discrete ill-posed problems: numerical aspects of linear inversion. In: SIAM (1998)
10. Hansen, P.C., Nagy, J.G., O'leary, D.P.: Deblurring images: matrices, spectra, and filtering. In: SIAM (2006)
11. Keller, M., Lenz, D., Wojciechowski, R.K.: Graphs and Discrete Dirichlet Spaces. Grundlehren der mathematischen Wissenschaften. Springer, Cham (2021). https://doi.org/10.1007/978-3-030-81459-5
12. Kingma, D.P., Ba, J.: Adam: a method for Stochastic Optimization. In: Bengio, Y., LeCun, Y. (eds.) 3rd International Conference on Learning Representations, ICLR 2015, San Diego, CA, USA, May 7–9, 2015, Conference Track Proceedings (2015). http://arxiv.org/abs/1412.6980
13. Li, H., Schwab, J., Antholzer, S., Haltmeier, M.: NETT: solving inverse problems with deep neural networks. Inverse Probl. **36**(6), 065005 (2020). https://doi.org/10.1088/1361-6420/ab6d57
14. Peyré, G., Bougleux, S., Cohen, L.: Non-local regularization of inverse problems. In: Forsyth, D., Torr, P., Zisserman, A. (eds.) ECCV 2008. LNCS, vol. 5304, pp. 57–68. Springer, Heidelberg (2008). https://doi.org/10.1007/978-3-540-88690-7_5
15. Ronneberger, O., Fischer, P., Brox, T.: U-net: convolutional networks for biomedical image segmentation. In: Navab, N., Hornegger, J., Wells, W.M., Frangi, A.F. (eds.) MICCAI 2015. LNCS, vol. 9351, pp. 234–241. Springer, Cham (2015). https://doi.org/10.1007/978-3-319-24574-4_28
16. Scherzer, O., Grasmair, M., Grossauer, H., Haltmeier, M., Lenzen, F.: Variational Methods in Imaging, 1st edn. Springer, New York (2009). https://doi.org/10.1007/978-0-387-69277-7
17. Tewodrose, D.: A survey on spectral embeddings and their application in data analysis. Séminaire de théorie spectrale et géométrie **35**, 197–244 (2021). https://doi.org/10.5802/tsg.369
18. Wang, Z., Bovik, A.C., Sheikh, H.R., Simoncelli, E.P.: Image quality assessment: from error visibility to structural similarity. IEEE Trans. Image Process. **13**(4), 600–612 (2004)

Learning Posterior Distributions in Underdetermined Inverse Problems

Christina Runkel[1(✉)], Michael Moeller[2], Carola-Bibiane Schönlieb[1], and Christian Etmann[1]

[1] Department of Applied Mathematics and Theoretical Physics, University of Cambridge, Cambridge, UK
cr661@cam.ac.uk

[2] Department of Electrical Engineering and Computer Science, University of Siegen, Siegen, Germany

Abstract. In recent years, classical knowledge-driven approaches for inverse problems have been complemented by data-driven methods exploiting the power of machine and especially deep learning. Purely data-driven methods, however, come with the drawback of disregarding prior knowledge of the problem even though it has shown to be beneficial to incorporate this knowledge into the problem-solving process.

We thus introduce an unpaired learning approach for learning posterior distributions of underdetermined inverse problems. It combines advantages of deep generative modeling with established ideas of knowledge-driven approaches by incorporating prior information about the inverse problem. We develop a new neural network architecture 'UnDimFlow' (short for Unequal Dimensionality Flow) consisting of two normalizing flows, one from the data to the latent, and one from the latent to the solution space. Additionally, we incorporate the forward operator to develop an unpaired learning method for the UnDimFlow architecture and propose a tailored point estimator to recover an optimal solution during inference. We evaluate our method on the two underdetermined inverse problems of image inpainting and super-resolution.

Keywords: Inverse Problems · Deep Learning · Normalizing Flows

1 Introduction

Machine learning has excelled in almost all fields of computational sciences over the past decade with the area of inverse imaging problems not being an exception: Classical knowledge-driven approaches have been replaced with or complemented by data-driven methods exploiting the expressive power of deep learning. Reasons for this are twofold. While knowledge-driven approaches provide mathematical guarantees like existence of a solution and convergence guarantees, they however lack the ability to capture bespoke structures in data and have the disadvantage of very high numerical complexity [3]. Data-driven methods tackle even this shortcomings by using machine and especially deep learning techniques.

L. Calatroni et al. (Eds.): SSVM 2023, LNCS 14009, pp. 187–209, 2023.
https://doi.org/10.1007/978-3-031-31975-4_15

Purely data-driven methods, however, come with the drawback of disregarding prior knowledge about the problem even though it has shown to be beneficial in order to obtain more stable and reliable reconstructions with less training data.

We therefore introduce an unpaired learning approach for learning posterior distributions of underdetermined linear inverse problems. It combines advantages of deep generative modeling with established ideas of knowledge-driven approaches by incorporating prior information about the inverse problem.

Our contributions are threefold. First, we develop a new neural network architecture '*UnDimFlow*' (short for Unequal Dimensionality Flow). It consists of two normalizing flows with unequal dimensionalities which are trained to learn the posterior distribution of the underlying inverse problem. Second, we incorporate the known forward operator to develop an unpaired learning method. In addition to learning the whole posterior, we third propose a point estimator to recover an optimal solution during inference. We evaluate our method for two underdetermined inverse problems in imaging – image inpainting and super-resolution.

Based on these examples, we show that the proposed approach is not only able to learn posteriors of general underdetermined inverse problems in an unpaired training setting but also captures the degree of uncertainty of every individual input during inference, to generate more or less diverse output samples.

Throughout the paper, we are considering underdetermined statistical linear inverse problems of the form

$$\mathbf{y} = \mathrm{A}\,\mathbf{x} + \mathbf{e} \tag{1}$$

with model parameters x of an X-valued random variable $\mathbf{x}$, measured data samples y of a Y-valued random variable $\mathbf{y}$, Gaussian noise $\mathbf{e} \in Y$ and a linear forward operator $\mathrm{A} : X \to Y$. We consider a discrete setting, i.e., $X = \mathbb{R}^m$ and $Y = \mathbb{R}^n$ for $m > n$. Unfortunately, most practically relevant problems of the form (1) are ill-posed. One typically aims to re-establish the well-posedness by incorporating prior knowledge (e.g., in the form of a regularizer) or by benefiting from large amounts of training data via machine learning techniques. The ideal type of data for the latter, i.e., a large number of exemplary pairs of real measurements and corresponding ground truth solutions (y, x), is often difficult to acquire. Thus, in this work we merely assume we are given two sets of data samples $\mathcal{Y} := \{y_1, \cdots, y_a\} \sim p_{\mathbf{y}}$ and model parameter samples $\mathcal{X} := \{x_1, \cdots, x_b\} \sim p_{\mathbf{x}}$ that are unpaired. Using these unpaired data points, we aim to approximate the intractable posterior $p^*_{\mathbf{x}|\mathbf{y}}(x \mid y)$ by a tractable model $p_{\mathbf{x}|\mathbf{y}}(x \mid y)$ which allows examining all of the four features of interest mentioned above.

2 Related Work

Inverse problems have been and still are a very important research direction in mathematics. Methods range from classical knowledge-driven approaches to more recent data-driven methods. For an in-depth overview of knowledge-driven approaches, the interested reader is referred to [5,9,12]. In the following, we give a broad overview of data-driven approaches for solving inverse problems with a focus on related work for solving inverse problems by generative modeling techniques. A more detailed overview of general data-driven approaches can be found in [3,23]. Data-driven approaches for solving inverse problems range from learning a regularizer (e.g., [7,15,16]), learning descent directions [18] and proximal operators [17,22] to deep image priors [28]. While all previously mentioned approaches make use of learning paradigms, in the following we further detail a specific subgroup of those – approaches using generative modeling techniques. Related work for solving inverse problems by generative modeling techniques can be classified into methods learning the whole posterior $p_{\mathbf{x}|\mathbf{y}}$ and approaches learning to generate samples x conditioned on data points y. Approaches that learn to approximate the posterior $p_{\mathbf{x}|\mathbf{y}}$ can be further grouped into supervised and unsupervised methods. *Supervised* methods like [1,2,20,24,31] come with the advantage of generating good results. They however are restricted by the amount of ground truth available for the specific inverse problem. As common for *unsupervised* and *unpaired* generative modeling approaches for inverse problems, Siahkoohi et al. [25] use a pre-trained generative model as a prior $p_{\mathbf{x}}$. Additionally, they train a conditional normalizing flow to find solutions x. Further unsupervised methods have been introduced e.g., by Whang et al. [29] and Asim et al. [4]. While both assume a pre-trained generative model as prior probability distribution $p_{\mathbf{x}}$, Whang et al. [29] introduce a pre-generator to learn structured noise for conditioning the input to the given pre-trained generative model. A major drawback of this approach, though, is the fact that the pre-generator needs to be retrained for every observation y. In [4], the authors combine a learning- and model-based approach by using a pre-trained generative model as prior for a variational method incorporating the forward operator A. Similar as in the previously mentioned approach, this method requires optimizing for every observation y. In contrast to all previous approaches, this method, however, is not capable of estimating the whole posterior probability distribution $p_{\mathbf{x}|\mathbf{y}}$ and rather computes a single deterministic solution for every observation y. Further methods using unsupervised or unpaired learning in combination with generative models to generate samples x conditioned on data points y without providing the whole posterior probability distribution are e.g., Daras et al. [8] and Sim et al. [26]. They both make use of variants of Generative adversarial networks (GANs) and show results for multiple inverse problems. Summing up, there exists a wide range of supervised learning approaches using generative models to generate single samples x conditioned on data points y. Methods learning the whole posterior probability distribution of the inverse problem and especially learning it in an unsupervised or unpaired manner, however, are extremely scarce. Existing unsupervised and unpaired approaches for learning the posterior almost always make use of a pretrained generative model which has been trained on data samples x of the model parameter space. In contrast, we introduce an approach to estimate the whole posterior distribution requiring unpaired data only.

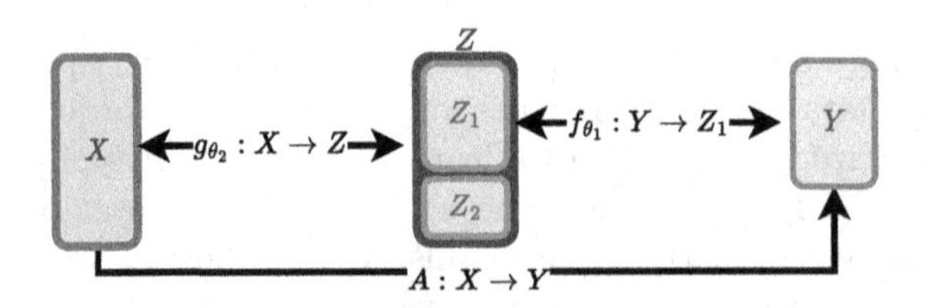

Fig. 1. Method overview – UnDimFlow, a composition of two normalizing flows f_{θ_1} and g_{θ_2} learns the posterior of the underdetermined inverse problem.

3 Unequal Dimensionality Flow

In contrast to other generative modeling techniques, normalizing flows [10,21] allow for explicit likelihood optimization and hence are capable of estimating true probability distributions. We thus propose a method to learn the posterior of an underdetermined inverse problem by making use of likelihood-based models and normalizing flows in particular. Normalizing flows, however, are diffeomorphisms and therefore not able to cope with the dimensionality reduction inherent in the forward mapping of the underdetermined inverse problem. We introduce a network architecture, UnDimFlow (short for **Un**equal **Dim**ensionality **Flow**; see Fig. 1), consisting of two jointly trained normalizing flows to overcome unequal dimensionalities of the measurement and model parameter space.
The **core ideas** of this method are the following:

1. **Learn two normalizing flows** – one learning the probability distribution of measurements $p_{\mathbf{y}}$, the other one learning the probability distribution of model parameters $p_{\mathbf{x}}$. We denote the flow learning the probability distribution of measurements $p_{\mathbf{y}}$ as $f_{\theta_1} : Y \to Z_1$, its inverse as $f_{\theta_1}^{-1} : Z_1 \to Y$ and the flow learning the probability distribution of model parameters as $g_{\theta_2} : X \to Z_1 \times Z_2$ as well as its inverse as $g_{\theta_2}^{-1} : Z_1 \times Z_2 \to X$. By definition, for flow f_{θ_1} it follows that $y = f_{\theta_1}^{-1}(z_1)$ and $z_1 = f_{\theta_1}(y)$, for flow g_{θ_2} it follows that $x = g_{\theta_2}^{-1}(z_1, z_2)$ and $(z_1, z_2) = g_{\theta_2}(x)$ for measured data samples y, model parameter samples x and latent space samples z_1 and z_2.
2. **Connect both normalizing flows by learning the same latent space** Z_1, with $\dim(Z_1) = \dim(Y)$. Additionally, the higher dimensional flow learning the probability distribution of model parameters makes up for the missing dimensions by adding a latent space Z_2 so that $\dim(Z) = \dim(Z_1) + \dim(Z_2) = \dim(X)$ for Z being the overall latent space. We further define a random variable $\mathbf{z}$ with samples $z \in Z$ and an overall probability distribution $p_{\mathbf{z}}(z)$. This probability distribution can be decomposed into $p_{\mathbf{z}}(z) = p_{\mathbf{z_1}}(z_1) \cdot p_{\mathbf{z_2}}(z_2)$, i.e., the random variables $\mathbf{z_1} \sim p_{\mathbf{z_1}}$ and $\mathbf{z_2} \sim p_{\mathbf{z_2}}$ for samples $z_1 \in Z_1$ and $z_2 \in Z_2$, respectively, are independent.
3. **Use the known forward operator** $A : X \to Y$ to connect model parameter and measurement space. Incorporating A into the energy function allows an unpaired training setting.

In comparison to existing methods, the proposed approach has the following **advantages**: It does not require pairs of training data and is thus able to learn the whole posterior of the inverse problem in an unpaired training setting. This is particularly useful as for most real-world inverse problems we only have little paired training data, if at all. Additionally, the proposed method is suitable for solving general underdetermined inverse problems and incorporates both data and prior knowledge of the problem. In complement to learning the posterior and in the style of the training procedure, we propose a point estimator that enables for explicit data discrepancy optimization in combination with a weighted regularization term consisting of the learned posterior. In the following, we introduce our approach. Due to the limited amount of space, a more detailed derivation of each of the results can be found in Appendix A.1.

3.1 Unpaired Training with UnDimFlow

Training the composed normalizing flows is realized by minimizing an energy function $E(\theta_1, \theta_2)$ consisting of four loss terms:

$$
\begin{aligned}
E(\theta_1, \theta_2) := & -\mathbb{E}_{\mathbf{y}}\left[\log p_{\mathbf{y}}(y)\right] - \lambda_1 \mathbb{E}_{\mathbf{x}}\left[\log p_{\mathbf{x}}(x)\right] \\
& + \lambda_2 \mathbb{E}_{\mathbf{y},\mathbf{z_2}}\left[\left\| \mathrm{A}\left(g_{\theta_2}^{-1}(f_{\theta_1}(y), z_2)\right) - y\right\|_2^2\right] \\
& + \lambda_3 \mathbb{E}_{\mathbf{x},\mathbf{z_2}}\left[\left\| g_{\theta_2}^{-1}(f_{\theta_1}(\mathrm{A}(x)), z_2) - x\right\|_2^2\right].
\end{aligned} \tag{2}
$$

The first and second loss term trains flow f_{θ_1} and g_{θ_2}, respectively, using a negative log likelihood loss (and the reparametrization trick of normalizing flows). The negative log likelihood loss plays an important role as it has regularizing effects and therefore also improves stability during training. As these loss terms minimize the true log-likelihood, them being minimized ensures that the network is not just learning a single best-fit solution but rather the whole high resolution probability distribution. Additionally, the third and fourth term of the energy function train the combination of both flows in an unpaired manner by making use of the forward operator.[1]

3.2 Computing the Posterior $p_{\mathbf{x}|\mathbf{y}}(x \mid y)$

In inverse problems, we are interested in computing the posterior $p_{\mathbf{x}|\mathbf{y}}(x \mid y)$. To compute $p_{\mathbf{x}|\mathbf{y}}(x \mid y)$ for our setting, we first need to compute $p_{\mathbf{x}|\mathbf{y},\mathbf{z_2}}(x \mid y, z_2)$ and second integrate over all $z_2 \in Z_2$.

Defining the Jacobians of g_{θ_2} and its inverse function $g_{\theta_2}^{-1}$ to be $J_{g_{\theta_2}}(x) := \frac{\partial g_{\theta_2}(x)}{\partial x}$ and $J_{g_{\theta_2}^{-1}}(z_1, z_2) := \frac{\partial g_{\theta_2}^{-1}(z_1, z_2)}{\partial (z_1, z_2)}$, respectively, the probability distribution of x given y and z_2 can be computed as follows:

[1] Note that although the third loss term might be sufficient to learn the correct mapping, in experiments the fourth loss term has shown to accelerate and stabilize training.

$$
\begin{aligned}
p_{\mathbf{x}|\mathbf{y},\mathbf{z_2}}(x \mid y, z_2) &= p_{\mathbf{z}}(g_{\theta_2}(x)) \cdot \left| \det\left(J_{g_{\theta_2}}(x)\right) \right| \\
&= p_{\mathbf{z_1}}(f_{\theta_1}(y)) \cdot p_{\mathbf{z_2}}(z_2) \cdot \left| \det\left(J_{g_{\theta_2}^{-1}}(f_{\theta_1}(y), z_2)\right) \right|^{-1}.
\end{aligned} \tag{3}
$$

The resulting term is easy to compute as y is given, z_2 is sampled from $p_{\mathbf{z_2}}$ and furthermore $p_{\mathbf{z_1}}(f_{\theta_1}(y))$ as well as $p_{\mathbf{z_2}}(z_2)$ are simple probability distributions, i.e., multivariate normal distributions, where we can sample from and compute probabilities of easily. The last part of the term involves computing the absolute value of the determinant of the Jacobian of the inverse of flow g_{θ_2}. As the most common invertible layers used in normalizing flows like affine coupling layers [11] allow for efficient computation of their Jacobian determinants, this part of the term is easily computable, too.

In the field of inverse problems, however, we are interested in the **posterior** $p_{\mathbf{x}|\mathbf{y}}$, independent of any sampled z_2. Taking Equation (3) into account, $p_{\mathbf{x}|\mathbf{y}}$ can be computed by integrating over z_2:

$$
\begin{aligned}
p_{\mathbf{x}|\mathbf{y}}(x \mid y) &= \int_{Z_2} p_{\mathbf{x}|\mathbf{y},\mathbf{z_2}}(x \mid y, z_2)\, dz_2 \\
&= \int_{Z_2} p_{\mathbf{z_1}}(f_{\theta_1}(y)) \cdot p_{\mathbf{z_2}}(z_2) \cdot \left| \det\left(J_{g_{\theta_2}^{-1}}(f_{\theta_1}(y), z_2)\right) \right|^{-1} dz_2.
\end{aligned} \tag{4}
$$

We have therefore shown that the proposed approach is able to approximate the true posterior of the inverse problem and thus to recover all possible solutions of the inverse problem as well as their probabilities. In practice, computing the posterior $p_{\mathbf{x}|\mathbf{y}}(x \mid y)$ can be done, e.g., by Monte Carlo Integration [19, pp. 30-43].

3.3 Recovering a Solution to the Inverse Problem During Inference

Besides computing the posterior, we are also interested in recovering a single solution to the inverse problem during inference. A solution $\bar{x}$ can be recovered either by generating a random solution, i.e., via posterior sampling or by optimizing for a solution under a point estimator. In the following, computing the Maximum a posteriori (MAP) estimator for the current setting is further detailed and we additionally introduce a new point estimator, the explicit data discrepancy optimization (EDDO), that explicitly optimizes for data discrepancy by making use of prior knowledge of the problem.

MAP Estimator: The goal of the MAP estimator is maximizing the probability of x given some y. As in our case x can be fully represented by $f_{\theta_1}(y)$ and z_2, i.e., by definition $x = g_{\theta_2}^{-1}(f_{\theta_1}(y), z_2)$, the value for x varies for changing z_2 (as we assume y to be fixed). We are able to rewrite the posterior using Bayes formula, the invertibility of both flows f_{θ_1} and g_{θ_2} and the fact that we assume y to be fixed, so that:

$$
p_{\mathbf{x}|\mathbf{y}}(x \mid y) = \frac{\delta\left(y - f_{\theta_1}^{-1}(g_{\theta_2}^{\langle z_1 \rangle}(x))\right) \cdot p_{\mathbf{x}}(x)}{p_{\mathbf{y}}(y)}. \tag{5}
$$

Using Eq. (5) and the fact that $p_{\mathbf{y}}(y)$ is a positive constant to formulate the optimization problem, we obtain

$$\bar{x}_{\text{MAP}} = \underset{x \in \{x \mid y = f_{\theta_1}^{-1}(g_{\theta_2}^{\langle z_1 \rangle}(x))\}}{\text{argmax}} \quad p_{\mathbf{x}}(x). \tag{6}$$

The constraint of the above problem means that there exists a z_2 such that $x = g_{\theta_2}^{-1}(f_{\theta_1}(y), z_2)$. This means optimizing over X can also be expressed as an optimization in latent space Z_2, which means computing

$$\begin{aligned} \bar{z}_{2_{\text{MAP}}} &= \underset{z_2 \in Z_2}{\text{argmax}}\, p_{\mathbf{x}}(g_{\theta_2}^{-1}(f_{\theta_1}(y), z_2)) \\ &= \underset{z_2 \in Z_2}{\text{argmax}} \left(p_{\mathbf{z_1}}(f_{\theta_1}(y)) \cdot p_{\mathbf{z_2}}(z_2) \cdot \left| \det \left(J_{g_{\theta_2}^{-1}}(f_{\theta_1}(y), z_2) \right) \right|^{-1} \right) \end{aligned} \tag{7}$$

and setting

$$\bar{x}_{\text{MAP}} = g_{\theta_2}^{-1}\big(f_{\theta_1}(y), \bar{z}_{2_{\text{MAP}}}\big) \tag{8}$$

yields a maximizer of Equation (7).

Explicit Data Discrepancy Optimization (EDDO): As the MAP estimator is not explicitly incorporating a data discrepancy term and therefore not using prior knowledge of the forward operator during inference, we propose yet another point estimator with explicit data discrepancy optimization. Similar to the proposed energy function, we minimize a weighted combination of data fidelity term $q_{\mathbf{y}|\mathbf{x}}(y \mid x)$ and regularization term $p_{\mathbf{x}|\mathbf{y}}(x \mid y)$ to recover an optimal solution

$$\bar{x}_{\text{disc}} = \underset{x \in X}{\text{argmax}} \left(q_{\mathbf{y}|\mathbf{x}}(y \mid x) \cdot p_{\mathbf{x}|\mathbf{y}}(x \mid y)^{\lambda} \right) \tag{9}$$

for a hyperparameter λ and model $q_{\mathbf{y}|\mathbf{x}}(y \mid x) := \exp\left(- \frac{\|Ax-y\|_2^2}{2\sigma^2} \right)$.[2] Using the fact that per definition $x = g_{\theta_2}^{-1}(z_1, z_2)$ and assuming y to be fixed, we optimize over latent space Z_2 so that $\bar{x}_{\text{disc}} = g_{\theta_2}^{-1}(f_{\theta_1}(y), \bar{z}_{2_{\text{disc}}})$ for $\bar{z}_{2_{\text{disc}}} = \text{argmax}_{z_2 \in Z_2}\, q_{\mathbf{y}|\mathbf{x}}\big(y \mid g_{\theta_2}^{-1}(z_1, z_2)\big) \cdot p_{\mathbf{x}}\big(g_{\theta_2}^{-1}(z_1, z_2)\big)^{\lambda}$. For optimization purposes we formulate the maximization problem as minimization problem, i.e.,

$$\begin{aligned} \bar{z}_{2_{\text{disc}}} = \underset{z_2 \in Z_2}{\text{argmin}} \Bigg(&\|A g_{\theta_2}^{-1}(f_{\theta_1}(y), z_2) - y\|_2^2 \\ &- \lambda\Big(\log p_{\mathbf{z_2}}(z_2) + \log \left| \det \left(J_{g_{\theta_2}^{-1}}(f_{\theta_1}(y), z_2) \right) \right| \Big) \Bigg). \end{aligned} \tag{10}$$

The proposed point estimator has the flexibility to directly scale the ratio between most likely solution under the learned posterior and a solution minimizing the squared L^2 distance $\|Ax - y\|_2^2$.

[2] Note that for clarification, we denote all learned probability distributions as p while denoting the modeled distributions as q.

4 Experimental Results

In the following, we experimentally verify our theoretical considerations on two underdetermined inverse problems – image super-resolution and inpainting. In all experiments, we use multivariate normal distributions as base distributions for both normalizing flows. Both flows use Glow-like building blocks (cf. [14]) as core building blocks. In contrast to standard Glow building blocks, we apply a random permutation as this has shown to improve stability during training of our experiments. Each block therefore computes

$$f_{\text{block}}(y) = P \cdot \sigma(s_{\text{an}}) \odot f_{\text{ac}}(y) + b_{\text{an}} \tag{11}$$

for an activation function σ, a permutation matrix P, a scaling parameter s_{an}, a bias parameter b_{an} (both parts of the ActNorm layer [14]) and the affine coupling function f_{ac} [11]. Further results and experimental details can be found in Appendix A.2 and A.3.

4.1 Super-Resolution Under High Uncertainty

Super-resolution imaging techniques aim to generate higher resolution images x from their lower resolution counterparts y [6]. To show that the proposed method is able to learn the super-resolution space X, we train an UnDimFlow network on the Fashion-MNIST dataset [30] for lower resolution images with a spacial resolution of 8×8 pixels, highlighting the benefits of the proposed approach for super-resolution under high uncertainty.

As downsampling to low resolution image sizes of only 8×8 pixels results in objects shown in the images that are hardly recognizable even for human observers, it introduces a high amount of uncertainty. In fact, Subfigure 2(b) highlights that the area of highest difference is exactly the part of the image that distinguishes trousers from dresses. This can additionally be verified by looking at the pixel-wise variance illustrated in Subfigure 2(c). Again, the area between possible trousers legs is an area of high variance just as the borders of the shown object. As the used bilinear downsampling operator introduces uncertainty especially in regions of edges, it is expectable that the pixel-wise variance between random samples is especially high in those parts of the image. Optimal solutions generated by the MAP estimator mirror the multi-modality of the underlying posterior. Since (7) is a non-convex problem, the MAP estimation can only be approximated by running local optimization algorithms from different initializations of z_2, which yield different estimates. In our example these estimates either end up being a dress (see Subfigure 2(d)) or trousers (see Subfigure 2(e)). We now compare the MAP estimator to the conditional mean and our proposed EDDO, highlighting their specific advantages as well as drawbacks. To compute the aforementioned point estimators, we choose the same example as detailed in Fig. 2 and compute point estimators for the same two different initializations of z_2 for both the MAP estimator and the EDDO. The results for the different point estimators are illustrated in Fig. 3. Comparing the point

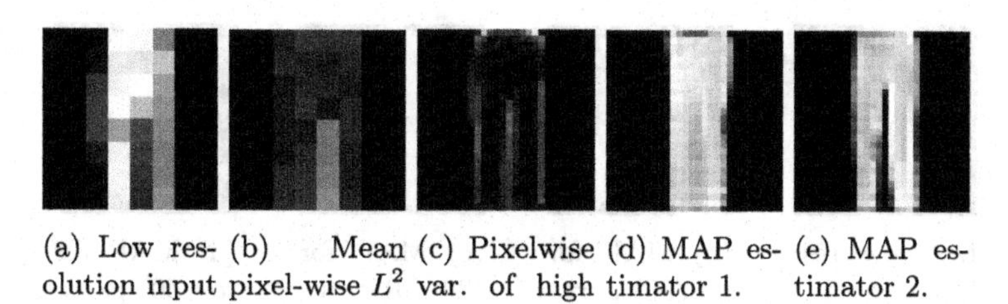

(a) Low resolution input image y. (b) Mean pixel-wise L^2 distance. (c) Pixelwise var. of high res. samples. (d) MAP estimator 1. (e) MAP estimator 2.

Fig. 2. Overview of Fashion-MNIST super-resolution results for 8×8 pixel low resolution images and $\lambda_2 = 200$.

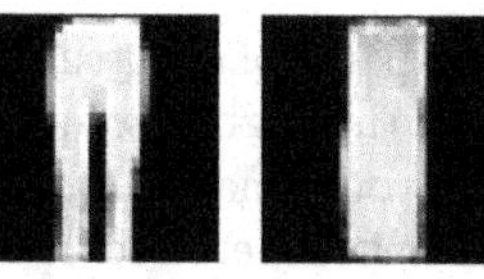

(a) Cond. mean over 1000 samples. (b) MAP estimators for two different init. of z_2. (c) EDDO for the same two initializations of z_2.

Fig. 3. Overview of point estimators for super-resolution on Fashion-MNIST.

estimators in terms of two factors – L^2 distance $\|A\bar{x}_{\text{opt}} - y\|_2$ for an optimal $\bar{x}_{\text{opt}}$ and how realistic the respective images look empirically – the conditional mean shows the worst visual result while having a medium L^2 distance of 1.33 and thus being closer to the input than each of the MAP estimators. They, however, show realistic and smooth results just as each of the samples generated by EDDO for $\lambda = 0.001$. With 1.0 and 1.55, the L^2 distance of the EDDO estimators is smaller than those for the results generated by the MAP estimators (2.1 and 1.85). All in all, the EDDO estimator provides the best results in regard to both factors.

4.2 Image Inpainting on Partitioned Training Dataset

After detailing the results of the proposed method on a super-resolution problem, in the following, we will on the one hand show that our approach is not tailored to a specific inverse problem but capable of solving underdetermined inverse problems in general. On the other hand, we furthermore highlight the ability of our approach to handle unbalanced training settings by training on different partitions and partition sizes of the training dataset.

Image inpainting is a general term for all methods concerned with the reconstruction of images $\mathbf{x}$ from degraded images $\mathbf{y}$ which have been destructed by some mask m, i.e., $\mathbf{y} = \mathbf{x} \odot m + \mathbf{e}$ [3, p. 9]. In order to apply the proposed method, we cut out the damaged parts of the image and thus reconstruct the original image x from a lower dimensional degraded image y. As we are assuming the forward operator A to be known, one might argue that having access to high resolution images x in combination with the forward operator, one would

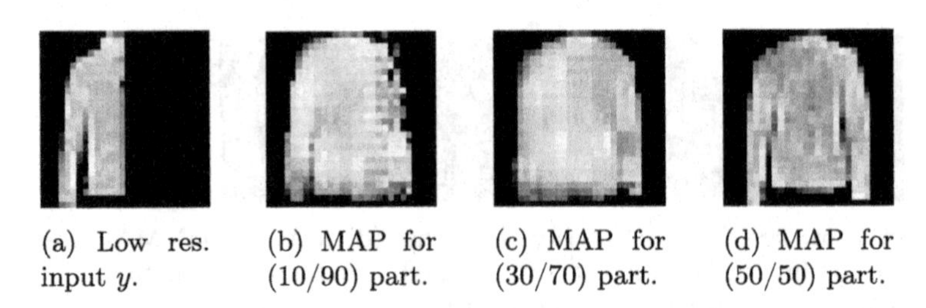

(a) Low res. input y. (b) MAP for (10/90) part. (c) MAP for (30/70) part. (d) MAP for (50/50) part.

Fig. 4. Overview of Fashion-MNIST inpainting results for partitioned datasets for partition ratios (50/50), (30/70) and (10/90) of HR/LR dataset.

be capable of constructing training pairs (y, x) by applying the forward operator A to the high resolution image x and adding noise e as $\mathbf{y} = A\mathbf{y} + \mathbf{e}$. This in fact would make an unpaired learning approach redundant and one might apply existing supervised learning approaches instead. However, a common issue in inverse problems is that while having access to a huge amount of lower dimensional data points y_i like degraded images for our particular inverse problem, the amount of high resolution data points x_j is often restricted. In fact, in some real-world scenarios the ratio between high resolution (HR) and low resolution (LR) data is $1 : 10$. We thus partition the Fashion-MNIST dataset randomly and test different partition sizes in conjunction with real-world settings where, e.g., only 10% of HR data are available. With regard to the first aspect, we train a (50/50) partitioning of HR/LR training data points and a (30/70) as well as (10/90) HR/LR training data split for analyzing the second aspect.

An overview of results can be found in Fig. 4. While the (50/50) split of HR and LR datasets results in a sharp, smooth, realistic and diverse MAP estimator, it can be reasoned that training on different partitions of a dataset still leads to good results. For a (30/70) split, however, missing information leads to more blurry images which are not that detailed and for an HR dataset of only 10%, training with the same parameters results in artifacts in parts of the image where information is missing due to the mask. These issues arise from the fact that using the same weighting of loss terms as during training on the whole Fashion-MNIST dataset, for unbalanced HR/LR dataset sizes this implicitly leads to weighting the L_y^2 and negative log-likelihood (NLL) loss term of flow f_{θ_1} more. Due to the higher amount of LR training data we optimize more often with respect to these two loss terms only. Taking into account the findings of the aforementioned partitioning experiments, we increase the weights of the remaining loss terms. We scale up λ_1 and λ_3 from $\lambda_1 = 2$, $\lambda_3 = 100$ to $\lambda_1 = 6$ and $\lambda_3 = 200$ so that the L_x^2 and NLL loss term of flow g_{θ_2} are weighted more. Additionally, we decrease the batch size of the HR training dataset from 128 to 64 to enable further optimization steps on HR data. Note that for a (10/90) partitioning of HR/LR data, decreasing the batch size by a factor of 2 still leads to a much smaller number of optimization steps than for the LR data partition. The results of this experiment, at the example of the more complex (10/90) partitioning, are illustrated in Fig. 5. The fine-tuning described above leads to smooth results

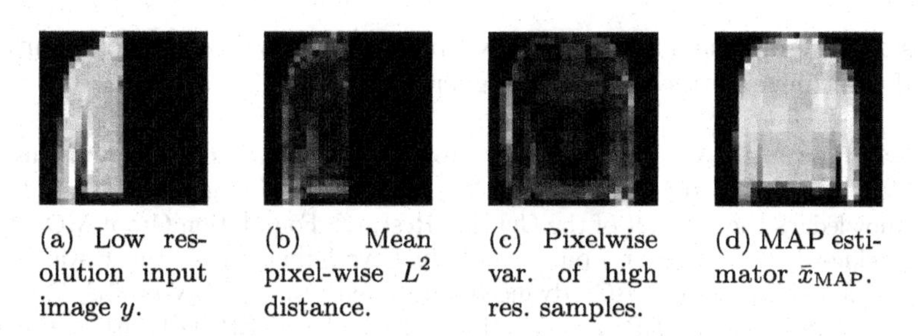

(a) Low resolution input image y. (b) Mean pixel-wise L^2 distance. (c) Pixelwise var. of high res. samples. (d) MAP estimator $\bar{x}_{\text{MAP}}$.

Fig. 5. Overview of Fashion-MNIST inpainting results for a (10/90) partitioning of HR/LR data with smaller batch size and adapted weightings of loss terms.

with much fewer artifacts than before. The results even seem more detailed than the results of the (30/70) partitioning of the previous experiment. The MAP estimator still shows a small amount of artifacts but improved strongly even for an unbalanced training setting like this. The proposed method thus exploits the advantages of unpaired learning and is capable of adopting to real-world training scenarios regarding HR/LR training data ratios.

5 Conclusion

In this work, we introduced a method for learning posterior probability distributions of underdetermined inverse problems by making use of normalizing flows. By additionally incorporating the known forward operator of the inverse problem, we overcame the necessity of paired training data and proposed a neural network architecture 'UnDimFlow' in combination with an unpaired learning method and a corresponding point estimator, simplifying finding a good solution under specified requirements like high similarity to the input. We evaluated our method on the Fashion-MNIST dataset for two underdetermined inverse problems in imaging – image inpainting and super-resolution. Based on these examples, we showed that the proposed approach is not only able to learn posterior probability distributions of general underdetermined inverse problems in an unpaired training setting but also captures the degree of uncertainty of every individual input during inference, to generate more or less diverse output samples. With EDDO, a further way to control the output during inference has been introduced as it weights a data discrepancy term against the regularization term $p_{\mathbf{x}|\mathbf{y}}(x \mid y)$. As the approach makes use of normalizing flows and it is a known issue that normalizing flows are hard to train on higher dimensional data, it however comes with the drawback of being difficult to use for higher dimensional data. This issue however has been recognized by the research community and recently there have been promising approaches like the method of Song et al. [27]. They use stochastic differential equations for score-based generative modeling, which shows promising results even for high dimensional images. As the proposed method is not restricted to normalizing flows and can be easily applied to other likelihood-based models, an interesting direction of future work is exchanging the normalizing flows by approaches like the aforementioned method to achieve better results on higher dimensional data.

Acknowledgements. CR acknowledges support from the Cantab Capital Institute for the Mathematics of Information (CCIMI) and the EPSRC grant EP/W524141/1. MM acknowledges the support of the German Research Foundation Grant MO 2962/7-1. CBS acknowledges support from the Philip Leverhulme Prize, the Royal Society Wolfson Fellowship, the EPSRC advanced career fellowship EP/V029428/1, EPSRC grants EP/S026045/1 and EP/T003553/1, EP/N014588/1, EP/T017961/1, the Wellcome Innovator Awards 215733/Z/19/Z and 221633/Z/20/Z, the European Union Horizon 2020 research and innovation programme under the Marie Skodowska-Curie grant agreement No. 777826 NoMADS, the CCIMI and the Alan Turing Institute. CE acknowledges support from the Wellcome Innovator Award RG98755.

A Appendix

In the following, we provide additional theoretical results as well as experimental details for all experiments described previously.

A.1 Additional Theoretical Results

This Subsection details some of the theoretical results of the main part of the paper.

Further Analysis of Third and Fourth Loss Term. To show that $(g_{\theta_2}^{-1} \circ f_{\theta_1})$ learns a correct mapping between the data measurement space Y and the model parameter space X when minimizing the energy function E in Eq. (2), we assume E to have fully converged so that $\hat{\theta}_1, \hat{\theta}_2 := \text{argmin}_{\theta_1,\theta_2} E(\theta_1, \theta_2)$. Explicitly, for the third loss term this means that $\mathbb{E}_{\mathbf{y},\mathbf{z_2}}\left[\|\, \mathrm{A}\left(g_{\hat{\theta}_2}^{-1}(f_{\hat{\theta}_1}(y), z_2)\right) - y\|_2^2\right]$ is minimal. Therefore, for the expectation value $\mathbb{E}_{\mathbf{y},\mathbf{z_2}}$ to be minimal, $\|\, \mathrm{A}\left(g_{\hat{\theta}_2}^{-1}(f_{\hat{\theta}_1}(y), z_2)\right) - y\|_2^2$ has to be minimal, as well. This means that the forward operator cancels out all values of x that map to Z_2 through flow g_{θ_2}. The remaining values of x, after applying the forward operator, need to equal those in y, up to a certain degree that is determined by the noise level. For this, there exist two cases: In the first case, $\|f_{\hat{\theta}_1}^{-1}(g_{\hat{\theta}_2}^{\langle z_1 \rangle}(x)) - y\|_2^2$ is minimal, as well, i.e., the inverse mapping between the corresponding part of x, mapping first to Z_1 and second to Y through flow $f_{\theta_1}^{-1}$, denoted as $f_{\hat{\theta}_1}^{-1}(g_{\hat{\theta}_2}^{\langle z_1 \rangle}(x))$, and y are equal. In the second case, considering the fact that still the expectation value in our loss term has to be minimal, parts of the sampled z_2 have to make up for these differences in the inverse mapping. Based on our design to sample z_2 randomly during training, the second case, however, may never lead to the expectation value in the loss term being minimal and thus the third loss term enforces the correct mapping.

For a fully converged energy function E, the fourth loss term $\mathbb{E}_{\mathbf{x},\mathbf{z_2}}\left[\|g_{\theta_2}^{-1}(f_{\theta_1}\left(\mathrm{A}(x)\right), z_2) - x\|_2^2\right]$ likewise is minimal. In contrast to the third loss term, the fourth loss term has an additional source of randomness involved as we are computing the expectation value for $\mathbf{x}$ living in the higher dimensional model parameter space X. Whenever computing the (squared) L^2 distance in X while applying the forward operator A with inherent information loss and *afterwards* making up for the missing information by randomly sampling z_2 as in the fourth loss term, minimizing $\mathbb{E}_{\mathbf{x},\mathbf{z_2}}$ may never reach zero because of the randomness involved. Still, similarly as for the third loss term, minimizing $\|g_{\theta_2}^{-1}(f_{\theta_1}(\mathrm{A}(x)), z_2) - x\|_2^2$ can be reached if either $f_{\hat{\theta}_1}^{-1}(g_{\hat{\theta}_2\langle z_1\rangle}(x))$ and y are equal or parts of the sampled z_2 make up for these differences which, due to the randomness involved, never minimizes $\mathbb{E}_{\mathbf{x},\mathbf{z_2}}$. Additionally, since we are computing the L^2 distance in the higher dimensional space, another possible minimum of the squared L^2 distance above might be a suboptimal mapping between Y and X whenever parts of z_2 by chance map to the correct values of the 'Z_2'-part of x. Due to random sampling of z_2, this however never minimizes the expectation value, either. In other words, whenever we are irrevocably throwing away information as in the case of applying the forward operator to the higher dimensional x and we are randomly sampling to make up for these information, the L_2 distance of our input x and the reconstructed solution x' is minimal if all values retained after applying the forward operator are mapped back to the corresponding values in X. Thus, the fourth loss term also enforces to learn the correct mapping between Y and X.

Detailed Derivation on Computing $p_{\mathbf{x}|\mathbf{y},\mathbf{z_2}}(x \mid y, z_2)$

$$
\begin{aligned}
p_{\mathbf{x}|\mathbf{y},\mathbf{z_2}}(x \mid y, z_2) &= p_{\mathbf{z}}(g_{\theta_2}(x)) \cdot \left|\det\left(J_{g_{\theta_2}}(x)\right)\right| && [1] \\
&= p_{\mathbf{z_1}}(z_1) \cdot p_{\mathbf{z_2}}(z_2) \cdot \left|\det\left(J_{g_{\theta_2}}(g_{\theta_2}^{-1}(z_1, z_2))\right)\right| && [2] \\
&= p_{\mathbf{z_1}}(z_1) \cdot p_{\mathbf{z_2}}(z_2) \cdot \left|\det\left(J^{-1}_{g_{\theta_2}^{-1}}(z_1, z_2)\right)\right| && [3] \\
&= p_{\mathbf{z_1}}(z_1) \cdot p_{\mathbf{z_2}}(z_2) \cdot \left|\det\left(J_{g_{\theta_2}^{-1}}(z_1, z_2)\right)^{-1}\right| && [4] \\
&= p_{\mathbf{z_1}}(z_1) \cdot p_{\mathbf{z_2}}(z_2) \cdot \left|\det\left(J_{g_{\theta_2}^{-1}}(z_1, z_2)\right)\right|^{-1} && [5] \\
&= p_{\mathbf{z_1}}(f_{\theta_1}(y)) \cdot p_{\mathbf{z_2}}(z_2) \cdot \left|\det\left(J_{g_{\theta_2}^{-1}}(f_{\theta_1}(y), z_2)\right)\right|^{-1} && [6]
\end{aligned}
\tag{12}
$$

In step [1], we apply the change-of-variables formula to then use the independence of $p_{\mathbf{z_1}}$ and $p_{\mathbf{z_2}}$ to substitute $p_{\mathbf{z}}$ by $p_{\mathbf{z_1}} \cdot p_{\mathbf{z_2}}$ and the fact that per definition $x = g_{\theta_2}^{-1}(z_1, z_2)$ (see step [2]). Making use of the inverse function theorem, we replace the Jacobian of g_{θ_2} by the inverse of the Jacobian of $g_{\theta_2}^{-1}$ as shown in step [3] and in step [4] pull out the inverse as the determinant of the inverse of a matrix equals the inverse determinant of the matrix, i.e., $\det(A^{-1}) = \det(A)^{-1}$. As the determinant is a scalar and the inverse of a scalar s is just $\frac{1}{s}$, the absolute value of this fraction can be further simplified so that $|s^{-1}| = |\frac{1}{s}| = \frac{|1|}{|s|} = \frac{1}{|s|} = |s|^{-1}$

(see step [5]). In the last step, we just replace z_1 with $f_{\theta_1}(y)$ as per definition $z_1 = f_{\theta_1}(y)$. The resulting term is easy to compute as y is given, z_2 is sampled from $p_{\mathbf{z_2}}$ and furthermore $p_{\mathbf{z_1}}(f_{\theta_1}(y))$ as well as $p_{\mathbf{z_2}}(z_2)$ are simple probability distributions where we can sample from and compute probabilities of easily. The last part of the term involves computing the absolute value of the determinant of the Jacobian of the inverse of flow g_{θ_2}. As invertible layers of normalizing flows allow for efficient computation of their Jacobian determinants, this part of the term is easily computable, as well.

Detailed Derivation of MAP Estimator. For optimization purposes, we formulate the optimization problem in Eq. (7) as a minimization problem, i.e.,

$$\bar{z}_{2\text{MAP}} = \underset{z_2 \in Z_2}{\operatorname{argmax}} \left(p_{\mathbf{x}}(g_{\theta_2}^{-1}(z_1, z_2))\right) \quad [1] \tag{13}$$

$$= \underset{z_2 \in Z_2}{\operatorname{argmin}} \left(-p_{\mathbf{x}}(g_{\theta_2}^{-1}(z_1, z_2))\right) \quad [2] \tag{14}$$

$$= \underset{z_2 \in Z_2}{\operatorname{argmin}} \left(-\log p_{\mathbf{x}}(g_{\theta_2}^{-1}(z_1, z_2))\right) \quad [3] \tag{15}$$

$$= \underset{z_2 \in Z_2}{\operatorname{argmin}} \left(-\log \left(p_{\mathbf{z_1}}(f_{\theta_1}(y)) \cdot p_{\mathbf{z_2}}(z_2) \cdot \left|\det\left(J_{g_{\theta_2}^{-1}}(f_{\theta_1}(y), z_2)\right)\right|^{-1}\right)\right) \quad [4] \tag{16}$$

$$= \underset{z_2 \in Z_2}{\operatorname{argmin}} \left(-\log p_{\mathbf{z_1}}(f_{\theta_1}(y)) - \log p_{\mathbf{z_2}}(z_2) - \log\left|\det\left(J_{g_{\theta_2}^{-1}}(f_{\theta_1}(y), z_2)\right)\right|^{-1}\right) \quad [5] \tag{17}$$

$$= \underset{z_2 \in Z_2}{\operatorname{argmin}} \left(-\log p_{\mathbf{z_2}}(z_2) - \log\left|\det\left(J_{g_{\theta_2}^{-1}}(f_{\theta_1}(y), z_2)\right)\right|^{-1}\right) \quad [6] \tag{18}$$

$$= \underset{z_2 \in Z_2}{\operatorname{argmin}} \left(-\log p_{\mathbf{z_2}}(z_2) - \log\left(\frac{1}{|\det(J_{g_{\theta_2}^{-1}}(f_{\theta_1}(y), z_2))|}\right)\right) \quad [7] \tag{19}$$

$$= \underset{z_2 \in Z_2}{\operatorname{argmin}} \left(-\log p_{\mathbf{z_2}}(z_2) + \log\left|\det\left(J_{g_{\theta_2}^{-1}}(f_{\theta_1}(y), z_2)\right)\right|\right) \quad [8] \tag{20}$$

While in step [2], we simply formulate the maximization problem as a minimization problem, in step [3], we use the fact that minimizing a function $d(x)$ due to monotonicity of the natural logarithm equals to minimizing $\log d(x)$. Substituting the results of Eq. (7) for $p_{\mathbf{x}}(g_{\theta_2}^{-1}(z_1, z_2))$ (see step [4]) and applying logarithmic laws (see step [5]), we simplify the original equation so that we are able to cancel $p_{\mathbf{z_1}}(f_{\theta_1}(y))$ in step [6] as $\mathbf{z_1}$ and $\mathbf{z_2}$ are independent random variables and we are solely optimizing over all $z_2 \in Z_2$. Applying logarithmic laws again (see step [7] and [8]), we end up with two terms that are easy to evaluate given that the first term computes the log probability of our simple base distribution and by construction, the log determinant of a normalizing flow is simple to calculate, as well.

Detailed Derivation of EDDO

$$
\begin{aligned}
\bar{z}_{2_{\text{disc}}} &= \underset{z_2 \in Z_2}{\operatorname{argmax}} \Big(q_{\mathbf{y}|\mathbf{x}}(y \mid g_{\theta_2}^{-1}(z_1, z_2)) \cdot \lambda^* \; p_{\mathbf{x}}(g_{\theta_2}^{-1}(z_1, z_2)) \Big) && [1] \\
&= \underset{z_2 \in Z_2}{\operatorname{argmax}} \left(\exp\left(- \frac{\|A g_{\theta_2}^{-1}(z_1, z_2) - y\|_2^2}{2\sigma^2} \right) \cdot \lambda^* \; p_{\mathbf{x}}(g_{\theta_2}^{-1}(z_1, z_2)) \right) && [2] \\
&= \underset{z_2 \in Z_2}{\operatorname{argmax}} \left(- \frac{1}{2\sigma^2} \|A g_{\theta_2}^{-1}(z_1, z_2) - y\|_2^2 + \lambda \log p_{\mathbf{x}}(g_{\theta_2}^{-1}(z_1, z_2)) \right) && [3] \\
&= \underset{z_2 \in Z_2}{\operatorname{argmax}} \Big(- \|A g_{\theta_2}^{-1}(z_1, z_2) - y\|_2^2 + \lambda \log p_{\mathbf{x}}(g_{\theta_2}^{-1}(z_1, z_2)) \Big) && [4] \\
&= \underset{z_2 \in Z_2}{\operatorname{argmin}} \Big(\|A g_{\theta_2}^{-1}(z_1, z_2) - y\|_2^2 - \lambda \log p_{\mathbf{x}}(g_{\theta_2}^{-1}(z_1, z_2)) \Big) && [5] \\
&= \underset{z_2 \in Z_2}{\operatorname{argmin}} \Big(\|A g_{\theta_2}^{-1}(f_{\theta_1}(y), z_2) - y\|_2^2 - \lambda \log p_{\mathbf{x}}(g_{\theta_2}^{-1}(f_{\theta_1}(y), z_2)) \Big) && [6] \\
&= \underset{z_2 \in Z_2}{\operatorname{argmin}} \Big(\|A g_{\theta_2}^{-1}(f_{\theta_1}(y), z_2) - y\|_2^2 && [7] \\
&\qquad - \lambda \Big(\log p_{\mathbf{z_2}}(z_2) + \log \Big| \det \Big(J_{g_{\theta_2}^{-1}}(f_{\theta_1}(y), z_2) \Big) \Big| \Big) \Big).
\end{aligned}
\tag{21}
$$

While we first insert the definition of $q_{\mathbf{y}|\mathbf{x}}$ (see step [2]) and use the fact that minimizing a function is equal to minimizing the natural logarithm of this function (see step [3]), in step [4] we eliminate $\frac{1}{2\sigma^2}$ as it is always a positive constant and thus can be omitted. Formulating the maximization as a minimization problem (see step [5]) and substituting z_1 by $f_{\theta_1}(y)$ in step [6], we are left with inserting the results of Eq. 13 for $p_{\mathbf{x}}(g_{\theta_2}^{-1}(f_{\theta_1}(y), z_2)))$.

Supervised Training with UnDimFlow. Additionally to the proposed unpaired learning method with which makes use of the forward operator A, the proposed approach is also capable of learning underdetermined inverse problems with unknown forward operator A, under the assumption that paired training data (y, x) is available. In the following, we will briefly describe the supervised training with UnDimFlow by highlighting and explaining the energy function $E(\theta_1, \theta_2)$.

Supervised training of the composed normalizing flows is realized by minimizing an energy function $E(\theta_1, \theta_2)$ consisting of four loss terms:

$$\begin{aligned} E(\theta_1, \theta_2) := &\, \mathbb{E}_{\mathbf{y},\mathbf{x}} \left[\| f_{\theta_1}^{-1}(g_{\theta_2}{}^{\langle z_1 \rangle}(x)) - y \|_2^2 \right] \\ &+ \lambda_1 \, \mathbb{E}_{\mathbf{y},\mathbf{x},\mathbf{z_2}} \left[\| g_{\theta_2}^{-1}(f_{\theta_1}(y), z_2) - x \|_2^2 \right] \\ &- \lambda_2 \, \mathbb{E}_{\mathbf{y}} \left[\log p_{\mathbf{y}}(y) \right] \\ &- \lambda_3 \, \mathbb{E}_{\mathbf{x}} \left[\log p_{\mathbf{x}}(x) \right]. \end{aligned} \tag{22}$$

The first two terms of the energy function train the combination of both flows in a supervised manner. In detail, in the first loss term we compute the squared L2 distance between data measurement samples $y \in Y$ and the output of the inverse function composition $h_{\theta_1,\theta_2}^{-1}$ of both flows for $h_{\theta_1,\theta_2}^{-1} := f_{\theta_1}^{-1}(g_{\theta_2 \langle z_1 \rangle}(x))$.

A.2 Further Results for Super-Resolution Under High Uncertainty

To show the variety of suitable solutions that our approach is able to learn, we additionally provide nine examples for seven different test images of the Fashion-MNIST dataset. While fixing the input image y, we randomly sample from the posterior. Figure 6 shows the results and the input image (rightmost column). All

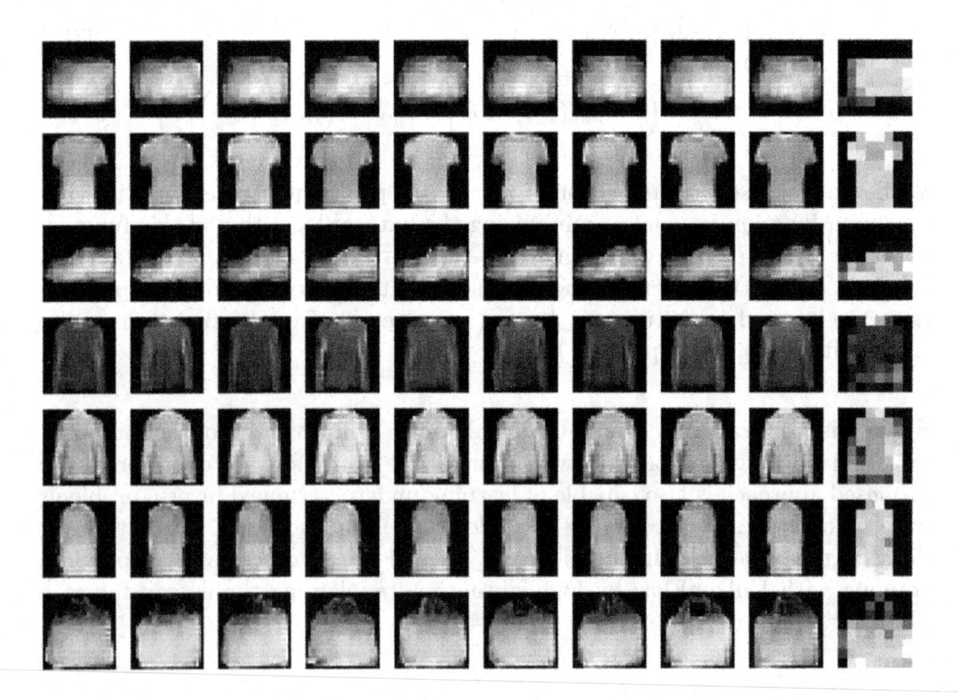

Fig. 6. Randomly sampled $\bar{x}_{\text{rnd}}$ for 8×8 pixel low resolution images with a weighting factor $\lambda_2 = 200$.

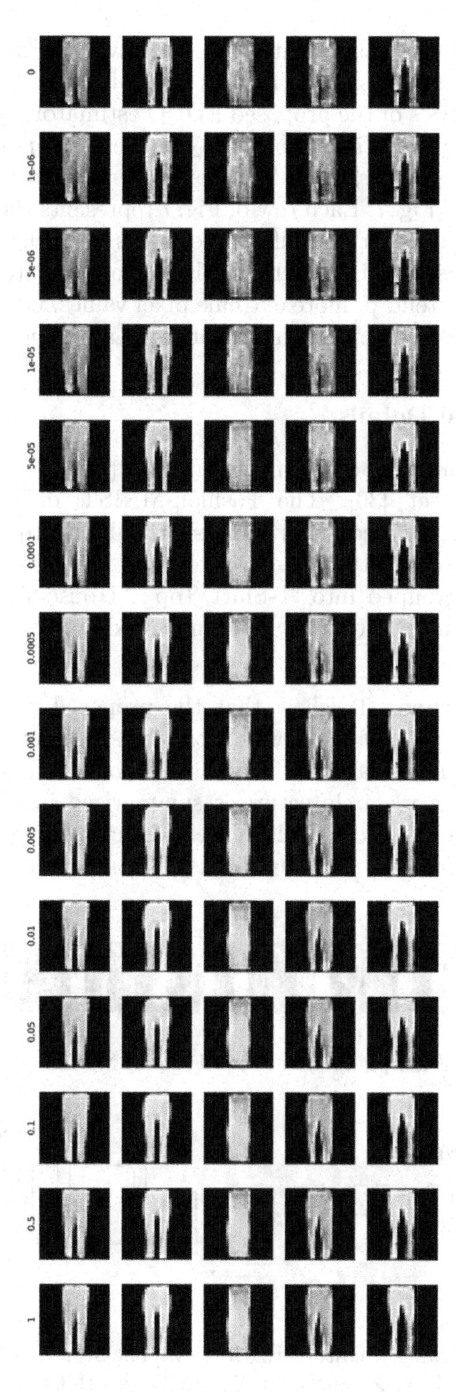

Fig. 7. Overview of five EDDO point estimators for different initializations of z_2 per weighting factor λ for super-resolution at the example of an image of the Fashion-MNIST dataset. Each row shows each of the five point estimators for $\lambda \in \{0, 10^{-6}, ..., 0.5, 1\}$. For each column, we initialize z_2 identically to ease comparability of results.

samples visually look like valid solutions to the inverse problem while showing some minor differences in uncertain regions of the images.

Detailing the analysis of the proposed EDDO estimator, we additionally conduct experiments for five initializations of z_2 and test different weightings of the regularization term $p_{\mathbf{x}|\mathbf{y}}(x \mid y)$ by scaling the hyperparameter λ, accordingly. The results are depicted in Fig. 7. Each row of Fig. 7 represents the results for each of the five initializations of z_2 for a specific value of λ as stated left to each row. It can be seen that for increasing weight of the regularization term, the images generated become smoother and tend to more extreme pixel values, i.e., pixels being either black or white which results in a reduction of grayscale values.

A.3 Experimental Details

Dataset. Throughout the experiments in this paper, we make use of the Fashion-MNIST dataset [30]. The Fashion-MNIST dataset introduced by Zalando Research in 2017 contains images of fashion objects. Images within the dataset have a resolution to 28 × 28 pixels and are grouped into ten classes, i.e., the images are grouped into 'T-Shirt/Top', 'Trousers', 'Pullover', 'Dress', 'Coat', 'Sandals', 'Shirt', 'Sneaker', 'Bag' and 'Ankle boots' (Fig. 8).

Network Architecture. To show that the proposed method is capable of learning multiple inverse problems, we choose the same network architecture of UnDimFlow on which we train and test the method for different inverse problems. The overall network architecture is summarized in Fig. 9. It consists of two normalizing flows f_{θ_1} and g_{θ_2} which are connected through the latent space Z_1. Both flows use Glow-like building blocks (cf. [14]) as core building blocks of

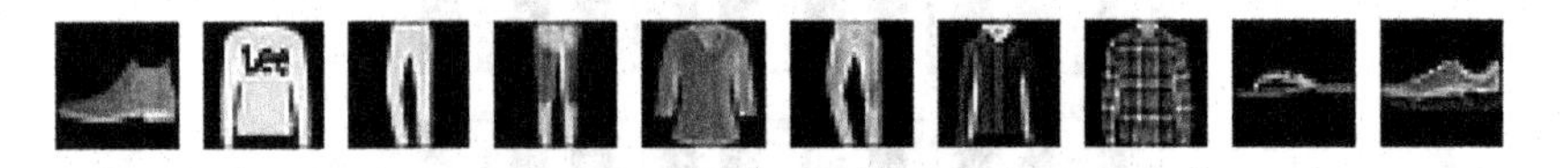

Fig. 8. Example images of the Fashion-MNIST dataset.

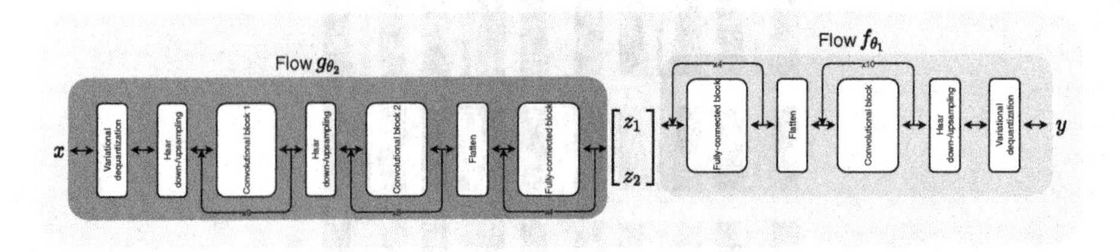

Fig. 9. UnDimFlow network architecture overview. The overall network consists of two normalizing flows $f_{\theta_1} : Y \rightarrow Z_1$ and $g_{\theta_2} : X \rightarrow Z_1 \times Z_2$. Both flows are connected via the same (part of the) latent space Z_1 and make use of Glow-like building blocks as core components. Additionally, a multi-scale architecture is realized by including invertible Haar downsampling layers.

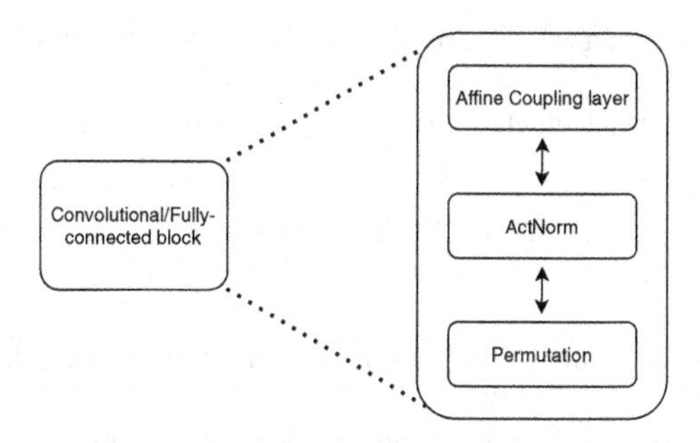

Fig. 10. Overview of the core component of the normalizing flows f_{θ_1} and g_{θ_2}. An affine coupling layer is applied before an ActNorm layer and a permutation. In contrast to a standard Glow building block, we use a random permutation. Depending of the type of the block, i.e., either convolutional or fully-connected, the subnetworks of the affine coupling layers are convolutional or fully-connected neural networks.

the normalizing flows. In contrast to standard Glow building blocks, we apply a random permutation as this has shown to improve stability during training of our experiments (Fig. 10). Each block therefore computes

$$f_{\text{block}}(y) = P \cdot \sigma(s_{\text{an}}) \odot f_{\text{ac}}(y) + b_{\text{an}} \tag{23}$$

for an activation function σ, a permutation matrix P, a scaling parameter s_{an}, a bias parameter b_{an} (both parts of the ActNorm layer [14]) and the affine coupling function f_{ac} [11].

We choose the subnetworks s and t of the affine coupling layers to either be shallow convolutional neural networks (denoted as 'Convolutional block' in Fig. 9) or fully-connected neural networks (denoted as 'Fully-connected block' in Fig. 9). The convolutional subnetworks consist of three convolutional layers with a kernel size of 3×3, 1×1 and again 3×3, accordingly. For fully-connected subnetworks we apply three fully-connected layers. Both types of subnetworks additionally make use of ReLU activations in between the linear layers.

Additionally, as an increasing number of channels has proven to be beneficial when using affine coupling layers, we incorporate downsampling operations via Haar transformations to invertibly increase the number of channels while decreasing spacial dimensionality.

For variational dequantization [13], we use a conditional normalizing flow and condition the aforementioned Glow-like building blocks on the image, i.e., either on y for the variational dequantization network of flow f_{θ_1} or on x for the variational dequantization network of flow g_{θ_2}. We use the most common technique to incorporate conditions into Glow-like building blocks and simply use the condition as further input to the affine coupling layer. As in our experiments we are working with grayscale images, we need to increase the number of channels to be able to use affine coupling layers. We thus apply a checkerboard mask, i.e., a pixel shuffling technique to increase the number of channels from one to two by moving every other pixel to the second channel. We apply this transformation to the

Table 1. Overview of structure of convolutional subnetworks of affine coupling layers.

Layer	# Input channels	# Output channels	Kernel size	Padding
1	N_{in}	64	3	1
2	64	128	1	0
3	128	N_{out}	3	1

Table 2. Overview of structure of fully-connected subnetworks of affine coupling layers.

Layer	# Input channels	# Output channels
1	N_{in}	128
2	128	128
3	128	N_{out}

condition image and to the input of the variational dequantization network after applying the logit function. Afterwards, we apply two Glow-like building blocks before transforming back via an inverse checkerboard mask and a sigmoid transformation.

Overview of Subnetworks for Affine Coupling Layers. In the context of affine coupling layers, we use shallow convolutional and fully-connected subnetworks. An overview of the details of these subnetworks are shown in Table 1 and 2.

General Training Settings. An overview of the general training settings can be found in Table 3.

Table 3. Overview of general training setting.

Parameter	
Network architecture	UnDimFlow
# Conv blocks flow f_{θ_1}	10
# Conv blocks flow g_{θ_2}	16
# FC blocks flow f_{θ_1}	4
# FC blocks flow g_{θ_2}	4
Layers	Glow-like blocks
Dataset	Fashion-MNIST
Loss function	as described in Sect. 3
Optimizer	Adam
Learning rate	10^{-4}
Learning rate schedule	Multiplication of learning rate with 0.99 every epoch
Epochs	200
Batch size	128
Gradient clipping	0.8

We use the Fashion-MNIST dataset with default train/test split and no data pre-processing steps during training and testing for all experiments but the partitioning experiments. For partitioning experiments we use random horizontal flipping during training.
All experiments were conducted on a NVIDIA Tesla V100 with 5120 CUDA Cores and 16 GB HBM2-Memory 3200 MHz MHz on the OMNI computing cluster of the University of Siegen.

Image Inpainting Experiments. For image inpainting, the general training settings have been extended by the following weightings of the individual loss terms of the energy function (Tables 4 and 5):

Table 4. Overview of weighting parameters for image inpainting experiments in Sect. 4.2 per individual loss term of the energy function.

Experiment	λ_1	λ_2	λ_3
Partitioning 1	2	200	100
Partitioning 2, i.e., (10/90) only	6	200	200

Table 5. Overview of weighting parameters for image super-resolution experiments in Sect. 4.1 per individual loss term of the energy function.

Experiment	λ_1	λ_2	λ_3
8×8	2	200	100
Comparison of point estimators	2	200	100

Image Super-Resolution Experiments. For image super-resolution, the general training settings have been extended by the following weightings of the individual loss terms of the energy function:

References

1. Ardizzone, L., Kruse, J., Rother, C., Köthe, U.: Analyzing inverse problems with invertible neural networks. In: International Conference on Learning Representations (2018)
2. Ardizzone, L., Lüth, C., Kruse, J., Rother, C., Köthe, U.: Guided image generation with conditional invertible neural networks. arXiv preprint arXiv:1907.02392 (2019)
3. Arridge, S., Maass, P., Öktem, O., Schönlieb, C.B.: Solving inverse problems using data-driven models. Acta Numerica **28**, 1–174 (2019)
4. Asim, M., Daniels, M., Leong, O., Ahmed, A., Hand, P.: Invertible generative models for inverse problems: mitigating representation error and dataset bias. In: International Conference on Machine Learning, pp. 399–409. PMLR (2020)

5. Benning, M., Burger, M.: Modern regularization methods for inverse problems. Acta Numerica **27**, 1–111 (2018)
6. Chaudhuri, S.: Super-Resolution Imaging, vol. 632. Springer Science, Cham (2001)
7. Chen, Y., Ranftl, R., Pock, T.: Insights into analysis operator learning: from patch-based sparse models to higher order MRFs. IEEE Trans. Image Process. **23**(3), 1060–1072 (2014)
8. Daras, G., Dean, J., Jalal, A., Dimakis, A.: Intermediate layer optimization for inverse problems using deep generative models. In: International Conference on Machine Learning, pp. 2421–2432. PMLR (2021)
9. Dashti, M., Stuart, A.M.: The Bayesian approach to inverse problems. In: Ghanem, R., Higdon, D., Owhadi, H. (eds.) Handbook of Uncertainty Quantification, pp. 311–428. Springer, Cham (2017). https://doi.org/10.1007/978-3-319-12385-1_7
10. Deco, G., Brauer, W.: Nonlinear higher-order statistical decorrelation by volume-conserving neural architectures. Neural Netw. **8**(4), 525–535 (1995)
11. Dinh, L., Sohl-Dickstein, J., Bengio, S.: Density estimation using Real NVP. In: International Conference on Learning Representations (2017)
12. Engl, H.W., Hanke, M., Neubauer, A.: Regularization of Inverse Problems, vol. 375. Springer, Cham (1996)
13. Ho, J., Chen, X., Srinivas, A., Duan, Y., Abbeel, P.: Flow++: improving flow-based generative models with variational dequantization and architecture design. In: International Conference on Machine Learning, pp. 2722–2730. PMLR (2019)
14. Kingma, D.P., Dhariwal, P.: Glow: generative flow with invertible 1× 1 convolutions. In: Proceedings of the 32nd International Conference on Neural Information Processing Systems, pp. 10236–10245 (2018)
15. Kobler, E., Effland, A., Kunisch, K., Pock, T.: Total deep variation for linear inverse problems. In: Proceedings of the IEEE/CVF Conference on Computer Vision and Pattern Recognition, pp. 7549–7558 (2020)
16. Mairal, J., Ponce, J., Sapiro, G., Zisserman, A., Bach, F.: Supervised dictionary learning. In: Advances in Neural Information Processing Systems, vol. 21 (2008)
17. Meinhardt, T., Moller, M., Hazirbas, C., Cremers, D.: Learning proximal operators: using denoising networks for regularizing inverse imaging problems. In: Proceedings of the IEEE International Conference on Computer Vision, pp. 1781–1790 (2017)
18. Moeller, M., Mollenhoff, T., Cremers, D.: Controlling neural networks via energy dissipation. In: Proceedings of the IEEE/CVF International Conference on Computer Vision, pp. 3256–3265 (2019)
19. Newman, M., Barkema, G.: Monte Carlo Methods in Statistical Physics, vol. 24. Oxford University Press, New York, USA (1999)
20. Padmanabha, G.A., Zabaras, N.: Solving inverse problems using conditional invertible neural networks. J. Comput. Phys. **433**, 110194 (2021)
21. Rezende, D., Mohamed, S.: Variational inference with normalizing flows. In: International Conference on Machine Learning, pp. 1530–1538. PMLR (2015)
22. Romano, Y., Elad, M., Milanfar, P.: The little engine that could: regularization by denoising (red). SIAM J. Imaging Sci. **10**(4), 1804–1844 (2017)
23. Scarlett, J., Heckel, R., Rodrigues, M.R., Hand, P., Eldar, Y.C.: Theoretical perspectives on deep learning methods in inverse problems. arXiv preprint arXiv:2206.14373 (2022)
24. Siahkoohi, A., Rizzuti, G., Louboutin, M., Witte, P., Herrmann, F.: Preconditioned training of normalizing flows for variational inference in inverse problems. In: Third Symposium on Advances in Approximate Bayesian Inference (2020)

25. Siahkoohi, A., Rizzuti, G., Witte, P.A., Herrmann, F.J.: Faster uncertainty quantification for inverse problems with conditional normalizing flows. arXiv preprint arXiv:2007.07985 (2020)
26. Sim, B., Oh, G., Kim, J., Jung, C., Ye, J.C.: Optimal transport driven CycleGAN for unsupervised learning in inverse problems. SIAM J. Imaging Sci. **13**(4), 2281–2306 (2020)
27. Song, Y., Sohl-Dickstein, J., Kingma, D.P., Kumar, A., Ermon, S., Poole, B.: Score-based generative modeling through stochastic differential equations. In: International Conference on Learning Representations (2021)
28. Ulyanov, D., Vedaldi, A., Lempitsky, V.: Deep image prior. In: Proceedings of the IEEE Conference on Computer Vision and Pattern Recognition, pp. 9446–9454 (2018)
29. Whang, J., Lindgren, E., Dimakis, A.: Approximate probabilistic inference with composed flows. In: NeurIPS 2020 Workshop on Deep Learning and Inverse Problems (2020)
30. Xiao, H., Rasul, K., Vollgraf, R.: Fashion-mnist: a novel image dataset for benchmarking machine learning algorithms. arXiv preprint arXiv:1708.07747 (2017)
31. Xiao, Z., Yan, Q., Amit, Y.: A method to model conditional distributions with normalizing flows. arXiv preprint arXiv:1911.02052 (2019)

Proximal Residual Flows for Bayesian Inverse Problems

Johannes Hertrich(✉)

Institute of Mathematics, TU Berlin, Straße des 17. Juni 136, 10623 Berlin, Germany
j.hertrich@math.tu-berlin.de

Abstract. Normalizing flows are a powerful tool for generative modelling, density estimation and posterior reconstruction in Bayesian inverse problems. In this paper, we introduce proximal residual flows, a new architecture of normalizing flows. Based on the fact, that proximal neural networks are by definition averaged operators, we ensure invertibility of certain residual blocks. Moreover, we extend the architecture to conditional proximal residual flows for posterior reconstruction within Bayesian inverse problems. We demonstrate the performance of proximal residual flows on numerical examples.

Keywords: Normalizing Flows · Proximal Neural Networks · Bayesian Inverse Problem

1 Introduction

Generative models for approximating complicated and high-dimensional probability distributions gained increasingly attention over the last years. One subclass of generative models are normalizing flows [13,40]. They are learned diffeomorphisms which push forward a complicated probability distribution to a simple one. More precisely, we learn a diffeomorphism $\mathcal{T}$ such that a distribution P_X can be approximately represented as $\mathcal{T}_{\#}^{-1} P_Z$ for a simple distribution P_Z. Several architectures of normalizing flows were proposed in the literature including Glow [33], real NVP [14], continuous normalizing flows [9,18] and autoregressive flows [15,31,37]. In this paper, we particularly focus on residual flows [5,8]. Here, the basic idea is that residual neural networks [23] are invertible as long as each subnetwork has a Lipschitz constant smaller than one. In [30], the authors figure out a relation between residual flows and Monge maps in optimal transport problems. For training residual flows, one needs to control the Lipschitz constant of the considered subnetworks. Training neural networks with a prescribed Lipschitz constant was addressed in several papers [17,35,38,42]. For example, the authors of [35] propose to rescale the transition matrices after each optimization step such that the spectral norm is smaller or equal than one. However, it is well known that enforcing a small Lipschitz constant within a neural network can lead to limited expressiveness.

In this paper, we propose to overcome these limitations by using proximal neural networks (PNNs). PNNs were introduced in [22,26] and are by

L. Calatroni et al. (Eds.): SSVM 2023, LNCS 14009, pp. 210–222, 2023.
https://doi.org/10.1007/978-3-031-31975-4_16

construction averaged operators. Using scaled PNNs as subnetworks, we prove that a residual neural network is invertible even if the scaled PNN has a Lipschitz constant larger than one. Further, we consider Bayesian inverse problems $Y = F(X) + \eta$, with an ill-posed forward operator F and some noise η. Here, we aim to reconstruct the posterior distributions $P_{X|Y=y}$ with using normalizing flows. To this end, we apply a conditional generative model [4,20,34,43]. For normalizing flows, this means that we aim to learn a mapping $\mathcal{T}(y, x)$ such that for any y it holds approximately $P_{X|Y=y} \approx \mathcal{T}(y, \cdot)^{-1}_{\#} P_Z$ for a simple distribution P_Z. Further, we show how proximal residual flows can be used for conditional generative modeling by constructing conditional proximal residual flows. Finally, we demonstrate the power of (conditional) proximal residual flows by numerical examples. First, we use proximal residual flows for sampling and density estimation of some complicated probability distributions including adversarial toy examples and molecular structures. Afterwards, we apply conditional proximal residual flows for reconstructing the posterior distribution in an inverse problem of scatterometry and for certain mixture models.

The paper is organized as follows. In Sect. 2, we first revisit residual flows and proximal neural networks. Afterwards we introduce proximal residual flows which combine both. Then, we extend proximal residual flows to Bayesian inverse problems in Sect. 3. We demonstrate the performance of proximal residual flows in Sect. 4. Conclusions are drawn in Sect. 5.

2 Proximal Residual Flows

Given i.i.d. samples $x_1, ..., x_N$ from an n-dimensional random variable X with unknown distribution P_X, a normalizing flow aims to learn a diffeomorphism $\mathcal{T}: \mathbb{R}^n \to \mathbb{R}^n$ such that $P_X \approx \mathcal{T}^{-1}_{\#} P_Z$. To this end, the diffeomorphism will be a neural network $\mathcal{T}_\theta$ with parameters θ, which is by construction invertible. For the training, we consider the maximum likelihood loss, i.e., we minimize $\mathcal{L}(\theta) = -\sum_{i=1}^N \log(p_{\mathcal{T}_{\theta\#}^{-1} P_Z}(x_i))$. Note, that using the change of variables formula for probability density functions, we have that $p_{\mathcal{T}_{\theta\#}^{-1} P_Z}(x)$ can be computed by $p_{\mathcal{T}_{\theta\#}^{-1} P_Z}(x) = p_Z(\mathcal{T}_\theta(x))|\nabla \mathcal{T}_\theta(x)|$. In this section, we propose a new architecture for normalizing flows based on residual flows [8] and proximal neural networks [22,26].

2.1 Residual Flows

Residual flows were introduced in [5,8]. The basic idea is to consider residual neural networks, where each subnetwork is constrained to be c-Lipschitz continuous for some $c < 1$, i.e., we have $\mathcal{T} = L_K \circ \cdots \circ L_1$, where each mapping $L_k: \mathbb{R}^n \to \mathbb{R}^n$ has the form $L(x) = x + g(x)$ with $\mathrm{Lip}(g) < 1$. Then, Banach's fixed point theorem yields that L is invertible and the inverse $L^{-1}(y)$ can be computed by the limit of the iteration $x^{(r+1)} = y - g(x^{(r)})$ starting at $x^{(0)} = y$.

For ensuring the Lipschitz continuity of g during the training of residual flows, the authors of [5,8] suggested to use spectral normalization [17,35]. Finally, to evaluate and differentiate $\log(|\nabla T(x)|)$, we have to evaluate and differentiate $\log(|\nabla L(x)|)$ for each residual block L. In small dimensions, this can be done by algorithmic differentiation, which is in high dimensions computationally intractable. Here we can apply the following theorem is from [8, Theorem 1, Theorem 2] which is based on an expansion of ∇L into a Neumann series.

Theorem 1. *Let Q be a random variable on $\mathbb{Z}_{>0}$ such that $P(Q = k) > 0$ for all $k \in \mathbb{Z}_{>0}$ and define $p_k = P(Q \geq k)$. Consider the function $L(x) = x + g(x)$, where $g\colon \mathbb{R}^n \to \mathbb{R}^n$ is differentiable and fulfills* $\mathrm{Lip}(g) < 1$. *Then, it holds*

$$\log(|\nabla L(x)|) = \mathbb{E}_{v\sim\mathcal{N}(0,I),q\sim P_Q}\left[\sum_{k=1}^{q} \frac{(-1)^{k+1}}{k} \frac{v^{\mathrm{T}}(\nabla g(x))^k v}{p_k}\right], \quad \textit{and}$$

$$\frac{\partial}{\partial\theta}\log(|\nabla L(x)|) = \mathbb{E}_{v\sim\mathcal{N}(0,I),q\sim P_Q}\left[\left(\sum_{k=0}^{q} \frac{(-1)^k}{p_k} v^{\mathrm{T}}(\nabla g(x))^k\right) \frac{\partial(\nabla g(x))}{\partial\theta} v\right].$$

2.2 Proximal Neural Networks

Averaged operators and in particular proximity operators received increasingly attention in deep learning over the last years [7,10,16]. For a proper, convex and lower semi-continuous function $f\colon \mathbb{R}^m \to \mathbb{R} \cup \{\infty\}$, the proximity operator of f is given by $\mathrm{prox}_{\lambda f}(x) = \arg\min_{y\in\mathbb{R}^n}\{\frac{1}{2\lambda}\|x-y\|^2 + f(y)\}$. Proximity operators are in particular $\frac{1}{2}$-averaged operators, i.e., $\mathrm{prox}_{\lambda f}(x) = \frac{1}{2}x + \frac{1}{2}R(x)$ for some R with $\mathrm{Lip}(R) \leq 1$.

In [11] the authors observed that most activation functions of neural networks are proximity operators. They proved that an activation function σ is a proximity operator with respect to some function g, which has 0 as a minimizer, if and only if σ is 1-Lipschitz continuous, monotone increasing and fulfills $\sigma(0) = 0$. They called the class of such activation functions *stable activation functions*. Using this result, Proximal Neural Networks (PNNs) were introduced in [22] as the concatenation of blocks of the form $B(\cdot; T, b, \alpha) := T^{\mathrm{T}}\sigma_\alpha(T\cdot +b)$, where $b \in \mathbb{R}^m$, T or T^{T} is in the Stiefel manifold $\mathrm{St}(n,m) = \{T \in \mathbb{R}^{n,m} : T^{\mathrm{T}}T = I\}$ and σ_α is a stable activation function which may depend on some additional parameter α. It can be shown that B is again a proximity operator of some proper, convex and lower semi-continuous function, see [22]. Now a PNN with K layers is defined as

$$\Phi(\cdot; u) = B_K(\cdot; T_K, b_K, \alpha_K) \circ \cdots \circ B_1(\cdot; T_1, b_1, \alpha_1),$$

where $u = ((T_k)_{k=1}^K, (b_k)_{k=1}^K, (\alpha_k)_{k=1}^K)$. Since Φ is the concatenation of K $\frac{1}{2}$-averaged operators, we obtain that Φ is $K/(K+1)$-averaged. From a numerical viewpoint, it was shown in [26] that (scaled) PNNs show a comparable performance as usual convolutional neural networks for denoising.

The training of PNNs is not straightforward due to the condition that T or T^{T} are contained in the Stiefel manifold. The authors of [22,26,27] propose a stochastic gradient descent on the manifold of the parameters, the minimization of a penalized functional and a stochastic variant of the inertial PALM algorithm [6,39]. However, to ensure the invertibility of proximal residual flows, it is important that the constraint $T_k \in \mathrm{St}(n,m)$ is fulfilled during the full training procedure. Therefore, we propose the following different training procedure.

Instead of training the matrices T_k directly, we define $T_k = P_{\mathrm{St}(n,m)}(\tilde{T}_k)$, where $P_{\mathrm{St}(n,m)}$ denotes the orthogonal projection onto the Stiefel manifold. For dense matrices and convolutions with full filter length, this projection is given by the U-factor of the polar decomposition, see [29, Sec. 7.3, Sec. 7.4], and can be computed by the iteration $Y_{n+1} = 2Y_n(I + Y_n^{\mathrm{T}} Y_n)^{-1}$ starting at $Y_0 = \tilde{T}_k$, see [28, Chap. 8]. Unfortunately, we are not aware of a similar iterative algorithm for convolutions with limited filter length. Finally, we optimize the matrices $\tilde{T}_k$ instead of the matrices T_k. In order to ensure numerical stability, we regularize the distance of $\tilde{T}_k$ to the Stiefel manifold by the penalizer $\|\tilde{T}_k^{\mathrm{T}}\tilde{T}_k - I\|_F^2$.

2.3 Proximal Residual Flows

Now, we propose proximal residual flows as the concatenation $\mathcal{T} = L_K \circ \cdots \circ L_1$ of residual blocks L_k of the form

$$L_k(x) = x + \gamma_k \Phi_k(x), \tag{1}$$

where $\gamma_k > 0$ is some constant and Φ_k is a PNN. The following proposition ensures the invertibility of L_k and $\mathcal{T}$.

Proposition 1. *Let Φ be a t-averaged operator with $\frac{1}{2} < t \le 1$ and let $0 < \gamma < \frac{1}{2t-1}$. Then, the function $L(x) = x + \gamma\Phi(x)$ is invertible and the inverse $L^{-1}(y)$ is given by the limit of the sequence*

$$x^{(r+1)} = \frac{1}{1+\gamma-\gamma t} y - \frac{\gamma t}{1+\gamma-\gamma t} R(x^{(r)}), \quad \textit{where} \quad R(x) := \frac{1}{t}\Phi(x) - \frac{1-t}{t}x.$$

Additionally, if Φ is t-averaged with $0 \le t \le \frac{1}{2}$, then the above statement is true for arbitrary $\gamma > 0$.

Note that $t = 1$ in the proposition exactly recovers the case considered in [5].

Proof. Since Φ is t-averaged, we get $\Phi = (1-t)I + tR$, where $R := \frac{1}{t}\Phi - \frac{1-t}{t}I$ is 1-Lipschitz continuous. Further, note that $\gamma < \frac{1}{2t-1}$ is equivalent to $\frac{\gamma t}{1+\gamma-\gamma t} < 1$. Therefore, Banach's fixed point theorem yields that the sequence $(x^{(r)})_r$ converges to the unique fixed point $x = \frac{1}{1+\gamma-\gamma t} y - \frac{\gamma t}{1+\gamma-\gamma t} R(x)$, which is equivalent to $y = x + \frac{\gamma}{2}(I+R)(x) = x + \gamma\Phi$. In particular, x is the unique solution of $L(x) = y$. In the case $t \le \frac{1}{2}$, we have that $\frac{\gamma t}{1+\gamma-\gamma t} < 1$ is true for any $\gamma > 0$ such that the same argumentation applies. □

Using the proposition, we obtain, that L_k from (1) is invertible, as long as $0 < \gamma_k < \frac{\kappa+1}{\kappa-1}$, where κ is the number of layers of Φ_k. In contrast to residual flows, the subnetworks $\gamma_k \Phi_k$ of proximal residual flows may have Lipschitz constants larger than 1. For instance, if $\kappa = 3$, then the upper bound on the Lipschitz constant of the subnetwork is 2 instead of 1.

Remark 1. Let $\Phi = B_K \circ \cdots \circ B_1 \colon \mathbb{R}^n \to \mathbb{R}^n$ be a PNN with layers $B_i(x) = T_i^{\mathrm{T}}\sigma(T_i x + b_i)$. Then, by definition, it holds $B_k \circ \cdots \circ B_1(x) \in \mathbb{R}^n$ for all $k = 1, ..., K$. In particular, each layer of the PNN has at most n neurons, which possibly limits the expressiveness. We overcome this issue with a small trick. Let $A = \frac{1}{\sqrt{p}}\big(I \cdots I\big)^{\mathrm{T}} \in \mathrm{St}(pn, n)$ and let $\Phi \colon \mathbb{R}^{pn} \to \mathbb{R}^{pn}$ be t-averaged. Then, a simple computation yields that also $\Psi = A^{\mathrm{T}}\Phi(A\cdot)$ is a t-averaged operator. In particular, we can use a PNN with pn neurons in each layer instead of a PNN with n neurons in each layer, which increases the expressiveness of the network a lot. □

For the evaluation of $\log(|\nabla T(x)|)$, we adapt Theorem 1 for proximal residual flows.

Corollary 1. *Let Q be a random variable on $\mathbb{Z}_{>0}$ such that $P(Q = k) > 0$ for all $k \in \mathbb{Z}_{>0}$ and define $p_k = P(Q \geq k)$. Consider the function $L(x) = x + \gamma\Phi(x)$, where $\Phi \colon \mathbb{R}^n \to \mathbb{R}^n$ is differentiable and t-averaged for $t \in (\frac{1}{2}, 1]$ and $0 < \gamma < \frac{1}{2t-1}$. Then, with $R(x) = \frac{1}{t}\Phi(x) - \frac{1-t}{t}x$ it holds*

$$\log(|\nabla L(x)|) = \mathbb{E}_{\substack{v \sim \mathcal{N}(0,I)\\ q \sim P_Q}}\left[\sum_{k=1}^{q} \frac{(-1)^{k+1}}{k} \frac{v^{\mathrm{T}}\big(\frac{\gamma t}{1+\gamma-\gamma t}\nabla R(x)\big)^k v}{p_k}\right] + n\log(1+\gamma-\gamma t),$$

$$\frac{\partial}{\partial\theta}\log(|\nabla L(x)|) = \mathbb{E}_{\substack{v \sim \mathcal{N}(0,I)\\ q \sim P_Q}}\left[\left(\sum_{k=0}^{q} \frac{(-1)^k}{p_k} v^{\mathrm{T}}\big(\tfrac{\gamma t}{1+\gamma-\gamma t}\nabla R(x)\big)^k\right) \frac{\partial\big(\frac{\gamma t}{1+\gamma-\gamma t}\nabla R(x)\big)}{\partial\theta} v\right].$$

Additionally, if Φ is t-averaged for $0 \leq t \leq \frac{1}{2}$, then the above statement holds true for any $\gamma > 0$.

Proof. Since Φ is t-averaged, it holds $\Phi = (1-t)I + tR$, where $R = \frac{1}{t}\Phi - \frac{1-t}{t}I$ is 1-Lipschitz continuous. Thus, we have

$$L = (1+\gamma-\gamma t)I + \gamma t R = (1+\gamma-\gamma t)\big(I + \tfrac{\gamma t}{1+\gamma-\gamma t}R\big),$$
$$\text{and}\quad \log(|\nabla L(x)|) = \log\big(|\nabla(I + \tfrac{\gamma t}{1+\gamma-\gamma t}R)(x)|\big) + n\log(1+\gamma-\gamma t).$$

Now, since $\frac{\gamma t}{1+\gamma-\gamma t} < 1$, applying Theorem 1 with $g = \frac{\gamma t}{1+\gamma-\gamma t}R$ gives the assertion. □

In the special case, that the PNN Φ consists of only one layer, we can derive the log-determinant explicitly by the following lemma.

Lemma 1. *Let $\Phi(x) = T^{\mathrm{T}}\sigma(Tx+b)$ for $T^{\mathrm{T}} \in \mathrm{St}(n,m)$, $n \geq m$ and a differentiable activation function $\sigma \colon \mathbb{R} \to \mathbb{R}$. Then, the log-determinant of the Jacobian of $L(x) = x + \gamma\Phi(x)$ is given by $\log(|\nabla L(x)|) = \sum_{i=1}^m \log(1 + \gamma\sigma_i'(Tx+b))$, where $\sigma_i'(Tx+b)$ is the ith component of $\sigma'(Tx+b)$.*

Proof. Let $\tilde{T}$ be a matrix, such that $S = (T^{\mathrm{T}}|\tilde{T}^{\mathrm{T}}) \in \mathbb{R}^{n\times n}$ is an orthogonal matrix. We have that

$$\nabla\Phi(x) = T^{\mathrm{T}}\sigma'(Tx+b)T, \quad \nabla L(x) = I_n + \gamma T^{\mathrm{T}}\sigma'(Tx+b)T.$$

Then, by orthogonality of S, it follows

$$|\nabla L(x)| = |S^{\mathrm{T}}||I_n + \gamma T^{\mathrm{T}} \sigma'(Tx+b)T||S| = |I_n + \gamma (TS)^{\mathrm{T}} \sigma'(Tx+b)(TS)|.$$

Since $T^{\mathrm{T}} \in \mathrm{St}(n,m)$, it holds by the definition of S that $TS = (I_m|0)$ such that

$$|\nabla L(x)| = \left| \begin{pmatrix} I_m + \gamma\sigma'(Tx+b) & 0 \\ 0 & I_{n-m} \end{pmatrix} \right| = \prod_{i=1}^{m} 1 + \gamma\sigma_i'(Tx+b).$$

Taking the logarithm proves the statement. □

3 Conditional Proximal Residual Flows

In the following, we consider for a random variable X the inverse problem $Y = F(X) + \eta$, where $F\colon \mathbb{R}^n \to \mathbb{R}^d$ is an ill-posed/ill-conditioned forward operator and η is some noise. Now, we aim to train a conditional normalizing flow model for reconstructing all posterior distributions $P_{X|Y=y}$, $y \in \mathbb{R}^d$. More precisely, we want to learn a mapping $\mathcal{T}\colon \mathbb{R}^d \times \mathbb{R}^n \to \mathbb{R}^n$ such that $\mathcal{T}(y,\cdot)$ is invertible for all $y \in \mathbb{R}^d$ and $P_{X|Y=y} = \mathcal{T}(y,\cdot)^{-1}_{\#} P_Z$. For this purpose, $\mathcal{T}_\theta$ will be a neural network with parameters θ. We learn $\mathcal{T}_\theta$ from i.i.d. samples $(x_1,y_1),\dots,(x_N,y_N)$ of (X,Y) using the maximum likelihood loss

$$\mathcal{L}(\theta) = \sum_{i=1}^{N} p_{\mathcal{T}_\theta(y_i,\cdot)^{-1}{}_{\#} P_Z}(x_i) \approx \mathbb{E}_{y\sim P_Y}[\mathrm{KL}(P_{X|Y=y}, P_{\mathcal{T}_\theta(y,\cdot)^{-1}{}_{\#} P_Z})] + \text{const}.$$

Note that for the real NVP architecture [14], such flows were considered in [1,4,12,20].

For using proximal residual flows as conditional normalizing flows, we need the following lemma.

Lemma 2. *Let $\Phi = (\Phi_1,\Phi_2)\colon \mathbb{R}^d \times \mathbb{R}^n \to \mathbb{R}^d \times \mathbb{R}^n$ be a t-averaged operator. Then, for any $y \in \mathbb{R}^d$, the operator $\Phi_2(y,\cdot)$ is t-averaged.*

Proof. Since Φ is t-averaged, we have that $\Phi(y,x) = (1-t)(y,x) + tR(y,x)$ for some 1-Lipschitz function $R = (R_1,R_2)\colon \mathbb{R}^n \times \mathbb{R}^d \to \mathbb{R}^n \times \mathbb{R}^d$. Now let $y \in \mathbb{R}^d$ be arbitrary fixed. Due to the Lipschitz continuity of R, it holds for $x_1, x_2 \in \mathbb{R}^n$ that

$$\|R(y,x_1) - R(y,x_2)\| \leq \|(y,x_1) - (y,x_2)\| = \|x_1 - x_2\|.$$

Thus, $R_2(\cdot,y)$ is 1-Lipschitz continuous and we get by definition that

$$\Phi_2(y,x) = (1-t)x + tR_2(y,x),$$

such that $\Phi_2(y,\cdot)$ is a t-averaged operator. □

Now, we define a conditional proximal residual flows as a mapping $\mathcal{T}\colon \mathbb{R}^d \times \mathbb{R}^n \to \mathbb{R}^n$ given by $\mathcal{T}(\cdot,y) = L_K(\cdot,y) \circ \cdots \circ L_1(\cdot,y)$ with

$$L_k(y,x) = x + \gamma_k \Phi_{k,2}(y,x), \tag{2}$$

where $\Phi_k = (\Phi_{k,1},\Phi_{k,2})\colon \mathbb{R}^n \times \mathbb{R}^d \to \mathbb{R}^n \times \mathbb{R}^d$ is a PNN. By definition, we have that $\mathcal{T}(y,\cdot)$ is a proximal residual flow for any fixed y such that the invertibility result in Proposition 1 applies.

4 Numerical Examples

In this section, we demonstrate the performance of proximal residual flows by numerical examples. First, in Subsect. 4.1 we apply proximal residual flows in an unconditional setting. Afterwards, in Subsect. 4.2, we consider a conditional setting with Bayesian inverse problems. In both cases, we compare our results with residual flows [5,8] and a variant of the real NVP architecture [3,14]. Within all architectures we use activation normalization [33] after every invertible block. For evaluating the quality of our results, we use the empirical Kullback-Leibler divergence and Wasserstein distance. All implementations are done in Python and Tensorflow[1]. We use fully connected PNNs with three layers as subnetworks and evaluate the log-determinant is done exactly by backprobagation. The following table contains for each experiment the dimension n, the dimension d of the condition, the number of residual blocks K, the parameter p from Remark 1, the hidden dimension h within the subnetwork, the parameter γ in the equations (1) and (2), the batch size b, the number of epochs e, the number of steps per epoch s and the learning rate τ.

Method	n	d	K	p	h	γ	b	e	s	τ
Toy examples	2	-	20	64	64	1.99	200	20	2000	10^{-3}
Alanine Dipeptide	66	-	20	2	100	1.99	200	20	2000	10^{-3}
Circle	2	1	20	64	64	1.99	800	20	2000	10^{-3}
Scatterometry	3	23	20	10	128	1.99	1600	20	2000	$5 \cdot 10^{-3}$
Mixture models	50	50	20	2	128	1.99	200	20	2000	$5 \cdot 10^{-3}$

4.1 Unconditional Examples

In the following, we apply proximal residual flows for density estimation, i.e., we are in the setting of Sect. 2.

Toy Densities. First, we train proximal residual flows onto some toy densities, namely 8 modes, two moons, two circles and checkboard. Samples from the training data and the reconstruction with proximal residual flows are given in Fig. 1. We observe that the proximal residual flow is able to learn all of the toy densities very well, even though it was shown in [21] that a diffeomorphism, which pushes forward a unimodal distribution to a multimodal one must have a large Lipschitz constant.

Alanine Dipeptide. Next, we evaluate proximal residual flows for an example from [2,44][2]. Here, we aim to estimate the density of molecular structures of alanine dipeptide molecules. The structure of such molecules is described by an

[1] https://github.com/johertrich/Proximal_Residual_Flows.

[2] For the data generation and evaluation of this example, we use the code of [44] available at https://github.com/noegroup/stochastic_normalizing_flows.

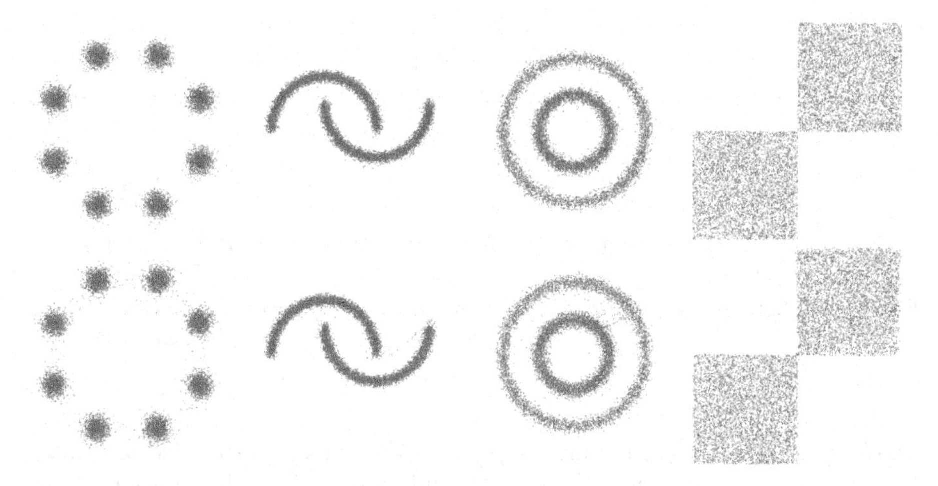

Fig. 1. Reconstruction of toy densities. Top: Ground truth, Bottom: Reconstructions with proximal residual flows.

66-dimensional vector. For evaluating the quality of the results, we follow [44] and consider the marginal distribution onto the torsion angles, as introduced in [36]. Afterwards, we consider the empirical Kullback Leibler divergence between these marginal distributions of the training data and the reconstruction by the proximal residual flows based on samples. We compare our results with a normalizing flow consisting of 20 real NVP blocks with subnetworks consisting of 3 fully connected layers and a hidden dimension of 128 and a residual flows with 20 residual blocks, where each subnetwork has three hidden layers with 128 neurons. The following table contains the empirical Kullback Leibler divergence between one-dimensional marginal distributions corresponding to the torsion angles ϕ, γ_1, ψ, γ_2 and γ_3. The results are averaged over five independent runs.

Method	ϕ	γ_1	ψ	γ_2	γ_3
Real NVP	0.12 ± 0.05	0.14 ± 0.04	0.06 ± 0.03	0.04 ± 0.01	0.07 ± 0.01
Residual Flows	0.12 ± 0.11	0.24 ± 0.29	0.10 ± 0.08	0.06 ± 0.03	0.07 ± 0.02
Proximal Residual Flow	0.05 ± 0.02	0.06 ± 0.01	0.03 ± 0.00	0.05 ± 0.01	0.05 ± 0.01

We observe that the proximal residual flow yields better results than the large real NVP network and the residual flow.

4.2 Posterior Reconstruction

Now, we aim to find a conditional proximal residual flow for reconstructing the posterior distribution $P_{X|Y=y}$ for all $y \in \mathbb{R}^d$ as in Sect. 3.

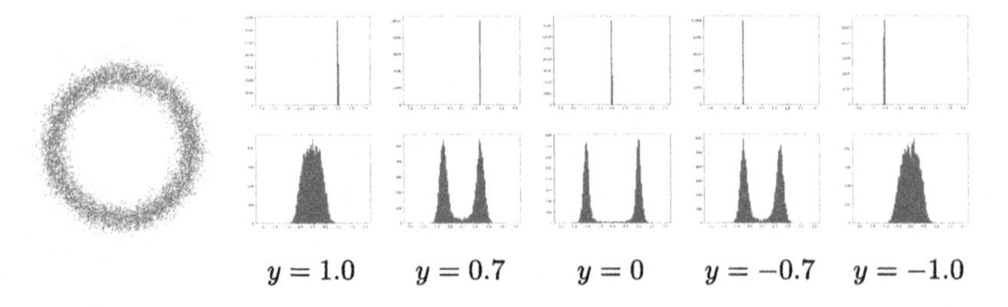

Fig. 2. Left: Samples from the P_X. Right: Histograms of the reconstructions of $P_{X|Y=y}$ for $y \in \{1, 0.7, 0, -0.7, -1\}$ within the circle example. Top: first coordinate, Bottom: second coordinate.

Circle. First, we consider the inverse problem $Y = F(X) + \eta$ specified as follows. Let the prior distribution P_X be the convolution of uniform distribution on the unit circle in $\mathbb{R}^2$ with the normal distribution $\epsilon \sim \mathrm{N}(0, 0.1^2 I_2)$. Further let the operator $F\colon \mathbb{R}^2 \to \mathbb{R}$ be given by $y := F(x_1, x_2) = x_1$ and define the noise distribution by $\eta \sim \mathcal{N}(0, 0.02^2 I)$. Figure 2 shows the prior distribution P_X and histograms of samples from the reconstructions of $P_{X|Y=y}$. As expected the estimation of $P_{X|Y=y}$ is unimodal for $y \in \{1, -1\}$ and bimodal otherwise.

Scatterometry. Next, we apply proximal residual flows to a Bayesian inverse problem in scatterometry with a nonlinear forward operator $F\colon \mathbb{R}^3 \to \mathbb{R}^{23}$. It describes the diffraction of monochromatic lights on line gratings which is a non-destructive technique to determine the structures of photo masks. For a detailed description, we refer to [24,25]. We use the code of [20] for the data generation, evaluation and the representation of the forward operator. As no prior information about the parameters x is given, we choose the prior distribution P_X to be the uniform distribution on $[-1, 1]^3$. Since we assume for normalizing flows that P_X has a strictly positive density p_X, we relax the probability density function of the uniform distribution for $x = (x_1, x_2, x_3) \in \mathbb{R}^3$ by

$$p_X(x) := q(x_1)q(x_2)q(x_3), \quad q(x) := \begin{cases} \frac{\alpha}{2\alpha+2}, & \text{for } x \in [-1, 1], \\ \frac{\alpha}{2\alpha+2}\exp(-\alpha|x-1|), & \text{for } x > 1, \end{cases}$$

where $\alpha \gg 0$ is some constant. In our numerical experiments, α is set to $\alpha = 1000$. We compare the proximal residual flow with the normalizing flow with real NVP architecture from [20] and with a residual flow of 20 residual blocks, where each subnetwork has three hidden layers with 128 neurons. As a quality measure, we use the empirical KL divergence of $P_{X|Y=y}$ and $\mathcal{T}(y, \cdot)^{-1}_{\#} P_Z$ for 100 independent samples with 540000 samples on a $75 \times 75 \times 75$ grid. As a ground truth, we use samples from $P_{X|Y=y}$ which are generated by the Metropolis Hastings algorithm, see [20]. The average empirical KL divergences of the reconstructions of $P_{X|Y=y}$ over 100 observations are given as follows.

	Real NVP	Residual Flows	Proximal Residual Flows
KL	0.773 ± 0.289	0.913 ± 0.407	0.637 ± 0.263

The proximal residual flow gives the best reconstructions.

Mixture Models. Next, we consider the Bayesian inverse problem $Y = F(X)+\eta$, where the forward operator $F\colon \mathbb{R}^{50} \to \mathbb{R}^{50}$ is linear and given by the diagonal matrix $A := 0.1\,\mathrm{diag}\big((\frac{1}{n})_{n=1}^{50}\big)$. Moreover, we add Gaussian noise with standard deviation 0.05. As prior distribution P_X, we choose a Gaussian mixture model with 5 components, where we draw the means uniformly from $[-1,1]^{50}$ and set the covariances to $0.01^2 I$. Note that in this setting, the posterior distribution can be computed analytically, see [20, Lem. 6.1]. We compare our results with a normalizing flow consisting of 20 real NVP blocks with subnetworks consisting of 3 fully connected layers and a hidden dimension of 128 and a residual flow with 20 residual blocks, where each subnetwork has three hidden layers with 128 neurons. Since the evaluation of the empirical KL divergence is intractable in high dimensions, we use the empirical Wasserstein distance as an error measure. The averaged errors over 100 observations is given in the follwoing table.

	Real NVP	Residual Flows	Proximal Residual Flows
Wasserstein-2 distance	2.122 ± 1.007	1.374 ± 0.050	1.028 ± 0.079

The proximal residual flow outperforms the comparisons significantly.

5 Conclusions

We introduced proximal residual flows, which improve the expressiveness of residual flows by the use of proximal neural networks. In particular, we proved that proximal residual flows are invertible, even though the Lipschitz constant of the subnetworks is larger than one. Afterwards, we extended the framework of proximal residual flows to the problem of posterior reconstruction within Bayesian inverse problems by using conditional generative modelling. Finally, we demonstrated the performance of proximal residual flows by numerical examples. This work can be extended in several directions. First, it is an open question, how to generalize the training procedure in this paper to convolutional networks. In particular, finding an efficient algorithm which computes the orthogonal projection onto the space of orthogonal convolutions with limited filter length is left for future research. Moreover, every invertible neural network architecture requires an exploding Lipschitz constant for reconstructing multimodal [21, 41] or heavy tailed distributions [32]. To overcome these topological constraints, the authors of [2, 20, 44] propose to combine normalizing flows with stochastic sampling methods. Finally, we could improve the expressiveness of proximal residual flows by combining them with other generative models, see e.g. [19].

Acknowledgements. Funding by the German Research Foundation (DFG) within the project STE 571/16-1 is gratefully acknowledged.

References

1. Altekrüger, F., Hertrich, J.: WPPNets and WPPFlows: The power of Wasserstein patch priors for superresolution. arXiv preprint arXiv:2201.08157 (2022)
2. Arbel, M., Matthews, A., Doucet, A.: Annealed flow transport Monte Carlo. In: International Conference on Machine Learning, pp. 318–330. PMLR (2021)
3. Ardizzone, L., Kruse, J., Rother, C., Köthe, U.: Analyzing inverse problems with invertible neural networks. In: International Conference on Learning Representations (2018)
4. Ardizzone, L., Lüth, C., Kruse, J., Rother, C., Köthe, U.: Guided image generation with conditional invertible neural networks. arXiv preprint arXiv:1907.02392 (2019)
5. Behrmann, J., Grathwohl, W., Chen, R.T., Duvenaud, D., Jacobsen, J.H.: Invertible residual networks. In: International Conference on Machine Learning, pp. 573–582 (2019)
6. Bolte, J., Sabach, S., Teboulle, M.: Proximal alternating linearized minimization for nonconvex and nonsmooth problems. Math. Program. **146**(1), 459–494 (2014)
7. Boyd, S., Parikh, N., Chu, E., Peleato, B., Eckstein, J.: Distributed optimization and statistical learning via the alternating direction method of multipliers. Found. Trends Mach. Learn. **3**(1), 1–122 (2011)
8. Chen, R.T.Q., Behrmann, J., Duvenaud, D.K., Jacobsen, J.H.: Residual flows for invertible generative modeling. In: Advances in Neural Information Processing Systems, vol. 32. Curran Associates, Inc. (2019)
9. Chen, R.T., Rubanova, Y., Bettencourt, J., Duvenaud, D.K.: Neural ordinary differential equations. In: Advances in Neural Information Processing Systems, vol. 31 (2018)
10. Combettes, P.L., Pesquet, J.C.: Proximal splitting methods in signal processing. In: Bauschke, H., Burachik, R., Combettes, P., Elser, V., Luke, D., Wolkowicz, H. (eds.) Fixed-point algorithms for inverse problems in science and engineering. Springer Optimization and Its Applications, vol. 49, pp. 185–212. Springer, New York (2011). https://doi.org/10.1007/978-1-4419-9569-8_10
11. Combettes, P.L., Pesquet, J.C.: Deep neural network structures solving variational inequalities. Set-Valued Variational Anal. **28**(3), 491–518 (2020)
12. Denker, A., Schmidt, M., Leuschner, J., Maass, P.: Conditional invertible neural networks for medical imaging. J. Imaging **7**(11), 243 (2021)
13. Dinh, L., Krueger, D., Bengio, Y.: NICE: non-linear independent components estimation. In: Bengio, Y., LeCun, Y. (eds.) 3rd International Conference on Learning Representations, Workshop Track Proceedings (2015)
14. Dinh, L., Sohl-Dickstein, J., Bengio, S.: Density estimation using real NVP. In: International Conference on Learning Representations (2017)
15. Durkan, C., Bekasov, A., Murray, I., Papamakarios, G.: Neural spline flows. In: Advances in Neural Information Processing Systems (2019)
16. Glowinski, R., Osher, S.J., Yin, W.: Splitting Methods in Communication, Imaging, Science, and Engineering. Springer, Cham (2017)
17. Gouk, H., Frank, E., Pfahringer, B., Cree, M.J.: Regularisation of neural networks by enforcing Lipschitz continuity. Mach. Learn. **110**(2), 393–416 (2021)

18. Grathwohl, W., Chen, R.T., Bettencourt, J., Sutskever, I., Duvenaud, D.: FFJORD: free-form continuous dynamics for scalable reversible generative models. In: International Conference on Learning Representations (2018)
19. Hagemann, P., Hertrich, J., Steidl, G.: Generalized normalizing flows via Markov Chains. arXiv preprint arXiv:2111.12506 (2021)
20. Hagemann, P., Hertrich, J., Steidl, G.: Stochastic normalizing flows for inverse problems: a Markov Chains viewpoint. SIAM/ASA J. Uncertainty Quantification **10**(3), 1162–1190 (2022)
21. Hagemann, P., Neumayer, S.: Stabilizing invertible neural networks using mixture models. Inverse Prob. **37**(8), 085002 (2021)
22. Hasannasab, M., Hertrich, J., Neumayer, S., Plonka, G., Setzer, S., Steidl, G.: Parseval proximal neural networks. J. Fourier Anal. Appl. **26**, 59 (2020)
23. He, K., Zhang, X., Ren, S., Sun, J.: Deep residual learning for image recognition. In: Proceedings of the IEEE Conference on Computer Vision and Pattern Recognition, pp. 770–778 (2016)
24. Heidenreich, S., Gross, H., Bär, M.: Bayesian approach to the statistical inverse problem of scatterometry: comparison of three surrogate models. Int. J. Uncertainty Quantification **5**(6) (2015)
25. Heidenreich, S., Gross, H., Bär, M.: Bayesian approach to determine critical dimensions from scatterometric measurements. Metrologia **55**(6), S201 (2018)
26. Hertrich, J., Neumayer, S., Steidl, G.: Convolutional proximal neural networks and plug-and-play algorithms. Linear Algebra Appl. **631**, 203–234 (2021)
27. Hertrich, J., Steidl, G.: Inertial stochastic PALM and applications in machine learning. Sampling Theory Signal Process. Data Anal. **20**(1), 4 (2022)
28. Higham, N.J.: Functions of Matrices: Theory and Computation. SIAM, Philadelphia (2008)
29. Horn, R.A., Johnson, C.R.: Matrix Analysis. Oxford University Press, Oxford (2013)
30. Huang, C.W., Chen, R.T., Tsirigotis, C., Courville, A.: Convex potential flows: universal probability distributions with optimal transport and convex optimization. In: International Conference on Learning Representations (2020)
31. Huang, C.W., Krueger, D., Lacoste, A., Courville, A.: Neural autoregressive flows. In: International Conference on Machine Learning, pp. 2078–2087 (2018)
32. Jaini, P., Kobyzev, I., Yu, Y., Brubaker, M.: Tails of Lipschitz triangular flows. In: International Conference on Machine Learning, pp. 4673–4681. PMLR (2020)
33. Kingma, D.P., Dhariwal, P.: Glow: Generative flow with invertible 1x1 convolutions. In: Advances in Neural Information Processing Systems, vol. 31 (2018)
34. Mirza, M., Osindero, S.: Conditional generative adversarial nets. arXiv preprint arXiv:1411.1784 (2014)
35. Miyato, T., Kataoka, T., Koyama, M., Yoshida, Y.: Spectral normalization for generative adversarial networks. In: International Conference on Learning Representations (2018)
36. Noé, F., Olsson, S., Köhler, J., Wu, H.: Boltzmann generators: sampling equilibrium states of many-body systems with deep learning. Science **365**(6457), 1147 (2019)
37. Papamakarios, G., Pavlakou, T., Murray, I.: Masked autoregressive flow for density estimation. In: Advances in Neural Information Processing Systems, pp. 2338–2347 (2017)
38. Pesquet, J.C., Repetti, A., Terris, M., Wiaux, Y.: Learning maximally monotone operators for image recovery. SIAM J. Imaging Sci. **14**(3), 1206–1237 (2021)

39. Pock, T., Sabach, S.: Inertial proximal alternating linearized minimization (iPALM) for nonconvex and nonsmooth problems. SIAM J. Imaging Sci. **9**(4), 1756–1787 (2016)
40. Rezende, D., Mohamed, S.: Variational inference with normalizing flows. In: International Conference on Machine Learning, pp. 1530–1538. PMLR (2015)
41. Salmona, A., De Bortoli, V., Delon, J., Desolneux, A.: Can push-forward generative models fit multimodal distributions? In: Advances in Neural Information Processing Systems (2022)
42. Sedghi, H., Gupta, V., Long, P.M.: The singular values of convolutional layers. In: International Conference on Learning Representations (2018)
43. Sohn, K., Lee, H., Yan, X.: Learning structured output representation using deep conditional generative models. In: Advances in Neural Information Processing Systems, vol. 28 (2015)
44. Wu, H., Köhler, J., Noé, F.: Stochastic normalizing flows. Adv. Neural. Inf. Process. Syst. **33**, 5933–5944 (2020)

A Model is Worth Tens of Thousands of Examples

Thomas Dagès[1(✉)], Laurent D. Cohen[2], and Alfred M. Bruckstein[1]

[1] Department of Computer Science, Technion Israel Institute of Technology, Haifa, Israel
{thomas.dages,freddy}@cs.technion.ac.il

[2] Ceremade, University Paris Dauphine, PSL Research University, UMR CNRS 7534, 75016 Paris, France
cohen@ceremade.dauphine.fr

Abstract. Traditional signal processing methods relying on mathematical data generation models have been cast aside in favour of deep neural networks, which require vast amounts of data. Since the theoretical sample complexity is nearly impossible to evaluate, these amounts of examples are usually estimated with crude rules of thumb. However, these rules only suggest when the networks should work, but do not relate to the traditional methods. In particular, an interesting question is: how much data is required for neural networks to be on par or outperform, if possible, the traditional model-based methods? In this work, we empirically investigate this question in two simple examples, where the data is generated according to precisely defined mathematical models, and where well-understood optimal or state-of-the-art mathematical data-agnostic solutions are known. A first problem is deconvolving one-dimensional Gaussian signals and a second one is estimating a circle's radius and location in random grayscale images of disks. By training various networks, either naive custom designed or well-established ones, with various amounts of training data, we find that networks require tens of thousands of examples in comparison to the traditional methods, whether the networks are trained from scratch or even with transfer-learning or finetuning.

Keywords: Deep learning · Model-based methods · Sample complexity

1 Introduction

Neural network-based machine learning has widely replaced the traditional methods for solving many signal and image processing tasks that relied on mathematical models for the data [10,14]. In some cases, the assumed models provided ways to optimally address the tasks at hand and resulted in well-performing estimation and prediction methods with theoretical guarantees [7,17,21]. Nowadays, gathering raw data and applying gradient descent-like processes to neural network structures [11,12,15,19] largely replaced modelling and mathematically developing provably optimal solutions.

L. Calatroni et al. (Eds.): SSVM 2023, LNCS 14009, pp. 223–235, 2023.
https://doi.org/10.1007/978-3-031-31975-4_17

It is commonly accepted that, if the networks are complex enough and when vast amounts of data are available, neural networks outperform traditionally designed methods [12,16] or even humans [4,8,9]. The required amount of data is called in statistical learning theory the sample complexity and is related to the VC-dimension of the problem [20], which is usually intractable for non trivial networks [2]. Instead, various rules of thumb have been used in the field to guess how many samples are needed: at least 10–50 times the number of parameters [1], at least 10 times per class in classification (and 50 times in regression) the data dimensionality [13] and at least 50–1000 times the output dimension [1].

However, these rules only suggest how much data is needed to get a "good" network, but they do not relate to the traditional data-generation model-based methods. A natural question hence arises: do the neural network-based solutions perform as well as, or even outperform, the processing methods based on traditional data-generation models when lots of data is available, and if so how much data is necessary? We address this question in two simple empirical examples, where the data is produced according to precisely defined models, and where well understood optimal or state-of-the-art mathematical solutions are available. The first is the deconvolution of Gaussian signals, optimally solved with the Wiener filter [21]. The second is the estimation of the radius and centre coordinates of a disk in an image, which can be elegantly solved using a Pointflow method [22]. This work aids engineers to decide when to use model-based classical methods or simply feed lots of data (if available) to deep neural networks.

Section 2 presents our comparison for the one-dimensional signal recovery, and Sect. 3 deals with estimation of disk characteristics in an image.

2 One-Dimensional Signal Recovery

We suggest to first analyse a simple and well-understood problem in the one-dimensional case where the optimal solution is provingly known.

2.1 Data Model and Optimal Solution

The original data consists of real random vectors φ of size D that are centred, i.e. $\mathbb{E}(\varphi) = 0$, and with known autocorrelation $R_\varphi = \mathbb{E}(\varphi\varphi^\top) \in \mathbb{R}^{D\times D}$. However, φ is degraded by blur and noise producing the observed data φ^{data} as follows:

$$\varphi^{data} = H\varphi + n, \tag{1}$$

where $H \in \mathbb{R}^{D\times D}$ is a known deterministic matrix and n is random additive noise independent from φ that is centred $\mathbb{E}(n) = 0$ and with known autocorrelation matrix $R_n \in \mathbb{R}^{D\times D}$. It is well-known [21] that the best linear recovery of φ in the L^2 sense, i.e. minimising the Expected Squared Error (ESE) $ESE(\hat{\varphi}, \varphi) = \mathbb{E}\left(\|\hat{\varphi} - \varphi\|_2^2\right)$ with respect to the matrix $M \in \mathbb{R}^{D\times D}$ such that $\hat{\varphi} = M\varphi^{data}$, is given by applying the Wiener filter $W = R_\varphi H^\top(HR_\varphi H^\top + R_n)^{-1}$, i.e. $\hat{\varphi}^* = W\varphi^{data}$. Moreover, if we further assume both $\varphi \sim \mathcal{N}(0, R_\varphi)$ and $n \sim \mathcal{N}(0, R_n)$

are Gaussian, then the Wiener filter W minimises the ESE over all possible recoveries including nonlinear ones. Furthermore, note that if R_φ is circulant, i.e. φ is cyclostationary, and so is n, e.g. if n has independent entries implying R_n is diagonal, and if H is circulant, then W is also circulant and Wiener filtering is a pointwise multiplication in the Fourier domain given by applying the unitary Discrete Fourier Transform $[DFT]$ with (k,l)-th entry $[DFT]_{k,l} = \frac{1}{\sqrt{D}} e^{-i\frac{2\pi kl}{D}}$.

In our tests, the dimensionality is $D = 32$ and the problem is circulant. We use an interpretable symmetric positive-definite autocorrelation matrix R_φ parameterised by a large number $\rho = 0.95$ to create high spatial correlation over a large support decaying with distance and H is a local smoothing convolution. The first lines of R_φ and H are $\left(1\; \rho\; \rho^2\; \rho^3 \cdots \rho^3\; \rho^2\; \rho\right)$ and $\left(1\; 1\; 0 — 0\; 1\right)$. The noise is i.i.d. $n \sim \mathcal{N}(0, \sigma_n^2 I)$ with $\sigma_n = 0.1$. We display example data in Fig. 1, the designed H, R_φ, and R_n along with their associated Wiener filter W in Fig. 2.

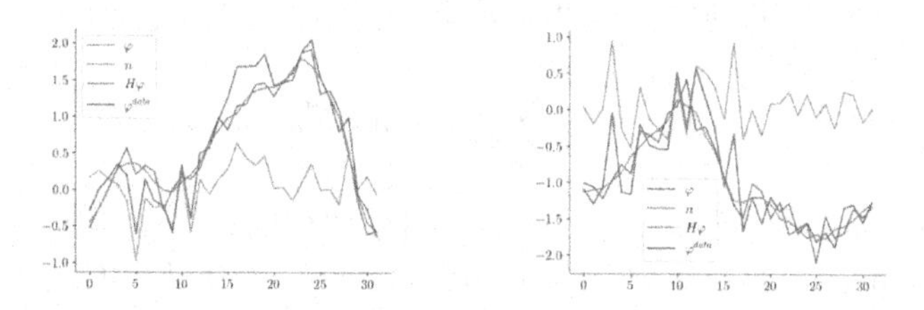

Fig. 1. Two example signals φ^{data}, with their associated blur $H\varphi$ and noise n.

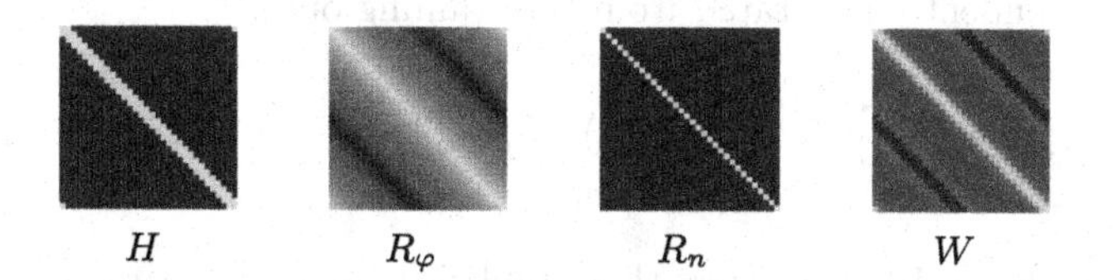

Fig. 2. Chosen model matrices and associated optimal Wiener filter.

2.2 Neural Models

We wish to evaluate the capabilities of neural networks by comparing them to humanly designed methods by classical experts using no training. Our criterion is the amount of random training samples N needed to reach or overtake human expertise. Working in the Gaussian case for the data model of Eq. (1), we create various random training datasets containing N data samples ranging in $N \in \{10, 100, 1000, 10000, 100000\}$. We train a variety of small Convolutional Neural Networks (CNNs) of various depths $K \in \{0, 1, 2, 3\}$. The depth of the network is measured as the number of successions of convolution-pointwise-nonlinearity layers. Each network ends with a final fully connected layer A (with bias b_A), i.e. a final unconstrained affine transformation. For simplicity, our CNNs will be single-channel only and without various architecture tricks, e.g. dropout, batch normalisation, or pooling. The network functions, denoted f_k for $k \in \{1, \ldots, K\}$ can thus be written as:

$$f_k(\varphi^{data}) = A\sigma \circ \tilde{C}_K \circ \sigma \circ \tilde{C}_{K-1} \circ \cdots \circ \sigma \circ \tilde{C}_1(\varphi^{data}) + b_A, \tag{2}$$

where $\tilde{C}_i(x) = C_i x + b_i$ is the i-th convolution layer comprising the circulant matrix C_i for the convolution and its additive unconstrained bias b_i and $\sigma =$ ReLU the standard pointwise nonlinearity in neural networks. Note that a CNN with depth 0 degenerates to an unconstrained affine transformation in $\mathbb{R}^D$ (no pointwise nonlinearity or convolution): $f_0(\varphi^{data}) = A\varphi^{data} + b_A$.

The networks are trained to minimise the Mean Squared Error (MSE)[1], a proxy for the ESE, using the N generated samples. Denoting $f_{k,N,\eta}$ the resulting networks (where η a hyperparameter of the optimisation algorithm), we have:

$$MSE_{train}(f_{k,N,\eta}) = \frac{1}{N}\sum_{i=1}^{N} \| f_{k,N,\eta}(\varphi^{data}_{train,i}) - \varphi_{train,i}\|_2^2, \tag{3}$$

where for a sample collection set, $\varphi_{set,i}$ and $\varphi^{data}_{set,i}$ denote the i-th original and degraded samples. This quantity is to be compared with $ESE(f_{k,N}(\varphi^{data}), \varphi)$, which evaluates the performance on all possible data of a network trained on N instances only. Naturally, this quantity cannot be computed by hand and is approximated by another MSE calculation on a large test set using N_t test samples independently generated from the training ones:

$$MSE_{test}(f_{k,N,\eta}) = \frac{1}{N_t}\sum_{i=1}^{N_t} \| f_{k,N,\eta}(\varphi^{data}_{test,i}) - \varphi_{test,i}\|_2^2 \xrightarrow[N_t \to \infty]{} ESE(f_{k,N,\eta}(\varphi^{data}), \varphi). \tag{4}$$

In our tests, $N_t = 100000$. Note that implicitly in $ESE(f_{k,N,\eta}(\varphi^{data}), \varphi)$ the network $f_{k,N,\eta}$ is the given result of a minimisation algorithm. For randomised algorithms, it is thus to be understood as the expectation conditional to the learned network $f_{k,N,\eta}$: $ESE(f_{k,N,\eta}(\varphi^{data}), \varphi) = \mathbb{E}(\| f_{k,N,\eta}(\varphi^{data}) - \varphi\|_2^2 \mid f_{k,N,\eta})$.

We train our networks using Stochastic Gradient Descent with Nesterov momentum parameter equal to 0.9. We train the networks using various learning rates $\eta \in \{0.0001, 0.0005, 0.001, 0.005, 0.01, 0.05, 0.1\}$ over 50 epochs, performing $N_r = 50$ independent training trials per learning rate, and compute the final median performance per learning rate on a validation set generated independently of the train and test data comprising $N_v = 100000$ validation samples:

$$MSE_{val}(f_{k,N,\eta}) = \frac{1}{N_v}\sum_{i=1}^{N_v} \| f_{k,N,\eta}(\varphi^{data}_{val,i}) - \varphi_{val,i}\|_2^2 \xrightarrow[N_v \to \infty]{} ESE(f_{k,N,\eta}(\varphi^{data}), \varphi). \tag{5}$$

[1] The loss function is actually scaled to $\frac{1}{D} MSE_{train}$ as is commonly done in practice.

The validation set is used to choose the best learning rate for each amount of training data $\eta^*(N)$ by taking:

$$\eta^*(N) = \underset{\eta}{\operatorname{argmin}}\, MEDIAN_r(MSE_{val}(f_{k,N,\eta})), \tag{6}$$

where $MEDIAN_r$ takes the median over the $r \leq N_r$ best independent runs on the validation set per η. Given that a significant amount of runs do not converge or get trapped early in a poor local minimum depending on the random initialisation, choosing $r \ll N_r$ ensures that only the networks finding a good local minimum are considered. The final performance of CNNs $SCORE_{k,r}(N)$ for each amount of data N is then the median of the test performance over those selected r trials[2] of the final test score at the chosen learning rate $\eta^*(N)$:

$$SCORE_{k,r}(N) = MEDIAN_r(MSE_{test}(f_{k,N,\eta^*(N)})). \tag{7}$$

We display the evolution of the networks' performance on the amount of training data N in Fig. 3 for each depth k, with detailed scores in Table 1, along with the performance of the Wiener filter. Regardless of N, the Wiener filter outperforms the neural models as expected by the theory, but their performance converges to the Wiener's one when a lot of data is available, with similar performance when at least 10000 training samples are available. We can thus consider this study as providing a criterion that a model would be preferable if data is limited to fewer than 10000 samples to train on.

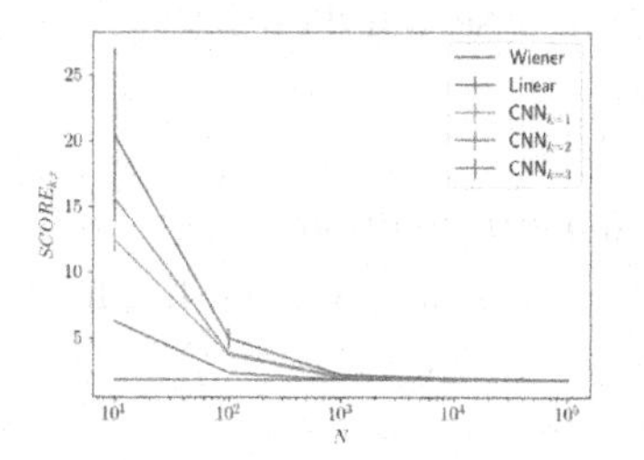

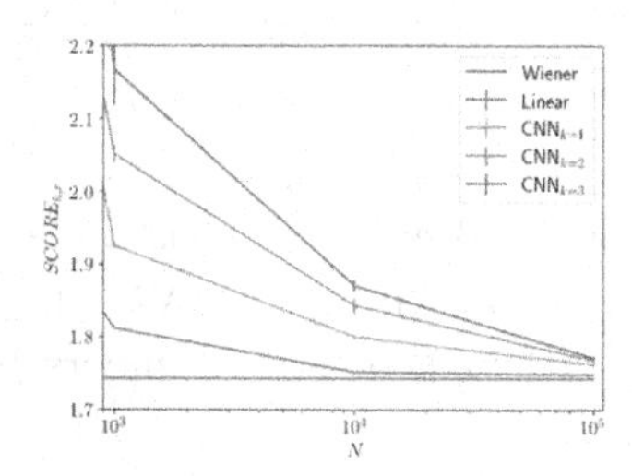

Fig. 3. Median test scores for CNNs with depth $k \in \{0, 1, 2, 3\}$ on $r = 10$ selected runs ($k = 0$ is just a linear layer). Vertical bars represent the standard deviation of the MSE of these runs. The right figure is a zoomed-in plot of the left one for large N.

3 Two-Dimensional Geometric Estimation

We next analyse a more complicated yet well-understood problem based on Euclidean geometry. The goal is to estimate basic geometric properties on simple data: the radius and centre location of a random disk in an image. It was shown in [6] that this seemingly trivial task is more complex than expected for neural models even when focusing on radius estimation of centred disks.

[2] Selected on the validation set.

Table 1. Median MSE scores $SCORE_{k,r}$ of the CNNs on $r = 10$ selected runs, compared to the theoretically optimal Wiener filter.

N	0	10	100	1000	10000	100000
Wiener	**1.743**	—	—	—	—	—
Linear ($k = 0$)	—	6.204	2.295	1.811	1.751	1.748
CNN ($k = 1$)	—	12.386	3.662	1.924	1.799	1.762
CNN ($k = 2$)	—	15.614	3.789	2.051	1.842	1.767
CNN ($k = 3$)	—	20.395	4.911	2.167	1.869	1.771

3.1 Data Model

The original data now consists of $D \times D$ random two-dimensional grayscale images of disks. Images are centred at $(0, 0)$, and for a pixel $x \in [-\frac{D-1}{2}, \frac{D-1}{2}]^2$:

$$\varphi(x) = \begin{cases} b & \text{if } \|x - c\|_2 > r \\ f & \text{if } \|x - c\|_2 \leq r, \end{cases} \tag{8}$$

where r is the circle's radius, $c = (c_x, c_y)$ its centre, and f (resp. b) is the foreground (resp. background) intensity. These parameters are independently[3] and uniformly chosen at random: $r \sim \mathcal{U}([\frac{\varepsilon_r}{2}\frac{D-1}{4}, (1 - \frac{\varepsilon_r}{2})\frac{D-1}{4}])$ with $\varepsilon_r = 0.4$, $c \sim \mathcal{U}([(D-1)\frac{\varepsilon_c}{2} - \frac{D-1}{2}, (D-1)(1 - \frac{\varepsilon_c}{2}) - \frac{D-1}{2}]^2)$ with $\varepsilon_c = 0.5$, $b \sim \mathcal{U}([0, 1])$, and $f \mid b \sim \mathcal{U}([0, 1] \setminus [b - \delta, b + \delta])$ with $\delta = \frac{50}{255}$ the minimum contrast[4]. However, φ is degraded with blur and noise giving the observed data φ^{data} as follows:

$$\varphi^{data} = g_{\sigma_b} * \varphi + n, \tag{9}$$

where $g_{\sigma_b}(x) = \frac{1}{2\pi}\exp(-\frac{\|x\|_2^2}{2\sigma_b^2})$ is a Gaussian convolution kernel, and n is i.i.d. white noise $n \sim \mathcal{N}(0, \sigma_n I_{D^2})$. We plot example data in Fig. 4. The task is to estimate the three geometric numbers $(r, c) = (r, c_x, c_y)$ from φ^{data}.

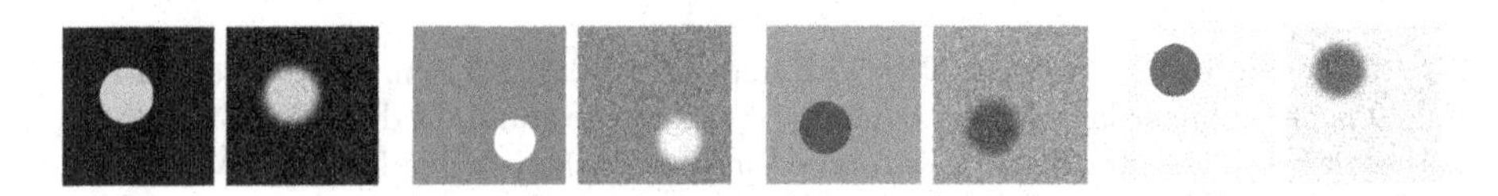

Fig. 4. Four examples of clean φ and degraded φ^{data} disk images.

3.2 Expert Engineer's Solution

Unlike in the Wiener case, the optimal estimator minimising the ESE is not so trivial to find. Instead, we choose a method called Pointflow designed by an expert engineer that perfectly tackles the problem at hand.

[3] Except f and b which are slightly correlated to ensure a minimal contrast $|f - b| > \delta$.
[4] In our tests, we take $D = 201$ implying that $r \sim \mathcal{U}([10, 40])$ and $c \sim \mathcal{U}([-50, 50]^2)$.

Pointflow [3,22] is an elegant subpixel level contour integrator and edge detector in images requiring no learning whatsoever. It consists in defining potential vector fields V along which random points P flow: $\frac{dP}{dt}(t) = V(P(t))$, such that end trajectories lie on edges of the image I. The vanilla Pointflow [22] uses two fields V_+ and V_- from the edge attraction V_a and rotating V_r fields based on the image gradients as follows:

$$V_a = \nabla \|\nabla I_b\|_2, \quad V_r = \nabla I_b^{\perp}, \quad V_{\pm} = \tfrac{1}{2}(V_a \pm V_r), \tag{10}$$

where $I_b = g_{\sigma_{Pf}} * I$ is a blurred version of I with a Gaussian kernel $g_{\sigma_{Pf}}$. Various stopping conditions and uses of $V_{\pm}$ exist to detect edges in natural images, however on our data containing a single circular edge per image, we need only consider the basic ones. Indeed, the possible cases for trajectories are: it loops (C_l), it leaves the image domain (C_o), or it is stuck in an area with small magnitude $\|V\|_2$ (C_s). Flowing initially from V_+, if we loop (C_l), then the point has reached the circle and it suffices to reflow along V_+ to extract just the circle's contour. If we end up outside the image domain (C_o), which is rare in our data, we reflow from the exit point along V_-. If we are in a low flow magnitude area (C_s), which is not rare, then we discard the trajectory. In total, $N_{Pf} = 200$ points are randomly uniformly sampled in the image domain and used for Pointflow, and a fraction of them end up flowing on the disk's edge with subpixel precision, as the other ones lead to discarded trajectories. For more details on our implementation of Pointflow see Appendix A. For some illustrations of Pointflow results on our data see Fig. 5.

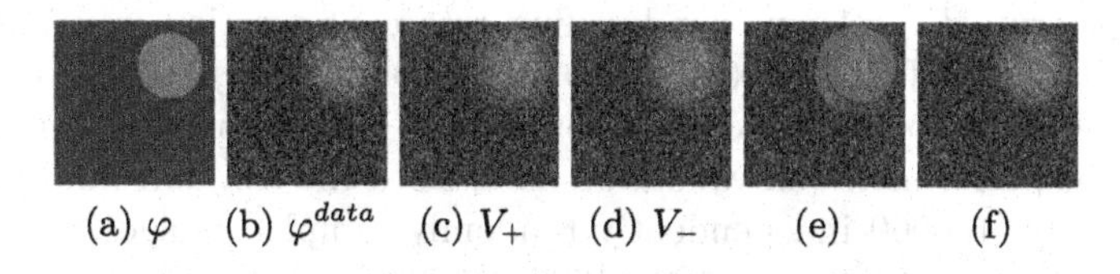

Fig. 5. Pointflow in practice. (5a): clean data. (5b): degraded data. (5c) and (5d): pointflow fields sampled every five pixels. (5e): initial flows of points without reflowing with groundtruth boundary in green. (5f): all final trajectories that have reflown in a closed loop (C_l) with groundtruth boundary in green and all regressed circles using least squares on each looped trajectory in blue (used for estimating the centre).

To estimate the disk's radius and centre from pointflow contours (C_l), we can simply compute the average length of the reflown closed contours and divide it by 2π. To compute the centre's coordinates, we could compute for each closed trajectory the average of its points, and then average over these estimations. However, this method empirically did not best perform on validation data, so we refined it by applying least-squares regression on the equation of a circle to estimate from it its location per trajectory and then average the estimations. Note that the least-squares regression did not provide a better estimation of the radius so we keep the crude length integration strategy for it.

3.3 Neural Models

As in the one-dimensional case, the expert's method is to be compared with a convolutional neural network. Although the learning problem seems trivial, it is actually harder than expected for networks, as has been shown in [6] even when the circles are centred. Empirically, we were not able to train correctly a small custom model similar to those previously used having just three layers and even many channels per layers and no further deep learning tricks. To overcome this limitation, we use famous networks in the deep learning literature: Alexnet [12], VGG [19] and Resnet[5] [11]. To adapt the model to our task, we change the final fully connected layer to have 3 outputs only.

For each architecture, we either train the networks from scratch (SC), or initialise the weights, except those of the final fully connected layers, to those publicly available obtained by classification on Imagenet [18], as is commonly done in the field. The pretrained weights can be either frozen for transfer-learning (TL) or retrained as well for finetuning (FT). Although the task and data are fundamentally different from ours, it is generally believed that the wide variety of natural images encourages the famous networks to learn features that generalise quite well to most reasonable tasks.

Once again, the *MSE* loss is used for training[6]. As it is significantly more expensive to train such networks compared to the tiny ones in the Wiener case, we only perform $N_r = 1$ run per learning rate configuration, ranging in $\eta \in \{0.000001, 0.00005, 0.00001, 0.0005, 0.0001, 0.005, 0.001, 0.05, 0.01, 0.5, 0.1\}$, with a batch size of 10 for 50 epochs. As previously, the optimal learning rate is chosen on the performance on validation data. Both the test and validation data use $N_v = N_t = 100000$ independently randomly sampled images, whereas the training sets have $N \in \{10, 100, 1000, 10000, 100000\}$ ones.

3.4 Results

We present the results in Fig. 6 and Table 2. Although the networks are simultaneously trained for both the radius and centre location estimation, we also present the MSE on the estimation of each geometric concepts separately.

First, the transfer-learning network Resnet-TL is not able to correctly estimate the radius or centre's location, meaning that its learned features on classification of Imagenet are not able to handle our simple data: they are not so general after all. Likewise, the prelearned features of Alexnet-TL and VGG-TL do not

[5] We use the simplest ones VGG11 and ResNet18, as larger ones are here unnecessary.

[6] To help the networks converge, the radius and centre coordinates are scaled to $[-1, 1]$ using $r_s = \frac{8}{D-1}(r - \frac{D-1}{8})$ and $c_s = \frac{2}{D-1}c$. In all plots and numbers provided in this paper, the results are rescaled to the original scale: r and c and not r_s and c_s.

generalise well to this toy problem, requiring significantly more data than the maximum available to compare with the simple data agnostic pointflow. However, when the networks are entirely trained, either finetuned or from scratch, they are either flatly beaten by pointflow when using small amounts of training data or on par or slightly outperform it when $N \geq 10000$. The only networks beating pointflow overall are VGG-SC and Resnet-FT when $N = 100000$, but more (VGG-FT, VGG-SC, Resnet-FT, and Resnet-SC) significantly outperform pointflow on the radius estimation when $N \geq 10000$. The difference between finetuning and training from scratch seems to only appear when small amounts of training data are available, and then finetuning is better. However, in these cases, both approaches pale in comparison to the reference data agnostic method.

From this experiment, we conclude that a realistic neural network is worth at least tens of thousands of examples on a fairly simple task (toy problem vs real world challenge) compared to an expert engineer. We thus provide a criterion that a data model is preferable if fewer than 10000 training samples are available.

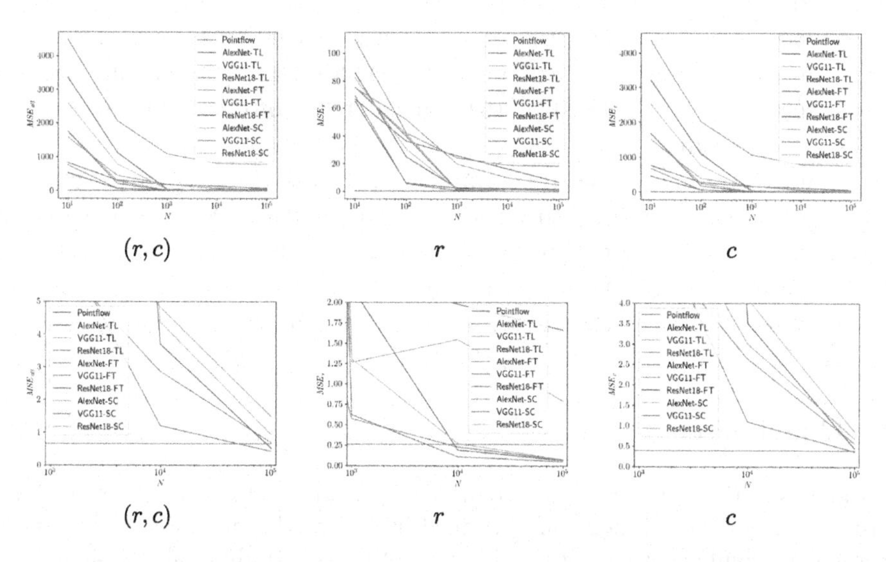

Fig. 6. Test scores of the data-agnostic Pointflow and of the best transfer-learned or finetuned networks. Left: MSE computed on both the radius and the centre's location estimation. Middle: same but only on the radius estimation. Right: same but only on the centre location's estimation. We zoom-in in the bottom set of figures.

Table 2. MSE scores of the networks compared to Pointflow on test data, computed on both r and c (top), just r (middle), or just c (bottom). Networks were trained on joint prediction of r and c. For the separate r (resp. c) scores, the selected networks were those providing the best r (resp. c) error on validation data.

	N	Pointflow	Alexnet			VGG			Resnet		
			TL	FT	SC	TL	FT	SC	TL	FT	SC
(r, c)	0	**0.66**									
	10		826	524	1748	1581	754	1748	5256	3358	2590
	100		314	66	291	448	69	215	2064	1124	793
	1000		183	23	9.0	189	6.5	7.6	1092	48	45
	10000		140	31	4.8	71	2.9	1.2	814	3.7	4.4
	100000		67	17	1.5	40	0.68	**0.42**	826	**0.51**	0.93
r	0	**0.26**									
	10		66	65	75	75	69	75	110	86	82
	100		36	5.9	42	52	5.6	40	42	25	31
	1000		26	2.4	1.3	20	0.57	0.62	24	2.2	1.3
	10000		16	2.0	1.5	9.9	**0.23**	**0.11**	19	**0.19**	0.27
	100000		6.8	1.7	0.69	4.9	**0.075**	**0.051**	19	**0.066**	**0.078**
c	0	**0.40**									
	10		759	454	1673	1506	675	1673	5142	3206	2507
	100		277	60	249	393	63	175	2020	1099	762
	1000		157	21	7.7	169	5.9	6.9	1067	46	43
	10000		118	29	3.0	61	2.6	1.1	795	3.5	4.1
	100000		57	6.4	0.79	36	0.57	**0.37**	807	0.45	0.86

4 Conclusion

We analysed the amount of data required by neural networks, either shallow custom ones or deep famous ones, trained from scratch, finetuned, or transfer-learned, to compete with optimal or state-of-the-art traditional data-agnostic methods based on mathematical data generation models. To do so, we mathematically generated data, and fed various amounts of samples to the networks for training. We found that tens of thousands of data examples are needed for the networks to be on par or beat the traditional methods, if they are able to. For mathematical accuracy, we did not investigate real-world problems, which are commonly harder with less accurate or non-existent mathematical data generation models, but more data should be needed in those complex tasks. We have empirically derived a simple criterion, enabling researchers working on tasks where data is not easily available, to choose whether to use model-based traditional methods, by using either preexisting or newly created data generation models, or simply feed data to deep neural networks.

Acknowledgements. This work is in part supported by the French government under management of Agence Nationale de la Recherche as part of the"Investissements d'avenir" program, reference ANR-19-P3IA-0001 (PRAIRIE 3IA Institute).

A Pointflow Implementation Details

The pointflow dynamics are implemented by discretising time and approximating the time derivative with a forward finite difference scheme, although it could be improved with a Runge-Kutta 4 implementation [5]. Given the small magnitudes of the fields, we found that a large time step $dt = 50$ works well. We define three thresholds, $\tau_l = 0.9$ for C_l, $\tau_s = 10^{-6}$ for C_s, and $\tau_{len} = 0.001$. we consider having looped C_l if a point reaches a previous point within squared Euclidean distance τ_l while having on the trajectory between the looping points at least one point with squared distance to them of at least τ_l. A trajectory is stuck if it reaches a point where the current flow V has small magnitude $\|V\|_2^2 \leq \tau_s$. Each flow is run for $N_i = 1000$ iterations, and trajectories shorter than τ_{len} are discarded, e.g. trajectories of type C_s. We used $\sigma_{Pf} = 5$ for blurring out the noise before computing the fields. The implemented pointflow algorithm for finding contours in our circle images is presented in Algorithm 1.

Algorithm 1. Contour integration with Pointflow on image I

Compute $I_b = g_{\sigma_{Pf}} * I$
Compute $V_a = \nabla \|\nabla I_b\|_2$ and $V_r = \nabla I_b^\top$
Compute $V_+ = \frac{1}{2}(V_a + V_r)$ and $V_- = \frac{1}{2}(V_a - V_r)$
Choose N_{Pf} random points independently and uniformly in the image domain $[-\frac{D-1}{2}, \frac{D-1}{2}]^2$
Let $\mathcal{C} = []$ be an empty list of computed contours
for $i = 1 \ldots N_{Pf}$ **do**
  Let $(traj_+, C)$ be the flow along V_+ starting from the i-th point
  if $C = C_l$ and length$(traj_+) \geq \tau_{len}$ **then**
    Let $(traj_l, C_l)$ be the reflow along V_+ starting from the endpoint of $traj_+$
    $\mathcal{C}$.append$(traj_l)$
  else if $C = C_o$ and length$(traj_+) \geq \tau_{len}$ **then**
    Let $(traj_-, C_-)$ be the reflow along V_- starting from the endpoint of $traj_+$
    if $C_- = C_l$ and length$(traj_-) \geq \tau_{len}$ **then**
      Let $(traj_l, C)$ be the reflow along V_- starting from the endpoing of $traj_-$
      $\mathcal{C}$.append$(traj_l)$
    end if
  end if
end for
Return $\mathcal{C}$

After computing the list of contours $\mathcal{C}$ in the image I, we estimate the radius using the average curve length $\hat{r} = \frac{1}{2\pi} \sum_{i=1}^{|\mathcal{C}|} \text{length}(\mathcal{C}_i)$. Since the average of the points did not yield the best estimation of the circle centre, we estimate it instead using least squares. The equation of a circle is naturally given by $(x - c_x)^2 + (y - c_y)^2 = r^2$, which can be written as $\theta_1 x + \theta_2 y + \theta_3 = x^2 + y^2$, where $\theta_1 = 2c_x$, $\theta_2 = 2c_y$, and $\theta_3 = r^2 - c_x^2 - c_y^2$. We can thus estimate for each contour $\theta = (\theta_1, \theta_2, \theta_3)^\top$ by least squares as $\hat{\theta} = A^\top(AA^\top)^{-1}B$, with $A_{i,:} = (x_i, y_i, 1)$ and $B_i = x_i^2 + y_i^2$ and i ranging in the number of computed points on the contour. From $\hat{\theta}$ we can estimate $\hat{c} = (\frac{\theta_1}{2}, \frac{\theta_2}{2})$. The final centre estimation is then given by the average of this estimation over all contours. Note that we can also estimate r using θ_3 but we found that it did not outperform the lenght strategy so we do not use it.

References

1. Alwosheel, A., van Cranenburgh, S., Chorus, C.G.: Is your dataset big enough? sample size requirements when using artificial neural networks for discrete choice analysis. J. Choice Model. **28**, 167–182 (2018)
2. Anthony, M., Bartlett, P.L.: Neural Network Learning: Theoretical Foundations. Cambridge University Press, Cambridge (1999)
3. Bai, B., Yang, F., Chai, L.: Point flow edge detection method based on phase congruency. In: 2019 Chinese Automation Congress (CAC), pp. 5853–5858. IEEE (2019)
4. Bengio, Y., Lecun, Y., Hinton, G.: Deep learning for AI. Commun. ACM **64**(7), 58–65 (2021)
5. Butcher, J.: Numerical Methods for Ordinary Differential Equations. Wiley, Hoboken (2008). https://books.google.co.il/books?id=opd2NkBmMxsC
6. Dagès, T., Lindenbaum, M., Bruckstein, A.M.: From compass and ruler to convolution and nonlinearity: On the surprising difficulty of understanding a simple CNN solving a simple geometric estimation task. arXiv preprint arXiv:2303.06638 (2023)
7. Elad, M.: Sparse and Redundant Representations: From Theory to Applications in Signal and Image Processing. Springer, Cham (2010)
8. Geirhos, R., Narayanappa, K., Mitzkus, B., Thieringer, T., Bethge, M., Wichmann, F.A., et al.: Partial success in closing the gap between human and machine vision. Adv. Neural. Inf. Process. Syst. **34**, 23885–23899 (2021)
9. Geirhos, R., Temme, C.R., Rauber, J., Schütt, H.H., Bethge, M., Wichmann, F.A.: Generalisation in humans and deep neural networks. In: Advances in Neural Information Processing Systems, vol. 31 (2018)
10. Goodfellow, I., Bengio, Y., Courville, A.: Deep Learning. MIT press, Cambridge (2016)
11. He, K., Zhang, X., Ren, S., Sun, J.: Deep residual learning for image recognition. In: Proceedings of the IEEE Conference on Computer Vision and Pattern Recognition, pp. 770–778 (2016)
12. Krizhevsky, A., Sutskever, I., Hinton, G.E.: Imagenet classification with deep convolutional neural networks. Commun. ACM **60**(6), 84–90 (2017)
13. Lakshmanan, V., Robinson, S., Munn, M.: Machine Learning Design Patterns. O'Reilly Media, Sebastopol (2020)
14. LeCun, Y., Bengio, Y., Hinton, G.: Deep learning. Nature **521**(7553), 436–444 (2015)

15. LeCun, Y., Boser, B., Denker, J., Henderson, D., Howard, R., Hubbard, W., et al.: Handwritten digit recognition with a back-propagation network. In: Advances in Neural Information Processing Systems, vol. 2 (1989)
16. Mohamed, A.R., Dahl, G., Hinton, G.: Deep belief networks for phone recognition. In: Nips Workshop on Deep Learning for Speech Recognition and Related Applications, vol. 1, p. 39 (2009)
17. Novikoff, A.B.: On convergence proofs for perceptrons. Technical report Stanford Research Institute, Menlo Park, CA, USA (1963)
18. Russakovsky, O., Deng, J., Su, H., Krause, J., Satheesh, S., Ma, S., et al.: Imagenet large scale visual recognition challenge. Int. J. Comput. Vis. **115**(3), 211–252 (2015)
19. Simonyan, K., Zisserman, A.: Very deep convolutional networks for large-scale image recognition. arXiv preprint arXiv:1409.1556 (2014)
20. Vapnik, V.N., Chervonenkis, A.Y.: On the uniform convergence of relative frequencies of events to their probabilities. In: Vovk, V., Papadopoulos, H., Gammerman, A. (eds.) Measures of Complexity, pp. 11–30. Springer, Cham (2015). https://doi.org/10.1007/978-3-319-21852-6_3
21. Wiener, N.: Extrapolation, Interpolation, and Smoothing of Stationary Time Series: With Engineering Applications, vol. 113. MIT press, Cambridge (1949)
22. Yang, F., Cohen, L.D., Bruckstein, A.M.: A model for automatically tracing object boundaries. In: 2017 IEEE International Conference on Image Processing (ICIP), pp. 2692–2696. IEEE (2017)

Resolution-Invariant Image Classification Based on Fourier Neural Operators

Samira Kabri[1(✉)], Tim Roith[1], Daniel Tenbrinck[1], and Martin Burger[2,3]

[1] Friedrich-Alexander-Universität Erlangen-Nürnberg, 91058 Erlangen, Germany
samira.kabri@fau.de
[2] Deutsches Elektronen-Synchrotron, 22607 Hamburg, Germany
[3] Universität Hamburg, Fachbereich Mathematik, 20146 Hamburg, Germany

Abstract. In this paper we investigate the use of Fourier Neural Operators (FNOs) for image classification in comparison to standard Convolutional Neural Networks (CNNs). Neural operators are a discretization-invariant generalization of neural networks to approximate operators between infinite dimensional function spaces. FNOs—which are neural operators with a specific parametrization—have been applied successfully in the context of parametric PDEs. We derive the FNO architecture as an example for continuous and Fréchet-differentiable neural operators on Lebesgue spaces. We further show how CNNs can be converted into FNOs and vice versa and propose an interpolation-equivariant adaptation of the architecture.

Keywords: neural operators · trigonometric interpolation · Fourier neural operators · convolutional neural networks · resolution invariance

1 Introduction

Neural networks, in particular CNNs, are a highly effective tool for image classification tasks. Substituting fully-connected layers by convolutional layers allows for efficient extraction of local features at different levels of detail with reasonably low complexity. However, neural networks in general are not resolution-invariant, meaning that they do not generalize well to unseen input resolutions. In addition to interpolation of inputs to the training resolution, various other approaches have been proposed to address this issue, see, e.g., [3,14,17]. In this work we focus on the interpretation of digital images as discretizations of functions. This allows to model the feature extractor as a mapping between infinite dimensional spaces with the help of so-called neural operators, see [13]. In Sect. 2, we use established results on Nemytskii operators to derive conditions for well-definedness,

This work was supported by the European Union's Horizon 2020 programme, Marie Skłodowska-Curie grant agreement No. 777826. TR and MB acknowledge the support of the BMBF, grant agreement No. 05M2020. SK and MB acknowledge the support of the DFG, project BU 2327/19-1. This work was carried out while MB was with the FAU Erlangen-Nürnberg.

L. Calatroni et al. (Eds.): SSVM 2023, LNCS 14009, pp. 236–249, 2023.
https://doi.org/10.1007/978-3-031-31975-4_18

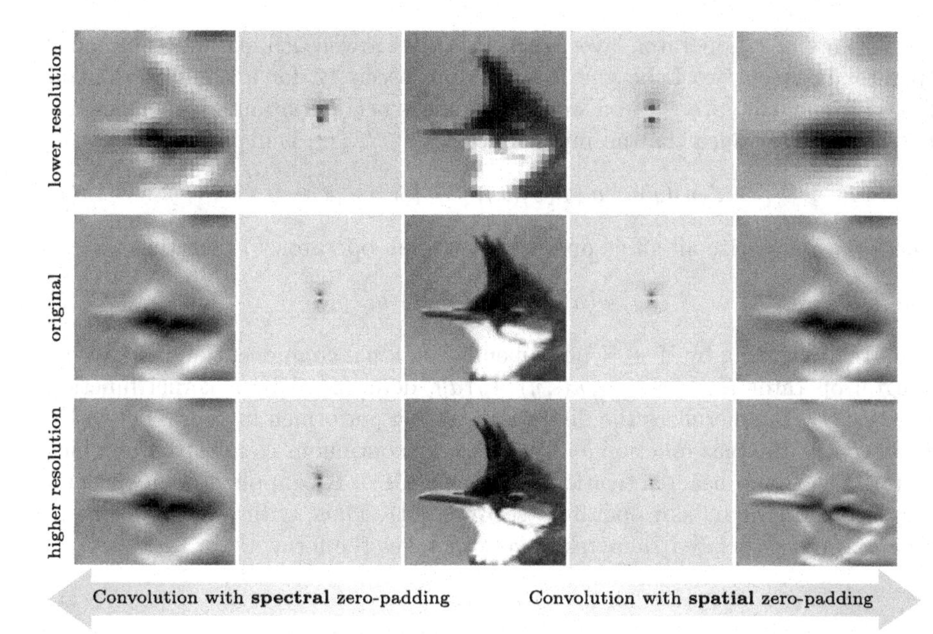

Fig. 1. Effects of applying a convolutional filter on the same image with different resolutions. Spatial zero-padding (standard CNN-implementation) changes the relation of kernel support to image domain, while spectral zero-padding (FNO-implementation) captures comparable features for all resolutions. The image depicts a red whiskered bulbul taken from the Birds500 dataset [18]. (Color figure online)

continuity, and Fréchet-differentiability of neural operators on Lebesgue spaces. We specifically show these properties for the class of FNOs proposed in [15] as a discretization-invariant generalization of CNNs.

The key idea of FNOs is to parametrize convolutional kernels by their Fourier coefficients, i.e., in the spectral domain. Using trainable filters in the Fourier domain to represent convolution kernels in the context of image processing with neural networks has been studied with respect to performance and robustness in recent works, see e.g., [4,19,25]. In Sect. 3 we analyze the interchangeability of CNNs and FNOs with respect to optimization, parameter complexity, and generalization to varying input resolutions. While we restrict our theoretical derivations to real-valued functions, we note that they can be naturally extended to vector-valued functions as well. Our findings are supported by numerical experiments on the FashionMNIST [24] and Birds500 [18] data sets in Sect. 4.

2 Construction of Neural Operators on Lebesgue Spaces

2.1 Well-Definedness and Continuity

A neural operator as defined in [13] is a composition of a finite but arbitrary number of so-called operator layers. In this section we derive conditions on the components of an operator layer, such that it is a well-defined and continuous operator between two Lebesgue spaces. More precisely, for a bounded domain $\Omega \subset \mathbb{R}^d$ and $1 \leq p, q \leq +\infty$ we aim to construct a continuous operator $\mathcal{L} : L^p(\Omega) \to L^q(\Omega)$, such that an input function $u \in L^p(\Omega)$ is mapped to

$$\mathcal{L}(u)(x) = \sigma\left(\Psi(u)(x)\right) \qquad \text{for a.e. } x \in \Omega, \tag{1}$$

where we summarize all affine operations with an operator Ψ, such that

$$\Psi(u) = Wu + \mathcal{K}u + b. \tag{2}$$

Here, the weighting by $W \in \mathbb{R}$ implements a residual component and the kernel integral operator $\mathcal{K} : u \mapsto \int_\Omega \kappa(\cdot, y)\, u(y)\, dy$, determined by a kernel function $\kappa : \Omega \times \Omega \to \mathbb{R}$ generalizes the discrete weighting performed in neural networks. Analogously, the bias function $b : \Omega \to \mathbb{R}$ is the continuous counterpart of a bias vector. The (non-linear) activation function $\sigma : \mathbb{R} \to \mathbb{R}$ is applied pointwise and thus acts as a Nemytskii operator (see e.g., [6]). Thus, with a slight abuse of notation, the associated Nemytskii operator takes the form

$$\sigma : v \mapsto \sigma(v(\cdot)), \tag{3}$$

where we assume σ to be a measurable function. In order to ensure that the associated Nemytskii operator defines a mapping $\sigma : L^p(\Omega) \to L^q(\Omega)$ for $1 \leq p, q \leq \infty$ we require the following conditions to hold:

$$\begin{aligned} \underline{p, q < \infty} &: |\sigma(x)| \leq K + \beta |x|^{\frac{p}{q}} \text{ for all } x \in \mathbb{R} \text{ and constants } \beta, K \in \mathbb{R}, \\ \underline{p = \infty} &: |\sigma(x)| \leq K(c) \text{ for every } c > 0 \text{ for all } x, |x| < c \\ &\quad \text{and a constant } K(c) \in \mathbb{R} \text{ depending on } c, \\ \underline{p < \infty, q = \infty} &: |\sigma(x)| \leq K \text{ for all } x \in \mathbb{R} \text{ and a constant } K \in \mathbb{R}, \end{aligned} \tag{4}$$

which were used in [6].

Lemma 1. *For $1 \leq p, q \leq \infty$ assume that σ fulfills* (4). *Then we have that the associated Nemytskii operator is a mapping $\sigma : L^p(\Omega) \to L^q(\Omega)$.*

Proof. Similar to [6, Th. 1], follows directly by employing the estimates in (4).

Since we are interested in continuity properties of the layer in (1) we consider the following continuity result for Nemytskii operators.

Lemma 2. *For $1 \leq p \leq \infty, 1 \leq q < \infty$ assume that the function $\sigma : \mathbb{R} \to \mathbb{R}$ is continuous and uniformly continuous in the case $q = \infty$. If the associated Nemytskii operator is a mapping $\sigma : L^p(\Omega) \to L^q(\Omega)$ then it is continuous.*

Proof. For $q < \infty$ the proof can be adapted from [23, p. 155–158]. For the case $q = \infty$ we refer to [6, Th. 5].

Remark 1. For $1 \leq p \leq q < \infty$ it is sufficient for σ to be p/q-Hölder continuous or locally Lipschitz continuous for $p, q = \infty$. In that case the Hölder and respectively the Lipschitz continuity transfers to the Nemytskii operator, see [22].

Example 1. The ReLU (Rectified Linear Unit, see [5]) function $\sigma(x) = \max(0, x)$ generates a continuous Nemytskii operator $\sigma : L^p(\Omega) \to L^q(\Omega)$ for any $p \geq q$. To show this, we note that the function σ is Lipschitz-continuous and with $p \geq q$ we have for all $x \in \mathbb{R}$ that $|\sigma(x)| \leq |x| \leq 1 + |x|^{\frac{p}{q}}$.

Proposition 1. *For $1 \leq p, q \leq \infty$ let $\mathcal{L}$ be an operator layer given by* (1) *with an activation function $\sigma : \mathbb{R} \to \mathbb{R}$. If there exists $r \geq 1$ such that*

(i) the affine part defines a mapping $\Psi : L^p(\Omega) \to L^r(\Omega)$,
(ii) the activation funtion σ generates a Nemytskii operator $\sigma : L^r(\Omega) \to L^q(\Omega)$,

then it holds that $\mathcal{L} : L^p(\Omega) \to L^q(\Omega)$. If additionally Ψ is a continuous operator on the specified spaces and the function σ is continuous, or uniformly continuous in the case $q = \infty$, the operator $\mathcal{L} : L^p(\Omega) \to L^q(\Omega)$ is also continuous.

Proof. With the assumptions on σ we directly have $\mathcal{L} = \sigma \circ \Psi : L^p(\Omega) \to L^q(\Omega)$. The continuity of $\mathcal{L}$ follows from Lemma 2.

Example 2. On the periodic domain $\Omega = \mathbb{R}/\mathbb{Z}$ consider an affine operator Ψ as defined in (2), where the integral operator is a convolution operator, i.e., $\kappa(x, y) = \kappa(x - y)$ with a slight abuse of notation. If for $1 \leq p, r, s \leq \infty$ we have that $\kappa \in L^s(\Omega)$ with $1/r + 1 = 1/p + 1/s$, it follows from Young's convolution inequality (see e.g., [8, Th. 1.2.12]) that $\mathcal{K} : L^p(\Omega) \to L^r(\Omega)$ is continuous. If further $b \in L^r(\Omega)$ and $W = 0$ in the case $r > p$, it follows directly that $\Psi : L^p(\Omega) \to L^r(\Omega)$ is continuous.

2.2 Differentiability

To analyze the differentiability of the neural operator layers we first transfer the result for general Nemytskii operators from [6, Th. 7] to our setting.

Theorem 1. *Let $1 \leq q < p < \infty$ or $q = p = \infty$ and $\sigma : \mathbb{R} \to \mathbb{R}$ a continuously differentiable function. Furthermore, let the Nemytskii operator associated to the derivative σ' be a continuous operator $\sigma' : L^p(\Omega) \to L^s(\Omega)$, with coefficient $s = pq/(p - q)$ for $q < p$ and $s = \infty$ for $q = p = \infty$. Then, the Nemytskii operator associated to σ is Fréchet-differentiable and its Fréchet-derivative $D\sigma(v) : L^p(\Omega) \to L^q(\Omega)$ in $v \in L^p(\Omega)$ is given by*

$$D\sigma(v)(h) = \sigma'(v) \cdot h, \qquad \text{for all } h \in L^p(\Omega).$$

Since the ReLU activation function from Example 1 is not differentiable, it does not fulfill the requirements of Theorem 1. An alternative is the so-called Gaussian Error Linear Unit (GELU), proposed in [10].

Example 3. The GELU function $\sigma(x) = x\,\Phi(x)$, where Φ denotes the cumulative distribution function of the standard normal distribution, generates a Fréchet-differentiable Nemytskii operator with derivative $D\sigma(v) : L^p(\Omega) \to L^q(\Omega)$ for any $p \geq q$ and $v \in L^p(\Omega)$. To show this, we compute $\sigma'(x) = \Phi(x) + x\phi(x)$, where $\phi(x) = \Phi'(x)$ is the standard normal distribution. We see that σ' is continuous and further $|\sigma'(x)| \leq 1 + |x|/\sqrt{2\pi} \leq 1 + 1/\sqrt{2\pi} + |x|^{\frac{p}{q}}/\sqrt{2\pi}$ for all $p \geq q$.

Proposition 2. *For $1 \leq p, q \leq \infty$, let $\mathcal{L}$ be an operator layer given by* (1) *with affine part Ψ as in* (2)*. If there exists $r > q$, or $r = q = \infty$ such that*

(i) the affine part is a continuous operator $\Psi : L^p(\Omega) \to L^r(\Omega)$,
(ii) the activation function $\sigma : \mathbb{R} \to \mathbb{R}$ is continuously differentiable
(iii) and the derivative of the activation function generates a Nemytskii operator $\sigma' : L^r(\Omega) \to [L^r(\Omega) \to L^s(\Omega)]$ with $s = rq/(r-q)$,

then it holds that $\mathcal{L} : L^p(\Omega) \to L^q(\Omega)$ is Fréchet-differentiable in any $v \in L^p(\Omega)$ with Fréchet-derivative $D\mathcal{L}(v) : L^p(\Omega) \to L^q(\Omega)$

$$D\mathcal{L}(v)(h) = \sigma'(\Psi(v)) \cdot \tilde{\Psi}(h),$$

where $\tilde{\Psi}$ denotes the linear part of Ψ, i.e., $\tilde{\Psi} = \Psi - b$.

Proof. Theorem 1 yields that $D\sigma(v) : L^r(\Omega) \to L^q(\Omega)$ is well defined and continuous for $v \in L^r(\Omega)$. Fréchet-differentiability of linear and continuous operators on Banach spaces (see e.g., [1, Ex. 1.3]) yields the continuity of $\Psi : L^p(\Omega) \to L^r(\Omega)$ in all $v \in L^p(\Omega)$ with $D\Psi(v)(h) = \tilde{\Psi}(h)$. The claim follows from the chain-rule for Fréchet-differentiable operators, see [1, Prop. 1.4 (ii)].

For $p < q$ Fréchet-differentiability of a Nemytskii operator implies that the generating function is constant, and respectively affine linear for $p = q < \infty$, see [6, Ch 3.1]. Therefore, unless $p = \infty$, Fréchet-differentiability of neural operators with non-affine linear activation functions is only achieved at the cost of mapping the output of the affine part into a less regular space.

Example 4. For a continuous convolutional neural operator layer as constructed in Example 2, we consider a parametrization of the kernel function by a set of parameters $\hat{\theta} = \{\hat{\theta}_k\}_{k \in I} \subset \mathbb{C}$, where I is a finite set of indices, such that

$$\kappa_{\hat{\theta}}(x) = \sum_{k \in I} \hat{\theta}_k \, b_k(x), \tag{5}$$

with Fourier basis functions $b_k(x) = \exp(2\pi i \, kx)$ for $x \in \Omega$. Effectively, this amounts to parametrizing the kernel function by a finite number of Fourier coefficients. The resulting linear operator and the operator layer are denoted by $\Psi_{\hat{\theta}}$ and $\mathcal{L}_{\hat{\theta}}$. We note that FNOs proposed in [15] are neural operators that consist of such layers. It is easily seen that the kernel function defined by (5) is bounded and thus $\kappa_{\hat{\theta}} \in L^\infty(\Omega)$. Therefore, for suitable activation functions, Proposition 2 yields Fréchet-differentiability of $\mathcal{L}_{\hat{\theta}}$ with respect to its input function v, which was similarly observed in [16]. Additionally, for fixed v we consider the operator $\mathcal{L}_{(\cdot)}(v) : \hat{\theta} \mapsto \mathcal{L}_{\hat{\theta}}(v)$ which maps a set of parameters to a function. With the arguments from Proposition 2 we derive the partial Fréchet-derivatives of an FNO-layer with respect to its parameters for $h_k = (1 + i)\, e_k$ as

$$D_{\hat{\theta}_k} \mathcal{L}_{\hat{\theta}}(v) := D\mathcal{L}_{\hat{\theta}}(v)(h_k) = \sigma'(\Psi_{\hat{\theta}}(v)) \, D\Psi_{\hat{\theta}}(v)(h_k),$$

where e_k denotes the k-th canonical basis vector. Computing the Fréchet-derivative of Ψ in the sense of Wirtinger calculus ([20, Ch. 1]), this can be rewritten as $D_{\hat{\theta}_k} \mathcal{L}_{\hat{\theta}}(v) = \sigma'(\Psi_{\hat{\theta}}(v))\, \bar{\hat{v}}_k\, \bar{b}_k$, where $\hat{v}_k$ denotes the k-th Fourier coefficient of v. Here, for a complex number $z \in \mathbb{C}$, we denote by $\bar{z}$ its complex conjugate.

3 Connections to Convolutional Neural Networks

In this section we analyze the connection between FNOs and CNNs. Thus, for the remainder of this work, we set the domain to be the d-dimensional torus, i.e., $\Omega = \mathbb{R}^d/\mathbb{Z}^d$. As described in Example 4, the main idea of FNOs is to parametrize the convolution kernel by a finite number of Fourier coefficients $\hat{\theta} = \{\hat{\theta}_k\}_{k \in I} \subset \mathbb{C}$, where $I \subset \mathbb{Z}^d$ is a finite set of indices. Making use of the convolution theorem, see e.g., [8, Prop. 3.1.2 (9)], the kernel integral operator can then be written as

$$\mathcal{K}_{\hat{\theta}} v = \mathcal{F}^{-1}\left(\hat{\theta} \cdot \mathcal{F} v\right), \tag{6}$$

where $\mathcal{F}\colon [\Omega \to \mathbb{C}] \to \left[\mathbb{Z}^d \to \mathbb{C}\right]$ denotes the Fourier transform on the torus (see e.g., [8, Ch. 3]) and $\cdot$ denotes elementwise multiplication in the sense that

$$\left(\hat{\theta} \cdot \mathcal{F} v\right)_k = \begin{cases} \hat{\theta}_k \, (\mathcal{F} v)_k & \text{for } k \in I, \\ 0 & \text{otherwise.} \end{cases} \tag{7}$$

We only consider parameters such that $\mathcal{K}$ maps real-valued functions to real-valued functions. This is equivalent to Hermitian symmetry, i.e., $\hat{\theta}_k = \overline{\hat{\theta}_{-k}}$ and in particular $\hat{\theta}_0 \in \mathbb{R}$. As proposed in [15], for $N \in \mathbb{N}$ we choose the set of multi-indices $I_N := \{-\lceil (N-1)/2 \rceil, \dots, 0, \dots, \lfloor (N-1)/2 \rfloor\}^d$, which corresponds to parametrizing the N lowest frequencies in each dimension. This is in accordance to the universal approximation result for FNOs derived in [12]. At this point, we assume N to be an odd number to avoid problems with the required symmetry and expand the approach to even choices of N in Sect. 3.3. Although an FNO is represented by a finite number of parameters, a discretization of (6) is needed to process discrete data, e.g., digital images. We therefore define the set of spatial multi-indices $J_N := \{0, \dots, N-1\}^d$ and write $v \in \mathbb{R}^{J_N}$ for mappings $v : J_N \to \mathbb{R}$. Furthermore, we discretize the Fourier transform for $v \in \mathbb{R}^{J_N}$ as

$$(Fv)_k = \frac{1}{\lambda} \sum_{j \in J_N} v_j \, e^{-2\pi i \left\langle k, \frac{j}{N} \right\rangle} \qquad \text{for all } k \in I_N$$

and its inverse for $\hat{v} \in \mathbb{C}^{I_N}$ as

$$\left(F^{-1} \hat{v}\right)_j = \frac{\lambda}{|J_N|} \sum_{k \in I_N} \hat{v}_k \, e^{2\pi i \left\langle k, \frac{j}{N} \right\rangle} \qquad \text{for all } j \in J_N,$$

where $\lambda \in \{1, \sqrt{|J_N|}, |J_N|\}$ determines the normalization factor. The discretized convolution operator parametrized by $\hat{\theta} \in \mathbb{C}^{I_N}_{\text{sym}} := F(\mathbb{R}^{J_N})$ is then defined by

$$K(\hat{\theta})(v) = F^{-1}\left(\hat{\theta} \cdot Fv\right) \qquad \text{for } v \in \mathbb{R}^{J_N}.$$

For the remainder of this work, we refer to the above implementation of convolution as the FNO-implementation. In the following we compare the FNO-implementation to the standard implementation of the convolution of θ and $v \in \mathbb{R}^{J_N}$ in a conventional CNN, which can be expressed as

$$C(\theta)(v)_j = \sum_{\tilde{j} \in J_N} \theta_{j-\tilde{j}} v_{\tilde{j}} \qquad \text{for all } j \in J_N.$$

For the sake of simplicity, we handle negative indices by assuming that the values can be perpetuated periodically, although this is usually not done in practice.

3.1 Extension to Higher Input-Dimensions by Zero-Padding

So far, the presented implementations of convolution require the dimensions of the parameters θ, or $\hat{\theta}$ and the input v to coincide. In accordance to (7), the authors of [15] propose to handle dimension mismatches by zero-padding of the spectral parameters. More precisely, a low-dimensional set of parameters $\hat{\theta} \in \mathbb{C}^{I_M}$ is adapted to an input $v \in \mathbb{R}^{J_N}$ with odd $N \in \mathbb{N}$ by setting

$$\hat{\theta}^{M \to N}_k = \begin{cases} \hat{\theta}_k & \text{for } k \in I_N \cap I_M, \\ 0 & \text{for } k \in I_N \backslash I_M. \end{cases}$$

Since we choose N to be odd, the required symmetry is not hurt by the above operation. The extended FNO-implementation of the convolution is then given for $\hat{\theta} \in \mathbb{C}^{I_M}$ and $v \in \mathbb{R}^{J_N}$ by

$$K(\hat{\theta})(v) := K(\hat{\theta}^{M \to N})(v).$$

Analogously, in the conventional CNN-implementation the convolution of parameters $\theta \in \mathbb{R}^{J_M}$ and $v \in \mathbb{R}^{J_N}$ with $N \geq M$ is computed as

$$C(\theta)(v) := C(\theta^{M \to N})(v),$$

where again, $\theta^{M \to N} \in \mathbb{R}^{J_M}$ denotes the zero-padded version of θ. We stress that, although the technique to generalize the implementations to higher input dimensions is the same, the outcome differs substantially. This was already mentioned in [13, Sec. 4] and is discussed further in Sect. 3.4.

3.2 Convertibility and Complexity

Deriving FNOs from convolutional neural operators using the convolution theorem suggests that there is a way to convert one implementation of convolution into the other as long as the input dimension is fixed. The following Lemma shows that this is indeed possible.

Lemma 3. *Let $M \leq N$ both be odd and let $T : \mathbb{R}^{J_N} \to \mathbb{C}^{I_N}$ be defined for $\theta \in \mathbb{R}^{J_N}$ as $T(\theta) = \lambda F(\theta)$. For any $\theta \in \mathbb{R}^{J_M}$ and $v \in \mathbb{R}^{J_N}$ it holds true that*

$$C(\theta)(v) = K(T(\theta^{M \to N}))(v)$$

and for any $\hat{\theta} \in \mathbb{C}^{I_M}_{sym}$ and $v \in \mathbb{R}^{J_N}$ it holds true that

$$K(\hat{\theta})(v) = C(T^{-1}(\hat{\theta}^{M \to N}))(v).$$

Proof. By the definition of the extension to higher input dimensions we can assume $M = N$. For $\theta, v \in \mathbb{R}^{J_N}$ we derive the discrete analogon of the convolution theorem by inserting the definitions of the discrete Fourier transform as

$$F(C(\theta)(v)) = \lambda F(\theta) \cdot F(v).$$

Employing that $F : \mathbb{R}^{J_N} \to \mathbb{C}^{I_N}_{\mathrm{sym}}$ is a bijection, it follows that

$$C(\theta)(v) = F^{-1}(\lambda F(\theta) F(v)) = K(T(\theta))(v).$$

The second statement can be proven analogously. We note that T^{-1} is well-defined since $\lambda \geq 1$ for odd N.

Although the above Lemma proves convertibility for a fixed set of parameters and fixed input dimensions, a conversion can increase the amount of required parameters, as in general, the dimension of the converted parameters has to match the input dimension. It becomes clear that spatial locality cannot be enforced with the proposed FNO-parametrization and spectral locality cannot be enforced with the CNN-parametrization. Therefore, different behavior during the training process is to be expected if the parameter size does not match the input size. Moreover, the following Lemma shows that even for matching dimensions, equivalent behavior for gradient-based optimization like steepest descent requires careful adaptation of the learning rate, since the computation of gradients is not equivariant with respect to the function T.

Lemma 4. *For odd $N \in \mathbb{N}$ and $v, \theta \in \mathbb{R}^{J_N}$ and $\hat{\theta} = T(\theta)$ it holds true that*

$$\nabla_{\hat{\theta}} K(\hat{\theta})(v) = \frac{1}{|J_N|} T\left(\nabla_\theta C(\theta)(v)\right).$$

Proof. Inserting $\hat{\theta} = T(\theta)$ it follows with the chain rule from Lemma 3 that

$$\frac{\partial K(\hat{\theta})(v)_l}{\partial \hat{\theta}_k} = \sum_{j \in J_N} \frac{\partial C(\theta)(v)_l}{\partial \theta_j} \frac{\partial T^{-1}(\hat{\theta})_j}{\partial \hat{\theta}_k} = \frac{1}{|J_N|} \sum_{j \in J_N} \frac{\partial C(\theta)(v)_l}{\partial \theta_j} e^{-i2\pi \langle k, \frac{j}{N} \rangle}$$

for $k \in I_N$. The claim now follows by inserting the definition of T.

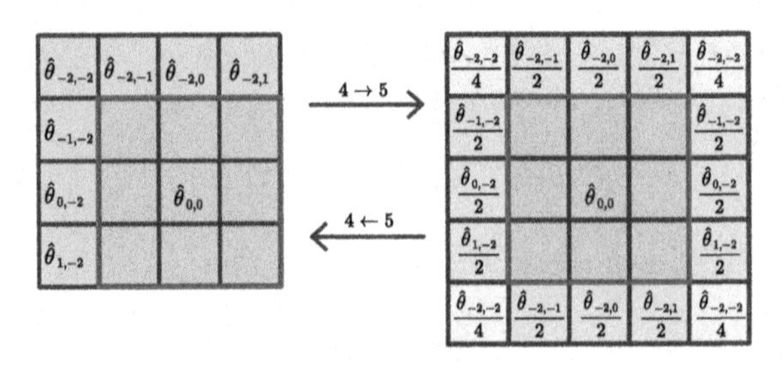

Fig. 2. Nyquist splitting for spectral parameters to extend real-valued trigonometric interpolation to even dimensions.

3.3 Adaptation to Even Dimensions

For the remainder of this paper we consider the special case $\Omega = \mathbb{R}^2/\mathbb{Z}^2$ and adapt the FNO-implementation to even dimensions. For odd dimensions M, N, zero-padding of a set of spectral coefficients does not violate the requirement $\hat{\theta}^{M \to N} \in \mathbb{C}^{I_N}_{\text{sym}}$. This property is lost in general for even dimensions. Since for odd dimensions, zero-padding in the spectral domain is equivalent to trigonometric interpolation, we perform the adaptation of dimensions such that $\hat{\theta}^{M \to N}$ is a trigonometric interpolator of a real-valued function (see [2] for an exhaustive study on this topic). In practice, this means splitting the coefficients corresponding to the Nyquist frequencies to interpolate from an even dimension to the next higher odd dimension, or to invert this splitting to interpolate from an odd dimension to the next lower even dimension (see Fig. 2). The real-valued trigonometric interpolation of $v \in \mathbb{R}^{J_M}$ to a dimension N is then given by

$$v^{M \xrightarrow{\Delta} N} := F^{-1}\left((Fv)^{M \to N}\right).$$

We extend the FNO-implementation to parameters $\hat{\theta} \in \mathbb{C}^{I_N}_{\text{sym}}$ and inputs $v \in \mathbb{R}^{J_N}$ with even $N \in \mathbb{N}$ by defining

$$K(\hat{\theta})(v) := \left(K(\hat{\theta}^{N \to \tilde{N}})(v^{N \xrightarrow{\Delta} \tilde{N}})\right)^{\tilde{N} \xrightarrow{\Delta} N}, \tag{8}$$

where $\tilde{N} = N + 1$. We note that by this choice we lose the direct convertibility to the CNN-implementation as in general for even dimensions

$$K(\hat{\theta})(v) \neq F^{-1}(T^{-1}(\hat{\theta})\, Fv),$$

as the right hand side corresponds to zero-padding of the spectral coefficients regardless of the oddity of the dimensions. However, we can still convert the FNO-implementation to the CNN-implementation and vice versa, by adapting the magnitude of coefficients to the effects of the Nyquist splitting.

3.4 Interpolation Equivariance

Our motivation to perform the adaptation to even dimension as proposed in the preceding section, is that the resulting implementation of convolution is equivariant with respect to (real-valued) trigonometric interpolation.

Corollary 1. *For $\hat{\theta} \in \mathbb{C}^{I_M}_{sym}$, $v \in \mathbb{R}^{J_N}, M \leq N$ it holds true for any $L \geq M$ that*

$$K(\hat{\theta})(v^{N \xrightarrow{\Delta} L}) = \Big(K(\hat{\theta})(v)\Big)^{N \xrightarrow{\Delta} L}.$$

Proof. We first note that it holds for any choice of $M \leq N, L$ that

$$K(\hat{\theta})(v^{N \xrightarrow{\Delta} L}) = \Big(K(\hat{\theta}^{M \to \tilde{M} \to \tilde{L}})(v^{N \xrightarrow{\Delta} \tilde{N} \xrightarrow{\Delta} \tilde{L}})\Big)^{\tilde{L} \xrightarrow{\Delta} L},$$

where $\tilde{L} = L + (1 - L\%2)$, $\tilde{M} = M + (1 - M\%2)$, $\tilde{N} = N + (1 - N\%2)$ and % denotes the modulo operation. Therefore, we can assume L, M and N to be odd without loss of generality and thus $\tilde{L} = L, \tilde{M} = M$ and $\tilde{N} = N$. Regarding the discrete Fourier coefficients then reveals that

$$F(K(\hat{\theta})(v^{N \xrightarrow{\Delta} L}))_k = \begin{Bmatrix} \hat{\theta}_k\, F(v)_k & \text{for } k \in I_M, \\ 0 & \text{otherwise} \end{Bmatrix} = F((K(\hat{\theta})(v))^{N \xrightarrow{\Delta} L})_k.$$

Applying the inverse Fourier transform completes the proof.

4 Numerical Examples

In this section we compare the discussed implementations of convolution numerically in the context of image classification.[1] Here, the task is to assign a label from $s \in \mathbb{N}$ possible classes to a given image $v : [0, 1]^2 \to \mathbb{R}^{n_c}$, with $n_c \in \mathbb{N}$ denoting the number of color channels. Solving this task numerically requires discrete input images of the form $v^N = v|_{J_N/N} \in \mathbb{R}^{J_N \times n_c}$, where $N \in \mathbb{N}$ denotes the dimension. We note that since we consider a fixed function domain the dimension is proportional to the resolution. If we assume N to be fixed the network is a function $f_\theta : \mathbb{R}^{J_N \times n_c} \to \mathbb{R}^s$. Given a finite training set $D \subset \mathbb{R}^{J_N \times nc} \times \mathbb{R}^s$ we optimize the parameters θ by minimizing the empirical loss based on the cross-entropy [7, Ch. 3]. The networks we use for our experiments consist of several convolutional layers for feature extraction followed by one fully connected classification layer. To make all architectures applicable to inputs of any resolution, we insert an adaptive average pooling layer between the feature extractor and the classifier.

[1] Our code is available online: github.com/samirak98/FourierImaging.

4.1 Expressivity for Varying Kernel Sizes

In the first experiment (see Fig. 3) we train a CNN without any residual components on the FashionMNIST[2] dataset. The network has two convolutional layers with periodic padding and without striding, followed by an adaptive pooling layer and a linear classifier. Since we do not observe major performance changes on the test set for different kernel sizes, we conclude that on this data set the expressivity of the small kernel architectures is comparable to large kernel architectures. We then convert the convolutional layers of the CNNs with 3×3- and 28×28-kernels to FNO-layers, employing varying numbers of spectral parameters. Here, we observe decreasing performance with smaller spectral kernel sizes, indicating that the learned spatial kernels cannot be expressed well by fewer frequencies. However, in this example, training an FNO with the same structure almost closes this performance gap. This implies the existence of low frequency kernels with sufficient expressivity. We refer to [11] for a study on training with a spectral parametrization.

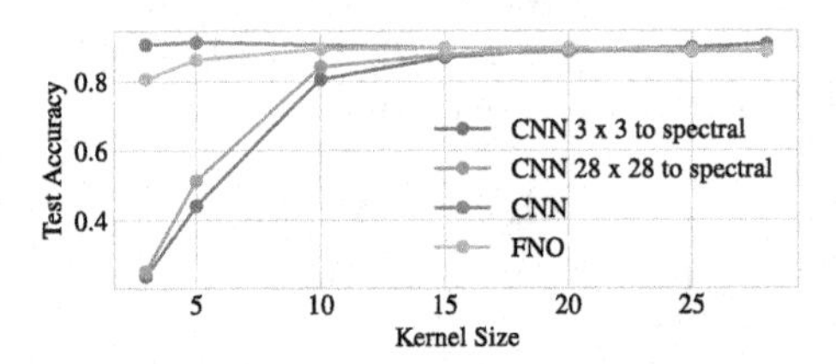

Fig. 3. Test accuracy of CNNs and FNOs for varying kernel sizes.

4.2 Resolution Invariance

In the second experiment, we investigate the resolution invariance of the different convolution implementations. In Fig. 4a we compare the accuracy on test data resized to different resolutions with trigonometric, or bilinear interpolation, respectively. Here, CNN refers to the conventional CNN-implementation with 5×5 kernel, where dimension mismatches are compensated for by spatial zero-padding of the kernel. FNO refers to the FNO-implementation, where the kernels are adapted to the input dimension by trigonometric interpolation. Additionally, we show the behavior of the CNN for inputs rescaled to the training resolution. Applying trigonometric interpolation before a convolutional layer can be interpreted as an FNO-layer with predetermined output dimensions.

The performance of the CNN varies drastically with the input dimension and peaks for the resolution it was trained on. This result is in accordance with the effect showcased in Fig. 1: Dimension adaption via spatial zero-padding modifies the locality of the kernel and consequently captures different features for different resolutions. While trigonometric interpolation performs best, we see

[2] This dataset consists of $60,000$ training and $10,000$ test 28×28 images (grayscale).

that the FNO adapts very well. In particular, the performance for higher input resolutions deters only slightly, which is not the case for the standard CNN.

Additionally (see Fig. 4b), we train a ResNet18 [9] on the Birds500 data set[3] with a reduced training size of 112×112. To regularize the generalization to different resolutions, especially for the FNO-implementation, we replace the standard striding operations by trigonometric downsampling. Compared to the first experiment it stands out that the FNO performs worse for inputs with resolutions below 112×112, but only slightly diminishes for higher resolutions. We attribute this fact to the dimension reduction operations in the architecture.

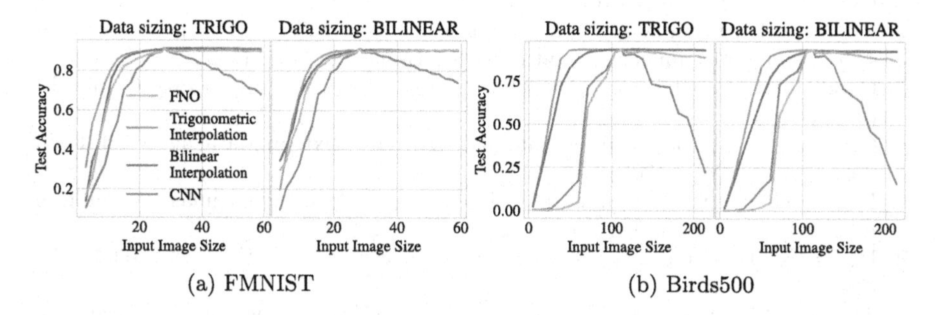

Fig. 4. Performance for different interpolation methods on test data that has been resized with the interpolation method denoted on top of the plots.

5 Conclusion and Outlook

In this work, we have studied the regularity of neural operators on Lebesgue spaces and investigated the effects of implementing convolutional layers in the sense of FNOs. Based on the theoretical derivation of the convertibility from standard CNNs to FNOs, our numerical experiments show that it is possible to convert a network that was trained with the standard CNN architecture into an FNO. By this, we could combine the benefits of both approaches: Enforced spatial locality with a small number of parameters during training and an implementation that generalizes well to higher input dimensions during the evaluation. However, we have seen that the trigonometric interpolation of inputs outperforms all other considered approaches. In future work, we want to investigate how the ideas of FNOs and trigonometric interpolation can be incorporated into image-to-image architectures like U-Nets as proposed in [21]. Additionally, we want to further explore the effects of training in the spectral domain, for example with respect to adversarial robustness.

[3] We employ a former version of the data set, which consists of 76, 262 RGB images for training and 2, 250 images for testing of size 224×224, where the task is to classify birds out of 450 possible classes.

References

1. Ambrosetti, A., Prodi, G.: A Primer of Nonlinear Analysis. Cambridge University Press, Cambridge (1993)
2. Briand, T.: Trigonometric polynomial interpolation of images. Image Process. On Line **9**, 291–316 (2019)
3. Cai, D., Chen, K., Qian, Y., Kämäräinen, J.K.: Convolutional low-resolution fine-grained classification. Pattern Recogn. Lett. **119**, 166–171 (2019)
4. Chi, L., Jiang, B., Mu, Y.: Fast Fourier convolution. Adv. Neural. Inf. Process. Syst. **33**, 4479–4488 (2020)
5. Fukushima, K.C.: Cognitron: a self-organizing multilayered neural network. Biol. Cybern. **20**, 121–136 (1975)
6. Goldberg, H., Kampowsky, W., Tröltzsch, F.: On NEMYTSKIJ operators in Lp-spaces of abstract functions. Math. Nachr. **155**(1), 127–140 (1992)
7. Goodfellow, I., Bengio, Y., Courville, A.: Deep Learning. MIT Press, Cambridge (2016)
8. Grafakos, L.: Classical Fourier Analysis. Graduate Texts in Mathematics, 3rd edn. Springer, New York (2014)
9. He, K., Zhang, X., Ren, S., Sun, J.: Deep residual learning for image recognition. In: Proceedings of the IEEE CVPR, pp. 770–778 (2016)
10. Hendrycks, D., Gimpel, K.: Gaussian error linear units (GELUs). arXiv:1606.08415 (2016)
11. Johnny, W., Brigido, H., Ladeira, M., Souza, J.C.F.: Fourier neural operator for image classification. In: 2022 17th Iberian Conference on Information Systems and Technologies (CISTI), pp. 1–6 (2022)
12. Kovachki, N.B., Lanthaler, S., Mishra, S.: On universal approximation and error bounds for Fourier neural operators. J. Mach. Learn. Res. **22**(1), 13237–13312 (2022)
13. Kovachki, N.B., et al.: Neural operator: Learning maps between function spaces. arXiv:2108.08481 (2021)
14. Koziarski, M., Cyganek, B.: Impact of low resolution on image recognition with deep neural networks: an experimental study. Int. J. Appl. Math. Comput. Sci. **28**(4), 735–744 (2018)
15. Li, Z., Kovachki, N.B., Azizzadenesheli, K., Liu, B., Bhattacharya, K., Stuart, A.M., Anandkumar, A.: Fourier neural operator for parametric partial differential equations. In: 9th International Conference on Learning Representations (ICLR) (2021)
16. Li, Z., et al.: Physics-informed neural operator for learning partial differential equations. arXiv preprint arXiv:2111.03794 (2021)
17. Peng, X., Hoffman, J., Stella, X.Y., Saenko, K.: Fine-to-coarse knowledge transfer for low-res image classification. In: 2016 IEEE International Conference on Image Processing (ICIP), pp. 3683–3687. IEEE (2016)
18. Piosenka, G.: Birds 500 - species image classification (2021). https://www.kaggle.com/datasets/gpiosenka/100-bird-species
19. Rao, Y., Zhao, W., Zhu, Z., Lu, J., Zhou, J.: Global filter networks for image classification. Adv. Neural. Inf. Process. Syst. **34**, 980–993 (2021)
20. Remmert, R.: Theory of Complex Functions. Springer, New York (1991)
21. Ronneberger, O., Fischer, P., Brox, T.: U-Net: convolutional networks for biomedical image segmentation. In: Navab, N., Hornegger, J., Wells, W.M., Frangi, A.F. (eds.) MICCAI 2015. LNCS, vol. 9351, pp. 234–241. Springer, Cham (2015). https://doi.org/10.1007/978-3-319-24574-4_28

22. Tröltzsch, F.: Optimal Control of Partial Differential Equations: Theory, Methods, and Applications, Graduate Studies in Mathematics, vol. 112. American Mathematical Society, Providence, Rhode Island (2010)
23. Vaĭnberg, M.M.: Variational method and method of monotone operators in the theory of nonlinear equations. No. 22090, John Wiley & Sons, Hoboken (1974)
24. Xiao, H., Rasul, K., Vollgraf, R.: Fashion-mnist: a novel image dataset for benchmarking machine learning algorithms. arXiv:1708.07747 (2017)
25. Zhou, M., et al.: Deep Fourier up-sampling. arxiv:2210.05171 (2022)

Graph Laplacian for Semi-supervised Learning

Or Streicher and Guy Gilboa(✉)

Technion - Israel Institute of Technology, 32000 Haifa, Israel
guy.gilboa@ee.technion.ac.il, orr.shtr@campus.technion.ac.il

Abstract. Semi-supervised learning is highly useful in common scenarios where labeled data is scarce but unlabeled data is abundant. The graph (or nonlocal) Laplacian is a fundamental smoothing operator for solving various learning tasks. For unsupervised clustering, a spectral embedding is often used, based on graph-Laplacian eigenvectors. For semi-supervised problems, the common approach is to solve a constrained optimization problem, regularized by a Dirichlet energy, based on the graph-Laplacian. However, as supervision decreases, Dirichlet optimization becomes suboptimal. We therefore would like to obtain a smooth transition between unsupervised clustering and low-supervised graph-based classification.

In this paper, we propose a new type of graph-Laplacian which is adapted for *Semi-Supervised Learning* (SSL) problems. It is based on both density and contrastive measures and allows the encoding of the labeled data directly in the operator. Thus, we can perform successfully semi-supervised learning using spectral clustering. The benefits of our approach are illustrated for several SSL problems.

Keywords: Graph Representation · Semi-Supervise Learning · Nonlocal Laplacian · Spectral Clustering

1 Introduction

Labeling information is a major challenge in modern learning techniques, which in many cases can be a long and expensive process. A possible solution to this problem is Semi-Supervised Learning (SSL). SSL methods can be thought of as the halfway between supervised and unsupervised learning. It uses large amounts of unlabeled data and a limited amount of labeled data, to improve the learning model. SSL techniques are usually used when one cannot employ supervised learning algorithms. The use of supervised learning when limited labels are available may result in a lack of generalization and overfitting. Those limited known labels, however, can significantly improve performance compared to unsupervised learning algorithms [11]. Intuitively, the purpose of SSL is to *generalize* the known labels to the unlabeled samples by an appropriate smoothing operator. The graph-Laplacian has shown to be highly effective for this purpose.

L. Calatroni et al. (Eds.): SSVM 2023, LNCS 14009, pp. 250–262, 2023.
https://doi.org/10.1007/978-3-031-31975-4_19

In this paper, we focus on *Graph-based methods* which are well-studied classical techniques. We note that the insights presented here can be further used by deep learning methods with spectral-graph modules, as in [1,6,18,21]. In classical graph methods, a weighted graph is constructed based on the affinities between data instances. These affinities are usually computed using a metric of the features representing each instance. The vast majority of graph-based learning methods use the graph-Laplacian as the smoothing operator for generalization. More advanced nonlinear methods use p-Laplacian operators [5,8]. In this study we limit the scope to the linear case (or quadratic energy), noting that our proposed operators can be further generalized. Data processing based on the Laplacian has shown to be effective for a wide range of problems including clustering [4,16,22], classification [9], segmentation [19], dimensionality reduction [2,7,17] and more. For the SSL setting, most graph-based learning techniques use the properties of the graph-Laplacian operator to define an optimization problem. The labeled information can be inserted as problem constraints, see e.g. [3,9,12–14,20].

In this paper, we propose a different approach to consider the labeled information, by inserting it into the affinity measure that defines the connectivity between the nodes of the graph. We first examine the work of [20] on the weighted nonlocal Laplacian. In this work, the density of the labeled data is essentially increased. This improves the solution of the constrained optimization problem. We found that it has a marginal effect on the spectral embedding. Based on contrastive arguments, we propose to increase connections between labeled and unlabeled data and to increase (remove) connections between labeled data of the same (different) clusters. This yields a considerably improved spectral embedding.

Our proposed method retains the following main qualities: 1) **Interpolation between unsupervised and semi-supervised learning.** The suggested approach enables learning for a changing range of labeled information. 2) **Low-label regime performance.** The proposed method was found most advantageous in the low-label regime, compared to competitive techniques. 3) **Different analysis tools.** Wide variety of analysis tools can be used for solving SSL problems, including spectral and functional analysis. We illustrate the advantages of using this new definition on toy examples and on real data sets.

2 Setting and Notation

Let $X = \{x_i\}_{i=1}^n$ be a finite set of instances in $\mathbb{R}^d$. These instances are represented as nodes on an undirected weighted graph $G = (V, E, W)$, where V is the vertices set, E is the edges set and W is the adjacency matrix. The adjacency matrix is symmetric and is usually defined by a distance measure between the nodes. For example, a common choice is a Gaussian kernel with Euclidean distance,

$$W_{ij} = \exp\left(-\frac{\|x_i - x_j\|_2^2}{2\sigma^2}\right), \tag{1}$$

where σ is a soft-threshold parameter.

The degree matrix D is a diagonal matrix where D_{ii} is the degree of the i-th vertex, i.e.,

$$D_{ii} = \sum_j W_{ij}. \tag{2}$$

The graph-Laplacian operator is defined by,

$$L := D - W. \tag{3}$$

The graph-Laplacian is a symmetric, positive semi-definite matrix, i.e., $\forall f \in \mathbb{R}^n$, $f^T Lf \geq 0$. For each vector $f \in \mathbb{R}^n$ it holds that

$$Lf \in \mathbb{R}^n \ , \ Lf(j) = \sum_{i=1}^{n} W_{ij}(f_j - f_i), \tag{4}$$

$$f^T Lf = \sum_{i=1}^{n} \sum_{j=1}^{n} W_{ij}(f_i - f_j)^2. \tag{5}$$

The eigenvalues of L are real and non-negative and sorted in ascending order $\lambda_1 \leq \lambda_2 \leq ... \leq \lambda_n$. The corresponding eigenvectors form an orthogonal basis, denoted by $u_1, u_2..., u_n$. The sample x_i can be represented in the spectral embedding space as the ith row of the matrix $U = \begin{bmatrix} u_1 \cdots u_K \end{bmatrix} \in \mathbb{R}^{n \times K}$, denoted as φ_i. More formally, the spectral embedding of an instance x_i can be formulated as

$$x_i \longmapsto \varphi_i = [u_1(i), u_2(i), ..., u_K(i)] \in \mathbb{R}^K, \tag{6}$$

where in most cases $K \ll d$.

For the SSL setting, let us define a discrete function $f \in \mathbb{R}^n$ over X and $S \subset X$, such that $|S| = m, m \leq n$, be a subset of X on which the values of f are known, i.e., $f(x) = g(x), \forall x \in S$, for a given function g. The purpose of the SSL model is to find the values of f of all data-points in X constrained by the values of the set S.

The main SSL problem this work focuses on is clustering. To evaluate the clustering performance we examined two common measures. The first one is *Normalized mutual information* (NMI) which is defined as,

$$NMI(c, \hat{c}) = \frac{I(c, \hat{c})}{\max\{H(c), H(\hat{c})\}}, \tag{7}$$

where $I(c, \hat{c})$ is the mutual information between the true labels c and the clustering result $\hat{c}$ and $H(\cdot)$ denotes entropy. We remind the definitions of entropy for a random variable U with distribution p_U, $H(U) := E[-\log p_U]$, where E is the expected value. For two random variables U, V with conditional probability $p_{V|U}$ the conditional entropy of V given U is $H(V|U) := E[-\log p_{V|U}]$. Mutual information measures the dependence between two random variable and admits the following identities $I(U, V) = H(U) - H(U|V) = H(V) - H(V|U)$. It is non-negative and equals zero when U and V are independent.

The second measure is *Unsupervised Clustering Accuracy* (ACC) which is defined as,

$$ACC(c, \hat{c}) = \frac{1}{n} \max_{\pi \in \Pi} \sum_{i=1}^{n} \mathbb{1}\{c_i = \pi(\hat{c}_i)\}, \tag{8}$$

where Π is the set of possible permutations of the clustering results. To choose the optimal permutation π we used the Kuhn-Munkres algorithm [15]. Both indicators are in the range $[0, 1]$, where high values indicate a better correspondence between the clustering result and the true labels.

3 Graph-Laplacian for SSL

3.1 Motivation

A common approach to solve SSL problems, based on the graph-Laplacian (GL), is to solve the Dirichlet problem

$$\min_f \frac{1}{2} \sum_{x_i, x_j \in X} W_{ij}(f(x_i) - f(x_j))^2, \tag{9}$$

$$\text{s.t. } f(x) = g(x), \; x \in S.$$

A solution obeys

$$Lf(x_i) = \sum_{x_j \in X} W_{ij}(f(x_i) - f(x_j)) = 0, x_i \in X \setminus S \tag{10}$$

$$f(x) = g(x), x \in S.$$

Note that Eq. (10) can be represented in matrix form. First, we define a constraints vector $b \in \mathbb{R}^n$ and a mask $M \in \mathbb{R}^{n \times n}$ such that

$$b_i = \begin{cases} g(x_i) & \text{if } x_i \in S \\ 0 & \text{if } x_i \in X \setminus S \end{cases}, \; M_{ij} = \begin{cases} \frac{1}{L_{ii}} & \text{if } x_i \in S \text{ and } i = j \\ 0 & \text{if } x_i \in S \text{ and } i \neq j \\ 1 & \text{if } x_i \in X \setminus S, \forall j \end{cases}, \tag{11}$$

where L_{ii} is the i-th element of the diagonal of L. Now, Eq. (10) can be introduced in matrix from by

$$(M \circ L)f = b, \tag{12}$$

where $\circ$ denotes element-wise multiplication.

A main problem with GL, as shown in [20], is that for a low sample rate of the labeled set, $|S|/|X|$, the solution is not continuous at the sample points. Thus the GL solution does not interpolate well the constraint values. In [20] the authors suggested solving the discontinuity problem by using Weighted Nonlocal Laplacian (WNLL), assigning a greater weight to the labeled set S compared to the unlabeled set $X \setminus S$. Formally, the optimization problem of WNLL is given by

$$\min_f \sum_{x_i \in X \setminus S} \sum_{x_j \in X} W_{ij}(f(x_i) - f(x_j))^2 + \mu \sum_{x_i \in S} \sum_{x_j \in X} W_{ij}(f(x_i) - f(x_j))^2 \quad (13)$$

$$\text{s.t. } f(x) = g(x), x \in S,$$

where μ is a regularization parameter. It was suggested to set

$$\mu = |X|/|S|, \quad (14)$$

the inverse of the sample rate. This can be interpreted as increasing the density (or measure) of the labeled instances. The solution of Eq. (13) is given by solving the following linear system,

$$\sum_{x_j \in X} (W_{ij} + W_{ji})(f(x_i) - f(x_j)) + (\mu - 1) \sum_{x_j \in S} W_{ji}(f(x_i) - f(x_j)) = 0, x_i \in X \setminus S$$

$$f(x) = g(x), x \in S. \quad (15)$$

Similarly to Eq. (12), one can define Eq. (15) in matrix form. Let us introduce the linear system as follows,

$$\sum_{x_j \in X} (W_{ij} + W_{ji})(f(x_i) - f(x_j)) + (\mu - 1) \sum_{x_j \in X} W_{ij}^{labeled}(f(x_i) - f(x_j)) = 0, x_i \in X \setminus S$$

$$f(x) = g(x), x \in S, \quad (16)$$

such that,

$$W_{ij}^{labeled} = \begin{cases} W_{ij} & x_i \in X \setminus S, x_j \in S \text{ or } x_i \in S, x_j \in X \setminus S \\ 0 & \text{otherwise} \end{cases} \quad (17)$$

or equivalently,

$$\sum_{x_j \in X} \left(W_{ij} + W_{ji} + (\mu - 1)W_{ij}^{labeled}\right)\left(f(x_i) - f(x_j)\right) = 0, x_i \in X \setminus S \quad (18)$$

$$f(x) = g(x), x \in S.$$

Now we can define,

$$\sum_{x_j \in X} [W_{WNLL}]_{ji}(f(x_i) - f(x_j)) = 0, x_i \in X \setminus S \quad (19)$$

$$f(x) = g(x), x \in S,$$

where

$$W_{WNLL} = 2W + (\mu - 1)W^{labeled}. \quad (20)$$

Based on W_{WNLL} one can define L_{WNLL}, following Eq. (3), such that Eq. (15) is equivalent to

$$(M \circ L_{WNLL})f = b. \tag{21}$$

Inspired by Eq. (20), we would like to define an affinity matrix that takes into account the known labels, such that it also distinguishes between labeled samples from the same and from different clusters.

3.2 Semi-supervised Laplacian Definition

The classical definition of the graph-Laplacian, Eq. (3), is based on the data features in an unsupervised manner. In this section, we introduce a novel definition of the graph affinity matrix for SSL problems. That means, the affinity measure takes into account not only the feature vectors but also the known information about the labels of a subset S of V. The proposed definition is intended to improve performance for SSL clustering problems. For K clusters, we denote by S_k the set of labeled nodes belonging to the k-th cluster, such that $S = \cup_{k=1}^{K} S_k$.

We suggest the following affinity measure

$$W_{SSL} = 2W + \alpha W^{labeled}, \tag{22}$$

where W is the known unsupervised affinity matrix, α is a scalar parameter, we set

$$\alpha = \mu - 1, \tag{23}$$

with μ as in (14) and $W^{labeled}$ is defined as follows,

$$W_{ij}^{labeled} = \begin{cases} \max(W) & x_i, x_j \in S_k \ , \forall k \in \{1, ..K\} \\ -\frac{2}{\alpha} W_{ij} & x_i \in S_k, x_j \in S_l \ , \forall k, l \in \{1, ..., K | k \neq l\} \\ W_{ij} & x_i \in S, x_j \in X \setminus S \text{ or } x_i \in X \setminus S, x_j \in S \\ 0 & x_i, x_j \in X \setminus S \end{cases}. \tag{24}$$

It can be observed that according to this definition, the connection between labeled nodes belonging to the same cluster is given the highest weight ($\max(W)$ is the maximum value of the unsupervised affinity matrix). This strong affinity ensures the nodes are well connected inducing high smoothness of the spectral solution at these regions. Labeled nodes of different clusters are disconnected. This increases the separation of these nodes, reducing smoothness and avoiding unnecessary regularity between nodes belonging to separate clusters. Note that although in the second line of (24) the weights are negative, the final respective weights of W_{SLL} are all non-negative. In addition, as in Eq. (17), we reinforce edges between labeled nodes and unlabeled nodes. Now we can define the SSL graph-Laplacian as follows,

$$L_{SSL} = D_{SSL} - W_{SSL}, \tag{25}$$

where W_{SSL} is defined in Eq. (22) and D_{SSL} is its associated degree matrix (see Eq. (2)). In a similar manner to Eq. (12), one can solve the following problem

$$(M \circ L_{SSL})f = b. \tag{26}$$

4 Analysis of L_{SSL}

In this section, we analyze the characteristics of L_{SSL} for different scenarios. First, we analyze the influence of each of the components in Eq. (24) on the spectral embedding. Let us define the following affinity matrices,

$$W_{ij}^1 = \begin{cases} \max(W) & x_i, x_j \in S_k \ , \forall k \in \{1, ..K\} \\ 0 & \text{otherwise} \end{cases} \tag{27}$$

$$W_{ij}^2 = \begin{cases} -\frac{2}{\alpha} W_{ij} & x_i \in S_k, x_j \in S_l \ , \forall k, l \in \{1, ..K | k \neq l\} \\ 0 & \text{otherwise} \end{cases} \tag{28}$$

$$W_{ij}^3 = \begin{cases} W_{ij} & x_i \in S, x_j \in X \setminus S \text{ or } x_i \in X \setminus S, x_j \in S \\ 0 & \text{otherwise} \end{cases} \tag{29}$$

Note that W^1 and W^2 can be interpreted as *Contrastive Affinities*, following insights of contrastive learning [10]. The contrastive paradigm aims at creating an embedding where instances of the same cluster are very close (the role of W^1), whereas instances of different clusters are distinctly separated (the role of W^2). On the other hand, W^3 can be thought of as *Density Affinity.* Its purpose is to increase the density of the graph in the vicinity of labeled nodes.

We would like to use those affinity matrices instead of $W^{labeled}$ in Eq. (22) and examine the resulting spectral representation based on the graph-Laplacians $\{L_{SSL}^i\}_{i=1}^3$ defined by $\{W_{SSL}^i\}_{i=1}^3$ such that

$$W_{SSL}^i = 2W + \alpha W^i. \tag{30}$$

The spectral embedding is examined for the 3-Moons dataset containing 900 nodes, of which 30 are labeled, as can be seen in Fig. 1.

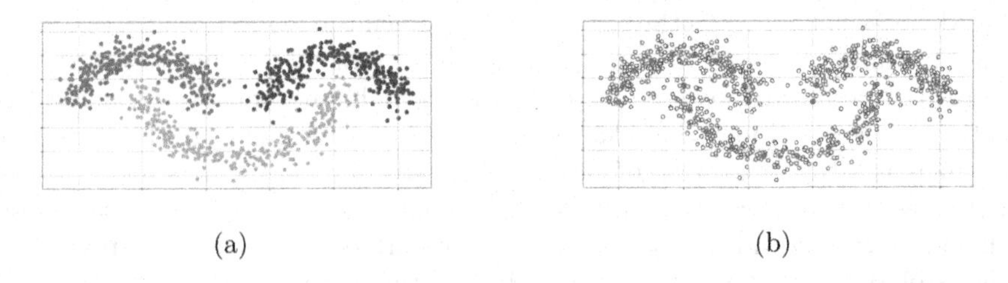

(a) (b)

Fig. 1. SSL Laplacian Illustration dataset. The 3-Moons dataset containing 900 nodes appears in Fig. 1a. The labeled nodes are shown in Fig. 1b.

We examine the spectral representation of the data, Eq. (6), spanned by the first two non-trivial leading eigenvectors of the unsupervised Laplacian L, $\{L_{SSL}^i\}_{i=1}^3$ and L_{SSL}. The spectral embedding for each case is shown in Fig. 2. We can observe that the spectral representation obtained by L_{SSL} produces the clearest division into clusters. Nodes of the same cluster ("moon") are grouped together, whereas nodes of different clusters are further apart. An interesting finding in this experiment is that the main effect on the spectral embedding is caused by the contrastive affinities, especially of W^1. We will see in the experimental part this trend is valid also for more complex data. Indeed, the Laplacian based on W^3, which is equivalent to L_{WNLL}, has virtually no contribution to the spectral embedding.

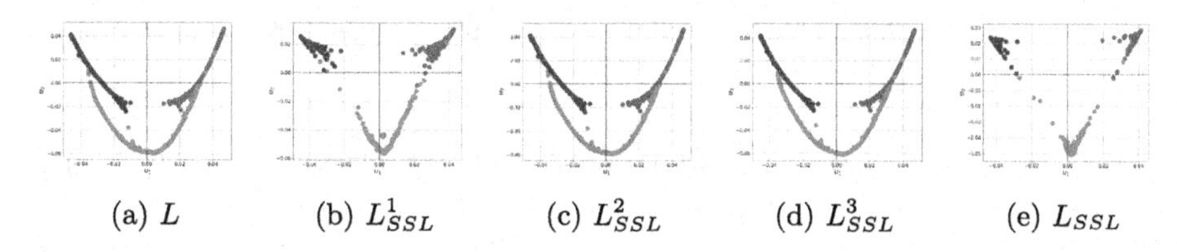

(a) L (b) L^1_{SSL} (c) L^2_{SSL} (d) L^3_{SSL} (e) L_{SSL}

Fig. 2. Spectral Embedding Illustration. The spectral embedding obtained on the 3-Moons dataset for L, $\{L^i_{SSL}\}^3_{i=1}$ and L_{SSL}.

Next, we would like to examine spectral processing compared to constrained optimization, both based on L_{SSL}. The performance of both approaches is tested over the 2-Moons dataset, which includes 1000 instances. Each node of the graph is defined by its Euclidean position. We analyze the spectral properties of the graph and the solution to the Dirichlet interpolation problem. The results are examined for different labeled sets S. In the spectral case, we find for each node of the graph its corresponding value according to the first non-trivial eigenvector of the graph-Laplacian, defined using W_{WNLL} or W_{SSL}. In order to find the solution for the Dirichlet problem, we set the value 1 over the labeled nodes of the first moon and -1 over the labeled nodes of the second moon. Then we find the solution to the interpolation problem using L, L_{WNLL} or L_{SSL}. To make a division into clusters, we perform K-Means over the resulting solution for each case. The obtained results are shown in Fig. 3.

Analyzing the results, it can be observed that when the amount of labeled samples is extremely small, the Dirichlet problem may not generalize that well the labels to the unlabeled data. This is especially true when the labels are not located near the cluster centers, as can be seen for the sets S_2 and S_3 (Fig. 3 bottom two rows). For the Dirichlet problem, the same performance is achieved for L_{WNLL} and for L_{SSL}. This is also valid in larger data sets. In more complex scenarios the advantages of using the above Laplacians are clear, compared to standard L. In this toy example, the differences are minor.

Intermediate conclusions for these toy examples are that in some cases spectral analysis of the data is preferred. In addition, the suggested definition of L_{SSL} allows us to get good performance for SSL problems both in the spectral case and for solving the Dirichlet problem. The reason for this is that L_{SSL}

includes the contrastive information, which is essential mainly for the spectral case, and the density information which is more significant in the optimization case. We will now test this in a more comprehensive manner.

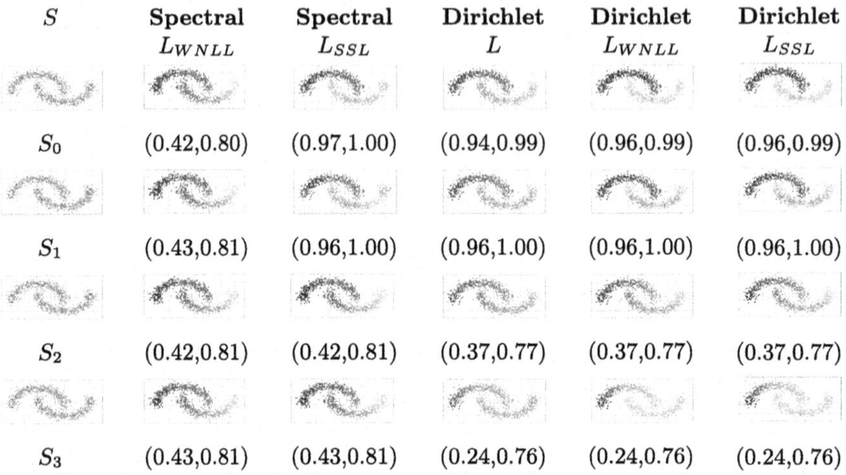

S	**Spectral** L_{WNLL}	**Spectral** L_{SSL}	**Dirichlet** L	**Dirichlet** L_{WNLL}	**Dirichlet** L_{SSL}
S_0	(0.42,0.80)	(0.97,1.00)	(0.94,0.99)	(0.96,0.99)	(0.96,0.99)
S_1	(0.43,0.81)	(0.96,1.00)	(0.96,1.00)	(0.96,1.00)	(0.96,1.00)
S_2	(0.42,0.81)	(0.42,0.81)	(0.37,0.77)	(0.37,0.77)	(0.37,0.77)
S_3	(0.43,0.81)	(0.43,0.81)	(0.24,0.76)	(0.24,0.76)	(0.24,0.76)

Fig. 3. SSL solutions of the 2 Moons dataset. The first column shows different labeled sets S. The 2nd and 3rd columns show the spectral clustering result for L_{WNLL} and L_{SSL}, respectively. The 4th, 5th and 6th columns show the Dirichlet problem result for L, L_{WNLL} and L_{SSL}, respectively. The clustering measures (NMI, ACC) are presented below each figure. Spectral L_{SSL} performs well in all configurations.

5 Experimental SSL Clustering Results

In this section, we examine the performance of the different definitions of the graph-Laplacian for the clustering problem. To perform clustering in the semi-supervised case, we examine two different methods. The first one is **Spectral Clustering** which is based on the division of the data into clusters by performing K-Means over the spectral embedding of the data, Eq. (6). The second method is based on **Dirichlet-form Clustering**. To adapt the Dirichlet interpolation problem to multiple clusters, we use the algorithm suggested in [20].

5.1 2-Moons Clustering

In this section, we present the statistical clustering performance for the 2-Moon dataset when the graph includes 500 nodes. We perform two experiments. In the first one, shown in Fig. 4, we examine the effect of changing the standard deviation of the noise of the data (that is, the deviation of each point from the position on the semicircle that defines the moon). In this case, 10 labeled nodes from each class are randomly defined. In the second experiment, we examine the effect of changing the size of the labeled set $|S|$ (for fixed noise standard deviation set to 0.1). The results of this experiment are summarized in Fig. 5. In both experiments, white Gaussian noise is used. The experiments show statistics of 100 trials, where a bold line represents the mean value (of NMI or ACC) and the lighter regions around each line depict the standard deviation of the measure.

The main conclusions from those experiments are that in the spectral case the results obtained for L_{SSL} are much better compared to the other Laplacians, where L_{WNLL} performance degenerates to the unsupervised case. For the Dirichlet problem, the performance of L_{SSL} and L_{WNLL} is similar and better than using the standard L, especially for the difficult cases, where the noise is significant and the amount of labeled information is small. We can conclude that the definition of L_{SSL} allows to get the best clustering performance when using the Dirichlet problem and especially for spectral analysis of the graph.

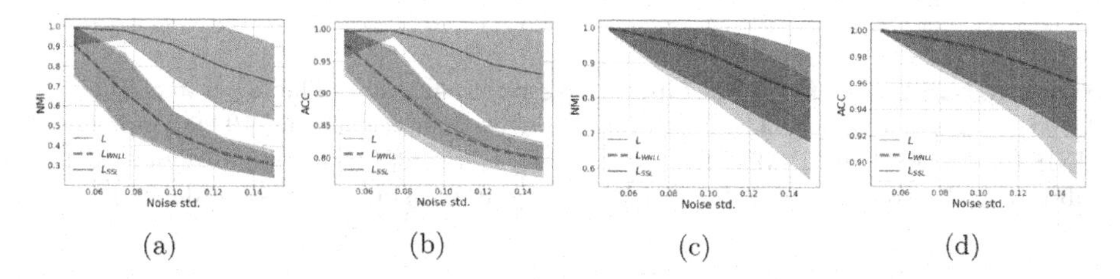

Fig. 4. 2 Moons clustering for different noise parameter. NMI and ACC measures over 100 different labeled set samples for different noise standard deviation. Figures 4a–4b are for Spectral Clustering. Figures 4c–4d are for Dirichlet Clustering.

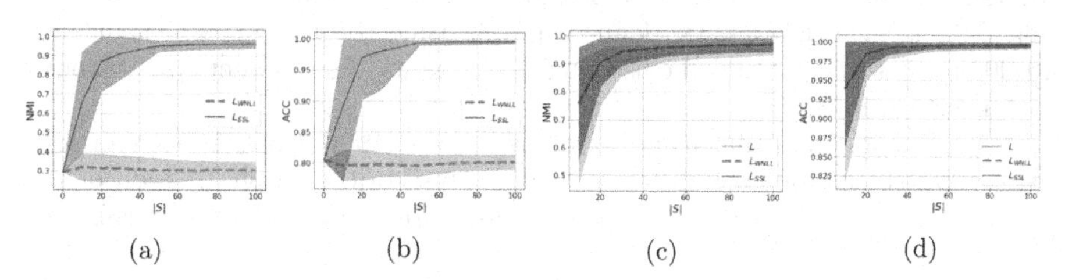

Fig. 5. 2 Moons clustering for different labeled set size. NMI and ACC measures over 100 different labeled set samples for different labeled set sizes. Figures 5a–5b are for Spectral Clustering. Figures 5c–5d are for Dirichlet Clustering.

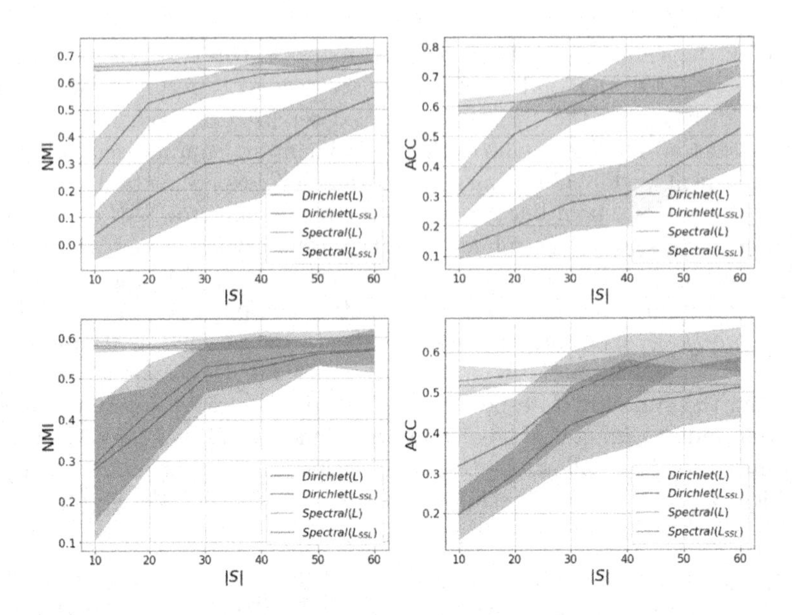

Fig. 6. MNIST and F-MIST Clustering performance. The mean and standard deviation of NMI and ACC of 10 different experiments over the 10,000 samples of MNIST (Top row) and F-MNIST (bottom row) test sets, for different size of a labeled subset $|S|$. We note that $Spectral(L)$ has no variability since it does not depend on S.

5.2 MNIST and F-MNIST

Now we examine the clustering performance over the MNIST and Fashion-MNIST datasets. Both of these well-known datasets include 28×28 gray-scale images. For each dataset, we define a graph using the test set which includes $10,000$ images. The obtained clustering performance, for different size of labeled sets, is shown in Fig. 6.

It can be observed that using L_{SSL} yields better performance, both for solving the Dirichlet problem and for spectral analysis of the graph. In addition, for small labeled set $|S|$, the performance obtained for spectral clustering is better.

6 Conclusions

In this paper, we propose a new definition for the graph-Laplacian designed to improve performance for SSL problems. The novel SSL Laplacian, which incorporates both contrastive and density affinities, yields improved spectral clustering and can be used also in constrained optimization problems. The proposed operator allows smooth interpolating between the unsupervised and the semi-supervised cases. The advantages are most prominent for an extremely low amount of labels or noisy data. In this work we have considered only the linear case, however, p-Laplacians may also be modified in a similar manner.

Acknowledgements. We acknowledge support by grant agreement No. 777826 (NoMADS), by the Israel Science Foundation (Grant No. 534/19), by the Ministry of Science and Technology (Grant No. 5074/22) and by the Ollendorff Minerva Center.

References

1. Aviles-Rivero, A.I., Sellars, P., Schönlieb, C.B., Papadakis, N.: Graphxcovid: explainable deep graph diffusion pseudo-labelling for identifying COVID-19 on chest x-rays. Pattern Recogn. **122**, 108274 (2022)
2. Belkin, M., Niyogi, P.: Laplacian eigenmaps for dimensionality reduction and data representation. Neural Comput. **15**(6), 1373–1396 (2003)
3. Belkin, M., Niyogi, P., Sindhwani, V.: Manifold regularization: a geometric framework for learning from labeled and unlabeled examples. J. Mach. Learn. Res. **7**(11) (2006)
4. Bresson, X., Laurent, T., Uminsky, D., Von Brecht, J.H.: Multiclass total variation clustering. arXiv preprint arXiv:1306.1185 (2013)
5. Calder, J.: The game theoretic p-laplacian and semi-supervised learning with few labels. Nonlinearity **32**(1), 301 (2018)
6. Chen, Z., Li, Y., Cheng, X.: Specnet2: orthogonalization-free spectral embedding by neural networks. arXiv preprint arXiv:2206.06644 (2022)
7. Coifman, R.R., Lafon, S.: Diffusion maps. Appl. Comput. Harmon. Anal. **21**(1), 5–30 (2006)
8. Elmoataz, A., Desquesnes, X., Toutain, M.: On the game p-laplacian on weighted graphs with applications in image processing and data clustering. Eur. J. Appl. Math. **28**(6), 922–948 (2017)
9. Garcia-Cardona, C., Merkurjev, E., Bertozzi, A.L., Flenner, A., Percus, A.G.: Multiclass data segmentation using diffuse interface methods on graphs. IEEE Trans. Pattern Anal. Mach. Intell. **36**(8), 1600–1613 (2014)
10. Hadsell, R., Chopra, S., LeCun, Y.: Dimensionality reduction by learning an invariant mapping. In: 2006 IEEE Computer Society Conference on Computer Vision and Pattern Recognition (CVPR 2006), vol. 2, pp. 1735–1742. IEEE (2006)
11. Hearty, J.: Advanced Machine Learning with Python. Packt Publishing (2016)
12. Joachims, T.: Transductive learning via spectral graph partitioning. In: Proceedings of the 20th International Conference on Machine Learning (ICML 2003), pp. 290–297 (2003)
13. Liu, W., He, J., Chang, S.F.: Large graph construction for scalable semi-supervised learning. In: ICML (2010)
14. Mao, Q., Tsang, I.W.: Parameter-free spectral kernel learning. arXiv preprint arXiv:1203.3495 (2012)
15. Munkres, J.: Algorithms for the assignment and transportation problems. J. Soc. Ind. Appl. Math. **5**(1), 32–38 (1957)
16. Ng, A., Jordan, M., Weiss, Y.: On spectral clustering: analysis and an algorithm. In: Advances in Neural Information Processing Systems, vol. 14 (2001)
17. Roweis, S.T., Saul, L.K.: Nonlinear dimensionality reduction by locally linear embedding. Science **290**(5500), 2323–2326 (2000)
18. Shaham, U., Stanton, K., Li, H., Nadler, B., Basri, R., Kluger, Y.: Spectralnet: spectral clustering using deep neural networks. In: Proceedings of the 6th International Conference on Learning Representations (2018)
19. Shi, J., Malik, J.: Normalized cuts and image segmentation. IEEE Trans. Pattern Anal. Mach. Intell. **22**(8), 888–905 (2000)

20. Shi, Z., Osher, S., Zhu, W.: Weighted nonlocal laplacian on interpolation from sparse data. J. Sci. Comput. **73**(2), 1164–1177 (2017)
21. Streicher, O., Cohen, I., Gilboa, G.: Basis: batch aligned spectral embedding space. arXiv preprint arXiv:2211.16960 (2022)
22. Zelnik-Manor, L., Perona, P.: Self-tuning spectral clustering. In: Advances in Neural Information Processing Systems, vol. 17 (2004)

A Geometrically Aware Auto-Encoder for Multi-texture Synthesis

Pierrick Chatillon[1,2](✉), Yann Gousseau[1], and Sidonie Lefebvre[2]

[1] LTCI, Télécom Paris, IP Paris, 19 place Marguerite Perey, 91120 Palaiseau, France
{pierrick.chatillon,Yann.Gousseau}@telecom-paris.fr
[2] DOTA & LMA2S, ONERA, Université Paris Saclay, 91123 Palaiseau, France
{pierrick.chatillon,sidonie.lefebvre}@onera.fr

Abstract. We propose an auto-encoder architecture for multi-texture synthesis. The approach relies on both a compact encoder accounting for second order neural statistics and a generator incorporating adaptive periodic content. Images are embedded in a compact and geometrically consistent latent space, where the texture representation and its spatial organisation are disentangled. Texture synthesis and interpolation tasks can be performed directly from these latent codes. Our experiments demonstrate that our model outperforms state-of-the-art feed-forward methods in terms of visual quality and various texture related metrics. The code is available online.

Keywords: texture synthesis · auto-encoder · scale/orientation models

1 Introduction

Texture synthesis, that is the process of synthesizing new image samples from a given exemplar, has experienced a clear breakthrough with the work of Gatys et al. [4]. Inheriting ideas from wavelet-based methods, it was first proposed in this work to synthesize textures by constraining second order statistics of the responses to classical classification neural networks. While this approach outperformed existing ones in term of visual fidelity, it relies on a relatively heavy optimization procedure that has to be carried out for each new exemplar. Therefore, methods have been proposed to perform synthesis using generative neural networks, such as networks learned for each new exemplar [17] or GANs [10]. More recently, such approaches have been extended to perform the synthesis from a whole set of textures, by using adaptive normalizations of generative networks [12,13].

The present work is in the continuation of these approaches and introduces an auto-encoder architecture enabling one to synthesize arbitrary textures from a compact latent representation. The autoencoder architecture avoids the training instability inherent to GAN-based approaches and is naturally suited to synthesis and editing tasks. The encoding step yields a latent code adapted to

Code available at: https://github.com/PierrickCh/TextureAutoEncoder.

L. Calatroni et al. (Eds.): SSVM 2023, LNCS 14009, pp. 263–275, 2023.
https://doi.org/10.1007/978-3-031-31975-4_20

an input texture, from which arbitrary samples can be synthesized. The resulting latent space is spatially agnostic, or in other words the network treats in the same way different translations of a given input, which is made possible by carefully designing both the encoder and the generator. Moreover, the proposed architecture deals with both stochastic and structured, periodic-like textures. This is achieved thanks to the inclusion of sine waves in the design of the generator. Contrarily to what is done in the literature, we deal with these periodic components in a way that allows arbitrary rotations and scalings of the exemplar textures, a property that is clearly desirable in order to achieve a generic representation of textures. To the best of our knowledge, the resulting architecture is the first approach proposing a generic latent representation for textures, including an encoder to embed texture images, from which new samples can be synthesized or in which editing operations can be performed.

2 Related Work

2.1 Periodic Texture Synthesis

Synthesizing periodic textures and textures with long range dependency is a challenging aspect of texture synthesis. In order to enable the synthesis of multiple stationary periodic textures, the authors of [2] improved the GAN-based synthesis method from [10] by providing additional long range periodic information to the generator. This concept has proven efficient and was reused in [13], coupled with the structure of a StyleGAN2 [11] network. Another approach is proposed in [5] by imposing some constraints on the Fourier power spectrum.

2.2 Universality and Latent Representations of Textures

Beyond the classical problem of exemplar-based texture synthesis, where the goal is to generate samples from a single input texture, a difficult problem is to develop a neural architecture that has the ability to synthesize arbitrary texture inputs, possibly not seen during the training phase. Incidentally, this also raises the question of how to efficiently encode textures, e.g. through latent spaces. The authors of [12] were the first to propose such an approach, leveraging the concept of adaptive instance normalization [9] to provide a flexible and universal style transfer and texture synthesis method, where each texture is represented by several thousands normalizing parameters related to VGG19 features. To allow for user control [1] proposes a modification of PSGAN [2] by introducing a convolutional encoder. They focus on very small latent spaces well suited to the representation of patches from a given image. Authors from [8] also use a convolutional encoder to generate textures, with exemplar-specific geometric transformations applied to the noise input of the generator. Building on attention mechanisms, similarity maps or Fourier representations, some works [7,14] also tackle the problem of universal neural texture synthesis but suffer from verbatim copy of structures, such as those usually observed in patch-based approaches. The method from [13] also has the ability to synthesize a variety of textures, but relies on a latent space which needs inverting the generator to represent given inputs.

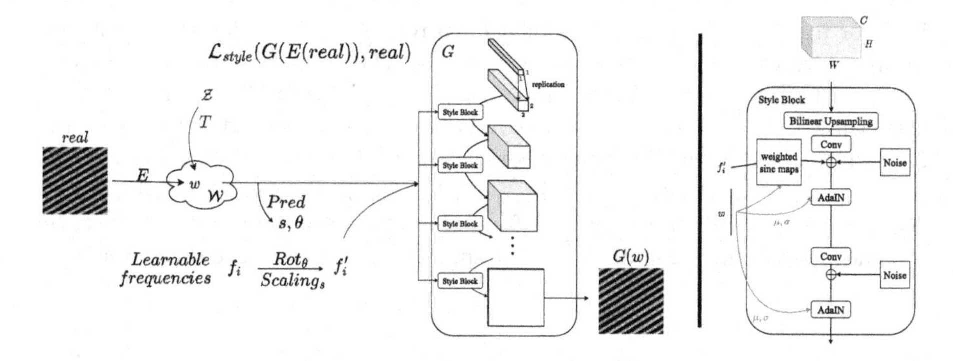

Fig. 1. Left: Texture Auto-encoder architecture. Right: Modified style block

3 Method

3.1 Architecture Overview

Our texture synthesis framework relies on an auto-encoder structure, that we call in short **TAE**. As explained in the introduction, we wish to develop a latent space that is spatially agnostic, deals with simple geometric transforms and with periodic textures.

The encoder E is used to map an image to a latent texture representation space $\mathcal{W}$. From this representation, new samples with different spatial arrangements can be generated with the generator G. For this generator, we rely on the classical StyleGAN architecture [11]. This architecture is augmented with spatially periodic information, namely sine maps as used in [2,13]. But unlike them, the periodic information is modulated using the latent representation and is thus adapted to each input sample. This adaptation of the periodic content is crucial and achieved by predicting scale and orientation parameters from the latent variable w.

The whole pipeline is mainly supervised by the texture reconstruction loss $\mathcal{L}_{style}$ [4] between the input and its reconstruction by the texture auto-encoder. Additionally, we introduce a network T designed for direct texture sampling, enabling us to map a noise distribution onto our latent space $\mathcal{W}$ in a meaningful way. This allows the exploration of the latent space and possibly the creation of new textures. An overview of our method is shown in Fig. 1.

3.2 Texture Encoder

In order to develop an encoder embedding texture images in the latent space of StyleGAN [11], we classically rely on the second order statistics of deep features, in the spirit of [4] and the numerous works that have followed. Given 5 different depths l, we retrieve features F^l from the VGG19 network, each one having dimension $(C_l, H_l W_l)$, to extract their second order statistics. The main difficulty to build a texture encoder from these statistics lies in their high dimensionality.

In order to extract a single low dimensional representation from these features, we first extract compact information from the VGG features at each scale (see Fig. 2, right), which we then combine into a single compact vector $w = E(I)$ using an MLP. At each scale, the second order features are extracted using the technique presented in [18]. Indeed, for each depth l, instead of computing whole (C, C) Gram matrices $G^l_{c,c'} = \sum_{k=1}^{H_l W_l} F^l_{c,k} F^l_{c',k}$ as in [3] and then reduce their dimensionality, we directly compute from G a representation vector $Q(G)$ of size n_w, the chosen dimension of our latent space. These vectors are defined as:

$$Q(G)_i = m_i^T \cdot G \cdot m_i \tag{1}$$

with $\{m_i\}_{1 \leq i \leq n}$ a set of trainable vectors of size $(C, 1)$. If we write $G = F^l \cdot F^{lT}$, the previous equation becomes:

$$Q(G)_i = m_i^T \cdot F^l \cdot F^{lT} \cdot m_i = (m_i^T \cdot F^l) \cdot (m_i^T \cdot F^l)^T \tag{2}$$

This allows the extraction of relevant second order features without having to compute the whole Gram matrix (as illustrated in Fig. 2), reducing the complexity from $n(2HWC^2 + 4C) \approx 2nHWC^2$ to $n(2 * CHW + 2HW) \approx 2nHWC$, i.e. by a factor C, which goes up to 512 for the feature maps used in the latest depths of the VGG network.

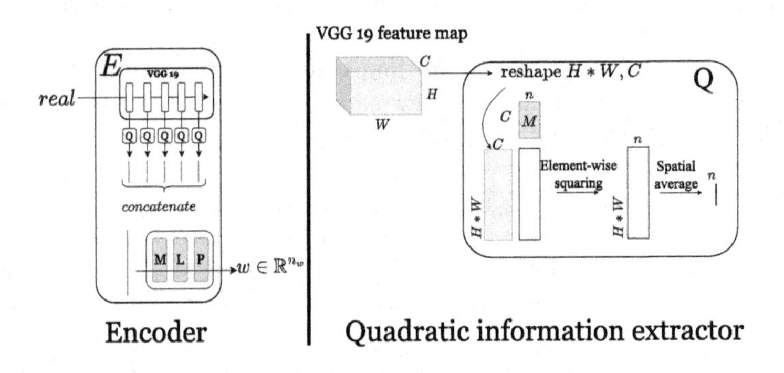

Fig. 2. Left: Encoder, Right: Quadratic information extraction module

This process results in a compact texture encoder, built only from second order statistics of some learned combination of features and carrying non localized textural information. This non-localized nature is of course a desirable property in view of texture synthesis and will be shared by the generator.

3.3 Texture Generator

We use a modified StyleGAN architecture as our generator G. To perform image synthesis, such an architecture uses a latent variable w in a non-localized way (contrarily to DCGAN-like architectures): w is used to predict and set the first and second moments of feature maps during the multi-scale synthesis process (through Adaptive Instance Normalization [9]).

Aiming to remove any systematic spatial information in the generation process, we modify the StyleGAN architecture so that all spatial information is due to the realizations of noise maps at each step of the generation process: instead of learning a tensor of size (512*4*4), we discard spatial dimensions by only learning an input tensor (n*1*1), which we then expand by replication at the start of the generation process. Besides, convolutions use no padding to avoid border artifacts.

Thanks to these changes, our auto-encoder achieves full disentanglement of texture information and pattern localization. Periodic information is also added at every level of the generator, as will be further described in Sects. 3.5–3.6.

3.4 Sine Waves

In order to synthesize periodic-like textures having long range dependency, we add sine waves to our generator, in a way similar to [2,13]. Given a set of n_{freq} frequencies f_i' (detailed in Sect. 3.6), we build $S \in \mathbb{R}^{n_{freq}*H*W}$ a volume where each channel S_i is a sine wave of frequency f_i': $S = (sin(f_i' \cdot \boldsymbol{x} + \phi))_i$, with $\boldsymbol{x} \in \mathbb{R}^{H*W}$ the spatial position, and ϕ a random phase. In contrast to the works in [2,13], we modulate elements S in the following ways:

- we weight each channel S_i of the sine maps S with the coefficient $weight_{level}(f_i)$ defined in Sect. 3.5. Its purpose is to inject the frequency f_i in the right level of the generator given its magnitude.
- we use the latent variable w to weight the use of every frequency according to the latent representation, using a fully connected layer $A \in \mathcal{M}_{n_{freq}, n_w}$, in the spirit of [11], to project w onto a weighting vector $A \cdot w$.

Finally, the addition of periodic content to a given feature map $F \in \mathbb{R}^{B*C*H*W}$ is performed by addition, after applying a 1 by 1 convolution filter to S to reach the number of feature maps C (see Fig. 1, right):

$$F \longleftarrow F + conv\Big(S * weight_{k_{level}} * A \cdot w)\Big) \tag{3}$$

3.5 Scale Independent Learnable Frequencies

In this section, we describe a novel way to generate periodic information independently from the architecture of the generator. We define a *level* of the network to be the set of operations delimited by a change of resolution. In our case each level is a StyleGAN block, delimited by upsampling by a factor 2.

In the literature, periodic content is usually incorporated at specific levels of generators. Sine maps are either used at the lowest resolution level of the network in [2] (not allowing for high-frequency content to be added), or at each level of the network as in [13] (leading to sudden discontinuity in the generation process due to the change of level). Instead of learning a different set of frequencies at each level, we choose to learn them independently of the architecture of the generator, and incorporate them in the generator at the adequate levels, depending on their

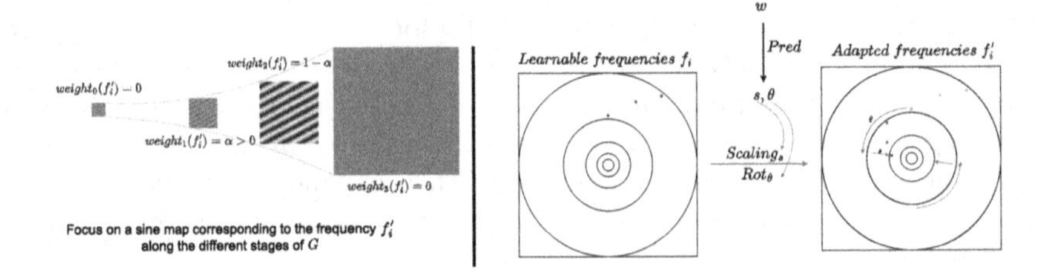

Fig. 3. Left: Use of a global frequency at different scales in the generator, depending on its magnitude. **Right:** Periodic information to incorporate to the network: learnable frequences are transformed using parameters depending on the input (through w).

magnitude. The idea is that a frequency f has to be added at the level of the network where its magnitude (relative to the resolution of the level) is not too low nor too high. The full procedure is illustrated in Fig. 3, left.

In short, we learn a set of n_{freq} frequencies f_i independently of the architecture of the generator. Depending on the magnitude of a given frequency, we use it to create sine maps in the two levels of the generator where its magnitude is adequate (relative to the resolution of the level). The use of a frequency f is then weighted based on $|f|$: for each level number k_{level} in the network, we define the weighting as:

$$weight_{k_{level}}(f) = \begin{cases} 1 - \left|1 - log_2(\frac{|f|}{f_0 \cdot 2^{k_{level}}})\right| & \text{if } \frac{f_0}{2} \leq \frac{|f|}{2^{k_{level}}} \leq 2f_0 \\ 0 \text{ otherwise} \end{cases} \quad (4)$$

where f_0 corresponds to a reference magnitude. This allows for a frequency's use level to smoothly move along the dyadic structure of the generator.

3.6 Image Specific Scale and Orientation Estimation

In this section, we describe a key component of our auto-encoder: each learned frequency f_i is rotated and scaled accordingly to the input, both at training time and for inference. This allows to generate a new sample with appropriate periodic content, aligned with the input.

For each input image, scale and orientation parameters are learned from the corresponding latent variable w. We directly infer these scale and rotation parameters with an auxiliary network *Pred*: $Pred(w) = (s, \theta)$. In order to define a scale and an orientation, one would usually need a reference, with scale 1 and orientation 0. We avoid this issue by forcing the network's predictions to be consistent with geometric transformations in the image space. This approach is reminiscent of self supervised learning (SSL) approaches.

Given an input image I, we geometrically transform I with two distinct scaling and rotation parameters $\hat{s}_i$ and $\hat{\theta}_i$ and then encode these images to get latent codes: $w_i = E(Scaling_{\hat{s}_i}(Rot_{\hat{\theta}_i}(I)))$. Finally, we predict transformation parameters from these latent codes: $s_i, \theta_i = Pred(w_i)$. The corresponding loss $\mathcal{L}_{SLL}$ reads:

$$\mathcal{L}_{SSL} = 1 - \left|\frac{\pi}{2} - \left((\hat{\theta}_1 - \hat{\theta}_0) - (\theta_1 - \theta_0)\right) mod_\pi\right| + \left(\frac{\hat{s}_1}{\hat{s}_0} - \frac{s_1}{s_0}\right)^2 \quad (5)$$

The first part of the loss takes into account the π-periodicity of the orientation of an image and is minimal when the difference in image rotation angles $\hat{\theta}_1 - \hat{\theta}_0$ is equal to the difference of predicted angles $\theta_1 - \theta_0 \quad mod\ \pi$. Similarly, the second part of the equation is minimal when the ratio between estimated scales matches the ratio of the scales used in the geometrical augmentation. Eventually, these input-dependent parameters are used to scale and rotate the set of global frequencies f_i, effectively adapting the frequency patterns to each specific image, as illustrated in Fig. 3, right. Writing $s_{pred}, \theta_{pred} = Pred(w)$, we get:

$$f_i' = Scaling_{s_{pred}}(Rot_{\theta_{pred}}(f_i)). \quad (6)$$

3.7 Direct Sampling

Using our architecture, the natural way to synthesize a texture is to encode and decode an image in the latent space, performing synthesis by example. Now, it may be desirable to synthesize new textures without exemplar input. In order to do so, we learn the distribution of the encoded training images $E(p_{data})$ in the $\mathcal{W}$ space. We choose an adversarial approach, where a network T is trained to map a noise distribution $p_{\mathcal{Z}}$ to $E(p_{data})$. We train T jointly with a discriminator $D_{\mathcal{W}}$, following the WGAN-GP framework [6], with a gradient penalty $\mathcal{L}_{GP}$:

$$\mathcal{L}_{adv}(T) = \mathbb{E}_z\left[D_{\mathcal{W}}(T(z))\right] \quad (7)$$

$$\mathcal{L}_{adv}(D_{\mathcal{W}}) = \mathbb{E}_{I\sim p_{data}}\left[D_{\mathcal{W}}(E(I))\right] - \mathbb{E}_z\left[D_{\mathcal{W}}(T(z))\right] + \lambda_{GP}\mathcal{L}_{GP} \quad (8)$$

3.8 Losses

We further constrain the texture synthesis process using the following losses. To compensate for low-frequency artifacts created by the optimization from [4], we add the spectral loss $\mathcal{L}_{Spe}$ from [5] (Equation (3.3)) to force the spectrum of the generated image to match the input's spectrum. This loss complements the texture loss, providing necessary information to the network to align the generated periodic content in $I_2 = G(E(I_1))$ with the input image I_1.

Using sliced histogram matching [15], we also implement a color histogram loss between I_1 and I_2:

$$\mathcal{L}_{Hist}(I_1, I_2) = \mathbb{E}_{x\in\mathbb{R}^3, \|v\|=1} \sum (sort_x(I_1) - sort_x(I_2))^2 \quad (9)$$

Here x denotes a randomly sampled color vector along which to perform color histogram matching, and $sort_x$ is the operation that sorts all pixels of an image accordingly to the value of the projection of the pixels's color value onto x. By randomly sampling x, we approximate a color histogram distance between I_1 and I_2.

Our final optimization objective reads:

$$\mathcal{L} = \mathcal{L}_{style} + \mathcal{L}_{Hist} + \mathcal{L}_{Spe} + \mathcal{L}_{adv} + \mathcal{L}_{SSL} \quad (10)$$

3.9 Training

The whole pipeline is trained in an end-to-end fashion, using the Adam optimizer with a learning rate of 10^{-4} for 600000 iterations with a batch size of 8. This amounts to a week of training on a single NVIDIA RTX 6000 GPU.

4 Experiments

4.1 Assessed Methods, Datasets and Evaluation Metrics

Among the many texture synthesis algorithms, we choose to compare ourselves to these methods:

- TextureCNN [4], as one of our goals is to shortcut this process with a feed-forward auto-encoder;
- PSGAN [2], as they introduced the idea of augmenting a generator with periodic content;
- Neural texture [8], since they also use a texture encoder;
- Whitening Coloring Transform (WCT) [12], because this method performs universal texture synthesis with no learning step.

We do not compare against [1] (unavailable code) nor [13] as no encoder is included in their architecture. For all methods requiring a training (ours, PSGAN and neural texture), we choose a latent space dimension of 32.

PSGAN does not come with an encoder to project an image into a global latent space variable z_g. As a baseline, we perform inversion of PSGAN as in [19]: given a data sample, randomly initialize a latent variable z_g and perform gradient descent in the latent space to minimize the texture loss $\mathcal{L}_{style}$ between the generated image $G(z_g)$ and the data sample. We keep the best result out of 10 random initializations. We call this resynthesis method PSGAN-GD.

We use the texture loss from [3] as a distance to measure performance, along with SIFID (introduced in [16]) and LPIPS [20]. We use the dataset Macro-Textures introduced in [13]. We strongly augment each dataset with geometric transformations: rotations of any angles and scalings to train our method, Neural texture, and PSGAN.

4.2 Visual and Quantitative Results

Synthesis results are simply obtained by first embedding an input exemplar into the latent space (using the encoder) and then generating a new sample from this latent code. Variability of the synthesized samples is classically obtained thanks to the noise maps in the generator. A comparison of syntheses obtained with the different methods is presented in Fig. 4. We observe that we indeed maintain appropriate long range dependencies, which is not the case in for TextureCNN nor WCT. We also get rid of low frequency artifacts mostly present in TextureCNN.

We show quantitative evaluations in Table 1. Our approach is very close to TextureCNN for the style metric, which leads to the best result as it directly optimizes $\mathcal{L}_{style}$ over the image.

Our method comes second best, notably beating WCT by a relatively big margin. Additionally, our method is orders of magnitude faster. We also obtain the best performance for the SIFID and LPIPS metrics, in particular outperforming the original TextureCNN approach. The results of PSGAN-GD are poor, which is essentially due to the instability of the inversion procedure in the latent space.

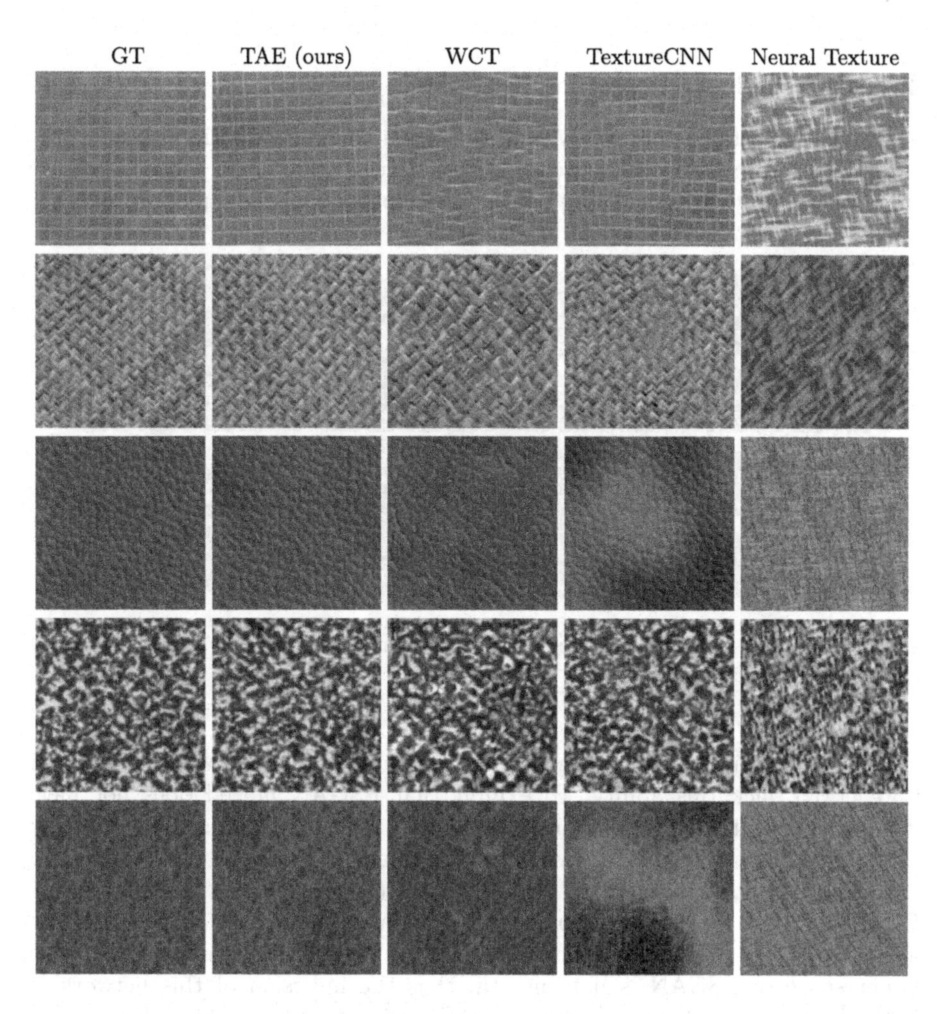

Fig. 4. Comparison of syntheses obtained from 4 different methods.

Table 1. Evaluation of the methods using $\mathcal{L}_{style}$, SIFID, LPIPS and runtime in seconds (all on one NVIDA RTX6000 GPU) on the MacroTextures dataset

	$\mathcal{L}_{style}$ $(\cdot 10^3)$	SIFID $(\cdot 10^{-6})$	LPIPS	runtime
TextureCNN [4]	$\mathbf{3.8 \pm 16.8}$	14 ± 14	0.51 ± 0.09	60
TAE (ours)	5.2 ± 11.7	$\mathbf{9 \pm 11}$	$\mathbf{0.35 \pm 0.09}$	**0.0072**
Neural texture [8]	135 ± 468	178 ± 190	0.56 ± 0.09	0.085
WCT [12]	29.8 ± 132.9	10 ± 18	0.44 ± 0.07	3.5
PSGAN-GD [2]	208 ± 611	381 ± 607	0.64 ± 0.11	24

4.3 Geometric Completeness

In order to illustrate the ability of the method to deal with simple geometric transforms, we perform auto-encoding on various scalings and rotations of the same input image, thus showing successful adaptation to the input in Fig. 5.

Fig. 5. Top row: texture image geometrically augmented, Bottom row: feed forward texture resynthesis $G(E(I))$

As a comparison we show that PSGAN, and thus the methods based upon it (e.g. [1]), lack this ability. To exhibit that PSGAN indeed does not have the capacity of handling different rotations of the same textures, we start by sampling a global variable z_g, generating an image $I = G(z_g)$.

Then, we create different rotations of I: $R_\theta(I)$ as shown in the first row of Fig. 6, and try to reproduce them with the generator G. Indeed, taking $I = G(z_g)$ avoids inversion problems. On the contrary, as observed in Fig. 6a) of [1], the latent space of PSGAN is not smooth, thus the inversion of this network is unstable.

The second row shows $G(z_{g,opt})$, where $z_{g,opt}$ is obtained by minimizing $\mathcal{L}_{style}(R_\theta(I), G(z_{g,opt}))$ gradient descent. We notice that the periodic structure is not rotated, while smaller details are correctly oriented. This is indeed not a satisfactory result. In the PSGAN architecture, the frequencies of sine maps are

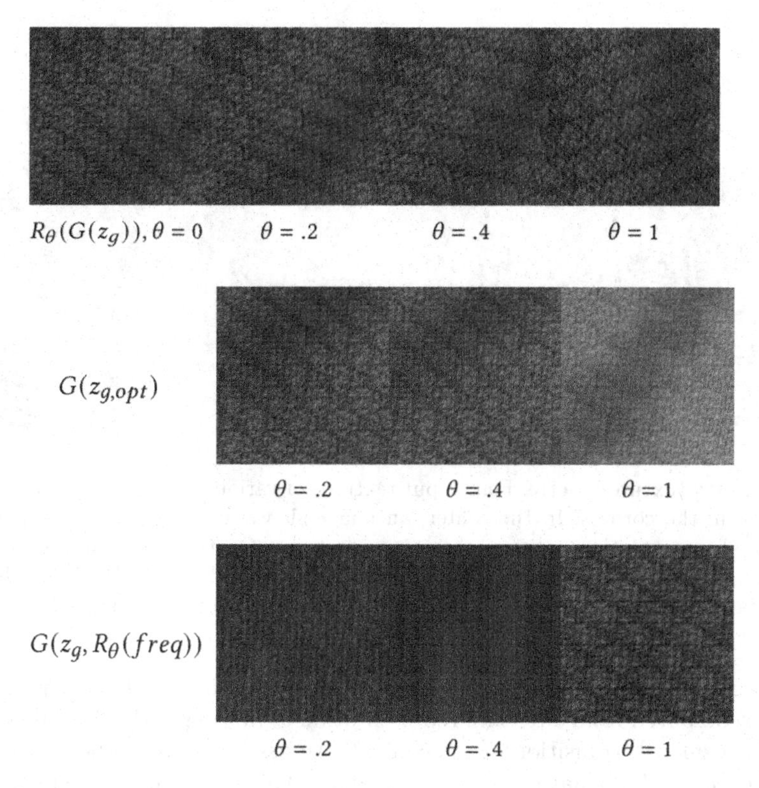

Fig. 6. Illustration of the difficulty of PSGAN to deal with arbitrary rotations **First row**: Rotations of input texture $R_\theta(I)$, $I = G(z_g)$ - **Second row**: PSGAN samples where $z_{g,opt}$ is obtained by gradient descent to match the texture $R_\theta(I)$ - **Third row**: Directly rotating the frequencies without changing z_g

inferred from z_g. We also tried to directly rotate these frequencies with the same angle θ as we rotated the image I with, thus aligning the periodic content with the image we want to reconstruct, $R_\theta(I)$. This operation yields poor pattern fidelity albeit somewhat aligned, as can be seen on the third row.

These two tests prove that, although trained on a dataset augmented with rotations, locally in the latent space of PSGAN, each texture is intrinsically binded to an orientation. There is no way to change the orientation without harming texture fidelity.

4.4 Spatial Interpolation

Our approach relies on a texture latent space and is therefore naturally adapted to texture interpolation, for instance by simply linearly interpolating latent variables. Spatial interpolation (building an image across which a texture is progressively varying), on the other hand, is more challenging. Indeed, AdaIN layers normalize features based on statistics computed across the whole image, therefore globally defining new features. We solve this problem by computing

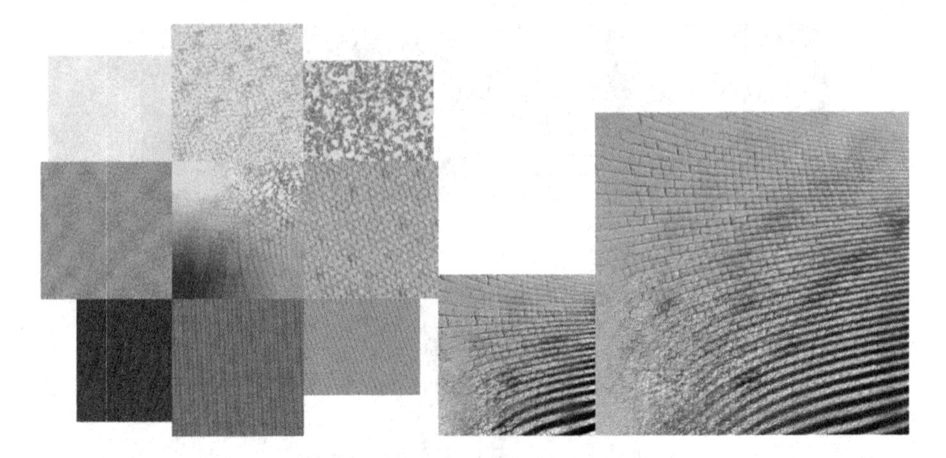

Fig. 7. Left: Texture Palette. Four input textures of various sizes and aspect ratio are displayed in the corners. In the center, an image shows a field obtained by performing spatial interpolation between the corner textures. On this field, the red dots are positions from which the images in the middle of each side are synthesized. **Right:** Interpolation between 4 textures without and with expansion by a factor 2

statistics locally. Additionally, the mean and variance imposed after normalization are computed from a texture representation w varying in the spatial domain, to allow a smooth transition of the content. An asset of this method is that the periodic content has a natural variation in space. We can also use it to create interpolated texture from a visual palette (Fig. 7, left). We show an example of texture interpolation and of interpolation combined with expansion in Fig. 7, right.

5 Conclusion

We introduced a feed forward auto-encoder network having the ability to learn a set of textures, yielding fast exemplar-based synthesis results. Thanks to adaptive periodic content, our method allows texture encoding and reconstruction of any orientation and scale. Its disentanglement of texture characteristics and spatial distribution allow for common texture manipulations such as interpolation and expansion. The proposed approach outperforms closely related methods for usual texture fidelity metrics.

Acknowledgements. This work was supported by the Defence Innovation Agency and the project MISTIC (ANR-19-CE40-005).

References

1. Alanov, A., Kochurov, M., et al.: User-controllable multi-texture synthesis with generative adversarial networks. In: Proceedings of the 15th International Joint Conference on Computer Vision, Imaging and Computer Graphics Theory and Applications (VISIGRAPP 2020), vol. 4, pp. 214–221 (2020)

2. Bergmann, U., Jetchev, N., et al.: Learning texture manifolds with the periodic spatial GAN. In: Proceedings of the 34th International Conference on Machine Learning, vol. 70, pp. 469–477 (2017)
3. Gatys, L.A., Ecker, A.S., et al.: Texture synthesis and the controlled generation of natural stimuli using convolutional neural networks (2015). http://arxiv.org/abs/1505.07376
4. Gatys, L.A., Ecker, A.S., et al.: Texture synthesis using convolutional neural networks. In: Proceedings of the 28th International Conference on Neural Information Processing Systems, vol. 1, pp. 262–270 (2015)
5. Gonthier, N., Gousseau, Y., et al.: High resolution neural texture synthesis with long range constraints. J. Math. Imaging Vis. **64**, 478–492 (2022)
6. Gulrajani, I., Ahmed, F., et al.: Improved training of Wasserstein GANs. In: Proceedings of the 31st International Conference on Neural Information Processing Systems, pp. 5769–5779 (2017)
7. Guo, S., Deschaintre, V., et al.: U-attention to textures: hierarchical hourglass vision transformer for universal texture synthesis. In: Proceedings of the 19th ACM SIGGRAPH European Conference on Visual Media Production (2022)
8. Henzler, P., Mitra, N.J., Ritschel, T.: Learning a neural 3D texture space from 2D exemplars. In: 2020 IEEE/CVF Conference on Computer Vision and Pattern Recognition (CVPR), pp. 8353–8361 (2020)
9. Huang, X., Belongie, S.: Arbitrary style transfer in real-time with adaptive instance normalization. In: Proceedings of Conference ICCV, pp. 1501–1510 (2017)
10. Jetchev, N., Bergmann, U., Vollgraf, R.: Texture synthesis with spatial generative adversarial networks (2016). http://arxiv.org/abs/1611.08207
11. Karras, T., Laine, S., et al.: Analyzing and improving the image quality of styleGAN. arXiv:1912.04958 (2019)
12. Li, Y., Fang, C., et al.: Universal style transfer via feature transforms. In: Proceedings of the 31st International Conference on Neural Information Processing Systems, pp. 385–395 (2017)
13. Lin, J., Sharma, G., Pappas, T.N.: Toward universal texture synthesis by combining texton broadcasting with noise injection in styleGAN-2. e-Prime Adv. Electr. Eng. Electron. Energy **3**, 100092 (2023)
14. Liu, G., Taori, R., et al.: Transposer: universal texture synthesis using feature maps as transposed convolution filter. arXiv:2007.07243 (2020)
15. Rabin, J., Delon, J., et al.: Regularization of transportation maps for color and contrast transfer. In: 2010 ICIP Conference, pp. 1933–1936 (2010)
16. Shaham, T.R., Dekel, T., et al.: SinGAN: learning a generative model from a single natural image. In: Proceedings of the IEEE/CVF International Conference on Computer Vision, pp. 4570–4580 (2019)
17. Ulyanov, D., Lebedev, V., et al.: Texture networks: feed-forward synthesis of textures and stylized images. In: Proceedings of the 33rd International Conference on International Conference on Machine Learning, vol. 48, pp. 1349–1357 (2016)
18. Yu, K., Salzmann, M.: Second-order convolutional neural networks. arXiv:1703.06817 (2017)
19. Yu, N., Barnes, C., et al.: Texture mixer: a network for controllable synthesis and interpolation of texture. In: 2019 IEEE/CVF Conference on Computer Vision and Pattern Recognition (CVPR), pp. 12156–12165 (2019)
20. Zhang, R., Isola, P., et al.: The unreasonable effectiveness of deep features as a perceptual metric. In: 2018 IEEE/CVF Conference on Computer Vision and Pattern Recognition (CVPR), pp. 586–595 (2018)

Fast Marching Energy CNN

Théo Bertrand(✉), Nicolas Makaroff, and Laurent D. Cohen

CEREMADE, UMR CNRS 7534, University Paris Dauphine,
PSL Research University, 75775 Paris, France
{bertrand,makaroff,cohen}@ceremade.dauphine.fr

Abstract. Leveraging geodesic distances and the geometrical information they convey is key for many data-oriented applications in imaging. Geodesic distance computation has been used for long for image segmentation using Image based metrics. We introduce a new method by generating isotropic Riemannian metrics adapted to a problem using CNN and give as illustrations an example of application. We then apply this idea to the segmentation of brain tumours as unit balls for the geodesic distance computed with the metric potential output by a CNN, thus imposing geometrical and topological constraints on the output mask. We show that geodesic distance modules work well in machine learning frameworks and can be used to achieve state-of-the-art performances while ensuring geometrical and/or topological properties.

Keywords: Geodesic Distance · Riemannian metric learning · Segmentation

1 Introduction

Geodesic curves and distances have been used to convey geometric properties in many different applications. The usual approach of those methods is to rely on prior knowledge of the task at hand to build a Riemannian metric g explicitly from data.

The approach presented in this work tries to get rid of the bias introduced in the choice of a metric tensor by generating it from data via a Neural Network architecture which parameters were previously optimized in a supervised learning approach with training data. Introducing such a bias is not a bad thing in itself, however, it requires an arbitrary decision from a user and parameter tuning, two issues that can be avoided by learning to generate a metric from data.

To demonstrate the effectiveness of this framework, we apply it to a segmentation task using a brain tumour MRI images dataset. By using our proposed method, we can obtain accurate results compared to traditional approaches, highlighting the capabilities of this approach. Furthermore, we also observe that our method has a remarkable ability to learn from data and somewhat generalize to unseen data.

T. Bertrand and N. Makaroff—Equal contribution.

L. Calatroni et al. (Eds.): SSVM 2023, LNCS 14009, pp. 276–287, 2023.
https://doi.org/10.1007/978-3-031-31975-4_21

The method introduced in this work offers a powerful and flexible way of using geodesic curves and distances in a wide range of applications in a holistic learning framework.

The rest of the paper is organised as follows. In Sect. 2 we present the computation of geodesic distances and their gradient. In Sect. 3 we introduce our experimental method for Fast Marching Energy CNN. In Sect. 4 we present the main results of our experiments and provide a discussion around our work.

Related Works

The use of geodesic distances in segmentation tasks has a long history. To the authors' knowledge, the first article to segment an image's region using a minimal path distance and fast marching is [9], with application on a 3D brain image. In the case of the segmentation of tubular tasks we can refer to [5] for instance, a method that segments the 3D vascular tree by propagating the front of the minimal path distance computation. Similarly, [6] segments vascular structures by introducing an anisotropic metric, determined dynamically by evaluating local orientation scores during the Fast Marching computations. Those 3 articles already use the level sets of the geodesic distance (or "geodesic balls") to provide the segmentation mask. We may also mention [3] that uses geodesic curves in an higher dimensional space to track vessels (as curves with an additional width component). These works generally aren't interested in treating the task in an holistic manner and focus on providing a good model for the structures to segment, whereas this work tries to treat the problem end-to-end and generalize to a large dataset of input images.

Only a few previous methods are interested in learning a metric from data. We may mention recent works such as [13] and [8] that try to find metric tensors that fit spatio-temporal data in order to capture the velocity fields and underlying geometry of the data. The first paper is modelling trajectories as the solutions of a dynamical system generated by a Neural Network and also taking into account the dynamics of the whole population by penalizing an optimal transport cost between two consecutive timestamps. However [8] tries to interpolate a sequence of histograms with Wasserstein barycenters by optimizing over the metric tensor appearing in the ground cost. Also, there are important links between the Wasserstein optimal transport, its dynamical formulation and geodesics, for further reading, we refer to [1]. These works propose interesting frameworks to work with, but they are not focused on generalizing the generation of the metric tensors.

[2] is an older article that is important for our work, as they laid the ground for the differentiation of the geodesic distance with respect to the metric in the Fast Marching algorithm. They then proceed to apply it in the setting of inverse problems to retrieve the metric from distance measurements. Its only concerns were to solve inverse problems involving the geodesic distance, whereas we go one step further by including a Fast Marching module in a deep learning segmentation procedure. The sub-gradient marching algorithm is briefly described

in Sect. 2 as it is essential to our framework to propagate through the Fast Marching module and carry the learning step.

In terms of Deep Learning, we might add a few references such as the classical [12] and [7] that respectively introduce the UNet and ResNet architectures, which are used for our method and as baseline comparisons. For a review of deep learning methods in medical imaging one might refer to [15]. The very general methods directly producing segmentation from medical images are already quite efficient, but they suffer from a lack of robustness and do not impose a lot of structure on the segmentation that comes out of the network. Contrary to this, our work allows to impose a lot of constraint on the topology of the segmented region (namely a set with trivial topology).

2 Computing Geodesic Distances and Their Gradient

The geodesic distance is a fundamental concept in the field of Riemannian geometry, and it is used to quantify the distance between two points on a (compact, path-connected) manifold $\mathcal{M}$. It is defined as the minimal length of all possible paths linking two points on the manifold.

Formally, the geodesic distance is given by the following:

$$d_g(x, y) = \inf_{\gamma \in \mathrm{Lip}([0,1], \mathcal{M}), \gamma(0)=x, \gamma(1)=y} \int_0^1 \sqrt{g_{\gamma(t)}(\gamma'(t), \gamma'(t))} \mathrm{d}t, \tag{1}$$

where $\mathrm{Lip}([0, 1], \mathcal{M})$ is the space of Lipschitz curves on the manifold $\mathcal{M}$ and parameterized by the interval $[0, 1]$. g is a metric tensor, which is a map defined at each point $x \in \mathcal{M}$ as $g_x : (u, v) \in \mathcal{T}_x\mathcal{M}^2 \mapsto g_x(u, v)$ is positive definite bilinear form. This means that $\sqrt{g_x}$ is a Euclidean norm on $\mathcal{T}_x\mathcal{M}$, the tangent space to $\mathcal{M}$ at point x.

In this work, we will consider a very simple mathematical framework, where $\mathcal{M}$ is simply a path-connected, open and bounded set Ω of $\mathbb{R}^d$ and $\mathcal{T}_x\mathcal{M}$ can be identified with $\mathbb{R}^d$. This simplification allows for a more straightforward implementation of the geodesic distance, while still maintaining its core properties and mathematical foundation. In the following we will have $g_x(u, v) = \phi(x)^2 \langle u, v \rangle_{\mathbb{R}^d}$.

Fast Marching Algorithm

Since the seminal work of [14], the Fast Marching algorithm has been one of the most widely used methods for computing geodesic distances on a manifold. The Fast Marching method computes the geodesic distance by front propagation.

The Eikonal equation, which has the geodesic distance as its unique positive viscosity solution, is the key component to the front propagation in Fast Marching.

The distance u from a set $S \subset \Omega$ satisfies the Eikonal equation:

$$\begin{cases} \forall x \in \Omega \setminus S, \ \|\nabla u(x)\| = \phi(x), \\ \forall x \in S, \quad\quad u(x) = 0, \end{cases} \tag{2}$$

It can be shown that the unique positive solution to the Eq. (2) in the sense of viscosity solutions is the geodesic distance from the set S, relative to the metric tensor field associated with the matrices $\phi(x)^2\mathbf{I}_d$.

The Eikonal equation is discretized using the upwind scheme:

$$\sum_{1\leq i\leq 2} \frac{1}{h^2} \max(u_p - u_{p+e_i}, u_p - u_{p-e_i}, 0)^2 = \phi_p^2, \tag{3}$$

with u_p and ϕ_p the geodesic distance and potential at point p in the discretized domain Ω, $p\pm e_i$ denote the adjacent points on the grid and h is the discretization parameter.

Fast Marching is an algorithm that iteratively visits each point on the grid from neighbour to neighbour. At each iteration we look at the neighbour points to those that have already been Accepted, and we accept the nearest point among the neighbours and we repeat by computing the new neighbourhood of the Accepted points. We initialise all values at $+\infty$ except the seed point at 0. Depending on the number of accepted points connected to p on the grid, Eq. (3) reduces either to a quadratic of affine equation to find u_p from the values of the parent points.

In practice, we use the python library *Hamiltonian Fast Marching* (*HFM*) that provides a fast and efficient implementation of the Fast Marching method and of the so-called Subgradient Marching Algorithm [10].

Differentiating Fast Marching. The ability to differentiate the geodesic distance with respect to the metric is an important tool in many applications, such as shape optimization and optimal control. The first work to propose a numerical method to differentiate the geodesic distance with respect to the metric is [2], and it has found few applications (see for instance [4]).

To differentiate the geodesic distance, we can use the update in the discretized Eikonal Eq. (4). By taking the Eikonal equation written in dimension 2, and using the setting of interest, i.e. an isotropic metric $g_x(v,w) = \phi(x)^2\langle v,w\rangle_{\mathbb{R}^2}$, we write the discretized version of the Eikonal equation, with h the discretization parameter, the discretized domain is simply a regular square grid :

$$\begin{cases} (u_p - u_{p\pm e_1})^2 + (u_p - u_{p\pm e_2})^2 = h^2\phi_p^2 \text{ if p has 2 parents,} \\ u_p = \min_i u_{p\pm e_i} + h\phi_p \text{ if p has only 1 parent or } h^2\phi_p^2 < (u_{p\pm e_1} - u_{p\pm e_2})^2. \end{cases} \tag{4}$$

Thus u_p is the value of the distance computed by fast marching at point p, and we define $D_\phi u_p \in \mathbb{R}^{n^2}$ the differential of u_p with respect to the potential ϕ.

Differentiating with respect to ϕ in the two cases of update, we get

$$\begin{cases} (u_p - u_{p\pm e_1})(D_\phi u_p - D_\phi u_{p\pm e_1}) + (u_p - u_{p\pm e_2})(D_\phi u_p - D_\phi u_{p\pm e_2}) = h^2\phi_p \\ \text{if p has 2 parents,} \\ D_\phi u_p = D_\phi u_{p\pm e_i} + h\mathbb{1}_p \text{ if p has only 1 parent or } h^2\phi_p^2 < (u_{p\pm e_1} - u_{p\pm e_2})^2, \end{cases} \tag{5}$$

with $\mathbb{1}_p \in \mathbb{R}^{n^2}$ the vector filled with zero except at coordinate p, which gives the update:

$$\begin{cases} D_\phi u_p = \frac{(u_p - u_{p\pm e_1})D_\phi u_{p\pm he_1} + (u_p - u_{p\pm e_2})D_\phi u_{p\pm e_2} + h^2\phi_p}{(u_p - u_{p\pm e_1}) + (u_p - u_{p\pm e_2})} \text{ if p has 2 parents,} \\ D_\phi u_p = D_\phi u_{p\pm e_i} + h\mathbb{1}_p \text{ if p has only 1 parent or } h^2\phi_p^2 < (u_{p\pm e_1} - u_{p\pm h_2})^2, \end{cases} \tag{6}$$

This update can then be used to compute the gradient of the geodesic distance with respect to the metric tensor during the Fast Marching iterations. In [2] it is named *Subgradient Marching Algorithm.* This method can be extended to higher dimensions as well as more general Finsler metrics.

3 Model

The proposed method presented in this study uses a neural network, specifically a modified version of the UNet architecture, to segment regions of an image as geodesic balls with respect to a metric. The metric is obtained by training a convolutional neural network (CNN) to provide both the metric and the center or seed of the geodesic ball. The framework, as shown in Fig. 1, processes the input image using the encoder component of the UNet, resulting in a vector representation of the image. This vector is then passed through two separate decoders to perform distinct tasks.

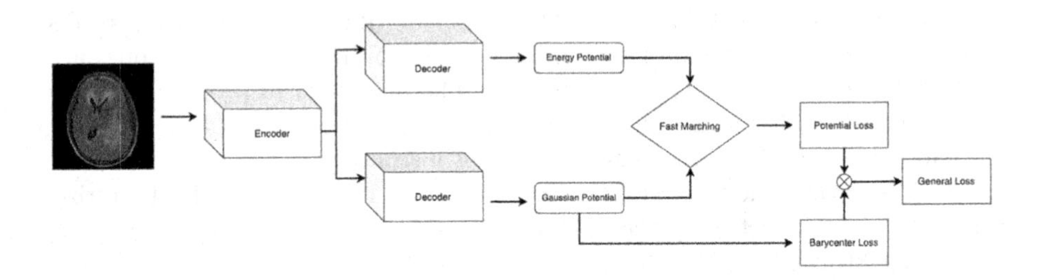

Fig. 1. Diagram of the framework from the input image to the loss.

The first decoder predicts the potential ϕ to be used by the fast marching module, which can be computed using the HFM library. The second decoder predicts a Gaussian potential that represents the probability of the presence of the region's barycenter in a given area, which is also provided as a seed to the fast marching module. The distance map generated by the fast marching procedure is then used to find a geodesic ball for segmentation. The expected segmentation is compared to the predicted segmentation, and the theoretical barycenter is compared to the predicted Gaussian potential to compute the error.

The distance computation module can be written as a function of both seed points and input metric. The metric ϕ is defined as the output of a CNN architecture, such as the widely used UNet, with θ being in the space of parameters. We enforce positive and non-zero properties of the metric by taking $\phi = f_\theta(u)^2 + \epsilon$, with u being the input image and let f_θ be a CNN, with $\theta \in \mathbb{R}^p$ the space of parameters. To avoid solutions that distribute a lot of mass everywhere, as noted in [2], we ensure that the total mass of the metric is reasonable by applying a transformation $\phi \mapsto \frac{\phi}{\max(\frac{1}{\lambda}\|\phi\|_1, 1)}$ that upper bounds the L^1 norm at a fixed level λ (We chose in this work to empirically bound the total mass at 5).

UNet

In this study, we focus on the task of potential generation and employ two different architectures commonly used for image segmentation: the UNet [12] and a combination of the UNet and ResNet [7]. The UNet is a fully convolutional neural network that is designed for image segmentation, comprising of a contracting path and an expansive path. The contracting path reduces the spatial resolution of feature maps while the expansive path increases it. The combination of these paths allows for the extraction of high-level features from the input image and recovery of the spatial resolution to provide a segmented output.

However, the depth of CNNs can cause the problem of vanishing gradients, which can affect model performance. To address this, we propose the use of ResNet-UNet, a combination of the UNet and ResNet-34 model in the encoder portion of the network. ResNet-34 benefits from deep residual learning and comprises of a 7×7 convolutional layer, a max pooling layer, and 16 residual blocks.

By combining these architectures, ResNet-UNet can capture fine and coarse features of input images and learn deeper and more complex representations. This results in a more accurate and robust model for image segmentation tasks, as demonstrated by our experimental results. Additionally, we introduced modifications to the expansive path of both networks, implementing a dual expansive path system to predict potential energy and a Gaussian potential for the prediction of barycenter. These modifications are illustrated in Fig. 1. Overall, our proposed model demonstrates promising results for potential generation tasks.

Generating Masks with Geodesic Balls

Applications may take advantage of topological priors on the label to reconstruct. For instance one may need to recover regions in an image that we know to be path-connected and of trivial topology. Such regions might be modelled as balls related to a specific distance and recovered as indicator function of such a ball. Formally, we expect for a set E to recover an indicator function as $\chi_{d_\phi(x_0,\cdot)\leq 1}$ for well chosen $x_0 \in \mathbb{R}^d$ and $\phi \in L^1(\Omega)$.

With this method of building masks for specific tasks, we can try to generalize using a neural network architecture and find good potential ϕ to segment interesting regions in images. To do this we would need to compute the gradient of a chosen loss function and thus would need to differentiate the mask, that is why we will replace the indicator function on the unit ball, that would yield zero gradients almost everywhere, by a sigmoid that will smoothly interpolate between the value 1 in the region inside the unit ball and 0 outside. Given the distance map $d_\phi(x_0, \cdot)$, our mask then becomes $\chi^\delta(d_\phi(x_0, \cdot)) = 1 - \frac{1}{1+\exp(-(d_\phi(x_0,\cdot)-1)/\delta)}$, which approaches characteristic function of the unit ball as the parameter δ approaches 0. δ will be taken typically of the order of the size of pixel, i.e. approximately the inverse of the image size.

Figure 2 shows how it is possible to approach the characteristic function of different sets with this formulation. This problem is not convex, so solutions may vary depending on the initialization for instance, but it seems that most of the time potentials converge to a solution that puts a lot of mass on the edges of the mask to recover. The seed here is fixed to x_0 the center of the balls to be fitted, and the potential ϕ is directly optimized using automatic differentiation and ADAM with a "learning rate" equal to 0.01. ϕ^2 is taken as input for the fast marching algorithm instead of ϕ as an easy way to smoothly enforce positivity of the potential.

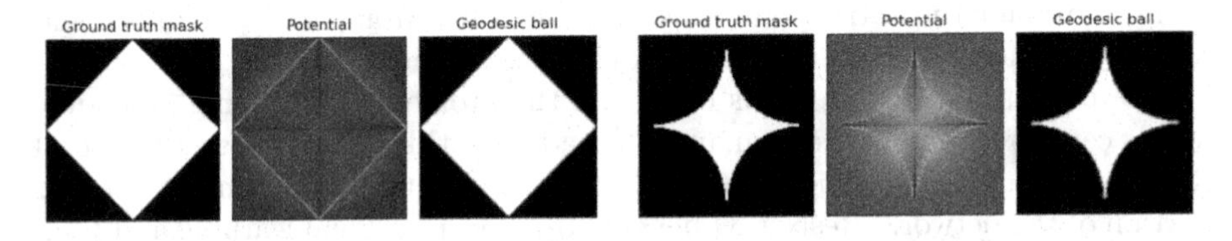

Fig. 2. Example of recovery of an isotropic metric fitting two regions by minimizing $\|\chi^\delta \circ d_{\phi^2} - y\|_2^2$ with respect to ϕ, where y is the ground truth mask, $\delta = 0.01$. x_0 is taken as the center of the mask to be recovered.

4 Experiments

As announced before, our experiments were led on tumour segmentation task.

Data

To conduct our experiments we have used a dataset of Brain MRI segmentation task that is the TCGA_LGG database openly available on the internet [11]. This database contains MRI scans of patients with brain tumours. They correspond to 110 patients (resulting in 1189 images) included in The Cancer Genome Atlas (TCGA) lower-grade glioma collection with at least fluid-attenuated inversion recovery (FLAIR) sequence and genomic cluster data available. We removed tumour with multiple connected components. This dataset is composed of the data of 110 patients. We have used the set of 2D MRI images as our learning and training datasets. We have set aside a 10 patients' data to form a test set as independent as possible (whereas two images from the same patient can be separated in the training and validation set, test data are always the result of a different acquisition from the training and validation set). We applied data augmentation on the training images to increase the diversity of the training set and improve the generalization of the model. The data augmentation techniques used were: horizontal flipping with probability p = 0.5, vertical flipping with probability p = 0.5, random 90-degree rotation with probability p = 0.5, transpose with probability p = 0.5, and a combination of shifting, scaling, and rotating with probability p = 0.25. We respectively set the shift limit, scale limit and rotation limit to 0.01, 0.04, and 0 (as we already perform rotation). We computed the tumour seed using a Euclidean barycenter of the mask region.

Model Training Procedures

The UNet architecture was employed for the task of image segmentation in this study. The model was initialized with Kaiming distribution and trained using the Adam optimizer, which has been widely used in literature due to its capability to adjust the learning rate during training. The learning rate was set to 1e-3, which is a commonly used value in CNNs, as it provides a balance between achieving convergence and avoiding overshooting the optimal solution. In order to optimize the model's performance, to penalize the error between the prediction mask and the groundtruth mask we used a combination of Dice loss and Binary Cross-Entropy (BCE) loss (Eq. (7)).

$$\mathcal{L}_S(x,y) = \frac{2 \times \sum_{i=1}^{N} x_i y_i}{\sum_{i=1}^{N} x_i + \sum_{i=1}^{N} y_i} + \frac{1}{N}\sum_{i=1}^{N} -(x_i \log(x_i) + (1-y_i)\log(1-y_i)) \quad (7)$$

To control the error on the seed prediction a Binary Cross-entropy loss was used.

$$\mathcal{L}_H(h^1,h^2) = \frac{1}{N}\sum_{i=1}^{N} -(h_i^1 \log(h_i^1) + (1-h_i^2)\log(1-h_i^2)) \quad (8)$$

The final loss is:

$$\mathcal{L}(x,y,h^1,h^2) = \mathcal{L}_S(x,y) + L_H(h^1,h^2) \quad (9)$$

The Dice loss function, which is known for its ability to handle imbalanced data, was combined with the BCE loss function, which provides stability during training.

In order to determine the distance between two barycenters, a transformation of the position coordinates into a Gaussian potential is used, based on the following formulation:

$$f(x,y) = \frac{1}{\sqrt{2\pi}\sigma} \exp(\frac{(x-b_i)^2 + (y-b_j)^2}{2\sigma^2}) \quad (10)$$

Here, (b_i, b_j) represent the coordinates of the barycenter. At inference time, the predicted potential is used to identify the maximum location, from which the barycenter coordinates can be extracted.

The model's architecture was initialized with 64 feature maps, which has been shown to be a suitable number for high resolution images, and a batch size of 16 was used during the training process. This combination of hyperparameters allowed the model to effectively use detailed information from the input image while maintaining a balance between generalization and overfitting, as demonstrated by the results presented in this paper. Perhaps it should be clarified that since the two decoders are different and predict two different things, these new parameters do not assist the segmentation compared to the direct method.

Potential Analysis

The potential generated by the neural network was analyzed with respect to the number of training epochs. Results show on Fig. 3 that the output distribution quickly converged towards the boundaries of the tumour to be segmented. However, as training progressed, the contour of the tumour sharpened and boundaries became more distinct and at the same time we can see the brain edges removed. The potential in the end only holds detailed information of the contours in a small area around the tumour.

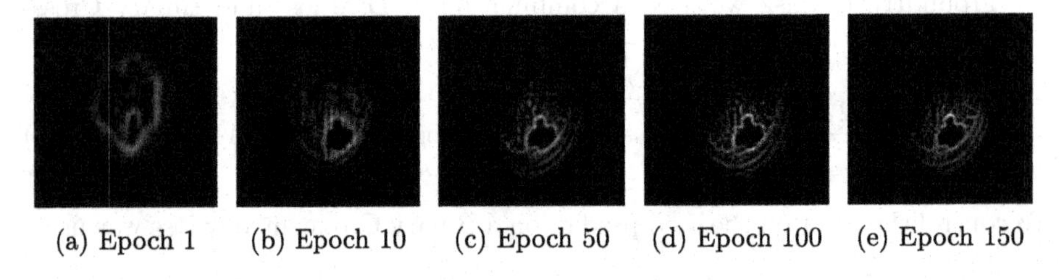

(a) Epoch 1 (b) Epoch 10 (c) Epoch 50 (d) Epoch 100 (e) Epoch 150

Fig. 3. Evolution of the predicted potential taken as input in the Fast Marching Module.

Segmentation Experiments

We compared our method to a standard UNet segmentation approach. As can be seen in the results plots Fig. 4, our method demonstrates clear edge detection. The well-defined contours produced by our method are a result of its ability to take into account the morphology of the image, which traditional filters are not able to do. Furthermore, the problem-specific nature of our method allows for improved performance in image segmentation. Classical metrics allows us to compare quantitatively the results of our segmentation. Overall we recover the same precision on the segmentation mask with minimal improvements of the symmetric Hausdorff distance. However the convergence towards an acceptable solution is faster when combined with the Fast Marching Module since with only a approximate potential the method converge to a relatively close segmentation. Time gives the neural network to more precisely learn the filter and sharpens the edge of the tumour. A general observation from the segmentation in Fig. 4 is that the method when failing to predict correctly a pixel tends to create false positive rather than true false. The Table 1 shows how our method has a high recall

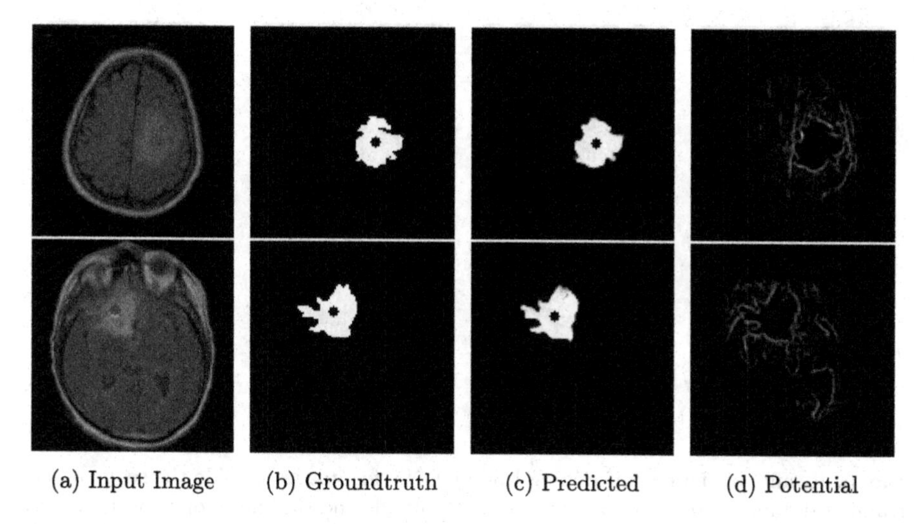

(a) Input Image (b) Groundtruth (c) Predicted (d) Potential

Fig. 4. Results of the segmentation on validation data. On the input image, the blue and green dots are respectively the groundtruth and predicted seed. (Color figure online)

controlling that there is a very low number of false negative. We performed the training with the library *HFM* and the heat method and recorded same results. Overall the UNet architecture shows difficulties to precisely learn the potential while from a metric point of view the ResNet-UNet performs comparatively as the classical segmentation technique using CNNs.

Table 1. Segmentation results (IOU) on the TGCA_LGG brain MRI database.

Name	Dice	IOU	Hausdorff	F1 Score	FPR	FNR
UNet	0.862	0.869	2.313	0.869	0.007	0.05
ResNet UNet	0.873	0.877	2.257	0.877	0.006	0.07
FM UNet (ours)	0.825	0.823	2.505	0.823	0.011	0.064
FM Resnet UNet (ours)	0.863	0.866	2.248	0.866	0.009	0.04

We further studied the properties of the generated potential of our CNN by testing it with dissimilar MRI images found randomly through an image search on Fig. 5 where activated areas correspond to the segmentation ranging from yellow to green for confidence. The results for the last two MRI images show that while the algorithm does not properly segment the tumour (as the predicted barycenter for initialization of the Fast Marching is not correctly placed), the learned filter detects small contours similar to tumours, focusing on the shape of the different objects.

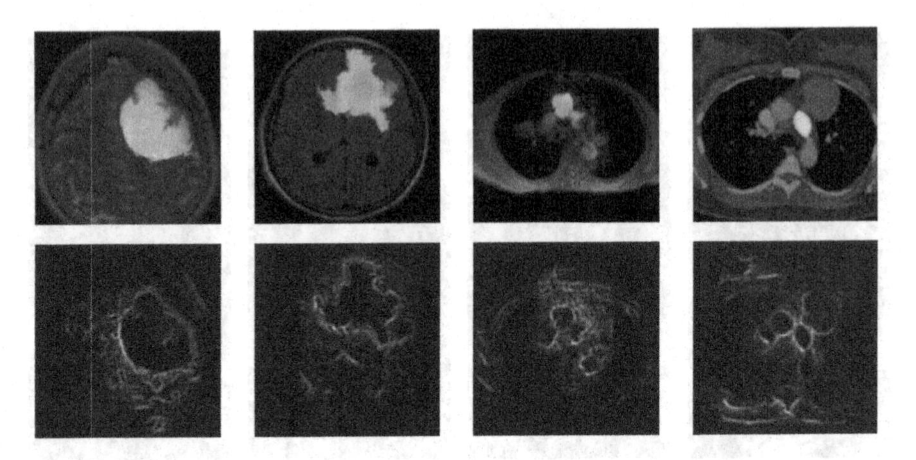

Fig. 5. Results of the Fast Marching Energy CNN for images outside the scope of the training database. Top row: segmentation of outside the training scope. Bottom row: Potential output by the CNN before fast marching.

5 Conclusion

Unlike traditional methods that focus solely on improving segmentation scores, our approach prioritizes the preservation of the tumour's geometrical structure. We have showed that it was possible to learn an interesting potential in order to segment brain tumours as unit balls of geodesic distances, and reach almost state-of-the-art performances on this task. By doing so, our method avoids the limitations of relying solely on convolutional operations, leading to more accurate and reliable results. Our approach offers an alternative path for tumour segmentation that considers both the quality of the segmentation and the preservation of the tumour's structure. This opens new possibilities in terms of geometrical and topological priors for all kinds of tasks.

Further works include extending this framework to general Riemannian metrics, evaluating the capabilities in terms of transfer learning of networks taught with our approach (see Fig. 5) and another possible direct extension of our framework is to try and include multiple connected components for the segmentation mask generated, for which one needs to know or find the number of components in the mask.

Acknowledgements. This work is in part supported by the French government under management of Agence Nationale de la Recherche as part of the "Investissements d'avenir" program, reference ANR-19-P3IA-0001 (PRAIRIE 3IA Institute).

References

1. Ambrosio, L., Brué, E., Semola, D.: Lectures on Optimal Transport. NITEXT. Springer, Cham (2021). https://books.google.fr/books?id=vcI5EAAAQBAJ

2. Benmansour, F., Carlier, G., Peyré, G., Santambrogio, F.: Derivatives with respect to metrics and applications: subgradient marching algorithm. Numerische Mathematik **116**(3), 357–381 (2010). https://doi.org/10.1007/s00211-010-0305-8. https://hal.archives-ouvertes.fr/hal-00360794
3. Benmansour, F., Cohen, L.D.: Tubular anisotropy segmentation. In: Tai, X.-C., Mørken, K., Lysaker, M., Lie, K.-A. (eds.) SSVM 2009. LNCS, vol. 5567, pp. 14–25. Springer, Heidelberg (2009). https://doi.org/10.1007/978-3-642-02256-2_2
4. Bonnivard, M., Bretin, E., Lemenant, A.: Numerical approximation of the Steiner problem in dimension 2 and 3. Math. Comput. (2019). https://doi.org/10.1090/mcom/3442. https://hal.science/hal-01791129
5. Chen, D., Cohen, L.D.: Vessel tree segmentation via front propagation and dynamic anisotropic riemannian metric. In: ISBI 2016, Prague, Czech Republic (2016). https://hal.science/hal-01415036
6. Cohen, L.D., Deschamps, T.: Segmentation of 3D tubular objects with adaptive front propagation and minimal tree extraction for 3D medical imaging. Comput. Methods Biomech. Biomed. Eng. **10**(4), 289–305 (2007). https://doi.org/10.1080/10255840701328239. pMID: 17671862
7. He, K., Zhang, X., Ren, S., Sun, J.: Deep residual learning for image recognition (2015). https://doi.org/10.48550/ARXIV.1512.03385. https://arxiv.org/abs/1512.03385
8. Heitz, M., Bonneel, N., Coeurjolly, D., Cuturi, M., Peyré, G.: Ground metric learning on graphs. J. Math. Imaging Vis. **63**(1), 89–107 (2021). https://doi.org/10.1007/s10851-020-00996-z. https://hal.science/hal-02989081
9. Malladi, R., Sethian, J.: A real-time algorithm for medical shape recovery. In: Sixth International Conference on Computer Vision (IEEE Cat. No. 98CH36271), pp. 304–310 (1998). https://doi.org/10.1109/ICCV.1998.710735
10. Mirebeau, J.M., Portegies, J.: Hamiltonian fast marching: a numerical solver for anisotropic and non-holonomic Eikonal PDEs. Image Process. On Line **9**, 47–93 (2019). https://doi.org/10.5201/ipol.2019.227
11. Pedano, N., et al.: The cancer genome atlas low grade glioma collection (TCGA-LGG) (version 3) (2016). https://doi.org/10.7937/K9/TCIA.2016.L4LTD3TK. https://www.kaggle.com/datasets/mateuszbuda/lgg-mri-segmentation?resource=download
12. Ronneberger, O., Fischer, P., Brox, T.: U-net: convolutional networks for biomedical image segmentation (2015). https://doi.org/10.48550/ARXIV.1505.04597. https://arxiv.org/abs/1505.04597
13. Scarvelis, C., Solomon, J.: Riemannian metric learning via optimal transport (2022). https://doi.org/10.48550/ARXIV.2205.09244. https://arxiv.org/abs/2205.09244
14. Sethian, J.A.: A fast marching level set method for monotonically advancing fronts. Proc. Natl. Acad. Sci. **93**(4), 1591–1595 (1996). https://doi.org/10.1073/pnas.93.4.1591. https://www.pnas.org/doi/abs/10.1073/pnas.93.4.1591
15. Zhou, S., et al.: A review of deep learning in medical imaging: imaging traits, technology trends, case studies with progress highlights, and future promises. Proc. Inst. Radio Eng. **109**(5), 820–838 (2021). https://doi.org/10.1109/JPROC.2021.3054390

Deep Accurate Solver for the Geodesic Problem

Saar Huberman(✉), Amit Bracha(✉), and Ron Kimmel(✉)

Technion - Israel Institute of Technology, Haifa, Israel
{saarhuberman,amit.bracha,ron}@cs.technion.ac.il

Abstract. A common approach to compute distances on continuous surfaces is by considering a discretized polygonal mesh approximating the surface and estimating distances on the polygon. We show that exact geodesic distances restricted to the polygon are at most second order accurate with respect to the distances on the corresponding continuous surface. Here, by *order of accuracy* we refer to the rate of convergence as a function of the average distance between sampled points. Next, a higher order accurate deep learning method for computing geodesic distances on surfaces is introduced. Traditionally, one considers two main components when computing distances on surfaces: a numerical solver that locally approximates the distance function, and an efficient causal ordering scheme by which surface points are updated. Classical minimal path methods often exploit a dynamic programming principle with quasi-linear computational complexity in the number of sampled points. The quality of the distance approximation is determined by the local solver that is revisited in this paper. To improve the accuracy, we consider a neural network based local solver which implicitly approximates the structure of the continuous surface. We supply numerical evidence that the proposed learned update scheme provides better accuracy compared to the best possible polyhedral approximations and previous learning-based methods. The result is a third order accurate solver with a bootstrapping-recipe for improvement.

Keywords: Distance map · multilayer perceptron · numerical convergence · latent space

1 Introduction

Geodesic distance is defined as the length of the shortest path connecting two points on a surface. It can be considered as a generalization of the Euclidean distance to curved manifolds. The approximation of geodesic distances is used as a building block in many applications. It can be found in robot navigation [8,11], and shape matching [4,7,17,19], to name just a few examples. Thus, for effective and reliable use, computation of geodesics is expected to be both *fast* and *accurate*.

Over the years, many methods have been proposed for computing distances on polygonal meshes that compromise between the accuracy of the distance

L. Calatroni et al. (Eds.): SSVM 2023, LNCS 14009, pp. 288–300, 2023.
https://doi.org/10.1007/978-3-031-31975-4_22

approximation and the complexity of the algorithm. One family of algorithms for computing distances in the domain of polyhedral meshes is based on solutions to the exact discrete geodesic problem introduced by Mitchell *et al.* [15]. The problem is defined as finding the exact distances on a polyhedron. Mitchell *et al.* proposed a $\mathcal{O}(N^2 \log(N))$ complexity algorithm, where N is the number of vertices, which was among the first method for computing exact distances on non-convex triangulated surfaces. That quadratic algorithm, known as MMP, is computationally demanding and challenging to code, with the first implementation introduced 18 years after its publication by Surazhsky *et al.* [20].

At the other end, a popular family of methods for efficient approximation of distances known as *fast marching*, involves quasi-linear computational complexity $\mathcal{O}(N \log(N))$. These methods are based on the solution of an *eikonal equation*. The fast marching method consists of two main components, a heap sorting strategy and a local numerical solver, often referred to as a numerical update procedure. Fast marching, originally introduced for regularly sampled grids [18,21], was extended to triangulated surfaces in [10]. While operating on curved surfaces approximated by triangulated meshes, the first proximity neighbors of a vertex in the mesh are used to locally approximate the solution of an eikonal equation, resulting in a *first-order-accurate scheme* in terms of a typical triangle's edge length denoted as $\mathcal{O}(h)$. Another prominent class of numerical solvers is the fast sweeping methods [9,12,22,23], iterative schemes that use alternating sweeping ordering. These methods have a worst case complexity of $\mathcal{O}(N^2)$, [6]. Finally, [2] proposed a Poisson solver that aligns the unit gradients of a distance function with the gradient directions of a short time heat kernel, ending up with a first order accurate solver.

We start our journey by proving that the exact geodesic distances computed on a polygonal mesh approximating a continuous surface would be at most a second order approximation of the corresponding distances on the surface. To overcome the second order accuracy limitation, we extend the numerical support about each vertex beyond the classical one ring approximation, and utilize the universal approximation properties of neural networks. The low complexity of the well-known dynamic programming update scheme [3], combined with our novel neural network-based solver, yields an *efficient* and *accurate* method.

In a related previous effort a neural network based local solver for the computation of geodesic distances was proposed [13]. We improve Lichtenstein's $\mathcal{O}(h^2)$ approach by extending the local neighborhood numerical support, and refining the network's architecture to obtain $\mathcal{O}(h^3)$ accuracy at similar linear complexity $\mathcal{O}(N \log(N))$. In fact, directly extending the local support in [13] to 3rd ring neighborhood, does not improve the accuracy of the method. It appears as if 90% of that model's latent space is inactive. Based on these observations we propose to add hidden layers, change the activation functions, and reduce the size of the latent space.

Similarly to [13], the suggested solver is trained in a supervised manner using ground truth examples. And yet, since geodesics can not be derived analytically except for a limited set of surfaces like spheres and planes, we propose a multi-hierarchy bootstrapping technique. Namely, we use distance approximations on high resolution sampled meshes to better approximate distances at low resolutions. We utilize our ability to apply given solvers at high resolution to generate higher order accurate training examples at low resolutions.

Contributions. We show that exact geodesics on polyhedrons are second order approximations. As a remedy, we develop a *fast* and *accurate* geodesic distance approximation method on surfaces. For fast computation, we use a distance update scheme (Algorithm 1) that guarantees quasi-linear computational complexity. For accuracy, by revisiting the ingredients of the solver suggested in [13], we propose a network based solver that operates directly on the sampled mesh vertices. Finally, to provide accurate ground truth distances required for training our solver, we propose a novel data generation bootstrapping procedure.

2 Exact Distances on Polyhedral Approximations

Theorem 1. *Let $\mathcal{M} : \Omega \in \mathbb{R}^2 \to \mathbb{R}^3$ be a Riemannian two dimensional manifold with effective Gaussian curvature a.e. Let $C(s) : [0, L] \to \mathcal{M}$ be a minimal geodesic connecting two surface points $C(0)$ and $C(L)$ on $\mathcal{M}$ with arclength parametrization s, and L the length of C. The length of C differs by $\mathcal{O}(h^2)$ from the sum of the lengths of the cords. These line segments, of length h each as measured in $\mathbb{R}^3$, are defined by a sequence of surface points $C(s_i)$ and $C(s_{i+1})$. That is, the length of the approximation γ defined by its vertices $\{C(0), C(s_2), \ldots, C(s_{n-1}), C(L)\}$, given by $L(\gamma) = \sum_{i=1}^{n-1} \|C(s_{i+1}) - C(s_i)\|_{\mathbb{R}^3} = nh$, differs by $\mathcal{O}(h^2)$ from $L(C) = \int_0^L ds$.*

Proof (Proof of Theorem 1). Consider the length parameterization along the line segment with end points $C(s_i)$ and $C(s_{i+1})$ be given by $t \in [-h/2, h/2]$, and assume w.l.o.g. the monotone increasing reparametrization $s(t)$ that would allow us to parametrize the surface geodesic segment between $C(s_i)$ and $C(s_{i+1})$. As t is the arclength along the cord connecting the two end points of the line segment, by freedom of parametrization, we could choose $|C_t(0)| = 1$.

Next, let us compute the length of $C(t)$ in the ith interval, $L_{\mathcal{M}}(C(s_i), C(s_{i+1})) = \int_{s_i}^{s_{i+1}} ds = \int_{-h/2}^{h/2} |C_t| dt$. Let us expand $|C_t|$ about 0, by which we have

$$\begin{aligned} |C_t(t)| &= |C_t(0)| + t\left(\frac{d}{dt}|C_t|\right)(0) + \frac{t^2}{2}\left(\frac{d^2}{dt^2}|C_t|\right)(0) + \cdots \\ &= |C_t(0)| + t\left(\frac{\langle C_t, C_{tt}\rangle}{|C_t|}\right)(0) + \frac{t^2}{2}\left(\frac{d}{dt}\frac{\langle C_t, C_{tt}\rangle}{|C_t|}\right)(0) + \cdots \end{aligned} \tag{1}$$

Let us focus on the second term, $\frac{\langle C_t, C_{tt}\rangle}{|C_t|} = \frac{\langle C_t, C_{tt}\rangle}{|C_t|^3}|C_t|^2 = \kappa |C_t|^2$, where κ is the curvature (normal curvature for a geodesic) of C at that point. The third term is given by $\frac{d}{dt}\kappa|C_t|^2 = \kappa_t|C_t|^2 + 2\kappa\langle C_t, C_{tt}\rangle = \kappa_s|C_t|^3 + 2\kappa^2|C_t|^3$.

We conclude with

$$L_i = \int_{s_i}^{s_{i+1}} ds = \int_{-h/2}^{h/2} |C_t| dt$$

$$= \int_{-h/2}^{h/2} \left(|C_t(0)| + t \left(\kappa |C_t|^2\right)(0) + \frac{t^2}{2} \left(\kappa_s |C_t|^3 + 2\kappa^2 |C_t|^3\right)(0) + \cdots \right) dt$$

$$= |C_t(0)| h + \left(\kappa_s |C_t|^3 + 2\kappa^2 |C_t|^3\right)(0) \frac{h^3}{24} + \mathcal{O}(h^5). \tag{2}$$

With our specific selection of $|C_t(0)| = 1$, we conclude with the overall error given by $\mathcal{E}rr = \sum_{i=1}^{n-1} |L_i - h| = \sum_{i=1}^{n-1} \left|(\kappa_s + 2\kappa^2)\frac{h^3}{24} + \mathcal{O}(h^5)\right| = \mathcal{O}(h^3)\mathcal{O}(n)$, where κ and κ_s are evaluated at $t = 0$ for each segment. Note, that κ and κ_s are geometric quantities and thus could be regarded as effective bounded constants. Then, assuming $h \approx \mathcal{O}(n^{-1})$ we have the convergence rate to be $\mathcal{O}(h^2)$.

3 Geodesics: $\mathcal{O}(h^3)$ Accuracy at $\mathcal{O}(N \log N)$ Complexity

We present a neural network based method for approximating accurate geodesic distances on surfaces. Similar to most dynamic programming methods, like the fast marching scheme, the proposed method consists of a numerical solver that locally approximates the distance function u at a surface point p, and an ordering scheme that defines the order of the visited points. Here, the N sampled surface points are divided into three disjoint sets.

1. *Visited*: points where the distance function $u(p)$ has already been computed and will not be changed.
2. *Wavefront*: points where the computation of $u(p)$ is in progress and is not yet fixed.
3. *Unvisited*: points where $u(p)$ has not yet been computed.

Algorithm 1. Distance Updating Scheme

1: **Definitions:**
  S - Set of all source points
  p - point on the surface
  $u(p)$ - minimal distance from sources to p
2: **Initialize:**
  $u(p) = 0$, tag p as *Visited*; $\forall p \in S$
  $u(p) = \infty$, tag p as *Unvisited*; $\forall p \notin S$
  Tag all *Unvisited* points adjacent to *Visited* points as *Wavefront*
3: **repeat**
4:   **for** $p \in$ *Wavefront* **do**
5:     Approximate $u(p)$ based on *Visited* points
6:   **end for**
7:   Tag the least distant *Wavefront* point p' as *Visited*
8:   Tag all *Unvisited* neighbors of p' as *Wavefront*
9: **until** all points are *Visited.*
10: **Return** u

The distances at the sampled surface points are computed according to Algorithm 1, where Step 5 of the scheme is performed by the proposed local solver. When applied to a target point $p \in$ *Wavefront*, the local solver uses a predefined maximum number of *Visited* points. These *Visited* points are chosen from the local neighborhood and are not related to the number of points on the mesh; hence, a single operation of our solver has constant complexity. Since, within our dynamic programming setting, the proposed method retains the heap sort ordering scheme, the overall computational complexity is $\mathcal{O}(N\log(N))$. Subsection 3.1 introduces the operation of the local solver, presents the required pre-processing and elaborates on the implementation of the neural network. Next, Subsect. 3.2 explains how the dataset is generated and the network weights are optimized. Finally, Subsect. 3.3 details how ground truth distances are calculated when no analytic closed form solution is available.

3.1 Local Solver

Here, we introduce a local neural network-based solver[1] for Step 5 of Algorithm 1. When the solver is applied to a given point $p \in$ *Wavefront*, it receives as input the coordinates and distance function values of its neighboring points. The neighboring points, denoted by $\mathcal{N}(p) = \{p_1, p_2, ..., p_M\}$, are defined by all vertices connected to p by a path of at most 3 edges, which is often referred to as third ring neighborhood. Based on the information from the *Visited* points in $\mathcal{N}(p)$, the local solver approximates the distance function $u(p)$. This way, we keep utilizing the order of updates that characterizes the construction of distance functions. As mentioned earlier, for a given target point p and neighboring points $\{p_i\}_{i=1}^{M} \subset \textit{Visited} \cap \mathcal{N}(p)$, the input to our solver is $\{(x_{p_i}, y_{p_i}, z_{p_i}, u(p_i))\}_{i=1}^{M} \cup \{(x_p, y_p, z_p)\}$. To address the solver's generalization capability and to handle diverse possible inputs, we transform the input to the neural network into a canonical representation. To this end, we design a preprocessing pipeline. The coordinates are centered with respect to the target point, resulting in relative coordinates $(x_{p_i} - x_p, y_{p_i} - y_p, z_{p_i} - z_p)$, and $\min_j\{u(p_j)\}$ is subtracted from the values of the distance function $\{u(p_i)\}_{i=1}^{M}$. After the input is centered, it is scaled so that the mean L2 norm of the coordinates is of unit size. Last, a $SO(3)$ rotation matrix is applied to the coordinates so that their first moment is aligned with a predefined principal directions. The processed input is fed into the neural network and the output is further processed to reverse the centering and scaling transformations.

The input neighborhood has no fixed order and can be viewed as a set. To properly handle our unstructured set of points, we train our neural network output to be permutation invariant. The network's architecture consists of three main components. A shared weight encoder that lifts the 4-dimensional input to 512 features using residual multi-layer perceptron (MLP) blocks [14]. A per-feature max pooling operation that results in a single 512 feature vector, and

[1] The code and further implementation consideration details of the neural network will be published upon acceptance.

a fully connected regression network of dimensions $(512, 1024, 512, 256, 1)$ that outputs the desired target distance.

3.2 Training the Local Solver

We use a customary supervised training procedure, using examples with corresponding ground-truth distances. These ground truth distances are obtained by applying a bi-level sampling strategy, as detailed in Sect. 3.3. Given an input $\{(x_{p_i}, y_{p_i}, z_{p_i}, u(p_i)\}_{i=1}^{M}$, our network is trained to minimize the difference between its output and its corresponding ground truth, denoted by $u_{gt}(p)$. To develop a reliable and robust solver, we create a diverse dataset that simulates a variety of scenarios. We construct this dataset by selecting various source points and sampling local neighborhoods at different random positions relative to the sources. According to the causal nature of our algorithm, we build our training examples, such that a neighboring point p' is defined as *Visited* and is allowed to participate in the prediction of $u(p)$ if $u_{gt}(p') < u_{gt}(p)$. The network's parameters Θ are optimized to minimize the Mean Square Error (MSE) loss

$$L(\Theta) = \frac{1}{K}\sum_{j=1}^{K}(f_{\Theta}(\{(x_{p_{i,j}}, y_{p_{i,j}}, z_{p_{i,j}}, u(p_{i,j}))\}_{i=1}^{M}) - u_{gt}(p_j))^2, \tag{3}$$

where K is the number of examples in the training set and $p_{i,j}$ corresponds to the ith neighbor of the target point p_j. The coordinates and distances used in our training procedure are in their canonical form, after being translated, rotated and scaled, as explained in Sect. 3.1.

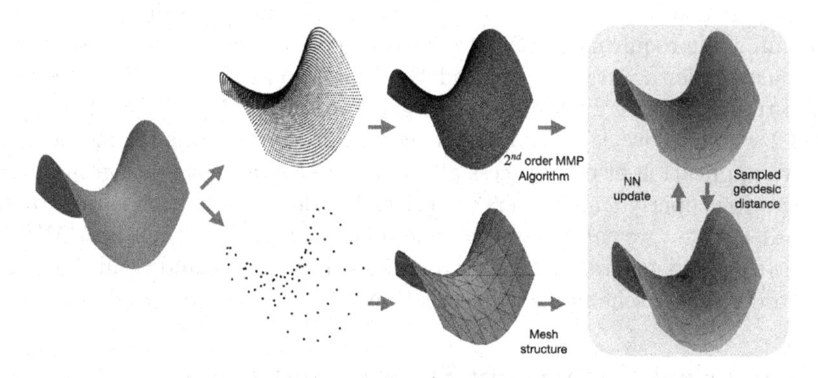

Fig. 1. Bootstrapping by training. Distance values computed for a high h^2-resolution sampled mesh of a continuous surface with an r accurate scheme yields $\mathcal{O}(h^{2r})$ accurate distances given at the mesh points. The mesh can then be sampled into a lower h-resolution mesh of the same continuous surface, while keeping the corresponding $\mathcal{O}(h^{2r})$ accurate distances at the vertices. See text for an elaborated discussion regarding data augmentation at high resolution and training more accurate update procedures at the low resolution.

3.3 Learning to Augment

Exact distances on continuous surfaces are given by analytic expressions for a very limited set of continuous surfaces; namely, for spheres and planes. Since our solver is trained on examples containing ground truth distances, an additional approximation algorithm must be considered to generate our training examples for general surfaces. Currently, the most accurate axiomatic method for distance computation is the MMP algorithm, which computes "exact" polyhedral distances. Considering polyhedral surfaces as sampled continuous ones, the "exact" distances on triangulated surfaces are 2nd order accurate with respect to the edge length h. Indeed, the MMP is an $\mathcal{O}(h^2)$ accurate method. In order to train our network with more accurate than $\mathcal{O}(h^2)$ distances for general smooth surfaces, we resort to the following bootstrapping idea.

We introduce a multi-resolution ground truth boosting generation technique that allows us to obtain ground truth distances of any desired order. The underlying idea is that distances computed on a mesh obtained from a denser sampling of the surface are a better approximation to the distances on the continuous surface. When generating examples from a given surface, two sampling resolutions of the surface are obtained and corresponding meshes are formed, denoted by $\mathcal{M}_{dense}$ and $\mathcal{M}_{sparse}$, respectively. Distances are computed on the high-resolution mesh $\mathcal{M}_{dense}$ and the obtained distance map is sampled at $\mathcal{M}_{sparse}$. Consider h_{dense}, h_{sparse} that correspond to the mean edge length of the polygons $\mathcal{M}_{dense}, \mathcal{M}_{sparse}$, so that, $h_{dense} = h^q_{sparse}$. The distances computed by an approximation method of order r on $\mathcal{M}_{dense}$ are r order accurate $\mathcal{O}(h^r_{dense})$. Therefore, the same approximated distances, sampled at the corresponding vertices of $\mathcal{M}_{sparse}$, have $\mathcal{O}(h^{qr}_{sparse})$ accuracy.

Using the distance samples of the polyhedral distances obtained by the MMP algorithm while requiring $q \geq 2$, allows us to generate distance maps that are at least fourth-order accurate. By considering these approximated distances as our ground truth, training examples can be generated from $\mathcal{M}_{sparse}$ as described in Sect. 3.2 which allow us to properly train a third-order accurate method. The iterative application of this process allows us to generate accurate ground truth distances to properly train solvers of arbitrary order. For example, after training a 3rd order solver, we can apply the same process while replacing the MMP with our new solver to generate a $\mathcal{O}(h^6)$ ground truth distances and train a solver up to 6th order. For a schematic representation of this technique, see Fig. 1.

4 Numerical Evaluation: Spheres and Beyond

Geodesic distances on spheres can be calculated analytically. Therefore, they are well suited for the evaluation of our method. For two given points $a = (x_a, y_a, z_a)$, $b = (x_b, y_b, z_b)$ lying on a sphere of radius r, the geodesic distance between them is defined by $u(a, b) = r \arccos\left((x_a x_b + y_a y_b + z_a z_b)/r^2\right)$.

To train our solver, we first randomly sampled spheres at different resolutions and obtained triangulated versions of them. Using the exact distances we created a data set with 100,000 examples and applied a training procedure as presented

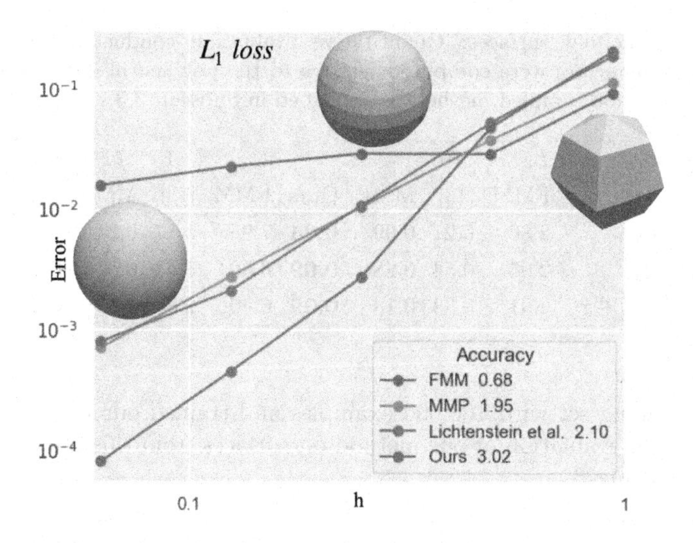

Fig. 2. Order of accuracy: Plots showing how the edge size effects the error. The accuracy of each scheme is defined by the corresponding slope. We used mesh approximations of a unit sphere with different edge size.

in Sect. 3.2. To evaluate our method, we constructed a hierarchy of spheres with various resolutions, see Fig. 2.

As described in [16], we assume that the exact solution $u(a, b)$ can be written as $u(a, b) = u_h(a, b) + Ch^R + \mathcal{O}(h^{R+1})$, where C is a constant, u_h is defined as the approximate solution on a mesh with a corresponding mean edge length of h, and R is the order of accuracy. For two given mesh resolutions of the same continuous surface $\mathcal{M}_1, \mathcal{M}_2$ with corresponding h_1, h_2, we can estimate our method's order of accuracy by

$$R = \log_{\frac{h_1}{h_2}} \left(\frac{u - u_{h_1}}{u - u_{h_2}} \right). \tag{4}$$

The evaluation of our method is shown in Fig. 2, where the slope of each graph indicates the order of accuracy R. It can be seen that our method has a higher order of accuracy than the classical fast marching (FMM), the MMP exact geodesic method, and the previous deep learning method proposed by Lichtenstein et al.

4.1 Generalization to Polynomial Surfaces

We evaluated our method on second order polynomial surfaces. In general, there is no closed form analytical expression for geodesic distances on such surfaces. To train our solver, we generated a wide variety of polynomial surfaces and obtained an accurate approximation of their geodesics for a range of sampling resolutions, as described in Sect. 3.3. After obtaining an accurate geodesic distance map, we

Table 1. Polynomial surfaces: Quantitative evaluation conducted on 2nd order paraboloids. The errors were computed relative to the polyhedral distance projected from high-resolution sampled meshes, as described in Subsect. 3.3.

Surface	L_1 $(\times 10^{-2})$				L_∞ $(\times 10^{-2})$			
	FMM	[13]	MMP	Ours	FMM	[13]	MMP	Ours
$x^2 - y^2$	2.86	0.21	0.09	**0.04**	7.95	1.17	0.24	**0.16**
$x^2 + y^2$	2.05	0.34	0.28	**0.09**	6.80	1.44	0.67	**0.26**
$x^2 - y^2 + xy$	2.91	0.33	0.18	**0.09**	6.40	2.60	0.63	**0.28**

created a training set with 100,000 examples and trained our model according to Sect. 3.2. An evaluation of our method on surfaces from this family is shown in Table 1 and Fig. 3.

4.2 Generalization to Arbitrary Surfaces

To better emphasize the generalization ability of our method, we conduct an additional experiment. We train our solver only on the three 2nd order polynomial surfaces shown in Table 1, and evaluate it on arbitrary shapes from the TOSCA dataset [1]. It can be seen in Fig. 4 and Table 2, that our method generalizes well and leads to significantly lower errors compared to the heat method, classical fast marching and the method presented by [13] when trained on the same polynomial surfaces. Errors are computed relative to the polyhedral distances, since they are the most accurate distances available to us for these shapes.

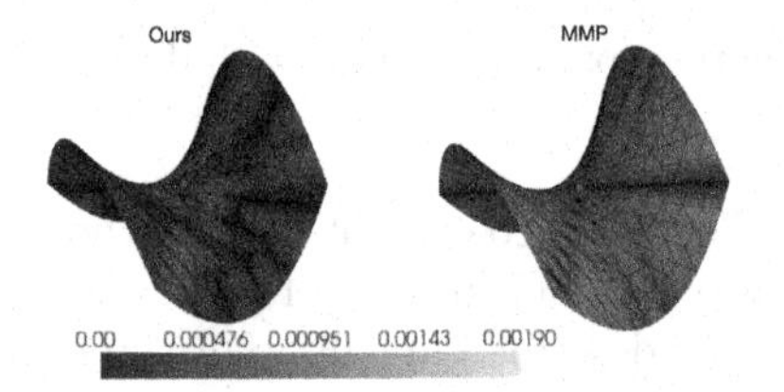

Fig. 3. Polynomial surfaces: Errors presented for the polyhedral scheme and the proposed method. Local errors, represented as colors on the surface, were computed relative to exact polyhedral distances computed at a high-resolution sampled mesh of the continuous surface, as described in Subsect. 3.3.

4.3 Ablation Study

To analyze the performance and robustness of our method, we conduct additional tests. These include modifying the local numerical support by which neighborhoods are defined and the precision point representation of the network weights.

Local Numerical Support. In the fast-marching method, the solver locally estimates a solution to the eikonal equation using a finite-difference approximation of the gradient. This approximation of the gradient is defined by a local stencil. For example, in the case of regularly sampled grids, the one-sided difference formula for a third order approximation requires a stencil with three

Fig. 4. Generalization to arbitrary surfaces. Top row: Iso-contours shown for our method. Bottom row: errors presented (left to right) for the heat kernel method, fast marching, Lichtenstein et al. and the proposed method. Local errors presented as colors on the surface (brighter color indicates higher error), were computed relative to the polyhedral distances. The evaluation was done on the TOSCA data base, whereas our solver and the solver proposed by Lichtenstein et al. were trained with only limited number of 2nd order polynomial surfaces, the three paraboloid surfaces presented in Table 1.

Table 2. Generalization to arbitrary surfaces: Quantitative evaluation tested on TOSCA. The errors are relative to the MMP scheme. Our solver and the one proposed in [13] were trained using our bootstrapping method Sect. 3.3 with a limited number of 2nd order polynomial surfaces, the three paraboloids presented in Table 1.

Shape	L_1 $(\times 10^{-1})$				L_∞ $(\times 10^{-1})$			
	Heat	FMM	[13]	Ours	Heat	FMM	[13]	Ours
Dog	0.728	0.110	0.123	**0.037**	8.688	1.514	1.318	**0.465**
Cat	1.596	0.136	0.386	**0.053**	6.541	0.631	3.450	**0.384**
Wolf	0.440	0.162	0.169	**0.072**	2.244	1.009	1.343	**0.422**
Horse	1.084	0.136	0.199	**0.068**	8.239	2.273	1.616	**0.896**
Michael	1.185	0.109	1.073	**0.054**	5.906	1.602	4.873	**0.881**
Victoria	1.211	0.076	0.479	**0.027**	4.538	1.075	2.289	**0.414**
Centaur	0.672	0.188	1.088	**0.062**	49.437	1.985	7.348	**1.417**

neighboring points [5]. In analogy to the stencil, our method uses a 3rd ring neighborhood. Our local solver does not explicitly solve an eikonal equation nor does it use an approximation of the gradient. Yet, the size of the numerical support is the underlying ingredient that allows our neural network to realize high order accuracy. We have empirically validated this, as depicted in Fig. 5 left.

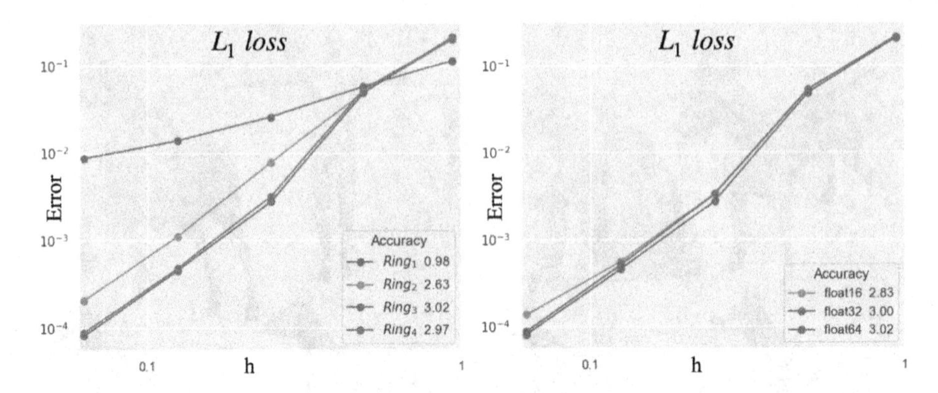

Fig. 5. Left: Local neighborhood: Evaluation of the proposed method on spheres with different local neighborhoods support. Ring_i corresponds to a neighborhood containing all vertices with at most i edges from the evaluated target. Right: Precision floating point representation: Evaluation of our method on spheres with different floating point precision representation of the neural network weights.

Numerical Precision. The choice of numerical representation is an important decision in the implementation of a neural network based solver. It leads to a trade-off between the accuracy of the solver and its execution time and memory footprint. In all our experiments, our main focus is on the accuracy of the method. Hence, we used double precision floating point for our neural network implementation. Figure 5 right shows a comparison between our implemented network with different precision, showing only a slight degradation when a single and half precision floating points are used.

5 Conclusions

A fast and accurate method for computing geodesic distances on surfaces was presented. Revisiting the method of Lichtenstein et al. [13] we modified the ingredients of a neural network based local solver. The solver proposed by Lichtenstein et al. was limited to second order accuracy. The suggested improvements involved extending the local solver's numerical support and redesigning the network's architecture. Next, we introduced a data generation mechanism that provides accurate high-order distances to train our solver by. It allowed us to use surfaces for which there is no analytic way to compute geodesic distances. We trained our solver using examples generated by a novel multi-resolution bootstrapping technique that projects distances computed at high resolutions to lower ones. We believe that the proposed bootstrapping idea could be utilized for training other numerical solvers while keeping in mind that the numerical support enables the required accuracy. We trained a neural network to locally extrapolate the values of the distance function on sampled surfaces. The result is the most accurate method that runs at the lowest (quasi-linear) computational complexity, compared to all existing methods.

Acknowledgements. The research was partially supported by the D. Dan and Betty Kahn Michigan-Israel Partnership for Research and Education, run by the Technion Autonomous Systems and Robotics Program. We are grateful to Mr. Alon Zvirin for all his efforts in improving the writeup and overall presentation of the ideas introduced in this paper.

References

1. Bronstein, A.M., Bronstein, M.M., Kimmel, R.: Numerical Geometry of Non-rigid Shapes. Springer, New York (2008). https://doi.org/10.1007/978-0-387-73301-2
2. Crane, K., Weischedel, C., Wardetzky, M.: Geodesics in heat: a new approach to computing distance based on heat flow. ACM Trans. Graph. (TOG) **32**(5), 1–11 (2013)
3. Dijkstra, E.W.: A note on two problems in connexion with graphs. Numer. Math. **1**(1), 269–271 (1959)
4. Elad, A., Kimmel, R.: Bending invariant representations for surfaces. In: Proceedings of the 2001 IEEE Computer Society Conference on Computer Vision and Pattern Recognition, CVPR 2001, vol. 1, pp. I-I. IEEE (2001)
5. Fornberg, B.: Generation of finite difference formulas on arbitrarily spaced grids. Math. Comput. **51**(184), 699–706 (1988)
6. Hysing, S.R., Turek, S.: The eikonal equation: numerical efficiency vs. algorithmic complexity on quadrilateral grids. In: Proceedings of ALGORITMY, vol. 22 (2005)
7. Ion, A., Artner, N.M., Peyré, G., Mármol, S.B.L., Kropatsch, W.G., Cohen, L.: 3D shape matching by geodesic eccentricity. In: 2008 IEEE Computer Society Conference on Computer Vision and Pattern Recognition Workshops, pp. 1–8. IEEE (2008)
8. Kimmel, R., Kiryati, N., Bruckstein, A.M.: Multivalued distance maps for motion planning on surfaces with moving obstacles. IEEE Trans. Robot. Autom. **14**(3), 427–436 (1998)
9. Kimmel, R., Maurer, R.: Method of computing sub-pixel euclidean distance maps, application filed Dec. 2000, US Patent 7,113,617 (2006)
10. Kimmel, R., Sethian, J.A.: Computing geodesic paths on manifolds. Proc. Natl. Acad. Sci. **95**(15), 8431–8435 (1998)
11. Kimmel, R., Sethian, J.A.: Optimal algorithm for shape from shading and path planning. J. Math. Imaging Vis. **14**(3), 237–244 (2001)
12. Li, F., Shu, C.W., Zhang, Y.T., Zhao, H.: A second order discontinuous galerkin fast sweeping method for eikonal equations. J. Comput. Phys. **227**(17), 8191–8208 (2008)
13. Lichtenstein, M., Pai, G., Kimmel, R.: Deep Eikonal solvers. In: Lellmann, J., Burger, M., Modersitzki, J. (eds.) SSVM 2019. LNCS, vol. 11603, pp. 38–50. Springer, Cham (2019). https://doi.org/10.1007/978-3-030-22368-7_4
14. Ma, X., Qin, C., You, H., Ran, H., Fu, Y.: Rethinking network design and local geometry in point cloud: a simple residual MLP framework. arXiv preprint arXiv:2202.07123 (2022)
15. Mitchell, J.S., Mount, D.M., Papadimitriou, C.H.: The discrete geodesic problem. SIAM J. Comput. **16**(4), 647–668 (1987)
16. Osher, S., Sethian, J.A.: Fronts propagating with curvature-dependent speed: algorithms based on Hamilton-Jacobi formulations. J. Comput. Phys. **79**(1), 12–49 (1988)

17. Panozzo, D., Baran, I., Diamanti, O., Sorkine-Hornung, O.: Weighted averages on surfaces. ACM Trans. Graph. (TOG) **32**(4), 1–12 (2013)
18. Sethian, J.A.: A fast marching level set method for monotonically advancing fronts. Proc. Natl. Acad. Sci. **93**(4), 1591–1595 (1996)
19. Shamai, G., Kimmel, R.: Geodesic distance descriptors. In: Proceedings of the IEEE Conference on Computer Vision and Pattern Recognition, pp. 6410–6418 (2017)
20. Surazhsky, V., Surazhsky, T., Kirsanov, D., Gortler, S.J., Hoppe, H.: Fast exact and approximate geodesics on meshes. ACM Trans. Graph. (TOG) **24**(3), 553–560 (2005)
21. Tsitsiklis, J.N.: Efficient algorithms for globally optimal trajectories. IEEE Trans. Autom. Control **40**(9), 1528–1538 (1995)
22. Weber, O., Devir, Y.S., Bronstein, A.M., Bronstein, M.M., Kimmel, R.: Parallel algorithms for approximation of distance maps on parametric surfaces. ACM Trans. Graph. (TOG) **27**(4), 1–16 (2008)
23. Zhao, H.: A fast sweeping method for eikonal equations. Math. Comput. **74**(250), 603–627 (2005)

Deep Image Prior Regularized by Coupled Total Variation for Image Colorization

Gaetano Agazzotti[1], Fabien Pierre[2(✉)], and Frédéric Sur[2]

[1] Mines Nancy and LORIA UMR 7503, Université de Lorraine, CNRS, INRIA, Nancy, France
[2] LORIA UMR 7503, Université de Lorraine, CNRS, INRIA, Nancy, France
fabien.pierre@univ-lorraine.fr

Abstract. Automatic image colorization is an old problem in image processing that has regained interest in the recent years with the emergence of deep-learning approaches with dramatic results. A careful examination shows that these methods often suffer from the so-called "color halos" or "color bleeding" effect: some colors are not well localized and may cross shape edges. This phenomenon is caused by the non-alignment of edges in the luminance and chrominance maps. We address this problem by regularizing the output of an efficient image colorization method with deep image prior and coupled total variation.

Keywords: Image colorization · Color halos · Deep image prior · Coupled total variation

1 Introduction

Image and video colorization is the process of adding colors to monochromatic (black-and-white or sepia) pictures and movies, for example legacy documents. This is an ill-posed problem since many solutions may match the human perception. Industrial applications require automatic methods to reduce costs. Historically, the first semi-automatic methods made use of a reference color image or of user-defined color scribbles, in particular through variational approaches. We refer the interested reader to the recent survey paper [14]. Since large datasets made of pairs of monochromatic and color images are easy to build (by simply desaturating a color image), it is possible to train neural networks to infer a color image from a monochromatic input image. Effective colorization algorithms based on deep learning have therefore emerged in the past few years. The present contribution focuses on automatic deep-learning-based approaches without user interaction (contrary to [20] or to the hybrid [6] for instance).

Most deep-learning models infer at each pixel either a color component [7,9,13,18,19] or a probability distribution over the color space [4,5,10,21]. To limit the computing time, these models are primarily able to infer a low-resolution color image, or are based on a super-pixel representation, or on an auto-encoder. This requires upsampling in a further stage. Upsampling the low-resolution

L. Calatroni et al. (Eds.): SSVM 2023, LNCS 14009, pp. 301–313, 2023.
https://doi.org/10.1007/978-3-031-31975-4_23

Fig. 1. Typical color halos in three outputs of Colorful Image Colorization [21]. The blue halo on the lion's rump, the orange shade on the building facade or in the sky on the right of the lighthouse, the yellow spots on the jellyfish image should be eliminated from colorization results. Input monochromatic images are shown in Fig. 8.

chrominance information or erroneously picking a color through the learnt probability distribution are, however, likely to give colors which are poorly localized and go across shape edges, resulting in unpleasant halos (also called color bleeding in [8]) across edges in some situations. Figure 1 shows typical outputs of a state-of-the-art method. Other examples are described in [16], a careful analysis of DeOldify [1] which is a recent colorization software with impressive results. DeOldify indeed renders color images at a lower resolution than the original monochromatic image, which may give in turn erroneous color halos. This halo effect is often quite subtle and cannot be seen in small-size images as shown in most papers, which mainly aim at getting images with vivid colors. A careful examination suggests, however, that color halos prevent the raw output of these models from being used in real industrial applications.

Several contributions try to circumvent this effect. For instance, the authors of [12] propose to adapt the variational method of [15] in order to restore the output of a deep network [21] to reduce these unwanted defects. They make use of the so-called coupled total variation whose minimization tends to align luminance and chrominance edges, so that color halos are reduced. In the context of color transfer, the authors of [2] use edge-aware texture descriptors with bilateral filtering to reduce halos. More recently, a deep-learning method based on user-defined scribbles was proposed [8]. However, these approaches do not explicitly consider the role of upsampling.

In this paper, we propose to investigate an automated restoration using deep image prior (DIP) [17] together with coupled total variation [15]. The goal is to obtain a colorization free from color halos, based on the hypothesized probability distribution given by [21]. DIP has shown good performances in solving super-resolution from a single image [17]. The expected benefit is thus a correct upsampling of color information to the resolution of the input monochromatic image. We think that this restoration process is of interest in most colorization methods.

Section 2 introduces deep image prior regularized by coupled total variation and its use in image colorization. Numerical experiments are presented in Sect. 3. We conclude with Sect. 4.

2 Deep Image Prior and Image Colorization

Deep image prior [17] has been recently introduced to solve ill-posed inverse problems in image processing. It is based on the observation that convolutional neural networks are good at producing images in many applications. The solution of an inverse problem is thus sought as the output of a given neural network f_θ, with a fixed random input z, whose weights θ are learnt to minimise some function based on data misfits between the observation x and the output $f_\theta(z)$. Minimization is performed with the standard optimization machinery based on the back-propagation of errors. DIP falls within the scope of unsupervised learning methods.

Although several neural architectures are possible, we use an encoder-decoder with skip connections all along this paper. This architecture is used in [17] for super-resolution and some other problems.

2.1 What Information Is Captured by Deep Image Prior?

Such a model encodes short and long range correlations between pixels, as illustrated by Fig. 2. This figure shows an experiment in which the preceding network is trained so that θ minimizes $\|f_\theta(z) - x\|$ where x is some image, the optimum weights being denoted by $\theta(x)$. New images are then generated as $f_{\theta(x)}(z')$ where the input z' is randomly drawn. We can see that the distribution of the colors is kept, fine textures and larger structures as well. These properties probably explain why DIP performs well as a Bayesian prior: contrary to the classic TV prior which tends to produce cartoon images, DIP is likely to produce textured images with geometrical shapes.

2.2 Deep Image Prior and Image Colorization

Of course, it is hopeless to try to solve the colorization problem by simply minimizing with respect to θ

$$\|L(f_\theta(z)) - y_0\|^2, \tag{1}$$

where $L(f_\theta(z))$ is the luminance channel of $f_\theta(z)$, y_0 is the monochromatic image to colorize, and $\|\cdot\|$ is the quadratic norm. The resulting $f_\theta(z)$ would just show random colors. Additional prior information is required.

From a monochromatic image, the "colorful image colorization" (CIC) model of [21] gives, at each pixel of a low-resolution 64×64 grid $\mathcal{G}$, the probability distribution over 313 samples covering the bidimensional chrominance space

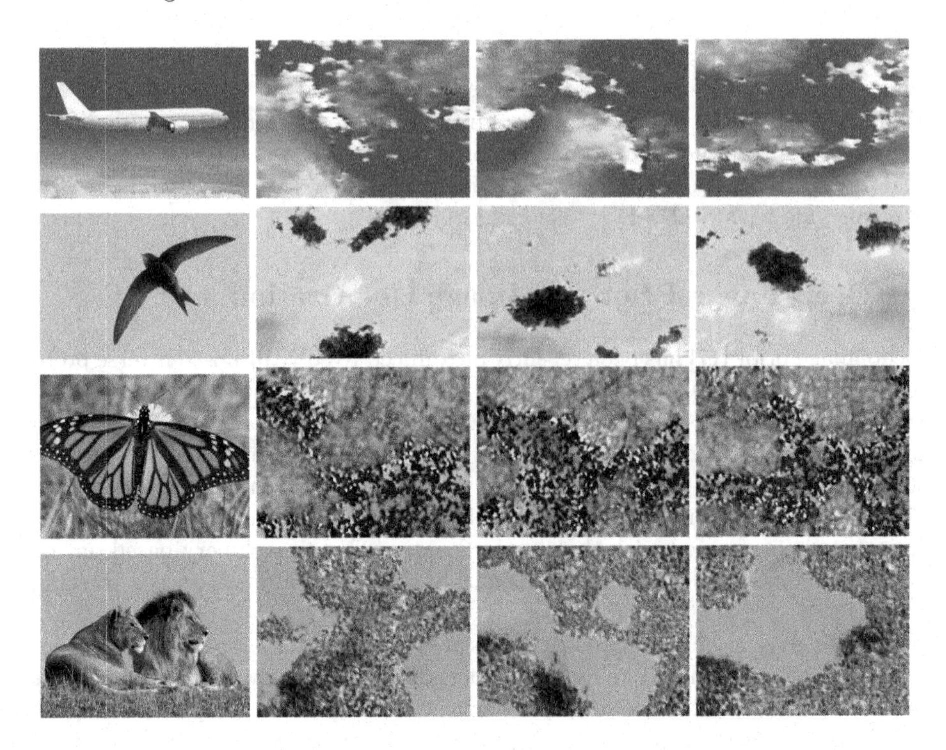

Fig. 2. Experimenting with deep image prior. First column: an image x. Second, third, and fourth columns: $f_{\theta(x)}(z')$ with different random inputs z'.

(see Fig. 3). In [21], a post-processing stage ensures a vivid output image whose resolution is the same as the input resolution. Although very good results are obtained, the probability distribution may not be able to clearly pick a consistent color, especially around the edges of the objects which are not accurately localized over the low-resolution grid. This is illustrated by Fig. 4. Besides, upsampling the chrominance channels is required to build the full-resolution output. This step is likely to produce interpolation artifacts: in the context of colorization, the high-resolution luminance is available but upsampling the chrominance independently from the luminance gives color halos around some objects of the image.

We propose to incorporate the probability distribution given by CIC over the low-resolution pixel grid in the framework of deep image prior to produce color images at the same resolution as the input monochromatic image.

2.3 From Probability Distributions over a Low-Resolution Grid to the Color Image

While an RGB output is chosen in the original DIP [17], in the present paper the output of the neural network is made of the three channels of the CIE Lab color representation, which is the color space used in [21]. Let us denote

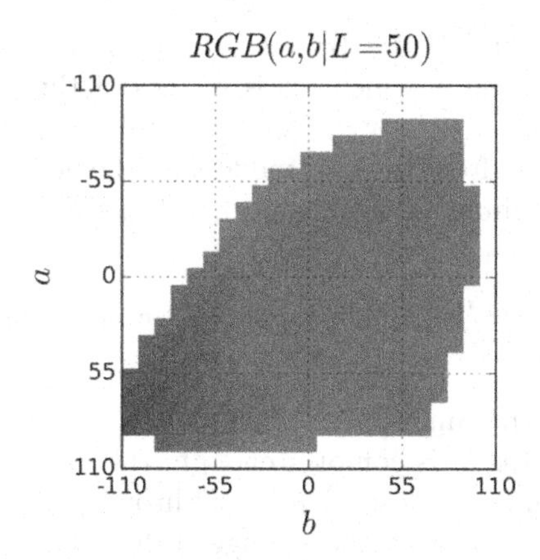

Fig. 3. The 313 color samples in (a, b) chrominance channels used in CIC [21]. Colors correspond to the luminance value $L = 50$.

Fig. 4. From left to right: a monochromatic image, the output of "colorful image colorization" [21] (default parameters), and entropy of the hypothesized probability distribution at each pixel of the 64×64 grid $\mathcal{G}$ given by CIC. The larger the entropy, the less sharp the distribution. Large entropy values can be noticed, especially around edges. Color halos can be seen, for example the orange halo on the grass at the bottom left of the butterfly, or the blue halo on the right lion's rump.

by L and C the mappings from an Lab image to the luminance and the (a, b) chrominance channels in CIE Lab, respectively. The original monochromatic image is denoted by y_0, and the neural network of DIP is f_θ. This network takes some fixed random z as input and produces an Lab image $f_\theta(z)$ with the same resolution as y_0. Here, the 313 samples in the bidimensional chrominance space are denoted by $(c_i)_{1 \leq i \leq 313}$, and $(w_i(x))_{1 \leq i \leq 313}$ is the probability distribution over the c_i's at any pixel x of the low-resolution grid $\mathcal{G}$, as given by CIC. Let d be the subsampling operator which maps images of the same resolution as y_0 and $f_\theta(z)$ to the grid $\mathcal{G}$.

Plugging information from the CIC model into the deep image prior of Eq. (1) can be achieved by minimizing with respect to θ the following quantity:

$$\sum_{x \in \mathcal{G}} \sum_{i=1}^{313} w_i(x) \left\| C\big(d(f_\theta(z))(x)\big) - c_i \right\|^2 + \alpha \left\| L(f_\theta(z)) - y_0 \right\|^2 \tag{2}$$

where $\alpha > 0$ is a hyperparameter.

If the neural network f_θ is complex enough, the optimum is attained for f_θ whose luminance channel is close to y_0 and chrominance channels are, at any pixel x of the low-resolution grid, the average of the (c_i) samples weighted by the probability distribution $(w_i(x))$. The DIP permits thus to upsample information from the low-resolution grid $\mathcal{G}$. The problem with this formulation is twofold. First, averaging chrominances gives dull colors, especially at pixels where the probability distribution is not sharp. Second, luminance and chrominance channels are coupled only through the hidden layers of f_θ, which still gives color halos.

We therefore adapt Eq. (2) in two ways:

1. We add a regularization term which explicitly enforces the coupling between luminance and chrominance channels. To this end, we use the coupled total variation (TV) introduced in [15]. If I is a color image of domain Ω, with luminance L and chrominance channels a and b, the coupled TV writes:

$$\mathrm{TV}_\gamma(I) = \int_\Omega \sqrt{\gamma \|\nabla L\|^2 + \|\nabla a\|^2 + \|\nabla b\|^2} \tag{3}$$

   where ∇ denotes the gradient and $\gamma > 0$ is a hyperparameter of the model. Minimizing the coupled TV makes the luminance and chrominance gradients to have small values at the same pixels [15]. This consequently aligns the edges in the chrominance and luminance channels, and reduces the color halo effect.
2. We allow the probability distribution $(w_i(x))$ to vary, in the same spirit as in [12]. Minimizing $\sum_{i=1}^{313} w_i(x) \|C\big(d(f_\theta(z))\big)(x) - c_i\|^2$ with respect to the non-negative $w_i(x)$ amounts to solving a simple linear program, the constraint being $\sum_{i=1}^{313} w_i(x) = 1$. The minimum is therefore obtained at a vertex of the polytope defined by the constraints: the optimum $w_i(x)$ is such that there exists i^* satisfying $w_{i^*}(x) = 1$ and for any $i \neq i^*, w_i(x) = 0$. For any x, the index i^* is thus simply $\mathrm{argmin}_i \|C\big(d(f_\theta(z))(x)\big) - c_i\|^2$. Such a sharp probability distribution has the advantage of giving vivid colors.

Input: a monochromatic image y_0, and probability distributions $(w_i(x))_{i=1...313}$ at every pixels x of a 64×64 grid $\mathcal{G}$ over samples $(c_i)_{i=1...313}$ spanning the bidimensional chromatic space (from [21]).

Repeat until convergence:

1. Minimize the loss (Eq. (5)) with respect to θ.
2. For any $x \in \mathcal{G}$, $i^*(x) = \text{argmin}_i \|C\big(d(f_\theta(z))\big)(x) - c_i\|$ and $w_i(x) = 1$ if $i = i^*(x)$, $w_i(x) = 0$ otherwise.

Output: the RGB image obtained from the luminance channel y_0 and the chrominance channels of $f_\theta(z)$.

Fig. 5. Algorithm for colorization from the low-resolution probability distributions of CIC [21] with deep image prior and coupled total variation.

As a consequence, we seek θ and w such that $\forall x \in \mathcal{G}, \sum_{i=1}^{313} w_i(x) = 1$, minimizing:

$$\sum_{x \in \mathcal{G}} \sum_{i=1}^{313} w_i(x) \left\|C\big(d(f_\theta(z))\big)(x) - c_i\right\|^2 + \alpha \left\|L(f_\theta(z)) - y_0\right\|^2 + \beta \text{TV}_\gamma(f_\theta(z)) \tag{4}$$

The optimal $f_\theta(z)$ is such that all three terms are small, that is, the luminance of f_θ is close to y_0, halo effects are reduced (small coupled TV), and the chrominances over the low-resolution grid are closed to ones picked from a 0-1 probability distribution.

Compared to the original DIP [17], it should be noted that the additional TV_γ regularisation term permits to get rid of overfitting. DIP is indeed known to overfit and requires an early-stopping strategy. The gradient descent optimization must be stopped after a certain number of iterations, which may critically depend on the inverse problem to solve. A classic way to prevent overfitting in neural network learning in general and in DIP in particular is to add a regularization term in the loss function, see [3,11] for total-variation based regularization of DIP. In the experiments of Sect. 3, we indeed simply minimize the loss function without any early-stopping process.

2.4 Optimization Algorithm

We use block coordinate descent to minimize Eq. (4): we alternate minimization with respect to θ, the parameters of the DIP neural network, and to w, the probability distributions. Minimizing with respect to θ is achieved with the classical back-propagation, and minimizing with respect to w amounts to solving a linear program as explained in Sect. 2.3.

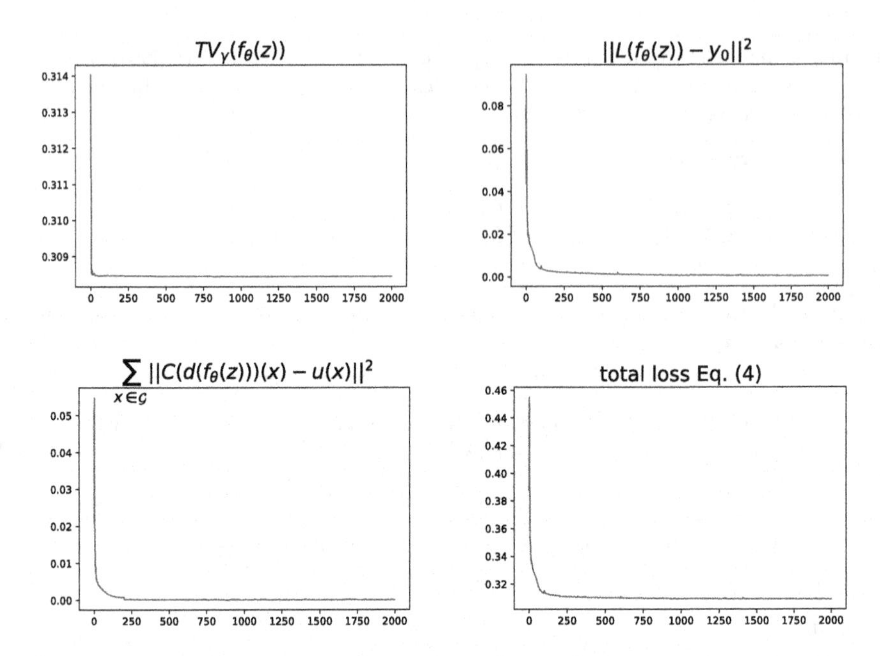

Fig. 6. Minimized loss (bottom right) and its components (related to coupled TV, luminance, and chrominance) against number of iteration in a typical case.

It can be noted that minimizing Eq. (4) with respect to θ by gradient descent is equivalent to minimizing:

$$\sum_{x \in \mathcal{G}} \left\| C\big(d(f_\theta(z))\big)(x) - u(x) \right\|^2 + \alpha \left\| L(f_\theta(z)) - y_0 \right\|^2 + \beta \, \mathrm{TV}_\gamma(f_\theta(z)) \tag{5}$$

with $u(x) = \sum_i w_i(x) c_i$. Indeed, with $v_\theta(x) = C\big(d(f_\theta(z))\big)(x)$,

$$\begin{aligned} \nabla_\theta \left(\sum_{x \in \mathcal{G}} \sum_{i=1}^{313} w_i(x) \left\| v_\theta(x) - c_i \right\|^2 \right) &= 2 \sum_{x \in \mathcal{G}} \sum_{i=1}^{313} w_i(x) \nabla_\theta v_\theta(x) \cdot (v_\theta(x) - c_i) \\ = 2 \sum_{x \in \mathcal{G}} \nabla_\theta v_\theta(x) \cdot \left(v_\theta(x) - \sum_i w_i(x) c_i \right) &= \nabla_\theta \left(\sum_{x \in \mathcal{G}} \left\| v_\theta(x) - u(x) \right\|^2 \right) \end{aligned} \tag{6}$$

where $\cdot$ denotes the dot product, since $\sum_i w_i(x) = 1$ for any $x \in \mathcal{G}$.

Our software implementation uses the equivalent Eq. (5) instead of Eq. (4) for the unsupervised learning step to reduce the computational burden of back-propagation. The resulting algorithm is given in Fig. 5. In practice, Step 2 is performed after repeating 200 gradient descent iterations of Step 1, which limits the number of Step 2 performed, this latter step being time-consuming in spite that only a few pixels are concerned by a change of their probability distribution.

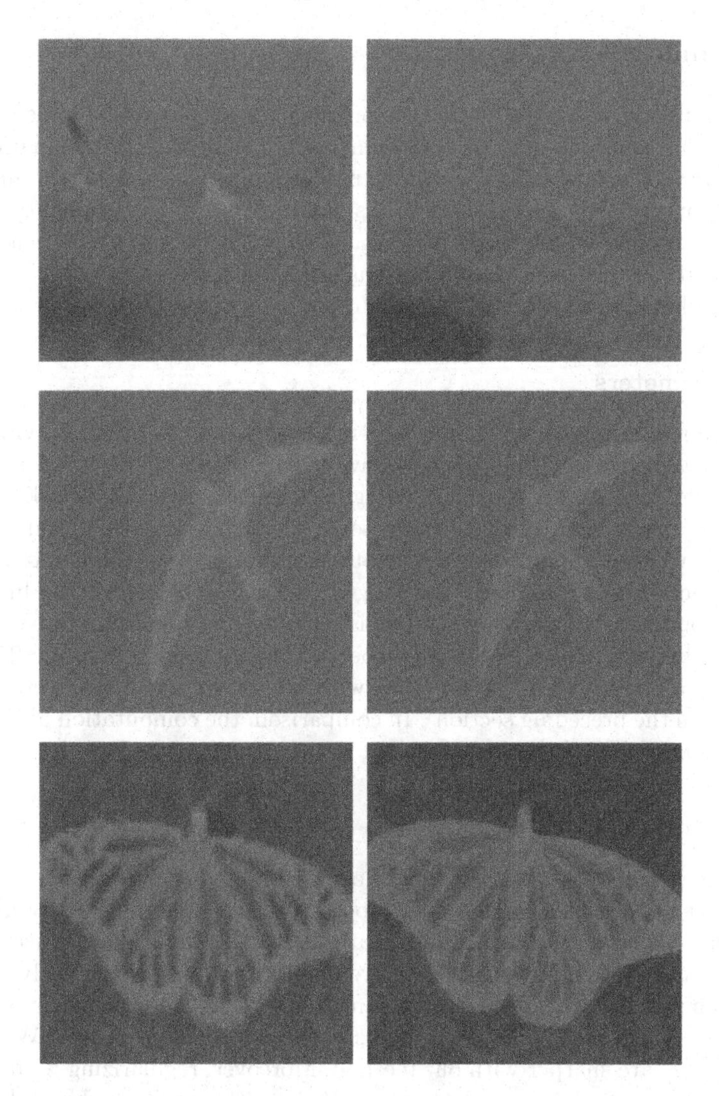

Fig. 7. Chrominance channels. Left: output of CIC [21]. Right: proposed approach (minimization of Eq. (5)). The proposed approach with regularized deep image prior gives a better localization of the chrominance information, which explains reduced color halos. The corresponding color images can be seen in Fig. 2. Best seen on screen.

We can see the effect of this alternating scheme in Fig. 6. The curve of the total loss is globally smooth but contains some small jumps. The function becoming asymptotically constant after 1,000 iterations, stopping after 2,000 iterations ensures the convergence of the iterative algorithm. In all the experiments of this paper, convergence curves are similar to Fig. 6. Contrary to the standard DIP approach [17], there is no need for early stopping.

3 Numerical Experiments

As mentioned in the introduction, colorization is an ill-posed problem as several different solutions may be consistent with a unique monochromatic image (for instance, the color of a car cannot be determined from its black-and-white image). Consequently, this section shows qualitative results: the goal is to assess whether the proposed approach gives plausible colorizations without color halos.

Our software programs are freely available at the following URL: https://gitlab.univ-lorraine.fr/pierre26/diptv

3.1 Parameters

The parameters have been chosen experimentally once and for all. On all the tested images, the same parameters have been used: $\alpha = 1$, $\beta = 5.10^{-7}$, and $\gamma = 80$. The β parameter controls the regularity of the image and works together with the γ parameter. β encourages some flat areas and small variations of the image whereas γ controls the smoothness/sharpness of the contours of the chrominance channels and the coupling of these channels with the luminance one. The optimizer minimizing the loss is ADAM with a learning rate of 2.10^{-2}.

The optimization of Eq. (4) with respect to θ takes about 162 sec for 256×256 images on a GeForce GTX 1060 GPU with 6GB memory (2,000 iterations, as explained in the preceding section). In comparison, the computation of the initial distribution by CIC [21] takes less than 2 s.

3.2 Experiments

In Fig. 7, chrominance channels are shown by applying a constant luminance channel equal to a 50% value. This process helps to qualitatively validate the results of a colorization algorithm by neutralising the effect of luminance. Indeed, colorized images often seem to be visually satisfactory, in spite of poorly localized chrominances. This is caused by the sensitivity of the human visual system to luminance, which hides potential defects of the colorization process. We can see that contours are sharper with our method. Moreover, regularizing with coupled total variation removes some defects on flat areas (for instance the bottom part of the sky below the plane). Sharpness of the chrominance channels shows some benefits in terms of colorization results: for instance, the orange halo on the left of the butterfly (which is the result of a wrong color assignment because of a flat probability distribution given by CIC, as proven by Fig. 4) is removed by our optimization scheme.

Figure 8 shows some comparisons between [12] and the results obtained with (4). It emphasizes the benefit of DIP with respect to a traditional variational method. It is important to note that the outputs of [12] are themselves significant improvements over the output of CIC [21]. It can be seen in the lion image that the contours are better respected. The blue halo over the lion's skin and the orange halo in the sky seen in the output of [12] are corrected by the proposed regularization step. The space between the two lions is not blue as one

Fig. 8. Comparison with [12]. Left: original monochromatic images. Center: results of [12] (from Figs. 5, 6 and supplementary material of [12]); right: proposed approach, minimization of Eq. (5). From top to bottom: lion, close-up, lighthouse, close-up, jellyfish. The proposed approach shows significant improvements over [12], which itself improves over CIC [21] (see Fig. 1).

would expect, but it turns out that, in this area, blue has a low probability in CIC on which our approach depends. In the lighthouse image, orange halo on the facade of the right-hand side building seen in the output of [12] disappears with the proposed approach. In the jellyfish image, the bottom right halo seen in [12] has been removed. Besides, the jellyfish is colorized with [12] through a brown blurry halo in the chrominance channels, whereas the proposed approach produces a structured colorization well-fitted to the body of the jellyfish. However, it can also be seen that the DIP approach produces a brown spot in the lower left of the image, not related to any structure of the luminance channel. It turns out that this phenomenon sometimes appears, depending on the random initialization z of DIP. It is quite rare: we show this particular output for the sake of completeness; most outputs are not affected by it.

4 Conclusion

This paper proposed a deep image prior approach to image colorization, based on the probability distribution given by a state-of-the-art neural network over a low resolution pixel grid. The proposed regularization scheme was able to produce full-resolution chrominance information through a deep image prior regularized by coupled total variation, which permitted to align the chrominance and luminance edges, the latter being available at full-resolution in the original monochromatic image. As a result, we have shown that the so-called color halos (or color bleeding) were reduced, which is a first step towards effective use of colorization.

Acknowledgments. This research was funded, in whole or in part, by l'Agence Nationale de la Recherche (ANR), project ANR-21-0008-01. For the purpose of open access, the authors have applied a CC-BY public copyright licence to any Author Accepted Manuscript (AAM) version arising from this submission.

References

1. Antic, J., Howard, J., Manor, U.: DeCrappification, DeOldification, and super resolution. Fast.ai course (2019). https://www.fast.ai/posts/2019-05-03-decrappify.html
2. Arbelot, B., Vergne, R., Hurtut, T., Thollot, J.: Local texture-based color transfer and colorization. Comput. Graph. **62**, 15–27 (2017)
3. Batard, T., Haro, G., Ballester, C.: DIP-VBTV: a color image restoration model combining a deep image prior and a vector bundle total variation. SIAM J. Imag. Sci. **14**(4), 1816–1847 (2021)
4. Deshpande, A., Lu, J., Yeh, M.C., Chong, M., Forsyth, D.: Learning diverse image colorization. In: Proceedings of the Conference on Computer Vision and Pattern Recognition (CVPR), pp. 2877–2885 (2017)
5. Deshpande, A., Rock, J., Forsyth, D.: Learning large-scale automatic image colorization. In: Proceedings of the International Conference on Computer Vision (ICCV), pp. 567–575 (2015)

6. Huang, Z., Zhao, N., Liao, J.: Unicolor: a unified framework for multi-modal colorization with transformer. ACM Trans. Graph. (Proc. SIGGRAPH2022) **41**(6) (2022)
7. Iizuka, S., Simo-Serra, E., Ishikawa, H.: Let there be color!: joint end-to-end learning of global and local image priors for automatic image colorization with simultaneous classification. ACM Trans. Graph. (Proc. SIGGRAPH'16) **35**(4), 1–11 (2016)
8. Kim, E., Lee, S., Park, J., Choi, S., Seo, C., Choo, J.: Deep edge-aware interactive colorization against color-bleeding effects. In: Proceedings of the International Conference on Computer Vision (ICCV), pp. 14667–14676 (2021)
9. Kim, G., et al.: Bigcolor: colorization using a generative color prior for natural images. In: Avidan, S., Brostow, G., Cisse, M., Farinella, G.M., Hassner, T. (eds.) Computer Vision— ECCV 2022. ECCV 2022. Lecture Notes in Computer Science, vol. 13667, pp. 350–366. Springer, Cham (2022). https://doi.org/10.1007/978-3-031-20071-7_21
10. Larsson, G., Maire, M., Shakhnarovich, G.: Learning representations for automatic colorization. In: Leibe, B., Matas, J., Sebe, N., Welling, M. (eds.) ECCV 2016. LNCS, vol. 9908, pp. 577–593. Springer, Cham (2016). https://doi.org/10.1007/978-3-319-46493-0_35
11. Liu, J., Sun, Y., Xu, X., Kamilov, U.: Image restoration using total variation regularized deep image prior. In: Proceedings of the International Conference on Acoustics, Speech and Signal Processing (ICASSP), pp. 7715–7719 (2019)
12. Mouzon, T., Pierre, F., Berger, M.-O.: Joint CNN and variational model for fully-automatic image colorization. In: Lellmann, J., Burger, M., Modersitzki, J. (eds.) SSVM 2019. LNCS, vol. 11603, pp. 535–546. Springer, Cham (2019). https://doi.org/10.1007/978-3-030-22368-7_42
13. Pan, X., Zhan, X., Dai, B., Lin, D., Loy, C., Luo, P.: Exploiting deep generative prior for versatile image restoration and manipulation. IEEE Trans. Pattern Anal. Mach. Intell. **44**(11), 7474–7489 (2022)
14. Pierre, F., Aujol, J.F.: Recent approaches for image colorization. In: Chen, K., Schönlieb, C.B., Tai, X.C., Younces, L. (eds.) Handbook of Mathematical Models and Algorithms in Computer Vision and Imaging: Mathematical Imaging and Vision. Springer (2021). https://doi.org/10.1007/978-3-030-03009-4_55-1
15. Pierre, F., Aujol, J.F., Bugeau, A., Papadakis, N., Ta, V.T.: Luminance-chrominance model for image colorization. SIAM J. Imag. Sci. **8**(1), 536–563 (2015)
16. Salmona, A., Bouza, L., Delon, J.: DeOldify: a review and implementation of an automatic colorization method. Image Process. Line **12**, 347–368 (2022)
17. Ulyanov, D., Vedaldi, A., Lempitsky, V.: Deep image prior. In: Proceedings of the Conference on Computer Vision and Pattern Recognition (CVPR) (2018)
18. Vitoria, P., Raad, L., Ballester, C.: ChromaGAN: adversarial picture colorization with semantic class distribution. In: Proceedings of the Winter Conference on Applications of Computer Vision, pp. 2445–2454 (2020)
19. Xia, M., Hu, W., Wong, T.T., Wang, J.: Disentangled image colorization via global anchors. ACM Trans. Graph. (Proc. SIGGRAPH2022) **41**(6) (2022)
20. Zhang, R., et al.: Real-time user-guided image colorization with learned deep priors. ACM Trans. Graph. (Proc. SIGGRAPH2017) **36**(4) (2017)
21. Zhang, R., Isola, P., Efros, A.A.: Colorful image colorization. In: Leibe, B., Matas, J., Sebe, N., Welling, M. (eds.) ECCV 2016. LNCS, vol. 9907, pp. 649–666. Springer, Cham (2016). https://doi.org/10.1007/978-3-319-46487-9_40

Hybrid Training of Denoising Networks to Improve the Texture Acutance of Digital Cameras

Raphaël Achddou(✉), Yann Gousseau, and Saïd Ladjal

LTCI, Telecom Paris, Institut Polytechnique de Paris, Palaiseau, France
raphael.achddou@telecom-paris.fr

Abstract. In order to evaluate the capacity of a camera to render textures properly, the standard practice, used by classical scoring protocols, is to compute the frequential response to a dead leaves image target, from which is built a *texture acutance metric*. In this work, we propose a mixed training procedure for image restoration neural networks, relying on both natural and synthetic images, that yields a strong improvement of this acutance metric without impairing fidelity terms. The feasibility of the approach is demonstrated both on the denoising of RGB images and the full development of RAW images, opening the path to a systematic improvement of the texture acutance of real imaging devices.

Keywords: Image denoising · Deep learning · Image quality assessment

1 Introduction

In order to correctly visualize a photograph, its corresponding RAW image undergoes a complex sequence of development operations including white balancing, demosaicking, tone mapping, and image restoration operations such as deblurring and denoising. Camera manufacturers implement proprietary algorithms fine-tuned for each setting of each camera. As a result, the overall image quality is a combination of hardware characteristics (quality of the lens, size of the sensor) and software performances. In order to fairly assess the quality of an imaging device, independent agencies have defined standard tests and ISO protocols. Each of these tests focus on a specific characteristic such as chromatic aberrations, noise reduction, or texture rendering.

Recently, with the increase in computational power and the advent of deep learning for image processing, more and more digital image processing stages can be replaced by learned neural networks [17]. Recent works already aim at completely replacing the full image development pipeline with a single neural network, producing impressive results in standard conditions [20] or extremely low-light conditions [13]. Moreover, light neural network architectures can now be integrated in embedded systems, e.g. on smartphone devices. Neural methods

L. Calatroni et al. (Eds.): SSVM 2023, LNCS 14009, pp. 314–325, 2023.
https://doi.org/10.1007/978-3-031-31975-4_24

present another key advantage: one can easily optimize their response to specific test images, by including them in training databases.

For the specific task of texture rendering evaluation, Cao et al. [10] first presented a protocol quantifying the ability of an imaging pipeline to preserve texture information. This is obtained through the frequential response of the system to dead leaves images with a specific perceptual metric called *texture acutance*. These images are known for their invariance properties, as well as statistical properties making them close to natural images (non Gaussianity, scaling property, distribution of the spectrum and gradient), as studied in [4,18,23]. This quality evaluation protocol later became an ISO standard to measure the preservation of textures [21] and is now used by classical camera scoring protocols. In a different direction, Achddou et al. [3] showed that image restoration networks could be trained from synthetic images only, using databases of dead leaves images.

Inspired by these results, we propose, in this paper, to train a denoising neural network on natural and dead leaves images, to jointly optimize a new metric derived from the texture acutance and the classic data fidelity metrics on natural images. After presenting some related works on image restoration in Sect. 2, we first introduce in Sect. 3 the texture acutance metric and the corresponding perceptual loss for image restoration networks. We then show in Sect. 4 that we can strongly improve the texture acutance metric without impairing performances on natural images, first for the task of Additive White Gaussian Noise removal (AWGN) and then for the development of RAW images. These results open the path to an automatic improvement of standard quality evaluation tests.

2 Related Works

The goal of image restoration is to retrieve a clean image from distorted observations. In many cases, the distortion process can be modeled as follows: $y = Ax + n$, where x is the theoretically perfect image, y the distorted observation, A is a linear operator and n is some noise.

In order to solve this problem, a first class of methods are based on prior hypotheses on the distribution of natural images. These methods try to impose regularity properties on the restored solutions. For instance, wavelet shrinkage methods [15,16] or DCT-filtering methods [31] reconstruct an image assuming that the targeted images can be well approximated by a sparse decomposition. In turn, variational methods based on the total variation [12,29] assume that the image gradient follows a Laplacian distribution. Based on the assumption of self-similarity, non-local methods leverage the redundancy in the image content. This is either done by weighted averaging (Non Local Means [9] Non Local Bayes [22]) or by collaborative filtering (BM3D [14]).

Over the past decade, learning-based approaches for image restoration have developed drastically. After the success of neural networks for high-level computer vision tasks [19], these methods have been adapted to image restoration through the use of generative models [32,33]. Rather than using prior hypotheses, the parameters of the neural networks are tuned in a long optimization process to directly minimize the reconstruction error in a black box manner. For the training to succeed, these methods require large databases of pairs of distorted and clean images. Even though they are hard to interpret, they surpassed prior-based methods on most image restoration benchmarks by a large margin for a wide variety of tasks such as image denoising [33], demosaicking [17] etc.

Following these initial works, recent papers extended the use of deep learning methods to real-world problems of image restoration such as RAW image denoising [5,24]. Ignatov et al. [20] and Chen et al. [13] also propose to fully replace the image development pipeline by a learned neural network, producing surprisingly good results. However, acquiring datasets of real-world pairs of distorted and clean RAW images is a cumbersome task [1,13], which often requires complex post-processing algorithms. In order to ease the training process, a more restrained approach consists in modeling the distortion process accurately, and to synthesize them accordingly [30].

Going further, Achddou et al. [2,3] proposed to train image restoration neural networks on generated dead leaves images in order to completely circumvent the data acquisition process, reaching performances close to the networks trained on real images, for various image restoration tasks. These images indeed exhibit statistical properties close to those of natural images [4,18,23] even though they depend from few parameters. Following [3], similar synthetic databases were also used to pre-train image classification networks [7] and disparity map estimators [25]. Prior to these works, dead leaves images were used to assess the capacity of cameras to render textures properly. This idea was first presented in 2009 by Cao et al. [10], which was later improved in the following references [6,11]. We will present in detail these works in the following section.

3 Texture Acutance: A Frequential Loss Assessing Texture Preservation

3.1 Dead Leaves Images

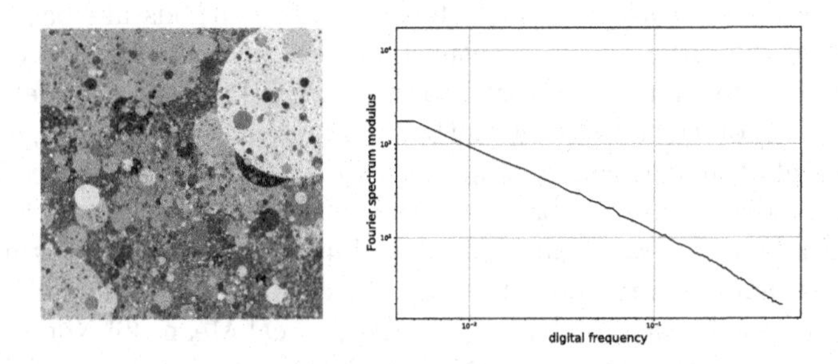

Fig. 1. Grey level dead leaves image and its associated digital spectrum in logarthmic scales. The theoretical value is a straight line. (Color figure online)

Dead leaves images were first introduced by Matheron in 1975 [26], with the aim of modeling porous media. It was later shown that if object sizes fulfill some scaling property, this model accounts for many statistics of natural images [4, 23]. To generate such images, shapes of random size, color and position are superimposed on top of each other until the whole image plan is covered. In the simplest set-up, these shapes are disks of random radius. An example of a dead leaves image is given in Fig. 1, along with its spectrum. A precise mathematical formulation of dead leaves images is given in [8].

Dead leaves images were first used for camera evaluation in 2009 by Cao et al. [10]. The proposed idea is to measure the response of a camera to a specific image target. Because of their invariances and statistical properties, the dead leaves model was chosen by the authors as the generation algorithm for the target. Among the desired properties, scale invariance is achieved when the disks radii follows a power law with $\alpha = 3$. The dead leaves target is therefore generated with this parameter. Note that to ensure the convergence of the algorithm, bounding parameters r_{min}, r_{max} are required [18].

3.2 Texture Acutance

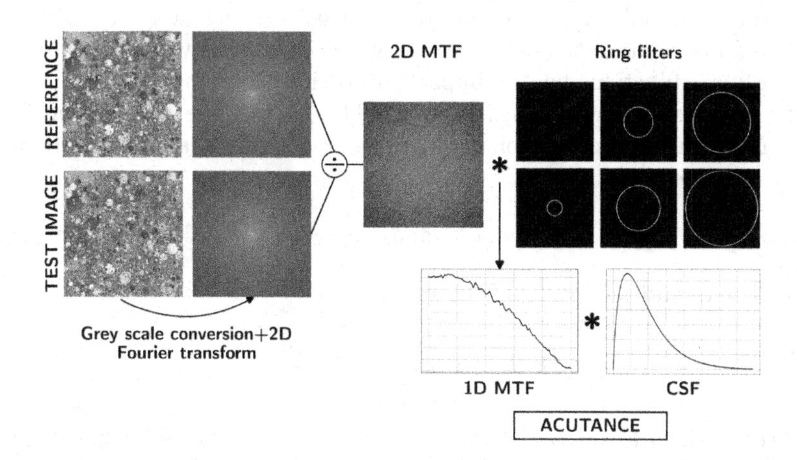

Fig. 2. Diagram explaining the computation of the acutance metric

In [10], the authors evaluate the response of a camera to the dead leaves target by computing the ratio of the power spectra, resulting in a Modulation Transfer Function (MTF). At each position (m, n) for an (N, N) image:

$$\mathrm{MTF}_{2D}(m,n) = \frac{|\hat{Y}(m,n)|}{|\hat{X}(m,n)|},$$

where $\hat{Y}$ is the spectrum of the obtained image and $\hat{X}$ is the ground truth spectrum. In all that follows, we compute the image spectra on a greyscale version of the color image, obtained by the standard linear combination

Grey $= 0.2126R + 0.7152G + 0.0722B$. The classical idea behind the MTF is that the ratio of the power spectra corresponds to the Fourier coefficients of the blur kernel induced by the camera, excluding non linear transforms often involved in the image development, as well as the impact of noise (Fig. 2).

In order to account for the impact of noise in the estimation of the MTF, Artmann first proposed a corrected version of the MTF, by subtracting an estimate of the noise spectrum. However, the latter was computed by taking a photograph of a uniform grey surface, assuming an additive and signal independent noise model, which is far from reality. Moreover, some image development pipelines include a nonlinear noise reduction operation, which affect the estimation of the real MTF.

In the same paper [6], Artmann proposes a new computation trying to correct these issues. Here, we consider the complex spectrum of a reference digital dead leaves target $\hat{X}$, rather than the estimate of the power spectrum $|\hat{X}|$ in the spatial domain. In the previous version, the phase information was lost. This is necessary in the context of camera calibration since phase information is reliable only if a registration algorithm is applied. In the context of training a denoising algorithm registration between noisy and restored image is supposed perfect and dealt with by the MSE-loss. Relying only on the amplitude of the spectrum meant that we could not differentiate frequencies which were already in the target and information that was added by the imaging device. Therefore noise and non linear functions had an impact in previous computations.

The proposed method, which we call MTF_{cross} uses the cross power density between the target and the obtained image $\phi_{XY}(m,n)$, and the auto power density $\phi_{XX}(m,n)$. More precisely,

$$\phi_{XY}(m,n) = \hat{Y}(m,n)\hat{X}^*(m,n) \quad \text{and} \quad \phi_{XX}(m,n) = \hat{X}(m,n)\hat{X}^*(m,n).$$

Given these quantities, the MTF becomes :

$$\text{MTF}_{cross}(m,n) = \left|\frac{\phi_{XY}(m,n)}{\phi_{XX}(m,n)}\right|. \tag{1}$$

Since the dead leaves target is rotationally invariant, so is its spectrum. We therefore express the MTF as a 1D function by averaging it on concentric rings of width 1. The MTF becomes :

$$\text{MTF}_{1D}(k) = \frac{\mathbb{1}_{C_k}}{\#C_k} \times \text{MTF}_{cross},$$

where $C_k = \left\{(i,j) \in [-N/2, N/2]^2 | (k-1)^2 \leq |i^2 + j^2| < k^2\right\}$ corresponds to a ring of radius k and $\#C_k$ is its cardinal.

Though the full MTF_{1D} is a good indicator of the camera's capacity to render textures, it is more helpful to compute a single score. To that end, the *texture acutance* [10] is defined as a weighted sum of the MTF_{1D}, with weights defined by a contrast sensitivity function (CSF), inspired by the slanted edge Spatial Frequency Response (SFR), used to evaluate the sharpness of a camera. Our visual system is indeed more sensitive to some frequencies than others. In that regard, the CSF models the sensitivity of the visual system to spatial frequencies expressed in cycle/degree.

Based on the physiological analysis of the contrast sensitivity of infants and monkeys led by Movshon and Kiorpes [27], the chosen formula to model the CSF is :

$\mathrm{CSF}(\nu) = a.\nu^{c}.e^{-b\nu}$, where ν is a spatial frequency expressed in cycles/degree, parameters are fixed as $b = 0.2$, $c = 0.8$, and a is a normalizing parameter so that $\int_0^{\mathrm{Nyquist}} \mathrm{CSF}(\nu)d\nu = 1$. Given this formula, the texture acutance score can be written as :

$$A = \int_0^{\mathrm{Nyquist}} \mathrm{CSF}(\nu).\mathrm{MTF}_{1D}(\nu)d\nu.$$

Note that we need to convert spatial frequencies in cycles/degree to a digital frequency in cycles/pixel for homogeneity. To do so we use the following formula: $f_{spatial} = \frac{1}{\alpha} f_{digital}$, where α is the viewing angle. The latter depends on viewing conditions with the equality $\alpha = \frac{180}{\pi}\arctan(\frac{P}{D})$, where P is the pixel size and D is the viewing distance, assumed to be equal to 0.2 mm and 1m respectively. This corresponds to a maximal spatial frequency of 40 cycles/degree which is approximately the limit of the human visual system.

The perfect MTF corresponds to a constant function equal to 1, meaning that the frequential content has been perfectly restored by the camera for every frequency. This leads to an acutance $A = 1$. An acutance greater than 1 indicates that some frequential content was added to the image, probably because of noise or sharpening. An acutance lower than 1 indicates that some frequencies have been lost.

3.3 Acutance Loss for Image Restoration CNNs

In [3], the authors showed that models trained on mixed databases (natural and synthetic images) perform on par with models trained on natural images only, while improving results on dead leaves image targets. We believe we can improve the frequential response of models trained on mixed sets, by using the acutance score in a loss function.

In the context of AWGN removal for color RGB images, the noisy image corresponds to $Y = X + n$ where X is a ground truth dead leaves image of size $(N, N, 3)$. The denoising network f_θ produces an estimate of the clean image $Z = f_\theta(Y)$. For our restoration problem, we can consider that the denoising network is analogous to the camera which acquires the dead leaves target. We can compute MTF_{cross} for the denoising network using Formula (1), based on the computation of the digital spectrum of both X and Z.

The obtained MTF_{cross} is turned into a 1D signal as described above. For faster computation, concentric ring masks are stored in GPU so that the computation of MTF_{1D} can be accelerated with parallel computing. Since the best possible acutance is 1, we define the acutance loss function as :

$$\mathcal{L}_{acutance}(Y, X) = |1 - A(f_\theta(Y), X)|,$$

which penalizes both adding or removing frequential information. In order to get a complete loss function, we add to it the $\mathcal{L}_2$ loss, the initial fidelity term of the network. Indeed, the acutance loss $\mathcal{L}_{acut}$ is computed solely on an aggregation of the Fourier spectrum and is therefore blind to the spatial organisation of the image and can not replace an MSE-loss. When training on dead leaves images the loss is therefore

$$\mathcal{L} = \mathcal{L}_2 + \lambda.\mathcal{L}_{acut},$$

where λ is a weighting parameter.

Since we train the image denoiser on both natural images and dead leaves images, we compute the acutance loss only on the dead leaves images in a mini-batch D of size K and the $\mathcal{L}_2$ loss for all images. The formation of minibatches during training indeed randomly samples images from the mixed set. Thus, the loss in a batch becomes:

$$\mathcal{L}_{batch} = \frac{1}{K}\sum_{i=0}^{K} ||x_i - f_\theta(x_i + n_i)||_2^2 + \frac{\lambda}{m^T\mathbf{1}}\sum_{i=0}^{K} m_i.\mathcal{L}_{acut}(f_\theta(x_i + n_i), x_i), \quad (2)$$

where m is a masking vector of size K such that $m_i = 1$ if x_i is a dead leaves image, or $m_i = 0$ otherwise. In order to count the number of dead leaves images we sum this masking vector which is given by $m^T\mathbf{1}$.

4 Image Denoising Results with FFDNet

We choose to train the FFDNet network [33] to illustrate the impact of the perceptual loss we presented in the previous section. We adapt the training scheme of the network to the present problem as follows. First, we increase the size of the training patches from $(50, 50, 3)$ to $(100, 100, 3)$. The reason for this is that the estimation of the 1D-MTF on a small patch is not sufficiently accurate. Keeping the same rings' width would result in fewer estimates for the 1D-MTF. On the other hand, decreasing the rings' width would lead to noisier estimates. Therefore, we perform the training with larger patches. Second, we reduce the batch size from 64 to 32 during training to decrease the memory footprint. We use 150000 samples, made of 100000 natural image patches and 50000 synthesized dead leaves patches. The other training hyper-parameters remain unchanged, such as the number of epochs or the learning rate decaying schedule.

4.1 Quantitative Evaluation

In order to show that the proposed scheme indeed has the potential to improve the texture acutance without impairing the usual PSNR evaluation of the performances on natural images, we compute both these metrics for various values of λ, the weighting parameter in Eq. (2). We consider values of $\lambda \in [0, 2, 5, 10, 20, 50, 100, 200, 500]$. Moreover, we also compute the classical SSIM metric and the perceptual metric PieAPP recently introduced in [28].

The models are evaluated numerically on two datasets. First, we evaluate the data fidelity by computing the PSNR, SSIM and PieAPP metrics on the Kodak24 dataset, a benchmark test set of 24 natural images. Second, we evaluated the acutance metric on a test set of synthesized dead leaves images.

We report, in Table 1, the numerical evaluation of the trained models. We observe a similar behaviour for the tested noise levels $\sigma = 25$ and $\sigma = 50$. In both cases, we notice that the standard evaluation metrics, i.e., the PSNR and SSIM, are not affected by the increase of the weighting parameter λ until $\lambda = 20$. For values greater than $\lambda = 100$ these metrics decrease rapidly. On the other hand, the acutance metric keeps improving until $\lambda = 100$ and then reaches a plateau. This table shows that we can optimize the texture acutance without impairing classic denoising evaluation. The perceptual evaluation with the PieAPP metric suggests that, for high noise values, the perceptual image quality is slightly enhanced by the addition of the acutance loss. Some results can be visualized in Fig. 3 (please zoom in the electronic version of this document). The result with and without using the acutance loss appear quite close, despite the strong improvement of the texture acutance measurement. Nonetheless, we can notice some improvements in the preservation of low-contrast details in the first row. Moreover, the contrast is also rendered better when training with the acutance loss. Finally, on the third row, details on the dead leaves images are better preserved using the acutance loss. On the second and third row, we see that the network trained with natural images sometimes hallucinates details, which are removed when training with dead leaves images.

Table 1. Denoising results of FFDNet trained with different weighting coefficients of the acutance loss for two noise levels. Each cell contains the PSNR, SSIM, PieAPP evaluated on Kodak 24 , and the Acutance metric evaluated on a test set of dead leaves images. Best results in blue, second results in red.

λ		0	2	5	10	20	50	100	200	500
$\sigma = 25$	PSNR ↑	31.88	31.87	31.88	31.87	31.88	31.85	31.77	31.65	31.56
	Acutance ↓	0.034	0.029	0.023	0.020	0.015	0.013	0.012	0.012	0.012
	SSIM ↑	0.877	0.876	0.875	0.875	0.877	0.876	0.873	0.872	0.869
	PieAPP ↓	0.568	0.596	0.598	0.602	0.586	0.587	0.591	0.62	0.612
$\sigma = 50$	PSNR ↑	28.81	28.79	28.80	28.80	28.80	28.76	28.66	28.58	28.42
	Acutance ↓	0.084	0.078	0.073	0.053	0.035	0.026	0.022	0.022	0.022
	SSIM ↑	0.791	0.788	0.789	0.789	0.790	0.789	0.786	0.783	0.779
	PieAPP ↓	0.932	0.956	0.952	0.948	0.912	0.925	0.934	0.940	0.953

4.2 Spectral Preservation

For mixed trainings of FFDNet, the texture acutance score is greatly improved when using the corresponding loss, which is expected. However, the acutance score only gives a partial information about the MTF of the trained network.

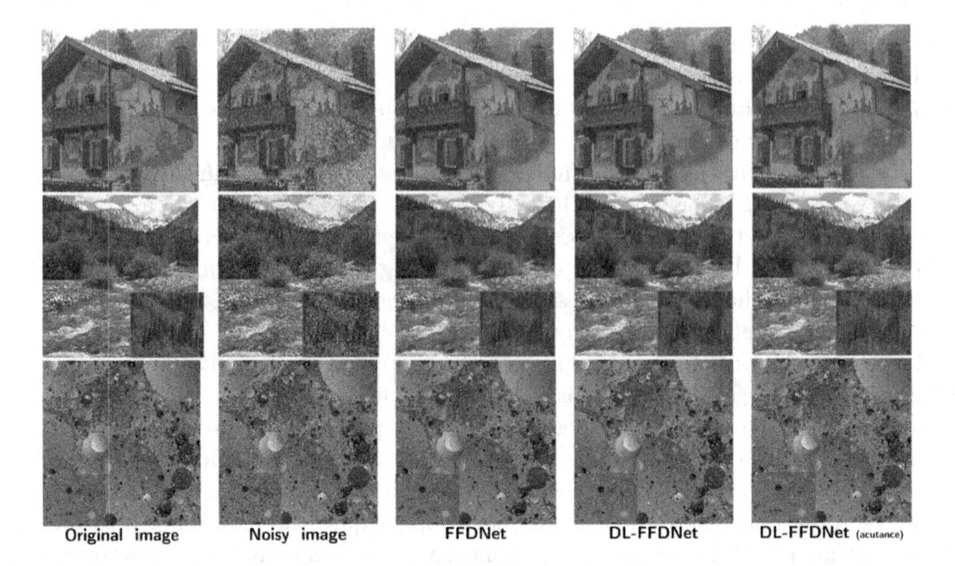

Fig. 3. Comparison of FFDNet results on two natural images and on a dead leaves image. From left to right: original image, noisy image, image denoised with standard FFDNet, image denoised with FFDNet trained on a mixed database without the acutance loss, and finally with the acutance loss.

In order to further understand the impact of the acutance loss on the spectral preservation ability of the network, we compute its MTF as described next. We compute the 1D-MTF from the denoised image and the original image for each dead leaves image of the synthetic test set.

Since the 1D-MTF depends on the image's content, which differs from image to image, we average the obtained MTF over the whole dataset. In Fig. 4, we report the MTF of FFDNet trained with and without the acutance loss (with $\lambda = 50$) for a noise level $\sigma = 25$. Recall that a perfect MTF should be equal to one. We can observe that for low to medium frequency, the MTF of the model trained with the accutance loss is much closer to one. Actually, the values for low frequency exceed one which is one way the system can improve the acutance and which indeed is a limitation of the approach. For high frequency, the gap between the two MTF is smaller, probably as a result of the profile of the CSF function, which quickly decreases for high frequency, see Fig. 1. This behavior, as well as the addition of low frequency, could be modified by considering alternative CSF functions and can be easily integrated into our frame-

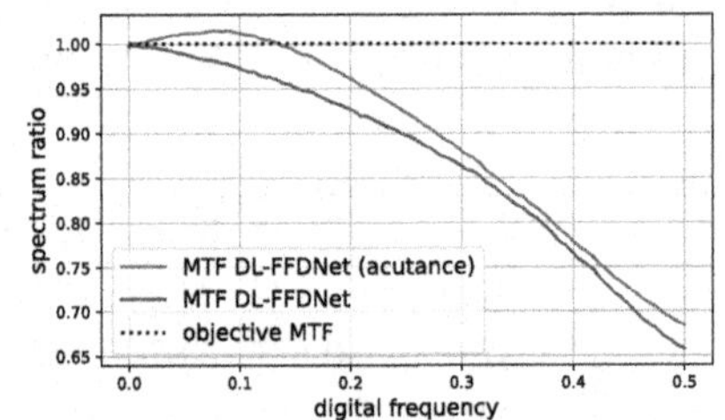

Fig. 4. Comparison of the MTF evaluated with FFDNet trained on a mixed database with or without the acutance loss, on the whole dead leaves image test set.

work. This could also further improve the preservation of details on examples such as those of Fig. 3. In this paper, we have decided to keep the original definition of the acutance, since our main goal is to show that this standard measure of the ability to preserve texture can be greatly improved without impairing the other aspects of image quality.

4.3 RAW Image Denoising

As a proof of concept, we extended our experiments to real-world image denoising on the SIDD benchmark [1] for cameraphones denoising. To that end, we trained the same denoising network with a U-Net architecture to denoise real RAW images and synthesized RAW dead leaves images. This network produces a RAW denoised image from a RAW input noisy image. To simulate RAW noise for dead leaves images, we used a Poisson-Gaussian model with realistic noise parameters. Unlike Gaussian noise removal, the loss is here a combination of the $\mathcal{L}_1$ loss and the acutance loss : $\mathcal{L} = \mathcal{L}_1 + \lambda\mathcal{L}_{acut}$. For RAW images, the acutance computation differs slightly. In order to convert a RAW image to a grey-scale image, we first pack the (H, W) image in a $(H/2, W/2, 4)$ RGGB tensor, then we average them in a single $(H/2, W/2)$ grey array, by weighting each channel with the white balance parameters. We ran the training for $\lambda \in [0, 10, 100]$. We report the numerical results obtained in Table 2. In comparison with $\lambda = 0$, the PSNR is still good for $\lambda = 10$, while the RAW acutance is largely improved. This improvement also translates in a better acutance in the RGB domain, which was not seen during training. This metric is computed on the denoised images developed with a standard ISP. This experiment shows that we can improve camera evaluation without impairing the image quality in the case of a full camera development pipeline. For $\lambda = 100$, the PSNR noticeably decreases while the RAW acutance reaches a plateau.

Table 2. Denoising results obtained by training a denoising Unet for real RAW images evaluated on the SIDD test set of cameraphone images [1]. We report the PSNR, Acutance RAW and acutance RGB metrics. Best results in **bold.**

λ	0	10	100
PSNR RAW	**51.31**	51.24	50.61
Acutance RAW	0.018	**0.011**	0.012
Acutance RGB	0.086	0.061	**0.049**

5 Conclusion

In this work, we have shown that a specific training of image restoration neural networks can greatly improve a standard evaluation metric quantifying the preservation of textures, without impairing classical performance evaluation criteria. As a proof of concept, we extended the use of the acutance loss for real-world image denoising networks, showing that the proposed framework can improve a complete RAW images development pipeline. Considering that the texture acutance metric is routinely used to evaluate digital camera, this founding has potential important practical applications.

References

1. Abdelhamed, A., Lin, S., Brown, M.S.: A high-quality denoising dataset for smartphone cameras. In: Proceedings of the IEEE Conference on Computer Vision and Pattern Recognition, pp. 1692–1700 (2018)
2. Achddou, R., Gousseau, Y., Ladjal, S.: Fully synthetic training for image restoration tasks, January 2023. https://hal.science/hal-03940525. working paper or preprint
3. Achddou, R., Gousseau, Y., Ladjal, S.: Synthetic images as a regularity prior for image restoration neural networks. In: Elmoataz, A., Fadili, J., Quéau, Y., Rabin, J., Simon, L. (eds.) SSVM 2021. LNCS, vol. 12679, pp. 333–345. Springer, Cham (2021). https://doi.org/10.1007/978-3-030-75549-2_27
4. Alvarez, L., Gousseau, Y., Morel, J.M.: The size of objects in natural and artificial images. In: Advances in Imaging and Electron Physics, vol. 111, pp. 167–242. Elsevier (1999)
5. Anwar, S., Barnes, N.: Real image denoising with feature attention. In: Proceedings of the IEEE/CVF International Conference on Computer Vision, pp. 3155–3164 (2019)
6. Artmann, U.: Image quality assessment using the dead leaves target: experience with the latest approach and further investigations. In: Digital Photography XI, vol. 9404, pp. 130–144. SPIE (2015)
7. Baradad, M., Wulff, J., Wang, T., Isola, P., Torralba, A.: Learning to see by looking at noise. In: Advances in Neural Information Processing Systems, vol. 34 (2021)
8. Bordenave, C., Gousseau, Y., Roueff, F.: The dead leaves model: a general tessellation modeling occlusion. Adv. Appl. Probab. **38**(1), 31–46 (2006)
9. Buades, A., Coll, B., Morel, J.M.: A non-local algorithm for image denoising. In: 2005 IEEE Computer Society Conference on Computer Vision and Pattern Recognition (CVPR2005), vol. 2, pp. 60–65. IEEE (2005)
10. Cao, F., Guichard, F., Hornung, H.: Measuring texture sharpness of a digital camera. In: Digital Photography V, vol. 7250, p. 72500H. International Society for Optics and Photonics (2009)
11. Cao, F., Guichard, F., Hornung, H.: Dead leaves model for measuring texture quality on a digital camera. In: Digital Photography VI, vol. 7537, p. 75370E. International Society for Optics and Photonics (2010)
12. Chambolle, A.: An algorithm for total variation minimization and applications. J. Math. Imag. Vis. **20**(1), 89–97 (2004)
13. Chen, C., Chen, Q., Xu, J., Koltun, V.: Learning to see in the dark. In: Proceedings of the IEEE Conference on Computer Vision and Pattern Recognition, pp. 3291–3300 (2018)
14. Dabov, K., Foi, A., Egiazarian, K.: Video denoising by sparse 3D transform-domain collaborative filtering. In: European Signal Processing Conference, vol. 16, no. 8, pp. 145–149 (2007)
15. Donoho, D., Johnstone, I.M.: Ideal spatial adaptation by wavelet shrinkage **81**(3), 425–455 (1994)

16. Donoho, D.L., Johnstone, I.M.: Minimax estimation via wavelet shrinkage. Ann. Stat. **26**(3), 879–921 (1998)
17. Gharbi, M., Chaurasia, G., Paris, S., Durand, F.: Deep joint demosaicking and denoising. ACM Trans. Graph. (ToG) **35**(6), 1–12 (2016)
18. Gousseau, Y., Roueff, F.: Modeling occlusion and scaling in natural images. Multiscale Model. Simulat. **6**(1), 105–134 (2007)
19. He, K., Zhang, X., Ren, S., Sun, J.: Deep residual learning for image recognition. In: Proceedings of the IEEE Conference on Computer Vision and Pattern Recognition, pp. 770–778 (2016)
20. Ignatov, A., Van Gool, L., Timofte, R.: Replacing mobile camera ISP with a single deep learning model. In: Proceedings of the IEEE/CVF Conference on Computer Vision and Pattern Recognition Workshops, pp. 536–537 (2020)
21. Photography – Digital cameras – Part 2: Texture analysis using stochastic pattern. Standard, International Organization for Standardization, Geneva, CH (2019)
22. Lebrun, M., Buades, A., Morel, J.M.: A nonlocal Bayesian image denoising algorithm. SIAM J. Imag. Sci. **6**(3), 1665–1688 (2013)
23. Lee, A.B., Mumford, D., Huang, J.: Occlusion models for natural images: a statistical study of a scale-invariant dead leaves model. Int. J. Comput. Vision **41**(1–2), 35–59 (2001)
24. Liu, Y., et al.: Invertible denoising network: a light solution for real noise removal. In: Proceedings of the IEEE/CVF Conference on Computer Vision and Pattern Recognition, pp. 13365–13374 (2021)
25. Madhusudana, P.C., Lee, S.J., Sheikh, H.R.: Revisiting dead leaves model: training with synthetic data. IEEE Signal Process. Lett. (2021)
26. Matheron, G.: Random sets and integral geometry (1975)
27. Movshon, J.A., Kiorpes, L.: Analysis of the development of spatial contrast sensitivity in monkey and human infants. JOSA A **5**(12), 2166–2172 (1988)
28. Prashnani, E., Cai, H., Mostofi, Y., Sen, P.: PieAPP: perceptual image-error assessment through pairwise preference. In: Proceedings of the IEEE Conference on Computer Vision and Pattern Recognition, pp. 1808–1817 (2018)
29. Rudin, L.I., Osher, S., Fatemi, E.: Nonlinear total variation based noise removal algorithms. Physica D **60**(1–4), 259–268 (1992)
30. Wei, K., Fu, Y., Zheng, Y., Yang, J.: Physics-based noise modeling for extreme low-light photography. IEEE Trans. Pattern Anal. Mach. Intell. (2021)
31. Yu, G., Sapiro, G.: DCT image denoising: a simple and effective image denoising algorithm. Image Process. Line **1**, 292–296 (2011). https://doi.org/10.5201/ipol.2011.ys-dct
32. Zhang, K., Zuo, W., Chen, Y., Meng, D., Zhang, L.: Beyond a Gaussian denoiser: residual learning of deep CNN for image denoising. IEEE Trans. Image Process. **26**(7), 3142–3155 (2017)
33. Zhang, K., Zuo, W., Zhang, L.: FFDNet: toward a fast and flexible solution for CNN-based image denoising. IEEE Trans. Image Process. **27**(9), 4608–4622 (2018)

Latent-Space Disentanglement with Untrained Generator Networks for the Isolation of Different Motion Types in Video Data

Abdullah Abdullah[1], Martin Holler[2(✉)], Karl Kunisch[2], and Malena Sabate Landman[3]

[1] Department of Mathematics, The Chinese University of Hong Kong, Hong Kong, Hong Kong
[2] Institute of Mathematics and Scientific Computing, University of Graz, Graz, Austria
{martin.holler,karl.kunisch}@uni-graz.at
[3] Department of Mathematics, Emory University, Atlanta, USA
malena.sabate.landman@emory.edu

Abstract. Isolating different types of motion in video data is a highly relevant problem in video analysis. Applications can be found, for example, in dynamic medical or biological imaging, where the analysis and further processing of the dynamics of interest is often complicated by additional, unwanted dynamics, such as motion of the measurement subject. In this work, it is empirically shown that a representation of video data via untrained generator networks, together with a specific technique for latent space disentanglement that uses minimal, one-dimensional information on some of the underlying dynamics, allows to efficiently isolate different, highly non-linear motion types. In particular, such a representation allows to freeze any selection of motion types, and to obtain accurate independent representations of other dynamics of interest. Obtaining such a representation does not require any pre-training on a training data set, i.e., all parameters of the generator network are learned directly from a single video.

Keywords: Isolation of motion · generator networks · deep image prior · latent-space disentanglement · magnetic resonance imaging

1 Introduction

Processing motion information in a time series of images is a classical but still very active research topic in computer vision and computational imaging, with a

Supplementary Information The online version contains supplementary material available at https://doi.org/10.1007/978-3-031-31975-4_25.

L. Calatroni et al. (Eds.): SSVM 2023, LNCS 14009, pp. 326–338, 2023.
https://doi.org/10.1007/978-3-031-31975-4_25

plethora of applications ranging from autonomous driving to biological and medical imaging. In this context, one can separate three (strongly interconnected) directions of research: i) Motion reconstruction, which aims at reconstructing dynamic image data from incomplete or indirect measurements, with applications for instance in dynamic magnetic resonance (MR) imaging or dynamic positron-emission tomography (PET), see [15,16] for examples. ii) Motion estimation, which aims to estimate and represent motion between different frames. This is one of the most classical problems in computer vision and often addressed, for instance, via optical flow estimation, see [4] for a review. iii) Motion correction, which aims to correct for motion, typically via registration techniques. The latter are again a well-established but still very active research topic, in particular in the context of medical imaging such as MR imaging, PET, computed tomography (CT), see for instance [12].

Naturally, close connections between these three direction of research exist: Having estimated motion fields available is crucial for motion correction, motion correction is strongly interconnected with image reconstruction, and adequately reconstructed time series data is required to carry out classical, image-based motion estimate techniques.

In the past years, deep learning based methods have enabled a significant progress in all of the above research direction connected to motion in time series of images, see [2,5,17] for recent review papers on deep learning techniques for reconstruction, motion estimation and motion correction, respectively.

A task that is related but still different to the above-described research directions is the *isolation* of different types of motion. By this, we mean the following general problem setting: Given a video with different types of motion, synthesize a new video that does not show all types of motion, but only a subset of motion-types that is relevant for further processing. A generic area of applications where this problem is relevant is medical imaging, e.g., MR imaging or PET. Here, a concrete example is a video showing cardiac motion together with additional motion resulting from breathing or patient movement, and the goal is to obtain a video showing only the isolated cardiac motion for further analysis.

While the task of isolating motion is strongly related to image registration techniques, the latter cannot directly be applied here since, even in the case where motion fields that register each frame of the time series to a representative template are available, it is still a highly non-trivial problem to decompose such motion fields into different components corresponding to different types of motion. In the specific context of measuring myocardial perfusion, there exist approaches that aim to overcome this by combining registration techniques, e.g., with independent- or principle component analysis [7,20] or sparsity priors on the perfusion dynamics [11]. Such approaches, however, require a specific structure of the dynamics of interest (i.e. perfusion) and an explicit modeling of deformations. The extension of such techniques, e.g. to allow for the isolation of two types of morphological motion such as cardiac and respiratory motion, is non-trivial.

A more generally applicable alternative to such approaches would be to use a non-linear extension of ICA or variational autoencodes, see for instance [9,10] or [6] and the references therein for more structured models. To the best knowledge of the authors, however, it is not clear to what extend non-linear ICA can be applied for isolating motion types based on a single video, since these approaches are typically applied in different contexts and for large datasets.

In this work, we present a new idea for motion isolation that we believe to have great potential for diverse applications, since it is generally applicable to different medical imaging modalities. Moreover, it can be directly incorporated in (variational) image reconstruction methods and, contrary to classical registration techniques, it does not rely on explicit modeling of different motion types. The main idea behind our approach is to employ untrained generator networks, together with a specific technique for latent space disentanglement, for motion isolation. More specifically, we consider the optimization of a generator network to represent a given time series of images, where different latent space variables are forced to independently explain the different types of motion. The latter is achieved by incorporating one-dimensional information on all but one of the different motion types present in the video.

The goal of this paper is to outline this new idea of motion isolation, and present a first numerical evaluation based on synthetic examples and semi-synthetic examples with real dynamic cardiac MR image data, where we isolate cardiac motion from respiratory motion. The one dimensional information on some of the underlying motion types in this case corresponds to a scalar describing the cardiac- or breathing state, a signal which in practice can easily be obtained for instance by simultaneous electrocardiogram (ECG) or chest displacement measurements.

The use of untrained generator networks for image representation was popularized under the name deep image prior by [19], which also partially inspired our work. Since the appearance of [19], many works employing the deep image prior for image reconstruction in various applications appeared. More recently, also works that employ the deep image prior for representation and reconstruction of dynamic MR data appeared, see for instance [8,21]. Existing works, however, focus on reconstruction and do not employ a specific latent-space disentanglement as proposed here, which is the main ingredient for not only representing but also isolating different motion types.

Latent space disentanglement is in turn an active research topic in the context of GANs, see for instance [3] for a seminal work on using latent space disentanglement to learn interpretable representations, and, e.g., [18] for a work on decomposing motion and content in videos. But again, also in the context of GANs, to the best knowledge of the authors, a method capable of isolating different motion types in video data based on a single video does not exist.

2 Method

We introduce the proposed approach in a general setting (including possibly indirect observations of the image data) first, and then specify its concrete application to isolating respiratory and cardiac motion in dynamic images that were obtained with MR imaging.

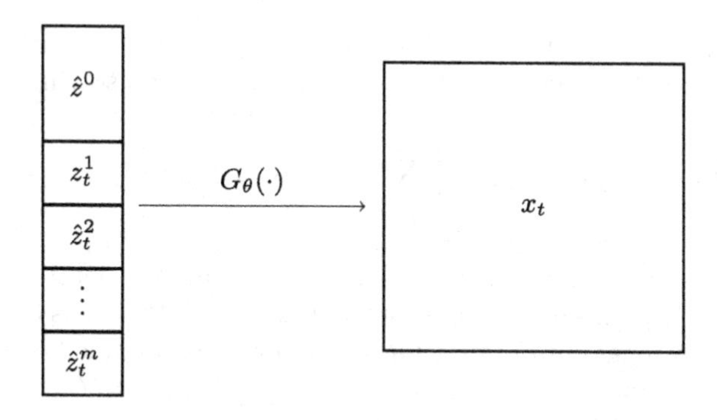

Fig. 1. Schematic representation of the image sequence generation: The latent space is split into a time-independent part $\hat{z}^0 \in \mathbb{R}^{q-m}$ and a time-dependent part $(z_t^1, \hat{z}_t^2, \ldots, \hat{z}_t^m) \in \mathbb{R}^m$. The network $G_\theta(\cdot)$ (whose parameters θ are time-independent) maps the latent space to the image domain. Variables with a hat $(\hat{z}^0, \hat{z}_t^2, \ldots, \hat{z}_t^m)$ are fixed, and the loss is optimized over z_t^1 and θ.

Consider a linear, discretized dynamic imaging inverse problem with data $(y_t)_{t=1}^T \subset \mathbb{R}^M$ and linear operators $A_t \in \mathcal{L}(\mathbb{R}^{N_1 \times N_2}, \mathbb{R}^M)$ of the form

$$y_t = A_t x_t, \quad t = 1, ..., T, \tag{1}$$

where the unknown is a sequence of images $(x_t)_{t=1}^T$ with $x_t \in \mathbb{R}^{N_1 \times N_2}$ for each t. Following the basic idea introduced in [19] for static images, we consider each frame $x_t \in \mathbb{R}^{N_1 \times N_2}$ to be the output of a generator $x_t = G_\theta(z_t)$, whose parameters we want to learn only from the given data. In particular, we consider generator networks $G_\theta : \mathbb{R}^q \to \mathbb{R}^{N_1 \times N_2}$ of the form

$$G_\theta(z_t) = \theta_L^1 * \sigma_{L-1}(\theta_{L-1}^1 * (...\sigma_1(\theta_1^1 * z_t + \theta_1^2)...) + \theta_{L-1}^2) + \theta_L^2, \tag{2}$$

with time-independent parameters $\theta_j^i \in \Theta$, for $1 \leq i \leq 2$ and $1 \leq j \leq L$ pointwise nonlinearities $\sigma_1, \ldots, \sigma_{L-1}$. Here, $*$ denotes a standard multi-channel convolution, convolving kernels defined by the parameters θ_j^1 with the output tensor of one layer to generate the input tensor of the subsequent layer, see Sect. 3 for details on the network architecture. The generator network maps the latent space $\mathbb{R}^q$ to the image (frame) space $\mathbb{R}^{N_1 \times N_2}$. Assuming the dynamic image sequence $(x_t)_{t=1}^T$ contains $m \in \{2, 3, \ldots\}$ independent types of motion, we split the latent variable $z \in \mathbb{R}^q$ into a time-independent part $z^0 \in \mathbb{R}^{q-m}$ and m time-dependent variables $(z_t^i)_{t=1}^T$ with $z_t^i \in \mathbb{R}$, $i = 1, \ldots, m$, see Fig. 1 for a schematic representation. We further assume that all the time dependent variables $(z_t^i)_{t=1}^T$ except $(z_t^1)_{t=1}^T$ are given as $(\hat{z}_t^i)_{t=1}^T$ from some one-dimensional a-priori information on the state of the respective types of motion. That is, we do not optimize over those latent variables but rather fix them according to some measurements (e.g. electrocardiograms or chest-displacement measurements) that we assume to be given and to encode the current state of the respective types of motion.

We then reconstruct the image sequence $(x_t)_{t=1}^T$ together with the network parameters θ and the time-dependent variable $(z_t^1)_{t=1}^T$ via solving

$$((\hat{z}_t^1)_{t=1}^T, \hat{\theta}) \in \arg\min_{(z_t^1)_{t=1}^T, \theta} \frac{1}{T} \sum_{t=1}^{T} \|y_t - A_t G_\theta(\hat{z}^0, z_t^1, \hat{z}_t^2 \ldots, \hat{z}_t^m)\|_2^2, \tag{3}$$

where $\|\cdot\|_2$ is the Euclidean norm (but can be a more general loss) and $\hat{z}^0 \in \mathbb{R}^{q-m}$ is a randomly initialized, fixed static latent variable (as is also the case in the original deep image prior [19]). Once a solution is obtained, we do not only obtain the reconstructed images sequence $(\hat{x}_t)_{t=1}^T$ via $\hat{x}_t = G_{\hat{\theta}}(\hat{z}^0, \hat{z}_t^1, \ldots, \hat{z}_t^m)$, but, more importantly, can generate image sequences $(\hat{x}_t^i)_{t=1}^T$ for $i = 1, \ldots, m$, which we expect to contain only the ith type of motion with all others being fixed, via

$$\hat{x}_t^i = G_{\hat{\theta}}(\hat{z}^0, \hat{z}_{h_1}^1, \ldots, \hat{z}_{h_{i-1}}^{i-1}, \hat{z}_t^i, \hat{z}_{h_{i+1}}^{i+1}, \ldots, \hat{z}_{h_m}^m), \tag{4}$$

where $h_1, \ldots, h_{i-1}, h_{i+1}, \ldots, h_m$ are fixed reference frames.

As a concrete example, in this paper we consider the application of this general approach to isolating cardiac motion from respiratory motion in dynamic images obtained from MRI, where one-dimensional information about the respiratory state (e.g. from measurements of the chest displacement) is available. In this case, $m = 2$, and the latent variable $z_t \in \mathbb{R}^q$ at time t is decomposed as $z_t = (z^0, z_t^1, z_t^2)$ with z_t^2 known (and given as $(\hat{z}_t^2)_{t=1}^T$). As our focus is on motion isolation rather than reconstruction, we further assume the reconstructed dynamic image sequence to be available, i.e., A_t is the identity, noting that a generalization of our approach to reconstruction does not pose any conceptual difficulties. In summary, this yields the following optimization problem

$$((\hat{z}_t^1)_{t=1}^T, \hat{\theta}) \in \arg\min_{(z_t^1)_{t=1}^T, \theta} \frac{1}{T} \sum_{t=1}^{T} \|y_t - G_\theta(\hat{z}^0, z_t^1, \hat{z}_t^2)\|_2^2. \tag{5}$$

Algorithmic Strategy. To solve the minimization problem (5), we use Pytorch [13] and the ADAM optimizer with default settings. To achieve a good minimization of the loss, and in particular stability w.r.t varying random initializations, we iteratively reduce the learning rate after a fixed number of epochs, track the network parameters and latent variables that achieve the minimal loss, and export those parameters and variables as the optimal solution (instead of the variables of the last iterate). Note that this does not cause much computational overhead due to the rather small dimensionality of our network (see Sect. 3).

Baseline Method. In order to have a baseline method for comparison, we also implemented an approach for motion isolation that is based on independent component analysis as follows: Any given video $(x_t)_{t=1}^T$ is first rearranged to a matrix $A \in \mathbb{R}^{(N_1 N_2) \times T}$, with the rows representing spatial dimension and columns representing the temporal dimension. This matrix is then decomposed as $A = SV$ with $S \in \mathbb{R}^{(N_1 N_2) \times 2}$ and $V \in \mathbb{R}^{2 \times T}$ using the FastICA implementation of [14], where the two rows of V correspond to the two independent components. Based on manual labeling, one row (say the first one) of V is then assigned to respiratory motion, and the other to cardiac motion. Given this, an approximation of the respiratory motion at any state t_1 and of the cardiac motion at any state t_2 can then obtained as $(S_{:,1}V_{1,:})_{:,t_1} + (S_{:,2}V_{2,:})_{:,t_2}$ (using standard matrix indexing). Fixing one of these states and letting the other one run from 1 to T, a video of only respiratory or cardiac motion can be obtained.

Note that, while this method is a simple, baseline approach to separate data into different, independent motion components, it has several limitations: Since the separation is linear, it can be expected to (approximately) work only for certain types of motions and images. Furthermore, the method can only be used for fully-sampled videos, but not for indirect measurements as in (3). Advantages of the method on the other hand are its simplicity and that it does not require any a-priori information on the motion.

3 Numerical Experiments

In this section the results of different experiments concerning dynamic images with respiratory and cardiac motion are presented to illustrate the behavior of the new method. In particular, we present a synthetic phantom example and four semi-synthetic examples where real dynamic cardiac MR images were enriched with synthetic respiratory motion. Note that a synthetization of one of the motion types is necessary in order to have a ground-truth with isolated motion.

For all experiments, we assume one-dimensional information, henceforth referred to as motion triggers, about the respiratory motion to be given. For the phantom images, the triggers for the respiratory motion are assumed to be given exactly, where for the real MR images we provide the method only with a perturbed version of the original respiratory motion triggers.

To assess the quality of the motion isolation, we compute the relative error norms ($\mathbf{E}_h^1$ and $\mathbf{E}_h^2$) of the dynamic images $\hat{x}^1 = (\hat{x}_t^1)_{t=1}^T$, $\hat{x}^2 = (\hat{x}_t^2)_{t=1}^T$ containing just one reconstructed kind of motion, where

$$\mathbf{E}_h^1 = \|\hat{x}^1 - x_{\text{true}}^1\|_2 / \|x_{\text{true}}^1\|_2, \quad \mathbf{E}_h^2 = \|\hat{x}^2 - x_{\text{true}}^2\|_2 / \|x_{\text{true}}^2\|_2, \tag{6}$$

and x_{true}^i is the ground truth showing only the ith type of motion, i.e., cardiac motion for $i = 1$ and respiratory motion for $i = 2$. Here, the subscript h refers to the frame at which the other motion state is fixed, see (4).

In principle, as described in Sect. 2, our method (as well as the ICA-based baseline approach) allows to freeze one kind of motion at any state, and generate images containing only the second kind of motion, as long as a sufficient mixing of motions was observed. In practice, the choice of the freezing frame h has an impact on the performance of the single motion reconstruction (though for the phantom at an overall rather low error regime). To allow for a fair comparison, for each method we always show the result for fixing the motion state that provides the best performance with respect to the ground truth.

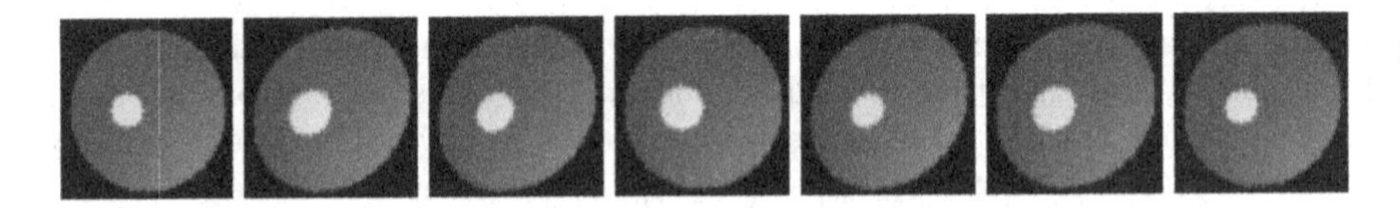

Fig. 2. Selected synthetic data frames displaying different phases of respiratory and cardiac movement.

For all experiments with our approach shown in the paper, we repeated the experiment 20 times with 20 different random seeds, and show the result whose performance w.r.t. the error measure $\mathbf{E}_h^1$ is closest to the median performance. Quantitative error measures for all experiments are provided in Table 1. Videos showing results for three different experiments are available in the supplementary material of this work. The source code to reproduce all experiments as well as videos showing all results of our method are available at [1]. All our experiments were conduced on a workstation with an AMD Ryzen 7 3800X 8-Core Processor and 32 GB of memory, using a Nvidia RTX 3090 GPU with 24 GB of memory. The smallest and largest experiment considered here (solving (5) with phantom and real cine MR data, respectively) took around 19 s and 3.8 min, respectively.

3.1 Synthetic Data

The first test problem, consisting of 80 frames with 64×64 pixels, corresponds to a synthetic example displaying two nested disks and is represented in Fig. 2. A more compact representation of this dynamic image can be observed in Fig. 3 (left), where a vertical slice of the image (marked in red on the left image), is displayed over time (and can be seen on the right image) clearly showing temporal changes. This representation will be used throughout the paper.

In this example, three cardiac motion cycles are simulated by dilation of the internal disk while approximately two simulated breathing motion cycles are represented by shearing of the whole image. Note that the size of the frames over time is maintained constant. The ground truth one-dimensional motion information, i.e., the motion triggers, that was used to parametrize the different types of motions is displayed in Fig. 3 (right). Note that the motion trigger for respiratory motion is provided to our method via the latent variable $(z^2(t))_{t=1}^T$, while the motion trigger for cardiac motion is an unknown optimization variable.

The generator modeling the solution for this example, as defined in Eq. (2), corresponds to a standard deep convolutional neural network (CNN) with 5 layers, where transpose two-dimensional convolutions are used for all convolutions in (2) and no biases are used. The network parameters are given as follows. Number of channels: [64, 128, 64, 32, 16, 1], (square) kernel size: [4, 4, 4, 4, 4], stride: [1, 2, 2, 2, 2], padding: [0, 1, 1, 1, 1], activation functions: [Tanh, LeakyReLU, Tanh, LeakyReLU, Tanh]. In total, the network has 3.03360×10^5 parameters. The architecture of the network, including the usage of different types of activation functions for different layers, is rather standard. While we could obtain reasonable

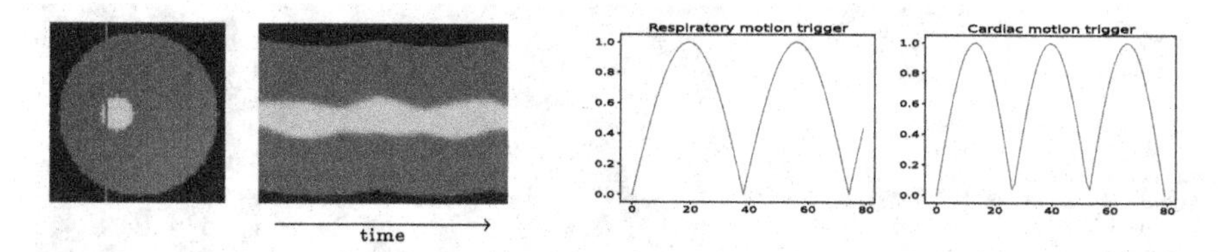

Fig. 3. Representation of the time evolution for the synthetic data. Left: Compact video representation, where the pixels on the red line of the left image are plotted over time to create the right image. Right: Ground truth motion triggers. (Color figure online)

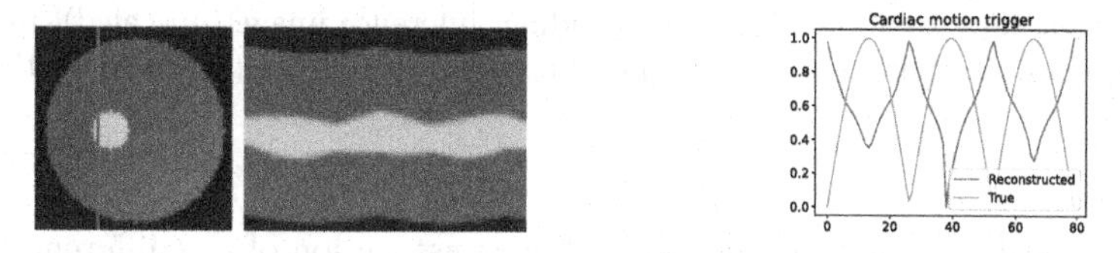

Fig. 4. Synthetic video reconstruction with full motion. Left: Video data, right: heart motion triggers (rescaled).

results with different architecture, in particular different combinations of activation functions, this one was chosen for giving the best results on average.

The latent space $Z \in \mathbb{R}^{64}$ is split into 62 static components and 2 dynamic components. Blocks of [4000, 4000, 4000, 4000] epochs with learning rates [0.01, 0.005, 0.001, 0.0005] are used for the Adam optimizer. The latent variables are initialized randomly from a uniform distribution on $[0, 1)$, and the network weights are initialized with Pytorch self-initialization.

The reconstruction of the dynamic image data resulting from the numerical solution of (5) is shown in Fig. 4. Even though a representation of the given data with mixed motion is not our primary goal, it can still be observed that resulting video is visually very similar to the ground truth, giving evidence that one-dimensional information on one of the movements is enough to disentangle the latent space without a significant degradation in representing the data. Regarding the reconstructed motion triggers for cardiac motion (the variable $(z_t^1)_{t=1}^T$, see Fig. 4 right), we do not expect an accurate reconstruction due to unavoidable ambiguities how such motion can be represented. Nevertheless, one would expect to see a periodic behavior in the reconstructed motion trigger, with the same period as the ground truth. Except for one larger deflection of the motion trigger during one period (which is probably canceled out by the nonlineraties in the network) this can indeed be observed.

After reconstruction, motion isolation based on the strategy described in Sect. 2 is straightforward. Figure 5 shows the isolated cardiac (top) and respiratory (bottom) motion reconstructions, together with results that were obtained with the ICA-based baseline approach. It can be seen that the proposed method indeed achieves a decomposition into the different types of motion, and that

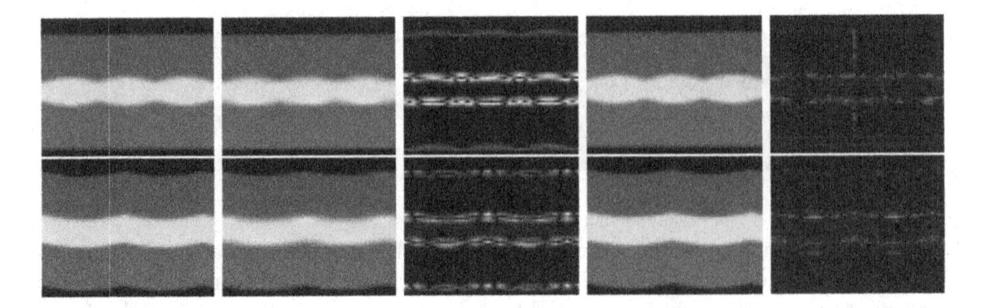

Fig. 5. Compact representation of the synthetic video reconstructed with only cardiac (top row) and only respiratory (bottom row) motion. From left to right: Ground truth, ICA-based motion isolation with corresponding difference image (upscaled), proposed motion isolation technique with corresponding difference image (upscaled by the same factor).

this decomposition yields a much sharper reconstruction of the different motion types compared to the ICA-based approach (see the error images).

3.2 Cardiac MR Images

We now test our method on cine MR images, comprising two images with a four-chamber view and two images with a short-axis view[1].

In all experiments considered here, the original videos were obtained via a sum-of-squares reconstruction from fully sampled MR data, and contain a 2D slice of the entire thorax showing the beating heart in one region. For our experiments, we simulated three heartbeats by concatenating the single-heartbeat-videos three times, and simulated two respiratory cycles with vertical (resp. horizontal) shearing motion for the four-chamber (resp. short-axis) view. After obtaining videos showing a slice of the entire thorax with three heartbeats and two respiratory cycles, we cropped the videos to a region of interest around the heart, see the the two left columns of Figs. 7 and 8. The final data consists of 99 resp. 90 frames with spatial resolution 100×100 for the two four-chamber view examples, and of 81 resp. 78 frames with spatial resolution 70×70 for the two short-axis view examples.

In order to account also for measurement errors in the motion trigger, the motion trigger for respiratory motion provided to our method was a perturbed one with the time-position where the motion trigger is sampled being perturbed by additive Gaussian noise with a standard deviation of 50% of the length of one timestep. To account for that, in our method we also allowed deviations from the perturbed trigger that are penalized with an L^2 discrepancy.

In all experiments with real data, the generator is a standard deep convolutional neural network (CNN) with 7 layers, ReLU activation functions in the first and last layer, LeakyReLUs in the middle layers and no biases. The latent

[1] Data from the ISMRM reconstruction challenge 2014 challenge.ismrm.org.

space $\mathbb{R}^{100}$ is split into 98 static components and 2 dynamic components. Network parameters shared by all experiments are given as: Number of channels: $[100, 640, 320, 160, 80, 40, 20, 1]$, stride: $[2, 2, 2, 2, 2, 2, 1]$. To account for the different image dimensions, the shape of the remaining parameters differs slightly: (Square) kernel size: $[4, 4, 4, 4, 4, 4, 3]$ (four-chamber), $[4, 4, 4, 4, 4, 5, 4]$ (short-axis), padding: $[0, 2, 0, 2, 2, 1, 1]$ (four-chamber), $[0, 2, 2, 2, 2, 1, 1]$ (short-axis). In total, the networks have 5.388980×10^6 (four-chamber) and 5.396320×10^6 (short-axis) parameters. Again this specific architecture was chosen in order to have a single architecture that gives good results for both four-chamber and short-axis data. Tuning the architecture to specific types of images might further improve the results.

The network is optimized according to Eq. (5), where the latent variables are randomly initialized from a uniform distribution on the interval $[0, 1)$, and the network weights are initialized with Pytorch self-initialization. For both experiments, blocks of $[4000, 4000, 4000, 4000, 4000]$ epochs with learning rates $[0.01, 0.008, 0.005, 0.003, 0.001]$ are used for the Adam optimizer. A comparison to the ICA-based approach was also carried out, but since we were not able to obtain reasonable results with the ICA-based approach, the comparison is not included here.

Motion isolation is performed on the MR images as explained in Sect. 2. The learned generator and the reconstructed dynamic latent space variables are used to freeze one type of motion while maintaining the dynamics of the other type of motion. The latent space variables associated to the cardiac motion are shown in Figs. 6. Note that these plots indeed seem to reassemble patterns associated to the heart's activity, and were completely unknown before performing the optimization.

The results of the motion isolation experiments are shown in Figs. 7 and 8 for the four-chamber view experiments and the short-axis view experiments, respectively. It can be observed that in all cases and for both types of motion, the isolation of motion works well, and the different structures of the motion are clearly visible in the slice-based visualization of the generated images. This is also confirmed by the quantitative values provided in Table 1.

We should note that some artifacts are visible in the breathing motion isolation. In our experience, those are mostly related to having obtained a sub-optimal solution of the minimization problem (recall that we provide results corresponding to the median of the performance of our method). In cases where a favorable random initialization leads to an improved convergence of the methods (in terms of minimizing the objective functional), these artifacts are reduced. Based on this observation, in order to further improve the results, one could in practice monitor the loss during iterations and, in case a suboptimal approximation of the data was achieved, restart the method with a different, random initialization.

Conclusions and Outlook. This paper introduces a new method for motion isolation based on the joint optimization of an untrained generator network over both the network parameters and the latent codes. Assuming one-dimensional information on all but one of the motions is known, motion isolation is achieved

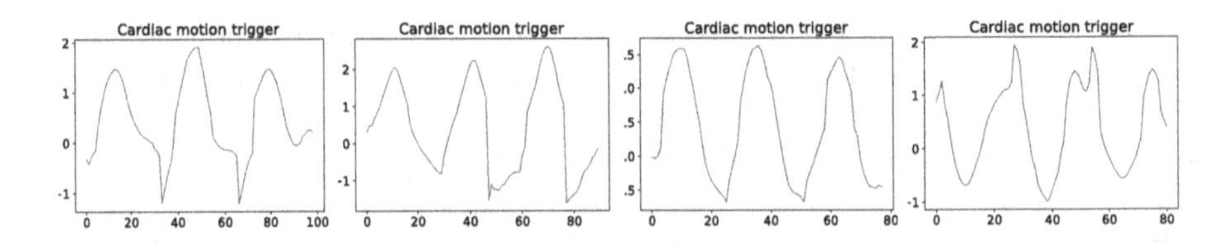

Fig. 6. Reconstructed motion triggers for the four-chamber examples (left two images) and the short-axis examples (right two images).

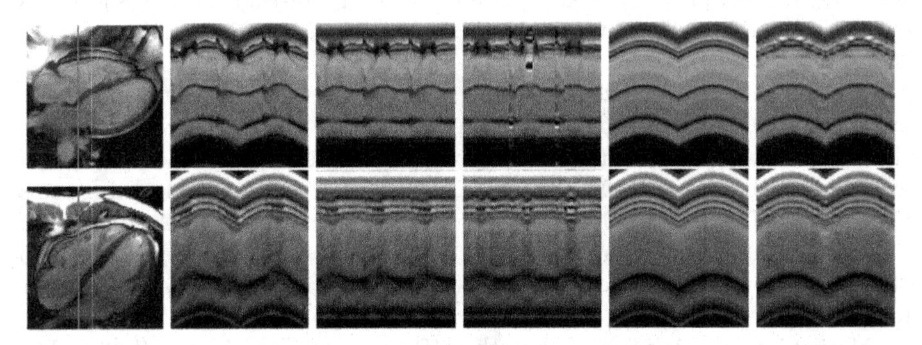

Fig. 7. Four-chamber examples. From left to right: Reference frame with marked slice, compact representation of full dynamics (ground truth), of cardiac dynamics (ground truth and reconstruction) and of respiratory dynamics (ground truth and reconstruction).

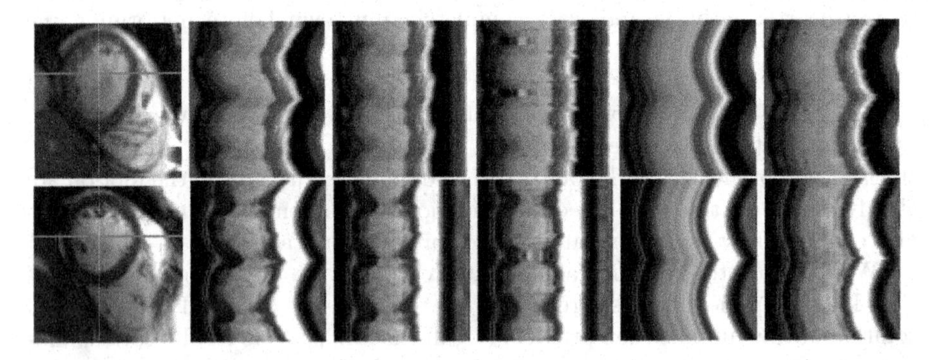

Fig. 8. Short-axis examples. From left to right: Reference frame with marked slice, compact representation of full dynamics (ground truth), of cardiac dynamics (ground truth and reconstruction) and of respiratory dynamics (ground truth and reconstruction).

through latent space disentanglement. Feasibility of this method was shown for isolating respiratory and cardiac motion in dynamic MR images, but the proposed method is general and can conceptually be used in many applications, e.g., in bio-medical imaging, biology, bio-mechanics or physics.

Table 1. Error in motion isolation (repeating each experiment for 20 different seeds). MAD denotes the median absolute deviation.

	Median	MAD	Mean	Std. dev
Phantom example, $\mathbf{E}_h^1$ (cardiac)	1.06e-02	2.87e-03	1.36e-02	8.32e-03
Phantom example, $\mathbf{E}_h^2$ (respiratory)	9.79e-03	1.41e-03	1.08e-02	2.97e-03
Four-chamber view, $\mathbf{E}_h^1$ (cardiac)	1.13e-01	2.89e-02	1.24e-01	3.93e-02
Four-chamber view, $\mathbf{E}_h^2$ (respiratory)	8.66e-02	1.69e-02	9.76e-02	2.75e-02
Four-chamber view, example 2, $\mathbf{E}_h^1$ (cardiac)	9.11e-02	6.71e-03	1.09e-01	5.70e-02
Four-chamber view, example 2, $\mathbf{E}_h^2$ (resp.)	7.64e-02	1.01e-02	8.98e-02	4.94e-02
Short-axis view, $\mathbf{E}_h^1$ (cardiac)	1.42e-01	1.88e-02	1.46e-01	3.38e-02
Short-axis view, $\mathbf{E}_h^2$ (respiratory)	1.02e-01	1.59e-02	1.10e-01	2.91e-02
Short-axis view, example 2, $\mathbf{E}_h^1$ (cardiac)	8.26e-02	6.87e-03	8.38e-02	1.22e-02
Short-axis view, example 2, $\mathbf{E}_h^2$ (respiratory)	6.91e-02	4.91e-03	6.94e-02	1.03e-02
Phantom example, ICA-method, $\mathbf{E}_h^1$ (cardiac)	3.08e-02	–	–	–
Phantom example, ICA-method, $\mathbf{E}_h^2$ (resp.)	3.87e-02	–	–	–

A limitation of the method, resulting from non-convexity, is its dependence on initializations, which is counteracted here via loss-based restarting strategies. Further, no explicit motion information is made available by our methods. While this might be considered as limitation, it also comes with the advantage that no explicit modeling of the underlying motion types is necessary. Our work shows a great potential of latent space disentanglement on untrained generators for video data. It opens the door to more advanced disentanglement schemes (e.g., based on modifications of the loss function or additional constraints on the latent space variables). Moreover, we expect the proposed method to be very well suited as image prior for dynamic inverse problems (e.g., in tomography, super-resolution) where the reconstructed solution displays different kinds of independent motion.

References

1. Abdullah, A., Holler, M., Kunisch, K., Landman, M.S.: Source code for: latent-space disentanglement with untrained generator networks for the isolation of different motion types in video data (2023). https://github.com/hollerm/generator_based_motion_isolation
2. Bustin, A., Fuin, N., Botnar, R.M., Prieto, C.: From compressed-sensing to artificial intelligence-based cardiac MRI reconstruction. Front. Cardiovasc. Med. **7**, 17 (2020). https://doi.org/10.3389/fcvm.2020.00017
3. Chen, X., Duan, Y., Houthooft, R., Schulman, J., Sutskever, I., Abbeel, P.: InfoGAN: interpretable representation learning by information maximizing generative adversarial nets. In: Advances in Neural Information Processing Systems, vol. 29 (2016)
4. Fortun, D., Bouthemy, P., Kervrann, C.: Optical flow modeling and computation: a survey. Comput. Vis. Image Underst. **134**, 1–21 (2015). https://doi.org/10.1016/j.cviu.2015.02.008

5. Fu, Y., Lei, Y., Wang, T., Curran, W.J., Liu, T., Yang, X.: Deep learning in medical image registration: a review. Phys. Med. Biol. **65**(20), 20TR01 (2020). https://doi.org/10.1088/1361-6560/ab843e
6. Hälvä, H., et al.: Disentangling identifiable features from noisy data with structured nonlinear ICA. In: Advances in Neural Information Processing Systems, vol. 34 (2021)
7. Hamy, V., et al.: Respiratory motion correction in dynamic MRI using robust data decomposition registration - application to DCE-MRI. Med. Image Anal. **18**(2), 301–313 (2014). https://doi.org/10.1016/j.media.2013.10.016
8. Hyder, R., Asif, M.S.: Generative models for low-dimensional video representation and reconstruction. IEEE Trans. Signal Process. **68**, 1688–1701 (2020). https://doi.org/10.1109/TSP.2020.2977256
9. Hyvärinen, A., Pajunen, P.: Nonlinear independent component analysis: existence and uniqueness results. Neural Netw. **12**(3), 429–439 (1999). https://doi.org/10.1016/S0893-6080(98)00140-3
10. Khemakhem, I., Kingma, D., Monti, R., Hyvarinen, A.: Variational autoencoders and nonlinear ICA: a unifying framework. In: International Conference on Artificial Intelligence and Statistics, pp. 2207–2217 (2020)
11. Lingala, S.G., DiBella, E., Jacob, M.: Deformation corrected compressed sensing (DC-CS): a novel framework for accelerated dynamic MRI. IEEE Trans. Med. Imaging **34**(1), 72–85 (2015). https://doi.org/10.1109/TMI.2014.2343953
12. Oliveira, F.P., Tavares, J.M.R.: Medical image registration: a review. Comput. Methods Biomech. Biomed. Engin. **17**(2), 73–93 (2014). https://doi.org/10.1080/10255842.2012.670855
13. Paszke, A., et al.: Automatic differentiation in PyTorch. In: Advances in Neural information processing systems (2017)
14. Pedregosa, F., et al.: Scikit-learn: machine learning in python. J. Mach. Learn. Res. **12**, 2825–2830 (2011)
15. Rahmim, A., Tang, J., Zaidi, H.: Four-dimensional (4D) image reconstruction strategies in dynamic PET: beyond conventional independent frame reconstruction. Med. Phys. **36**(8), 3654–3670 (2009). https://doi.org/10.1118/1.3160108
16. Schloegl, M., Holler, M., Schwarzl, A., Bredies, K., Stollberger, R.: Infimal convolution of total generalized variation functionals for dynamic MRI. Magn. Reson. Med. **78**(1), 142–155 (2017). https://doi.org/10.1002/mrm.26352
17. Tu, Z., et al.: A survey of variational and CNN-based optical flow techniques. Signal Process.: Image Commun. **72**, 9–24 (2019). https://doi.org/10.1016/j.image.2018.12.002
18. Tulyakov, S., Liu, M.Y., Yang, X., Kautz, J.: MoCoGAN: decomposing motion and content for video generation. In: IEEE Conference on Computer Vision and Pattern Recognition (CVPR), pp. 1526–1535 (2018)
19. Ulyanov, D., Vedaldi, A., Lempitsky, V.: Deep image prior. Int. J. Comput. Vision **128**(7), 1867–1888 (2020). https://doi.org/10.1007/s11263-020-01303-4
20. Wollny, G., Kellman, P., Santos, A., Ledesma-Carbayo, M.J.: Automatic motion compensation of free breathing acquired myocardial perfusion data by using independent component analysis. Med. Image Anal. **16**(5), 1015–1028 (2012). https://doi.org/10.1016/j.media.2012.02.004
21. Yoo, J., Jin, K.H., Gupta, H., Yerly, J., Stuber, M., Unser, M.: Time-dependent deep image prior for dynamic MRI. IEEE Trans. Med. Imaging **40**(12), 3337–3348 (2021). https://doi.org/10.1109/TMI.2021.3084288

Natural Numerical Networks on Directed Graphs in Satellite Image Classification

Karol Mikula[1,2], Michal Kollár[1,2], Aneta A. Ožvat[1,2](✉), Mária Šibíková[3], and Lucia Čahojová[3]

[1] Department of Mathematics, Slovak University of Technology in Bratislava, Radlinskeho 11, Bratislava 810 05, Slovakia
karol.mikula@gmail.com, aneta.ozvat@stuba.sk

[2] Algoritmy:SK s.r.o., Šulekova 6, 81106 Bratislava, Slovakia

[3] Plant Science and Biodiversity Center, Slovak Academy of Sciences, Institute of Botany, Dubravska cesta 9, Bratislava 845 23, Slovakia

Abstract. Natural numerical networks on directed graphs as a new supervised deep learning PDE-based classification algorithm are proposed in this work. The Natural numerical network (NatNet) is based on a forward-backward diffusion model, where the points of the given clusters are attracted together by the forward diffusion, and in contrast, the backward diffusion repulses points of different clusters from each other. First, the network is trained on the labelled data to achieve the highest possible accuracy on the learning dataset. Then, the method is applied to the classification of Sentinel-2 satellite optical data to automatically identify the protected oak habitat in Western Slovakia due to its threatened status. To that goal, the relevancy map, one of the outputs of the Natural numerical network, is created efficiently; its construction is significantly speed up thanks to the new NatNet formulation on directed graphs.

Keywords: Forward-backward diffusion · Partial differential equations on graph · Numerical methods · Data classification

1 Introduction

A new concept of Natural numerical network (NatNet) on directed graphs is introduced in this paper. The NatNet as a new supervised deep learning PDE-based classification method was presented in [13]. It introduces a forward-backward diffusion model on undirected graphs and its numerical discretisation to get the classification algorithm. The forward diffusion attracts the points of the given clusters together while the backward diffusion repulses the points of different clusters from each other. Such approach is inspired by the recent ODE and PDE-based deep learning methods from [4,10], and attraction-repulsion strategies are used also in other clustering applications such as high-dimensional data visualization [3]. For an interesting overview of PDE-based and variational approaches

Supported by grants APVV-19-0460, VEGA 1/0436/20 and ESA contract 4000140486/23/NL/SC/rp.

L. Calatroni et al. (Eds.): SSVM 2023, LNCS 14009, pp. 339–351, 2023.
https://doi.org/10.1007/978-3-031-31975-4_26

on graphs for high-dimensional data classification we refer also to [2]. In [10], the relation between a successful deep learning model, the so-called Residual Neural Network (ResNet) [4,11], and the numerical solution of the system of ordinary differential equations using the forward Euler method is shown. The authors then designed parabolic and hyperbolic networks for deep learning classification based on corresponding types of partial differential equations. The NatNet [13] uses another type of PDE, the nonlinear forward-backward diffusion equation, which is natural for supervised deep learning classification.

In [13], the NatNet supervised deep learning classification method was applied to the Sentinel-2 multispectral optical data to obtain in an automated way a spatial appearance of Natura 2000 [7] protected habitats in Slovakia and along the Danube river in Central and South-Eastern Europe. There exist various recent studies dealing with the classification of land cover classes by using the multispectral satellite images, and all standard classification algorithms, such as the Random Forest, k-Nearest Neighbour, and Support Vector Machine, were used, reported, and compared [5,16,18]. In general, the standard methods provide a meaningful classification of land cover classes, but a higher accuracy, exceeding 90%, is only reached, when simple categories are identified, e.g. water bodies, meadows, forests, fields and urbanised areas. They fail to classify accurately different forest types using purely the Sentinel-2 multispectral optical bands information, in such case, the classification accuracy drops to about 80% [18]. On the other hand, forest habitats defined in the Natura 2000 classification system are complex plant communities with variable species composition, and commonly one habitat cannot be defined by one dominant tree species. Providing classification of such detailed classes has been a challenging task with high demand on new reliable classification methods based on widely available multispectral satellite data. The NatNet designed in [13] provides a first successful approach to solve this task.

The trained NatNet can be represented by the forward-backward diffusion on undirected graphs [13], but for the classification of a large number of new observations, the directed graph concept is useful due to its computational efficiency. The NatNets classification output is given, together with the cluster membership of any pixel of the satellite image, by the so-called relevancy map. The relevancy map is a greyscale image giving information on the relevancy of cluster membership for every pixel and it has the same dimension as the satellite image. In order to create the relevancy map we have to let evolve the new observation by the dynamics of the trained network and classify it. In the undirected graph approach, we have to do it one by one solving as many times a small system of equations as there are pixels in the relevancy map. By using the directed graph concept, all pixel values in the relevancy map are computed at once by solving one system of equations which speeds up the computation many times, proportionally to the number of graph vertices in the learning dataset, and makes it possible to include the relevancy map computation directly into the NaturaSat software [15]. In this paper, we explain the NatNet on directed graphs in detail and apply it to the classification of protected oak habitats in Western Slovakia by Sentinel-2 satellite images.

2 Natural Numerical Network (NatNet)

2.1 Mathematical Model

A directed graph G consists of a non-empty finite set $V(G)$ of elements called vertices and a finite set $A(G)$ of ordered pairs of distinct vertices called arcs or directed edges [1]. We denote the number of vertices of the directed graph G by N_V. In general, for an arc $(u, v) = e_{uv}$, the first vertex u is its tail, and the second vertex v is its head, which means that the arc (u, v) leaves u and enters v. The head and tail of an arc are its end-vertices [1]. In the sequel, we will use a semi-complete directed graph G, where semi-complete means that there exists an arc between every pair of vertices $V(G)$.

Let us have the function $X : G \times [0, T] \to \mathbb{R}^k$ representing the Euclidean coordinates $X(v, t) = (x_1(v, t), \ldots, x_k(v, t))$ of the vertex $v \in V(G)$ in time $t \in [0, T]$. The index k represents a dimension of the feature space $\mathbb{R}^k$. A diffusion of the function $X(v, t)$ on the directed graph G is formulated as a partial differential equation (PDE)

$$\partial_t X(v, t) = \nabla \cdot (g \nabla X(v, t)), \qquad v \in V(G), \qquad t \in [0, T], \tag{1}$$

where g represents the diffusion coefficient, see also [9]. We consider Eq. (1) together with an initial condition $X(v, 0) = X^0(v)$, $v \in V(G)$. The boundary conditions are not necessary to prescribe because in our model diffusion occurs between all vertices of the semi-complete directed graph G.

We consider the diffusion coefficient g depending on the distance between two vertices v and u of the directed graph G. It will give a nonlinear diffusion model on the directed graph. We consider Eq. (1) with diffusion coefficient g in the form

$$g(e_{uv}) = \varepsilon(e_{uv}) \frac{1}{1 + \sum_{i=1}^{k} (K_i \; l_i^2(e_{uv}))}, \qquad K_i \geq 0, \quad i = 1, \ldots, k, \tag{2}$$

where K_i represents weights for each coordinate $l_i(e_{uv})$, $i = 1, \ldots, k$, of the vector

$$\begin{aligned} l(e_{uv}) &= (l_1(e_{uv}), \ldots, l_k(e_{uv}))^T = X(v, \cdot) - X(u, \cdot) = \\ &= (x_1(v, \cdot) - x_1(u, \cdot), \ldots, x_k(v, \cdot) - x_k(u, \cdot))^T, \qquad v, u \in V(G), \end{aligned} \tag{3}$$

and allow us to control the diffusion speed in each direction of the k-dimensional feature space. If the sum in the diffusion coefficient is large, the diffusion coefficient g is close to 0, which means that the diffusion process will be slow and the points do not diffuse by averaging. If the sum in the diffusion coefficient is small, the diffusion coefficient is close to 1, the diffusion process is faster, and the points are moving fast by diffusion.

The value of $\varepsilon(e_{uv})$ in the diffusion coefficient depends on the type of diffusion applied in each arc between each pair of vertices. For applying the forward diffusion, we choose $\varepsilon(e_{uv})$ as a positive constant, in all computations presented in this paper $\varepsilon(e_{uv}) = 1$. In our application, forward diffusion causes a moving, and thus clustering of points together. On the other hand, backward diffusion is represented by a negative diffusion coefficient, $\varepsilon(e_{uv})$ is a small negative value, and in our application it gives a repulsion of the points belonging to different clusters. Such a model with a combination of forward and backward diffusion is a suitable tool for supervised learning.

In the case of two directed edges between two vertices u and v, we consider $g(e_{uv}) = g(e_{vu})$.

In Fig. 1 we illustrate the behaviour of the model (1)–(2) on two given clusters where by light blue arrows we plot some of the links of forward diffusion and by red arrows some of the links of backward diffusion. This figure depicts the basic features and behaviour of the Natural network. The points inside a given cluster are attracted by forward diffusion, while there is a repulsion of points of different clusters by backward diffusion. The model allows the directed graph to have arcs with the same end-vertices, and in this figure all vertices have two directed edges, which means that the diffusion influence occurs in both directions. It is important to realise that the model does not allow pairs of arcs with the same tail and the same head (parallel arcs) or arcs whose heads and tails are the same vertex (loops).

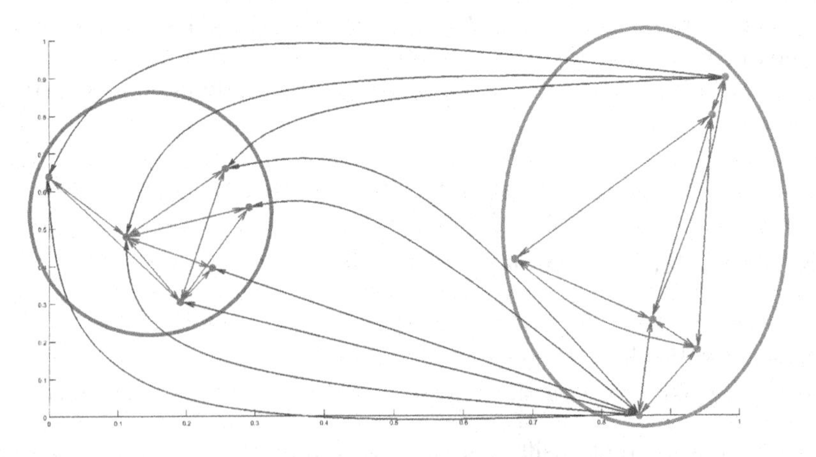

Fig. 1. Randomly generated 2D points in two clusters and some links of forward diffusion (light blue arrows) inside the clusters, and some links of backward diffusion (red arrows) between points from different clusters. (Color figure online)

Furthermore, in Fig. 2 we illustrate the situation that arises in the supervised learning and application phases of the classification method when a new observation is added to the network. Only forward diffusion is applied to all links of the vertex representing the new observation, which means that the new vertex is the head for directed edges connecting it with all other vertices. The forward diffusion links are depicted by the dark blue arrows connecting the new observation (black square) to every other point. Thus, this new observation is attracted by a certain diffusion speed to all existing clusters, which themselves are subject to the forward-backward diffusion as described before. The dynamics of the network decides on the cluster membership of the new observation.

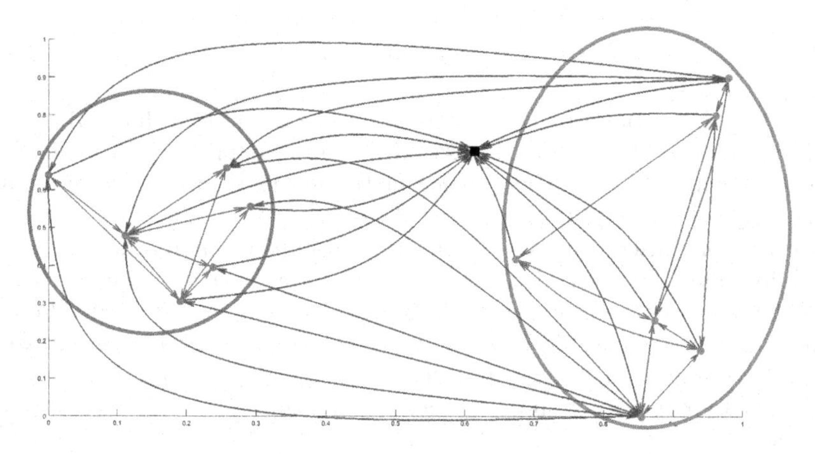

Fig. 2. Randomly generated 2D points in two clusters with one new observation (black square). The light blue arrows represent some of the forward diffusion links inside the clusters. The forward diffusion links from all other points to the new observation are represented by the dark blue arrows. The red arrows represent the links of backward diffusion between points from different clusters. (Color figure online)

We can reduce the influence of forward diffusion on the new observation vertex $w \in V(G)$ by using the diffusion coefficient in the form

$$g(e_{vw}) = \max(\varepsilon(e_{vw})\frac{1}{1+\sum_{i=1}^{k}(K_i\ l_i^2(e_{vw}))} - \delta,\ 0), \qquad \varepsilon(e_{vw}) > 0 \quad (4)$$

at all directed edges entering w, where δ is a parameter of the size of the "diffusion neighbourhood". The aforementioned modification causes that only the points for which the diffusion coefficient is larger than δ, attract new observation point w in the classification process.

2.2 Numerical Discretisation

To discretise the equation (1), we use *i)* the balance of diffusion fluxes (inflows and outflows) at each vertex $v \in V(G)$ and *ii)* the approximation of the diffusion flux to the vertex v along its arcs.

Let us define the diffusion flux approximation, which depends on the difference of the values of the function X at the vertices v and u, as

$$\mathcal{X}(v, e_{uv}, t) = g_{e_{uv}}(X(u,t) - X(v,t)), \tag{5}$$

for each directed edge (u, v), where $g_{e_{uv}}$ represents the diffusion coefficient on the directed edge (u, v). If $\mathcal{X}(v, e_{uv}, t) > 0$, it represents the diffusion inflow, while if $\mathcal{X}(v, e_{uv}, t) < 0$, it represents the diffusion outflow. Then the balance of diffusion fluxes at the vertex v is expressed by the equation

$$\partial_t X(v,t) = \sum_{\substack{u \in V(G)\,: \\ e_{uv} \in A(G)}} \mathcal{X}(v, e_{uv}, t). \tag{6}$$

By the substitution of the approximation of the diffusion flux (5) into the balance equation (6), we obtain the so-called "graph-Laplacian" with a measure equal to 1 which is a common choice in the graph theory. A more detailed description can be found in the paper [13].

For the time discretisation, we use the semi-implicit approach, see e.g. [14]. The finite difference method is used for the approximation of time derivative. Since the diffusion coefficient $g_{e_{uv}}$ at the directed edge e_{uv} can depend on the unknown quantity X, see (2) and (4), and thus can change over time, we take its value from the previous time step. In the case of classification of the data from the k-dimensional feature space, we get in each time step k systems of linear equations

$$\begin{gathered}(1 + \tau \sum_{\substack{u \in V(G)\,: \\ e_{uv} \in A(G)}} g_{e_{uv}}^{n-1}) x_i^n(v) - \tau \sum_{\substack{u \in V(G)\,: \\ e_{uv} \in A(G)}} g_{e_{uv}}^{n-1} x_i^n(u) = x_i^{n-1}(v), \\ i = 1, \dots, k, \quad v \in V(G),\end{gathered} \tag{7}$$

which are interconnected by the diffusion coefficient $g_{e_{uv}}^{n-1}$, which depends on all $x_i^{n-1}(v)$, $x_i^{n-1}(u)$, $i = 1, \dots, k$ and can be written in the form

$$g_{e_{uv}}^{n-1} = \varepsilon(e_{uv}^{n-1}) \frac{1}{1 + \sum_{i=1}^{k} (K_i \; l_i^2(e_{uv}^{n-1}))}, \qquad K_i \geq 0. \tag{8}$$

This system of equations is represented by a full matrix and, as we have said before, for considered semi-complete directed graphs, it is not necessary to define any boundary condition.

The Eqs. (7)–(8) represent our network, where the points inside the given clusters are moving together, and the clusters themselves are keeping away. The illustration of that dynamic can be found in the paper [13]. In the learning phase, and also in the application phase, the dynamics is modified in such a way that all other points are moving by (7)–(8) but for the new observation $w \notin C_i$, $i \in \{1, \dots, N_C\}$, the diffusion coefficient is set to

$$g_{e_{uw}}^{n-1} = \max(\varepsilon(e_{uw}^{n-1}) \frac{1}{1 + \sum_{i=1}^{k} (K_i \; l_i^2(e_{uw}^{n-1}))} - \delta, \; 0), \tag{9}$$

$\varepsilon(e_{uw}^{n-1}) \geq 0$, $K_i \geq 0$, $\delta > 0$ are given constants, see also (4). It is crucial to realise that the vertex w is the head of the directed edges connecting it with the neighbouring vertices, so the directed edges enter w, and there do not exist arcs leaving w.

To clearly conceive the role of the diffusion coefficient $g_{e_{uv}}$ in the classification algorithm, we describe and schematically show the system matrix of the semi-implicit scheme. As an example, we use the directed graph from Fig. 1, where the number of clusters is $N_C = 2$. For that directed graph, the matrix (10) is constructed by (7)–(8) representing the fundamental dynamics of the NatNet. The matrix contains four blocks. The first block B_1 corresponds to the vertices from the first cluster, and because the vertices are from the same cluster, only forward diffusion on their arcs is used. The vertices influence each other in the same way, so the first block is symmetric $B_1 = B_1^T$. Similarly, the fourth block B_4 corresponds to the vertices of the second cluster, and again the forward diffusion is applied to their arcs in a symmetric way, thus $B_4 = B_4^T$. Moreover, the diagonal of these blocks is positive while out of diagonal elements are negative. The entries in the second B_2 and the third B_3 block are given by the values of diffusion coefficient on the directed edges between the vertices of different clusters (in Fig. 1 red arrows). The backward diffusion with a small negative value of $\varepsilon(e_{uv}) = -10^{-2}$ is applied to these directed edges and cause the small positive values in the second and third block. The blocks are symmetric to each other $B_2 = B_3^T$, which means that the vertices from the different clusters affect themselves symmetrically.

$$\left(\begin{array}{c|c} B_1 & B_2 \\ \hline B_3 & B_4 \end{array}\right) \tag{10}$$

Modifying the diffusion coefficient for the new observation by using (9) in the directed graph approach changes the matrix of the system in the following way. Let us consider the same example as in the description of the matrix for fundamental network dynamics but with one new observation added to the directed graph, see Fig. 2. Adding the new observation in the directed graph enlarges the size of the matrix (10) by one row and column. Due to one-sided arcs between new observation and other vertices the system matrix is non-symmetric. Consequently, the matrix (11) for the directed graph enriched with a new observation differs only in the last row and column from the system matrix (10). The values in the row $\boldsymbol{N}$ for the new observation are calculated by (9), which means that if the new observation has some vertices in the "δ-diffusion neighbourhood", the non-zero value of diffusion coefficient is set in the intersection of the row of the new observation and the column of that vertex in the matrix. In the last column corresponding to the new observation there are only zero values because the new observation is not affecting any other vertex of the directed graph.

$$\left(\begin{array}{c|c|c} B_1 & B_2 & \multirow{2}{*}{$\mathbf{0}$} \\ \cline{1-2} B_3 & B_4 & \\ \hline \multicolumn{3}{c}{\boldsymbol{N}} \end{array}\right) \tag{11}$$

The aforementioned change in the Natural numerical network concept leads to a significant reduction in CPU time. In paper [13], the classification process is run for any new observation sequentially, which means that the new observations are added to the undirected graph one by one and dynamics of the network are run for every new observation independently. In theory, the CPU time for such an approach is proportional to $P(N_V + 1)$, where P are the number of new observations. With a new concept of the NatNets on directed graphs, we can classify all new observations simultaneously because the new observations do not affect any other vertex of the graph. In this case, theoretically, the CPU time is proportional to $N_V + P$, and the speed-up of the computation is thus proportional to N_V. To test the speed up in practice, we calculated the relevancy map of dimension 2000×2000, so we have $4 \cdot 10^6$ new observations to be classified by NatNet. For the approach from [13], we obtained CPU time 801 s for such large-scale classification task, while for the new concept presented in this paper, we obtained CPU time 25.2 seconds, which means the speed up 31.78 times. The computations were performed on the machine with AMD Ryzen Threadripper PRO 3975WX 32-Cores processor, RAM 256 GB DDR4 and using OpenMP parallelisation in both approaches.

A histogram stopping criterion is applied in the network dynamics. It is based on the calculation of the number of occurrences (frequency) of evolving points in prescribed spatial cells in every time step. For more details about the stopping criterion, see the paper [13].

2.3 Construction of Relevancy Maps

The relevancy map is a grayscale image with the same size as the images from satellite optical channels. The square $A(p, r)$ is created in every image pixel p with Chebyshev radius r. For each p of the square $A(p, r)$, the statistical characteristics (the mean, the standard deviation, the minimum value and the maximum value) are computed and considered and added in the directed graph G as new observation $w(p)$. Every new observation $w(p)$ is classified by the Natural network and its relevancy coefficient $R(w(p))$ is computed. Finally, depending on the Chebyshev radius r of the square $A(p, r)$, the relevancy map M_i^r, $i = 1, \ldots, N_C$, is defined for every cluster C_i, $i = 1, \ldots, N_C$, in every pixel p as follows

$$
\begin{aligned}
M_i^r(p) &= R(w(p)), \qquad \text{if } w(p) \text{ is classified into } C_i, \\
M_i^r(p) &= 0, \qquad\qquad\quad \text{if } w(p) \text{ is not classified into } C_i.
\end{aligned}
\tag{12}
$$

The definition of the relevancy coefficient $R(w(p))$ is given by

$$
R(w(p)) = \mathcal{L}(1 - \frac{l_1(w(p))}{l_1(w(p)) + l_2(w(p))}), \quad \mathcal{L}(x) = \frac{1}{1 + e^{\lambda(0.5-x)}} \tag{13}
$$

and it depends on the distance between the new observation and the centroid of the cluster $\mathcal{C}_a(w(p))$ to which it is assigned by the network dynamics,

$$
l_1(w(p)) = \mid X(w(p), 0) - \mathcal{C}_a(w(p)) \mid , \tag{14}
$$

and the average distance of the new observation to all other cluster centroids,

$$l_2(w(p)) = \frac{1}{N_c - 1} \sum_{\substack{i=1 \\ i \neq a}}^{N_c} \mid X(w(p), 0) - \mathcal{C}_i(w(p)) \mid \ . \tag{15}$$

The relevancy coefficient $R(w(p))$ has the nonlinear character and is assigning the values close to 1 for the new observation $X(w, 0)$ close to the centroid of the cluster to which it is classified, while in the other case it is low.

3 Natural Numerical Network in Nature Protection

The Natural numerical network is applied in ecology and nature conservation tasks. It is a suitable tool for identification and classification of protected forest habitats by using the remote sensing. In our experiments we are focused on the data from the Sentinel-2 satellite of the European Space Agency (ESA) [8]. The Sentinel-2 offers optical imagery in high spatial resolution with 17 channels. In addition to these 17 channels, we calculate one more, the normalized difference vegetation index (NDVI) [6] which quantify the vegetation. Thus, for the feature space construction, we use 18 channels in which we compute the statistical characteristics in a prescribed image subarea A. Therefore, the feature space is the 72-dimensional Euclidean space, i.e. $k = 72$.

The classification by NatNets is applied to two groups of forest habitats, QC forest - segments dominated by Quercus cerris (habitats 91M0 Pannonian-Balkanic turkey oak-sessile oak forests and 91I0 Euro-Siberian steppic woods with Quercus spp.) and QP forest - segments with the dominance of Quercus petraea (habitat 91G0 Pannonic woods with Quercus petraea and Carpinus betulus). The habitats 91M0, 91I0, and 91G0 are part of Natura 2000 protected network [7]. The motivation for the classification and identification of such oak forests is that the forests of 91M0 and 91I0 habitats are very endangered in Slovakia. The wood of the Quercus cerris (turkey oak), which is dominant in these habitats, is considered to be of lower quality compared to other oak species in Slovakia, and therefore the turkey oak was often eliminated in favour of other Quercus species.

3.1 Training of the Network

The vegetation scientists use automatic segmentation methods in NaturaSat software [15] to segment 42 areas of QC and QP forests in Western Slovakia, see Fig. 3 for areas examples of QC segmented areas. We denote the segmented areas as S_i, where $i = 1, \ldots, N_S$, $N_S = 42$, and in each segmented area, we choose randomly a square $A_i = A(p_i, r)$ centred in a pixel $p_i \in S_i$. The values for Chebyshev radius r are equal to 5 for large segmented areas, while for small areas, r can be smaller, equal to 4 or 3. The statistical characteristics are computed for every square A_i. The statistical characteristics of squares A_i represent the vertices of the initial network directed graph G. Since we have multi-dimensional feature space, we apply the Principal Component Analysis (PCA) [12,17] to reduce the data dimension but retain the maximum amount of information in the data. We observed experimentally, that using the first two principal components is sufficient, further coordinates do not help to differentiate clusters and can be omitted. Thus the dimension of the problem is reduced to $k = 2$ which gives at the same time the computationally tractable task and sufficiently accurate classification, exceeding 95%, in the training phase of the model. Because we have two types of forests, we have two clusters in the classification, $N_C = 2$, and each vertex in the directed graph is labelled by the cluster number to which its segmented area belongs.

Fig. 3. The subregion of Western Slovakia with segmented areas of protected QC forests (red curves). (Color figure online)

Network training aim is to tune the parameters of the model (7)–(9) to achieve the highest possible classification accuracy for the learning dataset. To attain that aim, we subsequently remove the cluster label from each vertex of the directed graph G, representing the learning dataset, and set it as the new observation. Then we classify it using the NatNet. We analyse the results and choose the model parameters with the highest success rate N_B/N_V, where N_B is a number of correctly classified observations.

Now let us consider the directed graph G having $N_V = 42$ vertices described above and denote it LDS42. We run the training of the network on the LDS42 dataset, the results are shown in the first row of Table 1. We achieved the success rate of 37/42. Since we have randomly chosen the squares $A(p_i, r)$, it is not the optimal approach. To increase the success rate, we adjust the learning dataset by a spatial shifting of representative squares inside the segmented areas S_i based on the relevancy maps computed using LDS42. We try to find the new square $A(p_i, r)$ for each segmented area S_i such that $M_a^r(p_i) >> 0$ for a new square centre $p_i \in S_i$ by analysing the relevancy map M_a^r of the cluster to which the segmented area S_i belongs. If we find a pixel p_i with high relevancy, we can construct the new square $A(p_i, r)$ and replace the randomly chosen square from LDS42 with the new one. In this manner, we construct the adjusted learning dataset LDS42adj and run the training of the network again. The result of the training is shown in the second row of Table 1. We achieve the success rate of $40/42 = 0.9524$, which is sufficiently high and allows us to use the trained NatNet for the identification of oak forests in Western Slovakia outside the areas used in the training of the network.

Table 1. The results of the learning phase on datasets LDS40 and LDS40adj.

Dataset name	Correctly classified	Incorrectly classified	Outliers	Success rate
LDS42	37	5	1	88.09%
LDS42adj	40	1	1	95.24%

3.2 Application of the Trained Network

By a successfully trained NatNet, we can classify satellite images and identify areas of target habitat in the examined territory. In our case, the target habitats are that dominated by Quercus cerris (turkey oak) due to their endangered status in Slovakia. The relevancy maps are computed for that purpose. We work with two habitats, thus we obtain two relevancy maps. Figure 4 depicts the area of Martinsky les special protected area with the segmented areas of QC forests (red curves). On the left part in Fig. 4, there is the image from the Sentinel-2 satellite, and on the right part, there is the relevancy map for the QC forests. The relevancy map shows bright colours in the interior of the segmented areas, which means high relevancy coefficient in the pixels and reflects the correct classification of the segmented area. We can notice that there is also apparent bright colour outside the segmented areas, which leads to field reviewal by vegetation scientists. The result of the field research is that the NatNet correctly classified a given territory. There are further areas of 91M0 and 91I0 habitats in such territory validated by fields visit and comparing them with forestry maps. The second relevancy map for QP forests is depicted in Fig. 5. This relevancy map shows the regions of appearance of QP forests in bright colour pixels, thus having a high relevancy coefficient. When we are focused on the segmented curves of the QC forests, we can conclude that the interior is quite dark, which expresses low

Fig. 4. The segmented areas of QC forests (red curve) plotted on the Sentinel-2 image (left) and on the relevancy map for QC forests (right). (Color figure online)

Fig. 5. The segmented areas of QC forests (red curve) plotted on the Sentinel-2 image (left) and on the relevancy map for QP forests (right). (Color figure online)

or no relevancy of the occurrence of the QP forest. The Natura 2000 habitats are complex compositions of various types of species, and it is impossible to have a homogeneous relevancy map for one habitat. Nevertheless, we can observe that the relevancy map for the QC forests and the QP forests complement each other.

References

1. Bang-Jensen, J., Gutin, G.R.: Digraphs: Theory, 2nd edn. Algorithms and Applications. Springer-Verlag, London (2010)
2. Bertozzi, A.L., Flenner, A.: Diffuse interface models on graphs for classification of high dimensional data. SIAM Rev. **58**(2), 293–328 (2016). https://doi.org/10.1137/16M1070426
3. Böhm, J.N., Berens, P., Kobak, D.: Attraction-repulsion spectrum in neighbor embeddings. J. Mach. Learn. Res. **23**(95), 1–32 (2022)
4. Chang, B., Meng, L., Haber, E., Ruthotto, L., Begert, D., Holtham, E.: Reversible architectures for arbitrarily deep residual neural networks. In: 32nd AAAI Conference on Artificial Intelligence, AAAI 2018. vol. 32, pp. 2811–2818 (April 2018), https://ojs.aaai.org/index.php/AAAI/article/view/11668
5. Cheng, K., Wang, J.: Forest type classification based on integrated spectral-spatial-temporal features and random forest algorithm-a case study in the qinling mountains. Forests **10**(7), 559 (2019). https://doi.org/10.3390/f10070559
6. Earth Observing System: Normalized difference vegetation index. https://eos.com/ndvi/ (2020)
7. European Environmental Agency: The natura 2000 protected areas network. https://www.eea.europa.eu/themes/biodiversity/natura-2000/the-natura-2000-protected-areas-network (2020)
8. European Space Agency: Sentinel 2. https://sentinel.esa.int/web/sentinel/missions/sentinel-2/data-products (2020)
9. Friedman, J., Tillich, J.P.: Calculus on graphs. CoRR cs.DM/0408028 (2004)
10. Haber, E., Ruthotto, L.: Stable architectures for deep neural networks. Inverse Probl. **34**(1) (2018). https://doi.org/10.1088/1361-6420/aa9a90
11. He, K., Zhang, X., Ren, S., Sun, J.: Deep residual learning for image recognition. In: CVPR, pp. 770–778 (June, 2016). https://doi.org/10.1109/CVPR.2016.90
12. Jolliffe, I.T.: Principal Component Analysis. Springer-Verlag, New York, 2nd edn. (2002). https://doi.org/10.1007/b98835
13. Mikula, K., Kollár, M., Ožvat, A.A., Ambroz, M., Čahojová, L., Jarolímek, I., Šibík, J., Šibíková, M.: Natural numerical networks for natura 2000 habitats classification by satellite images. Appl. Math. Model. **116**, 209–235 (2023). https://doi.org/10.1016/j.apm.2022.11.021
14. Mikula, K., Ramarosy, N.: Semi-implicit finite volume scheme for solving nonlinear diffusion equations in image processing. Numer. Math. **89**(3), 561–590 (2001)
15. Mikula, K., et al.: Naturasat-a software tool for identification, monitoring and evaluation of habitats by remote sensing techniques. Remote Sens. **13**(17) (2021). https://doi.org/10.3390/rs13173381
16. Noi, P.T., Kappas, M.: Comparison of random forest, k-nearest neighbor, and support vector machine classifiers for land cover classification using sentinel-2 imagery. Sensors (Switzerland) **18**(1), 18 (2018). https://doi.org/10.3390/s18010018
17. Rencher, A.: Methods of Multivariate Analysis. Wiley-Interscience (02 2002). https://doi.org/10.1002/0471271357
18. Waśniewski, A., Hoscilo, A., Zagajewski, B., Mouketou-Tarazewicz, D.: Assessment of sentinel-2 satellite images and random forest classifier for rainforest mapping in gabon. Forests **11**(9), 941 (2020). https://doi.org/10.3390/f11090941

Piece-wise Constant Image Segmentation with a Deep Image Prior Approach

Alessandro Benfenati[1], Ambra Catozzi[2,3], Giorgia Franchini[2](✉), and Federica Porta[2]

[1] Dipartimento di Scienze e Politiche Ambientali, Università di Milano, Milan, Italy
alessandro.benfenati@unimi.it

[2] Dipartimento di Scienze Fisiche, Informatiche e Matematiche, Università di Modena e Reggio Emilia, Modena, Italy
{ambra.catozzi,giorgia.franchini,federica.porta}@unimore.it

[3] Dipartimento di Scienze Matematiche, Fisiche e Informatiche, Università di Parma, Parma, Italy

Abstract. Image segmentation is a key topic in image processing and computer vision and several approaches have been proposed in the literature to address it. The formulation of the image segmentation problem as the minimization of the Mumford-Shah energy has been one of the most commonly used techniques in the last past decades. More recently, deep learning methods have yielded a new generation of image segmentation models with remarkable performance. In this paper we propose an unsupervised deep learning approach for piece-wise image segmentation based on the so called Deep Image Prior by parameterizing the Mumford-Shah functional in terms of the weights of a convolutional neural network. Several numerical experiments on both biomedical and natural images highlight the goodness of the suggested approach. The implicit regularization provided by the Deep Image Prior model allows to also consider noisy input images and to investigate the robustness of the proposed technique with respect to the level of noise.

Keywords: Image segmentation · Image segmentation with noise · Deep image prior · Neural Network · Unsupervised deep learning

1 Introduction

Image segmentation is a frequently used technique in signal processing to partition an image into multiple parts or regions, often based on the features (color, intensity, shape) of the image itself. Segmentation plays a central role in a broad range of applications, including medical image analysis [14], autonomous vehicles [25], video surveillance [11], and augmented reality to count a few. Several approaches have been proposed in the past for solving this problem, such as region growing [9], power watershed [20], color distances [2], k-means [10], Markov Random fields [12], among others.

Following a variational approach, the solution is theoretically formulated as a minimizer of the global energy obtained by properly approximating the

L. Calatroni et al. (Eds.): SSVM 2023, LNCS 14009, pp. 352–362, 2023.
https://doi.org/10.1007/978-3-031-31975-4_27

non-differentiable Mumford-Shah functional and its variants [3,19]. Gradient-like methods for the minimization of these models are usually computationally expensive, above all for large-scale data and real-time solutions and require High Parallel Computing techniques [28].

On the other hand, deep learning methods for image segmentation (see for example [1,8,17,18,21]) have become very popular over the last few years mainly due to rapid technology improvements, (faster GPUs) and to the availability of larger and larger datasets. The majority of the segmentation approaches are based on Convolutional Neural Networks (CNNs), which are usually trained in a supervised manner. This usually requires a complete dataset of appropriate dimension with the related ground truth labels, which typically can be rather expensive and/or time consuming to acquire. An alternative approach consists in considering unsupervised deep learning, where the networks are not trained on pre-labelled datasets. In this framework, several techniques have been proposed in recent years. A popular approach is to use generative models, such as Variational Autoencoders (VAEs) [15] or Generative Adversarial Networks (GANs) [27]. A further recent approach has been presented in [13], where the authors introduced a Mumford-Shah functional tailored for unsupervised image segmentation using neural networks. It is worth to remark that the approach in [13] still requires a training phase and a large dataset: indeed the training is pursued by minimizing the Mumford-Shah function (tailored for the neural networks) with respect to the outputs of the network and over mini-batches of the input images, however without using any ground truth label. Therefore, the technique suggested in [13] needs a training which is carried out in an unsupervised fashion.

The aim of this paper is to develop an unsupervised deep learning approach for piece-wise image segmentation coupling the Deep Image Prior (DIP) framework [26] and the version of the Mumford-Shah functional adapted for neural networks. DIP is among the most promising methods belonging to the class of unsupervised deep learning methods in the field of imaging inverse problems [6,16]. This novel framework relies on the implicit regularization provided by representing images as the output of generative CNN architectures. In more detail, in [26] the authors consider image restoration problems which can be modeled as

$$\text{find} \quad u \in \mathbb{R}^n \quad \text{s.t.} \quad Hu + \eta = g \tag{1}$$

where $H \in \mathbb{R}^{m \times n}$ is a known forward operator and $\eta \in \mathbb{R}^m$ is the noise corrupting the data $g \in \mathbb{R}^m$. Given a CNN generator $f : \mathbb{R}^s \times \mathbb{R}^t \to \mathbb{R}^n$ whose weights are denoted by $\theta \in \mathbb{R}^s$ and a random input vector $z \in \mathbb{R}^t$ sampled from a uniform distribution, the DIP approach estimates the restored image u^* as $f(\theta^*, z)$, where the set of weights θ^* are computed by solving the following minimization problem

$$\min_{\theta \in \mathbb{R}^s} \frac{1}{2} \|Hf(\theta, z) - g\|^2$$

through a gradient-based iterative scheme properly early stopped.

The property of DIP to induce implicit regularization by means of a CNN architecture has been exploited for image segmentation but into more traditional algorithms [4]. On the other hand, our approach generalizes the DIP paradigm to this task: indeed, we extend the original idea in [26], tailored for image restoration problems of the form (1), to image segmentation. Moreover, we investigate the robustness of the suggested technique with respect to noise affecting the images to be labelled, exploiting the regularizing behaviour provided by DIP architecture.

Several numerical experiments on image segmentation problems have been performed for both biomedical and natural images datasets.

2 Proposed Approach

Let $\Omega \subset \mathbb{R}^2$ be open and bounded, Γ be a closed subset in Ω consisting in a finite set of smooth curves, and Ω_i be the connected components of $\Omega \setminus \Gamma$ such that $\Omega = \cup_i \Omega_i \cup \Gamma$. Let $u_0 : \Omega \longrightarrow \mathbb{R}$ be a given input image. The image segmentation problem consists in finding a decomposition $\{\Omega_i\}_i$ of Ω and a piece-wise smooth function $u : \Omega \longrightarrow \mathbb{R}$ which approximate u_0 and such that u varies smoothly within each Ω_i, and discontinuously across the boundaries of Ω_i. In [19] the authors suggested to find u and Γ by solving the following optimization problem

$$\min_{u,\Gamma} \mathcal{E}(u,\Gamma) \equiv \lambda \int_{\Omega} (u(x) - u_0(x))^2 \, dx + \int_{\Omega \setminus \Gamma} |\nabla u(x)|^2 \, dx + \nu |\Gamma| \qquad (2)$$

where λ and ν are positive fixed parameters and $|\Gamma|$ denotes the sum of the lengths of curves in Γ itself. An interesting case of the above model arises when the segmented image u is equal to c_i on Ω_i, $i = 1, \dots, N$, i.e. u is piece-wise constant and hence $\nabla u(x) = 0$, $\forall x \in \Omega \setminus \Gamma$. We denote with c the vector of the constants $c = \{c_1, ..., c_N\}$. In this case, by assuming N piece-wise constant regions in the input image, the objective functional in (2) reduces to

$$\mathcal{E}(c,\Gamma) = \lambda \sum_{i=1}^{N} \int_{\Omega_i} (c_i - u_0(x))^2 \, dx + \nu |\Gamma| \qquad (3)$$

with $\Gamma = \cup_{i=1}^N \partial\Omega_i$ where $\partial\Omega_i$ is the boundary of the i-th region. It is easy to prove that, for a fixed Γ, the functional in (3) is minimized in the variables c_i by setting c_i as the intensity mean of the input image u_0 in Ω_i. By introducing the characteristic function of the i-th region, denoted by $\chi_i(x)$ and such that

we can rewrite (3) as

$$\mathcal{E}(\chi) = \lambda \sum_{i=1}^{N} \int_{\Omega} (c_i - u_0(x))^2 \chi_i(x) \, dx + \nu \sum_{i=1}^{N} \|\chi_i(x)\|_{\mathrm{TV}} \, dx, \qquad (4)$$

where $\|\cdot\|_{\mathrm{TV}}$ denotes the Total Variation functional [7]. The average pixel value c_i is given by

$$c_i(\chi) = \frac{\int_\Omega u_0(x)\chi_i(x)\,dx}{\int_\Omega \chi_i(x)\,dx}. \tag{5}$$

Before introducing the suggested approach, we consider a discretized version of (4). Let $u_0 : O \to \mathbb{R}$ be the vectorized input image, where $O \subset \mathbb{R}^d$ is the discretization of the continuous domain Ω, divided in the components O_i: $O = \cup O_i$. O_i refers to both Ω_i and its boundary $\partial\Omega_i$. Let $X_i : O \to \{0, 1\}$ be the discrete counterpart of the characteristic function of the i-th component O_i:

$$X_i(k) = \begin{cases} 1 & \text{if } k \in O_i \\ 0 & \text{otherwise} \end{cases}.$$

Hence, a discrete form of the first term of (4) can be written as

$$\lambda \sum_{i=1}^{N} \|\,(u_0 - c_i) \odot X_i\|_2^2,$$

where $\odot$ denotes the component-wise product of two vectors, and $\|\cdot\|_2$ denotes the classical Euclidean norm. The second term in (4) can be written in a discrete form as

$$\nu \sum_{i=1}^{N} \sum_{k=1}^{d} \|A_k X_i\|_2$$

where A_k is the 2D discrete first order difference operator at the element k of the vector-reshaped characteristic function X_i and the matrices $A_k \in \mathbb{R}^{2\times d}$, $k = 1, ..., d$ are sub-matrices of $A \in \mathbb{R}^{2d\times d}$, $A = \left(A_1^\top, \ldots, A_d^\top\right)^\top$. The discrete counterpart of (2) reads as

$$E(X) = \sum_{i=1}^{N} \left(\lambda\|\,(u_0 - c_i) \odot X_i\|_2^2 + \nu \sum_{k=1}^{d} \|A_k X_i\|_2 \right), \tag{6}$$

with $X \in \mathbb{R}^{d\times N}$, being $X^{(i)} \equiv X_i$, where $X^{(j)}$ denotes the j-th column of X; the values c_i, $i = 1, \ldots, N$ can be retrieved using

$$c_i = \frac{\langle u_0, X_i\rangle}{\|X_i\|_0} \tag{7}$$

By following the DIP paradigm, given a CNN generator

$$f : \mathbb{R}^s \times \mathbb{R}^t \to \mathbb{R}^{d\times N} \tag{8}$$
$$f : (\theta, z) \to f(\theta; z) \tag{9}$$

whose weights are denoted by θ and a random input vector z sampled from a uniform distribution, we parameterize X as $f(\theta; z)$ and then we optimize the weights by applying standard gradient-based iterative algorithms to the problem

$$\min_\theta E(f(\theta; z)). \tag{10}$$

It is worth to stress that $f(\theta, z)$ is the output of the considered CNN network. Inspired by [13], instead of a Sigmoid layer as a last layer of the DIP architecture, we consider a Softmax layer which mimics the role of the characteristic function in (6). Moreover, we remark that the i-th column of the output $f(\theta; z)$ corresponds thus to the i-th column of X, *i.e.* it is the approximation of the discrete characteristic function of the i-th region: $f(\theta; z)^{(i)} \sim X^{(i)} = X_i$. Hence, following (7), c_i can computed as

$$c_i = \frac{\langle u_0, f(\theta; z)^{(i)} \rangle}{\|f(\theta; z)^{(i)}\|_0}.$$

Remark 1. In (6), in order to avoid possible issue about differentiability of the objective function, we consider the smoothed version of the second term, namely

$$\sum_{k=1}^{d} \left\| \frac{A_k X_i}{\delta} \right\|_2$$

with $\delta \in \mathbb{R}^+$, $\delta \ll 1$.

Remark 2. In [13], as already noted, the unsupervised approach still requires a large dataset for the training. The proposed procedure, instead, does not require any dataset: solving (10) provides a neural network whose output is the segmentation of the image u_0.

3 Numerical Experiments

Several numerical experiments have been carried out to test the effectiveness of the proposed approach in addressing piece-wise constant image segmentation for both natural and biomedical images. All the numerical tests are carried on a Linux machine, equipped with Ubuntu 20.10 and with Matlab R2022b, using the Deep Learning Toolbox.

Datasets. For the experiments we use two datasets. The White Blood Cell (WBC) database [29] consists in leukocytes images and the labelled ground truths (for performance measurements): such ground truths specify the nucleus, the cytoplasm and the background. The second dataset employed is the GrabCut dataset [23], which includes 50 natural images (animals, people, objects), each of different dimensions, together with the ground truths.

We remark that the ground truths have been only used for the computation of the performance indices. Moreover we recall that the network is not trained neither on the pre-labelled datasets nor on the input datasets but learns its weights in a completely unsupervised fashion.

Architecture of the Network. The convolutional network architecture is inspired by the one in the seminal work [26]. Figure 1 presents a visual inspection of the network: it is an U-Net with skip connections, *i.e.* each stage of the "encoder"

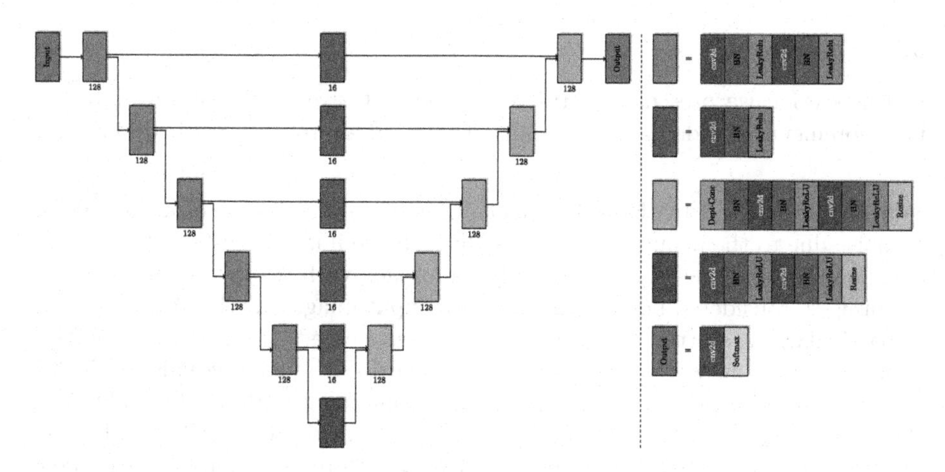

Fig. 1. Architecture of the network employed in the experiments on the Grab-Cut dataset. The structure of each stage is depicted on the right.

is connected with the corresponding stage of the "decoder"; with an abuse of notation, employing a parallelism with autoencoders, the layers on the left are referred to as the *encoder* part of the network, while the ones on the right part are referred to as the *decoder* part. Each stage of the encoder halves the spatial dimension of the input, whilst the upsampling layers of the decoder perform a bilinear interpolation, for doubling the dimensions. The number of filters employed in each convolutional layer is 128, whilst the number of filters in the skip connections is set to 4. The parameter of the LeakyReLU activation functions is set to 0.2 among the whole network. The last convolutional layer has a number of filters equal to the chosen number of classes among which we segment the image. A further difference with the network of [26] consists in having replaced the final layer, originally a Sigmoid layer, with Softmax activation function.

The input z is a 3D tensor having the same spatial dimension of the input image with 3 channels depending. In the experiments, we follow the common practice in the DIP framework of perturbing at each iteration the input z by a component sampled from a Gaussian distribution with zero mean and standard deviation equal to $\frac{1}{100}$.

Training and Hyperparameters Setting. Regarding the hyperparameters setting in the DIP approach several studies have been carried out on how the initialization [24] and the optimizer [5] affect its performance. All the results are obtained using Adam for the backpropagation phase with learning rate $\eta = 10^{-3}$. The regularization parameters λ and ν in (6) have been set equal to 1 and 10^{-2}, respectively. Moreover we considered the smoothed version of the second term in (6) by fixing $\delta = 10^{-4}$. We set the total number of iterations equal to 500.

3.1 Results

In this section we present qualitative and quantitative results obtained on the two aforementioned datasets by means of the DIP approach described in Sect.2.

White Blood Cell (WBC) Dataset. Biomedical images are often corrupted by noise due to the acquisition process through the imaging system. In order to evaluate the robustness of the suggested approach with respect to the presence of noise, we considered both clean and noisy input images for the WBC dataset. In particular, the starting noisy input images are created by considering additive white Gaussian noise of zero mean and standard deviation σ equals to $25/255$ and $50/255$. The presence of noise in the input images allows to evaluate if the implicit regularization provided by DIP in the restoration tasks can be also observed in this setting. The neural network architecture employed in these numerical tests is the same of Fig. 1 where the encoder and the decoder parts consist in two stages, instead of five.

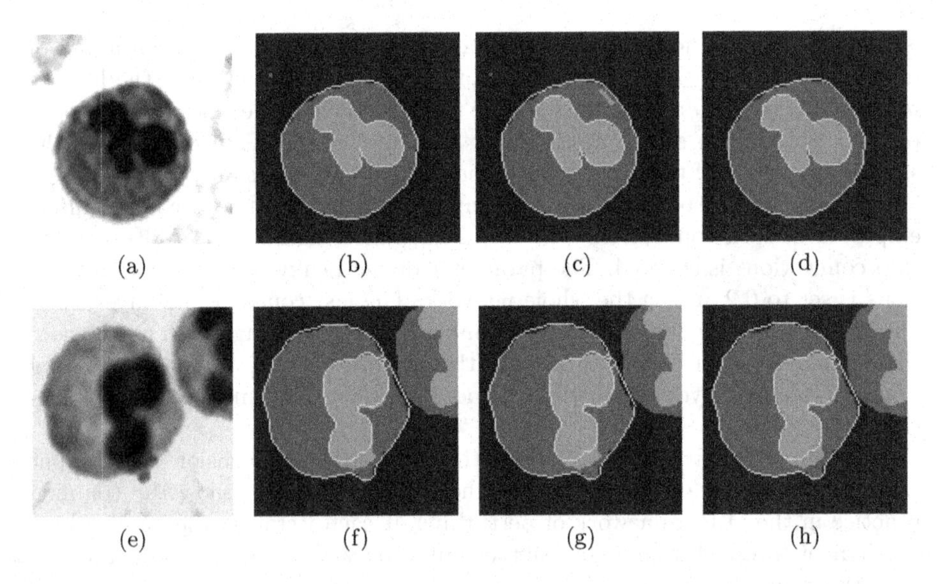

Fig. 2. Visual inspection of unsupervised segmentation on samples from WBC dataset. (a)–(e): true RGB image. (b)–(f) Segmentation result with $\sigma = 0$. (c)–(g) Segmentation result with $\sigma = 25$. (d)–(h) Segmentation result with $\sigma = 50$. The light lines depict the boundary of ground truth regions. The results in the second row show how the proposed method is able to segment regions of interest outside the given ground truth.

Among the measures for establishing the performance of a segmentation process (i.e. Jaccard index, precision, recall), we choose the Rand index which measures the similarity between two segmentations of the same image [22]. The Rand index ranges from 0 to 1, with 1 indicating that the two clusterings are identical and 0 indicating that they are completely different. Table 1 reports the mean and the standard deviation of the Rand indices obtained on 50 elements of the WBC dataset for different values of σ. It is possible to conclude that the averaged Rand index is high in all the three cases and the results are robust with respect to the noise. The well known regularization behaviour of DIP can be actually appreciated.

Table 1. Statistical results of Rand index based on 50 images from WBC dataset.

σ	Mean	St. Deviation
0	0.9369	0.0673
25	0.9333	0.0676
50	0.9403	0.0631

In Fig. 2 we show two examples of images image taken from the WBC dataset, together with the results obtained by the segmentation process with different level of noise. Particularly, in Fig. 2(a) the original RGB image is reported, while in Fig. 2(b), Fig. 2(c) and Fig. 2(d) we can appreciate the segmentation obtained by our method ($\sigma = 0, 25, 50$) compared to the ground truth. The corresponding Rand indexes are equal to 0.9878, 0.9867 and 0.9880, respectively. The suggested method results very stable with respect to noise.

The results achieved on the cell image of Fig. 2(e) shed some insights on the numerical performances showed in Table 1. Indeed, in this case the image Fig. 2(f) related to the clean case ($\sigma = 0$) achieves a Rand index of 0.7813, while the version Fig. 2(g) with corresponding to $\sigma = 25$ has a Rand index of 0.7808 and finally the version without the highest noise level ($\sigma = 50$) Fig. 2(h) has a Rand index of 0.7807. This relatively poor result is due to the fact that the ground truth considers only the cell in the center of the image, while the proposed procedure is able to segment even the regions of interest which have not been considered when the true labels where created.

GrabCut Dataset. This section is devoted to discuss the qualitative results of the described approach on five images of the Grab-Cut dataset. Figure 3 presents the result for the classification where the number of classes is set to 3. Figure 4 refers to the background/foreground segmentation task. The segmentation output for the plane is almost perfect, only a small portion of dark sky is recognized as belonging to the same class of the plane. The banana image presents some spots inside the fruit that are recognized as belonging to the background, due to the very similar color. The case of the portrait of the person in Fig. 4(c) presents some artifacts: indeed, some parts of the person (the hair and the t-shirt) have some common color components with the foliage in the background, hence they are classified as belonging to the same class. Nonetheless, the profile of the person is very well recognized.

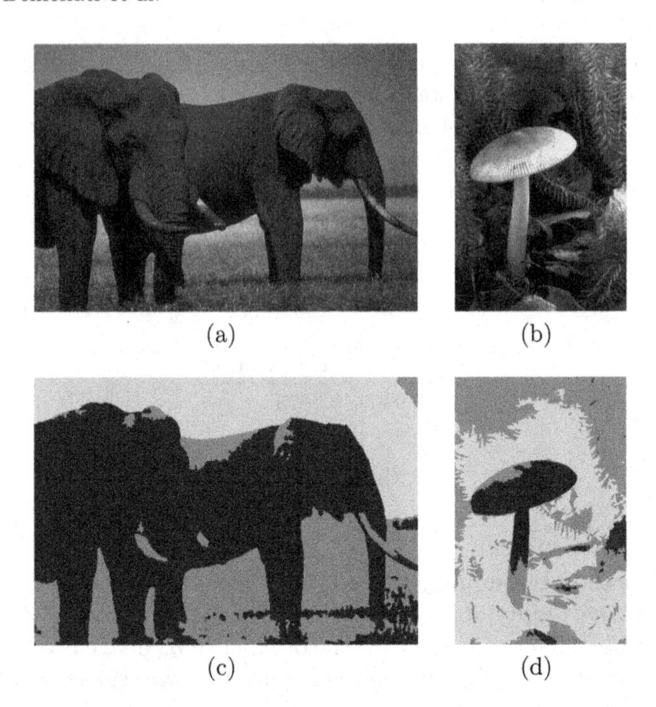

(a) (b)

(c) (d)

Fig. 3. 3-class unsupervised segmentation. (a) RGB image; (b) segmentation result.

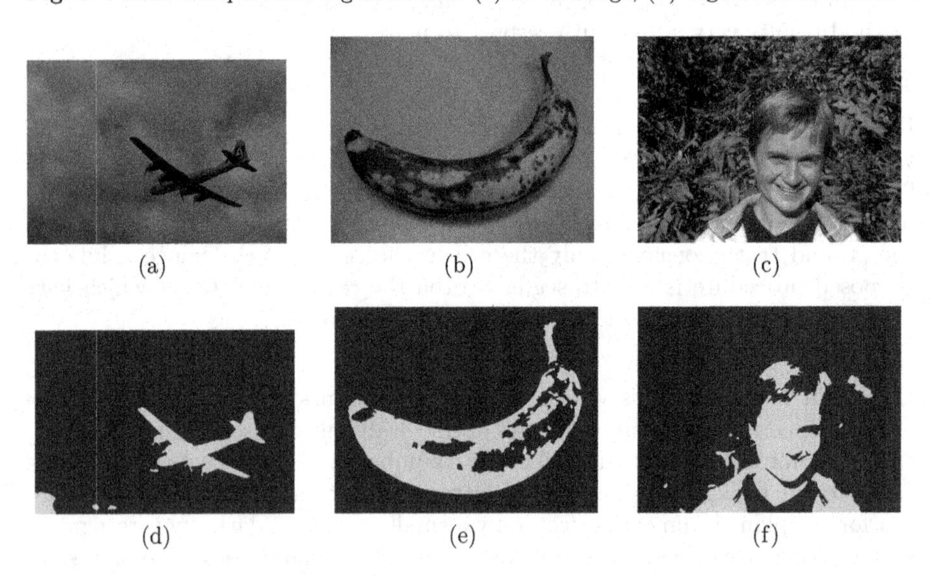

(a) (b) (c)

(d) (e) (f)

Fig. 4. Foreground-Background extraction. (a)–(c): Input RGB images; (d)–(f) relative unsupervised binary segmentation.

4 Conclusions

This paper proposes an unsupervised deep learning approach for piece-wise image segmentation that combines the Deep Image Prior and a tailored version of the Mumford-Shah functional for neural network. The numerical experiments show that the suggested approach is effective in segmenting both biomedical and natural images and demonstrates robustness to noise. The approach detailed in this work appears promising and it is worthy of further investigation.

References

1. Afshari, S., BenTaieb, A., Mirikharaji, Z., Hamarneh, G.: Weakly supervised fully convolutional network for pet lesion segmentation Med. Imag. 2019: Imag. Process. p. 109491K (2019)
2. Aletti, G., Benfenati, A., Naldi, G.: A semiautomatic multi-label color image segmentation coupling dirichlet problem and colour distances. J. Imaging **7**(10) (2021)
3. Ambrosio, L., Tortorelli, V.M.: Approximation of functional depending on jumps by elliptic functional via t-convergence. Commun. Pure Appl. Math. **43**(8), 999–1036 (1990). https://doi.org/10.1002/cpa.3160430805
4. Burrows, L., Chen, K., Torella, F.: Using deep image prior to assist variational selective segmentation deep learning algorithms. In: 17th International Symposium on Medical Information Processing and Analysis. vol. 12088,, pp. 243–252. SPIE (2021)
5. Cascarano, P., Franchini, G., Porta, F., Sebastiani, A.: On the First-Order Optimization Methods in Deep Image Prior. J. Verifi. Valid. Uncertain. Quant. **7**(4) (2023). https://doi.org/10.1115/1.4056470
6. Cascarano, P., Sebastiani, A., Comes, M.C., Franchini, G., Porta, F.: Combining weighted total variation and deep image prior for natural and medical image restoration via admm. In: 2021 21st International Conference on Computational Science and Its Applications (ICCSA). pp. 39–46 (2021). https://doi.org/10.1109/ICCSA54496.2021.00016
7. Chambolle, A., Caselles, V., Cremers, D., Novaga, M., Pock, T.: An introduction to total variation for image analysis. Theor. Found. Num. Methods Sparse Recov. **9**(263–340), 227 (2010)
8. Chen, L.C., Papandreou, G., Kokkinos, I., Murphy, K., Yuille, A.L.: DeepLab: Semantic image segmentation with deep convolutional nets, atrous convolution, and fully connected CRFs. In: IEEE Trans. Pattern Anal. Mach. Intell. **40**, 834–848 (2018)
9. Deshpande, A., Dahikar, P., Agrawal, P.: An experiment with statistical region merging and seeded region growing image segmentation techniques. In: Santosh, K.C., Hegadi, R.S. (eds.) RTIP2R 2018. CCIS, vol. 1035, pp. 493–506. Springer, Singapore (2019). https://doi.org/10.1007/978-981-13-9181-1_44
10. Dhanachandra, N., Manglem, K., Chanu, Y.J.: Image segmentation using k-means clustering algorithm and subtractive clustering algorithm. Procedia Compu.r Sci. **54**, 764–771 (2015)
11. Gruosso, M., Capece, N., Erra, U.: Human segmentation in surveillance video with deep learning. Multim. Tools Appl. **80**, 1175–1199 (2021)
12. Kato, Z., Zerubia, J.: Markov random fields in image segmentation. Found. Trends Signal Process. **5**(1–2), 1–155 (2012). https://doi.org/10.1561/2000000035

13. Kim, B., Ye, J.C.: Mumford-shah loss functional for image segmentation with deep learning. IEEE Trans. Image Process. **29**, 1856–1866 (2019)
14. Li, Y., Chouzenoux, E., Charmettant, B., Benatsou, B., Lamarque, J.P., Lassau, N.: Lightweight u-net for lesion segmentation in ultrasound images. In: 2021 IEEE 18th International Symposium on Biomedical Imaging (ISBI), pp. 611–615 (2021). https://doi.org/10.1109/ISBI48211.2021.9434086
15. Li, Z., Togo, R., Ogawa, T., Haseyama, M.: Variational autoencoder based unsupervised domain adaptation for semantic segmentation. In: 2020 IEEE International Conference on Image Processing (ICIP), pp. 2426–2430 (2020). https://doi.org/10.1109/ICIP40778.2020.9190973
16. Liu, J., Sun, Y., Xu, X., Kamilov, U.S.: Image restoration using total variation regularized deep image prior. In: ICASSP 2019–2019 IEEE International Conference on Acoustics, Speech and Signal Processing (ICASSP), pp. 7715–7719 (2019). https://doi.org/10.1109/ICASSP.2019.8682856
17. Long, J., Shelhamer, E., Darrell, T.: Fully convolutional networks for semantic segmentation. In: Proceedings of the IEEE Conference on Computer Vision and Pattern Recognition, pp. 3431–3340 (2015)
18. Minaee, S., Boykov, Y., Porikli, F., Plaza, A., Kehtarnavaz, N., Terzopoulos, D.: Image segmentation using deep learning: A survey. IEEE Trans. Pattern Anal. Mach. Intell. **44**(07), 3523–3542 (2022)
19. Mumford, D., Shah, J.: Optimal approximations by piecewise smooth functions and associated variational problems. Comm. Pure Appl. Math. **42**, 577–685 (1989)
20. Najman, L., Schmitt, M.: Watershed of a continuous function. Signal Process. **38**(1), 764–771 (1994)
21. Noh, H., Hong, S., Han, B.: Learning deconvolution network for semantic segmentation. In: IEEE International Conference on Computer Vision (ICCV). pp. 1520–1528 (2015)
22. Rand, W.M.: Objective criteria for the evaluation of clustering methods. J. Am. Stat. Assoc. **66**(336), 846–850 (1971). https://doi.org/10.1080/01621459.1971.10482356
23. Rother, C., Kolmogorov, V., Blake, A.: "grabcut" interactive foreground extraction using iterated graph cuts. ACM transactions on graphics (TOG) **23**(3), 309–314 (2004)
24. Sapienza, D., Franchini, G., Govi, E., Bertogna, M., Prato, M.: Deep image prior for medical image denoising, a study about parameter initialization. Front. Appl. Math. Stat. **8** (2022). https://doi.org/10.3389/fams.2022.995225
25. Scribano, C., Franchini, G., Olmedo, I.S., Bertogna, M.: Cerberus: Simple and effective all-in-one automotive perception model with multi task learning (2022). https://doi.org/10.48550/ARXIV.2210.00756
26. Ulyanov, D., Vedaldi, A., Lempitsky, V.: Deep image prior. In: Proceedings of the IEEE Conference on Computer Vision and Pattern Recognition, pp. 9446–9454 (2018)
27. Xun, S., et al.: Generative adversarial networks in medical image segmentation: A review. Comput. Biol. Med. **140**, 105063 (2022). https://doi.org/10.1016/j.compbiomed.2021.105063
28. Zanella, R., Porta, F., Ruggiero, V., Zanetti, M.: Serial and parallel approaches for image segmentation by numerical minimization of a second-order functional. Appl. Math. Comput. **318**, 153–175 (2018), recent Trends in Numerical Computations: Theory and Algorithms
29. Zheng, X., Wang, Y., Wang, G., Liu, J.: Fast and robust segmentation of white blood cell images by self-supervised learning. Micron **107**, 55–71 (2018)

On the Inclusion of Topological Requirements in CNNs for Semantic Segmentation Applied to Radiotherapy

Zoé Lambert[1,2], Carole Le Guyader[1(✉)], and Caroline Petitjean[3]

[1] INSA Rouen Normandie, Normandie Univ, LMI UR 3226, 76000 Rouen, France
{zoe.lambert,carole.le-guyader}@insa-rouen.fr
[2] CEREMA, ENDSUM, 76121 Le Grand Quevilly Cedex, France
[3] Normandie Univ, Université Le Havre Normandie, INSA Rouen Normandie, Normandie Univ, LITIS UR 4108, 76000 Rouen, France
caroline.petitjean@univ-rouen.fr

Abstract. The incorporation of prior knowledge into a medical segmentation task allows to compensate for the issue of weak boundary definition and to be more in line with anatomical reality even though the data do not explicitly show these characteristics. This motivation underlies the proposed contribution which aims to provide a unified variational framework involving topological requirements in the training of convolutional neural networks through the design of a suitable penalty in the loss function. More precisely, these topological constraints are implicitly enforced by viewing the segmentation assignment as a registration task between the considered image and its associated ground truth under incompressibility condition, making them homeomorphic. The application falls within the scope of organ-at-risk segmentation in CT (Computed Tomography) images, in the context of radiotherapy planning.

Keywords: Hybridisation variational methods/deep learning · Joint segmentation/registration · Nonlinear elasticity · Incompressibility · Splitting algorithm

1 Introduction

Image segmentation which aims to mirror the aptitude of human beings to pinpoint meaningful constituents and to merge elements sharing similarities, is an essential link in automatic image analysis. It is at the heart of many medical image processing chains requiring quantitative analysis such as radiotherapy

This project was co-financed by the European Union with the European regional development fund (ERDF, 18P03390/18E01750/18P02733), by the Haute-Normandie Régional Council via the M2SINUM project and by the French Research National Agency ANR via AAP CE23 MEDISEG ANR project. The authors would like to thank the CRIANN (Centre Régional Informatique et d'Applications Numériques de Normandie, France) for providing computational resources.

L. Calatroni et al. (Eds.): SSVM 2023, LNCS 14009, pp. 363–375, 2023.
https://doi.org/10.1007/978-3-031-31975-4_28

planning or computer-aided diagnosis. In this regard, it is generally agreed that deep convolutional neural networks (CNN) are now the state-of-the-art in medical image segmentation.
Training a CNN consists in learning its parameters through the optimisation of a differentiable data-driven loss function that forces the prediction to comply with the ground truth. To enhance the anatomical plausibility of the results, increase explicability of CNNs, mitigate their opacity and achieve significant qualitative and quantitative improvements [5], it seems natural to question the incorporation of prior topological knowledge in these networks so that the prediction no longer depends solely on the data, but also on known properties (*e.g.*, anatomical properties, contextual relations between objects, number of connected components, etc.). In particular, monitoring topology turns out to be relevant when the anatomical requirement is not reflected in the data/not in visual agreement with the data: for instance, in spite of its highly folded nature [7,15], the intrinsic unfolded structure of the cortex is the one of a 2D sheet. Two matters then naturally emerge:

(i) the first one is how to formalise mathematically topology requirements such as the number of connected components or the contextual relation between objects? This task is a difficult one, primarily due to the very nature of the concept of topology which is intrinsically dual in that it is both a global and local property [15]: small localised changes on a geometrical shape may modify its global connectivity. Moreover, topology is a continuous concept whose properties are difficult to transpose to the discrete case.
(ii) the second question is how to incorporate those topological requirements in a CNN not simply in a sequential prediction/correction pipeline but in a nested manner, either by designing suitable loss functions — note that in the supervised case, prior knowledge is thus not involved in the inference time — or by in-layer techniques achieved upstream of the loss function evaluation. In this latter case like in [13], activation functions are interpreted as dual variables of variational problems, yielding layer outputs theoretically fulfilling the prescribed constraints. Not only are these outputs fully involved in the gradient backpropagation process during training, but they still contribute during the testing phase. Nevertheless, hard constraints such as volume prescription must be slackened in this case, which requires the use of statistical estimations, thus approximate values.

Recent works attempt to provide a unifying response to both issues. In [10], Hu *et al.* investigate persistent homology [4] and in particular Betti numbers to design an objective function based on persistent diagrams in order to segment membranes on electronic microscope neuronal images. Clough *et al.* [3] also rely on persistent homology theory to measure the correspondence between the Betti numbers of the prediction and those known a priori via persistent bar-codes. This tool enables them to integrate a loss function, termed topological, into the training of a CNN to improve in particular the accuracy of segmentations of cardiac MRI images. Finally, the objective function clDice [16] by Shit et al. relies on the topological notion of skeletonisation to promote the respect of certain features such as connectivity and number of related components in the context of segmentation of tubular structures such as vessels or neurons. While the methods described above relate the discrete nature of digital images to their continuous representation via complex topological tools, the viewpoint we adopt in our approach is purely continuous, based on the joint treatment of the segmentation problem (assigning a label to each image pixel) and the registration one (consisting in matching the relevant features of the considered image onto their ground truth counterpart through a suitable deformation φ). As structure alignment and intensity distribution comparison rule registration, processing these tasks in a single framework might yield positive mutual influence and registration might be viewed as the incorporation of prior information. Indeed, the constraints prescribed on $\det \nabla\varphi$ (namely incompressibility constraint as will be seen later, reflecting topology and area/volume preservation) are now a substitute for the explicit monitoring of topology.

Note that some recent works related to the joint processing of these two tasks are emerging in the deep learning community without however focusing specifically on topology control, like [6,12,18]. To achieve our goal, we propose a unified variational framework blending segmentation and registration regarded as topological prior in the training process, through the design of an ad hoc loss function, implying that the network learning is dictated by these constraints. The mathematical formulation yields a non convex optimisation problem under equality constraint solved by introducing auxiliary variables to balance the numerical complexity. The original minimisation is thus split into subproblems for which, most of the time, closed form solutions are provided, in an alternating setting. Our contribution thus takes on several forms: (i) of a methodological nature through the design of a loss function inspired by image registration models in order to encode topological properties; (ii) of a theoretical nature with the existence of minimisers for the optimisation problem with respect to the deformation and an asymptotic analysis; (iii) of a numerical nature by developing an optimised implementation with explicit solutions; (iv) and finally of a more applied nature by evaluating the method in a multi-class case of thoracic scans. These elements are presented sequentially in the following.

Note that a longer version of this work is in preparation. It will include several elements complementary to what is presented: (i) of a methodological and theoretical aspect, with detailed proofs of the set of theoretical results related to the model and accurate description of the implemented algorithm (in particular, the closed-form solutions will be made explicit at each step), and of (ii) an experimental nature, with the evaluation of the generalisation ability of the method (validation on other datasets), the interpretability of the results by identifying the pixels that contributed the most to the decision, the exhaustive qualitative and quantitative analysis of the results (comparison of the training time with other classical loss functions, effectiveness when the amount of data is limited, comparison with other architectures, etc.).

2 Proposed Nested Framework Based on Deep Learning and Variational Approaches

2.1 CNN Formalism

In its most basic form, a CNN can be diagrammed as follows (refer to [13] for a deeper insight). Denoting by i^0 the input of the network — *an image* —, the output of the network i^T with values in $[0, 1]^L$, L being the number of classes, can be viewed as the resultant of the action of a parameterised nonlinear operator $\mathcal{F}_\theta$, $i^T = \mathcal{F}_\theta(i^0)$, θ being the unknown parameter set to be learnt. It represents a soft classification function whose component $i_l^T(x)$ returns the probability of pixel x to belong to the lth class.

Training the CNN requires to consider a dataset of K images (bidimensional in our case) and their associated ground truth maps denoted by $\{y^k\}_{k=1,\cdots,K}$ with $y^k \in \{1, \cdots, L\}$. Since the neural network optimisation problem is separable with respect to the variable k, we omit the dependence on k from now on. For each input image of the dataset, the CNN outputs a segmentation function parameterised by θ, denoted by $s(\theta)$, and desired to be as close as possible to y. To this end, learning the parameters θ of the network requires the design of a suitable and differentiable loss function $\mathcal{L}$ to make the backpropagation of the gradient error efficient. The next subsection is devoted to the design of a suitable loss involving implicit topological prescriptions. As our setting is the continuous one, some preliminary mathematical notations and assumptions are required. Moreover, the connection between segmentation and registration is made explicit by means of the deformation denoted by $\varphi(= \varphi^k)$ below.

2.2 Design of a Loss Function Blending Segmentation and Registration

Let Ω be a convex bounded open subset of $\mathbb{R}^2$ of class $\mathcal{C}^1$, therefore satisfying the cone property. This latter requirement is for technical purposes to ensure that Ball's theorems [1] apply. We assume that $y \in BV(\Omega, \{1, \cdots, L\})$. The connection between the segmentation and registration tasks is made through the introduction of the deformation. Let then $\varphi : \bar{\Omega} \to \mathbb{R}^2$ be the sought deformation allowing to warp y into $s(\theta) \in L^2(\Omega)$. Observe that the viewpoint is slightly changed in comparison to Subsect. 2.1 since $s(\theta)$ is no longer viewed as a tridimensional image. In practice, φ should be with values in $\bar{\Omega}$ but from a mathematical point of view, if we work with such spaces, we lose the structure of vector space. Nevertheless, based on Ball's results [1, Theorems 1 and 2], we prove that our model generates homeomorphisms from $\bar{\Omega}$ to $\bar{\Omega}$. The deformation gradient is $\nabla\varphi : \Omega \to M_2(\mathbb{R})$, with $M_2(\mathbb{R})$ the set of 2×2 matrices. Mechanically [2], a deformation is a smooth mapping that is orientation-preserving and injective except possibly on $\partial\Omega$ where self-contact is authorised. This translates

mathematically into the condition $\det \nabla\varphi > 0$ almost everywhere. The associated displacement field is denoted by u with $\varphi = \mathrm{Id} + u$ and $\nabla\varphi = I_2 + \nabla u$, Id being the identity mapping, and I_2, the 2×2 identity matrix. We also need the following notations: $A : B = \mathrm{tr}\, A^T B$, the inner product, and $\|A\| = \sqrt{A : A}$ the related norm (Frobenius norm).
The design of the functional to be minimised and that will be embedded in our loss function is dictated by several considerations: it appears to be a compromise between smoothness of the deformation to be recovered, and quality of the pairing between the deformed ground truth and the prediction. Thus the functional first comprises a term F_{id} quantifying how close the deformed ground truth $y \circ \varphi$ is from the prediction $s(\theta)$, *i.e.*, the degree of alignment, taken classically as the sum of squared differences $\mathrm{Fid}(\varphi) = \frac{1}{2}\,\|s(\theta) - y \circ \varphi\|^2_{L^2(\Omega)}$. Second, it includes a regularisation acting as a deformation model and based on mechanical concepts, the objects to be matched — the ground truth image in our setting — being viewed as bodies subjected to external forces. The point of view we embrace is the one of incompressible elasticity [9], which motivates the introduction of a first condition regarding the determinant $\det \nabla\varphi$. Indeed, the assumption of incompressibility means that the material does not undergo a change in volume when a deformation is applied. From the point of view of the segmentation problem, the incompressibility condition encodes two fine properties: (i) a first one of a topological nature, since requiring the prediction $s(\theta)$ to exhibit the same number of connected components as y; (ii) and a second one that ensures preservation of areas and thus more accurate segmentations. As will be seen later, in an underlying way, the recovered deformation φ is expected to be close to the identity mapping.
The requirement of incompressibility may then be stated in an algebraic form related to the deformation by $\det \nabla\varphi = 1$ almost everywhere ($a.e.$) on Ω. This local incompressibility condition is complemented by a more classical regularisation that can be seen, from the mechanical outlook, as a penalisation of the internal elastic energy $\pi(u)$ of the body, typically stated as $\pi(u) = \int_\Omega \sigma(x, \nabla u(x))\, dx$. Function σ is the stored energy function depending on the position x and on the displacement gradient ∇u. Its form characterises the properties of the material constituting the body and dictates the choice of the functional space on which π is defined. In our case, we propose introducing the following stored energy $W(\nabla\varphi) = W(I_2 + \nabla u) = \sigma(\nabla u) = \mu\,\left(\|\nabla\varphi\|^4 - 4\right)$, μ being a tuning parameter, resulting in the following regulariser

$$\mathrm{Reg}(\varphi) = \mu \int_\Omega \left(\|\nabla\varphi\|^4 - 4\right)\, dx. \tag{1}$$

The first term controls smoothness and penalises changes in length. The constant 4 is added to comply with the property $W(I_2) = 0$, I_2 being the Jacobian of the identity mapping. Combined with the local incompressibility condition, component (1) exhibits fine theoretical properties useful for the mathematical analysis conducted hereafter.

The proposed registration model in a variational setting thus reads as:

$$\inf_{\varphi \in \mathcal{W}} \left\{ \mathcal{F}(\varphi) = \frac{\nu}{2} \operatorname{Fid}(\varphi) + \operatorname{Reg}(\varphi), \right. \tag{$\mathcal{P}$}$$

$$\left. = \frac{\nu}{2} \|s(\theta) - y \circ \varphi\|^2_{L^2(\Omega)} + \int_\Omega W(\nabla \varphi)\, dx \right\},$$

under the constraint $\det \nabla\varphi = 1$ *a.e.*, with $\mathcal{W} = \{\psi \in \operatorname{Id} + W_0^{1,4}(\Omega, \mathbb{R}^2) \mid \det \nabla\psi = 1 \text{ } a.e.\}$. This functional is then embedded in a joint deep learning/variational-based framework through the following Loss Function $(\mathcal{LF})$

$$\inf_{\theta, \varphi \in \mathcal{W}} \mathcal{L}(\theta, \varphi) = \mathcal{F}_{DL}(s(\theta), y) + \frac{\nu}{2} \|s(\theta) - y \circ \varphi\|^2_{L^2(\Omega)} + \int_\Omega W(\nabla \varphi)\, dx, \tag{$\mathcal{LF}$}$$

reconciling both formalisms *deep learning-based approaches/variational models* in a single framework. Through the loss function $(\mathcal{LF})$, $s(\theta)$ is thus required to be aligned with the deformed ground truth $y \circ \varphi$, while being close, in terms of intensities, to the ground truth itself y. In an underlying fashion, the mapping φ which makes it possible to implicitly formulate topological prescriptions, is expected to be close to the identity mapping. Note that in a preliminary version of the model, we had considered explicitly including a quadratic penalty of the form $\|\varphi - \operatorname{Id}\|^2_{L^2(\Omega, \mathbb{R}^2)}$. The numerical simulations did not show a significant contribution to the results, which led us to dispense with this term. The processing chain is thus not a simple sequential pipeline but unifies the two methods. The first term $\mathcal{F}_{DL}(s(\theta), y)$ that composes our cost function $\mathcal{L}$ promotes the standard matching of intensity levels between the probability map $s(\theta)$ and the ground truth y. In practice, it is chosen to be the Dice metric.
In the next subsection, we present a result of existence of minimisers for the problem $(\mathcal{P})$ related to the deformation φ. Note that due to the data fidelity term involving a composition of a $BV(\Omega, \{1, \cdots, L\})$ function with a Sobolev mapping, the proof is not a straightforward application of the direct method in the calculus of variations. This result ensures the well-posed character of the partial minimisation problem in φ (whilst $s(\theta)$ is fixed) that naturally appears in the designed alternating algorithm (see Subsect. 3.1, Algorithm 1).

2.3 A First Theoretical Result

Theorem 1. *Problem $(\mathcal{P})$ admits at least one minimiser in $\mathcal{W}$.*

Proof. Due to the limited number of pages, we only sketch the proof here. It follows the arguments of the classical direct method of the calculus of variations. Clearly, $\mathcal{F}(\varphi) \geq \mu \|\nabla\varphi\|^4_{L^4(\Omega, M_2(\mathbb{R}))} - 4\mu \operatorname{meas}(\Omega)$, meas denoting the Lebesgue measure. The quantity $\mathcal{F}(\varphi)$ is thus bounded below by $-4\mu \operatorname{meas}(\Omega)$ and as for $\varphi = \operatorname{Id}$, $\mathcal{F}(\varphi) = \frac{\nu}{2} \|s(\theta) - y\|^2_{L^2(\Omega)}$ is finite (due to the embedding $BV(\Omega) \subset L^2(\Omega)$, or simply by the fact that $L^\infty(\Omega) \subset L^2(\Omega)$), the infimum is finite. Let then $(\varphi_k)_k \in \mathcal{W}$ be a minimising sequence, *i.e.* such that $\lim_{k \to +\infty} \mathcal{F}(\varphi_k) =$

$\inf\limits_{\varphi \in \mathcal{W}} \mathcal{F}(\varphi)$. Standard arguments from the theory of nonlinear elasticity enable one to conclude that there exists a subsequence of $(\varphi_k)_k$ — still denoted by $(\varphi_k)_k$ — and $\bar{\varphi} \in W^{1,4}(\Omega, \mathbb{R}^2)$ such that $\varphi_k \underset{k\to+\infty}{\rightharpoonup} \bar{\varphi}$ in $W^{1,4}(\Omega, \mathbb{R}^2)$, $\det \nabla\varphi_k \underset{k\to+\infty}{\overset{*}{\rightharpoonup}} \det \nabla\bar{\varphi}$, $\det \nabla\bar{\varphi} = 1$ *a.e.* and $\bar{\varphi} \in \mathrm{Id} + W_0^{1,4}(\Omega, \mathbb{R}^2)$. For k large enough,

$$\int_\Omega \| (\nabla\varphi_k)^{-1} \|^4 \det \nabla\varphi_k \, dx = \int_\Omega \frac{1}{(\det \nabla\varphi_k)^3} \|\nabla\varphi_k\|^4 \, dx \le C,$$

C being a positive constant depending only on Ω. The assumptions of Ball's theorems [1, Theorems 1 and 2] thus hold, yielding that φ_k is a homeomorphism of $\bar{\Omega}$ onto $\bar{\Omega}$ and $\varphi_k^{-1} \in W^{1,4}(\Omega, \mathbb{R}^2)$. The same reasoning applies to $\bar{\varphi}$ so that it inherits the same smoothness properties: it is a homeomorphism of $\bar{\Omega}$ onto $\bar{\Omega}$ and $\bar{\varphi}^{-1} \in W^{1,4}(\Omega, \mathbb{R}^2)$.
Obviously, $\int_\Omega W(\nabla\bar{\varphi})\, dx \le \liminf\limits_{k\to+\infty} \int_\Omega W(\nabla\varphi_k)\, dx$. It remains to study $\liminf\limits_{k\to+\infty} \|s(\theta) - y \circ \varphi_k\|^2_{L^2(\Omega)}$ — in fact, we handle $\lim\limits_{k\to+\infty} \|s(\theta) - y \circ \varphi_k\|^2_{L^2(\Omega)}$ —, inspired by prior work by Wirth [17]. We first prove that $\varphi_k \circ \bar{\varphi}^{-1} \underset{k\to+\infty}{\longrightarrow} \mathrm{Id}$ in $C^{0,\alpha}(\bar{\Omega}, \mathbb{R}^2)$ with $\alpha < \frac{1}{2}$. Let now $\mathcal{O}_i = \{x \in \Omega \,|\, y(x) = i\}$ with i any index in $\{1, \cdots, L\}$. We observe that $\Omega = \cup_{i=1}^L \mathcal{O}_i \cup \mathcal{N}$ with $\mathcal{N}$ such that $\mathrm{meas}(\mathcal{N}) = 0$. Due to the smoothness properties of $\bar{\varphi}$, $\Omega = \bar{\varphi}^{-1}(\Omega) = \cup_{i=1}^L \bar{\varphi}^{-1}(\mathcal{O}_i) \cup \bar{\varphi}^{-1}(\mathcal{N})$, which, combined with the fact that for any measurable set $\mathcal{S} \subset \Omega$ with Lebesgue measure $\mathrm{meas}(\mathcal{S}) \le \eta$, $\mathrm{meas}(\bar{\varphi}^{-1}(\mathcal{S})) \le \eta$, shows that to achieve our goal, it suffices to prove that $y \circ \varphi_k \underset{k\to+\infty}{\longrightarrow} y \circ \bar{\varphi}$ in $L^2(\bar{\varphi}^{-1}(\mathcal{O}_i))$.

3 Towards a Tractable Numerical Algorithm

3.1 Optimisation Strategy for the Nested Model

An alternating scheme which consists in splitting the original problem $(\mathcal{LF})$ into two subproblems is implemented: (i) the first one is smooth and non-convex (partial problem obtained when the deformation φ is fixed): it is the network parameter optimisation achieved by a stochastic gradient descent (SGD) method; (ii) the second one (partial problem obtained when the parameter set θ is fixed) is nonlinear and nonconvex, related to the updating of the deformation φ. It corresponds to problem $(\mathcal{P})$ with $\theta := \theta_n$. This latter problem is challenging and its resolution, based itself on a separation technique, is the subject of Subsect. 3.2. The sketch of the numerical method of resolution is summarised in Algorithm 1. To update the network parameters θ, we simply use a mini-batch gradient descent technique on the loss (the functional $\bar{\mathcal{L}}$ defined in Algorithm 1 to be minimised with respect to θ being smooth (non-convex) here). The parameters are then updated by taking a step opposite to the gradient, *i.e.*, $\theta_{n+1} = \theta_n - \eta \nabla_\theta \bar{\mathcal{L}}(\theta_n)$, where η is the learning rate and n represents the set of all successive steps of the mini-batch stochastic gradient descent technique performed during this epoch.

Algorithm 1. Nested Algorithm to update θ and φ

Initialise θ_0 randomly and set $\varphi = \text{Id}$
Fix μ, $\nu > 0$
for $n = 0, 1, \cdots$ **do**
  Compute $s(\theta_n)$
  $\varphi_n = \arg\min_{\varphi \in \mathcal{W}} \left[\frac{\nu}{2} \|s(\theta_n) - y \circ \varphi\|^2_{L^2(\Omega)} + \int_\Omega W(\nabla\varphi)\, dx \right]$
  Set $\bar{\mathcal{L}}(\cdot) = \mathcal{F}_{DL}(s(\cdot), y) + \frac{\nu}{2} \|s(\cdot) - y \circ \varphi_n\|^2$
  $\theta_{n+1} = \theta_n - \eta \nabla_\theta \bar{\mathcal{L}}(\theta_n)$
end for

3.2 Splitting Strategy for Problem ($\mathcal{P}$)

If problem ($\mathcal{P}$) exhibits fine theoretical properties (i) it admits at least one minimiser and (ii) this minimiser is a bi-Hölder homeomorphism from $\bar{\Omega}$ to $\bar{\Omega}$, it belongs, however, to the class of nonlinear non-convex problems, hard to solve from a numerical point of view. To balance the computational complexity, we introduce auxiliary variables in order to fragment the initial problem into a sequence of more easily solvable ones. In this line, inspired by the pioneering work of Negrón Marrero [14], two variables V and W both simulating $\nabla\varphi$ are added, the coupling being achieved through L^2-penalisations and the underlying idea being to transfer the non-linearity of the regulariser on V (while slackening the incompressibility constraint) and to constrain W to be such that $\det W = 1$ *a.e.*. The process thus breaks down the problem by separating the question of smoothness from the question of area preservation, yielding in the discrete setting, a series of problems with, for the most part, closed-form solutions. Precisely, the following decoupled problem ($\mathcal{DP}$) is now a substitute for the original one ($\mathcal{P}$):

$$\inf_{(\varphi,V,W)} E(\varphi, V, W) = \mu \|V\|^4_{L^4(\Omega, M_2(\mathbb{R}))} + \frac{\nu}{2} \|s(\theta) - y \circ \varphi\|^2_{L^2(\Omega)} \qquad (\mathcal{DP})$$
$$+ \frac{\mu\alpha}{2} \|\det V - 1\|^2_{L^2(\Omega)} + \frac{\gamma_1}{2} \|V - \nabla\varphi\|^2_{L^2(\Omega, M_2(\mathbb{R}))} + \frac{\gamma_2}{2} \|W - \nabla\varphi\|^2_{L^2(\Omega, M_2(\mathbb{R}))},$$

with $(\varphi, V, W) \in \left(\overline{\mathcal{W}} = \text{Id} + W_0^{1,2}(\Omega, \mathbb{R}^2)\right) \times L^4(\Omega, M_2(\mathbb{R}))$
$\times \left(\overline{\overline{\mathcal{W}}} = \{X \in L^2(\Omega, M_2(\mathbb{R})) \mid \det X = 1 \text{ a.e.}\}\right)$, and α, γ_1 and γ_2 being positive parameters. The role of parameter α will be made clearer next.

Remark 1. By considering smooth approximations of y — ensuring in particular that the composition $y \circ \varphi$ makes sense, since here φ is only $W^{1,2}(\Omega, \mathbb{R}^2)$ — and by replacing parameters γ's by γ_j, $(\gamma_j)_j$ being an increasing sequence of positive real numbers such that $\lim_{j\to+\infty} \gamma_j = +\infty$, an asymptotic result can be established. It is a result in itself and it will be the subject of a separate article in preparation.

We now place ourselves in the discrete setting and present the resulting alternating algorithm (see Algorithm 2). The problems in V and W are separable with

Algorithm 2. Splitting technique to solve $(\mathcal{DP})$

Initialise $\varphi_0 = \text{Id}$, $V_0 = I_2$ and $W_0 = I_2$
Fix γ_1, γ_2, μ, ν, $\alpha > 0$
for $n = 0, 1, \cdots$ **do**

$$V_{n+1} = \underset{V \in M_2(\mathbb{R}) \sim \mathbb{R}^4}{\arg\min} \ \mu \|V\|^4 + \frac{\mu\,\alpha}{2} (\det V - 1)^2 + \frac{\gamma_1}{2} \|V - \nabla\varphi^n\|^2$$

$$W_{n+1} = \underset{W \in M_2(\mathbb{R}) \sim \mathbb{R}^4}{\arg\min} \ \frac{\gamma_2}{2} \|W - \nabla\varphi^n\|^2 \text{ such that } \det W = 1$$

$$\varphi_{n+1} = \underset{\varphi}{\arg\min} \frac{\nu}{2} \|s(\theta) - y \circ \varphi\|^2_{L^2(\Omega)} + \frac{\gamma_1}{2} \|V^{n+1} - \nabla\varphi\|^2_{L^2(\Omega, M_2(\mathbb{R}))} + \frac{\gamma_2}{2} \|W^{n+1} - \nabla\varphi\|^2_{L^2(\Omega, M_2(\mathbb{R}))}$$

end for

respect to each pixel, yielding the following general formulations. Due to the limited number of pages, we can only give an outline of the proposed numerical resolutions :

(i) **Subproblem in V:** The problem is rephrased equivalently in terms of an auxiliary variable resulting from the multiplication of an orthogonal matrix and the vector $V \in M_2(\mathbb{R}) \sim \mathbb{R}^4$. It enables one to convert the rectangular terms involved in the determinant into square terms. Setting $\alpha = 8$ yields decoupled subproblems whose resolution amounts to finding the roots of cubic equations (closed-form solutions);

(ii) **Subproblem in W:** This part is inspired by prior related works by Glowinski *et al.* [9]. Again, the problem is rephrased equivalently in terms of an auxiliary variable resulting from the multiplication of an orthogonal matrix and the vector $W \in M_2(\mathbb{R}) \sim \mathbb{R}^4$. The theory of Lagrange multipliers allows to obtain the explicit expression of the — unique in the general case — solution of the problem;

(iii) **Subproblem in φ** solved through a Discrete Sine Transform method.

4 Experiments and Results

4.1 Dataset

Our new loss function is assessed in the context of multi-class segmentation of the SegTHOR dataset. It is composed of 60 thoracic CT-scans, acquired with or without intravenous contrast, of 60 patients diagnosed with lung cancer or Hodgkin's lymphoma and treated at the Henri Becquerel Center, Rouen, France. The images exhibit a size of $512 \times 512 \times (150\text{–}284)$ voxels (the number of slices changes depending on the patients). The healthy organs adjacent to the tumour that must be preserved from radiation are called Organs At Risk (OAR). In this case, 4 of them need to be segmented, among which the oesophagus, the heart, the trachea and the aorta. This dataset seems to have all the locks motivating the introduction of our model, *i.e.* class imbalance, inter-/intra-patient variability, absence of contours or slice-dependent topology. Indeed, the OAR represent considerably fewer pixels than the background, the lack of contrast makes the segmentation of the heart and aorta complex, and even more that of a small organ such as the oesophagus. This latter also shows a variable shape and position depending on the slice considered for the same patient and from one patient to another. Moreover, due to its cane shape, the aorta presents one or two connected components depending on the slice. Finally, the trachea being filled with air appears in black on the CT images and is easier to segment.

4.2 Implementation

In all our experiments, the network architecture that we used is a simplified version of U-Net, called sU-Net, previously introduced in [11], that has fewer dense connected layers than the original one. We evaluate the loss function $\mathcal{F}_{DL}$ as the Dice loss. The images are normalised and cropped from the centre to reduce their size to 304×240 pixels. In [11], the implemented data augmentation shows more accurate results so we used it to artificially triple the number of training images. The parameters of the network θ are randomly initialised using Glorot's technique [8] and updated by a stochastic gradient descent (SGD) whose initial learning rate is fixed at 10^{-2} and batch size at 24. We set the parameters related to the registration problem thanks to a grid search technique as follows: $\gamma_1 = 10000$, $\gamma_2 = 80000$ and $\mu = 125$. Furthermore, several values are evaluated for the weighting parameter ν, in particular we reported the segmentation results for this term set to 1 and 6. Finally, the calculation of our loss function takes 7 s longer than the Dice loss for a batch, but fewer iterations are needed. Thus, the training of our hybrid approach according to Algorithm 1 takes place over a maximum of 100 epochs, while in the results detailed below, the training of the classical DL approach (without topological constraint) occurs with a maximum of 120 epochs.

4.3 Protocol and Evaluation Metrics

Following the standardised protocol of the SegTHOR challenge, training is performed with the images of 40 patients and the last 20 are reserved for the test phase. To quantify the results of the automatic segmentations, two well-known metrics are used, namely the Dice score and the Hausdorff distance. The first one, which is an overlap measure, gives an overall idea of the segmentation, while the second one, which is a surface measure, highlights outliers and inconsistencies. These elements should be complemented by a qualitative analysis to reflect the strengths and weaknesses of the evaluated model.

4.4 Results

The automatic segmentations obtained with our nested method, which thus imbricates the two formalisms, can be analysed in the light of those obtained via the classical deep learning approach. Table 1 presents the quantitative segmentation results for the 4 organs at risk of the SegTHOR dataset (most significant

Table 1. Segmentation results (mean±stdev) obtained with a Dice loss function $\mathcal{F}_{DL}$ only and combined with the proposed variational model. $\mathcal{L}$ is the global loss function. Metrics used: Dice score and Hausdorff distance (HD). The most significant results are in bold.

$\mathcal{L}$	Metrics	Aorta	Trachea	Oesophagus	Heart
Dice	Dice score %	93.36 ± 1.62	90.60 ± 1.90	81.66 ± 5.45	92.30 ± 3.61
	HD (mm)	13.47 ± 8.55	18.75 ± 9.73	31.45 ± 19.02	45.23 ± 40.18
Dice + $\mathcal{F}(\varphi)$ with cont. (ν=1)	Dice score %	**94.61 ± 1.62**	90.69 ± 2.64	82.26 ± 4.98	**93.14 ± 3.04**
	HD (mm)	10.71 ± 7.36	17.96 ± 9.32	18.89 ± 12.44	18.99± 8.51
Dice + $\mathcal{F}(\varphi)$ with cont. (ν=6)	Dice score %	94.51 ± 1.88	91.26 ± 2.03	**82.56 ± 4.38**	93.05 ± 3.28
	HD (mm)	**8.95 ± 5.43**	17.85 ± 6.48	**15.83 ± 8.17**	**17.98 ± 7.96**

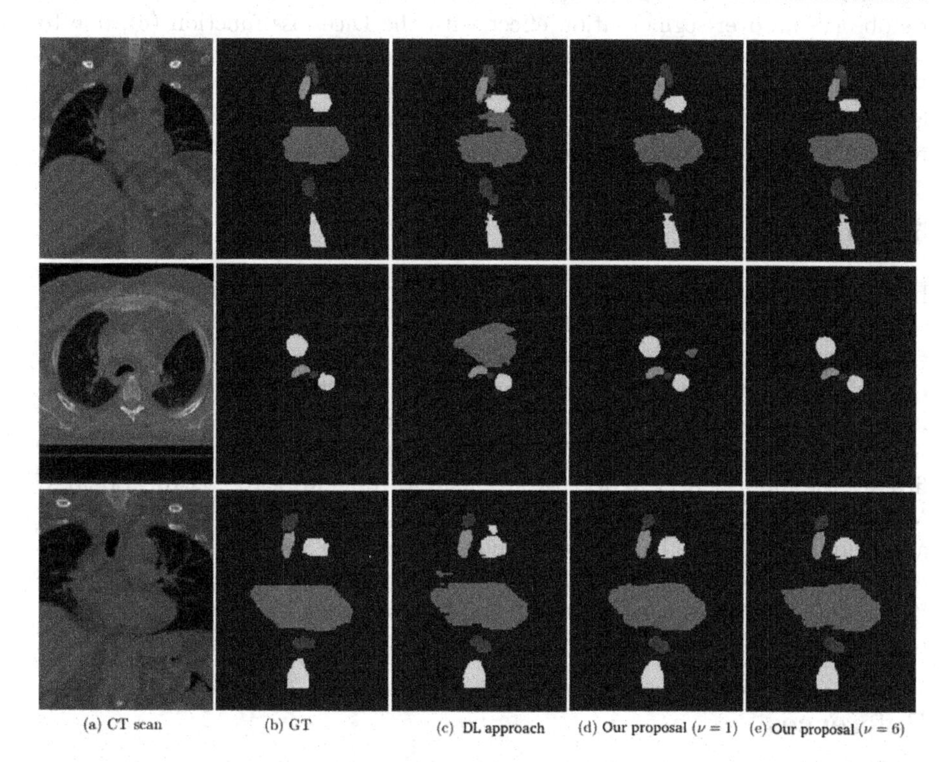

Fig. 1. Segmentation results on three patients of the aorta (yellow), the trachea (light green), the heart (dark green) and the oesophagus (blue) obtained with a classical Deep-Learning approach optimised with a Dice loss (c) and with our proposal (d-e). (Color figure online)

results compared to the classical deep learning approach in bold), and Fig. 1 the qualitative results. In particular, we show those obtained with two different values of ν. In view of this, we discuss each OAR.

As explained earlier, the trachea (in light green in Fig. 1) appears in black on the CT scan, making it easily distinguishable. Dice scores and Hausdorff distances are very similar for both methodologies. This translates visually into nearly identical shapes. For the aorta (shown in yellow in Fig. 1), adding our functional $\mathcal{F}$ and the constraint to the objective function saves about 1 pp of Dice score. Moreover, the Hausdorff distance decreases from 13.47 mm to 10.71 mm when ν is 1, and 8.95 for ν equal to 6. This decrease can be explained by a better respect of the topology as illustrated for example by the third line of Fig. 1. Concerning the Dice score for the oesophagus (in blue in Fig. 1) and the heart (in dark green in Fig. 1), the observation is equivalent to the one made in the case of the aorta with a gain of about 1 pp for these two OAR. On the other hand, our proposal significantly improves the HD metric by approximately dividing it by 2 for the oesophagus and 2.5 for the heart. For these two organs, we observe an over-segmentation effect with the Dice loss function (c), due to outliers and isolated pixel groups, which our method (d-e) tends to correct. Finally, if the increase of the weighting term ν gives globally similar Dice scores, it participates nevertheless to decrease a little more the Hausdorff distances by rectifying some remaining anomalies.

5 Conclusion

In this paper, we have introduced a new registration-based loss function in order to integrate topological knowledge in the learning of a 2D CNN. Indeed, here, the registration task under incompressibility constraint takes the form of a prior to guide the segmentation towards more realistic results. Quantitative and qualitative results show more accurate and anatomically faithful segmentations compared to a classical DL approach in the multi-class case of thoracic OAR, confirming the interest of a hybrid variational-DL framework to control topology. Moreover, this methodology has the advantage of being directly applicable to any neural networks. These benefits should be emphasised with other datasets, especially for those with limited data. Also, it would be interesting to include the process in the inference phase, for instance, by using an average template to be deformed.

References

1. Ball, J.M.: Global invertibility of Sobolev functions and the interpenetration of matter. P. Roy. Soc. Edin. A **88**(3–4), 315–328 (1981)
2. Ciarlet, P.: Three-Dimensional Elasticity. Elsevier Science, Mathematical Elasticity (1994)
3. Clough, J., Byrne, N., Oksuz, I., Zimmer, V.A., Schnabel, J.A., King, A.: Topological loss function for deep-learning based image segmentation using persistent homology. IEEE Trans. Pattern Anal. Mach. Intell. (2020). IEEE
4. Edelsbrunner, H., Harer, J.L.: Computational Topology: An Introduction. American Mathematical Society (2010)
5. El Jurdi, R., Petitjean, C., Honeine, P., Cheplygina, V., Abdallah, F.: High-level prior-based loss functions for medical image segmentation: A survey. Comput. Vis. Image Underst. **210**, 103248 (2021)

6. Estienne, T., et al.: U-ReSNet: ultimate coupling of registration and segmentation with deep nets. In: Shen, D., et al. (eds.) MICCAI 2019. LNCS, vol. 11766, pp. 310–319. Springer, Cham (2019). https://doi.org/10.1007/978-3-030-32248-9_35
7. Fischl, B., Liu, A., Dale, A.M.: Automated manifold surgery: constructing geometrically accurate and topologically correct models of the human cerebral cortex. IEEE Trans. Med. Imaging **20**(1), 70–80 (2001)
8. Glorot, X., Bengio, Y.: Understanding the difficulty of training deep feedforward neural networks. In: Proceedings of the Thirteenth International Conference on Artificial Intelligence and Statistics, pp. 249–256 (2010)
9. Glowinski, R., Le Tallec, P.: Numerical solution of problems in incompressible finite elasticity by augmented Lagrangian methods. I. Two-dimensional and axisymmetric problems. SIAM J. Appl. Math. **42**(2), 400–429 (1982)
10. Hu, X., Li, F., Samaras, D., Chen, C.: Topology-preserving deep image segmentation. Adv. Neural Inf. Process. Syst. **32** (2019)
11. Lambert, Z., Petitjean, C., Dubray, B., Ruan, S.: SegTHOR: Segmentation of Thoracic Organs at Risk in CT images. In: 2020 Tenth International Conference on Image Processing Theory, Tools and Applications (IPTA), pp. 1–6 (2020)
12. Li, B., et al.: A hybrid deep learning framework for integrated segmentation and registration: evaluation on longitudinal white matter tract changes. In: Shen, D., et al. (eds.) MICCAI 2019. LNCS, vol. 11766, pp. 645–653. Springer, Cham (2019). https://doi.org/10.1007/978-3-030-32248-9_72
13. Liu, J., Wang, X., Tai, X.C.: Deep Convolutional Neural Networks with Spatial Regularization, Volume and Star-shape Priori for Image Segmentation. J. Math. Imaging Vis. **64**(6), 625–645 (2022)
14. Negrón Marrero, P.: A numerical method for detecting singular minimizers of multidimensional problems in nonlinear elasticity. Numer. Math. **58**, 135–144 (1990)
15. Ségonne, F., Pacheco, J., Fischl, B.: Geometrically Accurate Topology-Correction of Cortical Surfaces Using Nonseparating loops. IEEE Trans. Med. Imaging **26**(4), 518–529 (2007)
16. Shit, S.: clDice-a novel topology-preserving loss function for tubular structure segmentation. In: Proceedings of the IEEE/CVF Conference on Computer Vision and Pattern Recognition, pp. 16560–16569 (2021)
17. Wirth, B.: On the Gamma-limit of joint image segmentation and registration functionals based on phase fields. Interfaces Free Bound. **18**(4), 441–477 (2016)
18. Xu, Z., Niethammer, M.: DeepAtlas: joint semi-supervised learning of image registration and segmentation. In: Shen, D., et al. (eds.) MICCAI 2019. LNCS, vol. 11765, pp. 420–429. Springer, Cham (2019). https://doi.org/10.1007/978-3-030-32245-8_47

Optimization for Imaging: Theory and Methods

A Relaxed Proximal Gradient Descent Algorithm for Convergent Plug-and-Play with Proximal Denoiser

Samuel Hurault[1(✉)], Antonin Chambolle[2], Arthur Leclaire[1], and Nicolas Papadakis[1]

[1] Univ. Bordeaux, CNRS, Bordeaux INP, IMB, UMR 5251, 33400 Talence, France
samuel.hurault@math.u-bordeaux.fr

[2] CEREMADE, CNRS, Université Paris-Dauphine, PSL, 75775 Paris, France

Abstract. This paper presents a new convergent Plug-and-Play (PnP) algorithm. PnP methods are efficient iterative algorithms for solving image inverse problems formulated as the minimization of the sum of a data-fidelity term and a regularization term. PnP methods perform regularization by plugging a pre-trained denoiser in a proximal algorithm, such as Proximal Gradient Descent (PGD). To ensure convergence of PnP schemes, many works study specific parametrizations of deep denoisers. However, existing results require either unverifiable or suboptimal hypotheses on the denoiser, or assume restrictive conditions on the parameters of the inverse problem. Observing that these limitations can be due to the proximal algorithm in use, we study a relaxed version of the PGD algorithm for minimizing the sum of a convex function and a weakly convex one. When plugged with a relaxed proximal denoiser, we show that the proposed PnP-αPGD algorithm converges for a wider range of regularization parameters, thus allowing more accurate image restoration.

Keywords: Plug-and-Play · nonconvex optimization · Inverse problems

1 Introduction

We focus on the convergence of plug-and-play methods associated to the class of inverse problems:

$$\hat{x} \in \arg\min_{x} \lambda f(x) + \phi(x), \tag{1}$$

where f is a L_f-Lipschitz gradient function acting as a data-fidelity term with respect to a degraded observation y, ϕ is a M-weakly convex regularization function (such that $\phi + \frac{M}{2}\left\|.\right\|^2$ is convex); and $\lambda > 0$ is a parameter weighting the influence of both terms. In our experimental setting, we consider degradation models $y = Ax^* + \nu \in \mathbb{R}^m$ for some groundtruth signal $x^* \in \mathbb{R}^n$, a linear operator $A \in \mathbb{R}^{n \times m}$ and a white Gaussian noise $\nu \in \mathbb{R}^n$. We thus deal with convex data-fidelity terms of the form $f(x) = \frac{1}{2}||Ax - y||^2$, with $L_f = ||A^T A||$. Our analysis can nevertheless apply to a broader class of convex or nonconvex functions f.

L. Calatroni et al. (Eds.): SSVM 2023, LNCS 14009, pp. 379–392, 2023.
https://doi.org/10.1007/978-3-031-31975-4_29

To find an adequate solution of the ill-posed problem of recovering x^* from y, the choice of the regularization ϕ is crucial. Convex [20] and nonconvex [27] handcrafted functions are now largely outperformed by learning approaches [19, 28], that may not even be associated to a closed-form regularization function ϕ.

1.1 Proximal Algorithms

Estimating a local or global optima of problem (1) is classically done using proximal splitting algorithms such as Proximal Gradient Descent (PGD) or Douglas-Rashford Splitting (DRS). Given an adequate stepsize $\tau > 0$, these methods alternate between explicit gradient descent, $\mathrm{Id} - \tau\nabla h$ for smooth functions h, and/or implicit gradient steps using the proximal operator $\mathrm{Prox}_{\tau h}(x) \in \arg\min_z \frac{1}{2\tau}||z - x||^2 + h(z)$, for proper lower semi-continuous functions h. Proximal algorithms are originally designed for convex functions, but under appropriate assumptions, PGD [2] and DRS [24] algorithms converge to a stationary point of problem (1) associated to nonconvex functions f and ϕ.

1.2 Plug-and-Play Algorithms

Plug-and-Play (PnP) [26] and Regularization by Denoising (RED) [19] methods consist in splitting algorithms in which the descent step on the regularization function is performed by an off-the-shelf image denoiser. They are respectively built from proximal splitting schemes by replacing the proximal operator (PnP) or the gradient operator (RED) of the regularization ϕ by an image denoiser. When used with a deep denoiser (*i.e* parameterized by a neural network) these approaches produce impressive results for various image restoration tasks [28].

Theoretical convergence of PnP and RED algorithms with deep denoisers has recently been addressed by a variety of studies [9,18,21,22]. Most of these works require specific constraints on the deep denoiser, such as nonexpansivity. However, imposing nonexpansivity of a denoiser can severely degrade its denoising performance.

Another line of works tries to address convergence by making PnP and RED algorithms be exact proximal algorithms. The idea is to replace the denoiser of RED algorithms by a gradient descent operator and the one of PnP algorithm by a proximal operator. Theoretical convergence then follows from known convergence results of proximal splitting algorithms. The authors of [6,11] thus propose to plug an explicit *gradient-step denoiser* of the form $D = \mathrm{Id} - \nabla g$, for a tractable and potentially nonconvex potential g parameterized by a neural network. As shown in [11], such a constrained parametrization does not harm denoising performance. The gradient-step denoiser guarantees convergence of RED methods without sacrificing performance, but it does not cover convergence of PnP algorithms. An extension to PnP has been addressed in [12]: following [8], when g is trained with contractive gradient, the gradient-step denoiser can be written as a proximal operator $D = \mathrm{Id} - \nabla g = \mathrm{Prox}_\phi$ of a nonconvex potential ϕ. A PnP scheme with this *proximal denoiser* becomes again a genuine proximal splitting algorithm associated to an explicit functional. Following existing convergence results of the PGD and DRS algorithms in the nonconvex setting, [12] proves convergence of PnP-PGD and PnP-DRS with proximal denoiser.

The main limitation of this approach is that the proximal denoiser $D = \mathrm{Prox}_\phi$ does not give tractability of $\mathrm{Prox}_{\tau\phi}$ for $\tau \neq 1$. Therefore, to be a provable converging proximal splitting algorithm, the stepsize of the overall PnP algorithm has to be fixed to $\tau = 1$. For instance, for the PGD algorithm with stepsize $\tau = 1$:

$$x_{k+1} \in \mathrm{Prox}_\phi(\mathrm{Id} - \lambda \nabla f)(x_k) \tag{2}$$

the convergence of x_k to a stationary point of (1) is only ensured for regularization parameters λ satisfying $L_f \lambda < 1$ [2]. This is an issue for low noise levels ν, for which relevant solutions are obtained with a dominant data-fidelity term in (1).

Our objective is to design a convergent PnP algorithm with a proximal denoiser, and with minimal restriction on the regularization parameter λ. Contrary to previous work on PnP convergence [6,9,11,21–23], we not only wish to adapt the denoiser but also the original optimization scheme of interest. We study a new proximal algorithm able to deal with a proximal operator that can only be computed for a predefined and fixed stepsize $\tau = 1$.

1.3 Contributions and Outline

In this paper, we propose a relaxation of the Proximal Gradient Descent algorithm, called *αPGD*, such that when used with a proximal denoiser, the corresponding PnP scheme *Prox-PnP-αPGD* can converge for any regularization parameter λ.

In Sect. 2, extending the result from [12], we show how building a denoiser D that corresponds to the proximal operator of a M-weakly convex function ϕ. Then we introduce a relaxation of the denoiser that allows to control M, the constant of weak convexity of ϕ.

In Sect. 3, we give new results on the convergence of Prox-PnP-PGD [12] with regularization constraint $\lambda(L_f + M) < 2$. In particular, using the convergence of PGD for nonconvex f and weakly convex ϕ given in Theorem 1, Corollary 1 improves previous PnP convergence results [12] for $M < L_f$.

In Sect. 4, we present *αPGD*, a relaxed version of the PGD algorithm[1] reminiscent of the accelerated PGD scheme from [25]. Its convergence is shown Theorem 2 for a smooth convex function f and a weakly convex one ϕ. Corollary 2 then illustrates how the relaxation parameter α can be tuned to make the proposed PnP-αPGD algorithm convergent for regularization parameters λ satisfying $\lambda L_f M < 1$. Having a multiplication, instead of an addition, between the constants L_f and M opens new perspectives. In particular, by plugging a relaxed denoiser with controllable weak convexity constant, Corollary 3 demonstrates that, for all regularization parameter λ, we can always decrease M such that $\lambda L_f M < 1$ i.e. such that the αPGD algorithm converges.

---

[1] There are two different notions of relaxation in this paper. One is for the relaxation of the proximal denoiser and the other for the relaxation of the optimization scheme.

In Sect. 5, we provide experiments for both image deblurring and image super-resolution applications. We demonstrate the effectiveness of our PnP-αPGD algorithm, which closes the performance gap between Prox-PnP-PGD and the state-of-the-art plug-and-play algorithms.

All proofs can be found in the extended version of the paper [10].

2 Relaxed Proximal Denoiser

This section introduces the denoiser used in our PnP algorithm. We first redefine the Gradient Step denoiser and show in Proposition 1 how it can be constrained to be a proximal denoiser; and finally introduced the relaxed proximal denoiser.

2.1 Gradient Step Denoiser

In this paper, we use the Gradient Step Denoiser introduced in [6,11].

It is defined as a gradient step over a differentiable potential g_σ parametrized by a neural network.

$$D_\sigma = \mathrm{Id} - \nabla g_\sigma. \tag{3}$$

This denoiser can then be trained to denoise white Gaussian noise ν_σ of various standard deviations σ by minimizing the ℓ_2 denoising loss $\mathbf{E}[||D_\sigma(x+\nu_\sigma)-x)||^2]$. When parametrized using a DRUNet architecture [28], it was shown in [11] that the Gradient Step Denoiser (3), despite being constrained to be a conservative vector field (as in [19]), achieves state-of-the-art denoising performance.

2.2 Proximal Denoiser

We propose here a new version of the result of [12] on the characterization of the gradient-step denoiser as a proximal operator of some potential ϕ. In particular, we state a new result regarding the weak convexity of the ϕ function. The proof of this result relies on the results from [8].

Proposition 1 (Proximal denoisers). *Let $g_\sigma : \mathbb{R}^n \to \mathbb{R}$ a $\mathcal{C}^2$ function with ∇g_σ L_{g_σ}-Lipschitz with $L_{g_\sigma} < 1$. Then, for $D_\sigma := \mathrm{Id} - \nabla g_\sigma$, there exists a potential $\phi_\sigma : \mathbb{R}^n \to \mathbb{R} \cup \{+\infty\}$ such that $\mathrm{Prox}_{\phi_\sigma}$ is one-to-one and*

$$D_\sigma = \mathrm{Prox}_{\phi_\sigma} \tag{4}$$

Moreover, ϕ_σ is $\frac{L_{g_\sigma}}{L_{g_\sigma}+1}$-weakly convex and it can be written $\phi_\sigma = \hat{\phi}_\sigma + K$ on $\mathrm{Im}(D_\sigma)$ (which is open) for some constant $K \in \mathbb{R}$, with $\hat{\phi}_\sigma : \mathcal{X} \to \mathbb{R} \cup \{+\infty\}$ defined by

$$\hat{\phi}_\sigma(x) := \begin{cases} g_\sigma(D_\sigma{}^{-1}(x))) - \frac{1}{2}\left|\left|D_\sigma{}^{-1}(x) - x\right|\right|^2 & \text{if } x \in \mathrm{Im}(D_\sigma), \\ +\infty & \text{otherwise.} \end{cases} \tag{5}$$

Additionally $\hat{\phi}_\sigma$ verifies $\forall x \in \mathbb{R}^n$, $\hat{\phi}_\sigma(x) \geq g_\sigma(x)$.

To get a proximal denoiser from the denoiser (3), the gradient of the learned potential g_σ must be contractive. In [12] the Lipschitz constant of ∇g_σ is softly constrained to satisfy $L_{g_\sigma} < 1$, by penalizing the spectral norm $\left|\left|\nabla^2 g_\sigma(x+\nu_\sigma)\right|\right|_S$ in the denoiser training loss.

2.3 Relaxed Denoiser

Once trained, the Gradient Step Denoiser $D_\sigma = \text{Id} - \nabla g_\sigma$ can be relaxed as in [11] with a parameter $\gamma \in [0, 1]$

$$D_\sigma^\gamma = \gamma D_\sigma + (1 - \gamma)\,\text{Id} = \text{Id} - \gamma \nabla g_\sigma. \tag{6}$$

Applying Proposition 1 with $g_\sigma^\gamma = \gamma g_\sigma$ which has a γL_{g_σ}-Lipschitz gradient, we get that if $\gamma L_g < 1$, there exists a $\frac{\gamma L_{g_\sigma}}{\gamma L_{g_\sigma}+1}$-weakly convex ϕ_σ^γ such that

$$D_\sigma^\gamma = \text{Prox}_{\phi_\sigma^\gamma}, \tag{7}$$

satisfying $\phi_\sigma^0 = 0$ and $\phi_\sigma^1 = \phi_\sigma$. Hence, one can control the weak convexity of the regularization function by relaxing the proximal denoising operator D_σ^γ.

3 Plug-and-Play Proximal Gradient Descent (PnP-PGD)

In this section, we give convergence results for the Prox-PnP-PGD algorithm, $x_{k+1} = D_\sigma \circ (\text{Id} - \lambda f)(x_k) = \text{Prox}_{\phi_\sigma} \circ (\text{Id} - \lambda f)(x_k)$, which is the PnP version of PGD, with plugged Proximal Denoiser (4). The authors of [12] proposed a suboptimal convergence result as the semiconvexity of ϕ_σ was not exploited. We here improve the condition on the regularization parameter λ for convergence.

We present properties of smooth functions and weakly convex functions in Sect. 3.1. Then we show in Sect. 3.2 the convergence of PGD in the smooth/weakly convex setting. We finally apply this result to PnP in Sect. 3.3.

3.1 Useful Inequalities

We present two results relative to weakly convex functions and smooth ones. We use the subdifferential of a proper, nonconvex function ϕ defined as $\partial\phi(x) = \left\{ v \in \mathbb{R}^n, \exists (x_k), \phi(x_k) \to \phi(x), v_k \to v, \underline{\lim}_{z \to x_k} \frac{\phi(z) - \phi(x_k) - \langle v_k, z - x_k \rangle}{\|z - x_k\|} \geq 0 \ \forall k \right\}$.

Proposition 2 (Properties of weakly convex functions). *For ϕ proper lsc and M-weakly convex with $M > 0$,*

(i) $\forall x, y$ *and* $t \in [0, 1]$,

$$\phi(tx + (1-t)y) \leq t\phi(x) + (1-t)\phi(y) + \frac{M}{2} t(1-t) \|x - y\|^2; \tag{8}$$

(ii) $\forall x, y,$ *we have* $\forall z \in \partial\phi(y)$,

$$\phi(x) \geq \phi(y) + \langle z, x - y \rangle - \frac{M}{2} \|x - y\|^2; \tag{9}$$

(iii) ***Three-points inequality.*** *For $z^+ \in \operatorname{Prox}_\phi(z)$, we have, $\forall x$*

$$\phi(x) + \frac{1}{2}\|x - z\|^2 \geq \phi(z^+) + \frac{1}{2}\|z^+ - z\|^2 + \frac{1-M}{2}\|x - z^+\|^2. \quad (10)$$

Lemma 1 (Descent Lemma for smooth functions). *For f proper differentiable and with a L_f-Lipschitz gradient, we have $\forall x, y$*

$$f(x) \leq f(y) + \langle \nabla f(y), x - y\rangle + \frac{L_f}{2}\|x - y\|^2. \quad (11)$$

3.2 Proximal Gradient Descent with a Weakly Convex Function

We consider the following minimization problem for a smooth nonconvex function f and a weakly convex function ϕ that are both bounded from below:

$$\min_x F(x) := \lambda f(x) + \phi(x). \quad (12)$$

We now show under which conditions the classical Proximal Gradient Descent

$$x_{k+1} \in \operatorname{Prox}_{\tau\phi} \circ (\operatorname{Id} - \tau\lambda\nabla f)(x_k) \quad (13)$$

converges to a stationary point of (12). We first show convergence of function values, and then convergence of the iterates, if F verifies the Kurdyka-Lojasiewicz (KL) property [2]. Large classes of functions, in particular all the proper, closed, semi-algebraic functions [1] satisfy this property, which is, in practice, the case of all the functions considered in this analysis.

Theorem 1 (Convergence of PGD algorithm (13)**).** *Assume f and ϕ proper lsc, bounded from below with f differentiable with L_f-Lipschitz gradient, and ϕ M-weakly convex. Then for $\tau < 2/(\lambda L_f + M)$, the iterates* (13) *verify*

(i) *$(F(x_k))$ monotonically decreases and converges.*
(ii) *$\|x_{k+1} - x_k\|$ converges to 0 at rate $\min_{k \leq K}\|x_{k+1} - x_k\| = \mathcal{O}(1/\sqrt{K})$*
(iii) *All cluster points of the sequence x_k are stationary points of F.*
(iv) *If the sequence x_k is bounded and if F verifies the KL property at the cluster points of x_k, then x_k converges, with finite length, to a stationary point of F.*

The proof follows standard arguments of the convergence analysis of the PGD in the nonconvex setting [2,3,17]. We only demonstrate here the first point of the theorem.

Proof (i). Relation (13) leads to $\frac{x_k - x_{k+1}}{\tau} - \lambda\nabla f(x_k) \in \partial\phi(x_{k+1})$, by definition of the proximal operator. As ϕ is M-weakly convex, Proposition 2 (ii) leads to

$$\phi(x_k) \geq \phi(x_{k+1}) + \frac{\|x_k - x_{k+1}\|^2}{\tau} + \lambda\langle \nabla f(x_k), x_{k+1} - x_k\rangle - \frac{M}{2}\|x_k - x_{k+1}\|^2. \quad (14)$$

The descent Lemma 1 gives for f:

$$f(x_{k+1}) \leq f(x_k) + \langle \nabla f(x_k), x_{k+1} - x_k \rangle + \frac{L_f}{2} ||x_k - x_{k+1}||^2 . \tag{15}$$

Combining both inequalities, for $F_{\lambda,\sigma} = \lambda f + \phi_\sigma$, we obtain

$$F(x_k) \geq F(x_{k+1}) + \left(\frac{1}{\tau} - \frac{M + \lambda L_f}{2} \right) ||x_k - x_{k+1}||^2 . \tag{16}$$

Therefore, if $\tau < 2/(M + \lambda L_f)$, $(F(x_k))$ is monotically deacreasing. As F is assumed lower-bounded, $(F(x_k))$ converges. □

3.3 Prox-PnP Proximal Gradient Descent (Prox-PnP-PGD)

Equipped with the convergence of PGD, we can now study the convergence of *Prox-PnP-PGD*, the PnP-PGD algorithm with plugged Proximal Denoiser (4):

$$x_{k+1} = D_\sigma(\mathrm{Id} - \lambda f)(x_k) = \mathrm{Prox}_{\phi_\sigma}(\mathrm{Id} - \lambda f)(x_k). \tag{17}$$

This algorithm targets stationary points of the functional $F_{\lambda,\sigma}$ defined as:

$$F_{\lambda,\sigma} := \lambda f + \phi_\sigma. \tag{18}$$

The following result, obtained from Theorem 1, improves [12] using the fact that the potential ϕ_σ is not any nonconvex function but a weakly convex one.

Corollary 1 (Convergence of Prox-PnP-PGD (17)**).** *Let $g_\sigma : \mathbb{R}^n \to \mathbb{R} \cup \{+\infty\}$ of class C^2, coercive, with L_{g_σ}-Lipschitz gradient, $L_{g_\sigma} < 1$, and $D_\sigma := \mathrm{Id} - \nabla g_\sigma$. Let ϕ_σ be defined from g_σ as in Proposition 1. Let $f : \mathbb{R}^n \to \mathbb{R} \cup \{+\infty\}$ differentiable with L_f-Lipschitz gradient. Assume f and D_σ respectively KL and semi-algebraic, and f and g_σ bounded from below. Then, for $\lambda L_f < (L_{g_\sigma} + 2)/(L_{g_\sigma} + 1)$, the iterates x_k given by the iterative scheme* (17) *verify the convergence properties (i)-(iv) of Theorem 1 for $F = F_{\lambda,\sigma}$.*

The proof of this result is a direct application of Theorem 1 using $\tau = 1$ and the fact that ϕ_σ defined in Proposition 1 is $M = L_{g_\sigma}/(L_{g_\sigma} + 1)$-weakly convex.

By exploiting the weak convexity of ϕ_σ, the convergence condition $\lambda L_f < 1$ of [12] is here replaced by $\lambda L_f < \frac{L_{g_\sigma}+2}{L_{g_\sigma}+1}$. Even if the bound is improved, we are still limited to regularization parameters satisfying $\lambda L_f < 2$. In the next section, we propose a modification of the PGD algorithm to relax this constraint.

4 PnP Relaxed Proximal Gradient Descent (PnP-αPGD)

In this section, we study the convergence of a relaxed PGD algorithm applied to problems (1) involving a smooth convex function f and a weakly convex function ϕ. Our objective is to design a convergent algorithm in which the proximal operator is only computable for $\tau = 1$ and the data-fidelity term constraint is less restrictive than the bound $\tau < 2/(M + \lambda L_f)$ of Theorem 1.

4.1 αPGD Algorithm

We present our main result which concerns, for weakly convex functions ϕ, the convergence of the following α-relaxed PGD algorithm, defined for $0 < \alpha < 1$ as

$$\begin{cases} q_{k+1} = (1-\alpha)y_k + \alpha x_k & \text{(19a)} \\ x_{k+1} = \text{Prox}_{\tau\phi}(x_k - \tau\lambda\nabla f(q_{k+1})) & \text{(19b)} \\ y_{k+1} = (1-\alpha)y_k + \alpha x_{k+1}. & \text{(19c)} \end{cases}$$

Algorithm (19) with $\alpha = 1$ exactly corresponds to the PGD algorithm (13). This scheme is reminiscent of [25] (taking $\alpha = \theta_k$ and $\tau = \frac{1}{\theta_k L_f}$ in Algorithm 1 of [25]), which generalizes Nesterov-like accelerated proximal gradient methods [3,16]. See also [7] for a variant with line-search. As shown in [13], there is a strong connection between the proposed algorithm (19) and the Primal-Dual algorithm [4] with Bregman proximal operator [5]. In the *convex* setting, one can show that ergodic convergence is obtained with $\tau\lambda L_f > 2$ and small values α. Convergence of a similar algorithm is also shown in [15] for a M-semi convex ϕ and a $c > M$-strongly convex f. However, ϕ_σ is here nonconvex while f is only convex, so that a new convergence result needs to be derived.

Theorem 2 (Convergence of αPGD (19)**).** *Assume f and ϕ proper lsc, bounded from below, f convex differentiable with L_f-Lipschitz gradient and ϕ M-weakly convex. Then*[2] *for $\alpha \in (0,1)$ and $\tau < \min\left(\frac{1}{\alpha\lambda L_f}, \frac{\alpha}{M}\right)$, the updates (19) verify*

(i) $F(y_k) + \frac{\alpha}{2\tau}\left(1 - \frac{1}{\alpha}\right)^2 ||y_k - y_{k-1}||^2$ *monotonically decreases and converges.*
(ii) $||y_{k+1} - y_k||$ *converges to 0 at rate* $\min_{k \leq K} ||y_{k+1} - y_k|| = \mathcal{O}(1/\sqrt{K})$
(iii) All cluster points of the sequence y_k are stationary points of F.

The proof relies on Lemma 1 and Proposition 2. It also requires the convexity of f.

With this theorem, αPGD is shown to verify convergence of the iterates and of the norm of the residual to 0. Note that we do not have here the analog of Theorem 1(iv) on the iterates' convergence using the KL hypothesis. Indeed, the nonconvex convergence analysis with KL functions from [2] or [17] do not extend to our case.

When $\alpha = 1$, Algorithms (19) and (13) are equivalent, but we get a slightly worse bound in Theorem 2 than in Theorem 1 ($\tau < \min\left(\frac{1}{\lambda L_f}, \frac{1}{M}\right) \leq \frac{2}{\lambda L_f + M}$). Nevertheless, when used with $\alpha < 1$, we next show that the relaxed algorithm is more relevant in the perspective of PnP with proximal denoiser.

[2] As shown in the proof, a better bound can be found, but with little numerical gain.

4.2 Prox-PnP-αPGD Algorithm

We can now study the Prox-PnP-αPGD algorithm obtained by taking the proximal denoiser (4) in the αPGD algorithm (19)

$$\begin{cases} q_{k+1} = (1-\alpha)y_k + \alpha x_k & \text{(20a)} \\ x_{k+1} = D_\sigma(x_k - \lambda \nabla f(q_{k+1})) & \text{(20b)} \\ y_{k+1} = (1-\alpha)y_k + \alpha x_{k+1} & \text{(20c)} \end{cases}$$

This scheme targets the minimization of the functional $F_{\lambda,\sigma}$ given in (18).

Corollary 2 (Convergence of Prox-PnP-αPGD (20)**).** *Let $g_\sigma : \mathbb{R}^n \to \mathbb{R} \cup \{+\infty\}$ of class $\mathcal{C}^2$, coercive, with $L_g < 1$-Lipschitz gradient and $D_\sigma := \mathrm{Id} - \nabla g_\sigma$. Let ϕ_σ be the $M = L_{g_\sigma}/(L_{g_\sigma}+1)$-weakly convex function defined from g_σ as in Proposition 1. Let f be proper, convex and differentiable with L_f-Lipschitz gradient. Assume f and g_σ bounded from below. Then, if $\lambda L_f M < 1$ and for any $\alpha \in [0,1]$*

$$M < \alpha < 1/(\lambda L_f) \tag{21}$$

the iterates x_k given by the iterative scheme (20) verify the convergence properties (i)–(iii) of Theorem 2 for $F = F_{\lambda,\sigma}$ defined in (18).

This PnP corollary is obtained by taking $\tau = 1$ in Theorem 2 and using the $M = (L_{g_\sigma})/(L_{g_\sigma}+1)$-weakly convex potential ϕ_σ defined in Proposition 1.

The existence of $\alpha \in [0,1]$ satisfying relation (21) is ensured as soon as $\lambda L_f M < 1$. As a consequence, when M gets small (*i.e* ϕ_σ gets "more convex") λL_f can get arbitrarily large. This is a major advance compared to Prox-PnP-PGD that was limited (Corollary 1) to $\lambda L_f < 2$ even for convex ϕ ($M = 0$). To further exploit this property, we now consider the relaxed denoiser D_σ^γ (6) that is associated to a function ϕ_σ^γ with a tunable weak convexity constant M^γ.

Corollary 3 (Convergence of Prox-PnP-αPGD with relaxed denoiser). *Let $F^\gamma_{\lambda,\sigma} := \lambda f + \phi_\sigma^\gamma$, with the $M^\gamma = \frac{\gamma L_g}{\gamma L_g + 1}$-weakly convex potential ϕ_σ^γ introduced in* (7) *and $L_g < 1$. Then, for $M^\gamma < \alpha < 1/(\lambda L_f)$, the iterates x_k given by the Prox-PnP-αPGD (20) with γ-relaxed denoiser D_σ^γ defined in* (6) *verify the convergence properties (i)–(iii) of Theorem 2 for $F = F^\gamma_{\lambda,\sigma}$.*

Therefore, using the γ-relaxed denoiser $D_\sigma^\gamma = \gamma D_\sigma + (1-\gamma)\,\mathrm{Id}$, the overall convergence condition on λ is now $\lambda < \frac{1}{L_f}\frac{\gamma L_g}{\gamma L_g + 1}$.

Provided γ gets small, λ can be arbitrarily large. Small γ means small amount of regularization brought by denoising at each step of the PnP algorithm. Moreover, for small γ, the targeted regularization function ϕ_σ^γ gets close to a convex function and it has already been observed that deep convex regularization can be sub-optimal compared to more flexible nonconvex ones [6]. Depending on the inverse problem, and on the necessary amount of regularization, the choice of the couple (γ, λ) will be of paramount importance for efficient restoration.

5 Experiments

The efficiency of the proposed Prox-PnP-αPGD algorithm (20) is now demonstrated on deblurring and super-resolution. For both applications, we consider a degraded observation $y = Ax^* + \nu \in \mathbb{R}^m$ of a clean image $x^* \in \mathbb{R}^n$ that is estimated by solving problem (1) with $f(x) = \frac{1}{2}\left|\left|Ax - y\right|\right|^2$. Its gradient $\nabla f = A^T(Ax-y)$ is thus Lipschitz with constant $L_f = \left|\left|A^T A\right|\right|_S$. We use for evaluation and comparison the 68 images from the CBSD68 dataset, center-cropped to $n = 256 \times 256$ and Gaussian noise with 3 noise levels $\nu \in \{0.01, 0.03, 0.05\}$.

For *deblurring*, the degradation operator $A = H$ is a convolution performed with circular boundary conditions. As in [11,18,28,29], we consider the 8 real-world camera shake kernels of [14], the 9×9 uniform kernel and the 25×25 Gaussian kernel with standard deviation 1.6.

For single image *super-resolution* (SR), the low-resolution image $y \in \mathbb{R}^m$ is obtained from the high-resolution one $x \in \mathbb{R}^n$ via $y = SHx + \nu$ where $H \in \mathbb{R}^{n \times n}$ is the convolution with anti-aliasing kernel. The matrix S is the standard s-fold downsampling matrix of size $m \times n$ and $n = s^2 \times m$. As in [28], we evaluate SR performance on 4 isotropic Gaussian blur kernels with standard deviations 0.7, 1.2, 1.6 and 2.0; and consider downsampled images at scale $s = 2$ and $s = 3$.

The proximal denoiser D_σ defined in Proposition 1 is trained following [12] with $L_g < 1$. For both Prox-PnP-PGD and Prox-PnP-αPGD algorithm, we use the γ-relaxed version of the denoiser (6). All the hypotheses on f and g_σ from Corollaries 1 and 3 are thus verified and convergence of Prox-PnP-PGD and Prox-PnP-αPGD are theoretically guaranteed provided the corresponding conditions on λ are satisfied. Hyper-parameters $\gamma \in [0,1]$, λ and σ are optimized via grid-search. In practice, we found that the same choice of parameters γ and σ are optimal for both PGD and αPGD, with values depending on the amount of noise ν in the input image. We thus choose $\lambda \in [0, \lambda_{lim}]$ where for Prox-PnP-PGD $\lambda_{lim}^{\text{PGD}} = \frac{1}{L_f}\frac{\gamma+2}{\gamma+1}$ and for Prox-PnP-αPGD $\lambda_{lim}^{\alpha\text{PGD}} = \frac{1}{L_f}\frac{\gamma+1}{\gamma} \geq \lambda_{lim}^{\text{PGD}}$. For both $\nu = 0.01$ and $\nu = 0.03$, λ is set to its maximal allowed value λ_{lim}. Prox-PnP-αPGD is expected to outperform Prox-PnP-PGD at these noise levels. Finally, for Prox-PnP-αPGD, α is set to its maximum possible value $1/(\lambda L_f)$.

We numerically compare in Table 1 the presented methods Prox-PnP-PGD (that improves [12]) and Prox-PnP-αPGD against three state-of-the-art deep PnP approaches: IRCNN [29], DPIR [28], and GS-PnP [11]. Among them, only GS-PnP has convergence guarantees. Both IRCNN and DPIR use PnP-HQS, the PnP version of the Half-Quadratic Splitting algorithm, with well-chosen varying stepsizes. GS-PnP uses the gradient-step denoiser (3) in PnP-HQS.

As expected, by allowing larger values for λ, we observe that Prox-PnP-αPGD outperforms Prox-PnP-PGD in PSNR at low noise level $\nu \in \{0.01, 0.03\}$. The performance gap is significant for deblurring and super-resolution with scale 2 and $\nu = 0.01$, in which case only a low amount of regularization is necessary, that is to say a large λ value. In these conditions, Prox-PnP-αPGD almost closes

Table 1. PSNR (dB) results on CBSD68 for deblurring (left) and super-resolution (right). PSNR are averaged over 10 blur kernels for deblurring (left) and 4 blur kernels along various scales s for super-resolution (right).

Noise level ν	Deblurring			Super-resolution					
	0.01	0.03	0.05	scale $s = 2$			scale $s = 3$		
				0.01	0.03	0.05	0.01	0.03	0.05
IRCNN [29]	31.42	28.01	26.40	26.97	25.86	25.45	25.60	24.72	24.38
DPIR [28]	**31.93**	**28.30**	26.82	27.79	26.58	25.83	**26.05**	25.27	24.66
GS-PnP [11]	31.70	28.28	**26.86**	27.88	**26.81**	**26.01**	25.97	**25.35**	**24.74**
Prox-PnP-PGD	30.91	27.97	26.66	27.68	26.57	25.81	25.94	25.20	24.62
Prox-PnP-αPGD	31.55	28.03	26.66	**27.92**	26.61	25.80	26.03	25.26	24.61

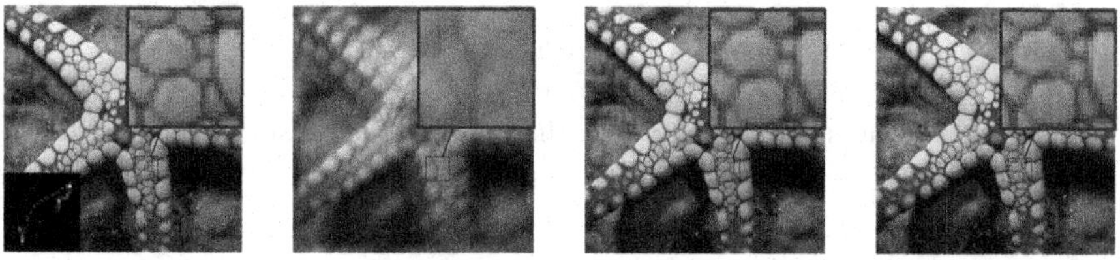

(a) Clean (b) Observed (c) Prox-PnP-PGD (33.32dB) (d) Prox-PnP-αPGD (33.62dB)

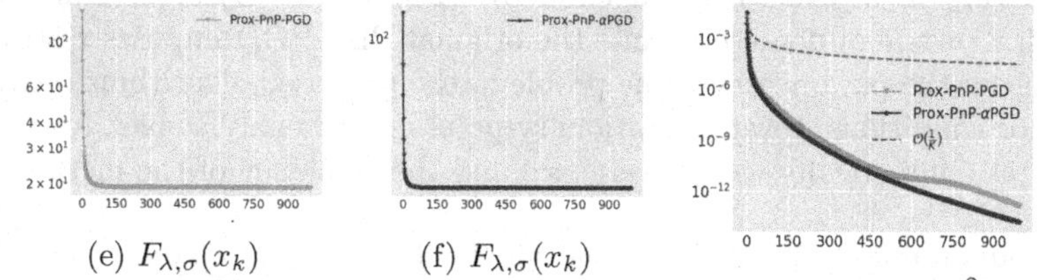

(e) $F_{\lambda,\sigma}(x_k)$ Prox-PnP-PGD (f) $F_{\lambda,\sigma}(x_k)$ Prox-PnP-αPGD (g) $||x_{i+1} - x_i||^2$

Fig. 1. Deblurring of "starfish" blurred with the shown kernel and noise $\nu = 0.01$.

the PSNR gap between Prox-PnP-PGD and the state-of-the-art PnP methods DPIR and GS-PnP that do not have any restriction on the choice of λ. We assume that the remaining difference of performance is due to the γ-relaxation of the denoising operation that affects the regularizer ϕ_γ. We qualitatively verify in Fig. 1 (deblurring) and Fig. 2 (super-resolution), on "starfish" and "leaves" images, the performance gain of Prox-PnP-αPGD over Prox-PnP-PGD. We also plot the evolution of $F_{\lambda,\sigma}(x_k)$ and $||x_{k+1} - x_k||^2$ to empirically validate the convergence of both algorithms.

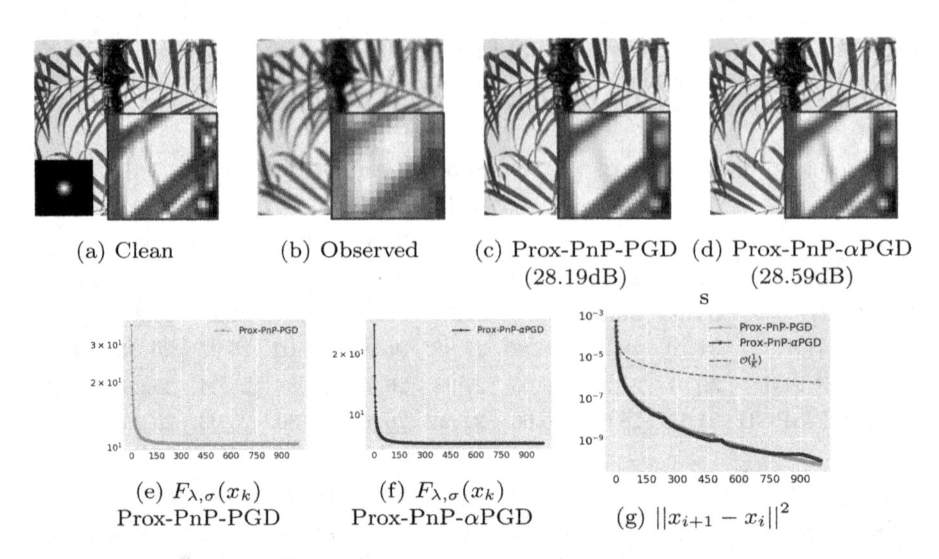

(a) Clean (b) Observed (c) Prox-PnP-PGD (28.19dB) (d) Prox-PnP-αPGD (28.59dB)

(e) $F_{\lambda,\sigma}(x_k)$ Prox-PnP-PGD (f) $F_{\lambda,\sigma}(x_k)$ Prox-PnP-αPGD (g) $||x_{i+1} - x_i||^2$

Fig. 2. SR of "leaves" downscaled with the shown kernel, scale 2 and $\nu = 0.01$.

6 Conclusion

In this paper, we propose a new convergent plug-and-play algorithm built from a relaxed version of the Proximal Gradient Descent (PGD) algorithm. When used with a proximal denoiser, while the original PnP-PGD imposes restrictive conditions on the parameters of the problem, the proposed algorithms converge, with minor conditions, towards stationary points of a weakly convex functional. We illustrate numerically the convergence and the efficiency of the method.

Acknowledgements. This work was partially funded through a CDSN grant of ENS Paris-Saclay. This study has also been carried out with financial support from the French Research Agency through the PostProdLEAP and Mistic projects (ANR-19-CE23-0027-01 and ANR-19-CE40-005). A.C. thanks Juan Pablo Contreras for many interesting discussions about nonlinear accelerated descent algorithms.

References

1. Attouch, H., Bolte, J., Redont, P., Soubeyran, A.: Proximal alternating minimization and projection methods for nonconvex problems: an approach based on the kurdyka-łojasiewicz inequality. Math. Oper. Res. **35**(2), 438–457 (2010)
2. Attouch, H., Bolte, J., Svaiter, B.F.: Convergence of descent methods for semi-algebraic and tame problems: proximal algorithms, forward-backward splitting, and regularized gauss-seidel methods. Math. Program. **137**(1–2), 91–129 (2013)
3. Beck, A., Teboulle, M.: Fast gradient-based algorithms for constrained total variation image denoising and deblurring problems. IEEE Trans. Image Process. **18**(11), 2419–2434 (2009)
4. Chambolle, A., Pock, T.: A first-order primal-dual algorithm for convex problems with applications to imaging. J. Math. Imaging Vis. **40**(1), 120–145 (2011)

5. Chambolle, A., Pock, T.: On the ergodic convergence rates of a first-order primal-dual algorithm. Math. Program. **159**(1), 253–287 (2016)
6. Cohen, R., Blau, Y., Freedman, D., Rivlin, E.: It has potential: gradient-driven denoisers for convergent solutions to inverse problems. In: Neural Information Processing Systems, vol. 34 (2021)
7. Dvurechensky, P., Gasnikov, A., Kroshnin, A.: Computational optimal transport: complexity by accelerated gradient descent is better than by Sinkhorn's algorithm. In: Proceedings of the 35th International Conference on Machine Learning, vol. 80, pp. 1367–1376. PMLR (2018)
8. Gribonval, R., Nikolova, M.: A characterization of proximity operators. J. Math. Imaging Vis. **62**(6), 773–789 (2020)
9. Hertrich, J., Neumayer, S., Steidl, G.: Convolutional proximal neural networks and plug-and-play algorithms. Linear Algebra Appl. **631**, 203–234 (2021)
10. Hurault, S., Chambolle, A., Leclaire, A., Papadakis, N.: A relaxed proximal gradient descent algorithm for convergent plug-and-play with proximal denoiser. arXiv preprint arXiv:2301.13731 (2023)
11. Hurault, S., Leclaire, A., Papadakis, N.: Gradient step denoiser for convergent plug-and-play. In: International Conference on Learning Representations (2022)
12. Hurault, S., Leclaire, A., Papadakis, N.: Proximal denoiser for convergent plug-and-play optimization with nonconvex regularization. In: International Conference on Machine Learning (2022)
13. Lan, G., Zhou, Y.: An optimal randomized incremental gradient method. Math. Program. **171**(1), 167–215 (2018)
14. Levin, A., Weiss, Y., Durand, F., Freeman, W.T.: Understanding and evaluating blind deconvolution algorithms. In: IEEE/CVF Conference on Computer Vision and Pattern Recognition, pp. 1964–1971 (2009)
15. Möllenhoff, T., Strekalovskiy, E., Moeller, M., Cremers, D.: The primal-dual hybrid gradient method for semiconvex splittings. SIAM J. Imaging Sci. **8**(2), 827–857 (2015)
16. Nesterov, Y.: Gradient methods for minimizing composite functions. Math. Program. **140**(1), 125–161 (2013)
17. Ochs, P., Chen, Y., Brox, T., Pock, T.: iPiano: inertial proximal algorithm for nonconvex optimization. SIAM J. Imaging Sci. **7**(2), 1388–1419 (2014)
18. Pesquet, J.C., Repetti, A., Terris, M., Wiaux, Y.: Learning maximally monotone operators for image recovery. SIAM J. Imaging Sci. **14**(3), 1206–1237 (2021)
19. Romano, Y., Elad, M., Milanfar, P.: The little engine that could: regularization by denoising (red). SIAM J. Imaging Sci. **10**(4), 1804–1844 (2017)
20. Rudin, L.I., Osher, S., Fatemi, E.: Nonlinear total variation based noise removal algorithms. Phys. D **60**, 259–268 (1992)
21. Ryu, E., Liu, J., Wang, S., Chen, X., Wang, Z., Yin, W.: Plug-and-play methods provably converge with properly trained denoisers. In: International Conference on Machine Learning, pp. 5546–5557 (2019)
22. Sun, Y., Wu, Z., Xu, X., Wohlberg, B., Kamilov, U.S.: Scalable plug-and-play ADMM with convergence guarantees. IEEE Trans. Comput. Imaging **7**, 849–863 (2021)
23. Terris, M., Repetti, A., Pesquet, J.C., Wiaux, Y.: Building firmly nonexpansive convolutional neural networks. In: IEEE International Conference on Acoustics, Speech, and Signal Processing, pp. 8658–8662 (2020)
24. Themelis, A., Patrinos, P.: Douglas-rachford splitting and ADMM for nonconvex optimization: tight convergence results. SIAM J. Optim. **30**(1), 149–181 (2020)

25. Tseng, P.: On accelerated proximal gradient methods for convex-concave optimization. SIAM J. Optim. **2**(3) (2008)
26. Venkatakrishnan, S.V., Bouman, C.A., Wohlberg, B.: Plug-and-play priors for model based reconstruction. In: IEEE Global Conference on Signal and Information Processing, pp. 945–948 (2013)
27. Wen, F., Chu, L., Liu, P., Qiu, R.C.: A survey on nonconvex regularization-based sparse and low-rank recovery in signal processing, statistics, and machine learning. IEEE Access **6**, 69883–69906 (2018)
28. Zhang, K., Li, Y., Zuo, W., Zhang, L., Van Gool, L., Timofte, R.: Plug-and-play image restoration with deep denoiser prior. IEEE Trans. Image Process. (2021)
29. Zhang, K., Zuo, W., Gu, S., Zhang, L.: Learning deep CNN denoiser prior for image restoration. In: IEEE/CVF Conference on Computer Vision and Pattern Recognition, pp. 3929–3938 (2017)

Off-the-Grid Charge Algorithm for Curve Reconstruction in Inverse Problems

Bastien Laville[1(✉)], Laure Blanc-Féraud[1], and Gilles Aubert[2]

[1] Morpheme Project: Université Côte d'Azur, Inria, CNRS, Nice, France
bastien.laville@inria.fr, laure.blanc_feraud@i3s.unice.fr
[2] Université Côte d'Azur, CNRS, LJAD, Nice, France
gilles.aubert@univ-cotedazur.fr

Abstract. Several numerical algorithms have been developed in the literature and employed for curves reconstruction. However, these techniques are developed within the discrete setting, namely the super-resolved image is defined on a finer grid than the observed images. Conversely, off-the-grid (or gridless) optimisation does not rely on a fine grid and offer a tractable theoretical and numerical framework. In this work, we present a gridless method accounting for the reconstruction of both open and closed curves, based on the latest theoretical development in off-the-grid curve reconstruction.

Keywords: Off-the-grid variational method · Inverse problem · Frank-Wolfe algorithm · Curve detection

1 Introduction

This work focuses on the numerical optimisation of the functional CROC designed for inverse problems in order to recover curves in an off-the-grid fashion, by considering the space of vector Radon measures with finite divergence.

Off-the-grid methods is a rather new field of research, introduced a decade ago to overcome some limitations of so-called discrete methods. Indeed, the source estimation problem amounts to recover the source with support lying in some set $\mathcal{X}$, thanks to an altered acquisition on a coarse grid: blurred, noisy, low-passed, *etc.* In a rather classical discrete framework, the recovered source lies on a refined grid *i.e.* it is a matrix. On the contrary, *off-the-grid* (or *gridless*) methods do not rely on a grid: for instance a spike source position is continuously estimated, and cannot be bound to a pixel in a fine grid, thus not bringing any discretisation error. The source is then encoded in a *measure*, lying in a

The work of Bastien Laville has been supported by the French government, through the UCA DS4H Investments in the Future project managed by the National Research Agency (ANR) with the reference number ANR-17-EURE-0004. The work of Laure Blanc-Féraud has been supported by the French government, through the 3IA Côte d'Azur Investments in the Future project managed by the National Research Agency (ANR) with the reference number ANR-19-P3IA-0002.

L. Calatroni et al. (Eds.): SSVM 2023, LNCS 14009, pp. 393–405, 2023.
https://doi.org/10.1007/978-3-031-31975-4_30

broader set of functions denoted $\mathscr{M}(\mathcal{X})$. Moreover, these gridless methods have several theoretical guarantees [8] and also add up structural information to the optimisation: geometrical information is then used to recover a certain object, on the contrary to discrete methods where it always yields a matrix. The off-the-grid literature built up around the spike reconstruction, but several other structures such as level sets [5], sinusoid [15], dynamic trajectories [3] *etc.* have likewise been explored.

However, to the best knowledge of the authors, the literature does not handle the off-the-grid curve reconstruction problem, let alone provides a tailored numerical algorithm for this specific task. The point/spike reconstruction problem, *i.e.* measure supported on a 0D set, was thoroughly explored [4,7] and the level set reconstruction problem, *i.e.* measure supported on a 2D set, has well-known theoretical results stemming from the geometrical measure theory and was lately successfully adapted to off-the-grid reconstruction [5]. Still, the numerical reconstruction of a measure supported on a 1D set, and more specifically a curve object, was yet to explore. It is all the more unfortunate as the curve structure naturally arise in many inverse problems, such as the super-resolution in biomedical imaging. In this paper, we propose a new algorithm for curve reconstruction in an off-the-grid fashion, based on the latest theoretical results [14] investigating a new regulariser to yield curve minima.

1.1 Notations

In the following, $\mathcal{X}$ is the ambient space where the positions of the objects (e.g. spikes, curves, *etc.*) live, it is a non-empty bounded open set of $\mathbb{R}^d$, hence a submanifold of dimension $d \in \mathbb{N}^*$. $\mathscr{H}_1$ denotes the 1-dimensional Hausdorff measure (see [9] for a definition).

1.2 Related Works

This paper makes an extensive use of the last theoretical results on the divergence vector field measure space brought by [14], based on some strong results of [12,17]. As we aim to close the gap between spike (sort of 0D, total variation) and set (somehow 2D, bounded variation) reconstruction, we relate to state-of-art Dirac [4,7] and level set [5] Frank-Wolfe algorithm.

2 Optimisation in the Space of Divergence Vector Fields

2.1 The Space of Charges

We give some useful definitions and properties from the off-the-grid literature, the interested reader can take a look at the review [13].

Definition 1 (Evanescent continuous function on $\mathcal{X}$). *We call $\mathscr{C}_0(\mathcal{X}, \mathcal{Y})$ the set of evanescent continuous functions from $\mathcal{X}$ to a normed vector space $\mathcal{Y}$, namely all the continuous map $\psi : \mathcal{X} \to \mathcal{Y}$ such that:*

$$\forall \varepsilon > 0, \exists K \subset \mathcal{X} \text{ compact}, \qquad \sup_{x \in \mathcal{X} \setminus K} \|\psi(x)\|_{\mathcal{Y}} \leq \varepsilon.$$

We write $\mathscr{C}_0(\mathcal{X})$ when $\mathcal{Y} = \mathbb{R}$. We now introduce:

Definition 2 (Set of Radon measures). *We denote by $\mathscr{M}(\mathcal{X})$ the set of real signed Radon measures on $\mathcal{X}$ of finite masses. It is the topological dual of $\mathscr{C}_0(\mathcal{X})$ endowed with* supremum *norm $\|\cdot\|_{\infty,\mathcal{X}}$ by the Riesz-Markov representation theorem [9]. Thus, a Radon measure $m \in \mathscr{M}(\mathcal{X})$ is a continuous linear form on functions $f \in \mathscr{C}_0(\mathcal{X})$, with the duality bracket denoted by $\langle f, m\rangle_{\mathscr{M}(\mathcal{X})} = \int_{\mathcal{X}} f \, \mathrm{d}m$.*

The space of integrable equivalent classes $\mathrm{L}^1(\mathcal{X})$ continuously injects into the Radon measure space $\mathrm{L}^1(\mathcal{X}) \hookrightarrow \mathscr{M}(\mathcal{X})$, measures are then a generalisation of functions to a broader set. An example of Radon measure is the Dirac measure δ_x, where $x \in \mathcal{X}$. Also, since $\mathscr{C}_0(\mathcal{X})$ is a normed vector space, $\mathscr{M}(\mathcal{X})$ is complete [4] if endowed with its dual norm called the total variation (TV) norm, defined for $m \in \mathscr{M}(\mathcal{X})$ by $\|m\|_{\mathrm{TV}}$.

Now, we use the latest developments in the off-the-grid literature concerning the curve reconstruction [14]. Consider the space of *vector* Radon measures:

Definition 3. *We define the set of vector Radon measures $\boldsymbol{\mathscr{M}}(\boldsymbol{\mathcal{X}})^{\mathbf{2}}$ as the topological dual of the space of continuous vector functions $\boldsymbol{\mathscr{C}}_{\mathbf{0}}(\boldsymbol{\mathcal{X}})^{\mathbf{2}} \stackrel{\text{def.}}{=} \boldsymbol{\mathscr{C}}_{\mathbf{0}}(\boldsymbol{\mathcal{X}}, \mathbb{R}^2)$. The properties of the scalar case hold for the vector one, indeed $\boldsymbol{\mathscr{M}}(\boldsymbol{\mathcal{X}})^{\mathbf{2}}$ has a natural TV-norm denoted by $\|\cdot\|_{\mathrm{TV}^2}$, a duality bracket $\langle\cdot,\cdot\rangle_{\mathscr{M}(\boldsymbol{\mathcal{X}})^2}$,* etc.

The following results in this section only hold in dimension $d = 2$, since some tumultuous pathological cases appear when $d > 2$, see [17, Section 1.3]. We denote by div the divergence operator, understood in the distributional sense. Indeed, for all $\boldsymbol{m} \in \boldsymbol{\mathscr{M}}(\boldsymbol{\mathcal{X}})^{\mathbf{2}}$ and $\mathscr{C}_0^\infty(\mathcal{X})$ the space of bump functions:

$$\forall \xi \in \mathscr{C}_0^\infty(\mathcal{X}), \quad \langle \operatorname{div} \boldsymbol{m}, \xi \rangle_{\mathscr{D}'(\mathcal{X}) \times \mathscr{C}_0^\infty(\mathcal{X})} = -\langle \boldsymbol{m}, \boldsymbol{\nabla} \xi \rangle_{\mathscr{M}(\boldsymbol{\mathcal{X}})^2}.$$

A measure $\boldsymbol{m}$ is *of finite divergence* if $\operatorname{div}(\boldsymbol{m}) \in \mathscr{M}(\mathcal{X})$. Let us now introduce the following useful space [16,17].

Definition 4 (Space of charges). *We denote by $\mathscr{V}$ the space of* divergence vector fields *or* charges, *namely the space of vector Radon measures with finite divergence:*

$$\mathscr{V} \stackrel{\text{def.}}{=} \left\{ \boldsymbol{m} \in \boldsymbol{\mathscr{M}}(\boldsymbol{\mathcal{X}})^{\mathbf{2}}, \operatorname{div}(\boldsymbol{m}) \in \mathscr{M}(\mathcal{X}) \right\}.$$

It is a Banach space with respect to the norm $\|\cdot\|_{\mathscr{V}} \stackrel{\text{def.}}{=} \|\cdot\|_{\mathrm{TV}^2} + \|\operatorname{div}(\cdot)\|_{\mathrm{TV}}$.

We define in the following the *curve measure* belonging to $\mathscr{V}$ *i.e.* a measure supported on a curve and defined through integration:

Definition 5 (Curve measure). *Let $\gamma : [0,1] \to \mathcal{X}$ a parametrised Lipschitz curve, we say that $\boldsymbol{\mu}_\gamma \in \mathscr{V}$ is a measure supported on the curve γ if:*

$$\forall \boldsymbol{g} \in \mathscr{C}_0(\mathcal{X})^2, \quad \langle \boldsymbol{\mu}_\gamma, \boldsymbol{g} \rangle_{\mathscr{M}(\mathcal{X})^2} \stackrel{\text{def.}}{=} \int_0^1 \boldsymbol{g}(\gamma(t)) \cdot \dot{\gamma}(t) \, \mathrm{d}t.$$

We denote by $\Gamma \stackrel{\text{def.}}{=} \gamma([0,1])$ the support of the curve.

The bracket w.r.t. to a curve measure is the circulation [17] of a test vector field function along the curve γ. The curve has a finite length since its parametrisation is Lipschitz, hence $\mathscr{H}_1(\Gamma) < +\infty$. Some properties that a curve might exhibit are introduced in the following.

Definition 6 (Several characterisation of curves). *A curve is called* simple *if the restriction of γ on $[0,1)$ is an injective mapping. A curve is* closed *if $\gamma(0) = \gamma(1)$, it is called a* loop *if it is simple and closed.*

Using Sard's theorem for Lipschitz functions, one can prove that $\boldsymbol{\mu}_\gamma$ does not depend on the way the curve is parametrized. Therefore, it is assumed that the curve γ has a constant speed parametrization[1], unless stated otherwise. Finally, we give the expression for the divergence of a curve:

Proposition 1 (Curve divergence). *Let $\boldsymbol{\mu}_\gamma$ be a measure supported on a curve γ, then $\operatorname{div}(\boldsymbol{\mu}_\gamma) = \delta_{\gamma(0)} - \delta_{\gamma(1)}$. In particular, $\operatorname{div}(\boldsymbol{\mu}_\gamma) = 0$ if γ is closed.*

2.2 The CROC Functional and Its Minimiser Structure

Similarly to the scalar case, one can define a variational problem on $\mathscr{V}$. The following functional (CROC) standing for *Curves Represented On Charges* [14] implements the curve reconstruction problem:

$$\underset{\boldsymbol{m} \in \mathscr{V}}{\operatorname{argmin}}\, T_\alpha(\boldsymbol{m}) \stackrel{\text{def.}}{=} \frac{1}{2} \|y - \Phi \boldsymbol{m}\|_{\mathcal{H}}^2 + \alpha \|\boldsymbol{m}\|_{\mathscr{V}}. \tag{CROC}$$

$\Phi : \mathscr{V} \to \mathcal{H}$ is linear and maps the divergence vector field set to the acquisition space $\mathcal{H}$ supposed to be Hilbert, where the data observation is $\boldsymbol{y} \in \mathcal{H}$. This functional exhibits existence of a solution, see [14] for a proof and a discussion on extremality conditions. The regulariser penalises the length of the curve and the number of curves. Now, consider the following:

Definition 7 (Curve measures set). *We denote by $\mathfrak{S}$ the space of curve measures, supported on either open or closed simple ones, endowed with weak-* topology:*

$$\mathfrak{S} \stackrel{\text{def.}}{=} \left\{ \frac{\boldsymbol{\mu}_\gamma}{\|\boldsymbol{\mu}_\gamma\|_{\mathscr{V}}}, \gamma \text{ is a simple oriented Lipschitz curve} \right\}.$$

It is a (non-complete) metric space for the weak- topology.*

[1] $\dot{\gamma}$ is a.e. equal to a constant.

We stress that the curve measures involved do not encode any variation of amplitude along the support, also that the elements of $\mathfrak{S}$ are normalised. This obviously affects the terms of the $\mathscr{V}$-norm: from now on, and to avoid any ambiguity, we denote by $\boldsymbol{\nu}_\gamma \stackrel{\text{def.}}{=} \boldsymbol{\mu}_\gamma / \|\boldsymbol{\mu}_\gamma\|_{\mathscr{V}}$ an element of $\mathfrak{S}$. The following result is a corollary of the main theorem of [14], which establishes that the extreme points of the unit ball of the $\mathscr{V}$-norm are precisely the elements of $\mathfrak{S}$, and the celebrated representer theorem [1,2]. The Hilbert acquisition space is now specifically $\mathcal{H} = \mathcal{H}_n \stackrel{\text{def.}}{=} \mathbb{R}^n$ a finite dimensional space.

Corollary 1 (Minimiser structure). *The problem* (CROC) *admits a minimiser denoted* $\overline{\boldsymbol{u}} \in \mathscr{V}$*:*

$$\overline{\boldsymbol{u}} = \sum_{i=1}^{p} \alpha_i \boldsymbol{u}_i$$

where $p \leq \dim \mathcal{H}_n$*,* $\boldsymbol{u} \in \mathfrak{S}^p$ *and* $\alpha_i > 0$ *for* $0 \leq i \leq p$*, while* $\sum_{i=1}^{p} \alpha_i = T_\alpha(\overline{\boldsymbol{u}})$*.*

The extreme points result of [14] and these latest corollary are the core component of our numerical implementation. Based on the compelling results obtained in [7], we use similarly a greedy algorithm namely the *Frank-Wolfe* algorithm [10], also known as the conditional gradient method. Hopefully, it consists in the iterative reconstruction of the solution with the regulariser atoms, *i.e.* level sets extreme points, yet precisely curve measures here. In the following sections, we present our main contribution, amounting to the numerical optimisation of (CROC) with an instance for a synthetic super-resolution.

3 The Charge (Sliding) Frank-Wolfe for Off-the-Grid Curve Reconstruction

The Frank-Wolfe algorithm performs the minimisation of a convex differentiable function over a weakly compact convex subset of a Banach space. It relies on the iterative minimisation of a linearised version of the objective function, benefiting from the fact that it uses the directional derivatives and that it does not require any Hilbertian structure, contrary to classical proximal algorithms. It has gained significant attention from data scientists as it produces iterates that are a combination of only a few atoms, specific to the chosen regulariser. Similarly to other off-the-grid implementation, this algorithm is not straightforwardly appliable to CROC: T_α is not differentiable, and the optimisation set $\mathscr{V}$ is not bounded. It is thus necessary to perform an *epigraphical lift* [7,11] to reach a differentiable functional that shares the same *minimum* measures as T_α. Our proposed algorithm is given in Algorithm 1.

We precise some notation: we denote $\boldsymbol{\nu}_\gamma{}^{[k]} \stackrel{\text{def.}}{=} (\boldsymbol{\nu}_{\gamma_1}{}^{[k]}, \dots, \boldsymbol{\nu}_{\gamma_{N^{[k]}}}{}^{[k]})$ the vector of reconstructed atoms at the k-th iteration, and $\Phi_{\boldsymbol{\nu}_\gamma{}^{[k]}}(a) \stackrel{\text{def.}}{=} \sum_{i=0}^{k} a_i \Phi \boldsymbol{\nu}_{\gamma_i}{}^{[k]}$

where $a \in \mathbb{R}^{N^{[k]}}$ is the vector of the estimated curves weights, see Algorithm 1. We denote by $\mathcal{S}(\mathcal{X})$ the set of curves spanning $\mathfrak{S}$, namely:

$$\mathcal{S}(\mathcal{X}) \stackrel{\text{def.}}{=} \{\gamma, \gamma \text{ is a simple oriented Lipschitz curve with support in } \mathcal{X}\}.$$

The length of γ is given by $\ell(\gamma) \stackrel{\text{def.}}{=} \mathscr{H}_1(\gamma([0,1]))$ and its $\mathscr{V}$-norm equivalent by $\ell_{\text{div}}(\gamma)$:

$$\ell_{\text{div}}(\gamma) = \begin{cases} \ell(\gamma) + 2 & \text{if } \gamma \text{ open.} \\ \ell(\gamma) & \text{otherwise.} \end{cases}$$

Algorithm 1: Charge Sliding Frank-Wolfe

Input: Acquisition $y \in \mathcal{H}$, number of iterations K, regularisation weight $\alpha > 0$.

1 Initialisation: $m^{[0]} = 0$, $N^{[k]} = 0$.

2 **for** k, $0 \leq k \leq K$ **do**

3 For $m^{[k]} = \sum_{i=1}^{N^{[k]}} a_i^{[k]} \boldsymbol{\nu}_{\gamma_i}{}^{[k]}$ such that $a_i^{[k]} \in \mathbb{R}$, $\boldsymbol{\nu}_{\gamma_i}{}^{[k]} \in \mathfrak{S}$, let

$$\boldsymbol{\eta}^{[k]}(x) \stackrel{\text{def.}}{=} \frac{1}{\alpha} \Phi^*(\Phi m^{[k]} - y).$$

4 Find $\boldsymbol{\gamma_*}^{[k]} \in \mathcal{S}(\mathcal{X})$ such that:

$$\gamma_* \in \underset{\gamma \in \mathcal{S}(\mathcal{X})}{\operatorname{argmax}} \frac{1}{\ell_{\text{div}}(\gamma)} \int_0^1 \boldsymbol{\eta}^{[k]}(\gamma(t)) \cdot \dot{\gamma}(t) \, dt.$$

if $\left|\eta^{[k]}(\gamma_*)\right| \leq 1$ **then**

5 $m^{[k]}$ is the solution of CROC. Stop.

6 **else**

7 Compute $m^{[k+1/2]} = \sum_{i=1}^{N^{[k]}} a_i^{[k+1/2]} \boldsymbol{\nu}_{\gamma_i} + a_{N^{[k]}+1}^{[k+1/2]} \boldsymbol{\nu}_{\gamma_*}{}^{[k+1/2]}$ such that:

$$a_i^{[k+1/2]} \in \underset{a \in \mathbb{R}^{N^{[k]}+1}}{\operatorname{argmin}} T_\alpha \left(\sum_{i=1}^{N^{[k]}} a_i \boldsymbol{\nu}_{\gamma_i}{}^{[k]} + a_{N^{[k]}+1} \boldsymbol{\nu}_{\gamma_*}{}^{[k+1/2]} \right)$$

8 Compute $m^{[k+1]} = \sum_{i=1}^{N^{[k+1]}} a_i^{[k+1]} \boldsymbol{\nu}_{\gamma_i}$, output of the optimisation initialised with $m^{[k+1/2]}$:

$$\left(a^{[k+1]}, \boldsymbol{\nu}_\gamma{}^{[k+1]}\right) \in \underset{(a,\gamma) \in \mathbb{R}^{N^{[k]}+1} \times \mathcal{S}(\mathcal{X})^{N^{[k]}+1}}{\operatorname{argmin}} T_\alpha \left(\sum_{i=1}^{N^{[k]}+1} a_i \boldsymbol{\nu}_{\gamma_i}{}^{[k]} \right).$$

9 Set $m^{[k+1]} = \sum_{i=1}^{N^{[k]}+1} a_i^{[k]} \boldsymbol{\nu}_{\gamma_i}{}^{[k]}$. Prune the low amplitude atoms.

10 **end**

11 **end**

Output: Discrete measure $m^{[k]}$ where k is the stopping iteration.

Our algorithm benefits from the *sliding* improvement, where the classic Frank-Wolfe is improved by the sliding step in line 8–9. Among others, this non-convex step allows a finite time convergence for off-the-grid spikes reconstruction, while a similar argument for off-the-grid level sets and curves is yet to be found, though observed in practice. The following property derived from Frank-Wolfe algorithm properties [6,7] holds:

Proposition 2. *Let* $(m^{[k]})_{k\geq 0}$ *a sequence produced by Algorithm 1. Then it has an accumulation point in weak-$*$ topology, the latter being a solution of* (CROC). *Also, there exists* $C > 0$ *such that for any minimiser* m^* *of* (CROC)*:*

$$\forall k \in \mathbb{N}^*, \quad T_\alpha(m^{[k]}) - T_\alpha(m^*) \leq \frac{C}{k}.$$

In practice, a curve measure $\boldsymbol{\nu}_\gamma$ is discretised by a polygonal curve with integer $n \geq 2$ segments[2]: it is the set of $x \in (\mathcal{X})^{2n}$ such that the list of vertices is simple. Such a choice of approximation is made for the sake of simplicity, in particular for the numerical implementation. A variant with splines or Bézier curves may be interesting to reach higher accuracy with fewer control points. In further works, we will pursue the theoretical (Γ-)convergence of the discrete approximation towards the continuous one. To help the reader to get a grasp on this numerical implementation, we illustrate the Charge Sliding Frank-Wolfe directly on a practical case of super-resolution.

4 A Numerical Illustration for Super-Resolution

The chosen application for these experiments is a super-resolution problem in the context of a Gaussian convolution operator. Let $g \in \mathcal{X}^n$ be the observed image with n pixels. Since our source $\boldsymbol{\nu}_\gamma$ is a *vector* measure, a natural choice of vector quantity for the fidelity term from this image would be the gradient of the image g, then we denote $\boldsymbol{y} = \boldsymbol{\nabla} g$. In further works, we plan to justify more thoroughly this choice. In practice, we exploit a trick enabled by the convolution derivative property. Indeed, consider formally the discrete image source I, then $\boldsymbol{y} = h * \boldsymbol{\nabla} I = \boldsymbol{\nabla} h * I$. Hence, $\Phi\, \boldsymbol{\nu}_\gamma$ will be computed as the support convolved with the gradient of the Gaussian kernel. Then we consider the *vector* forward operator Φ with kernel $\varphi(x)$ for $x \in \mathcal{X}$ writing down:

$$\forall x \in \mathcal{X}, \quad \varphi(x) \stackrel{\text{def.}}{=} \frac{1}{2\pi\sigma} \begin{pmatrix} -x_1\, e^{-(\frac{i-1}{K-1} - x_1)^2/2\sigma^2} \\ -x_2\, e^{-(\frac{i-1}{K-1} - x_2)^2/2\sigma^2} \end{pmatrix}_{1\leq i\leq n}.$$

Obviously, the gradient $\boldsymbol{y}$ ought to be smoothed up, since it is the vector image $\boldsymbol{y} = \boldsymbol{\nabla} g$ used in practice in the algorithm and all the more as the noise in g has even more impact on $\boldsymbol{y}$.

[2] and obviously $n \geq 3$ for closed curves.

In the following, we consider a synthetic example where one wants to recover curves from a classic image acquisition y, altered by a convolution with standard deviation $\sigma = 2 \times 10^{-2}$ and white additive noise with spread $\sigma_{\mathfrak{b}} = 4 \times 10^{-3}$. Let two measures $\boldsymbol{\mu_1}$ an open curve and $\boldsymbol{\mu_2}$ a closed one, both belonging to $\mathfrak{S}$; consider now the source charge $\boldsymbol{T} = 4\boldsymbol{\mu_1} + \boldsymbol{\mu_2}$, the latter 4 is chosen for the sake of visualisation. The image g and its source $\boldsymbol{T}$ are plotted on the Fig. 1. The gradient $\boldsymbol{y}$ of the image is plotted on the Fig. 2.

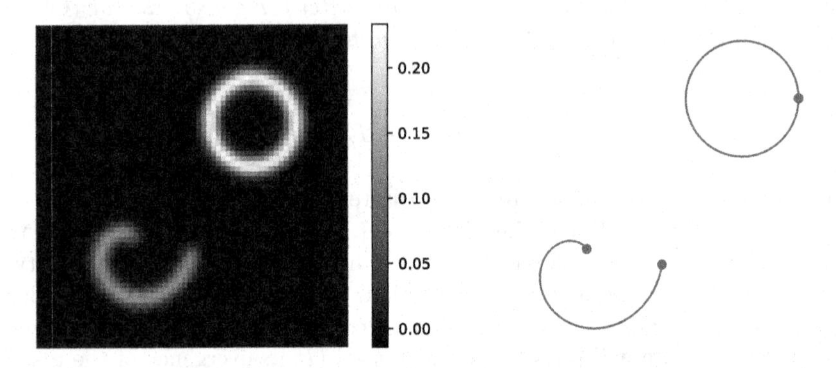

Fig. 1. Left: The observed image g, generated from a charge $\boldsymbol{T}$ composed of a spiral $\boldsymbol{\mu_1}$ and a loop $\boldsymbol{\mu_2}$, which have different intensities. Right: the two curves support. Note the smooth curvature of $\boldsymbol{\mu_1}$ and $\boldsymbol{\mu_2}$ we aim to recover.

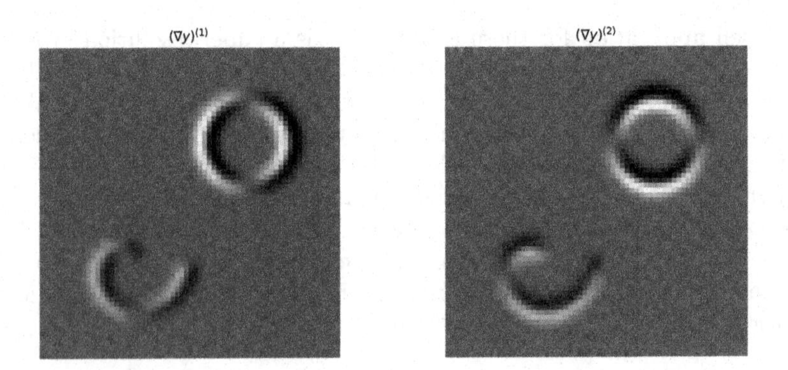

Fig. 2. The two components of the gradient $\boldsymbol{y} = \boldsymbol{\nabla} g$, is the relevant quantity for the fidelity term of (CROC). Note that the noise on the image has an even greater impact on its gradient, such that a gradient denoising strategy must be adopted, especially for experimental images.

As we stated before, the first step of the algorithm lies in the support estimation, this linear step bears an original approach in each off-the-grid regulariser.

The equivalent case for the classic TV-norm consists in a simple grid search of the greatest value pixel of $\eta^{[k]}$, similarly defined as in line 3 of Algorithm 1. Here we have to solve:

$$\underset{m \in \mathscr{V}}{\operatorname{argmax}} \langle \eta^{[k]}, m \rangle_{\mathscr{M}(\mathcal{X})^2}$$

equivalent to the following problem:

$$\underset{\gamma \in \mathcal{S}(\mathcal{X})}{\operatorname{argmax}} \frac{1}{\ell_{\mathrm{div}}(\gamma)} \int_0^1 \eta^{[k]}(\gamma(t)) \cdot \dot{\gamma}(t) \, \mathrm{d}t.$$

This can be interpreted as the length of the curve γ weighted by a 'metric' $\eta^{[k]}$. Since the kernel in Φ is the gradient of the 2D Gaussian kernel, the reader might be aware that the *certificate* $\eta^{[k]}$ in Algorithm 1, see [14] for more insights, is the Laplacian of the image g. Then, the support of the estimated curve γ_* appears naturally in the Laplacian, as it is the maximum of $|\eta^{[k]}|$. Concerning the heuristic to determine if γ_* is open or closed, note that our involved curves are simple. Therefore, we exploit that a loop separates $\mathcal{X}$ onto two connected components, on the contrary to an open curve whose support complement has only one connected set. The algorithm then selects a simple chain of pixels to give an approximate support estimation. The Fig. 3 shows the magnitude of the certificate at iteration 0, an initialisation support for the added atom is successfully found.

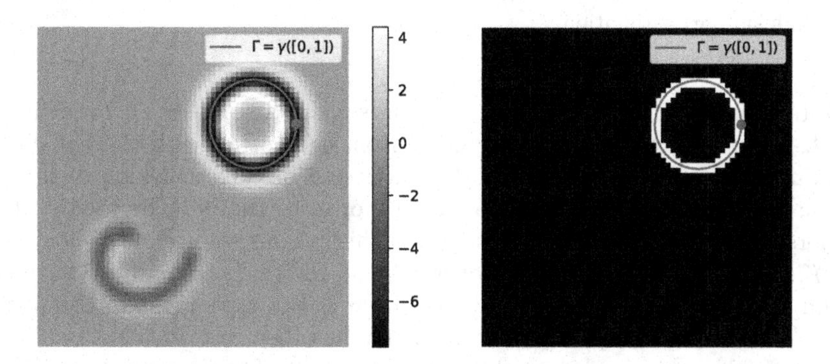

Fig. 3. Left: magnitude of $\Phi^* y$, the support of the estimated curve γ_* lies in the (near-)optima pixels. Right: threshold to reach a rough estimate of the support. The ground-truth curve γ with support Γ is traced in red. (Color figure online)

The convex step in line 7 is a fairly run-of-the-mill routine, the amplitudes of the atoms are estimated; this is a rather classical LASSO up to some tweaks, tackled in partice with a L-BFGS optimiser. The non-convex step in lines 8–9 on the contrary is way more challenging: the *sliding* performed here adjusts the atoms both in amplitudes and support. The chosen discretisation of curve measure here amounts to a chain, this optimisation can be understood then as a gradient descent over both amplitudes (common for one chain) and positions of numerous discretisation points. This is dealt with in our implementation with a flavour of stochastic gradient descent namely an ADAM optimiser, empirically shown here to outperform other non-convex solvers which have shown to be empirically. In Fig. 4 one can see the output of the convex step, then sharing the same estimated support as the crude one from the linear step; to be compared with the non-convex output on the right.

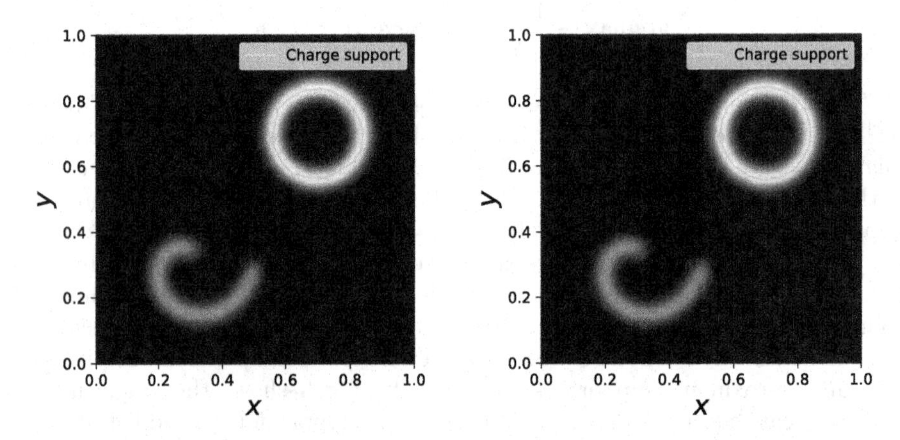

Fig. 4. Left: crude estimation from the oracle step. Right: the estimate at the end of the first iteration. Hopefully the sliding step smooths over the curvature and corrects the rough support estimation.

Interestingly enough, the loop endpoints *i.e.* $\gamma(0)$ and $\gamma(1)$, by definition equal, tends in fact to move a bit away from each other with the non-convex iteration. This is maybe related to the faces and broadly speaking to the geometric structure of the unit ball of the $\mathscr{V}$-norm: to the best knowledge of the authors there is no such work investigating this curious change of topology since the $\mathscr{V}$ space is a rather new concept for the off-the-grid community. Our observation is at this point purely empirical, and we only implemented a strategy to merge really closed endpoints. Our algorithm has then reconstructed one curve, and we loop over so forth to yield a reconstruction of the source charge $\boldsymbol{T}$, see Fig. 5.

Finally, the reconstruction is plotted in Fig. 6. The curves supports are greatly recovered[3], with of course some small differences due to the noise. A theoretical bound of this support estimation error is not trivial: it was investigated for spikes [8] and recently a similar result for level sets was advertised. Still, an equivalent

[3] 20 s on an *Intel Xeon E5-2687W v3* for a 64×64 image. A CUDA implementation is enabled for larger images, still ensuring reconstruction for a 512×512 image in less than 4 min on a *NVIDIA Quadro K2200*. See https://gitlab.inria.fr/blaville/amg.

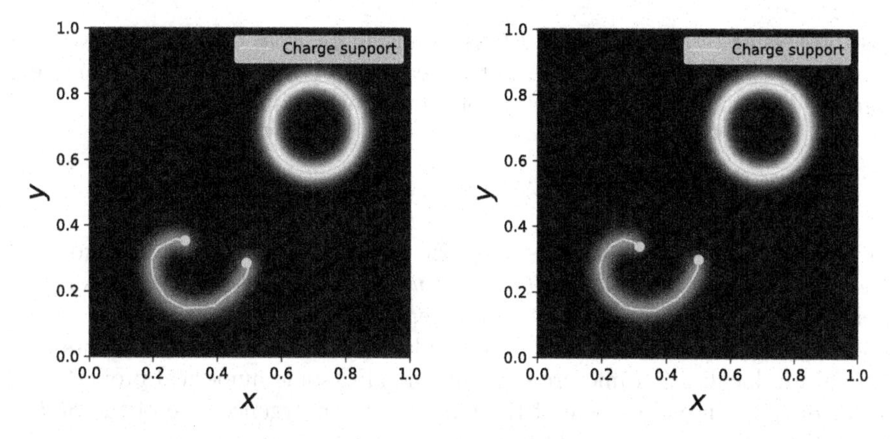

Fig. 5. Left: crude estimation from the oracle step. Right: the estimate at the end of the first iteration. Estimates matched the structure of the curves but exhibits some 'thorns': due to the noise, the reconstruction seems not as smoothed as the ground-truth. The non-convex step better recover the spiral folding.

for the off-the-grid curve reconstruction is yet to be explored and is clearly out of the scope of the present paper.

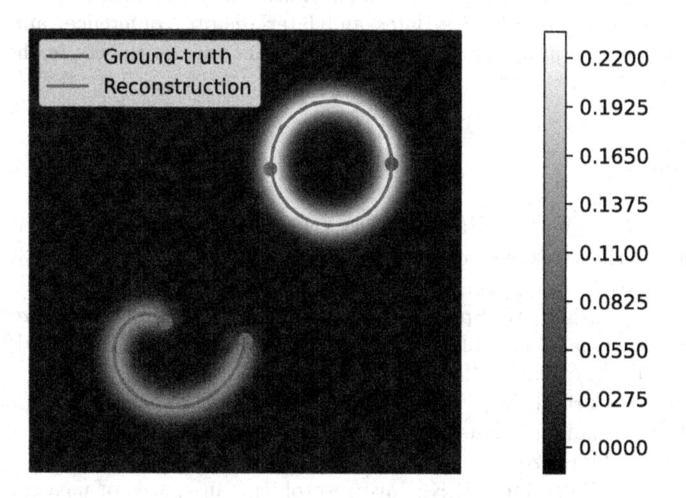

Fig. 6. The final reconstruction captures well the curvature of the source curves.

The proposed results might seem a bit simple. However, we emphasize that we can replace the convolution kernel with Fourier or Laplace measurements to handle more difficultly interpretable acquisition encountered in inverse problems. We wish to apply our algorithm to experimental data, but it requires a precise post-processing of the gradient such as denoising or a basic super-resolution, as the algorithm needs a (mediocre but at least usable) first guess to estimate a point of $\mathcal{V}$ in the basin of attraction of the global minimiser.

5 Conclusion

Based on the latest developments in off-the-grid theoretical results, we proposed a new algorithm called *Charge Sliding Frank-Wolfe* to perform a gridless reconstruction of curves while successfully implementing and reaching first results on a synthetic example. In further works, we plan to carry out the experiments with real data in localisation microscopy, while digging some numerical properties of our algorithm: an equivalence of the finite-time convergence rate of the *Sliding Frank-Wolfe* for instance would be a quite convenient property.

Acknowledgements. The work of Bastien Laville has been supported by the French government, through the UCA DS4H Investments in the Future project managed by the National Research Agency (ANR) with the reference number ANR-17-EURE-0004. The work of Laure Blanc-Féraud has been supported by the French government, through the 3IA Côte d'Azur Investments in the Future project managed by the National Research Agency (ANR) with the reference number ANR-19-P3IA-0002. This work was partially funded by the ANR Micro-Blind with the reference number ANR-21-CE48-0008. BL would like to warmly thanks Mr. Romain Petit for the genuinely constructive discussion they shared in Évian-les-Bains and later remote conference, and Mr. Théo Bertrand for his helpful advices on some of the numerous quirkiness in the numerical implementation.

References

1. Boyer, C., Chambolle, A., Castro, Y.D., Duval, V., de Gournay, F., Weiss, P.: On representer theorems and convex regularization. SIAM J. Optim. **29**(2), 1260–1281 (2019). https://doi.org/10.1137/18m1200750
2. Bredies, K., Carioni, M.: Sparsity of solutions for variational inverse problems with finite-dimensional data. Calc. Var. Partial. Differ. Equ. **59**(1), 1–26 (2019). https://doi.org/10.1007/s00526-019-1658-1
3. Bredies, K., Carioni, M., Fanzon, S., Romero, F.: On the extremal points of the ball of the Benamou-Brenier energy. Bull. London Math. Soc. (2021). https://doi.org/10.1112/blms.12509
4. Bredies, K., Pikkarainen, H.K.: Inverse problems in spaces of measures. ESAIM: Control Optim. Calc. Var. **19**(1), 190–218 (2012). https://doi.org/10.1051/cocv/2011205
5. De Castro, Y., Duval, V., Petit, R.: Towards off-the-grid algorithms for total variation regularized inverse problems. In: Elmoataz, A., Fadili, J., Quéau, Y., Rabin, J., Simon, L. (eds.) SSVM 2021. LNCS, vol. 12679, pp. 553–564. Springer, Cham (2021). https://doi.org/10.1007/978-3-030-75549-2_44
6. Demianov Vladimir Fedorovich, M.: Approximate Methods in Optimization Problems. American Elsevier Publishing Company, New York (1970)

7. Denoyelle, Q., Duval, V., Peyré, G., Soubies, E.: The sliding Frank-Wolfe algorithm and its application to super-resolution microscopy. Inverse Probl. **36**(1), 014001 (2019). https://doi.org/10.1088/1361-6420/ab2a29
8. Duval, V., Peyré, G.: Exact support recovery for sparse spikes deconvolution. Found. Comput. Math. **15**(5), 1315–1355 (2014). https://doi.org/10.1007/s10208-014-9228-6
9. Federer, H.: Geometric Measure Theory. Springer, Heidelberg (1996). https://doi.org/10.1007/978-3-642-62010-2
10. Frank, M., Wolfe, P.: An algorithm for quadratic programming. Nav. Res. Logist. Q. **3**(1–2), 95–110 (1956). https://doi.org/10.1002/nav.3800030109
11. Harchaoui, Z., Juditsky, A., Nemirovski, A.: Conditional gradient algorithms for norm-regularized smooth convex optimization. Math. Program. **152**(1–2), 75–112 (2014). https://doi.org/10.1007/s10107-014-0778-9
12. Khavin, V.P., Smirnov, S.K.: Approximation and extension problems for some classes of vector fields. St. Petersburg Dept. Steklov Inst. Math. Russian Acad. Sci. **10**(3), 507–528 (1998). https://mathscinet.ams.org/mathscinet-getitem?mr=1628034
13. Laville, B., Blanc-Féraud, L., Aubert, G.: Off-the-grid variational sparse spike recovery: methods and algorithms. J. Imaging **7**(12), 266 (2021). https://doi.org/10.3390/jimaging7120266. https://www.mdpi.com/2313-433X/7/12/266
14. Laville, B., Blanc-Féraud, L., Aubert, G.: Off-the-grid curve reconstruction through divergence regularisation: an extreme point result. Preprint (2023). https://hal.science/hal-03658949/document
15. Parhi, R., Nowak, R.D.: On continuous-domain inverse problems with sparse superpositions of decaying sinusoids as solutions. In: ICASSP 2022-2022 IEEE International Conference on Acoustics, Speech and Signal Processing (ICASSP). IEEE (2022). https://doi.org/10.1109/icassp43922.2022.9746165
16. Šilhavý, M.: Divergence measure vectorfields: their structure and the divergence theorem. Mathematical Modelling of Bodies with Complicated Bulk and Boundary Behavior, vol. 20, pp. 214–237 (2008)
17. Smirnov, S.K.: Decomposition of solenoidal vector charges into elementary solenoids, and the structure of normal one-dimensional flows. St. Petersburg Dept. Steklov Inst. Math. Russian Acad. Sci. **5**(4), 206–238 (1993). https://mathscinet.ams.org/mathscinet-getitem?mr=1246427

Convergence Guarantees of Overparametrized Wide Deep Inverse Prior

Nathan Buskulic(✉), Yvain Quéau, and Jalal Fadili

Normandie Univ., UNICAEN, ENSICAEN, CNRS, GREYC, Caen, France
nathan.buskulic@unicaen.fr

Abstract. Neural networks have become a prominent approach to solve inverse problems in recent years. Amongst the different existing methods, the Deep Image/Inverse Priors (DIPs) technique is an unsupervised approach that optimizes a highly overparametrized neural network to transform a random input into an object whose image under the forward model matches the observation. However, the level of overparametrization necessary for such methods remains an open problem. In this work, we aim to investigate this question for a two-layers neural network with a smooth activation function. We provide overparametrization bounds under which such network trained via continuous-time gradient descent will converge exponentially fast with high probability which allows to derive recovery prediction bounds. This work is thus a first step towards a theoretical understanding of overparametrized DIP networks, and more broadly it participates to the theoretical understanding of neural networks in inverse problem settings.

Keywords: Inverse problems · Deep Image/Inverse Prior · Overparameterization · Gradient flow

1 Introduction

1.1 Problem Statement

A linear inverse problem consists in reliably recovering an object $\overline{\mathbf{x}} \in \mathbb{R}^n$ from noisy indirect observations

$$\mathbf{y} = \mathbf{A}\overline{\mathbf{x}} + \epsilon, \tag{1}$$

where $\mathbf{y} \in \mathbb{R}^m$ is the observation, $\mathbf{A} \in \mathbb{R}^{m \times n}$ is a linear forward operator, and ϵ stands for some additive noise. Without loss of generality, we will assume throughout that $\mathbf{y} \in \operatorname{Im}(\mathbf{A})$.

In recent years, the use of sophisticated machine learning algorithms, including deep learning, to solve inverse problems has gained a lot of momentum and provides promising results, see e.g., reviews [2,12]. Most of these methods are supervised and require extensive datasets for training, which might not be available. An interesting unsupervised alternative [18] is known as Deep Image Prior,

L. Calatroni et al. (Eds.): SSVM 2023, LNCS 14009, pp. 406–417, 2023.
https://doi.org/10.1007/978-3-031-31975-4_31

which is also named Deep Inverse Prior (DIP) as it is not confined to images. In the DIP framework, a generator network $\mathbf{g} : (\mathbf{u}, \boldsymbol{\theta}) \in \mathbb{R}^d \times \mathbb{R}^p \mapsto \mathbf{x} \in \mathbb{R}^n$, with activation function ϕ, is optimized to transform some random input $\mathbf{u} \in \mathbb{R}^d$ into a vector in $\mathbf{x} \in \mathbb{R}^n$. The parameters $\boldsymbol{\theta}$ of the network are optimized via (possibly stochastic) gradient descent to minimize the squared Euclidean loss

$$\mathcal{L}(\mathbf{g}(\mathbf{u}, \boldsymbol{\theta})) = \frac{1}{2m} \|\mathbf{A}\mathbf{g}(\mathbf{u}, \boldsymbol{\theta}) - \mathbf{y}\|^2. \tag{2}$$

Theoretical understanding of recovery and establishing convergence guarantees for deep learning-based methods is of paramount importance to make their routine usage in critical applications reliable [11]. Our goal in this paper is to participate to this endeavour by explaining when gradient descent consistently and provably finds global minima of (2), and how this translates into recovery guarantees of (1). For this, we focus on a continuous-time gradient flow applied to (2):

$$\begin{cases} \dot{\boldsymbol{\theta}}(t) = -\nabla_{\boldsymbol{\theta}} \mathcal{L}(\mathbf{g}(\mathbf{u}, \boldsymbol{\theta}(t))), \\ \boldsymbol{\theta}(0) = \boldsymbol{\theta}_0. \end{cases} \tag{3}$$

This is an idealistic setting which makes the presentation simpler and it is expected to reflect the behavior of practical and commonly encountered first-order descent algorithms, as they are known to approximate gradient flows.

1.2 Contributions

We will deliver a first theoretical analysis of DIP models in the overparametrized regime. We will first analyze (3) by providing sufficient conditions for $\mathbf{y}(t) := \mathbf{A}\mathbf{g}(\mathbf{u}, \boldsymbol{\theta}(t))$ to converge exponentially fast to a globally optimal solution in the observation space. This result is then converted to a prediction error on $\overline{\mathbf{y}} := \mathbf{A}\overline{\mathbf{x}}$ through an early stopping strategy. Our conditions and bounds involve the conditioning of the forward operator, the minimum and maximum singular values of the Jacobian of the network, as well as its Lipschitz constant. We will then turn to evaluating these quantities for the case of a two-layer neural network

$$\mathbf{g}(\mathbf{u}, \boldsymbol{\theta}) = \frac{1}{\sqrt{k}} \mathbf{V}\phi(\mathbf{W}\mathbf{u}), \tag{4}$$

with $\mathbf{V} \in \mathbb{R}^{n \times k}$ and $\mathbf{W} \times \mathbb{R}^{k \times d}$, and ϕ an element-wise nonlinear activation function. The scaling by $\sqrt{k}$ will become clearer later. In this context, the network will be optimized with respect to the first layer (i.e., $\mathbf{W}$) while keeping the second (i.e., $\mathbf{V}$) fixed. Consequently, $\boldsymbol{\theta} = \mathbf{W}$. We show that for a proper random initialization $\mathbf{W}(0)$ and sufficient overparametrization, all our conditions are in force and the smallest eigenvalue of the Jacobian is indeed bounded away from zero independently of time. We provide a characterization of the overparametrization needed in terms of (k, d, n) and the conditioning of $\mathbf{A}$. Lastly, we show empirically that the behavior of real-world DIP networks is consistent with our theoretical bounds.

1.3 Relation to Prior Work

Data-Driven Methods to Solve Inverse Problems. Data-driven approaches to solve inverse problems come in various forms [2,12]. The first type trains an end-to-end network to directly map the observations to the signals for a specific problem. While they can provide impressive results, these methods can prove very unstable [11] as they do not use the physics of the problem which can be severely ill-posed. To cope with these problems, several hybrid models that mix model- and data-driven algorithms were developed in various ways. One can learn the regularizer of a variational problem [15] or use Plug-and-Play methods [19] for example. Another family of approaches, which takes inspiration from classical iterative optimization algorithms, is based on unrolling (see [10] for a review of these methods). Still, all these methods require extensive amount of training data, which may not always be available. Their theoretical recovery guarantees are also not well understood [11].

Deep Inverse Prior. The DIP model [18] (and its extensions that mitigate some of its empirical issues [8,9,16,20]) is an unsupervised alternative to the supervised approches briefly reviewed above. The empirical idea is that the architecture of the network acts as an implicit regularizer and will learn a more meaningful transformation before overfitting to artefacts or noise. With an early stopping strategy, one can get the network to generate a vector close to the sought signal. However, this remains purely empirical and there is no guarantee that a network trained in such manner converges in the observation space (and even less in the signal space). Our work aims at reducing this theoretical gap, by analyzing the behaviour of the network in the observation (prediction) space.

Theory of Overparametrized Networks. In parallel to empirical studies, there has been a lot of effort to develop some theoretical understanding of the optimization of overparametrized networks [3,6]. Amongst the theoretical models that emerged to analyze neural networks, the Neural Tangent Kernel (NTK) captures the behavior of neural networks in the infinite width limit during optimization via gradient descent. In the NTK framework, the neural network behaves as its linearization around the initialization, thus yielding a model equivalent to learning with a specific positive-definite kernel (so-called NTK). In [7], it was shown that in a highly overparametrized regime and random initialization, parameters $\boldsymbol{\theta}(t)$ stay near the initialization, and are well approximated by their linearized counterparts at all times (also called the "lazy" regime in [4]). With a similar aim, several works characterized the overparametrization necessary to obtain similar behaviour for shallow networks, see e.g., [1,5,13,14]. All these works provide lower bounds on the number of neurons from which they can prove convergence rates to a zero-loss solution. Despite some apparent similarities, our setting has important differences. On the one hand, we have indirect measurements through (fixed) $\mathbf{A}$, the output is not scalar, and there is no supervision. On the other hand, unlike all above works which deal with a supervised training setting, in the DIP model the dimension d of the input is a free parameter, while it is imposed in a supervised setting.

2 DIP Guarantees

2.1 Notations

For a matrix $\mathbf{M} \in \mathbb{R}^{a \times b}$ we denote, when dimension requirements are met, by $\lambda_{\min}(\mathbf{M})$ and $\lambda_{\max}(\mathbf{M})$ (resp. $\sigma_{\min}(\mathbf{M})$ and $\sigma_{\max}(\mathbf{M})$) its smallest and largest eigenvalues (resp. non-zero singular values), and by $\kappa(\mathbf{M}) = \frac{\sigma_{\max}(\mathbf{M})}{\sigma_{\min}(\mathbf{M})}$ its condition number. We also denote by $\|\cdot\|_F$ the Frobenius norm and $\|\cdot\|$ the Euclidean norm of a vector (or operator norm of a matrix). We use $\mathbf{M}^i$ (resp. $\mathbf{M}_i$) as the i-th row (resp. column) of $\mathbf{M}$. We represent a ball of radius r and center x by $\mathbb{B}(x, r)$. We also define $\mathbf{y}(t) = \mathbf{A}\mathbf{g}(\mathbf{u}, \boldsymbol{\theta}(t))$ and $\overline{\mathbf{y}} = \mathbf{A}\overline{\mathbf{x}}$. The Jacobian of the network is denoted $\mathcal{J}(\boldsymbol{\theta}(t))$. The Lipschitz constant of a mapping is denoted $\mathrm{Lip}(\cdot)$. We set $C_\phi = \sqrt{\mathbb{E}_{g \sim \mathcal{N}(0,1)}\left[\phi(g)^2\right]}$ and $C_{\phi'} = \sqrt{\mathbb{E}_{g \sim \mathcal{N}(0,1)}\left[\phi'(g)^2\right]}$ with $\mathbb{E}[X]$ the expected value of X.

2.2 Main Result

Standing Assumptions. In the rest of this work, we assume that:

A-1 $\mathbf{u}$ *is drawn uniformly on* $\mathbb{S}^{d-1}$;
A-2 $\mathbf{W}(0)$ *has iid entries from* $\mathcal{N}(0,1)$;
A-3 $\mathbf{V}$ *has iid columns with identity covariance and D-bounded entries;*
A-4 ϕ *is a twice differentiable function with B-bounded derivatives.*

Assumptions A-1, A-2 and A-3 are standard. Assumptions A-4 is met by many activations such as the softmax, sigmoid or hyperbolic tangent. Including the ReLU would require more technicalities that will be avoided here.

Well-posedness. In order for our analysis to hold, the Cauchy problem (3) needs to be well-defined. This is easy to prove upon observing that under A-4, the gradient of the loss is both Lipschitz and continuous. Thus, the Cauchy-Lipschitz theorem applies, ensuring that (3) has a unique global continuously differentiable solution trajectory.

Our main result establishes the prediction error for the DIP model.

Theorem 1. *Consider a network* $\mathbf{g}(\mathbf{u}, \boldsymbol{\theta})$, *with* ϕ *obeying A-4, optimized via the gradient flow* (3).

(i) Let $\sigma_{\mathbf{A}} = \inf_{\mathbf{z} \in \mathrm{Im}(\mathbf{A})} \left\|\mathbf{A}^\top \mathbf{z}\right\| / \|\mathbf{z}\| > 0$. *Suppose that*

$$\frac{\|\mathbf{y} - \mathbf{A}\mathbf{g}(\mathbf{u}, \boldsymbol{\theta}_0)\|}{\sigma_{\mathbf{A}}} < \frac{\sigma_{\min}(\mathcal{J}(\boldsymbol{\theta}_0))^2}{4\mathrm{Lip}(\mathcal{J})}. \tag{5}$$

Then for any $\epsilon > 0$

$$\|\mathbf{y}(t) - \overline{\mathbf{y}}\| \le 2\|\epsilon\| \qquad \textit{for all} \qquad t \ge \frac{4m \log\left(\|\mathbf{y} - \mathbf{A}\mathbf{g}(\mathbf{u}, \boldsymbol{\theta}_0)\| / \|\epsilon\|\right)}{\sigma_{\mathbf{A}}^2 \sigma_{\min}(\mathcal{J}(\boldsymbol{\theta}_0))^2}. \tag{6}$$

(ii) Let the one-hidden layer network (4) *with architecture parameters obeying*

$$k \geq C_1 \kappa(\mathbf{A})^2 n \left(\sqrt{n}\left(\sqrt{\log(d)}+1\right)+\sqrt{m}\right)^2$$

Then

$$\|\mathbf{y}(t)-\overline{\mathbf{y}}\| \leq 2\|\epsilon\| \qquad \textit{for all} \qquad t \geq \frac{C_2 m \log\left(\|\mathbf{y}-\mathbf{A}\mathbf{g}(\mathbf{u},\mathbf{W}(0))\|\right)}{\sigma_{\mathbf{A}}^2 C_{\phi'}^2}$$

with probability at least $1-n^{-1}-d^{-1}$, *where* C_i *are positive constants that depend only on the activation function and the bound* D.

Before proceeding with the proof, a few remarks are in order:

- We start with the scaling of the network architecture parameters required. First, the bound on k, the number of neurons of the hidden layer, scales quadratically in n^2 and linearly in m. We thus have the bound $k \gtrsim n^2 m$. The probability of success in our theorem is also dependent on the architecture parameters. More precisely, this probability increases with growing number of observations.
- The other scaling of the theorem is on the input size d and informs us that its influence is logarithmic. The bound is more demanding as $\mathbf{A}$ becomes more ill-conditioned. The latter dependency can be interpreted as follows: the more ill-conditioned the original problem is, the larger the network needs to be. Let us emphasize that, contrary to other learning settings in the literature, the size d of the random input $\mathbf{u}$ is free, and so far it has remained unclear how to choose it. Our result provides a first answer for shallow networks in the overparametrized setting: most of the overparametrization necessary for the optimization to converge is due to k.
- On our way to prove (1), we actually show that $\mathbf{y}(t)$ converges exponentially to $\mathbf{y}$, which is converted to a recovery of $\overline{\mathbf{y}}$ through an early stopping strategy. This ensures that the network does not overfit the noise and provides a solution in a ball around $\overline{\mathbf{y}}$ whose radius is linear in the noise level (so-called prediction linear convergence in the inverse problem literature). This provides a first result on convergence of wide DIP networks that ensures they behave well in the observation space.
- One has to keep in mind, however, that Theorem 1 does not say anything about the recovered vector generated by the network and its relation to $\overline{\mathbf{x}}$ (in absence of noise and at convergence, it might be any element of $\overline{\mathbf{x}}+\ker(\mathbf{A})$). Of course, when $\mathbf{A}$ is invertible, then we are done. In the general case, this is a much more challenging question which requires a more involved analysis and a restricted-type injectivity assumption. This will be the subject of a forthcoming paper.

3 Proof

The proof consists of two main steps. First, we prove that under (5), $\mathbf{y}(t)$ converges exponentially fast with a time-independent rate. We then use a triangle inequality and an early stopping criterion to show our result. The proof of the second claim will consist in verifying that (5) holds with high probability for our random model of the two-layer network under our scaling. Both proofs rely on several technical lemmas, which will be given later.

(i) The solution trajectory $\boldsymbol{\theta}(t)$, hence $\mathbf{y}(t)$, is continuously differentiable, and thus

$$\begin{aligned}
\frac{\mathrm{d}\frac{1}{2}\left\|\mathbf{y}(t)-\mathbf{y}\right\|^2}{\mathrm{d}t} &= (\mathbf{y}(t)-\mathbf{y})\dot{\mathbf{y}}(t) \\
&= (\mathbf{y}(t)-\mathbf{y})\mathbf{A}\mathcal{J}(\boldsymbol{\theta}(t))\dot{\boldsymbol{\theta}}(t) \\
&= -(\mathbf{y}(t)-\mathbf{y})\mathbf{A}\mathcal{J}(\boldsymbol{\theta}(t))\nabla_{\boldsymbol{\theta}}\mathcal{L}(\mathbf{g}(\mathbf{u},\boldsymbol{\theta}(t))) \\
&= -\frac{1}{m}(\mathbf{y}(t)-\mathbf{y})^\top\mathbf{A}\mathcal{J}(\boldsymbol{\theta}(t))\mathcal{J}(\boldsymbol{\theta}(t))^\top\mathbf{A}^\top(\mathbf{y}(t)-\mathbf{y}) \\
&= -\frac{1}{m}\left\|\mathcal{J}(\boldsymbol{\theta}(t))^\top\mathbf{A}^\top(\mathbf{y}(t)-\mathbf{y})\right\|^2 \le -\frac{\sigma_{\min}(\mathcal{J}(\boldsymbol{\theta}(t)))^2\sigma_{\mathbf{A}}^2}{m}\left\|\mathbf{y}(t)-\mathbf{y}\right\|^2, \quad (7)
\end{aligned}$$

where we used the fact that $\mathbf{y}(t)-\mathbf{y}\in\mathrm{Im}(\mathbf{A})$. In view of Lemma 1(iii), we have $\sigma_{\min}(\mathcal{J}(\boldsymbol{\theta}(t)))\ge\sigma_{\min}(\mathcal{J}(\boldsymbol{\theta}_0))/2$ for all $t\ge 0$ if the initialization error verifies (5), and in turn

$$\frac{\mathrm{d}\left\|\mathbf{y}(t)-\mathbf{y}\right\|^2}{\mathrm{d}t}\le-\frac{\sigma_{\min}(\mathcal{J}(\boldsymbol{\theta}_0))^2\sigma_{\mathbf{A}}^2}{2m}\left\|\mathbf{y}(t)-\mathbf{y}\right\|^2.$$

Integrating, we obtain

$$\left\|\mathbf{y}(t)-\mathbf{y}\right\|\le\left\|\mathbf{y}(0)-\mathbf{y}\right\|e^{-\frac{\sigma_{\min}(\mathcal{J}(\boldsymbol{\theta}_0))^2\sigma_{\mathbf{A}}^2}{4m}t}. \quad (8)$$

Using

$$\left\|\mathbf{y}(t)-\mathbf{y}\right\|\le\left\|\mathbf{y}(t)-\overline{\mathbf{y}}\right\|+\left\|\epsilon\right\|\le\left\|\mathbf{y}(0)-\mathbf{y}\right\|e^{-\frac{\sigma_{\min}(\mathcal{J}(\boldsymbol{\theta}_0))^2\sigma_{\mathbf{A}}^2}{4m}t}+\left\|\epsilon\right\|,$$

we get the early stopping bound by bounding the exponential term by $\|\epsilon\|$.

(ii) To show the statement, it is sufficient to check that (5) holds under our scaling. From Lemma 2, we have

$$\sigma_{\min}(\mathcal{J}(\boldsymbol{\theta}_0))\ge C_{\phi'}/2$$

with probability at least $1-n^{-1}$ provided $k\ge C_0 n\log(n)$ for $C_0>0$. Combining this with Lemma 3 and Lemma 4 and the union bound, it is sufficient for (5) to be fulfilled with probability at least $1-n^{-1}-d^{-1}$, that

$$C_1\kappa(\mathbf{A})\left(\sqrt{n}\left(\sqrt{\log(d)}+1\right)+\sqrt{m}\right)<\frac{C_{\phi'}^2\sqrt{k}}{16BD\sqrt{n}}.$$

We now prove the intermediate lemmas invoked in the proof.

Lemma 1. *(i) If* $\boldsymbol{\theta} \in \mathbb{B}(\boldsymbol{\theta}_0, R)$ *with* $R = \frac{\sigma_{\min}(\mathcal{J}(\boldsymbol{\theta}_0))}{2\mathrm{Lip}(\mathcal{J})}$, *then*

$$\sigma_{\min}(\mathcal{J}(\boldsymbol{\theta})) \geq \sigma_{\min}(\mathcal{J}(\boldsymbol{\theta}_0))/2.$$

(ii) If for all $s \in [0, t]$, $\sigma_{\min}(\mathcal{J}(\boldsymbol{\theta}(s))) \geq \frac{\sigma_{\min}(\mathcal{J}(\boldsymbol{\theta}_0)}{2}$, *then*

$$\boldsymbol{\theta}(t) \in \mathbb{B}(\boldsymbol{\theta}_0, R') \qquad \textit{with} \qquad R' = \frac{2}{\sigma_{\mathbf{A}} \sigma_{\min}(\mathcal{J}(\boldsymbol{\theta}_0))} \|\mathbf{y}(0) - \mathbf{y}\|.$$

(iii) If $R' < R$, *then for all* $t \geq 0$, $\sigma_{\min}(\mathcal{J}(\boldsymbol{\theta}(t))) \geq \sigma_{\min}(\mathcal{J}(\boldsymbol{\theta}_0))/2$.

Proof. (i) Since $\boldsymbol{\theta} \in \mathbb{B}(\boldsymbol{\theta}_0, R)$, we have

$$\|\mathcal{J}(\boldsymbol{\theta}) - \mathcal{J}(\boldsymbol{\theta}_0)\| \leq \mathrm{Lip}(\mathcal{J}) \|\boldsymbol{\theta} - \boldsymbol{\theta}_0\| \leq \mathrm{Lip}(\mathcal{J}) R.$$

By using that $\sigma_{\min}(A)$ is 1-Lipschitz, we obtain

$$\sigma_{\min}(\mathcal{J}(\boldsymbol{\theta})) \geq \sigma_{\min}(\mathcal{J}(\boldsymbol{\theta}_0)) - \|\mathcal{J}(\boldsymbol{\theta}) - \mathcal{J}(\boldsymbol{\theta}_0)\| \geq \frac{\sigma_{\min}(\mathcal{J}(\boldsymbol{\theta}_0)}{2}.$$

(ii) From (7), we have for all $s \in [0, t]$

$$\begin{aligned} \frac{\mathrm{d}\|\mathbf{y}(s) - \mathbf{y}\|}{\mathrm{d}s} &= -\frac{1}{m} \frac{\left\|\mathcal{J}(\boldsymbol{\theta}(s))^\top \mathbf{A}^\top (\mathbf{y}(s) - \mathbf{y})\right\|^2}{\|\mathbf{y}(s) - \mathbf{y}\|} \\ &\leq -\frac{\sigma_{\min}(\mathcal{J}(\boldsymbol{\theta}_0))\sigma_{\mathbf{A}}}{2m} \left\|\mathcal{J}(\boldsymbol{\theta}(s))^\top \mathbf{A}^\top (\mathbf{y}(s) - \mathbf{y})\right\|. \end{aligned}$$

The Cauchy-Schwarz inequality and (3) imply that

$$\frac{\mathrm{d}\|\boldsymbol{\theta}(s) - \boldsymbol{\theta}_0\|}{\mathrm{d}s} = \frac{\dot{\boldsymbol{\theta}}(s)^\top (\boldsymbol{\theta}(s) - \boldsymbol{\theta}_0)}{\|\boldsymbol{\theta}(s) - \boldsymbol{\theta}_0\|} \leq \left\|\dot{\boldsymbol{\theta}}(s)\right\| = \frac{1}{m} \left\|\mathcal{J}(\boldsymbol{\theta}(s))^\top \mathbf{A}^\top (\mathbf{y}(s) - \mathbf{y})\right\|.$$

We therefore get

$$\frac{\mathrm{d}\|\boldsymbol{\theta}(s) - \boldsymbol{\theta}_0\|}{\mathrm{d}s} + \frac{2}{\sigma_{\min}(\mathcal{J}(\boldsymbol{\theta}_0))\sigma_{\mathbf{A}}} \frac{\mathrm{d}\|\mathbf{y}(s) - \mathbf{y}\|}{\mathrm{d}s} \leq 0.$$

Integrating over $s \in [0, t]$, we get the claim.

(iii) Actually, we prove the stronger statement that $\boldsymbol{\theta}(t) \in \mathbb{B}(\boldsymbol{\theta}_0, R')$ for all $t \geq 0$, whence our claim will follow thanks to (i). Let us assume for contradiction that $R' < R$ and $\exists\, t < +\infty$ such that $\boldsymbol{\theta}(t) \notin \mathbb{B}(\boldsymbol{\theta}_0, R')$. By (ii), this means that $\exists\, s \leq t$ such that $\sigma_{\min}(\mathcal{J}(\boldsymbol{\theta}(s))) < \sigma_{\min}(\mathcal{J}(\boldsymbol{\theta}_0))/2$. In turn, (i) implies that $\boldsymbol{\theta}(s) \notin \mathbb{B}(\boldsymbol{\theta}_0, R)$. Let us define

$$t_0 = \inf\{\tau \geq 0 : \boldsymbol{\theta}(\tau) \notin \mathbb{B}(\boldsymbol{\theta}_0, R)\},$$

which is well-defined as it is at most s. Thus, for any small $\epsilon > 0$ and for all $t' \leq t_0 - \epsilon$, $\boldsymbol{\theta}(t') \in \mathbb{B}(\boldsymbol{\theta}_0, R)$ which, in view of (i) entails that $\sigma_{\min}(\mathcal{J}(\boldsymbol{\theta})(t')) \geq \sigma_{\min}(\mathcal{J}(\boldsymbol{\theta}_0))/2$. In turn, we get from (ii) that $\boldsymbol{\theta}(t_0 - \epsilon) \in \mathbb{B}(\boldsymbol{\theta}_0, R')$. Since ϵ is arbitrary and $\boldsymbol{\theta}$ is continuous, we pass to the limit as $\epsilon \to 0$ to deduce that $\boldsymbol{\theta}(t_0) \in \mathbb{B}(\boldsymbol{\theta}_0, R') \subsetneq \mathbb{B}(\boldsymbol{\theta}_0, R)$ hence contradicting the definition of t_0. □

Lemma 2 (Bound on $\sigma_{\min}(\mathcal{J}(\boldsymbol{\theta}_0))$). *For the one-hidden layer network* (4), *under assumptions A-1-A-4. We have*

$$\sigma_{\min}(\mathcal{J}(\boldsymbol{\theta}_0)) \geq C_{\phi'}/2$$

with probability at least $1 - n^{-1}$ *provided* $k \geq Cn\log(n)$ *for* $C > 0$ *large enough that depends only on* ϕ *and the bound on the entries of* $\mathbf{V}$.

Proof. Define the matrix $\mathbf{H} = \mathcal{J}(\boldsymbol{\theta}_0)\mathcal{J}(\boldsymbol{\theta}_0)^\top$. For the two-layer network, and since $\mathbf{u}$ is on the unit sphere, $\mathbf{H}$ reads

$$\mathbf{H} = \frac{1}{k}\sum_{i=1}^{k} \phi'(\mathbf{W}^i(0)\mathbf{u})^2 \mathbf{V}_i \mathbf{V}_i^\top.$$

It follows that

$$\mathbb{E}\left[\mathbf{H}\right] = \mathbb{E}_{g\sim\mathcal{N}(0,1)}\left[\phi'(g)^2\right]\frac{1}{k}\sum_{i=1}^{k}\mathbb{E}\left[\mathbf{V}_i\mathbf{V}_i^\top\right] = C_{\phi'}^2\mathbf{I}_n,$$

where we used A-1-A-2 and orthogonal invariance of the Gaussian distribution, hence $\mathbf{W}^i(0)\mathbf{u}$ are iid $\mathcal{N}(0,1)$, as well as A-3 and independence between $\mathbf{V}$ and $\mathbf{W}(0)$. Moreover,

$$\lambda_{\max}(\phi'(\mathbf{W}^i(0)\mathbf{u})^2\mathbf{V}_i\mathbf{V}_i^\top) \leq B^2D^2n.$$

We can then apply the matrix Chernoff inequality [17, Theorem 5.1.1] to get

$$\mathbb{P}\left(\sigma_{\min}(\mathcal{J}(\boldsymbol{\theta}_0)) \leq \delta C_{\phi'}\right) \leq ne^{-\frac{(1-\delta)^2 kC_{\phi'}^2}{2B^2D^2n}}.$$

Taking $\delta = 1/2$ and k as prescribed, we conclude. □

Lemma 3 (Lipschitz constant of the Jacobian). *For the one-hidden layer network* (4), *under assumptions A-1, A-2 and A-4, we have*

$$\mathrm{Lip}(\mathcal{J}) \leq BD\sqrt{\frac{n}{k}}.$$

Proof. We have for all $\mathbf{W}, \widetilde{\mathbf{W}} \in \mathbb{R}^{k\times d}$,

$$\begin{aligned}
\left\|\mathcal{J}(\mathbf{W}) - \mathcal{J}(\widetilde{\mathbf{W}})\right\|^2 &\leq \frac{1}{k}\sum_{i=1}^{k}|\phi'(\mathbf{W}^i\mathbf{u}) - \phi'(\widetilde{\mathbf{W}}^i\mathbf{u})|^2\left\|\mathbf{V}_i\mathbf{u}^\top\right\|_F^2 \\
&= \frac{1}{k}\sum_{i=1}^{k}|\phi'(\mathbf{W}^i\mathbf{u}) - \phi'(\widetilde{\mathbf{W}}^i\mathbf{u})|^2\left\|\mathbf{V}_i\right\|^2 \\
&\leq B^2D^2\frac{n}{k}\sum_{i=1}^{k}|\mathbf{W}^i\mathbf{u} - \widetilde{\mathbf{W}}^i\mathbf{u}|^2 \\
&\leq B^2D^2\frac{n}{k}\sum_{i=1}^{k}\left\|\mathbf{W}^i - \widetilde{\mathbf{W}}^i\right\|^2 = B^2D^2\frac{n}{k}\left\|\mathbf{W} - \widetilde{\mathbf{W}}\right\|_F^2.
\end{aligned}$$

□

Lemma 4 (Bound on the initial error). *Under the main assumptions, the initial error of the network is bounded by*

$$\|\mathbf{y}(0)-\mathbf{y}\| \leq \|\mathbf{A}\| \left(C\sqrt{n\log(d)} + \sqrt{n}\,\|\mathbf{x}_0\|_\infty + \sqrt{m}\,\|\epsilon\|_\infty\right),$$

with probability at least $1-d^{-1}$.

Proof. We first observe that

$$\|\mathbf{y}(0)-\mathbf{y}\| \leq \|\mathbf{A}\|\,\|\mathbf{g}(\mathbf{u},\mathbf{W}(0))\| + \|\mathbf{A}\|\left(\sqrt{n}\,\|\mathbf{x}_0\|_\infty + \sqrt{m}\,\|\epsilon\|_\infty\right),$$

where $\mathbf{g}(\mathbf{u},\mathbf{W}(0)) = \frac{1}{\sqrt{k}}\sum_{i=1}^{k}\phi(\mathbf{W}^i\mathbf{u})\mathbf{V}_i$. We now prove that this term concentrates around its expectation. First, we have by independence

$$\mathbb{E}\left[\|\mathbf{g}(\mathbf{u},\mathbf{W}(0))\|\right]^2 \leq \frac{1}{k}\mathbb{E}\left[\left\|\sum_{i=1}^{k}\phi(\mathbf{W}^i\mathbf{u})\mathbf{V}_i\right\|^2\right] = \mathbb{E}\left[\phi(\mathbf{W}^1\mathbf{u})^2\,\|\mathbf{V}_1\|^2\right] = C_\phi^2 n.$$

In addition,

$$\left|\|\mathbf{g}(\mathbf{u},\mathbf{W})\| - \|\mathbf{g}(\mathbf{u},\widetilde{\mathbf{W}})\|\right| \leq \frac{1}{\sqrt{k}}\left\|\sum_{i=1}^{k}\left(\phi(\mathbf{W}^i\mathbf{u}) - \phi(\widetilde{\mathbf{W}}^i\mathbf{u})\right)\mathbf{V}_i\right\|$$

$$\leq BD\sqrt{n}\left(\frac{1}{\sqrt{k}}\sum_{i=1}^{k}\left\|\mathbf{W}^i - \widetilde{\mathbf{W}}^i\right\|\right) \leq BD\sqrt{n}\left\|\mathbf{W} - \widetilde{\mathbf{W}}\right\|_F.$$

We then get

$$\mathbb{P}\left(\|\mathbf{g}(\mathbf{u},\mathbf{W}(0))\| \geq C_\phi\sqrt{n\log(d)} + \tau\right)$$
$$\leq \mathbb{P}\left(\|\mathbf{g}(\mathbf{u},\mathbf{W}(0))\| \geq \mathbb{E}\left[\|\mathbf{g}(\mathbf{u},\mathbf{W}(0))\|\right] + \tau\right)$$
$$\leq e^{-\frac{\tau^2}{2\mathrm{Lip}(\|\mathbf{g}(\mathbf{u},\mathbf{W}(0))\|)^2}} \leq e^{-\frac{\tau^2}{2nB^2D^2}}.$$

Taking $\tau = \sqrt{2}BD\sqrt{n\log(d)}$, we get the desired claim. □

4 Numerical Experiments

We conducted numerical experiments to verify our theoretical finding, by evaluating the convergence of networks with different architecture parameters in the noise-free context. Every network was initialized in accordance with the assumptions of our work and we used the sigmoid activation function. Both $\mathbf{A}$ and $\overline{\mathbf{x}}$ entries were drawn i.i.d from $\mathcal{N}(0,1)$. We used gradient descent to optimize the networks with a fixed step size of 1. A network was trained until it reached a loss of 10^{-7} or after 25000 optimization steps. For each set of architecture parameters, we did 50 runs and calculated the frequency at which the network arrived at the error threshold of 10^{-7}.

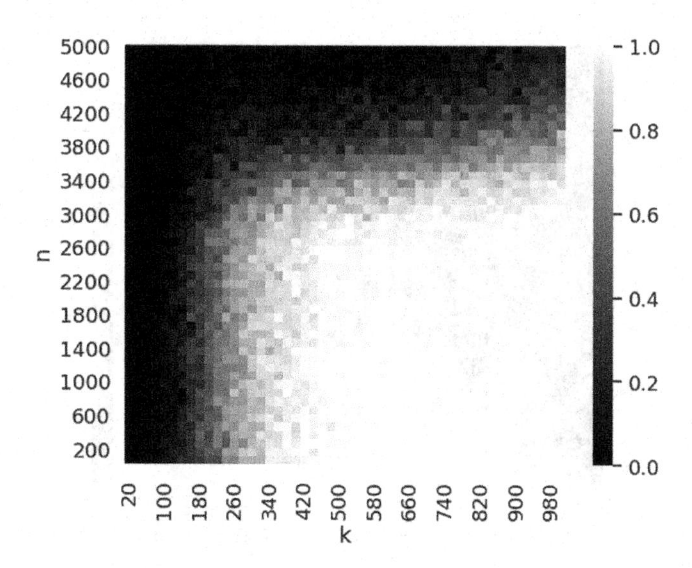

Fig. 1. Probability of arriving at a zero loss solution for networks with fixed number of observations m, yet varying number of neurons k and signal size n. This emphasizes that the required level of over-parametrization scales at least quadatically with n.

We present in Fig. 1 a first experiment where we fix the number of observations $m = 10$ and the input size $d = 500$, and we let the number of neurons k and the signal size n vary. It can be observed in this experiment that for any value of n, a zero-loss solution is reached with high probability as long as k is "large enough", where the phase transition seems to follow a quadratic law. Given that in this setup $n \gg m$ and $\mathbf{A}$ is Gaussian, this empirical observation is consistent with the theoretical quadratic relation $k \gtrsim n^2 m$ which is predicted by our main theorem. However, one may be surprised by the wide range of values of n which can be handled with a fixed k. Consider for instance the case $k = 900$: convergence is attained for values of n up to 3000, which includes cases where $k < n$. This goes against our intuition for such underparametrized cases.

Figure 2 presents a second experiment, where we now fix $n = 60$ (still with $d = 500$), while letting k vary with m. Therein, the expected linear relation between k and m clearly appears, which provides another empirical validation for our theoretical bound. Now, let us consider again a fixed level of over-parametrization, e.g., $k = 900$. Contrarily to the previous experiment on the signal size n, the range of observations number m which can be tackled is more restricted (here, convergence is observed for values of m up to 25). For problems where the ratio m/n largely deviates from zero, the level of required over-parametrization is thus much more important than for problems where $n \gg m$. Overall, these experiments validate the order of magnitude of our theoretical bounds, although they also emphasize that these bounds are not really tight.

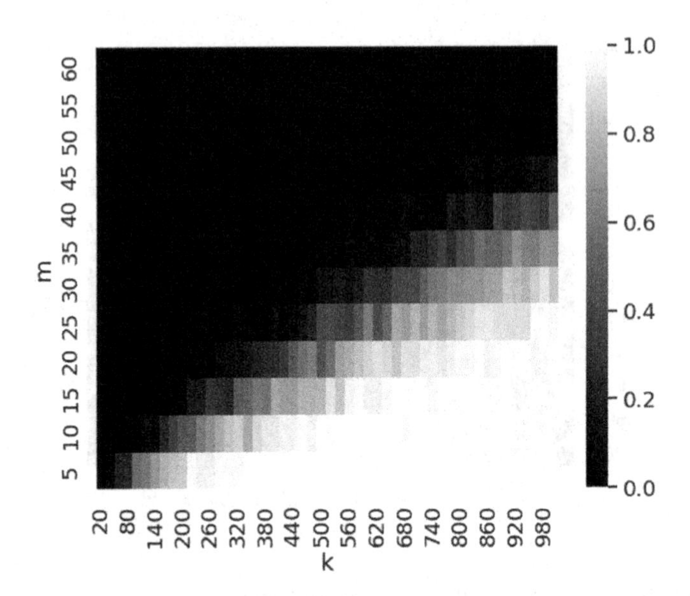

Fig. 2. Probability of arriving at a zero loss solution for networks with fixed signal size n, and varying number of neurons k and of observations m. This emphasizes that the required level of over-parametrization scales linearly with m.

5 Conclusion and Future Work

This paper studied the convergence of shallow DIP networks and provided bounds on the level of overparametrization, both in the input dimension and the hidden layer dimension, under which the method converges exponentially fast to a zero-loss solution. The proof relies on bounding the minimum singular values of the Jacobian of the network through an overparametrization that ensures a good initialization. These bounds are not tight, as demonstrated by the numerical experiments, but they provide an important step towards the theoretical understanding of DIP methods, and neural networks for inverse problems resolution in general. In the future, this work will be extended in several directions. First, we will study recovery guarantees of the signal $\overline{\mathbf{x}}$. Second, we will investigate the DIP model with unrestricted linear layers and possibly in the multilayer setting.

Acknowledgements. The authors thank the French National Research Agency (ANR) for funding the project ANR-19-CHIA-0017-01-DEEP-VISION.

References

1. Allen-Zhu, Z., Li, Y., Song, Z.: A Convergence theory for deep learning via over-parameterization. In: ICML, pp. 242–252 (2019)
2. Arridge, S., Maass, P., Ozan, Ö., Schönlieb, C.-B.: Solving inverse problems using data-driven models. Acta Numer. **28**, 1–174 (2019)

3. Bartlett, P.L., Montanari, A., Rakhlin, A.: Deep learning: a statistical viewpoint. Acta Numer. **30**, 87–201 (2021)
4. Chizat, L., Oyallon, E., Bach, F.: On Lazy training in differentiable programming. In: NeurIPS (2019)
5. Du, S.S., Zhai, X., Póczos, B., Singh, A.: Gradient descent provably optimizes over-parameterized neural networks. In: ICLR (2019)
6. Fang, C., Dong, H., Zhang, T.: Mathematical Models of Overparameterized Neural Networks. Proc. IEEE **109**(5), 683–703 (2021)
7. Jacot, A., Gabriel, F., and Hongler, C.: Neural tangent kernel: convergence and generalization in neural networks. In: NeurIPS (2018)
8. Liu, J., Sun, Y., Xu, X., Kamilov, U.S.: Image restoration using total variation regularized deep image prior. In: IEEE ICASSP, pp. 7715–7719 (2019)
9. Mataev, G., Milanfar, P., and Elad, M.: DeepRED: deep image prior powered by RED. In: ICCV, pp. 0–0 (2019)
10. Monga, V., Li, Y., Eldar, Y.C.: Algorithm unrolling: interpretable, efficient deep learning for signal and image processing. IEEE SPM **38**(2), 18–44 (2021)
11. Mukherjee, S., Hauptmann, A., Öktem, O., Pereyra, M., Schönlieb, C.-B.: Learned reconstruction methods with convergence guarantees (2022). arXiv:2206.05431 [cs]. Sept. 2022
12. Ongie, G., Jalal, A., Metzler, C.A., Baraniuk, R.G., Dimakis, A.G., Willett, R.: Deep learning techniques for inverse problems in imaging. IEEE J-SAIT **1**(1), 39–56 (2020)
13. Oymak, S., Soltanolkotabi, M.: Overparameterized nonlinear learning: gradient descent takes the shortest Path? In: ICML, pp. 4951–4960 (2019)
14. Oymak, S., Soltanolkotabi, M.: Toward moderate overparameterization: global convergence guarantees for training shallow neural networks. IEEE J-SAIT **1**, 84–105 (2020)
15. Prost, J., Houdard, A., Almansa, A., Papadakis, N.: Learning local regularization for variational image restoration. In: SSVM, pp. 358–370 (2021)
16. Shi, Z., Mettes, P., Maji, S., Snoek, C.G.M.: On measuring and controlling the spectral bias of the deep image prior. Int. J. Comput. Vis. **130**(4), 885–908 (2022). https://doi.org/10.1007/s11263-021-01572-7
17. Tropp, J.A.: An introduction to matrix concentration inequalities. arXiv:1501.01571 [cs, math, stat] (2015). arXiv: 1501.01571
18. Ulyanov, D., Vedaldi, A., and Lempitsky, V.: Deep image prior. Int. J. Comput. Vis. **128**(7), 1867–1888 (2020). arXiv: 1711.10925
19. Venkatakrishnan, S.V., Bouman, C.A., and Wohlberg, B.: Plug-and-Play priors for model based reconstruction. In: GlobalSIP, pp. 945–948 (2013)
20. Zukerman, J., Tirer, T., and Giryes, R.: BP-DIP: A Backprojection based deep image prior. In: EUSIPCO 2020, pp. 675–679 (2021)

On the Remarkable Efficiency of SMART

Max Kahl[1(✉)], Stefania Petra[2], Christoph Schnörr[3], Gabriele Steidl[4], and Matthias Zisler[2]

[1] Astroinformatics Group, HITS, Heidelberg, Germany
Max_Michael.Kahl@stud.uni-heidelberg.de
[2] Mathematical Imaging Group, Heidelberg University, Heidelberg, Germany
[3] Image and Pattern Analysis Group, Heidelberg University, Heidelberg, Germany
[4] Institute of Mathematics, TU Berlin, Berlin, Germany

Abstract. We consider the problem of minimizing the Kullback-Leibler divergence between two unnormalised positive measures, where the first measure lies in a finitely generated convex cone. We identify SMART (simultaneous multiplicative algebraic reconstruction technique) as a Riemannian gradient descent on the parameter manifold of the Poisson distribution. By comparing SMART to recent acceleration techniques from convex optimization that rely on Bregman geometry and first-order information, we demonstrate that it solves this problem very efficiently.

Keywords: KL divergence · accelerated mirror descent · relative smoothness · information geometry · Riemannian gradient descent

1 Introduction

This paper explores state-of-the-art first-order optimization methods for solving

$$\min_{x \in \mathbb{R}^n_+} f(x), \qquad f(x) = \mathrm{KL}(Ax, b), \qquad A \in \mathbb{R}^{m \times n}_+, \quad b \in \mathbb{R}^m_{++} \tag{1}$$

in order to recover a discretized *nonnegative* function x from linear *nonnegative* measurements $Ax \approx b$ by minimizing the Kullback-Leibler (KL) divergence, instead e.g., the usual least-squares norm, see e.g., [11] and references therein. We exploit the underlying Bregman geometry in a twofold way. First, *convex* optimization methods based on Bregman distances offer the possibility of matching the Bregman distance to the structure of the problem, leading to simple multiplicative gradient-like iterative schemes and enabling a reduced cost of the complexity per iteration. Secondly, by turning the interior of the feasible set into a Riemannian manifold, the geometry of the space allows smooth unconstrained optimization. We examine both aspects in a principled manner for problem (1) and consider discrete tomography as application scenario.

L. Calatroni et al. (Eds.): SSVM 2023, LNCS 14009, pp. 418–430, 2023.
https://doi.org/10.1007/978-3-031-31975-4_32

Related Work. A prototypical multiplicative iterative algorithm for (1) is *SMART (simultaneous multiplicative algebraic reconstruction technique)* [7]

$$x^{k+1} = x^k e^{-\tau_k \nabla f(x^k)}, \qquad x^0 \in \mathbb{R}^n_{++} \qquad \text{(SMART)} \tag{2a}$$

$$x_j^{k+1} = x_j^k \prod_{i=1}^{m} \Big(\frac{b_i}{(Ax^k)_i}\Big)^{\tau_k A_{ij}}, \qquad k = 0, 1, \dots \quad j = 1, \dots, n. \tag{2b}$$

In [20] SMART was identified as the classical mirror descent algorithm (MDA) [18] for *fixed* steplength τ_k for the particular objective (1), that converges at a (faster) $O(1/k)$ rate (as opposed to $O(1/\sqrt{k})$ for general MDA [5]) due to relative L-smoothness (see below). In addition, a computationally efficient acceleration scheme based on [22] was suggested, however, without a theoretical underpinning of a $O(1/k^2)$ rate. MDA, and thus SMART too, is a special instance of the Bregman proximal gradient (BPG) method [21]. Recent results concerning optimal complexity of Bregman first-order methods [12,14,15] including BPG, motivate us to explore the a-posteriori certification of accelerated rates for (1). In [17] the convergence of SMART was analyzed in the context of primal-dual methods. Hence, it is natural to ask how (1) can be solved by state-of-the-art (accelerated) primal dual splitting methods that employ generalized proximal operators defined in terms of a Bregman distance [8,9]. For a recent overview of Bregman divergences and proximity operators, see [13].

Contribution and Organization. Section 2 introduces essential concepts related to Bregman divergences. The acceleration of SMART is discussed from the viewpoint of BPG in Sect. 3. The Riemannian geometry of SMART is introduced in Sect. 4. In Sect. 5, we show in large scale experiments that SMART is on par with the state-of-the-art Bregman first order methods, and the $\mathcal{O}(1/k^2)$ rate of its accelerated version cannot be numerically certified.

Basic Notation. We denote the set of nonnegative real vectors by $\mathbb{R}^n_+$ and the set of positive ones by $\mathbb{R}^n_{++}$. Let $\langle \cdot, \cdot \rangle$ denote the standard inner product on $\mathbb{R}^n$, ∇h the gradient of a differentiable function $h : \mathbb{R}^n \to \mathbb{R}$ and h^* the Fenchel conjugate $h^*(p) = \sup_{x \in \mathbb{R}^n} \{\langle p, x \rangle - h(x)\}$. Given a sufficiently well-behaved convex function ϕ, we consider the so-called Bregman distance

$$D_\phi(x, y) = \phi(x) - \phi(y) - \langle \nabla \phi(y), x - y \rangle \tag{3}$$

between x and y. We frequently denote componentwise multiplication of vectors by $uv = (u_1 v_1, \dots, u_n v_n)^\top$ and, for strictly positive vectors $v \in \mathbb{R}^n_{++}$, componentwise division by $\frac{u}{v}$. Likewise, the functions $e^x, \log x$ apply componentwise to a vector x. For a smooth Riemannian manifold $(\mathcal{M}, g)$ with metric g, $T_x\mathcal{M}$ denotes the tangent space at $x \in \mathcal{M}$ and $\mathrm{d}_x h : T_x\mathcal{M} \to \mathbb{R}$ the differential of a smooth function $h : \mathcal{M} \to \mathbb{R}$. The Riemannian gradient $\operatorname{grad} h(x) \in T_x\mathcal{M}$ of h is uniquely defined by $\mathrm{d}_x h[\xi] = g_x(\operatorname{grad} h(x), \xi)$, $\forall \xi \in T_x\mathcal{M}$. The (squared) Riemannian norm is denoted by $\|v\|_x^2 = g_x(v, v)$, $\forall v \in T_x\mathcal{M}$.

2 Preliminaries

Throughout this paper, we assume $A \in \mathbb{R}_{+}^{m \times n}$ and $b \in \mathbb{R}_{++}^{m}$. We consider the KL divergence as the Bregman divergence

$$\mathrm{KL}(x,y) := D_{\varphi}(x,y) = \langle x, \log x - \log y\rangle - \langle \mathbb{1}, x - y\rangle, \tag{4}$$

defined on $\mathbb{R}_{+}^{n} \times \mathbb{R}_{++}^{n}$, which plays a distinguished role among all divergence functions [2, Section 3.4] and is induced by the Bregman kernel

$$\varphi(x) = \langle x, \log x\rangle - \langle \mathbb{1}, x\rangle, \qquad x \in \mathbb{R}_{+}^{n}. \tag{5}$$

The specific function φ in (5) is of Legendre type [4, Def. 2.8] which implies that both gradients $\nabla\varphi$ and $\nabla\varphi^{*}$ are one-to-one and inverses of each other. In particular, φ induces a dual structure induced by the Legendre transform

$$u := \nabla\varphi(x) = \log x, \qquad x = \nabla\varphi^{*}(u) = e^{u}, \qquad \varphi^{*}(u) = \langle \mathbb{1}, e^{u}\rangle. \tag{6}$$

and a corresponding dual divergence function due to (3) reads

$$D_{\varphi^{*}}(v,u) = \varphi^{*}(v) - \varphi^{*}(u) - \langle \nabla\varphi^{*}(u), v-u\rangle = D_{\varphi}(x,y)\big|_{x=e^{u},\, y=e^{v}}. \tag{7}$$

We now briefly discuss the attainment of minima in (1), that are related to the unique Bregman projection onto the cone $K_A := \{Ax \colon x \geq 0\}$, generated by the columns of A, that is a closed and convex set.

Theorem 1 (**[4, Thm. 3.12]**). *Suppose ϕ is closed proper convex and differentiable on* $\mathrm{int}(\mathrm{dom}\,\phi)$, *$C$ is closed convex with $C \cap \mathrm{int}(\mathrm{dom}\,\phi) \neq \emptyset$, and $b \in \mathrm{int}(\mathrm{dom}\,\phi)$. If ϕ is Legendre, then the Bregman projection $\overline{y}$ of b is unique and contained in* $\mathrm{int}(\mathrm{dom}\,\phi)$,

$$\operatorname*{argmin}_{y \in C \cap \mathrm{dom}\,\phi} D_{\phi}(y,b) = \{\overline{y}\}, \qquad \overline{y} \in \mathrm{int}(\mathrm{dom}\,\phi). \tag{8}$$

As K_A is nonempty, closed and convex, φ in (5) is Legendre with $\mathrm{dom}\,\varphi = \mathbb{R}_{+}^{m}$ and $K_A \cap \mathbb{R}_{++}^{m} \neq \emptyset$, in view of $A \in \mathbb{R}_{+}^{m \times n}$. Hence, the assumptions of the theorem above are satisfied. Hence $\overline{y}$ exists and is unique and all $\overline{x} \in \mathbb{R}_{+}^{n}$ with $\overline{y} = A\overline{x}$ are minimizers of (1). One can prove a similar result as in (8) for

$$\operatorname*{argmin}_{x \in C \cap \mathrm{dom}\,\phi} \left\{D_{\phi}(x,x^{0}) + \langle c, x\rangle\right\} = \{z\}, \qquad z \in \mathrm{int}(\mathrm{dom}\,\phi), \tag{9}$$

with $c \in \mathbb{R}^{n}$ arbitrary and $\|c\| \leq \infty$.

Lemma 1. *Let f and φ be given by* (1) *and* (5), *respectively. Then $D_f(x,y) = D_{\varphi}(Ax, Ay)$.*

Proof. From (5) and (3) it follows

$$\begin{aligned}
D_f(x,y) &= f(x) - f(y) - \langle \nabla f(y), x-y\rangle \\
&= D_\varphi(Ax,b) - D_\varphi(Ay,b) - \big\langle A^\top\big(\log(Ay) - \log b\big), x-y\big\rangle \\
&= \varphi(Ax) - \varphi(b) - \langle \nabla\varphi(b), Ax-b\rangle - \big(\varphi(Ay) - \varphi(b) - \langle \nabla\varphi(b), Ay-b\rangle\big) \\
&\qquad - \langle \log(Ay) - \log b, A(x-y)\rangle \\
&= \varphi(Ax) - \varphi(Ay) - \langle \nabla\varphi(b), A(x-y)\rangle - \langle \log(Ay) - \log b, A(x-y)\rangle \\
&\overset{\nabla\varphi(b)=\log b}{=} \varphi(Ax) - \varphi(Ay) - \langle \log(Ay), A(x-y)\rangle \\
&= D_\varphi(Ax,Ay).
\end{aligned} \tag{10}$$

□

3 SMART and Convex Acceleration

As mentioned, the SMART iteration (2) was studied in [20] in the context of mirror descent (aka Bregman proximal gradient), as investigated by Beck and Teboulle [5]. Specifically, using the Bregman divergence (3) as distance function, the update scheme with stepsize $\tau_k > 0$ reads

$$x^{k+1} = \underset{x\in\mathbb{R}^n_+}{\operatorname{argmin}}\ f(x^k) + \langle \nabla f(x^k), x\rangle + \frac{1}{\tau_k} D_\phi(x,x^k), \qquad x^0 \in \mathbb{R}^n_{++}, \tag{11}$$

which is well defined according to (9). For D_φ given by (5), one has

$$\nabla D_\varphi(x,x^k) = \nabla\varphi(x) - \nabla\varphi(x^k) \overset{(6)}{=} \log x - \log x^k, \tag{12}$$

so that evaluating the optimality condition with respect to (11) yields

$$0 = \nabla f(x^k) + \frac{1}{\tau_k}\big(\log x^{k+1} - \log x^k\big) \tag{13a}$$

$$\Leftrightarrow \qquad x^{k+1} = x^k e^{-\tau_k A^\top \log\frac{Ax^k}{b}}, \qquad x^0 \in \mathbb{R}^n_{++}, \tag{13b}$$

which is the SMART update (2). Below we summarize the main convergence results based on [7, Thm. 2] and [20, Thm. 2].

Theorem 2. *Let S be the solution set of* (1) *and $L = \|A\|_1$. For $(x^k)_{k\in\mathbb{N}}$ generated by (2) with starting point $x^0 \in \mathbb{R}^n_{++}$ and $\tau_k = \tau \leq 1/L$ we have*

(a) The sequence $(x^k)_{k\in\mathbb{N}}$ converges to a unique point in S, that is

$$\overline{x} = \arg\min_{x\in S} D_\varphi(x,x^0). \tag{14}$$

(b) For every k

$$f(x^k) - f(\overline{x}) \leq \frac{LD_\varphi(\overline{x},x^0)}{k}. \tag{15}$$

The following lemma, adapted from [20, Prop. 2], allows to establish the basic convergence rate $\mathcal{O}(1/k)$ of SMART without assuming that ∇f is L-Lipschitz.

Lemma 2. *Suppose $A \in \mathbb{R}^{m\times n}_+$ and consider φ in* (5). *Then*

$$\forall x, y \in \mathbb{R}^n_+, \qquad D_\varphi(Ax, Ay) \leq \|A\|_1 D_\varphi(x, y). \tag{16}$$

Acceleration in Bregman First-Order Convex Optimization. In [12] the $\mathcal{O}(1/k)$ rate is shown to be optimal for a broad class of Bregman proximal gradient (BPG) algorithms under *general* assumptions on the objective function f and the Bregman kernel ϕ. In particular, it is not required that ∇f is L-Lipschitz. Rather, f has merely to be L-smooth *relative* to ϕ, i.e.

$$D_f(x, y) \leq L D_\phi(x, y), \qquad \forall x, y \in \operatorname{dom} f \subset \operatorname{dom} \phi. \tag{17}$$

Accelerated Bregman proximal gradient (ABPG) algorithms can *only* be obtained under additional assumptions. The authors in [15] consider an assumption which yields a $\mathcal{O}(1/k^\gamma)$ rate with $\gamma \in [1, 2]$. In particular, they consider the *triangle-scaling property* with *uniform triangle-scaling exponent (TSE)* γ

$$D_\phi\big((1-\theta)x+\theta z, (1-\theta)x+\theta \tilde{z}\big) \leq \theta^\gamma D_\phi(z, \tilde{z}), \quad \forall \theta \in [0, 1], \quad \forall x, z, \tilde{z} \in \operatorname{rint} \operatorname{dom} \phi. \tag{18}$$

The focus is on *jointly* convex Bregman divergences D_ϕ since then (18) holds with $\gamma = 1$. Note that KL is jointly convex. Further, the *intrinsic TSE* of D_ϕ is defined by

$$\gamma_{\text{in}} = \limsup_{\theta \searrow 0} \frac{D_\phi\big((1-\theta)x + \theta z, (1-\theta)x + \theta \tilde{z}\big)}{\theta^\gamma} < \infty, \qquad \forall x, z, \tilde{z} \in \operatorname{rint} \operatorname{dom} \phi. \tag{19}$$

A broad class of Bregman divergences has $\gamma_{\text{in}} = 2$ which is the value the largest uniform TSE cannot exceed. The analysis in [15] rests upon the *triangle-scaling gain* $G(x, z, \tilde{z})$ defined by the relaxed triangle-scaling inequality

$$D_\phi\big((1-\theta)x + \theta z, (1-\theta)x + \theta \tilde{z}\big) \leq G(x, z, \tilde{z})\theta^\gamma D_\phi(z, \tilde{z}), \qquad \forall \theta \in [0, 1]. \tag{20}$$

$G(x, z, \tilde{z})$ is bounded based on the relative scaling of the Hessian of ϕ at different points. In particular, *adaptive* ABPG algorithms are proposed based on (20) for problems of the form

$$\min_{x \in C} F(x), \qquad F(x) = f(x) + \Psi(x), \tag{21}$$

with f being L-smooth relative to ϕ, C closed, and C, Ψ convex and simple, in the sense that the key step of the ABPG method

$$z_{k+1} = \operatorname{argmin}_{x \in C} \Big\{ f(y_k) + \langle \nabla f(y_k), x - y_k \rangle + \theta_k^{\gamma-1} L D_\phi(x, z_k) + \Psi(x) \Big\}, \tag{22}$$

can be solved efficiently. The convergence analysis of ABPG uses basic relations derived by [10] and [22] in order to relate two subsequent updates.

The ABPG-e method with *exponent adaption* starts with a large value $\gamma_k > 2$ and reduces it at each step by some fixed δ, until an inequality (the *local* triangle-scaling property, see below (24) for its specialization to our scenario) as stopping criterion is satisfied. The last value γ_k determines the convergence rate and serves as *empirical certificate.*

The ABPG-g method with *gain adaption* adapts the gain $G_k = G(x_k, z_k, \widetilde{z}_k)$ in an inner loop until a local triangle-scaling inequality is satisfied. In [15, Sect. 4.1], the authors discuss obstacles for proving a $\mathcal{O}(k^{-2})$ rate, which requires to bound the geometric mean $\overline{G} := (G_0^\gamma G_1 \cdots G_k)^{\frac{1}{k+\gamma}}$ of gains at each step, without additional assumptions. They argue that, in practical situations, one always works with a particular reference function ϕ that may have structural properties yielding fast convergence. Exploiting such a structure for φ (5) is subject to further research.

SMART and Acceleration. Combining Lemma 1 and Lemma 2, we conclude that f in (1) is L-smooth relative to φ (5) with $L = \|A\|_1$. The ABPG iteration (22) leads for $\gamma = 1$ and $\Psi \equiv 0$ to the F(ast)-SMART iteration [20],

$$y^k = (1-\theta_k)x^k + \theta_k z^k \tag{23a}$$

$$z^{k+1} = z^k \exp\left(-A^\top \log\left(\frac{Ay^k}{b}\right)/L\right) \tag{23b}$$

$$x^{k+1} = (1-\theta_k)x^k + \theta_k z^{k+1}, \tag{23c}$$

where $x^0 = z^0 \in \operatorname{int}(\operatorname{dom}\varphi)$ and $\theta_k \in (0,1]$ satisfies $\frac{1-\theta_{k+1}}{\theta_{k+1}^2} \leq \frac{1}{\theta_k^2}$. As the uniform TSE γ equals 1 for our choice φ in (5) only a $\mathcal{O}(1/k)$ rate can be guaranteed according to [15, Thm. 1]. Convergence of the sequence $(x^k)_{k\in\mathbb{N}}$ generated by FSMART, as it is guaranteed for SMART, remains an open issue. As discussed above ABPG-e uses a local triangle-scaling property that, in view of Lemma 1, takes the form

$$D_\varphi(Ax^{k+1}, Ay^{k+1}) < \theta_k^{\gamma_k} L D_\varphi(z^{k+1}, z^k), \tag{24}$$

when specialized to our scenario. Similarly, ABPG-g includes G_k in the r.h.s. above. Hence, satisfying such a condition brings extra cost for each iteration, similar to a line search.

4 SMART: A Geometric Perspective

Riemannian Geometry of the Positive Orthant. In this section, we represent the positive orthant $\mathbb{R}^n_{++}$ as a Riemannian manifold. To this end, we turn the open interval

$$\mathcal{P} := (0, +\infty), \qquad \mathcal{P}_n := \mathcal{P} \times \cdots \times \mathcal{P} = \mathbb{R}^n_{++} \tag{25}$$

into a manifold $(\mathcal{P}, g)$ with metric g and define $(\mathcal{P}_n, g)$ as the corresponding product manifold. In order to specify $(\mathcal{P}, g)$, we apply basic information geometry

[3]. Let points $\lambda \in \mathcal{P}$ parametrize the Poisson distribution $p(z;\lambda) = \frac{\lambda^z e^{-\lambda}}{z!}$ with rate parameter $\lambda > 0$ of a random variable $Z \in \mathbf{N}_0$. Then the metric tensor of the Fisher-Rao metric is a scalar function of λ given by

$$G(\lambda) = 4 \sum_{z \in \mathbf{N}_0} \Big(\frac{d}{d\lambda} \sqrt{p(z;\lambda)} \Big)^2 = \frac{e^{-\lambda}}{\lambda^2} \sum_{z \in \mathbf{N}_0} \frac{\lambda^z (z-\lambda)^2}{z!} = \frac{e^{-\lambda}}{\lambda^2} e^{\lambda} \lambda = \frac{1}{\lambda}. \tag{26}$$

In view of (25), this extends to the metric g on $T\mathcal{P}_n$ in terms of the diagonal matrix

$$G_n(x) = \operatorname{Diag}\Big(\frac{1}{x_1}, \dots, \frac{1}{x_n}\Big) = \operatorname{Diag}\Big(\frac{\mathbb{1}}{x}\Big) \tag{27}$$

as metric tensor. We naturally identify $T_x\mathcal{P}_n \cong \mathbb{R}^n$, $\forall x \in \mathcal{P}_n$ and denote this metric interchangeably by

$$g_x(v, v') = \langle v, v' \rangle_x = \langle v, G_n(x) v' \rangle, \qquad \forall v, v' \in T_x\mathcal{P}_n. \tag{28}$$

We point out that this geometry differs from the standard geometry of the positive orthant which underlies interior point methods [19].

Retraction. Retractions [1, Def. 4.1.1] are basic ingredients of first-order optimization algorithms on Riemannian manifolds. The main motivation is to replace the exponential map with respect to the metric (Levi Civita) connection by an approximation that can be efficiently evaluated or even computed in closed form. Below, we compute the exponential map with respect to the e-connection of information geometry [3] and show subsequently that it is a retraction.

Proposition 1. *The exponential maps on $\mathcal{P}$ resp. $\mathcal{P}_n$ with respect to the e-connection are given by*

$$\exp\colon \mathcal{P} \times T\mathcal{P} \to \mathcal{P}, \qquad \exp_\lambda(tv) = \lambda e^{t\frac{v}{\lambda}}, \quad t > 0, \tag{29a}$$

$$\exp\colon \mathcal{P}_n \times T\mathcal{P}_n \to \mathcal{P}_n, \qquad \exp_x(tv) = \big(\exp_{x_j}(tv_j)\big)_{j \in [n]}. \tag{29b}$$

Proof. By definition of the product manifold, it suffices to show (29a). A key concept of information geometry is to replace the metric connection by a pair of connections that are dual to each other with respect to the Riemannian metric g [3, Section 3.1]. In particular, under suitable assumptions, the parameter space of a probability distribution becomes a Riemannian manifold that is dually flat, i.e. two distinguished coordinate systems (the so-called m- and e-coordinates) exist with affine geodesics. We consider the case $(\mathcal{P}, g)$.

First, we rewrite the density of the Poisson distribution as distribution of the exponential family [6]

$$p(z;\theta) = h(z) \exp\big(z\theta - \psi(\theta)\big), \qquad \theta = \theta(\lambda) = \log \lambda \tag{30}$$

with exponential parameter θ, base measure $h(z) = \frac{1}{z!}$ and log-partition function $\psi(\theta) = e^\theta$ that is convex and of Legendre type. The aforementioned two coordinates are λ and θ with affine geodesics

$$t \mapsto \lambda_v(t) = \lambda + tv \in \mathcal{P}, \qquad t \mapsto \theta_u(t) = \theta + tu \in \mathbb{R}. \tag{31}$$

Note that unlike λ, v, the coordinate θ and the tangent u are unconstrained. Using (30), the e-geodesic reads

$$\lambda\big(\theta_u(t)\big) = e^{\theta_u(t)} = e^{\theta} e^{tu} \in \mathcal{P}. \tag{32}$$

We wish to express this curve in terms of λ and v using $\theta = \theta(\lambda)$ by (30) and the relation between the tangents u and v given by the differential

$$u = u(\lambda, v) = \frac{d}{dt}\theta\big(\lambda_v(t)\big)\big|_{t=0} = \frac{d}{dt}\theta(\lambda + tv)\big|_{t=0} = \frac{d}{dt}\log(\lambda + tv)\Big|_{t=0} = \frac{v}{\lambda}. \tag{33}$$

Substituting into (32) yields (29a)

$$\exp_{\lambda}(tv) := e^{\theta(\lambda)} e^{tu(\lambda,v)} = \lambda e^{t\frac{v}{\lambda}}. \tag{34}$$

□

Remark 1 (g is a Hessian metric). The dual nature of the exponential parametrization (30) is also highlighted by recovering the metric tensor (26) as Hessian metric from the potential $\varphi(\lambda)$ that is conjugate to the log-partition function $\psi(\theta) = e^{\theta}$, $\varphi(\lambda) = \psi^*(\lambda) = \lambda \log \lambda - \lambda$, to obtain $\varphi''(\lambda) = G(\lambda) = \frac{1}{\lambda}$. The dual coordinate chart and potential yield the inverse metric tensor $\psi''(\theta) = e^{\theta} = G(\lambda)^{-1}|_{\lambda=\lambda(\theta)}$.

Retractions provide a proper class of surrogate mappings for replacing the canonical exponential map corresponding to the metric connection.

Proposition 2 (exp **is a retraction).** *The mapping* $\exp \colon T\mathcal{P} \to \mathcal{P}$ *is a retraction in the sense of [1, Def. 4.1.1.].*

Proof. We check the two criteria that characterize retractions. First, by (29a) we have $\exp_{\lambda}(0) = \lambda$ for all $\lambda \in \mathcal{P}$. Second, the so-called local rigidity condition $d\exp_{\lambda}(0) = 1 = \mathrm{id}_{T_{\lambda}\mathcal{P}}$, $\forall \lambda \in \mathcal{P}$, holds as well, in view of the relation $d\exp_{\lambda}(u)v = e^{\frac{u}{\lambda}}v$ obtained from (29a). □

SMART as Riemannian Gradient Descent. The *Riemannian* gradient with respect to the metric g from (28), generally defined by [16, p. 89] here specifically reads

$$\operatorname{grad} f(x) := {G_n}^{-1}(x)\nabla f(x) \overset{(27)}{=} x\,\nabla f(x), \qquad x \in \mathcal{P}_n. \tag{35}$$

The retraction in Proposition 1 allows us to compute updates on the manifold based on numerical operations in the tangent space. Due to the simple structure of the constraints, this can be done in parallel for each coordinate. Furthermore, as a consequence of the choice (28) for g, the corresponding Riemannian gradient (35) exactly matches the exponent in the expression for (29b). Thus, applying $\exp_x$ to the Riemannian gradient simplifies to

$$\exp_x\big(-\tau \operatorname{grad} f(x)\big) = x e^{-\tau \frac{\operatorname{grad} f(x)}{x}} = x e^{-\tau \nabla f(x)}. \tag{36}$$

This results in the following representation of the SMART iteration.

Proposition 3. *Let* $(\mathcal{P}_n, g)$ *be endowed with the Riemannian metric* (28). *Then the SMART iteration (2) equals*

$$x^{k+1} = \exp_{x^k}\big(-\tau_k \operatorname{grad} f(x^k)\big). \tag{37}$$

The choice of τ_k can now be adapted to the current iterate by line search and we still obtain a global convergence result.

Theorem 3. *[1, Thm. 4.3.1] Let* $(x^k)_{k\in\mathbb{N}}$ *be a sequence generated by the iteration* (37) *with step-size* $\tau_k = \beta^m\alpha$ *and scalars* $\alpha > 0$, $\beta, \sigma \in (0,1)$, *where* m *is the smallest nonnegative integer such that*

$$f(x^k) - f\big(\exp_{x^k}\big(-\tau_k \operatorname{grad} f(x^k)\big)\big) \geq \sigma\tau_k \|\operatorname{grad} f(x^k)\|^2_{x^k}. \tag{38}$$

Then every cluster point $\hat{x}$ *is a critical point of* f, *i.e.* $\operatorname{grad} f(\hat{x}) = 0$.

The above statement assumes existence of a critical point. In view of (35) it is characterized by $\hat{x} > 0$ and $\nabla f(\hat{x}) = 0$. Clearly, such a critical point satisfies the optimality conditions $0 \leq \hat{x} \perp \nabla f(\hat{x}) \geq 0$ of (1). Hence, convergence of a subsequence of the iterates $(x_k)_{k\in\mathbb{N}}$ to a solution on the boundary, i.e., when $\hat{x}_i = 0$, is not covered by Theorem 3. In this paper, we are interested in assessing numerically the convergence speed of the iterates $(x_k)_{k\in\mathbb{N}}$ in the manifold $\mathcal{P}_n$ (25). An analysis of the behavior of these sequences close to the boundary of $\mathcal{P}_n$ will be reported in follow-up work.

5 Experiments

We compare SMART to the state-of-the-art accelerated Bregman proximal gradient methods in [15], which we adapt as described in Sect. 3, and to its geometric version employing Armijo line search and the retraction in (29). The latter version of SMART is denoted as Riemannian gradient (RG). In addition, we include a state-of-the-art primal dual Bregman method [9] in our comparison. For results and discussions we refer to Fig. 2, 3 and 4.

Problem and Data Setup. We consider large scale tomographic reconstruction as problem class, where we reconstruct the three phantoms shown in Fig. 1. We generated tomographic projection matrices A using the ASTRA-toolbox[1], with parallel beam geometry and equidistant angles in the range $[0, \pi]$. Each entry in A is nonnegative as it corresponds to the length of the intersection of a ray with a pixel. The undersampling rate was chosen to be 20%. None of the images in Fig. 1 is the unique nonnegative solution to $Ax = b$. Hence, different algorithms might converge to different nonnegative solutions. For the noisy setting we applied Poisson noise to b with a signal-to-noise ratio of SNR $= 20$ db.

Implementation Details. We implemented six different algorithms for solving (1) iteratively. In order to avoid numerical issues, we clip each component x_i to

[1] https://www.astra-toolbox.com.

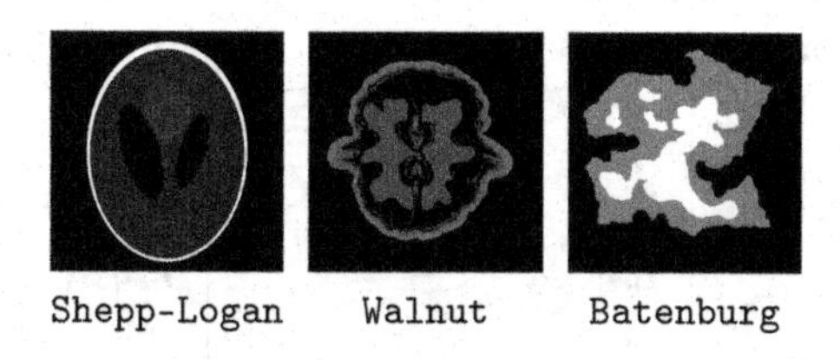

Fig. 1. The phantoms (1024 × 1024) used for the numerical evaluation.

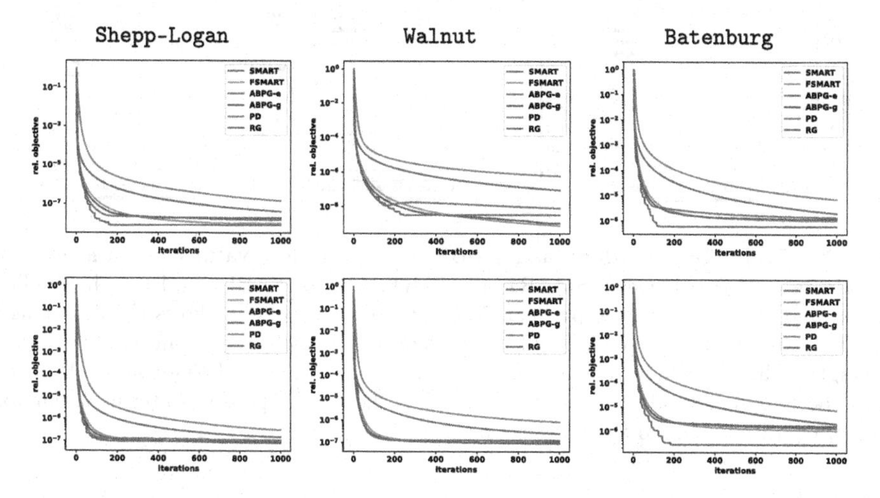

Fig. 2. Comparison of decreasing objective function values per iteration between SMART, FSMART, ABPG-e, ABPG-g, PD (Chambolle-Pock) and RG (Riemannian gradient descent). The i-th column shows the i-th image in Fig. 1, in noiseless (top row) and noisy scenarios (bottom row). Overall, RG (i.e. SMART with line search) aggressively minimizes the objective and in general outperforms the accelerated variants.

$\max\{x_i, \varepsilon\}$ with $\varepsilon = 10^{-10}$ before applying the logarithm. The maximum number of iterations was set to $n = 1000$, which also serves as a termination criterion. We always used the initialization $x^0 = \mathbb{1}$. For each algorithm the same set of parameters was used across all experiment instances. The algorithms and corresponding parameter choices are listed below:

SMART solely performs the multiplicative update specified in (2) with its stepsize fixed to $\tau_k = \frac{1}{L}$, where again f is L-smooth relative to φ.

FSMART is based on the iteration suggested in [20], where initially $\theta_0 = 1$ is chosen, which is then subsequently updated via $\theta_{k+1} = \frac{\sqrt{\theta_k^4 + 4\theta_k^2} - \theta_k^2}{2}$, as suggested in [22].

ABPG-e as described in [15, Algorithm 2] was applied to (1) with parameters $\gamma_{\min} = 1$, $\gamma_0 = 5$ and $\delta = 0.05$. The choices for γ_0 and δ deviate slightly from the recommendation in [15], but were chosen to facilitate fastest possible convergence on the selected problem instances. To ensure comparability

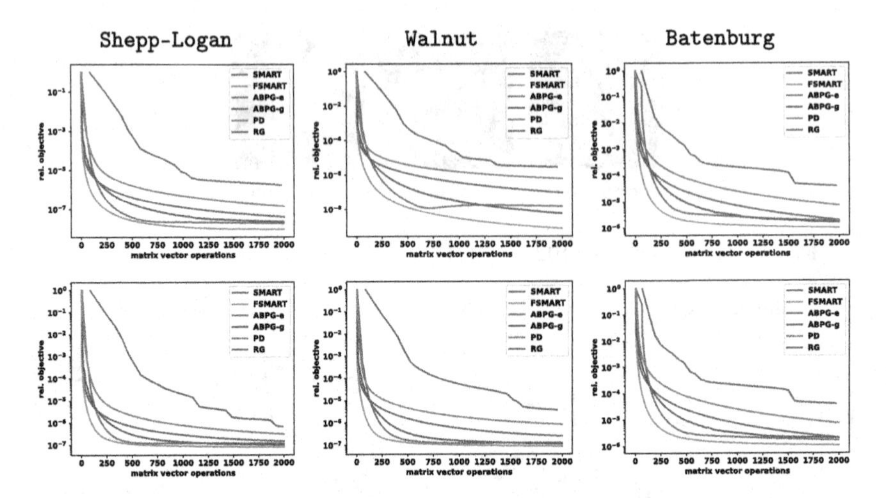

Fig. 3. Comparison of decreasing objective function values as function of costly operations between SMART, FSMART, ABPG-e, ABPG-g, PD (Chambolle-Pock) and RG (Riemannian gradient descent). The i-th column shows the i-th image in Fig. 1, in noiseless (top row) and noisy scenarios (bottom row). Checking the local triangle-scaling property incurs computational overhead for ABPG-e and ABPG-g. RG, the fastest method in terms of iterations, is the most expensive in terms of matrix vector operations due to the line search.

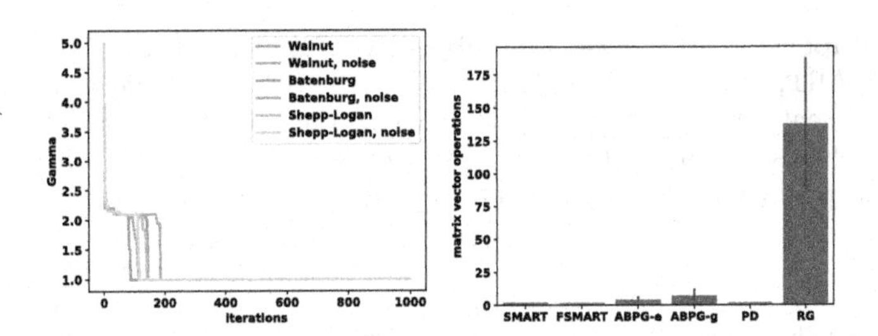

Fig. 4. A-posteriori certificates are obtained from ABPG-e by observing γ_k-values over iterations. These values are shown **left** for all problem instances. We observe that γ_k drops to 1 in all instances. We explored the drop-down-point for each instance and observed that it occurs when ABPG-e approaches the solution. Similar conclusions can be drawn from inspecting G_k in ABPG-g, which we omit here. **Average of matrix vector operations** are shown **right** for each algorithm over all iterates and tomography instances. By definition SMART, FSMART and PD always employ just two matrix vector operations per iteration. ABPG-e and ABPG-g require more such operations as they employ the local triangle scaling property, see (24), to guarantee sufficient decrease of the objective. Similarly, RG employs line search (38).

restarting mechanisms and stopping criteria based on the divergence of iterates were foregone. Updates for θ were conducted via Newton's method.

ABPG-g specified in [15, Algorithm 3] to (1) is used with parameters: $\rho = 1.2$, $\gamma = 2$ and $G_{\min} = 10^{-3}$. Restarting, stopping criteria and updating θ was handled analogously to ABPG-e.

RG is a SMART iteration with Armijo line search for choosing the step size τ_k via the retraction in (29) to iterate according to (37). The line search parameters are $\sigma = 0.5$, $\beta = 0.8$, $\alpha = 5.0$.

PD is the Chambolle-Pock primal dual algorithm [9, Algorithm 1] for solving convex composite structured optimization problems of the form $f(x) = g(x) + h(Ax)$. For $h(y) := \mathrm{KL}(y, b)$ with $y = Ax$ and $g \equiv 0$ we obtain

$$x^{k+1} = x^k e^{-\tau A^\top y^k} \qquad \text{(primal-step)} \tag{39}$$

$$y^{k+1} = \log\left(\frac{e^{y^k} + \sigma A(2x^{k+1} - x^k)}{\mathbb{1} + \sigma b}\right), \qquad \text{(dual-step)} \tag{40}$$

whereby we compute the primal step using the generalized proximal w.r.t. the KL divergence, defined in (4), and the dual step w.r.t. its dual divergence (7). The selected step size parameters were $\tau = \frac{1}{2L}$ and $\sigma = \frac{2}{L}$.

6 Conclusion

We explored recent acceleration techniques derived in the context of Bregman proximal methods (BPG) for SMART as well as the numerical a-posteriori certification of acceleration for a large scale problem. Even though the $\mathcal{O}(1/k^2)$ rate could not be certified in this way, the heuristically accelerated version FSMART turned out to be remarkably efficient. In addition, we characterized SMART as a Riemannian gradient descent scheme on the parameter manifold induced by the Fisher-Rao geometry which opens up possibilities for connecting the local triangle scaling property - employed by accelerated BPG for certifying convergence rates - with line search methods based on suitable retractions.

Acknowledgement. MK and MZ gratefully acknowledge the generous and invaluable support of the Klaus Tschira Foundation.

References

1. Absil, P.A., Mahony, R., Sepulchre, R.: Optimization Algorithms on Matrix Manifolds. Princeton University Press, Princeton (2008)
2. Amari, S.I., Cichocki, A.: Information geometry of divergence functions. Bull. Polish Acad. Sci **58**(1), 183–195 (2010)
3. Amari, S.I., Nagaoka, H.: Methods of Information Geometry. American Mathematical Society and Oxford University Press (2000)
4. Bauschke, H.G., Borwein, J.M.: Legendre functions and the method of random bregman projections. J. Convex Anal. **4**, 27–67 (1997)

5. Beck, A., Teboulle, M.: Mirror descent and nonlinear projected subgradient methods for convex optimization. Oper. Res. Lett. **31**(3), 167–175 (2003)
6. Brown, L.D.: Fundamentals of Statistical Exponential Families. Institute of Mathematical Statistics, Hayward (1986)
7. Byrne, C.L.: Iterative image reconstruction algorithms based on cross-entropy minimization. IEEE Trans. Image Process. **2**(1), 96–103 (1993)
8. Chambolle, A., Contreras, J.: Accelerated Bregman primal-dual methods applied to optimal transport and Wasserstein barycenter problems. SIAM J. Math. Data Sci. **4**(4), 1369–1395 (2022)
9. Chambolle, A., Pock, T.: On the ergodic convergence rates of a first-order primal-dual algorithm. Math. Program. **159**(1), 253–287 (2016)
10. Chen, G., Teboulle, M.: Convergence analysis of a proximal-like minimization algorithm using Bregman functions. SIAM J. Optim. **3**(3), 538–543 (1993)
11. Csiszár, I.: Why least squares and maximum entropy? An axiomatic approach to inference for linear inverse problems. Ann. Stat. **19**(4), 2032–2066 (1991)
12. Dragomir, R.A., Taylor, A.B., d'Aspremont, A., Bolte, J.: Optimal complexity and certification of Bregman first-order methods. Math. Program. **194**, 41–83 (2022)
13. El Gheche, M., Chierchia, G., Pesquet, J.C.: Proximity operators of discrete information divergences. IEEE Trans. Inf. Theory **64**(2), 1092–1104 (2017)
14. Gutman, D.H., Peña, J.F.: Perturbed Fenchel duality and first-order methods. Math. Program. **198**(1), 443–469 (2023)
15. Hanzely, F., Richtárik, P., Xiao, L.: Accelerated Bregman proximal gradient methods for relatively smooth convex optimization. Comput. Optim. Appl. **79**, 405–440 (2021)
16. Jost, J.: Riemannian Geometry and Geometric Analysis, 4th edn. Springer, Heidelberg (2005). https://doi.org/10.1007/3-540-28891-0
17. Lent, A., Censor, Y.: The primal-dual algorithm as a constraint-set-manipulation device. Math. Program. **50**(1–3), 343–357 (1991)
18. Nemirovski, A., Yudin, D.: Problem Complexity and Method Efficiency in Optimization. Wiley, Hoboken (1983)
19. Nesterov, Y.E., Todd, M.J.: On the Riemannian geometry defined by self-concordant barriers and interior-point methods. Found. Comput. Math. **2**(4), 333–361 (2002)
20. Petra, S., Schnörr, C., Becker, F., Lenzen, F.: B-SMART: Bregman-based first-order algorithms for non-negative compressed sensing problems. In: Kuijper, A., Bredies, K., Pock, T., Bischof, H. (eds.) SSVM 2013. LNCS, vol. 7893, pp. 110–124. Springer, Heidelberg (2013). https://doi.org/10.1007/978-3-642-38267-3_10
21. Teboulle, M.: A simplified view of first order methods for optimization. Math. Program. **170**(1), 67–96 (2018). https://doi.org/10.1007/s10107-018-1284-2
22. Tseng, P.: On Accelerated Proximal Gradient Methods for Convex-Concave Optimization (2008, unpublished manuscript)

Wasserstein Gradient Flows of the Discrepancy with Distance Kernel on the Line

Johannes Hertrich, Robert Beinert(✉), Manuel Gräf, and Gabriele Steidl

Institute of Mathematics, TU Berlin, Straße des 17. Juni 136, 10623 Berlin, Germany
{j.hertrich,beinert,graef,steidl}@math.tu-berlin.de
https://tu.berlin/imageanalysis/

Abstract. This paper provides results on Wasserstein gradient flows between measures on the real line. Utilizing the isometric embedding of the Wasserstein space $\mathcal{P}_2(\mathbb{R})$ into the Hilbert space $L_2((0,1))$, Wasserstein gradient flows of functionals on $\mathcal{P}_2(\mathbb{R})$ can be characterized as subgradient flows of associated functionals on $L_2((0,1))$. For the maximum mean discrepancy functional $\mathcal{F}_\nu$ with the non-smooth negative distance kernel $K(x,y) = -|x-y|$, we deduce a formula for the associated functional. This functional appears to be convex, and we show that $\mathcal{F}_\nu$ is convex along (generalized) geodesics. For the Dirac measure $\nu = \delta_q$, $q \in \mathbb{R}$ as end point of the flow, this enables us to determine the Wasserstein gradient flows analytically. Various examples of Wasserstein gradient flows are given for illustration.

Keywords: Maximum mean discrepancy · Wasserstein gradient flows · Riesz kernel

1 Introduction

Gradient flows provide a powerful tool for computing the minimizers of modeling functionals in certain applications. In particular, gradient flows on the Wasserstein space are an interesting field of research that combines optimization with (stochastic) dynamical systems and differential geometry. For a good overview on the theory, we refer to the books of Ambrosio, Gigli and Savaré [3], and Santambrogio [33]. Besides Wasserstein gradient flows of the Kullback–Leibler (KL) functional and the associated Fokker–Planck equation related to the overdamped Langevin dynamics, which were extensively examined in the literature, see, e.g., [21,28,30], flows of maximum mean discrepancy (MMD) functionals, i.e., of

$$\mathcal{F}_\nu(\mu) = \underbrace{\frac{1}{2}\iint_{\mathcal{X}\times\mathcal{X}} K(x,y)\,\mathrm{d}\mu(y)\,\mathrm{d}\mu(x)}_{\text{interaction energy } \mathcal{E}_K(\mu)} \underbrace{- \iint_{\mathcal{X}\times\mathcal{X}} K(x,y)\,\mathrm{d}\nu(y)\,\mathrm{d}\mu(x)}_{\text{potential energy } \mathcal{V}_{K,\nu}(\mu)} + \text{const}$$

Supported by the German Research Foundation (DFG) [grant numbers STE571/14-1, STE 571/16-1] and the Federal Ministry of Education and Research (BMBF, Germany) [grant number 13N15754].

L. Calatroni et al. (Eds.): SSVM 2023, LNCS 14009, pp. 431–443, 2023.
https://doi.org/10.1007/978-3-031-31975-4_33

with conditionally positive definite kernel $K\colon \mathcal{X}\times\mathcal{X}\to\mathbb{R}$ on a metric space $\mathcal{X}$, became popular in machine learning [4] and image processing [14]. Moreover, MMDs were used as loss functions in generative adversarial networks [6,13,24].

Wasserstein gradient flows of MMDs are not restricted to absolutely continuous measures and have a rich structure depending on the kernel. In the general setting, the analytic study of MMD flows is however very challenging. For example, it is unclear if particle flows—flows consisting of a fixed number of Dirac measures—are actually Wasserstein flows or not. If the kernel K is smooth, the authors of [4] answered this question and showed that particle flows of $\mathcal{F}_\nu$ are indeed Wasserstein gradient flows meaning that Wasserstein flows starting at an empirical measure remain empirical measures and coincide with usual gradient descent flows in $\mathbb{R}^d$. The situation changes for non-smooth kernels like the negative distance $K(x,y)=-\|x-y\|$ on $\mathcal{X}=\mathbb{R}^d$, which are of special interest in imaging applications like halftoning or dithering [14,16,18,34]. Restricting the study to the interaction energy $\mathcal{E}_K$, which is the repulsive part of the discrepancy and responsible for the proper spread of the measure, a theoretical study shows that empirical measures can become absolutely continuous ones and conversely, i.e. particles may explode. The concrete behavior of the flow depends also on the dimension, see [11,12,19,20], and can be determined using methods and results form potential theory [23,32]. Numerical studies in [2,20] indicate that this behaviour is also true for flows of the whole discrepancy $\mathcal{F}_\nu$.

In this paper, we take a first step to close the gap between the numerical and theoretical studies by restricting ourselves to the real line. More precisely, we give a analytic expression for flows of $\mathcal{F}_\nu$ starting or ending in a particle showing that the observed behaviour of the interaction energy part $\mathcal{E}_K$ dominates the potential energy part $\mathcal{V}_{K,\nu}$ and indeed governs the behaviour of the whole MMD flow, i.e., of $\mathcal{F}_\nu$. Optimal transport techniques that reduce the original transport to those on the line were successfully used in several applications [1,5,9,10,22,29]. When working on $\mathbb{R}$, we can exploit quantile functions of measures to embed the Wasserstein space $\mathcal{P}_2(\mathbb{R})$ into the Hilbert space of (equivalence classes) of square integrable functions $L_2((0,1))$. Then, instead of dealing with functionals on $\mathcal{P}_2(\mathbb{R})$, we can just work with associated functionals which are uniquely defined on a cone of $L_2((0,1))$. If the associated functional is convex, we will see that the original one is convex along (generalized) geodesics, which is a crucial property for the uniqueness of the Wasserstein gradient flow. Furthermore, we can characterize Wasserstein gradient flows using regular subdifferentials in $L_2((0,1))$. Note that the special case of Wasserstein gradient flows of the interaction energy was already considered in [7]. We will have a special look at the Wasserstein gradient flow of MMD for the negative distance kernel ending in δ_q. We will deduce an analytic formula for this flow and provide several examples to illustrate its behavior.

Outline of the Paper. In Sect. 2, we recall the basic notation on Wasserstein gradient flows in d dimensions. Then, in Sect. 3, we show how these flows can be simpler treated as gradient descent flows of an associated function on the Hilbert space $L_2((0,1))$. MMDs are introduced in Sect. 4. Then, in Sect. 5, we restrict our attention again to the real line and show how the associated functional looks for the MMD with negative distance kernel. In particular, this functional is convex. For the Dirac measure $\nu = \delta_q$, $q \in \mathbb{R}$, we give an explicit formula for the Wasserstein gradient flow of the MMD functional. Examples illustrating the behavior of the Wasserstein flows are provided in Sect. 6. Finally, conclusions are drawn in Sect. 7.

2 Wasserstein Gradient Flows

Let $\mathcal{M}(\mathbb{R}^d)$ denote the space of σ-additive, signed measures and $\mathcal{P}(\mathbb{R}^d)$ the set of probability measures. For $\mu \in \mathcal{M}(\mathbb{R}^d)$ and measurable $T\colon \mathbb{R}^d \to \mathbb{R}^n$, the *push-forward* of μ via T is given by $T_{\#}\mu := \mu \circ T^{-1}$. We consider the *Wasserstein space* $\mathcal{P}_2(\mathbb{R}^d) := \{\mu \in \mathcal{P}(\mathbb{R}^d) \colon \int_{\mathbb{R}^d} \|x\|_2^2 \, \mathrm{d}\mu(x) < \infty\}$ equipped with the *Wasserstein distance* $W_2\colon \mathcal{P}_2(\mathbb{R}^d) \times \mathcal{P}_2(\mathbb{R}^d) \to [0,\infty)$,

$$W_2^2(\mu,\nu) := \min_{\pi \in \Gamma(\mu,\nu)} \int_{\mathbb{R}^d \times \mathbb{R}^d} \|x-y\|_2^2 \, \mathrm{d}\pi(x,y), \qquad \mu,\nu \in \mathcal{P}_2(\mathbb{R}^d), \tag{1}$$

where $\Gamma(\mu,\nu) := \{\pi \in \mathcal{P}_2(\mathbb{R}^d \times \mathbb{R}^d) : (\pi_1)_{\#}\pi = \mu,\ (\pi_2)_{\#}\pi = \nu\}$ and $\pi_i(x) := x_i$, $i = 1,2$ for $x = (x_1,x_2)$. The set of optimal transport plans π realizing the minimum in (1) is denoted by $\Gamma^{\mathrm{opt}}(\mu,\nu)$. A curve $\gamma\colon I \to \mathcal{P}_2(\mathbb{R}^d)$ on an interval $I \subset \mathbb{R}$, is called a *geodesic* if there exists a constant $C \ge 0$ such that

$$W_2(\gamma(t_1),\gamma(t_2)) = C|t_2 - t_1|, \qquad \text{for all } t_1,t_2 \in I.$$

The Wasserstein space is a geodesic space, meaning that any two measures $\mu,\nu \in \mathcal{P}_2(\mathbb{R}^d)$ can be connected by a geodesic. The *regular tangent space* at $\mu \in \mathcal{P}_2(\mathbb{R}^d)$ is given by

$$\mathrm{T}_\mu\mathcal{P}_2(\mathbb{R}^d) := \overline{\{\lambda(T - \mathrm{Id}) : (\mathrm{Id},T)_{\#}\mu \in \Gamma^{\mathrm{opt}}(\mu,T_{\#}\mu),\ \lambda > 0\}}^{L_{2,\mu}}.$$

Here $L_{2,\mu}$ denotes the Bochner space of (equivalence classes of) functions $\xi : \mathbb{R}^d \to \mathbb{R}^d$ with finite $\|\xi\|_{L_{2,\mu}}^2 := \int_{\mathbb{R}^d} \|\xi(x)\|_2^2 \, \mathrm{d}\mu(x) < \infty$. Note that $\mathrm{T}_\mu\mathcal{P}_2(\mathbb{R}^d)$ is not a "classical" tangent space, in particular it is an infinite dimensional subspace of $L_{2,\mu}$ if μ is absolutely continuous and just $\mathbb{R}^d$ if $\mu = \delta_x$, $x \in \mathbb{R}^d$. In particular, this means that the Wasserstein space has only a "manifold-like" structure.

For $\lambda \in \mathbb{R}$, a function $\mathcal{F}\colon \mathcal{P}_2(\mathbb{R}^d) \to (-\infty,+\infty]$ is called *λ-convex along geodesics* if, for every $\mu,\nu \in \operatorname{dom}\mathcal{F} := \{\mu \in \mathcal{P}_2(\mathbb{R}^d) : \mathcal{F}(\mu) < \infty\}$, there exists at least one geodesic $\gamma\colon [0,1] \to \mathcal{P}_2(\mathbb{R}^d)$ between μ and ν such that

$$\mathcal{F}(\gamma(t)) \le (1-t)\,\mathcal{F}(\mu) + t\,\mathcal{F}(\nu) - \tfrac{\lambda}{2} t(1-t)\, W_2^2(\mu,\nu), \qquad t \in [0,1].$$

In the case $\lambda = 0$, we just speak about convex functions. For a proper and lower semi-continuous (lsc) function $\mathcal{F}\colon \mathcal{P}_2(\mathbb{R}^d) \to (-\infty,\infty]$ and $\mu \in \mathcal{P}_2(\mathbb{R}^d)$, the *reduced Fréchet subdifferential $\partial\mathcal{F}$ at μ* consists of all $\xi \in L_{2,\mu}$ satisfying

$$\mathcal{F}(\nu) - \mathcal{F}(\mu) \ge \inf_{\pi \in \Gamma^{\mathrm{opt}}(\mu,\nu)} \int_{\mathbb{R}^{2d}} \langle \xi(x), y - x\rangle \, \mathrm{d}\pi(x,y) + o(W_2(\mu,\nu)) \tag{2}$$

for all $\nu \in \mathcal{P}_2(\mathbb{R}^d)$. A curve $\gamma\colon I \to \mathcal{P}_2(\mathbb{R}^d)$ is *absolutely continuous*, if there exists a Borel velocity field $v_t\colon \mathbb{R}^d \to \mathbb{R}^d$ with $\int_I \|v_t\|_{L_{2,\gamma(t)}}\,\mathrm{d}t < +\infty$ such that

$$\partial_t \gamma(t) + \nabla_x \cdot (v_t\, \gamma(t)) = 0 \tag{3}$$

on $I \times \mathbb{R}^d$ in the distributive sense, i.e., for all $\varphi \in C_c^\infty(I \times \mathbb{R}^d)$ it holds

$$\int_I \int_{\mathbb{R}^d} \partial_t \varphi(t,x) + v_t(x) \cdot \nabla_x \varphi(t,x)\,\mathrm{d}\gamma(t)\,\mathrm{d}t = 0.$$

A locally absolutely continuous curve $\gamma\colon (0,+\infty) \to \mathcal{P}_2(\mathbb{R}^d)$ with velocity field $v_t \in \mathrm{T}_{\gamma(t)}\mathcal{P}_2(\mathbb{R}^d)$ is called *Wasserstein gradient flow with respect to* $\mathcal{F}\colon \mathcal{P}_2(\mathbb{R}^d) \to (-\infty,+\infty]$ if

$$v_t \in -\partial \mathcal{F}(\gamma(t)), \quad \text{for a.e. } t > 0. \tag{4}$$

3 Wasserstein Gradient Flows on the Line

Now we restrict our attention to $d = 1$, i.e., we work on the real line. We will see that the above notation simplifies since there is an isometric embedding of $\mathcal{P}_2(\mathbb{R})$ into $L_2((0,1))$. To this end, we consider the *cumulative distribution function* $R_\mu\colon \mathbb{R} \to [0,1]$ of $\mu \in \mathcal{P}_2(\mathbb{R})$, which is defined by $R_\mu(x) := \mu((-\infty,x])$, $x \in \mathbb{R}$. It is non-decreasing and right-continuous with $\lim_{x\to-\infty} R_\mu(x) = 0$ as well as $\lim_{x\to\infty} R_\mu(x) = 1$. The *quantile function* $Q_\mu\colon (0,1) \to \mathbb{R}$ is the generalized inverse of R_μ given by $Q_\mu(p) := \min\{x \in \mathbb{R}\colon R_\mu(x) \ge p\}$, $p \in (0,1)$. It is non-decreasing and left-continuous. The quantile functions form a convex cone $\mathcal{C}((0,1)) := \{Q \in L_2((0,1)) : Q \text{ nondecreasing}\}$ in $L_2((0,1))$. Note that both the distribution and quantile functions are continuous except for at most countably many jumps. For a good overview see [31, § 1.1]. By the following theorem, the mapping $\mu \mapsto Q_\mu$ is an isometric embedding of $\mathcal{P}_2(\mathbb{R})$ into $L_2((0,1))$. Here the Lebesgue measure and its restriction to intervals is denoted by $\Lambda_\bullet$.

Theorem 1 **([35, Thm 2.18]).** *For $\mu,\nu \in \mathcal{P}_2(\mathbb{R})$, the quantile function $Q_\mu \in \mathcal{C}((0,1))$ satisfies $\mu = (Q_\mu)_\#\Lambda_{(0,1)}$ and*

$$W_2^2(\mu,\nu) = \int_0^1 |Q_\mu(s) - Q_\nu(s)|^2 \mathrm{d}s.$$

Next we will see that instead of working with functionals $\mathcal{F} : \mathcal{P}_2(\mathbb{R}) \to (-\infty,+\infty]$, we can just deal with associated functionals $\mathrm{F}\colon L_2((0,1)) \to (-\infty,\infty]$ fulfilling $\mathrm{F}(Q_\mu) := \mathcal{F}(\mu)$. Note that F is defined in this way only on $\mathcal{C}((0,1))$, and there exist several continuous extensions to the whole linear space $L_2((0,1))$. Instead of the extended Fréchet subdifferential (2), we will use the *regular subdifferential* in $L_2((0,1))$ defined by

$$\partial G(f) := \big\{h \in L_2((0,1))\colon G(g) \ge G(f) + \langle h, g - f\rangle + o(\|g-f\|_{L_2})\ \forall g \in L_2((0,1))\big\}.$$

The following theorem characterizes Wasserstein gradient flows by this regular subdifferential and states a convexity relation between $\mathcal{F} : \mathcal{P}_2(\mathbb{R}) \to (-\infty,+\infty]$ and the associated functional F.

Theorem 2. i) *Let* $\gamma\colon (0,\infty) \to \mathcal{P}_2(\mathbb{R})$ *be a locally absolutely continuous curve and* $\mathrm{F}\colon L_2((0,1)) \to (-\infty,\infty]$ *such that the pointwise derivative* $\partial_t Q_{\gamma(t)}$ *exists and fulfills the* L_2 *subgradient equation*

$$\partial_t Q_{\gamma(t)} \in -\partial \mathrm{F}(Q_{\gamma(t)}), \quad \text{for almost every } t \in (0,+\infty).$$

Then γ *is a Wasserstein gradient flow with respect to the functional* $\mathcal{F} : \mathcal{P}_2(\mathbb{R}) \to (-\infty,+\infty]$ *defined by* $\mathcal{F}(\mu) := \mathrm{F}(Q_\mu)$.
ii) *If* $\mathrm{F} : \mathcal{C}((0,1)) \to (-\infty,\infty]$ *is convex, then* $\mathcal{F}(\mu) := \mathrm{F}(Q_\mu)$ *is convex along geodesics.*

Proof. i) Since γ is (locally) absolute continuous, the velocity field v_t from (3) fulfills by [3, Prop 8.4.6] for almost every $t \in (0,\infty)$ the relation

$$\begin{aligned}
0 &= \lim_{h\to 0} \frac{W_2(\gamma(t+h), (\mathrm{Id} + hv_t)_\# \gamma(t))}{|h|}\\
&= \lim_{h\to 0} \frac{W_2((Q_{\gamma(t+h)})_\# \Lambda_{(0,1)}, (Q_{\gamma(t)} + h(v_t \circ Q_{\gamma(t)}))_\# \Lambda_{(0,1)})}{|h|}\\
&= \lim_{h\to 0} \left\| \frac{Q_{\gamma(t+h)} - Q_{\gamma(t)}}{h} - v_t \circ Q_{\gamma(t)} \right\|_{L_2} = \| \partial_t Q_{\gamma(t)} - v_t \circ Q_{\gamma(t)} \|_{L_2}.
\end{aligned}$$

Thus, by assumption, $v_t \circ Q_{\gamma(t)} \in -\partial \mathrm{F}(Q_{\gamma(t)})$ a.e. In particular, for any $\mu \in \mathcal{P}_2(\mathbb{R})$, we obtain

$$\begin{aligned}
&\mathrm{F}(Q_\mu) - \mathrm{F}(Q_{\gamma(t)}) + \int_0^1 v_t(Q_{\gamma(t)}(s))\,(Q_\mu(s) - Q_{\gamma(t)}(s))\,\mathrm{d}s + o(\|Q_\mu - Q_{\gamma(t)}\|_{L_2})\\
&= \mathcal{F}(\mu) - \mathcal{F}(\gamma(t)) + \int_{\mathbb{R}\times\mathbb{R}} v_t(x)\,(y-x)\,\mathrm{d}\pi(x,y) + o\,(W_2(\mu,\gamma(t))) \geq 0,
\end{aligned}$$

where $\pi := (Q_{\gamma(t)}, Q_\mu)_\# \Lambda_{(0,1)}$. Since π the unique optimal transport plan between $\gamma(t)$ and μ, this yields by (2) that $v_t \in -\partial\mathcal{F}(\gamma(t))$ showing the assertion by (4).
ii) Let $\mathrm{F}\colon L_2((0,1)) \to \mathbb{R}$ be convex. For any geodesic $\gamma : [0,1] \to \mathcal{P}_2(\mathbb{R})$, since $\mu \mapsto Q_\mu$ is an isometry, the curve $t \mapsto Q_{\gamma(t)}$ is a geodesic in $L_2((0,1))$ too. Since $L_2((0,1))$ is a linear space, the convexity of $\mathrm{F}\colon L_2((0,1)) \to \mathbb{R}$ yields that $t \mapsto \mathrm{F}(Q_{\gamma(t)}) = \mathcal{F}(\gamma(t))$ is convex. Thus, $\mathcal{F}$ is convex along γ. □

Remark 1. If $\mathcal{F}\colon \mathcal{P}_2(\mathbb{R}) \to (-\infty,+\infty]$ is proper, lsc, coercive and λ-convex along so-called generalized geodesics, then the Wasserstein gradient flow starting at any $\mu_0 \in \overline{\mathrm{dom}\,\mathcal{F}}$ is uniquely determined and is the uniform limit of the miminizing movement scheme of Jordan, Kinderlehrer and Otto [21] when the time step size goes to zero, see [3, Thm 11.2.1]. In $\mathbb{R}$, but not in higher dimensions, λ-convex functions along geodesics fulfill also the stronger property that they are λ-convex along generalized geodesics, see [20].

4 Discrepancies

We consider symmetric and *conditionally positive definite kernels* $K\colon \mathbb{R}^d \times \mathbb{R}^d \to \mathbb{R}$ of order one, i.e., for any $n \in \mathbb{N}$, any pairwise different points $x^1, \ldots, x^n \in \mathbb{R}^d$ and any $a_1, \ldots, a_n \in \mathbb{R}$ with $\sum_{i=1}^n a_i = 0$ the relation $\sum_{i,j=1}^n a_i a_j K(x^i, x^j) \geq 0$ is satisfied. Typical examples are Riesz kernels

$$K(x,y) := -\|x-y\|^r, \quad r \in (0,2),$$

where we have strict inequality except for all a_j, $j = 1, \ldots, n$ being zero. The *maximum mean discrepancy* (MMD) $\mathcal{D}_K^2\colon \mathcal{P}(\mathbb{R}^d) \times \mathcal{P}(\mathbb{R}^d) \to \mathbb{R}$ between two measures $\mu, \nu \in \mathcal{P}(\mathbb{R}^d)$ is defined by

$$\mathcal{D}_K^2(\mu,\nu) := \mathcal{E}_K(\mu - \nu)$$

with the so-called *K-energy* or *interaction energy*

$$\mathcal{E}_K(\sigma) := \frac{1}{2} \int_{\mathbb{R}^d} \int_{\mathbb{R}^d} K(x,y) \,\mathrm{d}\sigma(x) \mathrm{d}\sigma(y), \qquad \sigma \in \mathcal{M}(\mathbb{R}^d).$$

The relation between discrepancies and Wasserstein distances is discussed in [15,26]. For fixed $\nu \in \mathcal{P}(\mathbb{R}^d)$, the MMD can be decomposed as

$$\mathcal{F}_\nu(\mu) = \mathcal{D}_K^2(\mu,\nu) = \mathcal{E}_K(\mu) + \mathcal{V}_{K,\nu}(\mu) + \text{const}, \tag{5}$$

where $\mathcal{V}_{K,\nu}(\mu) := -\int_{\mathbb{R}^d} \int_{\mathbb{R}^d} K(x,y) \,\mathrm{d}\nu(y) \,\mathrm{d}\mu(x)$ is the *potential energy* with respect to ν. In dimensions $d \geq 2$ neither $\mathcal{E}_K$ nor $\mathcal{D}_K^2$ with the Riesz kernel are λ-convex along geodesics, see [20], so that certain properties of Wasserstein gradient flows do not apply. We will see that this is different on the real line.

5 MMD Flows on the Line

In the rest of this paper, we restrict our attention to $d = 1$ and negative distance $K(x,y) = -|x-y|$, i.e. to Riesz kernels with $r = 1$.

Lemma 1. *For $\mathcal{F}_\nu$ in* (5) *with the negative distance kernel, the convex functional $\mathrm{F}_\nu\colon L_2((0,1)) \to \mathbb{R}$ defined by*

$$\mathrm{F}_\nu(f) := \int_0^1 \Big((1-2s)(f(s) + Q_\nu(s)) + \int_0^1 |f(s) - Q_\nu(t)| \,\mathrm{d}t\Big) \,\mathrm{d}s. \tag{6}$$

fulfills $\mathrm{F}_\nu(Q_\mu) = \mathcal{F}_\nu(\mu)$ for all $\mu \in \mathcal{P}_2(\mathbb{R})$. In particular, $\mathcal{F}_\nu$ is convex along (generalized) geodesics and there exists a unique Wasserstein gradient flow.

Proof. We reformulate $\mathcal{F}_\nu$ as

$$\begin{aligned}\mathcal{F}_\nu(\mu) &= -\frac{1}{2}\int_{\mathbb{R}\times\mathbb{R}} |x-y|(\mathrm{d}\mu(x)-\mathrm{d}\nu(x))(\mathrm{d}\mu(y)-\mathrm{d}\nu(y))\\ &= -\frac{1}{2}\int_0^1\int_0^1 |Q_\mu(s)-Q_\mu(t)| - 2|Q_\mu(s)-Q_\nu(t)| + |Q_\nu(s)-Q_\nu(t)|\,\mathrm{d}s\,\mathrm{d}t\\ &= \int_0^1\int_t^1 Q_\mu(t)-Q_\mu(s)+Q_\nu(t)-Q_\nu(s)\,\mathrm{d}s\,\mathrm{d}t + \int_0^1\int_0^1 |Q_\mu(s)-Q_\nu(t)|\,\mathrm{d}s\,\mathrm{d}t\\ &= \int_0^1\Big((1-2s)(Q_\mu(s)+Q_\nu(s)) + \int_0^1 |Q_\mu(s)-Q_\nu(t)|\,\mathrm{d}t\Big)\,\mathrm{d}s,\end{aligned}$$

which yields the first claim. The second one follows by Theorem 2ii) and Remark 1. □

Note that the lemma cannot be generalized to Riesz kernels with $r = (1,2)$. Finally, we derive for the special choice $\nu = \delta_q$ in $\mathcal{D}_K^2(\cdot,\nu)$ an analytic formula for its Wasserstein gradient flow.

Proposition 1. *Let $\mathcal{F}_{\delta_q} := \mathcal{D}_K^2(\cdot,\delta_q)$ with the negative distance kernel. Then the unique Wasserstein gradient flow of $\mathcal{F}_{\delta_q}$ starting at $\mu_0 = \gamma(0) \in \mathcal{P}_2(\mathbb{R})$ is $\gamma(t) = (g_t)_\# \Lambda_{(0,1)}$, where the function $g_t\colon (0,1)\to\mathbb{R}$ is given by*

$$g_t(s) := \begin{cases} \min\{Q_{\mu_0}(s)+2st, q\}, & Q_{\mu_0}(s) < q,\\ q, & Q_{\mu_0}(s) = q,\\ \max\{Q_{\mu_0}(s)+2st-2t, q\}, & Q_{\mu_0}(s) > q.\end{cases} \tag{7}$$

Proof. First, note that $g_t \in C((0,1))$ such that it holds $g_t = Q_{\gamma(t)}$. Since $Q_{\delta_q} \equiv q$, the subdifferential of F_{δ_q} in (6) at g_t consists of all functions

$$h(s) = \begin{cases} -2s, & Q_{\mu_0}(s) < q \text{ and } t < \frac{q-Q_{\mu_0}(s)}{2s},\\ 2-2s, & Q_{\mu_0}(s) > q \text{ and } t < \frac{Q_{\mu_0}(s)-q}{2-2s},\\ 1-2s+n(s), & \text{otherwise,}\end{cases}$$

with $-1 \le n(s) \le 1$ for $s \in (0,1)$. On the other hand, the pointwise derivative of g_t in (7) can be written as

$$\partial_t g_t(s) = \begin{cases} 2s, & Q_{\mu_0}(s) < q \text{ and } t < \frac{q-Q_{\mu_0}(s)}{2s},\\ 2s-2, & Q_{\mu_0}(s) > q \text{ and } t < \frac{Q_{\mu_0}(s)-q}{2-2s},\\ 0, & \text{otherwise,}\end{cases}$$

such that we obtain $\partial_t Q_{\gamma(t)} = \partial_t g_t \in -\partial \mathrm{F}_\nu(g_t) = -\partial\mathrm{F}_\nu(Q_{\gamma(t)})$. Thus, by Lemma 1 and Theorem 2, we obtain that γ is a Wasserstein gradient flow. It is unique since $\mathcal{F}_\nu$ is convex along geodesics by Theorem 2.ii, Lemma 1 and Remark 1. □

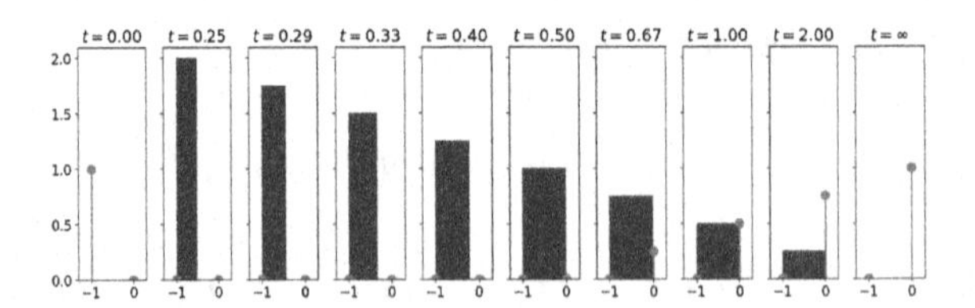

Fig. 1. Visualization of the Wasserstein gradient flow of $\mathcal{F}_{\delta_0}$ from δ_{-1} to δ_0. At various times t, the absolute continuous part is visualized by its density in blue (area equals mass) and the atomic part by the red dotted vertical line (height equals mass). The atomic part at the end point $x = 0$ starts to grow at time $t = 1/2$, where the support of the density touches this point for the first time.

6 Intuitive Examples

Finally, we provide some intuitive examples of Wasserstein gradient flows of $\mathcal{F}_\nu := \mathcal{D}_K^2(\cdot, \nu)$ with the negative distance kernel.

6.1 Flow Between Dirac Measures

We consider the flow of $\mathcal{F}_{\delta_0}$ starting at the initial measure $\gamma(0) = \mu_0 := \delta_{-1}$. Due to $Q_{\delta_0} \equiv 0$, Proposition 1 yields the gradient flow $\gamma(t) := (Q_t)_\# \Lambda_{(0,1)}$ given by

$$\gamma(t) = \begin{cases} \delta_{-1}, & t = 0, \\ \frac{1}{2t} \Lambda_{[-1,-1+2t]}, & 0 \le t \le \frac{1}{2}, \\ \frac{1}{2t} \Lambda_{[-1,0]} + \left(1 - \frac{1}{2t}\right) \delta_0, & \frac{1}{2} < t. \end{cases}$$

For $t \in (0, \frac{1}{2}]$, the initial Dirac measure becomes a uniform measure with increasing support, and for $t \in (\frac{1}{2}, \infty)$ it is the convex combination of a uniform measure and δ_0. A visualization of the flow is given in Fig. 1.

6.2 Flow on Restricted Sets

Next, we are interested in the Wasserstein gradient flows on the subsets $\mathcal{S}_i$, $i = 1, 2$, given by

(i) $\mathcal{S}_1 := \{\delta_x : x \in \mathbb{R}\}$,
(ii) $\mathcal{S}_2 := \{\mu_{m,\sigma} = \frac{1}{2\sqrt{3}\sigma} \Lambda_{[m-\sqrt{3}\sigma, m+\sqrt{3}\sigma]} : m \in \mathbb{R}, \sigma \in \mathbb{R}_{\ge 0}\}$.

Note that $\S_2$ is a special instance of sets of scaled and translated measures $\mu \in \mathcal{P}_2(\mathbb{R})$ defined by $\{T_{a,b\#}\mu : a \in \mathbb{R}_{\ge 0}, b \in \mathbb{R}\}$, where $T_{a,b}(x) := ax + b$. As mentioned in [17] the Wasserstein distance between measures μ_1, μ_2 from such sets has been already known to Fréchet:

$$W_2^2(\mu_1, \mu_2) = |m_1 - m_2|^2 + |\sigma_1 - \sigma_2|^2,$$

where m_i and σ_i are the mean value and standard deviation of μ_i, $i = 1, 2$. This provides an isometric embedding of $\mathbb{R} \times \mathbb{R}_{\ge 0}$ into $\mathcal{P}_2(\mathbb{R})$. The boundary of $\mathcal{S}_2$

is the set of Dirac measures $\mathcal{S}_1$ and is isometric to $\mathbb{R}$. The sets are convex in the sense that for $\mu, \nu \in \mathcal{S}_i$ all geodesics $\gamma : [0,1] \to \mathcal{P}(\mathbb{R})$ with $\gamma(0) = \mu$ and $\gamma(1) = \nu$ are in $\mathcal{S}_i$, $i \in \{1,2\}$. For $i = 1, 2$, we consider

$$\mathcal{F}_{i,\nu}(\mu) := \begin{cases} \mathcal{F}_\nu & \mu \in \mathcal{S}_i, \\ +\infty & \text{otherwise.} \end{cases}$$

Due to the convexity of $\mathcal{F}_\nu$ along geodesics and the convexity of the sets $\mathcal{S}_i$, we obtain that the functions $\mathcal{F}_{i,\nu}$ are convex along geodesics.

Flows of $\mathcal{F}_{1,\nu}$. We use the notation $f_x \equiv x$ for the constant function on $(0,1)$ with value x. It is straightforward to check that the function $\mathrm{F}\colon L^2((0,1)) \to (-\infty, \infty]$ given by

$$\mathrm{F}(f) = \begin{cases} F(x), & \text{if } f = f_x \text{ for some } x \in \mathbb{R}, \\ +\infty, & \text{otherwise,} \end{cases}$$

with

$$F(x) := \int_{\mathbb{R}} |x - y| \,\mathrm{d}\nu(y) - \frac{1}{2} \int_{\mathbb{R}\times\mathbb{R}} |y - z| \,\mathrm{d}\nu(y)\mathrm{d}\nu(z)$$

fulfills $\mathrm{F}(Q_\mu) = \mathcal{F}_{1,\nu}(\mu)$. In the following, we aim to find $x\colon [0,\infty) \to \mathbb{R}$ satisfying

$$\dot{x}(t) = -\partial F(x(t)).$$

Since the set $\{Q_\mu : \mu \in \mathcal{S}_1\}$ is a one-dimensional linear subspace of $L_2((0,1))$ spanned by f_1, this yields $f_{x(t)} \in -\partial \mathrm{F}(f_{x(t)})$ such that the Wasserstein gradient flow is by Theorem 2 given by $\gamma(t) = (f_{x(t)})_{\#}\Lambda_{(0,1)} = \delta_{x(t)}$.

In the special case $\nu = \delta_q$ for some $q \in \mathbb{R}$, we have

$$F(x) = |x - q|, \qquad \partial F(x) = \begin{cases} \{-1\}, & x < q, \\ [-1,1], & x = q, \\ \{1\}, & x > q. \end{cases}$$

Therefore, the Wasserstein gradient flow for $x(0) = x_0 \neq 0$ is given by

$$\gamma(t) = \delta_{x(t)}, \quad \text{with} \quad x(t) = \begin{cases} x_0 + t, & x_0 < q, \\ x_0 - t, & x_0 > q. \end{cases}, \qquad 0 \le t < |x_0 - q|$$

and $\gamma(t) = \delta_q$ for $t \ge |x_0 - q|$.

For $\nu = \frac{1}{2}\Lambda_{[-1,1]}$ the gradient flow starting at $x_0 \in [-1,1]$ is $x(t) = x_0 e^{-t}$, $t \ge 0$, and converges to the midpoint of the interval for $t \to \infty$. If it starts at $x_0 \in \mathbb{R} \setminus [-1,1]$ the gradient flow is

$$x(t) = \begin{cases} x_0 + t, & x_0 < -1, \\ x_0 - t, & x_0 > 1. \end{cases}, \qquad 0 \le t \le \min |x_0 - 1|, |x_0 + 1|,$$

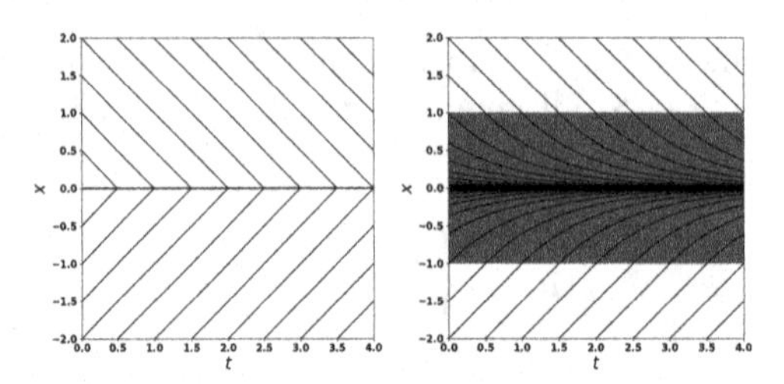

Fig. 2. Wasserstein gradient flow of $\mathcal{F}_{1,\nu}$ for $\nu = \delta_0$ (left) and $\nu = \frac{1}{2}\Lambda_{[-1,1]}$ (right) from various initial points δ_x, $x \in [-2, 2]$. The support of the right measure ν is depicted by the blue region. (Color figure online)

where it reaches the nearest interval end point in finite time. In Fig. 2, we plotted the $x(t)$ for different initial values $x(0)$. The examples show that gradient flows may reach the optimal points in finite or infinite time.

Flows of $\mathcal{F}_{2,\nu}$. We observe that $Q_{\mu_{m,\sigma}} = f_{m,\sigma}$, where $f_{m,\sigma}(x) = m+2\sqrt{3}\sigma(x-\frac{1}{2})$. By Lemma 1 we obtain that the function $\mathrm{F}\colon L_2((0,1)) \to (-\infty, \infty]$ given by

$$\mathrm{F}(f) = \begin{cases} F(m,\sigma), & \text{if } f = f_{m,\sigma} \text{ for } (m,\sigma) \in \mathbb{R} \times \mathbb{R}_{\geq 0}, \\ +\infty, & \text{otherwise}, \end{cases}$$

fulfills $\mathrm{F}(Q_\mu) = \mathcal{F}_{2,\nu}(\mu)$, where

$$F(m,\sigma) := \int_{(0,1)} (1-2s)(f_{m,\sigma}(s) + Q_\nu(s))\mathrm{d}s + \int_{(0,1)^2} |f_{m,\sigma}(s) - Q_\nu(t)|\mathrm{d}t\mathrm{d}s,$$

The set $\{f_{m,\sigma} : m, \sigma \in \mathbb{R}\}$ is a two dimensional linear subspace of $L_2((0,1))$ with orthonormal basis $\{f_{1,0}, f_{0,1}\}$. We aim to compute $m\colon [0,\infty) \to \mathbb{R}$ and $\sigma\colon [0,\infty) \to \mathbb{R}_{\geq 0}$ with

$$(\dot{m}(t), \dot{\sigma}(t)) = -\partial F(m(t), \sigma(t)), \qquad t \in I \subset \mathbb{R}, \tag{8}$$

yielding $f_{m(t),\sigma(t)} \in -\partial\mathrm{F}(f_{m(t),\sigma(t)})$ such that $\gamma(t) = (f_{m(t),\sigma(t)})_{\#}\Lambda_{(0,1)} = \mu_{m,\sigma}$ is by Theorem 2 the Wasserstein gradient flow.

In the following, we consider the special case $\nu = \delta_0 = \mu_{0,0}$. Then, the function F reduces to

$$\begin{aligned} F(m,\sigma) &= \int_{\mathbb{R}} (1-2s)(m + 2\sqrt{3}\sigma(s-\tfrac{1}{2})) + |m + 2\sqrt{3}\sigma(s-\tfrac{1}{2})|\mathrm{d}s \\ &= -\frac{\sigma}{\sqrt{3}} + \begin{cases} |m| & \text{if } |m| \geq \sqrt{3}\sigma, \\ \frac{m^2+3\sigma^2}{2\sqrt{3}\sigma^2} & \text{if } |m| < \sqrt{3}\sigma, \end{cases} \end{aligned}$$

and the subdifferential is given by

$$\partial F(m,\sigma) = \begin{cases} \operatorname{sgn}(m) \times \{-\frac{1}{\sqrt{3}}\}, & \text{if } |m| \geq \sqrt{3}\sigma, \\ \{(\frac{m}{\sqrt{3}\sigma^2}, \frac{-m^2}{\sqrt{3}\sigma^3} - \frac{1}{\sqrt{3}})\}, & \text{if } |m| < \sqrt{3}\sigma, \end{cases}$$

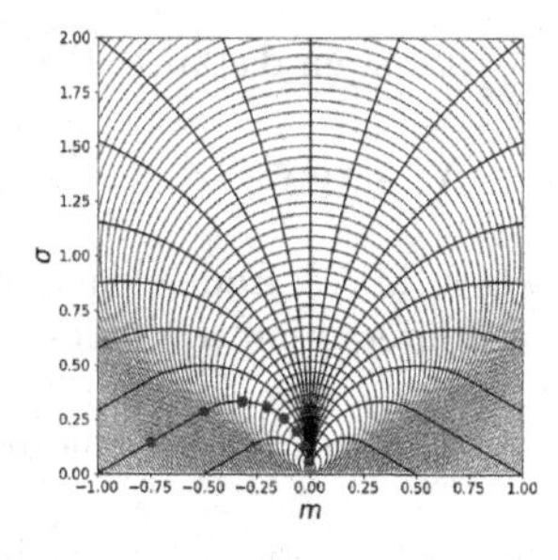

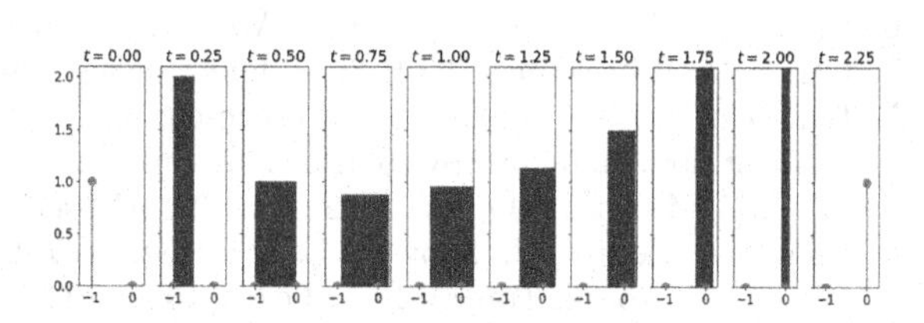

Fig. 3. Wasserstein gradient flow $\mathcal{F}_{2,\delta_0}$ from $(m(0), \sigma(0))$ to δ_0 (left) and from δ_{-1} to δ_0 (right). In contrast Fig. 1 it is a uniform measure for all $t \in (0, 1)$.

where $\mathrm{sgn}(m) = \{|m|/m\}$ if $m \neq 0$ and $\mathrm{sgn}(0) = [-1, 1]$. We observe that F is differentiable for $\sigma > 0$. Thus, for any initial intial value $(m(0), \sigma(0)) = (m_0, \sigma_0)$, we can compute the trajectory $(m(t), \sigma(t))$ solving (8) using an ODE solver. In Fig. 3 (left), we plotted the level sets of the function $F(m, \sigma)$ as well as the solution trajectory $(m(t), \sigma(t))$ for different initial values $(m(0), \sigma(0))$. For $(m(0), \sigma(0)) = (-1, 0)$, the resulting flow is illustrated in Fig. 3 (right).

7 Conclusions

We provided insight into Wasserstein gradient flows of MMD functionals with negative distance kernels and characterized in particular flows ending in a Dirac measure. We have seen that such flows are not simple particle flows, e.g. starting in another Dirac measure the flow becomes immediately uniformly distributed and after a certain time a mixture of a uniform and a Dirac measure. In our future work, we want to extend our considerations to empirical measures and incorporate deep learning techniques. Also the treatment of other functionals which incorporate an interaction energy part appears to be interesting. Further, we may combine univariate techniques with multivariate settings using Radon transform like techniques as in [8,25,27].

References

1. Abraham, I., Abraham, R., Bergounioux, M., Carlier, G.: Tomographic reconstruction from a few views: a multi-marginal optimal transport approach. Appl. Math. Optim. **75**(1), 55–73 (2017)
2. Altekrüger, F., Hertrich, J., Steidl, G.: Neural Wasserstein gradient flows for maximum mean discrepancies with Riesz kernels. arXiv:2301.11624 (2023)
3. Ambrosio, L., Gigli, N., Savare, G.: Gradient Flows. Lectures in Mathematics ETH Zürich, Birkhäuser, Basel (2005)
4. Arbel, M., Korba, A., Salim, A., Gretton, A.: Maximum mean discrepancy gradient flow. In: Wallach, H., Larochelle, H., Beygelzimer, A., d Alché-Buc, F., Fox, E., Garnett, R. (eds.) Advances in Neural Information Processing Systems, vol. 32, pp. 1–11. Curran Associates Inc., New York (2019)

5. Beier, F., Beinert, R., Steidl, G.: On a linear Gromov-Wasserstein distance. IEEE Trans. Image Process. **31**, 7292–7305 (2022)
6. Binkowski, M., Sutherland, D.J., Arbel, M., Gretton, A.: Demystifying MMD GANs. In: Proceedings ICLR 2018. OpenReview (2018)
7. Bonaschi, G.A., Carrillo, J.A., Francesco, M.D., Peletier, M.A.: Equivalence of gradient flows and entropy solutions for singular nonlocal interaction equations in 1D. ESAIM Control Optim. Calc. Var. **21**, 414–441 (2015)
8. Bonet, C., Courty, N., Septier, F., Drumetz, L.: Efficient gradient flows in sliced-Wasserstein space. Trans. Mach. Learn. Res. (2022)
9. Bonneel, N., Rabin, J., Peyré, G., Pfister, H.: Sliced and Radon Wasserstein barycenters of measures. J. Math. Imaging Vis. **1**(51), 22–45 (2015)
10. Cai, T., Cheng, J., Schmitzer, B., Thorpe, M.: The linearized Hellinger-Kantorovich distance. arXiv:2102.08807 (2021)
11. Carrillo, J.A., Huang, Y.: Explicit equilibrium solutions for the aggregation equation with power-law potentials. Kinetic Related Models **10**(1), 171–192 (2017)
12. Chafaï, D., Saff, E.B., Womersley, R.S.: Threshold condensation to singular support for a Riesz equilibrium problem. arXiv:2206.04956v1 (2022)
13. Dziugaite, G.K., Roy, D.M., Ghahramani, Z.: Training generative neural networks via maximum mean discrepancy optimization. In: Proceedings UAI 2015. UAI (2015)
14. Ehler, M., Gräf, M., Neumayer, S., Steidl, G.: Curve based approximation of measures on manifolds by discrepancy minimization. Found. Comput. Math. **21**(6), 1595–1642 (2021)
15. Feydy, J., Séjourné, T., Vialard, F.X., Amari, S., Trouvé, A., Peyré, G.: Interpolating between optimal transport and MMD using Sinkhorn divergences. In: Proceedings of Machine Learning Research, vol. 89, pp. 2681–2690. PMLR (2019)
16. Fornasier, M., Haskovec, J., Steidl, G.: Consistency of variational continuous-domain quantization via kinetic theory. Appl. Anal. **92**(6), 1283–1298 (2013)
17. Gelbrich, M.: On a formula for the L^2 Wasserstein metric between measures on Euclidean and Hilbert spaces. Math. Nachr. **147**(1), 185–203 (1990)
18. Gräf, M., Potts, M., Steidl, G.: Quadrature errors, discrepancies and their relations to halftoning on the torus and the sphere. SIAM J. Sci. Comput. **34**(5), 2760–2791 (2012)
19. Gutleb, T.S., Carrillo, J.A., Olver, S.: Computation of power law equilibrium measures on balls of arbitrary dimension. arXiv:2109.00843v1 (2021)
20. Hertrich, J., Gräf, M., Beinert, R., Steidl, G.: Wasserstein steepest descent flows of disrepancies with Riesz kernels. arXiv:2211.01804 v1) (2022)
21. Jordan, R., Kinderlehrer, D., Otto, F.: The variational formulation of the Fokker-Planck equation. SIAM J. Math. Anal. **29**(1), 1–17 (1998)
22. Kolouri, S., Park, S., Rohde, G.: The Radon cumulative distribution transform and its application to image classification. IEEE Trans. Image Process. **25**(2), 920–934 (2016)
23. Landkof, N.: Foundations of Modern Potential Theory. Grundlehren der mathematischen Wissenschaften. Springer, Berlin (1972)
24. Li, C.L., Chang, W.C., Cheng, Y., Yang, Y., Póczos, B.: MMD GAN: towards deeper understanding of moment matching network. arXiv:1705.08584 (2017)
25. Liutkus, A., Simsekli, U., Majewski, S., Durmus, A., Stöter, F.R.: Sliced-Wasserstein flows: nonparametric generative modeling via optimal transport and diffusions. In: Proceedings of Machine Learning Research, vol. 97. PMLR (2019)

26. Neumayer, S., Steidl, G.: From optimal transport to discrepancy. In: Chen, K., Schönlieb, C.B., Tai, X.C., Younes, L. (eds.) Handbook of Mathematical Models and Algorithms in Computer Vision and Imaging: Mathematical Imaging and Vision, pp. 1–36. Springer, Cham (2023). https://doi.org/10.1007/978-3-030-98661-2_95
27. Nguyen, K., Ho, N., Pham, T., Bui, H.: Distributional sliced-Wasserstein and applications to generative modeling. In: 9th International Conference on Learning Representations. IEEE (2021)
28. Otto, F.: The geometry of dissipative evolution equations: the porous medium equation. Comm. Partial Differ. Equ. **26**, 101–174 (2001)
29. Park, S., Kolouri, S., Kundu, S., Rohde, G.: The cumulative distribution transform and linear pattern classification. Appl. Comput. Harmonic Anal. **45**(3), 616–641 (2017)
30. Pavliotis, G.A.: Stochastic Processes and Applications: Diffusion Processes, the Fokker-Planck and Langevin Equations. Texts in Applied Mathematics, vol. 60. Springer, New York (2014). https://doi.org/10.1007/978-1-4939-1323-7
31. Rockafellar, R.T., Royset, J.O.: Random variables, monotone relations, and convex analysis. Math. Program. **148**, 297–331 (2014)
32. Saff, E., Totik, V.: Logarithmic Potentials with External Fields. Grundlehren der mathematischen Wissenschaften, Springer, Berlin (1997). https://doi.org/10.1007/978-3-662-03329-6
33. Santambrogio, F.: Optimal Transport for Applied Mathematicians, Progress in Nonlinear Differential Equations and their Applications, vol. 87. Birkhäuser, Basel (2015)
34. Teuber, T., Steidl, G., Gwosdek, P., Schmaltz, C., Weickert, J.: Dithering by differences of convex functions. SIAM J. Imag. Sci. **4**(1), 79–108 (2011)
35. Villani, C.: Topics in Optimal Transportation. Graduate Studies in Mathematics, vol. 58. American Mathematical Society, Providence (2003)

A Quasi-Newton Primal-Dual Algorithm with Line Search

Shida Wang[1(✉)], Jalal Fadili[2], and Peter Ochs[1]

[1] Department of Mathematics, University of Tübingen, Tübingen, Germany
shida.wang@math.uni-tuebingen.de

[2] Normandie Univ, ENSICAEN, CNRS, GREYC, Caen, France

Abstract. Quasi-Newton methods refer to a class of algorithms at the interface between first and second order methods. They aim to progress as substantially as second order methods per iteration, while maintaining the computational complexity of first order methods. The approximation of second order information by first order derivatives can be expressed as adopting a variable metric, which for (limited memory) quasi-Newton methods is of type "identity $\pm$ low rank". This paper continues the effort to make these powerful methods available for non-smooth systems occurring, for example, in large scale Machine Learning applications by exploiting this special structure. We develop a line search variant of a recently introduced quasi-Newton primal-dual algorithm, which adds significant flexibility, admits larger steps per iteration, and circumvents the complicated precalculation of a certain operator norm. We prove convergence, including convergence rates, for our proposed method and outperform related algorithms in a large scale image deblurring application.

Keywords: quasi-Newton · primal-dual algorithm · line search · saddle-point problems · large scale optimization

1 Introduction

In modern optimization, the datasets and dimensionality of the problems and parameters is vastly increasing. In the early stages of large scale optimization, the shift from second order to first order optimization could cope with the increasing dimensionality of the problems. Second order methods achieve a significant progress per iteration at the cost of a high computational load, since the computation of the second derivative (Hessian) and oftentimes its inverse are required, which is intractable in the large scale regime. In contrast, first order methods have a low computational effort per iteration but also less information about the

We acknowledge funding by the ANR-DFG joint project TRINOM-DS under the number DFG OC150/5-1.

Supplementary Information The online version contains supplementary material available at https://doi.org/10.1007/978-3-031-31975-4_34.

L. Calatroni et al. (Eds.): SSVM 2023, LNCS 14009, pp. 444–456, 2023.
https://doi.org/10.1007/978-3-031-31975-4_34

objective. The gradient cannot capture curvature information and hence may fail to provide directions that allow for large steps. Nevertheless, for the large scale regime this exchange pays off.

However, the ever increasing dimensionality of the considered problems and datasets asks for faster algorithms. Motivated by classical optimization and the discussion above, algorithms at the interface of first and second order methods are key to reach the next level. Tractability in the (nowadays extremely) large scale regime requires methods that are as cheap as first order methods, while progressing as substantially as second order algorithms: *Quasi-Newton methods*. While they are known for their outstanding performance in unconstrained smooth optimization, their development for non-smooth (or constrained smooth) problems is insufficiently understood. As we discuss in related work below, most algorithmic development is either too simplistic, in the sense that only a diagonal metric is admitted, which can hardly capture second order information of the objective, or too theoretical, in the sense that a good performance is proved in theory while the implementation cost is on a par with that of second order methods. Algorithmic subproblems (e.g. the evaluation of the proximal mapping) that are easy to solve with respect to the Euclidean metric may become intractable with respect to another metric.

We pursue the line of research initiated in [2,3] that considers both aspects as equally important. Key is the observation that quasi-Newton methods actually generate a metric of the specific type "identity $\pm$ low rank", which allows for an efficient proximal calculus (cf. Sect. 6) that unlocks the quasi-Newton power—well-known from classical optimization—in the area of optimization for Machine Learning. This idea was recently transferred to non-convex optimization in [15, 16] and to monotone inclusion problems in [27]. A special case of the latter setting comprises the extremely broad class of convex–concave saddle point problems, which has numerous applications in Machine Learning, Computer Vision, Image Processing and Statistics [1,7,8,14,25].

In this paper, we restrict our interest to saddle-point problems only. This focus allows us to design a quasi-Newton primal–dual algorithm that is tailored to this setting and therefore highly efficient and adaptable thanks to an additional line search procedure. This has several advantages as compared to a fixed step size: (i) the oftentimes expensive computation of the operator norm can be avoided, (ii) the choice of metric need not obey any static spectral restrictions, and (iii) in many situations larger steps and thus a faster convergence is observed.

Our main contribution is reduction of the gap between the outstanding performance of quasi-Newton methods in classical optimization and quasi-Newton methods for (non-smooth) convex–concave saddle point problems for modern optimization in Machine Learning. In detail, our contribution is the following:

1. We extend the line-search based primal–dual algorithm in [20] (extension of [6] by line search) to incorporate a quasi-Newton metric with efficiently implementable proximal mapping; thereby aiming equally at theoretical convergence guarantees (including convergence rates) as well as highly efficient implementation.

2. We unlock the use of multi-memory quasi-Newton metrics (L-BFGS and SR1 method) via a compact representation for primal–dual algorithms, including their efficient implementation via a semi-smooth Newton solver.
3. The proposed algorithm outperforms the line-search based primal–dual algorithm (with identity metric) on a challenging image deblurring problem.

1.1 Related Work

Due to page limitations, for an extended discussion of quasi-Newton approaches in non-smooth optimization and the vast literature on primal-dual algorithms, we refer to [27].

Non-smooth Quasi-Newton. For a class of non-smooth problems that are given as a composition of a smooth function h and a non-smooth function g, [21,24] combine quasi-Newton methods with forward–backward splitting via the forward–backward envelope. If g is an indicator function, [23] proposed a projected quasi-Newton method which requires either solving a complicated subproblem or is restricted to a diagonal metric. Later, their work was extended by [18] to a more general setting. [2,3,17] developed algorithms with efficient evaluation of the proximal operator with respect to a low-rank perturbed metric. It is worth to mention that in [2,3] the subproblem is a low dimensional root finding problem which can be solved efficiently. Inspired by [3], the work in [16] applied the limited-memory quasi-Newton method on non-convex problems.

PDHG. Primal-Dual Hybrid Gradient (PDHG) is widely used to solve saddle point problems [6,8]. However, in order to guarantee the convergence of PDHG, the computation of the norm of a operator K is required. To avoid this disadvantage, [20] combined line search with PDHG to get a new algorithm PDAL. For faster convergence, variable metric is being used [14]. It shows the potential of combining quasi-Newton methods and PDAL via a variable metric, however suffers again from the need to solve more complicated subproblems, which is remedied in [27] for the more general class of monotone inclusion problems and hence builds the grounding for our proposed line search variant.

2 Problem Setup: A Class of Saddle Point Problems

Let X and Y be finite dimensional real vector spaces with inner product $\langle \cdot, \cdot \rangle$ and induced norm $\| \cdot \| = \sqrt{\langle \cdot, \cdot \rangle}$. We consider saddle point problems

$$\min_{x \in X} \max_{y \in Y} \langle Kx, y \rangle + g(x) + h(x) - f^*(y) \,, \tag{1}$$

where $g, h \colon X \to \overline{\mathbb{R}}$ are proper, lower semi-continuous (lsc), convex functions with h, additionally, having an L-Lipschitz continuous gradient, $f^* \colon Y \to \overline{\mathbb{R}}$ is a proper, lsc, convex function; the convex conjugate (Legendre–Fenchel conjugate) of a function f, and $K \colon X \to Y$ is a bounded linear operator with operator norm

$L_K := \|K\|$ and adjoint K^*. Moreover, throughout this paper we assume that (1) has a saddle point. We remark that (1) is equivalent to the *primal problem*

$$\min_{x \in X} f(Kx) + g(x) + h(x), \tag{2}$$

and to the *dual problem*

$$\max_{y \in Y} -\Big(f^*(y) + (g^* \,\triangledown\, h^*)(-K^*y)\Big), \tag{3}$$

where we use the fact that the conjugate of a sum of two function $(g+h)^*$ equals the infimal convolution of the conjugate functions $g^* \triangledown h^*$.

3 Our Quasi-Newton Primal-Dual Algorithm with Line Search

The primal–dual algorithm that we develop in this paper is an extension of [20] to incorporate a variable metric of quasi-Newton type, which itself adds an efficient line search procedure to the primal–dual hybrid gradient (PDHG) algorithm [6] (aka Chambolle–Pock algorithm) and the extension in [19]. The handling of (a possibly) non-smooth functions essentially relies on evaluating the proximal mapping, which we define here for a function g and parameter τ with respect to an arbitrary metric $M \in \mathcal{S}_\alpha(X)$, $\alpha > 0$. Here, $\mathcal{S}_\alpha$ is the set of bounded self-adjoint linear operators from Hilbert space X to X such that $M - \alpha I$ is positive semi-definite for each $M \in \mathcal{S}_\alpha$. For simplicity, some notations are introduced:

$$\|x\|_M^2 = \langle Mx, x\rangle, \quad x \in X.$$

$$\text{prox}_{\tau g}^M(\bar{x}) := \operatorname*{argmin}_{x \in X} g(x) + \frac{1}{2\tau}\|x - \bar{x}\|_M^2, \quad \text{and set } \text{prox}_{\tau g} := \text{prox}_{\tau g}^I,$$

where I is the identity (Euclidean) metric.

Algorithm 1 presents the proposed algorithm. The algorithm alternates between updates of the dual variable (4) and the primal variable (5), where the line search is only implemented in the primal update. Let us discuss the algorithm for a fixed iteration k, i.e., we are given x^k, y^{k-1}, σ_{k-1}, θ_{k-1} and a monotone decreasing sequence of β_k. The discussion of the quasi-Newton type variable metric in Step (i) is deferred to Sect. 5. Step (ii) is a standard dual update step. In Step (iii), we perform the line search. We select a basic step size $\bar{\sigma}_k$ and, in the ith loop of the line search, we perform a trial step (5) with the current $\sigma_k = \bar{\sigma}_k \cdot \mu^i$ and check if the breaking condition (6) is satisfied. If 'yes', the current iteration k is completed. If 'no', the new trial step size is reduced to $\sigma_k = \bar{\sigma}_k \cdot \mu^{i+1}$ (in the subsequent $(i+1)$th line search step). Here, $M_k \in \mathcal{S}_\alpha(X)$ is symmetric positive definite and $\alpha \in (0, 1)$.If M_k is chosen as an identity, then we recover the breaking (stopping) criterion used in [20]. However, in this paper, we adopt a variable metric M_k which is generated by quasi-Newton methods to exploit the local geometry of the function h. As a result, it is more likely to obtain a larger step size σ_k and fewer inner loops for the line search procedure

Algorithm 1. Quasi-Newton PDHG with Line Search

Require: [initial data] $x^1 \in X$, [initial data] $y^0 \in Y$, [maximal iteration count] $N \geq 0$, [scaling of line search parameter] $\mu \in (0,1)$, [extrapolation parameter] θ_0, [initial dual step size] σ_0, [tolerance weight] $\delta \in (0,1)$, [primal-dual step ratio] $+\infty > \beta \geq \beta_{k+1} \geq \beta_k > 0$, $\forall k \in \mathbb{N}$.

Update for $k = 1, \ldots, N$:

(i) Compute M_k according to a quasi-Newton framework (cf. Section 5).

(ii) Compute dual update step:

$$y^k = \text{prox}_{\sigma_{k-1} f^*}(y^{k-1} + \sigma_{k-1} K x^k) . \tag{4}$$

(iii) Select $\bar{\sigma}_k \in [\frac{\beta_{k-1}}{\beta_k}\sigma_{k-1}, \sqrt{(1+\theta_{k-1})\frac{\beta_{k-1}}{\beta_k}}\sigma_{k-1}]$ and compute the quasi-Newton primal update step by:

Line search: Find the smallest power $i = 0, 1, 2, \ldots$ such that

$$\begin{aligned} \bar{y}^k &= y^k + \theta_k (y^k - y^{k-1}) , \\ x^{k+1} &= \text{prox}_{\tau_k g}^{M_k} (x^k - \tau_k M_k^{-1} K^* \bar{y}^k - \tau_k M_k^{-1} \nabla h(x^k)) \end{aligned} \tag{5}$$

with

$$\sigma_k = \bar{\sigma}_k \cdot \mu^i , \quad \theta_k = \frac{\sigma_k}{\sigma_{k-1}} , \quad \text{and} \quad \tau_k = \beta_k \sigma_k$$

satisfy

$$\tau_k \sigma_k \|K x^{k+1} - K x^k\|^2 + 2\tau_k \Big(h(x^{k+1}) - h(x^k) - \Big\langle \nabla h(x^k), x^{k+1} - x^k \Big\rangle \Big) \leq \delta \|x^{k+1} - x^k\|_{M_k}^2 . \tag{6}$$

End of for-loop

as compared to the Euclidean version ($M_k = I$). The employed metric is of type "identity $\pm$ low rank" for which the proximal mapping can be computed efficiently as shown in Sect. 6.

Remark 1. While the line search procedure is formulated for the primal problem, by duality, the primal problem can be interpreted as the dual of the dual problem and, thus, the dual problem as the primal problem. As a consequence, an equivalent algorithm with line search on the dual can be easily stated.

Discussion of Computational Cost for Line Search. In general, every loop of the line search procedure requires recomputing several quantities, including (5) and Kx^{k+1}, $h(x^{k+1})$ and $\|x^{k+1} - x^k\|_{M_k}^2$ in (6). While this seems to be expensive at first glance, often (6) is satisfied after 1–3 trial steps and hence large steps are taken with a low cost, as we underline in our experiments in Sect. 7. Nevertheless, the cost can be further reduced significantly in certain special cases, observed in [20] and generalized here to our setting, whenever $\text{prox}_{\tau_k g}^{M_k}$ is a linear or affine operator.

1. If $g(x) = \langle c, x\rangle$, then $\operatorname{prox}_{\tau_k g}^{M_k}(u) = u - \tau_k M_k^{-1} c$ and therefore, we obtain

$$\begin{aligned} x^{k+1} &= \operatorname{prox}_{\tau_k g}^{M_k}(x^k - \tau_k M_k^{-1} K^* \bar{y}^k - \tau_k M_k^{-1} \nabla h(x^k)) \\ &= x^k - \tau_k [M_k^{-1} K^* \bar{y}^k + M_k^{-1} \nabla h(x^k) + M_k^{-1} c] . \end{aligned}$$

$$Kx^{k+1} = Kx^k - \tau_k [K M_k^{-1} K^* \bar{y}^k + \tau_k K M_k^{-1} \nabla h(x^k) + \tau_k K M_k^{-1} c] .$$

2. If $g(x) = \frac{1}{2}\|x - b\|^2$, then $\operatorname{prox}_{\tau_k g}^{M_k}(u) = (I + \tau_k M_k^{-1})^{-1}[u + \tau_k M_k^{-1} b]$ and therefore, we obtain

$$\begin{aligned} x^{k+1} &= \operatorname{prox}_{\tau_k g}^{M_k}(x^k - \tau_k M_k^{-1} K^* \bar{y}^k - \tau_k M_k^{-1} \nabla h(x^k)) \\ &= (I + \tau_k M_k^{-1})^{-1}[x^k - \tau_k M_k^{-1} K^* \bar{y}^k - \tau_k M_k^{-1} \nabla h(x^k) + \tau_k M_k^{-1} b] , \end{aligned}$$

$$Kx^{k+1} = K(I + \tau_k M_k^{-1})^{-1}[x^k - \tau_k (M_k^{-1} K^* \bar{y}^k + M_k^{-1} \nabla h(x^k) - M_k^{-1} b)] .$$

3. Let $g(x) = \delta_H(x)$, where H refers to the hyperplane $H := \{u : \langle u, a\rangle = b\}$. Then $\operatorname{prox}_{\tau_k g}^{M_k}(u) = u + \frac{b - \langle u, a\rangle}{\|a\|^2_{M_k^{-1}}} M_k^{-1} a$ and therefore, we obtain

$$\begin{aligned} x^{k+1} &= \operatorname{prox}_{\tau_k g}^{M_k}(x^k - \tau_k M_k^{-1} K^* \bar{y}^k - \tau_k M_k^{-1} \nabla h(x^k)) \\ &= x^k - \tau_k [M_k^{-1} K^* \bar{y}^k + M_k^{-1} \nabla h(x^k)] \\ &+ \frac{b - \langle x^k - \tau_k [M_k^{-1} K^* \bar{y}^k + M_k^{-1} \nabla h(x^k)], a\rangle}{\|a\|^2_{M_k^{-1}}} M_k^{-1} a , \end{aligned}$$

$$\begin{aligned} Kx^{k+1} &= Kx^k - \tau_k [K M_k^{-1} K^* \bar{y}^k + K M_k^{-1} \nabla h(x^k)] \\ &+ \frac{b - \langle x^k - \tau_k [M_k^{-1} K^* \bar{y}^k + \tau_k M_k^{-1} \nabla h(x^k)], a\rangle}{\|a\|^2_{M_k^{-1}}} K M_k^{-1} a . \end{aligned}$$

4 Convergence Analysis of Algorithm 1

Let us now analyze the convergence of Algorithm 1. As for most variable metric primal–dual algorithms (cf. [10–12,27]), we require the following restriction for the change of the metric from one iteration to the next. Under this condition, we can generalize all convergence results from [20] by adapting their proofs.

Assumption 1. *Let* $\alpha \in (0, +\infty)$. $(M_k)_{k\in\mathbb{N}}$ *is a sequence in* $\mathcal{S}_\alpha(X)$ *such that*

$$\begin{cases} \exists C_M \in \mathbb{R}, \text{s.t.} \sup_{k\in\mathbb{N}} \|M_k\| \le C_M < \infty , \\ (\exists (\eta_k)_{k\in\mathbb{N}} \in \ell^1_+(\mathbb{N}))(\forall k \in \mathbb{N}): \quad (1 + \eta_k) M_k \succeq M_{k+1} . \end{cases} \tag{7}$$

Lemma 1.

(i) There exists some $\sigma > 0$ *such that* $\sigma_k \ge \sigma$ *for any* $k \in \mathbb{N}$.
(ii) The line search terminates.
(iii) If $\beta_k \equiv \beta$, θ_k *is bounded from above by some* θ *for any* $k \in \mathbb{N}$.

The proof is provided in Section B.1.

Theorem 1. *Consider Problem* (1) *and let the sequence* $(x^k, y^k)_{k\in\mathbb{N}}$ *be generated by Algorithm 1 with* $\beta_k \equiv \beta$ *where Assumption 1 holds. Then* $(x^k, y^k)_{k\in\mathbb{N}}$ *is a bounded sequence and its cluster points are solutions of* (1). *Furthermore, if* $f^*|_{\mathrm{dom} f^*}$ *is continuous and* σ_k *is bounded from above, then the whole sequence* $(x^k, y^k)_{k\in\mathbb{N}}$ *converges to a solution of* (1).

The proof is provided in Section B.2.

We obtain the same ergodic convergence rate as in [20], with respect to the primal–dual gap $\mathcal{G}_{\hat{x},\hat{y}}$ which is the difference (gap) between the optimal primal objective value in (2) and the optimal dual objective value (3).

Theorem 2. *Let the sequence* $(x^k, y^k)_{k\in\mathbb{N}}$ *be generated by Algorithm 1 with* $\beta_k \equiv \beta$ *where Assumption 1 holds and* $(\hat{x}, \hat{y})$ *be some saddle point of* (1). *Then it holds for a constant* $D = \Pi_{k\in\mathbb{N}}(1 + \eta_k) < +\infty$ *that*

$$\mathcal{G}_{\hat{x},\hat{y}}(\bar{X}^N, \bar{Y}^N) \leq \frac{D}{s_N}\Big(\frac{1}{2\beta}\|x^1 - \hat{x}\|^2_{M_1} + \frac{1}{2}\|y^1 - \hat{y}\|^2 + \sigma_1\theta_1 D_{\hat{x},\hat{y}}(y^0)\Big) = O\Big(\frac{1}{N}\Big), \tag{8}$$

where $s_N := \sum_{k=1}^N \sigma_k$, $\bar{X}^N := \frac{\sum_{k=1}^N \sigma_k x^{k+1}}{s_N}$ *and* $\bar{Y}^N := \frac{\sigma_1\theta_1 y^0 + \sum_{k=1}^N \sigma_k \bar{y}^k}{\sigma_1\theta_1 + s_N}$. *The last equality in* (8) *provides a simplified rate in terms of the big-O notation.*

The proof is provided in Section B.3.

Under the additional assumption that g is strongly convex, improved convergence rates can be derived when $(\beta_k)_{k\in\mathbb{N}}$ is varied appropriately.

Theorem 3. *Assume* g *is* γ*-strongly convex and* $(x^k, y^k)_{k\in\mathbb{N}}$ *is generated by Algorithm 1 with*

$$\beta_k = \frac{\beta_{k-1}}{\min\{1 + \frac{\gamma}{C_M}\beta_{k-1}\sigma_{k-1}, C_\theta\}}, \quad \forall k \in \mathbb{N}, \quad \text{and} \quad \beta_0 > 0, \tag{9}$$

where $C_\theta \in \mathbb{R}_+$ *is a constant, Assumption 1 holds and* $(\hat{x}, \hat{y})$ *be some saddle point of* (1). *Then, we have* $(\theta_k)_{k\in\mathbb{N}}$ *is bounded from above. Furthermore, we obtain*

$$\|x^N - \hat{x}\| = O(1/N) \quad \text{and} \quad \mathcal{G}_{\hat{x},\hat{y}}(\bar{X}^N, \bar{Y}^N) = O(1/N^2),$$

where $(\bar{X}^N, \bar{Y}^N)$ *are the ergodic sequences defined in Theorem 2.*

The proof is provided in Section B.4.

Remark 2. For the result in Theorem 3, $\delta = 1$ is also admitted.

5 Computing and Representing the Quasi-Newton Metric

In this section, we abuse notation in order to follow the conventions of quasi-Newton methods[1]. The metric M_k is expected to be an approximation of the

[1] For example, the variable y^k defined in (12) is not the dual variable in Algorithm 1.

Hessian $\nabla^2 h(x^k)$ at x^k for the k-th iteration. The most popular quasi-Newton methods are the SR1 and BFGS methods (and their low-memory variants), which update M_k by adding a rank-one modification (SR1 method)

$$M_{k+1} := M_{k+1}^{SR1} = M_k + \frac{(y^k - M_k s^k)(y^k - M_k s^k)^\top}{(y^k - M_k s^k)^\top s^k}, \tag{10}$$

or a rank-two modification (BFGS method)

$$M_{k+1} := M_{k+1}^{BFGS} = M_k + \frac{y^k (y^k)^\top}{(s^k)^\top y^k} - \frac{M_k s^k (s^k)^\top M_k}{(s^k)^\top M_k s_k}, \tag{11}$$

respectively, where

$$s^k := x^{k+1} - x^k \quad \text{and} \quad y^k := \nabla h(x^{k+1}) - \nabla h(x^k). \tag{12}$$

In order to apply quasi-Newton methods on large-scale problems, *m-limited memory quasi-Newton methods* are adopted [16], with the most popular version being L-BFGS [28]. It means that instead of generating M_k via all previous s^i and y^i for $i = 1, \ldots, k$ and M_0, for each k, the metric M_k is re-computed based on $M_{k,0}$ and the most recent m vectors s^i and y^i for $i = k - m + 1, \ldots, k$, if $k \geq m$. Usually, m is very small, such that only a small storage will be required. As pointed out by [5], the matrices of the m-limited memory version of quasi-Newton methods have a compact representation of the form

$$M_k = M_{k,0} + A_k Q_k^{-1} A_k^\top, \tag{13}$$

where $M_{k,0} \in \mathbb{R}^{n \times n}$, $n = \dim(X)$, is a symmetric positive definite matrix, $A_k \in \mathbb{R}^{n \times m}$, and a symmetric and non-singular matrix $Q_k \in \mathbb{R}^{m \times m}$ ($m \ll n$). For limited memory BFGS (known as L-BFGS), we have the following block-matrix representation

$$A_k = \begin{bmatrix} M_{k,0} S_k \ Y_k \end{bmatrix} \in \mathbb{R}^{n \times 2m} \quad \text{and} \quad Q_k = \begin{bmatrix} -S_k^* M_{k,0} S_k & -L_k \\ -L_k^* & D_k \end{bmatrix} \in \mathbb{R}^{2m \times 2m}, \tag{14}$$

where S_k and Y_k are matrices collecting the m most recent vectors in (12) as columns, $D_k := D(S_k^\top Y_k)$ and $L_k := L(S_k^\top Y_k)$ refer to the diagonal $D(\cdot)$ and the strict lower triangular $L(\cdot)$ part of the matrix $S_k^\top Y_k$, respectively. By using a spectral decomposition $Q^{-1} = V \Lambda V^\top$ with orthogonal $V \in \mathbb{R}^{s \times s}$ and diagonal $\Lambda \in \mathbb{R}^{s \times s}$, for some $s \in \mathbb{N}$, (13) is transformed into the compact representation

$$M_k = M_{k,0} + U_1 U_1^\top - U_2 U_2^\top, \tag{15}$$

for some $U_1 \in \mathbb{R}^{n \times m}$ and $U_2 \in \mathbb{R}^{n \times m}$. In detail, since Λ is a diagonal matrix with eigenvalues λ_i, $i = 1, 2, \ldots, s$ of Q_k^{-1} on the diagonal, we decompose $\Lambda = \Lambda_1 - \Lambda_2$ where Λ_1, given by $(\Lambda_1)_{i,i} = \max(\lambda_i, 0)$, $i = 1, 2, \ldots, s$, corresponds to the positive eigenvalues and Λ_2, given by $(\Lambda_2)_{i,i} = \max(-\lambda_i, 0)$, $i = 1, 2, \ldots, s$, corresponds to the negative eigenvalues. In this way, we obtain

$$U_1 := (A_k V) \Lambda_1^{1/2} \quad \text{and} \quad U_2 := (A_k V) \Lambda_2^{1/2}. \tag{16}$$

Theoretically, it is guaranteed that $M_k = M_{k,0} + U_1U_1^\top - U_2U_2^\top$ is positive definite [13] if $s^k y^k > 0$ for any $k \in \mathbb{N}$. However, in order to account for numerical rounding errors and the assumption that $M_k \in \mathcal{S}_\alpha$ is bounded from above by some C_M, we adopt a scaling version:

$$\begin{aligned} \tilde{M}_k &= M_{k,0} + \gamma_1 U_1U_1^\top - \gamma_2 U_2U_2^\top\,, \\ M_k &= \min\{\frac{C_M - \alpha}{\|\tilde{M}_k\|_2}, 1\}\tilde{M}_k + \alpha I\,, \end{aligned} \tag{17}$$

where $\|\tilde{M}_k\|_2$ denotes the l_2 norm of matrix $\tilde{M}_k$ and we set $\alpha = 0.01, \gamma_1 = 1, \gamma_2 = 1, C_M = 50$ in practice. There is an easy way to make sure that Assumption 1 is satisfied by setting $\gamma_1 = \frac{\eta_k}{\|U_1\|^2}$ and $\gamma_2 = \frac{\eta_k}{\|U_2\|^2}$ with arbitrary $\eta_k \in \ell_+^1$.

6 Proximal Calculus and Efficient Implementation

The transformation in Sect. 5 enables us to compute the proximal mapping with respect to the metric in the form of (15) via the proximal calculus developed in [3], which we state here for completeness.

Theorem 4. *Let $B = B_0 + U_1U_1^\top - U_2U_2^\top \in \mathcal{S}_\sigma(\mathbb{R}^n)$ with $\sigma > 0$, $B_0 \in \mathcal{S}_\sigma(\mathbb{R}^n)$ and $U_i \in \mathbb{R}^{n\times r_i}$ with rank r_i ($i = 1, 2$). Set $B_1 = B_0 + U_1U_1^\top$. Then, the following holds:*

$$\mathrm{prox}_g^B(x) = \mathrm{prox}_g^{B_0}(x + B_1^{-1}U_2\alpha_2^* - B_0^{-1}U_1\alpha_1^*)\,, \tag{18}$$

where $\alpha_i^*, i = 1, 2$, are the unique zeros of the coupled system $\mathcal{L}(\alpha) = \mathcal{L}(\alpha_1, \alpha_2) = 0$, where $\alpha = (\alpha_1, \alpha_2) \in \mathbb{R}^{r_1+r_2}$ and $\mathcal{L} = (\mathcal{L}_1, \mathcal{L}_2)$ is defined by

$$\begin{aligned} \mathcal{L}_1(\alpha_1, \alpha_2) &= U_1^\top(x + B_1^{-1}U_2\alpha_2 - \mathrm{prox}_g^{B_0}(x + B_1^{-1}U_2\alpha_2 - B_0^{-1}U_1\alpha_1)) + \alpha_1\,, \\ \mathcal{L}_2(\alpha_1, \alpha_2) &= U_2^\top(x - \mathrm{prox}_g^{B_0}(x + B_1^{-1}U_2\alpha_2 - B_0^{-1}U_1\alpha_1)) + \alpha_2\,. \end{aligned} \tag{19}$$

Here, $\mathcal{L}\colon \mathbb{R}^{r_1+r_2} \to \mathbb{R}^{r_1+r_2}$ is Lipschitz continuous.

The computation of a possibly complicated proximal mapping $\mathrm{prox}_g^B(x)$ is reduced to a simple (by assumption) proximal mapping $\mathrm{prox}_g^{B_0}$ and a low dimensional root finding problem in (19), which we tackle by the semi-smooth Newton solver proposed in [3], formulated in Algorithm 2. It requires to solve Newton-like equations where the classic Jacobian at the current iterate α_k is replaced by the Clarke Jacobian $\partial^c\mathcal{L}(\alpha_k)$ (see [9]), where we account for inexact solutions of these subproblems in terms of an error e_k. For completeness, we also state the convergence result of [3] for Algorithm 2.

Theorem 5. *If g is in addition a tame function, then the Lipschitz continuous function $\mathcal{L}$ is semi-smooth [4] and all elements of $\partial^C\mathcal{L}(\alpha^*)$ are non-singular [3]. Therefore, if $\rho_k \le \bar{\rho}$, $\forall k \in \mathbb{N}$, for some sufficiently small $\bar{\rho}$ and α_0 sufficiently close to α^*, then the sequence generated by the Algorithm 2 is well-defined and converges to α^* linearly. Additionally, if $\rho_k \to 0$, the convergence is superlinear.*

Algorithm 2. Semi-smooth Newton method to solve $\mathcal{L}(\alpha) = 0$ in (19)

Require: [initial data] $\alpha_0 \in \mathbb{R}^r$, [maximum iterations] N

Update for $k = 0, \cdots, N$:

(i) Select $G_k \in \partial^c \mathcal{L}(\alpha_k)$, compute α_{k+1} such that

$$\mathcal{L}(\alpha_k) + G_k(\alpha_{k+1} - \alpha_k) = e_k ,$$

and $e_k \in \mathbb{R}^r$ is an error term satisfying $\|e_k\| \leq \rho_k \|G_k\|$ and $\rho_k \geq 0$.

(ii) **if** $\mathcal{L}(\alpha_k) = 0$ **then terminate.**

End of for-loop

The tameness assumption is extremely mild, as it includes basically any function that occurs in practical applications, by excluding pathological special cases. For example this class of functions comprises all semi-algebraic functions [4].

7 Numerical Experiment

We apply our proposed algorithm for solving a challenging non-smooth image deblurring problem under a Poisson noise assumption [26]. Given the observation $b \in \mathbb{R}^{n_x \times n_y}$ as $n_x \times n_y$-sized image, the task is the following popular problem:

$$\min_{x \in \mathbb{R}_+^{n_x \times n_y}} D_{KL}(b, Ax) + \gamma \|\mathcal{D}x\|_{2,1} , \tag{20}$$

which involves the Kullback–Leibler divergence as data fidelity measure $h(x) := D_{\mathrm{KL}}(b, Ax) := \sum_{i,j} (Ax)_{i,j} - b_{i,j} \log((Ax)_{i,j})$ with respect to the blurred reconstruction Ax with known blur operator and a discrete total variation regularization term $f(x) = \gamma \|\mathcal{D}x\|_{2,1}$ that is steered by a weight $\gamma > 0$ where $\mathcal{D}$ implements discrete spatial finite differences. We recast (20) into the saddle point problem:

$$\min_{x \in \mathbb{R}^{n_x \times n_y}} \max_{y \in \mathbb{R}^{2 \times n_x \times n_y}} \langle \mathcal{D}x, y \rangle + \delta_{\mathbb{R}_+^{n_x \times n_y}}(x) + D_{KL}(b, Ax) - \delta_{\|\cdot\|_{2,\infty} \leq \gamma}(y) \tag{21}$$

and set $g(x) := \delta_{\mathbb{R}_+^{n_x \times n_y}}(x)$ and $K = \mathcal{D}$ in (1). Figure 1 compares several methods including PDHG with fixed stepsize (`PDHG`), PDHG with line search (`PDAL`), PDHG with fixed stepsize and variable metric (`VarPDHG`), PDHG with variable metric and line search (`VarPDAL`). The variable metric is generated by the limited memory BFGS method in Sect. 5. Figure 1 shows the primal gap where the optimal primal value was computed for 10000 iterations by running `PDHG`. For the update of the variable metric (17), we set $\gamma_1 = 1.0$ and $\gamma_2 = 0.99$. However, the Assumption 1 is not satisfied since it is not guaranteed by the construction of the metric that there is a sequence $(\eta_k)_k \in \ell_+^1(\mathbb{N})$ such that $M_{k+1} \preceq (1+\eta_k) M_k$. Fortunately, we still observe convergence of PDHG with this variable metric. Figure 1 shows that a variable metric (`VarPDHG`) improves the convergence vs only using line search. However, our algorithm `VarPDAL` that combines both features is even faster, with the best performance when $m = 9$.

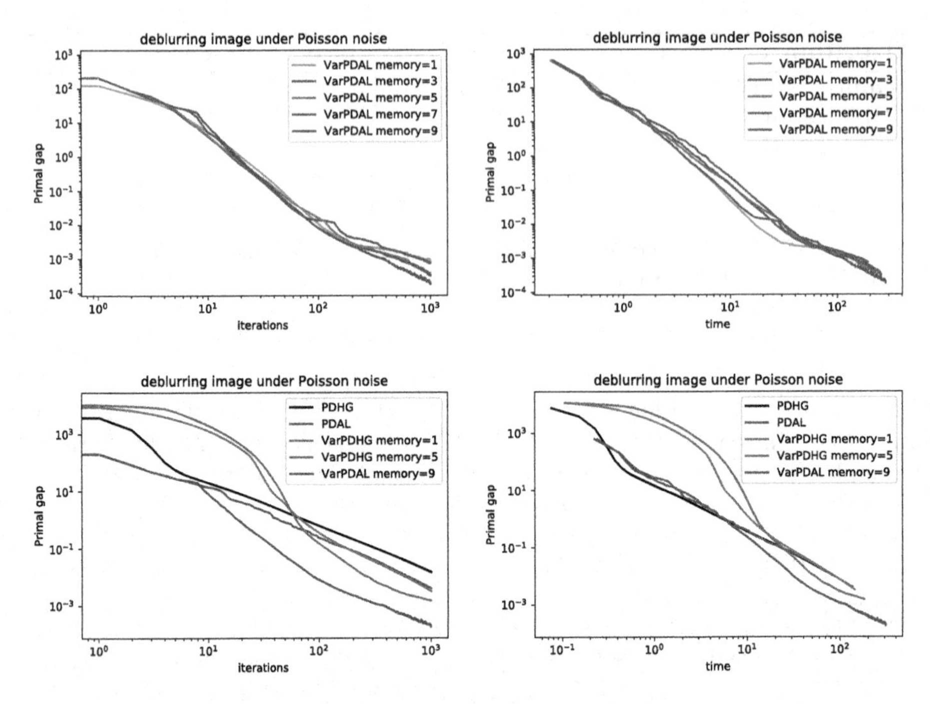

Fig. 1. Performance evaluation for the experiment in (21). Algorithms are described in the text. Our algorithms, which combine a quasi-Newton variable metric with line search, outperform all other algorithms (that either use a variable metric `VarPDHG` or line search `PDAL`; or none of the two `PDHG`).

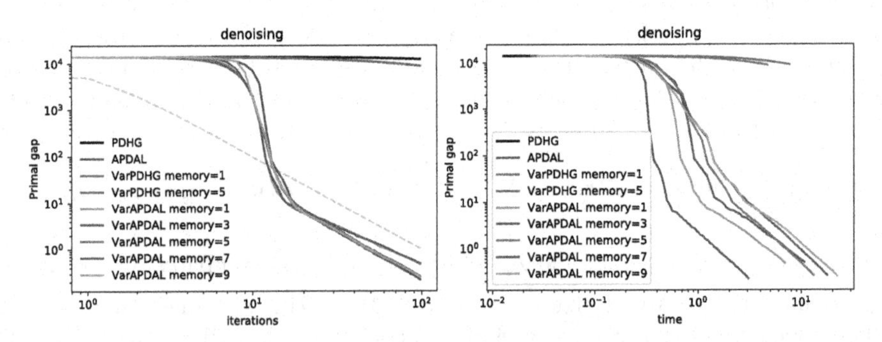

Fig. 2. Performance evaluation for the experiment for a strongly convex case. This figure reflects that our algorithms can retrieve $O(1/N^2)$ convergence rate as `PDAL`, which is theoretically guaranteed by Theorem 3.

As a second experiment to test out accelerated version, we consider $A = I$ and a constraint set $\mathcal{C} := \{x | x_{ij} \in [\epsilon, 255]\}$ since the grey value of each pixel should be less than 255 and be positive; We set $\epsilon = 0.1$. In this case $D_{KL}(b, x)$ restricted on $\mathcal{C}$ is a strongly convex function. We apply the accelerated version of Algorithm 1 in Theorem 3, and we use the notation `VarAPDAL`. Similarly, `APDAL` denotes the accelerated version of PDAL. The dashed line in Fig. 2 corresponds to $O(1/N^2)$. We can observe that VarPDAL, APDAL can converge faster than $O(1/N^2)$ as predicted by the convergence theorem.

In order to record the exact algorithmic details for our experiments, all code for experiments from this paper is available at https://github.com/wsdxiaohao/VarPDAL.git.

8 Conclusion

In this paper, we introduced a line search variant of a recently introduced quasi-Newton primal-dual algorithm. In contrast to related work, the employed quasi-Newton metric is of type "identity $\pm$ low rank", which captures significantly more second order information than a commonly used diagonal metric. We equally care for both, theoretical convergence guarantees including convergence rates as well as efficient practical implementation. The additional line search procedure usually leads to larger steps at a computational cost that pays off, which is confirmed by our numerical experiments.

References

1. Applegate, D., et al.: Practical large-scale linear programming using primal-dual hybrid gradient. In: Advances in Neural Information Processing Systems, vol. 34 (2021)
2. Becker, S., Fadili, J.: A quasi-Newton proximal splitting method. In: Advances in Neural Information Processing Systems, vol. 25 (2012)
3. Becker, S., Fadili, J., Ochs, P.: On quasi-Newton forward-backward splitting: proximal calculus and convergence. SIAM J. Optim. **29**(4), 2445–2481 (2019)
4. Bolte, J., Daniilidis, A., Lewis, A.: Tame functions are semismooth. Math. Program. **117**(1), 5–19 (2009)
5. Byrd, R.H., Nocedal, J., Schnabel, R.B.: Representations of quasi-newton matrices and their use in limited memory methods. Math. Program. **63**(1), 129–156 (1994)
6. Chambolle, A., Pock, T.: A first-order primal-dual algorithm for convex problems with applications to imaging. J. Math. Imaging Vis. **40**(1), 120–145 (2011)
7. Chambolle, A., Pock, T.: An introduction to continuous optimization for imaging. Acta Numer. **25**, 161–319 (2016)
8. Chambolle, A., Pock, T.: On the ergodic convergence rates of a first-order primal-dual algorithm. Math. Program. **159**(1), 253–287 (2016)
9. Clarke, F.H.: Optimization and Nonsmooth Analysis. Society for Industrial and Applied Mathematics (1990)
10. Combettes, P., Condat, L., Pesquet, J.C., Vu, B.: A forward-backward view of some primal-dual optimization methods in image recovery. In: IEEE International Conference on Image Processing (2014)
11. Combettes, P.L., Vũ, B.C.: Variable metric forward-backward splitting with applications to monotone inclusions in duality. Optimization **63**(9), 1289–1318 (2014)
12. Davis, D.: Convergence rate analysis of primal-dual splitting schemes. SIAM J. Optim. **25**(3), 1912–1943 (2015)
13. Fletcher, R.: Practical Methods of Optimization. Wiley, Hoboken (2013)

14. Goldstein, T., Li, M., Yuan, X., Esser, E., Baraniuk, R.: Adaptive primal-dual hybrid gradient methods for saddle-point problems. arXiv:1305.0546 (2013)
15. Kanzow, C., Lechner, T.: Globalized inexact proximal newton-type methods for nonconvex composite functions. Comput. Optim. Appl. **78**, 377–410 (2021)
16. Kanzow, C., Lechner, T.: Efficient regularized proximal quasi-Newton methods for large-scale nonconvex composite optimization problems. Technical report, University of Würzburg, Institute of Mathematics, January 2022
17. Karimi, S., Vavasis, S.: IMRO: a proximal quasi-Newton method for solving l1-regularized least squares problems. SIAM J. Optim. **27**(2), 583–615 (2017)
18. Lee, J.D., Sun, Y., Saunders, M.A.: Proximal Newton-type methods for minimizing composite functions. SIAM J. Optim. **24**(3), 1420–1443 (2014)
19. Lorenz, D.A., Pock, T.: An inertial forward-backward algorithm for monotone inclusions. J. Math. Imaging Vis. **51**(2), 311–325 (2015)
20. Malitsky, Y., Pock, T.: A first-order primal-dual algorithm with linesearch. SIAM J. Optim. **28**(1), 411–432 (2018)
21. Patrinos, P., Stella, L., Bemporad, A.: Forward-backward truncated Newton methods for convex composite optimization. arXiv:1402.6655 (2014)
22. Polyak, B.: Introduction to optimization. Optimization Software (1987)
23. Schmidt, M., Kim, D., Sra, S.: Projected Newton-type methods in machine learning. In: Optimization for Machine Learning, no. 1 (2012)
24. Stella, L., Themelis, A., Patrinos, P.: Forward-backward quasi-Newton methods for nonsmooth optimization problems. Comput. Optim. Appl. **67**(3), 443–487 (2017)
25. Valkonen, T.: A primal-dual hybrid gradient method for nonlinear operators with applications to MRI. Inverse Prob. **30**(5), 055012 (2014)
26. Vardi, Y., Shepp, L.A., Kaufman, L.: A statistical model for positron emission tomography. J. Am. Stat. Assoc. **80**(389), 8–20 (1985)
27. Wang, S., Fadili, J., Ochs, P.: Inertial quasi-newton methods for monotone inclusion: efficient resolvent calculus and primal-dual methods. arXiv:2209.14019 (2022)
28. Wright, S., Nocedal, J.: Numerical Optimization. Springer, New York (1999)

Stochastic Gradient Descent for Linear Inverse Problems in Variable Exponent Lebesgue Spaces

Marta Lazzaretti[1,3], Zeljko Kereta[2], Claudio Estatico[1], and Luca Calatroni[3(✉)]

[1] Dip. di Matematica, Università di Genova, Via Dodecaneso 35, 16146 Genoa, Italy
{lazzaretti,estatico}@dima.unige.it

[2] Department of Computer Science, University College London, London, UK
z.kereta@ucl.ac.uk

[3] CNRS, UCA, Inria, Laboratoire I3S, 06903 Sophia-Antipolis, France
calatroni@i3s.unice.fr

Abstract. We consider a stochastic gradient descent (SGD) algorithm for solving linear inverse problems (e.g., CT image reconstruction) in the Banach space framework of variable exponent Lebesgue spaces $\ell^{(p_n)}(\mathbb{R})$. Such non-standard spaces have been recently proved to be the appropriate functional framework to enforce pixel-adaptive regularisation in signal and image processing applications. Compared to its use in Hilbert settings, however, the application of SGD in the Banach setting of $\ell^{(p_n)}(\mathbb{R})$ is not straightforward, due, in particular to the lack of a closed-form expression and the non-separability property of the underlying norm. In this manuscript, we show that SGD iterations can effectively be performed using the associated modular function. Numerical validation on both simulated and real CT data show significant improvements in comparison to SGD solutions both in Hilbert and other Banach settings, in particular when non-Gaussian or mixed noise is observed in the data.

Keywords: Iterative regularisation · Stochastic gradient descent · Inverse problems in Banach spaces · Computed Tomography

1 Introduction

The literature on iterative regularisation methods for solving ill-posed linear inverse problems in finite/infinite-dimensional Hilbert or Banach settings is very vast, see, e.g., [7,21] for surveys. Given two normed vector spaces $(\mathcal{X}, \|\cdot\|_{\mathcal{X}})$ and $(\mathcal{Y}, \|\cdot\|_{\mathcal{Y}})$, we are interested in the inverse problem

$$\text{find} \quad x \in \mathcal{X} \quad \text{s.t.} \quad \mathcal{Y} \ni y = Ax + \eta, \tag{1}$$

where $A \in \mathcal{L}(\mathcal{X}; \mathcal{Y})$ is a bounded linear operator, and $\eta \in \mathcal{Y}$ denotes the (additive) noise perturbation of magnitude $\|\eta\|_{\mathcal{Y}} \le \delta$, $\delta > 0$, corrupting the measurements. Due to the ill-posedness, the standard strategy for solving (1) consists in computing $x^\star \in \operatorname{argmin}_{x\in\mathcal{X}} \, \Psi(x)$, where the functional $\Psi : \mathcal{X} \to \mathbb{R}_+ \cup \{+\infty\}$ quantifies the fidelity of a candidate reconstruction to the measurements, possibly combined

L. Calatroni et al. (Eds.): SSVM 2023, LNCS 14009, pp. 457–470, 2023.
https://doi.org/10.1007/978-3-031-31975-4_35

with a penalty or regularisation term enforcing prior assumptions on the sought quantity $x \in \mathcal{X}$. A popular strategy for promoting implicit regularisation through algorithmic optimisation consists in designing iterative schemes solving instances of the minimisation problem $\text{argmin}_{x\in\mathcal{X}} \|Ax - y\|_{\mathcal{Y}}$ or, more generally

$$\underset{x\in\mathcal{X}}{\text{argmin}}\ f(x) \qquad \text{with} \quad f(x) = \tilde{f}(Ax - y), \tag{P}$$

where, for $y \in \mathcal{Y}$, the function $f(\cdot) = \tilde{f}(A\cdot -y) : \mathcal{X} \to \mathbb{R}_{\geq 0}$ measures the discrepancy between the model observation Ax and y. The iterative scheme has to be endowed with a robust criterion for its early stopping in order to avoid that the computed reconstruction overfits the noise [16]. In this context, the role of the parameter tuning the amount of regularisation is thus played by nothing but the number of performed iterations. One-step gradient descent algorithms, such as the (accelerated) Landweber or the Conjugate Gradient, represent the main class of optimisation methods for the resolution of (P), see e.g. [6,18,19].

The most well-studied cases consider $\mathcal{X}$ and $\mathcal{Y}$ to be Hilbert spaces, e.g., $\mathcal{X} = \mathcal{Y} = \ell^2(\mathbb{R})$. In this setting, problem (P) takes the form $\text{argmin}_{x\in\ell^2(\mathbb{R})} \frac{1}{2}\|Ax - y\|^2_{\ell^2(\mathbb{R})}$ and can be solved by a standard Landweber iterative scheme

$$x^{k+1} = x^k - \mu_{k+1}A^*(Ax^k - y), \tag{2}$$

for $k \geq 0$, where $\mu_{k+1} > 0$ denotes the algorithmic step-sizes. However, many inverse problems require a more complex setting to retrieve solutions with specific features, such as sharp edges, piecewise constancy, sparsity patterns and/or to model non-standard (e.g., mixed) noise in the data. Either $\mathcal{X}$ or $\mathcal{Y}$, or both, can thus be modelled as more general Banach spaces. Notable examples are standard Lebesgue spaces $L^p(\Omega)$ and, in discrete settings, sequence spaces $\ell^p(\mathbb{R})$ with $p \in [1, +\infty] \setminus \{2\}$. While the solution space $\mathcal{X}$ affects the choice of the specific iterative scheme to be used, the measurement (or data) space $\mathcal{Y}$ is naturally connected to the norm appearing in (P). For example, for Hilbert $\mathcal{X} = \ell^2(\mathbb{R})$ and Banach $\mathcal{Y} = \ell^p(\mathbb{R})$, an instance of (P) reads as

$$\underset{x\in\ell^2(\mathbb{R})}{\text{argmin}}\ \frac{1}{q}\|Ax - y\|^q_{\ell^p}, \quad \text{with } q > 1,$$

for which a gradient descent-type scheme can still be used in the form $x^{k+1} = x^k - A^*\mathbf{J}^q_{\ell^p}(Ax^k - y)$, where $\mathbf{J}^q_{\ell^p} : \ell^p(\mathbb{R}) \to \ell^{p^*}(\mathbb{R})$ is the so-called q-duality map of $\ell^p(\mathbb{R})$, defined as $\mathbf{J}^q_{\ell^p}(\cdot) = \partial\left(\frac{1}{q}\|\cdot\|^q_{\ell^p(\mathbb{R})}\right)$. When both $\mathcal{X}$ and $\mathcal{Y}$ are Banach spaces, a popular algorithm for solving

$$\underset{x\in\mathcal{X}}{\text{argmin}}\ \frac{1}{q}\|Ax - y\|^q_{\mathcal{Y}}, \quad \text{with } q > 1$$

is the dual Landweber method [22]

$$x^{k+1} = \mathbf{J}^{p^*}_{\mathcal{X}^*}\left(\mathbf{J}^p_{\mathcal{X}}(x^k) - \mu_{k+1}A^*\mathbf{J}^q_{\mathcal{Y}}(Ax^k - y)\right) \tag{3}$$

where $\mathbf{J}^p_{\mathcal{X}} : \mathcal{X} \to \mathcal{X}^*$, is the p-duality map of $\mathcal{X}$, $\mathbf{J}^{p^*}_{\mathcal{X}^*} : \mathcal{X}^* \to \mathcal{X}$ is its inverse with p^* denoting the conjugate exponent of p, i.e. $1/p + 1/p^* = 1$. For other references of gradient-descent-type solvers in Banach settings, see, e.g. [11,21,22].

A non-standard Banach framework for solving linear inverse problems is the one of variable exponent Lebesgue spaces $L^{p(\cdot)}(\Omega)$ and $\ell^{(p_n)}(\mathbb{R})$ [5]. These Banach spaces are defined in terms of a Lebesgue measurable function $p(\cdot) : \Omega \to [1, +\infty]$, or a real sequence $(p_n)_n$, respectively, that assigns coordinate-wise exponents to all points in the domain. Variable exponent Lebesgue spaces have proven useful in the design of adaptive regularisation, suited to model heterogeneous data and complex noise settings. Iterative regularisation procedures in this setting have been recently studied [2] and also extended to composite optimisation problems involving non-smooth penalty terms [14].

While benefiting from several convergence properties, the use of such (deterministic) iterative algorithms may be prohibitively expensive in large-size applications as they require the use of all data at each iteration. In this work, we follow the strategy performed by the seminal work of Robbins and Monro [20] and adapt a stochastic gradient descent (SGD) strategy to the non-standard setting of variable exponent Lebesgue space, in order to reduce the per-iteration complexity costs. Roughly speaking, this is done by defining a suitable decomposition of the original problem and implementing an iterative scheme where only a batch of data, typically one, is used to compute the current update. Note that the use of SGD schemes has recently attracted the attention of the mathematical imaging community [10,13] due to its applicability in large-scale applications such as medical imaging [9,17,23]. However, its extension to variable exponent Lebesgue setting is not trivial due to some structural difficulties (e.g., non-separability of the norm), making the adaptation a challenging task.

Contribution. We consider an SGD-based iterative regularisation strategy for solving linear inverse problems in the non-standard Banach setting of variable exponent Lebesgue space $\ell^{(p_n)}(\mathbb{R})$. To overcome the non-separability of the norm in such space, we consider updates defined in terms of a separable function, the modular function. Numerical investigation of the methodology on CT image reconstruction are reported to show the advantages of considering such non-standard Banach setting in comparison to standard Hilbert scenarios. Comparisons between the modular-based deterministic and stochastic algorithms confirm improvements of the latter w.r.t. CPU times.

2 Optimisation in Banach Spaces

In this section we revise the main definitions and tools useful for solving a general instance of (P) in the general context of Banach spaces $\mathcal{X}$ and $\mathcal{Y}$. For a real Banach space $(\mathcal{X}, \|\cdot\|_{\mathcal{X}})$, we denote by $(\mathcal{X}^*, \|\cdot\|_{\mathcal{X}^*})$ its dual space and, for any $x \in \mathcal{X}$ and $x^* \in \mathcal{X}^*$, by $\langle x^*, x \rangle = x^*(x) \in \mathbb{R}$ its duality pairing.

The following definition is crucial for the development of algorithms solving (P) in Banach spaces. We recall that in Hilbert settings $\mathcal{H} \cong \mathcal{H}^*$ holds by the Riesz representation theorem, with $\cong$ denoting an isometric isomorphism.

Hence, for $x \in \mathcal{H}$, the element $\nabla f(x) \in \mathcal{H}^*$ can be implicitly identified with a unique element in $\mathcal{H}$ itself, up to the canonical isometric isomorphism, so that the design of gradient-type schemes is significantly simplified, as in (2). Since the same identification does not hold, in general, for a Banach space $\mathcal{X}$, we recall the notion of duality maps, which properly associate an element of $\mathcal{X}$ with an element (or a subset) of $\mathcal{X}^*$ [3].

Definition 1. *Let $\mathcal{X}$ be a Banach space and $p > 1$. The duality map $\mathbf{J}^p_{\mathcal{X}}$ with gauge function $t \mapsto t^{p-1}$ is the operator $\mathbf{J}^p_{\mathcal{X}} : \mathcal{X} \to 2^{\mathcal{X}^*}$ such that, for all $x \in \mathcal{X}$,*

$$\mathbf{J}^p_{\mathcal{X}}(x) = \left\{x^* \in \mathcal{X}^* : \langle x^*, x\rangle = \|x\|_{\mathcal{X}} \|x^*\|_{\mathcal{X}^*},\ \|x^*\|_{\mathcal{X}^*} = \|x\|^{p-1}_{\mathcal{X}}\right\}.$$

Under suitable smoothness assumptions on $\mathcal{X}$ [21], $\mathbf{J}^p_{\mathcal{X}}(x)$ is single valued at all $x \in \mathcal{X}$. For instance, for $\mathcal{X} = \ell^p(\mathbb{R})$, with $p > 1$, all duality maps are single-valued. The following Theorem (see [3]) provides an operative definition and a more intuitive interpretation of the duality maps.

Theorem 1 (Asplund's Theorem). *The duality map $\mathbf{J}^p_{\mathcal{X}}$ is the subdifferential of the convex functional $h : x \mapsto \frac{1}{p}\|x\|^p_{\mathcal{X}}$, that is, $\mathbf{J}^p_{\mathcal{X}}(\cdot) = \partial(\frac{1}{p}\|\cdot\|^p_{\mathcal{X}})$.*

The following result is needed for the invertibility of the duality map.

Proposition 1. *[21] Under suitable smoothness and convexity conditions on $\mathcal{X}$ and for $p > 1$, for all $x \in \mathcal{X}$ and all $x^* \in \mathcal{X}^*$, there holds*

$$\mathbf{J}^{p^*}_{\mathcal{X}^*}(\mathbf{J}^p_{\mathcal{X}}(x)) = x\,, \qquad \mathbf{J}^p_{\mathcal{X}}(\mathbf{J}^{p^*}_{\mathcal{X}^*}(x^*)) = x^*.$$

We notice that, if the gradient term $A^*\mathbf{J}^q_{\mathcal{Y}}(Ax^k - y)$ vanishes in iteration (3), then $x^{k+1} = \mathbf{J}^{p^*}_{\mathcal{X}^*}(\mathbf{J}^p_{\mathcal{X}}(x^k)) = x^k$ by Proposition 1.

For any $p, r > 1$ and for any $x, h \in \ell^p(\mathbb{R})$, the explicit formula for $\mathbf{J}^r_{\ell^p}$ is

$$\langle \mathbf{J}^r_{\ell^p}(x), h\rangle = \|x\|^{r-p}_p \sum_{n\in\mathbb{N}} \operatorname{sign}(x_n)|x_n|^{p-1} h_n. \tag{4}$$

Moreover, since $(\ell^p(\mathbb{R}))^* \cong \ell^{p^*}(\mathbb{R})$, then the inverse of the r-duality map $\mathbf{J}^r_{\ell^p}$ is nothing but $(\mathbf{J}^r_{\ell^p})^{-1} = \mathbf{J}^{r^*}_{(\ell^p)^*} = \mathbf{J}^{r^*}_{\ell^{p^*}}$. Hence, the explicit analytical expression of its inverse $(\mathbf{J}^r_{\ell^p})^{-1} = \mathbf{J}^{r^*}_{\ell^{p^*}}$ is also known [3].

2.1 Variable Exponent Lebesgue Spaces $\ell^{(p_n)}(\mathbb{R})$

In the following, we will introduce the main concepts and definitions on the variable exponent Lebesgue spaces in the discrete setting of $\ell^{(p_n)}(\mathbb{R})$. For surveys, we refer the reader to [4,5]. We define a family $\mathcal{P}$ of variable exponents as

$$\mathcal{P} := \left\{(p_n)_{n\in\mathbb{N}} \subset \mathbb{R} : 1 < p_- := \inf_{n\in\mathbb{N}} p_n \leq p_+ := \sup_{n\in\mathbb{N}} p_n < +\infty\right\}.$$

Definition 2. *For $(p_n)_{n\in\mathbb{N}} \in \mathcal{P}$ and any real sequence $x = (x_n)_{n\in\mathbb{N}}$,*

$$\rho_{(p_n)}(x) := \sum_{n\in\mathbb{N}} |x_n|^{p_n} \quad \text{and} \quad \bar{\rho}_{(p_n)}(x) := \sum_{n\in\mathbb{N}} \frac{1}{p_n}|x_n|^{p_n} \tag{5}$$

are called modular functions associated with the exponent map $(p_n)_{n\in\mathbb{N}}$.

Definition 3. *The Banach space $\ell^{(p_n)}(\mathbb{R})$ is the set of real sequences $x = (x_n)_{n\in\mathbb{N}}$ such that $\rho_{(p_n)}\left(\frac{x}{\lambda}\right) < 1$ for some $\lambda > 0$. For any $x = (x_n)_{n\in\mathbb{N}} \in \ell^{(p_n)}(\mathbb{R})$, the (Luxemburg) norm on $\ell^{(p_n)}(\mathbb{R})$ is defined as*

$$\|x\|_{\ell^{(p_n)}} := \inf\left\{\lambda > 0 : \ \rho_{(p_n)}\left(\frac{x}{\lambda}\right) \le 1\right\}. \tag{6}$$

We now report a result from [2] where a characterisation of the duality map $\mathbf{J}^r_{\ell^{(p_n)}}$ is given, in relation with (4).

Theorem 2. *Given $(p_n)_{n\in\mathbb{N}} \in \mathcal{P}$, then for each $x = (x_n)_{n\in\mathbb{N}} \in \ell^{(p_n)}(\mathbb{R})$ and for any $r > 1$, the duality map $\mathbf{J}^r_{\ell^{(p_n)}}(x) : \ell^{(p_n)}(\mathbb{R}) \to (\ell^{(p_n)})^*(\mathbb{R})$ is the linear operator defined, for all $h = (h_n)_{n\in\mathbb{N}} \in \ell^{(p_n)}(\mathbb{R})$ by:*

$$\langle \mathbf{J}^r_{\ell^{(p_n)}}(x), h\rangle = \frac{1}{\sum_{n\in\mathbb{N}} \frac{p_n|x_n|^{p_n}}{\|x\|^{p_n}_{\ell^{(p_n)}}}} \sum_{n\in\mathbb{N}} \frac{p_n \operatorname{sign}(x_n)|x_n|^{p_n-1}}{\|x\|^{p_n-r}_{\ell^{(p_n)}}} h_n. \tag{7}$$

By (6), we note that $\|\cdot\|_{\ell^{(p_n)}}$ is not separable as its computation requires the solution of a minimisation problem involving all elements x_n and p_n at the same time. As a consequence, the expression (7) is not suited to be used in a computational optimisation framework. The following result from [14] provides more flexible expressions associated to the modular functions (5).

Proposition 2. *The functions $\rho_{(p_n)}$ and $\bar{\rho}_{(p_n)}$ in (5) are Gateaux differentiable at any $x = (x_n)_{n\in\mathbb{N}} \in \ell^{(p_n)}(\mathbb{R})$. For $h = (h_n)_{n\in\mathbb{N}} \in \ell^{(p_n)}(\mathbb{R})$ their derivatives read*

$$\langle \mathbf{J}_{\rho_{(p_n)}}(x), h\rangle = \sum_{n\in\mathbb{N}} p_n \operatorname{sign}(x_n)|x_n|^{p_n-1}h_n, \quad \langle \mathbf{J}_{\bar{\rho}_{(p_n)}}(x), h\rangle = \sum_{n\in\mathbb{N}} \operatorname{sign}(x_n)|x_n|^{p_n-1}h_n. \tag{8}$$

Notice that, although $\mathbf{J}_{\rho_{(p_n)}}$ and $\mathbf{J}_{\bar{\rho}_{(p_n)}}$ are formally not duality maps, we adopt the same notation for the sake of consistency with Asplund Theorem 1.

3 Modular-Based Gradient Descent in $\ell^{(p_n)}(\mathbb{R})$

Given $(p_n)_{n\in\mathbb{N}}, (q_n)_{n\in\mathbb{N}} \in \mathcal{P}$, we now discuss how to implement a deterministic gradient-descent (GD) type algorithm for solving an instance of (P) with $\mathcal{X} = \ell^{(p_n)}(\mathbb{R})$ and $\mathcal{Y} = \ell^{(q_n)}(\mathbb{R})$. Recalling (3), GD iterations in this setting require knowing the duality map $\mathbf{J}^r_{\ell^{(p_n)}}$ and its inverse. However, as shown in [5, Corollary 3.2.14], such an inverse does not directly relate to the point-wise conjugate exponents of $(p_n)_{n\in\mathbb{N}}$ as the isomorphism between $(\ell^{(p_n)})^*(\mathbb{R})$ and $\ell^{(p_n^*)}(\mathbb{R})$

Algorithm 1: Modular-based Gradient Descent in $\ell^{(p_n)}(\mathbb{R})$

Parameters: $\{\mu_k\}_k$ s.t. $0 < \bar{\mu} \leq \mu_k \leq \frac{pc(1-\delta)}{K}$ with $0 < \delta < 1$, for all $k \geq 0$.
Initialisation: $x^0 \in \ell^{(p_n)}(\mathbb{R})$.
repeat

$$x^{k+1} = |\mathbf{J}_{\bar{\rho}_{(p_n)}}(x^k) - \mu_k \nabla f(x^k)|^{\frac{1}{p_n-1}} \operatorname{sign}(\mathbf{J}_{\bar{\rho}_{(p_n)}}(x^k) - \mu_k \nabla f(x^k)) \quad (9)$$

until convergence

-differing from the standard ℓ^p constant case- is not isometric. As discussed in [2], the approximation $\left(\mathbf{J}^r_{\ell^{(p_n)}}\right)^{-1} = \mathbf{J}^{r^*}_{(\ell^{(p_n)})^*} \approx \mathbf{J}^{r^*}_{\ell^{(p^*_n)}}$ can be used as an inexact (but explicit) formula for computing the duality map of $(\ell^{(p_n)})^*(\mathbb{R})$. Under this assumption, the dual Landweber method can thus be used to solve the minimisation problem $\operatorname{argmin}_{x\in\ell^{(p_n)}(\mathbb{R})} \frac{1}{q}\|Ax - y\|^q_{\ell^{(q_n)}}$, $q > 1$. Note, however, that the computation of the duality map $\mathbf{J}^p_{\ell^{(p_n)}}$ requires the computation of $\|x\|_{\ell^{(p_n)}}$ which, as previously discussed, makes the iterative scheme rather inefficient in terms of computational time. We thus follow [14] and define in Algorithm 1 a more efficient modular-based gradient descent iteration for the resolution of (P) in the general setting of variable exponent Lebesgue spaces. The following set of assumptions needs to hold:

A.1 $\nabla f : \ell^{(p_n)}(\mathbb{R}) \to (\ell^{(p_n)})^*(\mathbb{R})$ is $(\mathrm{p}-1)$–Hölder-continuous with exponent $1 < \mathrm{p} \leq 2$ and constant $K > 0$.
A.2 There exists $c > 0$ such that, for all $u, v \in \ell^{(p_n)}(\mathbb{R})$,

$$\langle \mathbf{J}_{\bar{\rho}_{(p_n)}}(u) - \mathbf{J}_{\bar{\rho}_{(p_n)}}(v), u - v\rangle \geq c \max\left\{\|u-v\|^{\mathrm{p}}_{\ell^{(p_n)}}, \|\mathbf{J}_{\bar{\rho}_{(p_n)}}(u) - \mathbf{J}_{\bar{\rho}_{(p_n)}}(v)\|^{\mathrm{p}^*}_{(\ell^{(p_n)})^*}\right\}.$$

The latter bound was previously used in [8,14]. It is a compatibility condition between the ambient space $\ell^{(p_n)}(\mathbb{R})$ and the Hölder smoothness properties of the residual function to minimise to achieve algorithmic convergence.

The minimisation of the specific function f of (P) is achieved solving at each iteration (9) the following minimisation problem:

$$x^{k+1} = \operatorname*{argmin}_{u\in\ell^{(p_n)}(\mathbb{R})} \bar{\rho}_{(p_n)}(u) - \langle \mathbf{J}_{\bar{\rho}_{(p_n)}}(x^k), u\rangle + \mu_k\langle \nabla f(x^k), u\rangle.$$

The following proof shows that the functional $\mathbf{J}_{\bar{\rho}_{(p_n)}}$ defined by (8) is invertible and gives a point-wise characterisation of its inverse.

Proposition 3. *The functional* $\mathbf{J}_{\bar{\rho}_{(p_n)}}$ *in* (8) *is invertible. For all* $v \in (\ell^{(p_n)})^*(\mathbb{R})$,

$$(\mathbf{J}_{\bar{\rho}_{(p_n)}})^{-1}(v) = \left(|v_n|^{\frac{1}{p_n-1}} \operatorname{sign}(v_n)\right)_{n\in\mathbb{N}} \in \ell^{(p_n)}(\mathbb{R}).$$

Proof. By straightforward componentwise computation, we have

$$|\mathbf{J}_{\bar{\rho}_{(p_n)}}(v_n)|^{\frac{1}{p_n-1}}\operatorname{sign}(\mathbf{J}_{\bar{\rho}_{(p_n)}}(v_n)) = |\mathbf{J}_{\bar{\rho}_{(p_n)}}(v_n)|^{\frac{1}{p_n-1}-1}\mathbf{J}_{\bar{\rho}_{(p_n)}}(v_n)$$
$$= |\mathbf{J}_{\bar{\rho}_{(p_n)}}(v_n)|^{\frac{2-p_n}{p_n-1}}\mathbf{J}_{\bar{\rho}_{(p_n)}}(v_n) = \big|\,|v_n|^{p_n-1}\operatorname{sign}(v_n)\big|^{\frac{2-p_n}{p_n-1}}|v_n|^{p_n-1}\operatorname{sign}(v_n) = v_n\,.$$

By the Proposition above, the update rule (9) of Algorithm 1, can be rewritten as

$$x^{k+1} = (\mathbf{J}_{\bar{\rho}_{(p_n)}})^{-1}\Big(\mathbf{J}_{\bar{\rho}_{(p_n)}}(x^k) - \mu_k\nabla f(x^k)\Big).$$

As a consequence, whenever $\nabla f(x_k) = 0$ at some $k \geq 0$, a stationary point $x^{k+1} = (\mathbf{J}_{\bar{\rho}_{(p_n)}})^{-1}\Big(\mathbf{J}_{\bar{\rho}_{(p_n)}}(x^k)\Big) = x^k$ is found, as expected.

The following convergence result is a special case of [14, Proposition 3.4] providing an explicit convergence rate for the iterates of Algorithm 1.

Proposition 4. *Let $x^* \in \ell^{(p_n)}(\mathbb{R})$ be a minimiser of f and let $(x^k)_k$ be the sequence generated by Algorithm 1. If (x^k) is bounded, then:*

$$f(x^k) - f(x^*) \leq \frac{\eta}{k^{p-1}},$$

*where $p > 1$ is defined in assumption **A.1** and $\eta = \eta(\bar{\mu}, \delta, p_-, x^0, x^*)$.*

Note that when the measurement space $\mathcal{Y}$ is a variable exponent Lebesgue space $\ell^{(q_n)}(\mathbb{R})$, a more effective and consistent choice for the objective function is the modular of the discrepancy between the model observation and the data, i.e. $f(x) = \bar{\rho}_{(q_n)}(Ax - y)$. In this way, the heavy computations of the $\|\cdot\|_{\ell^{(q_n)}}$ norm and of its gradient are not required, making the iteration scheme faster.

4 Stochastic Modular-Based Gradient-Descent in $\ell^{(p_n)}(\mathbb{R})$

The key challenge for the viability of many deterministic iterative methods for real-world image reconstruction problems is their scalability to data-size. For example, the highest per-iteration cost in emission tomography lies in the application of the entire forward operator at each iteration, whereas each image domain datum in computed tomography often requires several gigabytes of storage space. The same could thus be a bottleneck in the application of Algorithm 1. The stochastic gradient descent (SGD) paradigm addresses this issue [20].

We partition the forward operator A, and the forward model into a finite number of block-type operators $A_1, \ldots, A_{N_s}$, where $N_s \in \mathbb{N}$ is the number of subsets of data. The same partition is applied to the observations. Classical examples of this methodology include Kaczmarz methods in CT [9,17]. The SGD version of the iteration (3) in Banach spaces takes the form

$$x^{k+1} = \mathbf{J}^{p^*}_{\mathcal{X}^*}\left(\mathbf{J}^p_{\mathcal{X}}(x^k) - \mu_{k+1}A^*_{i_k}\mathbf{J}^q_{\mathcal{Y}}(A_{i_k}x^k - y)\right), \tag{10}$$

where the indices $i_k \in \{1, \ldots, N_s\}$ are sampled uniformly at random. Sampling reduces the per-iteration computational cost in $\mathcal{Y}$ by a factor of N_s. In [13] convergence of the iterates to a minimum norm solution is shown.

Algorithm 2: Stochastic Modular-based Gradient Descent in $\ell^{(p_n)}(\mathbb{R})$

Parameters: μ_0 s.t. $0 < \bar{\mu} \leq \mu_0 \leq \frac{pc(1-\delta)}{K}$, $0 < \delta < 1$, $N_s \geq 1$, $\gamma > 0$, $\eta > 0$.
Initialisation: $x^0 \in \ell^{(p_n)}(\mathbb{R})$.
repeat

Select uniformly at random $i_k \in \{1, \cdots, N_s\}$.
Set $\mu_k = \frac{\mu_0}{1+\eta(k/N_s)^\gamma}$
Compute

$$x^{k+1} = |\mathbf{J}_{\bar{\rho}_{(p_n)}}(x^k) - \mu_k \nabla f_{i_k}(x^k)|^{\frac{1}{p_n - 1}} \operatorname{sign}(\mathbf{J}_{\bar{\rho}_{(p_n)}}(x^k) - \mu_k \nabla f_{i_k}(x^k))$$

until convergence

Theorem 3. *Let* $\sum_{k=1}^{\infty} \mu_k = +\infty$ *and* $\sum_{k=1}^{\infty} \mu_k^{p^*} < +\infty$. *Then*

$$\mathbb{P}\Big(\lim_{k\to\infty} \inf_{\widetilde{x}\in\mathcal{X}_{\min}} \|x^{k+1} - \widetilde{x}\|_{\mathcal{X}} = 0\Big) = 1.$$

Let $\mathbf{J}_{\mathcal{X}}^p(x_0) \in \overline{\operatorname{range}(A^*)}$ *and let* $\mu_k^{p^*-1} \leq \frac{C}{L_{\max}^{p^*}}$ *for all* $k \geq 0$ *and some constant* $C > 0$, *where* $L_{\max} = \max_i \|A_i\|$. *Then* $\lim_{k\to\infty} \mathbb{E}[\|x^{k+1} - x^\dagger\|_{\mathcal{X}}] = 0$ $\lim_{k\to\infty} \mathbb{E}[\|\mathbf{J}_{\mathcal{X}}^p(x^{k+1}) - \mathbf{J}_{\mathcal{X}}^p(x^\dagger)\|^{p^*}] = 0$.

For noisy measurements, the regularising property of SGD should be established by defining suitable stopping criteria. However, robust stopping strategies are hard to use in practice and having methods that are less sensitive to data overfit is crucial for their practical use. Note that (10) is the standard form of SGD for separable objectives. Namely, for $f(x) = \|Ax - y\|_q^q$, we can choose $f_i(x; A, y) = \|A_i x - y_i\|_q^q$, so that $f(x) = \sum_{i=1}^{N_s} f_i(x)$. By Theorem 1, this decomposition shows that each step of (10) can thus be computed by simply taking a sub-differential of a single sum-function f_i.

To define a suitable SGD in variable exponent Lebesgue spaces, we take as objective function $f(x) = \bar{\rho}_{(q_n)}(Ax - y)$ and split it into $N_s \geq 1$ sub-objectives $f_i(x) := \bar{\rho}_{(q_n^i)}(A_i x - y_i)$, so that $\nabla f_i(x) = A_i^* \mathbf{J}_{\bar{\rho}_{(q_n^i)}}(A_i x - y_i)$. Exponents $(q_n^i)_n$ are obtained through the same partition of the exponents $(q_n)_n$ as the one used to split up the data. Then, at iteration k and a randomly sampled index $1 \leq i_k \leq N_s$, the corresponding stochastic iterates are given by

$$x^{k+1} = \underset{u\in\ell^{(p_n)}(\mathbb{R})}{\operatorname{argmin}} \ \bar{\rho}_{(p_n)}(u) - \langle \mathbf{J}_{\bar{\rho}_{(p_n)}}(x^k), u\rangle + \mu_k \langle \nabla f_{i_k}(x^k), u\rangle.$$

The pseudocode of the resulting stochastic modular-based gradient descent in $\ell^{(p_n)}(\mathbb{R})$ is reported in Algorithm 2. We expect that through minimal modifications an analogous convegence result as Theorem 3 can be proved in this setting too. A detailed convergence proof, however, is left for future research.

5 Numerical Results

We now present experimental results of the proposed Algorithm 2 on two exemplar problems in computed tomography (CT). The first set of experiments consider a simulated setting for quantitatively comparing the performance of Algorithm 2 with the corresponding Hilbert and Banach space versions (10). In the second set of experiments we consider the dataset of real-world CT scans of a walnut taken from doi:10.5281/zenodo.4279549, with a fan beam geometry. For these data, we utilise the insights from the first set of experiments and apply Algorithm 2 in a setting with different noise modalities across the sinogram space. The experiments were conducted in `python`, using the open source package [12] for the tomographic backend.

Hyper-Parameter Selection. In the following experiments, we employ a decaying stepsize regime such that it satisfies the conditions of Theorem 3 for the convergence of Banach space SGD, cf. [13]. A need for a decaying stepsize regime is common for stochastic gradient descent to mitigate the effects of inter-iterate variance. Specifically, we use $\mu_k = \frac{\mu_0}{1+c(k/N_s)^\gamma}$, where $\mu_0 > 0$ is the initial stepsize, and $\gamma > 0$ and $c > 0$ control the decay speed. For the Hilbert space setting, $\mathbf{SGD_2}$, initial stepsize μ_0 is given by the Lipschitz constant of the gradient of the objective function, namely $\mu_0 = 0.95/\max_i \|A_i\|^2$. For $\mathbf{SGD_p}$ and $\mathbf{SGD_{p_n,q_n}}$ the estimation of the respective Hölder continuity constant is more delicate and μ_0 has to be tuned to guarantee convergence. However, its tuning is rather easy and the employ of a decaying strategy makes the choice of μ_0 less critical.

As far as variable exponents are concerned, it is difficult (and somehow undesirable) to have a unified configuration as their selection is strictly problem-related. Parameters $(q_n)_n$ are related to the regularity of the measured sinograms as well as the different noise distributions considered. For instance, when impulsive noise is considered, values of q_- and q_+ closer to 1 are preferred while and for Gaussian noise values closer to 2 are more effective. Solution space parameters p_- and p_+ relate to the regularity of the solution to retrieve. As a consequence, their choice is intrinsically harder. We refer the reader to [2], where a comparison between different choices for p_- and p_+ and different interpolation strategies is carried out for image deblurring with gradient descent (3) in $\ell^{(p_n)}$.

Simulated Data. We considered (1) with A given by the discrete Radon transform. For its definition we use a 2D parallel beam geometry, with 180 projection angles on a 1 angle separation, 256 detector elements, and pixel size of 0.1. The synthetic phantom was provided by the CIL library, see Fig. 1(b). After applying the forward operator, a high level (15%) of salt-and-pepper noise is applied to the sinogram. The noisy sinogram is shown in Fig. 1(a).

To compute subset data A_i and y_i, the forward operator and the sinogram are pre-binned according to equally spaced views (w.r.t. the number of subsets) of the scanner geometry. Subsequent subset data are offset from one another by the subset index i. We consider $N_s = 30$ batches. We compare results obtained by solving (P) by:

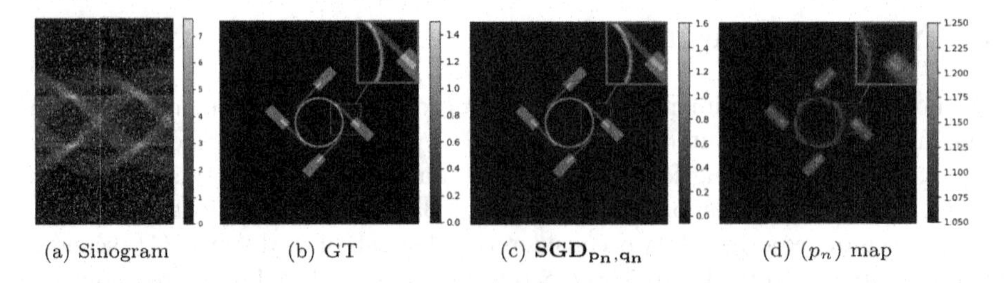

(a) Sinogram (b) GT (c) $\mathbf{SGD_{p_n,q_n}}$ (d) (p_n) map

Fig. 1. In (c) reconstruction of noisy sinogram (a) by $\mathbf{SGD_{p_n,q_n}}$, where $1.05 = p_- \leq (p_n) \leq p_+ = 1.25$ is shown in (d) and $1.05 = q_- \leq (q_n) \leq q_+ = 1.25$ is based on the model observation corresponding to (p_n).

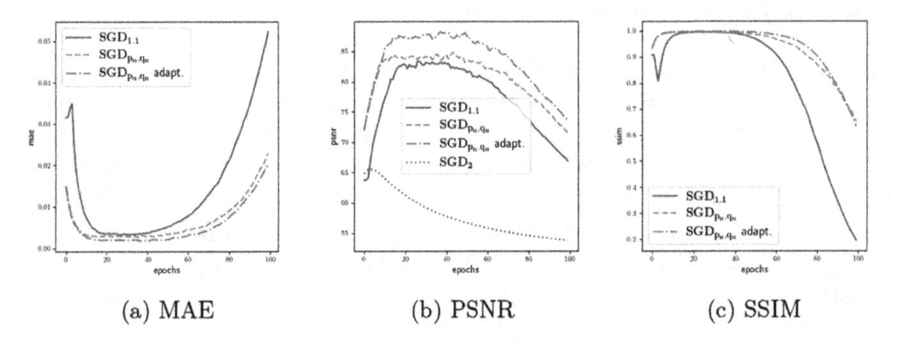

(a) MAE (b) PSNR (c) SSIM

Fig. 2. Quality metrics along the first 100 epochs of $\mathrm{SGD}_\mathbf{2}$; $\mathrm{SGD}_{\mathbf{1.1}}$; $\mathrm{SGD}_{\mathbf{p_n,q_n}}$ with and without adapting the exponent maps (p_n). $\mathrm{SGD}_\mathbf{2}$ is omitted from MAE and SSIM to improve the readability of the plots, due to its poor performance.

$\mathbf{SGD_2}$: $\mathcal{X} = \mathcal{Y} = \ell^2(\mathbb{R})$, $\mathcal{Y} = \ell^2(\mathbb{R})$, $f(x) = \frac{1}{2}\|Ax - y\|_2^2$ by SGD;
$\mathbf{SGD_p}$: $\mathcal{X} = \mathcal{Y} = \ell^p(\mathbb{R})$, $p = 1.1$, $f(x) = \frac{1}{p}\|Ax - y\|_p^p$ by Banach SGD (10);
$\mathbf{SGD_{p_n,q_n}}$: $\mathcal{X} = \ell^{(p_n)}(\mathbb{R})$, $\mathcal{Y} = \ell^{(q_n)}(\mathbb{R})$ for appropriately chosen exponent maps, $f(x) = \bar{\rho}_{(q_n)}(Ax - y)$ with modular-based SGD Algorithm 2.

We considered step-sizes $\mu_k = \frac{\mu_0}{1+0.1(k/N_s)^\gamma}$, with μ_0 and γ which depend on the algorithm.[1] Spaces $\ell^{(p_n)}(\mathbb{R})$ allow for variable exponent maps sensitive to local assumptions on both the solution and the measured data. A possible strategy for informed pixel-wise variable exponents consists in basing them on observed data (for (q_n)) and an approximation of the reconstruction (for (p_n)), as done in [1,2,14]. To this end, we first compute an approximate reconstruction $\tilde{x} \in \ell^{(p_n)}(\mathbb{R})$ by running $\mathbf{SGD_p}$ in $\ell^{1.1}(\mathbb{R})$ for 5 epochs with a constant stepsize regime. The map (p_n) is then computed via a linear interpolation of $\tilde{x}$ between $p_- = 1.05$ and $p_+ = 1.25$. The map (q_n) is chosen as the linear interpolation between $q_- = 1.05$ and $q_+ = 1.25$ of $A(p_n)$. The bounds p_-, p_+ and q_-, q_+ are chosen by prior assumptions on y (sparse phantom) and on the noise observed

[1] For $\mathbf{SGD_2}$ μ_0 is set as $0.95/\max_i \|A_i\|^2$ and $\gamma = 0.51$. For $\mathbf{SGD_p}$ and $\mathbf{SGD_{p_n,q_n}}$, we use $\mu_0 = 0.015$ with $\gamma = (p-1)/p + 0.01$ and $\gamma = (p_- - 1)/p_- + 0.01$ respectively.

Table 1. Comparison of per iteration cost and total CPU times after 3000 iterations for determistic algorithms and after 100 epochs for stochastic algorithm with $N_s = 30$. MAE, PSNR and SSIM values for stochastic algorithms are computed after 40 epochs (before noise overfitting).

Algorithm	Deterministic		Stochastic ($\cdot$ = **S**)					
	It.	Tot.	It.	Epoch	Tot.	MAE	PSNR	SSIM
$\cdot\mathbf{GD_2}$	0.44 s	1324 s	0.02 s	0.74 s	74.4 s	2.582e-1	57.89	0.0304
$\cdot\mathbf{GD_{1.1}}$	0.43 s	1297 s	0.03 s	0.81 s	81.3 s	3.671e-3	82.64	0.9897
$\cdot\mathbf{GD_{p_n,q_n}}$	0.47 s	1403 s	0.03 s	0.96 s	96.5 s	2.887e-3	84.05	0.9927
$\cdot\mathbf{GD_{p_n,q_n}}$ adapt	0.44 s	1317 s	0.03 s	0.91 s	91.2 s	1.777e-3	88.10	0.9965
Compute (p_n), (q_n)	0.45 s	16 s	0.03 s	0.8 s	4.0 s	-	-	-

(impulsive). We also tested an adaptive strategy by updating (p_n) based on the current solution estimate once every β_{updates} epochs to adapt the exponents along the iterations.

In Fig. 2, we report the mean absolute error (MAE), peak signal to noise ratio (PSNR) and structural similarity index (SSIM) of the iterates x^k w.r.t. the known ground-truth phantom along the first 100 epochs. Since PSNR favours smoothness, it is thus beneficial for $\mathbf{SGD_2}$, whereas MAE promotes sparsity hence is beneficial for both $\mathbf{SGD_p}$ and $\mathbf{SGD_{p_n,q_n}}$. Figure 2b shows that Banach space algorithms provide better performance than $\mathbf{SGD_2}$ in all three quality metrics. Note that all the results show the well-known semi-convergence behaviour with respect to the metrics considered. To avoid such behaviour an explicit regulariser or a sound early stopping criterion would be beneficial for reconstruction performance. We observe that the use variable exponents does not only improve all quality metrics, but also makes the algorithm more stable: the quality of the reconstructed solutions is significantly less sensitive to the number of epochs, making possible early stopping strategies more robust.

In Table 1, the CPU times for deterministic ($\mathbf{GD_2}$, $\mathbf{GD_p}$ and $\mathbf{GD_{p_n,q_n}}$) approaches and stochastic ones ($\mathbf{SGD_2}$, $\mathbf{SGD_p}$ and $\mathbf{SGD_{p_n,q_n}}$) are compared.

Real CT Datasets: Walnut. We consider a cone beam CT dataset of a walnut [15], from which we take a 2D fan beam sinograms from the centre plane of the cone. The cone beam data uses 0.5 angle separation over the range $[0, 360]$. The used sinogram is obtained by pre-binning the raw data by a factor of 8, resulting in 280 effective detector pixels. The measurements have been post-processed for dark current and flat-field compensation. As stepsize we used $\mu_k = \frac{\mu_0}{1+0.001(k/N_s)^\gamma}$, with $N_s = 10$ subsets, and suitable μ_0 and γ.[2] Initial images are computed by 5 epochs of $\mathbf{SGD_{1.4}}$ with a constant stepsize.

We consider a more delicate noise setting that requires exponential maps which vary in the acquisition domain. Here, we assume that noise has a different

[2] For $\mathbf{SGD}_2$, $\mu_0 = 0.95/\max_i \|A_i\|^2$, $\gamma = 0.51$. For $\mathbf{SGD}_{p_n,q_n}$ we $\mu_0 = 0.001$, $\gamma = 0.58$.

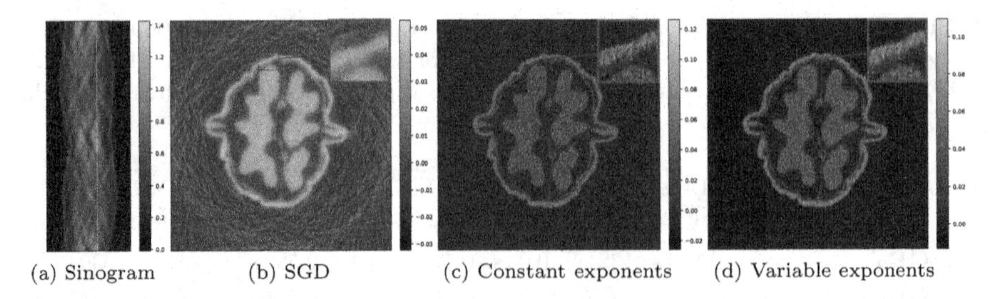

(a) Sinogram (b) SGD (c) Constant exponents (d) Variable exponents

Fig. 3. (a) Noisy sinogram with 10% salt & pepper (background) and speckle noise with 0 mean and variance 0.01 (foreground). (b) $\mathbf{SGD_2}$ result. (c) $\mathbf{SGD_{p_n,1.1}}$ result (d) $\mathbf{SGD_{p_n,q_n}}$ result. $p_- = 1.2$, $p_+ = 1.3$, $q_- = 1.1$ and $q_+ = 1.9$.

effect on the background (zero entries) and the foreground (non-zero entries) of the clean sinogram. Namely, we apply 10% salt and pepper noise to the background, and speckle noise with mean 0 and variance 0.01 to the foreground, cf. Fig. 3(a) for the resulting noisy sinogram. Notably, since this noise model has a non-uniform effect across the measurement data, Banach space methods favouring the adjustment of the Lebesgue exponents are expected to perform better than those making use of a constant value. Taking as a reference the result obtained by $\mathbf{SGD_2}$ (Fig. 3(b)), we compare here the effect of allowing variable exponents in the solution space only with the effect of allowing both maps (p_n) and (q_n) to be chosen. By choosing (p_n) based on the initial image and interpolating it between $p_- = 1.2$ and $p_+ = 1.3$ we then compare $\mathbf{SGD_{p_n,1.1}}$ (i.e., fixed exponent $q = 1.1$ in the measurement space), cf. Fig. 3(c), with $\mathbf{SGD_{p_n,q_n}}$ where (p_n) is as before while (q_n) is chosen from the sinogram by interpolating between $q_- = 1.1$ and $q_+ = 1.9$, cf. Fig. 3(d). The results show that a flexible framework where both maps (p_n) and (q_n) adapt to local contents are more suited for dealing with this challenging scenario.

6 Conclusions

We proposed a stochastic gradient descent algorithm for solving linear inverse problems in $\ell^{(p_n)}(\mathbb{R})$. After recalling its deterministic counterpart and the difficulties encountered due to the non-separability of the underlying norm, a modular-based stochastic algorithm enjoying fast scalability properties is proposed. Numerical results show improved performance in comparison to standard $\ell^2(\mathbb{R})$ and $\ell^p(\mathbb{R})$-based algorithms and significant computational gains. Future work should adapt the convergence result (Theorem 2) to this setting and consider proximal extensions for incorporating non-smooth regularisation terms.

Acknowledgement. CE and ML acknowledge the support of the Italian INdAM group on scientific calculus GNCS. LC acknowledges the support received by the ANR projects TASKABILE (ANR-22-CE48-0010) and MICROBLIND (ANR-21-CE48-0008), the H2020 RISE projects NoMADS (GA. 777826) and the GdR ISIS project SPLIN. ZK acknowledges support from EPSRC grants EP/T000864/1 and EP/X010740/1.

References

1. Alparone, M., Nunziata, F., Estatico, C., Lenti, F., Migliaccio, M.: An adaptive l^p -penalization method to enhance the spatial resolution of microwave radiometer measurements. IEEE Trans. Geosci. Remote Sens. **57**(9), 6782–6791 (2019)
2. Bonino, B., Estatico, C., Lazzaretti, M.: Dual descent regularization algorithms in variable exponent Lebesgue spaces for imaging. Numer. Algorithms **92**(6) (2023)
3. Cioranescu, I.: Geometry of Banach Spaces, Duality Mappings and Nonlinear Problems. Springer, Dordrecht (1990). https://doi.org/10.1007/978-94-009-2121-4
4. Cruz-Uribe, D.V., Fiorenza, A.: Variable Lebesgue Spaces. Springer Birkhäuser, Basel (2013). https://doi.org/10.1007/978-3-0348-0548-3
5. Diening, L., Harjulehto, P., Hästö, P., Ruzicka, M.: Lebesgue and Sobolev Spaces with Variable Exponents. Lecture Notes in Math, Springer, Heidelberg (2011). https://doi.org/10.1007/978-3-642-18363-8
6. Eicke, B.: Iteration methods for convexly constrained ill-posed problems in hilbert space. Numer. Funct. Anal. Optim. **13**(5–6), 413–429 (1992)
7. Engl, H.W., Hanke, A., Neubauer, M.: Regularization of Inverse Problems. Mathematics and Its Applications, Springer, Dordrecht (2000)
8. Guan, W.-B., Song, W.: The generalized forward-backward splitting method for the minimization of the sum of two functions in banach spaces. Numer. Funct. Anal. Optim. **36**(7), 867–886 (2015)
9. Herman, G.T., Meyer, L.B.: Algebraic reconstruction techniques can be made computationally efficient (positron emission tomography application). IEEE Trans. Med. Imaging **12**(3), 600–609 (1993)
10. Jin, Q., Lu, X., Zhang, L.: Stochastic mirror descent method for linear ill-posed problems in Banach spaces (2022). arXiv preprint https://arxiv.org/abs/2207.06584
11. Jin, Q., Stals, L.: Nonstationary iterated Tikhonov regularization for ill-posed problems in Banach spaces. Inverse Probl. **28**(10), 104011 (2012)
12. Jørgensen, J.S., et al.: Core imaging library - Part I: a versatile python framework for tomographic imaging. Phil. Trans. R. Soc. A **379**(2204), 20200192 (2021)
13. Kereta, Z., Jin, B.: On the convergence of stochastic gradient descent for linear inverse problems in Banach spaces. SIAM J. Imaging Sci. (2023, in press). arXiv preprint https://arxiv.org/abs/2302.05197
14. Lazzaretti, M., Calatroni, L., Estatico, C.: Modular-proximal gradient algorithms in variable exponent Lebesgue spaces. SIAM J. Sci. Comput. **44**(6), A3463–A3489 (2022)
15. Meaney, A.: X-ray dataset of walnut (2020-11-11), November 2020
16. Natterer, F.: The Mathematics of Computerized Tomography. Wiley, Hoboken (1986)
17. Needell, D., Zhao, R., Zouzias, A.: Randomized block Kaczmarz method with projection for solving least squares. Linear Algebra Appl. **484**, 322–343 (2015)
18. Neubauer, A.: Tikhonov-regularization of ill-posed linear operator equations on closed convex sets. J. Approx. Theory **53**(3), 304–320 (1988)
19. Piana, M., Bertero, M.: Projected Landweber method and preconditioning. Inverse Probl. **13**(2), 441–463 (1997)

20. Robbins, H., Monro, S.: A stochastic approximation method. Ann. Math. Stat. **22**(3), 400–407 (1951)
21. Schuster, T., Kaltenbacher, B., Hofmann, B., Kazimierski, K.S.: Regularization Methods in Banach Spaces. De Gruyter (2012)
22. Schöpfer, F., Louis, A.K., Schuster, T.: Nonlinear iterative methods for linear ill-posed problems in Banach spaces. Inverse Probl. **22**(1), 311–329 (2006)
23. Twyman, R., Arridge, S., et al.: An investigation of stochastic variance reduction algorithms for relative difference penalized 3D PET image reconstruction. IEEE Trans. Med. Imaging **42**(1), 29–41 (2023)

An Efficient Line Search for Sparse Reconstruction

Shima Shabani(✉) and Michael Breuß

Institute for Mathematics, Brandenburg Technical University, Platz der Deutschen Einheit 1, 03046 Cottbus, Germany
{shima.shabani,breuss}@b-tu.de

Abstract. A line search strategy is an iterative approach to finding a local minimizer of a nonlinear objective function. First, it finds a sensible direction and then an acceptable step size is computed that determines how far to move along that direction.

This paper introduces an efficient line search method for convex nonsmooth optimization problems. Like the Goldstein and Armijo conditions, the presented line search scheme uses only function values, apart from one gradient evaluation at each current iteration. For the novel line search method, we prove global convergence to a stationary point and R-linear convergence rate under some standard assumptions. Based on that, we propose an algorithm for the ℓ_1-minimization problem we call the iterative Shrinkage-Goldstein algorithm (ISGA). In numerical results, we report experiments that demonstrate the competitive performance of the ISGA approach for compressed sensing problems in comparison with some state-of-the-art algorithms.

Keywords: Nonsmooth minimization · Line search method · Goldstein quotient · Compressed sensing · Global convergence · R-linear convergence

1 Introduction

An important type of unconstrained nonsmooth convex optimization problems has the general form

$$\begin{array}{ll} \min & F(x) = f(x) + \mu c(x), \\ \text{s.t.} & x \in \mathbb{R}^n \end{array} \tag{1}$$

where $f(x)$ is a convex smooth function, the regularizer function $c(x)$ is convex and nonsmooth, and the regularization parameter $\mu \in \mathbb{R}^+$ is a trade-off parameter between the objective terms. The minimization problem (1) generalizes the well-known *basis pursuit denoising* problem [5,9–11,13], which is often formulated as

$$\begin{array}{ll} \min & F(x) = \frac{1}{2}\|Ax - b\|^2 + \mu\|x\|_1. \\ \text{s.t.} & x \in \mathbb{R}^n \end{array} \tag{2}$$

Here $\|\cdot\|$ stands for the standard Euclidean norm, $\|\cdot\|_1$ is the ℓ_1-norm, $A \in \mathbb{R}^{m\times n}$ ($m \ll n$) is the measurement matrix, and $b \in \mathbb{R}^m$ is the observation vector.

L. Calatroni et al. (Eds.): SSVM 2023, LNCS 14009, pp. 471–483, 2023.
https://doi.org/10.1007/978-3-031-31975-4_36

The model (2) is one of the standard robust models of sparse recovery, and it gives the sparsest solution of the underdetermined linear system $Ax = b$ when the measurements are noisy.

Popular areas in signal processing where the solution of (2) proves to be useful is compressed sensing [10,11,13]. Compressed sensing explores the use of sparse representations. It is a technique to reconstruct a signal from far fewer samples than required by the Nyquist-Shannon sampling theorem, by finding the solutions to the mentioned underdetermined linear systems.

Turning to the algorithmic solution of (2), a widely used method is an iterative procedure based on the shrinkage operator (also called soft thresholding, see Eq (11) below). The *iterative shrinkage-thresholding algorithm* (`ISTA`) [6,7] is a useful baseline algorithm to solve (2). However, it is a slow method with a sublinear convergence rate. Some algorithms have been proposed to accelerate `ISTA` in the past years, such as the fixed point continuation method with Barzilai-Borwein step size control (`FPC-BB`) [17], two-step `ISTA` (`TwIST`) [4], the fast iterative shrinkage thresholding algorithm (`FISTA`) [3], and sparse reconstruction by separable approximation (`SpaRSA`) [24], just to name a few.

Our Contribution. Inspired by [18], we propose a new nonsmooth strategy related to the Goldstein quotient to generate efficient step lengths. We show that the new strategy that steers the step size computation is easier to satisfy than the Goldstein condition, and it gives in some cases a more sharp decrease of the optimization objective than the Armijo condition. We study efficiency and convergence properties of the proposed method by a theoretical convergence analysis. Based on the presented nonsmooth line search methods, we propose an algorithm for the ℓ_1-minimization problem. We validate that the new method is competitive in quality and computational efficiency to some state-of-the art algorithms, for a number of tests in compressed sensing.

2 Related Work

In the following we briefly review some important topics and algorithms in unconstrained optimization methods related to this paper. First, we describe some standard conditions in the context of line search methods. Then we review some algorithms based on the shrinkage operator which are also employed later in the experimental section.

Line Searching. A *line search strategy* using the iterative scheme

$$x_{k+1} = x_k + \alpha_k d_k, \tag{3}$$

seeks the minimum of a nonlinear objective function by selecting a suitable direction vector d_k together with a step size α_k, so that in each iteration a function value closer to the minimum of the target function is obtained. The success and robustness of a line search method depend on suitable choices of both d_k and α_k. In robust minimization processes to design *globally convergent* algorithms, most line search algorithms require a *descent direction*. Ideally, the descent direction is also *efficient*.

Let us make precise what we mean. If a direction is a descent direction can be measured by $\Delta(x_k, d_k) < 0$ in

$$\Delta(x_k, d_k) = d_k^T \nabla f(x_k) + \mu\Big(c(x_k + d_k) - c(x_k)\Big), \tag{4}$$

which guarantees a reduction in the objective function, see e.g. [12,17,20]. For a line search method, *efficiency* in the sense of Warth & Werner [21] means, that in each iteration the step size α_k is chosen such that the following quotient has a fixed positive lower bound η for all $k \geq 0$,

$$\Big(F(x_k) - F(x_{k+1})\Big)\frac{\|d_k\|^2}{\Delta(x_k, d_k)^2} > \eta > 0. \tag{5}$$

Let us note that in [21], the condition (5) has been proposed for smooth optimization. As one of our contributions, we explore the use of (5) for *nonsmooth* optimization. We make this point precise in the first subsequent theorem.

Some nonsmooth optimization algorithms [12,17,20] employ a nonsmooth form of the *Armijo condition* [1] with backtracking to enforce sufficient decrease in the objective function with $0 < \nu_1 < 1$ as follows:

$$F(x_k + \alpha_k d_k) \leq F(x_k) + \nu_1 \alpha_k \Delta(x_k, d_k). \tag{6}$$

From the computational point of view, the Armijo condition just requires a function evaluation at the trial point $x_k + \alpha_k d_k$, which appears to make it an efficient choice for step size control. However, the Armijo condition by itself may not ensure that the step size is in practice satisfying. Since any small enough value of the step size will satisfy (6).

In [23], *Wolfe* used a curvature condition along with the smooth form of the Armijo formula (6) (setting $\mu = 0$ in (4), which is in turn used in (6)) to rule out very small steps. However, the Wolfe condition requires gradient evaluations at every trial point, which has a high computational cost and hinders its use in nonsmooth optimization.

In contrast to the Wolfe method, *Goldstein's condition* [15], which for a nonsmooth optimization strategy can be stated in the following way with $0 < \nu_2 < \nu_3 < 1$,

$$F(x_k) + \nu_3 \alpha_k \Delta(x_k, d_k) \leq F(x_k + \alpha_k d_k) \leq F(x_k) + \nu_2 \alpha_k \Delta(x_k, d_k), \tag{7}$$

only uses function values (apart from the $\Delta(x_k, d_k)$ evaluation at the current best point). This condition is designed so that the step size achieves a sufficient decrease and is at the same time not too short. But, in smooth optimization, Goldstein's condition may perform poorly in strongly nonconvex regions. It even may exclude all minimizers of the target function. To tackle this issue for smooth optimization, [18] proposes a line search scheme with a sufficient descent condition based on the Goldstein quotient [14]. This method produces efficient steps, in $(0, \infty)$, in the sense of Warth & Werner (5) with $\mu = 0$.

Now, we investigate the descent direction. As shown in [12], a descent direction for (1) can be found by

$$d_k = \text{argmin}_d \Big\{ \frac{1}{2} \|x_k + d - (x_k - \tau \nabla f(x_k))\|^2 + \mu\tau c(x_k + d) \Big\} - x_k, \quad (8)$$

where $\tau > 0$ is a suitable step length, chosen in different ways within some of the considered algorithms. According to the definition of a proximal operator [19] of the scaled function $\mu\tau c(x)$ at $x_k - \tau\nabla f_k$, (8) can be written as

$$d(x_k, \tau) = d_k = \mathbf{Prox}_{\mu\tau c}(x_k - \tau\nabla f(x_k)) - x_k. \quad (9)$$

It is of interest to find d_k easily. For instance, when fixing $c(x)$ in (1) to $c(x) = \|\cdot\|_1$, d_k can be found with a closed-form solution of (9), as

$$d_k = \mathcal{S}_{\mu\tau}\Big(x_k - \tau\nabla f(x_k)\Big) - x_k, \quad (10)$$

where

$$\mathcal{S}_\lambda(x) = sgn(x) \odot \max\Big\{|x| - \lambda, 0\Big\} \quad (11)$$

is the *shrinkage operator* [19] with factor $\lambda > 0$. Here, sgn$(\cdot)$ stands for the signum function and $\odot$ denotes the component-wise product, i.e., $(x \odot y)_i := x_i y_i$.

Review of Some Useful Methods. `ISTA` [6,7] uses (10) to solve (2) iteratively with $\alpha_k = 1$ for all k and employing a more conservative choice of τ, related to the Lipschitz constant of $\nabla f(x)$.

Accelerating `ISTA`, `TwIST` [4] and `FISTA` [3] use the idea of momentum term $\Theta_k(x_k - x_{k-1})$, through which the next step depends on the two previous steps x_k and x_{k-1} with $\Theta_k > 0$.

`SpaRSA` [24] solves (2) with separable structures. In each iteration, it solves an optimization subproblem involving a quadratic term with diagonal Hessian (i.e., separable in the unknown) along with the sparse regularizer term. It is an accelerated `ISTA` with improved practical performance resulting from a suitable variation of τ_k.

`FPC-BB` [17] uses an operator-splitting technique to solve a sequence of problem (2) and constructs the shrinkage step-size τ dynamically by Barzilai-Borwein [2] technique.

To be more precise at this point, one form of the Barzilai-Borwein methods produces τ_k dynamically with the safeguard parameters, $0 < \tau_{\min} \leq \tau_{\max} < \infty$, as

$$\tau_k = \max\left\{ \tau_{\min}, \min\left\{ \frac{\|s_{k-1}\|^2}{s_{k-1}^T y_{k-1}}, \tau_{\max} \right\} \right\}, \quad (12)$$

where $s_{k-1} = x_k - x_{k-1}$ and $y_{k-1} = \nabla f(x_k) - \nabla f(x_{k-1})$. This form is employed in [12,17] as well as in our proposed algorithm.

Remark 1. Based on Lemma 3.1 in [20], x_k is a stationary point for problem (1) if and only if in (9), $d_k = d(x_k, \tau_k) = 0$ for any $\tau_k > 0$ or if for the directional derivative of F at x_k along the all direction $d_k \in \mathbb{R}^n$, we have $F'(x_k, d_k) \geq 0$, [12,20,22]. Besides, Lemma 2.1 in [20] gives the descent property for $d_k \neq 0$ with $\tau_k > 0$ in (9) i.e.,

$$\Delta(x_k, d_k) \leq -\frac{1}{2\tau_k}\|d_k\|^2, \tag{13}$$

in (4). Considering the first-order approximation of the $f(x)$ for the directional derivative of the objective function in (1) and using the convexity of $c(x)$ and the relation (4), one can get

$$\begin{aligned} F'(x_k, d_k) &= \lim_{\alpha \downarrow 0} \frac{f(x_k + \alpha d_k) + \mu c(x_k + \alpha d_k) - f(x_k) - \mu c(x_k)}{\alpha} \\ &\leq \lim_{\alpha \downarrow 0} \frac{O(\alpha^2) + \alpha\Big(d_k^T \nabla f(x_k) + \mu c(x_k + d_k) - \mu c(x_k)\Big)}{\alpha} \\ &= \Delta(x_k, d_k). \end{aligned} \tag{14}$$

3 Efficient Step Sizes for Convex Nonsmooth Optimization

The idea presented in this segment extends the step size control strategy for the unconstrained smooth case in [18] to the unconstrained nonsmooth problem (1). The iterative scheme of our algorithm is like that in (3), where the descent direction d_k is obtained by (9). Now, estimating the nonzero parameter α_k, we present a nonsmooth version of the Goldstein quotient, as

$$\nu(\alpha_k) = \frac{F(x_k + \alpha_k d_k) - F(x_k)}{\alpha_k \Delta(x_k, d_k)}. \tag{15}$$

For a fixed $\theta > 0$, we consider the following *sufficient descent condition*

$$\nu(\alpha_k)\,|\nu(\alpha_k) - 1| \geq \theta, \tag{16}$$

that requires $\nu(\alpha_k)$ to be not only sufficiently positive but also sufficiently far from one to control the step size from below. Considering Goldstein's condition (7), we have

$$0 < \nu_2 \leq \nu(\alpha_k) \leq \nu_3 < 1, \tag{17}$$

which leads to (16) with

$$\theta = \nu_2(1 - \nu_3) > 0. \tag{18}$$

Conversely, (16) implies that (7) holds, in the case $\nu(\alpha_k) < 1$, with

$$\nu_2 = \frac{2\theta}{1 + \sqrt{1 - 4\theta}}, \quad \nu_3 = \frac{1 + \sqrt{1 - 4\theta}}{2}, \tag{19}$$

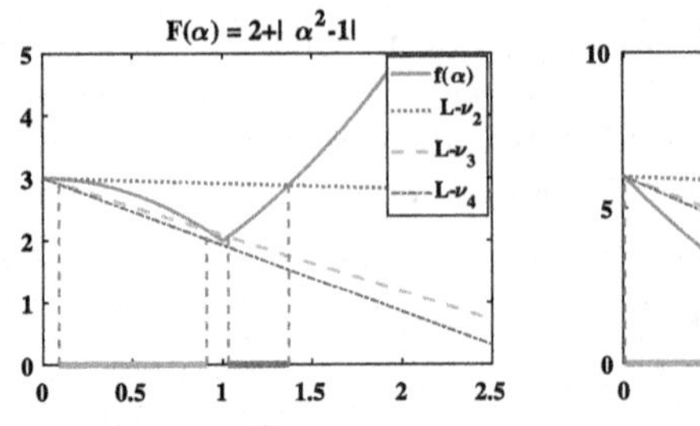

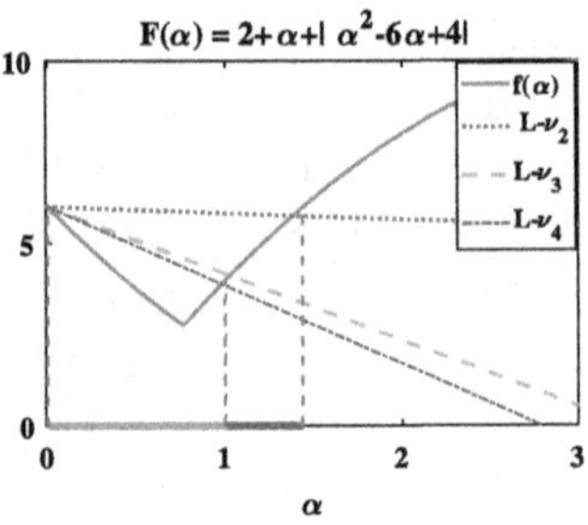

Fig. 1. The sufficient descent condition (16) with $\theta = 0.08$. Each subfigure presents the lines with slopes $\nu_i \Delta(0,1)$ for $i = 2,3,4$ and the resulting set of acceptable step sizes. The cone defined by the two lines with slops $\nu_i \Delta(0,1)$ for $i = 2,3$ cuts out a part of the graph that gives the acceptable step size. The other line gives another section of the graph, leading to the acceptable step size.

or the following alternative sharp descent condition holds for

$$\nu(\alpha_k) \geq \nu_4 = \frac{1+\sqrt{1+4\theta}}{2} > 1. \tag{20}$$

In this case, we have (6) with $\nu_4 > 1$ instead of $0 < \nu_1 < 1$, which makes a sharp decrease in objective function in comparison with the original Armijo condition (6). Therefore, condition (16) guarantees a sensible decrease in the objective function with a sizeable range for α_k, see Fig. 1 for some illustrative examples. As shown, we have descent directions for $\alpha_k \in (0, 1]$.

3.1 Convergence Analysis

We now aim to analyze the convergence of the proposed method. To investigate convergence analysis, we assume that $\nabla f(x)$ is Lipschitz continuous with constant $L > 0$. Also, to speed up convergence, we use dynamic τ_k as in (12).

The following theorem presents an explicit bound on the gain that is essential for the global and R-linear convergence statements and, based on (16), produces efficient steps as defined via (5).

Theorem 1. *If for the descent direction d_k, α_k satisfies the sufficient descent condition* (16), *then for any step size α' with $F(x_k + \alpha' d_k) \leq F(x_k + \alpha_k d_k)$ we have*

$$\Big(F(x_k) - F(x_k + \alpha' d_k)\Big)\frac{\|d_k\|^2}{\Delta(x_k, d_k)^2} \geq \frac{2\theta}{L}. \tag{21}$$

Proof. Lipschitz continuity of $\nabla f(x)$, convexity of $c(x)$, and the definition of $\Delta(x_k, d_k)$ in (4) result in

$$|F(x_k + \alpha_k d_k) - F(x_k) - \alpha_k \Delta(x_k, d_k)| \leq \Big|\alpha_k d_k^T \nabla f(x_k) + \frac{\alpha_k^2}{2} L\|d_k\|^2$$
$$+ \alpha_k \mu\Big(c(x_k + d_k) - c(x_k)\Big) - \alpha_k \Delta(x_k, d_k)\Big| = \frac{\alpha_k^2}{2} L\|d_k\|^2. \tag{22}$$

Therefore using (15), we have

$$\frac{\|d_k\|^2}{|\Delta(x_k, d_k)|} \geq \frac{2}{\alpha_k L} \left| \frac{F(x_k + \alpha_k d_k) - F(x_k) - \alpha_k \Delta(x_k, d_k)}{\alpha_k \Delta(x_k, d_k)} \right| = \frac{2}{\alpha_k L} |\nu(\alpha_k) - 1| . \tag{23}$$

Based on (13), $\Delta(x_k, d_k) < 0$. So,

$$\frac{F(x_k) - F(x_k + \alpha_k d_k)}{|\Delta(x_k, d_k)|} = \frac{F(x_k + \alpha_k d_k) - F(x_k)}{\Delta(x_k, d_k)} = \alpha_k \nu(\alpha_k). \tag{24}$$

Now, taking the product of terms from (23) and (24) and using (16), we get the result. □

Next theorem demonstrates that under some conditions, either we reach to an arbitrary large negative function value or we are arbitrary close to a stationary point of the problem.

Theorem 2. *Let* $\Delta_k = |\Delta(x_k, d_k)| > 0$. *Suppose that for every integer* $k \geq 1$

$$\sup_k \|d_k\|^2 \left(\frac{\|d_k\|^2}{\Delta_k^2} - \frac{\|d_{k-1}\|^2}{\Delta_{k-1}^2} \right) < \infty. \tag{25}$$

Then

$$\lim_{k \to \infty} \|d_k\| = 0 \quad \text{or} \quad \lim_{k \to \infty} F(x_k) = -\infty. \tag{26}$$

Proof. On the contrary suppose that the first case does not hold in (26). Then there is a constant $\epsilon > 0$ for which, $\|d_k\| \geq \epsilon$ for all k. Consider l for the supremum in (25), then

$$\frac{\|d_k\|^2}{\Delta_k^2} - \frac{\|d_{k-1}\|^2}{\Delta_{k-1}^2} \leq \frac{l}{\|d_k\|^2} \leq \frac{l}{\epsilon^2} \leq l' = \max\left(\frac{l}{\epsilon^2}, \frac{\|d_1\|^2}{\Delta_1^2} \right). \tag{27}$$

To explain the second term within the latter brackets, for $k \geq 1$, consider

$$\frac{\|d_k\|^2}{\Delta_k^2} \leq \frac{\|d_{k-1}\|^2}{\Delta_{k-1}^2} + l', \tag{28}$$

so that applying the corresponding relation recursively leads to

$$\frac{\|d_k\|^2}{\Delta_k^2} \leq kl'. \tag{29}$$

Now let δ be the infimum in the left hand side of (21) in Theorem 1. Then by (29),

$$F(x_k) - F(x_{k+1}) \geq \frac{\delta \Delta_k^2}{\|d_k\|^2} \geq \frac{\delta}{kl'}. \tag{30}$$

Recursive summation over all steps in (30) results in

$$F(x_k) \leq F(x_0) - \frac{\delta}{l'} \sum_{i=1}^{k-1} \frac{1}{i} \overset{k \to \infty}{\longrightarrow} -\infty, \tag{31}$$

which completes the proof. □

Now, we consider R-linear convergence of the method for the ℓ_1-minimization problem when $f(x)$ is a strongly convex function, and d_k is the particular closed-form solution (9). Let us recall some properties that we make use of for this purpose, see Lemma 2.1 in [22] for items (ii) and (iii):

(i) According to Theorem 1, the relation (12), and (13) for all k, we have

$$\frac{F(x_k)-F(x_{k+1})}{\|d_k\|^2} = \Big(F(x_k)-F(x_{k+1})\Big)\frac{\|d_k\|^2}{\Delta(x_k,d_k)^2}\Big(\frac{\Delta(x_k,d_k)}{\|d_k\|^2}\Big)^2 \geq \frac{\theta}{2L\tau_{\max}^2} > 0.$$

(ii) Suppose x^* is a stationary point for the ℓ_1-minimization problem. If $f(x)$ is strongly convex with $L'>0$ and $\nabla f(x)$ is Lipschitz continuous with $L>0$, then for every k,

$$\|x_k - x^*\| \leq \frac{1+\tau_k L}{L'\tau_k}\|d_k\|.$$

(iii) $\|\mathcal{S}_\lambda(x)-\mathcal{S}_\lambda(y)\| \leq \|x-y\|$ for all $x,y\in\mathbb{R}^n$ with $\lambda>0$.

Theorem 3. *Suppose that* $c(x)=\|x\|_1$, d_k *is produced by* (10), *the sequence* $\{x_k\}$ *converges to a minimizer* x^*, *and* $f(x)$ *is a strongly convex function, then there are constants* $q\in(0,1)$ *and* $\xi_1,\xi_2,\xi_3>0$ *such that, for all* k, *we have*

$$F(x_k)-F(x^*) \leq \xi_1 q^{2k}, \quad \|x_k-x^*\| \leq \xi_2 q^k, \quad \|d_k\| \leq \xi_3 q^k. \tag{32}$$

Proof. Since $\nabla f(x)$ is Lipschitz continuous with constant L, by Cauchy Schwarz inequality for all x and x', we get

$$\begin{aligned}
&F(x')-F(x)\\
&=\mu\big(c(x')-c(x)\big)+f(x')-f(x)\\
&=\mu\big(c(x')-c(x)\big)+(x'-x)^T\nabla f(x)\\
&\quad+\int_0^1\big(\nabla f(x+t(x'-x))-\nabla f(x)\big)^T(x'-x)dt\\
&\leq \mu\big(c(x')-c(x)\big)+(x'-x)^T\nabla f(x)\\
&\quad+\int_0^1\|\nabla f(x+t(x'-x))-\nabla f(x)\|\|x'-x\|dt\\
&\leq \mu\big(c(x')-c(x)\big)+(x'-x)^T\nabla f(x)\\
&\quad+\int_0^1 L\|x+t(x'-x)-x\|\|x'-x\|dt\\
&=\mu\big(c(x')-c(x)\big)+(x'-x)^T\nabla f(x)+\frac{L}{2}\|x'-x\|^2.
\end{aligned} \tag{33}$$

On the other hand, by strong convexity of $f(x)$, there exists constant L' such that for all x and x', we have

$$f(x') \geq f(x)+(x'-x)^T\nabla f(x)+\frac{L'}{2}\|x'-x\|^2. \tag{34}$$

So,

$$F(x') \geq F(x) + \mu\big(c(x') - c(x)\big) + (x' - x)^T \nabla f(x) + \frac{L'}{2}\|x' - x\|^2. \tag{35}$$

Now, (33) and (35) lead to

$$\frac{L'}{2}\|x' - x\|^2 \leq F(x') - F(x) - \Big((x' - x)^T \nabla f(x) + \mu(c(x') - c(x))\Big) \leq \frac{L}{2}\|x' - x\|^2. \tag{36}$$

As x^* is a minimizer, based on Remark 1, and (12)–(14), we get

$$0 \leq F'(x^*, x_k - x^*) \leq \Delta(x^*, x_k - x^*) \leq \frac{-1}{2\tau_{min}}\|x_k - x^*\|^2 \leq 0, \tag{37}$$

where

$$\Delta(x^*, x_k - x^*) = (x_k - x^*)^T \nabla f(x^*) + \mu\big(c(x_k) - c(x^*)\big). \tag{38}$$

Now by replacing x' by x_k and x by x^* in (36), the relations (36)–(38) lead to

$$\frac{L'}{2}\|x_k - x^*\|^2 \leq F(x_k) - F(x^*) \leq \frac{L}{2}\|x_k - x^*\|^2. \tag{39}$$

Property (ii) as mentioned before the theorem and (12) result in

$$\|x_k - x^*\| \leq \gamma \|d_k\|, \tag{40}$$

where, $\gamma = \frac{1+\tau_{\max}L}{L'\tau_{\min}}$. Now from (39) and (40), we conclude that

$$\frac{L'}{2}\|x_k - x^*\|^2 \leq F(x_k) - F(x^*) \leq l\|d_k\|^2, \tag{41}$$

with $l = \frac{\gamma^2 L}{2}$. Thus, we have

$$0 \leq \frac{F(x_{k+1}) - F(x^*)}{F(x_k) - F(x^*)} \leq 1 - \frac{F(x_k) - F(x_{k+1})}{l\|d_k\|^2} \leq q^2 = 1 - \inf_k \frac{F(x_k) - F(x_{k+1})}{l\|d_k\|^2}, \tag{42}$$

where based on property (i), $q^2 < 1$. Now, by recursive process derived from (42),

$$F(x_k) - F(x^*) \leq \xi_1 q^{2k}, \tag{43}$$

with $\xi_1 = F(x_0) - F(x^*)$. That gives the first relation in (32). Consequently, the first inequality in (41) along with (43) leads to

$$\|x_k - x^*\| \leq \xi_2 q^k, \tag{44}$$

and $\xi_2 = \sqrt{\frac{2}{L'}\xi_1}$. On the other hand, Lipschitz continuity of $\nabla f(x)$, Remark 1, the relation (10), and property (iii) result in

$$\begin{aligned}
\|d_k\| &= \|d(x_k, \tau_k) - d(x^*, \tau_k)\| \\
&= \|\mathcal{S}_{\mu\tau_k}\Big(x_k - \tau_k \nabla f_k\Big) - x_k - \mathcal{S}_{\mu\tau_k}\Big(x^* - \tau_k \nabla f(x^*)\Big) + x^*\| \\
&\leq \|x_k - x^*\| + \|(x_k - x^*) + \tau_k(\nabla f(x^*) - \nabla f(x_k))\| \\
&\leq (2 + \tau_{\max}L)\|x_k - x^*\|.
\end{aligned} \tag{45}$$

Thus,

$$\|d_k\| \leq \xi_3 q^k, \tag{46}$$

with $\xi_3 = (2 + \tau_{\max} L)\xi_2$, which completes the proof. □

3.2 Numerical Experiments

Based on the shrinkage descent direction (10) with dynamic parameter τ_k in (12) and the presented efficient line search in Section (3), we generate an iterative algorithm to solve (2). As already indicated, we call the new procedure the iterative Shrinkage-Goldstein algorithm (`ISGA`).

In this part, several numerical experiments are performed to compare `ISGA` with `FPC-BB`, `TwIST`, `FISTA`, and `SpaRSA` on compressed sensing problems. The experiments are run in `Matlab` R2022b on a LENOVO ThinkCenter M920t with a 12× Intel® Core^{TM} i7-8700 CPU @ 3.20 GHz processor and 15.5 GiB of RAM.

Experimental Setting. In(2), problem data A and b are generated by specifying the size and type of A matrix.

Six types of test matrices A (based on randomization of entries or composition) that comprise standard tests in the field of compressed sensing, cf. [12, 17], are evaluated: `Gaussian`, `scaled Gaussian`, `orthogonalized Gaussian`, `Bernoulli`, `partial Hadamard`, and `partial discrete cosine transform` matrices.

Having the dimension of the signal x as $n \in \{2^{10}, \ldots, 2^{14}\}$, in accordance with the given ρ and δ in $\{0.1, 0.2, 0.3\}$, we produce the dimension of the observation vector b by $m = \lfloor \delta n \rfloor$ as well as the number of nonzero elements in an exact solution xs by $k = \lfloor \rho m \rfloor$. In addition, we contaminated xs and b by Gaussian noise with values in

$$\{(10^{-1}, 10^{-1}), (10^{-3}, 10^{-3}), (10^{-5}, 10^{-5}), (10^{-7}, 10^{-7})\}, \tag{47}$$

because real problems are usually noisy. These choices lead to generating 840 random test problems in the compressed sensing area.

As for computational parameters, for all solvers, the initial point value is $x_0 = zeros(n, 1)$, and we set $\mu = 2^{-8}$ (chosen as in [12,17,22]). For $\texttt{FPC} - \texttt{BB}$, `SpaRSA` and `ISGA`, like that of [12], we select the safeguard parameters as $\tau_{\min} = 10^{-4}$, $\tau_{\max} = 10^4$, and $\theta = 10^{-10}$ specifically for `ISGA` based on some initial numerical results. All tested solvers are stopped whenever

$$\|F(x_{k+1}) - F(x_k)\| \leq \texttt{Ftol} \|F(x_k)\|, \tag{48}$$

with $\texttt{Ftol} = 10^{-10}$ or the number of iterations exceeds 10000.

Evaluation. In order to obtain a useful baseline for comparison, we use *performance profiles* `Matlab` code [8] to benchmark and compare solvers on the generated test problems in compressed sensing. Our cost metrics for comparison

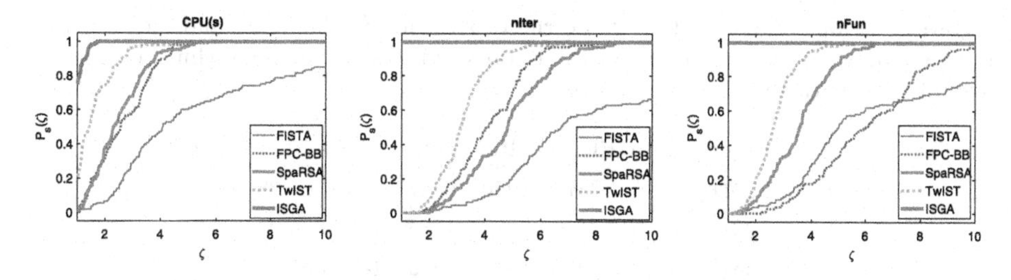

Fig. 2. Comparison among FISTA, FPC-BB, SpaRSA, TwIST, and ISGA with the performance cost metric CPU(s), nIter, and nFun, respectively.

are the total number of iterations (nIter), the number of function evaluations (nFun), and the time in seconds (CPU(s)).

Let us recall the definition of the performance profiles to make clear the meaning of results of this experiment. Assume that we have $\mathtt{n_s}$ solvers in S and $\mathtt{n_p}$ problems in P. We want for example to show computing time as a performance metric. Let $t_{\mathtt{p},\mathtt{s}}$ be computing time to solve p by s. We compare the performance on p by s with the best performance by any solver on p via $r_{\mathtt{p},\mathtt{s}}$ in the following:

$$r_{\mathtt{p},\mathtt{s}} = \frac{t_{\mathtt{p},\mathtt{s}}}{\min\{t_{\mathtt{p},\mathtt{s}} : \mathtt{s} \in \mathtt{S}\}}, \quad P_{\mathtt{s}}(\zeta) = \frac{1}{n_p} size\{\mathtt{p} \in \mathtt{P} : r_{\mathtt{p},\mathtt{s}} \leq \zeta\}. \tag{49}$$

In (49), $P_{\mathtt{s}}(\zeta)$ gives the probability for s that a performance ratio $r_{\mathtt{p},\mathtt{s}}$ is within a factor $\zeta \in \mathbb{R}$. So, if we are interested only in the number of wins, we need only consider $P_{\mathtt{s}}(1)$ for the all solvers.

Given the graphs depicted in Fig. 2, the proposed algorithm ISGA clearly outperforms the other methods with respect to all compared cost metrics. It attains the most wins concerning CPU(s) for around 75%, as well as nFun and nIter for 100% of the test problems.

Table 1 contains the arithmetic averages of considered cost metrics of the reconstructions concerning the original signal xs. These values are rounded (towards zero) to integers and show that, in solving problem (2), ISGA is faster than other compared solvers. Besides, we report the average value of the objective function in (2) at the end of the iterative process over all test problems named Fbest.

Further Discussion. All considered solvers are improved and accelerated forms of ISTA. Technically, the FPC-BB method is closest to the method we propose, as it relies on the same direction computation including the Barzilai-Borwein method, see (12). This is also reflected in the more similar performance of these two schemes compared to the other tested methods. However, FPC-BB uses the nonmonotone [16] Armijo condition in its iterative process with backtracking. The nonmonotone technique is for speeding up the algorithms. Comparing especially nIter and nFun metrics for FPC-BB and ISGA in the Table 1 shows the higher computational efficiency of the proposed new line search condition.

Table 1. Average values of the cost metrics `CPU(s)`, `nIter`, and `nFun`, and the objective function values `Fbest` at the end of the iterative process for all algorithms, computed over all test matrices

Algorithm	CPU(s)	nIter	nFun	Fbest
FISTA	0.32373	1956	1957	1.9578e+03
FPC-BB	0.17136	906	1776	1.7758e+03
SpaRSA	0.15925	1045	1046	1.0458e+03
TwIST	0.10062	697	780	1.0458e+03
ISGA	0.06081	217	292	292.1296

Let us remark that in very few test cases, the proposed algorithm needs relatively long convergence times, while in most cases it is extremely fast. By the very promising first results reported in Table 1, where these cases are taken into account, we conjecture that there is even some more potential to explore.

4 Conclusion and Future Work

To summarize our developments, we proposed an efficient line search condition in the sense of Warth & Werner for the convex nonsmooth optimization problem. Our strategy is an improved form of the Goldstein and Armijo strategies that leads to controlled step sizes that may have a faster decrease in the objective function in some iterations than with previous methods. This has been validated both theoretically and experimentally.

Future research in the theoretical part may include investigating R-linear convergence for a general form of the nonsmooth convex function $c(x)$ in (1) and extension to the nonmonotone line search. We expect better numerical results in nonmonotone cases for the $\ell_1 - \ell_2$ problem.

Further research in the experimental part may be devoted to make the proposed method even more efficient after considering a more detailed investigation of the numerical convergence process.

References

1. Armijo, L.: Minimization of functions having lipschitz continuous first partial derivatives. Pac. J. Math. **16**(1), 1–3 (1966)
2. Barzilai, J., Borwein, J.M.: Two-point step size gradient methods. IMA J. Numer. Anal. **8**, 141–148 (1988)
3. Beck, A., Teboulle, M.: A fast iterative shrinkage-thresholding algorithm for linear inverse problems. SIAM J. Imaging Sci. **2**(1), 183–202 (2009)
4. Bioucas-Dias, J.M., Figueiredo, M.A.T.: A new TwIST: two-step iterative shrinkage/thresholding algorithms for image restoration. IEEE Trans. Image Process. **16**, 2992–3004 (2007)

5. Candès, E.J., Romberg, J., Tao, T.: Robust uncertainty principles: exact signal reconstruction from highly incomplete frequency information. IEEE Trans. Inf. Theory **52**(2), 489–509 (2006)
6. Combettes, P.L., Wajs, V.R.: Signal recovery by proximal forward-backward splitting. Multiscale Model. Simul. **4**(4), 1168–1200 (2005)
7. Daubechies, I., Defrise, M., De Mol, C.: An iterative thresholding algorithm for linear inverse problems with a sparsity constraint. Commun. Pure Appl. Math. **57**(11), 1413–1457 (2004)
8. Dolan, E.D., Moré, J.J.: Benchmarking optimization software with performance profiles. Math. Program. **91**, 201–213 (2002)
9. Donoho, D.: Compressed sensing. IEEE Trans. Inf. Theory **52**(4), 1289–1306 (2006)
10. Elad, M.: Sparse and Redundant Representation from Theory to Application in Signal and Image Processing. Springer, New York (2010). https://doi.org/10.1007/978-1-4419-7011-4
11. Eldar, C.Y., Kutyniok, G.: Compressed Sensing: Theory and Application. Cambridge University Press, New York (2012)
12. Esmaeili, H., Shabani, S., Kimiaei, M.: A new generalized shrinkage conjugate gradient method for sparse recovery. Calcolo **56**(1), 1–38 (2019)
13. Foucart, S., Rauhut, H.: A Mathematical Introduction to Compressive Sensing. Springer, New York (2013). https://doi.org/10.1007/978-0-8176-4948-7
14. Goldstein, A.A.: On steepest descent. J. SIAM Ser. A: Control **3**, 147–151 (1965)
15. Goldstein, A.A., Price, J.F.: An effective algorithm for minimization. Nume. Math. **10**, 184–189 (1967)
16. Grippo, L., Lampariello, F., Lucidi, S.: A nonmonotone line search technique for Newton' method. SIAM J. Numer. Anal. **23**, 707–716 (1986)
17. Hale, E.T., Yin, W., Zhang, Y.: Fixed-point continuation applied to compressed sensing: implementation and numerical experiments. J. Comput. Math. **28**(2), 170–194 (2010)
18. Neumaier, A., Kimiaei, M.: An efficient gradient-free line search. Technical report, University of Vienna (2022). https://optimization-online.org/2022/11/an-efficient-gradient-free-line-search/
19. Parikh, N., Boyd, S.: Proximal algorithms. Found. Trend. Optim. **1**(3), 123–231 (2013)
20. Tseng, P., Yun, S.: A coordinate gradient descent method for nonsmooth separable minimization. Math. Program. **117**(1), 387–423 (2009)
21. Warth, W., Werner, J.: Effiziente Schrittweitenfunktionen bei unrestringierten Optimierungsaufgaben. Computing **19**, 59–72 (1977)
22. Wen, Z., Yin, W., Zhang, H., Goldfarb, D.: On the convergence of an active-set method for ℓ_1-minimization. Optim. Methods Softw. **27**(6), 1127–1146 (2012)
23. Wolfe, P.: Convergence conditions for ascent methods. SIAM Rev. **11**, 226–235 (1969)
24. Wright, S.J., Nowak, R.D., Figueiredo, M.A.T.: Sparse reconstruction by separable approximation. IEEE Trans. Signal Process. **57**(7), 2479–2493 (2009)

Learned Discretization Schemes for the Second-Order Total Generalized Variation

Lea Bogensperger[1(✉)], Antonin Chambolle[2], Alexander Effland[3], and Thomas Pock[1]

[1] Institute of Computer Graphics and Vision, Graz University of Technology, Graz, Austria
{lea.bogensperger,pock}@icg.tugraz.at
[2] CEREMADE, CNRS & Université Paris-Dauphine PSL, Paris, France
antonin.chambolle@ceremade.dauphine.fr
[3] Institute for Applied Mathematics, University of Bonn, Bonn, Germany
effland@iam.uni-bonn.de

Abstract. The total generalized variation extends the total variation by incorporating higher-order smoothness. Thus, it can also suffer from similar discretization issues related to isotropy. Inspired by the success of novel discretization schemes of the total variation, there has been recent work to improve the second-order total generalized variation discretization, based on the same design idea. In this work, we propose to extend this to a general discretization scheme based on interpolation filters, for which we prove variational consistency. We then describe how to learn these interpolation filters to optimize the discretization for various imaging applications. We illustrate the performance of the method on a synthetic data set as well as for natural image denoising.

Keywords: Total generalized variation · discretization · image denoising · bilevel optimization · piggyback algorithm · learning · primal-dual algorithms

1 Introduction

The total variation (TV) is a popular regularizer for many tasks in image reconstruction, yet it assumes as a prior that images/signals are essentially piecewise constant. The extension known as total generalized variation (TGV) [3] is a natural way to incorporate more complex signals (such as affine) in the prior, by combining the TV of different orders of derivatives into a global image. Like TV, TGV can be used in a plug-and-play style in various inverse problems [10,12,13,16]. In the continuous domain, it reads as

Supplementary Information The online version contains supplementary material available at https://doi.org/10.1007/978-3-031-31975-4_37.

L. Calatroni et al. (Eds.): SSVM 2023, LNCS 14009, pp. 484–497, 2023.
https://doi.org/10.1007/978-3-031-31975-4_37

$$\mathrm{TGV}_\alpha^2(u) = \sup_p \left\{ \int_\Omega u \, \mathrm{div}^2 p \, \mathrm{d}x : p \in \mathscr{C}^\infty(\Omega, \mathrm{Sym}^{2\times 2}), \|p\|_\infty \le \alpha_0, \|\mathrm{div}\, p\|_\infty \le \alpha_1 \right\}, \tag{1}$$

where $\mathrm{Sym}^{2\times 2}$ denotes the space of second-order symmetric tensors, $\alpha = (\alpha_0, \alpha_1)$ are positive parameters, div denotes the row-wise (or column-wise, as p is symmetric) divergence, and div^2 the divergence of the resulting vector. Note that one can similarly define regularizers combining higher orders of derivatives, yet the most common version used is TGV_α^2, therefore we stick to this case. Like TV, TGV is difficult to discretize while preserving isotropy and rotational invariance. For TV, improved discretization schemes have been studied in earlier works such as [5,6]. Recently, a discretization scheme was proposed in [9] to improve second-order TGV inspired by the work of Condat [6]. The idea is to impose the constraints in (1) on the dual variables in the discretized setting on n times denser grids using interpolations to deal with staggered pixel grids arising from the discretized finite difference operators.

We would like to extend on this work by expressing it in a more general framework for which we show consistency (see Theorem 1). This framework is based on local interpolation operations and requires only bounded filter kernels. Moreover, since it is not straightforward how to choose ideal filters, and this may depend on the underlying data and the context of the inverse problem, the question arises whether this can be further improved. An appealing idea is therefore to resort to learning such interpolation filters and subsequently investigate their performance, as recently done in [5] for TV.

2 Problem Setting

2.1 Notation

Let M, N be the dimension of the pixel grid. We usually denote an image $u \in \mathscr{M} := \mathbb{R}^{M\times N}$. For convenience, but with a slight abuse of notation, note that $\mathscr{M}$ determines the size of a pixel grid, whilst not assuming anything on the respective spatial locations within the grid.

If no specific norm is indicated, the $\|x\|_{1,2}$ norm is assumed, which is for $x \in \mathscr{M}^J$ the absolute sum of the 2-norm of its J components. Further, we set $\|x\|_Z = \|x\|_{1,1,2}$ for $x \in \mathscr{M}^{J\times I}$, which is the absolute sum consisting of I components of the 2-norm of its J components. Finally, let $\|\cdot\|_Z^*$ denote its corresponding dual norm.

2.2 Finite Difference Operators

On a standard Euclidean grid of size $hM \times hN$ with $M \times N$ pixels of size $h \times h$, we define the discrete forward operator $D : \mathscr{M} \to \mathscr{M}^2$ for $u \in \mathscr{M}$ via $Du = ((Du)^1, (Du)^2)$, where

$$(Du)^1_{i+\frac{1}{2},j} = \tfrac{1}{h}(u_{i+1,j} - u_{i,j}) \qquad i \le M-1, j \le N,$$

$$(Du)^2_{i,j+\frac{1}{2}} = \tfrac{1}{h}(u_{i,j+1} - u_{i,j}) \qquad i \le M, j \le N-1.$$

To ease the notation we set the values of the derivatives to 0 using Neumann boundary conditions if the index dies out before reaching M or N, which also implies that our resulting pixel grids remain of the same size. The tensor-valued symmetric counterpart is given by $E : \mathcal{M}^2 \to \mathcal{M}^3$, and its individual operator components also consist of forward differences[1]. Therefore, for $w = (w^1_{i+\frac{1}{2},j}, w^2_{i,j+\frac{1}{2}})$ one obtains the symmetrized tensor field $Ew = \begin{pmatrix} (Ew)^1 & (Ew)^2 \\ (Ew)^2 & (Ew)^3 \end{pmatrix}$ with

$$
\begin{aligned}
(Ew)^1_{i+1,j} &= \tfrac{1}{h}(w^1_{i+\frac{3}{2},j} - w^1_{i+\frac{1}{2},j}) && i \le M-1, j \le N,\\
(Ew)^2_{i+\frac{1}{2},j+\frac{1}{2}} &= \tfrac{1}{2h}(w^1_{i+\frac{1}{2},j+1} - w^1_{i+\frac{1}{2},j} + w^2_{i+1,j+\frac{1}{2}} - w^2_{i,j+\frac{1}{2}}) && i \le M-1, j \le N-1,\\
(Ew)^3_{i,j+1} &= \tfrac{1}{h}(w^2_{i,j+\frac{3}{2}} - w^2_{i,j+\frac{1}{2}}) && i \le M, j \le N-1.
\end{aligned}
$$

Again, the same handling of derivatives using Neumann boundary conditions is used. The symmetrized second-order finite difference operator is then given by $D^2 = ED$. The corresponding adjoint operators div and div^2 are directly given by the discrete Gauss-Green theorem.

2.3 Second-Order TGV Discretization

In the spirit of the recently proposed discretization [9] that builds upon the ideas of Condat's discretization [6], the aim is to state a more generalized definition of second-order TGV using interpolation filters K and L

$$K\colon \mathcal{M}^3 \to \mathcal{M}^{3\times n_K},\ L\colon \mathcal{M}^2 \to \mathcal{M}^{2\times n_L}, \tag{2}$$

with n_K and n_L denoting the number of filters, respectively. These filters can be chosen according to [9] as shown in Fig. 2, but also other choices exist (such as interpolating to arbitrary pixel grid locations), all being based on a staggered grid discretization. We start from the standard second-order TGV discretization in the primal domain:

$$\min_{w\in\mathcal{M}^2} \alpha_1 \|Du - w\| + \alpha_0 \|Ew\|. \tag{3}$$

Using interpolation filters from (2) this can be rewritten with $v_K \in \mathcal{M}^{3\times n_K}$ and $v_L \in \mathcal{M}^{2\times n_L}$ as

$$\min_{v_K,v_L,w} \alpha_1 \|v_L\|_Z + \alpha_0 \|v_K\|_Z,\ \text{s.t. } Du - w = L^* v_L,\ Ew = K^* v_K, \tag{4}$$

where w can be eliminated from the constraints such that we obtain

$$\min_{v_K,v_L} \alpha_1 \|v_L\|_Z + \alpha_0 \|v_K\|_Z,\ \text{s.t. } D^2 u = EL^* v_L + K^* v_K. \tag{5}$$

---

[1] Note that one could also resort to backward differences, however, for designing suitable interpolation operators on the dual variables this scheme is more convenient since it leads to the component consisting of mixed derivatives being located at the pixel corner.

In this sense, one possible interpretation is that we seek to learn a group-sparse coding for the symmetrized second-order discrete derivatives D^2 of u. For smooth regions, v_L will be close to 0 and the second-order gradients of the image will only be given by $K^* v_K$, whereas for discontinuities v_L contributes as well. Since $EL^* v_L$ will be symmetrized, it essentially leaves more freedom to the model as only the symmetric part of the second-order derivatives in the constraint must be fulfilled. Using convex conjugates and duality, the corresponding dual problem reads as

$$\sup_{p \in \mathcal{M}^3} \langle D^2 u, p \rangle, \text{ s.t. } \|L \operatorname{div} p\|_Z^* \leq \alpha_1, \ \|Kp\|_Z^* \leq \alpha_0. \tag{6}$$

Figure 1 shows the resulting pixel grids for the vector and tensor fields w and p arising from the finite difference operators D, E, and D^2. This basically suggests considering four different pixel locations for interpolation: the pixel center, the center of the horizontal and vertical edges, and the corner. All other pixel positions at this scale are contained in a superset of these four positions.

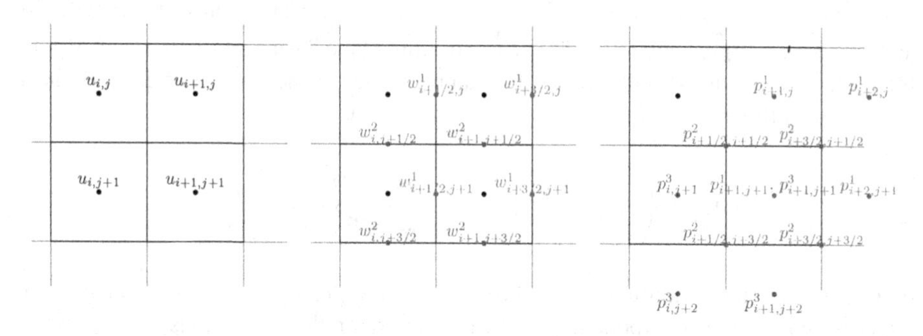

Fig. 1. Resulting pixel grids for w and p given an input u. Colors indicate different components of the vector/tensor fields for visualization purposes (best viewed on screen).

2.4 Interpolation Operators

Inspired by the improved discretization schemes on TV [6], the authors in [9] construct filters using $n_L = 3$ for the dual $\operatorname{div} p$ that is located at the same pixel grid positions as w for both vector field components. Thus both w^1, w^2 are interpolated to the three pixel grid positions $(i, j), (i + \frac{1}{2}, j), (i, j + \frac{1}{2})$. While the corresponding interpolation operations are given in detail by [9], a schematic of this is also shown in Fig. 2.

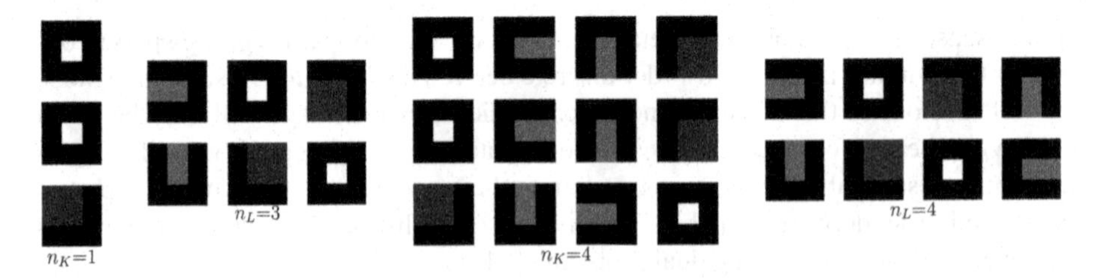

Fig. 2. Handcrafted interpolation filters where the intensity values are in $\{0, 0.25, 0.5, 1\}$ (ranging from dark to light) to ensure normalized filter coefficients. The authors in [9] use $n_K = 1$ for K and $n_L = 3$ for L ensuring that each component is interpolated from its resulting pixel grid location given in Fig. 1 to the pixel center and the horizontal and vertical edges, respectively. In case $n_K = 4$ and $n_L = 4$ this additionally interpolates to the pixel corner.

Thus, the filter $(L^{1,l}, L^{2,l})_{l=1}^{n_L}$ is applied using $L^{1,l}w^1 = (L^{1,1}w^1, L^{1,2}w^1, L^{1,3}w^1)$, and analogously for w^2. Naturally, this can be extended to also include the fourth position in the pixel corner $(i + \frac{1}{2}, j + \frac{1}{2})$ for $n_L = 4$. Using convolutions with filter kernels $(\eta^{1,l}_{m,n}, \eta^{2,l}_{m,n})$ and $(\xi^{1,r}_{m,n}, \xi^{2,r}_{m,n}, \xi^{3,r}_{m,n})$ of limited local support $(2\nu + 1) \times (2\nu + 1)$ for $\nu \in \mathbb{N}$ with bounded coefficients, this can be framed in the context of general interpolation operations. The filters can then be expressed as

$$(L^l w)_{i,j} = \begin{pmatrix} (L^{1,l}w^1)_{i,j} \\ (L^{2,l}w^2)_{i,j} \end{pmatrix} = \begin{pmatrix} \sum_{m,n=-\nu}^{\nu} \eta^{1,l}_{m,n} w^1_{i+\frac{1}{2}-m,j-n} \\ \sum_{m,n=-\nu}^{\nu} \eta^{2,l}_{m,n} w^2_{i-m,j+\frac{1}{2}-n} \end{pmatrix}. \tag{7}$$

In a similar manner, the dual p is interpolated to the pixel position (i, j) for $n_K = 1$ for each tensor field component [9]. Again, the other three positions at both pixel faces and at the corner can be included using $n_K = 4$. In the general setting using $(K^{1,r}, K^{2,r}, K^{3,r})_{r=1}^{n_K}$ this amounts to

$$(K^r p)_{i,j} = \begin{pmatrix} (K^{1,r}p^1)_{i,j} \\ (K^{2,r}p^2)_{i,j} \\ (K^{3,r}p^3)_{i,j} \end{pmatrix} = \begin{pmatrix} \sum_{m,n=-\nu}^{\nu} \xi^{1,r}_{m,n} p^1_{i-m,j-n} \\ \sum_{m,n=-\nu}^{\nu} \xi^{2,r}_{m,n} p^2_{i+\frac{1}{2}-m,j+\frac{1}{2}-n} \\ \sum_{m,n=-\nu}^{\nu} \xi^{3,r}_{m,n} p^3_{i-m,j-n} \end{pmatrix}. \tag{8}$$

In general, however, it is not straightforward how to select the interpolation points within the pixel grid of the dual variables with regards to an improved discretization. To gain a basic intuition, experiments on image denoising (see Sect. 5.1 on the respective data set and Sect. 4.1/Algorithm 1 on the optimization problem/reconstruction algorithm) have been conducted with varying n_K and n_L. In general, it seems that the choice of n_K does not impact the performance to a large extent, presumably due to the second-order finite differences which yield very smooth tensor fields. On the other hand, a larger n_L seems to be beneficial, resulting in a denser grid. Moreover, it is noteworthy that these tendencies exhibit small fluctuations depending on the parameters α_1 and α_0, the choice of data and the level of noise corruption. Due to this ambiguity of selecting the best set of suitable filters, it is tempting to directly learn the filters with the aim to obtain an even better discretization. This is also motivated by the success of learned discretization schemes for TV [5].

3 Γ-Convergence of the Discretization

For simplicity, we use a square grid of $N \times N$ pixels for the domain $\Omega = (0,1)^2$, where each pixel is of size $h \times h$, with $h = 1/N$. The operators and variables in the discrete setting are now marked with an h. We use both primal and dual definitions of the discretized second-order TGV

$$\begin{aligned}\mathrm{TGV}^2_{\alpha,h}(u^h) &= \min_{w^h, v^h_K, v^h_L} \left\{ h^2\alpha_1 \|v^h_L\|_Z + h^2\alpha_0 \|v^h_K\|_Z : L^*_h v^h_L = D_h u^h - w^h,\, K^*_h v^h_K = E_h w^h \right\} \\ &= \sup_{p^h} \left\{ h^2 \langle \mathrm{div}^2_h p^h, u^h \rangle : \|L_h \,\mathrm{div}_h p^h\|^*_Z \le \alpha_1, \|K_h p^h\|^*_Z \le \alpha_0 \right\}. \end{aligned} \tag{9}$$

In a slightly simpler setting where we assume that u is global affine plus periodic and w is periodic with periodic boundary conditions, the following theorem states the Γ-convergence of the discretized second-order TGV in (9) to the continuous second-order TGV in (1). The corresponding proof is given in the supplementary material. Note that the minimum in (9) is attained due to the finite-dimensional setting and the boundedness of (w, v_K, v_L).

Theorem 1. *We consider the setting where u is affine plus periodic with period 1 in $\mathbb{R}^2$, and w is 1-periodic. Then, for interpolation operators K and L that have local support and bounded filter coefficients,* $\mathrm{TGV}^2_{\alpha,h}(u^h)$ *Γ-converges to* $\mathrm{TGV}^2_\alpha(u)$.

The interpretation of this theorem is that minimizers of problems involving $\mathrm{TGV}^2_{\alpha,h}$ plus some continuous term (for instance, a quadratic penalization) will converge, when viewed as piecewise constant functions in the continuum, to minimizers of the corresponding continuous problem involving TGV^2_α defined in (1).

4 Numerical Methods

4.1 Image Reconstruction

Second-order TGV regularization is typically applied to image reconstruction problems being combined with a task-dependent convex data fidelity term $G(u, f)$, e.g. a typical use case is image denoising with $G(u,f) = \frac{1}{2}\|u - f\|_2^2$. Thus, given a corrupted image $f \in \mathcal{M}$ we obtain the following saddle point problem using the proposed discretization scheme from (5)

$$\min_{u, v_K, v_L} \max_p G(u,f) + \alpha_0 \|v_K\|_Z + \alpha_1 \|v_L\|_Z + \langle D^2 u - EL^* v_L - K^* v_K, p \rangle. \tag{10}$$

This can be solved with a primal-dual algorithm [4] as described in Algorithm 1, using diagonal block-preconditioning [14] to determine the step sizes. For details on how to compute the proximal maps, see [5].

Algorithm 1: Primal-dual algorithm to solve (10).

Input: initial values u^1, v_K^1, v_L^1 and p^1, block-preconditioned step size parameters $\tau_u, \tau_{v_K}, \tau_{v_L}, \sigma > 0$, $\theta \in [0,1]$, number of iterations J

Result: approximate saddle point (u^J, v_K^J, v_L^J, p^J)

for $j = 1, 2, \ldots, J$ **do**

$p^{j+1} = p^j + \sigma(D^2 u^j - EL^* v_L^j - K^* v_K^j)$;

$\bar{p}^{j+1} = p^{j+1} + \theta(p^{j+1} - p^j)$;

$u^{j+1} = \mathrm{prox}_{\tau_u G(\cdot, f)}(u^j - \tau_u D^{2^*} \bar{p}^{j+1})$;

$v_L^{j+1} = \mathrm{prox}_{\tau_{v_L} \|\cdot\|_Z}(v_L^j + \tau_{v_L} LE^* \bar{p}^{j+1})$;

$v_K^{j+1} = \mathrm{prox}_{\tau_{v_K} \|\cdot\|_Z}(v_K^j + \tau_{v_K} K \bar{p}^{j+1})$;

end

4.2 Learning Interpolation Filters

The interpolation filters can be learned with a bilevel approach, where the outer optimization problem enforces the similarity of the approximate reconstructions u^* from the inner problem to a known target data set t. To achieve this, a loss function is required (we use a quadratic loss $\ell(u^*, t) = \frac{1}{2}\|u^* - t\|_2^2$) with additional constraints on the learned interpolation filters

$$\min_{K,L} \frac{1}{S} \sum_{s=1}^{S} \ell(u^{s*}(K,L), t^s) + \mathcal{R}(K) + \mathcal{R}(L). \tag{11}$$

The constraints on the filters are given by $\mathcal{R}(K) = \delta_{(C_{\Sigma=1})^{3,n_K}}$ and $\mathcal{R}(L) = \delta_{(C_{\Sigma=1})^{2,n_L}}$, with $\delta_{C_{\Sigma=1}}$ the indicator function of the set $C_{\Sigma=1}$ per filter for each component, to ensure the boundedness of the filters such that for each the sum of the coefficients is 1 (also see Sect. 4.3 for more details).

As an alternative to an unrolling scheme, we resort to a piggyback-style algorithm for obtaining derivatives of the linear operators [2,5,8]. This bears the advantage of not being limited to the number of primal-dual iterations due to computational memory issues. While an estimate for a saddle point for (10) is obtained, the adjoint state of the corresponding bi-quadratic saddle point problem is simultaneously computed (see Algorithm 2). Using the resulting approximate saddle point (u^J, v_K^J, v_L^J, p^J) and its adjoint state (U^J, V_K^J, V_L^J, P^J), the gradients with respect to K and L can then be computed using automatic differentiation (see [5] for more details).

Algorithm 2: Piggyback primal-dual algorithm to solve (10) and its adjoint.

Input: initial values (u^1, v_K^1, v_L^1, p^1) and (U^1, V_K^1, V_L^1, P^1), block-preconditioned step size parameters $\tau_u, \tau_{v_K}, \tau_{v_L}, \sigma > 0$, $\theta \in [0,1]$, number of iterations J

Result: approximate saddle point (u^J, v_K^J, v_L^J, p^J) and its adjoint state (U^J, V_K^J, V_L^J, P^J)

for $j = 1, 2, \ldots, J$ **do**

$$
\begin{aligned}
p^{j+1} &= p^j + \sigma(D^2 u^j - EL^* v_L^j - K^* v_K^j), & P^{j+1} &= P^j + \sigma(D^2 U^j - EL^* V_L^j - K^* V_K^j);\\
\bar{p}^{j+1} &= p^{j+1} + \theta(p^{j+1} - p^j), & \bar{P}^{j+1} &= P^{j+1} + \theta(P^{j+1} - P^j);\\
\widetilde{u}^{j+1} &= u^j - \tau_u D^{2^*} \bar{p}^{j+1}, & \widetilde{U}^{j+1} &= U^j - \tau_u(D^{2^*} \bar{P}^{j+1} + \nabla \ell(u^j, t));\\
u^{j+1} &= \operatorname{prox}_{\tau_u G(\cdot, f)}(\widetilde{u}^{j+1}), & U^{j+1} &= \nabla \operatorname{prox}_{\tau_u G(\cdot, f)}(\widetilde{u}^{j+1}) \cdot \widetilde{U}^{j+1};\\
\widetilde{v}_L^{j+1} &= v_L^j + \tau_{v_L} LE^* \bar{p}^{j+1}, & \widetilde{V}_L^{j+1} &= V_L^j + \tau_{v_L} LE^* \bar{P}^{j+1};\\
v_L^{j+1} &= \operatorname{prox}_{\tau_{v_L} \|\cdot\|_Z}(\widetilde{v}_L^{j+1}), & V_L^{j+1} &= \nabla \operatorname{prox}_{\tau_{v_L} \|\cdot\|_Z}(\widetilde{v}_L^{j+1}) \cdot \widetilde{V}_L^{j+1};\\
\widetilde{v}_K^{j+1} &= v_K^j + \tau_{v_K} K \bar{p}^{j+1}, & \widetilde{V}_K^{j+1} &= V_K^j + \tau_{v_K} K \bar{P}^{j+1};\\
v_K^{j+1} &= \operatorname{prox}_{\tau_{v_K} \|\cdot\|_Z}(\widetilde{v}_K^{j+1}), & V_K^{j+1} &= \nabla \operatorname{prox}_{\tau_{v_K} \|\cdot\|_Z}(\widetilde{v}_K^{j+1}) \cdot \widetilde{V}_K^{j+1};
\end{aligned}
$$

end

The outer bilevel learning problem is solved using a block-wise Adam optimizer [11], whose block-wise structure is crucial due to the imposed constraints on the filters for the projections. This allows for individual adaptive learning rates for all groups of parameters subject to the same constraint by estimating the first and second gradient moments.

4.3 Filter Settings

Initialization. Since the underlying problem is of a non-convex nature due to its bilevel structure we lack any guarantee to obtain a global minimum. Initialization can thus make a huge difference. Experiments with different initialization schemes were conducted, comparing filter coefficients drawn from a uniform or normal distribution, using the recently proposed discretization [9], or using reference-style filters that introduce no sort of initial interpolation. We empirically found the initialization from [9] to work well for small filter kernel sizes of 3×3 and $n_K, n_L \leq 4$, whereas for larger filter kernels uniformly distributed filters $\sim \mathscr{U}(-1/\sqrt{b}, 1/\sqrt{b})$ (b depends on the number of input dimensions and the filter kernel size) yielded the most satisfactory results, which is inspired by the well-known Xavier initialization [7].

Constraints. As given in Sect. 4.2, the constraints are used to ensure the boundedness of the filter coefficients, such that the sum of each filter is constrained to be 1, i.e.

$$\sum_{m,n} \xi_{m,n}^{1,r} = \sum_{m,n} \xi_{m,n}^{2,r} = \sum_{m,n} \xi_{m,n}^{3,r} = 1, \qquad \sum_{m,n} \eta_{m,n}^{1,l} = \sum_{m,n} \eta_{m,n}^{2,l} = 1,$$

with $r = 1, \ldots, n_K$, $l = 1, \ldots, n_L$. The corresponding projection per filter is computed following [5]. The fact that second-order TGV requires choosing two hyperparameters α_0 and α_1 majorly influencing the resulting reconstructions, where proper tuning can be challenging especially for natural images. Therefore an option, in this case, is to implicitly include them in the aforementioned constraints of the learned filters, such that the filter coefficients sum up to the same values $\gamma_K, \gamma_L \in \mathbb{R}$, respectively.

Moreover, one can also attempt to include a symmetry constraint to construct filters with a 90° rotational invariance property on the filter coefficients of L [5]. This reduces the actual number of learnable filters, which can be seen as an additional form of regularization in the learning setting to reduce overfitting to the training data.

5 Numerical Results

5.1 Data Sets

Synthetic Data. A synthetic data set was generated which is inspired by the intrinsic nature of the second-order TGV regularizer that favors piecewise affine solutions. A train and test data set each with 32 images of size 128×128 was constructed by randomly drawing basic shapes such as triangles, rectangles, and circles with varying sizes, which were filled with piecewise affine intensity changes and embedded within different (piecewise affine) background scenes. Examples of such images can be seen in the first column in Fig. 3. Casting this as an inverse problem requires some sort of ground truth to compare the obtained reconstructions for specific (α_1, α_0). Although there exist special cases such as specific 1D functions where an actual solution for TGV_α^2 exists [15], there is no ground truth available for arbitrary 2D images. Thus, the idea is to upsample the images (we use a size of $8M \times 8N$) and to compute a pseudo ground truth using the Condat-inspired TGV [9] due to its rotational invariance, where the intuition is that this solution better approximates the ground truth. A downsampled version of this is used as a new ground truth to compare the effect of using different handcrafted and learned discretization versions of TGV_α^2. To enable a fair comparison, this was done for three distinct parameter settings $(\alpha_1, \alpha_0) \in \{(0.1, 0.2), (0.3, 0.6), (1.0, 2.0)\}$ leading to different levels of smoothing, which is shown in the last three columns in Fig. 3.

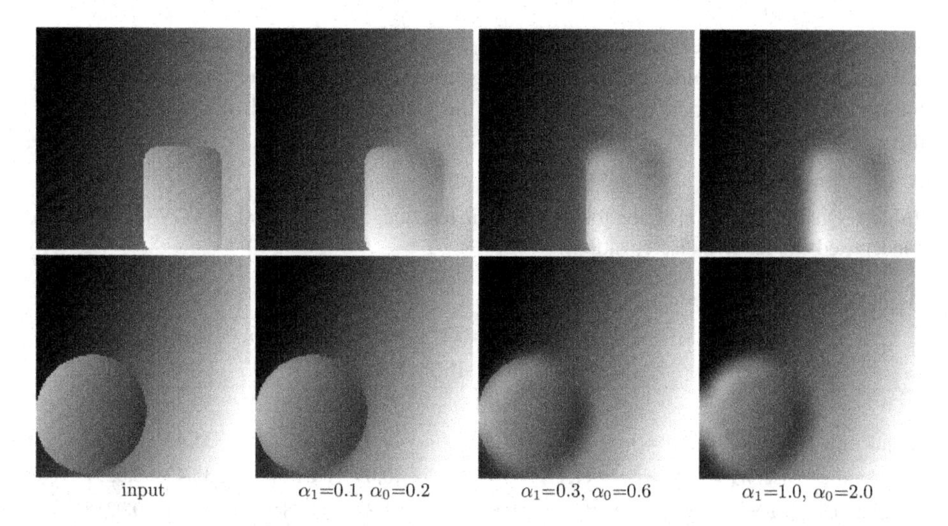

Fig. 3. Sample input images (left column) and respective "ground truth" reconstructions obtained from applying Condat's inspired TGV on upsampled images from the synthetic data set. This is done for three different combinations of (α_1, α_0).

Natural Images. Furthermore, a second, distinct data set was used. It is comprised of natural images where images were sampled from the well-known BSDS500 data set [1]. A train and a test data set each containing 32 images of size 128×128 were generated and all images were corrupted using zero-mean additive Gaussian noise $\sim \mathcal{N}(0, \sigma^2)$, which was sampled independently per pixel using noise levels $\sigma \in \{12.75, 25.5\}$ corresponding to 5% or 10% noise, respectively.

5.2 Results

Synthetic Data. The results presented here compare the solutions for the synthetic test data set using a "denoising" data term for different discretization schemes to the computed pseudo ground truth. In essence, the standard TGV and two scenarios of each handcrafted and learned discretizations are compared, where for both the two settings $n_K = 1$, $n_L = 3$ as in [9] and $n_K = 4$, $n_L = 4$ are used. For the learning setting mostly $3000 - 5000$ iterations in the outer learning problem were used or until the training loss function had stabilized, while the inner problem was solved with 100 piggyback primal-dual iterations using a warm-starting initialization scheme in each learning step. For the evaluation with Algorithm 1 the number of iterations was substantially increased to ensure absolute convergence.

Quantitative results reporting the mean peak signal-to-noise ratio (PSNR) and mean squared error (MSE) on the test set are presented in Table 1. They clearly show that both the handcrafted and the learned discretizations outperform the standard TGV, which is reflected for all three scenarios of (α_1, α_0). In each case using handcrafted filters gives some improvement, however directly learning the filters always outperforms these results by a larger margin. For $(\alpha_1, \alpha_0) = (1.0, 2.0)$, which introduces a substantial amount of smoothing and is, therefore, more challenging, the increase due to the handcrafted filters is barely present, while the learned filters still manage to improve the result further.

Table 1. Quantitative comparison for the synthetic test images of the standard TGV and different handcrafted and learned discretizations.

Method	Metric					
	$\alpha_1 = 0.1$, $\alpha_0 = 0.2$		$\alpha_1 = 0.3$, $\alpha_0 = 0.6$		$\alpha_1 = 1.0$, $\alpha_0 = 2.0$	
	PSNR	MSE $\cdot 10^{-2}$	PSNR	MSE $\cdot 10^{-2}$	PSNR	MSE $\cdot 10^{-2}$
TGV	40.36	0.0135	39.60	0.0151	35.36	0.0302
Handcrafted Disc. $n_K = 1$, $n_L = 3$	41.11	0.011	40.09	0.0128	35.43	0.0297
Handcrafted Disc. $n_K = 4$, $n_L = 4$	41.11	0.011	40.09	0.0128	35.42	0.0297
Learned Disc. $n_K = 1$, $n_L = 3$, 3×3	43.05	0.0065	41.37	0.0088	36.84	0.0218
Learned Disc. $n_K = 4$, $n_L = 4$, 3×3	**43.09**	**0.0064**	**41.45**	**0.0086**	**36.95**	**0.0211**

Natural Images. Moreover, results on natural image denoising for 5% and 10% additive Gaussian noise are presented, again comparing different handcrafted and learned discretizations. As for the previous task, the number of filters n_K and n_L is varied, however, due to the nature of natural images it is reasonable to also use a higher number of filters and larger kernel sizes. Thus, we extend the kernel size to 7×7, which is a good trade-off in terms of increased globality while maintaining moderate complexity. It is noteworthy that using the largest filter settings results in a computational time increase of up to twelve-fold, however, this trade-off is justified by the clear performance improvements. The learning setting remains similar to the previous experiments, and evaluations were conducted using 10^4 primal-dual iterations. The hyperparameters were set to $\alpha_1 = \{0.03, 0.0685\}$ per noise level, respectively, resulting from a prior grid search (and $\alpha_0 = 2\alpha_1$). Note that due to the constraint that the filter coefficients are allowed to sum up to γ_K, γ_L for the learning setting, the values of α_0, α_1 serve as an initialization, while their final values amount to $|\gamma_K|\alpha_0$ and $|\gamma_L|\alpha_1$ (whilst normalizing the learned filters with γ_K, γ_L).

Quantitative results are summarized in Table 2, clearly showing that learned discretizations with a higher number of filters (such as $n_K = 16$ and $n_L = 16$) and a filter kernel size of 7×7 yield the best results in terms of PSNR in dB and MSE (improvements up to approx. 0.6 dB). The same is confirmed when evaluating the structural similarity index measure (SSIM) [17]. This can be expected as these settings allow us to learn a more complex and rich discretization of natural images. The additional symmetry constraint on L – indicated by (sym.) – does not influence the results significantly. Further, an additional quantitative comparison using a TV regularizer with hand-tuned $\alpha = \{0.03, 0.0685\}$ for both noise levels confirms the well-established fact that the TV is a proper handcrafted regularizer especially considering its simplicity.

Table 2. Quantitative comparison of natural image denoising of the test set with 5% and 10% Gaussian noise for different handcrafted and learned discretizations.

Method	Metric					
	5% Gaussian noise			10% Gaussian noise		
	PSNR	MSE $\cdot 10^{-2}$	SSIM	PSNR	MSE$\cdot 10^{-2}$	SSIM
Corrupted f	26.04	0.2490	0.7885	20.02	0.9959	0.5382
TV	30.14	0.1049	0.9249	26.52	0.2445	0.8497
TGV	30.2	0.1043	0.9257	26.56	0.2431	0.8512
Handcrafted Disc. $n_K = 1$, $n_L = 3$	30.24	0.1046	0.9267	26.69	0.2394	0.8553
Handcrafted Disc. $n_K = 4$, $n_L = 4$	30.29	0.1030	0.9278	26.71	0.2370	0.8565
Learned Disc. $n_K = 1$, $n_L = 3$, 3×3	30.52	0.0935	0.9274	26.95	0.2172	0.8596
Learned Disc. $n_K = 4$, $n_L = 4$, 3×3	30.66	0.0906	0.9298	27.06	0.2123	0.8620
Learned Disc. $n_K = 8$, $n_L = 8$, 7×7	30.74	0.0896	0.9314	27.14	0.2090	0.8649
Learned Disc. $n_K = 8$, $n_L = 8$, 7×7, sym	30.72	0.0898	0.9311	27.15	0.2089	0.8649
Learned Disc. $n_K = 10$, $n_L = 10$, 7×7	30.73	0.0896	0.9313	27.17	0.2081	0.8657
Learned Disc. $n_K = 16$, $n_L = 16$, 7×7	30.77	0.0891	0.9319	27.16	0.2087	0.8654
Learned Disc. $n_K = 16$, $n_L = 16$, 7×7, sym	**30.77**	**0.0890**	**0.9320**	**27.18**	**0.2074**	**0.8659**

Exemplary learned filters K and L are displayed in Fig. 4 for the settings $n_K = 16$ and $n_L = 16$. Generally, it can be observed that different orientations are captured, while some of the filters in K introduce a bit of a smoothing effect. This can be associated with the fact that this filter operates on second-order finite difference arrays that are already very smooth, thus the discretization at this level will not contribute as much as opposed to the filters contained in L. Qualitative results on 10% Gaussian noise image denoising are shown in Fig. 5, where the handcrafted filters from [9] and the learned filters with $n_K = 16$ and $n_L = 16$ are compared in terms of reconstruction quality of two sample test images. Using learned filters tends to exhibit finer details and produces significantly more structure in the reconstructed images.

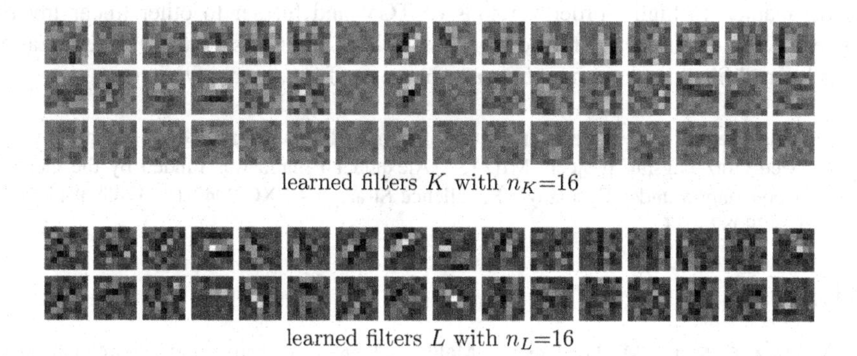

Fig. 4. Learned 7×7 filters using $n_L = 16$ and $n_K = 16$ for denoising (10% Gaussian noise). The row of a depicted filter denotes the component of the respective vector/tensor field that it acts upon, whereas the column refers to the specific filter r or l (with $r = 1, \cdots, n_K$ and $l = 1, \cdots, n_L$.)

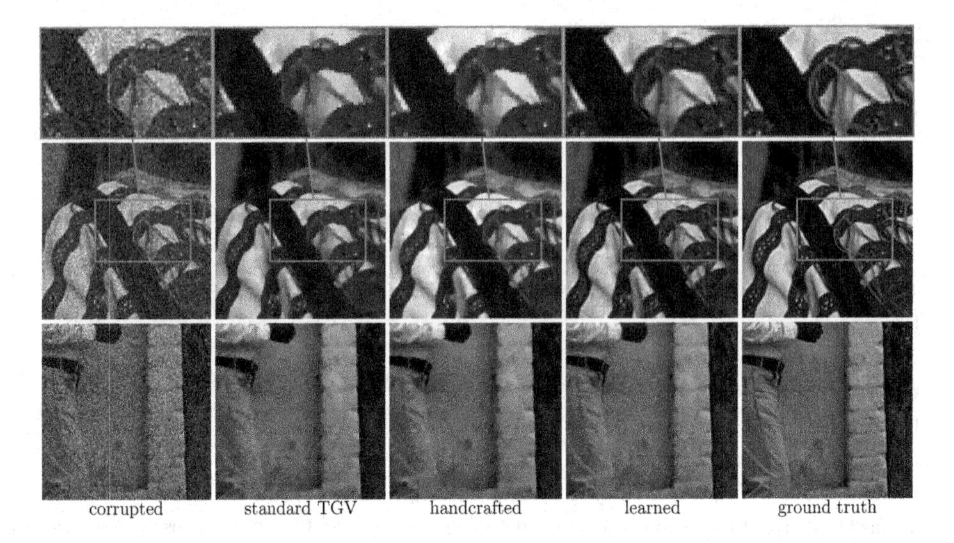

Fig. 5. Sample reconstructions from natural test images (10% Gaussian noise) comparing the standard TGV, the handcrafted discretization scheme with $n_K = 1$, $n_L = 3$ [9], and learned filters using $n_K = 16$, $n_L = 16$. For completeness, the ground truth images are also shown.

6 Conclusion

We proposed a general discretization scheme for the second-order TGV regularizer building upon the idea of [9]. This is supported by a proof of consistency by means of Γ-convergence of the newly discretized functional. Moreover, using a synthetic and a natural image data set we showcase that learning interpolation filters quantitatively and qualitatively improves the resulting image reconstruction in the setting of image denoising. It suggests that there might not be an ideal predefined set of interpolation filters applicable to all data sets and image reconstruction settings, but the most suited one can be obtained by learning within the respective setting. The proposed framework can be adapted to higher-order versions of TGV and further to other linear inverse problems. This is subject to future work, as well as an analysis on the generalization of the learned filters and on the robustness in terms of rotational invariance.

Acknowledgements. Lea Bogensperger and Thomas Pock acknowledge support by the BioTechMed Graz flagship project "MIDAS". Alexander Effland was funded by the German Research Foundation under Germany's Excellence Strategy - EXC-2047/1 - 390685813 and - EXC2151 - 390873048.

References

1. Arbeláez, P., Maire, M., Fowlkes, C., Malik, J.: Contour detection and hierarchical image segmentation. IEEE Trans. Pattern Anal. Mach. Intell. **33**(5), 898–916 (2011)
2. Bogensperger, L., Chambolle, A., Pock, T.: Convergence of a piggyback-style method for the differentiation of solutions of standard saddle-point problems. SIAM J. Math. Data Sci. **4**(3), 1003–1030 (2022). https://doi.org/10.1137/21M1455887

3. Bredies, K., Kunisch, K., Pock, T.: Total generalized variation. SIAM J. Imag. Sci. **3**(3), 492–526 (2010)
4. Chambolle, A., Pock, T.: A first-order primal-dual algorithm for convex problems with applications to imaging. J. Math. Imaging Vis. **40**(1), 120–145 (2011)
5. Chambolle, A., Pock, T.: Learning consistent discretizations of the total variation. SIAM J. Imag. Sci. **14**(2), 778–813 (2021)
6. Condat, L.: Discrete total variation: new definition and minimization. SIAM J. Imag. Sci. **10**(3), 1258–1290 (2017)
7. Glorot, X., Bengio, Y.: Understanding the difficulty of training deep feedforward neural networks. In: Proceedings of the Thirteenth International Conference on Artificial Intelligence and Statistics, pp. 249–256. JMLR Workshop and Conference Proceedings (2010)
8. Griewank, A., Faure, C.: Piggyback differentiation and optimization. In: Biegler, L.T., Heinkenschloss, M., Ghattas, O., van Bloemen Waanders, B. (eds.) Large-Scale PDE-Constrained Optimization, pp. 148–164. Springer, Heidelberg (2003). https://doi.org/10.1007/978-3-642-55508-4_9
9. Hosseini, A., Bredies, K.: A second-order TGV discretization with some invariance properties. arXiv preprint arXiv:2209.11450 (2022)
10. Huber, R., Haberfehlner, G., Holler, M., Kothleitner, G., Bredies, K.: Total generalized variation regularization for multi-modal electron tomography. Nanoscale **11**(12), 5617–5632 (2019)
11. Kingma, D., Ba, J.: Adam: a method for stochastic optimization. In: International Conference on Learning Representations (2014)
12. Knoll, F., Bredies, K., Pock, T., Stollberger, R.: Second order total generalized variation (TGV) for MRI. Magn. Reson. Med. **65**(2), 480–491 (2011)
13. Niu, S., et al.: Sparse-view X-ray CT reconstruction via total generalized variation regularization. Phys. Med. Biol. **59**(12), 2997 (2014)
14. Pock, T., Chambolle, A.: Diagonal preconditioning for first order primal-dual algorithms in convex optimization. In: 2011 International Conference on Computer Vision, pp. 1762–1769. IEEE (2011)
15. Pöschl, C., Scherzer, O.: Exact solutions of one-dimensional TGV. Commun. Math. Sci. **13**, 171–202 (2015)
16. Ranftl, R., Pock, T., Bischof, H.: Minimizing TGV-based variational models with non-convex data terms. In: Kuijper, A., Bredies, K., Pock, T., Bischof, H. (eds.) SSVM 2013. LNCS, vol. 7893, pp. 282–293. Springer, Heidelberg (2013). https://doi.org/10.1007/978-3-642-38267-3_24
17. Wang, Z., Bovik, A., Sheikh, H., Simoncelli, E.: Image quality assessment: from error visibility to structural similarity. IEEE Trans. Image Process. **13**(4), 600–612 (2004). https://doi.org/10.1109/TIP.2003.819861

Fluctuation-Based Deconvolution in Fluorescence Microscopy Using Plug-and-Play Denoisers

Vasiliki Stergiopoulou[1], Subhadip Mukherjee[2], Luca Calatroni[1](✉), and Laure Blanc-Féraud[1]

[1] UCA, CNRS, INRIA, Laboratoire I3S, 06903 Sophia-Antipolis, France
{vasiliki.stergiopoulou,calatroni,blancf}@i3s.unice.fr

[2] Department of Computer Science, University of Bath, Bath, UK
sm2467@cam.ac.uk

Abstract. The spatial resolution of images of living samples obtained by fluorescence microscopes is physically limited due to the diffraction of visible light, which makes the study of entities of size less than the diffraction barrier (around 200 nm in the x-y plane) very challenging. To overcome this limitation, several deconvolution and super-resolution techniques have been proposed. Within the framework of inverse problems, modern approaches in fluorescence microscopy reconstruct a super-resolved image from a temporal stack of frames by carefully designing suitable hand-crafted sparsity-promoting regularisers. Numerically, such approaches are solved by proximal gradient-based iterative schemes. Aiming at obtaining a reconstruction more adapted to sample geometries (e.g. thin filaments), we adopt a plug-and-play denoising approach with convergence guarantees and replace the proximity operator associated with the explicit image regulariser with an image denoiser (i.e. a pre-trained network) which, upon appropriate training, mimics the action of an implicit prior. To account for the independence of the fluctuations between molecules, the model relies on second-order statistics. The denoiser is then trained on covariance images coming from data representing sequences of fluctuating fluorescent molecules with filament structure. The method is evaluated on both simulated and real fluorescence microscopy images, showing its ability to correctly reconstruct filament structures with high values of peak signal-to-noise ratio (PSNR).

Keywords: Fluorescence microscopy · Image deconvolution · Variational regularisation · Proximal algorithms · Plug-and-Play regularisation

1 Introduction

In optical microscopy, the highest achievable spatial resolution is governed by some fundamental physical laws related to light propagation and is therefore limited. According to Rayleigh's criterion, the resolution of an optical microscope is defined as the smallest resolvable distance, i.e. the smallest distance

L. Calatroni et al. (Eds.): SSVM 2023, LNCS 14009, pp. 498–510, 2023.
https://doi.org/10.1007/978-3-031-31975-4_38

between two point sources so that they can be distinguished in the image. For conventional fluorescence microscopes, this distance is approximately equal to 200 nm in the lateral (x-y) plane. In order to resolve sub-cellular structures of size smaller than this barrier, several deconvolution and super-resolution techniques have emerged in the literature. Originally developed in the applied fields of chemistry, biology, and biophysics, such techniques can be naturally described in more mathematical terms as regularisation approaches for solving the ill-posed inverse problem considered. A big family of approaches achieving nanometric resolution (around 20 nm of lateral resolution) is known as *Single Molecule Localisation Microscopy* (SMLM) techniques (see [24] for a review). These methods rely on the use of sparse regularisation approaches for reconstructing frames of a temporal sequence of acquisitions where only a few molecules are active at a time. A more hardware-based super-resolution technique achieving a resolution of approximately 60–100 nm, is the *Stimulated Emission Depletion* (STED) microscopy approach [12] where the optical blur function (the microscope Point Spread Function, PSF) is depleted by means of suitable devices. While effective, these techniques show several drawbacks: for example, SMLM has long acquisition times while STED requires highly expensive commercial tools. Furthermore, both approaches require special (and expensive) fluorescent molecules able to support the high laser power required.

For overcoming such limitations, different types of approaches exploiting the independent stochastic temporal fluctuations of distinct fluorescent emitters became popular over the last decade. Such approaches rely on the use of both standard microscopes and fluorescent molecules and represent therefore a powerful class of approaches for applications. Some of these approaches are: the *Super-resolution Optical Fluctuation Imaging* (SOFI) approach [5] where the lack of correlation between distinct emitters is exploited by analyzing high-order statistics, the *Super-Resolution Radial Fluctuations* (SRRF) [11] microscopy, where super-resolution is achieved by calculating at each frame the degree of local symmetry and, finally, the *Sparsity-based Super-resolution Correlation Microscopy* (SPARCOM) [25] which models the sparse distribution of the fluorescent molecules via the use of a convex ℓ_1 regularisation applied on the emitters' covariance matrix. To improve the performance of SPARCOM, the Covariance-based ℓ_0 super-resolution microscopy with intensity estimation (COL0RME) method [27,29] has been proposed to estimate both molecule positions and intensities, which is a valuable piece of information in several applications, such as, e.g., 3D imaging [28], by means of a two-step procedure relying on hand-crafted sparsity promoting regularisers. The approach further estimates noise statistics and background terms (containing out-of-focus molecules). Both SPARCOM and COL0RME rely on the minimization of non-smooth (and possibly non-convex) functionals, for which tailored proximal optimization algorithms [19] based on soft- and hard-thresholding rules have thus been considered.

The applicability of these approaches to more complex geometries is limited due to the hand-crafted sparsity they enforce which creates biases (i.e., punctuated structures) in the reconstruction. This is particularly limiting when continuous curvilinear structures are desired, which is the case in several biological applications. For that, suitable regularisers can indeed be defined [18], with the major limitation of remaining tailored to particular shapes only. With the intent of developing a flexible regularisation approach suited to adapt to different geometries, we present in the following a data-driven optimization-inspired technique relying on the use of the so-called Plug-and-Play (PnP) approaches [31], which, over the last decade have been proved to represent an efficient framework for solving inverse image restoration problems, see [15] for a review. In this framework, the regulariser is parameterised by a deep neural network that can be trained on simulated data implicitly characterised by desired structures of interest, thus better capturing/promoting their shape after training. Our primary motivations behind using the PnP approach are three-fold: (i) Training the denoiser is independent of the imaging forward operator, which makes the pre-trained denoiser applicable even if the forward operator undergoes some changes, without having to retrain the denoiser from scratch. (ii) The training problem does not require pairs of measured and ground-truth images, unlike supervised machine learning approaches. To train the denoiser, one only needs high-quality ground-truth images and their noisy counterparts (with additive Gaussian noise). (iii) The PnP approaches are rooted in proximal point algorithms, so their convergence can be rigorously studied using results from fixed-point theory and/or convex analysis. This leads to better interpretability of the reconstruction algorithm and results in a principled way of combining knowledge about imaging physics with the available training data.

Contributions: In this work, we leverage the framework of PnP approaches with convergence guarantees [3,13,14] to show good empirical performance on the inverse problem of fluctuation-based image deconvolution presented, e.g., in [25, 27,29]. In Sect. 2, we review the recent advances in the field of PnP approaches for inverse problems, pointing out the convergent scheme we employ. In Sect. 3, we formulate the covariance-based deconvolution model and formulate its PnP extension, which we called PnP-COL0RME in the following. In Sect. 4 we report several numerical results on both simulated and real data where the advantages of using the PnP reconstruction model are shown in comparison to its model-based counterpart.

2 Plug-and-Play Approaches for Inverse Problems

A standard approach for solving ill-posed inverse problems in imaging consists in solving the optimisation problem:

$$\hat{\mathbf{x}} \in \arg\min_{\mathbf{x}\in\mathbb{R}^{n^2}} \mathcal{F}(\boldsymbol{\Psi}\mathbf{x};\mathbf{y}) + \lambda\mathcal{R}(\mathbf{x}), \quad \lambda > 0, \tag{1}$$

where, for observed data $\mathbf{y} \in \mathbb{R}^{n^2}$ (being the vectorisation of a 2D image of size $n \times n$) and model operator $\boldsymbol{\Psi} \in \mathbb{R}^{n^2 \times n^2}$, $\mathcal{F}$ denotes a (smooth) data fidelity term and $\mathcal{R}$ a regularisation term encoding prior knowledge on the solution $\hat{\mathbf{x}} \in \mathbb{R}^{n^2}$. Depending on the available prior information (such as sparsity, gradient smoothness, etc.), tailored hand-crafted functions $\mathcal{R}$ can be used. In most cases,

$\mathcal{R}$ is non-smooth, and proximal algorithms [19] can be used for solving (1). We recall that the proximity operator of parameter $\tau > 0$ of a proper, convex and non-smooth function $\mathcal{R}$ is defined by:

$$\text{prox}_{\tau\mathcal{R}}(\mathbf{z}) = \underset{\mathbf{x}\in\mathbb{R}^{n^2}}{\arg\min}\ \mathcal{R}(\mathbf{x}) + \frac{1}{2\tau}\|\mathbf{z}-\mathbf{x}\|_2^2, \quad \mathbf{z}\in\mathbb{R}^{n^2}. \tag{2}$$

Solving (2) corresponds to solving the problem of denoising an image $\mathbf{z}\in\mathbb{R}^{n^2}$ corrupted by an additive white Gaussian noise (AWGN) of constant variance equal to τ. Within a proximal gradient algorithm, (2) can thus be interpreted as a denoising step of the gradient descent iteration $\mathbf{z}^k = \mathbf{x}^k - \tau\nabla\mathcal{F}(\mathbf{\Psi}\mathbf{x}^\mathbf{k};\mathbf{y})$ at each iteration $k \geq 0$. This observation inspired the authors of [31] to develop the framework of PnP priors, whose main idea consists in replacing $\text{prox}_{\tau\mathcal{R}}(\cdot)$ with an off-the-shelf image denoiser $D_\sigma(\cdot)$ depending on a parameter $\sigma > 0$ corresponding to a regularisation functional $\mathcal{R}$ whose explicit definition is often not available. In a Bayesian framework, it is indeed possible to explicitly relate Gaussian minimum mean-squared error (MMSE) denoisers $D_\sigma(\cdot)$ with the (unknown) image prior $p(\cdot)$ one would like to model [17] using the Tweedie's identity: $\sigma^2\nabla\log p_\sigma(\mathbf{x}) = D_\sigma(\mathbf{x}) - \mathbf{x}$, where p_σ is the convolution of p with a Gaussian smoothing kernel of bandwith $\sigma > 0$, which makes p_σ smoother (namely, Lipschitz differentiable) than p under mild conditions. As observed in [20, Eq. 74–75], considering a (Gaussian) denoiser residual is in fact a good approximation to the score of the image prior. Along with their Bayesian interpretability, another advantage of PnP approaches is that they allow the use of advanced image denoising models, e.g. denoisers parameterised by convolutional neural networks (CNNs), within the iterative scheme, with impressive representational capabilities. In most cases, the CNN image denoiser D is trained to perform denoising on some pairs of clean-noisy images and can be used afterward for more-general inverse problems (e.g. deblurring, super-resolution, etc.), see [15] for a review. Some state-of-the-art denoisers include image-dependent filtering algorithms such as Block-Matching & 3D filtering (BM3D) [4], Denoising Convolutional Neural Networks (DnCNN) [34] and Dilated-Residual U-Net (DRUNET) deep learning network [33].

These denoisers are typically used in iterative proximal schemes (see, e.g., [16] for a FISTA-type PnP scheme), although they can be flexibly used in other algorithms such as, e.g., the Alternate Directions Method of Multipliers (ADMM) [2], the Douglas-Rachford Splitting (DRS) [6] and the Half-Quadratic Splitting (HQS) [8]. For all these algorithms, a corresponding PnP version can indeed simply be obtained as described above. PnP versions of proximal algorithms have been used to solve image restoration problems such as for example PnP-PGD in [30], PnP-ADMM and PnP-DRS in [21,22] and PnP-HQS in [3,33,34].

In [21] an explicit regularisation by denoising (RED) strategy was designed in terms of an explicit function $\mathcal{R}(\cdot)$ defined, for generic image denoiser D, by:

$$\mathcal{R}(\mathbf{x}) := \frac{1}{2}\mathbf{x}^T(\mathbf{x} - D(\mathbf{x})).$$

Under conditions of local homogeneity, non-expansiveness, and Jacobian symmetry, D was shown to be indeed equivalent to a gradient step on $\mathcal{R}$ [20], that is,

$D(\mathbf{x}) = \mathbf{x} - \nabla\mathcal{R}(\mathbf{x})$. However, as shown in [20], such requirements are unrealistic on the widely-used denoisers mentioned above, as they do not have symmetric Jacobians. In order to overcome this limitation, in [3,13,14], the authors proposed to formulate, similar to RED, a gradient step denoiser of the form:

$$D_\sigma(\mathbf{x}) = \mathbf{x} - \nabla\mathcal{R}_\sigma(\mathbf{x}), \tag{3}$$

where $\mathcal{R}_\sigma : \mathbb{R}^{n^2} \to \mathbb{R}$ is a scalar function parameterised by a neural network $N_\sigma : \mathbb{R}^{n^2} \to \mathbb{R}^{n^2}$. Interestingly, under mild structural assumption on D_σ, the authors are able to prove sound convergence guarantees for the underlying non-convex optimisation problem defined in terms of a non-trivial (but explicit) regularisation function $\mathcal{R}(\cdot)$. In the following section we specify the particular problem we are interested in and discuss its PnP extension based on the strategy discussed above.

3 Deconvolution via Sparse Auto-Covariance Analysis

We consider the following image formation model considered, e.g., in [25,27,29] to describe, for $t = 1, \ldots, T$, $T > 0$, a video of temporal acquisitions $\mathbf{y}_t \in \mathbb{R}^{n^2}$ by standard fluorescent microscopes of true images $\mathbf{x}_t \in \mathbb{R}^{n^2}$:

$$\mathbf{y}_t = \mathbf{\Psi}\mathbf{x}_t + \mathbf{b} + \mathbf{n}_t. \tag{4}$$

In (4), $\mathbf{\Psi} \in \mathbb{R}^{n^2 \times n^2}$ is a (known) convolution operator associated with the system point spread function (PSF), $\mathbf{b} \in \mathbb{R}^{n^2}$ is a background term and $\mathbf{n}_t \in \mathbb{R}^{n^2}$ is the realisation at time t of an i.i.d. Gausian noise random vector with unknown variance $s \geq 0$, i.e. $\mathbf{n} \sim \mathcal{N}(0, s\,\mathbf{Id})$. We look for a deconvolved image $\mathbf{x} \in \mathbb{R}^{n^2}$, defined as $\mathbf{x} = \frac{1}{T}\sum_{t=1}^{T} \mathbf{x_t}$. In [25,27], a reformulation of the model (4) was done in the covariance domain in order to exploit temporal information. In the following, we proceed similarly but consider a simplified modeling where only auto-covariance vectors are taken into account, thus neglecting cross-terms.

Considering the frames $\{\mathbf{y}_t\}_{t=1}^{T}$ as T realisations of a random variable $\mathbf{y}$, the sample auto-covariance (variance) vector $\tilde{\mathbf{r}}_\mathbf{y} \in \mathbb{R}^{n^2}$ of $\mathbf{y}$ can be estimated by:

$$\tilde{\mathbf{r}}_\mathbf{y} \approx \frac{1}{T-1}\sum_{t=1}^{T}(\mathbf{y}_t - \overline{\mathbf{y}})^2, \tag{5}$$

where $\overline{\mathbf{y}} = \frac{1}{T}\sum_{t=1}^{T} \mathbf{y}_t$ denotes the empirical mean. From (4) and (5), the following model thus holds between the auto-covariance vectors:

$$\tilde{\mathbf{r}}_\mathbf{y} = \mathbf{\Psi}^2\mathbf{r}_\mathbf{x} + \tilde{\mathbf{r}}_\mathbf{n}, \tag{6}$$

where $\mathbf{r}_\mathbf{x} \in \mathbb{R}^{n^2}$ and $\tilde{\mathbf{r}}_\mathbf{n} \in \mathbb{R}^{n^2}$ are the auto-covariance vectors associated to the samples $\{\mathbf{x}_t\}_{t=1}^{T}$ and $\{\mathbf{n}_t\}_{t=1}^{T}$, respectively, and where by $\mathbf{\Psi}^2 \in \mathbb{R}^{n^2 \times n^2}$ we denote the matrix $\mathbf{\Psi} \odot \mathbf{\Psi}$ where $\odot$ denotes the point-wise Hadamard product. Finally, note that by assumption $\tilde{\mathbf{r}}_\mathbf{n} = s\mathbf{1}$, where $\mathbf{1} = (1, \ldots, 1) \in \mathbb{R}^{n^2}$.

Algorithm 1. Model-based and PnP support estimation

Require: $\tilde{\mathbf{r}}_\mathbf{y}, {\mathbf{r}_\mathbf{x}}^0 \in \mathbb{R}^{n^2}$ and parameters ($\tau, \lambda > 0$ for model-based, σ for PnP)
repeat

$$s^{k+1} = \mathbf{1}^T(\tilde{\mathbf{r}}_\mathbf{y} - \boldsymbol{\Psi}^2 {\mathbf{r}_\mathbf{x}}^k)$$

$$\mathbf{z}^{k+1} = {\mathbf{r}_\mathbf{x}}^k - \tau\lambda(\boldsymbol{\Psi}^2)^T(\tilde{\mathbf{r}}_\mathbf{y} - \boldsymbol{\Psi}^2 {\mathbf{r}_\mathbf{x}}^k - s^{k+1}\mathbf{1})$$

$${\mathbf{r}_\mathbf{x}}^{k+1} = \begin{cases} \text{prox}_{\tau\lambda R}(\mathbf{z}^{k+1}) & \texttt{\% model-based (7)} \\ {\mathbf{r}_\mathbf{x}}^{k+1} = D_\sigma(\mathbf{z}^{k+1}) & \texttt{\% PnP} \end{cases}$$

until convergence
return $\boldsymbol{\Omega} := \{i \;:\; (\hat{\mathbf{r}}_\mathbf{x})_i \neq 0\}, \hat{s}$

Remark 1. Note that, upon reshaping, $\tilde{\mathbf{r}}_\mathbf{y} \in \mathbb{R}^{n^2}$ is in fact the second-order SOFI image associated to the stack $\{\mathbf{y}_t\}_{t=1}^T$, which, thanks to the 'squaring' of the underlying point spread function, enjoys better spatial resolution in comparison, e.g., to $\bar{\mathbf{y}}$, see [5] for details.

Remark 2. In comparison to the covariance-based modelling in SPARCOM [25] and COL0RME [27], model (6) is indeed a simplification. In those papers, the whole sample-covariance matrix $\mathbf{R}_\mathbf{y} \in \mathbb{R}^{n^2 \times n^2}$ with main diagonal $\tilde{\mathbf{r}}_\mathbf{y}$ was computed. Such a matrix is not diagonal, due to the correlation induced by the PSF. In order to deal with a simplified model and benefit from faster calculations, we consider in the following the relation (6) involving only auto-covariance terms and leave a complete modelling involving also cross terms for future work. The resulting observation model 6 is thus less rich but exact (not approximated).

Based on (6), we are now interested in finding the fluorescent molecule locations and estimate noise information. Namely, we are interested in finding the support of $\mathbf{r}_\mathbf{x}$, $\boldsymbol{\Omega} := \{i \;:\; (\mathbf{r}_\mathbf{x})_i \neq 0\}$ and the unknown noise variance $s \geq 0$. We do so by considering the following minimisation problem:

$$(\hat{\mathbf{r}}_\mathbf{x}, \hat{s}) \in \underset{\mathbf{r}_\mathbf{x} \geq 0,\; s \geq 0}{\arg\min} \; \frac{\lambda}{2}\|\tilde{\mathbf{r}}_\mathbf{y} - \boldsymbol{\Psi}^2\mathbf{r}_\mathbf{x} - s\mathbf{1}\|_2^2 + \mathcal{R}(\mathbf{r}_\mathbf{x}), \tag{7}$$

where $\lambda > 0$ is a regularization parameter and $\mathcal{R}(\cdot)$ is a regularisation term to be defined to enforce desirable properties (sparsity, for instance) of the solution. Problem (7) can be solved by Algorithm 1, where, to improve convergence speed, a global minimisation on s is performed followed by a proximal-gradient step on $\mathbf{r}_\mathbf{x}$. An analogous (a priori, slower) algorithm benefiting from theoretical convergence guarantees is the Proximal Alternating Linearized Minimisation (PALM) algorithm whose convergence is studied in [1].

Once $\boldsymbol{\Omega}$ and $\hat{s}$ have been computed, following [27,29] a second algorithmic step can be performed to estimate image intensities only in correspondence with the support points in $\boldsymbol{\Omega}$, i.e. by solving:

$$(\hat{\mathbf{x}}, \hat{\mathbf{b}}) \in \underset{\mathbf{x} \in \mathbb{R}_+^{|\Omega|},\; \mathbf{b} \in \mathbb{R}_+^{n^2}}{\arg\min} \; \frac{1}{2}\|\boldsymbol{\Psi}_\Omega \mathbf{x} - (\bar{\mathbf{y}} - \mathbf{b})\|_2^2 + \frac{\mu}{2}\|\nabla_\Omega \mathbf{x}\|_2^2 + \frac{\beta}{2}\|\nabla \mathbf{b}\|_2^2, \tag{8}$$

where the data term models the presence of Gaussian noise, $\overline{\mathbf{y}} = \sum_{t=1}^{T} \mathbf{y}_t$ and $\mu, \beta > 0$ are regularisation parameters. Moreover, $\mathbf{\Psi}_{\Omega} \in \mathbb{R}^{n^2 \times |\Omega|}$ is a matrix whose i-th column is extracted from $\mathbf{\Psi}$ for all indexes $i \in \Omega$ and ∇_Ω denotes the discrete gradient operator restricted to points in the support Ω.

A hand-crafted regularisation model $\mathcal{R}$ in (7) introduces reconstruction biases. For example, using the ℓ_1 norm [25] or the continuous exact relaxation of the ℓ_0 pseudo-norm [26]) enforces sparsity by promoting point reconstruction. Solutions thus appear dotted as reconstructed points have a given inter-distance which cannot be decreased [7]. For reconstructing filaments, a solution is to use a regularising term promoting curves. Such method is proposed, e.g., in [18] in an off-the-grid setting, but the numerical aspects are difficult and still under development. To overcome this limitation, in the following section, we propose a Plug-and-Play extension of the approach above where the proximal step is replaced by a denoiser D_σ trained on an appropriate dataset of covariance images representing the geometrical structures of interest.

3.1 Plug-and-Play Extension

The proximal step naturally appearing when solving problem (7) by proximal gradient algorithms, can be replaced by an off-the-shelf denoiser. To do so, we make use of a proximal gradient step denoiser as proposed by Hurault *et al.* in [14]. In their paper, the authors showed that this choice corresponds indeed to the proximal operator associated to a non-convex smooth function which allows the authors to derive convergence guarantees of the resulting proximal gradient scheme [14, Theorem 4.1]. Note, that differently to the setting proposed in [14], our algorithm processes auto-covariance images due to the model (6) and, along with $\hat{\mathbf{r}}_{\mathbf{x}}$, it provides an estimate $\hat{s}$ of the noise variance by alternate minimisation. We report the iterative scheme in Algorithm 1 and refer in the following to PnP-COL0RME to the case when a PnP regulariser is employed.

In [14] the authors considered a denoiser D_σ in the form of a gradient step (3) of a functional $\mathcal{R}_\sigma : \mathbb{R}^{n^2} \to \mathbb{R}$ with specific properties, e.g., bounded from below, and parameterised by a deep neural network N_σ. Recalling the characterisation of proximity operators [10] introduced by Gribonval & Nikolova, the authors proved in fact that D_σ can be written as proximal operator of a function ϕ_σ defined by:

$$\phi_\sigma(\mathbf{w}) := \mathcal{R}_\sigma(D_\sigma^{-1}(\mathbf{w})) - \frac{1}{2}\|D_\sigma^{-1}(\mathbf{w}) - \mathbf{w}\|_2^2, \quad \mathbf{w} \in \mathbb{R}^{n^2}.$$

The function minimised when employing PnP COL0RME reads: $F_\sigma(\mathbf{r_x}, s) := \frac{1}{2}\|\mathbf{r_y} - \mathbf{\Psi^2}\mathbf{r_x} - s\mathbf{1}\|_2^2 + \phi_\sigma(\mathbf{r_x})$, which, after recalling that $\mathbf{r}^k = D_\sigma(\mathbf{z}^k)$ at each k, can be written as:

$$F_\sigma({\mathbf{r_x}}^k, s^k) = \frac{1}{2}\|\mathbf{r_y} - \mathbf{\Psi^2}{\mathbf{r_x}}^k - s^k\mathbf{1}\|_2^2 + \mathcal{R}_\sigma(\mathbf{z}^k) - \frac{1}{2}\|\mathbf{z}^k - {\mathbf{r_x}}^k\|_2^2. \tag{9}$$

In [14, Theorem 4.1] the authors show that thanks to the structure of F_σ, the PnP proximal gradient scheme converges indeed to a stationary point of F_σ, whose decay can be indeed assessed throughout the iterations.

Note that, the regularisation parameter $\lambda > 0$ appearing in (1) to regulate the strength of the regularisation term $\mathcal{R}$ has been replaced by the hyperparameter σ in (9). Intuitively, the value of σ should correspond to the variance of AWGN appearing in the gradient steps of the proximal gradient Algorithm 1, hence its tuning is not straightforward. As discussed in [32], a possible remedy for avoiding a time-consuming parameter tuning consists in introducing a rescaling parameter whose setting is easier than σ.

4 Numerical Results

We now present some results obtained by using PnP-COL0RME on temporal sequences of blurred and noisy data. A natural extension to the actual problem of super-resolution where $\mathbf{\Psi} = \mathbf{M_q H} \in \mathbb{R}^{m^2 \times n^2}$ with $\mathbf{H} \in \mathbb{R}^{n^2 \times n^2}$ is PSF convolution matrix and $\mathbf{M_q} \in \mathbb{R}^{m^2 \times n^2}$ is a downsampling operator with $n = qm, q > 1$, is left for future work.

To train the denoiser D_σ we created a dataset composed of clean and noisy image pairs. The geometrical features of the images in this dataset should be the same as the one of the images to restore. Differently from other methods, the proposed algorithm works with a model formulated in the covariance domain, so that the denoiser takes as an input noisy sample auto-covariance matrices of a fluctuating temporal sequence of images. Hence, to create the dataset we first started by creating different spatial patterns (thin filaments) shown in Fig. 1 where the emitters have different positions in the continuous grid. Such patterns are the superposition, after rotations with different angles, of the ground truth spatial pattern provided in the MT0 microtubule training dataset uploaded for SMLM 2016 challenge[1]. Then, we used the fluctuation model discussed in [9] to simulate temporal fluctuations and create a temporal stack of $T = 500$ frames for each spatial pattern. Two exemplar frames of one temporal stack of images are reported in Figs. 2a and 2b. For each temporal stack of images, we could therefore calculate the temporal auto-covariance image (see Fig. 2c) corresponding to one instance of the clean images $\mathbf{r}_\mathbf{x}^{\mathrm{GT}}$ in our dataset. To create now its noisy version we added Gaussian noise $\boldsymbol{\eta}$ with constant variance σ^2, $\boldsymbol{\eta} \sim \mathcal{N}(\mathbf{0}, \sigma^2\mathbf{Id})$, with σ following a uniform distribution, $\sigma \sim \mathcal{U}(\sigma_1, \sigma_2)$. We remark that since the noise in the covariance data comes from additive Gaussian noise on the individual frames, its actual distribution is indeed χ^2. However, since the number of the degrees of freedom is high (as $T = 500$), the distribution can be approximated by a Gaussian distribution. In our experiments, after normalising $\mathbf{r}_\mathbf{x}^{\mathrm{GT}}$ with maximum value equal to 1, we select $\sigma_1 = \epsilon << 1$ and $\sigma_2 = 50/255$.

Training was performed following the procedure in [14] and using the code available on the authors' GitHub repository[2]. For the neural network $N_\sigma(\cdot)$ used to parameterise the denoiser (see (3)), we used DRUNet, a CNN proposed in [33]. For training, we used 500 pairs of clean-noise auto-covariance images and 100 for

[1] https://srm.epfl.ch/Challenge/ChallengeSimulatedData.
[2] https://github.com/samuro95/Prox-PnP.

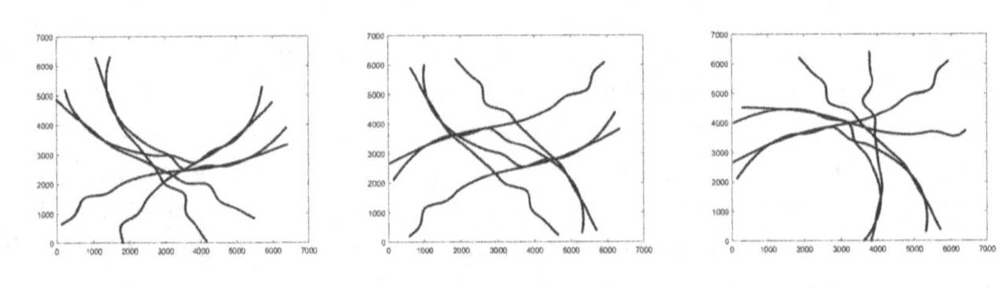

Fig. 1. Simulated spatial patterns

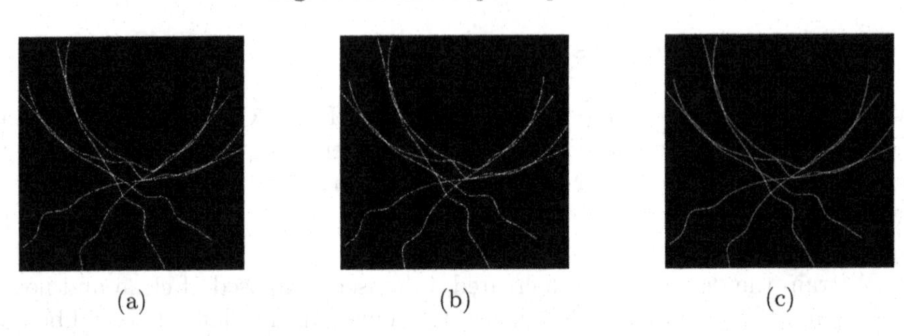

Fig. 2. (a–b) Two different frames of a simulated fluctuating stack made from the first spatial pattern from Fig. 1, (c) The auto-covariance image $\mathbf{r}_{\mathbf{x}}^{\mathrm{GT}}$ estimated from the whole temporal sequence.

validation. The network was trained using 1215 epochs via ADAM optimization and batch size equal to 16. In the following experiments, the choice $\tau = 1$ and $\lambda = 0.99$ in Algorithm 1 was performed to guarantee convergence, see [14, Section 4.1] for details.

4.1 Simulated Data

We first apply PnP COL0RME to simulated data presented in Fig. 3. The PSF used to generate the data has a FWHM equal to 176.6 nm, the pixel size is equal to 25 nm and the images have a size of 256 × 256 pixels.

Thanks to its training, we observe that the proposed approach is able to capture the filaments' geometry fairly well. We observe that in comparison to the ground truth support in Fig. 3c, the reconstruction in Fig. 3g is rather accurate. For the evaluation of the localization precision the Jaccard Index (JI) has been used. It is a quantity in $[0, 1]$ computed as the ratio between correct detections (CD) and the total (correct, false negatives false positive) decetions, i.e. $\mathrm{JI} := CD/(CD+FN+FP)$, up to a tolerance $\delta > 0$, measured in nm (see, e.g., [23]). For the reconstruction in Fig. 3g, the tolerance precision was chosen $\delta = 40$ nm. Moreover, by solving (8), intensities can also be estimated with high precision, see Fig. 3h. However, for the challenging dataset in Fig. 3, the appearance of small artefacts (e.g. incorrect duplication of filaments) due to the training dataset we built are observed. They could be potentially removed by retraining the model with more heterogeneous data.

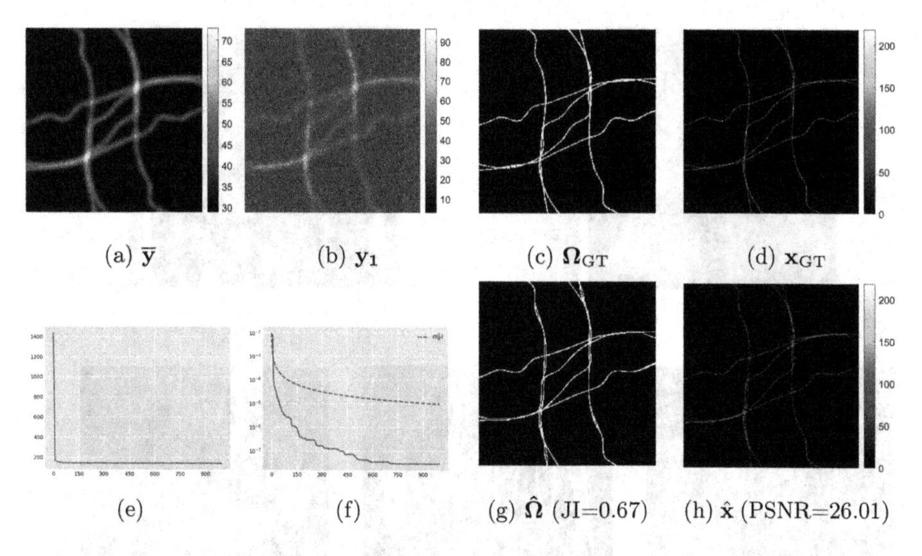

Fig. 3. (a) Mean of the acquired temporal sequence, (b) First frame (c) The ground truth support (d) The ground truth intensity image (e) Evolution of cost function F_σ in (9) (f) the evolution of $\min_{i\leq k} \|\mathbf{r_x}^{i+1} - \mathbf{r_x}^{i}\|^2/\|\mathbf{r_x}^{0}\|^2$, in logarithmic scale, (g) Reconstructed support (h) Reconstructed intensity image.

4.2 Real Data

We then applied the proposed approach to high-density SMLM acquisitions using a publicly available dataset created for the 2013 SMLM challenge[3], see Fig. 4. Although in SMLM the molecules do not have a blinking behaviour, but rather an on-to-off transition, we can consider as blinking the temporal behaviour of one pixel in high-density videos due to the presence of many molecules per pixel. The dataset contains $T = 500$ images, the PSF of the microscope used to acquire these data has a FWHM of 351.8 nm and the pixel size is equal to 100 nm. The support $\hat{\boldsymbol{\Omega}}$ computed by the model-based COL0RME approach in [27,29] based on the use of a relaxation of the ℓ_0 pseudo-norm is compared to the one PnP-COL0RME variant of Algorithm 1. Since no ground truth is available for these data, no quantitative assessment can be computed, however better continuation properties than COL0RME [27] are observed.

[3] https://srm.epfl.ch/Challenge/Challenge2013.

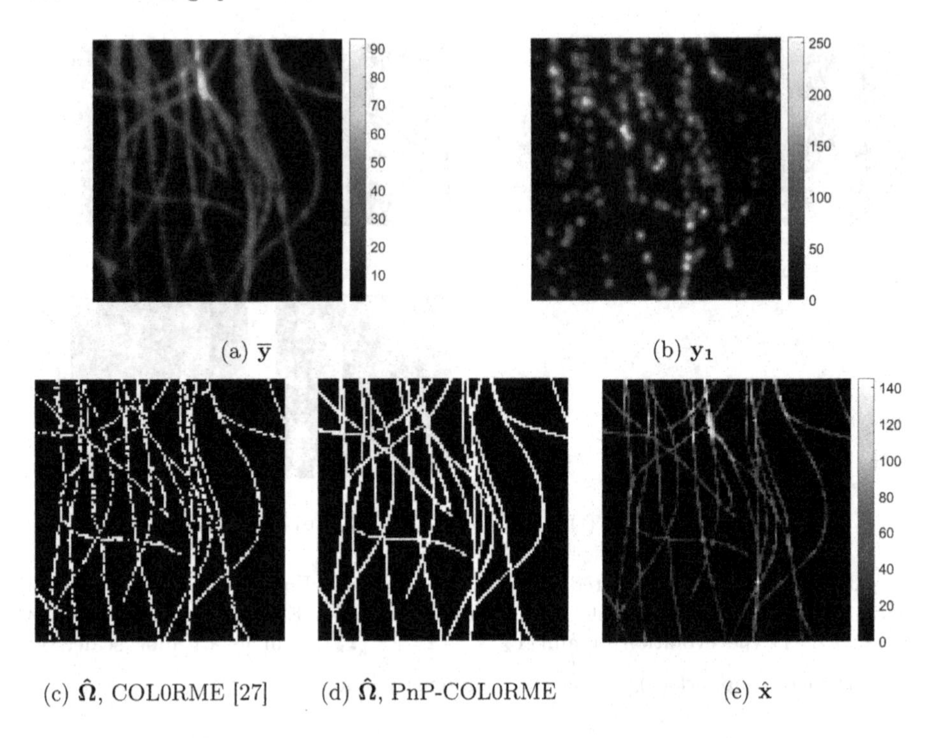

(a) $\overline{\mathbf{y}}$ (b) $\mathbf{y}_1$

(c) $\hat{\boldsymbol{\Omega}}$, COL0RME [27] (d) $\hat{\boldsymbol{\Omega}}$, PnP-COL0RME (e) $\hat{\mathbf{x}}$

Fig. 4. HD-SMLM data: (first row) The temporal mean and the first frame of the acquired temporal sequence (second row) Support (ℓ_0-based [27] VS. PnP) and intensity reconstruction, $\sigma = 10/255$.

5 Conclusions

We presented a PnP model for support localisation for the deconvolution of imaging data in fluorescence microscopy. PnP approaches rely on the use of off-the-shelf denoisers to model implicit prior regularisation functionals. They can be effectively used to replace proximal steps in proximal gradient algorithms. Following [14], we choose a denoiser with a particular structure to benefit from convergence guarantees. Our results show that the geometry of specific structures (thin filaments) can be captured by suitable training. Future work should take into account the presence of a downsampling operator in the image formation model and a more accurate modelling making use of also cross terms in the covariance data.

Acknowledgements. LC acknowledges the support received by the ANR project TASKABILE (ANR-22-CE48-0010) and the GdR ISIS project SPLIN. VS, LC and LBF acknowledge the support received by the ANR project MICROBLIND (ANR-21-CE48-0008). All authors acknowledge the support received by the H2020 RISE projects NoMADS (GA. 777826).

References

1. Bolte, J., Sabach, S., Teboulle, M.: Proximal alternating linearized minimization for nonconvex and nonsmooth problems. Math. Program. **146**(1), 459–494 (2014)
2. Boyd, S., Parikh, N., Chu, E., Peleato, B., Eckstein, J.: Distributed optimization and statistical learning via the alternating direction method of multipliers. Found. Trends Mach. Learn. **3**(1), 1–122 (2011)
3. Cohen, R., Blau, Y., Freedman, D., Rivlin, E.: It has potential: gradient-driven denoisers for convergent solutions to inverse problems. In: Advances in Neural Information Processing Systems, vol. 34, pp. 18152–18164. Curran Associates, Inc. (2021)
4. Dabov, K., Foi, A., Katkovnik, V., Egiazarian, K.: Image denoising by sparse 3-D transform-domain collaborative filtering. IEEE Trans. Image Process. **16**(8), 2080–2095 (2007)
5. Dertinger, T., Colyer, R., Iyer, G., Weiss, S., Enderlein, J.: Fast, background-free, 3D super-resolution optical fluctuation imaging (SOFI). PNAS **106**, 22287–22292 (2009)
6. Douglas, J., Rachford, H.H.: On the numerical solution of heat conduction problems in two and three space variables. Trans. Am. Math. Soc. **82**(2), 421–439 (1956)
7. Duval, V., Peyre, G.: Exact support recovery for sparse spikes. Found. Comput. Math. **15**, 1315–1355 (2015)
8. Geman, D., Yang, C.: Nonlinear image recovery with half-quadratic regularization. IEEE Trans. Image Process. **4**(7), 932–946 (1995)
9. Girsault, A., et al.: SOFI simulation tool: a software package for simulating and testing super-resolution optical fluctuation imaging. PLoS ONE **11**(9), 1–13 (2016)
10. Gribonval, R., Nikolova, M.: A characterization of proximity operators. J. Math. Imaging Vis. **62**, 773–789 (2020)
11. Gustafsson, N., Culley, S., Ashdown, G., Owen, D.M., Pereira, P.M., Henriques, R.: Fast live-cell conventional fluorophore nanoscopy with ImageJ through super-resolution radial fluctuations. Nature Commun. **7**(1), 12471 (2016)
12. Hell, S.W., Wichmann, J.: Breaking the diffraction resolution limit by stimulated emission: stimulated-emission-depletion fluorescence microscopy. Opt. Lett. **19**, 780–782 (1994)
13. Hurault, S., Leclaire, A., Papadakis, N.: Gradient step denoiser for convergent plug-and-play. In: International Conference on Learning Representations (2022)
14. Hurault, S., Leclaire, A., Papadakis, N.: Proximal denoiser for convergent plug-and-play optimization with nonconvex regularization. In: Proceedings of Machine Learning Research, vol. 162, pp. 9483–9505. PMLR (2022)
15. Kamilov, U.S., Bouman, C.A., Buzzard, G.T., Wohlberg, B.: Plug-and-play methods for integrating physical and learned models in computational imaging: theory, algorithms, and applications. IEEE Signal Process. Mag. **40**(1), 85–97 (2023)
16. Kamilov, U.S., Mansour, H., Wohlberg, B.: A plug-and-play priors approach for solving nonlinear imaging inverse problems. IEEE Signal Process. Lett. **24**(12), 1872–1876 (2017)
17. Laumont, R., Bortoli, V.D., Almansa, A., Delon, J., Durmus, A., Pereyra, M.: Bayesian imaging using plug & play priors: when Langevin meets tweedie. SIAM J. Imaging Sci. **15**(2), 701–737 (2022)
18. Laville, B., Blanc-Féraud, L., Aubert, G.: Off-the-grid curve reconstruction through divergence regularisation: an extreme point result (2022). HAL preprint

19. Parikh, N., Boyd, S.: Proximal algorithms. Found. Trends Optim. **1**(3), 123–231 (2014)
20. Reehorst, E.T., Schniter, P.: Regularization by denoising: clarifications and new interpretations. IEEE Trans. Comput. Imaging **5**(1), 52–67 (2019)
21. Romano, Y., Elad, M., Milanfar, P.: The little engine that could: regularization by denoising (red). SIAM J. Imaging Sci. **10**(4), 1804–1844 (2017)
22. Ryu, E., Liu, J., Wang, S., Chen, X., Wang, Z., Yin, W.: Plug-and-play methods provably converge with properly trained denoisers. In: Proceedings of the 36th International Conference on Machine Learning, Long Beach, California, USA, 09–15 June 2019, vol. 97, pp. 5546–5557. PMLR (2019)
23. Sage, D., et al.: Quantitative evaluation of software packages for single-molecule localization microscopy. Nat. Methods **12**, 717–724 (2015). https://doi.org/10.1038/nmeth.3442
24. Sage, D., et al.: Super-resolution fight club: assessment of 2D & 3D single-molecule localization microscopy software. Nat. Methods **16**, 387–395 (2019)
25. Solomon, O., Eldar, Y.C., Mutzafi, M., Segev, M.: SPARCOM: sparsity-based super-resolution correlation microscopy. SIAM J. Imaging Sci. **12**(1), 392–419 (2019)
26. Soubies, E., Blanc-Féraud, L., Aubert, G.: A continuous exact ℓ^0 penalty (CEL0) for least squares regularized problem. SIAM J. Imaging Sci. **8**(3), 1607–1639 (2015)
27. Stergiopoulou, V., Calatroni, L., de Morais Goulart, H., Schaub, S., Blanc-Féraud, L.: COL0RME: super-resolution microscopy based on sparse blinking/fluctuating fluorophore localization and intensity estimation. Biol. Imaging **2** (2022)
28. Stergiopoulou, V., Calatroni, L., Schaub, S., Blanc-Féraud, L.: 3D image super-resolution by fluorophore fluctuations and MA-TIRF microscopy reconstruction (3D-COL0RME). In: 2022 IEEE 19th International Symposium on Biomedical Imaging (ISBI), pp. 1–4 (2022)
29. Stergiopoulou, V., de M. Goulart, J.H., Schaub, S., Calatroni, L., Blanc-Féraud, L.: COL0RME: covariance-based ℓ_0 super-resolution microscopy with intensity estimation. In: 2021 IEEE 18th International Symposium on Biomedical Imaging (ISBI), pp. 349–352 (2021)
30. Terris, M., Repetti, A., Pesquet, J.C., Wiaux, Y.: Building firmly nonexpansive convolutional neural networks. In: ICASSP 2020-2020 IEEE International Conference on Acoustics, Speech and Signal Processing, pp. 8658–8662 (2020)
31. Venkatakrishnan, S.V., Bouman, C.A., Wohlberg, B.: Plug-and-play priors for model based reconstruction. In: 2013 IEEE Global Conference on Signal and Information Processing, pp. 945–948 (2013)
32. Xu, X., Liu, J., Sun, Y., Wohlberg, B., Kamilov, U.S.: Boosting the performance of plug-and-play priors via denoiser scaling. In: 2020 54th Asilomar Conference on Signals, Systems, and Computers, pp. 1305–1312. IEEE (2020)
33. Zhang, K., Li, Y., Zuo, W., Zhang, L., Van Gool, L., Timofte, R.: Plug-and-play image restoration with deep denoiser prior. IEEE Trans. Pattern Anal. Mach. Intell. **44**(10), 6360–6376 (2021)
34. Zhang, K., Zuo, W., Chen, Y., Meng, D., Zhang, L.: Beyond a Gaussian denoiser: residual learning of deep CNN for image denoising. IEEE Trans. Image Process. **26**(7), 3142–3155 (2017)

Segmenting MR Images Through Texture Extraction and Multiplicative Components Optimization

Laura Antonelli[1](✉), Valentina De Simone[2], and Marco Viola[3]

[1] Consiglio Nazionale delle Ricerche, Napoli, Italy
laura.antonelli@icar.cnr.it
[2] Università degli Studi della Campania "Luigi Vanvitelli", Caserta, Italy
valentina.desimone@unicampania.it
[3] University College Dublin, Dublin, Ireland
marco.viola@ucd.ie

Abstract. The segmentation of MRI is a challenging task due to artifacts introduced by the acquisition process, like bias field and noise. In this paper, using a cartoon-texture decomposition of the image, we present a strategy that segments the cartoon processed by simultaneous bias correction and denoising. Preliminary numerical tests show that our method is effective in segmenting MRI data corrupted by noise.

Keywords: MRI segmentation · cartoon-texture decomposition · Kullback-Leiber divergence · ADMM

1 Introduction

Magnetic Resonance Imaging (MRI) has been used for decades in medical diagnostic practice to analyze anatomical or functional information of human organs [14]. MRI technology is continuously improving; today, large and high-quality MRI scans point out stroke, traumatic injury, tumors, or malfunction by non-invasive and fast acquisitions. MRI combined with different diagnostic imaging data (computerized tomography, nuclear tomography, etc.) [15,25] or with omics data [4] may help to understand complex disease patterns better. Hence, ad-hoc numerical methods are required to improve accuracy and efficiency throughout the diagnosis process and analysis [21]. MRI segmentation task is essential in many clinical procedures since it influences the outcome of the entire analysis. The segmentation of brain MRI typically detects three regions of interest (ROIs): white matter (WM), gray matter (GM), and cerebrospinal fluid (CSF), as shown

This work was partially supported by Istituto Nazionale di Alta Matematica - Gruppo Nazionale per il Calcolo Scientifico (INdAM-GNCS, ICAR-CNR INdAM Research Unit) and by the Italian Ministry of University and Research under grant no. PON03PE_00060_5.

L. Calatroni et al. (Eds.): SSVM 2023, LNCS 14009, pp. 511–521, 2023.
https://doi.org/10.1007/978-3-031-31975-4_39

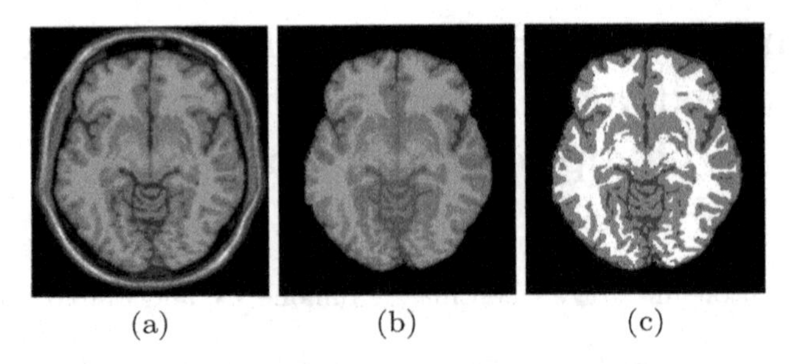

Fig. 1. (a) `brainweb64` original image of 3d T1-weighted MRI (slice n. 64). (b) Pre-processed `brainweb64` to be segmented. (c) The ground truth of the three ROIs: white matter (*white*), gray matter (*grey*), and cerebral fluid spinal (*dark grey*). Original images and ground truth are available from BrainWeb database. .

in Fig. 1. The segmentation result is a colorful map of labels identifying homogeneous regions or a set of contours drawing region boundaries. Such segmentation is essential for analyzing anatomical structure changes, lesion delineation, and disease identification. However, the segmentation process is nowadays challenging since MRI data are significantly affected by intensity inhomogeneity (referred to as *bias field*) and noise.

A bias field arises from the interaction between the magnetic field and the human body, introducing a spatially smooth artifact in intensity variation within the tissue of the same physical or biological properties. In clinical practice, typical magnetic field strength values for MR scanners are quantified from 1.5 and 3.0 Tesla (T), while research facilities use magnetic fields up to 11.7 T. In general, the higher is the magnetic field's magnitude, the more the bias field's smooth variation is affected. Consequently, well-known region-based methods based on a region uniformity assumption in intensity, color, or texture [2,9,16,20], may fail on MRI data, where the presence of MRI artifacts can often violate the uniformity model in intensity distribution. Early publications dealing with intensity inhomogeneity correction and noise go back to the '80s; since then, the literature has been enriched with plenty of methods able to segment MRI data affected by these MRI artifacts (see for example [1,13,17,23,24], and references therein).

Different models of MR image formation have been proposed in the literature, depending on how the inhomogeneity-free image and bias field interact [26]. Most frequently, the models assume that the inhomogeneity is multiplicative; two of the most common methods are the locally intensity clustering (LIC) [19] and the multiplicative intrinsic component optimization (MICO) model [18], in which the image intensity is decomposed into two multiplicative components, namely, the true image (piece-wise constant) and the bias field (smoothly varying). In addition to intensity inhomogeneity, the MR image contains a significant amount of noise which can be, in general, represented by means of a Rician distribution (based on the assumption that the noise on the real and imaginary channels is Gaussian). The presence of noise often requires additional preprocessing and/or postprocessing steps to improve the segmentation results. However, using linear or nonlinear denoising filters can degrade the details and the edges or the fine structure of the image, respectively [5].

In this work, we develop a MICO-based model that, using a cartoon-texture decomposition of the image, produces a multiplicative bias correction of the initial cartoon and progressively subtracts the remaining textural components, including the noise. The resulting corrected and denoised image can be then segmented by means of any standard segmentation method. Preliminary numerical test shows that our method, combined with the well-established K-means clustering criterion, is highly effective in segmenting MRI data corrupted by noise.

2 A CT Evolution Algorithm for MR Image Denoising and Bias Correction

In the multiplicative intrinsic components framework [18] an MR image, seen as a function $I : \Omega \subset \mathbb{R}^2 \to [0, \infty)$, is modeled as

$$I(x) = b(x)J(x) + n(x) \quad (x \in \Omega), \tag{1}$$

where $J(x)$ represents the true image, $b(x)$ the bias field, and $n(x)$ represents noise. By making the assumption that the original image is composed by N different tissues, coinciding with a region of the image characterized by a uniform color c_i, the original image can be seen as a piecewise constant function, i.e., one can write $J(x) = \sum_{i=1}^{N} c_i \cdot u_i(x)$ with $u_i(x) : \Omega \to \{0, 1\}$. As regards the bias field, which is in general considered as a smooth function, one can approximate it by linear combinations of smooth functions in a given basis $\{g_1(x), \ldots, g_M(x)\}$ (consisting, e.g., of polynomials), i.e., $b(x) = \sum_{j=1}^{M} w_j \cdot g_j(x)$. By exploiting the discrete MICO energy formulation (see eq. (6) in [18]) one can determine the components of this decomposition by solving the minimization problem

$$\begin{aligned} \min_{u,c,w} \;\; & F(u, c, w) = \sum_{p=1}^{P} \frac{1}{2} \left(I_p - w^\top G_p c^\top u_p\right)^2 \\ \text{s.t.} \;\; & 0 \leq (u_p)_i \leq 1, \quad \forall i = 1, \ldots, N, \; \forall p = 1, \ldots, P \\ & \sum_{i=1}^{N} (u_p)_i = 1, \quad \forall p = 1, \ldots, P, \end{aligned} \tag{2}$$

where P is the number of pixels of the original image I. In the model the bias field is represented pixel-wise by the term $w^\top G_p$, i.e., by a linear combination with coefficients $w \in \mathbb{R}^M$ of M vectors $G^1, \ldots, G^M \in \mathbb{R}^P$ consisting of the pixel-wise evaluations of the M polynomials $\{g_1(x), \ldots, g_M(x)\}$. For each pixel, the vector $u_p \in \mathbb{R}^N$ reflects the membership of the pixel to one of the uniform regions. Though, ideally, the membership should be a binary vector (as seen previously), here we consider "fuzzy" membership functions, i.e., for each pixel the vector $u_p \in \mathbb{R}^N$ has elements in $(0, 1)$ which sum to 1.

To properly handle the presence of noise in the image generation we propose to modify model (2) by following similar ideas to the ones introduced in [3]. Suppose a "rough" cartoon-texture decomposition of the original image I is available, i.e., suppose one can write

$$I = \bar{I} + \bar{v},$$

with $\bar{I}$ containing the cartoon part and $\bar{v}$ having information on the oscillating components of the original image (i.e., noise and texture). We define the MICO-CTD model (where CTD stands for Cartoon-Texture Decomposition) as

$$\begin{aligned} \min_{u,c,w,v} \quad & F(u,c,w) + \mu D_{KL}(v;\bar{v}) \\ \text{s.t.} \quad & 0 \leq (u_p)_i \leq 1, \quad \forall i = 1,\dots,N, \; \forall p = 1,\dots,P \\ & \sum_{i=1}^{N} (u_p)_i = 1, \quad \forall p = 1,\dots,P, \\ & w^\top G_p\, c^\top u_p + v_p = \bar{I}_p, \quad \forall p = 1,\dots,P, \end{aligned} \tag{3}$$

where μ is a given positive weighting parameter. The MICO-CTD model aims to construct the multiplicative components of the cartoon $\bar{I}$ while subtracting from it residual oscillatory components, namely v. By looking at the oscillatory components as a probability distribution, the latter task is performed by forcing similarity between v and $\bar{v}$ penalizing the Kullback-Leibler divergence

$$D_{KL}(v;\bar{v}) = \sum_p v_p \log\left(\frac{v_p}{\bar{v}_p}\right).$$

Although D_{KL} is not a distance it has good statistical properties and it has the nice property that v will be highly penalized to have large values where $\bar{v}$ is very small, thus forcing the two distributions to have similar support. As we will see later, the KL divergence also has beneficial computational properties, since its proximal operator is available in closed form.

By observing that, by the definition of F in (2) the quantity in the parentheses amounts to v_p, and that the first two constraints can be replaced by the requirement that each u_p lies in the standard simplex Δ in $\mathbb{R}^3$, we can rewrite (3) as

$$\begin{aligned} \min_{u,c,w,v} \quad & \chi_\Delta(u) + \frac{1}{2}\|v\|^2 + \mu D_{KL}(v;\bar{v}) \\ \text{s.t.} \quad & w^\top G_p\, c^\top u_p + v_p = \bar{I}_p, \quad \forall p = 1,\dots,P, \end{aligned} \tag{4}$$

where, with a slight abuse of notation, we denoted $\chi_\Delta(u) \equiv \sum_{p=1}^{P} \chi_\Delta(u_p)$, and χ_Δ is the indicator function of the standard 3D simplex. Problem (4) is a nonlinear optimization problem, characterized by a convex lower-semicontinuous objective function and subject to nonlinear constraints that, however, are linear with respect to each of the unknowns.

To find a solution to the optimization problem (4) we propose the use of a specialized version of the Alternate Directions Method of Multipliers (ADMM)

[7]. Given $\rho > 0$, and letting $\zeta \in \mathbb{R}^P$ representing a vector of Lagrange multipliers we can define the augmented Lagrangian of (4) as the function

$$\begin{aligned} \mathcal{L}_A(u, c, w, v, \zeta; \rho) = \quad & \chi_\Delta(u) + \frac{1}{2}\|v\|^2 + \mu D_{KL}(v; \bar{v}) + \\ & + \sum_{p=1}^{P} \left[\zeta_p \left(w^\top G_p c^\top u_p + v_p - \bar{I}_p \right) + \right. \\ & \left. + \frac{\rho}{2} \left\| w^\top G_p c^\top u_p + v_p - \bar{I}_p \right\|^2 \right]. \end{aligned} \tag{5}$$

By exploiting (5), one can define an ADMM scheme as follows: starting from an initialization $(u^0, c^0, w^0, v^0, \zeta^0)$ we propose to build the sequence $\{u^k, c^k, w^k, v^k, \zeta^k\}$ by

$$\begin{aligned} u^{k+1} &= \underset{u}{\operatorname{argmin}} \ \mathcal{L}_A(u, c^k, w^k, v^k, \zeta^k; \rho), \\ c^{k+1} &= \underset{c}{\operatorname{argmin}} \ \mathcal{L}_A(u^{k+1}, c, w^k, v^k, \zeta^k; \rho), \\ w^{k+1} &= \underset{w}{\operatorname{argmin}} \ \mathcal{L}_A(u^{k+1}, c^{k+1}, w, v^{k+1}, \zeta^k; \rho), \\ v^{k+1} &= \underset{v}{\operatorname{argmin}} \ \mathcal{L}_A(u^{k+1}, c^{k+1}, w^{k+1}, v, \zeta^k; \rho), \\ \zeta_p^{k+1} &= \zeta^k + \rho \left[(w^{k+1})^\top G_p \, (c^{k+1})^\top u_p^{k+1} + v_p^{k+1} - \bar{I}_p \right]. \end{aligned} \tag{6}$$

By introducing scaled Lagrange multipliers $\lambda_p^k = \frac{1}{\rho}\zeta_p^k$, one can rewrite the ADMM iterations as follows

$$u^{k+1} = \underset{u}{\operatorname{argmin}} \ \chi_\Delta(u) + \sum_{p=1}^{P} \frac{\rho}{2} \left\| (w^k)^\top G_p \, (c^k)^\top u_p + v_p^k - \bar{I}_p + \lambda_p^k \right\|^2, \tag{7}$$

$$c^{k+1} = \underset{c}{\operatorname{argmin}} \ \left\| A^k c + v^k - \bar{I} + \lambda^k \right\|^2, \tag{8}$$

$$w^{k+1} = \underset{w}{\operatorname{argmin}} \ \left\| B^k w + v^k - \bar{I} + \lambda^k \right\|^2, \tag{9}$$

$$v^{k+1} = \underset{v}{\operatorname{argmin}} \frac{1}{2}\|v\|^2 + \mu D_{KL}(v; \bar{v}) + \frac{\rho}{2} \left\| z^k + v \right\|^2, \tag{10}$$

$$\lambda_p^{k+1} = \lambda^k + (w^{k+1})^\top G_p \, (c^{k+1})^\top u_p^{k+1} + v_p^{k+1} - \bar{I}_p, \tag{11}$$

where

- $A^k \in \mathbb{R}^{P \times N}$ has rows $A_p^k = (w^k)^\top G_p (u_p^{k+1})^\top$,
- $B^k \in \mathbb{R}^{P \times M}$ has rows $B_p^k = (c^{k+1})^\top u_p^{k+1} (G_p)^\top$,
- $z^k \in \mathbb{R}^P$ has elements $z_p^k = (w^{k+1})^\top G_p \, (c^{k+1})^\top u_p^{k+1} - \bar{I} + \lambda^k$.

2.1 Implementation Details

We here analyze the solution of the four subproblems in (7)–(10).

Subproblem in u. Although the problem in (7) is separable into P independent problems of size N of the form

$$\min_{u \in \Delta} \left\| a^\top u + b \right\|^2 ,$$

they can easily be solved all at once as a single quadratic programming problem of size $N \times P$ subject to nonnegativity and linear equality constraints. In detail, we propose to solve the problem by the IP-PMM method[1] proposed in [12,22].

Subproblems in c and w. The update of c and w consists in the solution of small-scale least squares problems (usually $c = 3$ and $w = 10 \sim 20$). The two updates can be performed in a cheap and fast way by exact solution of the normal equations systems.

Subproblem in v. The update of v can be performed in closed form by observing that, by following [6, Theorem 6.1], the update can be rewritten as

$$v^{k+1} = \operatorname*{argmin}_{v} D_{KL}(v; \bar{v}) + \frac{1+\rho}{2\mu} \left\| \frac{\rho}{1+\rho} z^k + v \right\|^2 . \tag{12}$$

This allows one to update v by pixel-wise application of the Lambert W function, which corresponds to the proximal operator of the KL divergence.

3 Results on Brain MRI Segmentation

Our method was tested on several slices of brain T1-weighted MRIs coming from the BrainWeb dataset [10], containing a set of realistic MRI data volumes noise-free and corrupted by Rician distribution [11]. Here we show the results on two axial slices of a brain MRI, corrupted by two different percentages of Rician noise as displayed in Fig. 2. The proposed method was implemented in Matlab using the Image Processing toolbox to perform image operations. Moreover, the CTD was performed applying one iteration of the algorithm in [8]; further details can be found in [3]. To identify the three ROIs of the brain in MRIs, in all the experiments, we set $N = 3$. Dealing with a nonconvex problem, the choice of the starting point for the proposed ADMM scheme affects the quality of the solution. We decided to initialize the constants c_i for the three regions to 0.33, 0.66, 0.99, since the pixel intensities in all test images are normalized in $[0, 1]$. We then initialized each $u_p \in \mathbb{R}^3$ as

$$u_p = \begin{cases} (1,0,0), & \text{if } \bar{I}_p \in [0, 0.33], \\ (0,1,0), & \text{if } \bar{I}_p \in (0.33, 0.66], \\ (0,0,1), & \text{if } \bar{I}_p \in (0.66, 1]. \end{cases}$$

We set $M = 10$ and chose the vectors $G^1, \ldots, G^M \in \mathbb{R}^P$ to represent third-order Legendre polynomials. The initial value of w is set such that $w^\top G_p =$

[1] https://github.com/spougkakiotis/IP_PMM.

1 for each pixel p, and v is initialized to 0. We set $\mu = 10^{-2}$, $\rho = 10$ and empirically chose to run ADMM for 30 iterations. Figures 3 and 5 display the starting values for the product $c^\top u$, the bias field, and the extracted noise v, and their evolution after 10, 20, and 30 iterations of the ADMM scheme described in Sect. 2.1. As the ADMM iterations proceed, we note that the bias field is progressively reconstructed, enabling the correction of the cartoon. At each step, v is enforced to be close to $\bar{v}$, and the images show how the noise is progressively subtracted from the correct cartoon. Note that as the bias correction improves, the information in v increases, resulting in an improved denoising effect of the process. Figures 4 and 6 show the segmentation produced by our method (last column) compared with the segmentations produced by the K-means clustering algorithm applied to the noisy images (second column) and the cartoons (third column) respectively. Segmentations were also compared to MICO results on the noisy images (fourth column) and the cartoons (fifth column). As we can see in Fig. 4(b) and Fig. 4(d), K-means and MICO are sensitive to noise and other MRI artifacts producing the worst accuracy of the segmentation results. The two methods worked better on the cartoon as shown in Fig. 4(c) and 4(d) but the three ROI segmentation reveals several defects as the selected zoomed area pointed out. The same remarks can be made on the segmentation results of Fig. 6. We can conclude that our algorithm can efficiently correct the bias on the original image and extract the noise, resulting in a more accurate partitioning between the three brain ROIs.

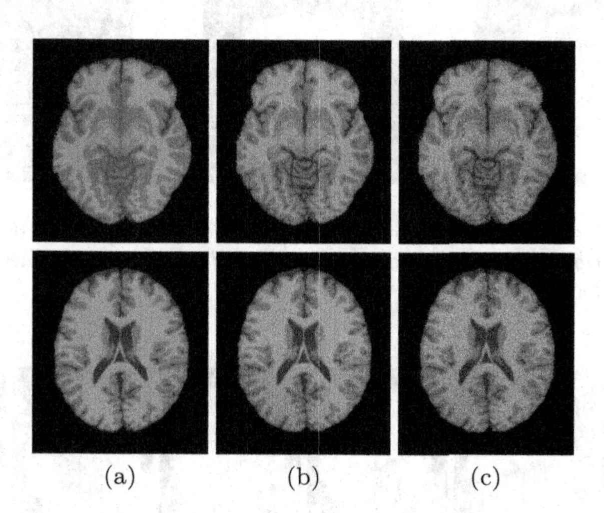

Fig. 2. (a) `brainweb64` (*top*) and `brainweb91` (*down*): original vs increased noise. Images were corrupted by Rician distribution at different percentages calculated with respect to the brightest tissue. (a) original image slices n. 64 and n. 91 from BRAINWEB. (b) Images exhibit 3% of Rician noise. (c) Images exhibit 7% of Rician noise.

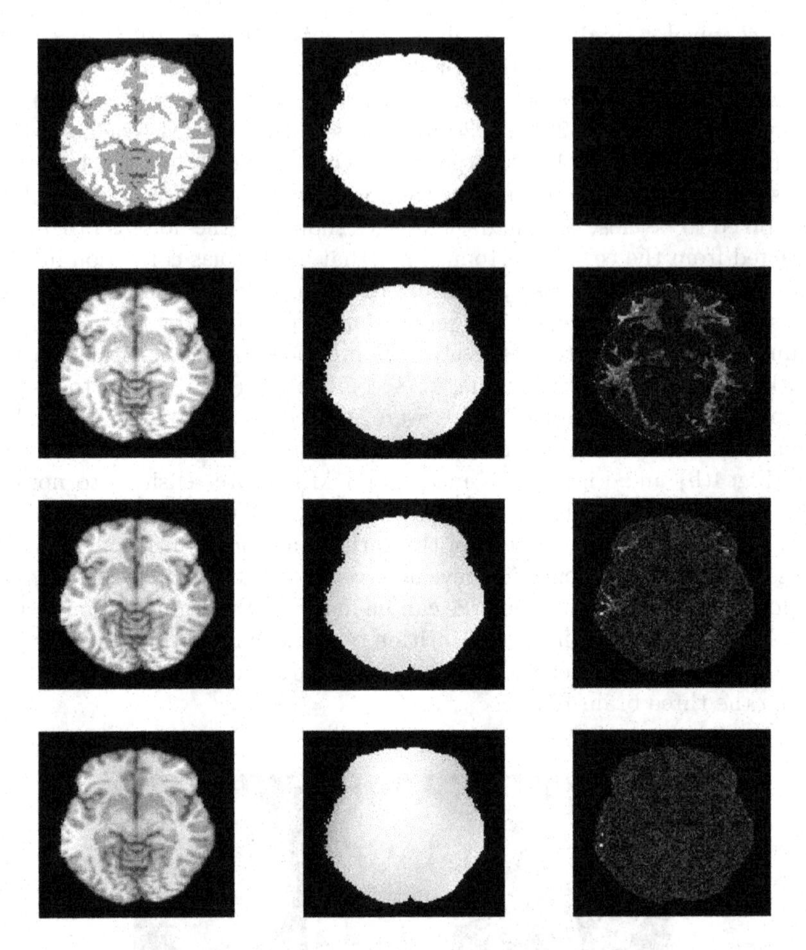

Fig. 3. History of convergence for the image `brainweb64` corrupted by 3% of Rician noise. *Left to right*: $c^\top u$, bias field, and v. *top to down*: starting values, iteration 10, iteration 20, final iteration.

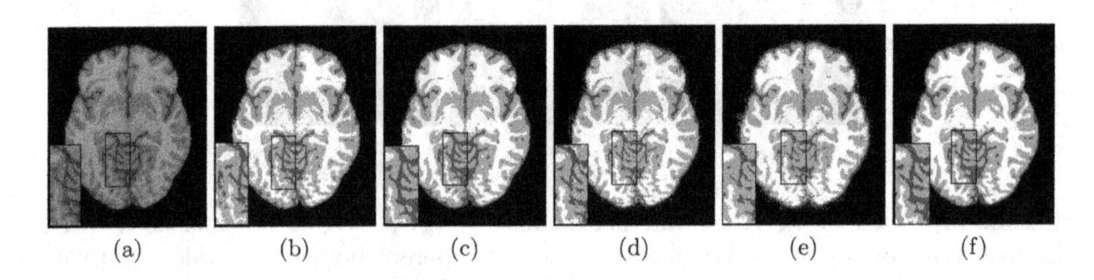

Fig. 4. Segmentation results obtained on the image `brainweb64` corrupted by 3% of Rician noise. *Left to right*: (a) noisy image, (b) K-means results, (c) K-means results on the cartoon, (d) MICO results, (e) MICO results on the cartoon, (f) our method results.

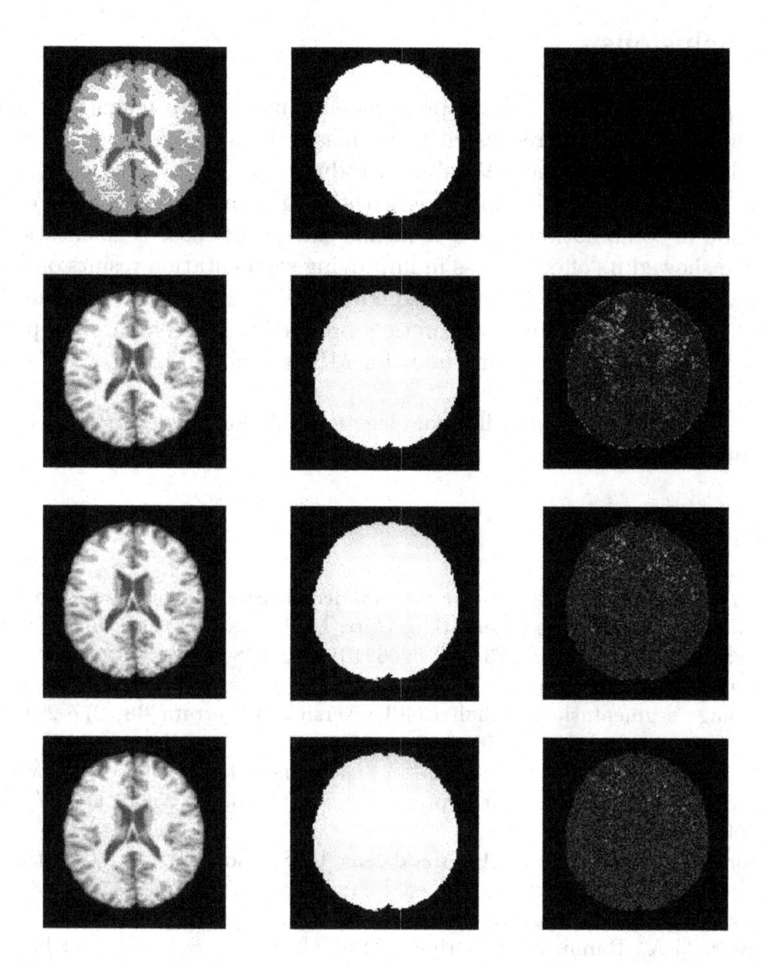

Fig. 5. History of convergence for the image `brainweb91` corrupted by 7% of Rician noise. Left to right: $c^\top u$, bias field, and v. *top to down*: starting values, iteration 10, iteration 20, final iteration.

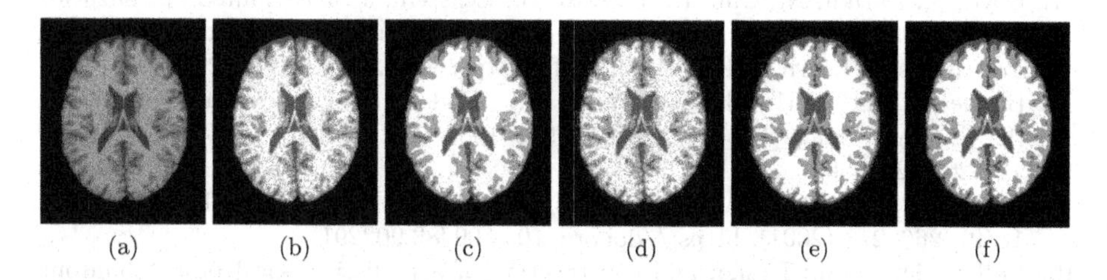

(a) (b) (c) (d) (e) (f)

Fig. 6. Segmentation results obtained on the image `brainweb91` corrupted by 7% of Rician noise. *Left to right*: (a) noisy image, (b) K-means results, (c) K-means results on the cartoon, (d) MICO results, (e) MICO results on the cartoon, (f) our method results.

4 Conclusions

We here presented MICO-CTD, a preprocessing method for magnetic resonance images which is able to reconstruct the bias field and, at the same time, to denoise the original image. We also introduced an ADMM scheme to minimize the nonlinear optimization problem arising from MICO-CTD. We tested our method in combination with a K-means-clustering-based segmentation algorithm. We showed its effectiveness in improving segmentation results on realistic MRI images corrupted by Rician noise. Future work will deal with the study of the proposed ADMM scheme's theoretical properties and compare the proposed strategy with state-of-the-art methods for MRI segmentation.

Acknowledgments. L. Antonelli would like to thank Simona Sada (ICAR-CNR) for her technical support.

References

1. Angulakshmi, M., Deepa, M.: A review on deep learning architecture and methods for MRI brain tumour segmentation. Curr. Med. Imaging **17**(6), 695–706 (2021). https://doi.org/10.2174/1573405616666210108122048
2. Antonelli, L., De Simone, V., di Serafino, D.: A view of computational models for image segmentation. Annali Dell'Università Di Ferrara **68**, 277–294 (2022). https://doi.org/10.1007/s11565-022-00417-6
3. Antonelli, L., De Simone, V., Viola, M.: Cartoon-texture evolution for two-region image segmentation. Comput. Optim. Appl. **84**(1), 5–26 (2023). https://doi.org/10.1007/s10589-022-00387-7
4. Antonelli, L., Guarracino, M.R., Maddalena, L., Sangiovanni, M.: Integrating imaging and omics data: a review. Biomed. Signal Process. Control **52**, 264–280 (2019). https://doi.org/10.1016/j.bspc.2019.04.032
5. Balafar, M.A., Ramli, A.R., Saripan, M.I., Mashohor, S.: Review of brain MRI image segmentation methods. Artif. Intell. Rev. **33**, 261–274 (2010). https://doi.org/10.1007/s10462-010-9155-0
6. Beck, A.: First-Order Methods in Optimization. Society for Industrial and Applied Mathematics, Philadelphia (2017). https://doi.org/10.1137/1.9781611974997
7. Boyd, S., Parikh, N., Chu, E., Peleato, B., Eckstein, J.: Distributed optimization and statistical learning via the alternating direction method of multipliers. Found. Trends Mach. Learn. **3**(1), 1–122 (2011). https://doi.org/10.1561/2200000016
8. Buades, A., Le, T.M., Morel, J., Vese, L.A.: Fast cartoon + texture image filters. IEEE Trans. Image Process. **19**(8), 1978–1986 (2010). https://doi.org/10.1109/TIP.2010.2046605
9. Chan, T.F., Vese, L.A.: Active contours without edges. IEEE Trans. Image Process. **10**(2), 266–277 (2001). https://doi.org/10.1109/83.902291
10. Collins, D., et al.: Design and construction of a realistic digital brain phantom. IEEE Trans. Med. Imaging **17**(3), 463–468 (1998). https://doi.org/10.1109/42.712135. https://brainweb.bic.mni.mcgill.ca/
11. Collins, D., et al.: Design and construction of a realistic digital brain phantom. IEEE Trans. Med. Imaging **17**(3), 463–468 (1998). https://doi.org/10.1109/42.712135

12. De Simone, V., di Serafino, D., Gondzio, J., Pougkakiotis, S., Viola, M.: Sparse approximations with interior point methods. SIAM Rev. **64**(4), 954–988 (2022). https://doi.org/10.1137/21M1401103
13. Dehdasht-Heydari, R., Gholami, S.: Automatic seeded region growing (ASRG) using genetic algorithm for brain MRI segmentation. Wireless Pers. Commun. **109**(2), 897–908 (2019). https://doi.org/10.1007/s11277-019-06596-4
14. Grover, V.P., Tognarelli, J.M., Crossey, M.M., Cox, I.J., Taylor-Robinson, S.D., McPhail, M.J.: Magnetic resonance imaging: principles and techniques: lessons for clinicians. J. Clin. Exp. Hepatol. **5**, 246–55 (2015). https://doi.org/10.1016/j.jceh.2015.08.001
15. Hao, X., et al.: Multimodal magnetic resonance imaging: the coordinated use of multiple, mutually informative probes to understand brain structure and function. Hum. Brain Mapp. **34**(2), 253–71 (2013). https://doi.org/10.1002/hbm.21440
16. Houhou, N., Thiran, J.P., Bresson, X.: Fast texture segmentation based on semi-local region descriptor and active contour. Numer. Math. Theory Methods Appl. **2**(4), 445–468 (2009). https://doi.org/10.4208/nmtma.2009.m9007s
17. Ji, Z., Liu, J., Cao, G., Sun, Q., Chen, Q.: Robust spatially constrained fuzzy C-means algorithm for brain MR image segmentation. Pattern Recogn. **47**(7), 2454–2466 (2014). https://doi.org/10.1016/j.patcog.2014.01.017
18. Li, C., Gore, J.C., Davatzikos, C.: Multiplicative intrinsic component optimization (MICO) for MRI bias field estimation and tissue segmentation. Magn. Reson. Imaging **32**(7), 913–923 (2014). https://doi.org/10.1016/j.mri.2014.03.010
19. Li, C., Huang, R., Ding, Z., Gatenby, J.C., Metaxas, D.N., Gore, J.C.: A level set method for image segmentation in the presence of intensity inhomogeneities with application to MRI. IEEE Trans. Image Process. **20**(7), 2007–2016 (2011). https://doi.org/10.1109/TIP.2011.2146190
20. Li, F., Ng, M.K., Zeng, T.Y., Shen, C.: A multiphase image segmentation method based on fuzzy region competition. SIAM J. Imag. Sci. **3**(3), 277–299 (2010). https://doi.org/10.1137/080736752
21. Luque-Baena, R.M., Despotović, I., Goossens, B., Philips, W.: MRI segmentation of the human brain: challenges, methods, and applications. Comput. Math. Methods Med. (2015). https://doi.org/10.1155/2015/450341
22. Pougkakiotis, S., Gondzio, J.: An interior point-proximal method of multipliers for convex quadratic programming. Comput. Optim. Appl. **78**(2), 307–351 (2021). https://doi.org/10.1007/s10589-020-00240-9
23. Song, J., Zhang, Z.: Magnetic resonance imaging segmentation via weighted level set model based on local kernel metric and spatial constraint. Entropy **23**(9), 1196 (2021). https://doi.org/10.3390/e23091196
24. Tarkhaneh, O., Shen, H.: An adaptive differential evolution algorithm to optimal multi-level thresholding for MRI brain image segmentation. Expert Syst. Appl. **138**, 1–18 (2019). https://doi.org/10.1016/j.eswa.2019.07.037
25. Vandenberghe, S., Marsden, P.K.: PET-MRI: a review of challenges and solutions in the development of integrated multimodality imaging. Phys. Med. Biol. **60**(4), R115 (2015). https://doi.org/10.1088/0031-9155/60/4/R115
26. Vovk, U., Pernus, F., Likar, B.: A review of methods for correction of intensity inhomogeneity in MRI. IEEE Trans. Med. Imaging **26**(3), 405–421 (2007). https://doi.org/10.1109/TMI.2006.891486

Scale Space, PDEs, Flow, Motion and Registration

Geodesic Tracking of Retinal Vascular Trees with Optical and TV-Flow Enhancement in SE(2)

Nicky J. van den Berg[1(✉)], Shuhe Zhang[2], Bart M. N. Smets[1], Tos T. J. M. Berendschot[2], and Remco Duits[1]

[1] CASA, Department of Mathematics and Computer Science, Eindhoven University of Technology, Eindhoven, The Netherlands
{n.j.v.d.berg,b.m.n.smets,r.duits}@tue.nl

[2] Ophthalmology Department, Maastricht University, Maastricht, The Netherlands
{shuhe.zhang,t.berendschot}@maastrichtuniversity.nl

Abstract. Retinal images are often used to examine the vascular system in a non-invasive way. Studying the behavior of the vasculature on the retina allows for noninvasive diagnosis of several diseases as these vessels and their behavior are representative of the behavior of vessels throughout the human body. For early diagnosis and analysis of diseases, it is important to compare and analyze the complex vasculature in retinal images automatically. In previous work, PDE-based geometric tracking and PDE-based enhancements in the homogeneous space of positions and orientations have been studied and turned out to be useful when dealing with complex structures (crossing of blood vessels in particular).

In this article, we propose a single new, more effective, Finsler function that integrates the strength of these two PDE-based approaches and additionally accounts for a number of optical effects (dehazing and illumination in particular). The results greatly improve both the previous left-invariant models and a recent data-driven model, when applied to real clinical and highly challenging images. Moreover, we show clear advantages of each module in our new single Finsler geometrical method.

Keywords: Geodesic Tracking · Optical Image Enhancement · TV-Flow Enhancement · Vascular Tree Tracking · Finsler Geometry

1 Introduction

The retina allows for noninvasive examination of the vascular system since the vessels in the eye, and their corresponding behavior, are representative of the behavior of vessels throughout the rest of the body. Therefore, studying the behavior of the vasculature on the retina allows for noninvasive diagnosis of several diseases, like diabetes, hypertension, and Alzheimer's disease [6,16,19]. Automatic vessel tracking algorithms help the efficient diagnosis of these diseases. Here, we rely on geodesic tracking methods which calculate the shortest

L. Calatroni et al. (Eds.): SSVM 2023, LNCS 14009, pp. 525–537, 2023.
https://doi.org/10.1007/978-3-031-31975-4_40

path connecting two points on the same blood vessel, following the biological structure. We will show that the single geodesic model will also allow for acceptable tracking of full vascular trees on realistic retinal images of limited quality.

Diseases such as cataract disease give rise to cloudy retinal images [20], while camera movements lead to motion artifacts [14] and uneven illumination [24]. This affects the clarity and visibility of the vasculature in the images we want to track. To cope with the limitations in the quality of ophthalmology images in practice, we must integrate both contrast enhancement from optical image processing [22] *and* crossing-preserving contextual TV-flows, in our correct geodesic tracking of the vasculature, as we will show.

Many approaches have been used when studying geodesic tracking methods on 2D images. However, in many methods, problems arise when dealing with difficult structures, like crossings and bifurcations, where standard geometric algorithms acting in the image domain $\mathbb{R}^2$ often take the wrong exit. Therefore one lifts the image to the homogeneous space of positions and orientations $\mathbb{M}_2$. In this lifted space, difficult structures are disentangled, cf. Fig. 2a. In previous works by various authors, it has been shown that PDE-based geometric tracking

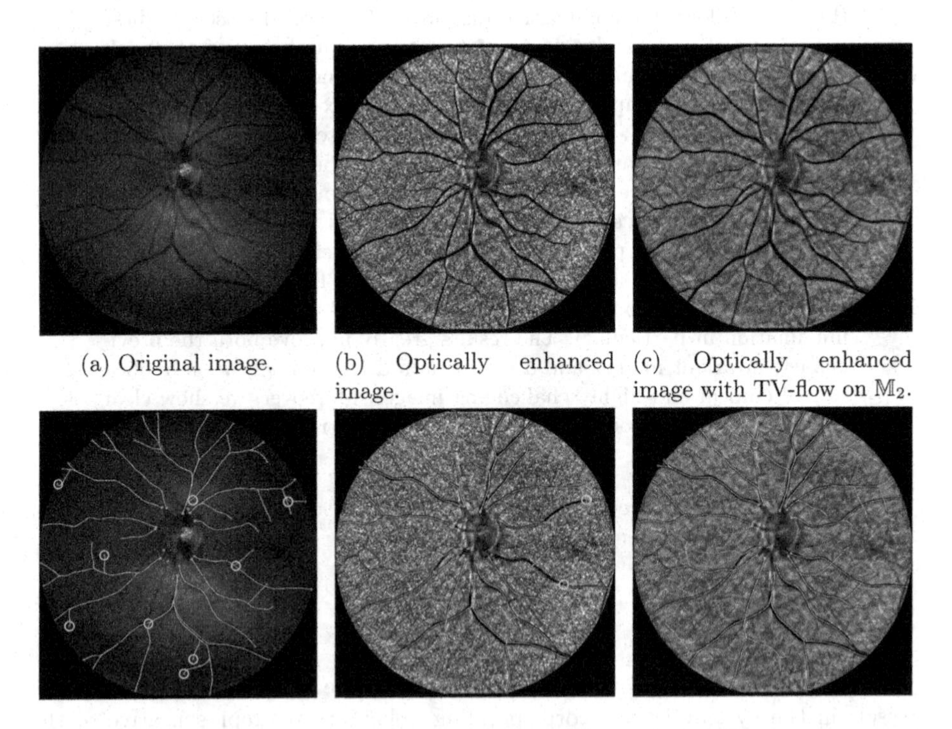

(a) Original image. (b) Optically enhanced image. (c) Optically enhanced image with TV-flow on $\mathbb{M}_2$.

Fig. 1. Geodesic tracking on the original image, (contrast-)enhanced image, and enhanced image after which TV-flow enhancement is done (left to right). The seeds and tips are indicated in resp. green and red. Yellow (/ red) circles indicate tracking mistakes that are (/ are *not*) fixed in the tracking of another column. (Color figure online)

algorithms [4,9,10,12] perform well in $\mathbb{M}_2$. Here, one first calculates a distance map which is based on the image data. After computing the distance map, the steepest descent algorithm is applied to find each shortest path from a tip (an endpoint) to the corresponding seed (a starting point).

In our tracking, we integrate PDE-enhancements, like crossing-preserving total variation flow (TV-flow) enhancement in $\mathbb{M}_2$ [18]. We will show this improves the results. Furthermore, optical enhancement [22] of limited-quality retinal images is required to keep equal contrast and intensity across the whole vasculature. This inevitably creates small noisy structures that are non-aligned with other structures in the data. Applying the TV-flow enhancement in $\mathbb{M}_2$ leads to crossing-preserving contextual denoising that preserves crossings, and line structures, and removes noisy non-aligned structures from the optical enhancement. Altogether the scheme results in a vascular tracking algorithm that provides better results as wavefronts follow the complex branching vasculature better than the approach in [4], see Fig. 1. Even a single geodesic front propagation (building the distance map initializing all seeds at the same time), where fronts follow the entire vasculature in one run produces good results, see Fig. 2.

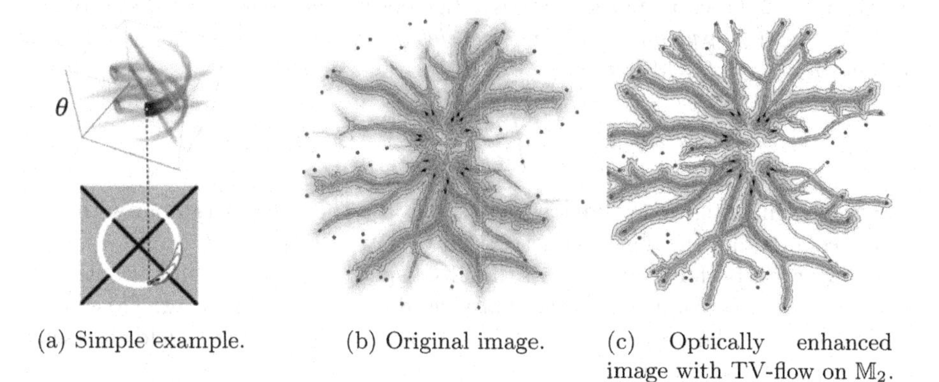

(a) Simple example. (b) Original image. (c) Optically enhanced image with TV-flow on $\mathbb{M}_2$.

Fig. 2. Distance map built in $\mathbb{M}_2$ (2a; top) and isocontours of the minimum projection over the orientations back onto $\mathbb{R}^2$ (2a; bottom). The isocontours of the minimum projection of the distance map are constructed based on the original image and optically enhanced image after which TV-flow enhancement is done (in Fig. 2b and 2c resp.).

The main contributions of this article are; **1)** the development of a new asymmetric, data-driven left-invariant Finsler geometric model that includes contextual contrast enhancement via TV-flows on $SE(2)$, **2)** experiments that show that application of this new Finsler geometric model reduces many tracking errors compared to previous left-invariant models [2,9] and the recent data-driven model [4], **3)** the new model performs very well on both realistic, unevenly illuminated retinal images and allows full vascular trees computations from a *single* distance map. The *inclusion* of the optical and TV-flow enhancements in the Finsler function no longer require a 2-step algorithm as in [4], but with the techniques in this work a single run of building the distance map suffices.

2 Lifted Space of Positions and Orientations $\mathbb{M}_2$

As explained in the introduction, it is beneficial to perform image processing (denoising, tracking) in the space of positions and orientations $\mathbb{M}_2$. This 3D space will offer us sufficient space to separate difficult (crossing) structures.

Definition 1 (Space of positions and orientations $\mathbb{M}_2$). *The space of two-dimensional positions and orientations $\mathbb{M}_2$ is defined as a smooth manifold $\mathbb{M}_2 := \mathbb{R}^2 \rtimes S^1$, where $S^1 \equiv \mathbb{R}/(2\pi\mathbb{Z}) \equiv SO(2)$ using the identification*

$$\mathbf{n} = (\cos\theta, \sin\theta) \longleftrightarrow \theta \longleftrightarrow R_\theta \in SO(2), \tag{1}$$

where R_θ is the counter-clockwise planar rotation over angle θ. Elements in $\mathbb{M}_2$ are denoted by $(\mathbf{x}, \theta) \in \mathbb{R}^2 \times \mathbb{R}/(2\pi\mathbb{Z})$. To stress the semidirect product of roto-translation group $SE(2) := \mathbb{R}^2 \rtimes SO(2)$ acting on $\mathbb{M}_2$, we write $\mathbb{M}_2 = \mathbb{R}^2 \rtimes S^1$.

We lift the image from $\mathbb{R}^2$ to $\mathbb{M}_2 \equiv SE(2)$ in order to separate crossing structures, using the orientation score transform.

Definition 2 (Orientation Score). *The orientation score transform $W_\psi f : \mathbb{L}_2(\mathbb{R}^2) \to \mathbb{L}_2(\mathbb{M}_2)$ using anisotropic wavelet ψ maps an image $f \in \mathbb{L}_2(\mathbb{R}^2)$ to an orientation score $U = W_\psi f$ and is given by*

$$(W_\psi f)(\mathbf{x}, \theta) := \int_{\mathbb{R}^2} \overline{\psi(R_\theta^{-1}(\mathbf{y} - \mathbf{x}))}\, f(\mathbf{y})\, \mathrm{d}\mathbf{y}.$$

In our experiments, we use cake wavelets [8,18] as they do not tamper with data evidence and allow for fast reconstruction by integration over θ.

In order to perform tracking in the lifted space of positions and orientations $\mathbb{M}_2$, we need to introduce a metric that is used to describe distances. This metric needs to satisfy the property that a roto-translation of the input yields the same roto-translation on the output.

Definition 3 (Left-Invariant Metric). *A metric tensor field $\mathcal{G}$ on $\mathbb{M}_2$ is called left invariant if*

$$\mathcal{G}_{g\cdot\mathbf{p}}\left((L_g)_*\dot{\mathbf{p}}, (L_g)_*\dot{\mathbf{p}}\right) = \mathcal{G}_{\mathbf{p}}\left(\dot{\mathbf{p}}, \dot{\mathbf{p}}\right),$$

for all $\mathbf{p} \in \mathbb{M}_2$, all $\dot{\mathbf{p}} \in T_{\mathbf{p}}(\mathbb{M}_2)$, the tangent space to $\mathbb{M}_2$ at point $\mathbf{p}$, and for all $g \in SE(2)$, where $L_g(\mathbf{p}) = g \cdot \mathbf{p} = (\mathbf{y}, R_\alpha) \cdot (\mathbf{x}, \theta) = (\mathbf{y} + R_\alpha \mathbf{x}, \alpha + \theta)$ with push-forward $(L_g)_*\dot{\mathbf{p}}(U) = \dot{\mathbf{p}}(U \circ L_g)$. More explicitly, $\dot{\mathbf{p}} = \sum_{i=1}^{3} \alpha^i\, \partial_{x^i}\big|_{\mathbf{p}}$ with $\alpha^i \in \mathbb{R}$, $\mathbf{p} = (x, y, \theta)$. Then

$$\begin{aligned}((L_g)_*\dot{\mathbf{p}}(U)) = \alpha^1(\cos\theta\partial_x + \sin\theta\partial_y)(U)(\mathbf{p}) + \alpha^2(-\sin\theta\partial_x + \cos\theta\partial_y)(U)(\mathbf{p}) \\ + \alpha^3\partial_\theta(U)(\mathbf{p}).\end{aligned}$$

More generally, a possibly asymmetric Finsler function defined on tangent-bundle $T(\mathbb{M}_2) = \{(\mathbf{p}, \dot{\mathbf{p}}) \mid \dot{\mathbf{p}} \in T_{\mathbf{p}}(\mathbb{M}_2)\}$ given by $\mathcal{F} : T(\mathbb{M}_2) \to \mathbb{R}^+$ is left invariant if $\mathcal{F}(\mathbf{p}, \dot{\mathbf{p}}) = \mathcal{F}(g \cdot \mathbf{p}, (L_g)_*\dot{\mathbf{p}})$ for all $(\mathbf{p}, \dot{\mathbf{p}}) \in T(\mathbb{M}_2), g \in SE(2)$.

Thereby the distance is left invariant:

$$d_{\mathcal{F}}(\mathbf{p}_1,\mathbf{p}_2)=\inf\left\{\int_0^1 \mathcal{F}(\gamma(t),\dot{\gamma}(t))\mathrm{d}t \,\middle|\, \gamma\in\Gamma_1,\gamma(0)=\mathbf{p}_1,\gamma(1)=\mathbf{p}_2\right\}=d_{\mathcal{F}}(g\cdot\mathbf{p}_1,g\cdot\mathbf{p}_2),$$

for all $g\in SE(2)$, optimizing over the set Γ_1 of piecewise $C^1([0,1],\mathbb{M}_2)$-curves.

In our application, the asymmetric Finsler function will restrict backward movement as we will see in Sect. 3, and consequently cusps are avoided [9].

3 Existing Reeds-Shepp Car Models

Over the years, many geometric control problems have been proposed for geodesic tracking of blood vessels or vehicles. The ones closest to our model are the symmetric and asymmetric Reeds-Shepp car models. The symmetric Reeds-Shepp Car model, proposed in [3,15], is given by

$$\mathcal{G}_{\mathbf{p}}(\dot{\mathbf{p}},\dot{\mathbf{p}})=C(\mathbf{p})^2\left(\xi^2|\dot{\mathbf{x}}\cdot\mathbf{n}|^2+\frac{\xi^2}{\zeta^2}\|\dot{\mathbf{x}}\wedge\mathbf{n}\|^2+\|\dot{\mathbf{n}}\|^2\right), \tag{2}$$

for all $\mathbf{p}=(\mathbf{x},\mathbf{n})\in\mathbb{M}_2, \dot{\mathbf{p}}=(\dot{\mathbf{x}},\dot{\mathbf{n}})\in T_{\mathbf{p}}(\mathbb{M}_2)$ with $\|\dot{\mathbf{x}}\wedge\mathbf{n}\|^2:=\|\dot{\mathbf{x}}\|^2-|\dot{\mathbf{x}}\cdot\mathbf{n}|^2$, where $\mathbf{n}$ is constructed using the identification in (1). The asymmetric Reeds-Shepp Car model, proposed in [9], is given by the asymmetric Finsler norm/function

$$|\mathcal{F}(\mathbf{p},\dot{\mathbf{p}})|^2=\mathcal{G}_{\mathbf{p}}(\dot{\mathbf{p}},\dot{\mathbf{p}})+C(\mathbf{p})^2(\varepsilon^{-2}-1)\xi^2|(\dot{\mathbf{x}}\cdot\mathbf{n})_-|^2, \tag{3}$$

for all $\mathbf{p}=(\mathbf{x},\mathbf{n})\in\mathbb{M}_2, \dot{\mathbf{p}}=(\dot{\mathbf{x}},\dot{\mathbf{n}})\in T_{\mathbf{p}}(\mathbb{M}_2)$ with $a_-:=\min\{0,a\}$. The parameter ξ influences the flexibility of the tracking, weighing between spatial and angular movement. The anisotropy parameter ζ penalizes sideways movement. When $\zeta\downarrow 0$, the classical sub-Riemannian model appears.

The cost function $C:\mathbb{M}_2\to[\delta,1]$, $\delta>0$, discourages movement outside the vascular structures via a crossing-preserving vesselness map $\mathcal{V}(\mathbf{p})$ [11],[4, App.D]. Typically [3, eq.5.1] one has $C(\mathbf{p})=(1+\lambda\mathcal{V}(\mathbf{p})^p)^{-1}$. The choice of the cost function is important and in this article (Sect. 4, 5, 6) we propose to include illumination enhancement and crossing preserving TV-flow (prior to the vesselness map computation) in the cost function as this will greatly improve tracking results.

In the asymmetric Reeds-Shepp car model, the parameter $\varepsilon\in(0,1]$ describes how strongly the model has to adhere to the forward gear. When $\varepsilon=1$, we are in the symmetric model. When $\varepsilon\downarrow 0$, backward motion becomes prohibited.

Computation of shortest paths (geodesics) connecting seeds and tips is done in 2 steps. First, the distances to all points in the domain are calculated, resulting in a distance map. Then, the shortest path is obtained by a steepest descent algorithm applied on this distance map. In all experiments we use the Anisotropic Fast Marching [4,13] by J.-M. Mirebeau for distance map computations.

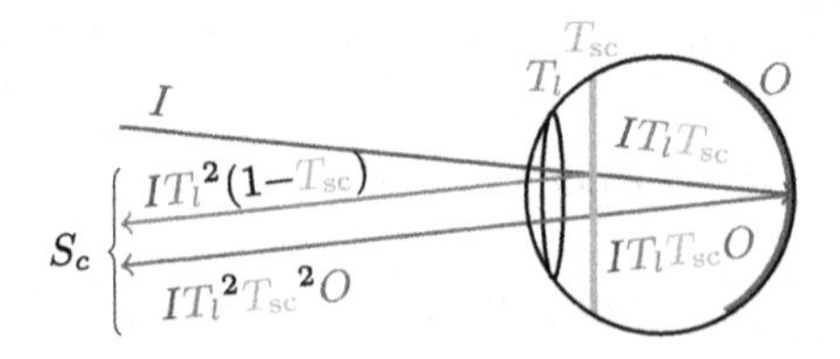

Fig. 3. Visualization of the physical model when imaging the retina; O is the actual image we would like to recover, S_c is the perceived image (sum of purple and red reflected light). I stands for input illumination, T_l and T_{sc} resp. denote the transmission ratio of the lens and the intraocular scattering (incl. cataract). (Color figure online)

4 Illumination Enhancement

Previous approaches in retinal vessel tracking typically consider the unprocessed picture S taken by the ophthalmologist, e.g. [2,9,11]. However, this may deviate from the actual retinal image O which we aim to recover, due to possible cataract and uneven illumination. The physical model of the construction of the output image is visualized in Fig. 3. This yields the following standard optics formula:

$$S_c(\mathbf{x}) = L(\mathbf{x})\left(T_{\mathrm{sc}}^2(\mathbf{x})O_c(\mathbf{x}) + 1 - T_{\mathrm{sc}}(\mathbf{x})\right) \text{ with } L = I \cdot T_l^2, \tag{4}$$

where c denotes the color channel in RGB or the luminance channel Y in YPbPr color space, cf. Wikipedia, and $L, T_{\mathrm{sc}} : \Omega \to [0,1]$ denote the illumination from outside the eye and transmission of the intraocular scattering respectively on domain $\mathbf{x} \in \Omega \subset \mathbb{R}^2$. The illumination from outside the eye L is composed of the input illumination I and the transmission ratio of the lens T_l. We apply an illumination correction, as done in [23]. After determining the illumination L, we re-express the Y channel (in YPbPr color space) of (4) to

$$\begin{aligned} &O_Y(\mathbf{x}) = \tfrac{L^{-1}(\mathbf{x})S_Y(\mathbf{x})-1+T_{\mathrm{sc}}(\mathbf{x})}{T_{\mathrm{sc}}^2(\mathbf{x})} \in [0,1], \text{ for all } \mathbf{x} \in \Omega, \\ &\text{with } T_{\mathrm{sc}} = 1 - A\left(L^{-1}(\cdot)S(\cdot) + \sum_{l=1}^{n} \tfrac{1}{n}\tfrac{1}{1+\exp(-\phi_l(\cdot))} G_{\sigma_l} * \left(L^{-1}(\cdot)S(\cdot)\right)\right), \end{aligned} \tag{5}$$

with Gaussian kernel standard deviations $\sigma_l = \text{pixelsize} \cdot 2^{(l-1)}$ of the retinal image at scale level $l \in \{1, \ldots, n\}$ where we took $n = 4$ and where the sigmoids on scale coefficients above are included to control the range in $[0,1]$ and to allow for stable optimization below. The Y channel of the actual image O is obtained by solving the Euler-Lagrange equation of the Tikhonov regularization problem via the Karush-Kuhn-Tucker conditions including the constraints $O_Y \in [0,1]$:

$$(\phi^{\min}, A^{\min}) = \operatorname{argmin}_{(\phi,A)} \{\|O_Y - \overline{O_Y}\|^2_{\mathbb{L}_2(\Omega)} + \lambda \|\nabla O_Y\|^2_{\mathbb{L}_2(\Omega)}\} \tag{6}$$

where we optimize w.r.t. $A > 0$ and $\phi = (\phi_l)_{l=1}^n \in \mathbb{R}^n$, not O_Y. Here, $\overline{O_Y} \in \mathbb{R}$ is an estimation of the desired intensity level [23, Sec.3.5], and λ regulates the smoothness of O_Y. After optimal non-constant $O_Y = O_Y(\phi^{\min}, A^{\min})$ is retrieved by (5), image O follows by linear conversion of YPbPr- to RGB-colors, via updated Pb- and Pr-channels.

5 TV-Flow Enhancement

TV-flow enhancement is a valuable technique to denoise surfaces, but at the same time preserve sharp edges. Recall that the metric intrinsic gradient is given by

$$\mathbb{M}_2 \ni \mathbf{p} \mapsto \nabla_{\mathcal{G}}\phi(\mathbf{p}) = \left((\mathcal{G})^{-1}\, \mathrm{d}\phi\right)(\mathbf{p}) \in T_{\mathbf{p}}(\mathbb{M}_2),$$

using $\mathcal{G}$ in (2) with $\zeta = C\xi = 1, C = 10$. Then TV-flow $U \mapsto W_0(\cdot, t)$ is given by

$$\begin{cases} \frac{\partial W_\varepsilon}{\partial t}(\mathbf{p}, t) = \mathrm{div}\left(\frac{\nabla_{\mathcal{G}} W_\varepsilon(\cdot,t)}{\varepsilon^2 + (\nabla_{\mathcal{G}} W(\cdot,t))^2}\right)(\mathbf{p}), & \mathbf{p} \in \mathbb{M}_2,\ t \geq 0, \\ W_\varepsilon(\mathbf{p}, 0) = U(\mathbf{p}) \end{cases}$$

and $W_0(\mathbf{p}, t) = \lim_{\varepsilon \downarrow 0} W_\varepsilon(\mathbf{p}, t)$. For proof of the $\mathbb{L}_2$-convergence, see [18]. Training of the end-time t of the TV-flow is not needed as for all lifted optically enhanced (cf. Sect. 4) images U of the STAR-dataset [21], end-time $t = 0.5$ is nearly optimal for subsequent tracking, and $\Delta t = 0.1$ always remains in the stability region [18]. The same settings provided optimal PSNR-ratios in [18].

6 A Finsler Metric on $\mathbb{M}_2$ that Includes the Enhancements

Our goal was to track vasculature accurately. In order to do so, one needs a metric tensor field that describes distances on the manifold. In some cases, it is beneficial to construct a metric tensor field $\mathcal{G}^U$ that depends explicitly on the underlying orientation score data U. This "data-driven" metric tensor field needs to be left invariant with respect to the roto-translation of the underlying data:

Definition 4 (Data-Driven Left-Invariant Metric (DDLIM)). *The metric tensor fields $\mathcal{G}^U$ and $\mathcal{F}^U$ on $\mathbb{M}_2$ are data-driven left invariant when they satisfy for all $(\mathbf{p}, \dot{\mathbf{p}}) \in T(\mathbb{M}_2)$ and all $g \in SE(2)$:*

$$\mathcal{G}^U_{\mathbf{p}}(\dot{\mathbf{p}}, \dot{\mathbf{p}}) = \mathcal{G}^{\mathcal{L}_g U}_{g\cdot\mathbf{p}}((L_g)_*\dot{\mathbf{p}}, (L_g)_*\dot{\mathbf{p}}), \text{ and } \mathcal{F}^U(\mathbf{p}, \dot{\mathbf{p}}) = \mathcal{F}^{\mathcal{L}_g U}(g \cdot \mathbf{p}, (L_g)_*\dot{\mathbf{p}}), \quad (7)$$

where $\mathcal{L}_g U(\mathbf{h}) := U(L_{g^{-1}}\mathbf{h}) := U(g^{-1} \cdot \mathbf{h})$ for all $\mathbf{h} \in \mathbb{M}_2$.

The considered data-driven left-invariant metric tensor fields are given by

$$\mathcal{G}^U_{\mathbf{p}}(\dot{\mathbf{p}}, \dot{\mathbf{p}}) = \mathcal{G}_{\mathbf{p}}(\dot{\mathbf{p}}, \dot{\mathbf{p}}) + \mu\, C^2(\mathbf{p}) \frac{\|HU|_{\mathbf{p}}(\dot{\mathbf{p}}, \cdot)\|^2_*}{\max\limits_{\|\dot{\mathbf{q}}\|=1} \|HU|_{\mathbf{p}}(\dot{\mathbf{q}}, \cdot)\|^2_*}, \quad (8)$$

$$\left|\mathcal{F}^U(\mathbf{p}, \dot{\mathbf{p}})\right|^2 = \left|\mathcal{F}(\mathbf{p}, \dot{\mathbf{p}})\right|^2 + \mu\, C^2(\mathbf{p}) \frac{\|HU|_{\mathbf{p}}(\dot{\mathbf{p}}, \cdot)\|^2_*}{\max\limits_{\|\dot{\mathbf{q}}\|=1} \|HU|_{\mathbf{p}}(\dot{\mathbf{q}}, \cdot)\|^2_*}, \quad (9)$$

for all $\mathbf{p} = (\mathbf{x}, \mathbf{n}), \dot{\mathbf{p}} = (\dot{\mathbf{x}}, \dot{\mathbf{n}})$, representing the symmetric and asymmetric metric tensor fields respectively. In (8) and (9), the metric tensor fields $\mathcal{G}$ and $\mathcal{F}$ were introduced in (2) and (3) respectively. The Hessian field HU is defined as $HU := \nabla(\mathrm{d}U)$, w.r.t. plus Cartan connection $\nabla^{[+]}$ for computational details see [7], [4, Rem.8], and $\|\cdot\|_*$ denotes the dual norm w.r.t. $\sqrt{\mathcal{G}_{\mathbf{p}}(\dot{\mathbf{p}}, \dot{\mathbf{p}})}$, where $\zeta = \xi = C = 1$.

The parameter $\mu > 0$ regulates the inclusion of the new data-driven term, and $C(\mathbf{p})$ denotes the cost function described in [4, App. D].

Remark 1. The construction of this cost now relies on the orientation score U of the optically enhanced image with TV-flow enhancement, whereas previously it relied on the orientation score of the unprocessed image. Akin to [4, App.C], one can show that the new Finsler/Riemannian metric tensor fields (8) are DDLIM.

The geodesics are calculated by steepest descent on distance maps using a metric that describes distances in $\mathbb{M}_2$.

Definition 5 (Data-Driven Riemannian Distance). *The data-driven Riemannian distance $d_{\mathcal{G}^U}$ from a point $\mathbf{p} \in \mathbb{M}_2$ to a point $\mathbf{q} \in \mathbb{M}_2$ is given by*

$$d_{\mathcal{G}^U}(\mathbf{p}, \mathbf{q}) = \inf_{\substack{\gamma \in \Gamma_1, \\ \gamma(0) = \mathbf{p}, \\ \gamma(1) = \mathbf{q}}} \int_0^1 \sqrt{\mathcal{G}^U_{\gamma(t)}(\dot{\gamma}(t), \dot{\gamma}(t))}\, \mathrm{d}t \tag{10}$$

where $\Gamma_1 := \{\gamma : [0,1] \to \mathbb{M}_2 | \gamma \in PC^1([0,1], \mathbb{M}_2)\}$ with PC^1 the space of piecewise continuously differentiable curves in $\mathbb{M}_2$, and $\dot{\gamma}(t) := \frac{\mathrm{d}}{\mathrm{d}t}\gamma(t)$. The quasi-distance that belongs to the asymmetric Finslerian model (9) is given by

$$d_{\mathcal{F}^U}(\mathbf{p}, \mathbf{q}) = \inf_{\substack{\gamma \in \Gamma_1, \\ \gamma(0) = \mathbf{p}, \\ \gamma(1) = \mathbf{q}}} \mathfrak{L}_{\mathcal{F}^U}(\gamma) := \inf_{\substack{\gamma \in \Gamma_1, \\ \gamma(0) = \mathbf{p}, \\ \gamma(1) = \mathbf{q}}} \int_0^1 \mathcal{F}^U(\gamma(t), \dot{\gamma}(t))\, \mathrm{d}t. \tag{11}$$

Lemma 1. *If $\mathcal{F}^U$ is DDLIM (Definition 4) then distance $d_{\mathcal{F}^U}$ satisfies:*

$$\forall_{g \in SE(2)} \forall_{\mathbf{p}_1, \mathbf{p}_2 \in \mathbb{M}_2} \; : \; d_{\mathcal{F}^{\mathcal{L}_g U}}(g \cdot \mathbf{p}_1, g \cdot \mathbf{p}_2) = d_{\mathcal{F}^U}(\mathbf{p}_1, \mathbf{p}_2). \tag{12}$$

Proof. One has $d_{\mathcal{F}^{\mathcal{L}_g U}}(g \cdot \mathbf{p}_1, g \cdot \mathbf{p}_2) = \inf\limits_{\substack{\gamma \in \Gamma_1, \\ \gamma(0) = g \cdot \mathbf{p}_1, \\ \gamma(1) = g \cdot \mathbf{p}_2}} \mathfrak{L}_{\mathcal{F}^{\mathcal{L}_g U}}(\gamma) \overset{(7)}{=} \inf\limits_{\substack{g^{-1} \cdot \gamma \in \Gamma_1, \\ g^{-1} \cdot \gamma(0) = \mathbf{p}_1, \\ g^{-1} \cdot \gamma(1) = \mathbf{p}_2}} \mathfrak{L}_{\mathcal{F}^U}(g^{-1} \cdot \gamma)$ $= d_{\mathcal{F}^U}(\mathbf{p}_1, \mathbf{p}_2)$, where $g^{-1} \cdot \gamma \in \Gamma_1 \Leftrightarrow \gamma \in \Gamma_1$, from which (12) follows. □

The shortest curves are computed using steepest descent on the distance map, departing from tip $\mathbf{p} \in \mathbb{M}_2$ towards seed $\mathbf{p}_0 \in \mathbb{M}_2$ as described in Theorem 1.

Theorem 1. *The shortest curve $\gamma : [0,1] \to \mathbb{M}_2$ with $\gamma(0) = \mathbf{p}$ and $\gamma(1) = \mathbf{p}_0$ can be computed by steepest descent tracking on distance map $W(\mathbf{p}) = d_{\mathcal{F}^U}(\mathbf{p}, \mathbf{p}_0)$*

$$\gamma(t) := \gamma^U_{\mathbf{p}, \mathbf{p}_0}(t) = Exp_{\mathbf{p}}(t\, v(W)), \qquad t \in [0,1], \tag{13}$$

where Exp integrates the following vector field on $\mathbb{M}_2$: $v(W) := -W(\mathbf{p}) \nabla_{\mathcal{F}^U} W$ and where W is the viscosity solution of the eikonal PDE system

$$\begin{cases} \mathcal{F}^*_U(\mathbf{p}, \mathrm{d}W(\mathbf{p})) = 1 & \mathbf{p} \in \mathbb{M}_2, \\ W(\mathbf{p}_0) = 0, \end{cases} \tag{14}$$

assuming $\mathbf{p}$ *is neither a 1st Maxwell-point nor a conjugate point, with dual Finsler function* $\mathcal{F}^*_U(\mathbf{p},\hat{\mathbf{p}}) := \max\{\langle\hat{\mathbf{p}},\dot{\mathbf{p}}\rangle \mid \dot{\mathbf{p}} \in T_{\mathbf{p}}(\mathbb{M}_2)$ *with* $\mathcal{F}^U(\mathbf{p},\dot{\mathbf{p}}) \leq 1\}$. *As* $v(W)$ *is data-driven left invariant, the geodesics carry the symmetry*

$$\gamma^{\mathcal{L}_g U}_{g\cdot\mathbf{p},g\cdot\mathbf{p}_0}(t) = g\,\gamma^{U}_{\mathbf{p},\mathbf{p}_0}(t) \text{ for all } g \in SE(2), \mathbf{p}, \mathbf{p}_0 \in \mathbb{M}_2, t \in [0,1]. \tag{15}$$

Proof. This is a special case of [4, Thm.1] with Lie group $SE(2) \equiv \mathbb{M}_2$. Then this yields the symmetric case $\|\nabla_{\mathcal{G}^U} W(\mathbf{p})\| = 1$ in the usual eikonal PDE form. Inclusion of the asymmetric front propagation (relying on asymmetric Finsler metric $\mathcal{F}^U$) requires a replacement of $\|\nabla_{\mathcal{G}^U} W(\mathbf{p})\| = 1$ with a dual norm expression $\mathcal{F}^*_U(\mathbf{p}, \mathrm{d}W(\mathbf{p})) = 1$, where one takes the positive part of the spatial momentum component in direction $\cos\theta \mathrm{d}x + \sin\theta \mathrm{d}y \in T^*(\mathbb{M}_2)$. This is similar to the technique in [9, Thm.4] but due to the data-driven behavior $\mathcal{F}^U$ this is subtle [4, Eq. 43, Lem. 3] and also directly applies to our model (16) using cost function C (incl. optical & TF-flow enhancement) as explained in Remark 1. Also, the backtracking requires a subtle adaptation: instead of ordinary intrinsic gradient descent in direction $\nabla_{\mathcal{G}^U} W = (\mathcal{G}^U)^{-1}\mathrm{d}W$ it now becomes more general descent in direction $\nabla_{\mathcal{F}^U} W(\cdot) := \mathrm{d}\mathcal{F}^*_U(\cdot, \mathrm{d}W(\cdot))$ as explained in [9, prop.4]. □

In the experimental section, we rely on the mixed metric tensor field, which is needed to avoid wrong exits at complex structures, see [4], and is given by:

$$\begin{aligned} &\mathcal{G}^M_{\mathbf{p}}(\dot{\mathbf{p}},\dot{\mathbf{p}}) = \kappa(\mathbf{x})\,\mathcal{G}_{\mathbf{p}}(\dot{\mathbf{p}},\dot{\mathbf{p}}) + (1-\kappa(\mathbf{x}))\,\mathcal{G}^U_{\mathbf{p}}(\dot{\mathbf{p}},\dot{\mathbf{p}}),\ \mathbf{p} = (\mathbf{x},\mathbf{n}) \in \mathbb{M}_2 \\ &\mathcal{F}^M(\mathbf{p},\dot{\mathbf{p}})^2 = \kappa(\mathbf{x})\,\mathcal{F}(\mathbf{p},\dot{\mathbf{p}})^2 + (1-\kappa(\mathbf{x}))\,\mathcal{F}^U(\mathbf{p},\dot{\mathbf{p}})^2, \end{aligned} \tag{16}$$

with $\kappa = \mathbb{1}_A * G_\sigma$ and A the crossing structure locations and Gaussian $G_{\sigma=1\text{pix}}$.

7 Experimental Results

We rely on the asymmetric metric tensor field (16) to calculate the geodesics of the 3 different models. These models are constructed based on a) the picture taken by the ophthalmologist (original image), b) the original image with illumination enhancement as explained in Sect. 4 (optically enhanced image), and c) the optically enhanced image with crossing-preserving TV-flow enhancement discussed in Sect. 5. In this section, we illustrate the results for a specific image and refer to Table 1 for an overview of the performance of the different models on the STAR dataset [1,21]. These results are consistent with the discussed example, and for reproducible code and all processed images see [5]. In all experiments we set standard parameter settings [3, eq.5.1], [4, eq.65] for cost-function C ($p = 3, \lambda = 1000$), for the metrics ($\xi = \zeta = 0.1$) in (2), for TV-flow ($t = 0.5$).

In prior research, the "original image" directly entered the metric tensor field on $\mathbb{M}_2$ when calculating the geodesics. These images are of varying quality, depending on the patient's condition and the used equipment. Applying a tracking algorithm, like Anisotropic Fast Marching [4], on the metric tensor field based on the original image, often results in tracking mistakes due to uneven illumination, both along vascular structures and on the background, making it

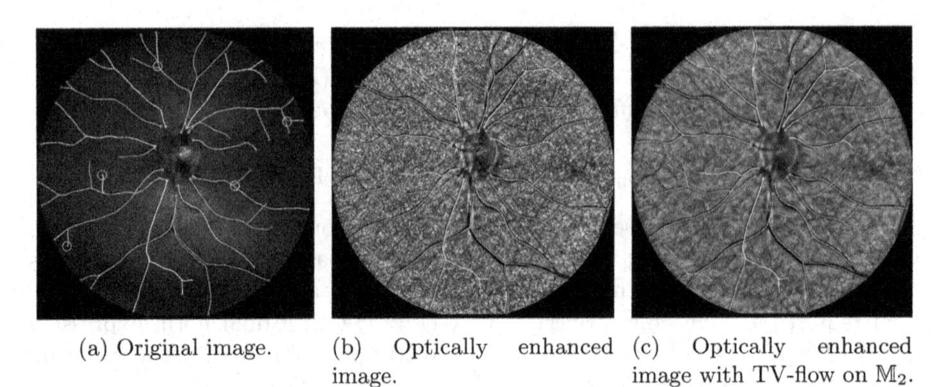

(a) Original image. (b) Optically enhanced image. (c) Optically enhanced image with TV-flow on $\mathbb{M}_2$.

Fig. 4. Tracking of Vascular Tree per Vessel Type: Tracking with the mixed model $(\mathbb{M}_2, \mathcal{F}^M)$ proposed in (16) with $\mu = 15$. Prior classification of vascular trees by type (artery/vein resp. white/cyan) only results in perfect tracking of the vessel tree on the enhanced images. Yellow circles indicate tracking mistakes.

hard to distinguish different structures. In Fig. 1a, one sees the tracking results on an original, unevenly illuminated, non-enhanced image, where all vessels were tracked in one single run. At a lot of locations (13), the tracking connects the seeds $\mathbf{p}_0$ and tips $\mathbf{p}$ incorrectly. The optical enhancement explained in Sect. 4 corrects for uneven illumination. Calculating the geodesics using the metric tensor field relying upon the optically enhanced image, reduces the number of tracking mistakes significantly (to 5), cf. Fig. 1b. Due to pointwise optimization in the optical enhancement, noise is generated. The crossing-preserving total variation flow enhancement suppresses this noise and indeed results in even fewer tracking mistakes (3), cf. Fig. 1c.

Calculating the tracking results in Fig. 1 uses no knowledge about the vasculature, apart from seed and tip locations. One might incorporate prior knowledge A) on vessel types (artery/vein), or B) on the connectivity of tips and seeds.

We start by investigating prior knowledge on A), where we first connect all tips on arteries to the seeds on arteries, and similarly for the tips and seeds on veins. Figure 4 shows that the tracking results improve significantly for all cases; the number of tracking mistakes at crossings reduces from (13, 5, 3) to (5, 0, 0) for resp. the original, optically enhanced image excl. and incl. TV-flow enhancement.

Second, we investigate the prior knowledge on B). In Fig. 5, the tracking results connecting the tips to their corresponding seed are presented. The number of correct tracks has improved once again, to only (3, 0, 0) mistakes for resp. the original, optically enhanced and optically enhanced with TV-flow image.

We report the tracking results for the three different approaches on images from STAR [1,21], in particular the example in Fig. 1, 4, and 5. We observe the same trend in performance for other images which we summarize in Table 1. We calculate the (weighted) percentage of incorrectly calculated geodesics by means of:

$$\epsilon := \frac{1}{|T|} \sum_{\mathbf{y} \in T} \left(1 - \frac{\|\mathbf{y} - \mathbf{x}_0(\mathbf{y})\|}{\sqrt{N_x^2 + N_y^2}}\right) C_0(\mathbf{x}(\mathbf{y}), \mathbf{y}) \geq 0, \tag{17}$$

Here S, T denote the sets of resp. seeds (near the optic disk) and tips. The image size is $N_x \times N_y$. The ground truth seed and calculated seed (first arriving front in the distance map, cf. Fig. 2) corresponding to the tip $\mathbf{y}$ are resp. denoted by $\mathbf{x}_0(\mathbf{y}) \in S$ and $\mathbf{x}(\mathbf{y}) \in S$. Function $C_0 : S \times T \to \{0, 1\}$ is given by $C_0(\mathbf{x}, \mathbf{y}) = 0$ if the tracking between $\mathbf{x}$ and $\mathbf{y}$ is correct and $C_0(\mathbf{x}, \mathbf{y}) = 1$ otherwise.

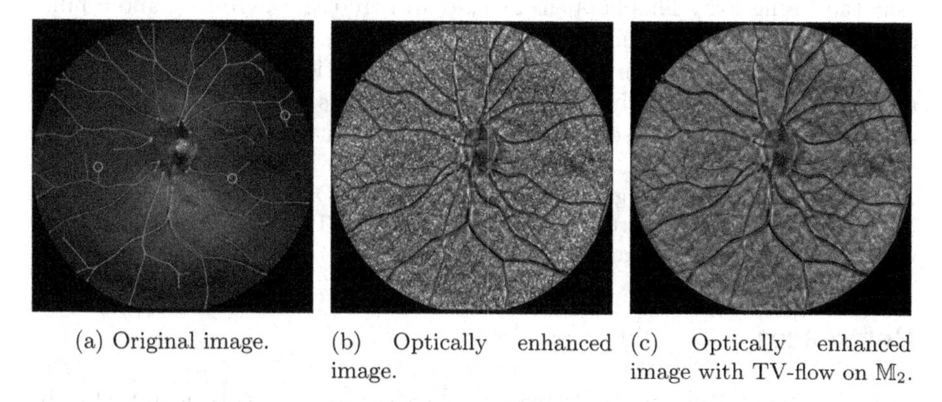

(a) Original image. (b) Optically enhanced image. (c) Optically enhanced image with TV-flow on $\mathbb{M}_2$.

Fig. 5. Tracking of Vascular Tree per Seed on the Optic Disk: Tracking with the mixed model $(\mathbb{M}_2, \mathcal{F}^M)$ proposed in (16) with $\mu = 15$. Prior grouping of tips (in red) and seeds (in green) only results in perfect tracking of the vessel tree on the enhanced images. Tracking mistakes are indicated by yellow circles. (Color figure online)

Table 1. Error measure ϵ of each tracking applied to STAR images in [5], calculated by (17). Highlighted: best results per tracking.

	Original image	Optically enhanced image	Optically enhanced image with crossing-preserving TV-flow
Single Run	0.34	0.23	0.20
Per Type (A/V)	0.25	0.12	0.10
Per Seed	0.23	0.09	0.09

We evaluate with a harsh error measure (17): one crossing mistake (indicated by a circle) often causes more errors, when the vessel bifurcates after the crossing.

The optical enhancement and TV-flow regularization applied on the original images, result in more accurate geodesics compared to those calculated directly on the original images, as can be seen in Table 1. The more prior information we use, the more accurate the geodesics follow the vasculature. Remarkably, tracking requiring artery-vein classification of seeds and tips performs similarly and is easier to automate than tracking with knowledge of seed-tip connectivity.

Conclusion

We developed a new asymmetric, data-driven left-invariant Finsler geometric model that includes contextual contrast enhancement via TV-flows on $SE(2)$. Experiments reveal that application of this new Finsler geometric model has benefits over previous left-invariant models [2,9] and the recent data-driven model [4]. The new model reduces many errors and performs very well on both realistic and challenging low-quality retinal images where full vascular trees are computed from a *single* asymmetric Finslerian distance map. Although we have shown that both the contrast enhancement and the TV-flow on $SE(2)$ in the new Finslerian model are highly beneficial, there are still exceptional cases where vessel tracts take the wrong exit. This happens at places where both a crossing and a bifurcation occur (cf. the red circles in Fig. 1c). Therefore, in future work, we aim to tackle these cases by automatic artery vein classification via PDE-G-CNNs [17], as our experiments show this allows us to obtain the same good practical results as with the 'tracking per seed' (that requires too costly user-knowledge).

Acknowledgements. We gratefully acknowledge the Dutch Foundation of Science NWO for its financial support by Talent Programme VICI 2020 Exact Sciences (Duits, Geometric learning for Image Analysis, VI.C 202-031).

References

1. Abbasi-Sureshjani, S., Smit-Ockeloen, I., Zhang, J., Ter Haar Romeny, B.: Biologically-inspired supervised vasculature segmentation in SLO retinal fundus images. In: Kamel, M., Campilho, A. (eds.) ICIAR 2015. LNCS, vol. 9164, pp. 325–334. Springer, Cham (2015). https://doi.org/10.1007/978-3-319-20801-5_35
2. Bekkers, E., Duits, R., Berendschot, T., ter Haar Romeny, B.: A multi-orientation analysis approach to retinal vessel tracking. JMIV **49**(3), 583–610 (2014)
3. Bekkers, E., Duits, R., Mashtakov, A., Sanguinetti, G.: A PDE approach to data-driven sub-Riemannian geodesics in SE(2). SIAM SIMS **8**(4), 2740–2770 (2015)
4. van den Berg, N., Smets, B., Pai, G., Mirebeau, J.M., Duits, R.: Geodesic tracking via new data-driven connections of Cartan type for vascular tree tracking (2022). arXiv preprint arXiv:2208.11004
5. van den Berg, N.: Data-driven left-invariant tracking on optically enhanced images with total variation flow in Mathematica (2023). https://github.com/NickyvdBerg/tracking_enhancement_total_variation_flow
6. Colligris, P., Perez de Lara, M.J., Colligris, B., Pintor, J.: Ocular manifestations of Alzheimer's and other neurodegenerative diseases: the prospect of the eye as a tool for the early diagnosis of Alzheimer's disease. J. Ophthalmol. **2018** (2018)
7. Duits, R., Smets, B.M.N., Wemmenhove, A.J., Portegies, J.W., Bekkers, E.J.: Recent geometric flows in multi-orientation image processing via a Cartan connection. In: Handbook of Mathematical Models and Algorithms in Computer Vision and Imaging: Mathematical Imaging and Vision, pp. 1–60 (2021). https://www.win.tue.nl/rduits/Bookchapter.pdf
8. Duits, R.: Perceptual organization in image analysis: a mathematical approach based on scale, orientation and curvature. Ph.D. thesis, Eindhoven University of Technology (2005)

9. Duits, R., Meesters, S.P.L., Mirebeau, J.M., Portegies, J.M.: Optimal paths for variants of the 2D and 3D Reeds-Shepp car with applications in image analysis. JMIV **60**, 1–33 (2018)
10. Franceschiello, B., Mashtakov, A., Citti, G., Sarti, A.: Geometrical optical illusion via sub-Riemannian geodesics in the roto-translation group. DGA **65**, 55–77 (2019)
11. Hannink, J., Duits, R., Bekkers, E.: Crossing-preserving multi-scale vesselness. In: Golland, P., Hata, N., Barillot, C., Hornegger, J., Howe, R. (eds.) MICCAI 2014. LNCS, vol. 8674, pp. 603–610. Springer, Cham (2014). https://doi.org/10.1007/978-3-319-10470-6_75
12. Liu, L., Wang, M., Zhou, S., Shu, M., Cohen, L.D., Chen, D.: Curvilinear structure tracking based on dynamic curvature-penalized geodesics. Pattern Recogn. **134**, 109079 (2023). https://doi.org/10.1016/j.patcog.2022.109079
13. Mirebeau, J.M.: Fast-marching methods for curvature penalized shortest paths. JMIV **60**(6), 784–815 (2018)
14. Mitra, A., Roy, S., Roy, S., Setua, S.K.: Enhancement and restoration of non-uniform illuminated Fundus Image of Retina obtained through thin layer of cataract. Comput. Methods Programs Biomed. **156**, 169–178 (2018)
15. Reeds, J., Shepp, L.: Optimal paths for a car that goes both forwards and backwards. Pac. J. Math. **145**(2), 367–393 (1990)
16. Sasongko, M.B., et al.: Retinal vessel tortuosity and its relation to traditional and novel vascular risk markers in persons with diabetes. Curr. Eye Res. **41**(4), 551–557 (2016)
17. Smets, B., Portegies, J., Bekkers, E., Duits, R.: PDE-based group equivariant convolutional neural networks. JMIV **65**(1), 209–239 (2022)
18. Smets, B.M., Portegies, J., St-Onge, E., Duits, R.: Total variation and mean curvature PDEs on the homogeneous space of positions and orientations. JMIV **63**(2), 237–262 (2021)
19. Weiler, D.L., Engelke, C.B., Moore, A.L., Harrison, W.W.: Arteriole tortuosity associated with diabetic retinopathy and cholesterol. OVS **92**(3), 384–391 (2015)
20. Xiong, L., Li, H., Xu, L.: An approach to evaluate blurriness in retinal images with vitreous opacity for cataract diagnosis. J. Healthc. Eng. (2017)
21. Zhang, J., Dashtbozorg, B., Bekkers, E.J., Pluim, J.P., Duits, R., ter Haar Romeny, B.M.: Robust retinal vessel segmentation via locally adaptive derivative frames in orientation scores. IEEE TMI **35**(12), 2631–2644 (2016)
22. Zhang, S., Berendschot, T.T., Webers, C.A.: Luminosity rectified blind Richardson-Lucy deconvolution for single retinal image restoration. SSRN (2023). https://doi.org/10.2139/ssrn.4132901
23. Zhang, S., Webers, C.A.B., Berendschot, T.T.J.M.: A double-pass fundus reflection model for efficient single retinal image enhancement. Signal Process. **192**, 108400 (2022)
24. Zhou, M., Jin, K., Wang, S., Ye, J., Qian, D.: Color retinal image enhancement based on luminosity and contrast adjustment. IEEE TBE **65**(3), 521–527 (2018)

Geometric Adaptations of PDE-G-CNNs

Gijs Bellaard[1(✉)], Gautam Pai[1], Javier Olivan Bescos[2], and Remco Duits[1]

[1] CASA: Center for Analysis, Scientific Computing and Applications, (Group: Geometric Learning & Differential Geometry), Department of Mathematics and Computer Science, Eindhoven University of Technology, Eindhoven, The Netherlands
{g.bellaard,g.pai,r.duits}@tue.nl
[2] Philips, Eindhoven, The Netherlands
javier.olivan.bescos@philips.com

Abstract. Group equivariant convolutional neural networks (G-CNNs) have been successfully applied in geometric deep learning. The recently introduced framework of PDE-based G-CNNs (PDE-G-CNNs) generalizes G-CNNs while simultaneously reducing network complexity and increasing performance. In PDE-G-CNNs the usual building blocks of neural networks are replaced with solvers for evolution PDEs, these PDEs being convection, diffusion, dilation, and erosion. We investigate three geometric adaptations of PDE-G-CNNs:

- We generalize the theory in [2] to a family of Lie groups in between roto-translation group $SE(2)$ and Heisenberg group $H(3)$. This geometric adaptation enables transferring training orientation score processing on $SE(2)$ to training processing of velocity scores, shearlet transforms, or frequency scores on $H(3)$.
- We theoretically prove that the trainable lifting layer in a PDE-G-CNN is interchangeable with a single fixed untrained lifting coupled with multiple trainable convections. We experimentally validate this theoretical insight and report identical performance. This fixing of the lifting layer makes PDE-G-CNNs more interpretable as they now solely train association fields from neurogeometry.
- We include curvature adaptation in PDE-G-CNNs. This curvature adaptation is beneficial within the convection part of PDE-G-CNNs as we show experimentally.

Keywords: Deep Learning · Lie Groups · PDE-G-CNN · Neurogeometry

1 Introduction

In [15] PDE-based group equivariant convolution neural networks (PDE-G-CNNs) are introduced. When compared to group equivariant convolutional neural networks (G-CNNs) [5] and CNNs it is shown in [15] and [2] that PDE-G-CNNs have reduced network complexity, better performance, and better geometric interpretability. They essentially train a sparse set of association fields [2, App.B] from neurogeometry.

L. Calatroni et al. (Eds.): SSVM 2023, LNCS 14009, pp. 538–550, 2023.
https://doi.org/10.1007/978-3-031-31975-4_41

In PDE-G-CNNs the usual components that make up CNNs, that being convolutions, max pooling, and non-linear activation functions, are replaced by solvers for evolution PDEs from geometric image analysis. This makes PDE-G-CNNs more geometrically interpretable than traditional networks.

Specifically, the PDEs are convection, diffusion, dilation, and erosion. These PDEs respectively correspond to shifting, smoothing, max pooling, and min pooling. They are solved by linear and morphological convolutions. Figure 1 illustrates the difference between a traditional CNN layer and a PDE layer.

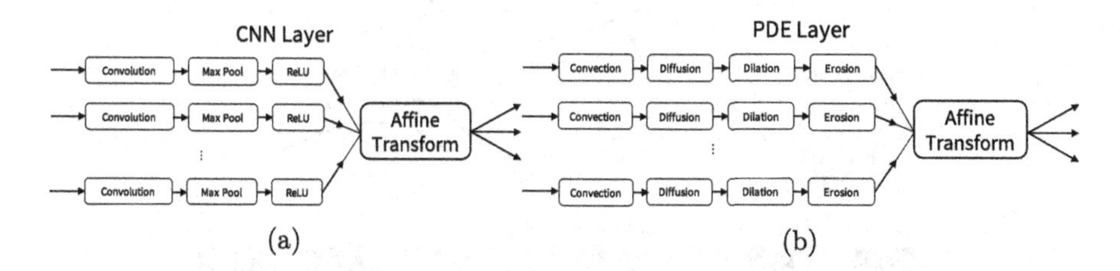

Fig. 1. The difference between a traditional CNN layer 1a, and a PDE layer 1b.

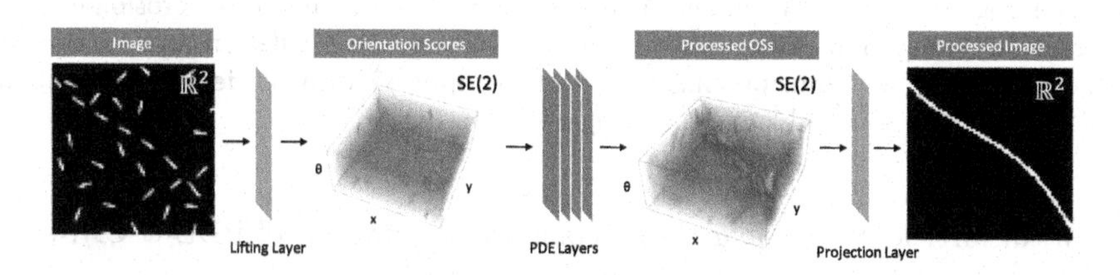

Fig. 2. An overview of a PDE-G-CNN performing line completion on the Lines dataset [2]. First, the image is lifted to an orientation score on $SE(2)$, then run through multiple PDE layers (Fig. 1b), after which the result is projected down to $\mathbb{R}^2$.

In this article we mainly discuss $SE(2)$ equivariant segmentation PDE-G-CNNs. Here $SE(2) = \mathbb{R}^2 \times \mathbb{R}/(2\pi\mathbb{Z})$ is the Lie group of roto-translations in the plane. The network being $SE(2)$ equivariant means that when an input image is roto-translated one can be *certain* that the segmentation is roto-translated accordingly.

To achieve inherent $SE(2)$ equivariance it is necessary to lift the input data, that being functions on $\mathbb{R}^2$, to functions on $SE(2)$. There are a multitude of ways of achieving this, but there is one very natural way using orientation score transforms [6]. Intuitively, this transform takes a kernel, roto-translates it across the input image, and "saves" the response as a function on $SE(2)$.

If an appropriate kernel is chosen to perform the lifting, such as cake wavelets [6], the lifted image has a meaningful interpretation: if the image contains a local orientation θ at position x, y the response at $(x, y, \theta) \in SE(2)$ is high.

After lifting the orientation scores are fed through the earlier discussed PDEs. This means that the PDEs are, in fact, defined on $SE(2)$, and the effect of diffusion, dilation, and erosion are completely determined by the Riemannian metric tensor field $\mathcal{G}$ that is chosen on $SE(2)$. The parameters that describe this metric are learned during training.

For an overview of a PDE-G-CNN performing line completion, as described in [2], see Fig. 2. A visualization of some of the feature maps of the network in Fig. 2 can be seen in Fig. 3, together with the geometric interpretation of a PDE layer.

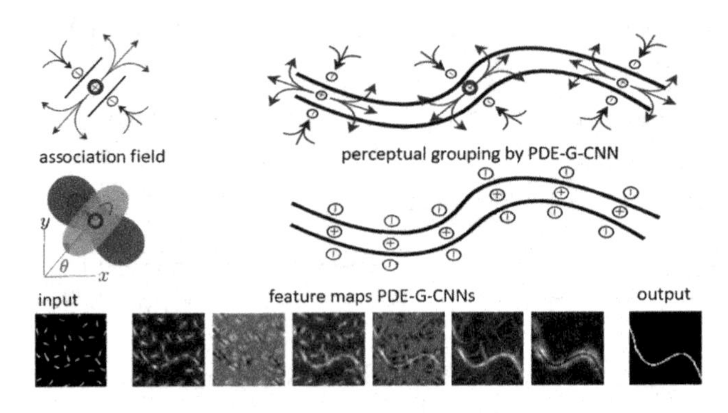

Fig. 3. Geometric Interpretation of a PDE layer, where a sparse set of association fields from neurogeometry is trained. Dilation (in green) is line extrapolation, erosion (in red) is line sharpening, and convection (in blue) is for centering. Dilation and erosion are respectively max and min pooling over trained (approximative) Riemannian balls in $SE(2)$ (as depicted in Table 1). (Color figure online)

In this article we consider three geometric adaptations of PDE-G-CNNs:

1.) The first geometric adaptation we investigate here is the generalization of the theoretical results in [2]. These results quantify the quality of the *logarithmic distance approximation* ρ_c, which acts as an approximation to the exact Riemannian distance on $SE(2)$. An approximation of the exact Riemannian distance is necessary to efficiently solve the diffusion, erosion, and dilation PDEs (as we shall see in Proposition 1). Specifically, we generalize the theory that assesses the quality of the approximations from Lie group $SE(2)$ to $SE(2)_\tau$, also described in [7, Sec.5.4]. The latter is a family of Lie groups that connects $SE(2)$ ($\tau = 1$) and the Heisenberg group $H(3)$ ($\tau \downarrow 0$) of positions and velocities. This generalization to $SE(2)_\tau$ enables us to transfer training orientation score processing on $SE(2)$ to training processing of velocity scores [4], shearlet transforms [11], or frequency scores [8] on $H(3)$. The first row of Table 1 visualizes the logarithmic distance approximation ρ_c for different τ's.

2.) The second geometric adaptation we consider is the fixing of the lifting layer. Currently in the LieTorch package [15] the lifting is done using learned kernels. We motivate theoretically that, within the PDE-G-CNN framework, a learned lifting layer can be replaced by a fixed lifting layer using cake wavelets [6]. We experimentally substantiate this finding by reporting that the performance of PDE-G-CNNs does not decrease when the lifting layer is fixed. Moreover, using orientation score kernels for lifting has the important benefit of increasing the geometric interpretation of PDE-G-CNNs (recall Fig. 3 and see [2, App.B] for more details) even further.

3.) The third geometric adaptation is the implementation of a more "free" Riemannian metric on $SE(2)$ than the one implemented in [15, Eq. 20] and analyzed in [2, Eq. 4]. Currently only diagonal metrics, with respect to a certain left-invariant frame (4), are implemented in the LieTorch package published in [15]. Using nondiagonal metrics has been beneficial in geometric image processing (contour perception [1], denoising [16], segmentation [17], tracking [12]) and we investigate it's importance within PDE-G-CNNs. The two rows in Table 1 show the difference between a diagonal and nondiagonal metric by looking at contours of the logarithmic distance approximation ρ_c. We see that using a nondiagonal metric has the effect of curving the contours, explaining why we call this 'curvature adaptation'.

Table 1. Contours $\rho_c = 0.5, 1.0, 1.5, 2.0, 2.5$ for a diagonal metric (5) (relative to the left-invariant frame (4)) and nondiagonal metric (allowing for curvature adaptation in erosion/dilation), for $\tau = 0, 0.5, 0.75, 1$. The domain of the plot is $(x, y, \theta) \in [-3, 3]^2 \times [-\pi, \pi)$, with the vertical axis being θ. On the bottom we have identically valued isocontours of the min-projection over θ.

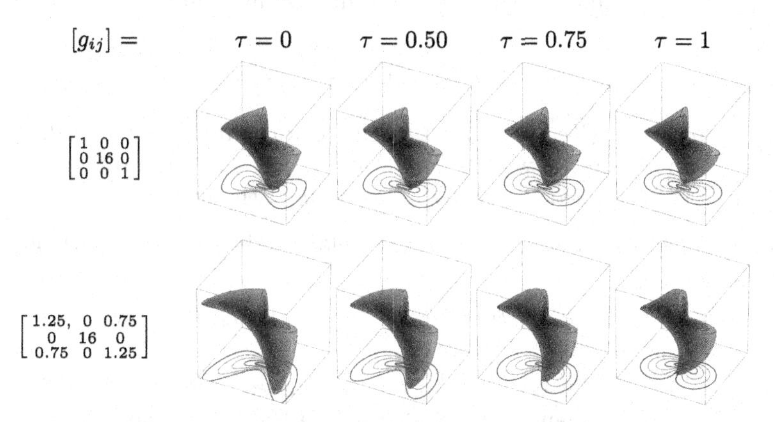

2 Preliminaries

Next we briefly list the basic and essential geometric tools that we will use.

Coordinates on $SE(2)$. Let $SE(2) = \mathbb{R}^2 \times \mathbb{R}/(2\pi\mathbb{Z})$ be the special Euclidean group of the two-dimensional plane, i.e. the group of two dimensional roto-translations. We always use the small angle identification: $\mathbb{R}/(2\pi\mathbb{Z}) = [-\pi, \pi)$. For all $g_1 = (x_1, y_1, \theta_1)$, $g_2 = (x_2, y_2, \theta_2) \in SE(2)$ we have the group product:

$$g_1 \cdot g_2 := (x_1 + x_2 \cos\theta_1 - y_2 \sin\theta_1, y_1 + x_2 \sin\theta_1 + y_2 \cos\theta_1, \theta_1 + \theta_2 \bmod 2\pi), \quad (1)$$

and the unit element is $e = (0, 0, 0)$.

Coordinates on $SE(2)_\tau$. Let $0 < \tau \leq 1$ be a scaling parameter. We define the Lie group $SE(2)_\tau = \mathbb{R}^2 \times [-\frac{\pi}{\tau}, \frac{\pi}{\tau}]$ as follows. Set scaling $s_\tau : SE(2)_\tau \to SE(2)$ defined as $s_\tau(x, y, \theta) := (x, \tau y, \tau\theta)$, and leverage the group product on $SE(2)$ to a group product $\cdot_\tau$ on $(SE(2)_\tau$ by defining $g_1 \cdot_\tau g_2 := s_\tau^{-1}(s_\tau(g_1) \cdot s_\tau(g_2))$, i.e. $(x_1, y_1, \theta_1) \cdot_\tau (x_2, y_2, \theta_2) =$

$$\left(x_1 + x_2 \cos\tau\theta_1 - \tau y_2 \sin\tau\theta_1, y_1 + \tfrac{1}{\tau} x_2 \sin\tau\theta + y_2 \cos\tau\theta_1, \theta_1 + \theta_2 \bmod \tfrac{2\pi}{\tau}\right). \quad (2)$$

Setting $\tau = 1$ we get the familiar group product of $SE(2)$. If we let $\tau \downarrow 0$ we get $(x_1, y_1, \theta_1) \cdot_\tau (x_2, y_2, \theta_2) = (x_1 + x_2, y_1 + y_2 + \theta_1 x_2, \theta_1 + \theta_2)$, which we indeed recognize as the product of the Heisenberg group $H(3)$.

Morphological Convolution. Given $f_1, f_2 : SE(2)_\tau \to \mathbb{R}$ we define:

$$(f_1 \,\square\, f_2)(g) = \inf_{h \in SE(2)_\tau} \left\{ f_1(h^{-1}g) + f_2(h) \right\}. \quad (3)$$

A basis of left-invariant (co-)vector fields is given by

$$\begin{aligned} \mathcal{A}_1 &= \cos\tau\theta \; \partial_x + \tfrac{1}{\tau} \sin\tau\theta \; \partial_y, \quad \mathcal{A}_2 = -\tau \sin\tau\theta \; \partial_x + \cos\tau\theta \; \partial_y, \quad \mathcal{A}_3 = \partial_\theta, \\ \omega^1 &= \cos\tau\theta \; \mathrm{d}x + \tau \sin\tau\theta \; \mathrm{d}y, \quad \omega^2 = -\tfrac{1}{\tau} \sin\tau\theta \; \mathrm{d}x + \cos\tau\theta \; \mathrm{d}y, \quad \omega^3 = \mathrm{d}\theta \,. \end{aligned} \quad (4)$$

Metric Tensor Field. We set the following left-invariant metric tensor field:

$$\mathcal{G} = \textstyle\sum_{i,j=1}^{3} g_{ij} \; \omega^i \otimes \omega^j, \quad (5)$$

and write $\|v\|_\mathcal{G} = \sqrt{\mathcal{G}(v, v)}$ for any tangent vector v. If the matrix is diagonal we write $g_{ii} = w_i^2$, where we will assume that $w_2 \geq w_1$, and the ratio $\zeta := \frac{w_2}{w_1} \geq 1$ will henceforth be called the *spatial anisotropy* of the metric.

Distances. Left-invariant metric tensor field $\mathcal{G}$ induces a left-invariant metric:

$$d_\mathcal{G}(g_1, g_2) = \inf_{\gamma \in \Gamma_t(g_1, g_2)} \int_0^t \|\dot{\gamma}(s)\|_\mathcal{G} \, \mathrm{d}s, \quad (6)$$

where $\Gamma_t(g_1, g_2)$ is the set of piecewise C^1-curves γ in $SE(2)_\tau$ with $\gamma(0) = g_1, \gamma(t) = g_2$. By invariance under re-parameterization we may set $t = 1$ above.

Exponential/Logarithm. One has $(x, y, \theta) = \exp(c^1 \; \partial_x|_e + c^2 \; \partial_y|_e + c^3 \; \partial_\theta|_e) =$

$$\left(\left(c^1 \cos\tfrac{\tau c^3}{2} - \tau c^2 \sin\tfrac{\tau c^3}{2}\right) \operatorname{sinc}\tfrac{\tau c^3}{2}, \left(\tfrac{1}{\tau} c^1 \sin\tfrac{\tau c^3}{2} + c^2 \cos\tfrac{\tau c^3}{2}\right) \operatorname{sinc}\tfrac{\tau c^3}{2}, c^3 \right) \quad (7)$$

for the exp map and for the log map: $\log(x, y, \theta) = c^1 \partial_x|_e + c^2 \partial_y|_e + c^3 \partial_\theta|_e \in T_e G$:

$$(c^1, c^2, c^3) = \left(\frac{x \cos\frac{\tau\theta}{2} + \tau y \sin\frac{\tau\theta}{2}}{\operatorname{sinc}\frac{\tau\theta}{2}}, \frac{-\frac{1}{\tau} x \sin\frac{\tau\theta}{2} + y \cos\frac{\tau\theta}{2}}{\operatorname{sinc}\frac{\tau\theta}{2}}, \theta \right). \quad (8)$$

Eqs. (7), (8) hold for $\tau \neq 0$. The case $\tau = 0$ follows by taking the limit. For $\tau > 0$ the θ-variable is $2\pi/\tau$-periodic. For $\tau \downarrow 0$ variable θ is a non-periodic velocity.

3 Analysis of Erosion and Dilation on $SE(2)_\tau$

The erosion and dilation PDEs on $G = SE(2)_\tau$ (for $\alpha > 1$) are defined as:

$$\frac{\partial W_\alpha}{\partial t} = \pm\frac{1}{\alpha}\left\|\nabla_{\mathcal{G}} W_\alpha\right\|^\alpha, \quad W_\alpha|_{t=0} = U, \tag{9}$$

where $W_\alpha : G \times \mathbb{R}_{\geq 0} \to \mathbb{R}$ is the (continuous) viscosity solution [9] obtained from the differentiable initial condition $U : G \to \mathbb{R}$. Here the +sign is a dilation scale space and the −sign is an erosion scale space [14]. Erosion and dilation correspond to min/max-pooling over Riemannian balls, both controlled by their own metric tensor field (5).

The following proposition expresses the solution of (9) in terms of a morphological convolution with a morphological kernel. The proof of this proposition can be found in [2, Prop.1].

Proposition 1 (Solution of Erosion and Dilation). *Let $\alpha > 1$. The viscosity solution W_α of the erosion PDE* (9) *is given by a morphological convolution* (3) *with a kernel $k_t^\alpha : G \to \mathbb{R}_{\geq 0}$:*

$$W_\alpha(g,t) = (k_t^\alpha \,\square\, U)(g), \text{ where } k_t^\alpha(g) := \frac{t}{\beta}\left(\frac{d_{\mathcal{G}}(e,g)}{t}\right)^\beta, \text{ with } \frac{1}{\alpha}+\frac{1}{\beta} = 1, \tag{10}$$

and the solution of the dilation PDE is $W_\alpha(g,t) = -(k_t^\alpha \square -U)(g)$, $g \in G$, $t \geq 0$.

The kernel k_t^α (10) depends on the exact Riemannian distance $d_{\mathcal{G}}(e,\cdot)$ (6), but calculating this is computationally demanding. Therefore we use approximations of the exact distance which are computationally cheap. We define the logarithmic distance approximation $\rho_c : G \to \mathbb{R}_{\geq 0}$, cf. [15, Def.19], by

$$\rho_c(g) = \|\log g\|_{\mathcal{G}} = \sqrt{(w_1 c^1(g))^2 + (w_2 c^2(g))^2 + (w_3 c^3(g))^2}, \tag{11}$$

where the 2nd equality holds because we for now only consider a diagonal metric. Also note that (11) is simpler than in the expression found in [15] as we only consider the Lie group $SE(2)_\tau$.

We also define the half-angle distance approximation ρ_b by replacing the logarithmic coordinates c^i by corresponding half-angle coordinates b^i:

$$b^1 = x\cos\frac{\tau\theta}{2} + \tau y \sin\frac{\tau\theta}{2},\; b^2 = -\frac{1}{\tau}x\sin\frac{\tau\theta}{2} + y\cos\frac{\tau\theta}{2},\; b^3 = \theta. \tag{12}$$

So ρ_b is defined as:

$$\rho_b(g) := \sqrt{(w_1 b^1(g))^2 + (w_2 b^2(g))^2 + (w_3 b^3(g))^2}. \tag{13}$$

We define an approximative kernel by replacing the exact distance in k_t^α (10) by an approximative distance. E.g. the half-angle distance approximation kernel:

$$k_{b,t}^\alpha(g) := \frac{t}{\beta}\left(\frac{\rho_b(g)}{t}\right)^\beta, \quad \text{for all } g \in G. \tag{14}$$

The theory in [2, Thm.1] discusses the quality of the half-angle distance approximation kernel $k_{b,t}^\alpha$ on Lie group $SE(2)$. We generalize this assessment to the Lie groups $SE(2)_\tau$.

Theorem 1 (Quality of the Half Angle Approximation). *Consider a metric tensor field* (5) *that is diagonal w.r.t. basis* $\{\omega^i\}$ (4), *let* $\zeta := \frac{w_2}{w_1}$ *be the spatial anisotropy,* k_t^α *be the exact kernel* (10), $k_{b,t}^\alpha$ *the half-angle approximation* (14), *and* β *be such that* $\frac{1}{\alpha} + \frac{1}{\beta} = 1$, *for* $\alpha > 1$ *fixed.*

- *The exact kernel* k_t^α *and the approximative kernel* $k_{b,t}^\alpha$ *are invariant under all the fundamental symmetries* ε_i, *see Table 2 and [13, Prop.4.3].*
- *Globally it holds that:* $(\frac{\zeta}{\tau})^{-\beta} k_t^\alpha \leq k_{b,t}^\alpha \leq (\frac{\zeta}{\tau})^{\beta} k_t^\alpha$.
- *Locally around the identity* $e = (0,0,0)$ *we have:* $k_{b,t}^\alpha(g) \leq (1+\epsilon(g))^{\beta/2} k_t^\alpha(g)$ *where* $\epsilon(g) := \frac{\zeta^2-\tau^2}{2w_3^2}(\frac{\zeta}{\tau})^4 \rho_b^2(g) + \mathcal{O}((\tau\theta)^3)$.

Proof. Theorem 1 in [2] is a summary of multiple corollaries and lemmas. To translate the theory to $SE(2)_\tau$ we will need to translate all that on which the theorem builds. Luckily, the strategy of every lemma and corollary stays exactly the same, and only the constants need to be adapted. We now highlight the important adaptations of the lemmas and corollaries in [2]:

- Lemma 4: We adapt $\tilde{\mathcal{G}}$ to $\tilde{\mathcal{G}} = w_1^2\, \omega^1 \otimes \omega^1 + (\tau w_1)^2\, \omega^2 \otimes \omega^2 + w_3^2\, \omega^3 \otimes \omega^3$ from which we get the lower bound $l := \sqrt{(w_1 x)^2 + (w_1 \tau y)^2 + (w_3\theta)^2}$. Similarly we get the upper bound $u := \sqrt{(w_2 x/\tau)^2 + (w_2 y)^2 + (w_3\theta)^2}$.
- Corollary 2: The relevant equation becomes $\frac{\tau}{\zeta} d \leq \rho_b \leq \frac{\zeta}{\tau} d$.
- Corollary 3: The global bounds of $k_{b,t}$ become $(\frac{\zeta}{\tau})^{-\beta} k_t^\alpha \leq k_{b,t}^\alpha \leq (\frac{\zeta}{\tau})^{\beta} k_t^\alpha$.
- Lemma 6: The bound on $\|d\rho_b\|$ becomes $\|d\rho_b\|^2 \leq 1 + \frac{\zeta^2-\tau^2}{2w_3^2}\rho_b^2 + \mathcal{O}((\tau\theta)^3)$.
- Lemma 7: The error between ρ_b and d becomes $\rho_b^2 \leq (1+\epsilon)d^2$, where $\epsilon := \frac{\zeta^2-\tau^2}{2w_3^2}(\frac{\zeta}{\tau})^4 \rho_b^2 + \mathcal{O}((\tau\theta)^3)$.

Furthermore, [2, Lem.3, Cor.1, Lem.5, and Cor.5] need no adaptations. □

Table 2. Overview of the symmetries in logarithmic coordinates c^i. For example $\varepsilon_3(c^1, c^2, c^3) = (-c^1, -c^2, c^3)$, which, when translated back to (x, y, θ) coordinates using (7) and (8), is equivalent to $\varepsilon_3(x, y, \theta) = (-x, -y, \theta)$. All symmetries in (x, y, θ) coordinates can be found in [13, Prop.4.3].

	ε_0	ε_1	ε_2	ε_3	ε_4	ε_5	ε_6	ε_7
c^1	+	+	−	−	−	−	+	+
c^2	+	−	+	−	+	−	+	−
c^3	+	+	+	+	−	−	−	−

4 Left-Invariant Convection on a Lie Group

The aim is to achieve the second geometric adaptation of PDE-G-CNNs: fixing the lifting layer. For this objective we need to study left-invariant convection PDEs on a Lie Group G and how they are solved.

Definition 1 (Left-Invariant Convection). *Let G be a Lie group and $c: G \to TG$ a left-invariant vector field. Left-invariant convection is defined as:*

$$\tfrac{\partial W}{\partial t} = -cW, \quad W|_{t=0} = U, \tag{15}$$

where $W : G \times \mathbb{R} \to \mathbb{R}$ is the solution obtained from the initial condition $U \in C^1(G)$.

Before presenting the solution of this convection PDE we define a specific group action of G.

Definition 2. *Let $\mathcal{R}_g : C^1(G) \to C^1(G)$ be given by $(\mathcal{R}_h U)(g) = U(gh)$, $h, g \in G$.*

Using this action we can succinctly state the solution to the convection PDE. Note that our solution is simpler than in the general homogeneous space setting of [15, Prop.5.1] as we are only considering convection on the full Lie group.

Proposition 2 (Solution of Convection). *The solution to left-invariant convection* (15) *is:*

$$W(\cdot, t) = \mathcal{R}_{e^{-tv}} U, \qquad \textit{where } v = c_e \in T_e(G). \tag{16}$$

Proof. We will show that (16) satisfies the convection Eq. (15):

$$\begin{aligned} \tfrac{\partial W}{\partial t}(g,t) &= \lim_{s \to 0} \frac{W(g, t+s) - W(g,t)}{s} = \lim_{s \to 0} \frac{\mathcal{R}_{e^{-sv}} W(g,t) - W(g,t)}{s} \\ &= \lim_{s \to 0} \frac{W(ge^{-sv}, t) - W(g,t)}{s} = (-cW)(g,t), \end{aligned} \tag{17}$$

where the second equality is due to $\mathcal{R}$ being an action together with properties of the exponential map, and the last equality from the fact that $s \mapsto ge^{-sv}$ is a curve corresponding to the tangent vector $-c_g \in T_g G$.

With this in mind we continue with the second geometric adaptation.

5 No Need to Train the Lifting Layer in a PDE-G-CNN

Within the PDE-G-CNN framework, a roto-translation group $G = SE(2)$ equivariant segmentation network $\Phi : \mathbb{L}_2(\mathbb{R}^2) \to \mathbb{L}_2(\mathbb{R}^2)$ is structured as follows:

$$\Phi = P \circ E^{\ell-1} \circ E^{\ell-2} \circ \cdots \circ E^2 \circ L, \tag{18}$$

where ℓ is total amount of layers, C_i denotes the amount of channels/features after the i'th layer, $L : \mathbb{L}_2(\mathbb{R}^2) \to \mathbb{L}_2(G)^{C_1}$ is the lifting layer, $E^i : \mathbb{L}_2(G)^{C_{i-1}} \to \mathbb{L}_2(G)^{C_i}$ are the PDE-based layers, and $P : \mathbb{L}_2(G)^{C_{\ell-1}} \to \mathbb{L}_2(\mathbb{R}^2)$ is the projection layer. Figure 4 shows a diagram of this architecture.

A lift, in this current setting, is any way to equivariantly map an image $f \in \mathbb{L}_2(\mathbb{R}^2)$ to an element of $\mathbb{L}_2(G)$. Conversely, a projection is any way to

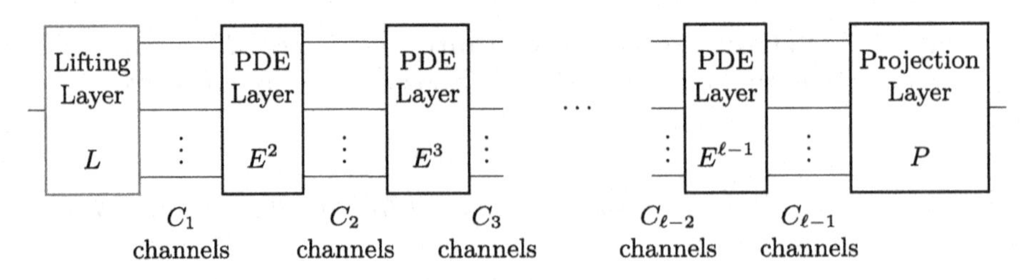

Fig. 4. Diagram of a PDE-G-CNN network (18). In a PDE layer every channel is convected (16), eroded (10), and dilated with $t = 1$.

equivariantly map an element of $\mathbb{L}_2(G)$ to $\mathbb{L}_2(\mathbb{R}^2)$. One natural, and in principle the only, linear way to lift is using *orientation score transforms*. The orientation score transform $\mathcal{W}_\psi : \mathbb{L}_2(\mathbb{R}^2) \to \mathbb{L}_2(G)$ w.r.t. a wavelet $\psi \in \mathbb{L}_2(\mathbb{R}^2)$, of an image $f \in \mathbb{L}_2(\mathbb{R}^2)$ is defined as:

$$(\mathcal{W}_\psi f)(g) := (\mathcal{U}_g \psi, f)_{\mathbb{L}_2(\mathbb{R}^2)} := \int_{\mathbb{R}^2} \psi(g^{-1} \triangleright \mathbf{y})\, f(\mathbf{y})\, \mathrm{d}\mathbf{y}, \tag{19}$$

where $\mathcal{U} : G \to B(\mathbb{L}_2(\mathbb{R}^2))$ is the representation defined by $(\mathcal{U}_g f)(\mathbf{y}) = f(g^{-1} \triangleright \mathbf{y}) = f(R_\theta^{-1}(\mathbf{x} - \mathbf{y}))$, where $\triangleright$ is the standard action of $SE(2)$ on $\mathbb{R}^2$, $g = (\mathbf{x}, \theta) \in SE(2)$, R_θ the corresponding rotation matrix, and $\mathbf{y} \in \mathbb{R}^2$.

As described in [15, Rem.4, Fig. 15], using orientation score transforms we can create a trained lifting layer L_{train} as follows:

$$(L_{\text{train}} f)_i = W_{\psi_i} f, \quad i = 1, 2, \ldots, C_1, \tag{20}$$

where ψ_i are learned kernels. Figure 5a shows a diagram of this layer.

In practice the real-valued kernels ψ_i are implemented as 2D arrays of a certain size. In [15, Fig. 15], for example, there are $C_1 = 16$ lifting kernels of 7×7. We can create a basis for this finite-dimensional space of $k \times k$ kernels using roto-translated versions of some "mother" wavelet ψ_m and write:

$$\psi_i = \sum_{j=1}^{k^2} A_{i,j}\, \mathcal{U}_{g_j} \psi_m, \text{ with } A \in \mathbb{R}^{C_1 \times k^2} \text{ and } g_j \in SE(2).$$

Lifting with a roto-translated version of the mother wavelet turns out to be equivalent to doing convection after lifting with the same mother wavelet:

Proposition 3. *Let $\psi \in \mathbb{L}_2(\mathbb{R}^2)$. For all $h \in SE(2)$ one has $\mathcal{R}_h \circ \mathcal{W}_\psi = \mathcal{W}_{\mathcal{U}_h \psi}$.*

Proof. Since $\mathcal{U}$ is a group representation we have for all $h, g \in G$ that

$$(\mathcal{R}_h \mathcal{W}_\psi f)(g) = (\mathcal{W}_\psi f)(gh) = (\mathcal{U}_{gh} \psi, f) = (\mathcal{U}_g(\mathcal{U}_h \psi), f) = (\mathcal{W}_{\mathcal{U}_h \psi} f)(g)$$

with $\mathbb{L}_2$ inner products $(\cdot,\cdot) = (\cdot,\cdot)_{\mathbb{L}_2(\mathbb{R}^2)}$, for all $g \in SE(2)$, all $f \in \mathbb{L}_2(\mathbb{R}^2)$. ☐

With this information we can show that the lifts W_{ψ_i} amount to:

$$W_{\psi_i} = W_{\sum_{j=1}^{k^2} A_{i,j}\, \mathcal{U}_{g_j}\psi_m} = \sum_{j=1}^{k^2} A_{i,j} W_{\mathcal{U}_{g_j}\psi_m} = \sum_{j=1}^{k^2} A_{i,j} \mathcal{R}_{g_j} W_{\psi_m},$$

where the second equality is due to linearity of the lift w.r.t. the wavelet, and the third due to Proposition 3. This motivates the creation of a *fixed lifting layer*:

$$(L_{\text{fix}} f)_i = \sum_{j=1}^{c} A_{i,j} \mathcal{R}_{g_j} W_{\psi_m} f, \quad i = 1, 2, \ldots, C_1, \tag{21}$$

where we perform $c \leq k^2$ convections, which we see as a new network hyperparameter. A diagram of the fixed lifting layer can be seen in Fig. 5.

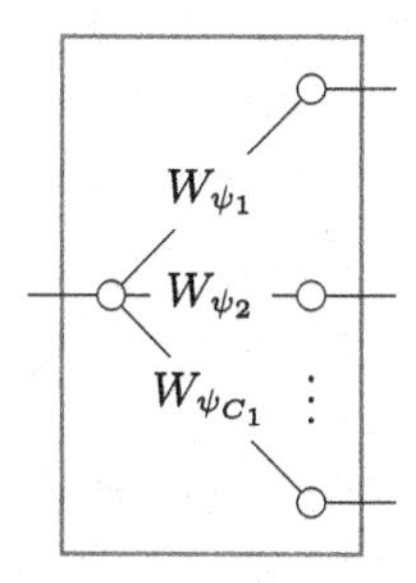

(a) A trained lifting layer L_{train} (20). We perform C_1 lifts W_{ψ_i} with learned kernels ψ_i.

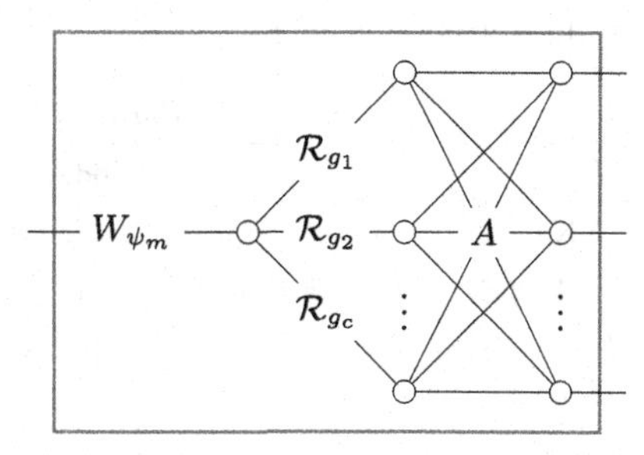

(b) A fixed lifting layer L_{fix} (21). We do a single fixed lift W_{ψ_m} with "mother" wavelet ψ_m, then c learned convections $\mathcal{R}_{g_i}$, and then apply a learned linear combination by $A \in \mathbb{R}^{C_1 \times c}$.

Fig. 5. Two diagrams illustrating a trained and fixed lifting layer, these being two possible ways to implement the first layer in Fig. 4. Both layers take an image in $\mathbb{L}_2(\mathbb{R}^2)$ as a input and return C_1 features in $\mathbb{L}_2(SE(2))$ as output.

In practice to calculate W_{ψ_m} in the fixed lifting layer L_{fix} we use so-called 'cake wavelets' [6] that do not tamper data-evidence (in view of the unitarity results in [6, Thm.19&20]). This data-evidence preserving fixing of the lifting layer makes PDE-G-CNNs and their geometric interpretation much more comprehensive. Recall to this end Fig. 2 depicting the PDE-G-CNN network with lifting and PDE-Layers, and Fig. 3 depicting a single PDE-Layer on $SE(2)$.

It can also be argued that the fixed lifting layer L_{fix} is more general than the trained lifting layer L_{train} in the following sense. We can reinterpret a trained lifting layer as a fixed lifting layer but with the convections constrained to pixel-step translations. This is in contrast to the fixed lifting layer in which the convections are allowed to take arbitrarily large sub-pixel roto-translations.

As we have shown above the fixed and trained lifting layers are theoretically equivalent. However, in general, functionally equivalent architectures do not necessarily yield comparable performance in practice (e.g. in view of sampling constraints). We therefore experimentally investigate these theoretical claims on the publicly available DCA1 dataset [3], following the method of [2, Sec.7.2].

We consider 4 networks, all with 6 layers: a standard CNN, a G-CNN with a trained lifting layer, a PDE-G-CNN with a trained lifting layer, and a PDE-G-CNN with a fixed lifting layer. Every network was trained 10 times for 80 epochs, and after every epoch the average Dice coefficient on the test set is stored. When training is finished the maximum of the average Dice coefficients over all 80 epochs is calculated. The result is 10 Dice coefficients for every architecture.

The results can be found in Table 3. We see that the PDE-G-CNN with a fixed lifting layer has identical performance to the one with a trained lifting layer, while having an effectively equal number of parameters.

Table 3. The architecture of different 6-layer networks, and their performance when applied to the DCA1 dataset. The crosses in the scatterplot indicate the mean and correspond with the last column of the table. The fixed lifting PDE-G-CNN was done with $c = 40$ convections (21).

Type	Lifting	C_1	**Params.**	**Dice Coeff.**
CNN	-	-	25662	0.756 $\pm$ 0.0051
G-CNN	Trained	8	24632	0.751 $\pm$ 0.0020
PDE-G-CNN	Trained	16	2560	0.775 $\pm$ 0.0042
PDE-G-CNN	Fixed	16	2536	0.775 $\pm$ 0.0052

0.70 0.80

6 Nondiagonal Metric for Erosion and Dilation

Next we address the third geometric adaptation of PDE-G-CNNs. In a PDE layer every channel is convected (16), eroded (10), and dilated. All of these 3 processes allow to include spatial curvature-adaptation. With convection, in the case of $\tau = 1$, the curvature of the spatially projected exponential curve $\gamma(t) = \exp(tv)$, where $v = c^1\partial_x|_e + c^2\partial_y|_e + c^3\partial_\theta|_e \in T_e(G)$, is constant and equals $\kappa(v) := c^3\text{sign}(c^1)/\sqrt{|c^1|^2 + |c^2|^2}$ as follows by (7) and [10, Eq. 32]. Thereby training for c^3 in the convection part is a curvature adaptation.

With dilation/erosion it is more subtle. Any left-invariant metric, diagonal or nondiagonal (5) w.r.t. the left-invariant frames (4), can always be diagonalized into an orthogonal basis of left-invariant vector fields $\mathcal{B}_i$. More concretely, one can decompose the metric as $[g_{ij}] = R^T D R$ where R is a rotation matrix and D a diagonal matrix with positive entries. We interpret R as the matrix that brings us from components w.r.t $\mathcal{A}_i$ to components w.r.t $\mathcal{B}_i$. This means that $\mathcal{B}_i = R_{i,1}\mathcal{A}_1 + R_{i,2}\mathcal{A}_2 + R_{i,3}\mathcal{A}_3$. By looking at the exponential curves $\exp(t\mathcal{B}_i|_e)$, just as we did for convection, we see that training a nondiagonal metric is also a curvature adaption. The effect of this adaptation can be seen in the second row of Table 1: the contours clearly curve towards the reader. Here we set rotation $R = R_{\mathbf{e}_2,-\pi/4}$ about axis $\mathbf{e}_2 = (0, 1, 0)$ making $\kappa(\mathcal{B}_1) = -1$.

To see if these curvature adaptations change the performance of the PDE-G-CNNs we perform some experiments on the Lines dataset from [2, Sec.7.3]. We consider 4 different 6-layer PDE-G-CNNs: one with diagonal metrics using the logarithmic distance approximation ρ_c, one with diagonal metrics and using ρ_b, one with nondiagonal metrics using ρ_b, and then another nondiagonal ρ_b where we fix $c^3 = 0$ for all convections in the PDE layers. Just as in [2] we increased the kernel size to [9, 9, 9]. Every network was trained 15 times for 60 epochs.

The results can be found in Table 4. We see that the curvature adaptation in the convection part is indeed beneficial, but for the dilation and erosion part we conclude that a nondiagonal metric has no significant impact on the performance of the PDE-G-CNNs on the Lines dataset. We hypothesize that the effect of nondiagonal dilation and erosion can be (sufficiently) emulated by diagonal dilation and erosion *after* convection.

Table 4. The architecture of different 6-layer PDE-G-CNN networks, and their performance when ran on the Lines dataset.

Approx.	Metric	b^3, c^3	Params.	Dice Coeff.
ρ_c	Diag.	Free	6018	0.9688 ± 0.0063
ρ_b	Diag.	Free	6018	0.9758 ± 0.0035
ρ_b	Nondiag.	Free	7458	0.9764 ± 0.0009
ρ_b	Nondiag.	0	7338	0.9718 ± 0.0028

0.95 1.00

7 Conclusion

In this article we have investigated 3 geometric adaptations of PDE-G-CNNs.

1.) In Theorem 1 we generalized the theoretical results of [2, Thm.1] to $SE(2)_\tau$, a family of Lie groups that lie in between the roto-translation group $SE(2)$ and the Heisenberg group $H(3)$. This enables us to transfer training orientation score processing on $SE(2)$ to training the processing of, for example, velocity scores on $H(3)$. Theorem 1 motivates that our newly proposed approximative kernels for PDE-G-CNNs are indeed qualitatively close to the exact kernels.
2.) We showed both theoretically and in practice (Table 3) that one does not need to lift with learned kernels in PDE-G-CNNs. A fixed lifting layer using cake wavelets gives identical performance on the DCA1 dataset, while keeping the network complexity effectively the same. Our experiments confirm that fixing the lifting layer preserves the performance while considerably increasing the geometric interpretation (Fig. 3) of PDE-G-CNNs.
3.) In Table 4 we showed on the Lines dataset that curvature adaptation is only useful within the convection part. Within the erosion and dilation parts curvature adaption, in the form of a nondiagonal metric w.r.t. the left-invariant frames (4), gives no significant increase in performance on the Lines dataset.

For a Git repository containing all code see: https://gitlab.com/bsmetsjr/lietorch.

Acknowledgements. We gratefully acknowledge the Dutch Foundation of Science NWO for funding of VICI 2020 Exact Sciences (Duits, Geometric learning for Image Analysis, VI.C. 202-031). We thank Andrii Kompanets for a Python implementation of the www.LieAnalysis.nl package, initially developed in Mathematica, which was used to construct the cake wavelets for the fixed lifting layer experiments.

References

1. August, J., Zucker, S.: Sketches with curvature: the curve indicator random field and Markov processes. IEEE PAMI **25**(4), 387–400 (2003)
2. Bellaard, G., Bon, D., Pai, G., Smets, B., Duits, R.: Analysis of (sub-)Riemannian PDE-G-CNNs (2022). https://doi.org/10.48550/ARXIV.2210.00935
3. Cervantes-Sanchez, F., Cruz-Aceves, I., Hernandez-Aguirre, A., Hernandez-Gonzalez, M.A., Solorio-Meza, S.E.: Automatic segmentation of coronary arteries in x-ray angiograms using multiscale analysis and artificial neural networks. Appl. Sci. **9**(24), 5507 (2019)
4. Cocci, G., Barbieri, D., Sarti, A.: Spatiotemporal receptive fields of cells in V1 are optimally shaped for stimulus velocity estimation. JOSA A **29**(1), 130–138 (2012)
5. Cohen, T.S., Welling, M.: Group equivariant convolutional networks. In: Proceedings of the 33rd International Conference on Machine Learning, vol. 48, pp. 1–12 (2016)
6. Duits, R.: Perceptual organization in image analysis: a mathematical approach based on scale, orientation and curvature. Ph.D. thesis, TU/e (2005)
7. Duits, R., Franken, E.M.: Left-invariant parabolic evolution equations on $SE(2)$ and contour enhancement via invertible orientation scores, part I: Linear left-invariant diffusion equations on $SE(2)$. QAM-AMS **68**, 255–292 (2010)
8. Duits, R., Führ, H., Janssen, B., Bruurmijn, M., Florack, L., van Assen, H.: Evolution equations on Gabor transforms & applications. ACHA **35**(3), 483–526 (2013)
9. Evans, L.C.: Partial differential equations, vol. 19. AMS (2010)
10. Franken, E.M., Duits, R.: Crossing-preserving coherence-enhancing diffusion on invertible orientation scores. IJCV **85**(3), 253–278 (2009)
11. Labate, D., Lim, W.Q., Kutyniok, G., Weiss, G.: Sparse multidimensional representation using shearlets. In: Papadakis, M., Laine, A.F., Unser, M.A. (eds.) Wavelets XI, vol. 5914, p. 59140U. SPIE (2005)
12. Liu, L., Wang, M., Zhou, S., Shu, M., Cohen, L.D., Chen, D.: Curvilinear structure tracking based on dynamic curvature-penalized geodesics. Pattern Recogn. **134**, 109079 (2023)
13. Moiseev, I., Sachkov, Y.L.: Maxwell strata in sub-Riemannian problem on the group of motions of a plane. ESAIM Control Optim. Calc. Var. **16**(2), 380–399 (2010)
14. Schmidt, M., Weickert, J.: Morphological counterparts of linear shift-invariant scale-spaces. JMIV **56**(2), 352–366 (2016)
15. Smets, B.M.N., Portegies, J.W., Bekkers, E.J., Duits, R.: PDE-based group equivariant convolutional neural networks. JMIV **65**(1), 209–239 (2022)
16. Smets, B.M.N., Portegies, J.W., St-Onge, E., Duits, R.: Total variation and mean curvature PDEs on the homogeneous space of positions and orientations. JMIV **63**(2), 237–262 (2021)
17. Zhang, J., Dashtbozorg, B., Bekkers, E., Pluim, J.P.W., Duits, R., ter Haar Romeny, B.M.: Robust retinal vessel segmentation via locally adaptive derivative frames in orientation scores. IEEE T-MI **35**(12), 2631–2644 (2016)

The Variational Approach to the Flow of Sobolev-Diffeomorphisms Model

Mara Guastini[1], Marko Rajković[1], Martin Rumpf[1], and Benedikt Wirth[2(✉)]

[1] Institute for Numerical Simulation, University of Bonn, Endenicher Allee 60, 53115 Bonn, Germany
{marko.rajkovic,martin.rumpf}@ins.uni-bonn.de

[2] Applied Mathematics Münster, University of Münster, Orléans-Ring 10, 48149 Münster, Germany
benedikt.wirth@uni-muenster.de

Abstract. The flow of diffeomorphisms, aka LDDMM, is a framework to define a group G of diffeomorphisms of chosen regularity with a Riemannian structure. If these diffeomorphisms are used to deform a template shape or image, they generate a space of shapes or images to which the Riemannian structure descends. For this reason LDDMM lies at the centre of computational anatomy, shape space theory, and image metamorphosis. Typically, to obtain a geodesic equation on G one formally applies the geodesic equation of Riemannian Lie groups, to which G has structural similarity. Then interpolation tasks between two given deformations, needed for all kinds of statistical analyses, are solved by shooting discretized geodesics within an optimal control approach. If G is chosen with Sobolev regularity, it is known to be a veritable infinite-dimensional Riemannian manifold. In this setting we derive the weak geodesic PDE, which in its strong form coincides with the formally derived one, and present a time discretization to compute a geodesic between given deformations by minimizing a time-discrete path energy, which Mosco-converges to the continuous path energy. This variational ansatz is a more natural alternative to shooting and to our knowledge the first numerical approach to LDDMM by minimization.

Keywords: LDDMM · flow of diffeomorphisms · discrete geodesic calculus · shape spaces

1 Introduction

Two competing, but of course tightly connected paradigms exist in the treatment of shape spaces and related notions: the dynamical systems approach, where paths in shape space are computed by solving Hamiltonian systems, and the variational methods approach, where minimizers of path energies are computed. The widely used flow of diffeomorphisms or LDDMM (Large Deformation Diffeomorphic Metric Mapping) model is traditionally treated in the former

L. Calatroni et al. (Eds.): SSVM 2023, LNCS 14009, pp. 551–564, 2023.
https://doi.org/10.1007/978-3-031-31975-4_42

framework, which readily allows to exploit its formal Lie group structure and thereby formally derive a geodesic equation. In this article we take the alternative, variational perspective on LDDMM:

1. We derive the weak geodesic PDE of Sobolev LDDMM as the optimality condition of the path energy. Not surprisingly, its strong form coincides with the known, formally derived equation, however, as in most PDE settings the latter can only be valid for special geodesics that exhibit sufficient regularity. In fact, for non-Sobolev LDDMM it is not even known whether the associated space of diffeomorphisms forms a manifold so that already the existence of a geodesic equation – weak or strong – is questionable. For Sobolev LDDMM, though, our result confirms the validity of the formally known geodesic equation (though in a weak form), even for geodesics of generic regularity.
2. We introduce a time-discrete path energy which Mosco-converges to the (time-continuous) path energy so that its minimizers, time-discrete geodesics, converge to (time-continuous) geodesics. This discretization is obviously based on the variational paradigm and therefore tailor-made to numerically solve the geodesic boundary value problem, while the dynamical viewpoint rather fits to the initial value problem.

2 The Flow of Diffeomorphisms Model

We here recapitulate the traditional viewpoint. The defining concept underlying LDDMM was developed in the '90s independently by different researchers, most notably Trouvé [13] and Christensen et al. [5]. Consider some open domain $D \subset \mathbb{R}^d$, e.g. the unit cube. As mentioned before, LDDMM brings forth a group of diffeomorphisms of D with a structure akin to Riemannian Lie groups. Fixing a Hilbert space $\mathcal{H}$ of vector fields ($\mathbb{R}^d$-valued functions) on D that vanish on ∂D, these diffeomorphisms, i.e. deformations of D into itself, are generated by temporally changing velocity fields $v : [0,1] \to \mathcal{H}$ via computing the associated *flow*

$$\dot{\psi}(t)(x) = v(t)(\psi(t)(x)), \qquad \psi(0)(x) = x, \tag{1}$$

for all $x \in D$, where the dot denotes differentiation with respect to time t. In the language of fluid mechanics, ψ is the Lagrangian and v the Eulerian description of the flow. One then defines the LDDMM group of diffeomorphisms as $G = \{\psi(1) \mid \psi \text{ solves (1) for} v \in L^2((0,1);\mathcal{H})\}$ with composition $(\phi, \theta) \mapsto \phi \circ \theta$ as the group product. As long as $\mathcal{H}$ continuously embeds into $C^1(\overline{D})^d$ this group turns out to be well-defined, and it can be equipped with the right-invariant geodesic distance

$$\mathrm{dist}(\phi,\theta) = \mathrm{dist}(\mathrm{id}, \theta\circ\phi^{-1}) = \inf\{\|v\|_{L^2((0,1);\mathcal{H})} \mid \theta\circ\phi^{-1} = \psi(1,\cdot) \text{for } \psi \text{ solving (1) }\}.$$

One can interpret G as a Riemannian Lie group (except in general it is unclear whether G actually is a manifold; also, multiplication will not be smooth): The tangent space at the identity deformation is $T_{\mathrm{id}}G = \mathcal{H}$, the space of admissible velocity fields, and it is equipped with the inner product $g_{\mathrm{id}} = (\cdot,\cdot)_{\mathcal{H}}$. This induces (by right action of G on itself) tangent spaces $T_\phi G = \mathcal{H}\phi$ with inner product $g_\phi(v,w) = g_{\text{id}}(v\phi^{-1}, w\phi^{-1}) = (v\circ\phi^{-1}, w\circ\phi^{-1})_\mathcal{H}$. Geodesics $\psi : [0,1] \to G$ between $\phi, \theta \in G$ are then minimizers of the path energy

$$\widetilde{\mathcal{E}}[\psi] = \int_0^1 g_{\psi(t)}(\dot\psi(t), \dot\psi(t))\mathrm{d}t = \int_0^1 (v(t), v(t))_\mathcal{H}\,\mathrm{d}t, \qquad \text{for } v(t) = \dot\psi(t)\circ\psi(t)^{-1}$$

over all paths ψ connecting ϕ with θ. Throughout the paper we will distinguish between different equivalent path energies that just differ in their argument: $\widetilde{\mathcal{E}}$ acts on time-continuous paths of diffeomorphisms, $\overline{\mathcal{E}}$ acts on the corresponding inverse diffeomorphisms, and $\mathcal{E}$ on the associated time-continuous velocity fields. Later, $\mathbf{E}$ will indicate the path energy of time-discrete velocity fields.

The geodesic equation is the optimality condition for ψ. It will typically be a PDE involving a differential or nonlocal operator $L = R^{-1}$, which represents the inverse of the Riesz isomorphism

$$R : \mathcal{H}' \to \mathcal{H}$$

and thus satisfies $\langle Lv, w\rangle = (v,w)_\mathcal{H}$ (it will be made more concrete later on).

On finite-dimensional Lie groups G with right-invariant Riemannian metric $g_{\text{id}}(v,w) = \langle Lv, w\rangle$ the geodesic equation, which is a second order ODE, has a well-known form: It is usually expressed as a Hamiltonian system of two first order ODEs in the state $\psi \in G$ and a *momentum* $p \in (T_\psi G)'$, where the (right-invariant) Hamiltonian associated with the (right-invariant) metric is $H(\psi, p) = H(\text{id}, p\psi^{-1}) = \frac{1}{2}\langle p\psi^{-1}, R(p\psi^{-1})\rangle$:

$$\dot\psi(t) = \partial_p H(\psi(t), p(t)), \qquad \dot p(t) = -\partial_\psi H(\psi(t), p(t)).$$

However, since the Hamiltonian is invariant under the action of G (in fact also under time shifts), by Noether's theorem the dynamical system conserves (some transformation of) the momentum and therefore can be reduced to a single first order ODE. This conservation can equivalently be expressed in three distinct momenta, $p \in (T_\psi G)'$ as well as its two pullbacks onto $(T_{\text{id}}G)'$ via $m = \psi^{-1}p$ or $\rho = p\psi^{-1} = \text{Ad}^*_{\psi^{-1}} m$ with $\text{Ad}_\psi : T_{\text{id}}G \to T_{\text{id}}G$, $v \mapsto \psi v \psi^{-1}$ denoting the *adjoint map* and * the dual operator. The second momentum m equals the so-called momentum map $\mathfrak{m}(\psi, p)$ defined by $\langle \mathfrak{m}(\psi,p), v\rangle = \langle p, \psi v\rangle$. It turns out to stay constant in time so that the desired first order ODE in ψ reads $\dot\psi(t) = \partial_p H(\psi(t), p(t))$ for $p(t) = \psi(t)m(t) = \psi(t)m(0) = \psi(t)\cdot\psi(0)^{-1}\cdot p(0)$. The third momentum ρ instead satisfies (using the relation $m = \text{Ad}^*_\psi \rho$)

$$\rho(t) = \text{Ad}^*_{\psi(t)^{-1}} \text{Ad}^*_{\psi(0)} \rho(0),$$

(see e.g. [15]) which is the integral of the *Euler–Poincaré equation*

$$\dot\rho = -(\text{ad}_{v(\rho)})^*\rho \qquad \text{for } v(\rho) = \dot\psi\psi^{-1} = \partial_\rho H(\text{id}, \rho),$$

where the last equality follows from a sequence of algebraic manipulations and $\text{ad}_v w = \partial_\psi \text{Ad}_\psi w|_{\psi=\text{id}}(v) = [v,w]$ equals the Lie bracket on $T_{\text{id}}G$. This equation describes the full dynamical system within the Lie algebra: Once it is solved, $\psi(t)$ is simply calculated by integrating $\dot{\psi} = (\dot{\psi}\psi^{-1})\psi = v(\rho)\psi$.

The Euler–Poincaré form of the geodesic equation is formally applied in LDDMM as well [10]: Since the momentum $p(t)$ is dual to $\dot{\psi}(t)$, the momentum $\rho(t) = p(t) \circ \psi(t)^{-1}$ is the one dual to $\dot{\psi}(t) \circ \psi^{-1}(t) = v(t) \in T_{\text{id}}G = \mathcal{H}$, thus $\rho(t) = Lv(t) \in \mathcal{H}'$. The Euler–Poincaré equation now formally reads

$$\begin{aligned}\langle \dot{\rho}(t), w \rangle = \langle \rho(t), \text{ad}_{v(t)} w \rangle &= \langle \rho(t), (Dw)v(t) - (Dv(t))w \rangle \\ &= \langle \rho(t), \text{div}(w \otimes v(t)) - w\text{div}v(t) - (Dv(t))w \rangle\end{aligned}$$

for all $w \in \mathcal{H}$. Using that the adjoints of $-D$ and $-\text{div}$ are the distributional divergence and gradient (since the velocity fields vanish on the boundary of D), an integration by parts turns this into

$$\dot{\rho}(t) = -[D\rho(t)v(t) + \rho(t)\text{div}v(t) + (Dv(t))^T \rho(t)] \quad \text{for} \quad v(t) = R\rho(t). \tag{2}$$

It is this equation which is usually solved numerically due to its simplicity (depending on the chosen Hilbert space, the Riesz operator R may be as simple as convolving with a mollifier). However, it only holds formally: On the right-hand side we differentiate $\rho \in \mathcal{H}'$ and $v \in \mathcal{H}$ and then multiply with v and ρ, respectively, which is not well-defined. Essentially, this issue originates from problems defining the adjoint map since left-multiplication (of a fixed element $\psi \in G$) is not differentiable (the derivative involves $\nabla\psi$ and thus a loss of regularity).

If $\mathcal{H}$ is chosen as the Sobolev space $W^{m,2}_{1,0}(D)^d := (W^{m,2}(D) \cap W^{1,2}_0(D))^d$ of functions with square integrable mth weak derivatives and vanishing function values on ∂D, $m > 1 + \frac{d}{2}$ (recall that the well-definedness of the flow requires $\mathcal{H} \hookrightarrow C^1(\overline{D})^d$), we denote the group G as the group $\mathcal{D}^m$ of Sobolev diffeomorphisms. In that specific setting it is actually known that $\mathcal{D}^m$ is the connected component of the identity of the open set

$$\{\phi = \text{id} + \xi \,|\, \xi \in W^{m,2}_{1,0}(D)^d, \ \det D\phi > 0 \text{ everywhere}\}$$

and thus a true Riemannian manifold, which furthermore is metrically and geodesically complete [4]. For the inner product on $\mathcal{H} = W^{m,2}_{1,0}(D)^d$ or equivalently the inverse Riesz operator $L : \mathcal{H} \to \mathcal{H}'$ there are many possible choices, a specific example being $L = (\text{Id} + \Delta)^m$. Geodesics ψ in $\mathcal{D}^m$ then correspond to minimizers of the path energy

$$\widetilde{\mathcal{E}}[\psi] = \int_0^1 \langle Lv, v \rangle \, \text{d}t \qquad \text{for } v(t) = \dot{\psi}(t) \circ \psi(t)^{-1} \tag{3}$$

for prescribed end points $\psi(0) = \psi_A, \psi(1) = \psi_B \in \mathcal{D}^m$. Equivalently, this can be expressed as an optimization problem in the velocity field $v \in L^2((0,1); \mathcal{H})$ by defining its flow ψ as

$$\dot{\psi}(t) = v(t) \circ \psi(t), \qquad \psi(0) = \psi_A \tag{4}$$

and minimizing the energy

$$\mathcal{E}[v] = \begin{cases} \int_0^1 \langle Lv, v \rangle \, \text{d}t & \text{if } \psi(1) = \psi_B \text{ for the solution of (4)}, \\ \infty & \text{else.} \end{cases} \tag{5}$$

3 The Weak Form of the Geodesic Equation

A direct formal derivation of the Euler–Poincaré Eq. (2) in the context of LDDMM is found in [10, p. 215] or [9, Thm. 3.1]. In the special case $G = \mathcal{D}^m$ of Sobolev diffeomorphisms, the same Euler–Poincaré equation can be derived rigorously, though it of course only holds in some weak form which is not obvious from (2). The validity of the Euler–Poincaré equation (or at least our derivation) depends on the metric chosen on $\mathcal{H} = W^{m,2}_{1,0}(D)^d$ (with $m > \frac{d}{2} + 1$) or equivalently the corresponding inverse Riesz isomorphism $L : \mathcal{H} \to \mathcal{H}'$. Abbreviating $N := d^{m+1}$, it can be expressed in the form

$$L = B^*B + \bar{L} \qquad \text{with } B : \mathcal{H} \to L^2(D)^N,\ Bv = AD^m v \tag{6}$$

for some $A \in L^\infty(D)^{N\times N}$ and compact, self-adjoint $\bar{L} : \mathcal{H} \to \mathcal{H}'$. We will assume

$$\bar{L} \text{ is continuous from } W^{m,2}_{1,0}(D)^d \text{ to } (W^{m-1,2}_{1,0}(D)^d)'. \tag{7}$$

The standard example would be $L = \mathrm{Id}+(-\Delta)^m$, but for instance also a nonlocal operator $L = (\mathrm{Id} - \Delta)^m + (\mathrm{Id} - \Delta)^{m-1/2}$ would be possible.

To derive the Euler–Poincaré equation, let $\psi \in W^{1,2}((0,1);\mathcal{D}^m)$ be a geodesic with velocity $v(t) = \dot{\psi}(t) \circ \psi(t)^{-1}$ and momentum $\rho(t) = Lv(t)$. Without changing the end points we perturb the geodesic to a curve $\psi_\epsilon(t) = (\mathrm{id}+\epsilon\eta(t))\circ\psi(t)$ using a smooth diffeomorphism $\mathrm{id} + \epsilon\eta(t)$ with $\eta \in C^\infty_c((0,1); C^\infty_c(D))$. Obviously we have $\psi_0 = \psi$ as well as $\psi_\epsilon \in \mathcal{D}^m$ for $|\epsilon|$ small enough (such that $\mathrm{id} + \epsilon\eta$ is invertible). The associated velocity field is given by

$$v_\epsilon(t) = \dot{\psi}_\epsilon(t) \circ \psi_\epsilon(t)^{-1} \text{ with momentum } \rho_\epsilon(t) = Lv_\epsilon(t).$$

A straightforward differentiation of the path energy $\widetilde{\mathcal{E}}[\psi_\epsilon] = \int_0^1 \langle Lv_\epsilon, v_\epsilon\rangle \,\mathrm{d}t$ is not feasible since in general the derivative of

$$v_\epsilon(t) = [(\mathbb{1} + \epsilon D\eta(t))v(t) + \epsilon\dot{\eta}(t)] \circ (\mathrm{id} + \epsilon\eta(t))^{-1}$$

with respect to ϵ does not lie in $W^{m,2}_{1,0}(D)^d$. Instead, we will first perform a change of variables via $\mathrm{id} + \epsilon\eta(t)$ to arrive at

$$\begin{aligned}\widetilde{\mathcal{E}}[\psi_\epsilon] &= \int_0^1 \langle \bar{L}v_\epsilon(t), v_\epsilon(t)\rangle \,\mathrm{d}t + \int_0^1\int_D Bv_\epsilon(t)\cdot Bv_\epsilon(t)\,\mathrm{d}x\,\mathrm{d}t\\ &= \int_0^1 \langle \bar{L}v_\epsilon(t), v_\epsilon(t)\rangle \,\mathrm{d}t + \int_0^1\int_D |(Bv_\epsilon(t))\circ(\mathrm{id}+\epsilon\eta(t))|^2 \det(\mathbb{1}+\epsilon D\eta(t))\,\mathrm{d}x\,\mathrm{d}t,\end{aligned}$$

which, with the help of the following lemma, can then be differentiated to arrive at the appropriate weak form of the Euler–Poincaré equation.

Lemma 1 (Regularity of velocity perturbation). *Under* (6)–(7) *the map*

$$t \mapsto \mathcal{O}(v(t), \eta(t), \epsilon) := \tfrac{\partial}{\partial\epsilon}\left[B(v \circ (\mathrm{id} + \epsilon\eta(t))^{-1}) \circ (\mathrm{id} + \epsilon\eta(t))\right]$$

is an element of $L^2((0,1); L^2(D)^N)$ *for* $|\epsilon|$ *small enough. At* $\epsilon = 0$ *it is given by*

$$t \mapsto -B(Dv\eta) + D(Bv)\eta,$$

which a priori only maps into $(W^{1,2}_{1,0}(D)^N)' = W^{-1,2}(D)^N$.

Essentially, the principal symbols of DB and BD coincide so that the problematic highest order derivatives cancel.

Proof. The explicit formula is straightforward to derive using $\frac{\partial}{\partial\epsilon}(\mathrm{id}+\epsilon\eta(t))^{-1} = -[(\mathbb{1}+\epsilon D\eta(t))^{-1}\eta(t)]\circ(\mathrm{id}+\epsilon\eta(t))^{-1}$, which comes from differentiating $(\mathrm{id}+\epsilon\eta(t))\circ(\mathrm{id}+\epsilon\eta(t))^{-1}=\mathrm{id}$ with respect to ϵ. It remains to prove the regularity of $\mathcal{O}(v(t),\eta(t),\epsilon)$. However, using $D(f\circ(\mathrm{id}+\epsilon\eta)^{-1})\circ(\mathrm{id}+\epsilon\eta) = Df(\mathbb{1}+\epsilon D\eta)^{-1}$ for any tensor-valued function f, we can explicitly calculate

$$B(v\circ(\mathrm{id}+\epsilon\eta(t))^{-1})\circ(\mathrm{id}+\epsilon\eta(t))=AD(...D(Dv(\mathbb{1}+\epsilon D\eta)^{-1})(\mathbb{1}+\epsilon D\eta)^{-1}..)(\mathbb{1}+\epsilon D\eta)^{-1}.$$

Obviously this equals $AD^m v((\mathbb{1}+\epsilon D\eta)^{-1}(\cdot),\dots,(\mathbb{1}+\epsilon D\eta)^{-1}(\cdot))$ plus a polynomial with bounded coefficients in derivatives of $(\mathbb{1}+\epsilon D\eta)^{-1}$ and derivatives of v up to order $m-1$, in which the polynomial is linear. Obviously, differentiating with respect to ϵ this becomes a linear polynomial in the derivatives of v up to order m with bounded coefficients. □

Theorem 1 (Euler–Poincaré equation for geodesics in $\mathcal{D}^m$). *Under* (6) *and* (7) *a geodesic ψ in $\mathcal{D}^m$ with Eulerian velocity v and momentum $\rho = Lv$ satisfies the weak Euler–Poincaré equation*

$$0=\int_0^1 -2\langle \bar{L}v, Dv\eta\rangle + 2\langle\rho,\dot\eta+D\eta v\rangle + \int_D |Bv|^2\mathrm{div}\eta + 2Bv\cdot\mathcal{O}(v,\eta,0)\,\mathrm{d}x\,\mathrm{d}t$$

for all $\eta\in C_c^\infty((0,1);C_c^\infty(D)^d)$, whose strong form is (2).

Proof. Since $\mathcal{D}^m$ is closed under composition and $\mathrm{id}+\epsilon\eta\in\mathcal{D}^m$ for $|\epsilon|$ small enough, $\eta(t)\circ\psi(t)$ is an admissible perturbation of the geodesic ψ. A necessary condition for ψ being geodesic thus is that the directional derivative of the path energy $\widetilde{\boldsymbol{\mathcal{E}}}$ at ψ in direction $\eta(t)\circ\psi(t)$ (if it exists) vanishes, thus $\frac{\mathrm{d}}{\mathrm{d}\epsilon}\widetilde{\boldsymbol{\mathcal{E}}}[\psi_\epsilon]_{\epsilon=0}=0$. However, using the above reformulation of the path energy and $\frac{\partial}{\partial\epsilon}[(Bv_\epsilon)\circ(\mathrm{id}+\epsilon\eta)]_{\epsilon=0} = B(\dot\eta+D\eta v)+\mathcal{O}(v,\eta,0)$ as well as $\frac{\partial}{\partial\epsilon}\langle\bar{L}v_\epsilon,v_\epsilon\rangle = 2\langle\bar{L}v_\epsilon,\frac{\partial}{\partial\epsilon}v_\epsilon\rangle$, this is exactly the stated weak form of the Euler–Poincaré equation. It is well-defined since η is smooth and $\mathcal{O}(v,\eta,0)\in L^2((0,1);L^2(D)^N)$ by Lemma 1. To obtain its strong form for sufficiently smooth ρ, v we first use the expression from Lemma 1 to arrive at

$$\begin{aligned}
0&=\int_0^1 -2\langle\bar{L}v,Dv\eta\rangle+2\langle\rho,\dot\eta+D\eta v\rangle+\int_D|Bv|^2\mathrm{div}\eta+2Bv\cdot(-B(Dv\eta)+D(Bv)\eta)\,\mathrm{d}x\,\mathrm{d}t\\
&=\int_0^1 -2\langle(B^*B+\bar{L})v,Dv\eta\rangle+\int_D 2\rho\cdot(\dot\eta+D\eta v)-D(|Bv|^2)\cdot\eta+2Bv\cdot(D(Bv)\eta)\,\mathrm{d}x\,\mathrm{d}t\\
&=-\int_0^1\int_D 2\eta(t)\cdot[\dot\rho(t)+D\rho(t)v(t)+\rho(t)\mathrm{div}v(t)+(Dv(t))^T\rho(t)]\,\mathrm{d}x\,\mathrm{d}t
\end{aligned}$$

after integrating by parts. Since η was arbitrary, this leads to the stated strong form. □

4 A First Variational Time Discretization

Recall that we aim to numerically find geodesics between two diffeomorphisms $\psi_A, \psi_B \in \mathcal{D}^m$ by minimizing (3) among all paths $\psi : [0,1] \to \mathcal{D}^m$ connecting ψ_A with ψ_B (which is in contrast to the shooting approach usually used in LDDMM, e.g. [1]) or equivalently by minimizing (5). To this end we discretize time $[0,1]$ by K steps of size $\frac{1}{K}$ and approximate the path ψ by a corresponding time-discrete path $\boldsymbol{\psi}_K = (\psi_{K,0}, \psi_{K,1}, \ldots, \psi_{K,K}) \in (\mathcal{D}^m)^{K+1}$ with $\psi_{K,0} = \psi_A$ and $\psi_{K,K} = \psi_B$ and the velocity field v by a time-discrete velocity $\boldsymbol{v}_K = (v_{K,1}, \ldots, v_{K,K}) \in \mathcal{H}^K$, so (4) turns into

$$\mathrm{id} + \frac{v_{K,k}}{K} = \psi_{K,k} \circ \psi_{K,k-1}^{-1} =: \phi_{K,k}, \qquad k = 1, \ldots, K.$$

A corresponding discrete energy could simply be defined as $\frac{1}{K}\sum_{k=1}^{K} \langle L v_{K,k}, v_{K,k} \rangle$. However, to illustrate that often better, physically motivated choices are possible we consider from now on a specific, exemplary inverse Riesz operator $L : \mathcal{H} \to \mathcal{H}'$ (which admits the decomposition (6)–(7) from the previous section), defined by

$$\langle Lv, w \rangle = \int_D \tfrac{\lambda}{2} \mathrm{tr}\varepsilon[v] \mathrm{tr}\varepsilon[w] + \mu \mathrm{tr}(\varepsilon[v]\varepsilon[w]) + \gamma D^m v \cdot D^m w \,\mathrm{d}x \quad \text{for all } v, w \in \mathcal{H},$$

where $\lambda, \mu, \gamma > 0$ and $\varepsilon[v] := \frac{1}{2}(Dv + Dv^T)$ the symmetrized gradient. The lower-order terms represent the viscous dissipation of an isotropic Newtonian fluid with shear and compression viscosity μ and λ, respectively. A time-discrete counterpart $\mathbf{E}_K$ of the time-continuous path energy $\boldsymbol{\mathcal{E}}$ is then given by

$$\mathbf{E}_K[\boldsymbol{v}_K] = K \sum_{k=1}^{K} \int_D \mathrm{W}(D(\mathrm{id} + \tfrac{v_{K,k}}{K})) + \gamma |\tfrac{1}{K} D^m v_{K,k}|^2 \,\mathrm{d}x. \tag{8}$$

Here, $\mathrm{W} : \mathbb{R}^{d,d} \to [0,\infty]$ is a C^4-function with $\mathrm{W}(\mathrm{id}) = 0$, $D\mathrm{W}(\mathrm{id}) = 0$, and $\frac{1}{2}D^2\mathrm{W}(\mathrm{id})(A,A) = \frac{\lambda}{2}(\mathrm{tr}A)^2 + \mu \mathrm{tr}\left((A^{\mathrm{sym}})^2\right)$ with $A^{\mathrm{sym}} = \frac{1}{2}(A + A^T)$, where in particular we think of rigid motion invariant stored elastic energy functions W that allow good accuracy already at coarse time discretizations [12]. For the purpose of our convergence analysis we shall assume there exists a constant $c_{\mathrm{W}} > 0$ such that $\mathrm{W}(A) \geq c_{\mathrm{W}} \min\{1, |A^{\mathrm{sym}} - \mathbb{1}|^2\}$, cf. [6].

We will constrain the incremental deformations $\phi_{K,k}$ to stay in the set

$$\mathcal{A} := \left\{ \phi \in W^{m,2}(D;D) \,\middle|\, |D\phi(x) - \mathbb{1}|_2 \leq \tfrac{1}{2} \text{ for all } x \in D, \phi = \mathrm{id} \text{ on } \partial D \right\} \subset \mathcal{D}^m$$

of admissible deformations (with $|\cdot|_2$ denoting the 2-operator norm) so that the sets of admissible vector fields and time-discrete velocity fields $\boldsymbol{v}$ become

$$\begin{aligned} \mathcal{A}_v &:= \left\{ v \in \mathcal{H} \,\middle|\, \mathrm{id} + \tfrac{v}{K} \in \mathcal{A} \right\}, \\ \mathcal{C}_K &:= \left\{ \boldsymbol{v}_K \in \mathcal{A}_v^K \,\middle|\, \psi_B = (\mathrm{id} + \tfrac{v_{K,K}}{K}) \circ \ldots (\mathrm{id} + \tfrac{v_{K,2}}{K}) \circ (\mathrm{id} + \tfrac{v_{K,1}}{K}) \circ \psi_A \right\}. \end{aligned} \tag{9}$$

Theorem 2 (Existence of discrete geodesic curves in $\mathcal{D}^m$). *Given $\psi_A, \psi_B \in \mathcal{D}^m$, the discrete path energy $\mathbf{E}_K$ has a minimizer $\boldsymbol{v}_K$ on $\mathcal{C}_K$ unless $\mathcal{C}_K = \emptyset$.*

Proof. We apply the direct method of the calculus of variations. By coercivity of $\mathbf{E}_K$ and reflexivity of $\mathcal{H}^K$, any minimizing sequence contains a subsequence $\boldsymbol{v}_K^n \in \mathcal{C}_K$, $n \in \mathbb{N}$, weakly converging in $\mathcal{H}^K$ to some $\boldsymbol{v}_K$ as $n \to \infty$. By Sobolev embedding we even get strong convergence in $(C^1(\overline{D})^d)^K$ so that $\boldsymbol{v}_K \in \mathcal{C}_K$. Finally, $\mathbf{E}_K[\boldsymbol{v}_K] \leq \liminf_{n\to\infty} \mathbf{E}_K[\boldsymbol{v}_K^n]$ by convexity of $\mathbf{E}_K[\boldsymbol{v}_K]$ in the mth derivatives of $\boldsymbol{v}_K$ and by Fatou's lemma applied to the lower order contributions to $\mathbf{E}_K$. □

To study convergence of discrete geodesics (or equivalently the associated velocity fields) to continuous ones we need to extend $\boldsymbol{v}_K \in \mathcal{C}_K$ to a velocity field in $L^2((0,1);\mathcal{H})$. It turns out convenient to use

$$\boldsymbol{v}_K^{\text{ext}}(t) = v_{K,k} \circ y_{K,k}(t)^{-1} \quad \text{with } y_{K,k}(t) = \text{id} + (t - \tfrac{k-1}{K})v_{K,k} \quad \text{if } t \in [\tfrac{k-1}{K}, \tfrac{k}{K}],$$

cf. [2], where invertibility of $y_{K,k}(t)$ follows from the assumptions on $v_{K,k}$. The discrete path energy can then be reformulated as $\boldsymbol{\mathcal{E}}_K : L^2((0,1);\mathcal{H}) \to [0,\infty]$ with

$$\boldsymbol{\mathcal{E}}_K[v] = \begin{cases} \mathbf{E}_K[\boldsymbol{v}_K] & \text{if } \boldsymbol{v}_K \in \mathcal{C}_K \text{ and } \boldsymbol{v}_K^{\text{ext}} = v, \\ \infty & \text{else.} \end{cases} \tag{10}$$

Theorem 3 (Mosco-convergence of the discrete path energy). *$\boldsymbol{\mathcal{E}}_K$ Mosco-converges in $L^2((0,1);\mathcal{H})$ to $\boldsymbol{\mathcal{E}}$ as $K \to \infty$. As a consequence, the minimizers of $\boldsymbol{\mathcal{E}}_K$ converge (up to a subsequence) to a minimizer of $\boldsymbol{\mathcal{E}}$ as $K \to \infty$.*

The proof is a direct adaptation of [2, Thm. 4.1 & 4.2], which is concerned with the convergence of an analogous time discretization of so-called image metamorphosis: In metamorphosis [14] one computes a flow of diffeomorphisms to deform an image over time, which can on top also change its color values. The optimal deformation and color change path is a geodesic in the space of images. Dropping the images actually simplifies the proof – the only necessary addition is the constraint that $\boldsymbol{v}_K^{\text{ext}}$ induces a flow from ψ_A to ψ_B. However, this is also the reason for the impracticality of this discretization, to be remedied next.

5 Relaxed Time Discretization

The above time-discrete path energy does not go well with a Galerkin-type spatial discretization of $\boldsymbol{v}_K$ and thus is impractical. Indeed, taking $v_{K,k}$ from a finite element or spline space V_h, compositions of deformations $\phi_{K,k} = \text{id} + \frac{v_{K,k}}{K}$ will no longer lie in V_h so that the constraint from $\mathcal{C}_K$ cannot be numerically implemented. In particular, the spatially continuous equation $\psi_{K,k} = \phi_{K,k} \circ \psi_{K,k-1}$ has no discrete counterpart in V_h.

For a remedy, consider approximations $\zeta_{K,k}$ of the inverse deformations $\psi_{K,k}^{-1}$ for $k = 1, \dots, K-1$ as new degrees of freedom (fixing $\zeta_{K,0} = \psi_A^{-1}$ and $\zeta_{K,K} = \psi_B^{-1}$). In these new variables the constraints read $\zeta_{K,k} \circ (\text{id} + \frac{v_{K,k}}{K}) =$

$\zeta_{K,k-1}$. We therefore suggest a relaxed discrete path energy in $\boldsymbol{v}_K$ and $\boldsymbol{\zeta}_K = (\zeta_{K,1}, \dots, \zeta_{K,K-1})$ as

$$\mathbf{E}^\delta_K[\boldsymbol{v}_K, \boldsymbol{\zeta}_K] = \mathbf{E}_K[\boldsymbol{v}_K] + \frac{K}{\delta}\sum_{k=1}^{K}\int_D |\zeta_{K,k}\circ(\mathrm{id}+\tfrac{v_{K,k}}{K})-\zeta_{K,k-1}|^2 \,\mathrm{d}x \tag{11}$$

for $\delta > 0$, to be minimized over $\mathcal{A}^K_v \times L^2(D;\mathbb{R}^d)^{K-1}$ using $\zeta_{K,0} = \psi_A^{-1}$ and $\zeta_{K,K} = \psi_B^{-1}$. The trick is that $\mathbf{E}^\delta_K$ coincides with the discrete path energy proposed in [2] for image metamorphosis, where $\boldsymbol{\zeta}_K$ plays the role of a vector of images with d color channels. Hence, the corresponding existence result transfers immediately. Even better, the optimal $\zeta_{K,k}$ is an explicit convex combination of $\zeta_{K,k-1}$ and $\zeta_{K,k+1}$,

$$\zeta_{K,k}(x) = \frac{\zeta_{K,k+1}\circ\phi_{K,k+1}(x)+(\zeta_{K,k-1}\circ\phi^{-1}_{K,k}(x))((\det D\phi_{K,k})^{-1}\circ\phi^{-1}_{K,k}(x))}{1+(\det D\phi_{K,k})^{-1}\circ\phi^{-1}_{K,k}(x)} \tag{12}$$

for $\phi_{K,k} = \mathrm{id} + \frac{v_{K,k}}{K} \in \mathcal{D}^m$ (cf. [2]), which implies together with the fact that $W^{m-1,2}(D;\mathbb{R})$ as a Banach algebra is closed under pointwise multiplication and under composition with functions from $\mathcal{D}^m$ [8] that the $\zeta_{K,k}$ are $W^{m-1,2}(D;\mathbb{R}^d)$ regular.

Theorem 4 (Existence of relaxed discrete geodesics). *Let $K \geq 2$ and $\psi_A, \psi_B \in \mathcal{D}^m$, then $\mathbf{E}^\delta_K$ has a minimizer $(\boldsymbol{v}_K, \boldsymbol{\zeta}_K)$ with $\boldsymbol{v}_K \in \mathcal{A}^K_v$ and $\zeta_{K,k} \in W^{m-1,2}(D;\mathbb{R}^d)$ for $k = 1, \dots, K-1$.*

Note that invertibility of $\zeta_{K,k}$ cannot be expected in general and will certainly depend on the interplay between $\delta > 0$, $K \in \mathbb{N}$, and the boundary data.

To investigate the convergence for $K \to \infty$ and simultaneously $\delta \to 0$ we once more extend the discrete path energy to the function space on which the continuous path energy is defined. To this end, we first extend the discrete inverse flow similar to the previous extension of the velocity field as

$$\zeta^{\mathrm{ext}}_K[\boldsymbol{\zeta}_K, \boldsymbol{v}_K](t) := \zeta_{K,k-1}\circ y_{K,k}(t)^{-1} + K(t-\tfrac{k-1}{K})(\zeta_{K,k}\circ\phi_{K,k}-\zeta_{K,k-1})\circ y_{K,k}(t)^{-1}$$

for $t \in [\frac{k-1}{K}, \frac{k}{K}]$, thus ζ^{ext}_K is an interpolation between $\zeta_{K,k-1}$ at time $\frac{k-1}{K}$ and $\zeta_{K,k}$ at time $\frac{k}{K}$ (recall that $\phi_{K,k} = \mathrm{id}+\frac{v_{K,k}}{K}$ and $y_{K,k}(t) = \mathrm{id}+(t-\frac{k-1}{K})v_{K,k}$). Following [7] we now define the extended discrete functional $\overline{\mathcal{E}}^\delta_K\colon L^2((0,1);L^2(D;\mathbb{R}^d)) \to [0,\infty]$ by

$$\overline{\mathcal{E}}^\delta_K[\zeta] := \inf_{\boldsymbol{v}_K\in\mathcal{A}^K_v,\, \boldsymbol{\zeta}_K\in L^2(D;\mathbb{R}^d)^{K-1}} \left\{ \mathbf{E}^\delta_K[\boldsymbol{v}_K, \boldsymbol{\zeta}_K] \,\middle|\, \zeta^{\mathrm{ext}}_K[\boldsymbol{\zeta}_K, \boldsymbol{v}_K] = \zeta \right\}$$

(the infimum over the empty set is infinity by convention). That the infimum on the right-hand side is attained is very similar to proving the existence result in Theorem 2. We can now reduce the convergence of relaxed discrete geodesics to the convergence results for metamorphosis from [7, Thms. 12, 14, 15]. To this end, we define the time-continuous relaxed energy

$$\overline{\mathcal{E}}^\delta[\zeta] := \inf_{(v,Z)\in\mathcal{C}(\zeta)} \int_0^1 \langle Lv, v\rangle + \frac{1}{\delta}\int_D |Z|^2 \,\mathrm{d}x\,\mathrm{d}t \quad \text{with}$$

$$\mathcal{C}(\zeta) := \{(v,Z) \in L^2((0,1);\mathcal{H}) \times L^2((0,1);L^2(D;\mathbb{R}^d)) \,|\, \forall x \in D, 0 \le t < s \le 1 :$$
$$\zeta(t)\circ\psi(t)(x) - \zeta(s)\circ\psi(s)(x) = \textstyle\int_t^s Z(r)\circ\psi(r)(x)\,\mathrm{d}r \text{ for the solution } \psi \text{ of (4) }\}.$$

Theorem 5 (Convergence of relaxed discrete inverse flows). $\overline{\mathcal{E}}^\delta_K$ *Mosco-converges in* $L^2((0,1);L^2(D;\mathbb{R}^d))$ *as* $K \to \infty$ *and* $\delta \to 0$ *to* $\overline{\mathcal{E}} : L^2((0,1);L^2(D;\mathbb{R}^d)) \to [0,\infty]$,

$$\overline{\mathcal{E}}[\zeta] = \begin{cases} \widetilde{\mathcal{E}}[(\psi(t))_{t\in[0,1]}] & \text{if } \psi(t) = \zeta(t)^{-1} \text{ is defined for a.e.t,} \\ \infty & \text{else.} \end{cases}$$

Further, as $K\to\infty$ *and* $\delta\to 0$, *the minimizers of* $\overline{\mathcal{E}}^\delta_K$ *converge up to a subsequence weakly in* $L^2((0,1);L^2(D;\mathbb{R}^d))$ *to* $t\mapsto\psi(t)^{-1}$ *for a geodesic* $\psi \in W^{1,2}((0,1),\mathcal{D}^m)$ *from* ψ_A *to* ψ_B.

Proof. In [7] (taking into account also Prop. 8 therein to arrive at the below formulation) it is proven that for $K \to \infty$ the energy $\overline{\mathcal{E}}^\delta_K$ Mosco-converges to $\overline{\mathcal{E}}^\delta$. Now let $K_n \to \infty$, $\delta_n \to 0$, $\zeta_n \overset{L^2}{\rightharpoonup} \zeta$ as $n \to \infty$, then by this Mosco-convergence

$$\overline{\mathcal{E}}^\delta[\zeta] \le \liminf_{n\to\infty} \overline{\mathcal{E}}^\delta_{K_n}[\zeta_n] \le \liminf_{n\to\infty} \overline{\mathcal{E}}^{\delta_n}_{K_n}[\zeta_n]$$

for any $\delta > 0$. The required lim inf inequality now follows from $\overline{\mathcal{E}}[\zeta] \le \liminf_{\delta\to 0} \overline{\mathcal{E}}^\delta[\zeta]$, which is obtained as follows: If $\overline{\mathcal{E}}^\delta[\zeta] \to \infty$ there is nothing to show, so we may assume $\overline{\mathcal{E}}^\delta[\zeta] < C < \infty$ uniformly in δ. Let $(v_\delta, Z_\delta) \in \mathcal{C}(\zeta)$ be δ-optimal in $\overline{\mathcal{E}}^\delta[\zeta]$, then by the uniform boundedness $\int_0^1 \langle Lv_\delta, v_\delta\rangle + \frac{1}{\delta}\int_D |Z_\delta|^2 \,\mathrm{d}x\,\mathrm{d}t \le C + \delta$ there exists a subsequence (v_δ, Z_δ) converging weakly in $L^2((0,1);\mathcal{H}) \times L^2((0,1);L^2(D;\mathbb{R}^d))$ to some $(v, 0)$ (convergence of Z_δ is even strong). Since $\mathcal{C}(\zeta)$ is closed under this convergence [7, proof of Prop. 13], $(v,0) \in \mathcal{C}(\zeta)$, and $\overline{\mathcal{E}}[\zeta] = \int_0^1 \langle Lv, v\rangle \,\mathrm{d}t \le \liminf_{\delta\to 0} \int_0^1 \langle Lv_\delta, v_\delta\rangle \,\mathrm{d}t \le \liminf_{\delta\to 0} \overline{\mathcal{E}}^\delta[\zeta]$, where the first inequality holds due to the convexity of the integral in v.

As for the lim sup inequality we may wlog. restrict to the case $\overline{\mathcal{E}}[\zeta] < \infty$ so that there exists a velocity field $v \in L^2((0,1);\mathcal{H})$ with $\frac{\mathrm{d}}{\mathrm{d}t}\zeta(t)^{-1} = v(t) \circ \zeta(t)^{-1}$. Given $K_n \to \infty$, $\delta_n \to 0$, we construct a recovery sequence $\zeta_n = \zeta^{\mathrm{ext}}_{K_n}[\boldsymbol{\zeta}_{K_n}, \boldsymbol{v}_{K_n}]$ for $\boldsymbol{\zeta}_{K_n} = (\zeta(0), \zeta(\frac{1}{K_n}), \ldots, \zeta(1))$, $\boldsymbol{v}_{K_n} = (K(\phi_{K_n,1} - \mathrm{id}), \ldots, K(\phi_{K_n,K_n} - \mathrm{id}))$ with $\phi_{K_n,k} = \zeta^{-1}(\frac{k}{K_n})\circ\zeta(\frac{k-1}{K_n})$. This is essentially the recovery sequence already proposed in [7], for which the authors prove

$$\int_0^1 \langle Lv, v\rangle \,\mathrm{d}t \ge \limsup_{n\to\infty} \mathbf{E}_{K_n}[\boldsymbol{v}_{K_n}].$$

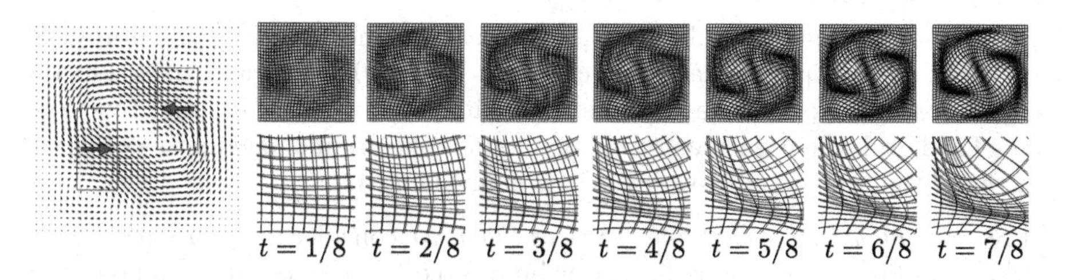

Fig. 1. Left: Velocity w of a stationary Stokes flow together with the inducing piecewise constant bulk force (red arrows in red rectangles). Right: We compare the flow $\tilde{\psi}$ (acting on a lattice of red lines), induced by the stationary velocity field w, with the LDDMM geodesic between $\psi_A = \tilde{\psi}(0) = \mathrm{id}$ and $\psi_B = \tilde{\psi}(1)$ (acting on black lines), approximated numerically via our relaxed time discretization with $\delta = 10^{-4}$. Times $t = 0, 1$ are omitted since the deformed lattices coincide by construction; the bottom images show a zoom. The significant differences illustrate that a stationary velocity field is in general not the most efficient way (in terms of the LDDMM path energy, i.e. the average squared $\mathcal{H}$-norm of the velocity field) to connect two diffeomorphisms. (Color figure online)

The left-hand side equals $\overline{\boldsymbol{\mathcal{E}}}[\zeta]$, and $\mathbf{E}_{K_n}[\boldsymbol{v}_{K_n}] = \mathbf{E}_{K_n}^{\delta_n}[\boldsymbol{\zeta}_{K_n}, \boldsymbol{v}_{K_n}] \geq \overline{\boldsymbol{\mathcal{E}}}_{K_n}^{\delta_n}[\zeta_{K_n}]$. This implies the required lim sup inequality and thus completes the proof of Mosco-convergence. The convergence of the minimizers to a minimizer of $\overline{\boldsymbol{\mathcal{E}}}$ (and thus to $\psi(t)^{-1}$ for a geodesic ψ) follows from the Mosco-convergence in combination with the equi-mild $L^2((0,1); L^2(D; \mathbb{R}^d))$-coercivity of the $\overline{\boldsymbol{\mathcal{E}}}_{K_n}^{\delta_n}$ [3], which in turn follows from the equi-mild coercivity of $\overline{\boldsymbol{\mathcal{E}}}_{K_n}^{\delta}$ for any fixed $\delta > 0$ proven in [7]. The regularity of ψ always holds for shortest geodesic paths. □

In fact, using the existing estimates from [2,7] one can also prove Mosco-convergence of the alternative energy

$$\boldsymbol{\mathcal{E}}_K^{\delta}[v] := \min_{\zeta_K \in (L^2(D;\mathbb{R}^d))^K, v_K \in \mathcal{A}_v^K} \left\{ \mathbf{E}_K^{\delta}[\boldsymbol{v}_K, \boldsymbol{\zeta}_K] \,\middle|\, \boldsymbol{v}_K^{\mathrm{ext}} = v \right\}$$

to $\boldsymbol{\mathcal{E}}$ in $L^2((0,1); \mathcal{H})$, however, the argument is substantially longer since we cannot directly apply the Mosco-convergence results from [2,7].

6 Numerical Experiments

We present two different numerical experiments as a proof of concept for the effectiveness of the variational time discretization.

For the implementation we restrict to $d = 2$ and $D = [0,1]^2$. All functions are discretized on a regular grid on D via cubic B-splines. To approximate the term $\zeta_{K,k} \circ (\mathrm{id} + \frac{v_{K,k}}{K})$ we use a warping operator defined by cubic B-splines as suggested in [6]. We make use of the iPALM algorithm [11] to optimize the fully discrete energy. We use $\mathrm{W}(A) = \frac{\lambda}{2}(e^{(\log \det A)^2} - 1) + \mu |A^{\mathrm{sym}} - \mathbb{1}|^2$ for $\lambda = \mu = 1$ in (8), which is consistent with our assumptions, while we use different (small) values of δ in (11) which allow stable computation and interpretation of results. However, for computational simplicity we set $\gamma = 0$ in (8) and thus skip the higher order term. Due to an inverse inequality in the space of B-spline functions on a fixed grid, the implemented energy bounds the original one up to a constant factor so that the qualitative behaviour is unchanged.

A Shorter Flow than a Stationary Stokes Flow. We consider the solution $w : D \to \mathbb{R}^2$ of the Stokes equations for given bulk forces and zero Dirichlet boundary conditions computed on a 257×257 grid of D using Matlab. Then we solve (1) with $v(t) = w$ (piecewise bilinear interpolated) via a second order Runge–Kutta scheme to obtain a flow $\tilde{\psi}$, and we set $\psi_B := \tilde{\psi}(1)$. Figure 1 shows the relaxed discrete geodesic from ψ_A to ψ_B and compares it with this longer Stokes flow $\tilde{\psi}$.

Comparison of Image Metamorphosis and the Flow of Diffeomorphisms Applied to Images. LDDMM is frequently applied to image matching. An extension is given by the metamorphosis model, which incorporates in addition changes of the colour along motion lines. Figure 2 compares the discrete image metamorphosis path between two binary images from [2] with the warp associated with the relaxed discrete geodesic path in $\mathcal{D}^m$. Even though the metamorphosis path

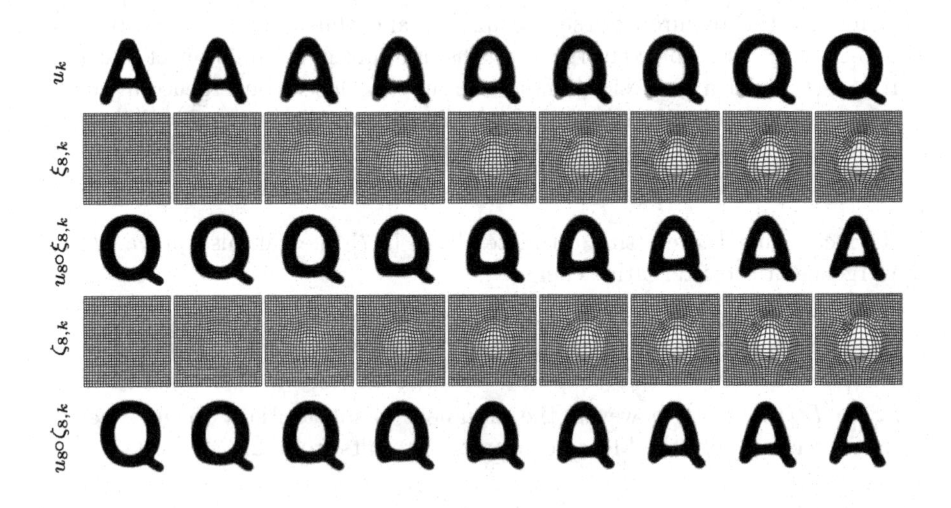

Fig. 2. Top row: Metamorphosis geodesic between letters 'A' and 'Q' for $\delta = 0.15$. Second row: Deformation of a regular grid by $\xi_{8,k} = \phi_{8,8} \circ \cdots \circ \phi_{8,9-k}$, $k = 1, \ldots, 8$, for the incremental deformations $\phi_{8,k}$ from the previous row (the $\xi_{8,k}$ represent the inverses of the time-reversed flow of diffeomorphisms). Third row: The pullback of 'Q' by $\xi_{8,k}$. Fourth row: Deformation of the grid along the LDDMM geodesic $(\zeta_{8,k})_{k=0,\ldots,8}$ between $\zeta_{8,0} = \mathbb{1}$ and $\zeta_{8,8} = \xi_{8,8}$, computed via our relaxed time discretization for $\delta = 0.01$. Bottom row: The pullback of 'Q' along $\zeta_{8,k}$. The experiment illustrates that the optimal deformation in metamorphosis in general slightly deviates from LDDMM geodesics (e.g. see the slightly different shape of the hole in the letters or their bottom left corner) due to exploitation of the possibility of changing also the colour values.

energy contains the LDDMM path energy for measuring the optimality of the diffeomorphism path, the interaction of deformations with grey value changes leads to slight differences in the diffeomorphisms compared to pure LDDMM.

7 Conclusions

We derived a weak formulation of the Sobolev LDDMM dynamics and proposed a variational time discretization with provable concergence to time-continuous LDDMM geodesics. As a proof of concept we showed results of a numerical implementation. Directions of future work include the application of the suggested discretization in areas such as computational anatomy, the development of a discrete exponential map based on our variational approach and its comparison to classical shooting methods.

Acknowledgements. This work was supported by the Deutsche Forschungsgemeinschaft (DFG, German Research Foundation) via project 211504053 – Collaborative Research Center 1060 and via project 431460824 – Collaborative Research Center 1450 and via Germany's Excellence Strategy project 390685813 – Hausdorff Center for Mathematics and project 390685587 – Mathematics Münster: Dynamics-Geometry-Structure.

References

1. Beg, M.F., Miller, M.I., Trouvé, A., Younes, L.: Computing large deformation metric mappings via geodesic flows of diffeomorphisms. Int. J. Comput. Vision **61**(2), 139–157 (2005). https://doi.org/10.1023/B:VISI.0000043755.93987.aa
2. Berkels, B., Effland, A., Rumpf, M.: Time discrete geodesic paths in the space of images. SIAM J. Imaging Sci. **8**(3), 1457–1488 (2015). https://doi.org/10.1137/140970719
3. Braides, A.: Γ-Convergence for Beginners. Oxford Lecture Series in Mathematics and its Applications, vol. 22. Oxford University Press, Oxford (2002)
4. Bruveris, M., Vialard, F.X.: On completeness of groups of diffeomorphisms. J. Eur. Math. Soc. (JEMS) **19**(5), 1507–1544 (2017). https://doi.org/10.4171/JEMS/698
5. Christensen, G., Miller, M., Rabbitt, R.: Deformable templates using large deformation kinematics. IEEE Trans. Image Process. **5**(10), 1435–1447 (1996)
6. Effland, A., Kobler, E., Pock, T., Rajković, M., Rumpf, M.: Image morphing in deep feature spaces: theory and applications. J. Math. Imaging Vis. **63**, 309–327 (2021). https://doi.org/10.1007/s10851-020-00974-5
7. Effland, A., Neumayer, S., Rumpf, M.: Convergence of the time discrete metamorphosis model on Hadamard manifolds. SIAM J. Imaging Sci. **13**(2), 557–588 (2020). https://doi.org/10.1137/19M1247073
8. Inci, H., Kappeler, T., Topalov, P.: On the regularity of the composition of diffeomorphisms. Mem. Am. Math. Soc. **226**(1062), vi+60 (2013). https://doi.org/10.1090/S0065-9266-2013-00676-4
9. Miller, M.I., Trouvé, A., Younes, L.: On the metrics and Euler-Lagrange equations of computational anatomy. Annu. Rev. Biomed. Eng. **4**(1), 375–405 (2002)

10. Miller, M.I., Trouvé, A., Younes, L.: Geodesic shooting for computational anatomy. J. Math. Imaging Vision **24**(2), 209–228 (2006). https://doi.org/10.1007/s10851-005-3624-0
11. Pock, T., Sabach, S.: Inertial proximal alternating linearized minimization (iPALM) for nonconvex and nonsmooth problems. SIAM J. Imaging Sci. **9**(4), 1756–1787 (2016). https://doi.org/10.1137/16M1064064
12. Rumpf, M., Wirth, B.: Discrete geodesic calculus in shape space and applications in the space of viscous fluidic objects. SIAM J. Imaging Sci. **6**(4), 2581–2602 (2013). https://doi.org/10.1137/120870864
13. Trouvé, A.: An infinite dimensional group approach for physics based models in pattern recognition. Technical report, ENS Cachan (1995). http://128.220.140.31/publications/papers_in_database/alain/trouve1995.pdf
14. Trouvé, A., Younes, L.: Local geometry of deformable templates. SIAM J. Math. Anal. **37**(1), 17–59 (2005). https://doi.org/10.1137/S0036141002404838
15. Trouvé, A., Younes, L.: Shape spaces. In: Scherzer, O. (ed.) Handbook of Mathematical Methods in Imaging, vol. 1, 2, 3, pp. 1759–1817. Springer, New York (2015). https://doi.org/10.1007/978-1-4939-0790-8_55

Image Comparison and Scaling via Nonlinear Elasticity

John M. Ball(✉) and Christopher L. Horner

Heriot-Watt University and Maxwell Institute for the Mathematical Sciences, Edinburgh, UK
jb101@hw.ac.uk

Abstract. A nonlinear elasticity model for comparing images is formulated and analyzed, in which optimal transformations between images are sought as minimizers of an integral functional. The existence of minimizers in a suitable class of homeomorphisms between image domains is established under natural hypotheses. We investigate whether for linearly related images the minimization algorithm delivers the linear transformation as the unique minimizer.

Keywords: Nonlinear elasticity · image registration · scaling

1 Introduction

In this paper we formulate and analyze a nonlinear elasticity model for comparing two images $P_1 = (\Omega_1, c_1),\ P_2 = (\Omega_2, c_2)$, regarded as bounded Lipschitz domains Ω_1, Ω_2 in $\mathbb{R}^n$ with corresponding intensity maps $c_1 : \Omega_1 \to \mathbb{R}^m, c_2 : \Omega_2 \to \mathbb{R}^m$. The model is based on an integral functional

$$I_{P_1,P_2}(y) = \int_{\Omega_1} \psi(c_1(x), c_2(y(x)), Dy(x))\, dx, \tag{1}$$

depending on c_1, c_2 and a map $y : \Omega_1 \to \Omega_2$ with gradient Dy, whose minimizers give optimal transformations y^* between images. The admissible transformations y between the images are orientation-preserving homeomorphisms with $y(\Omega_1) = \Omega_2$, and are not required to satisfy other boundary conditions.

The use of nonlinear elasticity, rather than models based on linear elasticity that are more commonly used in the computer vision literature, provides a conceptually clearer and more general framework. A key advantage is that nonlinear elasticity (of which linear elasticity is not a special case) respects rotational invariance, so that rigidly rotated and translated images can be identified as equivalent. Further, nonlinear elasticity is naturally suited for discussing the global invertibility of maps between images (see, for example, [5,29]), which in the context of mechanics describes non-interpenetration of matter.

Our work is closest in spirit to that of Droske & Rumpf [14], Rumpf [27] and Rumpf & Wirth [26], who like us make use of the existence theory for polyconvex

L. Calatroni et al. (Eds.): SSVM 2023, LNCS 14009, pp. 565–574, 2023.
https://doi.org/10.1007/978-3-031-31975-4_43

energies in [4], as also do Burger, Modersitski & Ruthotto [9], Debroux et al. [12], Iglesias, Rumpf & Scherzer [16] and Iglesias [15]. Other nonlinear elasticity approaches are due to Lin, Dinov, Toga & Vese [19], Ozeré, Gout & Le Guyader [22], Ozeré & Le Guyader [23], Simon, Sheorey, Jacobs & Basri [28] and Debroux & Le Guyader [13]. Key differences with these works are:

(i) that we minimize among homeomorphisms of the image domains rather than applying Dirichlet or other boundary conditions,
(ii) technical improvements as regards the regularity of the intensity maps, and
(iii) a novel analysis of linearly related images.

Our model is described in Sect. 2, in which it is shown (Proposition 1) that invariance of the integral (1) under rotation and translation requires that the integrand $\psi(c_1, c_2, \cdot)$ be isotropic. As described above, two images that are translated and rigidly rotated with respect to each other can reasonably be regarded as equivalent. In most applications the minimization algorithm should thus deliver this translation and rotation as the unique minimizer, and we give conditions on ψ under which this occurs. We also discuss symmetry with respect to interchange of images. Theorem 1 gives the existence of a minimizer for general pairs of images under polyconvexity and growth conditions on ψ, assuming only that the intensity maps are L^∞.

More generally we consider the case when two images are related by a linear transformation, and ask for which ψ the minimization algorithm delivers this linear transformation as the unique minimizer. We show (see Sect. 3) that ψ can be chosen such that for any pair of images related by a uniform magnification the unique minimizer is that magnification. However, for the functional to deliver as a minimizer the linear transformation between *any* linearly related pair of images the integrand must have a special form (see Theorem 2), in which the integrand depends on the gradient Dy as a convex function of $\det Dy$ alone. This degeneracy suggests that a better model might use an integrand depending also on the second gradient D^2y, and this is briefly discussed, together with other issues, in Sect. 4.

2 Nonlinear Elasticity Model

2.1 Comparing Images

We identify an image with a pair $P = (\Omega, c)$, where $\Omega \subset \mathbb{R}^n$ is a bounded Lipschitz domain, and $c : \Omega \to \mathbb{R}^m$ is an *intensity map* describing the greyscale intensity ($m = 1$), the intensity of colour channels ($m > 1$), and possibly other image characteristics. Our aim is to compare two images $P_1 = (\Omega_1, c_1)$, $P_2 = (\Omega_2, c_2)$ by means of a nonlinear elasticity based functional, whose minimizers give optimal transformation maps between the images.

To compare P_1, P_2 we minimize the functional

$$I_{P_1,P_2}(y) = \int_{\Omega_1} \psi(c_1(x), c_2(y(x)), Dy(x))\, dx, \tag{2}$$

over invertible maps $y : \Omega_1 \to \mathbb{R}^n$ such that $y(\Omega_1) = \Omega_2$, and which are *orientation-preserving*, that is $\det Dy(x) > 0$ for a.e. $x \in \Omega_1$. Here

$$\psi : \mathbb{R}^m \times \mathbb{R}^m \times M_+^{n\times n} \to [0,\infty),$$

where $M_+^{n\times n} = GL^+(n,\mathbb{R}) = \{\text{real } n \times n \text{ matrices } A \text{ with } \det A > 0\}$.

In (2), $Dy(x)$ denotes the distributional gradient of y at x. Throughout this section we assume that the maps y and their inverses y^{-1} have sufficient regularity; it is enough that $y \in W^{1,p}(\Omega_1,\mathbb{R}^n), y^{-1} \in W^{1,p}(\Omega_2,\mathbb{R}^n)$ for $p > n$ (for the definition of $W^{1,p}$ see Sect. 2.3), which is guaranteed by Theorem 1 below.

Note that we do not specify y on $\partial\Omega_1$, only that $y(\Omega_1) = \Omega_2$. Thus we allow 'sliding at the boundary', in order to better compare images with important boundary features. This is not typically done in the computer vision literature, but is considered in the context of elasticity by Iwaniec & Onninen [17], though for elasticity such a boundary condition would be difficult to realize mechanically.

2.2 Properties of ψ

We now list some desirable properties of the integrand ψ in (2).

(i) *Invariance under rotation and translation.* For two images $P = (\Omega, c)$ and $P' = (\Omega', c')$ write $P \sim P'$ if P, P' are related by a rigid translation and rotation, i.e.

$$\Omega' = E(\Omega),\ c'(E(x)) = c(x)$$

for some proper rigid transformation $E(x) = a + Rx$, $a \in \mathbb{R}^n$, $R \in SO(n)$.

If $P_1 \sim P_1'$, $P_2 \sim P_2'$, with corresponding rigid transformations $E_1(x) = a_1 + R_1x$, $E_2(x) = a_2 + R_2x$, we require that

$$I_{P_1,P_2}(y) = I_{P_1',P_2'}(E_2 \circ y \circ E_1^{-1}), \tag{3}$$

or, equivalently,

$$\int_{\Omega_1} \psi(c_1(x), c_2(y(x)), R_2 Dy(x) R_1^T)\, dx = \int_{\Omega_1} \psi(c_1(x), c_2(y(x)), Dy(x))\, dx. \tag{4}$$

Proposition 1. (4) *holds for all* P_1, P_2 *and orientation-preserving invertible* $y : \Omega_1 \to \Omega_2$ *with* $y(\Omega_1) = \Omega_2$ *iff* $\psi(c_1, c_2, \cdot)$ *is isotropic, i.e.*

$$\psi(c_1, c_2, QAR) = \psi(c_1, c_2, A) \tag{5}$$

for all $c_1, c_2 \in \mathbb{R}^n, A \in M_+^{n\times n}$, *and* $Q, R \in SO(n)$.

Proof. Setting c_1, c_2 constant, and $y(x) = Ax$, (4) implies (5), and the converse is obvious.

We denote by $v_i(A)$ the singular values of A (that is, the eigenvalues of $\sqrt{A^T A}$). A standard result of nonlinear elasticity (see, for example, [30, Theorem 8.5.1]) gives that $\psi(c_1, c_2, \cdot)$ is isotropic iff

$$\psi(c_1, c_2, A) = \Phi(c_1, c_2, v_1(A), ..., v_n(A))$$

with Φ symmetric with respect to permutations of the last n arguments.
(ii) *Matching of equivalent images.* We also require that the functional (2) is zero iff the two images are related by a rigid transformation, i.e. for invertible y with $y(\Omega_1) = \Omega_2$ we have

$$I_{P_1,P_2}(y) = 0 \text{ iff } P_1 \sim P_2 \text{ with corresponding rigid transformation } y. \tag{6}$$

Proposition 2. (6) *is equivalent to the condition*

$$\psi(c_1, c_2, A) = 0 \textit{ iff } c_1 = c_2 \textit{ and } A \in SO(n). \tag{7}$$

Proof. The only nontrivial part of the proof of equivalence is to show that if (7) holds and $I_{P_1,P_2}(y) = 0$ then $P_1 \sim P_2$ with corresponding rigid transformation y. But if $I_{P_1,P_2}(y) = 0$ then $c_2(y(x)) = c_1(x)$ and $Dy(x) = R(x) \in SO(n)$ for a.e. $x \in \Omega_1$. But this implies by [24] that $R(x) = R$ is constant, from which the conclusion follows.

(iii) *Symmetry with respect to interchanging images.* For applications in which both images are of the same type (but not, for example, when P_1 is a template image) it is reasonable to require that

$$I_{P_1,P_2}(y) = I_{P_2,P_1}(y^{-1}). \tag{8}$$

Equivalently

$$\begin{aligned} \int_{\Omega_1} \psi(c_1(x), c_2(y(x)), Dy(x))\, dx &= \int_{\Omega_2} \psi(c_2(y), c_1(x(y)), Dx(y))\, dy \\ &= \int_{\Omega_1} \psi(c_2(y(x)), c_1(x), Dy(x)^{-1}) \det Dy(x)\, dx. \end{aligned} \tag{9}$$

Taking c_1, c_2 constant and $y(x) = Ax$ this holds iff

$$\psi(c_1, c_2, A) = \psi(c_2, c_1, A^{-1}) \det A. \tag{10}$$

Such a symmetry condition was introduced by Cachier & Rey [10] and subsequently used by Kolouri, Slepčev & Rohde [18] and Iglesias [15]. It is also implicit in the work of Iwaniec & Onninen [17].

A class of integrands ψ satisfying the above conditions (5), (7), (10) is given by

$$\psi(c_1, c_2, A) = \Psi(A) + f(c_1, c_2, \det A), \tag{11}$$

where

(a) $\Psi \geqslant 0$ is isotropic, $\Psi(A) = \det A \cdot \Psi(A^{-1})$, $\Psi^{-1}(0) = SO(n)$,
(b) $f \geqslant 0$, $f(c_1, c_2, \delta) = \delta f(c_2, c_1, \delta^{-1})$, $f(c_1, c_2, 1) = 0$ iff $c_1 = c_2$.

In particular we can take

$$f(c_1, c_2, \delta) = (1 + \delta)|c_1 - c_2|^2, \tag{12}$$

or

$$f(c_1, c_2, \delta) = |c_1 - c_2\delta|^2 + \delta^{-1}|c_1\delta - c_2|^2, \tag{13}$$

which are both convex in δ.

2.3 Existence of Minimizers

Let $p > n$ and define the set of admissible maps

$$\mathcal{A} = \{y \in W^{1,p}(\Omega_1, \mathbb{R}^n) : y : \Omega_1 \to \Omega_2 \text{ an orientation-preserving homeomorphism}, y^{-1} \in W^{1,p}(\Omega_2, \mathbb{R}^n)\}. \tag{14}$$

Here, for a bounded domain $\Omega \subset \mathbb{R}^n$ and $1 < p < \infty$, $W^{1,p}(\Omega, \mathbb{R}^n)$ is the Sobolev space of maps $y : \Omega \to \mathbb{R}^n$ such that

$$\|y\|_{1,p} := \left(\int_\Omega (|y(x)|^p + |Dy(x)|^p) \, dx \right)^{\frac{1}{p}} < \infty.$$

We recall (see for example [1,21]) that if Ω is Lipschitz and $p > n$ then any $y \in W^{1,p}(\Omega, \mathbb{R}^n)$ has a representative that is continuous on the closure $\bar{\Omega}$ of Ω.

We now make some other technical hypotheses on ψ.

(H1) (*Continuity*) $\psi : \mathbb{R}^m \times \mathbb{R}^m \times M_+^{n\times n} \to [0, \infty)$ is continuous,

(H2) (*Coercivity*) $\psi(c, d, A) \geqslant C(|A|^p + \det A \cdot |A^{-1}|^p) - C_0$ for all $c, d \in \mathbb{R}^m, A \in M_+^{n\times n}$, where $C > 0$ and C_0 are constants,

(H3) (*Polyconvexity*) $\psi(c, d, \cdot)$ is polyconvex for each $c, d \in \mathbb{R}^s$, i.e. there is a function $g : \mathbb{R}^m \times \mathbb{R}^m \times \mathbb{R}^{\sigma(n)} \times (0, \infty) \to \mathbb{R}$ with $g(c, d, \cdot)$ convex, such that

$$\psi(c, d, A) = g(c, d, \mathbf{J}_{n-1}(A), \det A) \text{ for all } c, d \in \mathbb{R}^m, A \in M_+^{n\times n},$$

where $\mathbf{J}_{n-1}(A)$ is the list of all minors (i.e. subdeterminants) of A of order $\leqslant n-1$ and $\sigma(n)$ is the number of such minors,

(H4) (*Bounded intensities*) $c_1 \in L^\infty(\Omega_1, \mathbb{R}^s)$, $c_2 \in L^\infty(\Omega_2, \mathbb{R}^s)$.

We note that (H2) implies that $\psi(c, d, A) \geqslant C_1(\det A)^{1-\frac{p}{n}} - C_0$ for some constant $C_1 > 0$, so that $\psi(c, d, A) \to \infty$ as $\det A \to 0+$. This follows from the Hadamard inequality $|B|^n \geqslant n^{\frac{n}{2}} \det B$ for $B \in M_+^{n\times n}$ applied to $B = \text{cof} A$, noting that $\det \text{cof} A = (\det A)^{n-1}$.

Theorem 1. *Suppose that $\mathcal{A}$ is nonempty, and that the hypotheses* (H1)–(H4) *hold. Then there exists an absolute minimizer y^* in $\mathcal{A}$ of*

$$I_{P_1,P_2}(y) = \int_{\Omega_1} \psi(c_1(x), c_2(y(x)), Dy(x))\, dx.$$

The proof, which will appear in [8], follows the usual pattern for proving existence of minimizers in nonlinear elasticity for a polyconvex stored-energy function using the direct method of the calculus of variations (see [4,11,30]). However there are some extra issues. In particular, as observed by Rumpf [27] care has to be taken for intensity maps c_1, c_2 that are discontinuous, for which it is not even immediately obvious that $I_{P_1,P_2}(y)$ is well defined, and we are able to weaken his hypotheses by assuming only (H4). Note that the hypotheses on ψ discussed in Sect. 2.2 are not needed to prove the existence of minimizers.

3 Linear Scaling

Suppose that the images $P_1 = (\Omega_1, c_1)$ and $P_2 = (\Omega_2, c_2)$ are linearly related, i.e. for some $M \in M_+^{n\times n}$ we have

$$\Omega_2 = M\Omega_1, \quad c_2(Mx) = c_1(x). \tag{15}$$

Can we choose ψ such that the unique minimizer y of I_{P_1,P_2} is $y(x) = Mx$?

For simplicity consider ψ of the form (11), (12)

$$\psi(c_1, c_2, A) = \Psi(A) + (1 + \det A)|c_1 - c_2|^2.$$

Thus we require that for all orientation-preserving invertible y with $y(\Omega_1) = M\Omega_1$

$$\int_{\Omega_1} \left(\Psi(Dy(x)) + (1 + \det Dy(x))|c_1(x) - c_2(y(x))|^2\right)\, dx \geqslant \int_{\Omega_1} \Psi(M)\, dx, \tag{16}$$

with equality iff $y(x) = Mx$.

This holds for all c_1, c_2 iff

$$⨍_{\Omega_1} \Psi(Dy)\, dx \geqslant \Psi(M) \tag{17}$$

for y invertible with $y(\Omega_1) = M\Omega_1$, where $⨍_{\Omega_1} f\, dx := \frac{1}{|\Omega_1|}\int_{\Omega_1} f\, dx$ and $|\Omega_1|$ is the $n-$dimensional Lebesgue measure of Ω_1. The inequality (17) is a stronger version of *quasiconvexity at M*, the central convexity condition of the multi-dimensional calculus of variations implied by polyconvexity (see, e.g. [25]), in which the usual requirement that $y(x) = Mx$ for $x \in \partial\Omega_1$ is weakened.

We show that we can satisfy this condition if $M = \lambda\mathbf{1}$, $\lambda > 0$, where $\mathbf{1}$ denotes the identity matrix (or more generally if $M = \lambda R$, $R \in SO(n)$), so that P_2 is a uniform magnification (or reduction if $\lambda \leqslant 1$) of P_1. Let

$$\Psi(A) = \sum_{i=1}^{n} v_i^{\alpha} + (\det A)\sum_{i=1}^{n} v_i^{-\alpha} + h(\det A), \tag{18}$$

where $v_i = v_i(A)$ are the singular values of A, $\alpha > n$, and where $h : (0, \infty) \to \mathbb{R}$ is C^1, convex and bounded below with $h(\delta) = \delta h(\delta^{-1})$ and $h'(1) = -n$. Then Ψ is isotropic, $\Psi(A) = \det A \cdot \Psi(A^{-1})$, $\Psi \geqslant 0$, $\Psi^{-1}(0) = SO(n)$, and ψ satisfies (H1)–(H4).

Let y be invertible with $y(\Omega_1) = \lambda\Omega_1$. By the arithmetic mean – geometric mean inequality we have that, since $\det Dy = \prod_{i=1}^n v_i$,

$$\begin{aligned}
⨍_{\Omega_1} \Psi(Dy)\, dx &\geqslant ⨍_{\Omega_1} n\left((\det Dy)^{\frac{\alpha}{n}} + (\det Dy)^{1-\frac{\alpha}{n}}\right) + h(\det Dy)\, dx \\
&= ⨍_{\Omega_1} H(\det Dy(x))\, dx \\
&\geqslant H\left(⨍_{\Omega_1} \det Dy(x)\, dx\right) \\
&= H(\lambda^n) = \Psi(\lambda \mathbf{1}),
\end{aligned}$$

as required, where we have set $H(\delta) := n(\delta^{\frac{\alpha}{n}} + \delta^{1-\frac{\alpha}{n}}) + h(\delta)$ and used Jensen's inequality, noting that H is convex and that $\int_{\Omega_1} \det Dy(x)\, dx$ is the n-dimensional measure of $y(\Omega_1)$.

We have equality only when each $v_i = \lambda$, i.e. $Dy(x) = \lambda R(x)$ for $R(x) \in SO(n)$, which implies that $R(x) = R$ is constant and $a + \lambda R\Omega_1 = \lambda\Omega_1$, for some $a \in \mathbb{R}^n$, which for generic Ω_1 implies $a = 0$ and $R = \mathbf{1}$, hence $y(x) = \lambda x$.

However, for (17) to hold for general M implies that Ψ has a special form:

Theorem 2 (see [8]). *Let $\Psi \in C^1(M_+^{n\times n})$. Then*

$$⨍_{\Omega_1} \Psi(Dy)\, dx \geqslant \Psi(M)$$

for all orientation-preserving invertible y with $y(\Omega_1) = M\Omega_1$, and for every Ω_1 and $M \in M_+^{n\times n}$, iff

$$\Psi(A) = H(\det A)$$

for some convex H.

Sketch of proof. If $y = Mx$ is a minimizer, then we can construct a variation that slides at the boundary, so that the tangential component at the boundary of the 'Cauchy stress' is zero, i.e.

$$D\Psi(M)M^T = p(M)\mathbf{1},$$

for a scalar $p(M)$, from which it follows that Ψ corresponds to an elastic fluid, i.e. $\Psi(M) = H(\det M)$. But then $H(\det M)$ is quasiconvex, and so H is convex.

Conversely, if H is convex then

$$⨍_{\Omega_1} H(\det Dy(x))\, dx \geqslant H\left(⨍_{\Omega_1} \det Dy(x)\, dx\right) \tag{19}$$

$$= H(\det M). \tag{20}$$

4 Discussion

Theorem 1 gives conditions under which a minimizer y^* of I_{P_1,P_2} in $\mathcal{A}$ exists, but says nothing about the regularity properties of y^*. In the simpler problem of isotropic nonlinear elasticity essentially nothing is known. In particular it is an open question whether minimizers are smooth, or smooth outside some closed set of zero measure, or even if the usual weak form of the Euler-Lagrange equation holds (though some forms of the Euler-Lagrange equation can be established [7]). The presence of the (possibly discontinuous) lower order terms due to the intensity maps makes the problem for I_{P_1,P_2} even more challenging.

Theorem 2 shows that if the desirable property holds that for linearly related images y^* is the corresponding linear map, then ψ depends on Dy only through $\det Dy$, that is only on local volume changes, so that in particular the hypothesis (H2) of Theorem 1 does not hold. This suggests that a better model might be to minimize a functional such as

$$E_{P_1,P_2}(y) = \int_{\Omega_1} \left(\psi(c_1(x), c_2(y(x)), \det Dy(x)) + |D^2 y(x)|^2\right) dx, \tag{21}$$

for which existence of a minimizer can be proved for low dimensions n, and for which minimizers of linearly related images could be proved under suitable hypotheses to be linear. This idea is explored in [8].

We remark that it is straightforward to prove variants of Theorem 1 (a) for the case when y is required to map a finite number of landmark points in Ω_1 to corresponding points in Ω_2 (see e.g. [20]), and (b) for the case when P_1 is a template image that is to be compared to an unknown part of P_2 (such as in image registration). In the case (b), for example, one can minimize

$$I(y) = \int_{\Omega_1} \psi(c_1(x), c_2(y(x)), Dy(x))\, dx \tag{22}$$

subject to the constraint that $y : \Omega_1 \to \Omega$ is a homeomorphism for some (unknown) subdomain $\Omega = a + \lambda R \Omega_1 \subset \Omega_2$, where $a \in \mathbb{R}^n$, $R \in SO(n)$, $\alpha \leqslant \lambda \leqslant \beta$ and $0 < \alpha < \beta$ are given. Here we consider the case when the unknown part of P_2 is to be compared to a rigid transformation, rotation and uniform magnification of the template, but one can equally handle the case of more general affine transformations, which may be appropriate for images viewed in perspective. Such variants are also explored in [8].

Of course this work needs to be supplemented with appropriate numerical experiments on images. The numerical minimization of integrals such as (1) is not straightforward even without the presence of the (possibly discontinuous) intensity functions, and the fact that the minimization is to be carried out in the admissible set $\mathcal{A}$ of homeomorphisms, rather than, say, maps satisfying Dirichlet boundary conditions, presents additional difficulties. From a rigorous perspective, the numerical method should take into account the possible occurrence of the Lavrentiev phenomenon, whereby the infimum of the energy in $\mathcal{A}$ might be strictly less than the infimum among Lipschitz maps in $\mathcal{A}$ (such as those generated by a finite-element scheme). For discussions see [2,3,6].

Acknowledgements. We are grateful to Alexander Belyaev, Duvan Henao, José Iglesias, David Mumford, Ozan Öktem, Martin Rumpf, Carola Schonlieb and Benedikt Wirth for their interest and helpful suggestions. CLH was supported by EPSRC through grant EP/L016508/1.

References

1. Adams, R., Fournier, J.: Sobolev Spaces, 2nd edn. Academic Press, Cambridge (2003)
2. Bai, Y., Li, Z.: Numerical solution of nonlinear elasticity problems with Lavrentiev phenomenon. Math. Models Methods Appl. Sci. **17**(10), 1619–1640 (2007). https://doi.org/10.1142/S0218202507002406
3. Balci, A.K., Ortner, C., Storn, J.: Crouzeix-Raviart finite element method for non-autonomous variational problems with Lavrentiev gap. Numer. Math. **151**(4), 779–805 (2022)
4. Ball, J.M.: Convexity conditions and existence theorems in nonlinear elasticity. Arch. Ration. Mech. Anal. **63**, 337–403 (1977)
5. Ball, J.M.: Global invertibility of Sobolev functions and the interpenetration of matter. Proc. Royal Soc. Edinburgh **88A**, 315–328 (1981)
6. Ball, J.M.: Singularities and computation of minimizers for variational problems. In: DeVore, R., Iserles, A., Suli, E. (eds.) Foundations of Computational Mathematics. London Mathematical Society Lecture Note Series, vol. 284, pp. 1–20. Cambridge University Press, Cambridge (2001)
7. Ball, J.M.: Some open problems in elasticity. In: Newton, P., Holmes, P., Weinstein, A. (eds.) Geometry, Mechanics, and Dynamics, pp. 3–59. Springer, New York (2002). https://doi.org/10.1007/0-387-21791-6_1
8. Ball, J.M., Horner, C.L.: A nonlinear elasticity model in computer vision (in preparation)
9. Burger, M., Modersitzki, J., Ruthotto, L.: A hyperelastic regularization energy for image registration. SIAM J. Sci. Comput. **35**(1), B132–B148 (2013). https://doi.org/10.1137/110835955
10. Cachier, P., Rey, D.: Symmetrization of the non-rigid registration problem using inversion-invariant energies: application to multiple sclerosis. In: Delp, S.L., DiGoia, A.M., Jaramaz, B. (eds.) MICCAI 2000. LNCS, vol. 1935, pp. 472–481. Springer, Heidelberg (2000). https://doi.org/10.1007/978-3-540-40899-4_48
11. Ciarlet, P.G.: Mathematical Elasticity, Vol. I: Three-Dimensional Elasticity. North-Holland, Amsterdam (1988)
12. Debroux, N., et al.: A variational model dedicated to joint segmentation, registration, and atlas generation for shape analysis. SIAM J. Imag. Sci. **13**(1), 351–380 (2020)
13. Debroux, N., Le Guyader, C.: A joint segmentation/registration model based on a nonlocal characterization of weighted total variation and nonlocal shape descriptors. SIAM J. Imaging Sci. **11**(2), 957–990 (2018). https://doi.org/10.1137/17M1122906
14. Droske, M., Rumpf, M.: A variational approach to non-rigid morphological registration. SIAM J. Appl. Math. **64**(2), 668–687 (2004)
15. Iglesias, J.A.: Symmetry and scaling limits for matching of implicit surfaces based on thin shell energies. ESAIM Math. Model. Numer. Anal. **55**(3), 1133–1161 (2021)
16. Iglesias, J.A., Rumpf, M., Scherzer, O.: Shape-aware matching of implicit surfaces based on thin shell energies. Found. Comput. Math. **18**, 891–927 (2018)

17. Iwaniec, T., Onninen, J.: Hyperelastic deformations of smallest total energy. Arch. Ration. Mech. Anal. **194**(3), 927–986 (2008). https://doi.org/10.1007/s00205-008-0192-7
18. Kolouri, S., Slepčev, D., Rohde, G.K.: A symmetric deformation-based similarity measure for shape analysis. In: 2015 IEEE 12th International Symposium on Biomedical Imaging (ISBI), pp. 314–318. IEEE (2015)
19. Lin, T., Dinov, I., Toga, A., Vese, L.: Nonlinear elasticity registration and Sobolev gradients. In: Fischer, B., Dawant, B.M., Lorenz, C. (eds.) WBIR 2010. LNCS, vol. 6204, pp. 269–280. Springer, Heidelberg (2010). https://doi.org/10.1007/978-3-642-14366-3_24
20. Lin, T., Le Guyader, C., Dinov, I., Thompson, P., Toga, A., Vese, L.: Gene expression data to mouse atlas registration using a nonlinear elasticity smoother and landmark points constraints. J. Sci. Comput. **50**(3), 586–609 (2012). https://doi.org/10.1007/s10915-011-9563-6
21. Maz'ya, V.: Sobolev Spaces with Applications to Elliptic Partial Differential Equations. Grundlehren der Mathematischen Wissenschaften [Fundamental Principles of Mathematical Sciences], vol. 342. Springer, Heidelberg (2011). https://doi.org/10.1007/978-3-642-15564-2
22. Ozeré, S., Gout, C., Le Guyader, C.: Joint segmentation/registration model by shape alignment via weighted total variation minimization and nonlinear elasticity. SIAM J. Imaging Sci. **8**(3), 1981–2020 (2015). https://doi.org/10.1137/140990620
23. Ozeré, S., Le Guyader, C.: Topology preservation for image-registration-related deformation fields. Commun. Math. Sci. **13**(5), 1135–1161 (2015). https://doi.org/10.4310/CMS.2015.v13.n5.a4
24. Reshetnyak, Y.G.: Liouville's theorem on conformal mappings under minimal regularity assumptions. Siberian Math. J. **8**, 631–653 (1967)
25. Rindler, F.: Calculus of Variations. Universitext, Springer, Cham (2018). https://doi.org/10.1007/978-3-319-77637-8
26. Rumpf, M., Wirth, B.: An elasticity-based covariance analysis of shapes. Int. J. Comput. Vis. **92**(3), 281–295 (2011). https://doi.org/10.1007/s11263-010-0358-2
27. Rumpf, M.: Variational methods in image matching and motion extraction. In: Burger, M., Mennucci, A.C.G., Osher, S., Rumpf, M. (eds.) Level Set and PDE Based Reconstruction Methods in Imaging, pp. 143–204. Springer, Cham (2013). https://doi.org/10.1007/978-3-319-01712-9
28. Simon, K., Sheorey, S., Jacobs, D.W., Basri, R.: A hyperelastic two-scale optimization model for shape matching. SIAM J. Sci. Comput. **39**(1), B165–B189 (2017). https://doi.org/10.1137/15M1048562
29. Šverák, V.: Regularity properties of deformations with finite energy. Arch. Ration. Mech. Anal. **100**, 105–127 (1988)
30. Šilhavý, M.: The Mechanics and Thermodynamics of Continuous Media. Springer, Heidelberg (1997). https://doi.org/10.1007/978-3-662-03389-0

Learning Differential Invariants of Planar Curves

Roy Velich(✉) and Ron Kimmel

Technion - Israel Institute of Technology, Haifa, Israel
royve@campus.technion.ac.il, ron@cs.technion.ac.il

Abstract. We propose a learning paradigm for the numerical approximation of differential invariants of planar curves. Deep neural-networks' (DNNs) universal approximation properties are utilized to estimate geometric measures. The proposed framework is shown to be a preferable alternative to axiomatic constructions. Specifically, we show that DNNs can learn to overcome instabilities and sampling artifacts and produce consistent signatures for curves subject to a given group of transformations in the plane. We compare the proposed schemes to alternative state-of-the-art axiomatic constructions of differential invariants. We evaluate our models qualitatively and quantitatively and propose a benchmark dataset to evaluate approximation models of differential invariants of planar curves.

Keywords: Differential invariants · differential geometry · computer vision · shape analysis

1 Introduction

Differential Invariants and Signature Curves. According to an important theorem by É. Cartan [6], two curves are related by a group transformation $g \in G$, i.e., equivalent curves with respect to G, if and only if their signature curves, with respect to the transformation group G, are identical. This observation allows one to develop analytical tools to measure the equivalence of two planar curves, which has practical applications in various fields and tasks, such as computer vision, shape analysis, geometry processing, object detection, and more. Planar curves are often extracted from images as boundaries of objects, or level-sets of gray-scale images. The boundary of an object encodes vital information which can be further exploited by various computer vision and image analysis tools. A common computer vision task requires one to find common characteristics between two objects in two different image frames. This task can be naturally approached by representing image objects as signature curves of their boundaries. Since signature curves are proven to provide a full solution for the equivalence problem of planar curves [14, p. 183, Theorem 8.53], they can be further analyzed for finding correspondence between image objects. Signature curves are parametrized by differential invariants. Therefore, in order to generate the signature curve of a given planar curve, one has to evaluate the required differential

L. Calatroni et al. (Eds.): SSVM 2023, LNCS 14009, pp. 575–587, 2023.
https://doi.org/10.1007/978-3-031-31975-4_44

invariants at each point. In practice, planar curves are digitally represented as a discrete set of points, which implies that the computation of differential invariant quantities, such as the curvature at a point, can only be numerically approximated using finite differences techniques. Since many important and interesting differential invariants are expressed as functions of high-order derivatives, their approximations are prone to inherent numerical instabilities, and high sensitivity to sampling noise. For example, the equiaffine curvature, which is a differential invariant of the special-affine transformation group $\mathrm{SA}(2)$, is a fourth-order differential invariant, and its direct approximation using an axiomatic approach is practically infeasible for discrete curves, especially around inflection points.

Contribution. Inspired by the numerical approximation efforts of Calabi et al. [5], and the learning approaches introduced in [10,15], we present a complete learning environment for approximating differential invariants of planar curves with respect to various transformation groups, including the affine group, for which, very few attempts to approximate their differential invariants in the discrete setting have been made. Moreover, we introduce a shape-matching benchmark for evaluating differential invariants approximation models.

2 Related Work

Axiomatic Approximation of Differential Invariants. The axiomatic approach for approximating the differential invariants of discrete planar curves, as presented by Calabi et al. in [5], is based on the joint invariants between points of a discrete planar curve. As suggested by Calabi et al., given a transformation group G and a planar curve $\mathcal{C}$, one can approximate the fundamental differential invariant w.r.t. G at a point $\mathbf{x} \in \mathcal{C}$, which is denoted by $\kappa(\mathbf{x})$ and is known as the *G-invariant curvature* at $\mathbf{x}$, by first interpolating a set $\mathcal{P}_{\mathbf{x}} \subset \mathcal{C}$ of points from a small neighborhood around $\mathbf{x}$ with an auxiliary curve $\mathcal{C}_{\mathbf{x}}$, on which the differential invariant is constant by definition. Then, the constant differential invariant $\widetilde{\kappa}(\mathbf{x})$ of $\mathcal{C}_{\mathbf{x}}$ is evaluated using the joint invariants of the points in $\mathcal{P}_{\mathbf{x}}$. Finally, $\widetilde{\kappa}(\mathbf{x})$ is used as an approximation to $\kappa(\mathbf{x})$. Since the approximation is calculated by joint invariants of G, its evaluation is unaffected by the action of a group transformation $g \in G$, and therefore, as the size of the mesh of points $\mathcal{P}_{\mathbf{x}}$ tends to zero, the approximation of the differential invariant at $\mathbf{x}$ converges to the continuous value. For example, in the Euclidean case, the fundamental joint invariant of the groups $\mathrm{E}(2)$ and $\mathrm{SE}(2)$ is the Euclidean distance between a pair of points. By applying the approach suggested by Calabi et al., the Euclidean curvature $\kappa(\mathbf{x})$ can be approximated by first interpolating a circle through $\mathbf{x_{i-1}}$, $\mathbf{x_i}$, and $\mathbf{x_{i+1}}$, and then exploit the Euclidean distances between those three points to calculate the constant curvature $\widetilde{\kappa}(\mathbf{x})$ of the interpolated circle, using Heron's formula. The second differential invariant at a curve point $\mathbf{x}$, is defined as the derivative of the G-invariant curvature κ with respect to the *G-invariant arc-length element* s, and is denoted by $\kappa_s(\mathbf{x})$. According to Calabi et al., $\kappa_s(\mathbf{x})$ is approximated using finite differences, by calculating the ratio of the difference between $\widetilde{\kappa}(\mathbf{x}_{i-1})$ and $\widetilde{\kappa}(\mathbf{x}_{i+1})$ with respect to the distance between $\mathbf{x_{i-1}}$ and

$\mathbf{x}_{i+1}$. In their paper, Calabi et al. further explain how to approximate differential invariants of planar curves with respect to the equiaffine group SA (2). In this case, the fundamental joint invariant is the triangle area defined by a triplet of points, and the auxiliary curve is a conic section, which possesses a constant equiaffine curvature, and has to be interpolated through five curve points in the neighborhood of $\mathbf{x}$. The main limitation of their method, in the equiaffine case, is that it is practically applicable only to convex curves, since the equiaffine curvature is not defined at inflection points.

Learning-Based Approximation. A more recent approach was introduced in [15]. There, Pai et al. took a self-supervised learning approach and used a Siamese convolutional neural network to learn the euclidean curvature from a dataset of discrete planar curves. However, the scope of their work was lacking a module to approximate the Euclidean arc-length at each point and/or the derivative of the curvature w.r.t. arc-length, and therefore, a signature curve could not be generated. This means that their approach was not useful for evaluating the equivalence between planar shapes. Moreover, they have not extended their work to learning differential invariants of less restrictive transformation groups, such as the equiaffine and affine groups. In [10], Lichtenstein et al. introduced a deep learning approach to numerically approximate the solution to the Eikonal equation. They proposed to replace axiomatic local numerical solvers with a trained neural network that provides highly accurate estimates of local distances for a variety of different geometries and sampling conditions.

3 Mathematical Framework

We begin with the following theorem, as stated by [14, p. 176, Theorem 8.47],

Theorem 1 (The Fundamental Theorem of Differential Invariants). *A transformation group G acting on $\mathbb{R}^2$ admits the following properties. A unique differential invariant κ (up to a function of it), which is known as the G-invariant curvature. A unique G-invariant one-from $ds = p(x)\,dx$ (up to a constant multiple), which is known as the G-invariant arc-length element. Any other differential invariant of G is a function $I : (\kappa, \kappa_s, \kappa_{ss}, \ldots) \to \mathbb{R}$ of the G-invariant curvature κ and its derivatives with respect to the G-invariant arc length s.*

Definition 1 (Equivalent Planar Curves). *Let G be a transformation group acting on $\mathbb{R}^2$. Two planar curves $\mathcal{C}$ and $\hat{\mathcal{C}}$ are equivalent with respect to G if there exists $g \in G$ such that $\hat{\mathcal{C}} = g \cdot \mathcal{C}$.*

Definition 2 (Signature Curve). *The G-invariant signature curve associated with a parametrized planar curve $\mathcal{C} = \left\{ \left(x\left(t\right), y\left(t\right) \right) \right\}$, is a curve $S \subset \mathbb{R}^2$ parametrized by the G-invariant curvature of $\mathcal{C}$ at a point and its derivative with respect to the G-invariant arc-length, and is given by $S = \left\{ \left(\kappa\left(t\right), \kappa_s\left(t\right) \right) \right\}$.*

Signature curves can be used to solve the equivalence problem of planar curves for a general transformation group G, as stated by the next theorem, given in [14, p. 183, Theorem 8.53],

Theorem 2 (Equivalence of Planar Curves). *Two non-singular planar curves $\mathcal{C}$ and $\hat{\mathcal{C}}$ are equivalent w.r.t. a transformation group G if and only if their signature curves S and $\hat{S}$ are equal.*

For more information about differential invariants and signature curves, see [1–3,3,4,7–9,12–14].

4 Method

We introduce a simple neural-net architecture, along with an appropriate training scheme and loss function, for producing numerically stable approximations of the two fundamental differential invariants κ and κ_s, with respect to a group transformation G. Both quantities can be combined together to evaluate and plot the G-invariant signature curve of a planar curve. Given a planar curve $\mathcal{C}$ and a point $\mathbf{p} \in \mathcal{C}$ on the curve, the neural network receives as an input a discrete sample of the local neighborhood of $\mathbf{p}$, and outputs two scalars, of which we interpret as κ and κ_s at $\mathbf{p}$.

4.1 Learning Differential Invariants

There are two fundamental ideas that govern our training approach for learning a differential invariant w.r.t. a transformation group G.

Group and Reparametrization Invariance. The neural network should learn a representation that is both invariant to the action of a group transformation $g \in G$, and to a reparametrization of the input curve. That way, the network will learn a truly invariant representation, which is not biased towards any specific discrete sampling scheme, and that encapsulates the local geometric properties of the curve. Given a discrete planar curve $\mathcal{C}$, we denote by $\mathcal{C}(\mathcal{D})$ its down-sampled version, sampled non-uniformly with N points according to a random non-uniform probability mass function $\mathcal{D} : \mathcal{C} \to [0, 1]$. Given two different non-uniform probability mass functions $\mathcal{D}_1$ and $\mathcal{D}_2$, one can refer to the down-sampled curves $\mathcal{C}(\mathcal{D}_1)$ and $\mathcal{C}(\mathcal{D}_2)$, as two different reparametrizations of $\mathcal{C}$. This motivates us to require that if a point $\mathbf{p} \in \mathcal{C}$ was drawn both in $\mathcal{C}(\mathcal{D}_1)$ and $\mathcal{C}(\mathcal{D}_2)$, then, a valid prediction model should output the same curvature value for $\mathbf{p}$, whether it was fed with either a neighborhood of $g_1 \cdot \mathbf{p}$ or a neighborhood $g_2 \cdot \mathbf{p}$ as an input, for any two group transformations $g_1, g_2 \in G$.

Orthogonality of Differential Invariants. We expect κ and κ_s to be orthogonal quantities, since the latter is the derivative of the former w.r.t. to arc-length. So, for example, consider a collection of M points, such that each was sampled from an arbitrary planar curve. Moreover, let κ^i and κ_s^i denote the curvature and its derivative w.r.t. arc-length evaluated at the ith point. Then, if we consider $\left\{\kappa^i\right\}_{i=1}^{M}$ and $\left\{\kappa_s^i\right\}_{i=1}^{M}$ as M observations of random variables X_κ and X_{κ_s}, respectively, we expect the absolute value of the Pearson correlation coefficient $|\rho(X_\kappa, X_{\kappa_s})|$ to approach zero as $M \to \infty$. We empirically verify this assumption in the Euclidean case, by calculating the Pearson correlation coefficient of X_κ and X_{κ_s} on a collection of randomly sampled points, taken from a large collection of smooth curves. See Fig. 1.

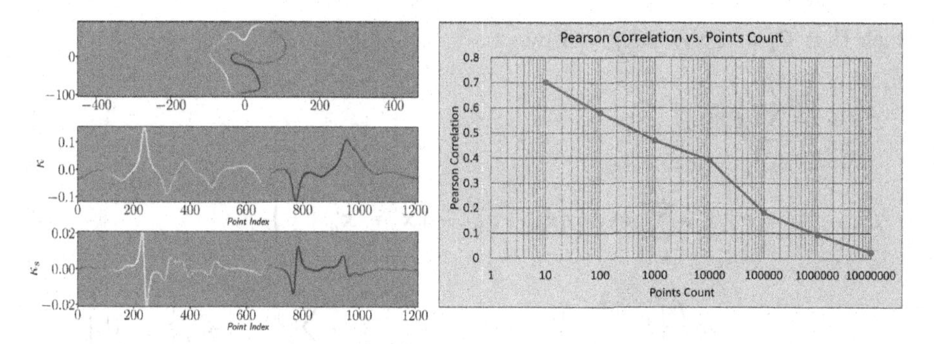

Fig. 1. Left: Plot of a smooth curve and its Euclidean κ and κ_s evaluated at each point. **Right**: A semi-log plot of the absolute value of the Pearson correlation coefficient of X_κ and X_{κ_s} as a function of points count.

4.2 Training Scheme

Given a dataset of n discrete planar curves $\{\mathcal{C}_i\}_{i=1}^n$, our training scheme generates a training batch on the fly, where each batch is made of a set of training tuplets. A training tuplet $\mathcal{T} = (\mathcal{E}_a, \mathcal{E}_p, \mathcal{E}_{n_1}, \ldots, \mathcal{E}_{n_m})$, is made up of a collection of sampled neighborhoods, which are referred to as the *anchor* ($\mathcal{E}_a$), *positive* ($\mathcal{E}_p$) and *negative* ($\{\mathcal{E}_{n_i}\}_{i=1}^m$) examples. The tuplet is generated by first drawing a random curve $\mathcal{C}$ and a random point $\mathbf{p} \in \mathcal{C}$ on it. Given two random group transformations $g_a, g_p \in G$ and two random non-uniform probability mass functions $\mathcal{D}_a$ and $\mathcal{D}_p$, both the anchor and positive examples, $\mathcal{E}_a$ and $\mathcal{E}_p$, are generated by selecting $2N$ adjacent points to $\mathbf{p}$ (N consecutive points that immediately precede $\mathbf{p}$, and additional N consecutive points that immediately succeed $\mathbf{p}$) from the transformed and down-sampled versions of $\mathcal{C}$ given by $g_a \cdot \mathcal{C}_a(\mathcal{D}_a)$ and $g_p \cdot \mathcal{C}_p(\mathcal{D}_p)$, respectively. In other words, both $\mathcal{E}_a$ and $\mathcal{E}_p$ represent a local neighborhood of $\mathbf{p}$ sampled under different reparametrizations, which were transformed by group transformations taken from the same transformation group G.

The ith negative example $\mathcal{E}_{n_i}$ is generated in a similar manner. We draw another random curve $\mathcal{C}_i$, and take a random point $\mathbf{p}_i \in \mathcal{C}_i$ on it. Then, we select again additional $2N$ adjacent points to $\mathbf{p}_i$ from a transformed and down-sampled version of $\mathcal{C}_i$ given by $g_{n_i} \cdot \mathcal{C}_i(\mathcal{D}_i)$, where $g_{n_i} \in G$ is a random group transformation, and $\mathcal{D}_i$ is a random probability mass function.

Given a batch of K training tuplets $\{\mathcal{T}_i\}_{i=1}^K$, we train a multi-head Siamese fully-connected neural network by minimizing the loss given by,

$$\mathcal{L} = \mathcal{L}_I + \mathcal{L}_O \tag{1}$$

Such that $\mathcal{L}_I$ and $\mathcal{L}_O$ are given by,

$$\mathcal{L}_I = \frac{1}{K}\sum_{i=1}^{K}\mathcal{L}^i_{\text{tuplet}} \tag{2}$$

$$\mathcal{L}^i_{\text{tuplet}} = \log\left(1+\sum_{j=1}^{m}\exp\left(\|\mathbf{x}^i_a-\mathbf{x}^i_p\|-\|\mathbf{x}^i_a-\mathbf{x}^i_{n_j}\|\right)\right) \tag{3}$$

$$\mathcal{L}_O = \frac{1}{m+2}\left(\left|\rho\left(\left\{\mathbf{x}^i_a\right\}_{i=1}^{K}\right)\right|+\left|\rho\left(\left\{\mathbf{x}^i_p\right\}_{i=1}^{K}\right)\right|+\sum_{j=1}^{m}\left|\rho\left(\left\{\mathbf{x}^i_{n_j}\right\}_{i=1}^{K}\right)\right|\right) \tag{4}$$

$$\rho\left(\left\{\mathbf{x}^i\right\}_{i=1}^{K}\right) = \frac{\sum_{i=1}^{K}\left(\mathbf{x}^i[0]-\bar{\mathbf{x}}[0]\right)\left(\mathbf{x}^i[1]-\bar{\mathbf{x}}[1]\right)}{\sqrt{\sum_{i=1}^{K}\left(\mathbf{x}^i[0]-\bar{\mathbf{x}}[0]\right)^2\sum_{i=1}^{K}\left(\mathbf{x}^i[1]-\bar{\mathbf{x}}[1]\right)^2}} \tag{5}$$

Where $\mathbf{x}^i_a$, $\mathbf{x}^i_p, \mathbf{x}^i_{n_1}, \ldots, \mathbf{x}^i_{n_m} \in \mathbb{R}^2$ are the neural-network's outputs for the ith input tuplet given by $\mathcal{T}_i = \left(\mathcal{E}^i_a, \mathcal{E}^i_p, \mathcal{E}^i_{n_1}, \ldots, \mathcal{E}^i_{n_m}\right)$, and where $\mathbf{x}[0]$ and $\mathbf{x}[1]$ are the first and second components of an output vector, respectively. Moreover, $\bar{\mathbf{x}}[0]$ and $\bar{\mathbf{x}}[1]$ are the mean values of the first and second components of a set of output vectors, respectively. Equation 3 is known as the *tuplet loss* [17] which is minimized when the distance between $\mathbf{x}^i_a$ and $\mathbf{x}^i_p$ is minimized, and the distance between $\mathbf{x}^i_a$ and $\mathbf{x}^i_{n_j}$ is maximized, for $j = 1, \ldots, m$. Equation 2 is named the *invariance loss* and it is equal to the mean of the tuplet loss over all the training tuplets in the training batch. Equation 4 is named the *orthogonality loss*, and it is minimized when the absolute value of the Pearson correlation coefficient is minimized w.r.t. the anchor, positive, and negative examples' outputs, over the whole batch. See Fig. 2 for an elaborate visual demonstration of the training scheme for a batch of training tuplets. Note, that we eliminate any translation and rotational ambiguity by transforming each input example into a canonical placement before it is fed into the neural-network. The transformation is done by translating each example such that its first point is located at the origin and then, rotating it such that the vector that starts at the example's first point and ends at the middle point is aligned with the positive x-axis.

5 Experiments and Results

5.1 Datasets and Training

We create three datasets of planar curves, by extracting level-curves from arbitrary images scrapped from the internet. The first two datasets are used for training and validation, and the third dataset is used for qualitative evaluation. Each dataset consists of tens of thousands of curves. We train a multi-head siamese neural-network based on a simple fully-connected architecture. The architecture is based on blocks of 3 fully-connected layers of the same size. The layers at the first block are of size 128, and the size of the layers in each subsequent block is

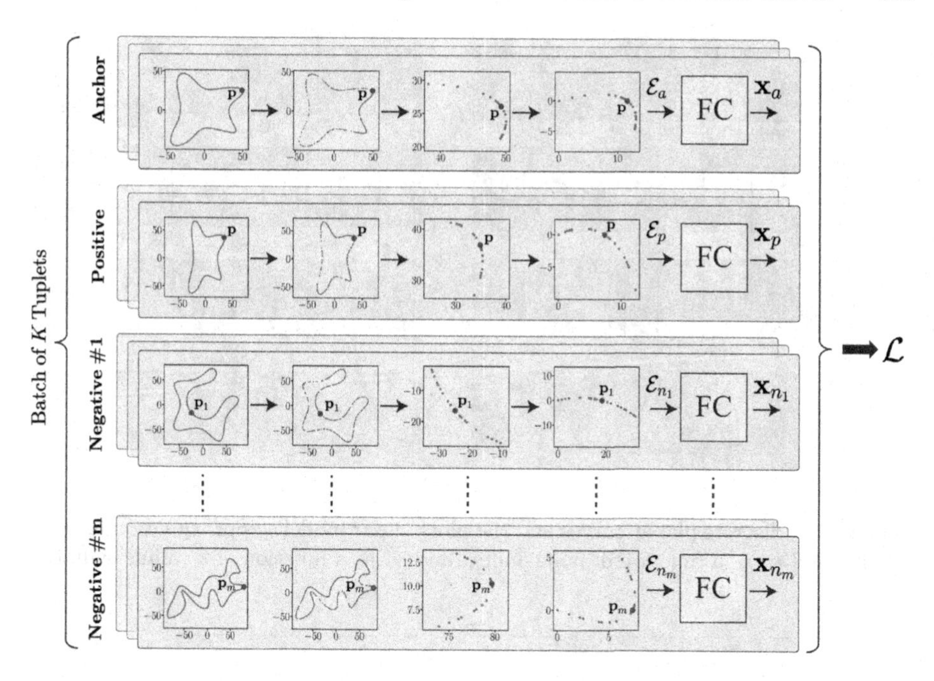

Fig. 2. A demonstration of the training scheme for a given input batch of training tuplets. **Left-to-right**: A random curve and a random point on it are drawn. The curve is down-sampled non-uniformly. A local neighborhood around the point is extracted. The extracted point neighborhood is transformed into a canonical placement, as described above. The canonical point neighborhood is fed into the neural-network, which outputs two scalars that we interpret as κ and κ_s. **Top-to-bottom**: The anchor and positive training examples, as well as the m negative training examples. Together, they form a training tuplet.

divided by two w.r.t. the size of layers in the previous block. We use a batch-norm layer after each linear layer, followed by a sine activation function. The reason for using a periodic activation function is due to the fact that its derivative can be expressed by a phase-shift. We believe that this fact makes it easier for our network to learn high-order derivatives [16].

5.2 Qualitative Evaluation

We evaluate our differential invariants approximation model qualitatively, w.r.t. the affine group. We focus on interpretability, group invariance, reparametrization invariance, and comparison against the Euclidean axiomatic signature.

Interpretability. As shown in Fig. 3, our model's outputs can be interpreted as κ and κ_s. The first output, which we interpret as κ, approximately reaches its extrema points when the second output, which we interpret as κ_s, crosses the zero line.

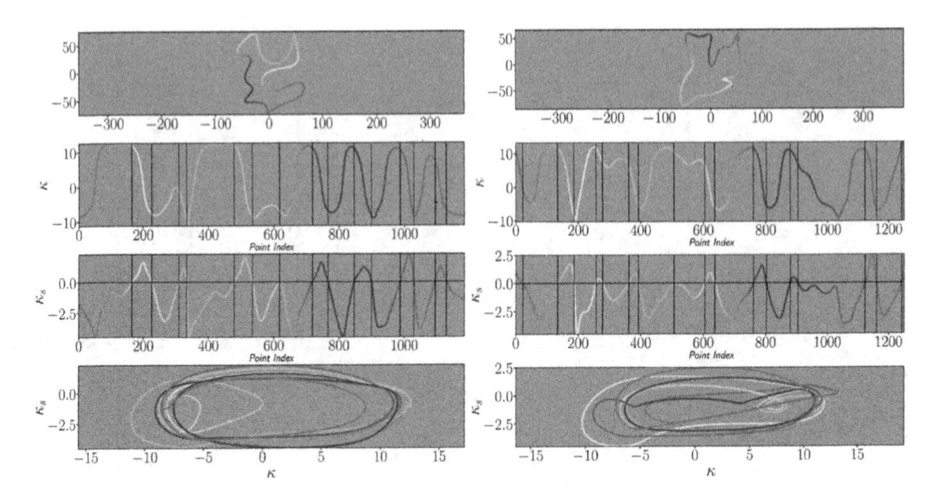

Fig. 3. Two discrete planar curves and the plots of our model's approximation for their affine κ and κ_s as a function of point-index, and κ_s as a function of κ (affine signature curve).

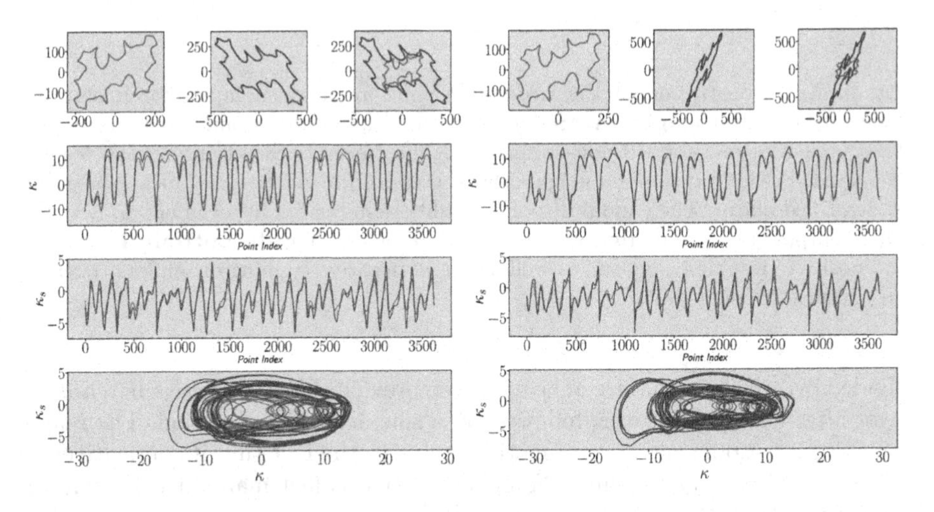

Fig. 4. A comparison of our model's approximation for the affine κ and κ_s as a function of point-index, of two pairs of equivalent curves. The blue curve was obtained by applying a linear operator on the red one. **Left**: Linear operator with condition number 2.5 and determinant 3.5. **Right**: Linear operator with condition number 5 and determinant 2. (Color figure online)

Group Action Invariance. As shown in Fig. 4, our approximation model exhibits clear group invariance w.r.t. the affine group. The peaks and valleys of the plots of κ and κ_s of the two equivalent curves (the red and blue curves) are aligned correctly, and the relative amplitude amplification is mild.

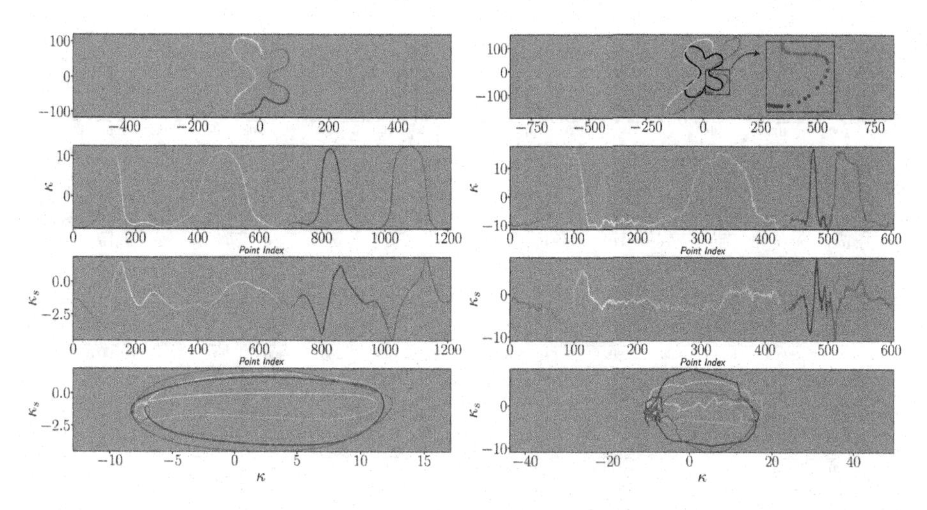

Fig. 5. A demonstration of our model's robustness for non-uniform down-sampling. **Left**: Our model's approximation for κ and κ_s of a reference planar curve. **Right**: Our model's approximation for κ and κ_s of the reference curve, after it was non-uniformly down-sampled by 50%, and an arbitrary affine transform was applied to it.

Reparametrization Invariance. As shown in Fig. 5, our approximation model is robust to reparametrization and non-uniform down-sampling. The peaks and valleys patterns of κ and κ_s w.r.t. the reference curve are clearly reproduced in its down-sampled and transformed version (matching colors).

Comparison with the Euclidean Axiomatic Approach. As shown in Fig. 6, our model's approximations for κ and κ_s w.r.t. the affine group are superior over the axiomatic Euclidean approach (which is based on finite differences). As expected, the axiomatic signature exhibits visible misalignment of κ and κ_s w.r.t. the two equivalent curves, as well as major amplitude amplification differences. On the other hand, the approximations made by our model are aligned while only a mild amplitude amplification is visible.

5.3 Quantitative Evaluation

We have compiled an evaluation benchmark of 34 collections of planar curves, which were extracted from the boundary of silhouettes of natural objects, which are not necessarily smooth (see Fig. 7). Each collection is comprised of 30 boundary curves of the same conceptual object in different poses and deformations. We perform a simple shape-matching evaluation which goes as follows. We iterate over all boundary curves in all collections. Given the jth curve in the ith collection, we deform it using an affine transformation and then down-sample it, in order to get what we call as a *query* curve. Then, we compare the query curve against all curves in its collection (which we refer to as *database* curves). A query curve is compared with a database curve by calculating the average Hausdorff distance of their corresponding signature curves, which were evaluated by our

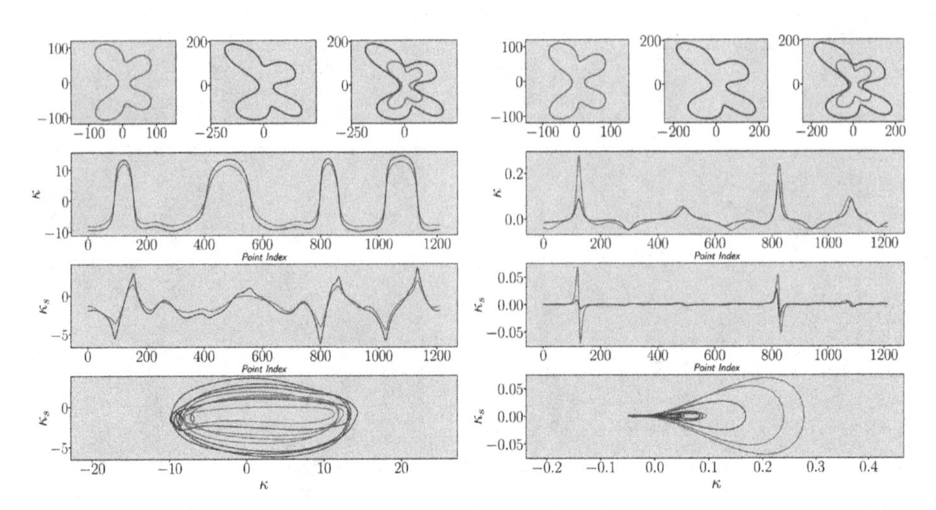

Fig. 6. Two affine equivalent curves (blue and red) and the approximation of their differential invariants. **Left**: The approximation of the affine κ and κ_s using our trained model. **Right**: The approximation of the Euclidean κ and κ_s using the axiomatic approach, which is based on finite differences. (Color figure online)

Table 1. Shape-matching evaluation (Bunnies collection)

	Success rate								
	Our model			Euclidean axiomatic			Equiaffine axiomatic		
Sampling rate	det = 2 cond = 2	det = 2 cond = 3	det = 3 cond = 2	det = 2 cond = 2	det = 2 cond = 3	det = 3 cond = 2	det = 2 cond = 2	det = 2 cond = 3	det = 3 cond = 2
100.00%	100.00%	100.00%	70.00%	3.33%	16.67%	10.00%	16.67%	10.00%	3.33%
90.00%	100.00%	100.00%	63.33%	13.33%	6.67%	10.00%	3.33%	0.00%	3.33%
80.00%	90.00%	90.00%	56.67%	10.00%	6.67%	10.00%	0.00%	0.00%	3.33%
70.00%	83.33%	90.00%	46.67%	6.67%	0.00%	3.33%	3.33%	0.00%	0.00%
60.00%	63.33%	73.33%	43.33%	13.33%	10.00%	6.67%	13.33%	3.33%	3.33%
50.00%	56.67%	46.67%	33.33%	3.33%	3.33%	3.33%	3.33%	3.33%	3.33%

model. If the minimum distance was acquired by comparing a curve with its own transformed and down-sampled version, then we call it a *successful match*. We score each collection by calculating the average number of successfully matched curves, which we denote by the *success rate* of the collection. We perform this experiment under various down-sampling ratios and affine transformations. We use this benchmark to evaluate our model as well as the axiomatic approximations for the Euclidean and equiaffine groups, as proposed by [5]. We run 3 different *flavors* of our benchmark. For the first flavor, we transform each curve by an affine transform $A \in \mathbb{R}^{2\times 2}$, where $\det(A) = 2$ and $\text{cond}(A) = 2$. For the second flavor we have $\det(A) = 2$ and $\text{cond}(A) = 3$, and for the third flavor we have $\det(A) = 3$ and $\text{cond}(A) = 2$. In all 3 flavors, we down-sample the query curves by increasing sampling ratios, where a sampling ratio of 100% means that we keep all points, a sampling rate of 90% means that we drop 10% of the points, and so on. Evaluation results for the *bunnies* collection are available in Table 1.

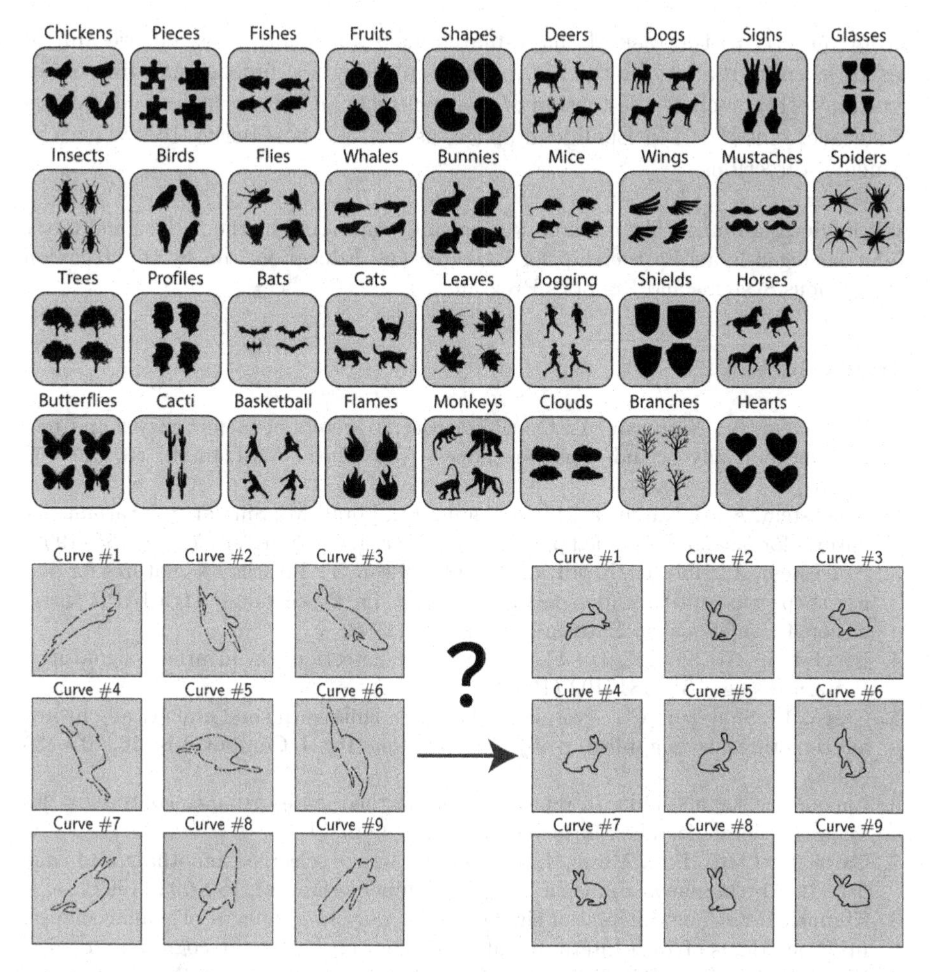

Fig. 7. **Top**: A sample of our benchmark's curve collections. We plot four samples (out of 30) from each collection. **Bottom**: A demonstration of our shape-matching benchmark for the *bunnies* collection. Each transformed and down-sampled curve on the left should be matched with its equivalent curve on the right. Although in this figure only 9 curves are shown, in the actual shape-matching benchmark each collection consists of 30 curves that need to be matched.

6 Conclusions and Future Research

We proposed a self-supervised approach, using multi-head Siamese neural networks, for the approximation of differential invariants of planar curves, with respect to a given transformation group G. We have shown qualitatively that our model is invariant w.r.t. the group actions, as well as reparametrization and non-uniform down-sampling. We have also demonstrated the applicability of our method to the task of shape-matching. As a future research, we aim to design learning models for the approximation of higher dimensional differential invariants [11], such as the torsion of space curves, and the principal curvatures of surfaces embedded in Euclidean space, as well as their derivatives w.r.t. the principal directions.

Acknowledgements. The research was partially supported by the D. Dan and Betty Kahn Michigan-Israel Partnership for Research and Education, run by the Technion Autonomous Systems and Robotics Program.

References

1. Bruckstein, A., Netravali, A.: On differential invariants of planar curves and recognizing partially occluded planar shapes. Ann. Math. Artif. Intell. **13**, 227–250 (1995)
2. Bruckstein, A.M., Katzir, N., Lindenbaum, M., Porat, M.: Similarity-invariant signatures for partially occluded planar shapes. Int. J. Comput. Vis. **7**, 271–285 (1992)
3. Bruckstein, A., Holt, R., Netravali, A., Richardson, T.: Invariant signatures for planar shape recognition under partial occlusion. In: Proceedings. 11th IAPR International Conference on Pattern Recognition (1992)
4. Bruckstein, A., Shaked, D.: Skew symmetry detection via invariant signatures. Pattern Recogn. **31**, 181–192 (1998)
5. Calabi, E., Shakiban, C., Olver, P.J., Haker, S.: Differential and numerically invariant signature curves applied to object recognition. Int. J, Comput. Vis. **26**, 107–135 (1998)
6. Cartan, E.: La méthode du repère mobile, la théorie des groupes continus et les espaces généralisés. The Mathematical Gazette (1935)
7. Catté, F., Lions, P.L., Morel, J.M., Coll, T.: Image selective smoothing and edge detection by nonlinear diffusion. SIAM J. Numer. Anal. **29**, 182–193 (1992)
8. Kimmel, R.: Affine differential signatures for gray level images of planar shapes. In: Proceedings of 13th International Conference on Pattern Recognition (1996)
9. Kimmel, R., Zhang, C., Bronstein, A.M., Bronstein, M.M.: Are MSER features really interesting? IEEE Trans. Pattern Anal. Mach. Intell. **33**(11), 2316–2320 (2011)
10. Lichtenstein, M., Pai, G., Kimmel, R.: Deep eikonal solvers. In: Lellmann, J., Burger, M., Modersitzki, J. (eds.) SSVM 2019. LNCS, vol. 11603, pp. 38–50. Springer, Cham (2019). https://doi.org/10.1007/978-3-030-22368-7_4
11. Olver, P.: Differential invariants of surfaces. Applications **27**, 230–239 (2009)
12. Olver, P., Sapiro, G., Tannenbaum, A.: Differential invariant signatures and flows in computer vision: a symmetry group approach. In: ter Haar Romeny, B.M. (ed.) Geometry-Driven Diffusion in Computer Vision. Computational Imaging and Vision, vol. 1. Springer, Dordrecht (1994). https://doi.org/10.1007/978-94-017-1699-4_11
13. Olver, P.J.: Equivalence. Invariants and Symmetry. Cambridge University Press (1995)
14. Olver, P.J.: Class. Invariant Theory. London Mathematical Society Student Texts, Cambridge University Press (1999)
15. Pai, G., Wetzler, A., Kimmel, R.: Learning invariant representations of planar curves. In: 5th International Conference on Learning Representations (2017)

16. Sitzmann, V., Martel, J.N., Bergman, A.W., Lindell, D.B., Wetzstein, G.: Implicit neural representations with periodic activation functions. In: Proceedings of NeurIPS (2020)
17. Yu, B., Tao, D.: Deep metric learning with tuplet margin loss. In: 2019 IEEE/CVF International Conference on Computer Vision (ICCV) (2019)

Diffusion–Shock Inpainting

Kristina Schaefer(✉) and Joachim Weickert

Mathematical Image Analysis Group, Department of Mathematics and Computer Science, E1.7, Saarland University, 66123 Saarbrücken, Germany
{schaefer,weickert}@mia.uni-saarland.de

Abstract. We propose diffusion–shock (DS) inpainting as a hitherto unexplored integrodifferential equation for filling in missing structures in images. It combines two carefully chosen components that have proven their usefulness in different applications: homogeneous diffusion inpainting and coherence-enhancing shock filtering. DS inpainting enjoys the complementary synergy of its building blocks: It offers a high degree of anisotropy along an eigendirection of the structure tensor. This enables it to connect interrupted structures over large distances. Moreover, it benefits from the sharp edge structure generated by the shock filter, and it exploits the efficient filling-in effect of homogeneous diffusion. The second order equation that underlies DS inpainting inherits a continuous maximum–minimum principle from its constituents. In contrast to other attractive second order inpainting equations such as edge-enhancing anisotropic diffusion, we can guarantee this property also for the proposed discrete algorithm. Our experiments show a performance that is comparable to or better than many linear or nonlinear, isotropic or anisotropic processes of second or fourth order. They include homogeneous diffusion, biharmonic interpolation, TV inpainting, edge-enhancing anisotropic diffusion, the methods of Tschumperlé and of Bornemann and März, Cahn–Hilliard inpainting, and Euler's elastica.

Keywords: Inpainting · Shock Filter · Diffusion · Mathematical Morphology · Image Processing

1 Introduction

Image inpainting [10,18] aims at restoring images with missing or damaged areas. Many popular methods use partial differential equations (PDEs), since they offer compact and transparent models with sound theoretical foundations. A particularly simple representative is homogeneous diffusion inpainting [6]. It is based on a linear second order PDE that satisfies a maximum–minimum principle and allows to design very efficient algorithms [13]. For inpainting-based compression,

This project has received funding from the European Research Council (ERC) under the European Union's Horizon 2020 research and innovation programme (grant agreement No. 741215, ERC Advanced Grant INCOVID).

L. Calatroni et al. (Eds.): SSVM 2023, LNCS 14009, pp. 588–600, 2023.
https://doi.org/10.1007/978-3-031-31975-4_45

it may give very good results, if the data are optimised carefully [17]. However, it cannot produce satisfactory continuations of sharp edges over large distances.

As a rememdy, higher order PDEs have been considered, such as the ones arising from the energy functionals of Euler's elastica [14,18,19] or the Cahn–Hilliard model [2,5]. From a theoretical perspective, these continuous models are attractive: They offer a low-curved continuation of level lines or the propagation of gradient information. Unfortunately, for such higher order PDEs it is fairly challenging to design good numerical algorithms that are computationally efficient and do not suffer from dissipative artefacts, which lead to a blurred continuation of edges.

An interesting alternative to higher order inpainting PDEs are second order anisotropic integrodifferential methods that implicitly exploit curvature information via a Gaussian convolution inside a diffusion tensor. This idea is pursued in edge-enhancing anisotropic diffusion (EED) [32]. It achieves state-of-the-art results for inpainting-based compression and can propagate structures over large distances [24]. However, current discretisations can violate a maximum–minimum principle. Moreover, although edges remain sharper than for most algorithms for higher order PDEs, they still show some dissipative artefacts.

Our Contribution. The goal of our work is to show that all the above mentioned problems can be addressed with a surprisingly simple combination of two processes with complementary qualities: homogeneous diffusion inpainting [6] and a coherence-enhancing shock filter [30]. While both techniques are well-established, their combination within an inpainting method is novel. The resulting *diffusion–shock (DS) inpainting* offers the best of two worlds: On the one hand, the coherence-enhancing shock filter is guided by the robust edge information from a structure tensor [11]. Its switch between the hyperbolic PDEs of dilation and erosion creates perfectly sharp shock fronts that can be propagated with basically no directional artefacts. This is illustrated in Fig. 1. On the other hand, the parabolic homogeneous diffusion PDE is a model with maximal simplicity that creates an efficient filling-in mechanism in flat areas. For both processes we use discretisations that offer a high degree of rotation invariance and satisfy a discrete maximum–minimum principle. These properties carry over to DS inpainting. In our experiments we compare DS inpainting with many inpainting PDEs and demonstrate its favourable performance in spite of its simplicity.

Related Work. With the goal of image sharpening, Kramer and Bruckner [16] proposed the first discrete definition of a shock filter, before Osher and Rudin gave the first PDE-based formulation [21]. In both cases, the morphological operations dilation and erosion with a disk-shaped structuring element are applied adaptively, depending on the sign of a second derivative operator. To make the process more robust against noise, Alvarez and Mazorra proposed to presmooth the image before computing this second derivative operator that guides the process [1]. Weickert's coherence-enhancing shock filter [30] uses the second derivative in direction of the dominant eigenvector of the structure tensor. Welk et al. [34] proved well-posedness results for 1-D semidiscrete and discrete shock

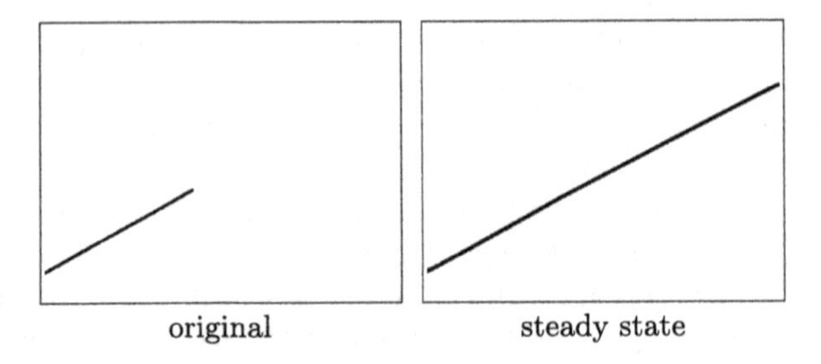

Fig. 1. Line elongation with the coherence-enhancing shock filter. **Left**: Original image, 512×384. **Right**: Steady state of the shock filter, $\sigma = 2$, $\rho = 5$.

filters. It is well-known that several classical image enhancement methods implicitly or explicitly combine a smoothing PDE with a shock filter; see e.g. [15,29]. Such combinations, however, rarely appear in inpainting applications.

Apart from the already mentioned PDE-based inpainting methods, there some additional ones that play a role in our performance evaluation. Biharmonic interpolation [8] is the fourth order counterpart to homogeneous diffusion inpainting and offers reasonable quality in inpainting-based compression [24]. Total variation (TV) inpainting [26] can be seen as a limit case of Perona–Malik [22] inpainting with a Charbonnier diffusivity [7]. Tschumperlé's method [28] involves a tensor-driven equation that uses the curvature of integral curves. This also qualifies it as candidate for bridging interrupted structures over large distances.

While many inpainting PDEs are elliptic or parabolic, hyperbolic ones such as the dilations and erosions in shock filters are rarely used within inpainting methods. A recent inpainting model by Novak and Reinić [20] combines a shock filter with the fourth order Cahn–Hilliard PDE. Our DS inpaiting is conceptually simpler: already a second order homogeneous diffusion PDE suffices to achieve the desired filling-in effect. The method of Bornemann and März [3] is closest in spirit to DS inpainting. It uses transport processes that are guided by a structure tensor. In contrast to our work, however, their paper follows a procedural–algorithmic approach without specifying a compact integrodifferential equation. In our experiments we will compare against this method.

Paper Structure. In Sect. 2, we review the concept of coherence-enhancing shock filtering since it is fundamental for our work. Section 3 introduces our proposed DS inpainting model. A corresponding algorithm is discussed in Sect. 4, followed by an experimental evaluation in Sect. 5. Section 6 concludes the paper and gives an outlook to future work.

2 Review of Coherence-Enhancing Shock Filtering

Since DS inpainting substantially relies on coherence-enhancing shock filters and their PDE-based definitions of dilation and erosion, we first review these concepts, in order to keep our paper self-contained.

2.1 PDE-Based Morphology

Let $f : \Omega \to \mathbb{R}$ denote a greyscale image on a rectangular image domain $\Omega \subset \mathbb{R}^2$. Mathematical morphology [27] considers a neighbourhood (structuring element) B. The dilation $\delta_B[f]$ replaces f by its supremum within B, and the erosion $\varepsilon_B[f]$ uses an infimum instead:

$$\delta_B[f](\boldsymbol{x}) = \sup\{f(\boldsymbol{x}-\boldsymbol{y}) \mid \boldsymbol{y} \in B\}, \tag{1}$$

$$\varepsilon_B[f](\boldsymbol{x}) = \inf\{f(\boldsymbol{x}+\boldsymbol{y}) \mid \boldsymbol{y} \in B\}. \tag{2}$$

If B is a disk of radius t, dilation and erosion $u(\boldsymbol{x},t)$ follow the PDE [4]

$$\partial_t u = \pm|\boldsymbol{\nabla} u| \tag{3}$$

with initial image $u(\boldsymbol{x},0) = f(\boldsymbol{x})$ and reflecting boundaries. The $+$ sign describes dilation, and the $-$ sign erosion. By $\boldsymbol{\nabla} = (\partial_x, \partial_y)^\top$ we denote the spatial nabla operator, and $|\cdot|$ is the Euclidean norm.

2.2 Coherence-Enhancing Shock Filtering

The dilation PDE propagates the grey values of local maxima, while erosion propagates the grey values of local minima. To enhance the sharpness of images, shock filters locally apply dilation in influence zones of maxima and erosion in influence zones of minima. The sign of a second order derivative operator determines these influence zones. Shocks are formed at its zero crossings. Individual shock filters differ in their second order derivative operator, which may also involve some smoothing to make the filter more robust under noise.

The coherence-enhancing shock filter [30] that we use is governed by the PDE

$$\partial_t u = -\operatorname{sgn}(\partial_{\boldsymbol{w}\boldsymbol{w}} u_\sigma)|\boldsymbol{\nabla} u| \tag{4}$$

with initial image $u(\boldsymbol{x},0) = f(\boldsymbol{x})$ and reflecting boundary conditions. It involves the second derivative along the dominant eigenvector $\boldsymbol{w}$ (with the larger eigenvalue) of the structure tensor [11]. Let $u_\sigma = K_\sigma * u$ denote the convolution of u with a Gaussian K_σ of standard deviation σ. Then the structure tensor is given by a componentwise convolution of the matrix $\boldsymbol{\nabla} u_\sigma \, \boldsymbol{\nabla} u_\sigma^\top$ with a Gaussian K_ρ:

$$\boldsymbol{J}_\rho(\boldsymbol{\nabla} u_\sigma) = K_\rho * \left(\boldsymbol{\nabla} u_\sigma \, \boldsymbol{\nabla} u_\sigma^\top\right). \tag{5}$$

Its eigenvalues describe the average quadratic contrast in the direction of the eigenvectors. Thus, steering the shock filter with $\partial_{\boldsymbol{w}\boldsymbol{w}} u_\sigma$ encourages the formation of coherent, flow-like structures with maximal contrast in direction $\boldsymbol{w}$. This justifies the name of the coherence-enhancing shock filter. For $t \to \infty$ it leads to a typically non-flat steady state that shows elongated image structures with very sharp edges. In contrast to [30], we use $\boldsymbol{J}_\rho(\boldsymbol{\nabla} u_\sigma)$ instead of $\boldsymbol{J}_\rho(\boldsymbol{\nabla} u)$, which yields better results in our case.

Figure 1 illustrates the performance of this shock filter. Without visible deviations, it elongates the black line over more than 200 pixels in a direction that is not grid aligned. Moreover, the result is extremely sharp without any dissipative artefacts. This quality is hardly ever seen in PDE-based algorithms.

3 Diffusion–Shock Inpainting

Coherence-enhancing shock filtering is designed to propagate information only in coherence direction. The scale of the created structures perpendicular to its propagation direction is determined by the presmoothing scale σ. Hence, it cannot fill in large homogeneous areas beyond that scale. Here, homogeneous diffusion inpainting [6] may serve as an ideal partner. It is simple, parameter-free, satisfies a maximum–minimum principle, and offers an efficient isotropic filling-in mechanism by solving $0 = \Delta u = \partial_{xx} u + \partial_{yy} u$ in inpainting regions.

Therefore, we design our *diffusion–shock (DS) inpainting* such that it performs a convex combination of both processes: Around edges, the coherence-enhancing shock filter from Subsect. 2.2 dominates, and homogeneous diffusion inpainting is activated in flat regions. Let the locations of the known values be given by a so-called *inpainting mask* $K \subset \Omega$. Here we do not alter the grey values. In the inpainting domain $\Omega \setminus K$, the reconstruction is computed as the steady state ($t \to \infty$) of the integrodifferential equation

$$\partial_t u \;=\; g\left(|\boldsymbol{\nabla} u_\nu|^2\right) \Delta u \;-\; \Big(1 - g\left(|\boldsymbol{\nabla} u_\nu|^2\right)\Big)\; \mathrm{sgn}\left(\partial_{\boldsymbol{ww}}\left(u_\sigma\right)\right) |\boldsymbol{\nabla} u| \tag{6}$$

with Dirichlet data at the boundaries ∂K of the inpainting mask, and reflecting boundary conditions on the image domain boundary $\partial \Omega$. By $u_\nu = K_\nu * u$ we denote a convolution of u with a Gaussian of standard deviation ν. The weight function $g : \mathbb{R} \to \mathbb{R}$ is a nonnegative decreasing function with $g(0) = 1$, and $g(s^2) \to 0$ for $s^2 \to \infty$. It has the same structure as a diffusivity function in nonlinear diffusion filters [22]. We use the Charbonnier diffusivity [7]

$$g\left(s^2\right) \;=\; \frac{1}{\sqrt{1 + s^2/\lambda^2}} \tag{7}$$

with a contrast parameter $\lambda > 0$. Thus, the weight $g\left(|\boldsymbol{\nabla} u_\nu|^2\right)$ secures a smooth transition between the shock term and the homogeneous diffusion term. Its Gaussian scale ν makes it robust under noise.

4 Numerical Algorithm

Continuous DS inpainting inherits a maximum–minimum principle from its diffusive and morphological components. To obtain an algorithm for DS inpainting that preserves this property, we discretise Eq. (6). We assume the same grid size $h > 0$ in x- and y-direction. Let $\tau > 0$ denote the time step size, and let $u_{i,j}^k$ be an equal discrete approximation of $u(\boldsymbol{x}, t)$ in pixel (i, j) at time $k\tau$. For simplicity, we use an explicit finite difference scheme. It approximates $\partial_t u$ at time level k with the forward difference

$$(\partial_t u)_{i,j}^k \;=\; \frac{u_{i,j}^{k+1} - u_{i,j}^k}{\tau} \tag{8}$$

and evaluates the right hand side of (6) at the old time level k. Thus, let us now focus on the space discretisation of the diffusion term and the shock term.

For the homogeneous diffusion term Δu classical central differences are suitable. Welk and Weickert show in [33] that a weighted combination of axial and diagonal central differences with weight $\delta = \sqrt{2} - 1$ results in a scheme with good rotation invariance. The corresponding stencil is given by

$$(\Delta u)_{i,j}^k = \left(\frac{1-\delta}{h^2} \begin{array}{|c|c|c|}\hline 0 & 1 & 0 \\ \hline 1 & -4 & 1 \\ \hline 0 & 1 & 0 \\ \hline \end{array} + \frac{\delta}{2h^2} \begin{array}{|c|c|c|}\hline 1 & 0 & 1 \\ \hline 0 & -4 & 0 \\ \hline 1 & 0 & 1 \\ \hline \end{array} \right) u_{i,j}^k . \tag{9}$$

With this stencil in space and the forward difference (8) in time we obtain an explicit scheme for the homogeneous diffusion equation $\partial_t u = \Delta u$. If

$$\tau \le \frac{h^2}{4 - 2\delta} =: \tau_D , \tag{10}$$

one easily verifies that $u_{i,j}^{k+1}$ is a convex combination of data at time level k. This implies stability in terms of the maximum–minimum principle

$$\min_{n,m} f_{n,m} \le u_{i,j}^k \le \max_{n,m} f_{n,m} \qquad \text{for all } i,\ j, \text{ and for } k \ge 1. \tag{11}$$

A space discretisation of morphological evolutions of type $\partial_t u = \pm|\boldsymbol{\nabla} u|$ is not as straightforward. Typically one uses upwind methods such as the Rouy–Tourin scheme [23]. In order to improve its rotation invariance, we follow Welk and Weickert [33] again, who propose a weighted combination of a Rouy–Tourin scheme in axial and diagonal direction. For the dilation term $|\boldsymbol{\nabla} u|$ this yields

$$\begin{aligned} |\boldsymbol{\nabla} u|_{i,j}^k = \ & \tfrac{1-\delta}{h} \Big(\max\{u_{i+1,j}^k - u_{i,j}^k,\ u_{i-1,j}^k - u_{i,j}^k,\ 0\}^2 \\ & \quad + \max\{u_{i,j+1}^k - u_{i,j}^k,\ u_{i,j-1}^k - u_{i,j}^k,\ 0\}^2 \Big)^{\frac{1}{2}} \\ & + \tfrac{\delta}{\sqrt{2}h} \Big(\max\{u_{i+1,j+1}^k - u_{i,j}^k,\ u_{i-1,j-1}^k - u_{i,j}^k,\ 0\}^2 \\ & \quad + \max\{u_{i-1,j+1}^k - u_{i,j}^k,\ u_{i+1,j-1}^k - u_{i,j}^k,\ 0\}^2 \Big)^{\frac{1}{2}} \end{aligned} \tag{12}$$

with some weight $\delta \in [0, 1]$. For the erosion term $-|\boldsymbol{\nabla} u|$, we use

$$\begin{aligned} -|\boldsymbol{\nabla} u|_{i,j}^k = \ & -\tfrac{1-\delta}{h} \Big(\max\{u_{i,j}^k - u_{i+1,j}^k,\ u_{i,j}^k - u_{i-1,j}^k,\ 0\}^2 \\ & \quad + \max\{u_{i,j}^k - u_{i,j+1}^k,\ u_{i,j}^k - u_{i,j-1}^k,\ 0\}^2 \Big)^{\frac{1}{2}} \\ & - \tfrac{\delta}{\sqrt{2}h} \Big(\max\{u_{i,j}^k - u_{i+1,j+1}^k,\ u_{i,j}^k - u_{i-1,j-1}^k,\ 0\}^2 \\ & \quad + \max\{u_{i,j}^k - u_{i-1,j+1}^k,\ u_{i,j}^k - u_{i+1,j-1}^k,\ 0\}^2 \Big)^{\frac{1}{2}} . \end{aligned} \tag{13}$$

It can be shown that an explicit scheme with time discretisation (8) and space discretisation (12) or (13) satisfies the maximum–minimum principle (11) if

$$\tau \le \frac{h}{\sqrt{2}\,(1-\delta) + \delta} =: \tau_M . \tag{14}$$

For approximating rotation invariance well, we follow [33] and set $\delta = \sqrt{2} - 1$.

To go from dilation and erosion to a coherence-enhancing shock filter, we have to discretise $\partial_{\boldsymbol{w}\boldsymbol{w}} u_\sigma$. We compute all Gaussian convolutions in the spatial domain with a sampled and renormalised Gaussian. To guarantee a high approximation quality, it is truncated at five times its standard deviation. In order to implement reflecting boundary conditions, we add an extra layer of mirrored dummy pixels around the image domain. For the first order derivatives within the structure tensor, we enforce this mirror symmetry by imposing zero values at the image boundaries. We approximate $\partial_x u$ and $\partial_y u$ in the structure tensor by means of Sobel operators [9], which offer a high degree of rotation invariance. Since the structure tensor is a symmetric 2×2 matrix, its normalised dominant eigenvector $\boldsymbol{w} = (c, s)^\top$ can be computed analytically. Moreover, we have

$$(\partial_{\boldsymbol{w}\boldsymbol{w}} v)_{i,j}^k \;=\; \left(c^2\, \partial_{xx} v + 2cs\, \partial_{xy} v + s^2\, \partial_{yy} v\right)_{i,j}^k . \tag{15}$$

All second order derivatives on the right hand side are approximated with their standard central differences.

Putting everything together yields the following explicit scheme for (6):

$$\frac{u_{i,j}^{k+1} - u_{i,j}^k}{\tau} \;=\; g_{i,j}^k \cdot (\Delta u)_{i,j}^k - (1 - g_{i,j}^k) \cdot \operatorname{sgn}\left((\partial_{\boldsymbol{w}\boldsymbol{w}} u_\sigma)_{i,j}^k\right) |\boldsymbol{\nabla} u|_{i,j}^k \tag{16}$$

with initial condition $u_{i,j}^0 = f_{i,j}$. It inherits its stability from the stability results of the schemes for diffusion and morphology:

Theorem 1 (Stability of the DS Inpainting Scheme). *Let the time step size τ of the scheme* (16) *be restricted by*

$$\tau \;\leq\; \min\{\tau_D, \tau_M\} \tag{17}$$

with τ_D and τ_M as in (10) and (14).
Then the scheme satisfies the discrete maximum–minimum principle

$$\min_{n,m} f_{n,m} \;\leq\; u_{i,j}^k \;\leq\; \max_{n,m} f_{n,m} \qquad \text{for all } i,\ j, \text{ and for } k \geq 1. \tag{18}$$

Proof. If $\tau \leq \min\{\tau_M, \tau_D\}$, it follows from the stability of the diffusion and morphological processes that

$$\begin{aligned} u_{i,j}^{k+1} &= u_{i,j}^k + \tau g_{i,j}^k \cdot (\Delta u)_{i,j}^k - (1 - g_{i,j}^k) \cdot \tau \operatorname{sgn}\left((\partial_{\boldsymbol{w}\boldsymbol{w}} u_\sigma)_{i,j}^k\right) |\boldsymbol{\nabla} u|_{i,j}^k \\ &\leq g_{i,j}^k \max_{n,m} f_{n,m} + (1 - g_{i,j}^k) \max_{n,m} f_{n,m} \\ &= \max_{n,m} f_{n,m} . \end{aligned}$$

Analogously, one can show the condition $\min_{n,m} f_{n,m} \leq u_{i,j}^k$. □

Thus, for $\delta = \sqrt{2} - 1$ and $h = 1$ our scheme is stable for $\tau \leq \tau_D \approx 0.31$. Theorem 1 shows an advantage of DS inpainting over EED inpainting [24,32], for which a discrete maximum–minimum principle cannot be guaranteed so far.

5 Experiments

Let us now evaluate DS inpainting experimentally. We mainly focus on binary images, since their high contrast is especially vulnerable to dissipative artefacts, but also present one experiment with greyscale images. Whenever the mask image is given, the white area denotes the inpainting mask, and the black area depicts the unknown regions. All methods in our evaluation use optimised parameters.

In Fig. 2, we apply DS inpainting to shape completion problems inspired by the experiments for Cahn–Hilliard inpainting in [2]. Our operator connects bars and restores a cross while maintaining the high contrast of the binary images. Compared to the results in [2], DS inpainting offers sharper reconstructions.

Figure 3 shows a more challenging experiment, inspired by [24]. It drives the sparsity of the inpainting data to the extreme by specifying only one or four dipoles. Nevertheless, DS inpainting shows a flawless performance: It creates two sharp half planes from one dipole, and a disk from four dipoles. This demonstrates its ability to bridge large distances and its curvature reducing effect.

Figure 4 shows an exhaustive comparison of DS inpainting with many results from Schmaltz et al. [24]. The goal is to reconstruct a Kaniza-like triangle. Homogeneous diffusion [6], biharmonic interpolation [8], and TV inpainting [26] are unable to produce sharp edges. Tschumperlé's algorithm [28] connects the corners, but fails to produce correctly oriented sharp edges. The Bornemann–März approach [3], edge-enhancing anisotropic diffusion (EED) [24,32], and DS inpainting reconstruct a satisfactory triangle. DS inpainting offers the best overall quality with high directional accuracy and no visible dissipative artefacts.

In Fig. 5, we consider the shape completion of a disk and a cat image, and compare DS inpainting with two advanced competitors: Euler's elastica [19] and EED inpainting [24]. The elastica results have been published in [25], while the cat data and its result for EED go back to [31]. While all methods accomplish their tasks, DS inpainting produces the sharpest edges.

Figure 6 shows that DS inpainting is also a powerful method for the reconstruction of greyscale images from sparse data. In both examples, the data were given by 10% randomly selected pixels of a 256×256 image. On a PC with an Intel© Core™i9-11900K CPU @ 3.50 GHz, the runtime was 3.32 seconds for the *peppers* image and 3.76 seconds for the *house* image.

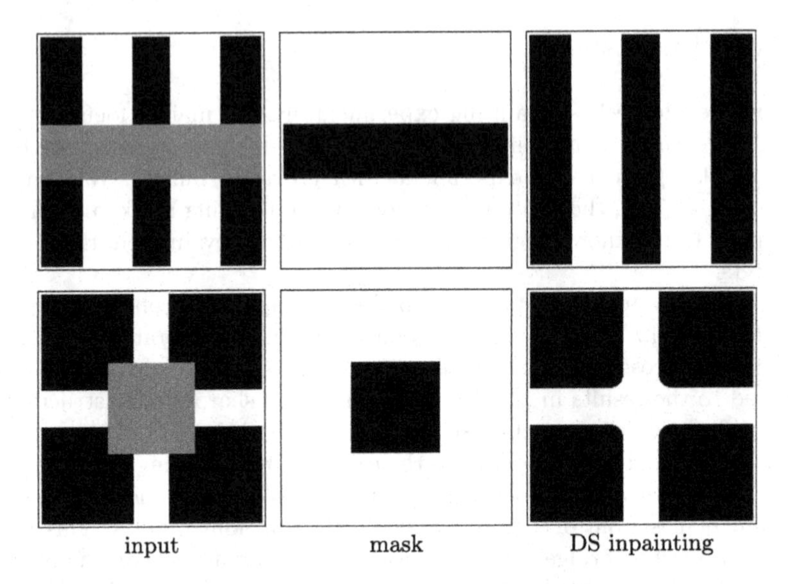

Fig. 2. DS inpainting of 256 × 256 test images. Parameters: **Top**: $\sigma = 2$, $\rho = 5$, $\nu = 3$, and $\lambda = 3$. **Bottom**: $\sigma = 2$, $\rho = 5$, $\nu = 2$, and $\lambda = 2$.

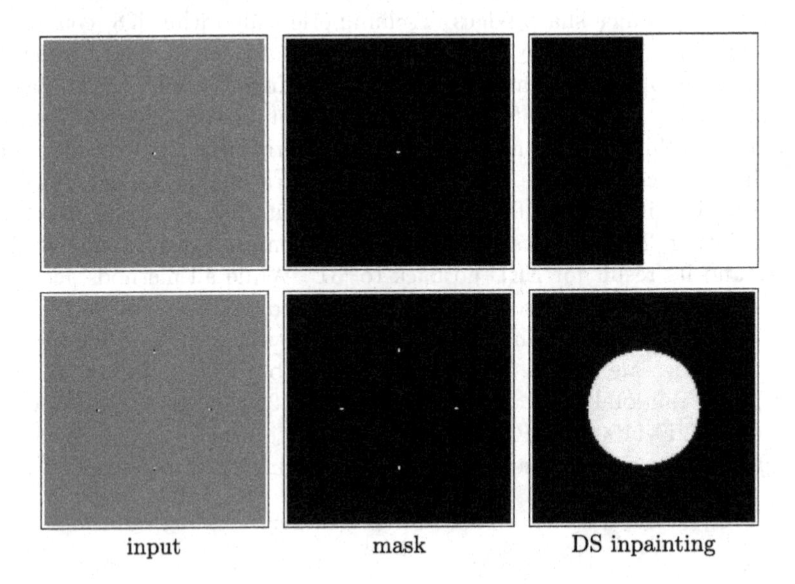

Fig. 3. DS inpainting from dipoles. **Top**: 128 × 128 image; $\sigma = 1$, $\rho = 2$, $\nu = 2$, $\lambda = 1$. **Bottom**: 127 × 127 image; $\sigma = 2.65$, $\rho = 4$, $\nu = 2$, $\lambda = 3$.

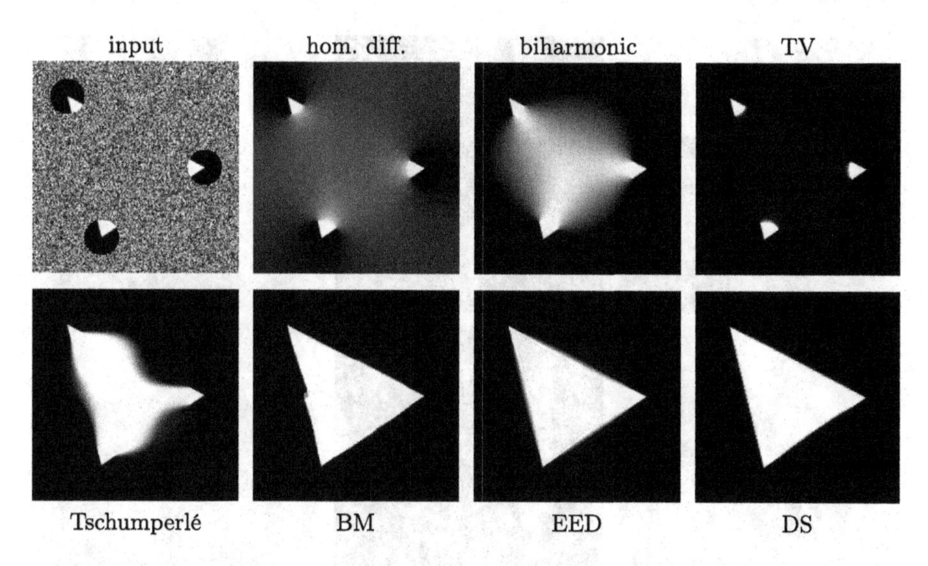

Fig. 4. Comparison of inpainting methods. **Top**: Input image with known data in the disks and noise in the unknown region, homogeneous diffusion, biharmonic interpolation, and TV inpainting. **Bottom**: Tschumperlé's approach, Bornemann–März (BM) method, EED inpainting, DS inpainting with $\sigma = 4.7$, $\rho = 6$, $\nu = 5.2$, and $\lambda = 3.4$.

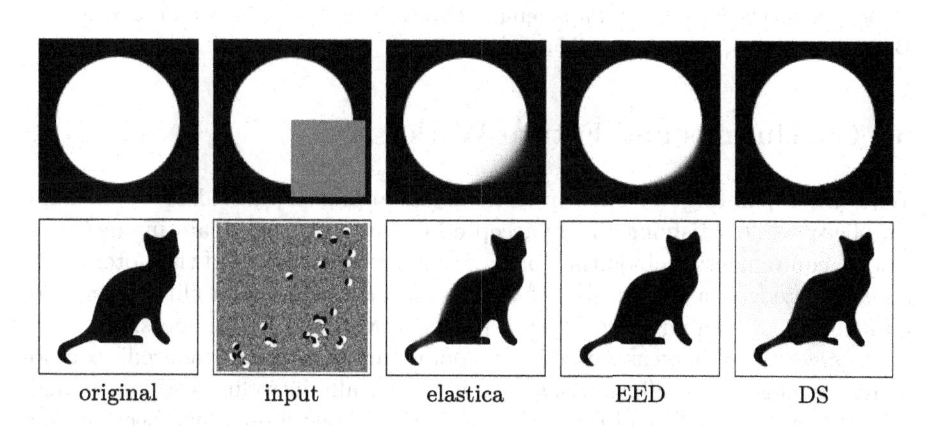

Fig. 5. Comparison of Euler's elastica, EED, and DS inpainting. The inpainting mask is shown in white. Parameters for DS inpainting: **Top**: $\sigma = 3.2$, $\rho = 3$, $\nu = 3$, and $\lambda = 2$; **Bottom:** $\sigma = 4.2$, $\rho = 4.8$, $\nu = 4.5$, and $\lambda = 7$.

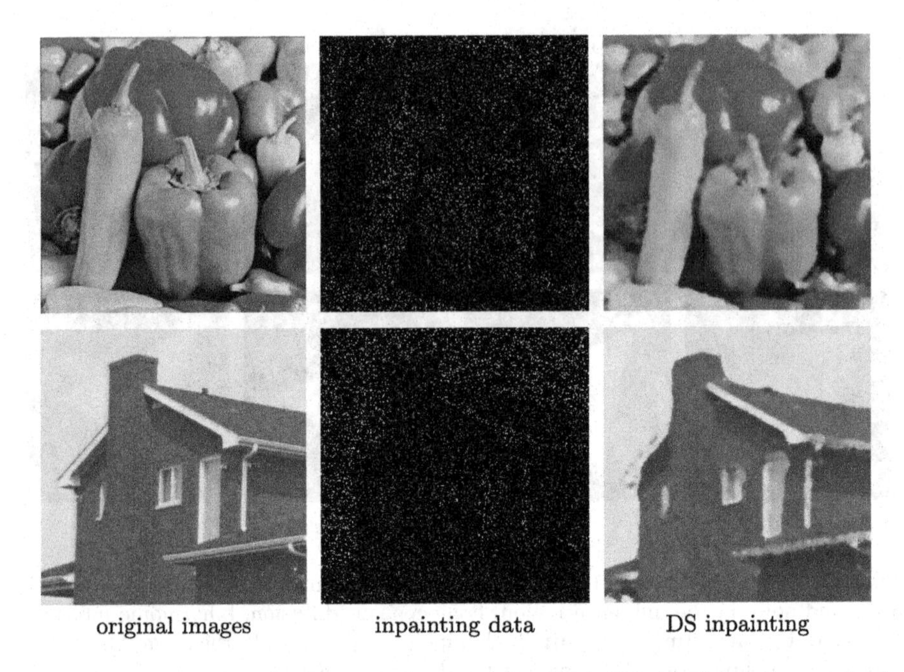

Fig. 6. Inpainting of sparse greyscale images (size 256×256, 10 % density, randomly chosen) with DS inpainting. Parameters: **Top**: $\sigma = 2$, $\rho = 1.5$, $\nu = 5$, and $\lambda = 3$; **Bottom**: $\sigma = 2.5$, $\rho = 1.8$, $\nu = 3.5$, and $\lambda = 3$.

6 Conclusions and Future Work

With DS inpainting we have introduced an approach that aims at maximal simplicity while satisfying widely accepted requirements on inpainting methods. These requirements include the ability to bridge large gaps and the potential to offer sharp edges, a high degree of rotation invariance, and stability in terms of a maximum–minimum principle that prevents over- and undershoots.

Interestingly, this was possible by combining two "time-honoured" components: Homogeneous diffusion has been axiomatically introduced to image analysis 61 years ago [12], and coherence-enhancing shock filters have been around for 20 years [30]. This demonstrates that classical methods still offer a huge potential that awaits being explored.

While most PDEs for inpainting are elliptic or parabolic, our results emphasise that the hyperbolic ones deserve far more attention. Hyperbolic PDEs are a natural concept for modelling discontinuities, and shock filters are a prototype for this. While their early representatives [16,21] produced oversegmentations and were highly sensitive w.r.t. to noise, more advanced variants such as coherence-enhancing shock filters have changed the game: Their performance reflects the high robustness of the structure tensor that guides them.

Last but not least, the fact that DS inpainting is a second order integrodifferential process that offers the full performance of higher order inpainting PDEs questions the necessity of higher order methods in practice. Without doubt, the latter ones are algorithmically far more challenging. As a general principle in science, Occam's razor suggests to prefer the simplest model that accomplishes a desired task. Our paper adheres to this principle.

In our ongoing work, we are aiming at a deeper understanding of such integrodifferential processes, and we are investigating alternative applications of this promising class of methods.

Acknowledgements. We thank Karl Schrader for providing us with the images and results from his publication [25].

References

1. Alvarez, L., Mazorra, L.: Signal and image restoration using shock filters and anisotropic diffusion. SIAM J. Numer. Anal. **31**, 590–605 (1994)
2. Bertozzi, A.L., Esedoglu, S., Gillette, A.: Inpainting of binary images using the Cahn-Hilliard equation. IEEE Trans. Image Process. **16**(1), 285–291 (2007)
3. Bornemann, F., März, T.: Fast image inpainting based on coherence transport. J. Math. Imaging Vis. **28**(3), 259–278 (2007)
4. Brockett, R.W., Maragos, P.: Evolution equations for continuous-scale morphology. In: Proceedings of IEEE International Conference on Acoustics, Speech and Signal Processing, San Francisco, CA, vol. 3, pp. 125–128 (1992)
5. Burger, M., He, L., Schönlieb, C.: Inpainting of binary images using the Cahn-Hilliard equation. SIAM J. Imag. Sci. **2**, 1129–11671 (2009)
6. Carlsson, S.: Sketch based coding of grey level images. Signal Process. **15**, 57–83 (1988)
7. Charbonnier, P., Blanc-Féraud, L., Aubert, G., Barlaud, M.: Deterministic edge-preserving regularization in computed imaging. IEEE Trans. Image Process. **6**(2), 298–311 (1997)
8. Duchon, J.: Interpolation des fonctions de deux variables suivant le principe de la flexion des plaques minces. RAIRO Analyse Numérique **10**, 5–12 (1976)
9. Duda, R.O., Hart, P.E.: Pattern Classification and Scene Analysis. Wiley, New York (1973)
10. Efros, A.A., Leung, T.: Texture synthesis by non-parametric sampling. In: Proceedings of Seventh International Conference on Computer Vision, Kerkyra, Greece, vol. 2, pp. 1033–1038. IEEE Computer Society Press (1999)
11. Förstner, W., Gülch, E.: A fast operator for detection and precise location of distinct points, corners and centres of circular features. In: Proceedings ISPRS Intercommission Conference on Fast Processing of Photogrammetric Data, Interlaken, Switzerland, pp. 281–305 (1987)
12. Iijima, T.: Basic theory on normalization of pattern (in case of typical one-dimensional pattern). Bull. Electrotech. Lab. **26**, 368–388 (1962). (in Japanese)
13. Kämper, N., Weickert, J.: Domain decomposition algorithms for real-time homogeneous diffusion inpainting in 4K. In: Proceedings of 2022 IEEE International Conference on Acoustics, Speech and Signal Processing, Singapore, pp. 1680–1684 (2022)
14. Kang, S., Tai, X.C., Zhu, W.: Survey of fast algorithms for Euler's elastica-based image segmentation. In: Kimmel, R., Tai, X.C. (eds.) Processing, Analyzing and Learning of Images, Shapes, and Forms: Part 2, Handbook of Numerical Analysis, vol. 20, pp. 533–552. Elsevier (2019)

15. Kornprobst, P., Deriche, R., Aubert, G.: Image coupling, restoration and enhancement via PDEs. In: Proceedings of 1997 IEEE International Conference on Image Processing, Washington, DC, vol. 4, pp. 458–461 (1997)
16. Kramer, H.P., Bruckner, J.B.: Iterations of a non-linear transformation for enhancement of digital images. Pattern Recogn. **7**, 53–58 (1975)
17. Mainberger, M., et al.: Optimising spatial and tonal data for homogeneous diffusion inpainting. In: Bruckstein, A.M., ter Haar Romeny, B.M., Bronstein, A.M., Bronstein, M.M. (eds.) SSVM 2011. LNCS, vol. 6667, pp. 26–37. Springer, Heidelberg (2012). https://doi.org/10.1007/978-3-642-24785-9_3
18. Masnou, S., Morel, J.M.: Level lines based disocclusion. In: Proceedings of 1998 IEEE International Conference on Image Processing, Chicago, IL, vol. 3, pp. 259–263 (1998)
19. Mumford, D.: Elastica and computer vision. In: Bajaj, C.L. (ed.) Algebraic Geometry and its Applications, vol. 5681, pp. 491–506. Springer, New York (1994). https://doi.org/10.1007/978-1-4612-2628-4_31
20. Novak, A., Reinić, N.: Shock filter as the classifier for image inpainting problem using the Cahn–Hilliard equation. Comput. Math. Appl. **123**, 105–114 (2022)
21. Osher, S., Rudin, L.I.: Feature-oriented image enhancement using shock filters. SIAM J. Numer. Anal. **27**, 919–940 (1990)
22. Perona, P., Malik, J.: Scale space and edge detection using anisotropic diffusion. IEEE Trans. Pattern Anal. Mach. Intell. **12**, 629–639 (1990)
23. Rouy, E., Tourin, A.: A viscosity solutions approach to shape-from-shading. SIAM J. Numer. Anal. **29**(3), 867–884 (1992)
24. Schmaltz, C., Peter, P., Mainberger, M., Ebel, F., Weickert, J., Bruhn, A.: Understanding, optimising, and extending data compression with anisotropic diffusion. Int. J. Comput. Vision **108**(3), 222–240 (2014)
25. Schrader, K., Alt, T., Weickert, J., Ertel, M.: CNN-based Euler's elastica inpainting with deep energy and deep image prior. In: 10th European Workshop on Visual Information Processing (EUVIP), Lisbon (2022)
26. Shen, J., Chan, T.F.: Mathematical models for local non-texture inpaintings. SIAM J. Numer. Anal. **62**(3), 1019–1043 (2002)
27. Soille, P.: Morphological Image Analysis, 2nd edn. Springer, Berlin (2004). https://doi.org/10.1007/978-3-662-05088-0
28. Tschumperlé, D.: Fast anisotropic smoothing of multi-valued images using curvature-preserving PDE's. Int. J. Comput. Vision **68**(1), 65–82 (2006)
29. van den Boomgaard, R.: Decomposition of the Kuwahara-Nagao operator in terms of linear smoothing and morphological sharpening. In: Talbot, H., Beare, R. (eds.) Mathematical Morphology: Proceedings of Sixth International Symposium, Sydney, Australia, pp. 283–292. CSIRO Publishing (2002)
30. Weickert, J.: Coherence-enhancing shock filters. In: Michaelis, B., Krell, G. (eds.) DAGM 2003. LNCS, vol. 2781, pp. 1–8. Springer, Heidelberg (2003). https://doi.org/10.1007/978-3-540-45243-0_1
31. Weickert, J.: Mathematische Bildverarbeitung mit Ideen aus der Natur. Mitteilungen der DMV **20**, 80–92 (2012)
32. Weickert, J., Welk, M.: Tensor field interpolation with PDEs. In: Weickert, J., Hagen, H. (eds.) Visualization and Processing of Tensor Fields, pp. 315–325. Springer, Heidelberg (2006). https://doi.org/10.1007/3-540-31272-2_19
33. Welk, M., Weickert, J.: PDE evolutions for M-smoothers in one, two, and three dimensions. J. Math. Imaging Vis. **63**, 157–185 (2021)
34. Welk, M., Weickert, J., Galić, I.: Theoretical foundations for spatially discrete 1-D shock filtering. Image Vis. Comput. **25**(4), 455–463 (2007)

Generalised Scale-Space Properties for Probabilistic Diffusion Models

Pascal Peter(✉)

Mathematical Image Analysis Group, Faculty of Mathematics and Computer Science, Campus E1.7, Saarland University, 66041 Saarbrücken, Germany
peter@mia.uni-saarland.de

Abstract. Probabilistic diffusion models enjoy increasing popularity in the deep learning community. They generate convincing samples from a learned distribution of input images with a wide field of practical applications. Originally, these approaches were motivated from drift-diffusion processes, but these origins find less attention in recent, practice-oriented publications.

We investigate probabilistic diffusion models from the viewpoint of scale-space research and show that they fulfil generalised scale-space properties on evolving probability distributions. Moreover, we discuss similarities and differences between interpretations of the physical core concept of drift-diffusion in the deep learning and model-based world. To this end, we examine relations of probabilistic diffusion to osmosis filters.

Keywords: probabilistic diffusion · scale-spaces · drift-diffusion · osmosis

1 Introduction

Probabilistic diffusion models introduced by Sohl-Dickstein et al. [26] are enjoying a rapid rise in popularity [9,12,21,27] fueled by the publicly available stable diffusion framework of Rombach et al. [21]. These deep learning models are generative in nature: Given a random seed, they can create new samples that fit to a given set of training data, for instance a certain class of images. Especially the numerous excellent text-to-image results based on stable diffusion [21] resonate not only with the scientific community, sparking many recent publications, but also with the general public.

Due to their tremendous success in practical applications, the roots of these approaches have received less attention than their efficient implementation by deep neural networks. While current interpretations often consider probabilistic diffusion as highly sophisticated versions of denoising autoencoders, its original roots lie in well-known physical processes, namely drift-diffusion. Diffusion processes have been closely investigated by the scale-space community [1,11,22,29], and drift-diffusion has model-based applications in the form of osmosis filtering

L. Calatroni et al. (Eds.): SSVM 2023, LNCS 14009, pp. 601–613, 2023.
https://doi.org/10.1007/978-3-031-31975-4_46

proposed by Weickert et al. [30]. We aim to bring the scale-space and deep learning communities closer together by showing that there are generalised scale-space concepts behind one of the most successful current paradigms in deep learning.

Our Contribution. We establish the first generalised scale-space interpretation for probabilistic diffusion. Compared to traditional scale-spaces, it describes the gradual simplification of probability distributions towards a non-flat steady state which does not carry information of the initial distribution anymore. We introduce entropy-based Lyapunov sequences and establish invariance statements. Moreover, we discuss relations to deterministic osmosis filters, highlighting similarities and differences.

Related Work. Probabilistic diffusion models were introduced by Sohl-Dickstein et al. [26] as an alternative to existing generative neural networks such as generative adversarial networks [7]. They use deep learning to invert a Markov process that gradually adds noise to an image. This allows to generate new samples from the distribution of the training data. Beyond the initial applications such as text-to-image, superresolution, and inpainting, many improvements and new practical uses have been proposed (e.g. [9,12,26,27]). With their publicly available stable diffusion model including trained weights, Rombach et al. [21] have unleashed a torrent of real-world applications for the initial concept. Due to the tremendous research interest in the topic, a full review is beyond the scope of this paper.

We investigate probabilistic diffusion from a new scale-space perspective. Classical scale-spaces embed images into a family of systematically simplified versions based on partial differential equations (PDEs) [1,11,14,22,29] or pseudo-differential operators [4,24].

Stochastic scale-spaces are fairly rare. Conceptually, the Ph.D. thesis of Majer [15] comes closest to our own considerations since it also deals with stochastic simplification and drift-diffusion. However, it is unrelated to deep learning and shuffles image pixels to remove information. A similar, local shuffling has been proposed by Koenderik and Van Doorn [13] under the name "locally orderless images". Stochastic considerations related to scale-spaces have been made w.r.t. the statistics of natural images reflected by Gaussian image models [19] and practical applications in stem cell differentiation [10].

Notably, probabilistic diffusion was originally motivated from drift-diffusion processes which can be described by the Fokker-Planck equation [20]. Osmosis filters for visual computing, a generalisation of diffusion filtering [29], have been derived from the same concept. The corresponding continuous theory was proposed by Weickert et al. [30], while Vogel et al. [28] provide results in the discrete setting, and additional properties were shown by Schmidt [23]. We discuss these in more detail in Sect. 4.1. Osmosis has proven particularly useful for image editing [3,28,30], shadow removal [3,18,30], and image fusion [17].

This implies connections to other fields of research. For instance, Sochen has established relationships between drift-diffusion and the Beltrami flow [25]. Hagemann et al. [8] have proposed a general framework that ties together many concepts, including probabilistic diffusion, under the common model of normalising flows.

Organisation of the Paper. In Sect. 2 we introduce probabilistic diffusion in its formulation as a Markov process and propose a corresponding generalised scale-space theory in Sect. 3. After a discussion of similarities and differences to osmosis filtering in Sect. 4, we draw conclusions and assess potential future benefits of this connection in Sect. 5.

2 Probabilistic Diffusion

Probabilistic diffusion [26] differs from most classical filters associated with scale-spaces. Instead of a single initial image, it considers a set of known images. These training data act as samples for common statistics that are not known directly. We can interpret discrete training images $\boldsymbol{f}_1, ..., \boldsymbol{f}_{n_t} \in \mathbb{R}^{n_x n_y n_c}$ with n_c colour channels of size $n_x \times n_y$ as realisations of a random variable $\boldsymbol{F}$. The unknown *target distribution* is expressed by its probability density function $p(\boldsymbol{F})$.

Probabilistic diffusion maps this unknown $p(\boldsymbol{F})$ to a simple, well-known distribution such as the standard normal distribution (i.e. Gaussian noise). Practical applications exploit that we can also map samples from the noise distribution back to the unknown distribution $p(\boldsymbol{F})$ (see Sect. 3.2). This makes the model generative, since it can create new images that resemble the training data.

However, we focus first on the *forward process* and show that it constitutes a scale-space in Sect. 3. Its evolution is described by a time-dependent random variable $\boldsymbol{U}(t)$. At time $t = 0$ it has the same distribution as our training data, i.e. $p(\boldsymbol{F})$. A sequence of m temporal realisations $\boldsymbol{u}_1, ..., \boldsymbol{u}_m$ at times $t_1 < t_2 < ... < t_m$ is referred to as one possible *trajectory* of $\boldsymbol{U}$. In a Markov process [6], $\boldsymbol{u}_i$ only depends on $\boldsymbol{u}_{i-1}$ and not on the previous trajectory $\boldsymbol{u}_0, ..., \boldsymbol{u}_{i-2}$. In terms of conditional transition probabilities, the Markov property is formulated as

$$p(\boldsymbol{u}_i|\boldsymbol{u}_{i-1}, ..., \boldsymbol{u}_0) \;=\; p(\boldsymbol{u}_i|\boldsymbol{u}_{i-1})\,. \tag{1}$$

This notation refers to the probability of the random variable $\boldsymbol{U}(t)$ assuming value $\boldsymbol{u}_i$ at time t_i, given that we observed $\boldsymbol{U}(t_{i-1}) = \boldsymbol{u}_{i-1}$. Due to the Markov property, the probability density of the whole trajectory can be successively traced back to the distribution $p(\boldsymbol{u}_0) = p(\boldsymbol{F})$ of the training data according to

$$p(\boldsymbol{u}_0, ..., \boldsymbol{u}_m) \;=\; p(\boldsymbol{u}_0)\prod_{i=1}^{m} p(\boldsymbol{u}_i|\boldsymbol{u}_{i-1})\,. \tag{2}$$

More concretely, we consider Gaussian transition probabilities

$$p(\boldsymbol{u}_i|\boldsymbol{u}_{i-1}) \;=\; \mathcal{N}\left(\sqrt{1-\beta_i}\,\boldsymbol{u}_{i-1},\, \beta_i \boldsymbol{I}\right)\,. \tag{3}$$

Here, $\mathcal{N}(\boldsymbol{\mu}, \boldsymbol{\sigma})$ denotes a multivariate Gaussian with unit matrix $\boldsymbol{I} \in \mathbb{R}^{n \times n}$ where $n = n_x n_y n_c$ is the number of pixels. Since the covariance matrix is diagonal, this corresponds to independent, identically distributed Gaussian noise with mean $\mu_{i,j} = \sqrt{1-\beta_i}u_{i-1,j}$ and standard deviation $\sigma_i = \sqrt{\beta_i}$ for each pixel j. The parameters $\beta_i \in (0,1)$ can either be user-specified or learned. For our following considerations, it is useful to express the random variable at a time t_i in terms of the random variable at the previous time t_{i-1} and Gaussian noise $\boldsymbol{G}$ from the standard normal distribution $\mathcal{N}(\boldsymbol{0}, \boldsymbol{I})$. Equation (2) directly implies that

$$\boldsymbol{U}_i = \sqrt{1-\beta_i}\,\boldsymbol{U}_{i-1} + \sqrt{\beta_i}\,\boldsymbol{G}. \tag{4}$$

With Eq. (3), a trajectory of images can be obtained from a starting image $\boldsymbol{u}_0 = \boldsymbol{f}$ by rescaling the image and adding a realisation of Gaussian noise in each step. As in [9], we can also specify the transition from time 0 to time t_i as

$$p(\boldsymbol{u}_i|\boldsymbol{u}_0) = \mathcal{N}\left(\sqrt{\prod_{j=0}^{m}(1-\beta_j)}\,\boldsymbol{u}_0,\ \boldsymbol{I} - \prod_{j=0}^{m}(1-\beta_j)\boldsymbol{I}\right). \tag{5}$$

With these insights into the forward process, we are suitably equipped to establish a scale-space theory for probabilistic diffusion.

3 Generalised Probabilistic Diffusion Scale-Space

In the following, we will consider a scale-space in the sense of probability distributions. As such, we do not specify properties of individual images, but of the marginal densities $p(\boldsymbol{u}_i)$ instead.

3.1 Generalised Scale-Space Properties

Property 1: Training Data Distribution as Initial State. By definition, the distribution $p(\boldsymbol{u}_0)$ at time $t_0 = 0$ is identical to the distribution $p(\boldsymbol{F})$ of the training data.

Property 2: Semigroup Property. We can acquire $p(\boldsymbol{u}_i)$ equivalently in i steps from $p(\boldsymbol{u}_0)$ or in ℓ steps from $p(\boldsymbol{u}_{i-\ell})$.

Proof. This follows from the recursive definition of the probability density (2). To find the distribution of an individual step in the trajectory, we integrate over all possible paths that lead to $\boldsymbol{u}_i$ and consider the marginal probability density

$$p(\boldsymbol{u}_i) = \int p(\boldsymbol{u}_0, ..., \boldsymbol{u}_i)\, d\boldsymbol{u}_0 \cdots d\boldsymbol{u}_{i-1}. \tag{6}$$

Using the definition of the joint probability density of the Markov process, we obtain the aforementioned two alternative ways to express $p(\boldsymbol{u}_i)$:

$$p(\boldsymbol{u}_i) = \int p(\boldsymbol{u}_0) \prod_{j=1}^{i} p(\boldsymbol{u}_j|\boldsymbol{u}_{j-1})\, d\boldsymbol{u}_0 \cdots d\boldsymbol{u}_{i-1} \tag{7}$$

$$= \int p(\boldsymbol{u}_{i-\ell}) \prod_{j=i-\ell+1}^{i-1} p(\boldsymbol{u}_j|\boldsymbol{u}_{j-1})\, d\boldsymbol{u}_{i-\ell} \cdots d\boldsymbol{u}_{i-1}. \tag{8}$$

Due to the Markov property (1), we can start the trajectory at any intermediate time $i - \ell$. □

Property 3: Lyapunov Sequences. In classical scale-spaces (e.g. with diffusion), Lyapunov sequences quantify the change in the evolving image with increasing scale parameter. They constitute a measure of image simplification [29] in terms of monotonic functions. In practice, they often represent the information content of an image at a given scale. Here, we define a Lyapunov sequence on the evolving probability density instead. Our first Lyapunov sequence, the differential entropy, indicates that the distribution of $\boldsymbol{U}$ gradually becomes more random with increasing time t_i.

Proposition 1 (Increasing Differential Entropy). *The differential entropy*

$$H(\boldsymbol{U}_i) := -\int p(\boldsymbol{u}_i) \ln p(\boldsymbol{u}_i) d\boldsymbol{u}_i \tag{9}$$

increases with t_i under the following assumptions for β_i:

$$\frac{1}{2} - \sqrt{\frac{1}{4} - \frac{1}{(2\pi e)^n}} \leq \beta_i \leq \frac{1}{2} + \sqrt{\frac{1}{4} - \frac{1}{(2\pi e)^n}} . \tag{10}$$

Here, $n = n_x n_y n_c$ denotes the total number of pixels.

Proof. According to Eq. (4), we can rewrite the entropy at $i+1$ as

$$\begin{aligned} H(\boldsymbol{U}_{i+1}) &= H\left(\sqrt{1-\beta_i}\,\boldsymbol{U}_i + \sqrt{\beta_i}\,\boldsymbol{G}\right) &(11)\\ &= H(\boldsymbol{U}_i) + \ln\left(\sqrt{1-\beta_i}\right) + H(\boldsymbol{G}) + \ln\left(\sqrt{\beta_i}\right) . &(12) \end{aligned}$$

Here we have used that the Gaussian noise does not depend on the images in the time step, and thus $\boldsymbol{U}_i$ and $\boldsymbol{G}$ are independent. Therefore, the entropy can be additively decomposed. Consequentially, the differential entropy is monotonously increasing if

$$H(\boldsymbol{G}) + \ln\left(\sqrt{(1-\beta_i)\beta_i}\right) \geq 0 . \tag{13}$$

This holds under restrictions for β_i:

$$\begin{aligned} \ln\left(\sqrt{\beta_i(1-\beta_i)}\right) \geq -H(\mathcal{N}(\boldsymbol{0}, \boldsymbol{I})) \quad &\Leftrightarrow \quad \sqrt{\beta_i(1-\beta_i)} \geq (2\pi e)^{-\frac{n}{2}} &(14)\\ \Leftrightarrow \quad \beta_i(1-\beta_i) \geq (2\pi e)^{-n} \quad &\Leftrightarrow \quad \beta_i^2 - \beta_i + (2\pi e)^{-n} \leq 0 . &(15) \end{aligned}$$

Standard rules for quadratic functions yield the conditions in Eq. (10) for which the inequality is fulfilled. Note that for $n \to \infty$, the lower limit goes to 0 and the upper limit to 1. In practice, this holds for reasonable choices of β_i that are not too close to the boundaries of its range $(0, 1)$. □

Alternatively, we can also consider the increasing conditional entropy given the distribution of the training data. Intuitively, this means that more information is needed to describe $\boldsymbol{U}_i$ given $\boldsymbol{U}_0$ with increasing t_i, which reflects that the initial information is gradually destroyed by noise.

Proposition 2 (Increasing Conditional Entropy). *The conditional entropy*

$$H(\boldsymbol{U}_i|\boldsymbol{U}_0) = -\int\int p(\boldsymbol{u}_i, \boldsymbol{u}_0) \ln p(\boldsymbol{u}_i|\boldsymbol{u}_0)\, d\boldsymbol{u}_0 d\boldsymbol{u}_i\,. \tag{16}$$

increases with t_i for all $\beta_i \in (0,1)$.

Proof.

$$H(\boldsymbol{U}_i|\boldsymbol{U}_0) = \int p(\boldsymbol{u}_0) - \int p(\boldsymbol{u}_i|\boldsymbol{u}_0) \ln p(\boldsymbol{u}_i|\boldsymbol{u}_0) d\boldsymbol{u}_i d\boldsymbol{u}_0 \tag{17}$$

$$= \int p(\boldsymbol{u}_0) \ln \left(2\pi e \prod_{j=1}^{i} (1-\beta_j) \right) d\boldsymbol{u}_0 \tag{18}$$

$$\geq \int p(\boldsymbol{u}_0) \ln \left(2\pi e \prod_{j=1}^{i-1} (1-\beta_j) \right) d\boldsymbol{u}_0 = H(\boldsymbol{U}_{i-1}|\boldsymbol{U}_0)\,. \tag{19}$$

According to Eq. (5), $p(\boldsymbol{u}_i|\boldsymbol{u}_0)$ is a normal distribution, and the inner integral is its entropy. It only depends on the covariance of the normal distribution and is thus independent of $\boldsymbol{u}_0$. □

Property 4: Permutation Invariance. Consider an arbitrary permutation function $P(\boldsymbol{f})$ that reorders the pixels of an image $\boldsymbol{f}$ from the initial distribution. Note that arbitrary permutations specifically also include translations and rotations by 90° increments. Probabilistic diffusion is invariant under permutations in the sense that trajectories are also permuted accordingly, which corresponds to classical invariances on individual images.

Proposition 3 (Permutation Invariant Trajectories). *Any trajectory $\boldsymbol{v}_1$, ...$\boldsymbol{v}_m$ starting from the permuted initial image $\boldsymbol{v}_0 := P(\boldsymbol{f})$ can be obtained by the same permutation P from a trajectory $\boldsymbol{u}_0, ...\boldsymbol{u}_m$ starting from the original image $\boldsymbol{f} =: \boldsymbol{u}_0$, i.e. for all i, we have $\boldsymbol{v}_i = P(\boldsymbol{u}_i)$, and vice versa.*

Proof. Let $\boldsymbol{g}_i$ denote the Gaussian noise realisation of $\boldsymbol{G}$ from Eq. (4) that occurs in the transition from $\boldsymbol{v}_{i-1}$ to $\boldsymbol{v}_i$. We define the transition noise from $\boldsymbol{u}_{i-1}$ to $\boldsymbol{u}_i$ as $\tilde{\boldsymbol{g}}_i := P^{-1}(\boldsymbol{g}_i)$, which is also from $\mathcal{N}(\boldsymbol{0}, \boldsymbol{I})$. Now we can prove the claim by induction, starting with $\boldsymbol{v}_0 = P(\boldsymbol{u}_0)$. Assuming that $\boldsymbol{v}_{i-1} = P(\boldsymbol{u}_{i-1})$, we obtain

$$\boldsymbol{v}_i = \sqrt{1-\beta_i}\boldsymbol{v}_{i-1} + \sqrt{\beta_i}\boldsymbol{g}_i = \sqrt{1-\beta_i}P(\boldsymbol{u}_{i-1}) + \sqrt{\beta_i}P(\tilde{\boldsymbol{g}}_i) = P(\boldsymbol{u}_i)\,. \tag{20}$$

Since P is a bijection, we have a one-to-one mapping between all possible trajectories from a permuted image and the permuted trajectories of the original. □

The proof above implies that permuting the initial data leads to the same permutation of the corresponding trajectories.

Property 5: Steady State. The steady state distribution for $i \to \infty$ is a multivariate Gaussian distribution $\mathcal{N}(\boldsymbol{0}, \boldsymbol{I})$ with mean $\boldsymbol{0}$ and standard deviation $\boldsymbol{I}$.

This is an effect that results directly from adding Gaussian noise in every step of the Markov process and has been used by Sohl-Dickstein et al. [26] without proof. In the following we provide a short formal argument for the sake of completeness.

Proposition 4 (Convergence to the Standard Normal Distribution). *With the assumptions on* β_i *from Property 3, the forward process described by Eq.* (2) *converges to the standard normal distribution for* $i \to \infty$.

Proof. The statement follows from Eq. (5) according to

$$\boldsymbol{u}_i = \underbrace{\sqrt{\prod_{j=0}^{i}(1-\beta_j)}}_{\overset{i\to\infty}{\to 0}}\, \boldsymbol{u}_0 + \underbrace{\sqrt{1-\prod_{j=0}^{i}(1-\beta_j)}}_{\overset{i\to\infty}{\to 1}}\, \boldsymbol{\xi} \tag{21}$$

with $\boldsymbol{\xi}$ from $\mathcal{N}(\boldsymbol{0}, \boldsymbol{I})$. □

Note that classical scale-spaces, e.g. those resulting from diffusion processes [29], converge to a flat image as the state of least information. However, in a generalised setting, probabilistic diffusion still follows the scale-space idea of gradual simplification by systematic removal of information.

3.2 PDE Formulation and Reverse Process

In Sect. 3.1, we have established that probabilistic diffusion in its Markov formulation constitutes a generalised scale-space. Like many existing scale-spaces, probabilistic diffusion can also be expressed in a PDE formulation. Feller [5] has shown a connection of a Markov process of type (2) if the stochastic moments

$$m_k(\boldsymbol{u}_t, t) = \lim_{h\to 0} \frac{1}{h} \int p(\boldsymbol{u}_{t+h}, \boldsymbol{u}_t)(\boldsymbol{u}_{t+h} - \boldsymbol{u}_t)^k d\boldsymbol{u}_{t+h} \tag{22}$$

exist for $k \in \{1, 2\}$. In this case the probability density $p(\boldsymbol{u}_\tau, \boldsymbol{u}_t)$ with $\tau < t$ is a solution to the partial differential equation

$$\frac{\partial}{\partial t} p = \frac{1}{2} \frac{\partial}{\partial \boldsymbol{u}_t \partial \boldsymbol{u}_t} \left(m_2(\boldsymbol{u}_t, t) p\right) + \frac{\partial}{\partial \boldsymbol{u}_t} \left(m_1(\boldsymbol{u}_t, t) p\right) . \tag{23}$$

This is a drift-diffusion equation with a drift coefficient m_1.

A crucial component for the success of probabilistic diffusion is the counterpart of the aforementioned forward process, the *backward* process. Feller [5] found that if a solution to the PDE (23) exists, then it also solves the backward equation

$$\frac{\partial}{\partial \tau} p = \frac{1}{2} m_2(\boldsymbol{u}_\tau, \tau) \frac{\partial}{\partial \boldsymbol{u}_\tau \partial \boldsymbol{u}_\tau} p + m_1(\boldsymbol{u}_\tau, \tau) \frac{\partial}{\partial \boldsymbol{u}_\tau} p \, . \tag{24}$$

Note that this equation is formulated w.r.t. the earlier time τ, thus yielding a backward perspective where transitions from t to τ are considered. Sohl-Dickstein et al. [26] use the fact that this reverse process has a very similar form compared to the forward process. It starts with the Gaussian noise distribution $\mathcal{N}(\mathbf{0}, \boldsymbol{I})$ and converges to $p(\boldsymbol{F})$. However, in contrast to the forward process, the mean and standard deviation of the Gaussian transition probabilities are unknown. These parameters are learned with a neural network such that the steady state minimises the cross entropy to $p(\boldsymbol{F})$. Details on how to find the reverse process have been the topic of many publications and have been refined considerably compared to the original publication of Sohl-Dickstein et al. [26]. Since this is not the focus of our work, we refer to [9,12,21,27] for more details.

The reverse process can sample new images from the distribution $p(\boldsymbol{F})$ that are not part of the training data. Additionally, this probability distribution can be conditioned with side information. Providing a textual description of the image content specifies the sampling, thus creating a text-to-image approach [21]. Similarly, inpainting can be implemented by using known image parts as side information [21,26].

4 Probabilistic Diffusion Models and Osmosis

After briefly reviewing osmosis filters, we establish connections to probabilistic diffusion via the PDE formulation, highlighting similarities and differences.

4.1 Continuous Osmosis Filtering

Consider an initial grey value image $f : \Omega \to \mathbb{R}_+$ that maps coordinates from the image domain $\Omega \subset \mathbb{R}^2$ to positive grey values. Osmosis describes the evolution of $u : \Omega \times [0, \infty) \to \mathbb{R}_+$ over time t, starting with f at $t = 0$. Colour images are covered by channel-wise processing. Besides the initial image, the so-called *drift vector field* $\boldsymbol{d} : \Omega \to \mathbb{R}^2$ has a decisive impact on the image evolution of $u(\boldsymbol{x}, t)$ which is determined by the PDE [30]

$$\partial_t u \;=\; \Delta u - \mathbf{div}\,(\boldsymbol{d}u) \qquad \text{on } \Omega \times (0, T]\,. \tag{25}$$

Reflecting boundary conditions prevent transport across the image boundaries. For $\boldsymbol{d} = \mathbf{0}$, Eq. (25) describes a homogeneous diffusion process [11], which smoothes the image over time. The drift component $-\mathbf{div}\,(\boldsymbol{d}u) = -\partial_x(d_1 u) - \partial_y(d_2 u)$ specifies local asymmetries in exchange of data between pixel cells. This makes it a valuable design tool for filters in visual computing.

The PDE formulation of Feller [5] for probabilistic diffusion from Eq. (23) closely resembles the osmosis equation (25). Consider the 1-D version of the osmosis equation (25):

$$\partial_t u \;=\; \partial_{xx} u - \partial_x(du) \qquad \text{on } \Omega \subset \mathbb{R} \times (0, T]\,. \tag{26}$$

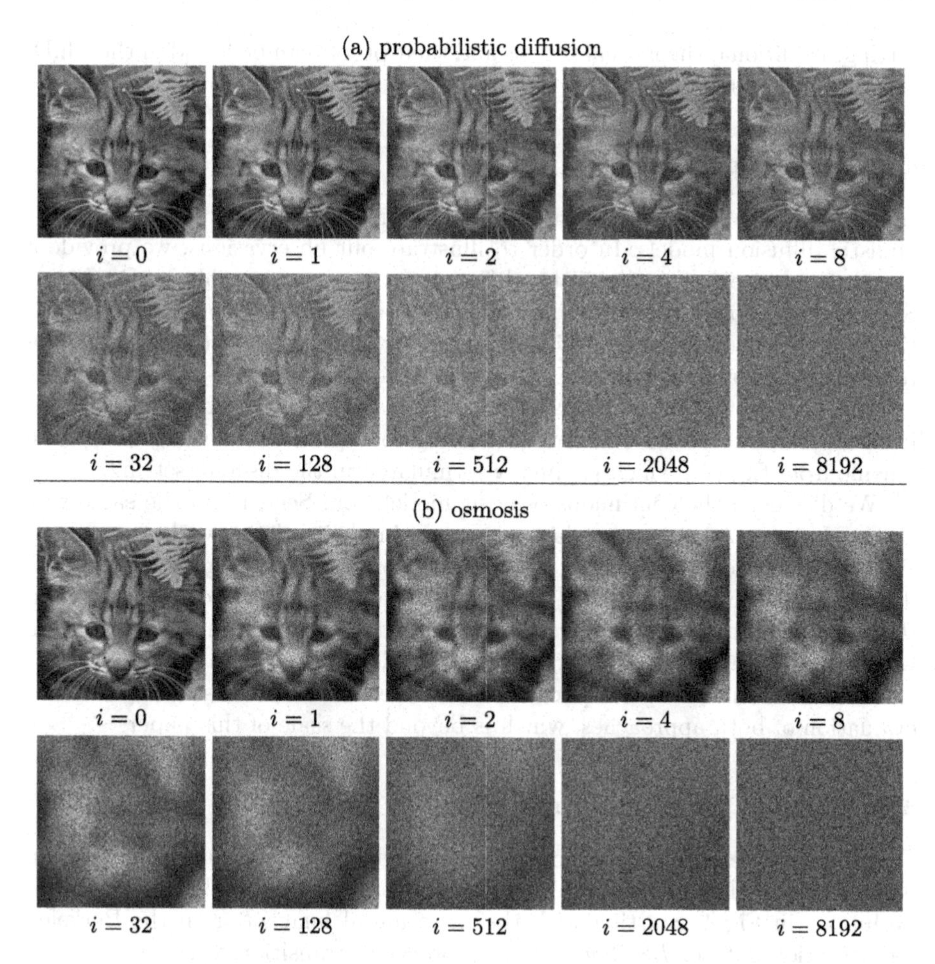

Fig. 1. The probabilistic diffusion trajectory (a) was obtained with $\beta_t = 0.02$ and i indicates the number of steps. The osmosis evolution (b) uses time step size $\tau = 1$ and a canonical drift vector field of a Gaussian noise image.

A structural comparison to Eq. (23) reveals that the evolving image u corresponds to the evolving probability density p. In both cases, ∂_t is the derivative w.r.t. the time variable of the image evolution. The spatial derivative ∂_x corresponds to the derivative ∂_{u_t} w.r.t. positions of individual particles in the probabilistic diffusion model (see Sect. 4.4). Hence, we can identify m_2 with a diffusivity which is set to one in the linear osmosis equation. The moment m_1 corresponds to the scalar drift term d in Eq. (26). Overall, we can interpret probabilistic diffusion as a 1-D osmosis PDE on the probability density p. Thus, there is a close conceptual connection between both methods.

In this paper, we consider only those osmosis properties that are most relevant for our comparison to probabilistic diffusion. For the more theoretical details we refer to Weickert et al. [30] in the continuous and Vogel et al. [28] in the discrete setting. Additional theoretical results and their proofs can be found in the Ph.D. thesis of Schmidt [23].

4.2 Visual Comparison

In the following, we compare structural properties of osmosis filtering and probabilistic diffusion models. In order to illustrate our observations, we provide a visual comparison of probabilistic diffusion to an osmosis evolution in Fig. 1. Since the probabilistic diffusion model defines an evolution of probability densities, the visual comparison considers a single, exemplary trajectory. It acts as a representative for the effects on the level of individual images.

Such a trajectory of the probabilistic diffusion model can be directly obtained from the forward process. It is straightforward to implement following the update scheme from Eq. (4), which is already formulated in the discrete setting.

We discretise the continuous osmosis model from Sect. 4.1 in the same way as Vogel et al. [28]. In particular, we use an implicit scheme with a stabilised BiCGSTAB solver [16]. In order to obtain the osmosis evolution, we use a standard normal noise sample v to define the so-called canonical drift vector field $\boldsymbol{d} = \frac{\boldsymbol{\nabla} v}{v}$. Weickert et al. [30] have shown that this yields a steady state $w = \frac{\mu_f}{\mu_v} v$ where μ_u denotes the average grey value of an image u.

Note that our visual comparison is not intended as a full-scale, systematic evaluation of both approaches, which is beyond the scale of this paper.

4.3 Common Structural Properties

Due to the observation that both osmosis and probabilistic diffusion rely on drift-diffusion, they share some theoretical properties and yield similar image evolutions in Fig. 1. Starting with the same initial image f from the Berkeley segmentation dataset *BSDS500* [2] both processes transition to noise.

We observe that a comparable amount of information of the initial image is removed over time. Schmidt [23] has shown that Lyapunov functionals can be defined for osmosis. In particular, the relative entropy of u w.r.t. w,

$$L(t) := - \int_{\Omega} u(\boldsymbol{x}, t) \ln \left(\frac{u(\boldsymbol{x}, t)}{w(\boldsymbol{x})} \right) d\boldsymbol{x} , \tag{27}$$

is increasing in t. It indicates that the information w.r.t. to the steady state is increasing. This conceptually resembles the conditional entropy from Sect. 3, which is formulated from the point of view of the initial distribution instead.

Finally, the backward process for probabilistic diffusion has a conceptual counterpart in osmosis. By swapping the roles of initial and guidance image, osmosis can also transition from noise to an image. However, this would not yield the same intermediate scales as the osmosis evolution in Fig. 1.

4.4 Differences

In contrast to probabilistic diffusion, osmosis is deterministic and thus has different applications. Osmosis is applied to individual images only, but also does not require any training data to perform tasks like image editing [3,28,30] or shadow removal [3,18,30]. Due to its stochastic nature, probabilistic diffusion is a natural choice for generating new images from given user prompts such as text. However, it can also be used for the restoration of individual images, e.g. restoring large missing image parts with matching generated content [21,27].

The second major difference lies in the physical interpretation of images. Osmosis directly models the macroscopic aspects of propagation in the 2-D image domain Ω, where pixel values can be seen as concentrations in the respective area. In contrast, the implementation of probabilistic diffusion via a Markov process models individual particles with 1-D Brownian motion. The pixel values are thus positional data instead, and propagation occurs in the co-domain. Effects of these conceptual differences are visible in Fig. 1. Osmosis gradually reduces the features of the cat by smoothing in the two-dimensional image domain. In contrast, the independent pixel-wise Brownian motion of probabilistic diffusion keeps edges of the original image intact until they are drowned out by noise.

Considering the steady state, we see that osmosis preserves the average colour value of the initial image, while samples from the standard normal distribution in probabilistic diffusion always have zero mean. Note that for osmosis, we only receive a noise steady state since we specified the guidance image accordingly – we can also use arbitrary other images to guide osmosis to receive non-trivial steady states. In contrast, probabilistic diffusion in the sense of Sohl-Dickstein et al. [26] always converges to a tractable (noise) distribution, e.g. the standard normal distribution.

5 Conclusions and Outlook

Investigating probabilistic diffusion from the point of view of scale-space research yields surprising results: Probabilistic diffusion defines an evolution that resembles traditional scale-spaces in important aspects such as causality and gradual simplification. However, probabilistic diffusion acts on distributions rather than on individual images and removes information by creating chaos instead of uniformity. Thus, it does not converge to a flat steady state, but to a noise distribution. Theoretical and practical results allow bi-directional traversal of this scale-space, which is rare in deterministic scale-spaces.

Interestingly, probabilistic diffusion can be seen as the stochastic counterpart to classical PDE-based, deterministic osmosis filtering. In the future, we plan to investigate this connection in more detail. Moreover, recognising probabilistic diffusion as a scale-space implies potential applications that make use of intermediate results of the evolution instead of only relying on steady states.

Acknowledgements. I thank Joachim Weickert, Karl Schrader, and Kristina Schaefer for fruitful discussions and advice.

References

1. Alvarez, L., Guichard, F., Lions, P.L., Morel, J.M.: Axioms and fundamental equations in image processing. Arch. Ration. Mech. Anal. **123**, 199–257 (1993)
2. Arbelaez, P., Maire, M., Fowlkes, C., Malik, J.: Contour detection and hierarchical image segmentation. IEEE Trans. Pattern Anal. Mach. Intell. **33**(5), 898–916 (2011)
3. d'Autume, M., Morel, J.M., Meinhardt-Llopis, E.: A flexible solution to the osmosis equation for seamless cloning and shadow removal. In: Proceedings of 2018 IEEE International Conference on Image Processing, pp. 2147–2151. Athens, Greece, October 2018
4. Duits, R., Florack, L., de Graaf, J., ter Haar Romeny, B.: On the axioms of scale space theory. J. Math. Imaging Vision **20**, 267–298 (2004)
5. Feller, W.: On the theory of stochastic processes, with particular reference to applications. In: First Berkeley Symposium on Mathematical Statistics and Probability, pp. 403–432. Berkeley, CA, January 1949
6. Gardiner, C.W.: Handbook of Stochastic Methods for Physics, Chemistry and the Natural Sciences, Springer Series in Synergetics, vol. 13. Springer, Berlin (1985) https://link.springer.com/book/9783540156079
7. Goodfellow, I.J., et al.: Generative adversarial nets. In: Ghahramani, Z., Welling, M., Cortes, C., Lawrence, N.D., Weinberger, K.Q. (eds.) Proceedings of 28th International Conference on Neural Information Processing Systems. Advances in Neural Information Processing Systems, vol. 27, pp. 2672–2680. Montréal, Canada, December 2014
8. Hagemann, P.L., Hertrich, J., Steidl, G.: Generalized normalizing flows via Markov chains. In: Elements in Non-local Data Interactions: Foundations and Applications. Cambridge University Press (2023). (in press)
9. Ho, J., Jain, A., Abbeel, P.: Denoising diffusion probabilistic models. In: Advances in Neural Information Processing Systems, vol. 33, pp. 6840–6851. NeurIPS Foundation, San Diego, CA (2020)
10. Huckemann, S., Kim, K.R., Munk, A., Rehfeldt, F., Sommerfeld, M., Weickert, J., Wollnik, C.: The circular sizer, inferred persistence of shape parameters and application to early stem cell differentiation. Bernoulli **22**(4), 2113–2142 (2016)
11. Iijima, T.: Basic theory on normalization of pattern (in case of typical one-dimensional pattern). Bull. Electrotechn. Lab. **26**, 368–388 (1962). in Japanese
12. Kingma, D., Salimans, T., Poole, B., Ho, J.: Variational diffusion models. Adv. Neural. Inf. Process. Syst. **34**, 21696–21707 (2021)
13. Koenderink, J.J., Van Doorn, A.J.: The structure of locally orderless images. Int. J. Comput. Vision **31**(2), 159–168 (1999)
14. Lindeberg, T.: Generalized Gaussian scale-space axiomatics comprising linear scale-space, affine scale-space and spatio-temporal scale-space. J. Math. Imaging Vision **40**, 36–81 (2011)
15. Majer, P.: A statistical approach to feature detection and scale selection in images. Ph.D. thesis, Department of Mathematics, Saarland University, Göttingen, Germany (2000)
16. Meister, A.: Numerik linearer Gleichungssysteme, 5th edn. Vieweg, Braunschweig (2015)
17. Parisotto, S., Calatroni, L., Bugeau, A., Papadakis, N., Schönlieb, C.B.: Variational osmosis for non-linear image fusion. IEEE Trans. Image Process. **29**, 5507–5516 (2020)

18. Parisotto, S., Calatroni, L., Caliari, M., Schönlieb, C.B., Weickert, J.: Anisotropic osmosis filtering for shadow removal in images. Inverse Prob. **35**(5), 054001 (2019)
19. Pedersen, K.S.: Properties of Brownian image models in scale-space. In: Griffin, L.D., Lillholm, M. (eds.) Scale-Space 2003. LNCS, vol. 2695, pp. 281–296. Springer, Heidelberg (2003). https://doi.org/10.1007/3-540-44935-3_20
20. Risken, H.: The Fokker-Planck Equation. Springer, New York (1984). https://doi.org/10.1007/978-3-642-61544-3
21. Rombach, R., Blattmann, A., Lorenz, D., Esser, P., Ommer, B.: High-resolution image synthesis with latent diffusion models. In: Proceedings of 2022 IEEE/CVF Conference on Computer Vision and Pattern Recognition. pp. 10684–10695. New Orleans, LA, June 2022
22. Scherzer, O., Weickert, J.: Relations between regularization and diffusion filtering. J. Math. Imaging Vision **12**(1), 43–63 (2000)
23. Schmidt, M.: Linear Scale-Spaces in Image Processing: Drift-Diffusion and Connections to Mathematical Morphology. Ph.D. thesis, Department of Mathematics, Saarland University, Saarbrücken, Germany (2018)
24. Schmidt, M., Weickert, J.: Morphological counterparts of linear shift-invariant scale-spaces. J. Math. Imaging Vision **56**(2), 352–366 (2016)
25. Sochen, N.A.: Stochastic processes in vision: From Langevin to Beltrami. In: Proceedings of Eighth International Conference on Computer Vision, vol. 1, pp. 288–293. Vancouver, Canada, July 2001
26. Sohl-Dickstein, J., Weiss, E., Maheswaranathan, N., Ganguli, S.: Deep unsupervised learning using nonequilibrium thermodynamics. In: Bach, F., Blei, D. (eds.) Proceedings 32nd International Conference on Machine Learning. Proceedings of Machine Learning Research, vol. 37, pp. 2256–2265. Lille, France, July 2015
27. Song, Y., Durkan, C., Murray, I., Ermon, S.: Maximum likelihood training of score-based diffusion models. Adv. Neural. Inf. Process. Syst. **34**, 1415–1428 (2021)
28. Vogel, O., Hagenburg, K., Weickert, J., Setzer, S.: A fully discrete theory for linear osmosis filtering. In: Kuijper, A., Bredies, K., Pock, T., Bischof, H. (eds.) SSVM 2013. LNCS, vol. 7893, pp. 368–379. Springer, Heidelberg (2013). https://doi.org/10.1007/978-3-642-38267-3_31
29. Weickert, J.: Anisotropic Diffusion in Image Processing. Teubner, Stuttgart (1998)
30. Weickert, J., Hagenburg, K., Breuß, M., Vogel, O.: Linear osmosis models for visual computing. In: Heyden, A., Kahl, F., Olsson, C., Oskarsson, M., Tai, X.-C. (eds.) EMMCVPR 2013. LNCS, vol. 8081, pp. 26–39. Springer, Heidelberg (2013). https://doi.org/10.1007/978-3-642-40395-8_3

Gromov–Wasserstein Transfer Operators

Florian Beier(✉)

Institute of Mathematics, Technische Universität Berlin, Straße des 17. Juni 136, 10623 Berlin, Germany
f.beier@tu-berlin.de

Abstract. Gromov–Wasserstein (GW) transport is inherently invariant under isometric transformations of the data. Having this property in mind, we propose to estimate dynamical systems by transfer operators derived from GW transport plans, when merely the initial and final states are known. We focus on entropy regularized GW transport, which allows to utilize the fast Sinkhorn algorithm and a spectral clustering procedure to extract coherent structures. Moreover, the GW framework provides a natural quantitative assessment on the shape-coherence of the extracted structures. We discuss fused and unbalanced variants of GW transport for labelled and noisy data, respectively. Our models are verified by three numerical examples of dynamical systems with governing rotational forces.

Keywords: optimal transport · Gromov–Wasserstein transport · Perron–Frobenius transfer operators · dynamical systems · coherent structures

1 Introduction

Optimal transport (OT) aims to find an optimal mass transport between two input (marginal) measures according to an underlying cost function. To improve the speed of the numerical computation, Cuturi [6] introduced a regularized OT version which can be solved by the fast and parallelizable Sinkhorn algorithm. Further effort has been made to generalize the OT for different settings as, e.g., unbalanced optimal transport [12], which relaxes the hard matching of the marginal measures. Another line of work pioneered by Mémoli [13] focuses on so-called Gromov–Wasserstein (GW) distances. Here, the inputs have additional structure in the sense of intrinsic (dis-)similarities. The difference to OT is that a meaningful cost function on the product space of the inputs might not be available. Instead, the mass is transported so that pairwise (dis-)similarities are preserved. GW distances are invariant under isometric transformations, making them a valuable tool for e.g. shape classification [3], word alignment [1] or graph matching [21]. For certain applications, a transport which simultaneously takes structural data in the GW sense as well as labelled data in the OT sense into account, is desirable. This is possible in the framework of fused GW transport

Supported by the German Research Foundation (DFG) within the RTG 2433 DAEDALUS.

L. Calatroni et al. (Eds.): SSVM 2023, LNCS 14009, pp. 614–626, 2023.
https://doi.org/10.1007/978-3-031-31975-4_47

[19]. Moreover, (fused) GW transport allows for a similar entropic regularization and unbalanced relaxation as OT [4,16,18]. In [2] the authors propose a framework which extends OT to be invariant to various classes of linear transformations such as e.g. orthogonal transformations. Compared to GW this method is numerically more appealing but has the drawback that it requires the inputs to be embedded in a common space and centered.

Recently, Koltai et al. [11] examined OT-based estimations of dynamical systems from observed initial and final states. More precisely, the authors leveraged solutions to regularized (unbalanced) optimal transport to estimate so-called transfer operators. These are linear operators that characterize dynamical systems in the form of density flows. Furthermore, a clustering procedure based on the spectral information of the estimator was used to extract so-called coherent structures of the dynamical system. Although no unified definition of such structures is available, it is understood that they are persistent in time and space. Coherent structures are of particular interest e.g. in fluid dynamics, since they capture important flow dynamics. This makes precise knowledge of the formation of coherent structures very appealing, since it may lead to a deeper understanding of the dynamics or computational advancements. In [10], the authors assumed instead that the exact transfer operator is known on a finite subset of the full state space. Then, using regularized OT, a finite-dimensional approximation is constructed which limit is a regularized version of the ground truth and exhibits desirable properties, such as retention of the spectral information.

In this paper, we build on the work in [11], but use entropic GW transport plans for constructing transfer operators. This is motivated by the fact that GW transport is readily able to detect isometric transformations such as rotation. Additionally, data labels can be incorporated. We will see that our proposed model includes a quantitative assessment of shape-coherence of the extracted structures.

Outline of the Paper. In Sect. 2, we briefly recall regularized (unbalanced) OT, associated transfer operators and related spectral clustering procedures. GW transport and its (unbalanced) regularized and fused variants are introduced in Sect. 3. Then, we expand the derivation of transfer operators and spectral clustering towards GW transport plans. In Sect. 4, we present numerical examples which indicate the potential of our method.

2 Optimal Transport and Transfer Operators

We consider (unbalanced) entropic OT, show how transfer operators can be derived from OT plans, and elaborate on spectral clustering. The derivation of transfer operators will be generalized to GW plans in the next section.

Optimal Transport. Let $X, Y \subset \mathbb{R}^d$ be compact sets equipped with the Euclidean distance d_{E}. By $\mathcal{M}^+(X)$, we denote the set of non-negative (Borel) measures and by $\mathcal{P}(X) \subset \mathcal{M}^+(X)$ the set of probability measures on X. Furthermore, let $L^2_\mu(X)$ be the Hilbert space of (equivalence classes) of square integrable functions with respect to the finite measure $\mu \in \mathcal{M}^+(X)$ equipped with the inner product $\langle \cdot, \cdot \rangle_\mu$. By 1_A, we denote the characteristic function on A. For

$\mu, \nu \in \mathcal{M}^+(X)$, the *Kullback–Leibler divergence* is defined by

$$\mathrm{KL}(\mu, \nu) := \int_X \log\left(\tfrac{\mathrm{d}\mu}{\mathrm{d}\nu}\right) \mathrm{d}\mu + \nu(X) - \mu(X),$$

if the Radon–Nikodym derivative $\frac{\mathrm{d}\mu}{\mathrm{d}\nu}$ exists, and by $\mathrm{KL}(\mu, \nu) := \infty$ otherwise. For $\mu \in \mathcal{P}(X)$, $\nu \in \mathcal{P}(Y)$, a lower semi-continuous cost function $c : X \times Y \to [0, \infty)$ and $\varepsilon > 0$, the *regularized OT problem* is given by

$$\mathrm{OT}_\varepsilon(\mu, \nu) := \min_{\pi \in \Pi(\mu,\nu)} \underbrace{\int_{X\times Y} c(x, y) \,\mathrm{d}\pi + \varepsilon \,\mathrm{KL}(\pi, \mu \otimes \nu)}_{=:F_\varepsilon^{\mathrm{OT}}(\pi)}, \tag{1}$$

where $\Pi(\mu, \nu) := \{\pi \in \mathcal{P}(X \times Y) : P_{1\#}\pi = \mu, P_{2\#}\pi = \nu\}$ with $P_i(x_1, x_2) := x_i$ and push forward measures $P_{i\#}\pi = \pi \circ P_i^{-1}$, $i = 1, 2$. Elements of $\Pi(\mu, \nu)$ are called *transport plans*. For $\varepsilon = 0$, we obtain the unregularized optimal transport $\mathrm{OT}(\mu, \nu)$. The minimizer in (1) is called (entropic) optimal transport plan $\hat{\pi}_\varepsilon$. In the following, we will mainly use $c = d_{\mathrm{E}}^2$ which leads to the *Wasserstein distance* $\mathrm{OT}(\mu, \nu)^{\frac{1}{2}}$. The dual problem of OT_ε is

$$\begin{aligned}\mathrm{OT}_\varepsilon(\mu, \nu) = \max_{(f,g)\in L^\infty_\mu(X)\times L^\infty_\nu(Y)} \Big\{ &\int_X f \,\mathrm{d}\mu + \int_Y g \,\mathrm{d}\nu \\ &- \varepsilon \int_{X\times Y} \exp\Big(\frac{f(x) + g(y) - c(x, y)}{\varepsilon}\Big) - 1 \,\mathrm{d}(\mu \otimes \nu)\Big\}\end{aligned}$$

Optimal potentials $\hat{f}_\varepsilon \in L^\infty_\mu(X)$, $\hat{g}_\varepsilon \in L^\infty_\nu(Y)$ exist and are unique on $\mathrm{supp}(\mu)$ and $\mathrm{supp}(\nu)$ up to an additive constant. They are related to $\hat{\pi}_\varepsilon$ by

$$\hat{\pi}_\varepsilon = \exp\Big(\frac{\hat{f}_\varepsilon(x) + \hat{g}_\varepsilon(y) - c(x, y)}{\varepsilon}\Big) (\mu \otimes \nu) =: k_\varepsilon(\mu \otimes \nu). \tag{2}$$

For atomic measures, the solution can be approximated efficiently by Sinkhorn's algorithm. In some applications, it is useful to deal with *regularized unbalanced OT*

$$\mathrm{UOT}_{\varepsilon,\kappa}(\mu, \nu) := \min_{\pi\in\mathcal{M}^+(X\times Y)} F_\varepsilon^{\mathrm{OT}}(\pi) + \kappa \left(\mathrm{KL}(P_{1\#}\pi, \mu) + \mathrm{KL}(P_{2\#}\pi, \nu)\right), \ \kappa > 0,$$

which relaxes the hard marginal constraints on the objective to penalizing the KL divergence of its marginals with respect to the inputs. Unbalanced optimal transport is treated in detail in [12] and its regularized version in [15]. Similarly as in the balanced case, there is a dual problem formulation with optimal potentials $(\hat{f}_{\varepsilon,\kappa}, \hat{g}_{\varepsilon,\kappa}) \in L^\infty_\mu(X) \times L^\infty_\nu(Y)$ and the optimal transport plan is given by

$$\hat{\pi}_{\varepsilon,\kappa} = \exp\Big(\frac{\hat{f}_{\varepsilon,\kappa}(x) + \hat{g}_{\varepsilon,\kappa}(y) - c(x, y)}{\varepsilon}\Big) (\mu \otimes \nu) =: k_{\varepsilon,\kappa} (\mu \otimes \nu). \tag{3}$$

A generalization of Sinkhorn's algorithm can be used to solve the corresponding discrete problem, see [15].

Transfer Operators. *Transfer operators*, also known as *Perron-Frobenius operators*, are linear operators which characterize dynamical systems in the form of density flows [9]. We consider transfer operators derived from entropic transport plans as in [11]. Here we restrict ourselves to the balanced setting as the unbalanced case follows in a similar way. To this end, we associate $\hat{\pi}_\varepsilon$ in (2) with the transfer operator $K_\varepsilon : L^2_\mu(X) \to L^2_\nu(Y)$ given by

$$(K_\varepsilon \psi)(y) := \int_X k_\varepsilon(x, y)\psi(x)\,\mathrm{d}\mu(x).$$

Figuratively, K_ε captures the structure of the transport of $\hat{\pi}_\varepsilon$ independent of the marginal masses. Since $\hat{\pi}_\varepsilon \in \Pi(\mu, \nu)$, it holds

$$\int_X k_\varepsilon(x, y)\,\mathrm{d}\mu(x) = 1_Y \;\; \nu\text{-a.e.} \quad \text{and} \quad \int_Y k_\varepsilon(x, y)\,\mathrm{d}\nu(y) = 1_X \;\; \mu\text{-a.e..}$$

In particular, for atomic measures $\mu = \sum_{i=1}^m \mu(i)\delta_{x_i}$ and $\nu = \sum_{j=1}^n \nu(j)\delta_{y_j}$ and an optimal transport plan $\hat{\pi}_\varepsilon = \sum_{i,j=1}^{m,n} \hat{\pi}_\varepsilon(i, j)\delta_{x_i,y_j}$ using the matrix-vector notation $\mu := (\mu(i))_{i=1}^m$, $\nu := (\nu(j))_{j=1}^n$, $D_\mu := \operatorname{diag}(\mu)$ and $\hat{\pi}_\varepsilon := (\hat{\pi}_\varepsilon(i, j))_{i,j=1}^{m,n}$, the transfer kernel and operator are given by

$$k_\varepsilon = D_\mu^{-1}\, \hat{\pi}_\varepsilon\, D_\nu^{-1} \quad \text{and} \quad K_\varepsilon = D_\nu^{-1}\, \hat{\pi}_\varepsilon^{\mathrm{T}}.$$

Spectral Clustering. In order to extract coherent structures in dynamical systems, we can apply a spectral clustering procedure on K_ε, see [11]. The clustering premise is just the knowledge of two observations from $\mu \in \mathcal{P}(X)$ and $\nu \in \mathcal{P}(Y)$ in a dynamical system without any knowledge of the true dynamics. The goal is to find measurable partitions $X = X_1 \dot{\cup} X_2$, $Y = Y_1 \dot{\cup} Y_2$ fulfilling ideally

$$K_\varepsilon 1_{X_k} = 1_{Y_k} \quad \text{and} \quad \mu(X_k) = \nu(Y_k), \quad k = 1, 2. \tag{4}$$

These conditions may be interpreted as coherence and mass preservation of the partitions. One way to tackle this problem is to consider the following optimization problem

$$\max_{X_1 \dot{\cup} X_2 = X, Y_1 \dot{\cup} Y_2 = Y} \left\{ \frac{\langle K_\varepsilon 1_{X_1}, 1_{Y_1} \rangle_\nu}{\mu(X_1)} + \frac{\langle K_\varepsilon 1_{X_2}, 1_{Y_2} \rangle_\nu}{\mu(X_2)} \right\},$$

which is usually relaxed to

$$\max_{(\varphi, \psi) \in L^2_\mu(X) \times L^2_\nu(Y)} \left\{ \frac{\langle K_\varepsilon \varphi, \psi \rangle_\nu}{\|\varphi\|_\mu \|\psi\|_\nu} : \langle \varphi, 1_X \rangle_\mu = \langle \psi, 1_Y \rangle_\nu = 0 \right\}. \tag{5}$$

Since K_ε is bounded and non-negative $(\mu \otimes \nu)$-a.s., it follows that the largest singular value of $K_\varepsilon^* K_\varepsilon$ is simple [9, Lem. 3]. Moreover, the largest singular value of K_ε is 1 and the corresponding left and right singular functions are 1_X and 1_Y, respectively. Notably, $(1_X, 1_Y)$ are not included by the constraints in (5). Hence, a maximizing pair $(\hat{\varphi}, \hat{\psi})$ in (5) is given by the right and left singular functions of K_ε associated to the second largest singular value of K_ε. The desired partitioning is then readily obtained by thresholding $(\hat{\varphi}, \hat{\psi})$ at zero. Solving (5) in practice amounts to computing a (truncated) singular value decomposition of K_ε.

3 Transfer Operators from GW Transport Plans

In both references [10,11], the assumption on the underlying dynamics is that they are compliant with an optimal transport. For certain situations this might not be the case. Consider e.g. particles on the two-dimensional unit disk with a driving rotational force. If the rotation angle between two observations is large, OT will not be able to recover this dynamic, see the first Example in Sect. 4. A transport setting which naturally handles isometric transforms such as rotation is given by the framework of GW transport [13]. As before, we consider compact state spaces $X, Y \subset \mathbb{R}^d$ and measures $\mu \in \mathcal{P}(X)$, $\nu \in \mathcal{P}(Y)$. In the contrast to classic OT, a cost function on the product space $X \times Y$ is not required. Instead we seek the preservation of the internal structure of the spaces. Here we focus on the Euclidean metrics, for generalizations see [17]. We set $d_X := d_{\mathrm{E}}|_{X \times X}$. Then the triples $\mathbb{X} := (X, d_X, \mu)$, $\mathbb{Y} := (Y, d_Y, \nu)$ are called *metric measure (mm-) spaces.* We introduce the notation $\mu^\otimes := \mu \otimes \mu$. For $\varepsilon > 0$, the *regularized GW transport* between two mm-spaces $\mathbb{X}$ and $\mathbb{Y}$ is defined by

$$\mathrm{GW}_\varepsilon(\mathbb{X}, \mathbb{Y}) := \inf_{\pi \in \Pi(\mu,\nu)} F_\varepsilon^{\mathrm{GW}}(\pi), \tag{6}$$

$$F_\varepsilon^{\mathrm{GW}}(\pi) := \int_{(X \times Y)^2} (d_X(x,x') - d_Y(y,y'))^2 \,\mathrm{d}\pi(x,y)\,\mathrm{d}\pi(x',y') + \varepsilon\,\mathrm{KL}\big(\pi^\otimes, (\mu \otimes \nu)^\otimes\big). \tag{7}$$

In contrast to OT, we regularize with the quadratic KL divergence as in [16]. For $\varepsilon = 0$, we obtain the unregularized GW transport GW which was originally introduced in [13]. Notably, $\mathrm{GW}^{\frac{1}{2}}$ defines a metric on the space of mm-spaces up to identification by measure-preserving isometries. More precisely, $\mathrm{GW}(\mathbb{X}, \mathbb{Y}) = 0$ if and only if there exists an isometry $I : X \to Y$ with $\nu = I_\# \mu$. In this case, $(\mathrm{id}, I)_\# \mu$ is an optimal GW plan. In particular, this shows the invariance of GW with respect to isometric transformations. Figuratively, optimal GW plans are such that whenever they transport (infinitesemal) mass from x to y and x' to y' one has $d_X(x,x') \approx d_Y(y,y')$ which favors a near-isometric transport.

Similar to OT_ε, GW_ε admits unbalanced versions [16], we focus on marginal penalization using KL. For $\varepsilon, \kappa > 0$, the *unbalanced regularized GW transport* is defined by

$$\mathrm{UGW}_{\varepsilon,\kappa}(\mathbb{X}, \mathbb{Y}) = \inf_{\pi \in \mathcal{M}^+(X \times Y)} F_\varepsilon^{\mathrm{GW}}(\pi) + \kappa\left(\mathrm{KL}((P_{1\#}\pi)^\otimes, \mu^\otimes) + \mathrm{KL}((P_{2\#}\pi)^\otimes, \nu^\otimes)\right).$$

Here the marginals of optimal plans differ from the inputs whenever an exact matching results in large values under the functional $F_\varepsilon^{\mathrm{GW}}$. This can make $\mathrm{UGW}_{\varepsilon,\kappa}$ somewhat robust to outliers.

When working with labelled data, we might be interested in a transport plan which preserves the internal geometrical information in the form of metrics as well as feature information in the form of labels. This leads to a fused version of the GW and the Wasserstein distance. To incorporate label information, we introduce an additional set $A \subset \mathbb{R}^m$ endowed with $d_A := d_{\mathrm{E}}|_{A \times A}$. We assume that each point in X, Y admits only one label, which we characterize by label functions $l_X : X \to A$, $l_Y : Y \to A$, respectively. Clearly, a more general treatment would be to consider distributions in the label space as in e.g. [19]. In our case, the *regularized fused GW distance* is defined by

$$\mathrm{FGW}_\varepsilon((\mathbb{X}, l_X), (\mathbb{Y}, l_Y)) := \inf_{\pi \in \Pi(\mu,\nu)} F_\varepsilon^{\mathrm{GW}}(\pi) + \int_{X \times Y} d_A\big(l_X(x), l_Y(y)\big)\, \mathrm{d}\pi(x, y).$$

As with the original formulation, the marginal constraints may be relaxed in the same way which leads to an unbalanced, fused variant $\mathrm{UFGW}_\varepsilon^\kappa$ which was discussed in [4,18].

The previously discussed GW formulations are quadratic with respect to the objective plan which renders them numerically challenging. For our numerical experiments below, we rely on a class of simple iterative algorithms which are based on block-coordinate relaxations. The main idea consists of alternately fixing one plan while minimizing with respect to the other. The problem that is then minimized in each iteration step can be written as an entropic OT problem for which Sinkhorn's algorithm can be leveraged. Details regarding this procedure can be found in [14] (balanced GW), [16] (unbalanced GW) and [4,18] (unbalanced, fused GW). Solutions $\hat{\pi}_\varepsilon$ obtained with this procedure are also solutions to an entropic (unbalanced) OT problem and thus have the form (2), i.e. it holds $\hat{\pi}_\varepsilon = k_\varepsilon\,(\mu \otimes \nu)$. Ultimately, this allows us to apply the spectral clustering procedure on the associated transfer operator K_ε as described in Sect. 2. The next remark highlights another benefit of GW over OT transfer operators for extracting coherent structures.

Remark 1 (Quantitative assessment of shape-coherence). Let $\hat{\pi}_\varepsilon$ be an optimal GW plan between $\mathbb{X}$ and $\mathbb{Y}$ with associated transfer operator K_ε and X_i, Y_i, $i = 1, 2$ the spectral clustering partition. Even if the partitions satisfy $K_\varepsilon 1_{X_i} \approx 1_{Y_i}$ and $\mu(X_i) \approx \nu(Y_i)$, it may be that the intrinsic shapes of X_i and Y_i, $i = 1, 2$ differs significantly. It depends on the application, if these structures should be considered coherent or not. The GW framework readily gives us the possibility for a quantitative assessment of shape-coherence by evaluating the GW functional at $\hat{\pi}_\varepsilon$ restricted to $X_i \times Y_i$, $i = 1, 2$. The closer the evaluation is to 0, the more the associated partitions can be considered shape-coherent or isometric under the transfer operator K_ε. We apply this for Example 3 in Sect. 4.

4 Numerical Examples

In this section we provide three examples of our proposed GW transfer method. In OT comparisons we use the quadratic Euclidean cost function. We partly rely on the Python Optimal Transport library [8]. For our experiments we aim to set the entropic regularization parameter $\varepsilon > 0$ as small as possible while avoiding numerical overflow.

1. Particles on a rotating disk. First, we are interested in the ability of OT plans to recover the dynamics of a rotating system and compare it with a GW

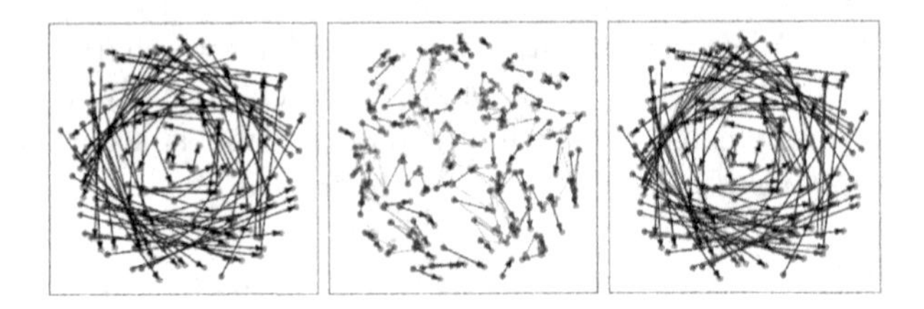

Fig. 1. An example of X (blue), $R_{\frac{\pi}{2}}(X)$ (orange), as well as the associated transfer kernels k_{true} (left), k_ε^{W} (middle) and $k_\varepsilon^{\text{GW}}$ (right). The arrows are drawn from x to y with opacity proportional to the respective kernel at (x, y). (Color figure online)

based approach. We consider $n = 50$ uniformly sampled particles on the 2D unit disk $D := \{x : \|x\|_2 \le 1\} \subset \mathbb{R}^2$. Let $X \subset D$ be the set of particles. We consider a counter-clockwise rotation of degree $\theta \in (0, 2\pi)$. More precisely, the true transfer operator is characterized by the bijective map $R_\theta : D \to D$ given by

$$(r\cos(\phi), r\sin(\phi)) \mapsto (r\cos(\phi+\theta), r\sin(\phi+\theta)), \quad r \in [0,1], \theta \in [0, 2\pi).$$

We focus on the transfer associated to kernel $k_{\text{true}}(x, y) = \delta_{\{R_\theta(x)=y\}}$. An illustration of X and R_θ is shown in Fig. 1. We investigate how well the GW transfer operator estimates the true transfer operator for $\theta = \frac{\pi}{30}, \frac{2\pi}{30}, \dots, \pi$. To this end, we sample the initial state X and compute the GW transport plan with $\varepsilon = 0.0008$ between between the uniform distributions on X and $R_\theta(X)$, respectively. As discussed, all plans admit the form (2) for respective kernels $k_\varepsilon^{\text{W}}, k_\varepsilon^{\text{GW}}$. To compare the performance we consider the error measure

$$\mathrm{e}(k_\varepsilon^\bullet) := \frac{1}{n^2} \sum_{x \in X} \sum_{y \in R_\theta(X)} k_\varepsilon^\bullet(x, y) d_{\mathrm{E}}(R_\theta(x), y), \quad \bullet \in \{\mathrm{W}, \mathrm{GW}\}.$$

Intuitively, this gives us the mean Euclidean distance when comparing the transfer operator associated to the kernel against the true transfer.

The right-hand side of Fig. 1 shows the qualitative difference between the OT and GW-based approaches for one example with $\theta = \frac{\pi}{2}$. A quantitative comparison is given on the left-hand side of Fig. 2. More precisely, we sampled 10 independent choices of X to obtain 10 OT plans $\pi_\varepsilon^{\text{W},1}, \dots, \pi_\varepsilon^{\text{W},10}$ and 10 GW plans $\pi_\varepsilon^{\text{GW},1}, \dots, \pi_\varepsilon^{\text{GW},10}$ for each angle θ. We plot the mean errors $\frac{1}{10}\sum_{i=1}^{10} \mathrm{e}(k_\varepsilon^{\bullet,i})$, $\bullet \in \{\mathrm{W}, \mathrm{GW}\}$ as a function of the angle θ. As expected, for large values of θ, the OT-based transfer operator is a poor estimator. This is evident since, e.g. for a 90° rotation, points are transferred far distances which is sub-optimal in the OT sense. Even for smaller angles such as 18°, we observe a mean error of 0.15. On the other hand, the GW based approach recovers R_θ nearly exactly in all cases. In the previous example, $Y = R_\theta(X)$ was given by an exact rotation of X. However, in practice the observed end state Y of the dynamical system might be a noisy version of $R_\theta(X)$. Hence, we repeat the previous experiment, where this time $Y = R_\theta(X) + m\eta$ with $\eta \sim \mathcal{U}([-0.1, 0.1]^2)$ and $m = 1$. To make this comparable to the previous experiment, we consider the same sampled initial

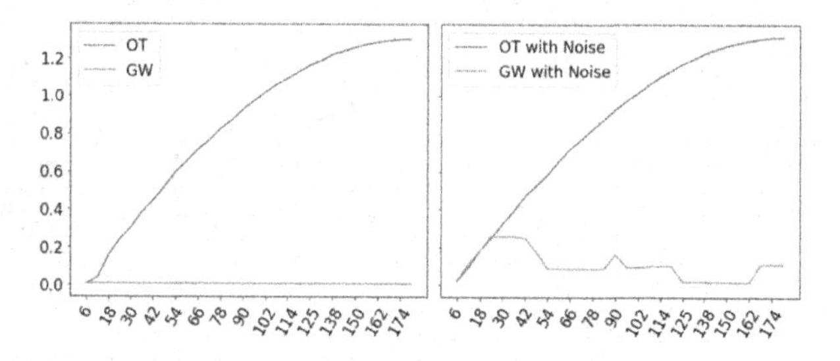

Fig. 2. The mean errors plotted against rotation angle θ in degrees without noise (left) and with noise (right).

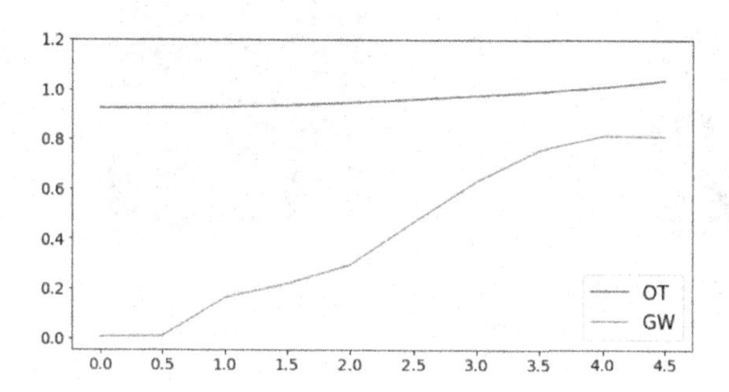

Fig. 3. The mean errors for fixed $\theta = \frac{\pi}{2}$ plotted against the noise magnitude m.

states X as above. We proceed as before and plot the error of the OT and GW-based approach on the right-hand side of Fig. 2. For small angles, GW remains comparable to OT whereas for large angles a better estimation is achieved by utilizing GW. Finally, we repeat the procedure this time for a fixed angle $\theta = \frac{\pi}{2}$ and for varying $m = 0.5, 1, 1.5, \dots, 4.5$. The result is plotted in Fig. 3.

2. Multiple rotating disks. In our next example, let $\theta = \pi/2$, and D, R_θ as above. In addition, for $i = 1, 2$, we consider

$$D_i = \{x \in D : \|x - x^{(i)}\| \leq 1/2\}, \quad x^{(i)} = (-1/2, 0), \quad x^{(i)} = (1/2, 0).$$

We set $F = (R^{(1)} + R^{(2)}) \circ R_\theta$, where $R^{(i)}$ constitutes a rotation of $-\pi/4$ around $x^{(i)}$, restricted to D_i, $i = 1, 2$. Let $n = 80$, we uniformly sample $n/2$ points of D_1 and D_2, respectively. Denote the entire set of n points by $X \subset D$. Let $Y = F(X)$ and equip X and Y with the uniform distribution. Figure 4 illustrates X, Y and F. We focus on the estimation of the transfer operator associated to $k_{\text{true}}(x, y) = \delta_{\{F(x)=y\}}$. We compute an OT plan $\pi_\varepsilon^{\text{W}}$ and an GW plan $\pi_\varepsilon^{\text{GW}}$ both with $\varepsilon = 0.001$. Illustrations of the matrices $\pi_\varepsilon^{\text{W}}, \pi_\varepsilon^{\text{GW}}$ as well as a visualizations of the transfer operators $K_\varepsilon^\bullet$, associated to respective kernels $k_\varepsilon^\bullet$, $\bullet \in \{\text{W}, \text{GW}\}$ are provided in Fig. 5. Clearly, neither approach is able to recover the ground truth. However, the figure indicates that $K_\varepsilon^{\text{GW}}$ transfers most of the mass from D_i to $F(D_i)$, $i = 1, 2$, while the OT-based approach does not. We apply the spectral clustering procedure, i.e. we compute the left and right eigenvectors

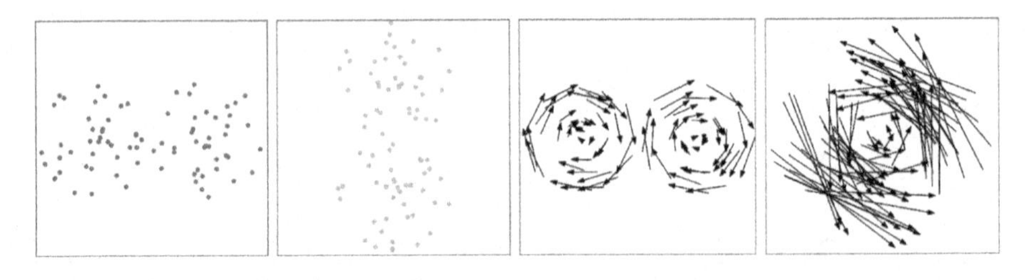

Fig. 4. Left to right: X, Y, $(R^{(1)} + R^{(2)})$ and R_θ.

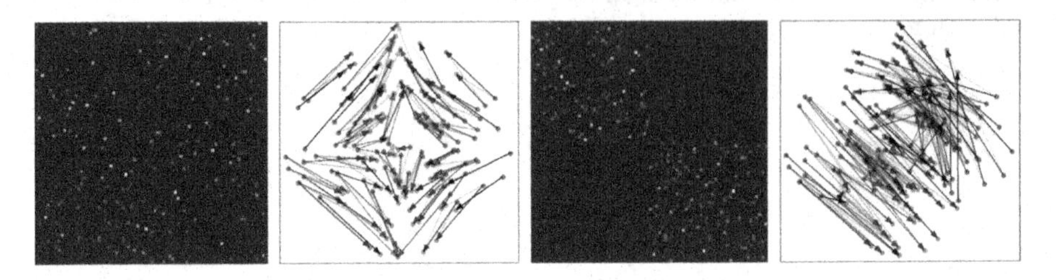

Fig. 5. Left to right: Matrix $\pi_\varepsilon^{\mathrm{W}}$, transfer kernel $k_\varepsilon^{\mathrm{W}}$, matrix $\pi_\varepsilon^{\mathrm{GW}}$, transfer kernel $k_\varepsilon^{\mathrm{GW}}$.

associated to the second largest eigenvalue of $K_\varepsilon^\bullet$, $\bullet \in \{\mathrm{W}, \mathrm{GW}\}$ and present them in Fig. 6. As expected, the partitioning according to $K_\varepsilon^{\mathrm{GW}}$ is able to find both coherent disks. Now, the local dynamics within the partitions can readily be obtained by computing the GW transport of the partitioned subspaces.

We conclude this example by remarking that the correct identification of the discs may also fail and is not stable with respect to noise. This is due to the fact that e.g. $\mathbb{X}$ is almost isometric to a 180° rotation as well as a reflection along the vertical axis. If the inputs are subjected to noise, an optimal GW plan might match D_1 with $F(D_2)$ and D_2 with $F(D_1)$.

3. Vorticity field of the 2D Navier–Stokes equation. Finally, we consider a two-dimensional flow in time which behaves according to the 2D Navier–Stokes equations on the square $[0, 2\pi]^2$ (periodic boundary conditions)

$$\begin{aligned} \partial_t u + (u \cdot \nabla)u &= -\nabla p + v\nabla^2 u \\ \nabla \cdot u &= 0, \end{aligned}$$

where $u : [0, T] \times [0, 2\pi]^2 \to \mathbb{R}^2$ is the velocity, $p : [0, T] \times [0, 2\pi]^2 \to \mathbb{R}^2$ the pressure and $v \in \mathbb{R}$ the kinematic viscosity. Numerically, it is more efficient to solve the scalar advection-diffusion equation

$$\partial_t \omega + (u \cdot \nabla)\omega = v\nabla^2 \omega, \tag{8}$$

where $\omega = \partial_x u_y - \partial_y u_x$ is the vorticity of u. Following [20, Sec IV], the equation is solved in the Fourier domain after a adding a small-scale forcing term and a large-scale damping function on a 4096×4096 grid. Ultimately, we obtain two

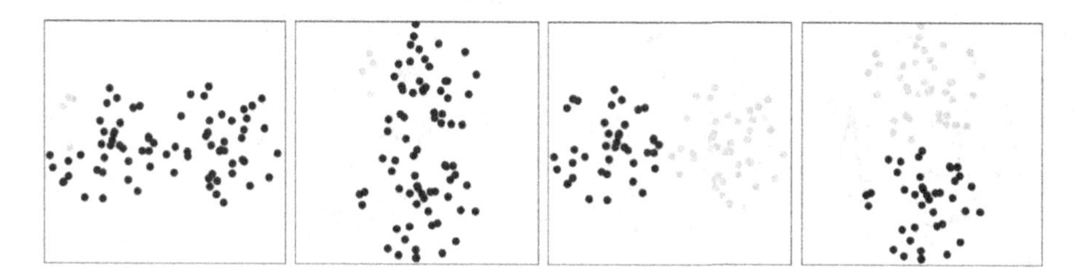

Fig. 6. The spaces X, Y coloured according to the sign of the left and right eigenvector of $K_\varepsilon^{\mathrm{W}}$ (left) and $K_\varepsilon^{\mathrm{GW}}$ (right) corresponding to the second largest eigenvalue.

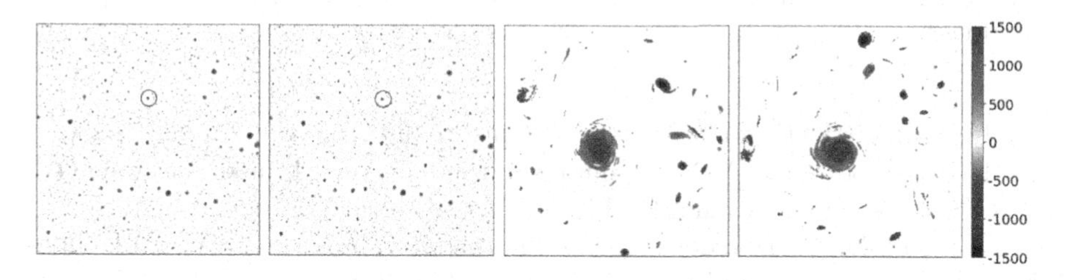

Fig. 7. Left: Two thresholded snapshots ω_0, ω_1 of a direct numerical simulation of (8) on a 4096×4096 pixel grid. The images on the right-hand side show the marked circular patch for both time-steps.

time snapshots ω_0, ω_1 of the vorticity field, which we restrict to $|\omega_i| \geq 600$. The snapshots as well as a zoom into a circular patch with a 290 pixel diameter is shown in Fig. 7. As we can see, the flow exhibits coherent structures in the form of vortices on large and small scales. Large vortices essentially determine most of the local dynamics. This can be seen for instance in the in selected patch, where smaller vortices are rotating around the large center vortex.

We proceed to estimate the dynamics of the extracted patch. Similarly to our motivating example, we compare the OT and GW transfer operators. From ω_0 and ω_1 we extract the mm-spaces $\mathbb{X} = (X, d_X, \mu)$ and $\mathbb{Y} = (Y, d_Y, \nu)$, respectively. More precisely, X and Y are the sets of patch points in $\mathbb{R}^2$ where $|\omega_0| \geq 600$ and $|\omega_1| \geq 600$. Furthermore, d_X and d_Y are the normalized Euclidean metrics on X and Y, respectively. Finally, μ, ν are the (fully supported) probability measures proportional to the absolute value of the vorticity field. For our model we want to prohibit the transport between positive and negative vorticity. To this end we label our data in the following way. Let l_X, l_Y be the label function on X, Y given by 0, 1 for negative, positive vorticity, respectively. Additionally, due to possible dissipation of vorticity, we focus on unbalanced approaches for the estimation of the transfer operator. We proceed to solve the entropic unbalanced OT problem between μ and ν with respect to the cost function $c(x, y) = d_{\mathrm{E}}(x, y)^2 + d_{\mathrm{E}}^2(l_X(x), l_Y(y))$, regularization parameter $\varepsilon = 0.0003$ and marginal relaxation parameter $\kappa = 0.1$. This can be understood as the (entropic and unbalanced) Wasserstein distance with an additional penalty on transport-

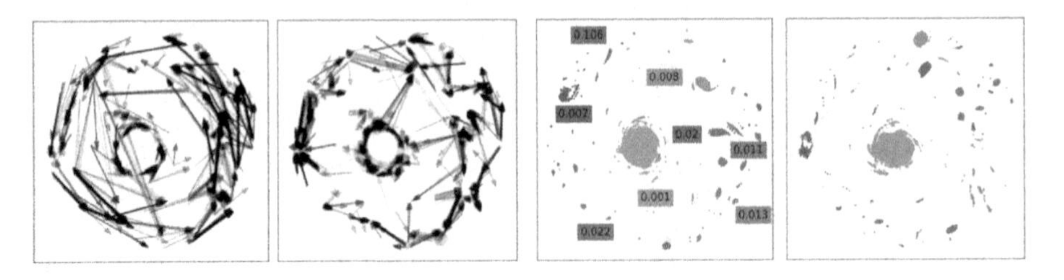

Fig. 8. Left to right: GW transfer kernel, OT transfer kernel. The last two images show distinctly coloured partitions according to the spectral clustering of $K^{\mathrm{GW}}_{\varepsilon,\kappa}$. The plotted numbers are the evaluations of the GW functional at π_{GW} restricted to the respective partitions.

ing between distinctly signed vorticity. Thus we obtain a solution denoted $\hat{\pi}^{\mathrm{W}}_{\varepsilon,\kappa}$. In the same way, let $\hat{\pi}^{\mathrm{GW}}_{\varepsilon,\kappa}$ be a solution to the unbalanced, fused, entropic GW problem between $(\mathbb{X}, l_X)$ and $(\mathbb{Y}, l_Y)$ and ε, κ as above. On the left-hand side of Fig. 8, we illustrate the associated transfer operators $K^{\bullet}_{\varepsilon,\kappa}$, $\bullet \in \{\mathrm{W}, \mathrm{GW}\}$. Similar to the previous examples, the OT transfer operator is not able to recover the underlying rotation. On the other hand, by favoring the preservation of intrinsic distances, the GW transport nicely reflects a counter-clockwise rotation. Finally, we apply the discussed spectral clustering procedure, where we focus on $K^{\mathrm{GW}}_{\varepsilon,\kappa}$. To obtain more than two coherent structures, we proceed in a nested manner. More precisely, applying the clustering procedure once yields two partitions of each mm-space X_1, X_2, Y_1, Y_2, respectively. Then we apply the procedure with respect to the associated (labelled) sub mm-spaces

$$\left(\left(X_i, d_X|_{X_i}, \frac{\mu(\cdot \cap X_i)}{\mu(X_i)}\right), l_X|_{X_i}\right), \qquad \left(\left(Y_i, d_Y|_{Y_i}, \frac{\nu(\cdot \cap Y_i)}{\nu(Y_i)}\right), l_Y|_{Y_i}\right),$$

and the restricted transfer operator $K^{\mathrm{GW}}_{\varepsilon,\kappa}|_{X_i \times Y_i}$, $i = 1, 2$. This yields two sub-partitions per partition. We repeat this three times so that we obtain 8 partitions in total. The right-hand side of Fig. 8 shows the mm-spaces $\mathbb{X}, \mathbb{Y}$, where points of the same partition are coloured equally. Additionally, we evaluate the GW functional of $\hat{\pi}^{\mathrm{GW}}_{\varepsilon,\kappa}$ restricted to the partitions as explained in Remark 1 and add the evaluation in the plot of $\mathbb{X}$. As expected, the center vortex is clearly identified. Additionally, we are able to identify even smaller structures such as the coherent structures in brown, pink, grey and red. The orange partition attains the smallest GW evaluation by far. This is followed by brown and pink which represent smaller coherent structures and highlights their shape preservation under the transfer $K^{\mathrm{GW}}_{\varepsilon,\kappa}$.

5 Conclusions

In this paper, we proposed a novel approach to estimate dynamical systems based on (unbalanced, fused) GW transport plans. Moreover, we demonstrated that the obtained transport plans can be leveraged for a spectral clustering procedure to extract coherent structures. The resulting method is convenient as it can be quickly implemented by using out-of-the-box methods for GW and the singular value decomposition. We verified its potency on three numerical examples.

As future work we leave a direct comparison with the method proposed in [2]. The latter provides a numerically more appealing framework for obtaining transport plans which minimize the OT functional under additional invariances such as orthogonal transformations. Moreover, we are interested in applying our method on non-Euclidean data such as e.g. graphs.

Acknowledgements. This work is supported by funds from the German Research Foundation (DFG) within the RTG 2433 DAEDALUS. The author thanks Jiahan Wang for fruitful discussions and his support regarding numerical implementations as well as Gabriele Steidl for valuable discussions.

References

1. Alvarez-Melis, D., Jaakkola, T.: Gromov-Wasserstein alignment of word embedding spaces. In: Proceedings of the 2018 Conference on EMNLP, Brussels, Belgium, pp. 1881–1890. ACL (2018)
2. Alvarez-Melis, D., Jegelka, S., Jaakkola, T.: Towards optimal transport with global invariances. In: Proceedings of the 22nd AISTATS, Naha, Okinawa, Japan, pp. 1870–1879. PMLR (2019)
3. Beier, F., Beinert, R., Steidl, G.: On a linear Gromov-Wasserstein distance. IEEE Trans. Image Process. **31**, 7292–7305 (2022)
4. Beier, F., Beinert, R., Steidl, G.: Multi-marginal Gromov-Wasserstein transport and barycenters. arXiv:2205.06725 (2022)
5. Beier, F., von Lindheim, J., Neumayer, S., Steidl, G.: Unbalanced multi-marginal optimal transport. J. Math. Imaging Vis. (2022)
6. Cuturi, M.: Sinkhorn distances: lightspeed computation of optimal transport. In: Advances in Neural Information Processing Systems, Lake Tahoe, Nevada, United States, vol. 26, pp. 2292–2300. Curran Associates Inc. (2013)
7. Elvander, F., Haasler, I., Jakobsson, A., Karlsson, J.: Multi-marginal optimal transport using partial information with applications in robust localization and sensor fusion. Signal Process. **171**, 107474 (2020)
8. Flamary, R., Courty, N.: POT Python Optimal Transport library (2017). https://github.com/PythonOT/POT. Accessed 19 Jan 2023
9. Froyland, G.: An analytic framework for identifying finite-time coherent structures in time-dependent dynamical systems. Phys. D **250**, 1–19 (2013)
10. Junge, O., Matthes, D., Schmitzer, B.: Entropic transfer operators. arXiv:2204.04901 (2022)
11. Koltai, P., von Lindheim, J., Neumayer, S., Steidl, G.: Transfer operators from optimal transport plans for coherent set detection. Phys. D **426**, 132980 (2021)
12. Liero, M., Mielke, A., Savaré, G.: Optimal entropy-transport problems and a new Hellinger-Kantorovich distance between positive measures. Invent. Math. **211**(3), 969–1117 (2018)
13. Mémoli, F.: Gromov-Wasserstein distances and the metric approach to object matching. Found. Comput. Math. **11**(4), 417–487 (2011)

14. Peyré, G., Cuturi, M., Solomon, J.: Gromov-Wasserstein averaging of kernel and distance matrices. In: Proceedings of the 33rd ICML, New York, NY, United States, pp. 2664–2672. PMLR (2016)
15. Séjourné, T., Feydy, J., Vialard, F.-X., Trouvé, A., Peyré, G.: Sinkhorn divergences for unbalanced optimal transport. arXiv:1910.12958 (2019)
16. Séjourné, Th., Vialard, F.-X., Peyré, G.: The unbalanced Gromov Wasserstein distance: conic formulation and relaxation. In: Advances in Neural Information Processing Systems, Virtual Only, vol. 34, pp. 8766–8779. Curran Associates Inc. (2021)
17. Sturm, K.-T.: The space of spaces: curvature bounds and gradient flows on the space of metric measure spaces. arXiv:1208.0434 (2012)
18. Thual, A., et al.: Aligning individual brains with fused unbalanced Gromov-Wasserstein. arXiv:2206.09398 (2022)
19. Vayer, T., Chapel, L., Flamary, R., Tavenard, R., Courty, N.: Fused Gromov-Wasserstein distance for structured objects. Algorithms **13**(9), 212 (2020)
20. Wang, J., Sesterhenn, J., Müller, W.-C.: Coherent structure detection and the inverse cascade mechanism in two-dimensional Navier-Stokes turbulence. arXiv:2203.11336 (2022)
21. Xu, H., Luo, D., Zha, H., Carin, L.: Gromov-Wasserstein learning for graph matching and node embedding. In: Proceedings of the 36th ICML, Long Beach, California, USA, pp. 6932–6941. PMLR (2019)

Optimal Transport Between GMM for Multiscale Texture Synthesis

Julie Delon[1], Agnès Desolneux[2], Laurent Facq[3], and Arthur Leclaire[3](✉)

[1] Université Paris Cité, CNRS, MAP5 UMR 8145, 75006 Paris, France
julie.delon@u-paris.fr

[2] Centre Borelli, CNRS and ENS Paris-Saclay, 91190 Gif-sur-Yvette, France
agnes.desolneux@ens-paris-saclay.fr

[3] Univ. Bordeaux, Bordeaux INP, CNRS, IMB, UMR 5251, 33400 Talence, France
{laurent.facq,arthur.leclaire}@math.u-bordeaux.fr

Abstract. Using optimal transport in image processing tasks has become very popular. However, it still faces difficult computational issues when dealing with high-dimensional distributions. We propose here to use the recently introduced GMM-OT formulation, which consists in restricting the optimal transport problem to the set of Gaussian mixture models. As a proof of concept, we use it to improve the texture model Texto based on optimal transport between distributions of image patches. Using GMM-OT in this texture model allows to deal with larger patches, hence providing results with better geometric details. This new model allows for synthesis, mixing, and style transfer.

Keywords: Optimal transport · Gaussian mixture models · texture synthesis

1 Introduction

Numerical optimal transport (OT) has undergone spectacular progress in the last ten years, and is now used in a large variety of applications [1,4,7,11,12,15]. Important advances have been made in the numerical approximations of optimal transport, with the emergence of efficient tools like regularized optimal transport [5] or the sliced optimal transport [2]. Nevertheless, it remains complex to compute optimal transport distances between empirical distributions when the dimension (the number of samples n or the space dimension d) of the problem increases too much.

In this context, several questions were raised concerning the ability to compute numerical solutions of high-dimensional problems or the sample complexity of the different transport approximations [3,10,17]. However, for several applications, targeting exact solutions of optimal transport might not be desirable, whereas proxy formulations sharing similar properties might deliver more relevant solutions in practice. Among these alternative formulations, an OT-like distance between Gaussian Mixture Models (GMM) has been introduced in [6]. It consists in restricting the set of possible couplings to GMM in the product

L. Calatroni et al. (Eds.): SSVM 2023, LNCS 14009, pp. 627–638, 2023.
https://doi.org/10.1007/978-3-031-31975-4_48

space. Solutions of this formulation are easy to compute and merely require to calculate Bures distances between Gaussian measures and solve a small-scale discrete OT problem. When applied to discrete data, the dependency on the dimension d and the number of samples n lies only in the GMM fitting step on the data, and in the computation of Bures distances. This makes the approach very versatile and robust to dimension in practice. In this paper we explore the use of GMM-OT for texture modeling, as a proof of concept. To this aim, we improve the texture model Texto [8], which is based on semi-discrete OT on image patches. We end up with a lighter and simpler formulation of the same problem. More precisely, in this context, the computing time of GMM-OT is at least one order of magnitude faster than the ones of semi-discrete OT or regularized OT, for similar (or even better) quality of synthesized images. This permits to use larger patch dimensions, and much more patches than the original Texto model.

2 Reminders on Optimal Transport Between Gaussian Mixture Models

This section recalls the main results on OT between GMM [6]. The quadratic Wasserstein distance between two probability measures μ_0, μ_1 on $\mathbb{R}^d$ with finite second moments is defined as

$$W_2^2(\mu_0, \mu_1) := \inf_{\gamma \in \Pi(\mu_0, \mu_1)} \int_{\mathbb{R}^d \times \mathbb{R}^d} \|y_0 - y_1\|^2 d\gamma(y_0, y_1), \tag{1}$$

where $\Pi(\mu_0, \mu_1)$ is the set of probability measures on $\mathbb{R}^d \times \mathbb{R}^d$ with marginals μ_0, μ_1. A solution γ^* of (1) is called an OT plan between μ_0 and μ_1. This distance has been extensively used for various applications in data science, and especially to define Wasserstein barycenters of probability measures, which are defined similarly to Euclidean barycenters, replacing the Euclidean distance by W_2.

2.1 Definition of MW_2

Let us denote by GMM_d the set of probability distributions which can be written as finite GMM on $\mathbb{R}^d$. OT plans and Wasserstein barycenters between GMM are usually not GMM themselves, which can be troublesome if we rely on this modeling to analyse or generate data. For this reason, the authors of [6] propose to modify the formulation of the classical Wasserstein distance by restricting the set of possible coupling measures to GMM on $\mathbb{R}^d \times \mathbb{R}^d$. More precisely, for $\mu_0, \mu_1 \in GMM_d$, we define

$$MW_2^2(\mu_0, \mu_1) := \inf_{\gamma \in \Pi^{\mathrm{GMM}}(\mu_0, \mu_1)} \int_{\mathbb{R}^d \times \mathbb{R}^d} \|y_0 - y_1\|^2 d\gamma(y_0, y_1), \tag{2}$$

where $\Pi^{\mathrm{GMM}}(\mu_0, \mu_1)$ is the set of probability measures in GMM_{2d} with marginals μ_0 and μ_1. It is shown in [6] that MW_2 defines a distance on GMM_d.

Moreover, if $\mu_0 = \sum_{k=1}^{K_0} \pi_0^k \mu_0^k$ and $\mu_1 = \sum_{l=1}^{K_1} \pi_1^l \mu_1^l$, where the $\pi_0^k \geq 0$, $\pi_1^l \geq 0$ are scalars with $\sum_k \pi_0^k = \sum_l \pi_1^l = 1$ and where the μ_0^k, μ_1^l are Gaussian probability measures, it can be shown [6] that

$$MW_2^2(\mu_0, \mu_1) = \min_{w \in \Pi(\pi_0, \pi_1)} \sum_{k,l} w_{kl} W_2^2(\mu_0^k, \mu_1^l), \tag{3}$$

where $\Pi(\pi_0, \pi_1)$ is the of $K_0 \times K_1$ matrices with non-negative entries and discrete marginals π_0 and π_1. This discrete expression makes MW_2 easy to compute in practice, even in large dimension. Indeed, the distance W_2 between two Gaussian measures $\mu = \mathcal{N}(m, \Sigma)$ and $\tilde{\mu} = \mathcal{N}(\tilde{m}, \tilde{\Sigma})$ has a closed-form expression:

$$W_2^2(\mu, \tilde{\mu}) = \|m - \tilde{m}\|^2 + \operatorname{tr}\left(\Sigma + \tilde{\Sigma} - 2\left(\Sigma^{\frac{1}{2}} \tilde{\Sigma} \Sigma^{\frac{1}{2}}\right)^{\frac{1}{2}}\right), \tag{4}$$

where we denote by $M^{\frac{1}{2}}$ the unique semi-definite positive square-root of a symmetric semi-definite positive matrix M. If the different parameters of the GMM μ_0 and μ_1 are known, computing (3) boils down to computing $K_0 \times K_1$ Wasserstein distances between Gaussian measures and then to solve a $K_0 \times K_1$ discrete OT problem. Similarly to the Wasserstein distance, it is possible to define barycenters for MW_2, and this also gives rise to a simple discrete formulation.

2.2 Using MW_2 in Practice

Optimal plans for MW_2 are not supported on the graph of a function and hence do not directly yield a transport map between the mixtures μ_0 and μ_1. In order to define a transport map from the optimal plan γ^* we can for instance use

$$T(x) = \mathbb{E}_{(X,Y)\sim\gamma^*}[Y|X = x]. \tag{5}$$

As shown in [6], the closed-form formula for T is given by

$$T(x) = \frac{\sum_{k,l} w_{k,l}^* g_{m_0^k, \Sigma_0^k}(x) T_{k,l}(x)}{\sum_k \pi_0^k g_{m_0^k, \Sigma_0^k}(x)}, \tag{6}$$

where w^* is the optimal solution of the discrete problem (3), $T_{k,l}$ are the optimal affine maps between Gaussians μ_0^k and μ_1^l and $g_{m,\Sigma}$ is the density of $\mathcal{N}(m, \Sigma)$. In the following, this map (6) will be called the GMM-OT map.

As shown in [6], this allows to express also the MW_2-barycenters between μ_0 and μ_1 with a closed-form formula:

$$\forall \alpha \in [0, 1] \quad \mu_\alpha = \sum_{k,l} w_{k,l}^* \big((1 - \alpha)\mathrm{Id} + \alpha T_{k,l}\big)\sharp\mu_0, \tag{7}$$

where $T\sharp\mu$ is the pushforward measure of the measure μ by the map T.

3 TextoGMM, a Multiscale Texture Synthesis Approach with Optimal Transport Between Patches

Let $u : \Omega \to \mathbb{R}^d$ be a texture defined on a rectangle $\Omega \subset \mathbb{Z}^2$. Let U be an initialization for the synthesized texture. The idea of TextoGMM, inspired by [8], is to use OT to force the patch distribution of U to look like the one of u at several scales. The synthesized texture is then recomposed from this patch distribution by simple local averages. In what follows, we denote by $\omega = \{0, \ldots, \mathsf{w}-1\}^2$ the patch domain with patch size w, and by $u_{|a+\omega} \in \mathbb{R}^\omega$ the patch at position a in u.

3.1 Monoscale Model

We start by describing the synthesis at a single scale. To initialize the process, we generate a stationary Gaussian random field U with the same mean and covariance as the example u, defined by

$$\forall a \in \mathbb{Z}^2, \quad U(a) = \bar{u} + \frac{1}{\sqrt{|\Omega|}} \sum_{b \in \Omega} (u(b) - \bar{u}) W(a-b) \tag{8}$$

where $\bar{u} = \frac{1}{|\Omega|} \sum_{a \in \Omega} u(a)$ and where W is a random field on $\mathbb{Z}^2$ whose pixel values are i.i.d. with distribution $\mathcal{N}(0,1)$.

The random field U has some global features of u, but not its details. The distribution $\hat{\mu}$ of patches of U is then sent by optimal transport towards the distribution $\hat{\nu}$ of patches of u in order to reimpose these details on the synthesized image. To make this transport calculation fast, we use here GMM-OT as described in the previous section, approximating the discrete distributions $\hat{\mu}, \hat{\nu}$ by GMMs μ and ν using the EM algorithm[1]. The formula (5) allows us to deduce a transport map $T : \mathbb{R}^\omega \to \mathbb{R}^\omega$ in the patch space from the GMM-OT plan that solves $MW_2(\mu, \nu)$. Finally, the new synthesized image V is computed by averaging all the transported patches:

$$\forall a \in \mathbb{Z}^2, \quad \tilde{U}(a) = \frac{1}{|\omega|} \sum_{h \in \omega} T(U_{|a-h+\omega})(h). \tag{9}$$

Note that because of averaging, the distribution of patches of $\tilde{U}$ is not quite that of the transported patches. Imposing more precisely the distribution of patches of the synthesized image would require more sophisticated techniques.

3.2 Multiscale Model

Let us now extend the previous model to several scales. For $0 \leq s \leq S-1$, consider a subsampled version u_s of u defined on the subdomain $\Omega_s \subset 2^s\mathbb{Z}^2$. At the coarse scale, U_{S-1} is initialized as the Gaussian field (8) estimated from u_{S-1}.

[1] It would be interesting here to have a GMM estimation method that directly minimizes a transport cost between the GMM and the discrete patch distribution.

Now assume that U_s at scale $s \in \{1, \ldots, S-1\}$ is given. Again, we estimate GMMs μ_s and ν_s from the patch distributions of U_s and u_s, and derive a GMM-OT map T_s between μ_s and ν_s, and then recompute a synthesized image $\tilde{U}_s$ at scale s from the transported patches:

$$\forall a \in 2^s\mathbb{Z}^2, \quad \tilde{U}_s(a) = \frac{1}{|\omega|} \sum_{b \in 2^s\omega} T_s(U_{s|a-b+2^s\omega})(b). \tag{10}$$

However, for the need of the upcoming upsampling step, we need to compose with a L^2 nearest-neighbor (NN) projection on the exemplar patches at scale s. Therefore, we set

$$\forall a \in 2^s\mathbb{Z}^2, \quad V_s(a) = \frac{1}{|\omega|} \sum_{b \in 2^s\omega} u_s\big(C_s(a-b)+b\big), \tag{11}$$

where the coordinate map C_s is defined by

$$\forall a \in 2^s\mathbb{Z}^2, \quad C_s(a) = \underset{a' \text{ st. } a'+2^s\omega \subset \Omega_s}{\operatorname{Argmin}} \|T_s(U_{s|a+2^s\omega}) - u_{s|a'+2^s\omega}\|^2. \tag{12}$$

This mechanism makes it easy to initialize synthesis at the next scale, by taking patches twice as large at the same positions. More precisely, U_{s-1} is initialized by setting for all $a \in 2^s\mathbb{Z}^2$ and all $k \in \{0, 2^{s-1}\}^2$,

$$U_{s-1}(a+k) = \frac{1}{|\omega|} \sum_{b \in 2^s\omega} u_{s-1}\big(C_s(a-b)+b+k\big). \tag{13}$$

At the end of the process, we obtain the synthesized image V_0. This *coarse-to-fine* synthesis process is illustrated on Fig. 1. Let us mention that, once the model estimated (i.e. all GMMs and transport plans computed from one synthesis), it can be used off-line to do image synthesis on demand (and of arbitrary size).

3.3 Adaptation to Style Transfer and Texture Mixing

The TextoGMM model can be easily adapted for style transfer and texture mixing. The adaptation to style transfer is a straightforward extension of the technique explained in [13] where the texture information can be treated with GMM-OT maps and then blended with the geometric features.

The adaptation to texture mixing requires more explanation, because it relies on a benefit of the GMM-OT cost, which is to have closed-form barycenters. For mixing, we exploit the explicit formula (7) that gives the expression of the MW_2 barycenter between μ_0 and μ_1. Let us fix a parameter $\alpha \in [0,1]$ that controls the mixing between the texture models associated with two source images u^0, u^1. For the initialization at the coarse scale, we can rely on the Gaussian model U^{S-1}

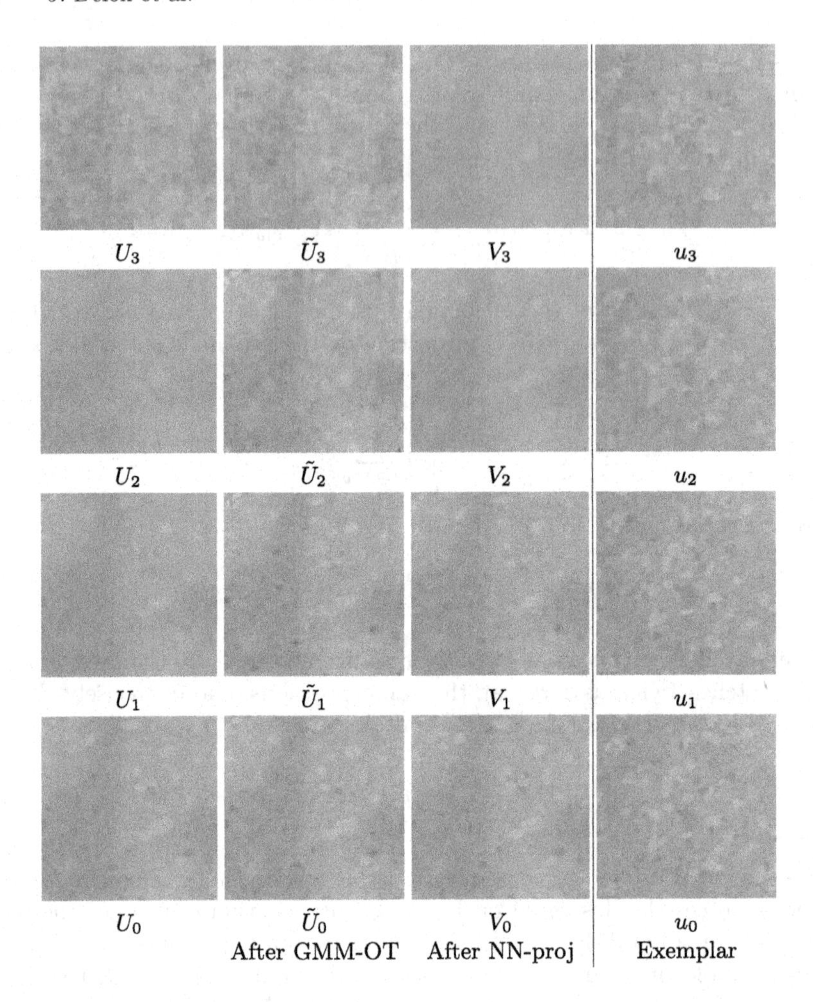

Fig. 1. Coarse-to-fine process ($\mathrm{w} = 5, S = 4$). On this figure, we illustrate the coarse-to-fine synthesis process of one exemplar texture (shown on the 4th column). It must be read from left to right and then top to bottom. At the coarse scale (here $S - 1 = 3$), the synthesis is initialized with the Gaussian field U_3. At each scale, from the current synthesis (U_s, 1st column), the patches are first transported with GMM-OT ($\tilde{U}_s$, 2nd column) and then projected back on exemplar patches with a NN projection (V_s, 3rd column). One can notice here the effect of the patch GMM-OT maps that will help to better cover the exemplar patch distributions, thus counter-acting the effects of NN projections that may sometimes flatten the image dynamic (especially for coarse scales with fewer patches).

obtained as the W_2 barycenter of the Gaussian models associated to u^0 and u^1, which can still be expressed as a convolution of a Gaussian white noise with an explicit image as in (8) [18]. For the patch transport at each scale s we apply a

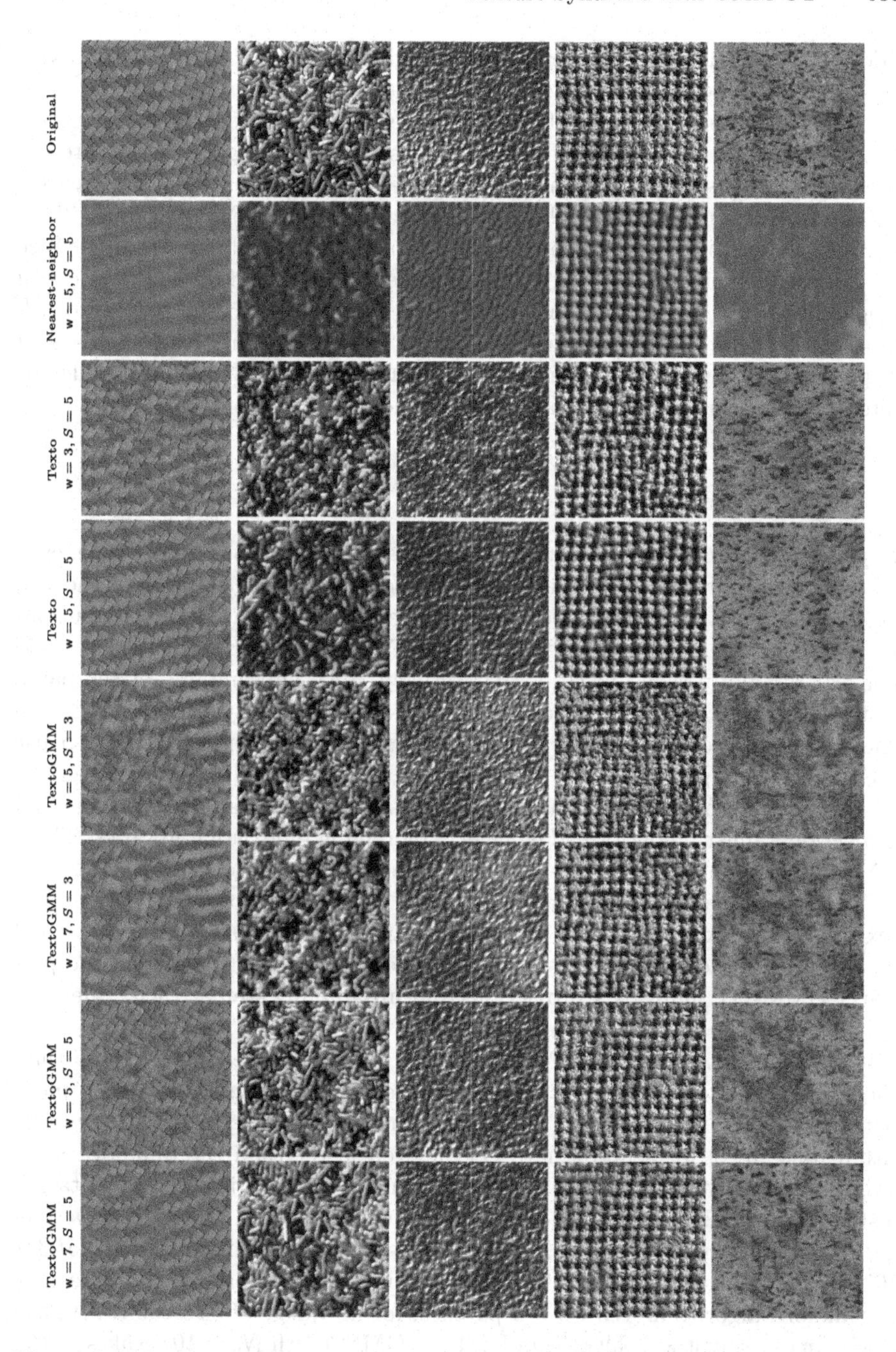

Fig. 2. Synthesis results. In this figure, we display several 512×512 original texture images (row 1), and synthesized images obtained with different models explained in the paper, with different parameters w (patch size) and S (number of scales). In row 2, the synthesis is obtained by using only patch nearest-neighbor projections at each scale. And then, we display the results obtained with the previous Texto model [8] (rows 3,4) and the TextoGMM model proposed here (rows 5–8).

GMM-OT map that targets the mixed patch distribution ν_s^α obtained by mixing the GMM patch distributions ν_s^0, ν_s^1 associated with u^0, u^1.

But a delicate point is that, for mixing, one cannot use a direct NN projection on exemplar patches to perform exemplar-based upsampling, because there is a priori no "mixed exemplar" that is available. We solve this issue by creating a collection of mixed patches by relying, again, on the GMM-OT map T (6) between ν_s^0, ν_s^1: once this transport map obtained, the mixed patches are defined as a linear interpolation $(1-\alpha)p + \alpha q$ between a patch p of ν_s^0 and the nearest neighbor $q = P_{NN}(T(p))$ of the transported patch $T(p)$ in the patches of ν_s^1. Since p and q are both patches taken from the original images at scale s, it is possible to compute the corresponding mixed patch at the next scale, by interpolating twice-larger patches taken at the same positions.

4 Experiments

Implementation Details. In order to keep a reasonable computational time, even for very large images and large patches, we do not use all patches in the EM algorithms and the NN projection steps. More precisely, the GMM distributions μ_s, ν_s at each scale are estimated using only $N_p = 10^4$ patches (or less for small images) randomly taken in U_s, u_s respectively. Such subsampling of the patch distribution was observed to be harmless for the texture synthesis application and speed-up these two steps. For the NN projection step (12), we also use the same set of patches taken in u_s. Notice also that, in order to fasten the whole synthesis process, the patch operations (extraction, aggregation and NN projections) must be performed efficiently by relying on existing libraries. These NN projections may even be accelerated with dedicated algorithms [14].

Let us emphasize on the fact that the TextoGMM model associated to a texture can be computed with one pass of analysis-synthesis. Once the model has been estimated, it can be sampled on-the-fly by directly applying the pre-learnt GMM-OT maps, NN projections and upsampling step at each scale.

Computation Time. TextoGMM allows for a much faster estimation step than the original Texto model [8], even though TextoGMM handles 10 times more patches. Indeed, the GMM-OT algorithm runs faster than the stochastic algorithm used in [8] for semi-discrete OT or even faster than the Sinkhorn algorithm [5]. For instance, on an OT problem with 10^4 points in source and target distributions on $\mathbb{R}^{147}$ (for 7×7 color patches), with a modern laptop with parallel CPU computations, solving GMM-OT takes $\approx 1'$, to be compared with $2.5h$ for 10^5 iterations of the stochastic OT algorithm, or $40'$ to perform 10^3 iterations of Sinkhorn algorithm. With 5×5 patches, for an image of size 256×256, the whole analysis-synthesis algorithm for TextoGMM (with $N_p = 10^4$) takes ≈ 15" while the analysis-synthesis algorithm for Texto (with $N_p = 10^3$) takes more than $1h$. In the same setting, once the model estimated, one synthesis takes ≈ 5". Therefore, a major benefit of the TextoGMM model is to considerably fasten the model estimation, while allowing for larger patches.

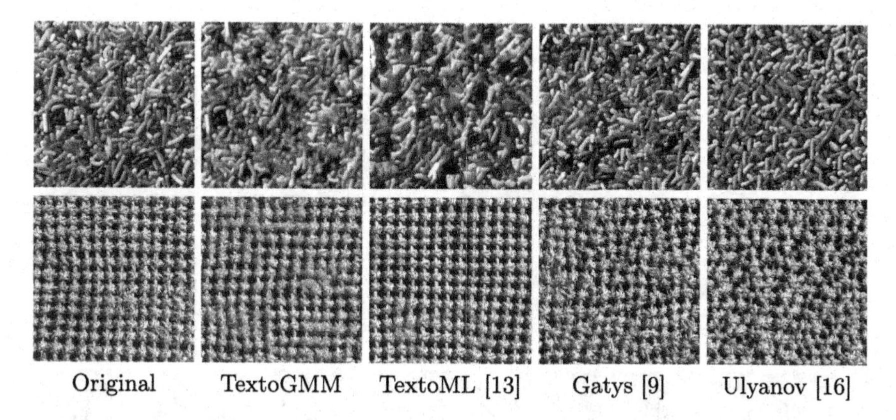

Fig. 3. Comparison with other synthesis algorithms. We display synthesis results obtained with several recent models: the here proposed TextoGMM model, the TextoML model from [13] also based on OT (both with patch size $\mathsf{w} = 5$ and $S = 5$ scales), and the neural-network based techniques from [9] and [16]. One can see that the synthesis quality with TextoGMM gets close to the one attained by TextoML, while keeping a considerably simpler model estimation step. One can also observe that TextoGMM is able to recover some complex and large structures (since it works on large patches) as [9] and [16], but produces blurrier results due to patch aggregation.

Visual Comments. Several synthesis examples are displayed on Fig. 1, Fig. 2, Fig. 3. In the captions, the output of the multiscale process explained in Sect. 3.2 is referred to as TextoGMM. In all the results of TextoGMM shown here, the estimated GMM models have 4 components, except in Fig. 4.

GMM-OT allows to partially cope with the curse of dimensionality, and thus permits to use much larger patches in TextoGMM than Texto (also because GMM-OT can handle distributions with much more points). This leads to more faithful synthesis of structured textures, with a better preservation of the sharp details, as illustrated in Fig. 2. The second row of Fig. 2 confirms the importance of using OT for patch transformation and not only simple NN matching. Also, the third and fourth rows confirm that the previous Texto model works only with small patches. In contrast, TextoGMM produces remarkable results on these textures with the proper choice of parameters w and S. In Fig. 3, we observe that the TextoGMM attains a visual quality similar to the model of [13] based on a multi-layer approximation of OT, while allowing for much faster estimation. Also, compared to recent texture synthesis methods, TextoGMM is able to reproduce large structures in a coherent way, while allowing for much faster estimation. In Fig. 4, we vary the number of number of Gaussian components K used in the GMM. Even if the visual details are more precisely retrieved when using more components, one can see that the syntheses obtained with very few Gaussian components (even 1 or 2!) appear already convincing. For $K = 1$, this illustrates the capacity of a very light texture synthesis algorithm based only on affine transformations and NN projections.

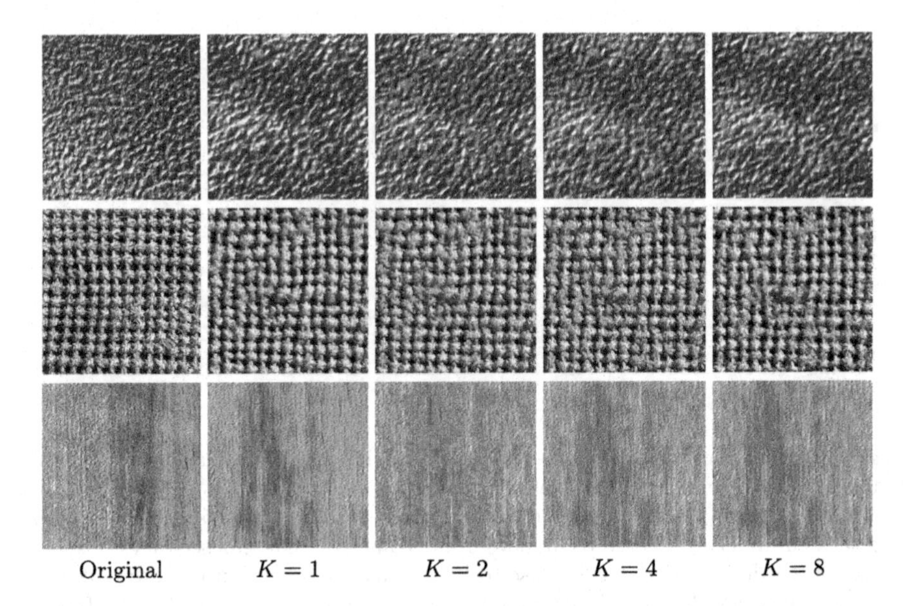

Fig. 4. Number of Gaussian components with ($\mathsf{w} = 5, S = 4$). We display synthesis results obtained with the TextoGMM model with varying number of components in the GMM. Even if results appear surprisingly good for $K = 1$, the fine details of the exemplar are better recovered with larger values of K (the reader is invited to examine fine details by comparing the images after a strong zoom-in).

Fig. 5. Style transfer with TextoGMM. Adapting the style transfer technique from [13] to the TextoGMM model produces convincing style transfer results.

Finally, on Fig. 5 and Fig. 6, we display some results of style transfer and texture mixing, respectively, with the TextoGMM model. For both these applications, the visual results appear convincing. In particular, for texture mixing, it is interesting to notice how the intermediate patch models are able to mix some structures seen in the exemplar textures u^0, u^1. To confirm the relevance of this approach, it would be interesting to compare the output distribution of the intermediate images to the true W_2-barycenter obtained with the patch distributions of u^0, u^1.

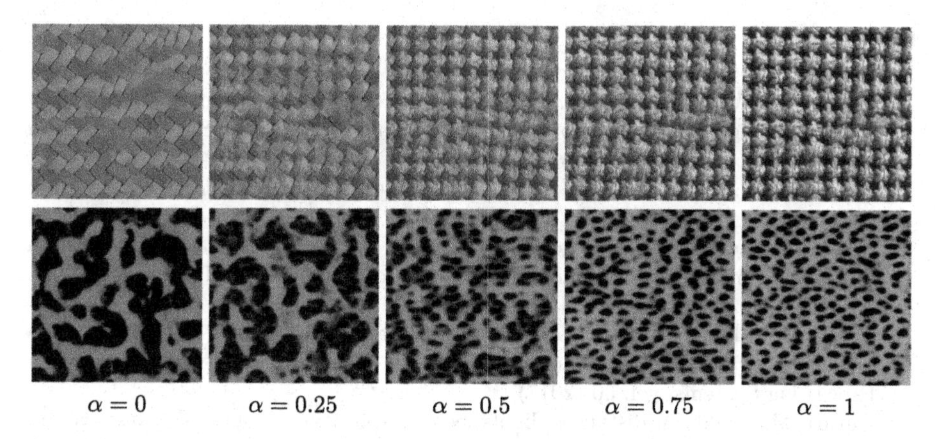

Fig. 6. Texture mixing. We display here examples of texture mixing computed from two source images u_0, u_1 (not shown in this figure). From left to right, we display a sample of the mixed texture model with mixing parameter $\alpha \in [0, 1]$. For $\alpha = 0, 1$, we get a sample of the TextoGMM model associated with u_0, u_1. It is interesting to see how the mixed TextoGMM model is able to combine the geometric structures of u_0, u_1. Parameters are $(\mathsf{w}, S) = (7, 5)$ for the 1st row and $(5, 4)$ for the 2nd row.

5 Conclusion

In this paper we proposed to exploit a new formulation of optimal transport specific to Gaussian mixture models, in order to improve the texture model Texto which is based on optimal transport in patch space. This new formulation allows to work with larger patches using a very simple parameterization of the transportation maps. Compared to Texto, it thus brings a clear improvement on the visual quality of generated textures, while considerably reducing the computational time required for estimating the model. Also, this model allows for fast on-the-fly synthesis because it only needs, from coarse to fine scales, a few affine transformations of the patches composed with a nearest neighbor search. Finally, this texture model can also be used for style transfer or texture mixing, exploiting the closed-form barycenters for the GMM-OT cost.

The main limitation of this new model is now the practical complexity of the EM algorithm used to approximate empirical patch distributions by GMM. Even if the number of used components can be set small (between 1 and 10), the practical behavior of the EM algorithm applied to very rich empirical distributions in high dimension remains problematic. It would be interesting to examine more thoroughly the impact of the obtained GMM approximation on the quality of synthesized textures, and to see if another GMM learning algorithm could be used in order to better scale up to the dimension of the patch space. Also, it would be interesting to compare more thoroughly the GMM-OT approximation of the OT cost with the semi-discrete multilayer approximation of [13]. These methods are respectively based on a soft and hard clustering of a target distribution, and it would be useful to draw a theoretical connection between them.

References

1. Bonneel, N., Van De Panne, M., Paris, S., Heidrich, W.: Displacement interpolation using Lagrangian mass transport. In: Proceedings of the 2011 SIGGRAPH Asia Conference, pp. 1–12 (2011)
2. Bonnotte, N.: Unidimensional and evolution methods for optimal transportation. Ph.D. thesis, Paris 11 (2013)
3. Chizat, L., Roussillon, P., Léger, F., Vialard, F.X., Peyré, G.: Faster Wasserstein distance estimation with the Sinkhorn divergence. Adv. Neural. Inf. Process. Syst. **33**, 2257–2269 (2020)
4. Courty, N., Flamary, R., Habrard, A., Rakotomamonjy, A.: Joint distribution optimal transportation for domain adaptation. In: Advances in Neural Information Processing Systems, vol. 30 (2017)
5. Cuturi, M.: Sinkhorn distances: lightspeed computation of optimal transport. In: Advances in Neural Information Processing Systems, pp. 2292–2300 (2013)
6. Delon, J., Desolneux, A.: A Wasserstein-type distance in the space of Gaussian mixture models. SIAM J. Imag. Sci. **13**(2), 936–970 (2020)
7. Feydy, J., Roussillon, P., Trouvé, A., Gori, P.: Fast and scalable optimal transport for brain tractograms. In: Shen, D., et al. (eds.) MICCAI 2019. LNCS, vol. 11766, pp. 636–644. Springer, Cham (2019). https://doi.org/10.1007/978-3-030-32248-9_71
8. Galerne, B., Leclaire, A., Rabin, J.: A texture synthesis model based on semi-discrete optimal transport in patch space. SIAM J. Imag. Sci. **11**(4), 2456–2493 (2018)
9. Gatys, L.A., Ecker, A.S., Bethge, M.: Image style transfer using convolutional neural networks. In: Proceedings of the IEEE Conference on Computer Vision and Pattern Recognition, pp. 2414–2423 (2016)
10. Genevay, A., Chizat, L., Bach, F., Cuturi, M., Peyré, G.: Sample complexity of Sinkhorn divergences. In: The 22nd International Conference on Artificial Intelligence and Statistics, pp. 1574–1583. PMLR (2019)
11. Hertrich, J., Houdard, A., Redenbach, C.: Wasserstein patch prior for image super-resolution. IEEE Trans. Comput. Imaging **8**, 693–704 (2022)
12. Karras, T., Laine, S., Aila, T.: A style-based generator architecture for generative adversarial networks. In: Proceedings of the IEEE Conference on Computer Vision and Pattern Recognition, pp. 4401–4410 (2019)
13. Leclaire, A., Rabin, J.: A stochastic multi-layer algorithm for semi-discrete optimal transport with applications to texture synthesis and style transfer. J. Math. Imaging Vis. **63**(2), 282–308 (2021)
14. Liang, L., Liu, C., Xu, Y.Q., Guo, B., Shum, H.Y.: Real-time texture synthesis by patch-based sampling. ACM Trans. Graph. **20**(3), 127–150 (2001)
15. Mignon, S., Galerne, B., Hidane, M., Louchet, C., Mille, J.: Semi-unbalanced regularized optimal transport for image restoration. In: Actes du GRETSI (2022)
16. Ulyanov, D., Lebedev, V., Vedaldi, A., Lempitsky, V.: Texture networks: feed-forward synthesis of textures and stylized images. In: Proceedings of the International Conference on Machine Learning, vol. 48, pp. 1349–1357 (2016)
17. Weed, J., Bach, F.: Sharp asymptotic and finite-sample rates of convergence of empirical measures in Wasserstein distance. Bernoulli **25**(4A), 2620–2648 (2019)
18. Xia, G., Ferradans, S., Peyré, G., Aujol, J.: Synthesizing and mixing stationary Gaussian texture models. SIAM J. Imag. Sci. **7**(1), 476–508 (2014)

Asymptotic Result for a Decoupled Nonlinear Elasticity-Based Multiscale Registration Model

Noémie Debroux[1](✉) and Carole Le Guyader[2]

[1] Université Clermont Auvergne, CNRS, SIGMA Clermont, Institut Pascal, Aubière, France
noemie.debroux@uca.fr

[2] INSA Rouen Normandie, Normandie Univ, LMI UR 3226, 76000 Rouen, France
carole.le-guyader@insa-rouen.fr

Abstract. In this paper, a theoretical asymptotic result in relation to the nonlinear elasticity-based multiscale registration model [Debroux *et al.* 2023] is proved. Specifically, it establishes a link between the original minimisation problem comprising high non linearity and non convexity, and the one derived from splitting techniques by means of auxiliary variables and L^p-penalisations which exhibits increased numerical manageability. This latter problem thus appears to be a good compromise between mechanical/physical realism and practical feasibility.

Keywords: Registration · Multiscale decomposition · Ogden materials · Bi-Lipschitz homeomorphisms · Asymptotic analysis

1 Introduction

1.1 Original Mathematical Model

In the preliminary paper [8], the authors propose extending Tadmor *et al.*'s work [20] devoted to multiscale image representation and achieved through hierarchical (BV, L^2) decompositions to the case of registration. The rationale underpinning this study is to build a hierarchical decomposition of the deformation matching two images in the form of a composition of intermediate deformations $\varphi_0 \circ \cdots \circ \varphi_k \circ \cdots \circ \varphi_n$, that is, to hierarchise the information carried by the deformation. The coarser one, φ_0, is computed from rough versions of the two involved images obtained from [20], and encodes the main structural/geometrical deformation. Iterating the process and injecting more and more details into both images yields more accurate deformations of the type $\varphi_0 \circ \cdots \circ \varphi_k$ mapping faithfully small-scale features. The underlying aim is (i): to dissociate the main deformation from the more localised

This project was co-financed by the European Union with the European regional development fund (ERDF, 18P03390/18E01750/18P02733), by the Haute-Normandie Régional Council via the M2SINUM project and by the French Research National Agency ANR via AAP CE23 MEDISEG ANR project.

L. Calatroni et al. (Eds.): SSVM 2023, LNCS 14009, pp. 639–651, 2023.
https://doi.org/10.1007/978-3-031-31975-4_49

displacements. In this respect, a parallel can be drawn with multiresolution techniques [13] even though those latter refer primarily to the size of the images being processed, while we focus more on the scale of the features; (ii) to open the way to a posteriori analyses like deriving statistics.

Mathematically, the resulting iterative refinement scheme that takes into account the granulometry of the deformation is broken down as follows:

$$(\mathcal{P}_k) \qquad \begin{cases} \varphi_0 = & \underset{\varphi\in\mathcal{W}}{\arg\min}\{\mathcal{F}(\varphi, T_0, R_0)\}, \\ \varphi_k = \underset{\varphi\in\mathcal{X}_k}{\arg\min}\{\mathcal{F}(\varphi_0\circ\varphi_1\circ\ldots\circ\varphi_{k-1}\circ\varphi, \sum_{j=0}^{k} T_j, \sum_{j=0}^{k} R_j)\}, \end{cases}$$

the quantity $\bar{R}_k := \sum_{j=0}^{k} R_j$ (*resp.* $\bar{T}_k := \sum_{j=0}^{k} T_j$) denoting the hierarchical decomposition of the Reference image $R : \bar{\Omega} \to \mathbb{R}$ (*resp.* the Template image $T : \bar{\Omega} \to \mathbb{R}$), and Ω being a convex bounded open subset of $\mathbb{R}^2$ of class $\mathcal{C}^1$, $\mathcal{W}$ and $\mathcal{X}_k$ being defined later. Images R and T are supposed to be elements of $BV(\Omega)$. Also, for theoretical purposes, we assume that T is such that its essential support ess supp(T) is included in $\Omega' \subset\subset \Omega$, Ω' being a bounded open set of Ω. Functional $\mathcal{F}$ can be viewed as a parent functional from which the successive minimisation subproblems are derived. Arguments borrowed from the theory of mechanics (hyperelasticity [6, Part A, Chap. 4 and Part B, Chap. 7]) motivate the way this functional is designed, the objects to be matched being seen as bodies subjected to external forces. Precisely, it is defined by

$$\begin{aligned} \mathcal{F}(\varphi, T, R) =& \frac{\lambda}{2}\|T\circ\varphi - R\|^2_{L^2(\Omega)} + \int_\Omega \mathcal{W}_{Op}(\nabla\varphi, \det(\nabla\varphi))\,dx \\ &+ \mathbb{1}_{\{\|\cdot\|_{L^\infty(\Omega, M_2(\mathbb{R}))}\leq\alpha\}}(\nabla\varphi) + \mathbb{1}_{\{\|\cdot\|_{L^\infty(\Omega, M_2(\mathbb{R}))}\leq\beta\}}((\nabla\varphi)^{-1}), \end{aligned}$$

and reflects a balance (modulated by the λ parameter) between the classical L^2-fidelity term and smoothness requirements. Regularity is first encoded through the component $\int_\Omega \mathcal{W}_{Op}(\nabla\varphi, \det(\nabla\varphi))\,dx$, where

$$\mathcal{W}_{Op}(F, \delta) = \begin{cases} a_1\|F\|^4 + a_2(\delta-1)^2 + \frac{a_3}{(\delta)^{10}} - 4a_1 - a_3 & \text{if} \quad \delta > 0 \\ \infty & \text{otherwise} \end{cases},$$

is the stored energy of an Ogden material [6]—*Op* standing for '*Ogden particular*' and $\|\cdot\|$ being the Frobenius norm—. The first term $\|\cdot\|^4$ controls the smoothness of the deformation, namely changes in length, the second one restricts changes in area, while the third one ensures orientation preservation, the last two components being added to comply with the property $\mathcal{W}_{Op}(I, \det I) = 0$, where I stands for the identity matrix. The choice of the power 4 in $\|\cdot\|^4$ ($4 > 2$ the dimension of the ambient space) combined with the components phrased in terms of the convex characteristic functions and the constraint $\det F > 0$ *a.e.* ensures that the obtained deformations are bi-Lipschitz homeomorphisms. In the first step of the algorithm, the unknown φ is searched in the functional space $\mathcal{W} = \{\psi \in \text{Id} + W_0^{1,\infty}(\Omega, \mathbb{R}^2) \,|\, \|\nabla\psi\|_{L^\infty(\Omega, M_2(\mathbb{R}))} \leq \alpha,\ \|(\nabla\psi)^{-1}\|_{L^\infty(\Omega, M_2(\mathbb{R}))} \leq$

β, $\det \nabla\psi > 0$ a.e. in $\Omega\}$, α and β being two prescribed positive parameters, $M_2(\mathbb{R})$ being the set of 2×2 real matrices. Then to pass from step $k-1$ to k, the minimisation problem is formulated on a functional space depending on k (emphasising the fact that the stages are correlated to each other), namely

$$\mathcal{X}_k = \Big\{ \psi \,|\, \Phi := \varphi_0 \circ \varphi_1 \circ \ldots \circ \varphi_{k-1} \circ \psi \in \mathrm{Id} + W_0^{1,\infty}(\Omega, \mathbb{R}^2),$$

$$\det \nabla\Phi > 0 \; a.e., \; \|\nabla\Phi\|_{L^\infty(\Omega, M_2(\mathbb{R}))} \leq \alpha, \; \|(\nabla\Phi)^{-1}\|_{L^\infty(\Omega, M_2(\mathbb{R}))} \leq \beta \Big\}.$$

To make things more concrete, Fig. 1 illustrates the versatility of the model [8]. In particular, it demonstrates its ability to model deformations capturing increasingly fine details as more subtle features appear in the images.

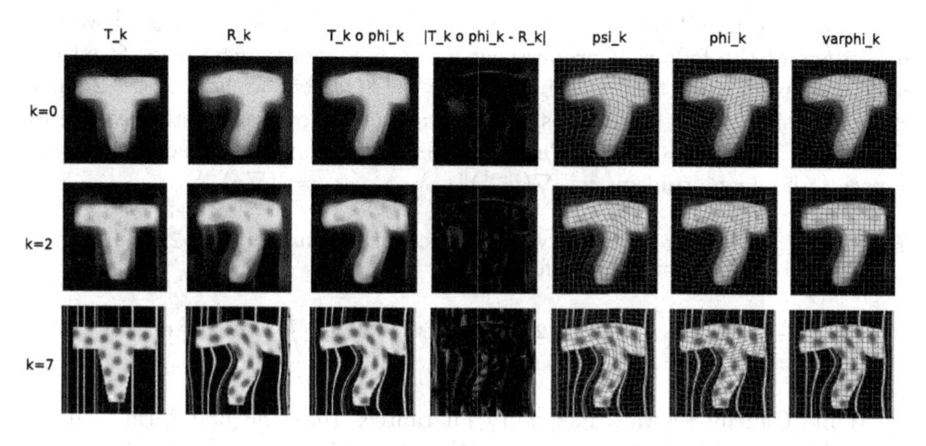

Fig. 1. Multiscale registration results on synthetic T-shape images with texture.

For the sake of completeness, we would like to add that several previous works [2,11,12,14–19] have suggested promoting multiscale decomposition in the context of registration, whether this step is involved in the decomposition of the images to be matched themselves, in the decomposition of the deformation pairing the two images, or in both. Nevertheless, the work [8] differs from the existing literature either on the paradigm used or on methodological aspects. At last, we would like to point out that this contribution is a continuation of the companion papers [7,8]. In particular, [7] contains the upstream motivation and description of the modelling, justification of the functional spaces involved, theoretical results highlighting the mathematical soundness among which the existence of minimisers and an asymptotic result, along with exhaustive numerical simulations demonstrating the ability of the model to produce accurate hierarchical representations of deformations. Here, the focus is on the derivation of an algorithm allowing to alleviate the computational burden of the proposed highly nonlinear nonconvex model.

1.2 Towards a Decoupled Problem

If problem $(\mathcal{P}_k)$ exhibits fine theoretical properties ((i) it admits at least one minimiser and this minimiser is a bi-Lipschitz homeomorphism from $\bar{\Omega}$ to $\bar{\Omega}$, (ii) an asymptotic result can be established when k tends to ∞, etc.), it is nevertheless part of the class of non convex, non differentiable problems, hard to solve from a numerical point of view. To balance the numerical complexity, we introduce several auxiliary variables, and the original problem $(\mathcal{P}_k)$ is split into subproblems that are more easily tractable. These additional variables are: (i) ϕ that simulates the composition of deformations $\varphi_0 \circ \cdots \circ \varphi_{k-1} \circ \varphi$ to lighten the non linearity stemming from the composition $\bar{T}_k \circ \varphi_0 \circ \cdots \circ \varphi_{k-1} \circ \varphi$, (ii) Ψ that mimics ϕ^{-1}, (iii) V that approximates $\nabla \phi$ and that inherits a part of the non convexity and non linearity, (iv) W that reproduces $\nabla \Psi$. Note at last that two considerations rule the design of the decoupled problem:

Remark 1. To handle the L^∞ penalty on $(\nabla \varphi_0 \circ \varphi_1 \circ \ldots \circ \varphi_{k-1} \circ \varphi)^{-1}$ we take advantage of the following property: if u is a homeomorphism from Ω into Ω, and the inverse function u^{-1} belongs to $W^{1,q}(\Omega, \mathbb{R}^2)$ with $q > 2$—2 being the dimension of the ambient space—, the matrix of weak derivatives reads $\nabla(u^{-1}) = (\nabla u)^{-1}(u^{-1})$ ([3, Theorem 2]) and it can be proved that for bi-Lipschitz homeomorphisms u, $\|\nabla(u^{-1})\|_{L^\infty(\Omega, M_2(\mathbb{R}))} = \|(\nabla u)^{-1}\|_{L^\infty(\Omega, M_2(\mathbb{R}))}$.

Remark 2. Little more regularity is assumed on $\bar{T}_k$, namely $\bar{T}_k \in L^4(\Omega)$. With the prescribed functional space for Ψ,

$$\int_\Omega (\bar{T}_k - \bar{R}_k \circ \Psi)^2 \det \nabla \Psi \, dx \leq 2 \int_\Omega |\bar{T}_k|^2 \det \nabla \Psi \, dx + 2 \int_\Omega (\bar{R}_k \circ \Psi)^2 \det \nabla \Psi \, dx.$$

While Cauchy-Schwarz inequality guarantees that the first term is finite, Theorem 1 of [3] holds, ensuring that the classical change of variable formula applies to the second term which is thus well-defined.

Equipped with these theoretical precautions, the derived decoupled problem $(\mathcal{DP}_{k,i})$ is in the end stated as follows, using L^p-type penalisations: let $(\gamma_{k,i})_{i \in \mathbb{N}}$ be an increasing sequence of positive real numbers such that $\lim_{i \to \infty} \gamma_{k,i} = \infty$ for fixed k. Then one denotes by $(\mathcal{DP}_{k,i})$ the problem

$$\begin{aligned} &\inf_{\varphi,\phi,\psi,V,W} \Big\{ \bar{\mathcal{F}}_{k,i}(\varphi, \phi, \psi, V, W) \qquad (\mathcal{DP}_{k,i}) \\ &= \frac{\lambda}{2} \int_\Omega (\bar{T}_k - \bar{R}_k \circ \psi)^2 \det \nabla \psi \, dx + \int_\Omega W_{Op}(V, \det V) \, dx \\ &+ \mathbb{1}_{\{\|\cdot\|_{L^\infty(\Omega, M_2(\mathbb{R}))} \leq \alpha\}}(V) + \mathbb{1}_{\{\|\cdot\|_{L^\infty(\Omega, M_2(\mathbb{R}))} \leq \beta\}}(W) + \frac{\gamma_{k,i}}{4} \|V \\ &- \nabla \phi\|^4_{L^4(\Omega, M_2(\mathbb{R}))} + \frac{\gamma_{k,i}}{4} \|W - \nabla \psi\|^4_{L^4(\Omega, M_2(\mathbb{R}))} \\ &+ \frac{\gamma_{k,i}}{2} \|\zeta_{k-1}^{-1} \circ \phi - \varphi\|^2_{L^2(\Omega, \mathbb{R}^2)} + \frac{\gamma_{k,i}}{2} \|\psi \circ \phi - \mathrm{Id}\|^2_{L^2(\Omega, \mathbb{R}^2)} \Big\}, \end{aligned}$$

where one has set $\zeta_{k-1} = \varphi_0 \circ \varphi_1 \circ \ldots \circ \varphi_{k-1}$, bi-Lipschitz homeomorphism from $\bar{\Omega}$ to $\bar{\Omega}$ with $\zeta_{k-1} \in \mathrm{Id} + W_0^{1,\infty}(\Omega, \mathbb{R}^2)$, $\det \zeta_{k-1} > 0$ *a.e.*, $\|\nabla \zeta_{k-1}\|_{L^\infty(\Omega, M_2(\mathbb{R}))} \leq \alpha$ and $\|(\nabla \zeta_{k-1})^{-1}\|_{L^\infty(\Omega, M_2(\mathbb{R}))} \leq \beta$, and with $\varphi \in L^2(\Omega, \mathbb{R}^2)$, $\phi \in \{u \in \mathrm{Id} + W_0^{1,4}(\Omega, \mathbb{R}^2)\}$, $\psi \in \{u \in \mathrm{Id} + W_0^{1,4}(\Omega, \mathbb{R}^2), \det \nabla u > 0 \; a.e.\}$, $V \in \{u \in L^4(\Omega, M_2(\mathbb{R})), (\det u)^{-1} \in L^{10}(\Omega), \det u > 0 \, a.e., \|u\|_{L^\infty(\Omega, M_2(\mathbb{R}))} \leq \alpha\}$ and $W \in \{u \in L^\infty(\Omega, M_2(\mathbb{R})), \|u\|_{L^\infty(\Omega, M_2(\mathbb{R}))} \leq \beta\}$, $\det u$ being the determinant of the u matrix belonging to $M_2(\mathbb{R})$. Note that we will give a precise meaning to $\zeta_{k-1}^{-1} \circ \phi$ and $\psi \circ \phi$ afterwards.

We are now ready to state the main theoretical result. It claims that for fixed k and for sufficiently large parameters weighting the L^p-penalisations ensuring the coupling between the auxiliary variables and the variables they are supposed to simulate, the $(\mathcal{DP}_{k,i})$ problem constitutes a good approximation to $(\mathcal{P}_k)$.

2 Main Result

We work at fixed k so in order to lighten the notation, we remove the dependency in k when unnecessary.

Theorem 1 (Asymptotic result). *Let* $(\varphi_{i,n}, \phi_{i,n}, \Psi_{i,n}, V_{i,n}, W_{i,n})_n$ *be a minimising sequence of* $\bar{\mathcal{F}}_{k,i}$. *Then there exists a subsequence denoted by* $(\varphi_{\tau(i),N(\tau(i))}, \phi_{\tau(i),N(\tau(i))}, \Psi_{\tau(i),N(\tau(i))}, V_{\tau(i),N(\tau(i))}, W_{\tau(i),N(\tau(i))})$ *that we now write* $(\varphi_i, \phi_i, \Psi_i, V_i, W_i)$ *for the sake of simplicity, and a minimiser* $\bar{\varphi}$ *of* $\mathcal{F}_k$ *such that*

$$\lim_{i \to \infty} \bar{\mathcal{F}}_{k,i}(\varphi_i, \phi_i, \Psi_i, V_i, W_i) = \mathcal{F}_k(\bar{\varphi}),$$

with $\phi_i \underset{i\to\infty}{\rightharpoonup} \bar{\phi}$ *in* $W^{1,4}(\Omega, \mathbb{R}^2)$, $\varphi_i \underset{i\to\infty}{\rightharpoonup} \bar{\varphi} = (\varphi_0 \circ \cdots \circ \varphi_{k-1})^{-1} \circ \bar{\phi}$ *in* $L^2(\Omega, \mathbb{R}^2)$, $V_i \underset{i\to\infty}{\overset{*}{\rightharpoonup}} \nabla \bar{\phi}$ *in* $L^\infty(\Omega, M_2(\mathbb{R}))$, $\det V_i \underset{i\to\infty}{\rightharpoonup} \det \nabla \bar{\phi}$ *in* $L^2(\Omega)$, $W_i \underset{i\to\infty}{\rightharpoonup} \nabla \bar{\phi}^{-1}$ *in* $L^\infty(\Omega, M_2(\mathbb{R}))$ *and* $\Psi_i \underset{i\to\infty}{\rightharpoonup} \bar{\phi}^{-1}$ *in* $W^{1,4}(\Omega, \mathbb{R}^2)$.

Proof. The guidelines for the proof are as follows: (i) the involved sequences are proved to be uniformly bounded in the respective functional spaces, leading to weakly converging subsequences; (ii) the fact that the sequence $(\gamma_i)_{i\in\mathbb{N}}$ diverges to $+\infty$ makes it possible to establish relations on the limit functions; (iii) classical weak lower semi-continuity arguments combined with [3, Theorems 1 and 2] enable one to obtain the desired smoothness on the deformations; (iv) the last part concentrates the main technical difficulties and address the treatment of the data fidelity terms.

For $(\hat{\varphi}, \hat{\phi}, \hat{\psi}, \hat{V}, \hat{W}) = (\zeta_{k-1}^{-1}, \mathrm{Id}, \mathrm{Id}, I_2, I_2)$, $\bar{\mathcal{F}}_{k,i}(\hat{\varphi}, \hat{\phi}, \hat{\psi}, \hat{V}, \hat{W}) < \infty$, bounded uniformly with respect to k and i since $\bar{T}_k \underset{k\to\infty}{\longrightarrow} T$ strongly in $L^2(\Omega)$, respectively, $\bar{R}_k \underset{k\to\infty}{\longrightarrow} R$, and ψ_{k-1}^{-1}, is uniformly bounded in $L^2(\Omega, \mathbb{R}^2)$. The derivation of a standard coercivity inequality, combined with the existence of the uniform bound shows that the infimum is finite.

Let $(\varphi_{i,n}, \phi_{i,n}, \psi_{i,n}, V_{i,n}, W_{i,n})$ be a minimising sequence. By definition of the infimum, there exists $\varphi_l \in \mathcal{X}_k$—meaning that φ_l is a bi-Lipschitz homeomorphism from $\bar{\Omega}$ to $\bar{\Omega}$ with $\varphi_l \in \text{Id} + W_0^{1,\infty}(\Omega, \mathbb{R}^2)$, $\det \nabla\varphi_l > 0$ *a.e.*, $\|\nabla(\zeta_{k-1} \circ \varphi_l)\|_{L^\infty(\Omega, M_2(\mathbb{R}))} \leq \alpha$ and $\|(\nabla(\zeta_{k-1} \circ \varphi_l))^{-1}\|_{L^\infty(\Omega, M_2(\mathbb{R}))} \leq \beta$, so that in particular, $\|(\nabla(\zeta_{k-1}\circ\varphi_l))^{-1}\|_{L^\infty(\Omega, M_2(\mathbb{R}))} = \|\nabla(\zeta_{k-1} \circ \varphi_l)^{-1}\|_{L^\infty(\Omega, M_2(\mathbb{R}))}$—such that $\mathcal{F}_k(\varphi_l) = \bar{\mathcal{F}}_{k,i}(\varphi_l, \zeta_{k-1}\circ\varphi_l, (\zeta_{k-1}\circ\varphi_l)^{-1}, \nabla(\zeta_{k-1}\circ\varphi_l), \nabla(\zeta_{k-1}\circ\varphi_l)^{-1}) \leq \inf_{\varphi\in\mathcal{X}_k} \mathcal{F}_k(\varphi) + \frac{1}{l} < \infty$. Next, by definition of a minimising sequence, $\forall\epsilon > 0$, $\exists N(\epsilon, \gamma_{k,i}) \in \mathbb{N}$, $\forall n \in \mathbb{N}$, $\big(n \geq N(\epsilon, \gamma_i := \gamma_{k,i}) \implies$

$$\bar{\mathcal{F}}_{k,i}(\varphi_{i,n}, \phi_{i,n}, \psi_{i,n}, V_{i,n}, W_{i,n}) \leq \epsilon + \inf \bar{\mathcal{F}}_{k,i} \leq \inf_{\varphi\in\mathcal{X}_k} \mathcal{F}_k(\varphi) + \frac{1}{l} + \epsilon < \infty\big).$$

Let us set in particular $\epsilon = \frac{1}{\gamma_i}$. Then there exists $N_i \in \mathbb{N}$ such that for all $n \in \mathbb{N}$, $n \geq N_i \Rightarrow \bar{\mathcal{F}}_{k,i}(\varphi_{i,n}, \phi_{i,n}, \psi_{i,n}, V_{i,n}, W_{i,n}) \leq \inf_{\varphi\in\mathcal{X}} \mathcal{F}_k + \frac{1}{l} + \frac{1}{\gamma_i} \leq \inf_{\varphi\in\mathcal{X}_k} \mathcal{F}_k + \frac{1}{l} + \frac{1}{\gamma_0} < \infty$. For the sake of conciseness, from now on, we denote by $\varphi_i := \varphi_{i,N_i}$ and similarly for the other components.

According to standard coercivity arguments, we get that:

(i) (V_i) is uniformly bounded in $L^\infty(\Omega, M_2(\mathbb{R}))$, so in $L^4(\Omega, M_2(\mathbb{R}))$,
(ii) $(\det V_i)$ is uniformly bounded in $L^2(\Omega)$,
(iii) (W_i) is uniformly bounded in $L^\infty(\Omega, M_2(\mathbb{R}))$, so in $L^4(\Omega, M_2(\mathbb{R}))$,
(iv) (ϕ_i) is uniformly bounded in $W^{1,4}(\Omega, \mathbb{R}^2)$ thanks to the generalised Poincaré inequality,
(v) (ψ_i) is uniformly bounded in $W^{1,4}(\Omega, \mathbb{R}^2)$, again thanks to the generalised Poincaré inequality,
(vi) and (φ_i) is uniformly bounded in $L^2(\Omega, \mathbb{R}^2)$.

Thus there exist a subsequence still denoted by (V_i), $\bar{V}$ and $\bar{\delta}$ such that $V_i \underset{i\to\infty}{\rightharpoonup} \bar{V}$ in $L^4(\Omega, M_2(\mathbb{R}))$, $V_i \underset{i\to\infty}{\overset{*}{\rightharpoonup}} \bar{V}$ in $L^\infty(\Omega, M_2(\mathbb{R}))$ and $\det V_i \underset{i\to\infty}{\rightharpoonup} \bar{\delta}$ in $L^2(\Omega)$. By the weak-$*$ lower semi-continuity of $\|\cdot\|_{L^\infty(\Omega, M_2(\mathbb{R}))}$, we deduce that $\|\bar{V}\|_{L^\infty(\Omega, M_2(\mathbb{R}))} \leq \liminf_{i\to\infty} \|V_i\|_{L^\infty(\Omega, M_2(\mathbb{R}))} \leq \alpha$ and $\mathbb{1}_{\{\|\cdot\|_{L^\infty(\Omega, M_2(\mathbb{R}))}\leq\alpha\}}(\bar{V}) \leq \liminf_{i\to\infty} \mathbb{1}_{\{\|\cdot\|_{L^\infty(\Omega, M_2(\mathbb{R}))}\leq\alpha\}}(V_i)$.

Additionally, there exist a (common) subsequence still denoted by (ϕ_i) and $\bar{\phi}$ in $W^{1,4}(\Omega, \mathbb{R}^2)$ such that $\phi_i \underset{i\to\infty}{\rightharpoonup} \bar{\phi}$ in $W^{1,4}(\Omega, \mathbb{R}^2)$ and $\det \nabla\phi_i \underset{i\to\infty}{\rightharpoonup} \det \nabla\bar{\phi}$ in $L^2(\Omega)$.

Hence, let us set $x_i = V_i - \nabla\phi_i$. Obviously $x_i \underset{i\to\infty}{\longrightarrow} 0$ in $L^4(\Omega, M_2(\mathbb{R}))$. Strong convergence implying weak convergence, we have (denoting by : the standard matrix inner product, *i.e.*, $\forall(A, B) \in M_2(\mathbb{R})^2$, $A : B = \text{tr}\,(A^T B)$), $\forall\xi \in L^{\frac{4}{3}}(\Omega, M_2(\mathbb{R}))$,

$$\int_\Omega x_i : \xi\, dx = \int_\Omega (V_i - \nabla\phi_i) : \xi\, dx \underset{i\to\infty}{\longrightarrow} 0.$$

But $\int_\Omega V_i : \xi\, dx \underset{i\to\infty}{\longrightarrow} \int_\Omega \bar{V} : \xi\, dx$, resulting in $\nabla\phi_i \underset{i\to\infty}{\rightharpoonup} \bar{V}$ in $L^4(\Omega, M_2(\mathbb{R}))$. By uniqueness of the weak limit in $L^4(\Omega, M_2(\mathbb{R}))$, we get that $\nabla\bar{\phi} = \bar{V} \in L^\infty(\Omega, M_2(\mathbb{R}))$.

Now, omitting the index i for the sake of readability,

$$\det V_i = \det x_i + \det \nabla\phi_i + x_{11}\frac{\partial \phi_2}{\partial x_2} + x_{22}\frac{\partial \phi_1}{\partial x_1} - x_{12}\frac{\partial \phi_2}{\partial x_1} - x_{21}\frac{\partial \phi_1}{\partial x_2}.$$

Let us then set $d_i = \det x_i + x_{11}\frac{\partial \phi_2}{\partial x_2} + x_{22}\frac{\partial \phi_1}{\partial x_1} - x_{12}\frac{\partial \phi_2}{\partial x_1} - x_{21}\frac{\partial \phi_1}{\partial x_2}$, which is an element of $L^2(\Omega)$ from the generalised Hölder's inequality. By invoking this latter inequality combined with the property of equivalence of norms in finite dimension (in particular, there exist $(c_1, c_2) \in \mathbb{R}^2$ such that $\forall x \in M_2(\mathbb{R}) \sim \mathbb{R}^4$, $c_1\|x\|_2 \leq \|x\|_4 \leq c_2\|x\|_2$. Similarly, there exist $(c_3, c_4) \in \mathbb{R}^2$ such that $\forall x \in \mathbb{R}^4$, $c_3\|x\|_4 \leq \|x\|_{\frac{4}{3}} \leq c_4\|x\|_4$), it yields:

$$\begin{aligned}
\|d_i\|_{L^2(\Omega)} &\leq \|x_{11}\|_{L^4(\Omega)}\|\frac{\partial \phi_2}{\partial x_2}\|_{L^4(\Omega)} + \|x_{22}\|_{L^4(\Omega)}\|\frac{\partial \phi_1}{\partial x_1}\|_{L^4(\Omega)} + \|x_{12}\|_{L^4(\Omega)}\|\frac{\partial \phi_2}{\partial x_1}\|_{L^4(\Omega)} \\
&\quad + \|x_{21}\|_{L^4(\Omega)}\|\frac{\partial \phi_1}{\partial x_2}\|_{L^4(\Omega)} + \|x_{11}\|_{L^4(\Omega)}\|x_{22}\|_{L^4(\Omega)} + \|x_{12}\|_{L^4(\Omega)}\|x_{21}\|_{L^4(\Omega)}, \\
&\leq (\|x_{11}\|^4_{L^4(\Omega)} + \|x_{22}\|^4_{L^4(\Omega)} + \|x_{12}\|^4_{L^4(\Omega)} + \|x_{21}\|^4_{L^4(\Omega)})^{\frac{1}{4}} \\
&\quad (\|\frac{\partial \phi_1}{\partial x_1}\|^{\frac{4}{3}}_{L^4(\Omega)} + \|\frac{\partial \phi_1}{\partial x_2}\|^{\frac{4}{3}}_{L^4(\Omega)} + \|\frac{\partial \phi_2}{\partial x_1}\|^{\frac{4}{3}}_{L^4(\Omega)} + \|\frac{\partial \phi_2}{\partial x_2}\|^{\frac{4}{3}}_{L^4(\Omega)})^{\frac{3}{4}} \\
&\quad + (\|x_{11}\|^4_{L^4(\Omega)} + \|x_{12}\|^4_{L^4(\Omega)})^{\frac{1}{4}}(\|x_{22}\|^{\frac{4}{3}}_{L^4(\Omega)} + \|x_{21}\|^{\frac{4}{3}}_{L^4(\Omega)})^{\frac{3}{4}}, \\
&\leq c_4c_2^2\|x_i\|_{L^4(\Omega,M_2(\mathbb{R}))}\|\nabla\phi_i\|_{L^4(\Omega,M_2(\mathbb{R}))} + c_4c_2^2\|x_i\|^2_{L^4(\Omega,M_2(\mathbb{R}))},
\end{aligned}$$

$\|\nabla\phi_i\|_{L^4(\Omega,M_2(\mathbb{R}))}$ being uniformly bounded with respect to i, yielding $d_i \underset{i\to\infty}{\longrightarrow} 0$ strongly in $L^2(\Omega)$.

Coming back to the study of the weak limit of $\det V_i \in L^2(\Omega)$, $\forall \xi \in L^2(\Omega)$,

$$\int_\Omega \det V_i\, \xi\, dx = \int_\Omega \det \nabla\phi_i\, \xi\, dx + \int_\Omega d_i\, \xi\, dx,$$

with $|\int_\Omega d_i\, \xi\, dx| \leq \|d_i\|_{L^2(\Omega)}\|\xi\|_{L^2(\Omega)}$ from Cauchy-Schwarz inequality and $\det\nabla\phi_i \underset{i\to\infty}{\rightharpoonup} \det\nabla\bar{\phi}$ in $L^2(\Omega)$, leading to $\forall \xi \in L^2(\Omega)$, $\int_\Omega \bar{\delta}\, \xi\, dx = \int_\Omega \det \nabla\bar{\phi}\, \xi\, dx$, and subsequently, $\bar{\delta} = \det \nabla\bar{\phi}$ by uniqueness of the weak limit in $L^2(\Omega)$. Function $\mathcal{W}_{Op}$ is continuous and convex. If $V_n \to \bar{V}$ in $L^4(\Omega, M_2(\mathbb{R}))$, one can extract a subsequence still denoted by V_n such that $V_n \underset{n\to\infty}{\longrightarrow} \bar{V}$ almost everywhere in Ω, and if $\kappa_n \underset{n\to\infty}{\longrightarrow} \bar{\kappa}$ strongly in $L^2(\Omega)$, one can extract a (common) subsequence still denoted by (κ_n) such that $\kappa_n \underset{n\to\infty}{\longrightarrow} \bar{\kappa}$ almost everywhere in Ω. Then by applying Fatou's lemma and taking a common subsequence, one gets that

$$\int_\Omega \mathcal{W}_{Op}(\bar{V}, \bar{\kappa}) \leq \liminf_{n\to,\infty} \int_\Omega \mathcal{W}_{Op}(V_n, \kappa_n)\, dx.$$

As $\mathcal{W}_{Op}$ is convex, so is $(\cdot,\cdot) \mapsto \int_\Omega \mathcal{W}_{Op}(\cdot,\cdot)\, dx$, and [4, Corollary III.8] holds so that $\int_\Omega \mathcal{W}_{Op}(\cdot,\cdot)\, dx$ is also weakly lower semi-continuous on $L^4(\Omega, M_2(\mathbb{R})) \times$

$L^2(\Omega)$ and $\int_\Omega W_{Op}(\nabla\bar{\phi}, \det\nabla\bar{\phi})\,dx \leq \liminf_{i\to\infty} \int_\Omega W_{Op}(V_i, \det V_i)\,dx$, resulting in particular in $\det\nabla\bar{\phi} \in L^\infty(\Omega) > 0$ a.e. and $\frac{1}{\det\nabla\bar{\phi}} \in L^{10}(\Omega)$. Since $\bar{\phi} \in \mathrm{Id} + W_0^{1,\infty}(\Omega,\mathbb{R}^2)$ by continuity of the trace operator (Gagliardo-Nirenberg inequality ([4, p.197, Example 3]) enables one to show that $\bar{\phi} \in W^{1,\infty}(\Omega,\mathbb{R}^2)$), $\det\nabla\bar{\phi} > 0$ a.e. and $\frac{1}{\det\nabla\bar{\phi}} \in L^{10}(\Omega)$, we have $\forall q \in]2,11]$,

$$\int_\Omega \|(\nabla\bar{\phi})^{-1}\|^q\,dx = \int_\Omega \|\nabla\bar{\phi}\|^q(\det\nabla\bar{\phi})^{1-q}\,dx \leq \alpha^q \|\frac{1}{\det\nabla\bar{\phi}}\|^{q-1}_{L^{q-1}(\Omega)} < \infty.$$

By Ball's results [3], we deduce that $\bar{\phi}$ is a homeomorphism of $\bar{\Omega}$ onto $\bar{\Omega}$ and $\bar{\phi}^{-1}$ belongs to $W^{1,q}(\Omega,\mathbb{R}^2)$. The matrix of weak derivatives of $\bar{\phi}^{-1}$ is given by $\nabla(\bar{\phi}^{-1}) = (\nabla\bar{\phi})^{-1}(\bar{\phi}^{-1})$ almost everywhere in Ω. (In fact, $\bar{\phi}$ is a homeomorphism of $\bar{\Omega}$ onto $\bar{\Omega}$). At this stage, $\bar{\phi} \in \mathrm{Id} + W_0^{1,\infty}(\Omega,\mathbb{R}^2)$ with $\|\nabla\bar{\phi}\|_{L^\infty(\Omega,M_2(\mathbb{R}))} \leq \alpha$ is such that $\det\nabla\bar{\phi} > 0$ a.e., $\frac{1}{\det\nabla\bar{\phi}} \in L^{10}(\Omega)$ and $\bar{\phi}^{-1} \in W^{1,q}(\Omega,\mathbb{R}^2)$, $\forall q \in]2,11]$.

By invoking similar arguments to those previously used, there exist a subsequence still denoted by (W_i) and $\bar{W}$ such that $W_i \underset{i\to\infty}{\overset{*}{\rightharpoonup}} \bar{W}$ in $L^\infty(\Omega, M_2(\mathbb{R}))$ and $W_i \underset{i\to\infty}{\rightharpoonup} \bar{W}$ in $L^4(\Omega, M_2(\mathbb{R}))$.

Also, $\mathbb{1}_{\{\|\cdot\|_{L^\infty(\Omega,M_2(\mathbb{R}))}\leq\beta\}}(\bar{W}) \leq \liminf_{i\to\infty} \mathbb{1}_{\{\|\cdot\|_{L^\infty(\Omega,M_2(\mathbb{R}))}\leq\beta\}}(W_i)$ thanks to the w.l.s.c. of the L^∞-norm. As $y_i = W_i - \nabla\psi_i$ strongly converges to 0 in $L^4(\Omega, M_2(\mathbb{R}))$ and as $\psi_i \underset{i\to\infty}{\rightharpoonup} \bar{\psi}$ in $L^4(\Omega, M_2(\mathbb{R}))$, it yields $\nabla\bar{\psi} = \bar{W} \in L^\infty(\Omega, M_2(\mathbb{R}))$ a.e..

Now, recall that according to Gagliardo-Nirenberg inequalities ([4, p.197, Example 3]), whenever $1 \leq q \leq p \leq \infty$ and $r > N$ (in the general case where $\Omega \subset \mathbb{R}^N$ is a bounded open set with smooth boundary),

$$\|u\|_{L^p} \leq C\,\|u\|_{L^q}^{1-a}\,\|u\|_{W^{1,r}}^a,\ \forall u \in W^{1,r}(\Omega),$$

with $a = \frac{\frac{1}{q}-\frac{1}{p}}{\frac{1}{q}+\frac{1}{N}-\frac{1}{r}}$. In our case, $\bar{\psi}$ being an element of $\mathrm{Id} + W_0^{1,4}(\Omega,\mathbb{R}^2)$, taking $\begin{cases} p = \infty \\ q = 4 \\ r = 4 \end{cases}$ (thus yielding $a = \frac{1}{2}$), $\|\bar{\psi}\|_{L^\infty(\Omega,\mathbb{R}^2)} \leq C\,\|\bar{\psi}\|^{\frac{1}{2}}_{L^4(\Omega,\mathbb{R}^2)}\,\|\bar{\psi}\|^{\frac{1}{2}}_{W^{1,4}(\Omega,\mathbb{R}^2)}$, showing in the end that $\bar{\psi}$ is an element of $\mathrm{Id} + W_0^{1,\infty}(\Omega,\mathbb{R}^2)$ with $\|\nabla\bar{\psi}\|_{L^\infty(\Omega,M_2(\mathbb{R}))} \leq \beta$.

The next part of the proof is devoted to the handling of the term $\|\psi \circ \phi - \mathrm{Id}\|^2_{L^2(\Omega,\mathbb{R}^2)}$ and in particular, to the meaning given to it to ensure that the composition makes sense. In that purpose, we use the embedding result [4, Theorem IX.2] and reformulate the L^2-coupling $\|\psi \circ \phi - \mathrm{Id}\|^2_{L^2(\Omega,\mathbb{R}^2)}$ into $\|P\psi \circ \phi - \mathrm{Id}\|^2_{L^2(\Omega,\mathbb{R}^2)}$, P being the operator defined in [4, Theorem IX.2], guaranteeing thus the well-posed nature of this component. We also need to introduce some suitable functional spaces (we restrict ourselves to their presentation in dimension 2): Ω being a bounded open set of $\mathbb{R}^2$, $\mathcal{C}^0(\Omega)$ denotes the space of continuous functions on Ω, while $\mathcal{C}^0_b(\Omega)$ is the subset of $\mathcal{C}^0(\Omega)$ consisting of those functions which are bounded and uniformly continuous on Ω. By endowing this subspace with the norm $\|\varphi\|_{\mathcal{C}^0_b(\Omega)} = \sup_{x\in\Omega} |\varphi(x)|$, it makes it a Banach space. The definition of the next functional space, $\mathcal{C}_b^{0,\lambda}(\Omega)$, can be found in [9] and is complemented by a Sobolev continuous embedding theorem. Theorem 2.31 from [9] then states that $0 < \lambda \leq \frac{1}{2}$ implies that $W^{1,4}(\mathbb{R}^2) \hookrightarrow \mathcal{C}_b^{0,\lambda}(\mathbb{R}^2)$ and in particular, $W^{1,4}(\mathbb{R}^2) \hookrightarrow \mathcal{C}_b^{0,\frac{1}{2}}(\mathbb{R}^2)$ Returning to the matter at hand, we observe first that $P\psi_i \circ \phi_i$ strongly converges to Id in $L^2(\Omega, \mathbb{R}^2)$, so up to a subsequence pointwise almost everywhere. Classical Sobolev embedding theorems claim that $W^{1,4}(\Omega, \mathbb{R}^2) \hookrightarrow \mathcal{C}^0(\bar{\Omega}, \mathbb{R}^2)$ is compact and the injection $W^{1,4}(\Omega, \mathbb{R}^2) \hookrightarrow \mathcal{C}^{0,\frac{1}{2}}(\bar{\Omega}, \mathbb{R}^2)$ is continuous. For all $x \in \Omega$, ($|\cdot|$ denoting the Euclidean norm in $\mathbb{R}^2$)

$$\begin{aligned}
|P\psi_i \circ \phi_i(x) - P\bar{\psi} \circ \bar{\phi}(x)| &= |P\psi_i \circ \phi_i(x) - P\psi_i \circ \bar{\phi}(x) + P\psi_i \circ \bar{\phi}(x) - P\bar{\psi} \circ \bar{\phi}(x)|, \\
&\leq |P\psi_i \circ \phi_i(x) - P\psi_i \circ \bar{\phi}(x)| + |P\psi_i \circ \bar{\phi}(x) - P\bar{\psi} \circ \bar{\phi}(x)|, \\
&\leq \|P\psi_i\|_{0,\frac{1}{2}} \|\phi_i - \bar{\phi}\|^{\frac{1}{2}}_{\mathcal{C}^0(\bar{\Omega},\mathbb{R}^2)} + \sup_{x \in \Omega} |\psi_i(x) - \bar{\psi}(x)|,
\end{aligned}$$

since $\bar{\phi}$ is a homeomorphism from Ω onto Ω and $P\psi_i|_\Omega = \psi_i$. Then

$$\begin{aligned}
&|P\psi_i \circ \phi_i(x) - P\bar{\psi} \circ \bar{\phi}(x)| \\
&\leq C\|P\psi_i\|_{W^{1,4}(\mathbb{R}^2,\mathbb{R}^2)} \|\phi_i - \bar{\phi}\|^{\frac{1}{2}}_{\mathcal{C}^0(\bar{\Omega},\mathbb{R}^2)} + \sup_{x \in \Omega} |\psi_i(x) - \bar{\psi}(x)|, \\
&\leq C\|\psi_i\|_{W^{1,4}(\Omega,\mathbb{R}^2)} \|\phi_i - \bar{\phi}\|^{\frac{1}{2}}_{\mathcal{C}^0(\bar{\Omega},\mathbb{R}^2)} + \|\psi_i - \bar{\psi}\|_{\mathcal{C}^0(\bar{\Omega},\mathbb{R}^2)}
\end{aligned}$$

from [4, Theorem IX.2], with $C > 0$ a constant that may change line to line, and ψ_i being uniformly bounded with respect to i. By virtue of Sobolev compact embedding injections, the right quantity tends to 0 when i tends to ∞, yielding $\left| \begin{array}{ll} P\psi_i \circ \phi_i \underset{i \to \infty}{\longrightarrow} \text{Id in } L^2(\Omega, \mathbb{R}^2) & \text{, on the one hand} \\ \forall x \in \Omega,\ P\psi_i \circ \phi_i(x) \underset{i \to \infty}{\longrightarrow} P\bar{\psi} \circ \bar{\phi}(x) & \text{, on the other hand} \end{array} \right.$. In the end, $\forall x \in \Omega$, $P\bar{\psi} \circ \bar{\phi}(x) = x$ and so $\forall y \in \Omega$, $P\bar{\psi}(y) = \bar{\phi}^{-1}(y)$ since $\bar{\phi}$ is a bi-Hölder homeomorphism of Ω to Ω and thus $\forall y \in \Omega$, $\bar{\psi} = \bar{\phi}^{-1}$ since $P\bar{\psi} = \bar{\psi}$ on Ω. Since $\bar{\phi}^{-1} \in W^{1,q}(\Omega, \mathbb{R}^2)$ with $q \in]2, 11]$, we have then $\nabla\bar{\psi} = \bar{W} = \nabla\bar{\phi}^{-1} \in L^\infty(\Omega, M_2(\mathbb{R}))$. The generalised Poincaré inequality combined with Gagliardo-Nirenberg inequalities enables one to conclude that $\bar{\phi}^{-1} \in W^{1,\infty}(\Omega, \mathbb{R}^2)$. Additionally, for almost every x in Ω, $\nabla\bar{\phi}^{-1} = (\nabla\bar{\phi})^{-1}(\bar{\phi}^{-1})$. From the above, $\det(\nabla\bar{\phi})^{-1} > 0$ *a.e.*. Let $\mathcal{N} \subset \Omega$ be such that $\text{meas}(\mathcal{N}) = 0$ and for all $x \in \Omega \setminus \mathcal{N}$, $\det(\nabla\bar{\phi})^{-1}(x) > 0$. Let now $\mathcal{N}'$ be such that $\mathcal{N}' = \bar{\phi}(\mathcal{N})$. Then $\text{meas}(\mathcal{N}') = 0$ since $\bar{\phi}$ is a Lipschitz map (and thus for every measurable set E, $\text{meas}(\bar{\phi}(E)) \leq C'\text{meas}(E)$, C' being a constant depending only on the dimension and on the Lipschitz constant of $\bar{\varphi}$) and for every $y \notin \mathcal{N}'$,

$$\det(\nabla\bar{\phi})^{-1}(\bar{\phi}^{-1})(y) = \det(\nabla\bar{\phi})^{-1}(x)$$

with $x \in \Omega \setminus \mathcal{N}$, yielding $\det \nabla\bar{\psi} = \det \nabla\bar{\phi}^{-1} > 0$ *a.e.*. In the end, $\bar{\psi}$ is an element of $\text{Id} + W_0^{1,\infty}(\Omega, \mathbb{R}^2)$ with $\det \nabla\bar{\psi} > 0$ *a.e.*.

The same kind of reasoning applies to the term $\|P\zeta_{k-1}^{-1} \circ \phi_i - \varphi_i\|^2_{L^2(\Omega,\mathbb{R}^2)}$ (the definition of ζ_{k-1} can be found p. 4).

First, $P\zeta_{k-1}^{-1} \circ \phi_i - \varphi_i$ strongly converges to 0 in $L^2(\Omega, \mathbb{R}^2)$, so weakly in $L^2(\Omega, \mathbb{R}^2)$. Let us now show that $P\zeta_{k-1}^{-1} \circ \phi_i$ strongly converges to $P\zeta_{k-1}^{-1} \circ \bar{\phi}$ in $L^2(\Omega, \mathbb{R}^2)$ so pointwise almost everywhere up to a subsequence.

$$\int_\Omega |P\zeta_{k-1}^{-1} \circ \phi_i - P\zeta_{k-1}^{-1} \circ \bar{\phi}|^2 \, dx \leq \|P\zeta_{k-1}^{-1}\|_{0,\frac{1}{2}}^2 \, \|\phi_i - \bar{\phi}\|_{L^2(\Omega,\mathbb{R}^2)} \underset{i \to \infty}{\longrightarrow} 0.$$

Strong convergence in $L^2(\Omega, \mathbb{R}^2)$ implying weak convergence in $L^2(\Omega, \mathbb{R}^2)$, it yields, $\forall u \in L^2(\Omega, \mathbb{R}^2)$ and denoting by $\langle \cdot, \cdot \rangle$ the standard Euclidean scalar product in $\mathbb{R}^2$,

$$\int_\Omega \langle \varphi_i, u \rangle \, dx = \int_\Omega \langle \varphi_i - P\zeta_{k-1}^{-1} \circ \phi_i, u \rangle \, dx + \int_\Omega \langle P\zeta_{k-1}^{-1} \circ \phi_i, u \rangle \, dx,$$

$$\underset{i \to \infty}{\longrightarrow} 0 + \int_\Omega \langle P\zeta_{k-1}^{-1} \circ \bar{\phi}, u \rangle \, dx.$$

By uniqueness of the weak limit in $L^2(\Omega, \mathbb{R}^2)$, $\bar{\varphi} = P\zeta_{k-1}^{-1} \circ \bar{\phi}$ a.e. in Ω. As $\bar{\phi}$ is a bi-Lipschitz homeomorphism from Ω onto Ω, $\bar{\varphi} = \zeta_{k-1}^{-1} \circ \bar{\phi} \in W^{1,\infty}(\Omega, \mathbb{R}^2)$ a.e. in Ω. By taking the continuous representative of $\bar{\varphi}$, we have for all $x \in \Omega$, $\bar{\phi}(x) = \zeta_{k-1} \circ \bar{\varphi}(x)$.

It remains to handle the fidelity component. We assume slightly more smoothness on $\bar{T}_k$ as previously mentioned, *i.e.* $\bar{T}_k \in L^4(\Omega)$, which is not too restrictive in the context of image processing, and $\bar{T}_k \geq 0$ *a.e.*. For theoretical purposes, we assume that $\bar{R}_k$ is such that its essential support is included in $\tilde{\Omega} \subset\subset \Omega \subset\subset \Omega_0$, Ω_0 being a bigger open bounded set containing Ω and $\tilde{\Omega}$ being a bounded open set of Ω. We then consider an extension by 0 of $\bar{R}_k$ on $\mathbb{R}^2$ following [10, Theorem 1, p. 183]. Also, it is assumed that $\bar{R}_k \geq 0$ *a.e.*.

Recall that $\bar{T}_k$ is assumed to be an element of $L^4(\Omega)$ while $\bar{R}_k$ is in $L^2(\Omega)$. Let us show that $(\bar{T}_k - \bar{R}_k \circ \psi_i)^2 \det \nabla \psi_i \in L^1(\Omega) \underset{i \to \infty}{\rightharpoonup} (\bar{T}_k - \bar{R}_k \circ \bar{\psi})^2 \det \nabla \bar{\psi}$ in $L^1(\Omega)$ or equivalently, (see [4, Theorem IV.14]), since $(L^1)'(\Omega) = L^\infty(\Omega)$, that $\forall \xi \in L^\infty(\Omega)$, $\int_\Omega ((\bar{T}_k - \bar{R}_k \circ \psi_i)^2 \det \nabla \psi_i - (\bar{T}_k - \bar{R}_k \circ \bar{\psi})^2 \det \nabla \bar{\psi}) \xi \, dx \underset{i \to \infty}{\longrightarrow} 0$. Note that $(\bar{T}_k - \bar{R}_k \circ \bar{\psi})^2 \det \nabla \bar{\psi} \in L^1(\Omega)$ since $\det \nabla \bar{\psi} \in L^\infty(\Omega)$, $\bar{T}_k \in L^2(\Omega)$ and $\bar{R}_k \circ \bar{\psi} \in BV(\Omega) \subset L^2(\Omega)$, $\bar{\psi}$ being a bi-Lipschitz homeomorphism from Ω to Ω (even from $\tilde{\Omega}$ to $\tilde{\Omega}$) (see [1, Theorem 3.16]).

For the sake of simplicity, we denote by $d_i = \det \nabla \psi_i$, $d = \det \nabla \bar{\psi}$, $\rho = \bar{R}_k$.

First, $\int_\Omega \bar{T}_k^2 (d_i - \bar{d}) \xi \, dx \underset{i \to \infty}{\longrightarrow} 0$ since $d_i \underset{i \to \infty}{\rightharpoonup} \bar{d}$ in $L^2(\Omega)$, $\bar{T}_k^2 \in L^2(\Omega)$ and $\xi \in L^\infty(\Omega)$. Let us now consider $\int_\Omega ((\rho(\psi_i))^2 d_i - (\rho(\bar{\psi}))^2 \bar{d}) \xi \, dx$. ρ being $L^2(\Omega_0)$, there exists a sequence $\zeta_n \in \mathcal{D}(\Omega_0)$ such that (ζ_n) converges strongly to ρ in $L^2(\Omega_0)$. Let $\epsilon > 0$. We can pick a N such that $\|\zeta_N\|_{L^2(\Omega_0)} \leq 1 + \|\rho\|_{L^2(\Omega_0)}$ and $\|\rho - \zeta_N\|_{L^2(\Omega_0)} \leq \frac{\epsilon}{4\|\xi\|_{L^\infty(\Omega)}(2\|\rho\|_{L^2(\Omega_0)}+1)}$.

$$\int_\Omega ((\rho(\psi_i))^2 d_i - (\rho(\bar{\psi}))^2 \bar{d}) \xi \, dx = \int_\Omega \big[(((\rho(\psi_i))^2 - (\zeta_N(\psi_i))^2 + (\zeta_N(\psi_i))^2$$
$$- (\zeta_N(\bar{\psi}))^2) \, d_i \, \xi + (\zeta_N(\bar{\psi}))^2 d_i \, \xi - (\zeta_N(\bar{\psi}))^2 \bar{d} \, \xi + (\zeta_N(\bar{\psi}))^2 \bar{d} \, \xi - (\rho(\bar{\psi}))^2 \bar{d} \, \xi \big] \, dx.$$

We now detail each component.

- First component:

$$\int_\Omega ((\rho(\psi_i))^2 - (\zeta_N(\psi_i))^2) d_i \xi \, dx \leq \|\xi\|_{L^\infty(\Omega)} \int_\Omega (\rho(\psi_i)^2 - \zeta_N(\psi_i)^2) d_i \, dx,$$
$$\leq \|\xi\|_{L^\infty(\Omega)} \int_{\mathbb{R}^2} |\rho(y)^2 - \zeta_N(y)^2| \, dy,$$

using [5, Theorem 6, Remark 8 and Proposition 9], thus yielding

$$\int_\Omega ((\rho(\psi_i))^2 - (\zeta_N(\psi_i))^2) d_i \xi \, dx \leq \|\xi\|_{L^\infty(\Omega)} \|\rho - \zeta_N\|_{L^2(\Omega_0)} \|\rho + \zeta_N\|_{L^2(\Omega_0)},$$
$$\leq \|\xi\|_{L^\infty(\Omega)} (2\|\rho\|_{L^2(\Omega_0)} + 1) \|\rho - \zeta_N\|_{L^2(\Omega_0)},$$

bounded by $\frac{\epsilon}{4}$.
- Second component: $\int_\Omega ((\zeta_N(\psi_i))^2 - (\zeta_N(\bar{\psi}))^2) \, d_i \, \xi \, dx = \int_\Omega ((\zeta_N(\psi_i)) - (\zeta_N(\bar{\psi})))(\zeta_N(\psi_i) + \zeta_N(\bar{\psi})) \, d_i \, \xi \, dx$. ζ_N is Lipschitz continuous with some constant L depending on $N = N(\epsilon)$ and ζ_N is bounded (we denote by $L' = L'(N(\epsilon))$ such a bound) while both ψ_i and $\bar{\psi}$ are continuous due to the (compact) embedding $W^{1,4}(\Omega, \mathbb{R}^2) \hookrightarrow \mathcal{C}^0(\bar{\Omega}, \mathbb{R}^2)$. Consequently,

$$\int_\Omega ((\zeta_N(\psi_i))^2 - (\zeta_N(\bar{\psi}))^2) \, d_i \, \xi \, dx \leq 2\|\xi\|_{L^\infty(\Omega)} \int_\Omega LL' |\psi_i - \bar{\psi}| \, d_i \, dx,$$
$$\leq 2LL' \|\xi\|_{L^\infty(\Omega)} \|\psi_i - \bar{\psi}\|_{L^2(\Omega, \mathbb{R}^2)} \|d_i\|_{L^2(\Omega)},$$
$$\leq 2ELL' \, \|\xi\|_{L^\infty(\Omega)} \|\psi_i - \bar{\psi}\|_{L^2(\Omega, \mathbb{R}^2)},$$

by applying Cauchy-Schwarz inequality and using the fact that d_i is uniformly bounded in $L^2(\Omega)$ by some E. Due to the strong convergence of ψ_i to $\bar{\psi}$ in $L^2(\Omega, \mathbb{R}^2)$, one can always find I_1 such that $(i \geq I_1 \Rightarrow \|\psi_i - \bar{\psi}\|_{L^2(\Omega, \mathbb{R}^2)} \leq \frac{\epsilon}{8ELL' \|\xi\|_{L^\infty(\Omega)}})$.
- Third component: $\zeta_N(\bar{\psi})$ is bounded by $L' = (L(N(\epsilon))$ and $\xi \in L^\infty(\Omega)$ so that $(\zeta_N(\bar{\psi}))^2 \xi \in L^2(\Omega)$. As $d_i \underset{i \to \infty}{\rightharpoonup} \bar{d}$ in $L^2(\Omega)$, one can always find I_2 such that $(i \geq I_2 \Rightarrow \int_\Omega (\zeta_N(\bar{\psi}))^2 (d_i - \bar{d}) \, \xi \, dx \leq \frac{\epsilon}{4})$.
- Last component: $\int_\Omega ((\zeta_N(\bar{\psi}))^2 - (\rho(\bar{\psi}))^2 \bar{d} \, \xi \, dx \leq \|\xi\|_{L^\infty(\Omega)} \int_\Omega |(\zeta_N(y))^2 - (\rho(y))^2| \, dy$, since $\bar{\psi}$ is a bi-Lipschitz homeomorphism from Ω to Ω. Then $\int_\Omega ((\zeta_N(\bar{\psi}))^2 - (\rho(\bar{\psi}))^2 \bar{d} \, \xi \, dx \leq \|\xi\|_{L^\infty(\Omega)} \|\zeta^N - \rho\|_{L^2(\Omega_0)} (2\|\rho\|_{L^2(\Omega_0)} + 1)$, bounded by $\frac{\epsilon}{4}$.

By combining all the above inequalities, we get the expected intermediate result.

It remains to study $\int_\Omega (\bar{T}_k \, (\rho(\psi_i) d_i - \rho(\bar{\psi}) \bar{d}) \, \xi) \, dx = \int_\Omega (\bar{T}_k (\rho(\psi_i) - \zeta_N(\psi_i) + \zeta_N(\psi_i) - \zeta_N(\bar{\psi})) d_i \, \xi + \bar{T}_k \zeta_N(\bar{\psi}) d_i \, \xi - \bar{T}_k \zeta_N(\bar{\psi}) \bar{d} \, \xi + \bar{T}_k \zeta_N(\bar{\psi}) \bar{d} \, \xi - \bar{T}_k \rho(\bar{\psi}) \bar{d} \, \xi) \, dx$.

- Last component: Using Cauchy-Schwarz inequality and the fact that $L^\infty(\Omega) \ni \bar{d} > 0$ *a.e*,

$$\begin{aligned}
&\int_\Omega (\zeta_N(\bar{\psi}) - \rho(\bar{\psi}))\bar{T}_k \bar{d}\,\xi\, dx \\
&\leq \|\xi\|_{L^\infty(\Omega)} \|\sqrt{\bar{d}}\|_{L^\infty(\Omega)} \|\bar{T}_k\|_{L^2(\Omega)} \left(\int_\Omega |\zeta_N(\bar{\psi}) - \rho(\bar{\psi})|^2 \bar{d}\, dx\right)^{\frac{1}{2}}, \\
&\leq \|\xi\|_{L^\infty(\Omega)} \|\sqrt{\bar{d}}\|_{L^\infty(\Omega)} \|\bar{T}_k\|_{L^2(\Omega)} \|\zeta_N - \rho\|_{L^2(\Omega_0)},
\end{aligned}$$

since $\bar{\psi}$ is a bi-Lipschitz homeomorphism from Ω toΩ.

- Third component: $\int_\Omega \bar{T}_k \zeta_N(\bar{\psi})\xi(d_i - \bar{d})\, dx \underset{i\to\infty}{\longrightarrow} 0$ since $d_i \underset{i\to\infty}{\rightharpoonup} \bar{d}$ in $L^2(\Omega)$.
- Second component:

$$\begin{aligned}
\int_\Omega \bar{T}_k(\zeta_N(\psi_i) - \zeta_N(\bar{\psi}))\, d_i\, \xi\, dx &\leq \int_\Omega \bar{T}_k\, L|\psi_i - \bar{\psi}| d_i\, \xi\, dx, \\
&\leq L\|\xi\|_{L^\infty(\Omega)} \|\psi_i - \bar{\psi}\|_{L^4(\Omega,\mathbb{R}^2)} \|\bar{T}_k d_i\|_{L^{\frac{4}{3}}(\Omega)},
\end{aligned}$$

from Hölder's inequality and since ψ_i strongly converges to $\bar{\psi}$ in $L^4(\Omega, \mathbb{R}^2)$.
- First component:

$$\begin{aligned}
&\int_\Omega \bar{T}_k(\rho(\psi_i) - \zeta_N(\psi_i)) d_i\, \xi\, dx \\
&\leq \|\xi\|_{L^\infty(\Omega)}\, \|\bar{T}_k \sqrt{d_i}\|_{L^2(\Omega)} \int_\Omega |\rho(\psi_i) - \zeta_N(\psi_i)|^2\, d_i\, dx, \\
&\leq \|\xi\|_{L^\infty(\Omega)}\, \|\bar{T}_k \sqrt{d_i}\|_{L^2(\Omega)} \|\rho - \zeta_N\|_{L^2(\Omega_0)},
\end{aligned}$$

still from [5, Theorem 6] and the intermediate conclusion is straightforward. We thus have proved that $\forall \xi \in L^\infty(\Omega)$, $\int_\Omega (\bar{T}_k - \bar{R}_k \circ \psi_i)^2 \mathrm{det}\nabla\psi_i\, \xi\, dx \underset{i\to\infty}{\longrightarrow} \int_\Omega (\bar{T}_k - \bar{R}_k \circ \bar{\psi})^2 \mathrm{det}\nabla\bar{\psi}\, \xi\, dx$. In particular, for $\xi = 1$ on Ω, we deduce that $\int_\Omega (\bar{T}_k - \bar{R}_k \circ \psi_i)^2 \mathrm{det}\nabla\psi_i\, dx \underset{i\to\infty}{\longrightarrow} \int_\Omega (\bar{T}_k - \bar{R}_k \circ \bar{\psi})^2 \mathrm{det}\nabla\bar{\psi}\, dx$, meaning that for i sufficiently large, $\|(\bar{T}_k - \bar{R}_k \circ \psi_i)^2 \mathrm{det}\nabla\psi_i\|_{L^1(\Omega)}$ is uniformly bounded, and subsequently that for i sufficiently large, $(\bar{T}_k - \bar{R}_k \circ \psi_i)^2 \mathrm{det}\nabla\psi_i$ is $L^1(\Omega)$. It yields $\int_\Omega (\bar{T}_k - \bar{R}_k \circ \bar{\psi})^2 \mathrm{det}\nabla\bar{\psi}\, dx = \int_\Omega (\bar{T}_k \circ \bar{\phi} - \bar{R}_k)^2\, dy \leq \liminf_{i\to\infty} \|(\bar{T}_k - \bar{R}_k \circ \psi_i)^2 \mathrm{det}\nabla\psi_i\|_{L^1(\Omega)}$ since $\bar{\psi}$ is a bi-Lipschitz homeomorphism and $\bar{\phi} = \bar{\psi}^{-1}$.

3 Conclusion

In this paper, we have proposed a theoretical asymptotic result sustaining the idea that for such non convex non linear minimisation problems, introducing a decoupled problem involving splitting variables constitutes a good compromise between compliance with the original problem and numerical tractability.

References

1. Ambrosio, L., Fusco, N., Pallara, D.: Functions of Bounded Variation and Free Discontinuity Problems. Oxford University Press, Oxford (2000)
2. Athavale, P., Xu, R., Radau, P., Nachman, A., Wright, G.A.: Multiscale properties of weighted total variation flow with applications to denoising and registration. Med. Image Anal. **23**(1), 28–42 (2015)
3. Ball, J.M.: Global invertibility of Sobolev functions and the interpenetration of matter. P. Roy. Soc. Edin. A **88**(3–4), 315–328 (1981)
4. Brezis, H.: Analyse fonctionnelle. Dunod Paris (2005)
5. Burger, M., Modersitzki, J., Suhr, S.: A Nonlinear Variational Approach to Motion-Corrected Reconstruction of Density Images. arXiv e-prints arXiv:1511.09048 (2015)
6. Ciarlet, P.: Three-Dimensional Elasticity. Mathematical Elasticity. Elsevier Science (1994)
7. Debroux, N., Le Guyader, C., Vese, L.: A Multiscale Deformation Representation. SIAM J. Imaging Sci. (2023, accepted for publication)
8. Debroux, N., Le Guyader, C., Vese, L.A.: Multiscale registration. In: Elmoataz, A., Fadili, J., Quéau, Y., Rabin, J., Simon, L. (eds.) Scale Space and Variational Methods in Computer Vision, pp. 115–127. Springer, Cham (2021). https://doi.org/10.1007/978-3-030-75549-2_10
9. Demengel, F., Demengel, G., Erné, R.: Functional Spaces for the Theory of Elliptic Partial Differential Equations. Universitext, Springer, London (2012). https://doi.org/10.1007/978-1-4471-2807-6
10. Evans, L., Gariepy, R.: Measure Theory and Fine Properties of Functions. CRC Press, Boca Raton (1992)
11. Gris, B., Durrleman, S., Trouvé, A.: A sub-Riemannian modular framework for diffeomorphism-based analysis of shape ensembles. SIAM J. Imaging Sci. **11**(1), 802–833 (2018)
12. Lam, K.C., Ng, T.C., Lui, L.M.: Multiscale representation of deformation via Beltrami coefficients. Multiscale Model. Simul. **15**(2), 864–891 (2017)
13. Modersitzki, J.: FAIR: Flexible Algorithms for Image Registration. SIAM (2009)
14. Modin, K., Nachman, A., Rondi, L.: A multiscale theory for image registration and nonlinear inverse problems. Adv. Math. **346**, 1009–1066 (2019)
15. Paquin, D., Levy, D., Schreibmann, E., Xing, L.: Multiscale image registration. Math. Biosci. Eng. **3**(2), 389–418 (2006)
16. Paquin, D., Levy, D., Xing, L.: Hybrid multiscale landmark and deformable image registration. Math. Biosci. Eng. **4**(4), 711–737 (2007)
17. Paquin, D., Levy, D., Xing, L.: Multiscale deformable registration of noisy medical images. Math. Biosci. Eng. **5**(1), 125–144 (2008)
18. Risser, L., Vialard, F.X., Wolz, R., Murgasova, M., Holm, D.D., Rueckert, D.: Simultaneous multi-scale registration using large deformation diffeomorphic metric mapping. IEEE T. Med. Imaging **30**(10), 1746–1759 (2011)
19. Sommer, S., Lauze, F., Nielsen, M., Pennec, X.: Sparse multi-scale diffeomorphic registration: the kernel bundle framework. J. Math. Imaging Vis. **46**, 292–308 (2013)
20. Tadmor, E., Nezzar, S., Vese, L.: A multiscale image representation using hierarchical (BV, L^2) decompositions. Multiscale Model. Simul. **2**(4), 554–579 (2004)

Image Blending with Osmosis

Paul Bungert, Pascal Peter(✉), and Joachim Weickert

Mathematical Image Analysis Group, Faculty of Mathematics and Computer Science, Campus E1.7, Saarland University, 66041Saarbrücken, Germany
{bungert,peter,weickert}@mia.uni-saarland.de

Abstract. Image blending is an integral part of many multi-image applications such as panorama stitching or remote image acquisition processes. In such scenarios, multiple images are connected at predefined boundaries to form a larger image. A convincing transition between these boundaries may be challenging, since each image might have been acquired under different conditions or even by different devices.

We propose the first blending approach based on osmosis filters. These drift-diffusion processes define an image evolution with a non-trivial steady state. For our blending purposes, we explore several ways to compose drift vector fields based on the derivatives of our input images. These vector fields guide the evolution such that the steady state yields a convincing blended result. Our method benefits from the well-founded theoretical results for osmosis, which include useful invariances under multiplicative changes of the colour values. Experiments on real-world data show that this yields better quality than traditional gradient domain blending, especially under challenging illumination conditions.

Keywords: osmosis · blending · drift-diffusion · gradient domain methods

1 Introduction

Image stitching refers to the task of merging multiple images that depict different areas of the same scene or object. It has many practical applications, in particular for panorama photography [12] or various forms of remote image acquisition that create mosaic problems [4,8]. In practice, differing lighting conditions in the individual input images may yield large differences in brightness and contrast. Therefore, naïve stitching as a mosaic of the individual input images often creates undesirable boundary artefacts.

Stitching results without visible seams are the objective of so-called blending methods. Similar problems exist in image editing, which also requires seamless merging of content from different sources. In this related field, osmosis has yielded excellent results [2,23,24]. This class of filters is inspired by physical processes. In its drift–diffusion formulation for visual computing applications, a continuous [19,24] and a discrete [23] theory have been established. Moreover, it has also been successfully used for shadow removal [2,15,24], which suggests

L. Calatroni et al. (Eds.): SSVM 2023, LNCS 14009, pp. 652–664, 2023.
https://doi.org/10.1007/978-3-031-31975-4_50

that it performs well under large brightness differences. Despite these indications that osmosis filtering might also perform well for blending problems, its potential has not been investigated so far. Our goal is to address this problem.

1.1 Our Contributions

We propose an image blending approach that merges several images seamlessly by osmosis. To this end, we use derivative data of the input images to define a suitable drift vector field. It guides the osmosis process such that it converges to the blending result as a steady state.

We introduce multiple ways to construct these drift vector field and demonstrate that even straightforward approaches yield visually highly convincing results. Our evaluations on real world and synthetic data reveal that osmosis blending benefits from its multiplicative invariance, especially on challenging image sets with high contrast differences.

1.2 Related Work

Osmosis models in visual computing go back to Weickert et al. [24]. They have presented a continuous theory along with applications for compact image representations, shadow removal, and seamless image fusion. The continuous theory has been extended by Schmidt [19]. Vogel et al. [23] have established a discrete framework and have proposed a fast implicit solver. These considerations refer to linear osmosis models, which already allow a large degree of flexibility.

On the application side, Parisotto et al. [15] have advocated a nonlinear variant of osmosis for shadow removal. As a new application field for osmosis, Parisotto et al. [16] proposed the fusion of spectral images with linear osmosis, and they also considered nonlinear fusion approaches [14].

Osmosis models have several interesting predecessors and share some conceptual similarities with other methods. A lattice Boltzmann model for halftoning by Hagenburg et al. [9] can be seen as an early nonlinear drift–diffusion process in visual computing. Outside the field of visual computing, Hagenburg et al. [10] advocated osmosis models for improving numerical methods for hyperbolic conservation laws. They used a Markov chain formulation rather than a drift–diffusion model. In statistical physics, drift–diffusion has close ties to the Fokker-Planck equation [18] and by that connections to Langevin formulations and the Beltrami flow [21].

With its ability to "integrate" nonintegrable derivative information in a seamless way, osmosis resembles gradient domain methods. They have been introduced for shape from shading by Frankot and Chelappa [6] and have found numerous applications in computer graphics. These include tone mapping [5] and image editing [17], but also image blending [12], as we will discuss below.

In contrast to osmosis models that are invariant under multiplicative rescalings of the pixel values, gradient domain methods are invariant under additive rescalings. Georgiev's covariant derivative approach for image editing [7] is invariant under multiplicative changes, but has not been applied to image blending so far. Illner and Neunzert [11] have proposed directed diffusion which shares conceptual similarities with osmosis in that it converges to a so-called background image. However, they did not use it for any visual computing application.

Regarding image blending, traditional methods either attempted to find optimal boundaries between the input images in order to minimise artefacts [3] or directly average information between overlapping regions, e.g. with so-called feathering [22] or pyramid blending [1]. The gradient domain method of Levin et al. [12] not only yielded a significant gain in quality over the image domain method, but also has the closest relation to our own osmosis blending. This connection is discussed in detail in Sect. 2.3.

While a full review of blending is beyond the scope of this paper, there are also approaches with watershed segmentation and graph cuts [8] or wavelets [20]. Others focus on compensating colour differences [4], and there are also blending methods based on deep learning [25]. Due to its direct focus on the blending problem itself and conceptual relations, we will focus our comparative evaluation on the gradient domain method of Levin et al. [12].

1.3 Organisation of the Paper

In Sect. 2 we review the theoretical background of osmosis process which constitutes the foundation of our blending. Combining this basic model with multiple different ways to guide the osmosis-driven propagation leads to the blending approaches described in Sect. 3. We compare these methods against each other and gradient domain blending in Sect. 4 and conclude with a discussion and outlook on future work in Sect. 5.

2 Osmosis

2.1 Continuous Model

In our blending setting, we consider nonnegative colour images $\boldsymbol{f} : \Omega \to \mathbb{R}_+^{n_c}$. They map the image domain $\Omega \subset \mathbb{R}^2$ to a positive colour domain $\mathbb{R}_+^{n_c}$ with n_c colour channels ($n_c = 3$ for RGB images, $n_c = 1$ for greyscale). We denote individual colour channels as f_i, i.e. $f_i : \Omega \to \mathbb{R}_+$.

In addition to the initial image $\boldsymbol{f}$, the osmosis process also relies on a given multi-channel *drift vector field* $\boldsymbol{d} : \Omega \to \mathbb{R}^{2n_c}$. For each channel i of the image $\boldsymbol{u}(\boldsymbol{x}, t)$, the osmosis evolution over time t is described by the initial boundary value problem

$$\partial_t u_i = \Delta u_i - \mathbf{div}\,(\boldsymbol{d}_i u_i) \qquad \text{on } \Omega \times (0, \infty), \tag{1}$$

$$u_i(\boldsymbol{x}, 0) = f_i(\boldsymbol{x}) \qquad \text{on } \Omega, \tag{2}$$

$$\boldsymbol{n}^\top (\boldsymbol{\nabla} u_i - \boldsymbol{d}_i u_i) = 0 \qquad \text{on } \partial\Omega \times (0, \infty). \tag{3}$$

The linear partial differential equation (PDE) (1) describes the propagation of grey levels over time t.

The evolution Eq. (1) consists of a diffusion part defined by the Laplace operator $\Delta u_i = \partial_{xx} u_i + \partial_{yy} u_i$ and the drift component $-\mathbf{div}\,(\boldsymbol{d}_i u_i) = -\partial_x d_{i,1} u_i - \partial_y d_{i,2} u_i$. The drift vector field $\boldsymbol{d}_i \in \mathbb{R}^2$ contains two-dimensional drift vectors for the colour channel i, and $d_{i,1}$ and $d_{i,2}$ denote its first and second components. It distinguishes osmosis from a pure diffusion process, which would lead to a flat steady state for $t \to \infty$. Instead, for osmosis, non-flat steady states are possible.

In addition to the image $\boldsymbol{f}$ as an initial condition at time 0 in Eq. (2), we define homogeneous Neumann boundary conditions in Eq. (3). They prevent transport across the image boundaries $\partial\Omega$ with outer normal vector $\boldsymbol{n}$.

2.2 Theoretical Properties

Weickert et al. [24] have shown several characteristic properties of osmosis processes. Let the average colour value μ_{f_i} of a channel f_i be described by

$$\mu_{f_i} = \frac{1}{|\Omega|} \int_\Omega f_i(\boldsymbol{x})\, d\boldsymbol{x}\,. \tag{4}$$

As for diffusion processes, this average is preserved by osmosis for all intermediate results $\boldsymbol{u}(\cdot, t)$, yielding $\mu_{f_i} = \mu_{u_i(\cdot,t)}$. Furthermore, all intermediate results retain their nonnegativity, i.e. $u_i(\boldsymbol{x}, t) \geq 0$ for all $\boldsymbol{x} \in \Omega$, $i \in \{1, ..., n_c\}$, and $t > 0$.

For visual computing purposes, statements on the steady-state for $t \to \infty$ are of particular interest. Let us first consider the so-called *compatible* case for osmosis. Here, the drift vector field $\boldsymbol{d}$ fulfills

$$\boldsymbol{d}_i = \frac{\boldsymbol{\nabla} v_i}{v_i} \tag{5}$$

for a so-called *guidance image* $\boldsymbol{v} : \Omega \to \mathbb{R}_+^{n_c}$. The corresponding drift vector field $\boldsymbol{d}$ is referred to as the *canonical drift vector field* of $\boldsymbol{v}$. The steady state $\boldsymbol{w}(\boldsymbol{x})$ of the osmosis process is given by

$$w_i(\boldsymbol{x}) = \frac{\mu_{f_i}}{\mu_{v_i}} v_i(\boldsymbol{x}) \tag{6}$$

for all channels i. Thus, the osmosis process converges to the guidance image rescaled to the average grey value of the initial image.

Note here that the definition of the canonical drift vector field in Eq. 5 implies a multiplicative invariance w.r.t. the guidance image. This is a very useful property for visual computing applications that deal with brightness differences. Therefore, osmosis seems well-suited for blending.

In our blending scenario, we want to obtain the blended image as the steady state of an osmosis process. Since this result is unknown, we also do not have access to its canonical drift vector field. Therefore, we need to consider the so-called *incompatible case* of osmosis.

2.3 Incompatible Drift Vector Fields and Gradient Domain Methods

Weickert et al. [24] also consider the steady state of the osmosis evolution on the image $\boldsymbol{u}(\boldsymbol{x}, t)$, which implies $\partial_t u_i = 0$ for all channels i. Plugging this into Eq. (1), the steady state result $\boldsymbol{w}$ fulfills

$$\Delta w_i = \mathbf{div}\,(\boldsymbol{d}_i w_i)\,. \tag{7}$$

This closely resembles the Poisson equation

$$\Delta w_i = \mathbf{div}\,\boldsymbol{g}_i\,. \tag{8}$$

It arises for so-called gradient domain methods [6,17] as a necessary condition for finding a minimiser of the energy

$$E(u_i) = \int_{\Omega} |\boldsymbol{\nabla} u_i - \boldsymbol{g}_i|^2\, d\boldsymbol{x}\,. \tag{9}$$

For a gradient field $\boldsymbol{g}_i = \boldsymbol{\nabla} v_i$ this yields an exact integration, and thus $\boldsymbol{v}$ as a result, up to an additive constant. This corresponds to the compatible osmosis case. For our blending purposes, it will however be vital that osmosis exhibits multiplicative invariances instead.

Minimising the energy from Eq. (9) for a non-integrable vector field $\boldsymbol{g}_i$ yields an approximate integration result $\boldsymbol{u}_i$. Similarly, for osmosis, we can also calculate a steady state that fulfills Eq. (7) for a drift vector field $\boldsymbol{d}_i$ which does not correspond to a guidance image. This *incompatible* case allows us to design drift vector fields that can be used for blending in Sect. 3.

2.4 Discrete Osmosis

In order to apply osmosis to digital images, we use the discrete implementation proposed by Vogel et al. [23]. They have shown that the theoretical results carry over into the discrete setting and can be implemented with appropriate finite difference discretisations.

We are only interested in the steady state, i.e. the solution of the elliptic Eq. (7). Therefore, we use an implicit scheme with a stabilised BiCGSTAB solver as described by Meister [13], since this yields faster results compared to an explicit formulation [23].

Note that in the scheme from [23], the drift vector field is discretised on a grid shifted by $h/2$ for a grid size h. Therefore, the discrete samples coincide with pixel boundaries. This is of particular importance for our seam removal approach in Sect. 3.2.

3 Blending with Osmosis

In the following we define multiple blending approaches based on the osmosis model from Eq. (1)–(3). Our blending differs only in terms of the drift vector

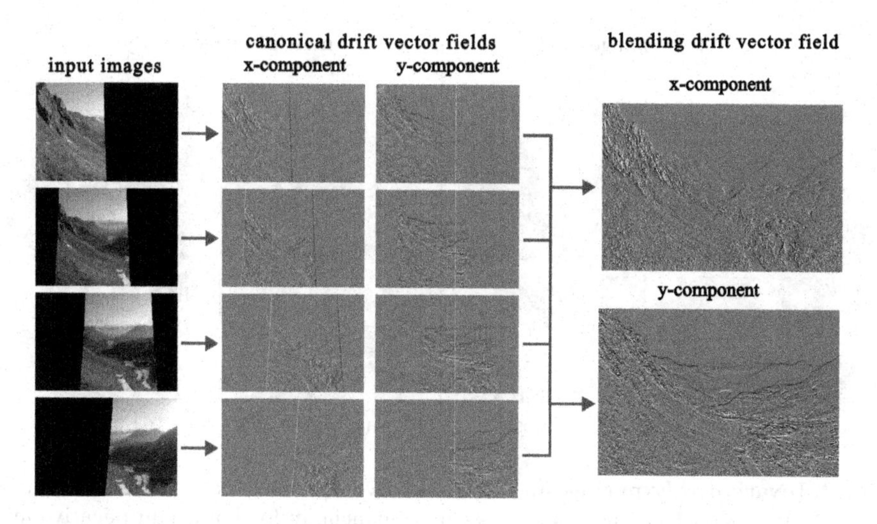

Fig. 1. Drift Vector Blending. The canonical drift vector fields for all input images are stitched together at the predetermined seams. The steady state of an osmosis process based on this joint blending drift vector field yields the blended image.

field $\boldsymbol{d}$. For all approaches, we assume that the panoramic or mosaic images are already aligned manually or by a suitable algorithm.

All following approaches share the same basic problem formulation. Given are n aligned, partially overlapping images $\boldsymbol{v}_1, ..., \boldsymbol{v}_n \in \mathbb{R}_+^{n_x \times n_y \times n_c}$ with spatial resolution $n_x \times n_y$ and n_c colour channels. Our approach merges these images into a result $\boldsymbol{w} \in \mathbb{R}_+^{n_x \times n_y \times n_c}$, the steady state of an osmosis process. As illustrated by Fig. 1, the aligned input images are padded to the target resolution $n_x \times n_y$.

Note that according to the properties of osmosis discussed in Sect. 2, the steady state is almost completely independent of the initial image. Only the average colour value in each channel carries over, otherwise the steady state is fully determined by the drift vector field. The osmosis process can thus be initialised with a naïvely stitched image or a flat image with the same average colour value. Both variants will lead to the same results. Moreover, osmosis preserves positivity, but not necessarily the maximum colour values. Therefore, we clip the osmosis steady state to the original image range $[0, 255]$.

3.1 Drift Vector Blending

The most straightforward approach to osmosis blending is to build a composite drift vector field by stitching at a hard seam. To this end, we first compute the canonical drift vector fields $\boldsymbol{d}_1, ..., \boldsymbol{d}_n$ that correspond to the input images $\boldsymbol{v}_1, ..., \boldsymbol{v}_n$ according to Eq. (5).

We split overlapping image parts in the middle, thus partitioning the image into n parts. The corresponding drift vector fields $\boldsymbol{d}_i$ form a partition of the joint drift vector field $\boldsymbol{d}$ as depicted in Fig. 1. At partition boundaries, the drift vectors

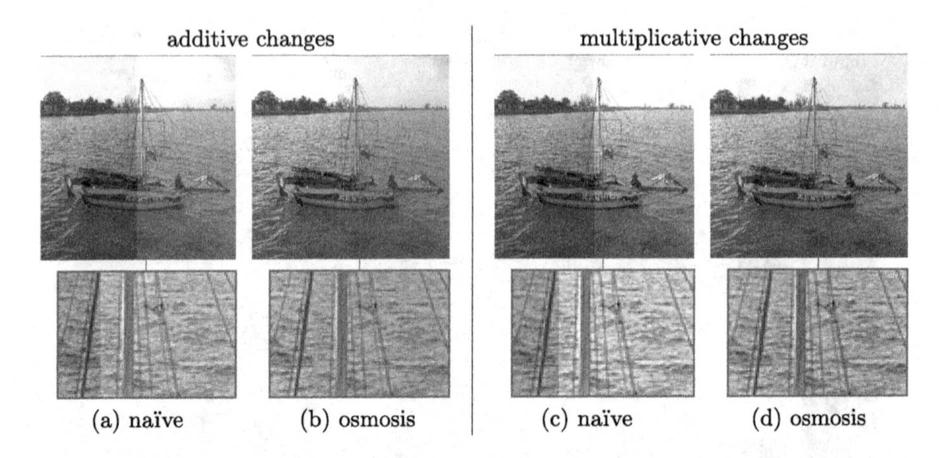

Fig. 2. Invariance Experiments for Osmosis. Experiments on synthetic examples show that osmosis blending removes seams convincingly for both multiplicative and additive changes.

at the pixel boundary overlap (see Sect. 2.4). At such locations, we average the values of both adjacent drift vector fields.

Instead of splitting images in the middle, one can also attempt to minimise brightness differences between neighbouring pixels with an optimal seam algorithm. For our experiments we follow Levin et al. [12] and use the minimum error boundary cut [3]. It computes a seam between two image areas by minimising the overlapping error in terms of the Euclidean distance. The path along the pixel boundaries with the lowest overall error is computed with dynamic programming.

3.2 Seam Removal

The blending problem can be also interpreted as a generalisation of shadow removal. In both cases, there are multiple regions with different illumination and the goal is to fuse them seamlessly into a single image.

Therefore, we also investigate if established methods for osmosis-based shadow removal yield better results than the drift vector stitching. Weickert et al. [24] compute the canonical drift vector field of the image to be edited and modify it by setting the drift vector field to zero at the shadow boundaries.

After acquiring seams as in the previous section, we first stitch the images together directly in the image domain without performing any blending. This allows us to compute the canonical drift vector field of this preliminary composite image. The seams coincide with pixel boundaries, which also correspond to the locations of the drift vector field in our discretisation (see Sect. 2.4). Therefore, we can simply edit the drift vector field by setting it to zero at seam locations.

(a) drift vector blending (b) seam removal (c) alpha blending

Fig. 3. Comparison of Osmosis Blending Approaches. The simple direct blending yields results that are just as convincing as seam removal. Alpha blending leads to blurry seams close to the camera, where alignment is imperfect.

3.3 Alpha Blending with Osmosis

Levin et al. [12] also considered soft seam methods [22] for gradient domain blending. Therefore, we also investigate if this concept can be useful in the case of osmosis blending.

Soft seam approaches do not simply stitch together input data at the seams as we did in Sect. 3.1. Instead, they perform weighted averaging in overlapping regions. Let $\boldsymbol{d}_\ell$ denote the left and $\boldsymbol{d}_r$ the right canonical drift vector field in a vertically split overlapping region.

We blend the drift vector fields by location adaptive weighting, i.e. at location i, j, we get

$$\boldsymbol{d}_{i,j} = \alpha(i,j)\, \boldsymbol{d}_{\ell,i,j} + (1 - \alpha(i,j))\, \boldsymbol{d}_{r,i,j}\,.$$

The weight α changes according to its distance to the seam. For vertical seams, the shortest distance can be simplified to a description via the horizontal index i. For our experiments, we define the weight α according to

$$\alpha(i,j) = \begin{cases} 1 & \text{for } i < s_j - w, \\ \frac{s_j + w - i}{2w} & \text{for } s_j - w \leq i \leq s_j + w, \\ 0 & \text{for } i > s_j + w. \end{cases} \tag{10}$$

Here, s_j denotes the horizontal position of the seam at the vertical position j. We weight both fields equally at the seam and linearly blend the two drift vector fields inside of the window $[s_j - w, s_j + w]$ with size $2w$.

(a) naïve blending (b) gradient domain (c) osmosis

Fig. 4. Plausibility Check: Low Brightness Differences. As expected, on simple sequences with moderate brightness differences, osmosis and gradient domain blending both perform well. But already here, a slightly better preservance of contrast hints at the robustness of osmosis blending.

4 Experiments

All real-world images in the evaluation have been aligned and warped with the open source tool Hugin 2021.0.0[1]. Osmosis was implemented with an implicit scheme and a BiCGSTAB solver as proposed by Vogel et al. [23]. For each step with a time step size of $\tau = 10^5$ this solver is stopped if the relative Euclidean norm of the residual drops below 10^{-9}. We also determine the number of time steps on a relative Euclidean norm on the steady state Eq. (7), requiring a decay of $\|\Delta u_i - \mathbf{div}\,(\boldsymbol{d}_i u_i)\|_2$ by more than a factor 10^{-6} compared to the initialisation.

In addition to evaluating different osmosis blending variants against each other, we compare against gradient domain blending. Here, gradient fields of the input images are stitched together with the same seams as our osmosis method to allow a direct comparison between the properties of these approximate integration methods (see Sect. 2.3). This corresponds to algorithm GIST2 of Levin et al. [12], a widely known approach that is conceptually closest to our own.

4.1 Invariances of Osmosis Blending in Practice

In Sect. 2 we discussed the multiplicative invariance for the compatible osmosis case. This is the dominating type of illumination changes in blending, and we verify that this carries over to the blending application. Moreover, we also investigate the impact of additive changes and compare against naïve stitching. These are less common, but in practice, more complex mixtures of lighting

[1] https://hugin.sourceforge.io/.

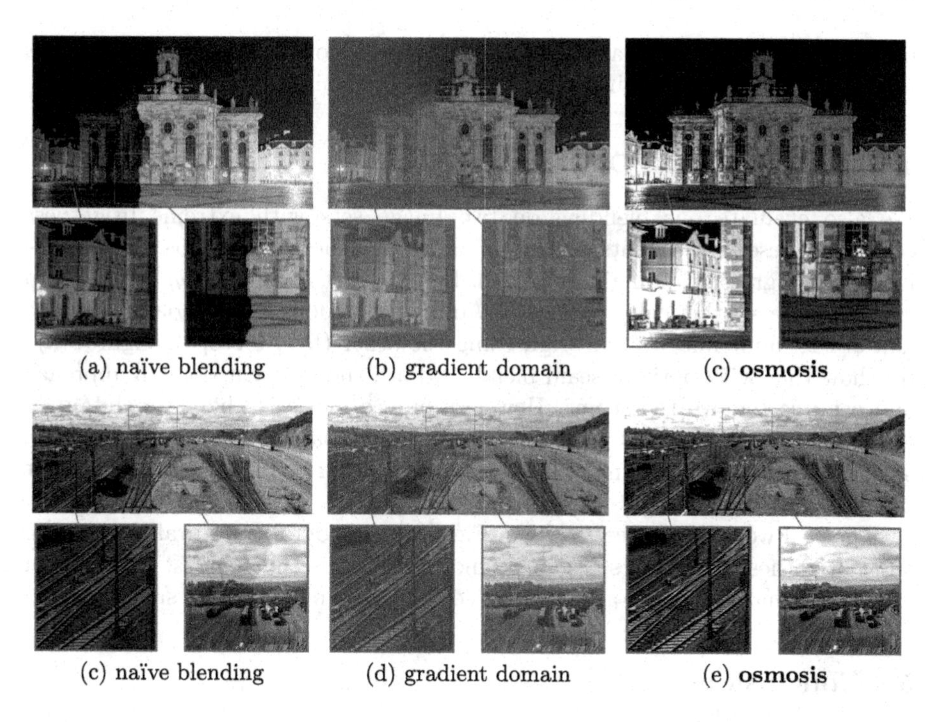

Fig. 5. High Brigthness Differences. We perform optimal seam blending of *building* and *tracks* sequences. On these challenging sequences, osmosis preserves the image contrast significantly more convincingly due to its multiplicative invariance.

changes than pure multiplicative ones can occur. Therefore we created two synthetic test cases for the image *boat* in Fig. 2. For the additive change in Fig. 2(a), each pixel value in the left image half was increased by 30, and in Fig. 2(c), each pixel was multiplied by 1.3. In both cases, the results were clipped at 255. Osmosis blending is robust under both types of brightness changes.

4.2 Comparing Variants of Osmosis Blending

On the *mountain* sequence from Fig. 1 we compare our three osmosis blending approaches from Sect. 3. In Fig. 3, all three approaches equilibrate brightness differences well. Drift vector blending and seam removal also produce no visible artefacts at the seams. Only alpha blending leads to blur at stitching boundaries. Due to its simplicity and quality, drift vector blending is our method of choice.

4.3 Plausibility Check: Low Brightness Differences

As a plausibility check, we first investigate the *mountain* sequence with visible, but fairly low brightness differences for naïve blending in Fig. 4. As expected, gradient domain blending yields good results without visible seams. Osmosis offers similar quality with a slightly better preservation of high contrast, such as the shadow of the mountains in the right half of the image.

4.4 High Brightness Differences

We also evaluate our blending on two image sets with extreme brightness changes. These experiments reveal the differences between our osmosis blending and the gradient domain approach. For the first set, *building*, illumination changes were synthetically created in Adobe Lightroom v5.5 by darkening one half of the input images and brightening the other (by 1.5 stops). Figure 5(a)–(c) show that with optimal seam blending, both the gradient domain approach and osmosis remove the seams. However, only osmosis is able to maintain or even enhance the contrast in the darkened image parts on the left-hand side. This highlights how the multiplicative invariance of osmosis has a significant practical impact.

The real world sequence *tracks* in Fig. 5(d)–(e) contains naturally occurring large brightness differences. Again, osmosis preserves the contrast better than gradient domain blending and yields a considerably more vivid result.

5 Conclusions

We have proposed the first osmosis model for image blending. Our investigation has shown that already simple stitching of drift vector fields yields excellent results without any visible seams. The natural invariances of the osmosis filter take care of multiplicative brightness differences without the need of any further processing such as alpha blending.

In particular for challenging image sequences with large brightness changes, osmosis clearly produces superior results compared to gradient domain methods.

This application highlights the practical value of the theoretical properties provided by osmosis filtering. In the future, we intend to leverage these strengths for additional applications in computer vision and computer graphics.

References

1. Burt, P.J., Adelson, E.H.: A multiresolution spline with application to image mosaics. ACM Trans. Graph. **2**(4), 217–236 (1983)
2. d'Autume, M., Morel, J.M., Meinhardt-Llopis, E.: A flexible solution to the osmosis equation for seamless cloning and shadow removal. In: Proceedings of 2018 IEEE International Conference on Image Processing, Athens, Greece, pp. 2147–2151 (2018)
3. Efros, A., Freeman, W.T.: Image quilting for texture synthesis and transfer. In: Proceedings of 28th Annual Conference on Computer Graphics and Interactive Techniques, Los Angeles, CA, pp. 341–346 (2001)
4. Fang, F., Wang, T., Fang, Y., Zhang, G.: Fast color blending for seamless image stitching. IEEE Geosci. Remote Sens. Lett. **16**(7), 1115–1119 (2019)

5. Fattal, R., Lischinski, D., Werman, M.: Gradient domain high dynamic range compression. In: Proceedings of SIGGRAPH 2002, San Antonio, TX, pp. 249–256 (2002)
6. Frankot, R., Chellappa, R.: A method for enforcing integrability in shape from shading algorithms. IEEE Trans. Pattern Anal. Mach. Intell. **10**(4), 439–451 (1988)
7. Georgiev, T.: Covariant derivatives and vision. In: Leonardis, A., Bischof, H., Pinz, A. (eds.) ECCV 2006. LNCS, vol. 3954, pp. 56–69. Springer, Heidelberg (2006). https://doi.org/10.1007/11744085_5
8. Gracias, N., Mahoor, M., Negahdaripour, S., Gleason, A.: Fast image blending using watersheds and graph cuts. Image Vis. Comput. **27**(5), 597–607 (2009)
9. Hagenburg, K., Breuß, M., Vogel, O., Weickert, J., Welk, M.: A lattice Boltzmann model for rotationally invariant dithering. In: Bebis, G., et al. (eds.) ISVC 2009. LNCS, vol. 5876, pp. 949–959. Springer, Heidelberg (2009). https://doi.org/10.1007/978-3-642-10520-3_91
10. Hagenburg, K., Breuß, M., Weickert, J., Vogel, O.: Novel schemes for hyperbolic PDEs using osmosis filters from visual computing. In: Bruckstein, A.M., ter Haar Romeny, B.M., Bronstein, A.M., Bronstein, M.M. (eds.) SSVM 2011. LNCS, vol. 6667, pp. 532–543. Springer, Heidelberg (2012). https://doi.org/10.1007/978-3-642-24785-9_45
11. Illner, R., Neunzert, H.: Relative entropy maximization and directed diffusion equations. Math. Methods Appl. Sci. **16**, 545–554 (1993)
12. Levin, A., Zomet, A., Peleg, S., Weiss, Y.: Seamless image stitching in the gradient domain. In: Pajdla, T., Matas, J. (eds.) ECCV 2004. LNCS, vol. 3024, pp. 377–389. Springer, Heidelberg (2004). https://doi.org/10.1007/978-3-540-24673-2_31
13. Meister, A.: Numerik linearer Gleichungssysteme, 5th edn. Vieweg, Braunschweig (2015)
14. Parisotto, S., Calatroni, L., Bugeau, A., Papadakis, N., Schönlieb, C.B.: Variational osmosis for non-linear image fusion. IEEE Trans. Image Process. **29**, 5507–5516 (2020)
15. Parisotto, S., Calatroni, L., Caliari, M., Schönlieb, C.B., Weickert, J.: Anisotropic osmosis filtering for shadow removal in images. Inverse Probl. **35**(5), Article 054001 (2019)
16. Parisotto, S., Calatroni, L., Daffara, C.: Digital cultural heritage imaging via osmosis filtering. In: Mansouri, A., El Moataz, A., Nouboud, F., Mammass, D. (eds.) ICISP 2018. LNCS, vol. 10884, pp. 407–415. Springer, Cham (2018). https://doi.org/10.1007/978-3-319-94211-7_44
17. Pérez, P., Gagnet, M., Blake, A.: Poisson image editing. ACM Trans. Graph. **22**(3), 313–318 (2003)
18. Risken, H.: The Fokker-Planck Equation. Springer, New York (1984)
19. Schmidt, M.: Linear scale-spaces in image processing: drift-diffusion and connections to mathematical morphology. Ph.D. thesis, Department of Mathematics, Saarland University, Saarbrücken, Germany (2018)
20. Sevcenco, I.S., Hampton, P.J., Agathoklis, P.: Seamless stitching of images based on a Haar wavelet 2D integration method. In: Proceedings of 17th International Conference on Digital Signal Processing, Kanoni, Greece (2011)
21. Sochen, N.A.: Stochastic processes in vision: from Langevin to Beltrami. In: Proceedings of Eighth International Conference on Computer Vision, Vancouver, Canada, vol. 1, pp. 288–293. IEEE Computer Society Press (2001)
22. Uyttendaele, M., Eden, A., Skeliski, R.: Eliminating ghosting and exposure artifacts in image mosaics. In: Proceedings of 2001 IEEE Computer Society Conference on Computer Vision and Pattern Recognition, Kauai, HI, pp. 509–516 (2001)

23. Vogel, O., Hagenburg, K., Weickert, J., Setzer, S.: A fully discrete theory for linear osmosis filtering. In: Kuijper, A., Bredies, K., Pock, T., Bischof, H. (eds.) SSVM 2013. LNCS, vol. 7893, pp. 368–379. Springer, Heidelberg (2013). https://doi.org/10.1007/978-3-642-38267-3_31
24. Weickert, J., Hagenburg, K., Breuß, M., Vogel, O.: Linear osmosis models for visual computing. In: Heyden, A., Kahl, F., Olsson, C., Oskarsson, M., Tai, X.-C. (eds.) EMMCVPR 2013. LNCS, vol. 8081, pp. 26–39. Springer, Heidelberg (2013). https://doi.org/10.1007/978-3-642-40395-8_3
25. Wu, H., Zheng, S., Zhang, J., Huang, K.: GP-GAN: towards realistic high-resolution image blending. In: Proceedings of 27th ACM International Conference on Multimedia, Nice, France, pp. 2487–2495 (2019)

α-Pixels for Hierarchical Analysis of Digital Objects

Kouki Tosaka[1] and Atsushi Imiya[2](✉)

[1] School of Science and Engineering, Chiba University, Yayoi-cho 1-33, Inage-ku, Chiba 263-8522, Japan

[2] Institute of Management and Information Technologies, Chiba University, Yayoi-cho 1-33, Inage-ku, Chiba 263-8522, Japan

imiya@faculty.chiba-u.jp

Abstract. In this paper, we apply a scale space analysis method to digital geometry. We deal with the reconstruction of Euclidean polylines controlling the size of pixels in discrete space. Numerical examples show that the size of pixels of digital geometry processes the scale for the hierarchical reconstruction of Euclidean objects from digital objects.

1 Introduction

In this paper, we apply a scale space analysis method to digital geometry [1–3]. Linear digital geometry on a plane mainly deals with two problems: (1) the generation of a string of pixels from polylines and (2) the reconstruction of polylines from a string of pixels. The string generated in the first problem is called a digital object. The second problem is called Euclidean reconstruction [1–3] in digital geometry. Therefore, digital and discrete geometry deals with mathematical relations between the generation of digital objects and the reconstruction of Euclidean objects from digital objects.

We deal with Euclidean reconstruction if the size of pixels varies. We numerically show that the size of pixels in digital expression governs the scale in the hierarchical expression for the reconstruction of Euclidean objects. In digital geometry, there are three types of local structure of strings: naive, standard and supercover [1–3]. We deal with supercover [3,4] for the string expression constructed from Euclidean polylines.

Two-dimensional digital geometry deals with geometric objects defined on the equi-spaced grid system Furthermore, objects in digital geometry are collections of pixels centred at grid points. Therefore, since digital geometry is an application of interval analysis to geometry, each point is dealt with as a square centred at a grid point. On the other hand, the α-shape method [5] is dealt with a point as a small circle centred at the point for geometric processing with rounding errors [6,7], As an analogue of the α-shape method to digital geometry, we introduce α-digital geometry, which deals with various sizes of pixels. An extension of the geometric properties of pixels in digital geometry is the isothetic irregular grid system geometry. In the irregular grid system geometry, digital images are expressed by isothetic rectangles [4,9,10]. The irregular grid system

L. Calatroni et al. (Eds.): SSVM 2023, LNCS 14009, pp. 665–676, 2023.
https://doi.org/10.1007/978-3-031-31975-4_51

geometry has advantages for the expression and analysis of digital images, since for smooth parts on the boundary of an object, the less numbers of the isothetic rectangles are expected for image expressions.

The curve shorting flow [11–13] is used for shape retrievals since the flow extracts dominant local features from the boundary curves of objects. The proposing method extracts the local features by reducing the number of edges in polylines extracted from the digital boundary of digitised objects on a plane.

2 Pixels and Discrete Object

Setting $\boldsymbol{n} = (m, n)^\top$ to be a point in $\mathbf{Z}^2$, a square,

$$\mathbf{p}(\boldsymbol{n}) = \left\{ \boldsymbol{x} | \, |\boldsymbol{x} - \boldsymbol{n}|_\infty \leq \frac{1}{2} \right\}, \tag{1}$$

for $\boldsymbol{x} = (x, y)^\top$ is called a pixel of $\mathbf{R}^2$ centred at point $\boldsymbol{n}$, where $|\boldsymbol{x}|_\infty = \max(|x|, |y|)$ is the infinity norm ℓ_∞ of $\mathbf{R}^2$.

Definition 1. *For* $\boldsymbol{n} = (m, n)^\top$,

$$\mathbf{p}^\alpha(\boldsymbol{n}) = \left\{ \boldsymbol{x} | \, |\boldsymbol{x} - \boldsymbol{n}|_\infty \leq \alpha \frac{1}{2}, \alpha > 0 \right\}, \tag{2}$$

is an α*-pixel of* $\mathbf{R}^2$ *centred at point* $\boldsymbol{n} \in \mathbf{Z}^2$.

Therefore, the pixels are 1-pixels. We call the 1-pixels simply pixels. Furthermore, we set $\lim_{\alpha \to 0} \mathbf{p}^\alpha(\boldsymbol{x}) = \boldsymbol{x}$

We define the connectivity of points in $\mathbf{Z}^2$.

Definition 2. *For* $\boldsymbol{x} = (x, y)^\top \in \mathbf{Z}^2$, *the set* $\mathbf{N}(\boldsymbol{x}) = \{(x \pm 1, y)^\top, (x, y \pm 1)^\top\}$ *is the four neighbourhoods of* $\boldsymbol{x}$.

Definition 3. *If* $\boldsymbol{y} \in \mathbf{N}(\boldsymbol{x})$, $\boldsymbol{y}$ *is four connected to* $\boldsymbol{x}$ *and described as* $\boldsymbol{x} \sim \boldsymbol{y}$.

Therefore, a pair points $\boldsymbol{x}$ and $\boldsymbol{y}$ in $\mathbf{Z}^2$ are four connected, the pair of pixels $\mathbf{p}(\boldsymbol{x})$ and $\mathbf{p}(\boldsymbol{y})$ shares an edge. We describe $\mathbf{P}(\boldsymbol{x}) \sim \mathbf{P}(\boldsymbol{y})$ if $\boldsymbol{x} \sim \boldsymbol{y}$. Therefore, for the collection of centre points $\mathbf{p} = \{\boldsymbol{x}_1, \boldsymbol{x}_2, \cdots, \boldsymbol{x}_n\}$, if the connection relations $\boldsymbol{x}_1 \sim \boldsymbol{x}_2 \sim \cdots \sim \boldsymbol{x}_n$ are satisfied elements in a collection of pixels $\mathbf{P} = \{\mathbf{p}_1, \mathbf{p}_2, \cdots, \mathbf{p}_n\}$ satisfy the connection property such that $\mathbf{p}_1 \sim \mathbf{p}_2 \sim \cdots \sim \mathbf{p}_n$, We call the collection of pixels $\mathbf{P}$ four-connected objects.

Definition 4. *If pixels in the collection of four-connected pixels* $\mathbf{P} = \{\mathbf{p}\}_{i=1}^n$ *are replaced with* α*-pixels, we call the object an* α*-object and express it as* $\mathbf{P}^\alpha$.

Definition 5. *If the* α*-object* $\mathbf{P}^\alpha$ *is generated from a string of* α*-pixels, we call this object an* α*-string.*

Definition 6. *We express the* α*-string generated from string* $\mathbf{S}$ *as* $\mathbf{S}^\alpha$.

Next, we define the supercover of sample points.

Definition 7. *For* $a, b, \mu \in \mathbf{Z}$ *and* $\omega = |a| + |b|$, *points* $(x, y)^\top \in \mathbf{Z}^2$ *on the supercover satisfy the double Diophantus inequality*

$$0 \leq ax + by + \mu + \frac{\omega}{2} \leq \omega \tag{3}$$

for $\gcd(a, b) = 1$ *and* $\gcd(\mu, 1) = 1$.

Furthermore, we define the supercover of sample points.

Definition 8. *For* $a, b, \mu \in \mathbf{Z}$ *and* $\omega = |a| + |b|$, *points* $(x, y)^\top \in \mathbf{Z}^2$ *on the* α*-supercover satisfy the double inequality*

$$0 \leq ax + by + \mu + \alpha\frac{\omega}{2} \leq \alpha\omega \tag{4}$$

for $\gcd(a, b) = 1$ *and* $\gcd(\mu, 1) = 1$.

The α-pixels satisfy the following properties.

Property 1. *For* $\alpha_1 < \alpha_2$, *the inclusive relation* $\mathbf{p}^{\alpha_1}(\boldsymbol{x}) \subset \mathbf{p}^{\alpha_2}(\boldsymbol{x})$ *is satisfied.*

Property 2. *For the four-connected point set* $\mathbf{s} = \{\boldsymbol{x}_i\}_{i=1}^n$, *the collection of* α*-pixels* $\mathbf{S}^\alpha = \{p_\alpha(\boldsymbol{x}_i)\}_{i=1}^n$ *satisfies the inclusive relation* $\mathbf{S}^{\alpha_1} \subset \mathbf{S}^{\alpha_2}$ *if* $1 \leq \alpha_1 < \alpha_2$.

For $\alpha = 1$, the supercover of line $ax + b + \mu = 0$ is the four connected pixel string $\mathbf{P}_{(a,b,\mu)}$. The line $ax + b + \mu = 0$ crosses with all elements of the string $\mathbf{P}_{(a,b,\mu)}$. Figure 1(a) shows the configuration pixels in the supercover of a line. In Fig. 1(b) and 1(c), dark squares are α-pixels for $\alpha = 2$ and $\alpha = 3/4$, respectively. Furthermore, in Fig. 1(b) and 1(c), light gray squares are α-supercovers for $\alpha = 2$ and $\alpha = 3/4$, respectively. Figures 1(b) and 1(c) the local structures of 2-supercove and 3/4-supercover for a line, respectively.

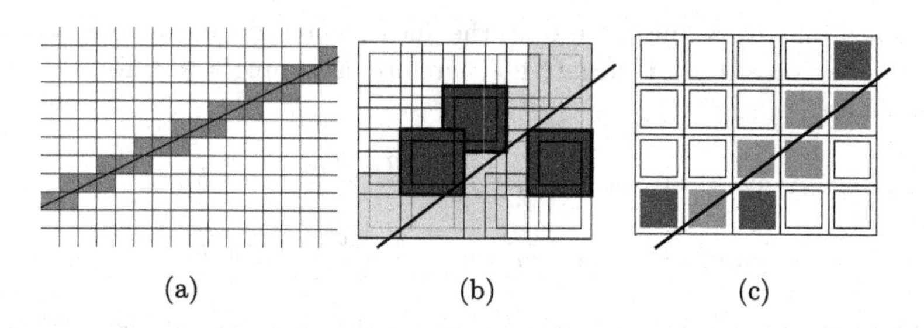

Fig. 1. Supercover and α-supercovers (a) Supercover of an Euclidean line. (b) The local structure of 2-supercover. (c) The local structure of 3/4-supercover. In (b) and (c) dark squares are α-pixels of sample points and light grey squares are elements of α-supercovers.

If $\mathbf{S}$ is a string of four-connected pixels, Property 2 implies the inclusive relations of regions in which Euclidean reconstructions exist. Figure 2 shows the geometrical expressions of this inclusive relation for digital circles $\mathbf{C}^\alpha$ for $\alpha = 1, 2^2, 2^3$, and, 2^4 in (a), (b), (c) and (d), respectively.

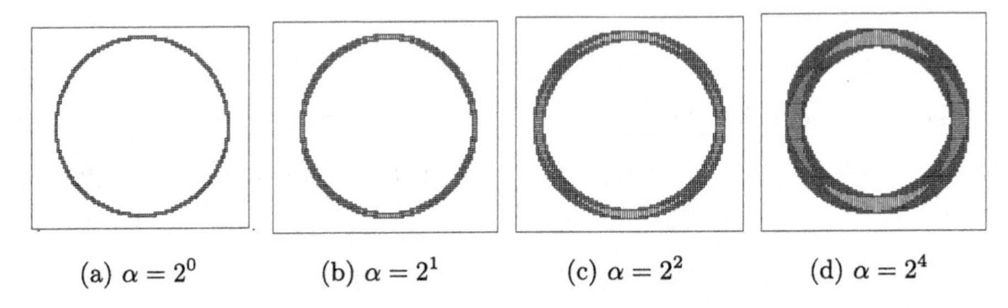

(a) $\alpha = 2^0$ (b) $\alpha = 2^1$ (c) $\alpha = 2^2$ (d) $\alpha = 2^4$

Fig. 2. α-pixel strings of circles. (a) Circle of pixels $\mathbf{C} = \mathbf{C}_{2^0}$. (b) Circle of 2-pixels $\mathbf{C}^2$ (c) Circle of 4-pixels $\mathbf{C}^4 = \mathbf{C}^{2^2}$. (d) Circle of 4-pixels $\mathbf{C}^8 = \mathbf{C}^{2^3}$. These figures show that digital circles satisfy the inclusive relation $\mathbf{C}^{2^0} \subset \mathbf{C}^{2^1} \subset \mathbf{C}^{2^2} \subset \mathbf{C}^{2^3}$.

3 Reconstruction of Line

Definition 9. *Recognition of the α-supercover from sample pixels $\{\boldsymbol{x}_i = (x_i, y_i)^\top\}_{i=1}^n$ is to find integers a, b and μ which minimise*

$$M = |a| + |b| + |\mu| \tag{5}$$

with the conditions

$$0 \leq ax_i + by_i + \mu + \alpha\frac{\omega}{2} \leq \alpha\omega, \ i = 1, 2, \cdots, n, \tag{6}$$

for $\gcd(a, b) = 1$, $\gcd(\mu, 1) = 1$ *and* $\omega = |a| + |b|$.

It is possible to assume $a \geq 0$ since the line $ax+by+\mu = 0$ can be transformed to $-ax - by - \mu = 0$ if a is negative. Therefore, assuming $b \geq 0$, Eq. (3) can be expressed as

$$\text{case1}: 0 \leq ax + by + \mu + \alpha\frac{a+b}{2} \leq \alpha(a+b), \tag{7}$$

$$\text{case2}: 0 \leq ax - by + \mu + \alpha\frac{a+b}{2} \leq \alpha(a+b). \tag{8}$$

Theorem 1. *For these two cases, the vector forms of inequalities of Eqs. (7) and (8) for $\boldsymbol{a} = (a, b)^\top$ are*

$$\begin{cases} -\boldsymbol{x}_i^\top \boldsymbol{a} - \alpha\frac{1}{2}\omega \leq \mu, \ \mu \leq -\boldsymbol{x}_i^\top \boldsymbol{a} + \alpha\frac{1}{2}\omega, \\ \boldsymbol{x}_{ij}^\top \boldsymbol{a} > 0, \\ \boldsymbol{a} > 0, \end{cases} \quad \text{for} \begin{cases} \boldsymbol{x}_{ij} = \boldsymbol{x}_i - \boldsymbol{x}_j + \alpha \boldsymbol{e}, \\ \boldsymbol{e} = (1, 1)^\top, \\ \omega = a + b \end{cases} \tag{9}$$

and

$$\begin{cases} -\boldsymbol{x}_i^\top \boldsymbol{M}\boldsymbol{a} - \alpha\frac{1}{2}\omega \le \mu,\ \mu \le -\boldsymbol{x}_i^\top \boldsymbol{M}\boldsymbol{a} + \alpha\frac{1}{2}\omega, \\ \boldsymbol{x}_{ij}^\top \boldsymbol{a} > 0, \\ \boldsymbol{a} > 0, \end{cases} \quad \text{for} \begin{cases} \boldsymbol{x}_{ij} = \boldsymbol{M}(\boldsymbol{x}_i - \boldsymbol{x}_j) + \alpha\boldsymbol{e}, \\ \boldsymbol{e} = (1,1)^\top, \\ \omega = a + b, \end{cases} \tag{10}$$

where $\boldsymbol{M} = \begin{pmatrix} 1, & 0 \\ 0, & -1 \end{pmatrix}$.

(Proof) For case 1, using the system of inequalites

$$\begin{cases} ax_i + by_i + \mu + \alpha\frac{\omega}{2} \ge 0, \\ -ax_j - by_j - \mu - \alpha\frac{\omega}{2} \ge -\alpha(a+b), \end{cases} \tag{11}$$

Eq. (6) implies the system of inequalities

$$\begin{cases} -(x_i a + y_i b) - \alpha\frac{a+b}{2} \le \mu, \\ -(x_j a + y_j b) + \alpha\frac{a+b}{2} \ge \mu, \\ X_{ij}a + Y_{ij}b \ge 0, \\ a, b > 0, \end{cases} \tag{12}$$

for $i \ne j$, $i, j = 1, 2, \cdots, n$, where $X_{ij} = x_i - x_j + \alpha$ and $Y_{ij} = y_i - y_j + \alpha$. Therefore, $\boldsymbol{x}_{ij} = (X_{ij}, Y_{ij})^\top = \boldsymbol{x}_i - \boldsymbol{x}_j + \alpha\boldsymbol{e}$ for $\boldsymbol{e} = (1,1)^\top$.

For case 2, since

$$\begin{cases} -(ax_i - by_i) - \alpha\frac{1}{2}\omega \le \mu, \\ -(ax_j - by_j) + \alpha\frac{1}{2}\omega \ge \mu, \\ (ax_i - by_i) + \mu + \alpha\frac{\omega}{2} \ge 0, \\ -(ax_j - by_j) - \mu - \alpha\frac{\omega}{2} \ge -\alpha(a+b). \end{cases} \tag{13}$$

we have the relations

$$\boldsymbol{x}_{ij} = \boldsymbol{M}(\boldsymbol{x}_i - \boldsymbol{x}_j) + \alpha\boldsymbol{e} \tag{14}$$

and

$$-\boldsymbol{x}_i^\top \boldsymbol{M}\boldsymbol{a} - \alpha\frac{1}{2}\omega < \mu,\ \mu < -\boldsymbol{x}_i^\top \boldsymbol{M}\boldsymbol{a} + \alpha\frac{1}{2}\omega. \tag{15}$$

(Q.E.D)

For case 1, setting $\boldsymbol{a}^* = (a^*, b^*)^\top = \arg\min_k(|\boldsymbol{a}|_1 = k,\ \boldsymbol{x}_{ij}^\top \boldsymbol{a} \ge 0)$, where $|\boldsymbol{a}|_1 = a + b = \boldsymbol{e}^\top \boldsymbol{a}$, $\mu^* \in \mathbf{Z}$ is computed from the system of inequalies

$$\begin{cases} -\boldsymbol{x}_i^\top \boldsymbol{a}^* - \alpha\frac{1}{2}\boldsymbol{e}^\top \boldsymbol{a}^* \le \mu^*, \\ -\boldsymbol{x}_j^\top \boldsymbol{a}^* + \alpha\frac{1}{2}\boldsymbol{e}^\top \boldsymbol{a}^* \ge \mu^* \end{cases} \tag{16}$$

for $i \neq j, i = 1, 2, \cdots, n$. These relations imply that the computation of a positive integer pair $(a, b)^\top$ and an integer μ is decomposed into two steps. Furthermore, this decomposition strategy leads to the following lemma.

Lemma 1. *There is no solution for* $(a, b)^\top$, *if* $\mathbf{A}_2 = \{\boldsymbol{a} = (a, b)^\top | a > 0, b > 0, a, b \in \mathbf{Z}\}$ *is empty.*

Definition 10. *We call the region defined by* $\boldsymbol{x}_{ij}^\top \boldsymbol{a} \geq 0$ *the feasible region.*

For α-supercover recognition, we analyse the relation between the geometrical configuration of sample points $\boldsymbol{x}_i = (x_i, y_i)^\top$ for $i = 1, 2, \cdots n$ and the feasible region defined by the sample points. From the geometric configurations of regions

$$\begin{aligned}
H &= \{\boldsymbol{x}_{ij} | i \neq j, i, j = 1, 2, \cdots, n\}, \\
Q_{++} &= \{\boldsymbol{x}_{ij} | \boldsymbol{x}_{ij} \in H, X_{ij} > 0, Y_{ij} > 0, i \neq j\}, \\
Q_{--} &= \{\boldsymbol{x}_{ij} | \boldsymbol{x}_{ij} \in H, X_{ij} < 0, Y_{ij} < 0, i \neq j\}, \\
Q_{+-} &= \{\boldsymbol{x}_{ij} | \boldsymbol{x}_{ij} \in H, X_{ij} > 0, Y_{ij} < 0, i \neq j\}, \\
Q_{-+} &= \{\boldsymbol{x}_{ij} | \boldsymbol{x}_{ij} \in H, X_{ij} < 0, Y_{ij} > 0, i \neq j\}, \\
Q_{0X} &= \{\boldsymbol{x}_{ij} | \boldsymbol{x}_{ij} \in H, X_{ij} = 0, i \neq j\}, \\
Q_{0Y} &= \{\boldsymbol{x}_{ij} | \boldsymbol{x}_{ij} \in H, Y_{ij} = 0, i \neq j\},
\end{aligned}$$

and the regions defined by Eq. (6), Eq. (6) has no solution if all the following conditions

1. $Q_{--} \neq \emptyset$,
2. $\forall \boldsymbol{x}_{ij} \in Q_{0X}, Y_{ij} \leq 0$,
3. $\forall \boldsymbol{x}_{ij} \in Q_{0Y}, X_{ij} \leq 0$,
4. $Q_{+-} \neq \emptyset$, $Q_{-+} \neq \emptyset$ and

$$\min\left(-X_{ij}/Y_{ij}\right)_{|\boldsymbol{x}_{ij} \in Q_{+-}} < \max\left(-X_{nm}/Y_{nm}\right)_{|\boldsymbol{x}_{nm} \in Q_{-+}}$$

are satisfied. Therefore, we have the following theorem for recognition of the supercover from sample points in $\mathbf{Z}^2$.

Theorem 2. *If all relations*

$$\begin{cases}
Q_{--} = \emptyset, \\
\forall \boldsymbol{x}_{ij} \in Q_{0X}, Y_{ij} > 0, \\
\forall V \in Q_{0Y}, X_{ij} > 0, \\
\min(-\frac{X_{ij}}{Y_{ij}})_{|\boldsymbol{x}_{ij} \in Q_{+-}} \geq \max(-\frac{X_{nm}}{Y_{nm}})_{|\boldsymbol{x}_{nm} \in Q_{-+}},
\end{cases} \tag{17}$$

are satisfied, the corresponding pixels $\{\mathbf{p}(\boldsymbol{x}_i)\}_{i=1}^n$ *of sample points* $\{\boldsymbol{x}_i\}_{i=1}^n$ *on* $\mathbf{Z}^2$ *define a supercover of the line* $ax + by + \mu = 0$ *for* $a > 0$, $b > 0$ *and* $\gcd(a, b) = 1$.

For the computation of a Euclidean line from sample points $\{\boldsymbol{x}_i\}_{i=1}^n$, we define

$$(\widehat{X}_{ij}, \widehat{Y}_{ij}) = \min(-X_{ij}/Y_{ij})_{|(X_{ij},Y_{ij})\in Q_{+-}}, \tag{18}$$

$$(\widetilde{X}_{nm}, \widetilde{Y}_{nm}) = \max(-X_{nm}/Y_{nm})_{|(X_{nm},Y_{nm})\in Q_{-+}} \tag{19}$$

Then, Eq. (17) implies the following theorem.

Theorem 3. *There exists a line if the region defined by the system of inequalities*

$$\begin{cases} \widehat{X}_{ij}a + \widehat{Y}_{ij}b \geq 0, \\ \widetilde{X}_{mn}a + \widetilde{Y}_{mn}b \geq, 0 \end{cases} \tag{20}$$

is not empty.

Assuming the feasible region is the positive cone in (a, b)-space defined by the system of inequalities

$$\begin{cases} u_1 a + v_1 b \leq 0, \\ u_2 a + v_2 b \geq 0, \\ a > 0, \\ b > 0, \end{cases} \tag{21}$$

the solutions of the system of linear equations

$$\begin{cases} a + b = k, \\ u_1 a + v_1 b = 0, \end{cases} \quad \begin{cases} a + b = k, \\ u_2 a + v_2 b = 0, \end{cases} \tag{22}$$

are

$$\begin{cases} a_1 = \dfrac{v_1}{v_1 - u_1} k, \\ b_1 = k - a_1, \end{cases} \quad \begin{cases} a_2 = \dfrac{v_2}{v_2 - u_2} k, \\ b_2 = k - a_2. \end{cases} \tag{23}$$

These solutions are points on the boundary of the positive cone. By incrementing k until both a_1 and b_2 are integers, we compute the minimiser of $a + b$. Once $a > 0$ and $b > 0$ are computed from Eq. (20), Eq. (12) implies

$$\max\left\{-(x_i + \frac{1}{2})a - (y_i + \frac{1}{2})b\right\} \leq \mu, \quad \mu \leq \min\left\{(\frac{1}{2} - x_i)a + (\frac{1}{2} - y_i)b\right\}. \tag{24}$$

Therefore,

- If $\max\{-(x_i + \frac{1}{2})a - (y_i + \frac{1}{2})b\} \geq 0$, then $\mu = \max\{-(x_i + \frac{1}{2})a - (y_i + \frac{1}{2})b\}$.
- If $\min\{(\frac{1}{2} - x_i)a + (\frac{1}{2} - y_i)b\} \leq 0$, then $\mu = \min\{(\frac{1}{2} - x_i)a + (\frac{1}{2} - y_i)b\}$.
- If $\max\{-(x_i + \frac{1}{2})a - (y_i + \frac{1}{2})b\} < 0$ and $\min\{(\frac{1}{2} - x_i)a + (\frac{1}{2} - y_i)b\} > 0$, then $\mu = 0$.

4 Polygonalisation of α-String

For the computation of a Euclidean polyline from the 1-string, at least three connected sample points are requested. Even if there is no solution for four-connected sample points $\{\boldsymbol{x}_i = (x_i, y_i)^\top\}_{i=1}^n$, there might be two supercovers for each connected $\{\boldsymbol{x}_i = (x_i, y_i)^\top\}_{i=1}^m$ and $\{\boldsymbol{x}_i = (x_i, y_i)^\top\}_{m+1}^n$ for $3 \leq m \leq n-3$. This geometrical property of connected sample points implies the computation of open and closed polygonal curves from connected-sample points. This problem automatically requests the computation of the partition of connected sample points and the supercover to each partition of sample points.

We assume that strings of pixels $\mathbf{S}$ form the four-connected digital boundary of segment $\mathbf{S}$ on $\mathbf{Z}^2$. This boundary curve is extracted using appropriate morphological operations [14]. For the sample points $\{\boldsymbol{x}_i\}_{i=1}^n$, $\boldsymbol{x}_i \neq \boldsymbol{x}_j$ if $i \neq j$. Furthermore, we can set $\boldsymbol{x}_{i+m} = \boldsymbol{x}_i$. Then, the polygonalisation of $\mathbf{S}$ is described bellow.

Problem 1. *Let $\mathbf{S}^\alpha$ be the digital boundary curve of an object on the square grid system. Compute a partition of $\mathbf{S}^\alpha$ such that $\mathbf{S}^\alpha = \cup_{i=1}^n \mathbf{S}_i^\alpha$, $\mathbf{S}_i^\alpha = \{\mathbf{p}_{ij}^\alpha\}_{j=1}^{n(i)}$ $|\mathbf{S}_i^\alpha \cap \mathbf{S}_{i+1}^\alpha| = \varepsilon$ for an appropriate small integer ε.*

The polygonalisation of $\mathbf{S}^\alpha$ is described as following.

Problem 2. *Compute the number of polygonal edges n and triplets of parameters $\{(a_i, b_i, \mu_i)\}_{i=1}^n$ for edges that minimise the criterion*

$$z = \sum_{i=1}^{n} (|a_i| + |b_i| + |\mu_i) \tag{25}$$

with respect to the system of inequalities

$$0 \leq a_i x_{ij} + b y_{ij} + \mu_i + \alpha \frac{\omega_i}{2} \leq \alpha \omega_i, \tag{26}$$

for $i = 1, 2, \cdots n$ and $j = 1, 2, \cdots, n(i)$, where $\omega_i = |a_i| + |b_i|$, $\gcd(a_i, b_i) = 1$ and $\gcd(\mu_i, 1) = 1$.

The solution of Problem 2 yields a collection of curves

$$\mathcal{L}_i = \left\{ (x, y)^\top \mid |a_i x + b_i y + |\mu_i| \leq \alpha \frac{1}{2}(|a_i| + |b_i|) \right\} \tag{27}$$

$i = 1, 2, \cdots, n$ for an appropriate integer n.

Since we reconstruct polygonal edges from the α-string generated from four-connected pixels, the minimum number of connected pixels for the initialisation of the algorithm is three. Then, incrementing four connected pixels on the string, the algorithm computes feasible regions from a series of the local string of α-pixels whose centre points are four-connected. If no feasible region exists upon incrementing the α-pixel, the algorithm fixes the parameters (a_i, b_i) and μ_i from a partial string one step before using the algorithm for supercover recognition. The algorithm restarts the parameter computation pixel by pixel. Finally, if all α-pixels in the string are used for parameter computation, the algorithm stops and outputs polygonal edges.

5 Numerical Examples

Figure 3 shows the hierarchical expression of contour curves. Figures 3 (a), (b), (c), (d), (e) and (f) show the four-connected pixel contour, the Euclidean contour reconstructed 2-pixels, the Euclidean contour reconstructed 4-pixels, the Euclidean contour reconstructed 1/60-pixels, the Euclidean contour reconstructed 1-pixels and the Euclidean contour reconstructed 10-pixels, respectively.

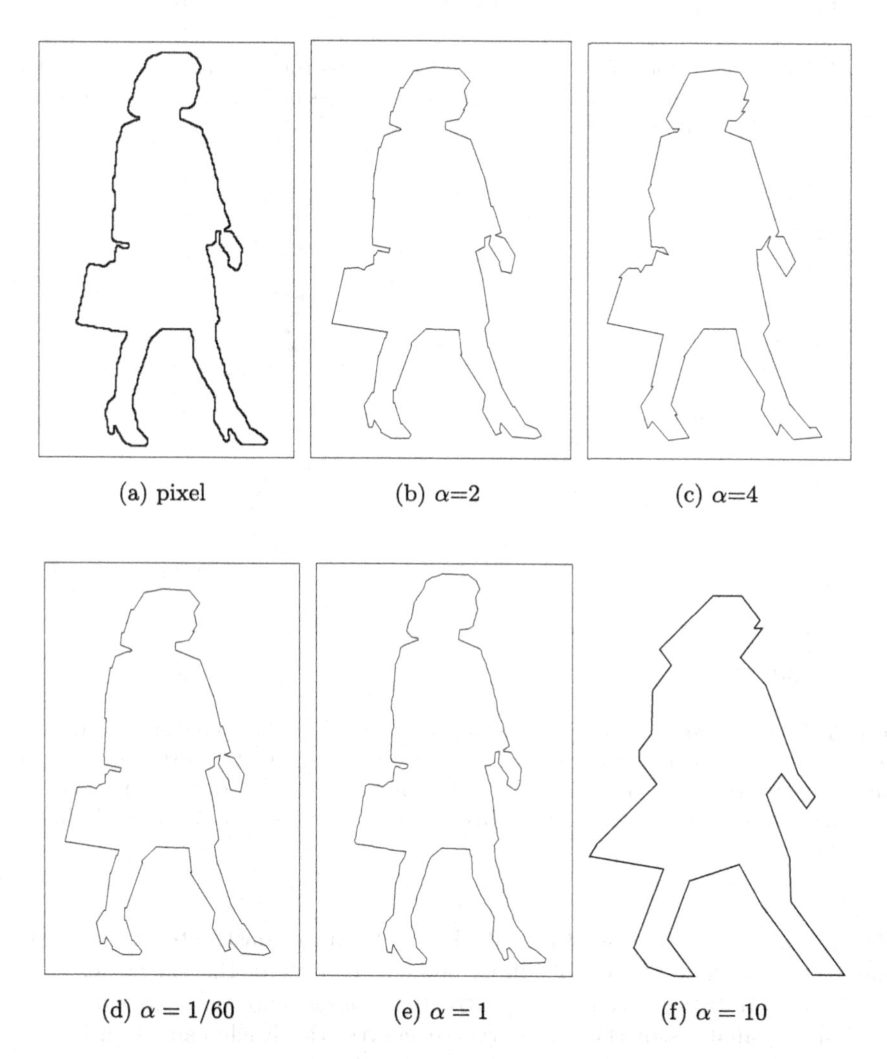

Fig. 3. Euclidean reconstruction from α-pixels. (a), (b), (c), (d), (e) and (f) show the four-connected pixel contour, the Euclidean contour reconstructed 2-pixels, the Euclidean contour reconstructed 4-pixels, the Euclidean contour reconstructed 1/60-pixels, the Euclidean contour reconstructed 1-pixels and the Euclidean contour reconstructed 10-pixels, respectively.

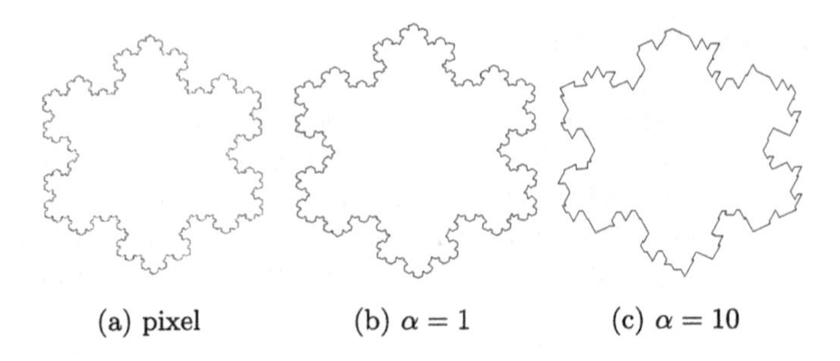

Fig. 4. Koch curves for evaluation of Euclidean reconstruction. (a) is the pixel curve. The Euclidean reconstruction from digital contour curves in (b) and (c) is reconstructed for $\alpha = 1$ and $\alpha = 10$, respectively.

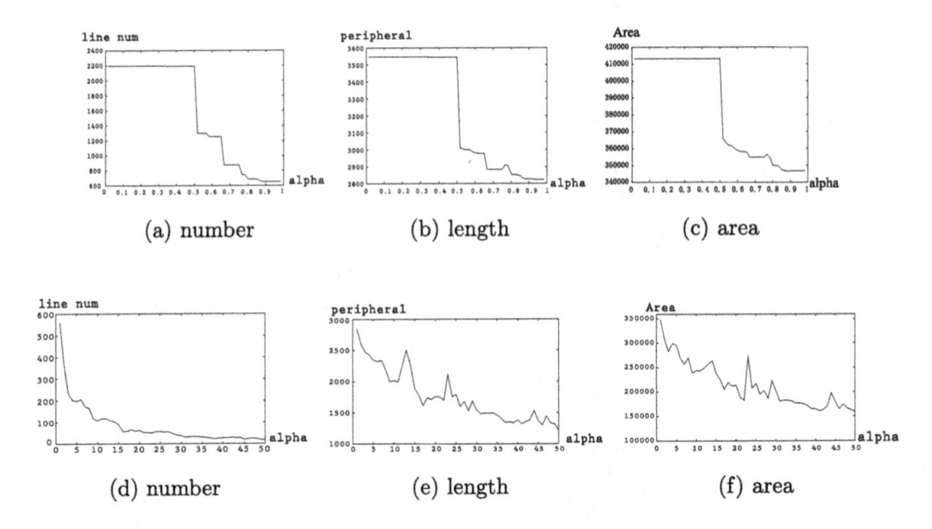

Fig. 5. Hierarchical analysis of Euclidean reconstruction. The numbers of edges, the lengths of contour curves and the areas encircled by reconstructed curves for $\alpha < 1$ are shown in (a), (b) and (c), respectively, The numbers of edges, the lengths of contour curves and the areas encircled by reconstructed curves for $\alpha < 1$ are shown in (d), (e) and (f), respectively.

These examples show that if $\alpha < 0$, the Euclidean reconstructed curve is similar to the polyline fitting to sample points. For $\alpha > 0$, the Euclidean curves reconstructed form α-pixel strings extracts absolute shapes if α increases.

For the analysis of the reconstructed curves, the Euclidean reconstruction curves are computed for various α values for the Koch curve in Fig. 4(a). Figures 4(b) and (c) show the Euclidean reconstruction from digital contour curves of $\alpha = 1$ and $\alpha = 10$, respectively.

Figure 5 shows evaluations of the numbers of edges, the lengths of contour curves and the areas encircled by reconstructed curves, respectively. These results show that for $\alpha < 1/2$ the numbers of edges, the lengths of contour curves and the areas encircled by reconstructed curves are almost stational. On the other hand, for $\alpha > 1$, α acts as a scale for the hierarchical expression of shapes.

6 Conclusions

The Euclidean reconstruction problem of a line from samples $\{\boldsymbol{x}_i = (x_i, y_i)^\top\}_{i=1}^n$ for $\alpha \to 0$ implies the minimisation criterion

$$J(a, b, \mu) = \sum_{i=1}^{n} (|ax_i + b_i + |\mu|). \tag{28}$$

Equation (28) is the minimiser for the ℓ_1 line fitting [15,16]. Setting

$$\boldsymbol{s} = (s_1, s_2, \cdots, s_n)^\top, \ \ \boldsymbol{e} = (1, 1, \cdots, 1)^\top, \tag{29}$$

the ℓ_1 line fitting is established by minimising $z = \boldsymbol{e}^\top \boldsymbol{s}$ subject to

$$\boldsymbol{s} \geq 0, \ \ -\boldsymbol{s} \leq (\boldsymbol{X}\boldsymbol{a} + \mu\boldsymbol{e}) \leq \boldsymbol{s}, \ \ \boldsymbol{X}^\top = \left(\boldsymbol{x}_1, \boldsymbol{x}_2, \ldots, \boldsymbol{x}_n\right), \tag{30}$$

where $\boldsymbol{a} = (a, b)^\top$ [16,17], using linear programming [17].

Therefore, our methods for Euclidean line reconstruction and polyline reconstruction from a collection of α-pixels, respectively, are modifications of ℓ_1 line fitting [16] and ℓ_1-polyline estimation [18,19] optimisation to interval analysis [6–8].

Acknowledgement. This research was supported by the Grants-in-Aid for Scientific Research funded by the Japan Society for the Promotion of Science, under 20K11881.

References

1. Rosenfeld, A., Klette, R.: Digital Geometry: Geometric Methods for Digital Image Analysis (The Morgan Kaufmann Series in Computer Graphics). Morgan Kaufmann, San Diego (2004)
2. Bruckstein, A.M.: Digital geometry in image-based metrology. In: Brimkov, V., Barneva, R. (eds.) Digital Geometry Algorithms. Lecture Notes in Computational Vision and Biomechanics, vol. 2. Springer, Cham (2012). https://doi.org/10.1007/978-94-007-4174-4_1
3. Imiya, A., Sato, K.: Shape from silhouettes in discrete space. In: Brimkov, V., Barneva, R. (eds.) Digital Geometry Algorithms. Lecture Notes in Computational Vision and Biomechanics, vol. 2. Springer, Dordrecht (2012). https://doi.org/10.1007/978-94-007-4174-4_11
4. Coeurjolly, D., Zerarga, L.: Supercover model, digital straight line recognition and curve reconstruction on the irregular isothetic grids. Comput. Graph. **30**, 46–53 (2006). https://doi.org/10.1016/j.cag.2005.10.009

5. Amenta, N., Choi, S., Dey, T.K., Leekha, N.: A simple algorithm for homeomorphic surface reconstruction. In: SCG 2000: Proceedings of the Sixteenth Annual Symposium on Computational Geometry, pp. 213–222 (2000). https://doi.org/10.1145/336154.336207
6. Karasick, M., Lieber, D., Nackman, L.: Efficient Delaunay triangulation using rational arithmetic. ACM Trans. Graph. **10**, 71–91 (1991)
7. Alefeld, G., Mayer, G.: Interval analysis: theory and applications. J. Comput. Appl. Math. **121**, 421–464 (2000). https://doi.org/10.1016/S0377-0427(00)00342-3
8. Ratschek, H.J., Rokne, J.: Geometric Computations with Interval and New Robust Methods Applications in Computer Graphics, GIS and Computational Geometry. Woodhead Publishing Limited (2003)
9. Vacavant, A., Roussillon, T., Kerautret, B., Lachaud, J.-O.: A combined multiscale/irregular algorithm for the vectorization of noisy digital contours. CVIU **117**, 438–450 (2013). https://doi.org/10.1016/j.cviu.2012.07.006
10. Vacavant, A.: Fast distance transformation on two-dimensional irregular grids. Pattern Recogn. **43**, 3348–3358 (2010). https://doi.org/10.1016/j.patcog.2010.04.018
11. Gage, M., Hamilton, R.S.: The heat equation shrinking convex plane curves. J. Differ. Geom. **23**, 69–96 (1986). https://doi.org/10.4310/jdg/1214439902
12. Grayson, M.A.: The heat equation shrinks embedded plane curves to round points. J. Differ. Geom. **26**, 285–314 (1987). https://doi.org/10.4310/jdg/1214441371
13. Bruckstein, A.M., Sapiro, G., Shaked, D.: Evolutions of planar polygons. IJPRAI **9**, 991–1014 (1995). https://doi.org/10.1142/S0218001495000407
14. Serra, J.: Image Analysis and Mathematical Morphology. Academic Press (1983)
15. Barrodale, I., Roberts, F.D.K.: An improved algorithm for discrete l_1 linear approximation. SIAM J. Numer. Anal. **10**, 839–848 (1973). https://www.jstor.org/stable/2156318
16. Wstson, G.A.: Approximation in normed linear spaces. J. Comput. Appl. Math. **121**, 1–36 (2000). https://doi.org/10.1016/S0377-0427(00)00333-2
17. Matouăek, J., Gärtner, B.: Understanding and using linear programming. Springer (2007). https://doi.org/10.1007/978-3-540-30717-4
18. Imai, H., Kato, K., Yamamoto, P.: A linear-time algorithm for linear L_1 approximation of points. Algorithmica **4**, 77–96 (1989). https://doi.org/10.1007/BF01553880
19. Aronov, B., Asano, T., Katoh, N., Mehlhorn, K., Tokuyama, T.: Polyline fitting of planar points under min-sum criteria. In: Fleischer, R., Trippen, G. (eds.) ISAAC 2004. LNCS, vol. 3341, pp. 77–88. Springer, Heidelberg (2004). https://doi.org/10.1007/978-3-540-30551-4_9

Hypergraph p-Laplacians, Scale Spaces, and Information Flow in Networks

Ariane Fazeny[1(✉)], Daniel Tenbrinck[1], and Martin Burger[2,3]

[1] Friedrich-Alexander-Universität Erlangen-Nürnberg, 91058 Erlangen, Germany
ariane.fazeny@fau.de

[2] Deutsches Elektronen-Synchrotron, 22607 Hamburg, Germany

[3] Fachbereich Mathematik, Universität Hamburg, 20146 Hamburg, Germany

Abstract. The aim of this paper is to revisit the definition of differential operators on hypergraphs, which are a natural extension of graphs in systems based on interactions beyond pairs. In particular we focus on the definition of Laplacian and p-Laplace operators, their basic spectral properties, variational structure, and their scale spaces.

We shall see that the corresponding gradient flows, i.e., diffusion equations on hypergraphs, are possible models for the information flow on social networks, e.g., in opinion formation based on group discussion. Moreover, the spectral analysis and scale spaces induced by these operators provide a potential method to further analyze complex networks and their multiscale structure.

The quest for spectral analysis on hypergraphs motivates in particular a definition of differential operators with trivial first eigenfunction and thus more interpretable second eigenfunctions. This property is not automatically satisfied in existing definitions of hypergraph p-Laplacians and we hence provide a novel axiomatic approach that extends previous definitions and can be specialized to satisfy such (or other) desired properties.

Keywords: Hypergraphs · PDEs on (hyper)graphs · Diffusion models · Social networks · Information flow · Hypergraph spectral clustering

1 Introduction

Methods for data analysis and simulation of information propagation have strongly benefited from using graph structures in the past, and the modeling with PDEs on graphs including graph p-Laplacians and associated flow became a standard tool for analyzing graph structures and dynamics on such. Traditional

The authors acknowledge financial support by the European Unions Horizon 2020 research and innovation programme under the Marie Skodowska-Curie grant agreement No. 777826 (NoMADS) and the German Science Foundation (DFG) through CRC TR 154 "Mathematical Modelling, Simulation and Optimization Using the Example of Gas Networks", subproject C06. This work was carried out while M. Burger was with the FAU Erlangen-Nürnberg.

L. Calatroni et al. (Eds.): SSVM 2023, LNCS 14009, pp. 677–690, 2023.
https://doi.org/10.1007/978-3-031-31975-4_52

graphs can however capture only *pairwise interactions* of individuals and thus are unable to directly model group relationships, which are relevant e.g. in social networks. In order to mitigate for this problem we propose to apply a more general structure, namely a **hypergraph** with which it is straightforward to encode group interactions. Here we adopt the definition of oriented hypergraphs, whose hyperarcs (generalizing edges) can have more than one ingoing and more than one outgoing vertex.

1.1 Motivation

The hypergraph structure gives additional flexibility when modeling social phenomena, e.g., by connecting opinion leaders to all their followers directly and hence representing a community within a social network. One field of study in analyzing information flow in social networks is *opinion formation*, an interesting phenomenon that can be observed for a group of individuals which interact and have complex relationships with each other. Some individuals of the social network, so-called *opinion leaders* or *social media influencers*, with a large group of followers (up to half a billion people) have a strong influence on the opinion of many others and can even make profit by leveraging their impact on large groups of social media users (see, e.g., [12]). Modeling information flow in social networks mathematically is typically performed by using traditional graphs. With such graphs it is possible to link two social media users with a pairwise connection, if they are online friends or follow each other (see, e.g., [9]). The information flow in the social network can then be modeled in terms of diffusion processes on the graph, e.g., by solving a partial differential equation (PDE) (see, e.g., [1,3]). However, recent work suggest that interactions beyond pairs are of particular relevance (cf. [13]). Structures reminiscent of a Laplacian on hypergraphs can be found in the model of [11].

A similar question arises in the analysis of community structures, where graph spectral clustering is a standard technique. In order to understand networks including group connections, a more general structure such as hypergraphs seems to be more appropriate. The success of PDE-based methods on graphs motivates a further study on hypergraphs in order to explore the potential of PDEs on such objects. For this sake we need appropriate definitions of hypergraph gradients and Laplacians, which we revisit in this paper. Moreover, the study of scales on hypergraphs is a relevant topic, which could naturally be defined by the evolution of diffusion type processes we hence study here.

1.2 Related Work

There already exists extensive literature about traditional graph theory and its application to social networks. In [9], an overview of social network modeling with traditional graphs is given, including community clustering, similarity analysis and community-based event detection. It indicates how the versatile structure of a graph can be applied to real world problems. [1] introduces the so-called ego network, a graph which focuses on one specific social media user in the center and their surrounding concentric layers of followers, sorted hierarchically depending on their contact frequency.

[4] introduces first-order differential operators and a family of p-Laplacian operators for traditional oriented graphs. The proposed partial difference, adjoint, divergence and anisotropic p-Laplacian for traditional graphs are a special case of our vertex gradient, adjoint, divergence and p-Laplacian operators for hypergraphs, which are introduced in Sect. 4 and Sect. 5. The theoretical results of [4] are applied for mathematical image analysis, such as filtering, segmentation, clustering, and inpainting, but not for social network modeling.

[6] generalizes the already known p-Laplacian operators for normal graphs to the hypergraph setting and performs spectral analysis with a specific focus on the 1-Laplacian. The spectral properties are then applied to common (hyper)graph problems, for instance vertex partitioning, cuts in graphs, coloring of vertices and hyperarc partitioning, but the paper does not include any numerical experiments or the modeling of social networks with hypergraphs. In comparison, our gradient, adjoint, and p-Laplacian definitions are more general and also have the property of the gradient null space including constant vertex functions. Additionally, they are also more flexible with respect to their adaptability for application tasks.

[11] uses unoriented hypergraphs to model different sociological phenomena of cliques, such as peer pressure, with consensus models. Diffusion processes in multi-way interactions with convergence to one united group consensus are modeled with a simple 2-Laplacian inspired by the traditional graph setting. Due to the lack of orientation in the hypergraphs, the described consensus models are not able to capture the effects of a one-sided connection through following someone (e.g., Twitter, Instagram), but only mutual connection through being friends (e.g., Facebook).

Furthermore, [14] uses unoriented hypergraphs in machine learning and shows how hypergraph modelling of data relationships can outperform normal graphs in spectral clustering tasks. Similarly, [8] compares two different algorithms for submodular hypergraph clustering, for not oriented hypergraphs with positive vertex weights and a normalized positive hyperedge weight function, namely the Inverse Power Method (IPM) and the clique expansion method (CEM).

1.3 Main Contributions

The contributions of this paper are manifold. First, we introduce generalized vertex p-Laplacian operators for oriented hypergraphs, which generalize the normal graph definitions described in [4] and the simple hypergraph definitions introduced in [6]. However, in comparison to the definitions in [6] we include two different vertex weight functions and hyperarc weight functions respectively, which can be chosen individually to model specific properties of vertices and hyperarcs. Furthermore, the presented vertex gradient definition leading to the vertex p-Laplacian fulfills the expected properties of the continuum setting (antisymmetry and the gradient of a constant function being equal to zero), based on less strict assumptions compared to the implicit gradient of [6].

Moreover, to the best of our knowledge, oriented hypergraphs have not been used to model opinion formation in social networks before. Even though [11] describe group dynamics with unoriented hypergraphs, their models are not applicable to the one-sided information flow of social media networks such as Twitter or Instagram, where opinion leaders influence a large group of followers, but generally do not follow their followers back and hence are only partly (or not at all impacted) by them. With the help of our generalized p-Laplacian operators we are able to mimic the existence of opinion leaders with a large group of followers. Furthermore, due to the included weight functions we can model the level of influence of an individual on their followers and the users they are following as well as individually adapting the strength of the bond between a user and their followers. By summarizing the relationships in the social network as hyperarcs, instead of looking at pairwise connections, we are able to catch group dynamics between one user and their followers. Furthermore, storing the structure of the social network as hyperarcs, instead of regular arcs, lead to more efficient computations of information flow, since the number of hyperarcs is in many cases significantly smaller than the number of arcs.

2 Introduction of Oriented Hypergraphs

The definition of the oriented hypergraph is a generalization of oriented normal graphs. Given a finite set of vertices $\mathcal{V} = \{v_1, v_2, \dots v_N\}$, then a hypergraph does not only capture pairwise connection between two vertices, but higher-order relationships within any subset of all vertices.

Definition 1 (Oriented hypergraph OH). *[6] An oriented hypergraph $OH = (\mathcal{V}, \mathcal{A}_H)$ consists of a finite set of vertices $\mathcal{V}$, and a set of so-called hyperarcs $\mathcal{A}_H$. Each hyperarc $a_q \in \mathcal{A}_H$ is made of two disjoint subsets of vertices*

$$a_q = \left(a_q^{out}, a_q^{in}\right) \tag{1}$$

with $\emptyset \subset a_q^{out}, a_q^{in} \subset \mathcal{V}$, $a_q^{out} \cap a_q^{in} = \emptyset$, a_q^{out} being the set of all output vertices and a_q^{in} being the set of all input vertices of the hyperarc a_q.

Remark 1. As proposed in [10], we differentiate between oriented and unoriented hypergraphs instead of directed and undirected hypergraphs, because for every oriented hyperarc there is only one orientation but two possible directions: the direction along the orientation and the direction against the orientation.

Remark 2. Furthermore, for clarity reasons we assume that each hyperarc in the set of hyperarcs $\mathcal{A}_H$ is unique and thus occurs only once. This automatically implies that the cardinality of the hyperarc set is finite due to set of vertices $\mathcal{V}$ being finite. More precisely the number of hyperarcs $OH = (\mathcal{V}, \mathcal{A}_H)$ is limited by $|\mathcal{A}_H| \leq N^N$.

3 Functions on Oriented Hypergraphs

We now define different functions on oriented hypergraphs, which are used in Sect. 4 to introduce differential operators inspired by the continuum setting.

In order to check if a vertex is part of an hyperarc as an output or an input vertex, we use two kinds of vertex-hyperarc characteristic functions.

Definition 2 (Vertex-hyperarc characteristic functions δ_{out}, δ_{in}). *For an oriented hypergraph $OH = (\mathcal{V}, \mathcal{A}_H)$, we define the output vertex-hyperarc characteristic function δ_{out} as:*

$$\delta_{out} : \mathcal{V} \times \mathcal{A}_H \longrightarrow \{0,1\} \qquad (v_i, a_q) \longmapsto \delta_{out}(v_i, a_q) = \begin{cases} 1 & v_i \in a_q^{out} \\ 0 & otherwise \end{cases}. \tag{2}$$

Respectively, the input vertex-hyperarc characteristic function δ_{in} is given by:

$$\delta_{in} : \mathcal{V} \times \mathcal{A}_H \longrightarrow \{0,1\} \qquad (v_i, a_q) \longmapsto \delta_{in}(v_i, a_q) = \begin{cases} 1 & v_i \in a_q^{in} \\ 0 & otherwise \end{cases}. \tag{3}$$

Real valued functions can be defined both on the set of vertices $\mathcal{V}$ and the set of hyperarcs $\mathcal{A}_H$ to link any data to the oriented hypergraph.

Definition 3 (Vertex functions f and hyperarc functions F). *For an oriented hypergraph $OH = (\mathcal{V}, \mathcal{A}_H)$ vertex functions are defined on the set of vertices:*

$$f : \mathcal{V} \longrightarrow \mathbb{R} \qquad v_i \longmapsto f(v_i). \tag{4}$$

Hyperarc functions are consequently defined on the domain of the set of hyperarcs:

$$F : \mathcal{A}_H \longrightarrow \mathbb{R} \qquad a_q \longmapsto F(a_q). \tag{5}$$

The most important examples for functions defined on oriented hypergraphs are vertex weight functions

$$w : \mathcal{V} \longrightarrow \mathbb{R}_{>0} \qquad v_i \longmapsto w(v_i) \tag{6}$$

and hyperarc weight functions

$$W : \mathcal{A}_H \longrightarrow \mathbb{R}_{>0} \qquad a_q \longmapsto W(a_q). \tag{7}$$

Definition 4 (Symmetric hyperarc functions F). *A hyperarc function F defined on an oriented hypergraph $OH = (\mathcal{V}, \mathcal{A}_H)$ is called symmetric, if for all hyperarcs $a_q = \left(a_q^{out}, a_q^{in}\right) \in \mathcal{A}_H$ it holds true that:*

$$F\left(a_q^{in}, a_q^{out}\right) = \begin{cases} F\left(a_q^{out}, a_q^{in}\right) & \left(a_q^{in}, a_q^{out}\right) \in \mathcal{A}_H \\ 0 & otherwise \end{cases} \tag{8}$$

Note: $a_q = \left(a_q^{out}, a_q^{in}\right) \in \mathcal{A}_H$ generally does not imply $\left(a_q^{in}, a_q^{out}\right) \in \mathcal{A}_H$.

The space of all vertex functions and all hyperarc functions defined on an oriented hypergraph can be identified with an N- or an at most N^N-dimensional Hilbert space, respectively.

Definition 5 (Space of vertex functions $\mathcal{H}(\mathcal{V})$ and space of hyperarc functions $\mathcal{H}(\mathcal{A}_H)$). *The space of all vertex functions f for an oriented hypergraph $OH = (\mathcal{V}, \mathcal{A}_H)$ is given by*

$$\mathcal{H}(\mathcal{V}) = \{f \mid f: \mathcal{V} \longrightarrow \mathbb{R}\} \tag{9}$$

with the inner product $\langle f, g\rangle_{\mathcal{H}(\mathcal{V})} = \sum_{v_i \in \mathcal{V}} w_I(v_i)^{\alpha} f(v_i) g(v_i)$ for any two vertex functions $f, g \in \mathcal{H}(\mathcal{V})$, vertex weight function w_I, and parameter $\alpha \in \mathbb{R}$.

Similarly, the space of all hyperarc functions F is defined as

$$\mathcal{H}(\mathcal{A}_H) = \{F \mid F: \mathcal{A}_H \longrightarrow \mathbb{R}\} \tag{10}$$

with the corresponding inner product $\langle F, G\rangle_{\mathcal{H}(\mathcal{A}_H)} = \sum_{a_q \in \mathcal{A}_H} W_I(a_q)^{\beta} F(a_q) G(a_q)$ for any two hyperarc functions $F, G \in \mathcal{H}(\mathcal{A}_H)$, hyperarc weight function W_I, and parameter $\beta \in \mathbb{R}$.

4 First-Order Differential Operators on Hypergraphs

Utilizing the introduced definitions for hypergraphs we can now generalize the definitions of the vertex gradient, the vertex adjoint, and the vertex p-Laplacian for normal graphs, which have already been discussed in a simplified form with less weight functions and parameters in [4].

Definition 6 (Vertex gradient operator ∇_v). *For an oriented hypergraph $OH = (\mathcal{V}, \mathcal{A}_H)$ with vertex weight functions w_I and w_G and a symmetric hyperarc weight function W_G, we define the vertex gradient operator ∇_v with parameters $\alpha, \gamma, \epsilon, \eta \in \mathbb{R}$ as:*

$$\nabla_v : \mathcal{H}(\mathcal{V}) \longrightarrow \mathcal{H}(\mathcal{A}_H) \quad f \longmapsto \nabla_v f$$
$$\nabla_v f : \mathcal{A}_H \longrightarrow \mathbb{R} \quad a_q \longmapsto \nabla_v f(a_q) =$$

$$W_G(a_q)^{\gamma} \sum_{v_i \in \mathcal{V}} \left(\delta_{in}(v_i, a_q) \frac{w_I(v_i)^{\alpha} w_G(v_i)^{\epsilon}}{|a_q^{in}|} - \delta_{out}(v_i, a_q) \frac{w_I(v_i)^{\alpha} w_G(v_i)^{\eta}}{|a_q^{out}|} \right) f(v_i). \tag{11}$$

The introduced vertex gradient fulfills two expected properties from the continuum setting, namely antisymmetry and the gradient of a constant function being equal to zero.

Theorem 1 (Vertex gradient operator properties). *The vertex gradient ∇_v defined on an oriented hypergraphs $OH = (\mathcal{V}, \mathcal{A}_H)$ with vertex weight functions w_I and w_G and a symmetric hyperarc weight function W_G, satisfies the following properties:*

- **Vanishing gradient of a constant vertex function:** *Let the weight condition*

$$w_I(v_k)^{\alpha} w_G(v_k)^{\epsilon} = w_I(v_j)^{\alpha} w_G(v_j)^{\eta}$$

hold for all vertex combinations $v_j, v_k \in \mathcal{V}$ with $v_j \in a_q^{out}$ and $v_k \in a_q^{in}$ for a hyperarc $a_q \in \mathcal{A}_H$, Then for every constant function f, i.e. $f(v_i) \equiv \overline{f}$ for all vertices $v_i \in \mathcal{V}$, we have $\nabla_v f(a_q) = 0$ for all hyperarcs $a_q \in \mathcal{A}_H$

- **Antisymmetry:** *Let* $\epsilon = \eta$. *Then the identity* $\nabla_v f\left(a_q^{out}, a_q^{in}\right) = -\nabla_v f\left(a_q^{in}, a_q^{out}\right)$ *holds for all hyperarcs* $a_q \in \mathcal{A}_H$.

Proof. See [5] Theorem 9.2 (Vertex gradient operator properties).

Let us mention one additional complication compared to the traditional graph case: While it is trivial to see that for a connected graph constant functions are the only elements in the nullspace of the gradient, this is not apparent for hypergraphs.

By computing the adjoint ∇_v^* of the vertex gradient we can introduce the definition of a divergence operator on the hypergraph in analogy to traditional calculus. For the sake of brevity, we skip the detailed computation based on the relation

$$\langle G, \nabla_v f\rangle_{\mathcal{H}(\mathcal{A}_H)} = \langle f, \nabla_v^* G\rangle_{\mathcal{H}(\mathcal{V})} \tag{12}$$

for all vertex functions $f \in \mathcal{H}(\mathcal{V})$ and all hyperarc functions $G \in \mathcal{H}(\mathcal{A}_H)$.

Definition 7 (Vertex adjoint operator ∇_v^*). *For an oriented hypergraph* $OH = (\mathcal{V}, \mathcal{A}_H)$ *with vertex weight function* w_G *and symmetric hyperarc weight functions* W_I *and* W_G, *the vertex adjoint operator* ∇_v^* *with parameters* $\beta, \gamma, \epsilon, \eta \in \mathbb{R}$ *is given by:*

$$\nabla_v^* : \mathcal{H}(\mathcal{A}_H) \longrightarrow \mathcal{H}(\mathcal{V}) \quad F \longmapsto \nabla_v^* F$$

$$\nabla_v^* F : \mathcal{V} \longrightarrow \mathbb{R} \quad v_i \longmapsto \nabla_v^* F(v_i) =$$

$$\sum_{a_q \in \mathcal{A}_H} \left(\delta_{in}(v_i, a_q) \frac{w_G(v_i)^{\epsilon}}{\left|a_q^{in}\right|} - \delta_{out}(v_i, a_q) \frac{w_G(v_i)^{\eta}}{\left|a_q^{out}\right|}\right) W_I(a_q)^{\beta} W_G(a_q)^{\gamma} F(a_q). \tag{13}$$

Definition 8 (Vertex divergence operator div_v). *For an oriented hypergraph* $OH = (\mathcal{V}, \mathcal{A}_H)$ *with vertex weight function* w_G *and symmetric hyperarc weight functions* W_I *and* W_G, *the vertex divergence operator* div_v *with parameters* $\beta, \gamma, \epsilon, \eta \in \mathbb{R}$ *is given by:*

$$\mathrm{div}_v : \mathcal{H}(\mathcal{A}_H) \longrightarrow \mathcal{H}(\mathcal{V}) \quad F \longmapsto \mathrm{div}_v F$$

$$\mathrm{div}_v F : \mathcal{V} \longrightarrow \mathbb{R} \quad v_i \longmapsto \mathrm{div}_v F(v_i) =$$

$$-\nabla_v^* F(v_i) = \sum_{a_q \in \mathcal{A}_H} \left(\delta_{out}(v_i, a_q) \frac{w_G(v_i)^{\eta}}{\left|a_q^{out}\right|} - \delta_{in}(v_i, a_q) \frac{w_G(v_i)^{\epsilon}}{\left|a_q^{in}\right|}\right) W_I(a_q)^{\beta} W_G(a_q)^{\gamma} F(a_q). \tag{14}$$

5 p-Laplacian Operators on Hypergraphs

Based on the previous definitions we introduce a generalized vertex p-Laplacian inspired by the continuum setting, which implies that for all $p \in (1, \infty)$ and all vertex functions $f \in \mathcal{H}(\mathcal{V})$ it holds true that:

$$\Delta_v^p f = \mathrm{div}_v \left(|\nabla_v f|^{p-2} \nabla_v f\right).$$

Note that from the definition of the divergence as a negative adjoint of the gradient it becomes clear the hypergraph p-Laplacian is the negative variation of the p-norm of the gradient, which allows to apply the full theory of eigenvalues of p-homogeneous functionals (see, [2]). In particular, the hypergraph Laplacian is a negative semidefinite linear operator and has a spectrum on the negative real line.

Definition 9 (Vertex p-Laplacian operator Δ_v^p). *For an oriented hypergraph $OH = (\mathcal{V}, \mathcal{A}_H)$ with vertex weight functions w_I and w_G and with symmetric hyperarc weight functions W_I and W_G, the vertex p-Laplacian operator Δ_v^p with parameters $\alpha, \beta, \gamma, \epsilon, \eta \in \mathbb{R}$ is given by:*

$$\Delta_v^p : \mathcal{H}(\mathcal{V}) \longrightarrow \mathcal{H}(\mathcal{V}) \quad f \longmapsto \Delta_v^p f \qquad \Delta_v^p f : \mathcal{V} \longrightarrow \mathbb{R} \quad v_i \longmapsto \Delta_v^p f(v_i) =$$

$$\sum_{a_q \in \mathcal{A}_H} \left(\delta_{out}(v_i, a_q) \frac{w_G(v_i)^\eta}{|a_q^{out}|} - \delta_{in}(v_i, a_q) \frac{w_G(v_i)^\epsilon}{|a_q^{in}|} \right) W_I(a_q)^\beta W_G(a_q)^{p\gamma}$$

$$\left| \sum_{v_j \in \mathcal{V}} \left(\delta_{in}(v_j, a_q) \frac{w_I(v_j)^\alpha w_G(v_j)^\epsilon}{|a_q^{in}|} - \delta_{out}(v_j, a_q) \frac{w_I(v_j)^\alpha w_G(v_j)^\eta}{|a_q^{out}|} \right) f(v_j) \right|^{p-2}$$

$$\sum_{v_k \in \mathcal{V}} \left(\delta_{in}(v_k, a_q) \frac{w_I(v_k)^\alpha w_G(v_k)^\epsilon}{|a_q^{in}|} - \delta_{out}(v_k, a_q) \frac{w_I(v_k)^\alpha w_G(v_k)^\eta}{|a_q^{out}|} \right) f(v_k). \quad (15)$$

The following theorem states that the vertex p-Laplacian is well-defined.

Theorem 2 (Connection vertex gradient ∇_v, vertex divergence div_v, and vertex p-Laplacian Δ_v^p). *For an oriented hypergraph $OH = (\mathcal{V}, \mathcal{A}_H)$ with vertex weight functions w_I and w_G and symmetric hyperarc weight functions W_I and W_G, the vertex p-Laplacian Δ_v^p fulfills the equality*

$$\Delta_v^p f = \text{div}_v \left(|\nabla_v f|^{p-2} \nabla_v f \right) \quad (16)$$

for all vertex functions $f \in \mathcal{H}(\mathcal{V})$.

Proof. See [5] Theorem 10.13 (Connection vertex divergence div_v, vertex gradient ∇_v, and vertex p-Laplacian Δ_v^p).

Moreover, our vertex p-Laplacian definition is a valid generalization of the definition introduced in [6].

Remark 3 (Parameter choice for the vertex p-Laplacian operator). The simplified definition of the vertex p-Laplacian introduced in [6] for any vertex function $f \in \mathcal{H}(\mathcal{V})$ and for any vertex $v_i \in \mathcal{V}$ can be written in our notation as:

$$\Delta_p f(v_i) = \frac{1}{\deg(v_i)} \sum_{\substack{a_q \in \mathcal{A}_H:\ \delta_{out}(v_i,a_q)=1 \\ \text{or } \delta_{in}(v_i,a_q)=1}} \left| \sum_{v_j \in a_q^{in}} f(v_j) - \sum_{v_j \in a_q^{out}} f(v_j) \right|^{p-2}$$

$$\left(\sum_{v_k \in \mathcal{V}} \left(\delta_{out}(v_i,a_q)\,\delta_{out}(v_k,a_q) + \delta_{in}(v_i,a_q)\,\delta_{in}(v_k,a_q)\right) f(v_k) - \right.$$

$$\left. \sum_{v_k \in \mathcal{V}} \left(\delta_{out}(v_i,a_q)\,\delta_{in}(v_k,a_q) + \delta_{in}(v_i,a_q)\,\delta_{out}(v_k,a_q)\right) f(v_k) \right). \quad (17)$$

The factor $\left(\delta_{out}(v_i,a_q)\,\delta_{out}(v_k,a_q) + \delta_{in}(v_i,a_q)\,\delta_{in}(v_k,a_q)\right)$ is always equal to zero, unless $v_i, v_k \in a_q^{out}$ or $v_i, v_k \in a_q^{in}$, which means that the vertices v_i and v_k are co-oriented. Similarly, the factor $\big(\delta_{out}(v_i,a_q)\,\delta_{in}(v_k,a_q) + \delta_{in}(v_i,a_q)\,\delta_{out}(v_k,a_q)\big)$ ensures to only consider vertices $v_k \in \mathcal{V}$ which are anti-oriented compared to vertex v_i and hence either $v_i \in a_q^{out}, v_k \in a_q^{in}$ or $v_i \in a_q^{in}, v_k \in a_q^{out}$.

Thus, choosing the parameters of the vertex p-Laplacian Δ_v^p as $\alpha = 0$, $\beta = 0$, $\gamma = 0$, $\epsilon = 0$ and $\eta = 0$ together with excluding the $\frac{1}{|a_q^{out}|}$ and $\frac{1}{|a_q^{in}|}$ multiplicative factors and including a new $-\frac{1}{\deg(v_i)}$ factor in the vertex adjoint and the vertex divergence, results in the simplified vertex p-Laplacian introduced in [6].

Furthermore, applying these parameter choices to the vertex gradient, the vertex adjoint and the vertex divergence lead to the following definitions for all vertex functions $f \in \mathcal{H}(\mathcal{V})$, all hyperarc functions $f \in \mathcal{H}(\mathcal{A}_H)$, for all hyperarcs $a_q \in \mathcal{A}_H$ and all vertices $v_i \in \mathcal{V}$:

- $\nabla_v f(a_q) = \sum_{v_i \in \mathcal{V}} \left(\delta_{in}(v_i,a_q) - \delta_{out}(v_i,a_q)\right) f(v_i)$
- $\nabla_v^* F(v_i) = -\frac{1}{\deg(v_i)} \sum_{a_q \in \mathcal{A}_H} \left(\delta_{in}(v_i,a_q) - \delta_{out}(v_i,a_q)\right) F(a_q)$
- $\mathrm{div}_v(F)(v_i) = -\frac{1}{\deg(v_i)} \sum_{a_q \in \mathcal{A}_H} \left(\delta_{out}(v_i,a_q) - \delta_{in}(v_i,a_q)\right) F(a_q)$.

Proof. See [5] Theorem 10.12 (Parameter choice for the vertex p-Laplacian operator).

6 Scale Spaces Based on Hypergraph p-Laplacians

In the following we discuss PDEs based on the family of p-Laplace operators on hypergraphs introduced in Sect. 5, which can be used for modeling and analyzing information flow in social networks. The obvious starting point is to investigate the scale space for the p-Laplacian operator, i.e. the gradient flow of the p-Laplacian energy:

$$\begin{aligned} \frac{\partial f}{\partial t}(v_i,t) &= \Delta_v^p f(v_i,t), && v_i \in \mathcal{V}, t \in (0,\infty) \\ f(v_i,0) &= f_0(v_i), && v_i \in \mathcal{V}. \end{aligned} \quad (18)$$

Solving (18) for every time step $t \in (0,\infty)$ amounts in computing the information flow between vertices of the hypergraph along the respective hyperarcs.

Note that although there are no explicit boundaries in hypergraphs, we can interpret the above problem as the homogeneous Neumann boundary problem. Due to the properties of the proposed family of hypergraph p-Laplace operators it is easy to see that the mean-value of f is conserved in time and we can naturally interpret the evolution as a scale space towards coarser and coarser scales on the graph. Moreover, the general asymptotic of gradient flows for p-homogeneous energies (cf. [2]) yields that $f \to \overline{f}$ as $t \to \infty$ with $\overline{f}$ being the mean value of f_0. Moreover, the rescaled quantity $g = \frac{f-\overline{f}}{\|f-\overline{f}\|}$ converges to a multiple of a second eigenfunction for generic initial values.

Similar to the Neumann boundary problem we can also introduce a Dirichlet type problem, where the Dirichlet boundary $\partial\mathcal{V} \subset \mathcal{V}$ denotes a subset of the vertex set $\mathcal{V}$, for which we introduce boundary values and keep them fixed over time. The corresponding stationary solution is not necessarily constant

$$\begin{aligned} \Delta_v^p f(v_i) &= 0, \qquad v_i \in \mathring{\mathcal{V}}, \\ f(v_j) &= F_j, \qquad v_j \in \partial\mathcal{V}. \end{aligned} \tag{19}$$

Then, we aim to solve the p-Laplace equation on the complementary vertex set $\mathring{\mathcal{V}} := \mathcal{V} \setminus \partial\mathcal{V}$ of the hypergraph. Instead of solving (19) directly, we solve the hyperbolic PDE model (18) on the vertex set $\mathring{\mathcal{V}}$, while keeping the vertex function $f \in \mathcal{H}(\mathcal{V})$ fixed on the boundary set $\partial\mathcal{V}$. The reason for this approach is that any stationary solution of (18) on $\mathring{\mathcal{V}}$ with fixed boundary values is also a solution to the p-Laplace equation in (19). To solve the two proposed PDE models discussed above we numerically have to solve the initial value problem in (18). For this sake we employ a forward-Euler time discretization with fixed time step size $\tau > 0$ and use the renormalized variable g to observe convergence to a nontrivial eigenfunction. This leads to the following explicit iteration scheme

$$f_{n+1}(v_i) \;=\; f_n(v_i) + \tau \cdot \Delta_v^p f_n(v_i). \tag{20}$$

7 Numerical Experiments

In the following we present the results of our numerical experiments in which we solve the two PDEs (18) and (19) by using the explicit forward Euler discretization scheme until the relative change between two iterations is smaller than $\epsilon := 10^{-6}$. We choose τ in (20) small enough to fulfill the CFL condition for numerical stability. This leads to very small time steps for the iteration scheme for the cases $1 \leq p < 2$.

The source code of our method can be downloaded from the repository available at https://gitlab.com/arianefazeny/hypergraph_p-laplace.

For our numerical experiments we use the Twitter data set provided by Stanford University [7]. It consists of 41,652,230 vertices (users) and 1,468,364,884 arcs (oriented pairwise connections indicating that one person follows another). Due to the size of the data set, we restrict our numerical experiments to a comparatively small sub-network within the first 1,000,000 lines of the Twitter input

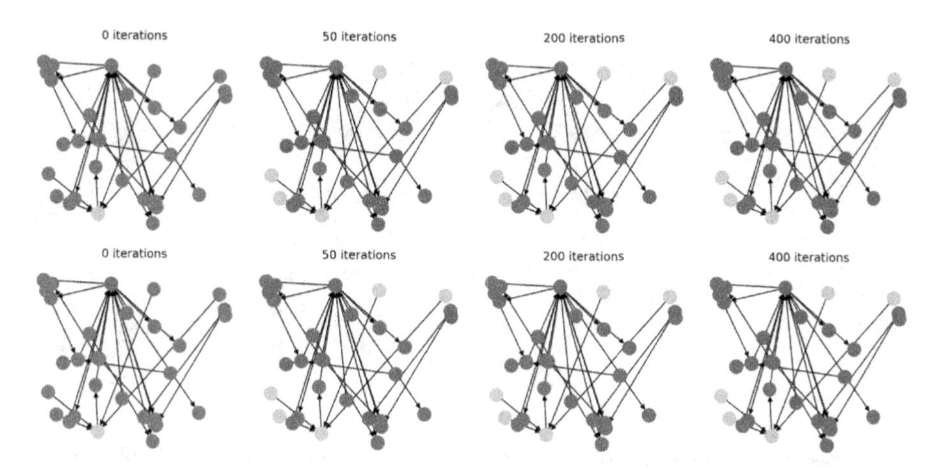

Fig. 1. Solution of the boundary value problem of graph (top) and hypergraph (bottom) p-Laplace operator for $p = 2$.

data. We chose the sub-network of individuals such that all users are directly or indirectly linked to each other to avoid cliques of individuals, which are not connected to the rest of the sub-network and thus also not influenced by users outside their small circle. Furthermore, we ensure that each sub-network includes an opinion leader with a large number of followers in the sub-network. Therefore, we can observe how one influential user impacts the opinion of the rest of the network. In order to generate hyperarcs from the given arcs, we put one Twitter user as a singleton output vertex set and summarize all followers of this user as the set of input vertices. This especially allows highlighting the effect of opinion leaders, for instance famous people with a large group of followers on Twitter.

We simulate the opinion of all individuals in the social network towards an imaginary hypothesis by a vertex function $f\colon \mathcal{V} \times [0, \infty) \to [-1, 1]$, which can be interpreted as the following. If an individual believes the hypothesis the corresponding value of the vertex function is positive (with 1 being the strongest level of trust), while for an individual that opposes the hypothesis the corresponding value of the vertex function is negative (with -1 being the strongest level of distrust).

For the **boundary value problem** (19) we initialize the opinion of all individuals in a social network by setting the vertex function f to zero, which can be interpreted as having no opinion towards an imaginary hypothesis. We now simulate information flow in the social network by giving two opinion leaders (i.e., vertices with many followers) two opposing opinions towards this hypothesis and setting the respective values of the vertex function to -1 and 1. We keep these values fixed to simulate Dirichlet boundary conditions. By using the explicit forward Euler discretization scheme to solve the boundary value problem for $p = 2$ the opinion of the two dedicated individuals is propagated in the social network as can be seen in Fig. 1. We initialize the vertex function equally

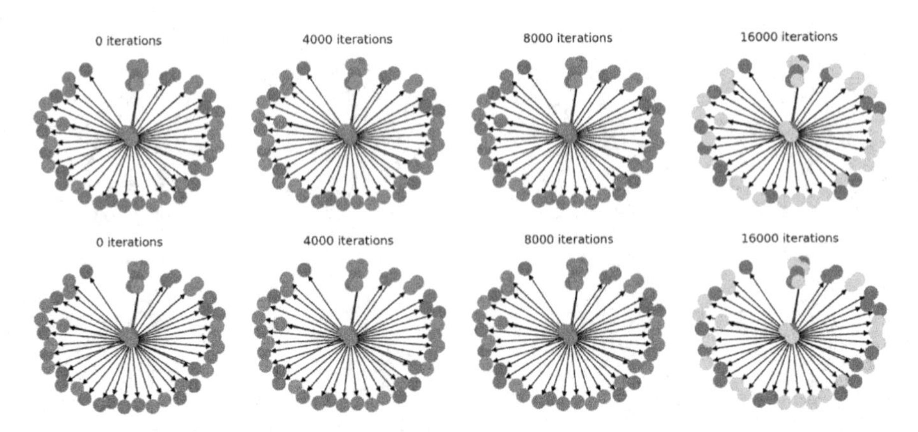

Fig. 2. Second eigenfunction of graph (top) and hypergraph (bottom) p-Laplace operator for $p = 1$ with thresholding at 0 after 16.000 iterations.

for the normal graph (top row) and the hypergraph (bottom row) and calculate the diffusion process until convergence. As can be seen, in both cases the opinion is propagated in the social network based on the underlying network topology and the final state is equivalent for both the normal graph and the hypergraph experiment. However, as can be observed information within the hypergraph is distributed at a higher rate compared to the normal graph and thus converging faster. This is due to the fact that opinion leaders in a normal graph have a less direct impact on their followers compared to the hypergraph case, where the follower's believe $f(v_i)$ is scaled with $\frac{1}{|a_q^{in}|}$, where $|a_q^{in}|$ is the number of followers of the individual user. This can be seen in (15) since in our modeling for this application the parameter a_q^{out} is set to 1.

For the **initial value problem** (18) we choose $p = 1$. We initialize each individual's opinion $f_0(v_i)$ randomly with a uniform distribution in the interval $[-1, 1]$. Additionally, we make sure that the vertex function f is initialized with average 0 and normalized. As can be observed in Fig. 2 the information flow in the social network converges to a second eigenfunction of both the graph p-Laplacian (top row) and the hypergraph p-Laplacian. For both cases we thresholded at 0 after 16.000 iterations to induce a spectral clustering of the opposing opinions and separating the social network into smaller communities based on the topology of the network (i.e., the relationship of following an individual). As can be seen the resulting second eigenfunctions differ significantly with respect to the underlying topology of the normal graph and the hypergraph. This yields potential for further analysis and experiments in other applications, e.g., segmentation of images via spectral clustering.

8 Conclusion

In this paper we derived a general version of differential operators and a family of p-Laplacian operators on hypergraphs, which generalizes known graph and hypergraph operators from literature. The resulting operators and the associated scale space flows can be employed for modelling information flows or perform spectral analysis in social networks, where we can directly incorporate group relationships via hyperarcs. Preliminary results indicate a great potential for future research. Interesting further questions in addition to a more detailed study of spectral clustering are e.g. the relation between hypergraph gradients and higher-order methods for partial differential equations or the definition of distance functions via eigenfunctions of the infinity-Laplacian. Also, we aim to investigate the potential of the proposed diffusion models on hypergraphs for problems in mathematical image processing and computer vision, e.g., for non-local image denoising.

References

1. Arnaboldi, V., Conti, M., Passarella, A., Dunbar, R.: Online social networks and information diffusion: the role of ego networks. Online Soc. Netw. Media **1**, 44–55 (2017)
2. Bungert, L., Burger, M.: Asymptotic profiles of nonlinear homogeneous evolution equations of gradient flow type. J. Evol. Equ. **20**, 1061–1092 (2020)
3. Chamley, C., Scaglione, A., Li, L.: Models for the diffusion of beliefs in social networks: an overview. IEEE Signal Process. Mag. **30**(3), 16–29 (2013)
4. Elmoataz, A., Toutain, M., Tenbrinck, D.: On the p-Laplacian and infinity-Laplacian on graphs with applications in image and data processing. SIAM J. Imag. Sci. **8**(4), 2412–2451 (2015)
5. Fazeny, A.: p-Laplacian Operators on Hypergraphs. Master thesis at FAU Erlangen-Nürnberg. https://arxiv.org/abs/2304.06468 (2023)
6. Jost, J., Mulas, R., Zhang, D.: p-laplace operators for oriented hypergraphs. Vietnam J. Math. **50**(2), 323–358 (2021)
7. Leskovec, J., Krevl, A.: SNAP Datasets: Stanford Large Network Dataset Collection. http://snap.stanford.edu/data. Accessed 5 Oct 2022
8. Li, P., Milenkovic, O.: Submodular hypergraphs: p-laplacians, cheeger inequalities and spectral clustering. In: International Conference on Machine Learning, pp. 3014–3023 (2018)
9. Majeed, A., Rauf, I.: Graph theory: a comprehensive survey about graph theory applications in computer science and social networks. Inventions **5**(1), 10 (2020)
10. Mulas, R., Kuehn, C., Böhle, T., Jost, J.: Random walks and Laplacians on hypergraphs. Discret. Appl. Math. **317**, 26–41 (2022)
11. Neuhäuser, L., Lambiotte, R., Schaub, M.: Consensus dynamics and opinion formation on hypergraphs. In: Battiston, F., Petri, G. (eds.) Higher-Order Systems, pp. 347–376. Springer, Cham (2022). https://doi.org/10.1007/978-3-030-91374-8_14
12. Turcotte, J., York, C., Irving, J., Scholl, R., Pingree, R.: News recommendations from social media opinion leaders: effects on media trust and information seeking. J. Comput.-Mediat. Commun. **20**(5), 520–535 (2015)

13. Zanette, D.: H: Beyond networks: opinion formation in triplet-based populations. Philos. Trans. Royal Soc. A Math. Phys. Eng. Sci. **367**, 3311–3319 (2009)
14. Zhou, D., Huang, J., Schölkopf, B.: Learning with hypergraphs: clustering, classification, and embedding. In: Advances in Neural Information Processing Systems, vol. 19 (2006)

On Photometric Stereo in the Presence of a Refractive Interface

Yvain Quéau[1(✉)], Robin Bruneau[2,3], Jean Mélou[2], Jean-Denis Durou[2], and François Lauze[3]

[1] Normandie Univ, UNICAEN, ENSICAEN, CNRS, GREYC, 14000 Caen, France
yvain.queau@ensicaen.fr
[2] IRIT, UMR CNRS 5505, Toulouse, France
[3] DIKU, Copenhagen University, Copenhagen, Denmark

Abstract. We conduct a discussion on the problem of 3D-reconstruction by calibrated photometric stereo, when the surface of interest is embedded in a refractive medium. We explore the changes refraction induces on the problem geometry (surface and normal parameterization), and we put forward a complete image formation model accounting for refracted lighting directions, change of light density and Fresnel coefficients. We further show that as long as the camera is orthographic, lighting is directional and the interface is planar, it is easy to adapt classic methods to take into account the geometric and photometric changes induced by refraction. Moreover, we show on both simulated and real-world experiments that incorporating these modifications of PS methods drastically improves the accuracy of the 3D-reconstruction.

1 Introduction

Photometric stereo (PS) is a 3D computer vision technique which was pioneered by Woodham in the late 70s [27]. It aims at inferring the shape of an opaque surface from a series of images captured under the same viewing angle, but varying illumination. Compared to other 3D-reconstruction techniques, PS excels at recovering the thinnest geometric variations (high-frequency information given by surface normals), and it is the only photographic 3D-reconstruction method which is also able to infer the reflectance of the surface. Such properties are essential in applications such as relighting or cultural heritage artifacts digitization.

However, a fundamental assumption in PS is that the light sources, the camera and the pictured surface all lie in the same homogeneous medium – usually the air. In the present paper, we revisit PS in the presence of a refractive interface i.e., when the camera and the light sources both lie in one homogeneous medium, while the surface is immersed in another homogoneous medium with a different index of refraction (pure water, glass, alcohol, etc.). This particular setting finds applications, for instance, in underwater imaging (Fig. 1a) or in the digitization of natural historic museal objects preserved in amber or alcohol.

L. Calatroni et al. (Eds.): SSVM 2023, LNCS 14009, pp. 691–703, 2023.
https://doi.org/10.1007/978-3-031-31975-4_53

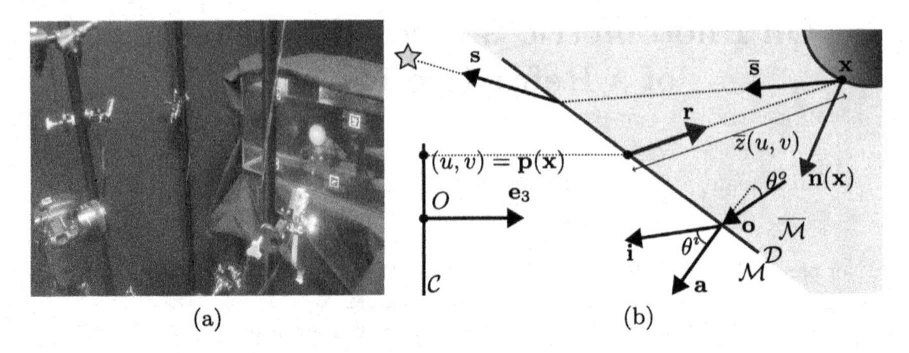

Fig. 1. We discuss the problem of recovering through photometric stereo the 3D-shape and the albedo of a surface immersed in a refractive medium, as in (a) where a white sphere is immersed in an aquarium filled with pure water. In particular, we show how to adapt classic PS methods when the lighting is directional, the camera is orthographic and the interface is planar, as illustrated in the sketch (b) which summarizes our notations. Therein, given a plane $\mathcal{D}$ with normal $\mathbf{a}$, Snell's law (2.3) gives the relation between an incident ray $\mathbf{i}$ in medium $\mathcal{M}$, and the refracted one $\mathbf{o}$ in medium $\overline{\mathcal{M}}$. Even if the camera is orthographic, a point $\mathbf{x}$ on the surface projects non-orthogonally onto the image plane $\mathcal{C}$: a pixel (u, v) first deprojects onto $\mathcal{D}$ along the viewing direction $\mathbf{e}_3$, and then travels the distance $\overline{z}(u, v)$ along the refracted viewing direction $\mathbf{r}$. Besides, the effective lighting direction $\overline{\mathbf{s}}$ differs from the direction $\mathbf{s}$ which is calibrated outside the refractive medium.

The difference with classic PS lies in the presence of an interface between the two media, which have different indices of refraction. Refraction will have profound consequences in 3D shape recovery techniques, as it modifies the geometry of image acquisition, and light direction and density will be changed as well (this is, after all, the principle behind a lot of lensing effects). While these points are well-understood by designers of optical systems, either to use them or for limiting some of their undesirable consequences, they have seldom been investigated from the photometric shape recovery side.

Assumptions and Contributions. We address the PS problem, in the presence of a *Lambertian* surface (specularities are viewed as outliers) embedded in a *homogeneous* refractive medium with *known geometry*, imaged in the *visible spectrum*. After reviewing related works in Sect. 2, we explore the impact of a *planar* (but not necessarily fronto-parallel) refractive interface on the geometry of PS under *orthographic projection* in Sect. 3. In Sect. 4, we derive a complete image formation model for this case, under *directional lighting calibrated outside the refractive medium*. This model accounts for refraction of lighting directions, attenuation of lighting densities, and Fresnel coefficients. Then, we discuss in Sect. 5 the inversion of this model by adapting classic PS algorithms. Eventually, in Sect. 6 we draw our conclusions, and mention possible extensions of our work to more complicated setups (pinhole camera, non-directional lighting, non-planar interface, and light absorption).

2 Background

Photometric Stereo. In the traditional PS setup, the pictured surface $\mathcal{S}$ is assumed Lambertian i.e., it reflects light diffusively, as the reflectance at $\mathbf{x} \in \mathcal{S}$ is characterized by the albedo $\rho(\mathbf{x}) \in [0,1]$. Let us consider a surface lit by a single, known point light source at infinity (calibrated directional lighting) represented by unit direction $\mathbf{s} \in \mathbb{R}^3$ and density $\varphi > 0$, and denote $\mathbf{n}(\mathbf{x}) \in \mathbb{R}^3$ the unit outward normal to the surface at $\mathbf{x}$. Then, the measured brightness at pixel $(u,v) = \mathbf{p}(\mathbf{x})$, which is the projection of the surface point $\mathbf{x}$ onto the camera image plane, is $I(\mathbf{p}) \propto \varphi \max\{0, \rho(\mathbf{x})\mathbf{n}(\mathbf{x})^\top \mathbf{s}\}$, with the proportionality constant independent of $\mathbf{x}$. Omitting the max{} operator, which models self-shadows (they are usually dealt with robust estimators), integrating the proportionality coefficient into the albedo (which can be normalized a posteriori), and considering $k \geq 3$ light sources yields the following image formation model:

$$I_i(\mathbf{p}) = \varphi_i \, \rho(\mathbf{x}) \, \mathbf{n}(\mathbf{x})^\top \mathbf{s}_i, \quad i \in \{1, \dots, k\}. \tag{2.1}$$

This model can be inverted as long as the light directions $\mathbf{s}_i$ are non-coplanar, so as to compute the Lambertian reflectance $\rho(\mathbf{x})$ and the surface normal $\mathbf{n}(\mathbf{x})$ for each $\mathbf{p}$. This approach can also be extended to non-Lambertian reflectance and uncalibrated lighting, for instance by resorting to deep neural networks [7].

Surface Parameterization. The surface $\mathcal{S}$ is parameterized as $(u,v) \mapsto \mathbf{x}(u,v) = \mathcal{S}(u,v)$ and its normal at $\mathbf{x}$ is written as

$$\mathbf{n}(\mathbf{x}) = \pm \frac{\mathcal{S}_u \times \mathcal{S}_v}{|\mathcal{S}_u \times \mathcal{S}_v|}, \tag{2.2}$$

where $\mathcal{S}_u$ and $\mathcal{S}_v$ are the partial derivatives of $\mathcal{S}$, and where the $\pm$ ambiguity is resolved by taking arbitrarily the normal oriented towards the camera. Once the normal field $\mathbf{n}(\mathbf{x})$ is estimated, retrieving $\mathcal{S}$ then comes down to a 2D integration problem, for which various solutions exist [20]. The parameterization $\mathcal{S}$ is a right-inverse to the projection: $\mathbf{p}(\mathcal{S}(u,v)) = (u,v)$. It is constrained by the form that $\mathbf{p}$ takes (orthographic projection, perspective projection, etc.), and this has of course important consequences on the integration process.

Refraction. The index of refraction (IoR) n of a material is the ratio c/v of the speed of light in vacuum and the velocity in the medium. Snell's laws assert that 1) the normal $\mathbf{a}$ to the interface, the incident light direction $\mathbf{i}$ and the refracted light direction $\mathbf{o}$ are coplanar; and 2) the refracted and incident angles satisfy the relation $n \sin\theta^i = \overline{n} \sin\theta^o$, with n the IoR of the first medium, $\overline{n}$ the IoR of the second one, θ^i the angle between $\mathbf{i}$ and $\mathbf{a}$, and θ^o the angle between $\mathbf{a}$ and $\mathbf{o}$ (see Fig. 1b). In vectorial form [14], defining $\mu = n/\overline{n}$:

$$\mathbf{o} = \mathrm{Snell}_\mu^{\mathbf{a}}(\mathbf{i}) = \mu \, \mathbf{i} + \left(\sqrt{1 - \mu^2 \left(1 - (\mathbf{i}^\top \mathbf{a})^2\right)} - \mu \, (\mathbf{i}^\top \mathbf{a}) \right) \mathbf{a}. \tag{2.3}$$

Refractive 3D-Vision. Snell's law (2.3) of refraction has been considered in few 3D-vision contexts. For instance, the epipolar geometry theory has been extended to the case where the camera and the surface are separated by a refractive plane [5]. This constitutes the basis for the development of refractive structure-from-motion algorithms [4,13]. Multi-view stereo in the presence of a refractive interface has also been recently explored [3,6,11]. In the photometric stereo context, underwater imaging has attracted some attention [10,16,17,25,26]. These works focus mostly on light absorption, which occurs when scattering is involved (inhomogeneous medium such as murky water) or in near-infrared imaging. Yet, other refraction effects (e.g., change of incident light direction and density, and of surface parameterization) are neglected. For instance, it is usually assumed that all the sources have the same relative intensity, and that their directions can be obtained using a calibration target immersed in the medium. Yet, even if all the sources outside the refractive medium have exactly the same intensity, the refractive interface will induce luminous fluxes with different densities (see Sect. 4). Therefore, it would be more convenient to calibrate light directions and densities outside the refractive medium, and account for refraction within the image formation model. This has been achieved in [18] but only for a fronto-parallel interface and with a somehow naive numerical solution, and in [9] but by relying on laser triangulation. Instead, the present paper aims at modeling and evaluating the effects of a refractive interface with arbitrary orientation on shape recovery by pure PS, and at providing an efficient numerical solution by adapting state-of-the-art algorithms.

3 Geometry of Refractive PS

Notations. As illustrated in Fig. 1b, we work in $\mathbb{R}^3$ with its canonical frame $(O, \mathbf{e}_1, \mathbf{e}_2, \mathbf{e}_3)$, where O is the camera's principal point, $\mathbf{e}_3$ is the optical axis direction and $\mathcal{C} := \mathbf{e}_3^\perp$ is the image plane. A generic point in $\mathbb{R}^3$ is denoted by $\mathbf{x}$, while $\mathbf{p}(\mathbf{x}) = (u, v)^\top$ denote the 2D coordinates of its conjugate pixel in the frame $(O, \mathbf{e}_1, \mathbf{e}_2)$. The projection from $\mathbb{R}^3 \to \mathbb{R}^2$ keeping the first two coordinates is represented by the matrix $\mathbf{\Pi} = \begin{pmatrix} 1\,0\,0 \\ 0\,1\,0 \end{pmatrix}$, whose transpose is the canonical injection $\mathbb{R}^2 \to \mathbb{R}^3$. The interface plane $\mathcal{D}$ is given by the equation $\mathbf{a}^\top \mathbf{x} + \alpha = 0$, $\alpha \in \mathbb{R}$, where $\mathbf{a} = (a_1, a_2, a_3)^\top \in \mathbb{S}^2$ is a known unit normal vector to $\mathcal{D}$ ($\mathbb{S}^2$ being the unit sphere of $\mathbb{R}^3$), oriented towards the camera ($\mathbf{a}^\top \mathbf{e}_3 \leq 0$). We assume that $\mathbf{a}^\top \mathbf{e}_3 \neq 0$. The medium containing the camera is located in $\mathcal{M} = \{\mathbf{x} \in \mathbb{R}^3, \mathbf{a}^\top \mathbf{x} + \alpha > 0\}$ and has IoR n, while the medium containing the object under scrutiny is located in $\overline{\mathcal{M}} = \{\mathbf{x} \in \mathbb{R}^3, \mathbf{a}^\top \mathbf{x} + \alpha \leq 0\}$ and has IoR $\overline{n}$, and we denote $\mu = n/\overline{n} < 1$. Lastly, for a plane $\mathcal{P}$ of equation $\mathbf{v}^\top \mathbf{x} + \beta = 0$ and $\mathbf{w} \in \mathbb{R}^3$ with $\mathbf{v}^\top \mathbf{w} \neq 0$, we define the projection on plane $\mathcal{P}$ along direction $\mathbf{w}$ as

$$\boldsymbol{P}_{\mathcal{P}}^{\mathbf{w}}(\mathbf{x}) = \mathbf{x} - \frac{\mathbf{v}^\top \mathbf{x} + \beta}{\mathbf{v}^\top \mathbf{w}} \mathbf{w} = \left(\mathrm{id} - \frac{\mathbf{w} \mathbf{v}^\top}{\mathbf{w}^\top \mathbf{v}} \right) \mathbf{x} - \frac{\beta \mathbf{w}}{\mathbf{v}^\top \mathbf{w}}. \tag{3.1}$$

The orthogonal projection on $\mathbf{v}^\top \mathbf{x} = 0$ is simply denoted by $\boldsymbol{P}_{\mathbf{v}^\perp}$.

Depth from the Interface. In the orthographic case, all the light rays reaching the camera are orthogonal to the image plane, i.e., parallel to $\mathbf{e}_3$. In the absence of a refractive interface (or when the interface is fronto-parallel as in [16,18]), the projection is simply $\mathbf{p}(\mathbf{x}) = \boldsymbol{\Pi}\mathbf{x}$ and the surface parameterization, as its right inverse, is $\mathcal{S}(u,v) = (u, v, z(u,v))^\top$ with z the depth map. However, when a non fronto-parallel refractive interface comes into play, the projection becomes non-orthogonal (see Fig. 1b). In this case, the rays reaching the camera are parallel to direction $\mathbf{e}_3$, and come from parallel incident rays with common direction $\mathbf{r} \in \mathbb{S}^2$ within the refractive medium $\overline{\mathcal{M}}$ as the interface is planar. Vector $\mathbf{r}$ is fully determined by Snell's law (2.3) as it must refract to viewing direction $\mathbf{e}_3$:

$$\mathbf{r} = \mathrm{Snell}_\mu^{-\mathbf{a}}(\mathbf{e}_3), \tag{3.2}$$

where the sign before $\mathbf{a}$ comes from the fact that $\mathbf{a}$ is oriented towards the camera, while $\mathbf{e}_3$ and $\mathbf{r}$ are oriented towards the surface (see Fig. 1b).

Therefore, a point $\mathbf{x} \in \overline{\mathcal{M}}$ on the immersed surface first projects non-orthogonally onto $\mathcal{D}$ along the refracted viewing direction $\mathbf{r}$, before being orthogonally projected onto the camera image plane along the viewing direction $\mathbf{e}_3$:

$$\mathbf{p}(\mathbf{x}) = \boldsymbol{\Pi}\boldsymbol{P}_{e_3^\perp}(\boldsymbol{P}_\mathcal{D}^\mathbf{r}(\mathbf{x})). \tag{3.3}$$

This leads to a straightforward model where we deproject pixel $(u,v)^\top$ in the image plane to a point on the refractive interface $\mathcal{D}$, and then follow the incident ray up to the object: $\mathcal{S}(u,v) = \boldsymbol{P}_\mathcal{D}^{e_3}(u,v,0)^\top + \bar{z}(u,v)\mathbf{r}$, with $\bar{z}$ the pseudo-depth (distance travelled along the refracted ray $\mathbf{r}$). One readily checks that $\mathbf{p}(\mathcal{S}(u,v)) = (u,v)^\top$. Using (3.1), we can write

$$\mathcal{S}(u,v) = \boldsymbol{P}_\mathcal{D}^{\mathbf{e}_3}(u,v,0)^\top + \bar{z}(u,v)\mathbf{r} = \mathbf{A}(u,v,0)^\top + \mathbf{t} + \bar{z}(u,v)\mathbf{r}, \tag{3.4}$$

with known quantities

$$\mathbf{A} = \begin{pmatrix} 1 & 0 & 0 \\ 0 & 1 & 0 \\ -\frac{a_1}{a_3} & -\frac{a_2}{a_3} & 0 \end{pmatrix}, \quad \mathbf{t} = -\frac{\alpha}{a_3}\mathbf{e}_3. \tag{3.5}$$

Surface Normals. Now, let us establish the link between the pseudo-depth $\bar{z}$ from the interface, and the normal $\mathbf{n}$ to the surface. To do this, let us consider the partial derivatives of the parameterization. They are given by $\mathcal{S}_u = \mathbf{A}\mathbf{e}_1 + \bar{z}_u\mathbf{r}$ and $\mathcal{S}_v = \mathbf{A}\mathbf{e}_2 + \bar{z}_v\mathbf{r}$. An (unnormalized) normal to the surface $\mathcal{S}(u,v)$ is $\mathcal{S}_u \times \mathcal{S}_v = (\mathbf{A}\mathbf{e}_1 + \bar{z}_u\mathbf{r}) \times (\mathbf{A}\mathbf{e}_2 + \bar{z}_v\mathbf{r})$. Set $\mathbf{b}^1 = \mathbf{r} \times \mathbf{A}\mathbf{e}_2$, $\mathbf{b}^2 = \mathbf{A}\mathbf{e}_1 \times \mathbf{r}$ and $\mathbf{b}^3 = \mathbf{A}\mathbf{e}_2 \times \mathbf{A}\mathbf{e}_1$. Then $\mathcal{S}_u \times \mathcal{S}_v = \bar{z}_u\mathbf{b}^1 + \bar{z}_v\mathbf{b}^2 - \mathbf{b}^3$. By letting $\mathbf{B}$ be the matrix $-(\mathbf{b}^1, \mathbf{b}^2, \mathbf{b}^3)$,

$$\mathbf{n}(u,v) = \mathbf{n}(\mathcal{S}(u,v)) \propto \mathbf{B}\begin{pmatrix} \nabla\bar{z}(u,v) \\ -1 \end{pmatrix}, \quad \mathbf{B} = \begin{pmatrix} \frac{a_2 r_2}{a_3} + r_3 & -\frac{a_1 r_2}{a_3} & \frac{a_1}{a_3} \\ -\frac{a_2 r_1}{a_3} & \frac{a_1 r_1}{a_3} + r_3 & \frac{a_2}{a_3} \\ -r_1 & -r_2 & 1 \end{pmatrix}. \tag{3.6}$$

Equation (3.6) relates the surface normals to the underlying gradient of the pseudo-depth from the interface. When the interface is fronto-parallel, $\mathbf{a} = -\mathbf{e}_3$, $\mathbf{r} = \mathbf{e}_3$ and $\bar{z} = z - \beta$. Hence, $\mathbf{B} = \mathbf{I}_3$ and the formula matches the classic one obtained in the absence of refraction: $\mathbf{n}(u,v) \propto \left(\nabla z(u,v)^\top, -1\right)^\top$.

4 Image Formation Model Under Directional Lighting

Now, we turn our attention to extending the image formation model (2.1) to the refractive case. We assume to have at hand a series of k images $I_1, \dots, I_k$, with lighting directions $\mathbf{s}_1, \dots, \mathbf{s}_k$ and densities $\varphi_1, \dots, \varphi_k$ calibrated inside $\mathcal{M}$. The effective lighting inside $\overline{\mathcal{M}}$ will however be different from the calibrated one, due to refraction.

Effective Lighting Directions and Densities. We assume that light directions and densities are known in the camera medium $\mathcal{M}$ thanks to calibration, and we denote these calibrated parameters by $\mathbf{s}_i$ and φ_i. However, the directions $\overline{\mathbf{s}}_i$ and densities $\overline{\varphi}_i$ of the effective light beams reaching the surface differ from calibrated values, see Fig. 2.

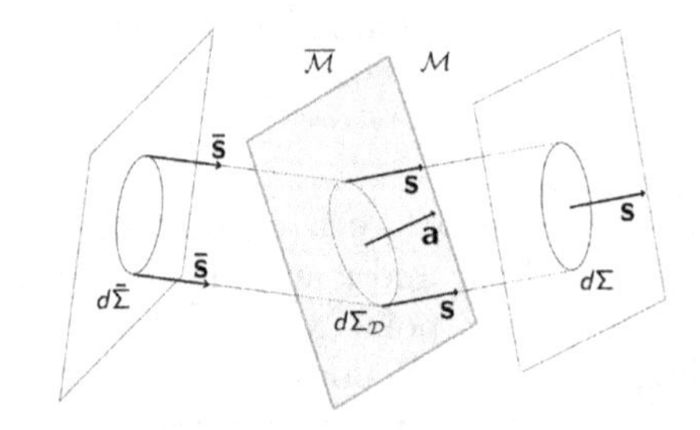

Fig. 2. Light refraction by a planar interface with normal $\mathbf{a}$, with $\mathbf{s}$ the light direction calibrated outside the refractive medium, and $\overline{\mathbf{s}}$ the effective refracted direction (in this drawing, the light source is on the right). Light direction is changed according to Snell's law, while its density is multiplied by $\frac{\mathbf{a}^\top \mathbf{s}}{\mathbf{a}^\top \overline{\mathbf{s}}}$.

After crossing the refractive interface, the incident light beams obviously remain parallel, yet their directions become, according to Snell's law (2.3):

$$\overline{\mathbf{s}}_i = -\text{Snell}_{\mu}^{-\mathbf{a}}(-\mathbf{s}_i), \quad i \in \{1, \dots, k\}. \tag{4.1}$$

Moreover, the size of the surface elements orthogonal to the rays also changes, according to $\frac{d\overline{\Sigma}_i}{\mathbf{a}^\top \overline{\mathbf{s}}_i} = d\Sigma_{\mathcal{D}} = \frac{d\Sigma_i}{\mathbf{a}^\top \mathbf{s}_i}$. Then:

$$\overline{\varphi}_i = \frac{\mathbf{a}^\top \mathbf{s}_i}{\mathbf{a}^\top \overline{\mathbf{s}}_i} \varphi_i, \quad i \in \{1, \dots, k\}. \tag{4.2}$$

Let us emphasize that, even if all the sources have exactly the same intensity i.e., $\varphi_i = \varphi_j, \forall i \neq j$, the effective densities will be different. For instance, when $n = 1$, $\overline{n} = 1.5$, and $\varphi_1 = \varphi_2 = 1$, a lighting orthogonal to the interface yields $\overline{\varphi}_1 = 1$, while an incident angle of 30° yields $\overline{\varphi}_2 = 0.91$. This effect is thus far from negligible in a calibrated PS setup.

Fresnel Coefficients. The interface may act partially as a mirror, with the amount of transmitted light being a function of the incident angle. This happens twice in the process: first when going from the light source in $\mathcal{M}$ to the surface embedded in $\overline{\mathcal{M}}$, and then when going from the latter to the camera, back in $\mathcal{M}$.

The incident and outgoing angles when going from $\mathcal{M}$ towards $\overline{\mathcal{M}}$ will vary depending on the incident direction $\mathbf{s}_i$, $i \in \{1, \dots, k\}$: each light source will thus induce a different transmission rate. This rate is however the same whatever the point $\mathbf{x}$, since lighting is assumed directional - this would not be the case for instance under a near point light source. Assuming that all the light beams are unpolarized, each transmission rate is given by the Fresnel coefficient

$$T_i^{\mathcal{M}\to\overline{\mathcal{M}}} = 1 - \frac{1}{2}\left(\frac{\left(\mu\,\mathbf{a}^\top\mathbf{s}_i - \mathbf{a}^\top\overline{\mathbf{s}}_i\right)^2}{\left(\mu\,\mathbf{a}^\top\mathbf{s}_i + \mathbf{a}^\top\overline{\mathbf{s}}_i\right)^2} + \frac{\left(\mu\,\mathbf{a}^\top\overline{\mathbf{s}}_i - \mathbf{a}^\top\mathbf{s}_i\right)^2}{\left(\mu\,\mathbf{a}^\top\overline{\mathbf{s}}_i + \mathbf{a}^\top\mathbf{s}_i\right)^2}\right),\ i \in \{1,\ldots,k\}. \tag{4.3}$$

Taking again as an example the case $n = 1$, $\overline{n} = 1.5$, an incident lighting orthogonal to the interface yields $T_1^{\mathcal{M}\to\overline{\mathcal{M}}} = 0.9600$, while an incident angle of 30° yields $T_2^{\mathcal{M}\to\overline{\mathcal{M}}} = 0.9585$. This shows that this Fresnel coefficient is non-negligible, although less dramatic than the change in the incident densities.

When going from $\overline{\mathcal{M}}$ to $\mathcal{M}$, the incident and outgoing angles are the same for all images $I_1, \ldots, I_k$ (viewing direction is independent from the incident lighting directions), therefore the transmission rate simply scales all the brightness values at pixel $\mathbf{p}$ conjugate to $\mathbf{x}$ by the same coefficient $T^{\overline{\mathcal{M}}\to\mathcal{M}}(\mathbf{x})$, $\forall i \in \{1,\ldots,k\}$. Besides, since we assume orthographic viewing, these angles are the same for all pixels, hence $T^{\overline{\mathcal{M}}\to\mathcal{M}}$ is independent from $\mathbf{x}$ as well - this would not be the case under pinhole projection. The Fresnel coefficient is then written as

$$T^{\overline{\mathcal{M}}\to\mathcal{M}} = 1 - \frac{1}{2}\left(\frac{\left(-\mathbf{a}^\top\mathbf{r} + \mu\,\mathbf{a}^\top\mathbf{e}_3\right)^2}{\left(-\mathbf{a}^\top\mathbf{r} - \mu\,\mathbf{a}^\top\mathbf{e}_3\right)^2} + \frac{\left(-\mathbf{a}^\top\mathbf{e}_3 + \mu\,\mathbf{a}^\top\mathbf{r}\right)^2}{\left(-\mathbf{a}^\top\mathbf{e}_3 - \mu\,\mathbf{a}^\top\mathbf{r}\right)^2}\right). \tag{4.4}$$

Note that this second Fresnel coefficient simply scales all the observations by the same constant, hence it can be taken into account by normalization.

Forward Model for Refractive PS. To summarize the effects described above, in the presence of refraction the classic image formation model (2.1) becomes

$$I_i(\mathbf{p}) = \underbrace{\left(\overline{\varphi}_i T_i^{\mathcal{M}\to\overline{\mathcal{M}}}\right)}_{:=\overline{\psi}_i} \underbrace{\left(T^{\overline{\mathcal{M}}\to\mathcal{M}}\rho(\mathbf{x})\right)}_{:=\varrho(\mathbf{x})} \mathbf{n}(\mathbf{x})^\top \underbrace{\left(-\mathrm{Snell}_\mu^{-\mathbf{a}}(-\mathbf{s}_i)\right)}_{:=\overline{\mathbf{s}}_i},\ i \in \{1,\ldots,k\}, \tag{4.5}$$

where:

- the effective lighting directions $\overline{\mathbf{s}}_i$ must be deduced from the calibrated ones $\mathbf{s}_i$ according to Snell's law (4.1);
- the effective lighting densities $\overline{\psi}_i$ must be deduced from the calibrated ones φ_i using (4.2) (density attenuation) and (4.3) (Fresnel coefficients);
- the Fresnel-scaled albedo $\varrho(\mathbf{x})$ (see (4.4)) and the surface normal $\mathbf{n}(\mathbf{x})$ (see (3.6)) constitute the unknowns of the PS problem.

To summarize, we have established the geometric parameterization of the surface, and shown how to deduce the effective lighting directions and densities from the ones calibrated outside the refractive medium. In the next section, we turn our attention to the numerical resolution of the system of Eqs. (4.5), by adapting state-of-the-art strategies.

5 Solving Refractive PS

To invert the image formation model (4.5), it is possible to either sequentially estimate normals and the 3D-shape, or to directly compute the 3D-shape.

Normal and Albedo Estimation. Estimating the surface normals and albedo comes down to solving the system of Eqs. (4.5) with known effective lighting densities $\overline{\psi}_i$ and effective incident lighting directions $\overline{\mathbf{s}}_i$. This system of equations admits a unique approximate solution as long as $k \geq 3$ and the effective directions $\overline{\mathbf{s}}_i$ are non-coplanar (which is the case if the incident directions $\mathbf{s}_i$ are themselves non-coplanar). Any calibrated PS method can be applied for this task, simply changing the light directions and densities so as to take refraction into account. For instance, defining $\mathbf{m} := \varrho\, \mathbf{n}$, one may consider the following pixelwise linear least-squares solution, $\forall \mathbf{p}$:

$$\mathbf{m}(\mathbf{p}) = \underset{\mathbf{m}\in\mathbb{R}^3}{\operatorname{argmin}} \sum_{i=1}^{k} \left(\overline{\psi}_i \overline{\mathbf{s}}_i^\top \mathbf{m} - I_i(\mathbf{p})\right)^2, \; \varrho(\mathbf{x}) = |\mathbf{m}(\mathbf{p})|, \; \mathbf{n}(\mathbf{x}) = \frac{\mathbf{m}(\mathbf{p})}{|\mathbf{m}(\mathbf{p})|}, \quad (5.1)$$

which can be computed in closed-form by using the pseudo-inverse. If robustness (e.g., to shadows or specularities) needs to be addressed, more evolved solutions based on deep neural networks [7] can be considered. Semi-calibrated algorithms [8] could also be employed for automatically inferring the coefficients $\overline{\psi}_i$. Provided that the integrability constraint [28] is adapted to the refractive case, uncalibrated algorithms [12] would even provide the $\overline{\mathbf{s}}_i$ up to a generalized bas-relief ambiguity [2], which could be resolved a posteriori using one of the methods discussed in [24].

Normal Integration. The next stage consists in obtaining the surface from its normals. Equation (3.6) tells us that once $\mathbf{n}(u, v)$ is estimated, computing $\mathbf{B}^{-1}\mathbf{n}(u, v)$ using the definition in (3.6) of $\mathbf{B}$, and then normalizing both its first components by the third one provides an estimate for $\nabla \overline{z}(u, v)$. Given these gradient estimates, the pseudo-depth map from the interface can be obtained by integration. Any approach designed for the classic case can be employed at this stage, just changing the input gradient estimates (see [20]). Once the pseudo-depth has been computed, one simply has to apply Eq. (3.4) to obtain the 3D-surface.

Direct Differential Approach. To avoid bias accumulation due to the sequential estimation of normals and shape, it is also possible to follow a direct differential approach. Plugging (3.6) into (4.5), we get, $\forall (i, \mathbf{p})$:

$$I^i(\mathbf{p}) = \overline{\psi}_i \underbrace{\frac{\varrho(\mathbf{x})}{\left|\mathbf{B}\begin{pmatrix}\nabla \overline{z}(\mathbf{p}) \\ -1\end{pmatrix}\right|}}_{:=\tilde{\varrho}(\mathbf{p})} \underbrace{\left(\mathbf{B}^\top \overline{\mathbf{s}}_i\right)^\top}_{:=\tilde{\mathbf{s}}_i^\top} \begin{pmatrix}\nabla \overline{z}(\mathbf{p}) \\ -1\end{pmatrix}, \quad (5.2)$$

which is a system of nonlinear PDEs. Therein, $\tilde{\varrho}$ will be considered as the unknown "pseudo-albedo" and vectors $\tilde{\mathbf{s}}_i$ as known "pseudo light vectors". The direct joint estimation of the pseudo-albedo and the pseudo-depth from the interface can then be written as a variational problem:

$$\min_{\overline{z},\tilde{\varrho}} \sum_{\mathbf{p}} \sum_{i} \Phi\left(\overline{\psi}_i \tilde{\varrho}(\mathbf{p}) \tilde{\mathbf{s}}_i^\top \begin{pmatrix} \nabla \overline{z}(\mathbf{p}) \\ -1 \end{pmatrix} - I_i(\mathbf{p})\right), \tag{5.3}$$

using some robust estimator Φ and a finite differences approximation of the gradient operator. Once the depth from the interface and the pseudo-albedo have been estimated, it only remains to deduce the true Fresnel-scaled albedo ϱ from $\tilde{\varrho}$ and $\nabla \overline{z}$ using the definition in Eq. (5.2), and eventually the 3D-surface by using Eq. (3.4). Again, such a differential approach can be extended to the semi-calibrated scenario [21], or even to refine the pseudo light vectors [22].

Validation on Synthetic and Real-World Data. In order to empirically validate our forward model and its inversion, we first generated synthetic PS images using Blender [1]. The Lambertian surfaces were placed inside glass ($\overline{n} = 1.5$) while the light sources and the orthographic camera were placed inside air ($n = 1$). In each experiment, 12 images were rendered under varying parallel lighting, whose direction and relative density are provided by the engine.

We first considered images of a perfect sphere. We used the sequential approach (5.1) followed by DCT integration [20], as well as the differential approach (5.3) with Cauchy estimator [22]. In both cases, we carried out 3D-reconstruction first neglecting all refraction effects, and then with refraction considered. To quantitatively evaluate the results, we fit a sphere to the 3D-reconstruction using least-squares, and compute the normalized RMSE between the 3D-reconstruction and the spherical fit. Results are shown in Table 1. It can be seen that for both approaches, considering refraction drastically improves the 3D-reconstruction, even when the interface is not rotated. Indeed, as can be seen in Fig. 3, neglecting refraction causes the 3D-reconstruction to "flatten".

Table 1. Normalized root mean square error between the estimated surface and a least-squares spherical fit, for a planar refractive interface (the angles stand for the rotations around the horizontal and vertical axes, respectively). Considering refraction systematically improves performance, for both the sequential and the differential approaches.

	No interface	$(0°, 0°)$	$(11.5°, 0°)$	$(11.5°, 22.5°)$
Sequential w/o refraction	0.0035	0.0195	0.0232	0.0403
Sequential w/ refraction	0.0035	0.0116	0.0129	0.0261
Differential w/o refraction	0.0020	0.0202	0.0239	0.0396
Differential w/ refraction	**0.0020**	**0.0114**	**0.0127**	**0.0254**

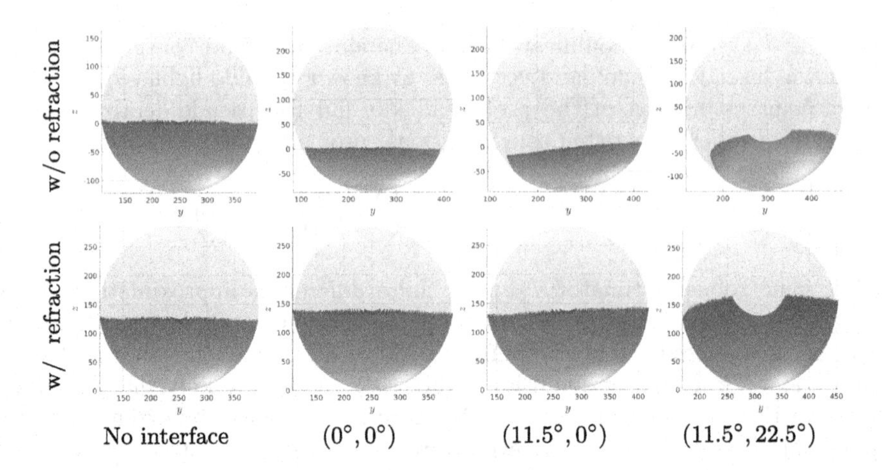

Fig. 3. 3D-reconstructions of the spheres from Table 1 using the differential approach, neglecting (top) or considering (bottom) refraction. The light grey spheres are the least-squares spherical fits to the estimated surfaces used for the quantitative evaluation in Table 1. Neglecting refraction induces a severe "flattening".

Then, we replaced the sphere by two objects with a more complex shape: an insect (imaged with the interface rotated by 5° around the horizontal axis) and a skull (imaged with the interface rotated by 20° around the horizontal axis, and by 11.25° around the vertical one). The results in Fig. 4, obtained with the differential approach, show that it is possible to achieve a 3D-reconstruction which is indistinguishable from the one obtained in the absence of refraction. In particular, the "flattening" effect is corrected.

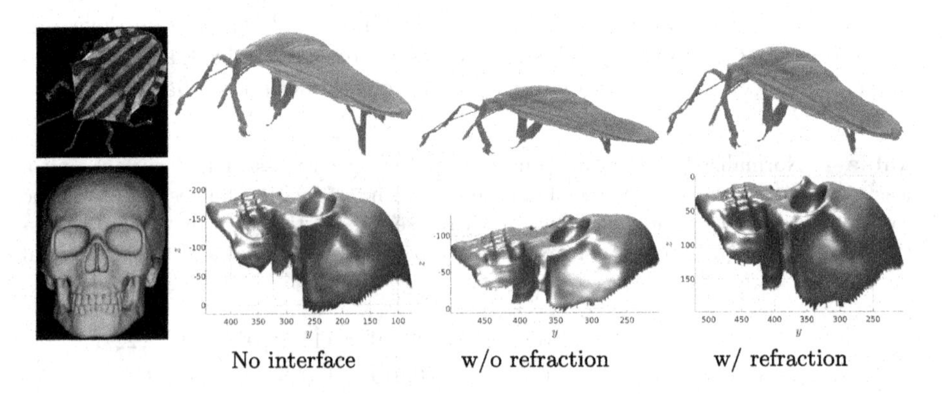

Fig. 4. 3D-reconstruction of an insect (top) and a skull (bottom). In each row, the first image represents one of the input images (out of 12); the second one shows the 3D-reconstruction obtained in the absence of the interface (for reference); and the other ones show the 3D-reconstruction in the presence of the interface, while neglecting or considering refraction.

Lastly, we conducted experiments on a real-world dataset. Our acquisition setup, illustrated in Fig. 1a, consists of 8 calibrated directional light sources. A diffuse sphere was imaged in the air, and then immersed in an aquarium filled with pure water (see Fig. 5). We performed the 3D-reconstruction using (5.3), and compared the results neglecting or considering refraction effects. For both a fronto-parallel and a rotated interface, considering refraction largely reduces the flattening and distortion effects, which empirically validates our method.

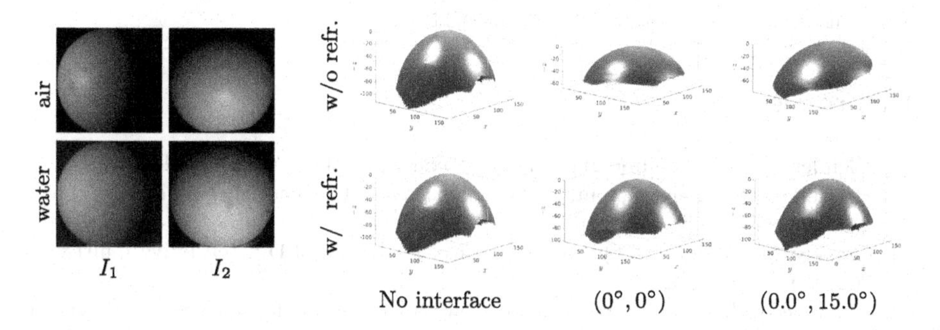

Fig. 5. 3D-reconstruction of a real-world sphere. On the left, we show two of the input images, in the absence ("air") and in presence ("water") of a refractive interface which is rotated by 15.0° around the vertical axis. On the right, we show the 3D-reconstruction of the sphere, taking into account (top) or not (bottom) refraction effects, in three cases: in the air, with a fronto-parallel interface, and with a rotated interface. Neglecting refraction leads to flattened and distorded 3D-reconstructions, while these effects are much attenuated with the proposed approach.

6 Conclusion and Future Work

In this paper, we have explored the impact of the presence of a refractive interface on the modeling of the photometric stereo problem, both in terms of geometry and of photometric image formation model. We further showed how to adapt existing solutions so as to take into account geometric deformation, refraction of incident directions, attenuation of densities and Fresnel coefficients. We showed that taking into account such phenomena largely improves the accuracy of the 3D-reconstruction. However, the explicit modeling of refraction effects was eased by a few simplifying assumptions: orthographic viewing, directional lighting, planar interface and absence of light absorption. In the future, we plan to explore the changes induced by the relaxation of these assumptions. This can partially be achieved by making the forward more realistic through the incorporation of, e.g., a refractive near-field illumination model [23] or distance-dependent light attenuation [16,26]. However, we believe that differentiable inverse rendering frameworks may constitute an even more promising track for solving nonstandard photometric 3D-reconstruction problems in a somehow generic manner.

Such approaches have recently been successfully employed for solving complex multi-view 3D-reconstruction problems [15], yet for now they remain limited to cases where the surface projection onto the camera comes down to a simple rasterization. To cope with evolved refractive effects, one could thus imagine combining differentiable inverse rendering with powerful renderers such as Mitsuba 2 [19].

Acknowledgements. This project was partially supported by the KU Data+ Project Phylorama, the ALICIA-Vision LabCom (ANR-19-LCV1-0002), and the Inclusive Museum Guide project (ANR-20-CE38-0007).

References

1. Blender - a 3D modelling and rendering package. http://www.blender.org
2. Belhumeur, P.N., Kriegman, D.J., Yuille, A.L.: The bas-relief ambiguity. IJCV **35**(1), 33–44 (1999)
3. Cassidy, M., Mélou, J., Quéau, Y., Lauze, F., Durou, J.D.: Refractive multi-view stereo. In: 3DV (2020)
4. Chadebecq, F., et al.: Refractive two-view reconstruction for underwater 3D vision. IJCV **128**(5), 1101–1117 (2020)
5. Chari, V., Sturm, P.: Multiple-view geometry of the refractive plane. In: BMVC (2009)
6. Chen, C., Wang, H., Zhang, Z., Gao, F.: Three-dimensional reconstruction from a fringe projection system through a planar transparent medium. OptEx **30**(19), 34824–34834 (2022)
7. Chen, G., Han, K., Shi, B., Matsushita, Y., Wong, K.Y.K.: Deep photometric stereo for non-Lambertian surfaces. PAMI **44**(1), 129–142 (2020)
8. Cho, D., Matsushita, Y., Tai, Y.W., Kweon, I.S.: Semi-calibrated photometric stereo. PAMI **42**(1), 232–245 (2018)
9. Fan, H., et al.: Underwater optical 3-D reconstruction of photometric stereo considering light refraction and attenuation. IEEE J. Ocean. Eng. **47**(1), 46–58 (2021)
10. Fujimura, Y., Iiyama, M., Hashimoto, A., Minoh, M.: Photometric stereo in participating media considering shape-dependent forward scatter. In: CVPR (2018)
11. Fujitomi, T., Sakurada, K., Hamaguchi, R., Shishido, H., Onishi, M., Kameda, Y.: LB-NERF: light bending neural radiance fields for transparent medium. In: ICIP (2022)
12. Hayakawa, H.: Photometric stereo under a light source with arbitrary motion. JOSA A **11**(11), 3079–3089 (1994)
13. Hu, X., Lauze, F., Pedersen, K.S.: Refractive pose refinement. IJCV (2023). https://doi.org/10.1007/s11263-023-01763-4
14. Mikš, A., Novák, P.: Determination of unit normal vectors of aspherical surfaces given unit directional vectors of incoming and outgoing rays: comment. JOSA A **29**(7), 1356–1357 (2012)
15. Munkberg, J., et al.: Extracting triangular 3D models, materials, and lighting from images. In: Proceedings of the IEEE/CVF Conference on Computer Vision and Pattern Recognition, pp. 8280–8290 (2022)
16. Murai, S., Kuo, M.Y.J., Kawahara, R., Nobuhara, S., Nishino, K.: Surface normals and shape from water. In: CVPR (2019)

17. Murez, Z., Treibitz, T., Ramamoorthi, R., Kriegman, D.: Photometric stereo in a scattering medium. In: CVPR (2015)
18. Narasimhan, S.G., Nayar, S.K.: Structured light methods for underwater imaging: light stripe scanning and photometric stereo. In: OCEANS (2005)
19. Nimier-David, M., Vicini, D., Zeltner, T., Jakob, W.: Mitsuba 2: a retargetable forward and inverse renderer. ACM Trans. Graph. (TOG) **38**(6), 1–17 (2019)
20. Quéau, Y., Durou, J.D., Aujol, J.F.: Normal integration: a survey. JMIV **60**(4), 576–593 (2018)
21. Quéau, Y., Wu, T., Cremers, D.: Semi-calibrated near-light photometric stereo. In: Lauze, F., Dong, Y., Dahl, A.B. (eds.) SSVM 2017. LNCS, vol. 10302, pp. 656–668. Springer, Cham (2017). https://doi.org/10.1007/978-3-319-58771-4_52
22. Quéau, Y., Wu, T., Lauze, F., Durou, J.D., Cremers, D.: A non-convex variational approach to photometric stereo under inaccurate lighting. In: CVPR (2017)
23. Sanao, H., Yingjie, S., Ming, L., Jingwei, Q., Ke, X.: Underwater 3D reconstruction using a photometric stereo with illuminance estimation. Appl. Opt. **62**(3), 612–619 (2023)
24. Shi, B., Wu, Z., Mo, Z., Duan, D., Yeung, S.K., Tan, P.: A benchmark dataset and evaluation for non-Lambertian and uncalibrated photometric stereo. In: CVPR (2016)
25. Tsiotsios, C., Angelopoulou, M.E., Kim, T.K., Davison, A.J.: Backscatter compensated photometric stereo with 3 sources. In: CVPR (2014)
26. Tsiotsios, C., Davison, A.J., Kim, T.K.: Near-lighting photometric stereo for unknown scene distance and medium attenuation. IVC **57**, 44–57 (2017)
27. Woodham, R.J.: Photometric stereo: a reflectance map technique for determining surface orientation from image intensity. In: Image understanding systems and industrial applications I, vol. 155, pp. 136–143 (1979)
28. Yuille, A.L., Snow, D., Epstein, R., Belhumeur, P.N.: Determining generative models of objects under varying illumination: shape and albedo from multiple images using SVD and integrability. IJCV **35**(3), 203–222 (1999)

Multi-view Normal Estimation – Application to Slanted Plane-Sweeping

Lilian Calvet[1,3](✉), Nicolas Maignan[2], Baptiste Brument[3], Jean Mélou[3], Silvia Tozza[4], Jean-Denis Durou[3], and Yvain Quéau[5]

[1] Research in Orthopedic Computer Science, Balgrist University Hospital, University of Zurich, 8008 Zurich, Switzerland
lilian.calvet@balgrist.ch
[2] Université de Lorraine CNRS Inria LORIA, 54000 Nancy, France
[3] IRIT, UMR CNRS 5505, 31000 Toulouse, France
[4] Department of Mathematics, Università di Bologna, 40126 Bologna, Italy
[5] Normandie Univ, UNICAEN, ENSICAEN, CNRS, GREYC, 14000 Caen, France

Abstract. In this paper, we show how to estimate the normals of a 3D surface from a minimum of two views, assuming that the poses of a calibrated camera are perfectly known. For each pair of image points, the normal at the corresponding 3D point is expressed in function of the local gradients of the grey level, whatever the type of image formation (orthogonal or perspective projection). As an application, this allows us to fully estimate the inter-image homography, which not only depends on the relative pose between views, but also on the local orientation of the surface. Hence, the photo-consistency between patches from two images, which is the basis of the so-called "plane-sweeping" method, is improved. Experiments on synthetic and real data validate our approach.

Keywords: Normal Estimation · Multi-view Stereo · Plane-sweeping

1 Introduction

Image-based 3D reconstruction pipelines usually comprise three steps: 1) feature extraction and matching across the images; 2) structure-from-motion to estimate the camera poses and a sparse 3D point cloud; 3) multi-view stereo (MVS), which reconstructs a dense 3D geometry. A common approach to MVS is to match the pixels between the different views by maximizing the photo-consistency of a specific image, called *reference image*, with the others, called *control images*. To measure photo-consistency, the reference image is warped to the control images, assuming known poses and making a guess on the depth.

Assuming the surface is locally flat, the plane-sweeping method allows for a more robust comparison than pixel-to-pixel, as it allows for patch comparison. The change of point of view implies a distortion of a patch from one image to another, which takes the form of a homography depending on the normal to the tangent plane of the surface. However, this dependency is usually ignored, due

Lilian Calvet and Nicolas Maignan contributed equally to this work.

L. Calatroni et al. (Eds.): SSVM 2023, LNCS 14009, pp. 704–716, 2023.
https://doi.org/10.1007/978-3-031-31975-4_54

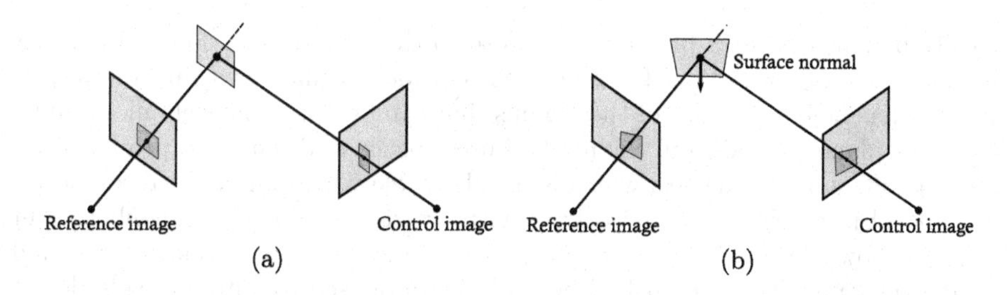

Fig. 1. (a) Standard plane-sweeping assumes that the scene is locally fronto-parallel to the reference image. (b) We propose to estimate the surface normal from image gradients and integrate this knowledge into a slanted plane-sweeping method.

to the difficulty of estimating this normal, which is considered to be collinear with the optical axis of the reference camera (see Fig. 1-a).

In this paper, we present a new method for estimating the normal to a surface from two images, using the gradients of the grey levels in a pair of conjugate image points, which allows us to characterise the inter-image homography, and thus to improve the comparison between conjugate patches (see Fig. 1-b). After a brief review of related approaches in Sect. 2, we present our method for normal estimation in Sect. 3. Preliminary experiments on both synthetic and real data are conducted in Sect. 4, which present the strengths and weaknesses of our approach. Our work is eventually summarised in Sect. 5.

2 Related Work

MVS methods can be divided in three main approaches. The first one exploits inter-image correspondences for multi-view 3D reconstruction [1,8,9,25]. These methods typically estimate depth maps, fuse them into point clouds and optionally generate meshes [16]. The second one uses implicit representations and leverages differentiable rendering to reconstruct 3D geometry with appearance from image collections [15,22,24]. NeRF [22] and most of its follow-ups use volumetric representations and compute radiance by ray marching through a neurally encoded 5D light field. The third approach uses explicit surface representations and estimates an explicit 3D mesh from images [5,6,17,19,21,26]. Most methods based on this approach assume a given, fixed mesh topology [5,6,21], but this assumption has been relaxed recently [17,23,26].

In this work, we focus on the method based on the first approach using inter-image correspondences. These methods commonly assume a fronto-parallel scene structure (see Fig. 1-a). Gallup et al. [10] observed the distortion of the cost function caused by structures that deviate from this prior and combated it by using multiple sweeping directions deduced from the sparse reconstruction. Earlier approaches [3,4,28] similarly account for the surface normal in stereo matching. Bleyer et al. [27] use PatchMatch to estimate per-pixel normals to compensate for the distortion of the cost function. They initialize each pixel with a random plane, hoping that at least one pixel of the region, supposedly locally planar, carries a plane that is close to the correct one. The method has no guarantee of converging to the correct normal estimate and, in practice, its success depends on the size of the regions that can be approximately modeled by the same plane. In addition, the method uses spatial and temporal propagations from "good" normal guesses, which is not desirable when aiming at reconstructing fine objects' details and working without video sequence. Schönberger et al. [25] follow the same approach while considering a variety of photometric and geometric priors. More recently, Liu et al. [20] proposed to automatically detect piecewise planar regions in the input images, and to compute the associated planes' parameters. They use a slanted plane-sweeping strategy. A set of three-dimensional slanted plane hypotheses is generated over both normal and depth values following uniform distributions of learned ranges.

In contrast to these approaches, we propose to estimate pixel-level normals from image gradients given camera parameters. We integrate this knowledge into a slanted plane-sweeping strategy (see Fig. 1-b) in order to overcome distortions induced by deviation of the surface to the fronto-parallel scene structure assumption. The problem of surface normal estimation from image gradients was also tackled by Lindeberg et al. in [11,18]. Therein, it is shown that the surface normal can be obtained from the transformation that relates the second moment matrices (computed from image gradients), in a scale-space framework. However, integrating this approach in a plane-sweeping strategy would require either solving a difficult two-parameters optimization problem, or resorting to a keypoint-based procedure which is not suitable when texture is lacking. On the contrary, the proposed approach requires solving a simpler one-parameter optimization problem, and it is suitable for poorly textured surfaces.

3 Slanted Plane-Sweeping

3.1 Photo-Consistency-Based MVS

Let us first recall the principle of MVS, which estimates the depth map associated with the reference camera by browsing, for each pixel, a set of possible depths.

Let us consider an opaque surface observed by $n+1$ identical cameras providing a reference image I and n control images I_i, $i \in \{1,\dots,n\}$. The poses of these cameras are assumed to be known and expressed in a world frame aligned with the reference camera frame. Cameras intrinsics are also supposed known. Let $\mathbf{Q} = [x, y, z]^\top$ be a 3D point expressed in the reference camera frame. Since the camera parameters are known, we can note $\mathbf{q} = \pi(\mathbf{Q})$ the projection of $\mathbf{Q}$ in the reference image, whose coordinates $\mathbf{q} = [u, v]^\top$ are expressed in the reference image frame. We define the same way $\{\pi_i\}_{i \in \{1,\dots,n\}}$, the projections from 3D points to pixels in the control cameras.

The central projection π is invertible if the depth function z is known. In this case, there is a bijection between the visible 3D points of the scene and their images, which is written $\pi_z^{-1}(u, v)$, where the subscript z is used to indicate that, without knowledge of the function z, this writing would be ambiguous. Then, for a 3D point $\mathbf{Q}$ on the surface which is visible from all cameras, the Lambertian assumption gives:

$$I_i \circ \pi_i \circ \pi_z^{-1}(u, v) = I(u, v), \qquad i \in \{1, \ldots, n\} \tag{1}$$

MVS searches for the depth function z corresponding to the reference image that maximises its **photo-consistency** with the n control images. Equations (1) are turned into a least squares problem, which has to be reformulated in discrete form (we do not know the grey level functions, but only their values in each pixel). The problem can then be solved separately in each pixel $\mathbf{q} = [u, v]^\top$:

$$\widehat{z}(u, v) = \underset{z \in \mathbb{R}}{\operatorname{argmin}} \; \frac{1}{n} \sum_{i=1}^{n} \left[I_i \circ \pi_i \circ \pi_z^{-1}(u, v) - I(u, v) \right]^2 \tag{2}$$

where $I_i \circ \pi_i \circ \pi_z^{-1}(u, v)$ has to be computed by interpolation.

For now, photo-consistency is reduced to the least squares comparison of two grey levels. In practice, photo-consistency is more complex [7, chapter 2]. The problem may then become nonlinear, non-smooth and non-convex, and the optimization tedious. Therefore, minimization is usually carried out using brute-force grid-search over the sampled depth space. This "winner-takes-all" strategy was first advocated in [14]. Despite its simplicity, it is remarkably efficient, and impressive depth map reconstructions of highly textured scenes have long been demonstrated [12].

3.2 Plane-Sweeping

In practice, comparing pixel signals over a single pixel value, as shown in Equation (2), works very poorly due to the lack of information. To overcome this limitation, the photo-consistency is minimised over an image patch, namely over $\mathbf{v}(u, v)$ representing the pixel intensities in the vicinity of a pixel $\mathbf{q} = [u, v]^\top$. Considering the i-th control camera, $i \in \{1, \ldots, n\}$, the photo-consistency then measures the difference between vectors $I_i \circ \mathbf{v}_i \circ \pi_i \circ \pi_z^{-1}(u, v)$ and $I \circ \mathbf{v}(u, v)$, in the sense of a function ρ. Problem (2) then becomes:

$$\widehat{z}(u, v) = \underset{z \in \mathbb{R}}{\operatorname{argmin}} \; \frac{1}{n} \sum_{i=1}^{n} \rho \left(I_i \circ \mathbf{v}_i \circ \pi_i \circ \pi_z^{-1}(u, v), I \circ \mathbf{v}(u, v) \right) \tag{3}$$

As far as the reference camera is concerned, it seems natural to use the pixel grid to define a neighbourhood, and thus the $\mathbf{v}$ function. Now, we need to define the patch used in the control images. To do this, we are interested in the homography transforming $\mathbf{v}(u, v)$ into $\mathbf{v}_i \circ \pi_i \circ \pi_z^{-1}(u, v)$.

To go further, we need to explicit the coordinates $[u_i, v_i, 1]^\top$ of the projection on image plane i of a 3D point $\mathbf{Q} = [x, y, z]^\top$. Since the cameras are supposed to be identical, the calibration matrix $\mathbf{K}$ is independent of index i. The projection formula gives us:

$$[u_i, v_i, 1]^\top = \frac{1}{z_i} \mathbf{K} \left(\mathbf{R}_{0 \to i} \, \mathbf{Q} + \mathbf{t}_{0 \to i} \right) \tag{4}$$

where the rotation matrix $\mathbf{R}_{0\to i}$ and the translation vector $\mathbf{t}_{0\to i}$ charaterize the pose of camera i. Denoting $\mathbf{C}_i = -\mathbf{R}_{0\to i}^{-1}\,\mathbf{t}_{0\to i}$ the location of the optical center of camera i, (4) becomes:

$$[u_i, v_i, 1]^\top = \frac{1}{z_i}\,\mathbf{K}\,\mathbf{R}_{0\to i}\,(\mathbf{Q} - \mathbf{C}_i) \tag{5}$$

which provides us with the following expression for the coordinates of the 3D point $\mathbf{Q}$:

$$\mathbf{Q} = z_i\,\mathbf{R}_{0\to i}^{-1}\,\mathbf{K}^{-1}[u_i, v_i, 1]^\top + \mathbf{C}_i \tag{6}$$

Putting (5) and (6) together, the movement of a point from camera i to camera j can be written:

$$[u_j, v_j, 1]^\top = \frac{1}{z_j}\,\mathbf{K}\,\mathbf{R}_{0\to j}\left(z_i\,\mathbf{R}_{0\to i}^{-1}\,\mathbf{K}^{-1}\,[u_i, v_i, 1]^\top + \mathbf{C}_i - \mathbf{C}_j\right)$$

which can be condensed in the following equation:

$$[u_j, v_j, 1]^\top = \frac{z_i}{z_j}\left(\mathbf{H}_{i,j}^{\infty}\,[u_i, v_i, 1]^\top + \frac{\mathbf{e}_{i,j}}{z_i}\right) \tag{7}$$

where $\mathbf{H}_{i,j}^{\infty} = \mathbf{K}\,\mathbf{R}_{i\to j}\,\mathbf{K}^{-1}$ is the homography which maps points at infinity from image i to image j, and $\mathbf{e}_{i,j} = \mathbf{K}\,\mathbf{R}_{0\to j}\,(\mathbf{C}_i - \mathbf{C}_j)$ is the *epipole* in image j.

The plane-sweeping method consists in assuming the surface to be locally flat during the exhaustive search for the depth z. The homography is then supported by the tangent plane to the surface, characterized by a unit-length normal $\mathbf{n}$ and located at a distance d_i from the optical centre of camera i, whose Cartesian equation is written:

$$\mathbf{n}^\top\mathbf{Q} = d_i \tag{8}$$

Plugging the expression (6) of $\mathbf{Q}$ in (8), we obtain:

$$\frac{1}{z_i} = \frac{\mathbf{n}^\top\mathbf{R}_{0\to i}^{-1}\,\mathbf{K}^{-1}\,[u_i, v_i, 1]^\top}{d_i - \mathbf{n}^\top\mathbf{C}_i} \tag{9}$$

and finally, combining (7) and (9):

$$[u_j, v_j, 1]^\top = \frac{z_i}{z_j}\left(\mathbf{H}_{i,j}^{\infty} + \frac{\mathbf{e}_{i,j}\,\mathbf{n}^\top\mathbf{R}_{0\to i}^{-1}\,\mathbf{K}^{-1}}{d_i - \mathbf{n}^\top\mathbf{C}_i}\right)[u_i, v_i, 1]^\top \tag{10}$$

Thus, the *inter-image homography* depends not only on the camera movement between two poses, but also on the normal vector $\mathbf{n}$ of the tangent plane. Facing the difficulty of estimating the normal, it is usual to assume that this plane is fronto-parallel to the image plane of the first camera i.e., to arbitrarily impose $\mathbf{n}$ colinear to its optical axis. In the next subsection, we show how to estimate this normal from the depth and the gradients of the grey level of a pair of images, which will allow us to use the inter-image homography expressed in (10).

3.3 Surface Normal Estimation

In this subsection, we establish the expression of the normal as a function of the depth and the gradients of the grey levels from two views of known poses. Whatever the type of camera projection, the normal is written, in the world frame:

$$\mathbf{n}(\mathbf{Q}) = \frac{1}{\sqrt{|\nabla z|^2 + 1}} \begin{bmatrix} \nabla z \\ -1 \end{bmatrix} \tag{11}$$

Thus, estimating the gradient $\nabla z = [\frac{\partial z}{\partial x}, \frac{\partial z}{\partial y}]^\top$ suffices to characterize the surface normal. By derivation of (1) along the axes of the world frame, and introducing the notation $I_i(u_i, v_i) = I_i \circ \pi_i \circ \pi_z^{-1}(u, v)$, we get:

$$\nabla_{x,y} I_i(u_i, v_i) = \nabla_{x,y} I(u, v), \quad i \in \{1, \ldots, n\} \tag{12}$$

According to the chain rule:

$$\nabla_{x,y} I_i(u_i, v_i) = \mathbf{J}_i^\top \, \nabla I_i(u_i, v_i) \tag{13}$$

where:

$$\mathbf{J}_i^\top = \begin{bmatrix} \frac{\partial u_i}{\partial x} & \frac{\partial v_i}{\partial x} \\ \frac{\partial u_i}{\partial y} & \frac{\partial v_i}{\partial y} \end{bmatrix} \tag{14}$$

On the other hand, since the reference camera is aligned with the world reference frame, and denoting α the number of pixels per meter, we have:

$$\nabla_{x,y} I(u, v) = \alpha \, \nabla I(u, v) \tag{15}$$

From (12), (13) and (15), we get:

$$\mathbf{J}_i^\top \, \nabla I_i(u_i, v_i) = \alpha \, \nabla I(u, v) \tag{16}$$

As we shall see next, the depth gradient can be deduced from this equation, for both orthogonal and perspective projections.

Case of Orthogonal Projection – The change of coordinates of a 3D point $\mathbf{Q}$ from the world frame to the camera frame $\mathcal{R}_i$ writes:

$$[x_i, y_i, z_i]^\top = \mathbf{R}_{0\to i} \, \mathbf{Q} + \mathbf{t}_{0\to i} \tag{17}$$

where, as already said, $\mathbf{R}_{0\to i}$ and $\mathbf{t}_{0\to i}$ characterize the pose of camera i.

Under the assumption of orthogonal projection, it is easy to deduce from (17):

$$\begin{cases} u_i = \alpha \left[r_i^{1,1} \, x + r_i^{1,2} \, y + r_i^{1,3} \, z + t_i^1 \right] + u_i^0 \\ v_i = \alpha \left[r_i^{2,1} \, x + r_i^{2,2} \, y + r_i^{2,3} \, z + t_i^2 \right] + v_i^0 \end{cases} \tag{18}$$

where $r_i^{j,k}$, $(j, k) \in \{1, 2, 3\}^2$, and t_i^j, $j \in \{1, 2, 3\}$, designate the current elements of matrix $\mathbf{R}_{0\to i}$ and of vector $\mathbf{t}_{0\to i}$, respectively, and $[u_i^0, v_i^0]^\top$ are the coordinates of the principal point in image i. By derivation of (18), we get:

$$\mathbf{J}_i^\top = \alpha \begin{bmatrix} r_i^{1,1} + r_i^{1,3} \frac{\partial z}{\partial x} & r_i^{2,1} + r_i^{2,3} \frac{\partial z}{\partial x} \\ r_i^{1,2} + r_i^{1,3} \frac{\partial z}{\partial y} & r_i^{2,2} + r_i^{2,3} \frac{\partial z}{\partial y} \end{bmatrix} \tag{19}$$

From (16) and (19), we finally obtain:

$$\left\{ \begin{bmatrix} r_i^{1,1} & r_i^{2,1} \\ r_i^{1,2} & r_i^{2,2} \end{bmatrix} + \begin{bmatrix} \frac{\partial z}{\partial x} \\ \frac{\partial z}{\partial y} \end{bmatrix} \begin{bmatrix} r_i^{1,3} & r_i^{2,3} \end{bmatrix} \right\} \nabla I_i(u_i, v_i) = \nabla I(u, v) \tag{20}$$

From this equation, we deduce the following expression for the depth gradient $\nabla z = [\frac{\partial z}{\partial x}, \frac{\partial z}{\partial y}]^\top$:

$$\nabla z = \frac{\nabla I(u, v) - \begin{bmatrix} r_i^{1,1} & r_i^{2,1} \\ r_i^{1,2} & r_i^{2,2} \end{bmatrix} \nabla I_i(u_i, v_i)}{\begin{bmatrix} r_i^{1,3} & r_i^{2,3} \end{bmatrix} \nabla I_i(u_i, v_i)} \tag{21}$$

Equation (21) provides us with a closed-form expression for depth gradient, and hence the surface normal, at every point where the denominator does not vanish. This can happen in the following three cases:

- The vector $\begin{bmatrix} r_i^{1,3} & r_i^{2,3} \end{bmatrix}$ is null in the case of a pure rotation around the optical axis of the camera. In this case, the normal cannot be evaluated in any point on the surface. However, this type of camera movement is to be avoided in the context of multi-view stereo.
- The gradient $\nabla I_i(u_i, v_i)$ may be null at certain points in control image i, particularly if the surface is not sufficiently textured. This will happen in the case of a flat, untextured surface that is uniformly illuminated.
- Finally, it is possible that none of these vectors is null, but that this is the case for their scalar product. However, if we have several control images, it is very unlikely that this scalar product cancels for all of them.

Case of Perspective Projection – For the vast majority of real images, the projection onto the camera is no longer orthogonal but perspective. Extending the previous rationale to perspective projection is not difficult, however we skip the proof for space limitation reasons. We therefore content ourselves with giving the new expressions of $\frac{\partial z}{\partial x}$ and $\frac{\partial z}{\partial y}$, which still involve $\nabla I(u, v) = [\frac{\partial I}{\partial u}, \frac{\partial I}{\partial v}]^\top$ and $\nabla I_i(u_i, v_i) = [\frac{\partial I_i}{\partial u_i}, \frac{\partial I_i}{\partial v_i}]^\top$, but also z and z_i:

$$\begin{cases} \dfrac{\partial z}{\partial x} = \dfrac{z_i^2\, z\, \frac{\partial I}{\partial u} + z^2 \left(w_i^{2,1} \frac{\partial I_i}{\partial u_i} - w_i^{1,1} \frac{\partial I_i}{\partial v_i} \right)}{z^2 \left(-w_i^{2,3} \frac{\partial I_i}{\partial u_i} + w_i^{1,3} \frac{\partial I_i}{\partial v_i} \right) + z_i^2 \left(x \frac{\partial I}{\partial u} + y \frac{\partial I}{\partial v} \right)} \\[2ex] \dfrac{\partial z}{\partial y} = \dfrac{z_i^2\, z\, \frac{\partial I}{\partial v} + z^2 \left(w_i^{2,2} \frac{\partial I_i}{\partial u_i} - w_i^{1,2} \frac{\partial I_i}{\partial v_i} \right)}{z^2 \left(-w_i^{2,3} \frac{\partial I_i}{\partial u_i} + w_i^{1,3} \frac{\partial I_i}{\partial v_i} \right) + z_i^2 \left(x \frac{\partial I}{\partial u} + y \frac{\partial I}{\partial v} \right)} \end{cases} \tag{22}$$

In these expressions, the coefficients $w_i^{1,k}$, $w_i^{2,k}$ and $w_i^{3,k}$, $k \in \{1, 2, 3\}$, are defined by the following cross products, where (x_i, y_i, z_i) are already defined in (17):

$$\begin{bmatrix} w_i^{1,k} \\ w_i^{2,k} \\ w_i^{3,k} \end{bmatrix} = \begin{bmatrix} r_i^{1,k} \\ r_i^{2,k} \\ r_i^{3,k} \end{bmatrix} \wedge \begin{bmatrix} x_i \\ y_i \\ z_i \end{bmatrix} = \begin{bmatrix} z_i\, r_i^{2,k} - y_i\, r_i^{3,k} \\ x_i\, r_i^{3,k} - z_i\, r_i^{1,k} \\ y_i\, r_i^{1,k} - x_i\, r_i^{2,k} \end{bmatrix}$$

4 Experiments

We conducted experiments with synthetic and real images to quantify the performance of the proposed method.

4.1 Implementation Details

In the following experiments, the size of the patch used for photo-consistency computation is fixed empirically to 9×9 pixels and the number of control views to four. As shown in Sect. 3, the surface normal can be computed from two views. In practice, we use the median of the set of normals estimated over all the pairs between the reference view and one of the four control views, which is the normal that minimizes the sum of the geodesic distances to these normals.

4.2 Synthetic Data

A synthetic sphere cap of radius 1, whose center is located at $(0, 0, 0)$, cut at the height $\sqrt{1 - 0.7^2}$, is viewed from five poses of the same camera with focal length equal to 1000. In each pose, the camera points towards the cap summit, at a distance equal to 4. In Fig. 2-a, the reference camera is displayed in blue, while the four control cameras are displayed in red. The reference view and the four control views, which are generated using a uniform directional lighting parallel to the z-axis, are displayed in Figs. 2-b and 2-c.

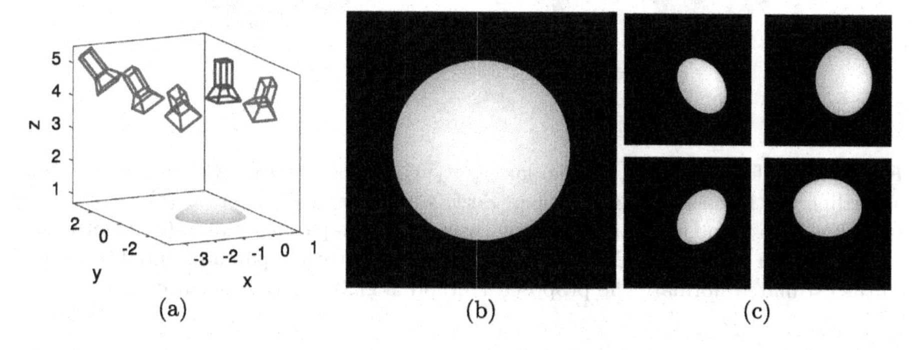

Fig. 2. (a) Synthetic dataset comprising one reference camera (in blue) and four control cameras (in red). The scene is a sphere cap illuminated by a uniform directional lighting parallel to the z-axis. (b) Reference view, whose depth is to be reconstructed. (c) Four control views. (Color figure online)

The method described in Sect. 3 is applied to estimate the surface normal in each pixel of the reference view (see Fig. 3-a). The angular errors are shown in Fig. 3-b. The normal estimation method shows to perform well, except at the edge of the cap, where the grey level gradient may become infinite.

The slanted plane-sweeping algorithm, which makes use of the estimated normal, is applied to estimate the depth map. The depth errors of plane-sweeping under the fronto-parallel assumption, and of the slanted plane-sweeping method, are shown in Figs. 3-c and 3-d. The proposed method is globally more accurate, but this highly depends on the local orientation of the surface.

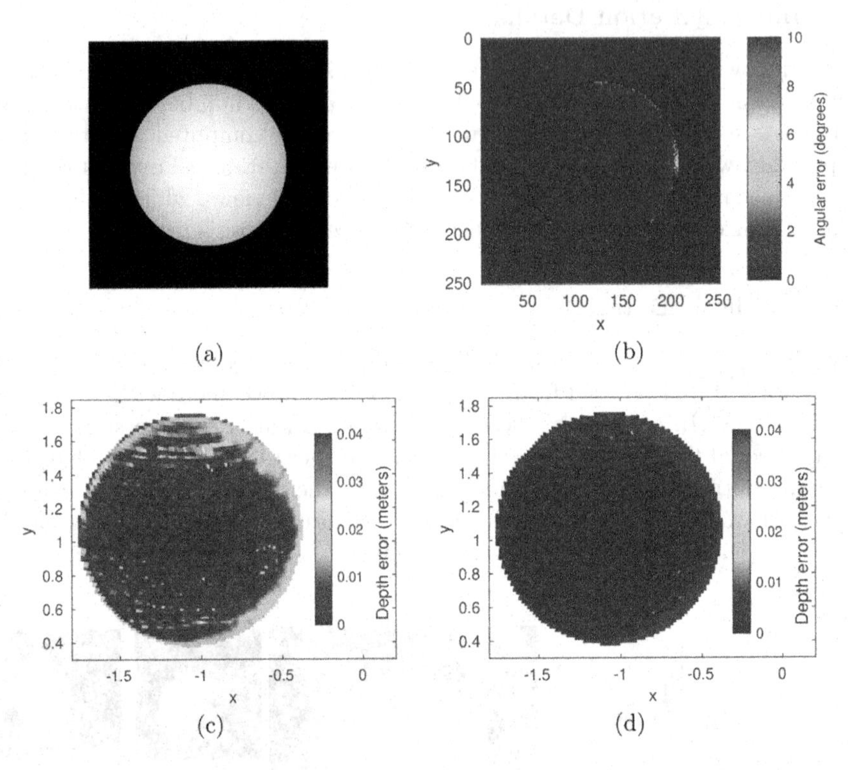

Fig. 3. (a) Reference view. (b) Angular error (in degrees) between the normal estimated according to the method described in Sect. 3 and the ground truth. (c) Depth error resulting from standard plane-sweeping, using a fronto-parallel patch. (d) Depth error resulting from the proposed slanted plane-sweeping, using a patch oriented according to the estimated normal. The proposed method is globally more accurate.

The spherical cap is then segmented into five regions, based on the angle between the surface normal and the optical axis of the reference camera, which are highlighted in color in Fig. 4-a. The mean and median depth errors per region are shown in Fig. 4-b, in the absence of noise in the images, and in Fig. 4-c (resp. 4-d), by adding a uniform noise in the range $[-3, 3]$ (resp. $[-6, 6]$) to the images, whose grey level values are between 0 and 255. Depth errors shown by the proposed slanted plane-sweeping are overall lower than the ones obtained under the fronto-parallel assumption, but this is particularly true for areas nearby the cap edge, associated to the highest out-of-plane rotations of the tangent plane, strongly violating the fronto-parallel patch assumption.

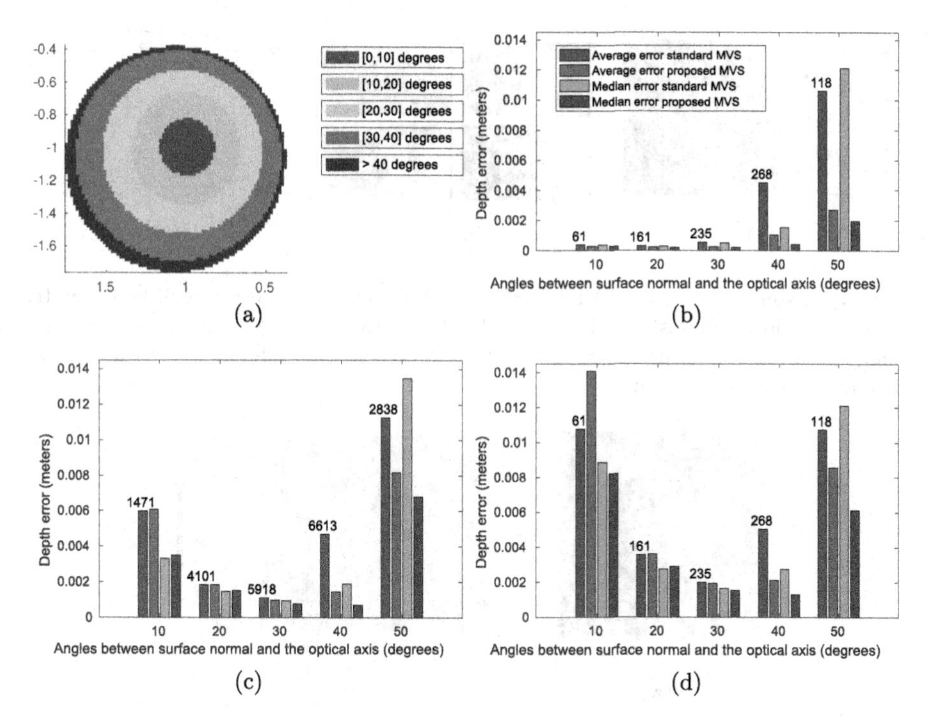

Fig. 4. (a) Segmentation of the reference view in five regions, based on the orientation of the surface normal. Histograms of the average error and of the median error on the estimated depth, using standard or slanted plane-sweeping methods: (b) in the absence of noise in the images; (c) adding to the images a uniform noise in the range $[-3, 3]$; (d) adding to the images a uniform noise in the range $[-6, 6]$.

4.3 Real Data

Then, the method has been evaluated on a real dataset, which is a set of five views of a plane. ArUco markers have been glued on the plane, in order to obtain its 3D reconstruction, along with the camera poses, using the structure-from-motion pipeline AliceVision Meshroom [13]. The normals of the plane are obtained by standard plane fitting from the marker locations, and used as ground truth. The input views are shown in Figs. 5-a and 5-b, while the results of structure-from-motion are shown in Fig. 5-c.

The angular errors, which are shown in Fig. 6-a, show a mean of 13.9^o and a standard deviation of 6.75^o. In this real scenario, the errors obtained on the estimated normals are too large to be exploited by the proposed slanted plane-sweeping. This may partly be explained by a gradient computation very sensitive to noise or by inaccuracies in camera pose estimation.

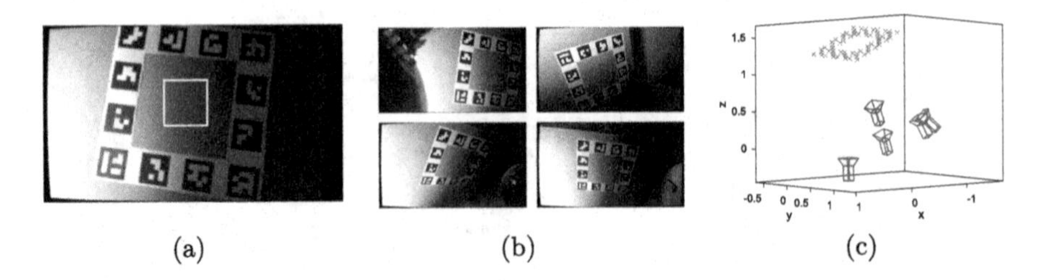

Fig. 5. (a) Reference view. (b) Four control views. The depth errors will be evaluated within the white frame shown in (a). (c) Real camera and marker positions, computed using a standard structure-from-motion technique [13]. The reference camera is shown in blue, the control ones in red.

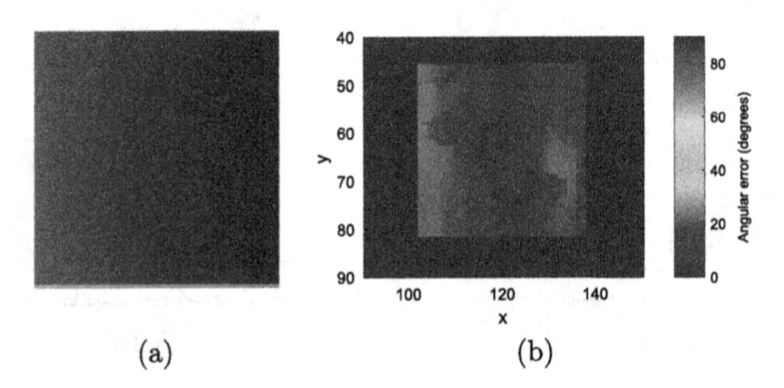

Fig. 6. (a) Image area over which normal estimation is performed (highlighted in white in Fig. 5-a). (b) Angular errors on the estimated normals, which show a mean and standard deviation of 13.9^o and 6.75^o, respectively, computed over 1296 pixels.

5 Conclusion and Perspectives

In this paper, we show how to estimate the normals of a surface from two views, provided that the poses of the camera, assumed calibrated, are perfectly known. For a pair of homologous points, we show that the normal at the corresponding 3D point can be computed unambiguously, as a function of the grey level gradient at each of these points, mentioning three cases for which this estimation is impossible. We detail the estimation method in the case of an orthogonal projection, and give its generalization to the perspective case.

Among the various applications of this new method of normal estimation, it makes it possible to estimate the inter-image homography of a surface portion, assumed to be locally planar, which depends not only on the change of pose, but also on the local orientation of the plane. This is therefore of interest for plane-sweeping matching, which is usually based on an approximate estimate of the inter-image homography. The tests carried out on synthetic images validate the theoretical part of our approach, and show that it is indeed worthwhile to take into account the local orientation of the surface in the criterion of photo-consistency. The tests on real data are less convincing, as the estimation of the normal by our approach gives too high angular errors, of the order of 10°, to be able to claim an improvement of plane-sweeping matching.

Among the follow-ups to this work, we should first make the normal estimation method more robust. In addition to a possible inaccurate estimation of the camera poses (we used only five images in the real dataset), it is likely that the computation of the grey level gradients, as we did it, is grossly lacking in robustness. Moreover, a purely local estimate of the normal may be inherently too sensitive to noise. Another perspective is therefore to estimate the normals for a set of neighboring points, assumed to belong to a common tangent plane. It will then be appropriate to make the link with the work by Bartoli et al. on the estimation of normals from the deformations of a template [2].

References

1. Agarwal, S., et al.: Building Rome in a day. Commun. ACM **54**(10), 105–112 (2011)
2. Bartoli, A., Gérard, Y., Chadebecq, F., Collins, T., Pizarro, D.: Shape-from-template. PAMI **37**(10), 2099–2118 (2015)
3. Birchfield, S., Tomasi, C.: Multiway cut for stereo and motion with slanted surfaces. In: ICCV (1999)
4. Burt, P.J., Wixson, L.E., Salgian, G.: Electronically directed "focal" stereo. In: ICCV (1995)
5. Chen, W., et al.: Learning to predict 3D objects with an interpolation-based differentiable renderer. In: NeurIPS (2019)
6. Chen, W., et al.: DIB-R++: learning to predict lighting and material with a hybrid differentiable renderer. In: NeurIPS (2021)
7. Furukawa, Y., Hernàndez, C.: Multi-view stereo: a tutorial. Found. Trends Comput. Graph. Vis. **9**(1–2), 1–148 (2013)
8. Furukawa, Y., Ponce, J.: Accurate, dense, and robust multiview stereopsis. PAMI **32**(8), 1362–1376 (2010)
9. Galliani, S., Lasinger, K., Schindler, K.: Massively parallel multiview stereopsis by surface normal diffusion. In: ICCV, pp. 873–881 (2015)
10. Gallup, D., Frahm, J., Mordohai, P., Yang, Q., Pollefeys, M.: Real-time plane-sweeping stereo with multiple sweeping directions. In: CVPR (2007)
11. Gårding, J., Lindeberg, T.: Direct computation of shape cues using scale-adapted spatial derivative operators. IJCV **17**(2), 163–191 (1996)
12. Goesele, M., Curless, B., Seitz, S.M.: Multi-view stereo revisited. In: CVPR, vol. 2, pp. 2402–2409 (2006)
13. Griwodz, C., et al.: AliceVision Meshroom: an open-source 3D reconstruction pipeline. In: ACM Multimedia Systems Conference, pp. 241–247 (2021)
14. Hernández, C., Schmitt, F.: Silhouette and stereo fusion for 3D object modeling. CVIU **96**(3), 367–392 (2004)
15. Jiang, Y., Ji, D., Han, Z., Zwicker, M.: SDFDiff: differentiable rendering of signed distance fields for 3D shape optimization. In: CVPR (2020)
16. Kazhdan, M.M., Hoppe, H.: Screened poisson surface reconstruction. ACM Trans. Graph. **32**(3), 1–13 (2013)
17. Liao, Y., Donné, S., Geiger, A.: Deep marching cubes: learning explicit surface representations. In: CVPR (2018)
18. Lindeberg, T., Gårding, J.: Shape-adapted smoothing in estimation of 3-D depth cues from affine distortions of local 2-D brightness structure. In: Eklundh, J.-O. (ed.) ECCV 1994. LNCS, vol. 800, pp. 389–400. Springer, Heidelberg (1994). https://doi.org/10.1007/3-540-57956-7_42

19. Liu, H.D., Williams, F., Jacobson, A., Fidler, S., Litany, O.: Learning smooth neural functions via Lipschitz regularization. In: SIGGRAPH (2022)
20. Liu, J., et al.: PlaneMVS: 3D plane reconstruction from multi-view stereo. In: CVPR (2022)
21. Liu, S., Chen, W., Li, T., Li, H.: Soft rasterizer: a differentiable renderer for image-based 3D reasoning. In: ICCV (2019)
22. Mildenhall, B., Srinivasan, P.P., Tancik, M., Barron, J.T., Ramamoorthi, R., Ng, R.: NeRF: representing scenes as neural radiance fields for view synthesis. Commun. ACM **65**(1), 99–106 (2022)
23. Munkberg, J., et al.: Extracting triangular 3D models, materials, and lighting from images. In: CVPR (2022)
24. Niemeyer, M., Mescheder, L.M., Oechsle, M., Geiger, A.: Differentiable volumetric rendering: learning implicit 3D representations without 3D supervision. In: CVPR (2020)
25. Schönberger, J.L., Zheng, E., Frahm, J.-M., Pollefeys, M.: Pixelwise view selection for unstructured multi-view stereo. In: Leibe, B., Matas, J., Sebe, N., Welling, M. (eds.) ECCV 2016. LNCS, vol. 9907, pp. 501–518. Springer, Cham (2016). https://doi.org/10.1007/978-3-319-46487-9_31
26. Shen, T., Gao, J., Yin, K., Liu, M., Fidler, S.: Deep marching tetrahedra: a hybrid representation for high-resolution 3D shape synthesis. In: NeurIPS (2021)
27. Strecha, C., Fransens, R., Gool, L.V.: Wide-baseline stereo from multiple views: a probabilistic account. In: CVPR (2004)
28. Zabulis, X., Daniilidis, K.: Multi-camera reconstruction based on surface normal estimation and best viewpoint selection. In: 3DPVT (2004)

Partial Shape Similarity by Multi-metric Hamiltonian Spectra Matching

David Bensaïd(✉), Amit Bracha, and Ron Kimmel

Technion - Israel Institute of Technology, Haifa, Israel
{dben-said,amit.bracha}@campus.technion.ac.il, ron@cs.technion.ac.il

Abstract. Estimating the similarity of non-rigid shapes and parts thereof plays an important role in numerous geometry analysis applications. We propose a method for evaluating the similarity and matching of shapes describing articulated objects that gracefully handles partiality. The correspondence between a part and a whole is formulated as the alignment of spectra of operators closely related to the Laplace-Beltrami operator (LBO). The proposed approach considers multiple metrics defined on the same surface, which provide a compact description of the underlying geometric structure from different perspectives. Specifically, we study the scale-invariant metric and the corresponding scale-invariant Laplace-Beltrami operator (SI-LBO) together with the regular metric and the regular LBO. We demonstrate that, unlike the regular LBO, the low pass part of the SI-LBO eigen-structure is sensitive to regions with high Gaussian curvature which are of semantic importance in articulated objects. Thus, the low part of the SI-LBO's spectrum better captures curved regions and complements the information encapsulated in the lower part of the regular LBO's spectrum. A two spectra matching loss lends itself to a method that outperforms state of the art axiomatic and learning based techniques when evaluated on the task of partial matching on well established benchmarks (Code and results are available at: https://github.com/davidgip74/DualSpectraAlignmnent).

Keywords: Laplace-Beltrami operator · Gaussian curvature · metric tensor · shape analysis · partial shape matching

1 Introduction

Non-rigid shape matching is a fundamental task in many computer vision applications such as augmented reality, medical image analysis, and face recognition. Aligning shapes captured in real-world scenarios, where occlusions and partial views are common, is particularly difficult when considering the missing information. Some recent papers [3,15,16,18,21,23] address this challenge by introducing methods for *partial shape matching.*

Supplementary Information The online version contains supplementary material available at https://doi.org/10.1007/978-3-031-31975-4_55.

L. Calatroni et al. (Eds.): SSVM 2023, LNCS 14009, pp. 717–729, 2023.
https://doi.org/10.1007/978-3-031-31975-4_55

In 2019, Rampini *et al.* [21] addressed the problem by matching a part of a given shape to the whole. The matching region in the full shape is found by aligning the eigenvalues of differential operators closely related to the LBO. Unlike methods that rely on local descriptors that produce dense point-to-point correspondences, the region matching formulation adopts a global perspective that makes it inherently resilient to local discrepancies [14]. Such spectral-based region matching can be adapted to challenging settings such as matching approximately isometric shapes or shapes with missing parts. In this paper we continue the *shape-DNA* [22] line of thought, and the more recent efforts [8,17,20], that argue that the eigenvalues of the Laplace-Beltrami operator (LBO), known as *spectrum*, could be used as shape descriptors. Exploiting this idea, we adopt and adapt the potential alignment method proposed in [21] and define a combination of two Hamiltonian operators [7] by which the shape matching by spectrum alignment is performed.

Roughly speaking, the LBO spectrum captures the surface structure as a whole. It therefore has a *global* flavor that limits its ability to describe fine details [19,21] that are associated with *local* geometric structures of regions with high Gaussian curvature. We demonstrate that the spectrum of the scale-invariant LBO [2] depends mainly on curved regions which usually contain meaningful details, like joints and fingertips, that are essential when considering articulated objects. The complementary structures captured by each of the two spectra, that share the same support, motivates the proposed multi-metric approach. To demonstrate the advantage of this approach, we first show that comparing the dual spectra of full shapes results in improved separation of closely related classes of shapes, such as dogs and wolves. We then study the problem of matching a part of a shape to its whole. We demonstrate that matching two spectra allows to lock onto fine details that are missed when aligning a single spectrum.
The main contributions of this paper include,

- Matching part of a shape to its whole using more than a single metric defined on the same surface.
- A theoretical interpretation of the scale-invariant LBO spectrum as an indicator of regions with high Gaussian curvature that are semantically important when describing articulated objects.
- Outperforming state of the art learning and axiomatic methods for partial shape matching tested on challenging benchmarks like SHREC'16 CUTS [9] and PFARM [3,13].

2 Background

Shapes as Riemannian Manifolds. We model a shape as a Riemannian manifold $\mathcal{M} = (S, g)$, where S is a smooth two-dimensional surface embedded in $\mathbb{R}^3$ and g a metric tensor, also referred to as *first fundamental form.* The metric tensor defines *geometric* quantities on the surface, such as lengths of curves and angles between vectors. Note, that the same surface S with a different metric $\tilde{g}$ can be used to define a different manifold $\tilde{\mathcal{M}} = (S, \tilde{g})$.

2.1 The Laplace-Beltrami Operator

The Laplace-Beltrami operator (LBO) Δ_g is an ubiquitous operator in shape analysis. It generalizes the Laplacian operator to Riemannian manifolds,

$$\Delta_g f \triangleq \frac{1}{\sqrt{|g|}} \mathrm{div}(\sqrt{|g|} g^{-1} \nabla f), \quad f \in \mathcal{L}^2(\mathcal{M}), \tag{1}$$

where $|g|$ is the determinant of g, and $\mathcal{L}^2(\mathcal{M})$ stands for the Hilbert space of square-integrable scalar functions defined on $\mathcal{M}$.

The LBO admits a spectral decomposition under homogeneous Dirichlet boundary conditions,

$$\begin{aligned} -\Delta_g \phi_i(x) &= \lambda_i \phi_i(x), & x &\in \mathcal{M} \setminus \partial\mathcal{M} \\ \phi_i(x) &= 0, & x &\in \partial\mathcal{M} \end{aligned} \tag{2}$$

where $\partial\mathcal{M}$ stands for the boundary of manifold $\mathcal{M}$. The set $\{\phi_i\}_{i\geq 0}$ constitutes a basis invariant to isometric deformations. It can be regarded as a generalization of the Fourier basis [26].

2.2 The Hamiltonian Operator in Shape Analysis

In 2018, Choukroun *et al.* [7] adapted the well-known Hamiltonian operator H_g from quantum mechanics to shape analysis,

$$\mathrm{H}_g \triangleq -\Delta_g + v, \tag{3}$$

with $v : S \to \mathbb{R}^+$. H_g is a semi-positive definite operator that admits a spectral decomposition under homogeneous Dirichlet boundary conditions.

The Hamiltonian operator was first introduced to shape analysis in [7] where a comprehensive survey in this context could be found, see also [12]. For our discussion, the most important property of the Hamiltonian is,

Property 1 *Let $\mathcal{M} = (S, g)$ be a Riemmanian manifold and $v : S \to \mathbb{R}^+$ a potential function. The absolute value of eigenfunctions ϕ_i of the Hamiltonian exponentially decay in every $\hat{s} \in S$ for which $v(\hat{s}) > \lambda_i$.*[1]

Figure 1 illustrates that the potential can be considered as a mask determining the domain at which the LBO embedded in the Hamiltonian is effective. A second key property of the Hamiltonian is the differentiability of its eigenvalues with respect to its potential function [1].

Property 2 *The eigenvalues $\{\lambda_i\}_{i\geq 0}$ of the discretized Hamiltonian operator* **H** *are differentiable with respect to the potential v.*

[1] See [4] page 403 for a sketch of proof.

Fig. 1. Top: First eigenvectors of the LBO of the *partial* shape $\mathcal{N}$. **Bottom**: First eigenvectors of the Hamiltonian of the *full* shape $\mathcal{M}$. The Hamiltonian is defined with a step potential v (in black) corresponding to the effective support of $\mathcal{N}$. With the potential v, the eigenfunctions of the LBO and the Hamiltonian are similar up to a sign.

3 A Single Surface Treated as Two Manifolds

Shape properties, including the notion of similarity itself, are affected by the choice of a metric. Two surfaces can be isometric w.r.t. one metric and non-isometric w.r.t. another, see examples in [11]. Considering multiple metrics for the same surface can therefore be viewed as considering alternative perspectives of the same shape, each being sensitive to different types of deformations. Specifically, we use the *regular* and the scale-invariant metrics. The leading eigenvalues in the spectrum of the SI-LBO are influenced by regions with high Gaussian curvature like joints and fingertips which are essential in representing articulated shapes. The LBO leading eigenvalues, at the other end, treat all surface points alike, and are thus less sensitive to these geometric structures.

3.1 Scale Invariance as a Measure of Choice

Scale Invariant Metric. One version of a scale-invariant metric for surfaces utilizes the Gaussian curvature K [10] as local scaling of the regular metric elements. A scale invariant pseudo-metric $\tilde{g}$ can be defined by its elements as,

$$\tilde{g}_{ij} \triangleq |K|\, g_{ij}\,, \tag{4}$$

where g_{ij} are the elements of the *regular* metric tensor. Adding a small positive constant $\epsilon \in \mathbb{R}^+$ to the Gaussian curvature, so that, $\tilde{g}_{ij} = \sqrt{\epsilon + K^2}\, g_{ij}$, defines a metric. The modulation by a Gaussian curvature conformally shrinks intrinsically flat regions into points. The LBO of a Riemmanian manifold $\tilde{\mathcal{M}}$, defined by the surface S equipped with the scale-invariant metric $\tilde{g}$ is called the *scale-invariant Laplace-Beltrami Operator* (SI-LBO) [2],

$$\Delta_{\tilde{g}} = |K|^{-1} \Delta_g. \tag{5}$$

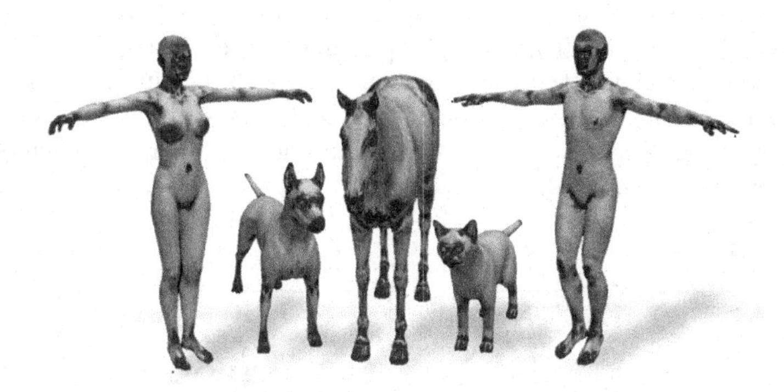

Fig. 2. Absolute value of the Gaussian curvature texture mapped to shapes from TOSCA [6]. Regions with large curvature magnitudes are darker, and have a larger influence on the spectrum of the SI-LBO, according to Theorem 1. These regions contain important details that can be used to classify an object.

Scale-Invariant Metric and Articulated Objects. The scale-invariant metric inflates semantically important regions in articulated objects. Figure 2 shows the curvature magnitude for some shapes from SHREC'16 [9]. In the human case, for instance, the scale invariant metric *accentuates* the head, the hands and the feet, at the expense of flat regions such as the back and the legs. Intuitively, the scale-invariant metric shrinks intrinsically flat regions into points and inflates intrinsically curved regions. We formalize this intuition in the spectral domain with a direct generalization of the Weyl law for the SI-LBO.

Lemma 1 *Let $\tilde{\mathcal{M}} = (S : \Omega \subseteq \mathbb{R}^2 \to \mathbb{R}^3, \tilde{g})$ be a Riemmanian manifold defined with the scale-invariant metric tensor $\tilde{g}$ and $\{\tilde{\lambda}_i\}_{i\geq 1}$ the spectrum of the scale-invariant LBO of $\tilde{\mathcal{M}}$. It holds, $\tilde{\lambda}_i \sim \frac{2\pi i}{\int_{\omega\in\Omega} |K(\omega)| da(\omega)}$, where $\sim$ stands for asymptotic equality.*

Using Lemma 1, a simple perturbation analysis expresses the influence of the curvature on the spectrum of the SI-LBO.

Theorem 1. *Consider a perturbation $\epsilon\,\delta_p$ of the curvature at $p \in \tilde{\mathcal{M}}$ where $\epsilon > 0$ and δ is a Dirac delta function on $\tilde{\mathcal{M}}$. Denote by $\delta K \triangleq \frac{\int_\Omega \epsilon\,\delta_p(\omega) da(\omega)}{\int_\Omega |K(\omega)| da(\omega)}$ the relative perturbation of the curvature and by $\delta\tilde{\lambda}_i \triangleq \frac{\tilde{\lambda}_i - \tilde{\mu}_i}{\tilde{\lambda}_i}$ the relative perturbation of $\tilde{\lambda}_i$, where $\tilde{\mu}_i$ is the ith eigenvalue of the perturbed manifold. $\delta\tilde{\lambda}_i$ respects $\delta\tilde{\lambda}_i \sim \delta K$.*

Theorem 1[2] implies that the SI-LBO spectrum is mostly determined by regions with effective Gaussian curvature. To illustrate the benefits of the proposed multi-metric approach and the scale-invariant metric for matching articulated shapes, we apply the ShapeDNA representation by using the spectra of both

[2] Sketch of proof: $\delta\tilde{\lambda}_i \triangleq \frac{\tilde{\lambda}_i - \tilde{\mu}_i}{\tilde{\lambda}_i} \sim \frac{1}{\tilde{\lambda}_i} \frac{2\pi i \epsilon}{(\int_\Omega |\mathrm{K}| da)(\int_\Omega |\mathrm{K}| da + \epsilon)} \sim \frac{1}{\tilde{\lambda}_i} \frac{2\pi i \epsilon}{(\int_\Omega |\mathrm{K}| da)^2} \sim \delta\mathrm{K}$.

the LBO $\{\lambda_i\}_{i=1}^k$ and the SI-LBO $\{\tilde{\lambda}_i\}_{i=1}^k$, defined for each shape. We normalize each spectrum to balance the two. As shown in Fig. 3, the incorporation of the SI-LBO results in a clear separation of shape classes such as quadrupeds like horses, cats, dogs, wolves, and humans, that can not be separated when only the LBO spectrum is considered.

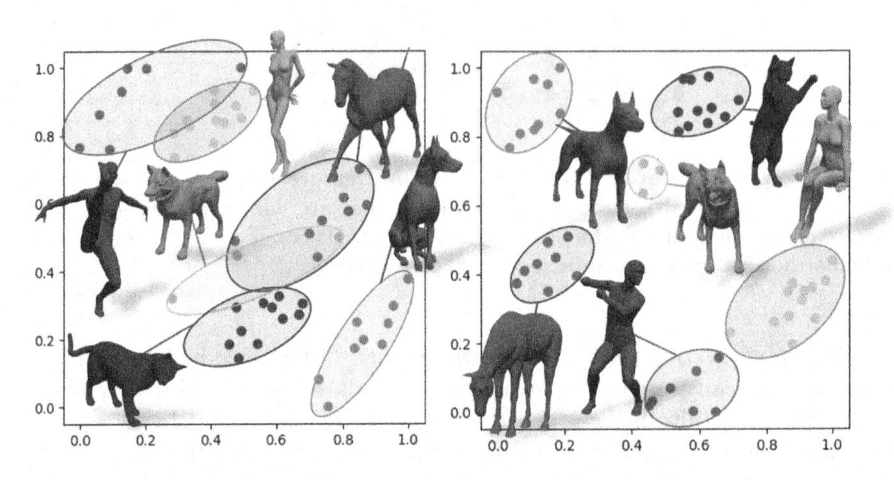

Fig. 3. Multidimensional scaling mapping to the plane applied to the distances between truncated spectra of the LBO (left) and that of the LBO together with the SI-LBO (right) for shapes from TOSCA [6]. Points that share the same color represent shapes of the same object in different poses. (Color figure online)

4 Dual Spectra Alignment for Region Localization

Overview. In this section we introduce a framework that finds the effective support of a given part of a shape in a shape given as a whole. Following [21], we use Property 1 to reduce the localization of a *partial* shape within a *full* shape to a search for a Hamiltonian's potential v that represents the effective support of the partial shape. The search for the proper potential function is formulated by an alignment cost of the spectra of Hamiltonian operators defined on the full shape with the spectra of the Laplace-Beltrami operators defined on the partial shape. Finally, Property 2 allows the minimization of the cost function with a first-order optimization algorithm.

Notations. The full shape equipped with the *regular* metric is denoted by $\mathcal{M} = (S_f, g)$ and the spectrum of the Hamiltonian defined over $\mathcal{M}$ by $\{\lambda_i\}_{i=1}^k$ with $\lambda_1 \leq ... \leq \lambda_k$. $\tilde{\mathcal{M}} = (S_f, \tilde{g})$ stands for the full shape defined with the scale-invariant metric and $\{\tilde{\lambda}_i\}_{i=1}^k$, with $\tilde{\lambda}_1 \leq ... \leq \tilde{\lambda}_k$, for the spectrum of the *scale-invariant Hamiltonian*. We denote by $\Phi \in \mathbb{R}^{n \times k}$ and $\tilde{\Phi} \in \mathbb{R}^{n \times k}$ the k first eigenfunctions of the discrete versions of the Hamiltonian and of the scale-invariant Hamiltonian, respectively. In the same way, the partial shape equipped with the *regular* and the scale-invariant metrics are referred to as $\mathcal{N} = (S_p, g)$, $\tilde{\mathcal{N}} = (S_p, \tilde{g})$. The spectra of the discrete versions of the LBO and of the SI-LBO are respectively denoted by $\{\mu_i\}_{i=1}^k$ and by $\{\tilde{\mu}_i\}_{i=1}^k$, with $\mu_1 \leq ... \leq \mu_k$ and $\tilde{\mu}_1 \leq ... \leq \tilde{\mu}_k$.

Cost Function. We consider a cost function that measures the alignment of the LBO spectrum of $\mathcal{N}$ and the SI-LBO of $\tilde{\mathcal{N}}$ with the spectra of the regular and the scale-invariant Hamiltonians of $\mathcal{M}$ and $\tilde{\mathcal{M}}$. Namely,

$$f(v) = \|\lambda(v) - \mu\|_w^2 + \|\tilde{\lambda}(v) - \tilde{\mu}\|_w^2 \,. \tag{6}$$

Following [21], the weighted L2 norm $\|.\|_w$ is defined as,

$$\|a - b\|_w^2 = \sum_{i=1}^{k} \frac{1}{b_i^2}(a_i - b_i)^2 \,, \tag{7}$$

to mitigate the weight given to high frequencies.

Optimization. The cost function Eq. (6) induces a constrained optimization problem,

$$\arg\min_{\mathrm{v} \geq 0} f(v) \,. \tag{8}$$

According to Property 2 and [1], the gradient of the last equation with respect to v is,

$$\nabla_v f = 2\mathbf{A}\,(\Phi \otimes \Phi)((\lambda - \mu) \oslash \mu^2) + 2\,\mathbf{A}\text{—}\mathbf{K}\text{—}(\tilde{\Phi} \otimes \tilde{\Phi})((\tilde{\lambda} - \tilde{\mu}) \oslash \tilde{\mu}^2) \,. \tag{9}$$

where $\oslash$ stands for point-wise division, $\otimes$ for the point-wise multiplication and $\mathbf{A}$ for the mass matrix of the discretization of $\mathcal{M}$. To simplify the optimization process, we minimize an unconstrained relaxation of Eq. (8) instead,

$$\arg\min_{v} f(q(v)) \,, \tag{10}$$

with $q : \mathbb{R} \to \mathbb{R}^+$ a smooth function operating element-wisely. We consider the saturation function $q(x) = \mathrm{c}(\tanh(x) + 1)$ with $\mathrm{c} \gg \mu_k$. By promoting high step potentials, q limits the eigenfunctions that can be considered within the region where $v \approx 0$. Equation (10) is finally minimized with a trust-region procedure [28], a first order optimization algorithm.

Initialization Strategy for the Dual Spectra Alignment. We follow the initialization procedure described in [21], where the proposed method is performed over multiple initial potentials. The final solution is selected by comparing the projections of the SHOT descriptors [24] onto the first eigenfunctions of the *regular* and the scale-invariant Hamiltonian of the full shape, with the projection of SHOT descriptors [24] onto the first eigenfunctions of the LBO and the SI-LBO of the partial shape.

Gaussian Curvature Estimation. The Gaussian curvature K is approximated with the Gauss-Bonnet formula. We smooth the result to overcome discretization artifacts. For each vertex, we take the average of the approximated Gaussian curvatures obtained at the first ring neighbors. Moreover, following [2,5,11], we use the metric,

$$\tilde{g}_{ij} = (\sqrt{\epsilon + K^2})^\alpha g_{ij}, \tag{11}$$

with $\alpha \in [0, 1]$, which interpolates between the regular, for $\alpha = 0$, and the scale-invariant metric, for $\alpha = 1$. Interestingly, the introduction of the parameter α is meaningful in light of the interpretation of the scale-invariant metric proposed in Sect. 3.1. α regulates the influence of the *shape prior* and quantifies the importance given to features found in curved regions. In all our experiments we used $\alpha = 0.33$ and $\epsilon = 10^{-8}$.

Table 1. Quantitative analysis of the proposed method compared to state-of-the-art techniques applied to SHREC'16 CUTS [9] and PFARM [3,13]. The proposed framework achieved state-of-the-art results compared to both learning and axiomatic competing methods.

	SHREC'16 [9]			PFARM [13]		
Method	**Precision**	**Recall**	**IoU**	**Precision**	**Recall**	**IoU**
Bag-of-words of SHOT descriptors [24]	0.653	0.589	0.430	0.475	0.454	0.310
PFC [23]	0.938	0.573	0.564	0.333	0.067	0.060
Single spectra alignment (Rampini *et al.*) [21]	0.775	0.738	0.668	**0.850**	0.763	0.701
DPFM [3]	**0.975**	0.576	0.569	0.642	0.275	0.248
Proposed dual spectra alignment	0.859	**0.838**	**0.751**	0.846	**0.787**	**0.710**

5 Experiments

5.1 Datasets

We evaluate the dual spectra alignment method on two databases. The first is SHREC'16 Partial Matching Benchmark (CUTS) [9], the standard benchmark to train and assess partial non-rigid shape matching frameworks. It contains 120 partial shapes from 8 classes: dog, horse, wolf, cat, centaur, and 3 human subjects. The partial shapes are obtained by cutting full shapes that have undergone various non-rigid transformations with random planes. The second is PFARM [3,13], an extension of the FARM test set [13] proposed in [3]. PFARM contains 27 test pairs of humans with significantly different connectivity and vertex density. Shapes that make up a test pair are taken from different human individuals and are therefore only approximately isometric, see Fig. 6. PFARM allows to evaluate the generalization ability of the models in challenging setups which are closer to real-life applications.

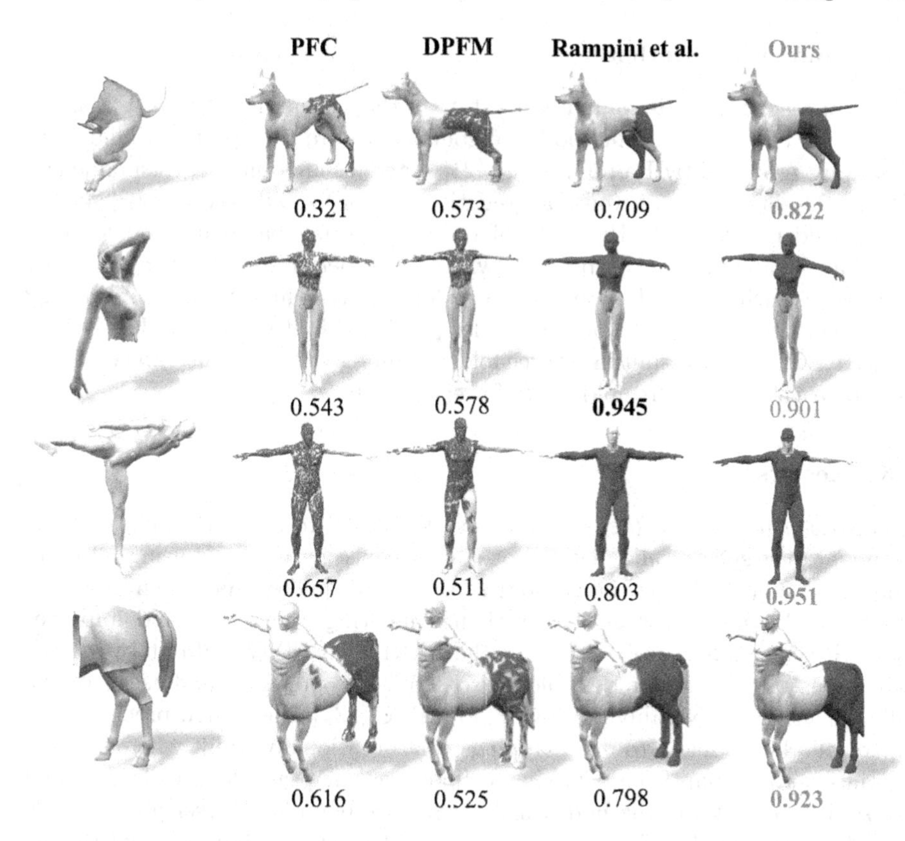

Fig. 4. Region localization. Comparison of the proposed approach for partial shape similarity with competing state of the art methods applied to SHREC'16. Red regions correspond to parts on the full shapes that should match the query parts (left column). IoU is indicated below each mask.

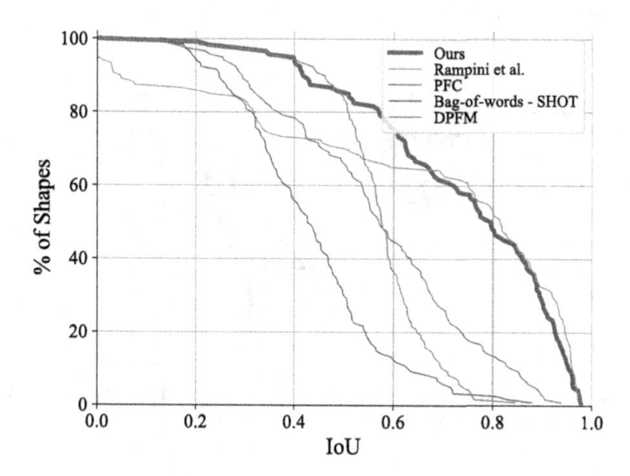

Fig. 5. Quantitative analysis. Cumulative IoU of each method on SHREC'16 CUTS. Areas under the curves are the mean IoUs reported in Table 1.

5.2 Competing Methods

The proposed method is compared to state of the art methods for partial shape matching. The axiomatic methods we compare to are the spectrum alignment procedure proposed by Rampini *et al.* [21], Partial Functional Correspondences (PFC) [23], and a bag-of-words aggregation [27] of SHOT descriptors [24]. We also consider DPFM [3], the state of the art learning based method for partial shape matching. It has been shown to be superior to competing learning approaches such as [25]. Please refer to [3] for further comparisons with learning based methods. We use the original codes published by the authors. For DPFM [3], SHREC'16 [9] was split into three folds. Training was performed on two folds, keeping each time a third fold apart for evaluation.

5.3 Results

Comparison to State of the Art Methods. We compare the performance of the proposed dual spectra alignment procedure to existing methods using intersection over union (IoU) as the standard measure of quality. As shown in Fig. 5, our method achieves better results than competing approaches and improves upon the current state of the art in the SHREC'16 CUTS dataset by 12.4%, as demonstrated in Table 1. Table 1 also shows the precision and recall of the different methods. Notably, DPFM [3] and PFC [23] achieve high precision but low recall, due to many-to-one mappings that occasionally result when deriving a point-to-point map from a predicted functional map and lead to fragmented masks. In contrast, our method, which does not rely on local descriptors, avoids such discrepancies. Qualitative comparisons of the proposed method with alternatives on shapes from the SHREC'16 dataset are shown in Fig. 4. This figure illustrates the benefits of the multi-metric approach in detecting important parts such as the head of the third object or the tail of the centaur, which are missed by the single metric approach [21].

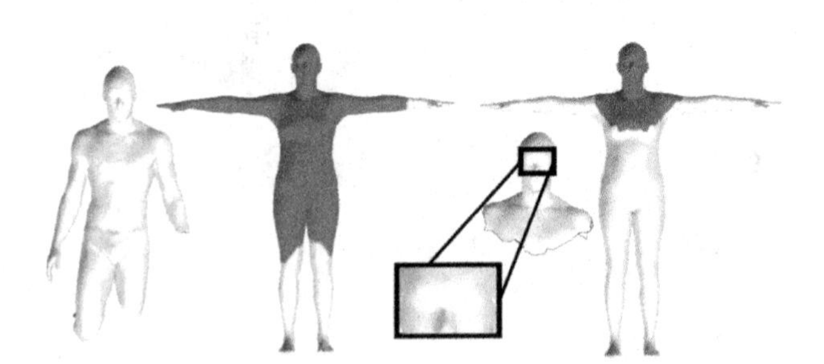

Fig. 6. The proposed region localization procedure applied to approximately isometric, low resolution shapes from PFARM [3,13].

Generalization Tested on Challenging Datasets. When comparing to the state of the art learning methods, generalization could become an issue. To that end, we trained DPFM on SHREC'16 and evaluated on PFARM. Table 1 shows that the proposed method significantly outperforms DPFM [3] and has an advantage when processing new databases with unknown shapes and challenging setups such as approximate isometries and inconsistent discretizations.

5.4 Ablation Study

Multi-metric vs Single Metric. To demonstrate the benefits of the multi-metric approach for region localization, we compare the alignment of single and multiple spectra considering *the same* number of eigenvalues. Figure 7 shows the cumulative IoU of the proposed framework applied to SHREC'16 [9] while using 20 eigenvalues of the LBO and 20 eigenvalues of the SI-LBO. It is compared to one setup with 40 eigenvalues of the LBO (without SI-LBO) and to a second setup with 40 eigenvalues of the SI-LBO (without LBO). The region alignment problems are thereby solved with the same number of constraints for each problem. Figure 7 shows that the multi-metric approach clearly outperforms single metric approaches.

6 Future Research Directions

We proposed a novel approach for shape similarity that leverages the complementary perspectives offered by the regular and scale-invariant metrics to

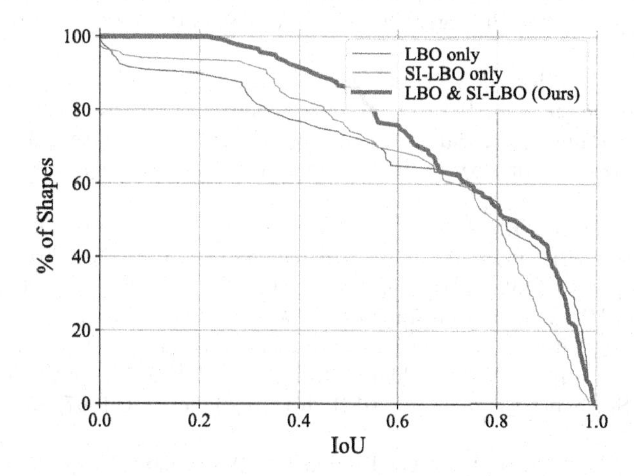

Fig. 7. Ablation study. Comparison of the proposed multi-metric approach, which includes the spectra of both the LBO and SI-LBO, with a method based only on the spectrum of the LBO and the SI-LBO spectrum. The plot shows the cumulative score of each setup tested on SHREC'16 [9]. The multi-metric approach reaches a mean IoU of 0.75. The single metric approach, involving only the LBO, mean IoU is 0.71. While using only the SI-LBO the mean IoU is 0.7.

achieve performance improvements compared to state of the art methods. In future research, we plan to extend the proposed spectra alignment procedure to more challenging datasets such as SHREC'16 Partial Matching (HOLES) [9]. It could be done by exploring new metric spaces and differentiable shape representations that also adopt a multi-metric perspective, such as self-functional maps [11]. Finally, the proposed method is fully differentiable and can potentially serve as an unsupervised loss to train a learning framework for partial shape matching.

Acknowledgement. We are extremely grateful to Alon Zvirin for his help in improving the readability and exposition of ideas in the paper. The research was partially supported by the D. Dan and Betty Kahn Michigan-Israel Partnership for Research and Education, run by the Technion Autonomous Systems and Robotics Program.

References

1. Abou-Moustafa, K.T.: Differentiating eigenvalues and eigenvectors. McGill Technical Report MTR No. TR-CIM-10-09(2009)
2. Aflalo, Y., Kimmel, R., Raviv, D.: Scale invariant geometry for nonrigid shapes. SIAM J. Imag. Sci. **6**, 1579–1597 (2013)
3. Attaiki, S., Pai, G., Ovsjanikov, M.: DPFM: deep partial functional maps. In: 3DV (2021)
4. Berger, M.: A Panoramic View of Riemannian Geometry. Springer, Heidelberg (2003). https://doi.org/10.1007/978-3-642-18245-7
5. Bracha, A., Halimi, O., Kimmel, R.: Shape correspondence by aligning scale-invariant LBO eigenfunctions. In: 3DOR (2020)
6. Bronstein, A.M., Bronstein, M.M., Kimmel, R.: Numerical Geometry of Non-Rigid Shapes. MCS, Springer, New York (2009). https://doi.org/10.1007/978-0-387-73301-2
7. Choukroun, Y., Shtern, A., Bronstein, A., Kimmel, R.: Hamiltonian operator for spectral shape analysis. IEEE Trans. Vis. Comp. Graph. **26**, 1320–1331 (2018)
8. Cosmo, L., Panine, M., Rampini, A., Ovsjanikov, M., Bronstein, M., Rodolà, E.: Isospectralization, or how to hear shape, style, and correspondence. In: CVPR (2019)
9. Cosmo, L., Rodola, E., Bronstein, M.M., Torsello, A., Cremers, D., Sahillioglu,Y.: SHREC'16: partial matching of deformable shapes. In: 3DOR (2016)
10. Do Carmo, M.P.: Differential Geometry of Curves And Surfaces (2016)
11. Halimi, O., Kimmel, R.: Self functional maps. In: 3DV (2018)
12. Iglesias, J.A., Kimmel, R.: Schrödinger diffusion for shape analysis with texture. In: Fusiello, A., Murino, V., Cucchiara, R. (eds.) ECCV 2012. LNCS, vol. 7583, pp. 123–132. Springer, Heidelberg (2012). https://doi.org/10.1007/978-3-642-33863-2_13
13. Kirgo, M., Melzi, S., Patane, G., Rodola, E., Ovsjanikov, M.: Wavelet-based heat kernel derivatives: towards informative localized shape analysis. In: Computer Graphics Forum (2021)
14. Kleiman, Y., Ovsjanikov, M.: Robust structure-based shape correspondence. In: Computer Graphics Forum (2019)
15. Litany, O., Rodolà, E., Bronstein, A., Bronstein, M.: Fully spectral partial shape matching. In: Computer Graphics Forum (2017)

16. Litany, O., Rodolà, E., Bronstein, A., Bronstein, M., Cremers, D.: Partial single- and multishape dense correspondence using functional maps. In: Handbook of Numerical Analysis (2018)
17. Marin, R., Rampini, A., Castellani, U., Rodola, E., Ovsjanikov, M., Melzi, S.: Instant recovery of shape from spectrum via latent space connections. In: 3DV (2020)
18. Melzi, S., Ren, J., Rodola, E., Sharma, A., Wonka, P., Ovsjanikov, M.: Zoomout: spectral upsampling for efficient shape correspondence. ACM Trans. Graph. (2019)
19. Melzi, S., Rodolà, E., Castellani, U., Bronstein, M.M.: Localized manifold harmonics for spectral shape analysis. In: Computer Graphics Forum (2018)
20. Moschella, L., et al.: Learning spectral unions of partial deformable 3D shapes. In: Computer Graphics Forum (2022)
21. Rampini, A., Tallini, I., Ovsjanikov, M., Bronstein, A.M., Rodola, E.: Correspondence-free region localization for partial shape similarity via hamiltonian spectrum alignment. In: 3DV (2019)
22. Reuter, M., Wolter, F.E., Peinecke, N.: Laplace-Beltrami spectra as 'Shape-DNA' of surfaces and solids. Comput.-Aided Des. **38**, 342–366 (2006)
23. Rodolà, E., Cosmo, L., Bronstein, M., Torsello, A., Cremers, D.: Partial functional correspondence. In: Computer Graphics Forum (2017)
24. Salti, S., Tombari, F., Di Stefano, L.: SHOT: unique signatures of histograms for surface and texture description. CVIU **125**, 251–264 (2014)
25. Sharma, A., Ovsjanikov, M.: Weakly supervised deep functional maps for shape matching. In: NeurIPS (2020)
26. Taubin, G.: A signal processing approach to fair surface design. In: Conference on Computer Graphics and Interactive Techniques (1995)
27. Toldo, R., Castellani, U., Fusiello, A.: A bag of words approach for 3D object categorization. In: ICCV (2009)
28. Yuan, Y.X.: A review of trust region algorithms for optimization. In: ICIAM (2000)

Modeling Large-Scale Joint Distributions and Inference by Randomized Assignment

Bastian Boll[1(✉)], Jonathan Schwarz[1], Daniel Gonzalez-Alvarado[1], Dmitrij Sitenko[1], Stefania Petra[2], and Christoph Schnörr[1]

[1] Image and Pattern Analysis Group, Heidelberg University, Heidelberg, Germany
bastian.boll@iwr.uni-heidelberg.de

[2] Mathematical Imaging Group, Heidelberg University, Heidelberg, Germany

Abstract. We propose a novel way of approximating energy-based models by randomizing the parameters of assignment flows, a class of smooth dynamical data labeling systems. Our approach builds on averaging flow limit points within the combinatorially large simplex of joint distributions. In an initial learning stage, the distribution of flow parameters is selected to match a given energy-based model. This entails the difficult problem of estimating model entropy which we address by differentiable approximation of a bias-corrected estimator. The model subsequently allows to perform probabilistic inference by computationally efficient draws of structured integer samples which are approximately governed by the energy-based target Gibbs measure in the low-temperature regime. We conduct a rigorous quantitative assessment by approximating a small two-dimensional Ising model and find close approximation of the combinatorial solution in terms of relative entropy which outperforms a mean-field approximation baseline.

Keywords: Probabilistic Inference · Assignment Flows · Energy-based Models · Structured Prediction

1 Introduction

Probabilistic models for context-sensitive decision making and structured prediction have been a focal point of research during the last decades, devoted to data modeling and analysis, imaging science and machine learning. Major paradigms for representing mathematically complex probability distributions include Gibbs distributions, probabilistic graphical models [17,32] and measure transport using push-forward maps parameterized by neural networks [16,26]. A range of variational approximations [3] have been developed for the – typically intractable – problems of inference and parameter learning.

The most basic one, the so-called ('naive') mean-field approximation [32, Section 5], minimizes the Kullback-Leibler distance of a fully factorized distribution and the intractable target distribution. More advanced structured mean-field approaches include the well-known Bethe- and Kikuchi approximations and

L. Calatroni et al. (Eds.): SSVM 2023, LNCS 14009, pp. 730–742, 2023.
https://doi.org/10.1007/978-3-031-31975-4_56

related algorithms for approximating marginals of the target distribution by belief propagation [20], [32, Section 4], convexified Bethe approximations [11,31] and related methods in statistical physics, like the cavity method [25].

This paper presents a preliminary step for approaching the problem from quite a different angle. Specifically, we consider a *randomized* dynamical system, the assignment flow approach, proposed by [1] for data and image labeling. Using learned *deterministic* parameters, this approach provides a continuous-time model for deep networks whose layers emerge by geometric integration. Key differences to established image labeling methods based on minimizing energies over discrete variables, like Maximum A-Posterior (MAP) inference using Markov Random Fields [13], include (i) inherent smoothness, (ii) efficient inference by geometric integration and (iii) amenability to learning parameters directly from data. Our goal is to achieve *probabilistic* inference, beyond *deterministic* MAP *point* estimates, by efficient evaluations of *randomized* assignment flows.

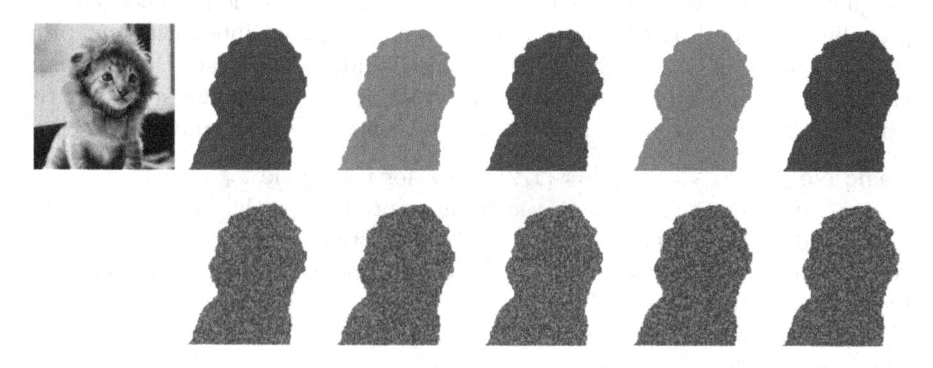

Fig. 1. Image segmentation of an ambiguous subject (This image was created by DALL-E 2 [24], cat or lion with equal probability). Our model of the joint distribution (samples in first row) captures the structure of the data by coupling subject pixels. In contrast, the marginal distribution (samples in second row) makes a pixelwise independence assumption and therefore fails to represent spatial context.

The rationale behind our approach appears natural when considering the embedding of assignment flows into a meta-simplex, as recently proposed by [4]. *Any* probability distribution over discrete variables can be represented as a point in a combinatorially large probability simplex, each vertex of which represents a *single* labeling corresponding to a single *joint* configuration of *all* involved discrete assignment variables. The embedding of assignment flows then ranges over a submanifold in this simplex corresponding to *factorized* distributions, akin to the basic mean-field approach mentioned above.

In this situation, we utilize (i) that assignment flows converge to labelings under mild conditions, i.e. they approach a vertex in the meta-simplex and, consequently, (ii) that *randomized* assignment flows define *implicitly* via (i) a *probability distribution* on the set of all vertices of the meta-simplex. This distribution, in turn, defines a *barycenter* in the meta-simplex which generally lies *outside* the submanifold corresponding to embedded assignment flows. In other words, using *randomized* assignment flows, we achieve approximate probabilistic representations and inference that are *more* expressive than *any* (naive) mean-field model. Figure 1 provides an example which illustrates this difference.

We point out that our approach to probabilistic modeling and inference, by convex combination of extreme points of compact convex sets of probability distributions, is not at all new in mathematics, but in fact extends far beyond the scenarios with discrete random variables considered here [8]. Our approach to constructing these representations, using randomized assignment flows, is novel however. Randomized assignment flows were also used in [5], yet within the different context of PAC-Bayes risk certification.

Organization. Section 2 briefly presents a specific parameterization of assignment flows introduced in [27] and the embedding approach of [4]. Randomized assignment flows are introduced in Sect. 3. The approximation of energy-based probability target distributions by our approach using the Gibbs variational principle is detailed in Sect. 4. Experiments which validate quantitatively our claims are presented in Sect. 5, using problems which are small enough such that the results of *exact* probabilistic inference can be computed as unequivocal baseline.

Basic Notation. We set $[n] = \{1, 2, \dots, n\}$ for $n \in \mathbb{N}$ and $\mathbb{1}_n = (1, 1, \dots, 1)^\top \in \mathbb{R}^n$. $\langle\cdot,\cdot\rangle$ denotes the Euclidean vector inner product or the Frobenius matrix inner product. The canonical basis of $\mathbb{R}^n$ is denoted by $(e_1, \dots, e_n) = I_n$. $\mathbb{R}^n_{++}$ denotes the set of vectors in $\mathbb{R}^n$ with strictly positive entries. $\otimes$ denotes the Kronecker product.

2 (S-)Assignment Flows

2.1 Definition

Let $\mathcal{G} = (\mathcal{V}, \mathcal{E}, \omega)$ denote an undirected weighted graph with $n = |\mathcal{V}|$ vertices and a nonnegative weight function $\omega\colon \mathcal{E} \to [0, \infty)$ on graph edges $\mathcal{E} \subseteq \mathcal{V} \times \mathcal{V}$. Let $S\colon \mathcal{V} \to \mathcal{S}_c$ denote a function that takes values $S_i \in \mathcal{S}_c$, $i \in \mathcal{V}$ in the relative interior $\mathcal{S}_c = \{s \in \mathbb{R}^c_{++}\colon \langle \mathbb{1}_c, s\rangle = 1\}$ of the probability simplex. $(\mathcal{S}_c, g)$ is a Riemannian manifold with the trivial tangent bundle $T_p\mathcal{S}_c \cong \mathcal{S}_c \times T_0\mathcal{S}_c$, with tangent space $T_0\mathcal{S}_c = \{v \in \mathbb{R}^c\colon \langle \mathbb{1}_c, v\rangle = 0\}$ and with the Fisher-Rao metric

$$g\colon T_p\mathcal{S}_c \times T_p\mathcal{S}_c \to \mathbb{R}, \qquad (v_1, v_2) \mapsto \langle v_1, v_2\rangle_g = \Big\langle \frac{v_1}{p}, v_2 \Big\rangle. \tag{1}$$

This metric can be seen as an infinitesimal version of relative entropy on $\mathcal{S}_c$. The assignment manifold $(\mathcal{W}, g)$ is the product manifold $\mathcal{W} = \mathcal{S}_c \times \cdots \times \mathcal{S}_c$ with $n = |\mathcal{V}|$ factors and the natural corresponding extension of the Fisher-Rao metric g. It defines the set of feasible assignment matrices $S \in \mathcal{W} \subset \mathbb{R}^{n\times c}_{++}$.

We consider a version of the assignment flow approach [1,28] introduced in [27],

$$\dot{S} = R_S(\Omega S),\ S(0) = S_0, \qquad \Omega_{ij} = \omega_{ij}, \quad ij \in \mathcal{E}, \tag{2a}$$

$$\big(R_S(\Omega S)\big)_i = R_{S_i}(\Omega S)_i, \qquad R_{S_i} = \mathrm{Diag}(S_i) - S_i S_i^\top,\ i \in \mathcal{V}. \tag{2b}$$

Determining $S(t)$ by geometric numerical integration [34] converges for $t \to \infty$ under mild conditions towards unit vectors $S_i^* = e_{j(i)}$, $i \in \mathcal{V}$ that assign the label j to the data point at vertex i represented by the initial point $S_{0;i}$ [35]. Using the row-stacking operator $s(t) = \mathrm{vec}_r(S(t))$ and extending Ω from $\mathrm{vec}_r(\Omega S) = (\Omega \otimes I_c)\mathrm{vec}_r(S)$ to $\Omega^{\mathfrak{v}}\mathrm{vec}_r(S)$, as done in [5] in order to take both spatial and label interaction into account, vectorization of (2) yields

$$\dot{s}(t) = R^{\mathfrak{v}}_{s(t)} \Omega^{\mathfrak{v}} s(t), \qquad s(0) = s_0 \tag{3}$$

with $R^{\mathfrak{v}}_s \Omega^{\mathfrak{v}} s = \mathrm{Diag}(R_{S_1}, \dots, R_{S_n}) \Omega^{\mathfrak{v}} \mathrm{vec}_r(S)$. We further define the *lifting map* $\exp_S(V) = \mathrm{softmax}(V + \log S)$ where softmax is applied along the second dimension (c) and log applies componentwise. To simplify notation, we will in the following re-use the symbol Ω to mean the extended operator $\Omega^{\mathfrak{v}} \in \mathbb{R}^{nc \times nc}$.

2.2 Embedding

In [4], a formal reduction of assignment flows to replicator dynamics has been proposed. To this end, the authors regard the joint state of n nodes each carrying a distribution over c classes as point on a single meta-simplex $\mathcal{S}_N$ with $N = c^n$ vertices (extreme points). As in this work, we will use multi-index notation $\alpha \in [c]^n$ instead of single indices $i \in [N]$ to refer to entries of $\mathcal{S}_N$ and its tangent space. We will use the embedding result [4, Theorem 1] as well as the following associated definitions.

Definition 1 (Embedding Maps). *The maps*

$$T\colon \mathcal{W} \to \mathcal{S}_N, \qquad T(S)_\alpha := \prod_{i \in [n]} S_{i,\alpha_i}, \quad N := c^n, \tag{4a}$$

$$Q\colon T_0\mathcal{W} \to T_0\mathcal{S}_N, \qquad Q(V)_\alpha := \sum_{l \in [n]} V_{l,\alpha_l}. \tag{4b}$$

are diffeomorphisms between their domain and a subset of their range. In the case of (4a), *the range* $\mathrm{img}\, T$ *is the set of joint distributions in* $\mathcal{S}_N$ *which factorize into marginals* S_i.

With abuse of notation, we will use the same symbol Q to denote the linear map $\mathbb{R}^{n \times c} \to \mathbb{R}^N$ defined analogously by (4b). From this perspective, the adjoint linear map $Q^\top$ was shown in [4, Lemma 2] to perform marginalization of distributions in $\mathcal{S}_N$ which is the inverse map of T on its range. In addition, we will analogously apply the maps T and Q to vectorized arguments $s = \mathrm{vec}(S)$ and $v = \mathrm{vec}(V)$. With these definitions, the central result of [4] is that for $s(t)$ with dynamics (3), the quantity $p(t) = T(s(t)) \in \mathcal{S}_N$ has the dynamics

$$\dot{p}(t) = R_{p(t)}[Q\Omega Q^\top p(t)], \qquad p(0) = T(s_0). \tag{5}$$

3 Randomized Assignment Flow

We regard the interaction Ω as a random variable with distribution μ. This in turn makes $p(t) = p(t, \Omega, S_0)$ defined by the dynamics (5) a random variable whose distribution $\nu(t)$ in $\mathcal{S}_N$ varies over time. We will use the first moment

$$P(t) = \mathbb{E}_{p\sim\nu(t)}[p] = \mathbb{E}_{\Omega\sim\mu}[p(t, \Omega, S_0)] \tag{6}$$

to represent a joint distribution of random variables on the graph $\mathcal{G}$. In this section, we examine properties of $P(t)$ and its limit P^∞ over time.

First note that $P(t)$ lies in $\mathcal{S}_N$ because every $p(t, \Omega, S_0)$ lies in $\mathcal{S}_N$ and $\mathcal{S}_N$ is a convex set. Thus, the limit P^∞ lies in the closure $\overline{\mathcal{S}}_N$ which contains all joint distributions of random variables on $\mathcal{G}$ with possibly non-full support. Suppose that (5) converges to a unit vector $e_{\gamma(\Omega)} \in \overline{\mathcal{S}}_N$ for (almost) every Ω drawn from μ. Then

$$P^\infty_\alpha = \lim_{t\to\infty} P(t)_\alpha \to \mathbb{E}_{\Omega\sim\mu}[e_{\gamma(\Omega)}]_\alpha = \mathbb{E}_{\Omega\sim\mu}[\![\alpha = \gamma(\Omega)]\!] = \mathbb{P}_{\Omega\sim\mu}(\alpha = \gamma(\Omega)). \tag{7}$$

Thus, we can draw samples of P^∞ *efficiently* by numerical integration of (3) and these samples *will be integer distributions*, i.e. hard node-label assignments. A distribution μ governing Ω, which meets these requirements is specified next.

Theorem 1 ([35, Thm. 2]). *Let* $\Omega = \max(Z + Z^\top, 0) + \epsilon\mathbb{I}_n$, $Z \in \mathbb{R}^{n\times n}$, $\epsilon > 0$, *where the entries of Z follow a multivariate normal distribution and maximization is componentwise. Then the embedded S-flow* (5) *converges to a unit vector for every draw of Ω and almost every $S_0 \in \mathcal{W}$.*

Proof. For the given shape of Ω, the assumptions of [35, Thm. 2] are met, which guarantees that the solution $S(t)$ of (2) converges to an integral solution for almost every initialization $S_0 \in \mathcal{W}$. Because T bijectively maps the corners of $\mathcal{W}$ to the corners of $\mathcal{S}_N$, the assertion is immediate from the embedding theorem [4, Theorem 1].

Crucially, even though every $p(t, \Omega, S_0)$ lies in the image of T and thus factorizes into node marginals, the expected value $P(t)$ typically *does not factorize*. This is due to the fact that $\operatorname{img} T$, the set of rank-1 tensors, is a relatively low-dimensional, curved, non-convex subset of $\overline{\mathcal{S}}_N$. To see this, consider the following Lemma.

Lemma 1 (Lifting Map Lemma [4]**).** *For any $S \in \mathcal{W}$ and $V \in \mathbb{R}^{n\times c}$ it holds*

$$T(\exp_S(V)) = \exp_{T(S)}(QV) \tag{8}$$

By using Lemma 1, we can show the following properties of $\operatorname{img} T$.

Lemma 2 (Properties of $\operatorname{img} T$). $\operatorname{img} T$ *is a curved, non-convex submanifold of $\mathcal{S}_N$ with dimension at most $n(c-1)$.*

Proof. By choosing $S \in \mathcal{W}$ as the uniform distribution on every node, the statement of Lemma 1 becomes

$$T(\mathrm{softmax}(V)) = \mathrm{softmax}(QV) \tag{9}$$

Since $\mathrm{softmax}\colon T_0\mathcal{W} \to \mathcal{W}$ is surjective onto $\mathcal{W}$, (9) characterizes $\mathrm{img}\, T$ as the image of the linear subspace $\mathrm{img}\, Q \subseteq T_0\mathcal{S}_N$ under $\mathrm{softmax}\colon T_0\mathcal{S}_N \to \mathcal{S}_N$. Thus, $\mathrm{img}\, T$ is flat in e-coordinates on $\mathcal{S}_N$, making it curved in m-coordinates. The subspace $\mathrm{img}\, Q$ has dimension at most $n(c-1)$ because Q is linear and $T_0\mathcal{W}$ has dimension $n(c-1)$. To see that $\mathrm{img}\, T$ is not convex, note that the extreme points of $\overline{\mathcal{W}}$ are bijectively mapped to the extreme points of $\overline{\mathcal{S}}_N$ by T. Suppose $\mathrm{img}\, T$ was convex. Then $\mathrm{img}\, T$ contains the convex hull of every subset of $\mathrm{img}\, T$. But the convex hull of the extreme points of $\mathcal{S}_N$ is just all of $\mathcal{S}_N$, contradicting the fact that $\mathrm{img}\, T$ has lower dimension than $\mathcal{S}_N$. □

4 Approximation of Energy-Based Models

Here we consider the approximation of energy-based models, i.e. models in which the probability of configuration α is given by

$$P^*_\alpha = \frac{1}{Z^*}\exp(-E_\alpha), \qquad Z^* = \sum_{\alpha\in[c]^n}\exp(-E_\alpha) \tag{10}$$

where energy E_α of each individual configuration is tractable but the *partition function* Z^* is intractable, because it contains a combinatorially large number of summands. We enumerate the energies of all configurations and collect them in the single vector $E \in \mathbb{R}^N$. As an instance of (10), consider the class of pairwise graphical models with energy

$$E_\alpha = \sum_{i\in[n]}\langle\theta^i, e_{\alpha_i}\rangle + \sum_{ij\in\mathcal{E}}\langle e_{\alpha_j}, \theta^{ij} e_{\alpha_i}\rangle \qquad \theta^i \in \mathbb{R}^c, \quad \theta^{ij} \in \mathbb{R}^{c\times c} \tag{11}$$

Tying back to the notation of earlier sections, we transform the pairwise term in (11) to

$$\sum_{ij\in\mathcal{E}}\langle e_{\alpha_j}, \theta^{ij} e_{\alpha_i}\rangle = \sum_{ij\in\mathcal{E}}\langle (Q^\top e_\alpha)_j, \theta^{ij}(Q^\top e_\alpha)_i\rangle = \langle Q^\top e_\alpha, \theta^{(p)} Q^\top e_\alpha\rangle \tag{12}$$

with matrix $\theta^{(p)} \in \mathbb{R}^{nc\times nc}$ built from blocks $\theta^{ij} \in \mathbb{R}^{c\times c}$, $i,j \in [n]$. Similarly, combining unary parameters θ^i into a single vector $\theta^{(u)} \in \mathbb{R}^{nc}$ yields the vectorized form of (11)

$$E = Q\theta^{(u)} + \mathrm{diag}(Q\theta^{(p)}Q^\top). \tag{13}$$

Suppose one approximates P^* by a tractable $P \in \mathcal{S}_N$. This entails minimization of

$$\mathrm{KL}[P\colon P^*] = \left\langle P, \log\frac{P}{P^*}\right\rangle = \langle P, \log P\rangle - \langle P, \log P^*\rangle \tag{14a}$$

$$= -H(P) - \langle P, E\rangle + \underbrace{\log Z^*}_{\text{const}}\underbrace{\langle P, \mathbb{1}_N\rangle}_{=1} \tag{14b}$$

which mirrors the well-known conjugacy relation [7, Lemma 1.1.3]

$$\log \left\langle \frac{1}{N}\mathbb{1}_N, \exp(-E) \right\rangle = \sup_P \; -\langle E, P\rangle - \mathrm{KL}[P\colon \frac{1}{N}\mathbb{1}_N]. \tag{15}$$

Thus, in order to learn P efficiently, we need to be able to compute its entropy and expected energy as well as their respective gradients. Since the energy of each individual configuration is tractable, the expected energy of a tractable model is typically easy to estimate. However, estimating entropy from samples is generally a difficult problem, which makes tractable entropy a key design criterium for surrogate models P. Along this line of reasoning, the basic *mean-field approach* is to approximate P^* by a factorizing distribution $T(M)$. The model entropy in (14) then simplifies to

$$-H(T(M)) = \langle T(M), \log T(M)\rangle = \langle T(M), Q\log M\rangle = \langle M, \log M\rangle \tag{16}$$

by [4, Lemma 2]. Because both the uniform distribution in $\mathcal{S}_N$ and every extreme point of $\mathcal{S}_N$ is a factorizing distribution, the mean-field approximation generally works best if either (a) all configurations have close to the same probability (high temperature regime) or (b) P^* is close to an integer distribution (the system is essentially deterministic). It is the challenging medium or low temperature regime in which a more sophisticated model is typically required – entailing the problem of entropy estimation.

Here we propose to approximate P^* by P as defined in (6). This goes beyond mean field approaches because, as discussed in Sect. 3, P typically lies outside $\operatorname{img} T$ i.e. does not factorize. Expected model energy reads

$$\langle P, E\rangle = \langle \mathbb{E}_\Omega T(S(\Omega)), E\rangle = \mathbb{E}_\Omega \langle T(S(\Omega)), E\rangle \tag{17}$$

which amounts to an expected value of mean field energies. Thus, if mean field energy is tractable, the empirical energy over samples of Ω is an unbiased estimator of model energy.

We turn to the more challenging problem of entropy estimation. Typically, estimating model entropy $H(P)$ from samples is difficult because the support $|\operatorname{supp} P| = s$ of P is large compared to the number m of available samples. The support of P^* can be arbitrarily large in principle. In fact, as a prerequisite for the Hammersley-Clifford theorem [6, Thm. 9.1.10], full support has formal merit in Markov random fields. On the other hand, many situations of practical interest do not benefit from a model with very large support. For instance, in image segmentation, most configurations of classes on the pixels of an image will have very little semantic content. In statistical mechanics, full support is beneficial to model high temperature systems. However, the behavior of these systems is dominated by randomness and they are well-described by a mean field approximation. More sophisticated models are beneficial in the challenging medium or low temperature regime and in this case small support can suffice.

Suppose the support size s is small compared to the number m of available samples $\{\alpha(i)\}_{i\in[m]} \subseteq [c]^n$ drawn from P^∞. Denote by

$$\widehat{p} = \frac{1}{m} \sum_{i\in[m]} e_{\alpha(s)} \in \mathcal{S}_N \tag{18}$$

the empirical distribution of samples. A classical analysis by [18] shows that the *plugin estimator*

$$H(P) \approx H(\widehat{p}) = - \sum_{\alpha\in\mathrm{supp}(\widehat{p})} \widehat{p}_\alpha \log \widehat{p}_\alpha \tag{19}$$

has bias

$$\mathbb{E}[H(\widehat{p})] - H(P) = -\frac{s-1}{2m} + \mathcal{O}\Big(\frac{1}{m^2}\Big) \tag{20}$$

which leads to the Miller-Maddows bias correction for known support s. It was shown that this only achieves a consistent estimator if $m \gg s$ [21] which is far from the optimal rate of $m \gg s/\log s$ [12] and thus motivates the use of more advanced approaches such as [12,29,30,33].

The support of P as defined in (6) is typically not known. However, in our experiments (Sect. 5) we observe that the empirical distribution is only supported on relatively few configurations. For this reason, we judge (19) with bias correction (20) to be sufficient for the case at hand. Note that an unbiased estimator of entropy from samples exists [19] but is not practical for our use case, because it entails drawing a potentially large number of samples which is not known a priori.

As a key issue it remains to find a differentiable approximation of $-H(\widehat{p})$. Under the assumption of integer samples, we find

$$-H(\widehat{p}) = \Big\langle \frac{1}{m} \sum_{k\in[m]} T(S^k), \log \frac{1}{m} \sum_{k\in[m]} T(S^k) \Big\rangle \tag{21a}$$

$$= \Big\langle \frac{1}{m} \sum_{k\in[m]} e_{\alpha(k)}, \log \frac{1}{m} \sum_{k\in[m]} e_{\alpha(k)} \Big\rangle = \frac{1}{m} \sum_{k\in[m]} \log \Big(\frac{1}{m} \sum_{l\in[m]} e_{\alpha(l)}\Big)_{\alpha(k)} \tag{21b}$$

$$= -\log m + \frac{1}{m} \sum_{k\in[m]} \log \Big(\sum_{l\in[m]} e_{\alpha(l)}\Big)_{\alpha(k)}. \tag{21c}$$

This motivates the approximation

$$-H(\widehat{p}) \approx -\log m + \frac{1}{m} \sum_{k\in[m]} \log \Big(\sum_{l\in[m]} T(S^l)\Big)_{\alpha(k)} \tag{22}$$

for non-integer samples S^l. The fact that assignment flows converge to integer labelings is crucial to this construction, because quantities in $\mathcal{S}_N$ can not be explicitly represented in numerical computations and integrality of S allows to reduce the sparse sum in (21b) to a tractable quantity. Note that $T(S^l)_{\alpha(k)}$ above is a product of n numbers in $(0,1)$. We thus rewrite the summands in (22) as

$$\log\Big(\sum_{l\in[m]} T(S^l)\Big)_{\alpha(k)} \overset{(4a)}{=} \log\Big(\sum_{l\in[m]}\prod_{i\in[n]} S^l_{i,\alpha(k)_i}\Big)_{\alpha(k)} \tag{23a}$$

$$= \log\sum_{l\in[m]}\exp\Big(\sum_{i\in[n]}\log S^l_{i,\alpha(k)_i}\Big) \tag{23b}$$

to avoid numerical underflow problems by leveraging a stabilized implementation of $\operatorname{logexp}_{\epsilon=1}$. Note that the right-hand side is differentiable.

Once a suitable approximation of P^* is found by minimizing (14) with respect to parameters governing μ (Sect. 5), the model P can be used for probabilistic inference. Marginal distributions are easily estimated via

$$Q^\top P = \mathbb{E}_{\Omega\sim\mu}[Q^\top T(S(\Omega))] = \mathbb{E}_{\Omega\sim\mu}[S(\Omega)] \tag{24}$$

and more generally, any quantity $Q\phi$ with $\phi \in \mathbb{R}^{nc}$ can be inferred by

$$\mathbb{E}_P[Q\phi] = \langle \mathbb{E}_{\Omega\sim\mu}[T(S(\Omega))], Q\phi\rangle = \mathbb{E}_{\Omega\sim\mu}[\langle S(\Omega),\phi\rangle]. \tag{25}$$

5 Experiments

The introductory example in Fig. 1 was produced by approximating a large Potts model on the grid graph of image pixels[1]. This was achieved by randomizing the generalized S-flow (3), giving $\Omega \in \mathbb{R}^{nc\times nc}$ the structure of multi-channel convolution with weights following a multivariate normal distribution. Suitable moments for this normal distribution together with a suitable flow initialization s_0 are the result of a *training procedure* which minimizes (14). To this end, a reparameterization trick [15] is applied in conjunction with the approximation (22) and bias correction (20) where the unknown support s is replaced by the empirical support $\widehat{s} = |\operatorname{supp}\widehat{p}|$ smoothed by the mean entropy of nodewise assignment. Numerical integration of (3) via the simple geometric Euler scheme [34] (step size 0.1, end time 1.0) is unwound and automatically differentiated by PyTorch [22] which allows to find a local optimum of parameters by employing the Adam optimizer [14] with step length 0.01.

In this section, we further demonstrate the approximation of energy-based models on a small *two-dimensional Ising model*, i.e. a system of binary random variables with nearest-neighbor interaction on a grid graph, governed by a Gibbs distribution of the form (10) with a corresponding energy function E_α.

These systems are classical ones in statistical mechanics [23]. They prototypically represent a combinatorially large configuration space and *long range* correlation at low temperatures. As a consequence, in the presence of an 'external field' [2], i.e. data defining unary potentials in E_α, minimizing E_α and probabilistic inference become NP-hard even for moderate problem sizes. Such models

[1] All experiments were run on a single NVIDIA RTX 2080ti graphics card.

initiated research on image segmentation and Bayesian inference [9,10] and have been stimulating research on variational approximations for many years [17,32]. As a consequence, they define an ideal testbed for evaluating our approach and validating also experimentally our claims.

$\mathcal{G}$ is chosen as a 3×8 grid graph such that the combinatorial partition function and true marginals can still be computed by brute force. This allows to give numerical values for the distance to the combinatorial model in terms of relative entropy via (14). The number of classes is $c = 2$. Unary energy is chosen as -3.0 for the 0-configuration of nodes on the left boundary and as 3.0 on the right boundary. All other unary energies are zero. Pairwise energy is set to $\theta^{ij} = \frac{7}{10} \cdot (\mathbb{1}_c \mathbb{1}_c^\top - \mathbb{I}_c)$ for each edge.

We approximate this model by the same training procedure as above with reduced learning rate $5 \cdot 10^{-3}$ over 5k iterations. This takes around 21 min on a single desktop graphics card. To guarantee S-flow convergence via Theorem 1, we omit label interaction as afforded by the generalization (3) and instead use (2) with symmetric matrix $\Omega \in \mathbb{R}^{n \times n}$ parameterized as $\Omega = \max(Z + Z^\top, 0) + 10^{-3}\mathbb{I}_n$ with entries of $Z \in \mathbb{R}^{n \times n}$ following a multivariate normal distribution. We initialize the distribution of Z centered at $\frac{1}{20}\mathbb{1}_n \mathbb{1}_n^\top$ and with componentwise variance 10^{-1}. In the early stages of optimization, samples are not integral due to the finite time horizon, but we observe that the sample entropy gradually decreases over the course of optimization, making the approximation (22) already close to exact for finite time. Once a model is learnt through convergence to a local minimum, integrality of samples is guaranteed by Theorem 1 for $t \to \infty$. In fact, it was shown in [35] that the same integer limit is also found by rounding after sufficiently large but finite time t which is relevant for numerical implementation.

As a baseline, we compute a mean field approximation $M \in \mathcal{W}$ by parameterizing $M = \text{softmax}(V)$ and using the Adam optimizer to learn V by minimizing the tractable form of (14). This procedure is repeated for 1k initializations drawn randomly from a standard normal distribution of $V \in \mathbb{R}^{n \times c}$ and a model with minimal KL distance is selected from resulting local optima. The true distribution has multiple modes, of which mean field approximation can only represent just one. In contrast, our model is able to capture the multimodality as is apparent from samples (Fig. 2), close approximation of marginals (Fig. 3) and low relative entropy (Table 1).

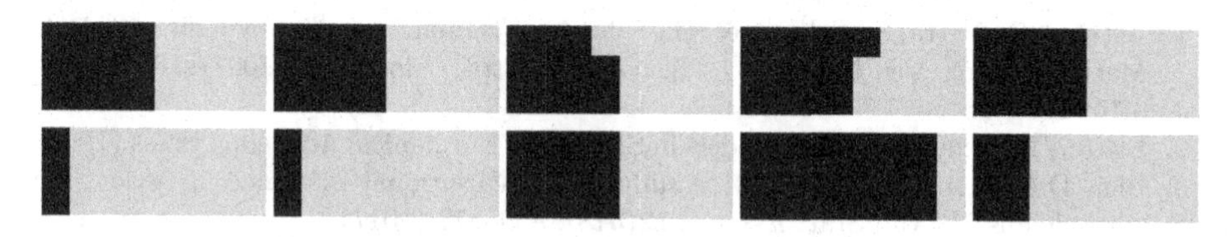

Fig. 2. Samples of Ising model from the mean-field baseline (first row) and from our model (second row). This demonstrates that, unlike the mean-field approximation, our approach can explore multiple modes in the low-temperature regime.

Fig. 3. Marginals of the true distribution (left), our approximation via randomized assignment (middle) and the baseline mean-field approximation (right).

Table 1. Summary of Ising model approximation. Relative entropy to the true distribution is computed by brute-force evaluation of the combinatorial partition function. Entropy of our model is closely approximated by (19) with bias correction (20) using $m = 1$M integer samples.

Model	KL	Energy	Entropy	Marginal Difference
AF (ours)	**0.599**	−1.98	2.56	**0.090**
Mean Field	1.974	−1.57	1.60	0.198

6 Discussion and Conclusion

In the low temperature regime (E large), the mass of P^* concentrates around its modes. For this reason, the proposed model – for which small support is computationally beneficial – actually becomes more effective at lower temperature. This unusual performance characteristic makes our approach promising in challenging structured prediction scenarios where mean-field approximation fails to capture the structure of interest. A natural direction of future work is to construct differentiable approximations such as (22) for more advanced entropy estimators.

Acknowledgements. This work is funded by the Deutsche Forschungsgemeinschaft (DFG), grant SCHN 457/17-1, within the priority programme SPP 2298: "Theoretical Foundations of Deep Learning". This work is funded by the Deutsche Forschungsgemeinschaft (DFG) under Germany's Excellence Strategy EXC-2181/1 - 390900948 (the Heidelberg STRUCTURES Excellence Cluster).

References

1. Åström, F., Petra, S., Schmitzer, B., Schnörr, C.: Image labeling by assignment. J. Math. Imaging Vis. **58**(2), 211–238 (2017). https://doi.org/10.1007/s10851-016-0702-4
2. Baxter, R.: Exactly Solved Models in Statistical Mechanics. Academic Press (1982)
3. Blei, D.M., Kucukelbir, A., McAuliffe, J.D.: Variational inference: a review for statisticians. J. Am. Stat. Assoc. **112**(518), 859–877 (2017)
4. Boll, B., Schwarz, J., Schnörr, C.: On the correspondence between replicator dynamics and assignment flows. In: Elmoataz, A., Fadili, J., Quéau, Y., Rabin, J., Simon, L. (eds.) SSVM 2021. LNCS, vol. 12679, pp. 373–384. Springer, Cham (2021). https://doi.org/10.1007/978-3-030-75549-2_30

5. Boll, B., Zeilmann, A., Petra, S., Schnörr, C.: Self-certifying classification by linearized deep assignment. preprint arXiv:2201.11162 (2022)
6. Brèmaud, P.: Discrete Probability Models and Methods. Springer, Cham (2017). https://doi.org/10.1007/978-3-319-43476-6
7. Catoni, O.: PAC-Bayesian Supervised Classification: The Thermodynamics of Statistical Learning. Institute of Mathematical Statistics (2007)
8. Dynkin, E.B.: Sufficient statistics and extreme points. Ann. Probab. **6**(5), 705–730 (1978)
9. Geman, S., Geman, D.: Stochastic relaxation, Gibbs distributions, and the Bayesian restoration of images. IEEE Trans. Pattern Anal. Mach. Intell. **6**(6), 721–741 (1984)
10. Gidas, B.: A renormalization group approach to image processing problems. IEEE Trans. Pattern Anal. Mach. Intell. **11**(11), 164–180 (1989)
11. Heskes, T.: Convexity arguments for efficient minimization of the Bethe and Kikuchi free energies. J. Artif. Intell. Res. **26**, 153–190 (2006)
12. Jiao, J., Venkat, K., Han, Y., Weissman, T.: Minimax estimation of functionals of discrete distributions. IEEE Trans. Inf. Theory **61**(5), 2835–2885 (2015)
13. Kappes, J., et al.: A comparative study of modern inference techniques for structured discrete energy minimization problems. Int. J. Comput. Vis. **115**(2), 155–184 (2015). https://doi.org/10.1007/s11263-015-0809-x
14. Kingma, D.P., Ba, J.: Adam: a method for stochastic optimization. preprint arXiv:1412.6980 (2014)
15. Kingma, D.P., Welling, M.: Auto-encoding variational bayes. preprint arXiv:1312.6114 (2013)
16. Kobyzev, I., Prince, S.D., Brubaker, M.A.: Normalizing flows: an introduction and review of current methods. IEEE Trans. Pattern Anal. Mach. Intell. **43**(11), 3964–3979 (2021)
17. Mézard, M., Montanari, A.: Information, Physics, and Computation. Oxford University Press, Oxford (2009)
18. Miller, G.: Note on the Bias of Information Estimates. Information Theory in Psychology: Problems and Methods (1955)
19. Montgomery-Smith, S., Schürmann, T.: Unbiased estimators for entropy and class number. arXiv preprint arXiv:1410.5002 (2014)
20. Pakzad, P., Anantharam, V.: Estimation and marginalization using Kikuchi approximation methods. Neural Comput. **17**(8), 1836–1873 (2005)
21. Paninski, L.: Estimation of entropy and mutual information. Neural Comput. **15**(6), 1191–1253 (2003)
22. Paszke, A., et al.: Pytorch: an imperative style, high-performance deep learning library. In: NIPS (2019)
23. Pathria, R.K., Beale, P.D.: Statistical Mechanics, 3rd edn. Academic Press (2011)
24. Ramesh, A., Dhariwal, P., Nichol, A., Chu, C., Chen, M.: Hierarchical text-conditional image generation with CLIP latents (2022)
25. Rizzo, T., Wemmenhove, B., Kappen, H.J.: Cavity approximation for graphical models. Phys. Rev. E **76**(1), 011102 (2007)
26. Ruthotto, L., Haber, E.: An introduction to deep generative modeling. GAMM Mitt. **44**(2), 24 (2021)
27. Savarino, F., Schnörr, C.: Continuous-domain assignment flows. Eur. J. Appl. Math. **32**(3), 570–597 (2021)
28. Schnörr, C.: Assignment flows. In: Grohs, P., Holler, M., Weinmann, A. (eds.) Handbook of Variational Methods for Nonlinear Geometric Data, pp. 235–260. Springer, Cham (2020). https://doi.org/10.1007/978-3-030-31351-7_8

29. Valiant, G., Valiant, P.: Estimating the unseen: an n/log (n)-sample estimator for entropy and support size, shown optimal via new CLTs. In: Proceedings of the 43th ACM Symposium on Theory of Computing, pp. 685–694 (2011)
30. Valiant, G., Valiant, P.: Estimating the unseen: improved estimators for entropy and other properties. J. ACM **64**(6), 1–41 (2017)
31. Wainwright, M.J., Jaakola, T.S., Willsky, A.S.: Tree-based reparameterization framework for analysis of sum-product and related algorithms. IEEE Trans. Inf. Theory **49**(5), 1120–1146 (2003)
32. Wainwright, M., Jordan, M.: Graphical models, exponential families, and variational inference. Found. Trends Mach. Learn. **1**(1–2), 1–305 (2008)
33. Wu, Y., Yang, P.: Minimax rates of entropy estimation on large alphabets via best polynomial approximation. IEEE Trans. Inf. Theory **62**(6), 3702–3720 (2016)
34. Zeilmann, A., Savarino, F., Petra, S., Schnörr, C.: Geometric numerical integration of the assignment flow. Inverse Probl. **36**(3), 034004 (33pp) (2020)
35. Zern, A., Zeilmann, A., Schnörr, C.: Assignment flows for data labeling on graphs: convergence and stability. Inf. Geom. **5**, 355–404 (2022). https://doi.org/10.1007/s41884-021-00060-8

Quantum State Assignment Flows

Jonathan Schwarz[3](✉), Bastian Boll[3], Daniel Gonzalez-Alvarado[3], Dmitrij Sitenko[3], Martin Gärttner[1], Peter Albers[2], and Christoph Schnörr[3]

[1] Kichhoff Institute for Physics, Heidelberg University, Heidelberg, Germany
[2] Mathematical Institute, Heidelberg University, Heidelberg, Germany
[3] Institute for Applied Mathematics, Heidelberg University, Heidelberg, Germany
jonathan.schwarz@iwr.uni-heidelberg.de

Abstract. This paper extends the assignment flow approach from categorial distributions to complex-valued Hermitian density matrices, used as state spaces for representing and analyzing data associated with vertices of an underlying graph. Determining the flow of the resulting dynamical system by geometric integration causes an interaction of these non-commuting states across the graph, and the assignment of a pure (rank-one) state to each vertex after convergence. Experiments with toy systems indicate the potential of the novel approach for data representation and analysis.

Keywords: Assignment Flows · Density Matrix · Information Geometry

1 Introduction

The *assignment flow approach* provides data models in terms of state spaces that interact geometrically across an underlying graph. In principle, any open convex set can serve as a state space which becomes a Riemannian manifold when endowed with a Riemannian metric. The canonical cases are parameter spaces of distributions of the exponential family and the Fisher-Rao metric, which is the subject of *information geometry* [2].

The assignment flow approach has been introduced by [3] for the basic family of *categorial distributions*, in order to assign a unique element of a finite set of labels set to each data point observed in a metric space. We refer to [17] for more details and a review of related work.

Contribution. In this paper, we apply the assignment flow approach to a novel class of state spaces, the class of *complex valued, Hermitian positive semidefinite matrices*, known as density matrices in quantum mechanics, where they represent a physical system [5]. Even though this extension appears to be straightforward from an abstract mathematical viewpoint, details matter with regards to both the components of the approach and the scope of applications.

Specifically, a *key difference* is the *non-commutative* interaction of density matrices, opposed to the multiplicative interaction of discrete probability vectors.

L. Calatroni et al. (Eds.): SSVM 2023, LNCS 14009, pp. 743–756, 2023.
https://doi.org/10.1007/978-3-031-31975-4_57

Furthermore, regarding the objects to be assigned to data, the finite set of labels is replaced by the uncountable set of rank-one density matrices, i.e. the set of orthogonal projectors onto one-dimensional subspaces. We show that the original assignment flow can be recovered by restriction to the submanifold of *diagonal* density matrices.

Approach Component	Discrete Labeling Assignment Flow	Quantum State Assignment Flow
state space	product manifold of categorial distributions	product manifold of density matrices
state evolution	$\dot{S} = R_S[\Omega S]$	$\dot{\rho} = R_\rho[\Omega[\rho]]$
assigned limit states	unit vector at each vertex of $\mathcal{G}$	pure (rank one) state at each vertex of $\mathcal{G}$

Scope. Regarding applications, we have in mind data modeling and analysis as well as applications in quantum mechanics. Regarding the former aspect, we confine ourselves to a few proof of concept examples that demonstrate the different character and indicate the enhanced flexibility for data modeling of the novel assignment flow approach. Regarding the latter aspect, we point out that the *representation* of image data for quantum computing is an active field of research [7]. This paper contributes an approach for data *modeling and analysis* based on concepts of information geometry and quantum mechanics.

Organization. Section 2 collects concepts of information geometry. The novel approach is introduced in Sect. 3. Its properties are illustrated and discussed in Sect. 4. We conclude and point out further work in Sect. 5.

Due to the page limit, we have to omit many details and almost all proofs, and refer to [18].

2 Information Geometry

Information geometry [1,12] is concerned with the representation of parametric probability distributions like, e.g., the exponential family of distributions [6], from a geometric viewpoint. Specifically, an open convex set $\mathcal{M}$ of parameters of a probability distribution becomes a Riemannian manifold $(\mathcal{M}, g)$ when equipped with a Riemannian metric g. The Fisher-Rao metric is the canonical choice due to its invariance properties with respect to reparametrization [19].

A key ingredient of information geometry is the so-called α-family of affine connections introduced by Amari, which comprises the so-called e-connection ∇ and m-connection ∇^* as special cases. These connections are torsion-free and dual to each other in the sense that they jointly satisfy the equation which uniquely characterizes the Levi-Civita connection as metric connection [1, Def. 3.1, Thm. 3.1]. Regarding numerical computations, working with the exponential map induced by the e-connection is particularly convenient since its domain is the entire tangent space. We refer to [2,4,8] for further reading and to [13], [2, Ch. 7] for the specific case of the state spaces of quantum mechanics.

In this paper, we are concerned with two classes of convex sets, the relative interior of *probability simplices*, each of which represents the categorial (discrete) distributions of the corresponding dimension, and *density matrices*, i.e. the set of positive-definite Hermitian matrices with trace equal to one. Sections 2.1 and 2.2 introduce the information geometry for the former and the latter class of sets, respectively.

2.1 Categorial Distributions

We set $[c] := \{1, 2, \dots, c\}$ and $\mathbb{1}_c := (1, \dots, 1)^\top \in \mathbb{R}^c$ for $c \in \mathbb{N}$. The probability simplex of distributions on $[c]$ is denoted by $\Delta_c := \{p \in \mathbb{R}^c_+ : \langle \mathbb{1}_c, p\rangle = \sum_{i\in[c]} p_i = 1\}$. Its relative interior equipped with the Fisher-Rao metric g becomes the Riemannian manifold $(\mathcal{S}_c, g)$, where

$$\mathcal{S}_c := \operatorname{rint} \Delta_c = \{p \in \Delta_c : p_i > 0,\ i \in [c]\}, \tag{2.1a}$$

$$g_p(u, v) = \langle u, \operatorname{Diag}(p)^{-1} v\rangle, \qquad \forall u, v \in T_{c,0}, \quad p \in \mathcal{S}_c \tag{2.1b}$$

with the tangent space (with the barycenter of $\mathcal{S}_c$ denoted by $\mathbb{1}_{\mathcal{S}_c} = \frac{1}{c}\mathbb{1}_c$)

$$T_{c,0} := T_{\mathbb{1}_{\mathcal{S}_c}} \mathcal{S}_c = \{v \in \mathbb{R}^c : \langle \mathbb{1}_c, v\rangle = 0\} \tag{2.2}$$

and the trivial tangent bundle $T\mathcal{S}_c \cong \mathcal{S}_c \times T_{c,0}$. The orthogonal projection onto $T_{c,0}$ reads

$$\pi_0 : \mathbb{R}^c \to T_{c,0}, \qquad \pi_0 v = (I_c - \mathbb{1}_c \mathbb{1}_{\mathcal{S}_c}^\top) v, \tag{2.3}$$

A key role plays the *replicator mapping*

$$R : \mathcal{S}_c \times T_{c,0} \to T_{c,0}, \qquad R_p v := (\operatorname{Diag}(p) - pp^\top) v, \tag{2.4}$$

which is parametrized by $p \in \mathcal{S}_c$ and has the properties

$$R_p \mathbb{1}_c = 0 \qquad \text{and} \qquad \pi_0 R_p = R_p \pi_0 = R_p, \qquad \forall p \in \mathcal{S}_c. \tag{2.5}$$

In particular, R_p is the inverse metric tensor expressed in the ambient coordinates p and its restriction to the tangent space $T_{c,0}$ is a linear isomorphism [16, Lemma 3.1]. Accordingly, given a smooth function $f : \mathcal{S}_c \to \mathbb{R}$, its Riemannian gradient with respect to the Fisher-Rao metric (2.1b) is given by

$$\operatorname{grad} f(p) = R_p \partial f(p). \tag{2.6}$$

We list two further mappings required below. The exponential map induced by the e-connection is defined on the entire space $T_{c,0}$ and reads [4]

$$\operatorname{Exp} : \mathcal{S}_c \times T_{c,0} \to \mathcal{S}_c, \qquad \operatorname{Exp}_p(v) := \langle p, e^{\frac{v}{p}}\rangle^{-1} (p \cdot e^{\frac{v}{p}}), \tag{2.7}$$

where $\cdot$ denotes *componentwise* multiplication of vectors (Hadamard product). The so-called *lifting map* introduced in [3] reads

$$\exp : \mathcal{S}_c \times T_{c,0} \to \mathcal{S}_c, \qquad \exp_p(v) := \operatorname{Exp}_p \circ R_p(v) = \langle p, e^v\rangle^{-1} (p \cdot e^v). \tag{2.8}$$

The subscript of $\exp_p$ disambiguates its meaning in view of the ordinary exponential *function* e^v written without subscripts, and from the symbol $\exp_{\mathrm{m}}$ which always means the *matrix exponential* function.

2.2 Density Matrices

We denote by $\rho^* = \overline{\rho}^\top$ the conjugate transpose of a matrix $\rho \in \mathbb{C}^{c \times c}$. The inner products on $\mathbb{C}^c$ and $\mathbb{C}^{c\times c}$, respectively, are denoted by $\langle a, b\rangle = a^* b$ and $\langle A, B\rangle = \operatorname{tr}(A^* B)$. We denote the open convex cone of positive definite Hermitian matrices by $\mathcal{P}_c := \{\rho \in \mathbb{C}^{c\times c} \colon \rho = \rho^*,\ \rho \succ 0\}$ and its intersection with the hyperplane defined by constraint $\operatorname{tr}\rho = 1$, the space of *density matrices*, by

$$\mathcal{D}_c := \{\rho \in \mathcal{P}_c \colon \operatorname{tr}\rho = 1\}. \tag{2.9}$$

We refer to [5] for the physical background and to [14] for mathematical aspects related to quantum information theory. Denoting the vector space of Hermitian matrices by $\mathcal{H}_c := \{X \in \mathbb{C}^{c\times c} \colon X^* = X\}$, we have analogous to (2.2) the tangent space (with $\mathbb{1}_{\mathcal{D}_c} := \operatorname{Diag}(\mathbb{1}_{\mathcal{S}_c})$)

$$\mathcal{H}_{c,0} := T_{\mathbb{1}_{\mathcal{D}_c}}\mathcal{D}_c = \mathcal{H}_c \cap \{X \in \mathbb{C}^{c\times c} \colon \operatorname{tr} X = 0\} \tag{2.10}$$

and the trivial tangent bundle $T\mathcal{D}_c \cong \mathcal{D}_c \times \mathcal{H}_{c,0}$. The corresponding orthogonal projection reads[1]

$$\pi_0 \colon \mathcal{H}_c \to \mathcal{H}_{c,0}, \qquad \pi_0 X := X - (\operatorname{tr} X) I_{\mathcal{D}_c}. \tag{2.11}$$

The metric g is the *Bogoliubov-Kubo-Mori (BKM) metric* [15]

$$g_\rho(X,Y) := \int_0^\infty \operatorname{tr}\big(X(\rho+\lambda I)^{-1} Y (\rho+\lambda I)^{-1}\big) d\lambda,\ X, Y \in \mathcal{H}_{c,0},\ \rho \in \mathcal{D}_c. \tag{2.12}$$

This metric uniquely ensures that the e-connection ∇ induced on $\mathcal{D}_c$ is symmetric and the connections ∇, ∇^* are mutually dual to each other in the sense of information geometry [9], [2, Thm. 7.1].

The following map and its inverse, defined in terms of the matrix exponential $\exp_{\mathrm{m}}$ and its inverse $\log_{\mathrm{m}} = \exp_{\mathrm{m}}^{-1}$ will be convenient: $\mathbb{T} \colon \mathcal{D}_c \times \mathcal{H}_c \to \mathcal{H}_c$, with

$$\mathbb{T}_\rho[X] := \frac{d}{dt}\log_{\mathrm{m}}(\rho + tX)\big|_{t=0} = \int_0^\infty (\rho+\lambda I)^{-1} X (\rho+\lambda I)^{-1} d\lambda, \tag{2.13a}$$

$$\mathbb{T}_\rho^{-1}[X] = \frac{d}{dt}\exp_{\mathrm{m}}(H + tX)\big|_{t=0} = \int_0^1 \rho^{1-\lambda} X \rho^\lambda d\lambda, \qquad \rho = \exp_m(H). \tag{2.13b}$$

The inner product (2.12) may now be written in the form $g_\rho(X,Y) = \langle X, \mathbb{T}_\rho[Y]\rangle$ since the trace is invariant with respect to cyclic permutations of a matrix product as argument. Likewise, the relation $\langle \rho, X\rangle = \operatorname{tr}(\rho X) = \operatorname{tr}\mathbb{T}_\rho^{-1}[X]$ holds.

3 Quantum State Assignment Flows

This section summarizes our results regarding the extension of the assignment flow on the manifold of categorial distributions to the manifold of density matrices. The extension significantly generalizes the state space and the corresponding

[1] We keep the symbol π_0 for notational simplicity. The argument disambiguates the projections (2.3) and (2.11), respectively.

assignment flow. The positivity condition imposed on discrete probability vectors is replaced by the positive definiteness condition imposed on Hermitian matrices, and mass conservation of categorial distributions is replaced by trace normalization. The assignment flow has to be generalized accordingly, which is accomplished within the framework of information geometry. Rather than assigning a single label from a *finite* set of labels to each data point, the resulting quantum-state assignment flow (QSAF) assigns a pure (rank-one) state from an *uncountable* set to each data point. And analogous to the encoding of labels by unit vectors on the boundary of the product of probability simplices (assignment manifold), the QSAF converges towards the boundary of the corresponding density matrix product manifold.

This section is organized as follows. A few basic relations are collected in Sect. 3.1. Section 3.2 introduces the flow for a single state space which is generalized in Sect. 3.3 to the case of multiple states whose evolutions interact across a graph. This approach generalizes the assignment flow approach introduced by [3]. Section 3.4 generalizes accordingly the reparametrization introduced by [16] that enables to characterize the approach as a Riemannian gradient flow with respect to a nonconvex potential. Finally, Sect. 3.5 elucidates that the original assignment flow can be recovered as special case by restricting the quantum state assignment flow to diagonal density matrices.

Due to the page limit, we omit the proofs, except for the short one of Proposition 6, and refer to the journal version of this paper [18].

3.1 Basic Relations

A parametrization of the manifold $\mathcal{D}_c$ is given by

$$\Gamma\colon \mathcal{H}_{c,0} \to \mathcal{D}_c, \qquad \Gamma(X) := \frac{\exp_{\mathrm{m}}(X)}{\operatorname{tr} \exp_{\mathrm{m}}(X)}. \tag{3.1}$$

Lemma 1. *The mapping* (3.1) *is bijective with inverse*

$$\Gamma^{-1}\colon \mathcal{D}_c \to \mathcal{H}_{c,0}, \qquad \Gamma^{-1}(\rho) = \pi_0 \log_{\mathrm{m}} \rho, \tag{3.2}$$

Furthermore, for $H, X \in \mathcal{H}_{c,0}$ with $\Gamma(H) = \rho$ and $Y \in T\mathcal{H}_{c,0} \cong \mathcal{H}_{c,0}$, the respective differential mappings are given by

$$d\Gamma(H)[Y] = \mathbb{T}_\rho^{-1}\big[Y - \langle \rho, Y\rangle I\big], \qquad \rho = \Gamma(H) \tag{3.3a}$$

$$d\Gamma^{-1}(\rho)[X] = \pi_0 \circ \mathbb{T}_\rho[X]. \tag{3.3b}$$

A key concept of information geometry is the affine e-connection and the corresponding exponential map. It rarely occurs that pairs of points on a Riemannian manifold can be connected in closed form by a corresponding geodesic.

Proposition 1. *The e-geodesic emanating at $\rho \in \mathcal{D}_c$ in the direction $X \in \mathcal{H}_{c,0}$ and the corresponding exponential map are given by*

$$\gamma^{(e)}_{\rho,X}(t) := \mathrm{Exp}^{(e)}_{\rho}(tX), \quad t \geq 0 \tag{3.4a}$$

$$\mathrm{Exp}^{(e)}_{\rho}(X) := \Gamma\big(\Gamma^{-1}(\rho) + d\Gamma^{-1}(\rho)[X]\big) \tag{3.4b}$$

$$= \Gamma\big(\Gamma^{-1}(\rho) + \pi_0 \circ \mathbb{T}_{\rho}[X]\big). \tag{3.4c}$$

We next consider the evaluation of Riemannian gradients.

Proposition 2. *The Riemannian gradient of a smooth function $f\colon \mathcal{D}_c \to \mathbb{R}$ with respect to the BKM-metric* (2.12) *is given by*

$$\mathrm{grad}_{\rho} f = \mathbb{T}^{-1}_{\rho}[\partial f] - \langle \rho, \partial f\rangle \rho, \tag{3.5}$$

where $\mathbb{T}^{-1}_{\rho}$ is given by (2.13b) *and ∂f is the ordinary gradient with respect to the Euclidean structure of the ambient space $\mathbb{R}^{c\times c}$.*

The general defining formula $g_{\rho}(\mathrm{grad}\, f, X) = df_{\rho}X,\ \forall X \in \mathcal{H}_{c,0}$ for the Riemannian gradient [11, pp. 337] generally yields an expression of the form $\mathrm{grad}\, f_{\rho} = G(\rho)^{-1}\partial f(\rho)$ given in local coordinates. The role of the inverse metric tensor G^{-1} is played by the replicator map (2.4) which, in the present context and in view of the result (3.5), takes the more general form

$$R_{\rho}\colon \mathcal{H}_c \to \mathcal{H}_{c,0}, \quad R_{\rho}[X] := \mathbb{T}^{-1}_{\rho}[X] - \langle \rho, X\rangle \rho \qquad \textbf{(replicator map)} \tag{3.6}$$

where $\rho \in \mathcal{D}_c$ and $X \in \mathcal{H}_{c,0}$.

For a given graph $\mathcal{G} = (\mathcal{V}, \mathcal{E})$, we finally introduce the product spaces

$$\mathcal{H} := \mathcal{H}_c \times \cdots \times \mathcal{H}_c, \qquad \mathcal{H}_0 := \mathcal{H}_{c,0} \times \cdots \times \mathcal{H}_{c,0}, \tag{3.7}$$

each with $|\mathcal{V}|$ factors.

3.2 Single-Vertex Quantum State Assignment Flow

Let $D \in \mathcal{H}_c$ denote a given Hermitian matrix. Then we define the corresponding *likelihood matrix* by

$$L_{\rho}\colon \mathcal{H}_c \to \mathcal{D}_c, \qquad L_{\rho}(D) := \exp_{\rho}(-\pi_0 D), \qquad \rho \in \mathcal{D}_c. \tag{3.8}$$

The *single-vertex quantum state assignment flow* equation reads

$$\dot{\rho} = R_{\rho}[L_{\rho}(D)], \qquad \rho(0) = \mathbb{1}_{\mathcal{D}_c}, \tag{3.9}$$

where R_{ρ} is given by (3.6). The evolution of $\rho(t)$ solving (3.9) behaves as follows.

Proposition 3. *Let $D = Q\Lambda_D Q^{\top}$ be the spectral decomposition of D with eigenvalues $\lambda_1 \geq \cdots \geq \lambda_c$ and orthonormal eigenvectors $Q = (q_1, \ldots, q_c)$. Assume the minimal eigenvalue λ_c is unique. Then the solution $\rho(t)$ to* (3.9) *satisfies*

$$\lim_{t\to\infty} \rho(t) = \Pi_{q_c} := q_c q_c^{\top}. \tag{3.10}$$

We refer to [18] for a proof. It relies on a decomposition of $\mathcal{H}_{c,0}$ [2, Section 7.1] that allows for a reduction of the single-vertex quantum state assignment flow to the standard assignment flow [3]. The convergence of the latter is discussed in detail in [21].

A natural question is how multiple states which evolve in this way, can interact so as to influence their limit points, but not their property of being pure (rank one) states. Such a dynamical system is provided next.

3.3 Quantum State Assignment Flow

Let $\mathcal{G} = (\mathcal{V}, \mathcal{E}, \omega)$ be a given graph with nonnegative weight function $\omega \colon \mathcal{E} \to \mathbb{R}_+$, satisfying $\sum_{k \in \mathcal{N}_i} \omega_{ik} = 1$ with respect to the neighborhood system $\mathcal{N}_i := \{i\} \cup \{k \in \mathcal{V} \colon k \sim i\}$, $i \in \mathcal{V}$, induced by the adjacency relation $\mathcal{E}$. The motivation as well as one possible choice will be further clarified in the experimental section. Based on (2.9), we define the product manifold $(\mathcal{Q}_c, g)$ where

$$\rho = (\dots, \rho_i, \dots) \in \mathcal{Q}_c := \underbrace{\mathcal{D}_c \times \dots \times \mathcal{D}_c}_{|\mathcal{V}| \text{ factors}} \tag{3.11}$$

and the Riemannian metric in view of (2.12) and (3.7) is given by

$$g_\rho(X, Y) := \sum_{i \in \mathcal{V}} g_{\rho_i}(X_i, Y_i), \qquad X, Y \in T_\rho \mathcal{Q}_c := \mathcal{H}_0, \quad \forall \rho. \tag{3.12}$$

Let $\mathbb{1}_{\mathcal{Q}_c}$ denote the barycenter of $\mathcal{Q}_c$ given by $(\mathbb{1}_{\mathcal{Q}_c})_i = \mathbb{1}_{\mathcal{D}_c}$ for all $i \in \mathcal{V}$. Then the geometric interaction of likelihood matrices of the form (3.8) is defined by the *similarity mapping*

$$S \colon \mathcal{V} \times \mathcal{Q}_c \to \mathcal{D}_c, \qquad S_i(\rho) := \mathrm{Exp}^{(e)}_{\rho_i}\Big(\sum_{k \in \mathcal{N}_i} \omega_{ik} \big(\mathrm{Exp}^{(e)}_{\rho_i}\big)^{-1} \big(L_{\rho_k}(D_k)\big)\Big). \tag{3.13}$$

A characterization of the similarity map and a formula for evaluating it conveniently follow. They illustrate the benefit of using information geometry and the e-connection, rather than the Riemannian connection, from the computational viewpoint.

Proposition 4. *An equivalent expression of the similarity mapping* (3.13) *is given by*

$$S_i(\rho) = \Gamma\Big(\sum_{k \in \mathcal{N}_i} \omega_{ik} (\log_{\mathrm{m}} \rho_k - D_k) \Big), \qquad i \in \mathcal{V}. \tag{3.14}$$

Furthermore, if $\overline{\rho} \in \mathcal{D}_c$ solves the equation $0 = \sum_{k \in \mathcal{N}_i} \omega_{ik} \big(\mathrm{Exp}^{(e)}_{\overline{\rho}}\big)^{-1} \big(L_{\rho_k}(D_k)\big)$, which corresponds to the optimality condition for Riemannian centers of mass [10, Lemma 6.9.4], except for using a different exponential map, then

$$\overline{\rho} = S_i(\rho), \tag{3.15}$$

with $S_i(\rho)$ given by (3.13) *and* (3.14), *respectively.*

We are now in the position to define the

$$\dot{\rho} = R_\rho[S(\rho)], \quad \rho(0) = \mathbb{1}_{\mathcal{Q}_c} \qquad \textbf{(quantum state assignment flow)} \tag{3.16a}$$

where both the replicator map R_ρ and the similarity map $S(\rho)$ apply factorwise,

$$S(\rho)_i = S_i(\rho), \qquad R_\rho[S(\rho)]_i = R_{\rho_i}[S_i(\rho)], \qquad i \in \mathcal{V}. \tag{3.16b}$$

3.4 Riemannian Gradient Flow Parametrization

The proposition below provides a reparametrization of the quantum state flow equation (3.16a) that constitutes a Riemannian gradient flow with respect to a nonconvex potential.

Based on the weight function $\omega\colon \mathcal{E} \to \mathbb{R}_+$ of a given graph $\mathcal{G} = (\mathcal{V}, \mathcal{E}, w)$, we define the linear mapping

$$\Omega\colon \mathcal{Q}_c \to \mathcal{Q}_c, \qquad \Omega[\rho]_i := \sum_{k \in \mathcal{N}_i} \omega_{ik}\rho_k \in \mathcal{D}_c, \qquad i \in \mathcal{V}. \tag{3.17}$$

In addition, we adopt the *symmetry assumption*

$$\omega_{ij} = \omega_{ji}, \qquad j \in \mathcal{N}_i \Leftrightarrow i \in \mathcal{N}_j, \qquad \forall i, j \in \mathcal{V} \tag{3.18}$$

which makes the mapping (3.17) self-adjoint: $\langle \mu, \Omega[\rho] \rangle := \sum_{i \in \mathcal{V}} \langle \mu_i, \Omega[\rho]_i \rangle = \langle \Omega[\mu], \rho \rangle$ for all $\mu, \rho \in \mathcal{Q}_c$. Then the following holds.

Proposition 5. *The flow Eq.* (3.16a) *is equivalent to the system*

$$\dot{\rho} = R_\rho[\mu], \qquad \rho(0) = \mathbb{1}_{\mathcal{Q}_c}, \tag{3.19a}$$

$$\dot{\mu} = R_\mu\big[\Omega[\mu]\big], \qquad \mu(0) = S(\mathbb{1}_{\mathcal{Q}_c}). \tag{3.19b}$$

Furthermore, (3.19b) *is the Riemannian gradient flow*

$$\dot{\mu} = -\mathrm{grad}_\mu J(\mu) \tag{3.20a}$$

with respect to the potential $J(\mu)$ given by

$$J(\mu) := -\frac{1}{2}\langle \mu, \Omega[\mu] \rangle = \frac{1}{2}\big(\langle \mu, L_{\mathcal{G}}[\mu] \rangle - \|\mu\|^2\big), \tag{3.20b}$$

and with the 'Laplacian' operator $L_{\mathcal{G}}\colon \mathcal{Q}_c \to \mathcal{Q}_c$, $L_{\mathcal{G}} := \mathrm{id} - \Omega$.

The crucial point of Proposition (5) is that the evolution of $\mu(t)$ described by (3.19b) represents the 'essential' part of the quantum state flow equation (3.16a), since $\rho(t)$ solving (3.19a) is a function of $\mu(t)$ but *not* vice versa. Hence the geometric potential flow (3.20) provides a suitable basis for analyzing 'deep' quantum state assignment flows that result from the geometric integration of $\mu(t)$ (where each time step defines a 'layer') and a task-dependent choice of a time-variant mapping $\Omega(t)$, typically to be learnt from data.

3.5 Recovering the Assignment Flow for Categorial Distributions

In this section, we show that the quantum state assignment flow on a product manifold of density matrices contains as special case the assignment flow for categorial distributions, when the former flow is restricted to diagonal density matrices. This is quite natural because $\rho \succ 0$ implies positive diagonal elements and $\operatorname{tr}\rho = 1$ implies $\operatorname{diag}(\rho) \in \mathcal{S}_c$.

We confine ourselves to the formulation of quantum state assignment flow provided by Proposition 5 whose flow corresponds one-to-one to the flow generated by Eq. (3.16a).

Proposition 6. *Let*

$$\mathcal{Q}_c^d := \mathcal{D}_c^d \times \cdots \times \mathcal{D}_c^d \subset \mathcal{Q}_c, \qquad \mathcal{D}_c^d := \{\operatorname{Diag}(p) \colon p \in \mathcal{S}_c\} \tag{3.21}$$

denote the product submanifold of diagonal density matrices of the manifold $\mathcal{Q}_c$ *given by* (3.11)*. Then the quantum state flow equation in the form* (3.19b) *reduces to the dynamical system*[2]

$$\dot{S} = R_S[\Omega S], \qquad S(0) = S(\mathbb{1}_{\mathcal{W}_c}) \tag{3.22}$$

called 'S-flow' in [16, Prop. 3.6], where

$$S \in \mathbb{R}_+^{|\mathcal{V}|\times c}, \quad S_i = \operatorname{diag}(\mu_i), \quad i \in \mathcal{V} \tag{3.23a}$$

$$\Omega \in \mathbb{R}_+^{n\times n}, \quad \Omega_{ij} = \omega(ij), \quad ij \in \mathcal{E} \tag{3.23b}$$

$$R_S[\Omega S]_i = R_{S_i}(\Omega S)_i, \tag{3.23c}$$

with R_{S_i} *given by* (2.4)*, with* $\mathbb{1}_{\mathcal{W}_c}$ *denoting the barycenter of the assignment manifold and with the initial point* $S(0)$ *defined as in [16, Prop. 3.6].*

Proof. The proof basically reduces to identifying (i) the restriction of product states of the form (3.11) to diagonal density matrices as factors and (ii) matrices $S \in \mathbb{R}_{|V|\times c}$ with row vectors

$$S_i = \operatorname{diag}(\mu_i), \qquad i \in \mathcal{V}. \tag{3.24}$$

Then the mapping (3.17) takes the form ΩS with Ω given by (3.23b), whereas the right-hand side in (3.19b) takes for *diagonal* – and hence *commuting* – density matrices μ_i, $i \in \mathcal{V}$, the form

$$R_\mu[\Omega\mu]_i \overset{(3.16b)}{=} R_{\mu_i}[\Omega[\mu]_i] = R_{\mu_i}\Big[\sum_{k\in\mathcal{N}_i} \omega_{ik}\rho_k\Big] \tag{3.25a}$$

$$\overset{\substack{(3.6)\\(2.13b)}}{=} \sum_{k\in\mathcal{N}_i} \omega_{ik}\Big(\int_0^1 \mu_i^{1-\lambda}\mu_k\mu_i^{\lambda}d\lambda - \operatorname{tr}(\mu_i\mu_k)\mu_i\Big) \tag{3.25b}$$

[2] The use of the symbol S in the present context should *not* be confused with the similarity mapping (3.13). We just adhere to the notation used in prior work in order to reference clearly.

$$\overset{(3.24)}{=} \sum_{k\in\mathcal{N}_i} \omega_{ik}\big(S_i \cdot S_k - \langle S_i, S_k\rangle S_i\big) = R_{S_i}\Big(\sum_{k\in\mathcal{N}_i} \omega_{ik} S_k\Big) \tag{3.25c}$$

$$= R_{S_i}(\Omega S)_i \overset{(3.23c)}{=} R_S[\Omega S]_i. \tag{3.25d}$$

□

Proposition 6 basically says that the quantum assignment flow introduced in this paper is the *natural* generalization of the assignment flow approach introduced by [3,17] to *non-commutative* state spaces.

4 Experiments and Discussion

This section presents few academical results which illustrate properties of the novel ***QSAF*** approach (3.16). *We do not intend to consider and discuss any serious and fully worked out application in this paper* (see Sect. 5). Rather, the focus is on properties of the novel approach that cannot be achieved with the original assignment flow of categorial distributions. All computations were done using a geometric numerical Euler scheme adapted to the QSAF, after generalizing the approach presented in [20] accordingly.

Basic Patch Smoothing. Figure 1 shows an application of the QSAF to a *random* spatial arrangement (grid graph) of patches, where each vertex represents a patch, not a pixel. We refer to the caption for a description. The result demonstrates

- that *geometric* smoothing of image data at the *patch level* can preserve spatial image structure;
- that after convergence the final state constitutes a piecewise constant labeling with pure (rank-one) states.

Thus the QSAF, directly applied to the raw data, performs image partitioning which is *not* piecewise *constant* at the pixel level.

Noise Separation at the Patch Level. Figure 2 shows an application of the QSAF to a spatial collection of patches, each of which is pixelwise the mean of a randomly oriented patch and a patch with a fixed orientation. The result demonstrates that the QSAF effectively separates and removes 'random patch noise' at the *patch level without* any prior information or accessing the pixel level.

Patch Smoothing Using Harmonic Frames. Any matrix ensemble of the form $\{M_j\}_{j\in[c]} \subset \overline{\mathcal{P}}_c\colon \sum_{j\in[c]} M_j = I_c$ induces the categorial probability distribution $p \in \Delta_c$ on $[c]$ by taking inner products: $p_j = \langle M_j, \rho\rangle = \operatorname{tr}(M_j\rho)$, $j \in [c]$, for any $\rho \in \mathcal{D}_c$. A simple instance are the projection operators $M_j = F_2^j (F_2^j)^*$ formed by the columns F_2^j of the unitary discrete Fourier matrix $F_2 = F \otimes F \in \mathbb{C}^{c\times c}$ which performs the 2D discrete Fourier transform when applied to a vectorized patch with c pixels. Subtracting the mean of a vectorized patch followed by normalization using the $\|\cdot\|_2$ norm, the patch at vertex $i \in \mathcal{V}$ was

Fig. 1. **Left pair:** A random collection of patches with oriented image structure. The colored image displays for each patch its orientation using the color code depicted by the rightmost panel. Each patch is represented by a rank-one matrix D in (3.8), obtained by vectorizing the patch and taking the tensor product. **Center pair:** The final state of the QSAF obtained by geometric integration with uniform weighting $\omega_{ik} = \frac{1}{|\mathcal{N}_i|}$, $\forall k \in \mathcal{N}_i$, $\forall i \in \mathcal{V}$, of the nearest neighbors states. It represents an image partition but preserves image structure, due to geometric smoothing of patches encoded by non-commutative state spaces.

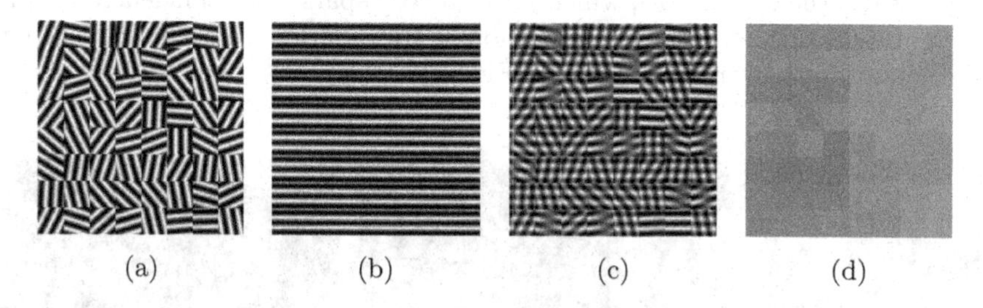

Fig. 2. **(a)** A random collection of patches with oriented image structure. **(b)** A collection of patches with the same oriented image structure. **(c)** Pixelwise mean of the patches (a) (b) at each location. **(d)** The QSAF recovers a close approximation of (b) (color code: see Fig. 1) by iteratively smoothing the states ρ_k, $k \in \mathcal{N}_i$ corresponding to (c) through geometric integration.

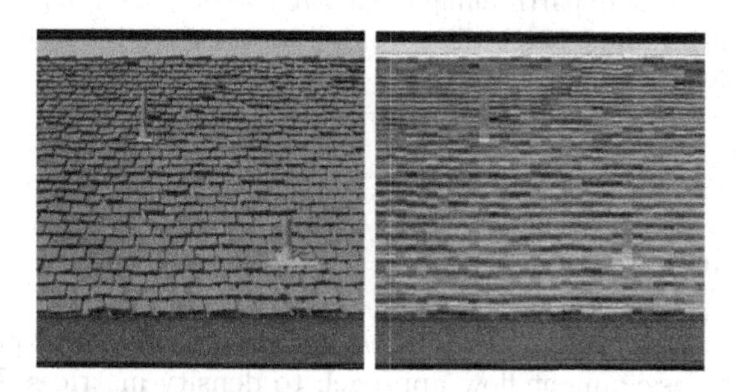

Fig. 3. **Left:** A real roof texture. **Right:** 8×8 patches were encoded using the discrete Fourier frame (see text). Integrating the QSAF yields the same effect as shown by Fig. 2, here with respect to the Fourier frame, however.

encoded by a matrix D_i in (3.8) of the form $D_i = F_2\mathrm{Diag}(-|\widehat{p}_i|^2)F_2^*$ with $|\widehat{p}_i|_j = |(F_2\mathrm{vec}(P_i))_j|$, $\forall j$, for patch P_i.

Integrating the QSAF yields a denoising effect at the patch level similar to Fig. 2, here in the Fourier domain, however. After convergence, each state has the form $\rho_i = q_i q_i^\top$ for some unit vector q_i which was used to filter the Fourier-transformed patch vector using the Hadamard product, followed by decoding the patch. Accordingly, "assignment" here means scale and orientation in a spatial context, as encoded by the harmonic Fourier frame; see Fig. 3.

Translation-Invariant Patch Smoothing in a Harmonic Frame on a Non-grid Graph. The scenario of Fig. 3 was extended: rather than using nearest-neighborhoods, the 8 most similar patches in the *entire* collection of image patches were selected for each patch, to define corresponding non-grid edges and irregular neighborhoods. Similarity was defined in terms of the distance between the orbits of patches, generated by 2D cyclic translation. The resulting unitary 'registration operator' was attached to each edge. Such elementary preprocessing changes the data encoding in terms of the matrix D in (3.8), but not the QSAF flow, which yields a very sparse but sufficiently detailed representation of image structure, although the Fourier frame only represents two orientations at various scales; see Fig. 4.

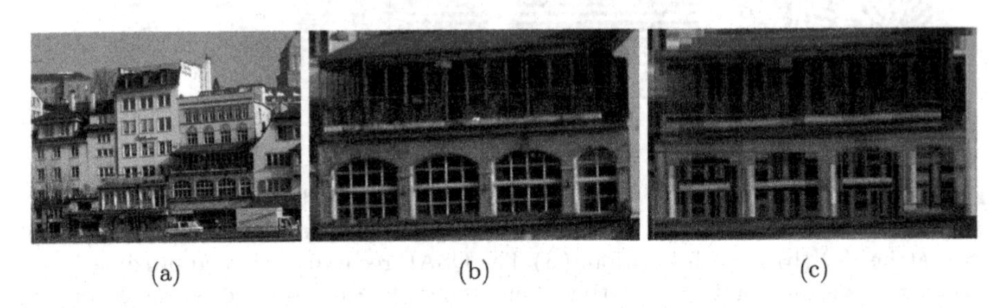

Fig. 4. **(a)** A real scene. **(b)** A section of (a). **(c)** QSAF-filtered patches using a single harmonic frame, irregular non-local neighborhoods and translation invariant patch encoding (see text). Due to partitioning the image into patches, using a single harmonic frame and a shift-invariant patch distance, image structure is encoded very sparsely but sufficiently detailed.

5 Conclusion

We introduced a novel dynamical system for data representation and analysis, by extending the assignment flow approach to density matrices. Few numerical examples illustrated context-sensitive patch smoothing by geometric averaging of the non-commuting state spaces.

The model expressivity of the approach which performs the assignment of rank-one density matrices as 'labels', is larger than our preliminary experiments indicate. For instance, *latent* states may be used to parametrize Parseval frames which in turn transform the primary states. Moreover, by learning the parameters Ω in (3.13) and (3.17), respectively, from data, our approach may be seen as a novel 'neural ODE' from the viewpoint of machine learning.

Acknowledgements. This work is funded by the Deutsche Forschungsgemeinschaft (DFG) under Germany's Excellence Strategy EXC-2181/1 - 390900948 (the Heidelberg STRUCTURES Excellence Cluster). This work is funded by the Deutsche Forschungsgemeinschaft (DFG), grant SCHN 457/17-1, within the priority programme SPP 2298: "Theoretical Foundations of Deep Learning".

References

1. Amari, S.: Differential-Geometrical Methods in Statistics. Lecture Notes in Statistics, vol. 28. Springer, New York (1985). https://doi.org/10.1007/978-1-4612-5056-2
2. Amari, S.I., Nagaoka, H.: Methods of Information Geometry. American Mathematical Society and Oxford University Press (2000)
3. Åström, F., Petra, S., Schmitzer, B., Schnörr, C.: Image labeling by assignment. J. Math. Imaging Vis. **58**(2), 211–238 (2017). https://doi.org/10.1007/s10851-016-0702-4
4. Ay, N., Jost, J., Lê, H.V., Schwachhöfer, L.: Information Geometry. Springer, Cham (2017). https://doi.org/10.1007/978-3-319-56478-4
5. Bengtsson, I., Zyczkowski, K.: Geometry of Quantum States: An Introduction to Quantum Entanglement, 2nd edn. Cambridge University Press, Cambridge (2017)
6. Brown, L.D.: Fundamentals of Statistical Exponential Families. Institute of Mathematical Statistics, Hayward (1986)
7. Cai, Y., Lu, X., Jiang, N.: A survey on quantum image processing. Chin. J. Electron. **27**(4), 667–888 (2018)
8. Calin, O., Udriste, C.: Geometric Modeling in Probability and Statistics. Springer, Cham (2014). https://doi.org/10.1007/978-3-319-07779-6
9. Grasselli, M., Streater, R.: On the uniqueness of the Chentsov metric in quantum information geometry. Infinite Dimensional Anal. Quantum Probab. Relat. Top. **4**(2), 173–182 (2001)
10. Jost, J.: Riemannian Geometry and Geometric Analysis, 7th edn. Springer, Heidelberg (2017). https://doi.org/10.1007/978-3-319-61860-9
11. Kobayashi, S., Nomizu, K.: Foundations of Differential Geometry, vol. II. Wiley, Hoboken (1969)
12. Lauritzen, S.L.: Statistical manifolds. In: Gupta, S.S., Amari, S.I., Barndorff-Nielsen, O.E., Kass, R.E., Lauritzen, S.L., Rao, C.R. (eds.) Differential Geometry in Statistical Inference, pp. 163–216. IMS, Hayward (1987)
13. Petz, D.: Geometry of canonical correlation on the state space of a quantum system. J. Math. Phys. **35**(2), 780–795 (1994)
14. Petz, D.: Quantum Information Theory and Quantum Statistics. Springer, Heidelberg (2008). https://doi.org/10.1007/978-3-540-74636-2
15. Petz, D., Toth, G.: The Bogoliubov inner product in quantum statistics. Lett. Math. Phys. **27**(3), 205–216 (1993). https://doi.org/10.1007/BF00739578
16. Savarino, F., Schnörr, C.: Continuous-domain assignment flows. Eur. J. Appl. Math. **32**(3), 570–597 (2021)

17. Schnörr, C.: Assignment flows. In: Grohs, P., Holler, M., Weinmann, A. (eds.) Handbook of Variational Methods for Nonlinear Geometric Data, pp. 235–260. Springer, Cham (2020). https://doi.org/10.1007/978-3-030-31351-7_8
18. Schwarz, J., et al.: The present paper. Journal version, in preparation (2023)
19. Čencov, N.N.: Statistical Decision Rules and Optimal Inference. AMS (1981)
20. Zeilmann, A., Savarino, F., Petra, S., Schnörr, C.: Geometric numerical integration of the assignment flow. Inverse Probl. **36**(3), 034004 (33pp) (2020)
21. Zern, A., Zeilmann, A., Schnörr, C.: Assignment flows for data labeling on graphs: convergence and stability. Inf. Geom. **5**, 355–404 (2022). https://doi.org/10.1007/s41884-021-00060-8

Author Index

L. Calatroni et al. (Eds.): SSVM 2023, LNCS 14009, pp. 757–759, 2023.
https://doi.org/10.1007/978-3-031-31975-4